THE

INFRARED SPECTRA

HANDBOOK OF

PRIORITY POLLUTANTS AND TOXIC CHEMICALS

THE INFRARED SPECTRA HANDBOOK OF

PRIORITY POLLUTANTS AND TOXIC CHEMICALS

SADTLER
HEYDEN

Published by
Sadtler Research Laboratories
Division of Bio-Rad Laboratories, Inc.
3316 Spring Garden Street
Philadelphia, Pennsylvania 19104

ISBN 0-8456-0077-X

Library of Congress Catalog Card Number: 81-52795

Co-published for exclusive distribution in Europe by
Heyden & Son Ltd.
Spectrum House, Hillview Gardens
London N.W. 4, England 2JQ

ISBN 0-85501-552-7

Printed in the United States of America

ORGANIZATION OF THE HANDBOOK

INTRODUCTION

In recent years a tremendous increase in the development and application of synthetic organic chemicals has occurred. The use of inorganics has also expanded and intensified, and there is now an urgent need for society to be aware of and concerned with the toxic effects and other risks associated with many compounds from both of these groups. Thus the purpose of this handbook is to serve as a convenient and practical spectral reference for research, industry and all others who are engaged in analyzing, monitoring, controlling or studying environmental, physiological or occupational pollutants and toxic substances.

PRESENTATION

The Sadtler Infrared Spectra Handbook of Priority Pollutants and Toxic Chemicals includes 564 infrared reference spectra representing 363 chemical substances which have appeared in the EPA Priority Pollutants List, the Occupational Safety and Health Administration (OSHA) Category I List of Carcinogenic Substances, and a list of hazardous compounds common to industry and of concern during interstate transportation. In addition, an expanded appendix is included which shows detailed GC/FT-IR analysis of polychlorinated biphenyls (PCBs) that appear on the EPA Priority Pollutant list. The first section of this handbook contains 201 compounds that are represented by two types of spectra, infrared condensed phase and infrared vapor phase. The condensed phase spectra are typical of those obtained using standard laboratory sample handling techniques for condensed phase spectroscopy. The vapor phase spectra are representative of those one would obtain using GC/FT-IR analysis where the spectra are measured above ambient temperature in a heated optical cell. Condensed phase spectra for compounds that are gases at room temperature are not included; instead, a spectrum of the gas at 25° C is presented in addition to the vapor phase spectrum measured above ambient temperature. Also, some of the vapor phase spectra presented in this section of the handbook are for mixtures of compounds, such as the PCBs, dieldrin, chlordan, etc. These spectra are presented for band correlation purposes and are not spectra of pure compounds. The GC/FT-IR analyses of PCBs are presented in the Appendix to illustrate the type of analysis that is required to obtain pure component spectra of these complicated mixtures. A separate introduction to the Appendix has been given for the GC/FT-IR techniques used to evaluate these substances.

The second section of the handbook is composed of compounds represented by a single spectrum. In these cases, the compounds are generally not suitable for gas chromatography analysis, i.e. inorganics. All spectra in the handbook are arranged in a large, easy-to-read format, with one spectrum presented on each page. The spectra are presented in a transmittance vs. frequency (wavenumber) format over the spectral region 4000 to 400 cm^{-1}. Each compound is listed by its Chemical Abstracts name together with frequently used common names. The molecular formula, the structure, the molecular weight, the source of sample, the CAS Registry number and the NIOSH number are all given when the information is available. (The NIOSH numbers are the sequence numbers assigned as permanent descriptors to the chemicals listed in the Registry of Toxic Effects of Chemical Substances by the National Institute for Occupational Safety and Health.) The sample preparation method is specified with the condensed phase spectra, and pertinent physical and instrumental data including cell temperature are presented with vapor phase spectra. Traces of impurities were found in a small number of the samples used to compile this collection. Rather than remove these spectra from the data compilation and present no data for these important compounds, we have elected to include them and mark the impurity bands with a small dot or star. In addition, nujol bands present in certain inorganic spectra are indicated in a similar manner.

Condensed Phase Spectra

Condensed phase spectra were prepared using either a Perkin-Elmer 521 or 621 high resolution grating spectrometer or a Digilab FTS-14 or 15 FT-IR spectrometer. With the latter instrument, an optical retardation of 0.25 was used corresponding to 4 cm^{-1} spectral resolution. Triangular apodization was used for all interferometrically measured spectra. Standard techniques, developed in Sadtler's laboratories, insure that the published spectra are of the best possible quality, and are reproducible for comparison and identification purposes. The preferred sample preparation methods are the capillary cell for liquids and low-melting solids and the KBr wafer for solids. The KBr method is used for solids since it is a standard technique and requires only a small amount of sample for the preparation of good spectra, leaving the remainder for further analytical investigations. Since the KBr method cannot be used for solids which undergo ionic exchange with the KBr, these samples were mulled in perfluorinated hydrocarbon and measured over the 4000 - 1350 cm^{-1} region and mulled in mineral oil for a measurement over the 1350 - 400 cm^{-1} region. Thus a complete spectrum of the compound is provided. Each spectrum is clearly labeled with the sample preparation technique used.

Spectra of Gases at 25° C

All spectra were measured on a Digilab FTS-14 Fourier transform spectrometer. For some compounds which have very narrow band structure a nominal resolution of 2 cm^{-1} across the complete spectral range was used. Most materials have been measured in a 10 cm path-length cell, with KBr windows. Occasionally a 1 meter long-path cell has been used, generally when the compound has either weak absorption bands (such as HI, H_2S) or a low vapor pressure (such as dimethyl sulfoxide, p-xylene). Certain samples where these criteria do not apply have also been measured in the long-path cell. This has been for the purpose of eliminating water vapor interferences from the spectra, since it was found that when liquid samples could be injected into the cell through a septum, the rigorous drying process to which all the vapors were subjected became less necessary. Liquid samples were dried with anhydrous calcium chloride, magnesium sulfate or a molecular sieve. The vapor was passed through a column of silica gel before entering the gas-cell.

Vapor Phase Spectra

All spectra were determined at Sadtler Research Laboratories using a Digilab FTS-14 Fourier transform spectrometer in the spectral region 4000 cm^{-1} to 450 cm^{-1} with a nominal resolution of 4 cm^{-1} being maintained across the entire spectral region. A Sadtler CIRA 101 chromatographic infrared analyzer was used to vaporize and trap the compounds for examination. The carrier gas was helium. The compounds were examined at temperatures ranging from 100° C to 280° C; the temperature used for each compound was based on the boiling point or melting point of the material, and its chemical stability. Thermally sensitive compounds were always run at lower vaporization temperatures.

EXPERIMENTAL

One microliter of liquid samples was injected into the CIRA 101 with the helium flow rate operating at 20 ml/min. at the selected vaporization temperature. For samples that were solids at ambient temperature, 1 mg of sample was introduced into the system using the CIRA 101 Solid Sampling Accessory. For samples which were gases at room temperature, or low boiling liquids, 500 μl of the vapor at atmospheric pressure was injected into the CIRA 101. The sample was trapped within the CIRA cell when the interferogram, monitored in real time with an oscilliscope, decreased in magnitude from the maximum to minimum level, therefore indicating maximum concentration of the compound inside the cell. The reference spectrum was then measured. After the spectrum was measured, the gas flow was restarted for 3 - 5 minutes to purge the sample from the system before proceeding to the next sample.

Gas Chromatography/FT-IR

Gas chromatography/FT-IR analysis of six PCBs and chlordan are presented in the Appendix of this book. The analyses were completed using a Digilab Model FTS-20 spectrometer with a GC/C capillary accessory. A detailed description of the experimental conditions for obtaining the GC/IR analysis is presented in the Appendix preface.

All spectra in this handbook were indexed alphabetically, by molecular formula and by strongest band to provide rapid location and identification.

ACENAPHTHENE
1,2–DIHYDROACENAPHTHYLENE

$C_{12}H_{10}$

MOLECULAR WEIGHT: 154.21

SOURCE OF SAMPLE: Aldrich Chemical Company, Inc.
Milwaukee, Wisconsin

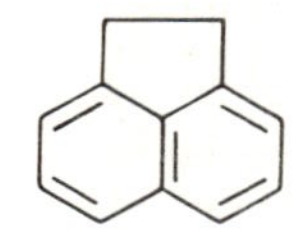

CAS REGISTRY NO.: 83-32-9

NIOSH NUMBER:

KBr Wafer

CONDENSED PHASE

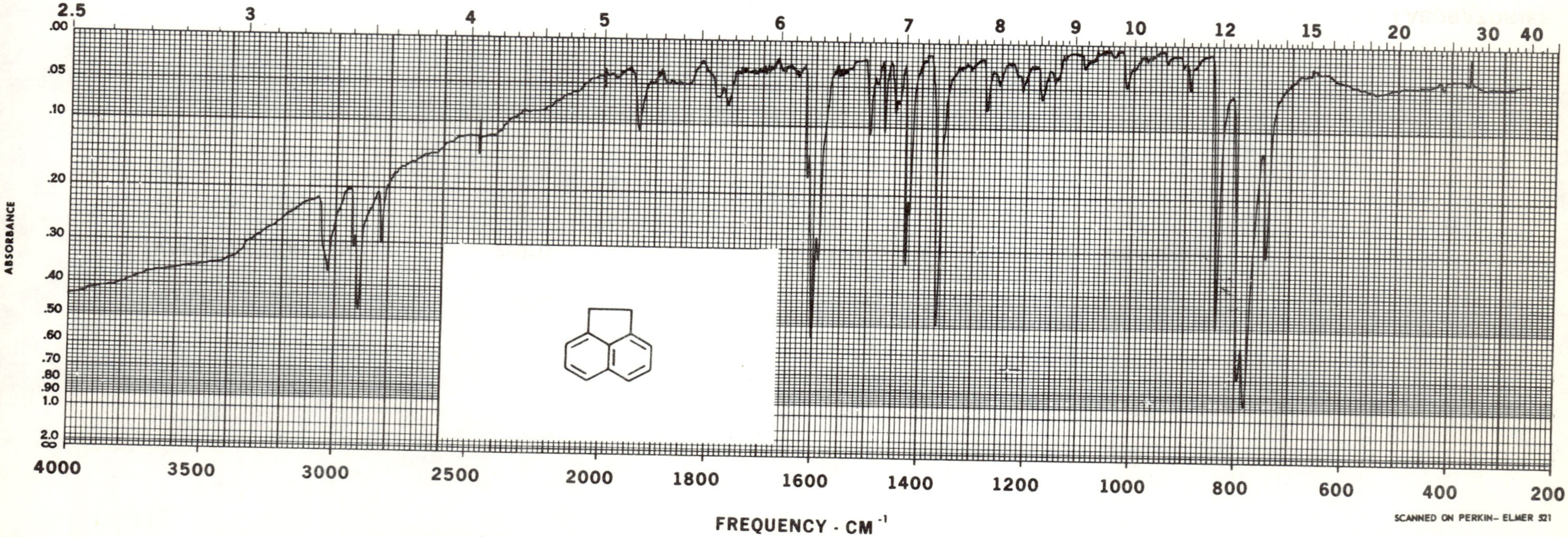

BOILING POINT:	279°C	INJECTION AMT.:	0.69 mg
VAPORIZATION TEMP.:	280°C	PRESSURE:	Atmospheric
CELL VOLUME:	7.5 cc	MAX./MIN.:	101.7/46.2
PATH LENGTH:	7.6 cm	SCALE EXPANSION:	1.66

VAPOR PHASE

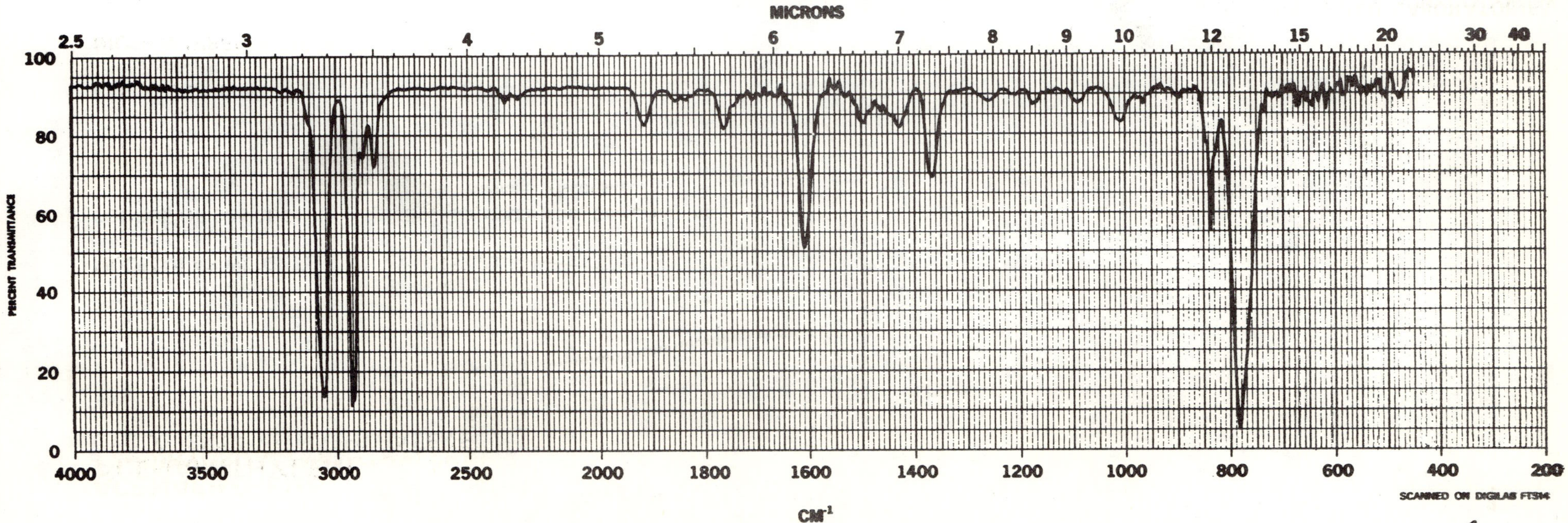

1

ACENAPHTHYLENE
CYCLOPENTA[de] NAPHTHALENE

$C_{12}H_8$

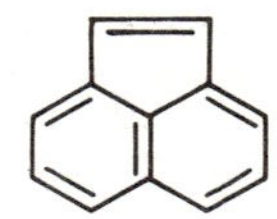

MOLECULAR WEIGHT: 152.20

SOURCE OF SAMPLE: Aldrich Chemical Company, Inc.
Milwaukee, Wisconsin

CAS REGISTRY NO.: 208-96-8

NIOSH NUMBER: —

KBr Wafer

CONDENSED PHASE

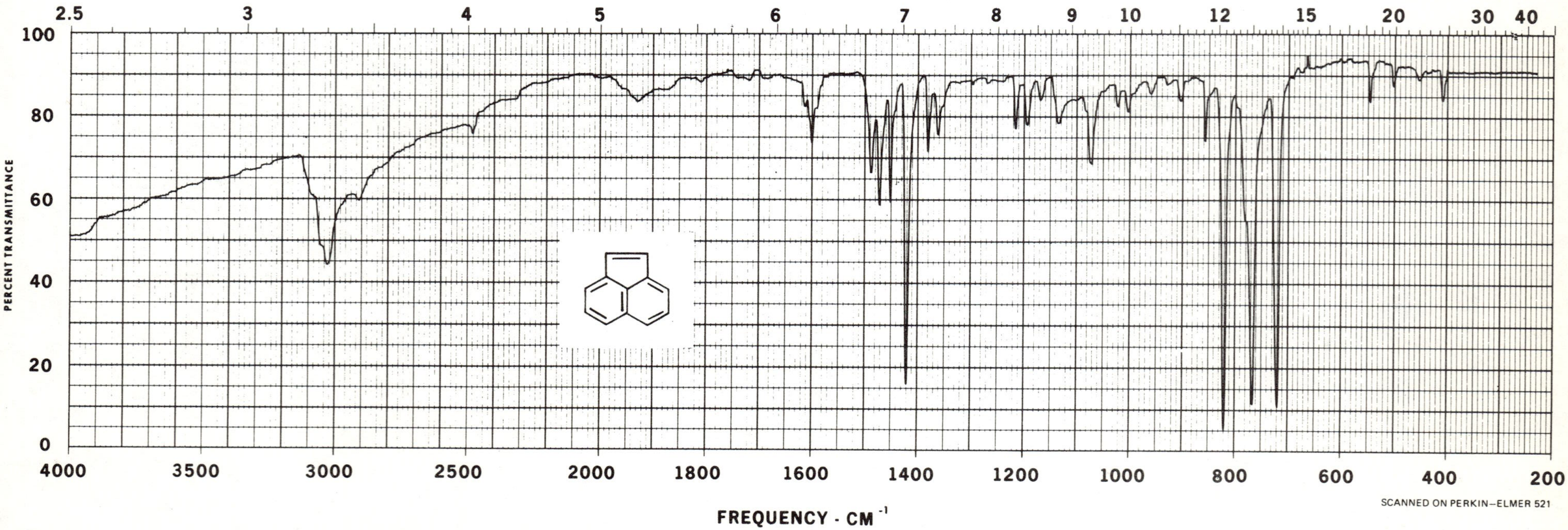

SCANNED ON PERKIN–ELMER 521

BOILING POINT:	280°C	INJECTION AMT.:	0.80 mg
VAPORIZATION TEMP.:	280°C	PRESSURE:	Atmospheric
CELL VOLUME:	7.5 cc	MAX./MIN.:	108.3/58.4
PATH LENGTH:	7.6 cm	SCALE EXPANSION:	1.84

VAPOR PHASE

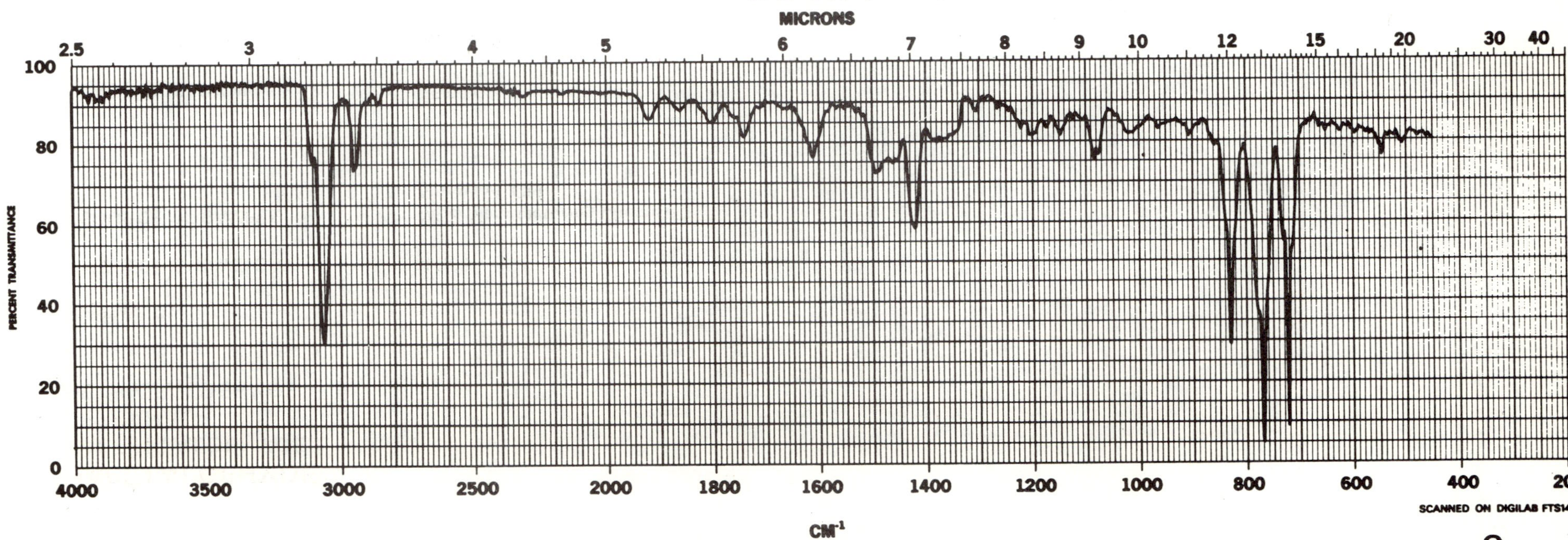

SCANNED ON DIGILAB FTS14

ACETALDEHYDE
ACETIC ALDEHYDE
ETHANAL
ETHYL ALDEHYDE

C_2H_4O

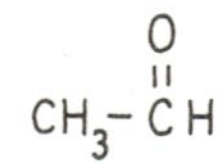

MOLECULAR WEIGHT:	44.05
SOURCE OF SAMPLE:	MCB Manufacturing Chemists Norwood, Ohio
CAS REGISTRY NO.:	75-07-0
NIOSH NUMBER:	AB19250
Path Length:	10 cm
Complete Scan:	20 mm Hg

A GAS AT 25°C

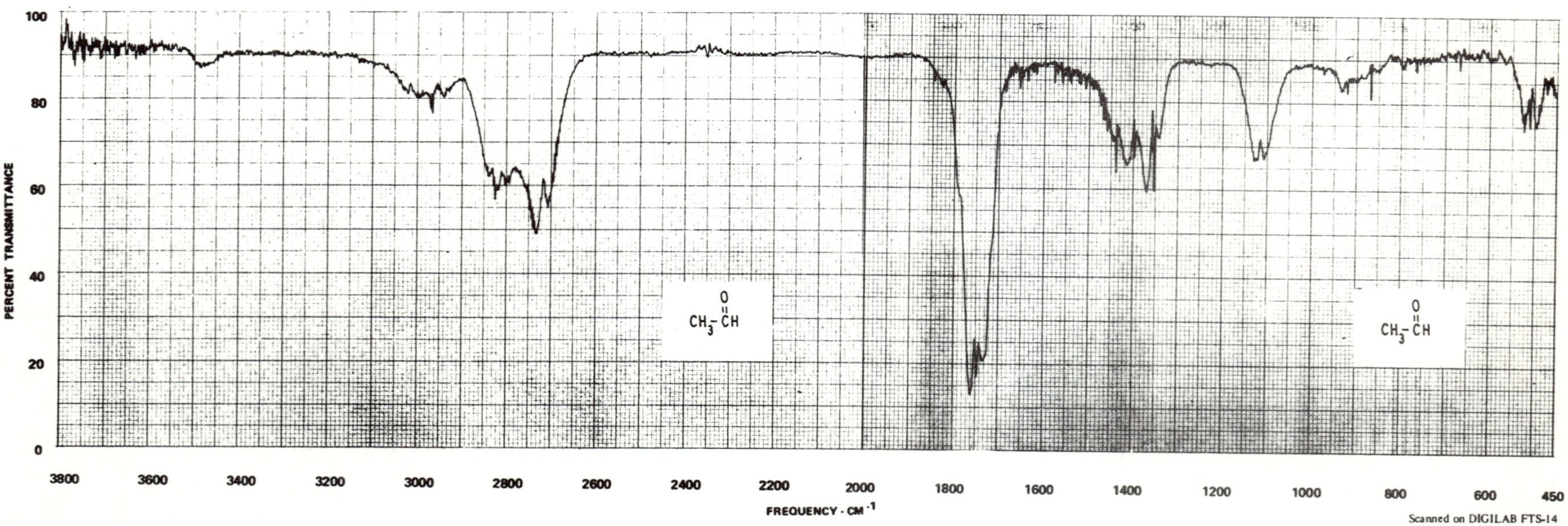

BOILING POINT:	20.8°C	INJECTION AMT.:	–
VAPORIZATION TEMP.:	160°C	PRESSURE:	Atmospheric
CELL VOLUME:	7.5 cc	MAX./MIN.:	99.4/17.3
PATH LENGTH:	7.6 cm	SCALE EXPANSION:	1.12

VAPOR PHASE

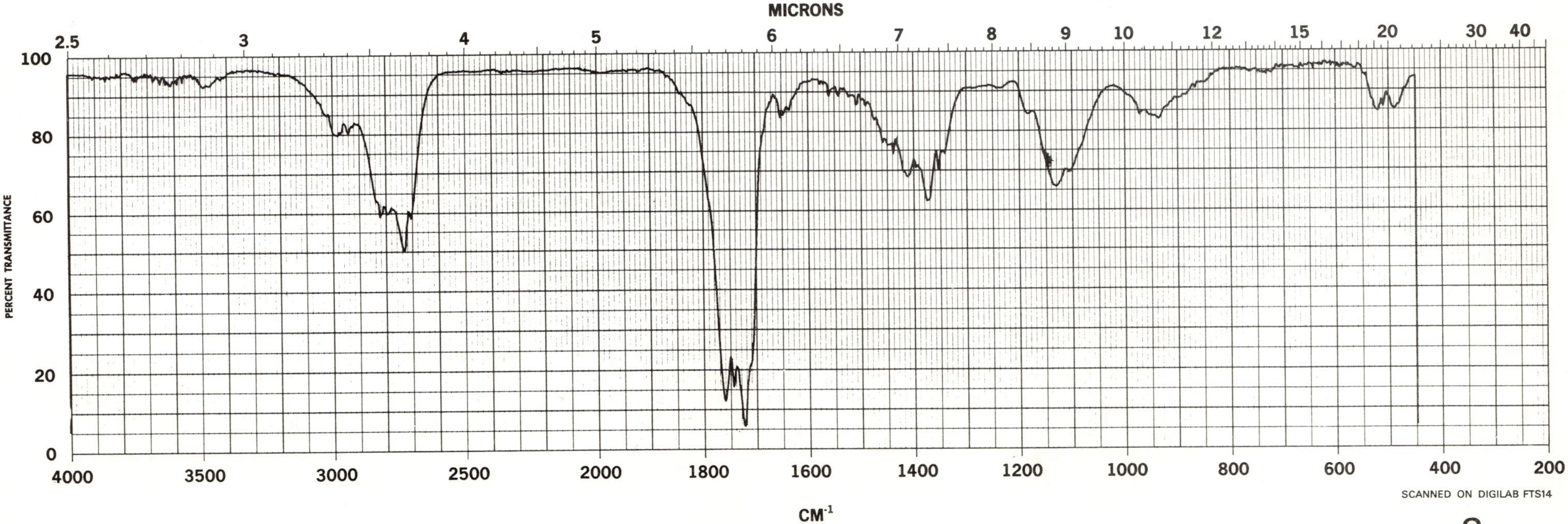

SCANNED ON DIGILAB FTS14

ACETAMIDE
ACETIC ACID AMIDE
ETHANAMIDE
METHANECARBOXAMIDE

C_2H_5NO

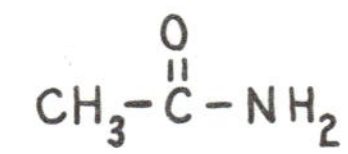

MOLECULAR WEIGHT:	59.07
SOURCE OF SAMPLE:	MCB Manufacturing Chemists Norwood, Ohio
CAS REGISTRY NO.:	60-35-5
NIOSH NUMBER:	AB40250

KBr Wafer

CONDENSED PHASE

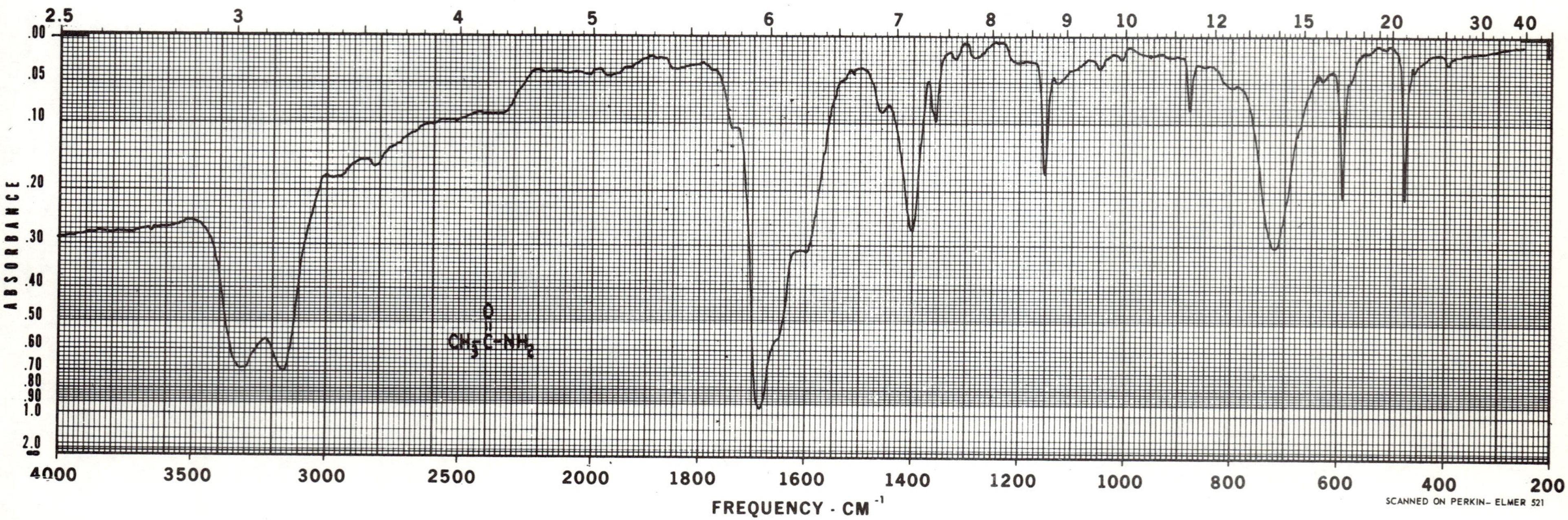

BOILING POINT:	221.2°C	INJECTION AMT.:	0.43 mg
VAPORIZATION TEMP.:	280°C	PRESSURE:	Atmospheric
CELL VOLUME:	7.5 cc	MAX./MIN.:	101.2/13.1
PATH LENGTH:	7.6 cm	SCALE EXPANSION:	1.04

VAPOR PHASE

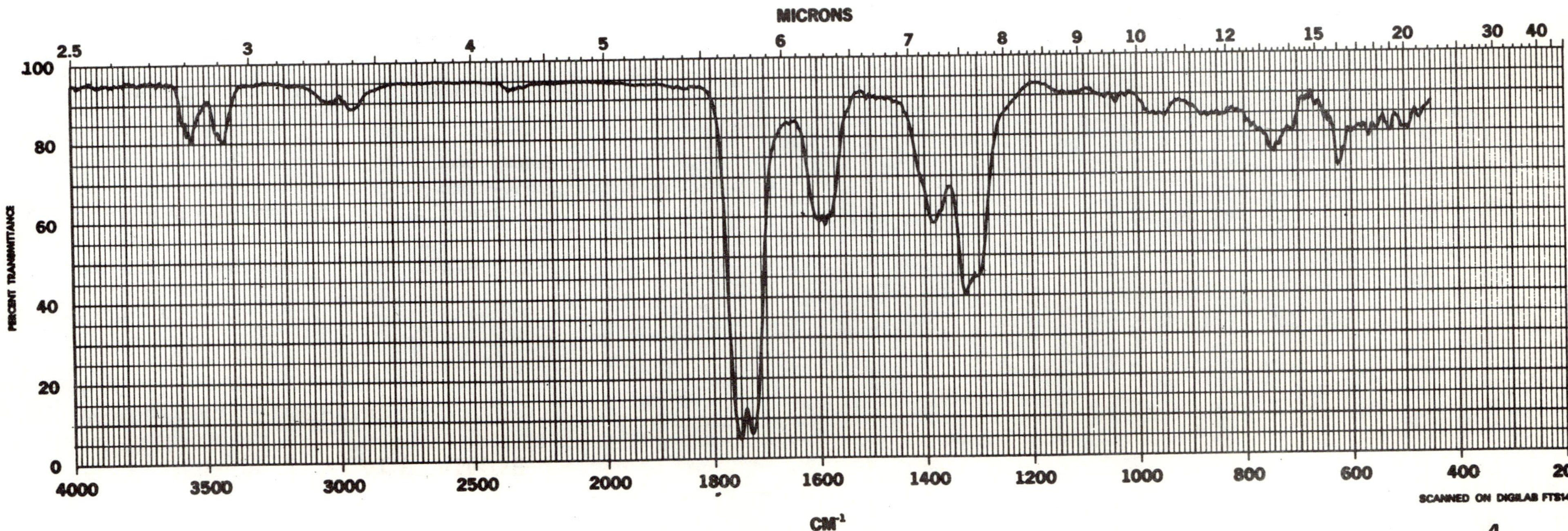

N—FLUOREN—2—YLACETAMIDE
AAF
2—ACETAMIDOFLUORENE
FAA

$C_{15}H_{13}NO$

MOLECULAR WEIGHT: 223.28

SOURCE OF SAMPLE: S. Laham
Department of National Health & Welfare
Ottawa, Canada

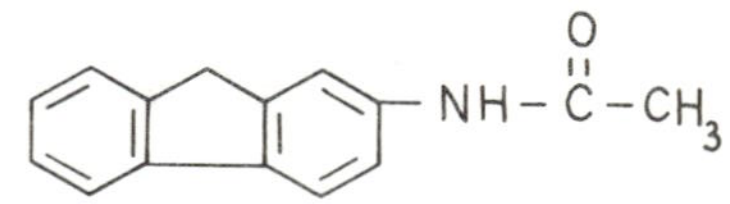

CAS REGISTRY NO.: 53-96-3

NIOSH NUMBER: AB94500

KBr Wafer

CONDENSED PHASE

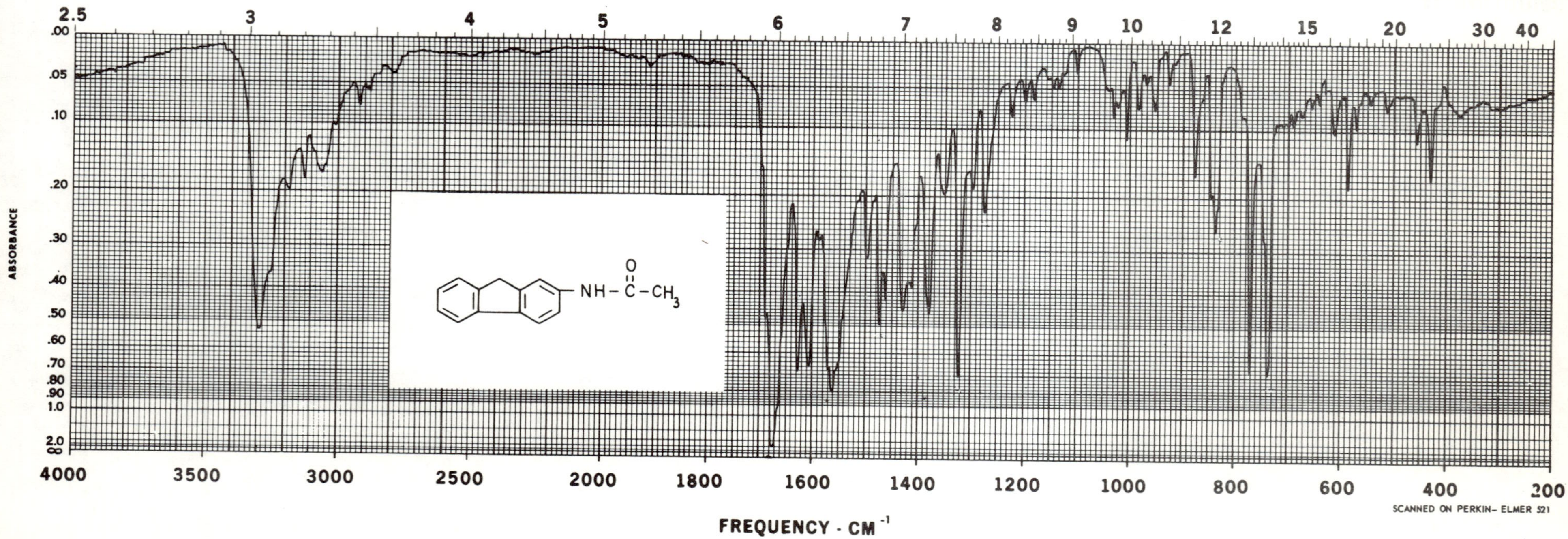

MELTING POINT:	194°C	INJECTION AMT.:	–
VAPORIZATION TEMP.:	280°C	PRESSURE:	Atmospheric
CELL VOLUME:	7.5 cc	MAX./MIN.:	98.0/52.6
PATH LENGTH:	7.6 cm	SCALE EXPANSION:	2.11

VAPOR PHASE

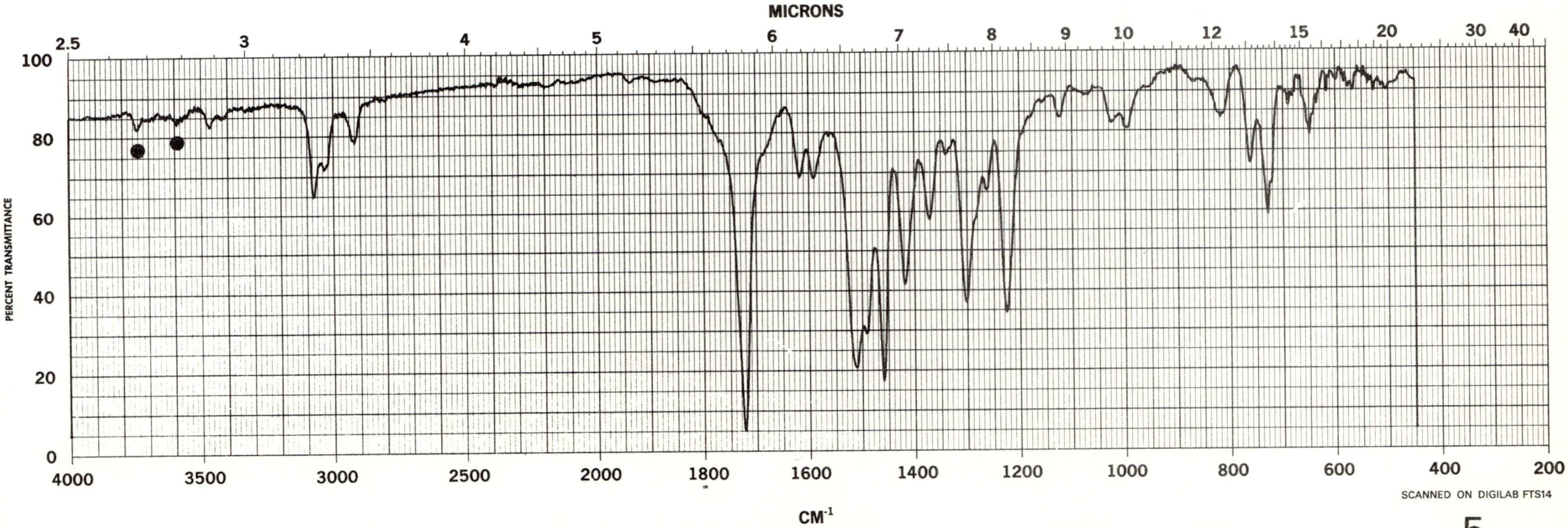

SCANNED ON DIGILAB FTS14

THIOACETAMIDE
ACETOTHIOAMIDE
ETHANETHIOAMIDE
TAA

C_2H_5NS

MOLECULAR WEIGHT: 75.13

SOURCE OF SAMPLE: MCB Manufacturing Chemists
Norwood, Ohio

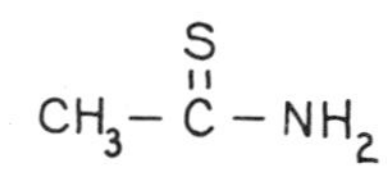

CAS REGISTRY NO.: 62-55-5

NIOSH NUMBER: AC89250

KBr Wafer

CONDENSED PHASE

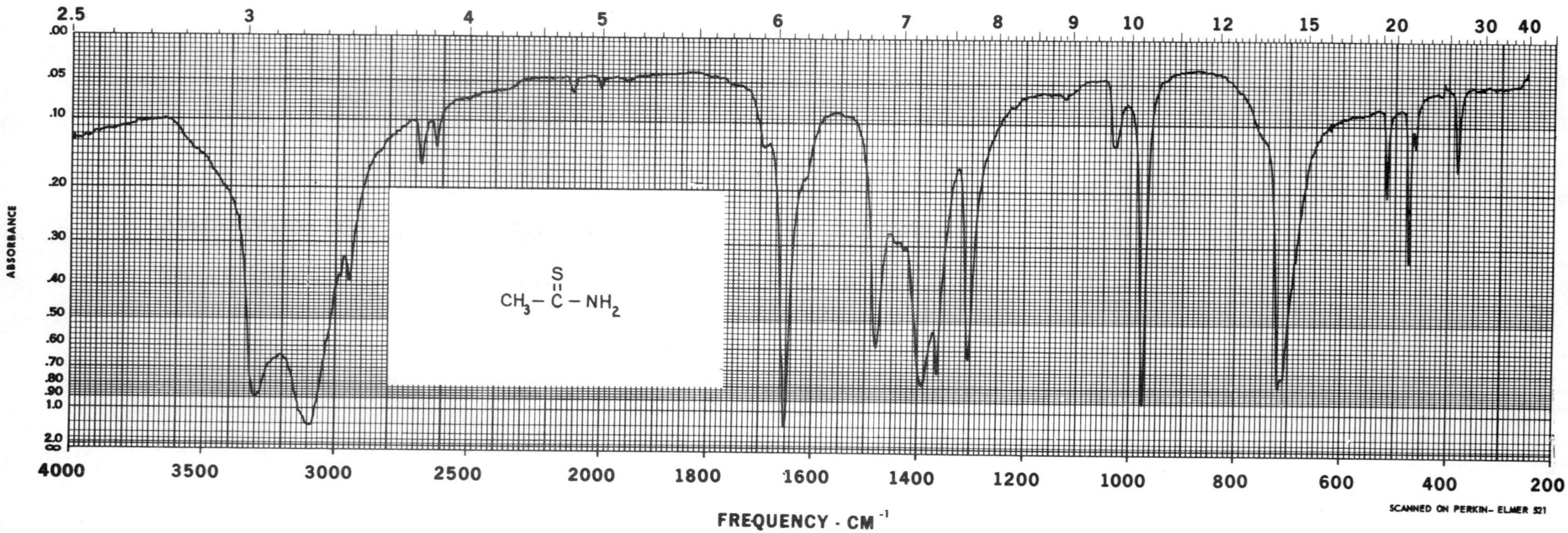

MELTING POINT:	110-113°C	INJECTION AMT.:	—
VAPORIZATION TEMP.:	200°C	PRESSURE:	Atmospheric
CELL VOLUME:	7.5 cc	MAX./MIN.:	99.7/18.9
PATH LENGTH:	7.6 cm	SCALE EXPANSION:	1.14

VAPOR PHASE

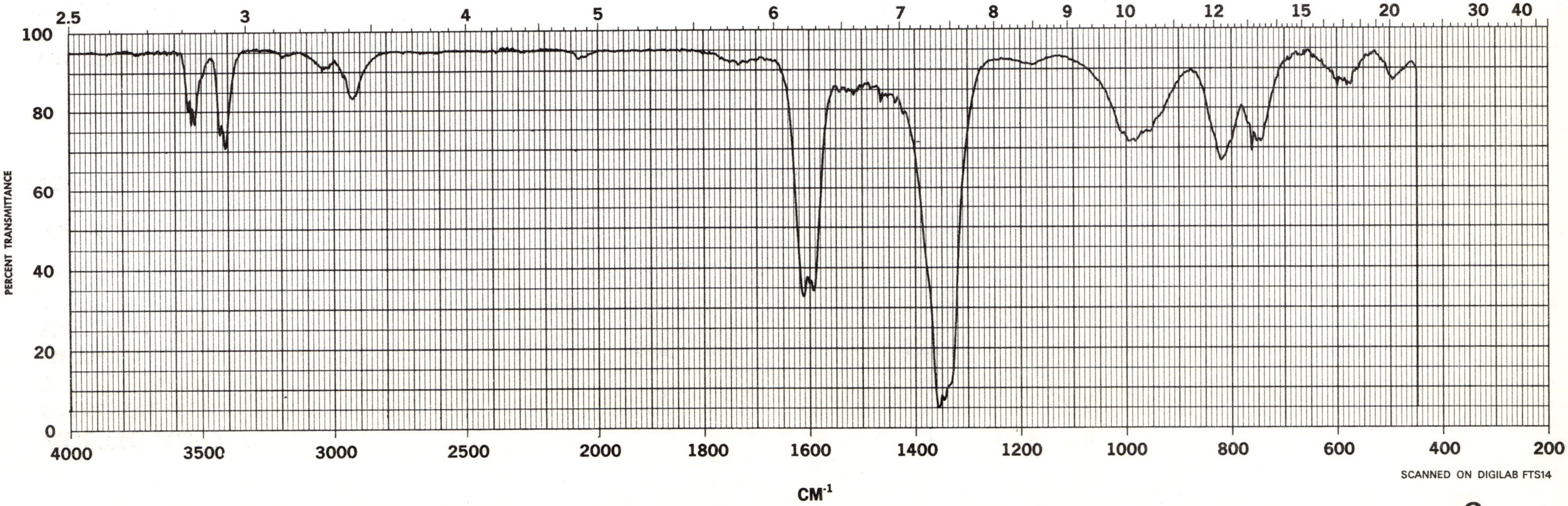

SCANNED ON DIGILAB FTS14

ACETIC ACID
ETHANOIC ACID
ETHYLIC ACID
METHANECARBOXYLIC ACID
VINEGAR ACID

$C_2H_4O_2$

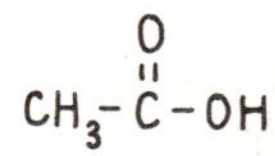

MOLECULAR WEIGHT: 60.05

SOURCE OF SAMPLE: Eastman Organic Chemicals
Rochester, New York

CAS REGISTRY NO.: 64-19-7

NIOSH NUMBER: AF12250

CAPILLARY CELL: Neat

CONDENSED PHASE

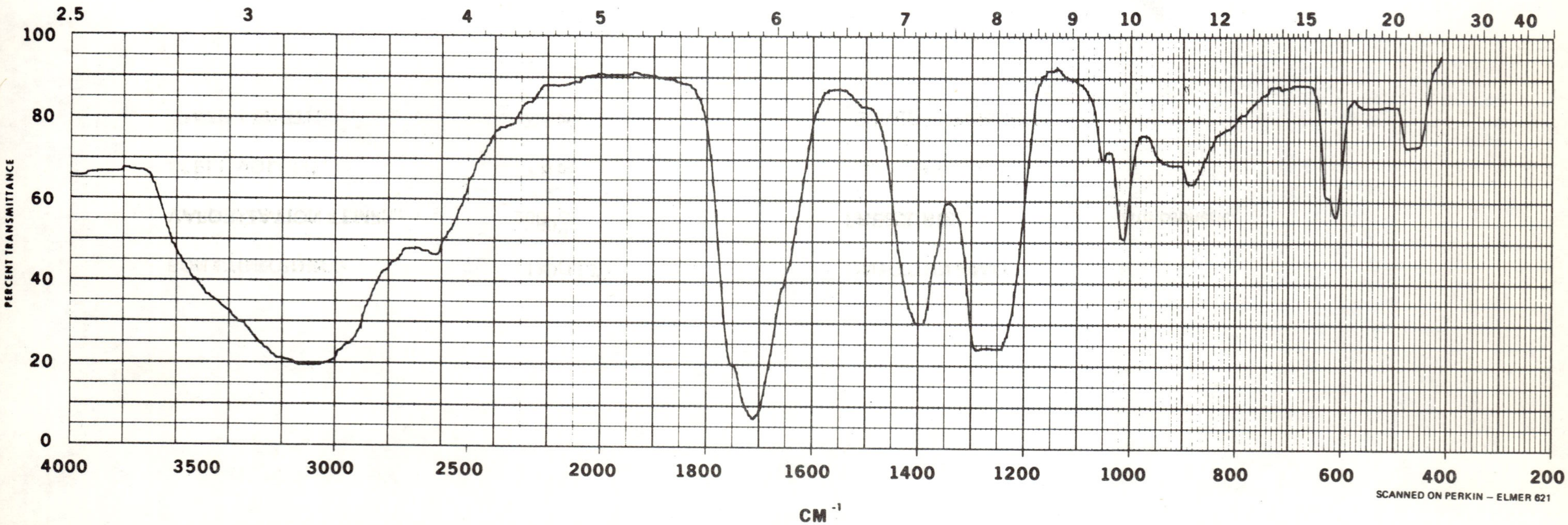

BOILING POINT:	$118^{\circ}C$	INJECTION AMT.:	0.2 µl
VAPORIZATION TEMP.:	$200^{\circ}C$	PRESSURE:	Atmospheric
CELL VOLUME:	7.5 cc	MAX./MIN.:	—
PATH LENGTH:	7.6 cm	SCALE EXPANSION:	0.91

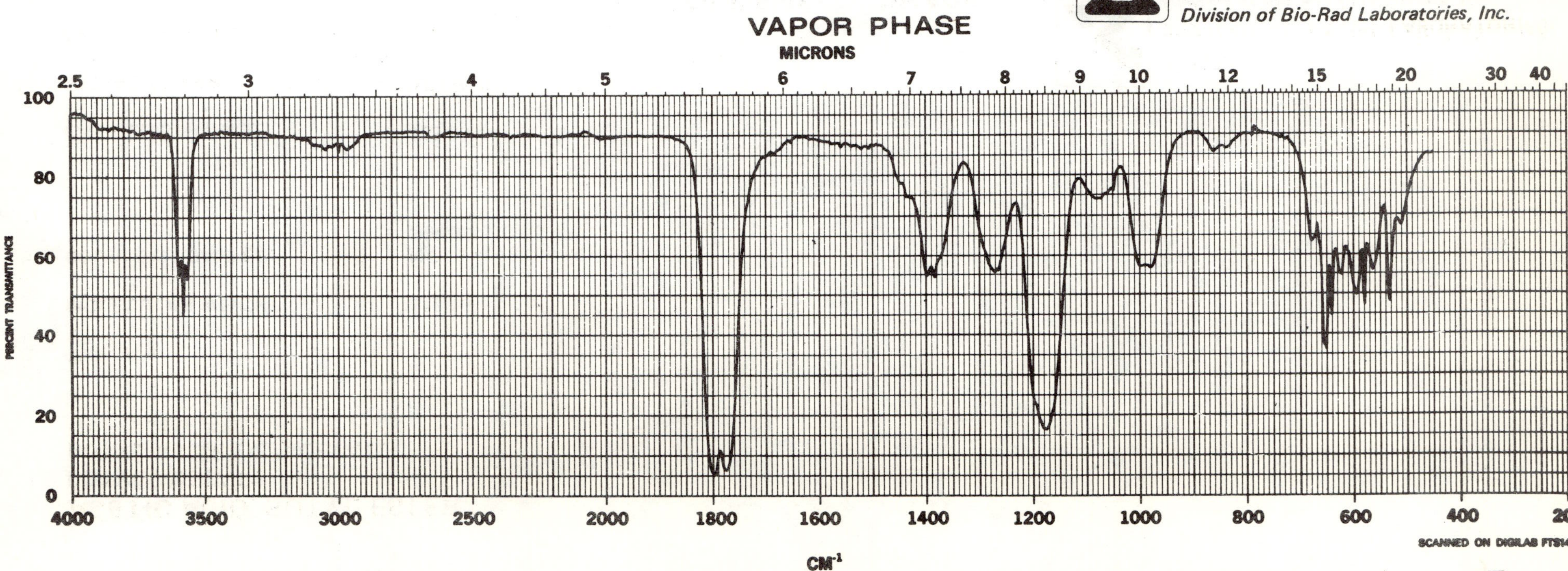

ACETIC ACID, BUTYL ESTER
BUTYL ACETATE
BUTYL ALCOHOL, ACETATE

$C_6H_{12}O_2$

$CH_3-C(=O)-OCH_2CH_2CH_2CH_3$

MOLECULAR WEIGHT:	116.16
SOURCE OF SAMPLE:	MCB Manufacturing Chemists Norwood, Ohio
CAS REGISTRY NO.:	123-86-4
NIOSH NUMBER:	AF73500
CAPILLARY CELL:	Neat

CONDENSED PHASE

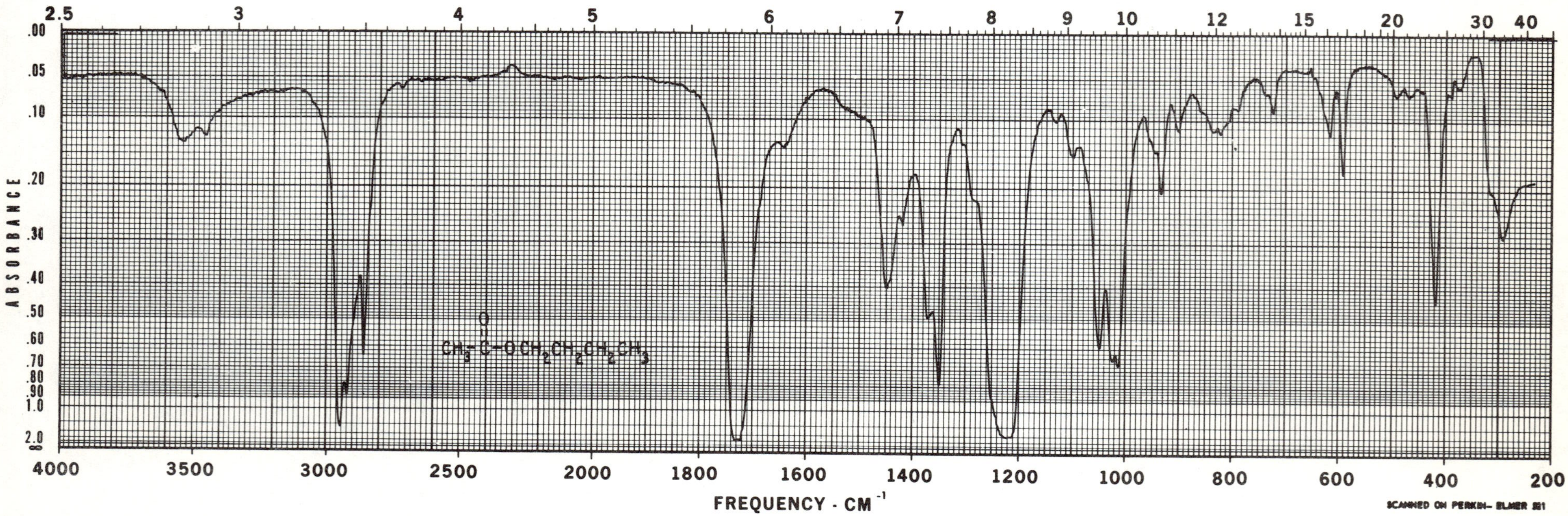

BOILING POINT:	126°C	INJECTION AMT.:	0.5 µl
VAPORIZATION TEMP.:	200°C	PRESSURE:	Atmospheric
CELL VOLUME:	7.5 cc	MAX./MIN.:	–
PATH LENGTH:	7.6 cm	SCALE EXPANSION:	0.93

VAPOR PHASE

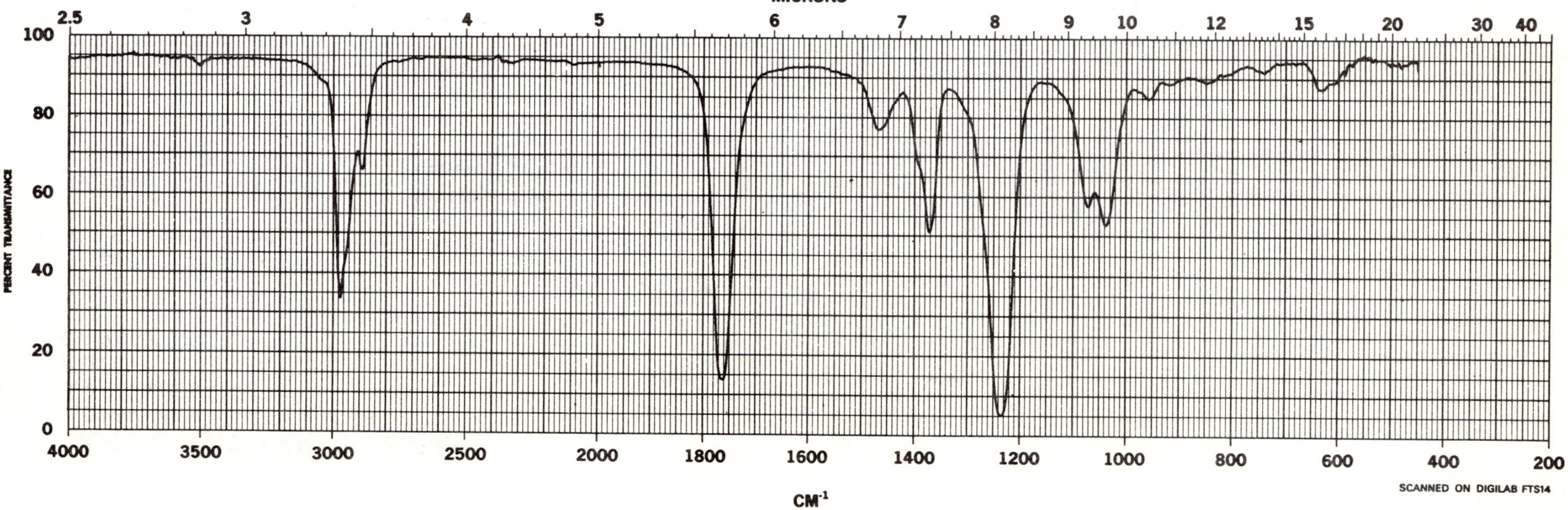

ACETIC ACID, PENTYL ESTER

ACETIC ACID, AMYL ESTER
AMYL ACETATE
PENTYL ACETATE
BANANA OIL

$C_7H_{14}O_2$

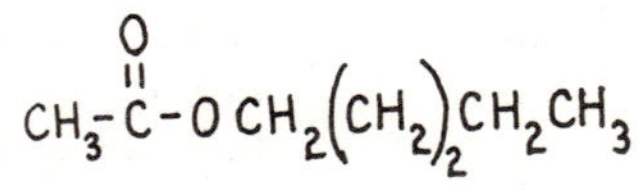

MOLECULAR WEIGHT: 130.19

SOURCE OF SAMPLE: MCB Manufacturing Chemists, Norwood, Ohio

CAS REGISTRY NO.: 628-63-7

NIOSH NUMBER: AJ19250

CAPILLARY CELL: Neat

CONDENSED PHASE

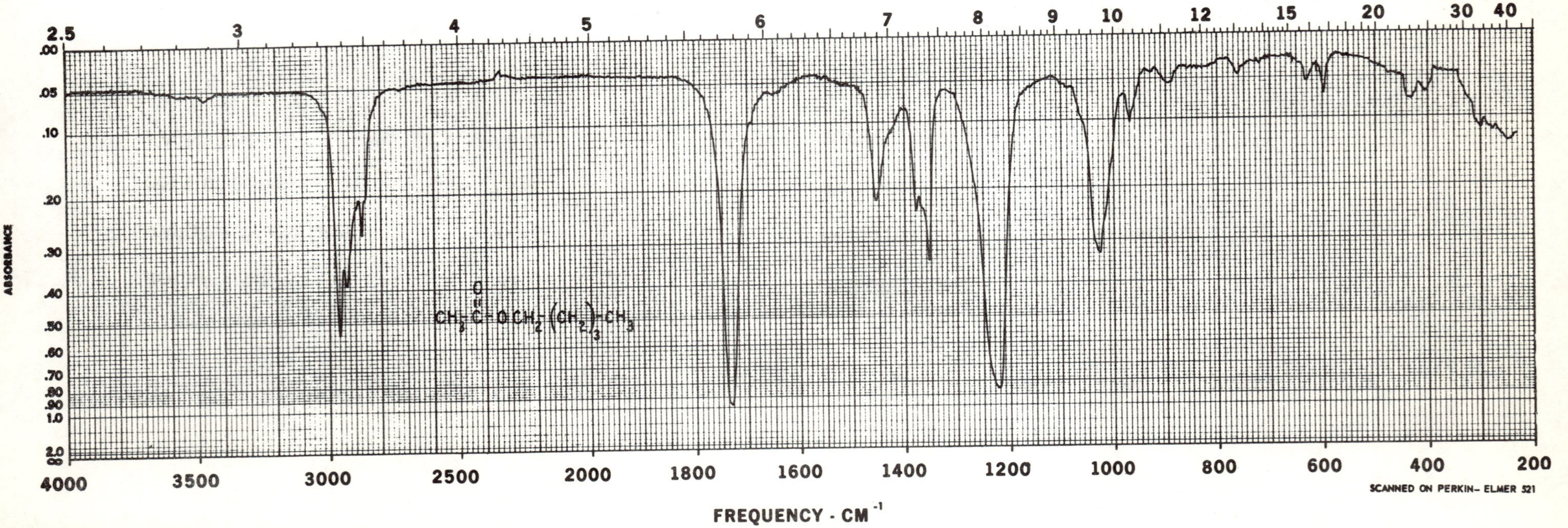

BOILING POINT:	147-149°C	INJECTION AMT.:	0.7 µl
VAPORIZATION TEMP.:	200°C	PRESSURE:	Atmospheric
CELL VOLUME:	7.5 cc	MAX./MIN.:	—
PATH LENGTH:	7.6 cm	SCALE EXPANSION:	0.91

VAPOR PHASE

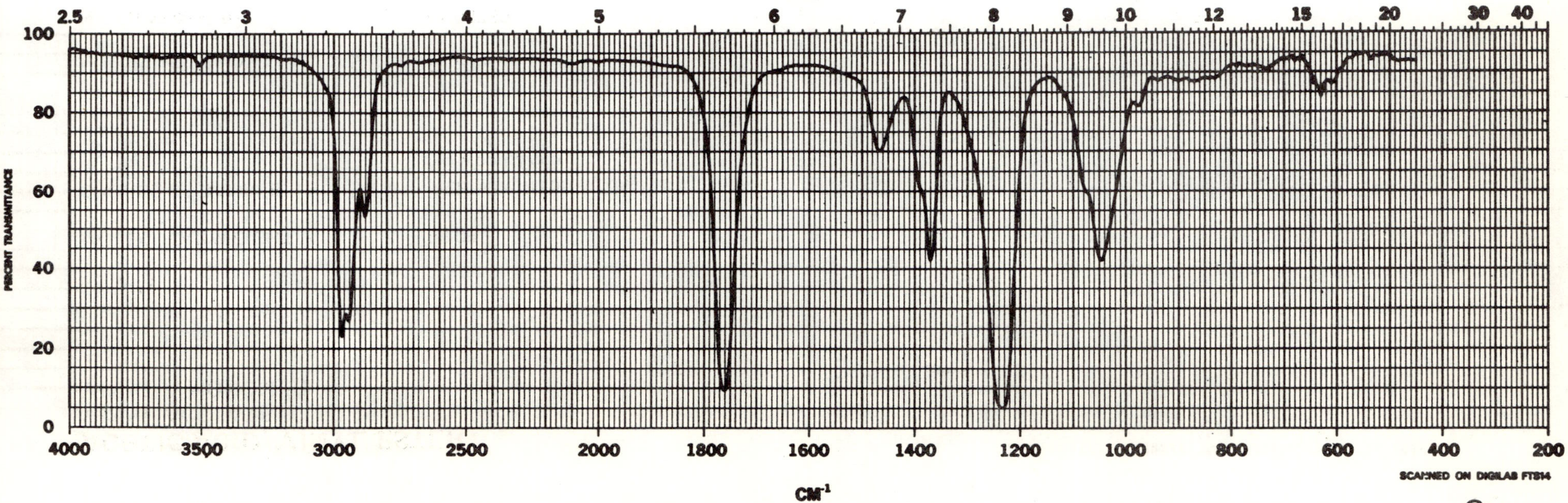

ACETIC ACID, VINYL ESTER
VINYL ACETATE
VINYL A MONOMER

$C_4H_6O_2$

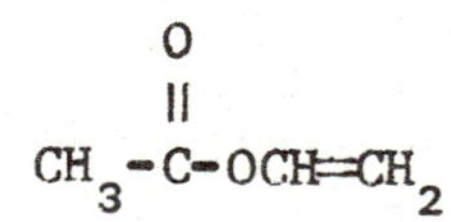

MOLECULAR WEIGHT:	86.09
SOURCE OF SAMPLE:	MCB Manufacturing Chemists Norwood, Ohio
CAS REGISTRY NO.:	108-05-4
NIOSH NUMBER:	AK08750
0.025mm CsBr Cell:	Neat

CONDENSED PHASE

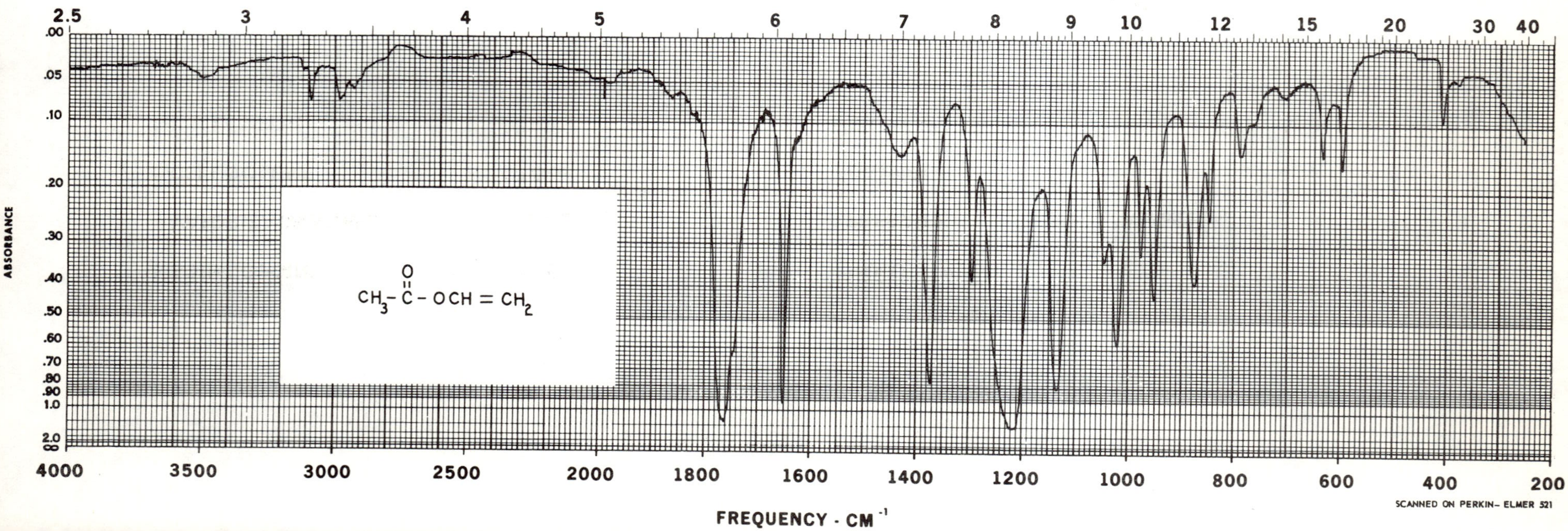

BOILING POINT:	73°C	INJECTION AMT.:	0.2 μl
VAPORIZATION TEMP.:	200°C	PRESSURE:	Atmospheric
CELL VOLUME:	7.5 cc	MAX./MIN.:	–
PATH LENGTH:	7.6 cm	SCALE EXPANSION:	0.90

VAPOR PHASE

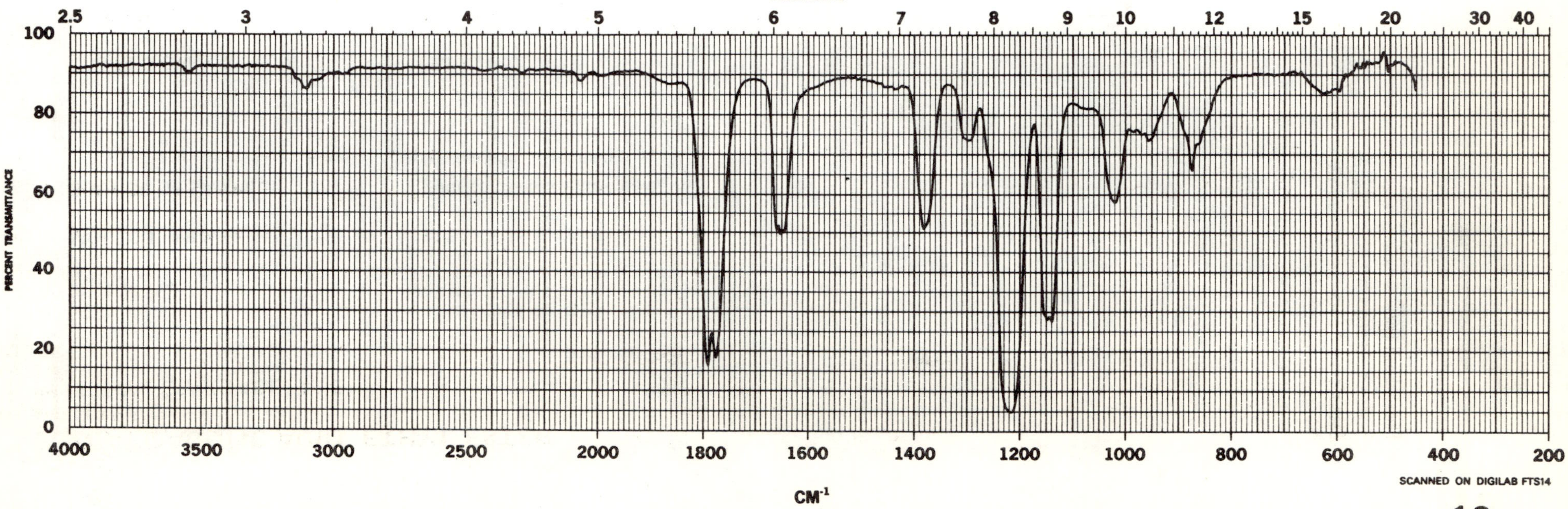

SCANNED ON DIGILAB FTS14

DIAZOACETIC ACID, ETHYL ESTER
DAEE
ETHYL DIAZOACETATE

$C_4H_6N_2O_2$

MOLECULAR WEIGHT: 114.11

SOURCE OF SAMPLE: Aldrich Chemical Company, Inc.
Milwaukee, Wisconsin

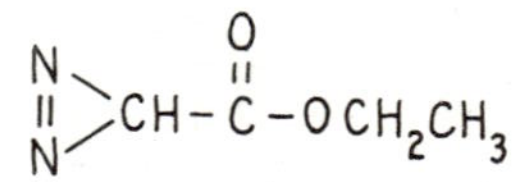

CAS REGISTRY NO.: 623-73-4

NIOSH NUMBER: AG57750

CAPILLARY CELL: Neat

CONDENSED PHASE

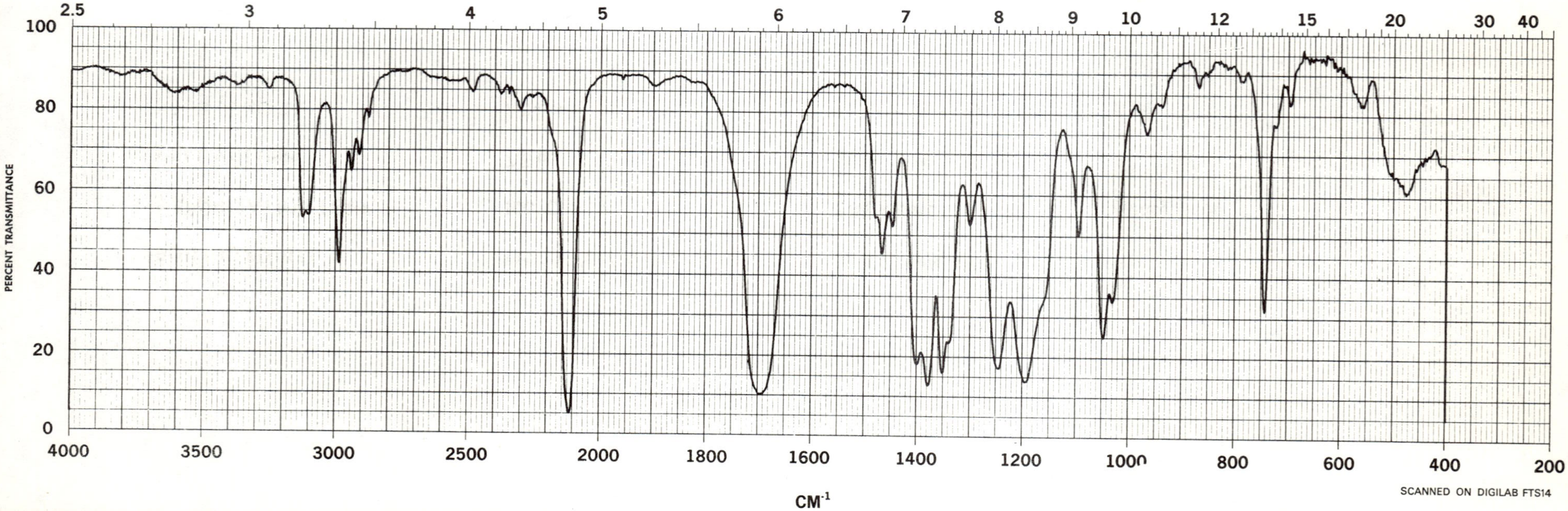

BOILING POINT:	140-141°C/720 mm	INJECTION AMT.:	–
VAPORIZATION TEMP.:	100°C	PRESSURE:	Atmospheric
CELL VOLUME:	7.5 cc	MAX./MIN.:	98.3/11.2
PATH LENGTH:	7.6 cm	SCALE EXPANSION:	1.06

VAPOR PHASE

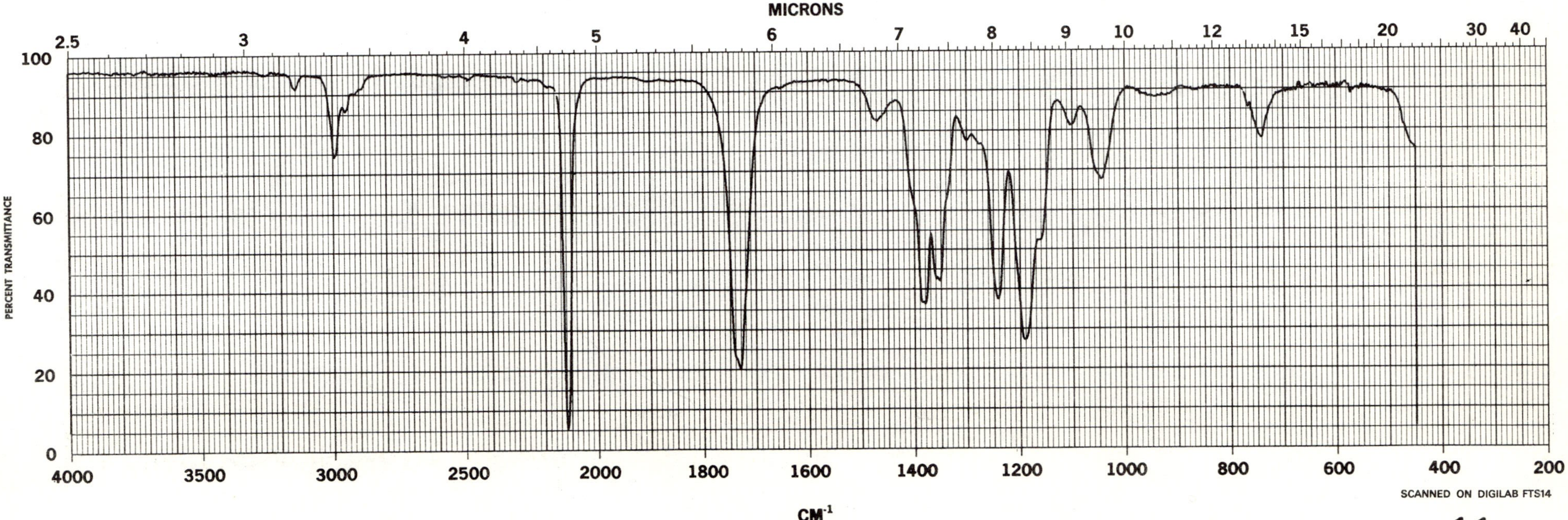

SCANNED ON DIGILAB FTS14

(2,4–DICHLOROPHENOXY)ACETIC ACID

2,4–D
AGROTECT
DED WEED
ESTERON

$C_8H_6Cl_2O_3$

MOLECULAR WEIGHT: 221.04

SOURCE OF SAMPLE: MCB Manufacturing Chemists
Norwood, Ohio

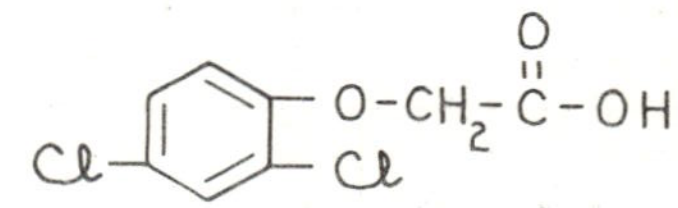

CAS REGISTRY NO.: 94-75-7

NIOSH NUMBER: AG68250

KBr Wafer

CONDENSED PHASE

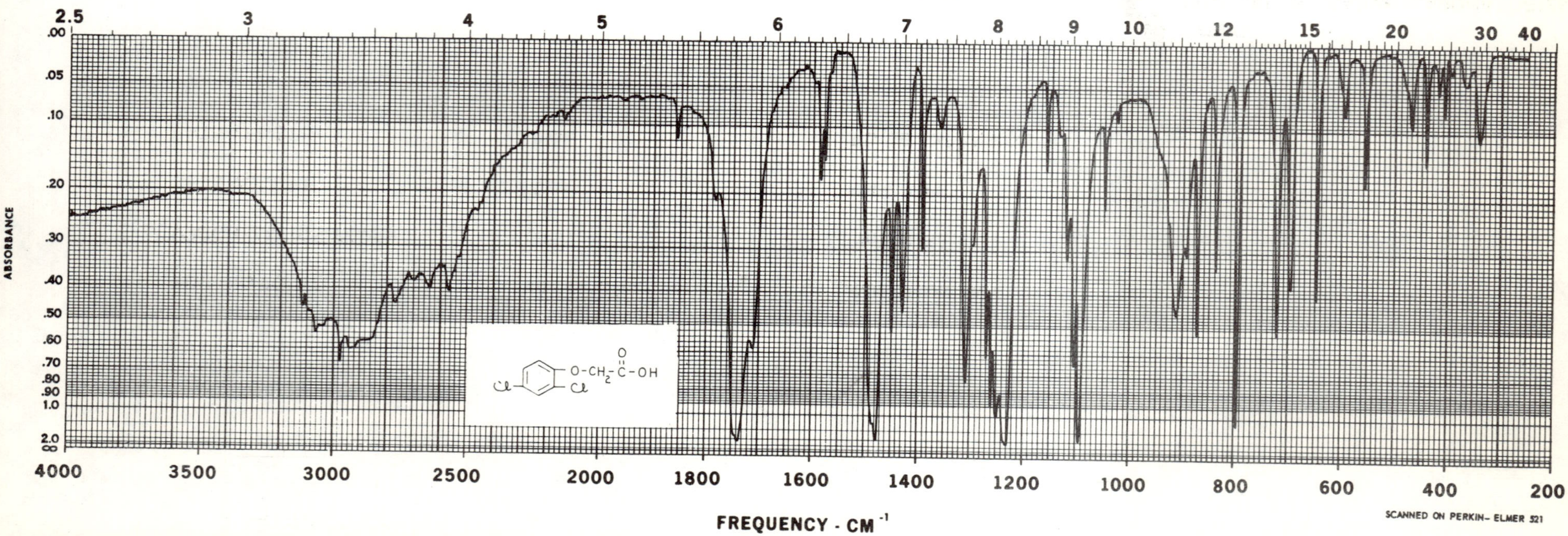

MELTING POINT:	136-140°C	INJECTION AMT.:	0.88 mg
VAPORIZATION TEMP.:	240°C	PRESSURE:	Atmospheric
CELL VOLUME:	7.5 cc	MAX./MIN.:	91.9/32.2
PATH LENGTH:	7.6 cm	SCALE EXPANSION:	1.54

VAPOR PHASE

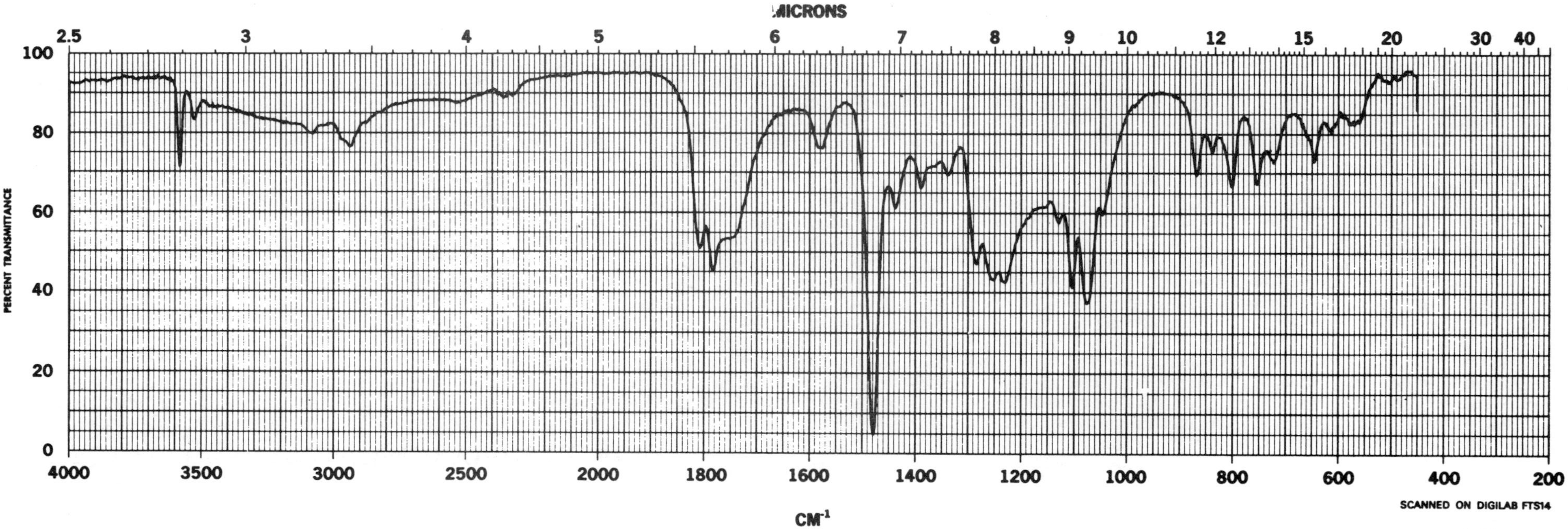

SCANNED ON DIGILAB FTS14

(2,4,5–TRICHLOROPHENOXY)ACETIC ACID

ESTERON
REDDOX
2,4,5–T
WEEDONE

$C_8H_5Cl_3O_3$

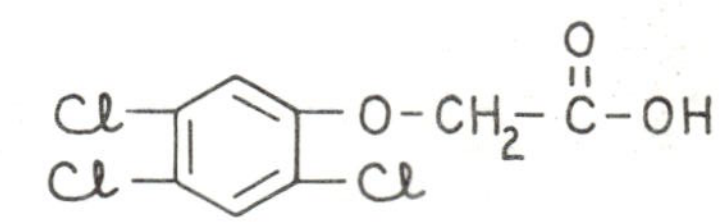

MOLECULAR WEIGHT:	255.49
SOURCE OF SAMPLE:	Fluka AG Buchs, Switzerland
CAS REGISTRY NO.:	93-76-5
NIOSH NUMBER:	AJ84000

KBr Wafer

CONDENSED PHASE

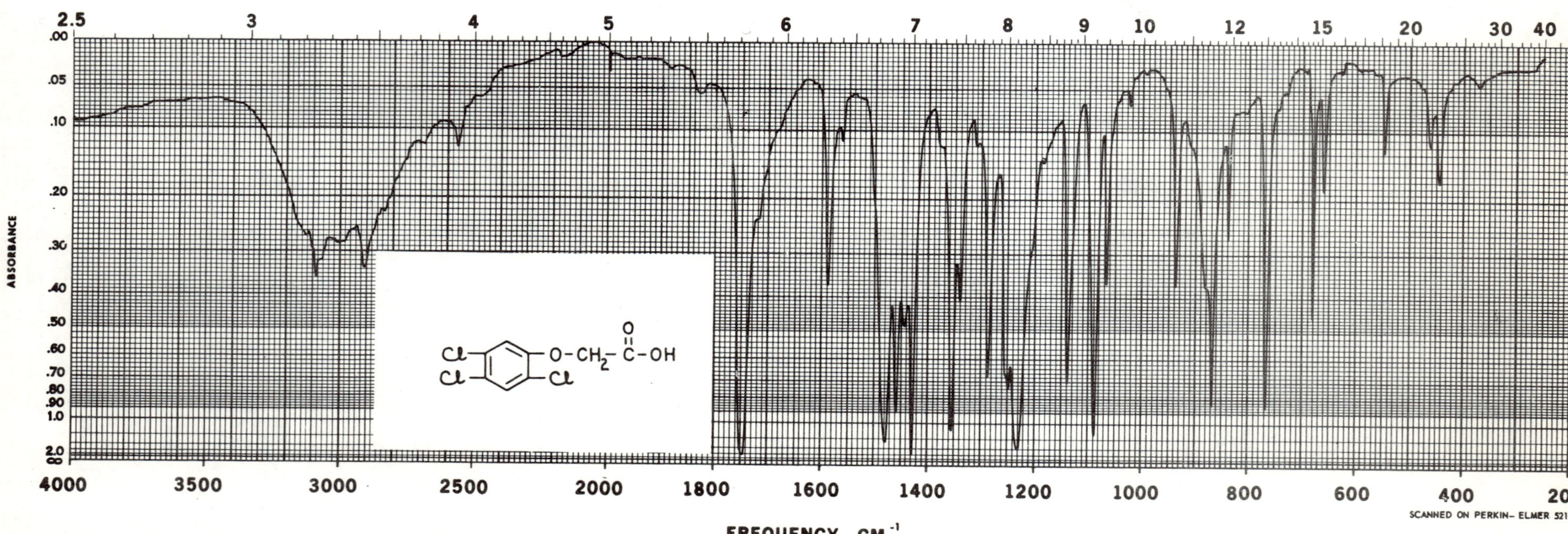

MELTING POINT:	158-159°C	INJECTION AMT.:	—
VAPORIZATION TEMP.:	280°C	PRESSURE:	Atmospheric
CELL VOLUME:	7.5 cc	MAX./MIN.:	96.5/11.2
PATH LENGTH:	7.6 cm	SCALE EXPANSION:	1.08

VAPOR PHASE

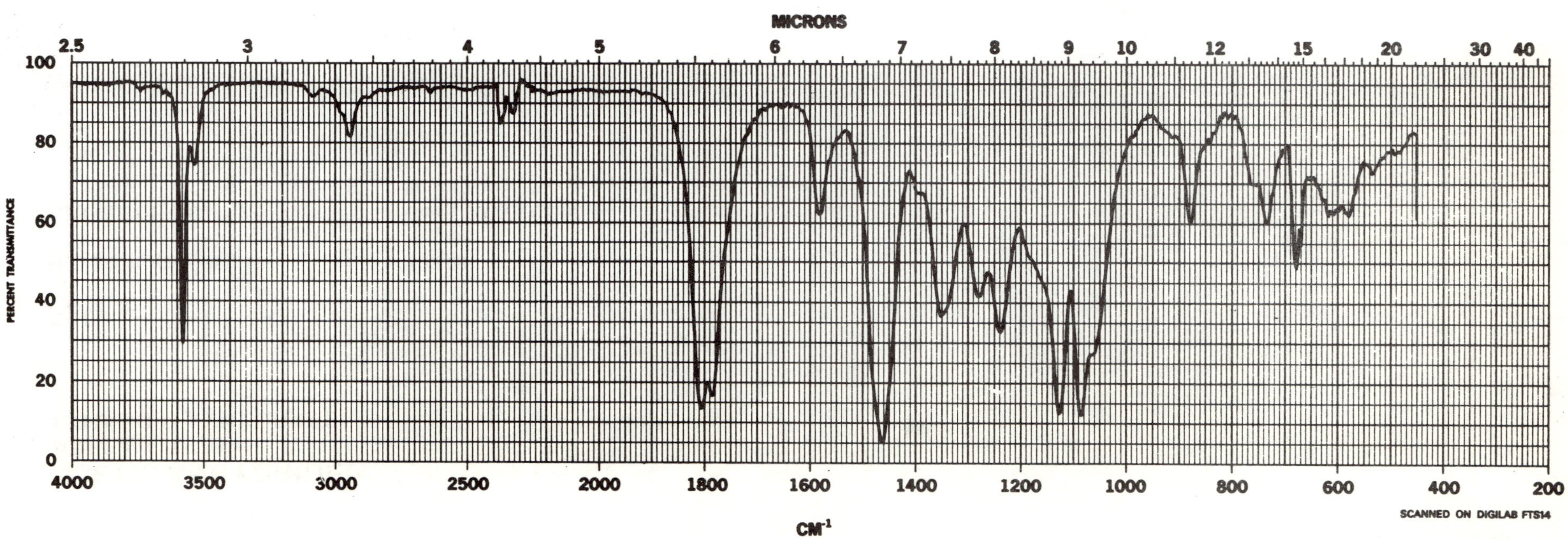

ACETIC ANHYDRIDE
ACETIC OXIDE
ETHANOIC ANHYDRIDE

$C_4H_6O_3$

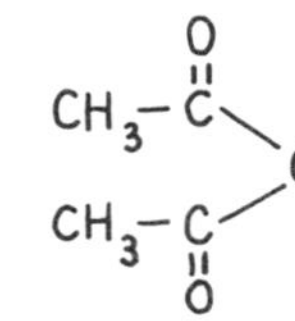

MOLECULAR WEIGHT:	102.09
SOURCE OF SAMPLE:	Chem Service Company West Chester, Pennsylvania
CAS REGISTRY NO.:	108-24-7
NIOSH NUMBER:	AK19250
CAPILLARY CELL:	Neat

CONDENSED PHASE

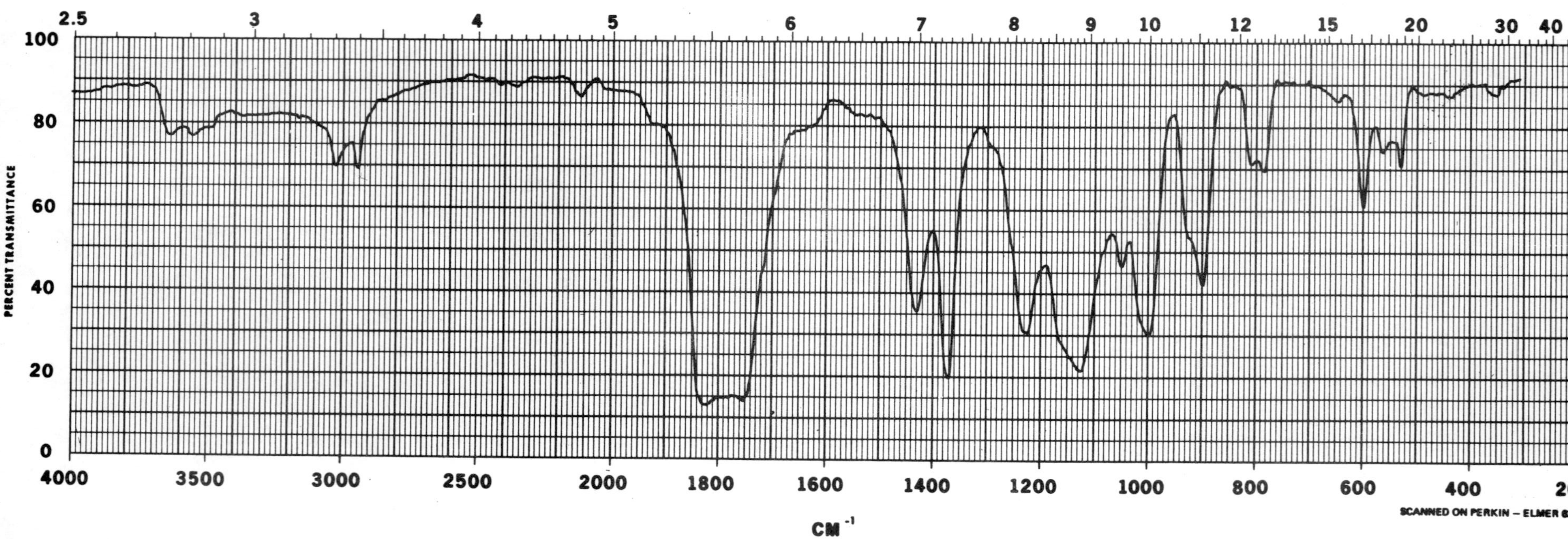

BOILING POINT:	139.6°C	INJECTION AMT.:	0.2 µl
VAPORIZATION TEMP.:	200°C	PRESSURE:	Atmospheric
CELL VOLUME:	7.5 cc	MAX./MIN.:	104.8/1.5
PATH LENGTH:	7.6 cm	SCALE EXPANSION:	0.89

VAPOR PHASE

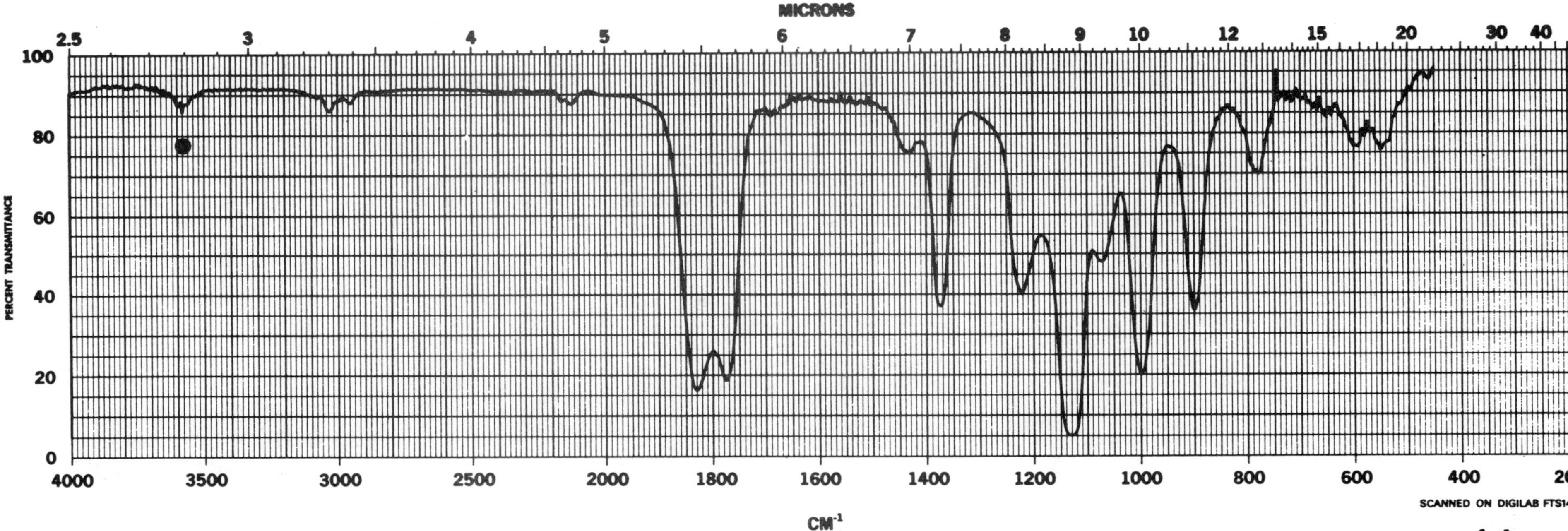

SCANNED ON DIGILAB FTS14

ACETYL CHLORIDE
ACETIC ACID, CHLORIDE
ETHANOYL CHLORIDE

C_2H_3ClO

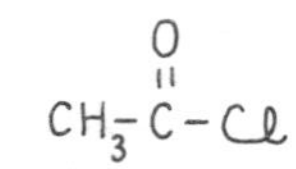

MOLECULAR WEIGHT: 78.50

SOURCE OF SAMPLE: J. T. Baker Chemical Company
Phillipsburg, New Jersey

CAS REGISTRY NO.: 75-36-5

NIOSH NUMBER: AO63900

CAPILLARY CELL: 0.015 mm

CONDENSED PHASE

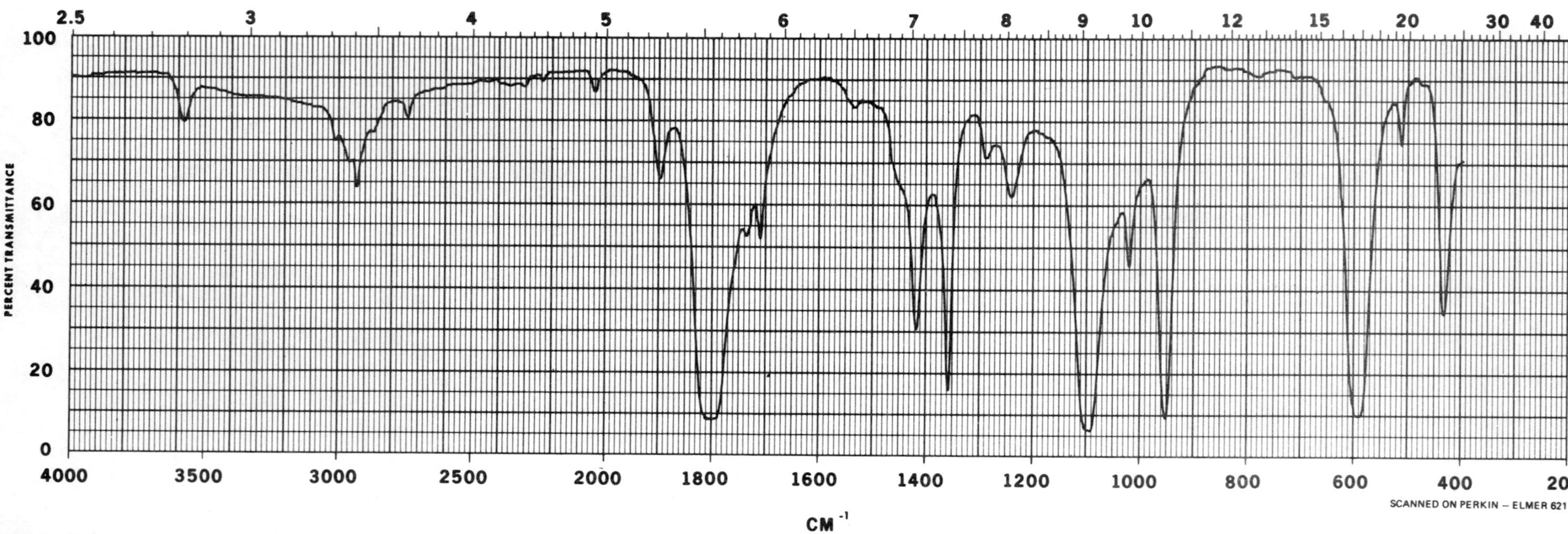

BOILING POINT:	51-52°C (lit.)	INJECTION AMT.:	0.1 μl
VAPORIZATION TEMP.:	160°C	PRESSURE:	Atmospheric
CELL VOLUME:	7.5 cc	MAX./MIN.:	104.0/10.5
PATH LENGTH:	7.6 cm	SCALE EXPANSION:	0.98

VAPOR PHASE

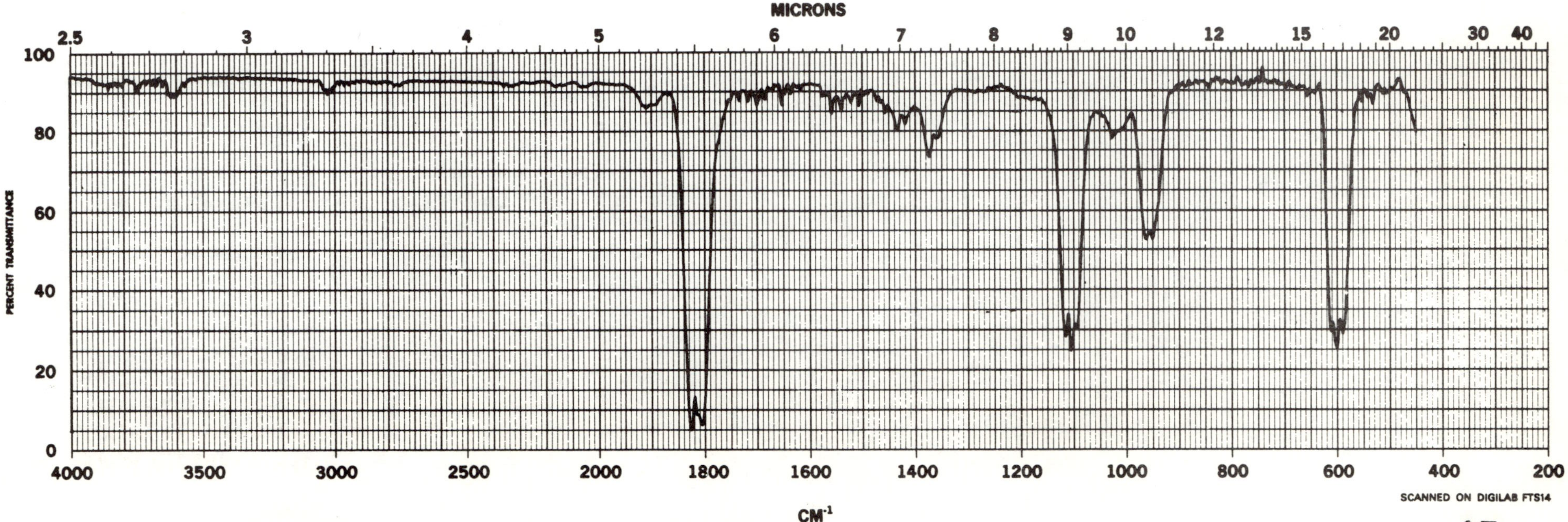

SCANNED ON DIGILAB FTS14

ACROLEIN
ACRYLALDEHYDE
PROPENAL
ALLYL ALDEHYDE
ETHYLENE ALDEHYDE

C_3H_4O

$CH_2=CH-\overset{O}{\overset{\|}{C}}H$

MOLECULAR WEIGHT:	56.06
SOURCE OF SAMPLE:	Fluka AG Buchs, Switzerland
CAS REGISTRY NO.:	107-02-8
NIOSH NUMBER:	AS10500
CAPILLARY CELL:	0.015 mm

CONDENSED PHASE

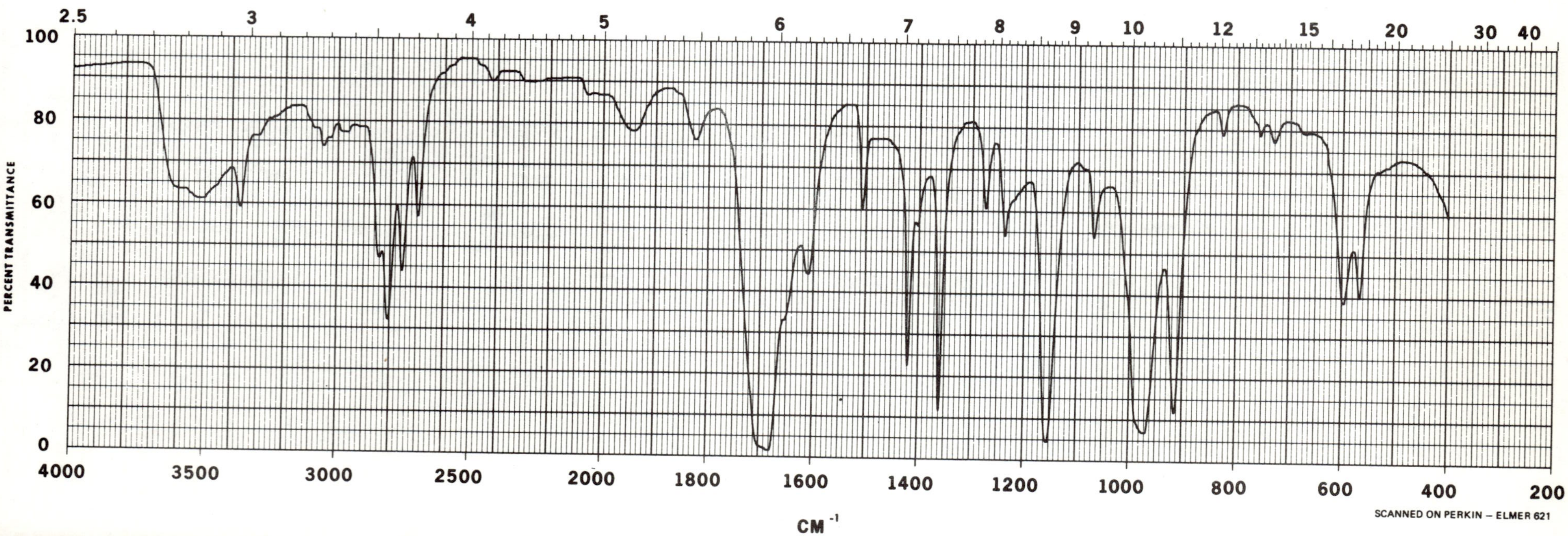

BOILING POINT:	51-53°C	INJECTION AMT.:	0.5 μl
VAPORIZATION TEMP.:	200°C	PRESSURE:	Atmospheric
CELL VOLUME:	7.5 cc	MAX./MIN.:	—
PATH LENGTH:	7.6 cm	SCALE EXPANSION:	0.92

VAPOR PHASE

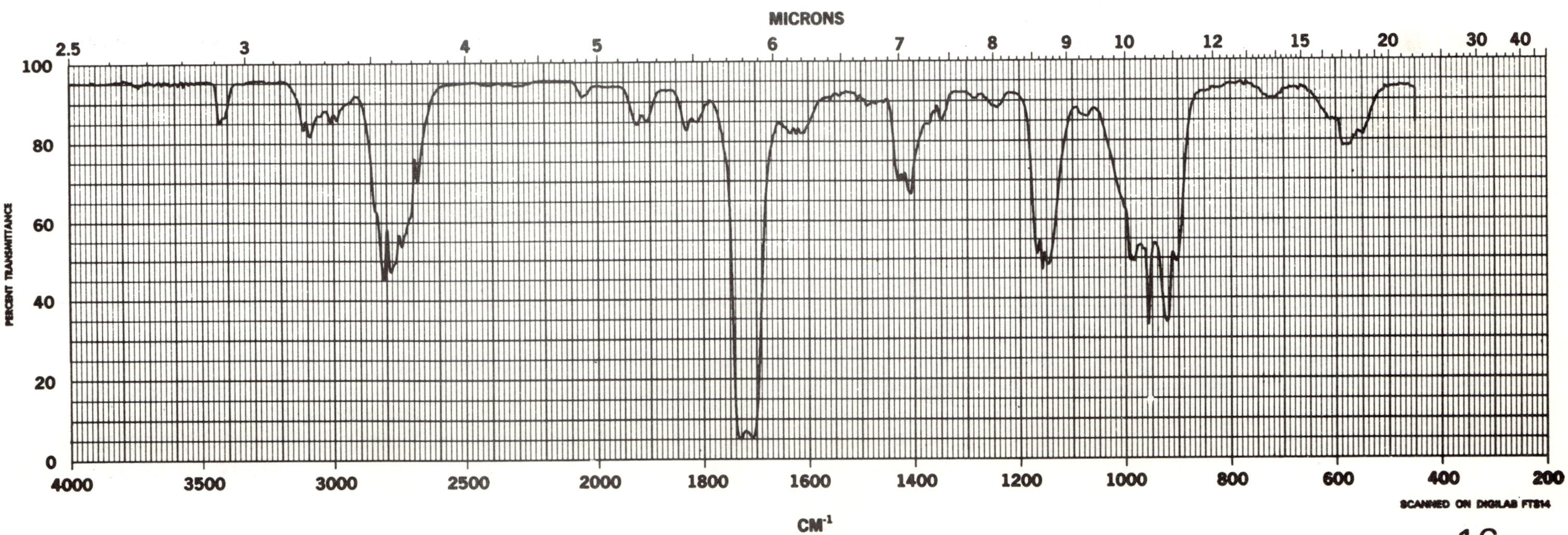

ACRYLIC ACID
ACROLEIC ACID
ETHYLENECARBOXYLIC ACID
PROPENOIC ACID

$C_3H_4O_2$

MOLECULAR WEIGHT: 72.06

SOURCE OF SAMPLE: MCB Manufacturing Chemists
Norwood, Ohio

$CH_2=CH-C(=O)-OH$

CAS REGISTRY NO.: 79-10-7

NIOSH NUMBER: AS43750

A = CAPILLARY CELL: Neat
B = CAPILLARY CELL: 0.015 mm

CONDENSED PHASE

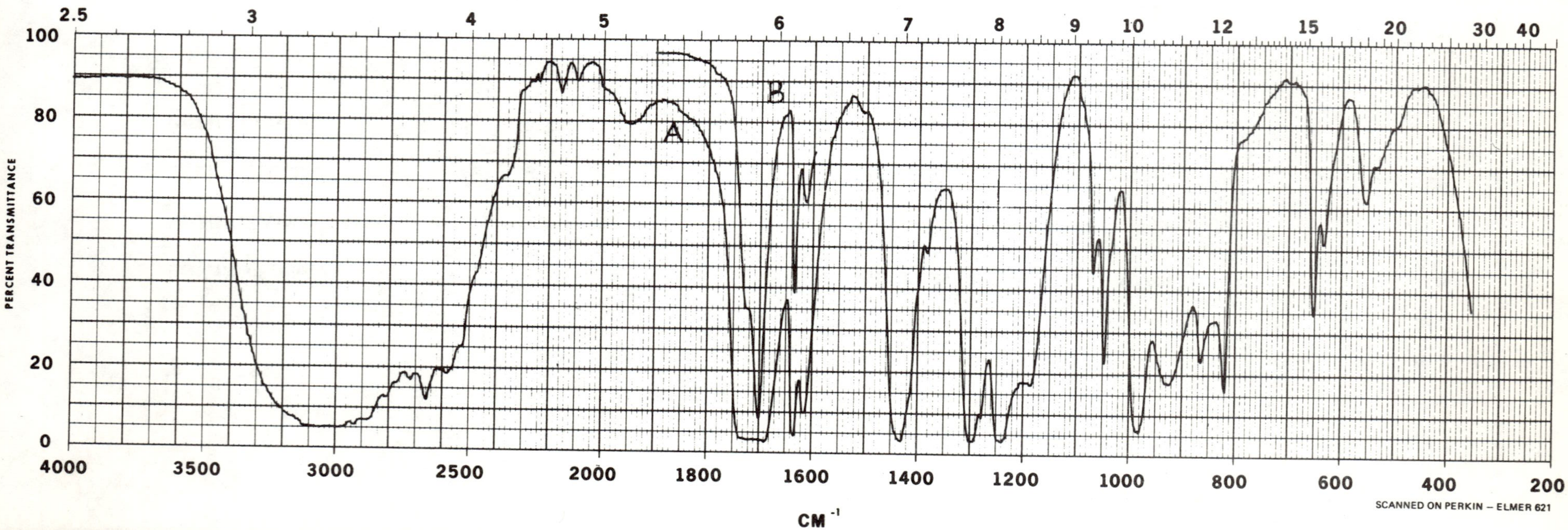

BOILING POINT:	142°C	INJECTION AMT.:	0.4 µl
VAPORIZATION TEMP.:	200°C	PRESSURE:	Atmospheric
CELL VOLUME:	7.5 cc	MAX./MIN.:	101.3/31.9
PATH LENGTH:	7.6 cm	SCALE EXPANSION:	1.33

VAPOR PHASE

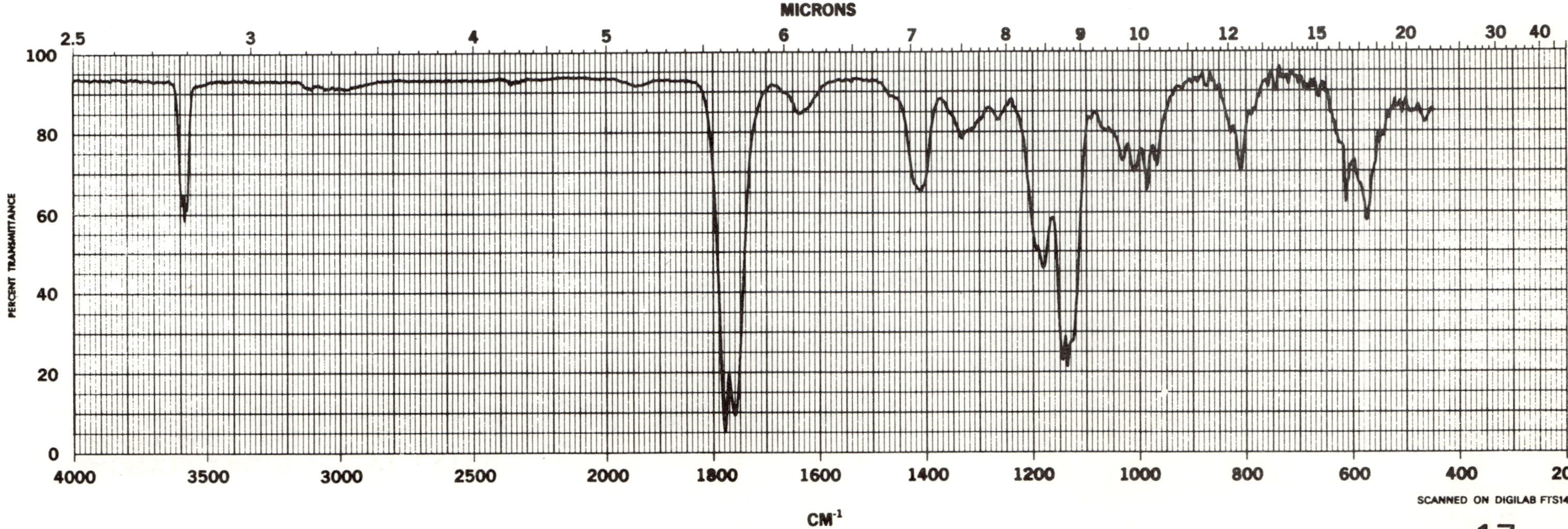

ACRYLIC ACID, BUTYL ESTER
BUTYL ACRYLATE
BUTYL ALCOHOL, ACRYLATE

$C_7H_{12}O_2$

MOLECULAR WEIGHT: 128.17

SOURCE OF SAMPLE: MCB Manufacturing Chemists
Norwood, Ohio

$CH_2=CH-\overset{O}{\overset{\|}{C}}-OCH_2CH_2CH_2CH_3$

CAS REGISTRY NO.: 141-32-2

NIOSH NUMBER: UD31500

CAPILLARY CELL: Neat

CONDENSED PHASE

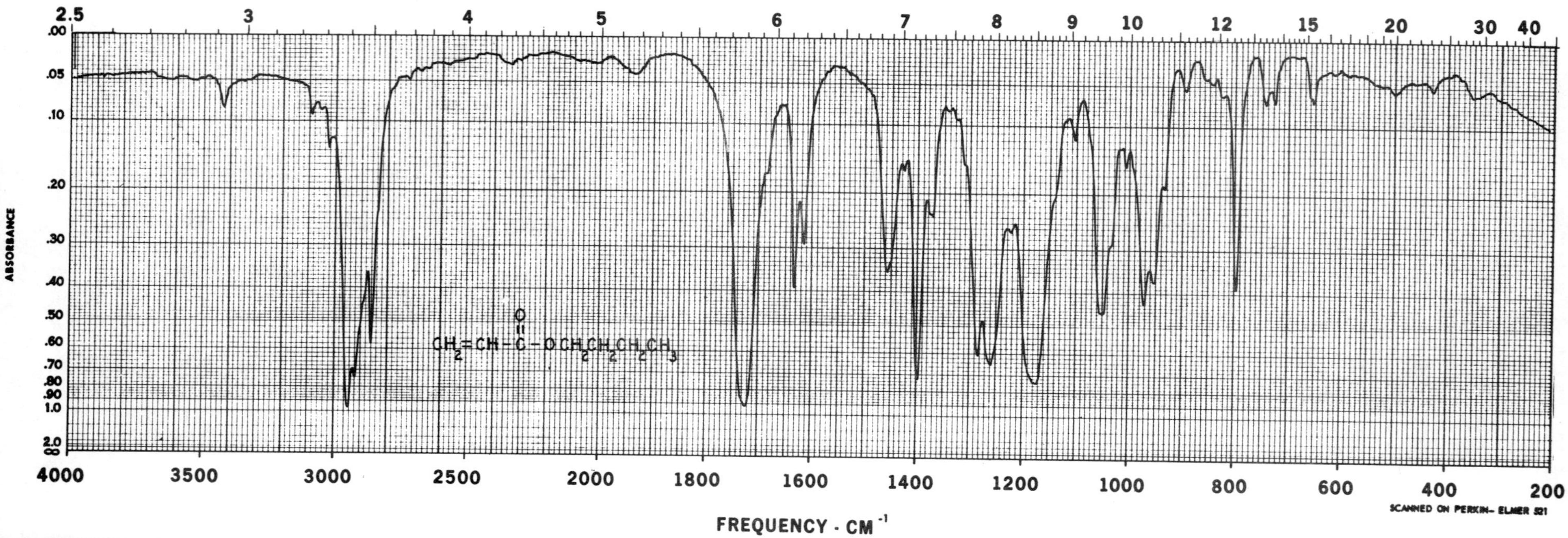

BOILING POINT:	145-148°C	INJECTION AMT.:	0.4 μl
VAPORIZATION TEMP.:	160°C	PRESSURE:	Atmospheric
CELL VOLUME:	7.5 cc	MAX./MIN.:	109.8/5.0
PATH LENGTH:	7.6 cm	SCALE EXPANSION:	0.88

VAPOR PHASE

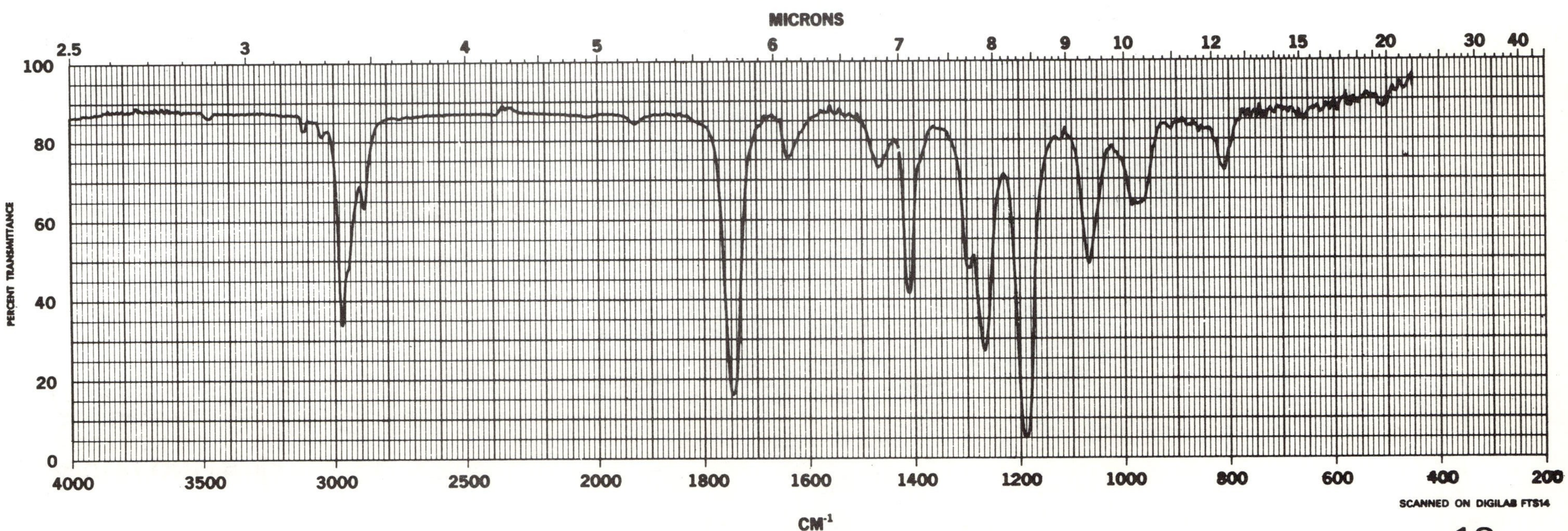

ACRYLONITRILE
ACRYLON
CARBACRYL
PROPENENITRILE

C_3H_3N

$CH_2=CH-C\equiv N$

MOLECULAR WEIGHT:	53.06
SOURCE OF SAMPLE:	MCB Manufacturing Chemists Norwood, Ohio
CAS REGISTRY NO.:	107-31-1
NIOSH NUMBER:	AT52500
CAPILLARY CELL:	Neat

CONDENSED PHASE

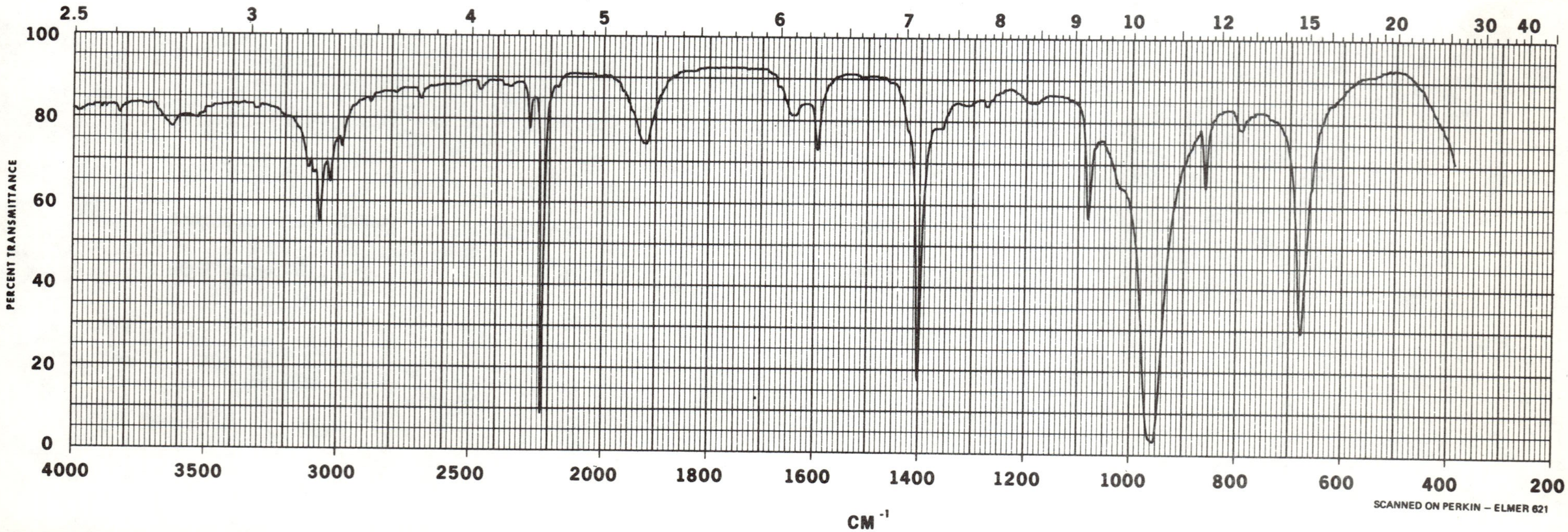

BOILING POINT:	78.5°C	INJECTION AMT.:	2.0 µl
VAPORIZATION TEMP.:	200°C	PRESSURE:	Atmospheric
CELL VOLUME:	7.5 cc	MAX./MIN.:	–
PATH LENGTH:	7.6 cm	SCALE EXPANSION:	1.07

VAPOR PHASE

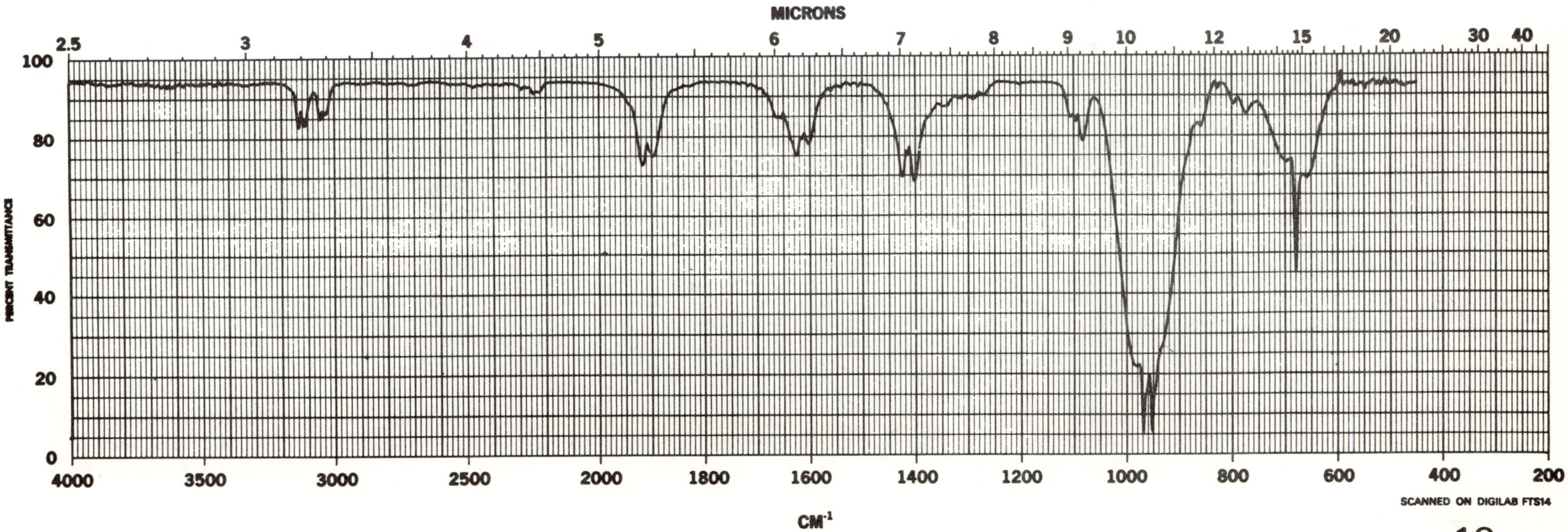

SCANNED ON DIGILAB FTS14

ALLYL ALCOHOL
2–PROPEN–1–OL
VINYLCARBINOL

C_3H_6O

MOLECULAR WEIGHT: 58.08

SOURCE OF SAMPLE: MCB Manufacturing Chemists
Norwood, Ohio

$CH_2=CHCH_2-OH$

CAS REGISTRY NO.: 107-18-6

NIOSH NUMBER: BA50750

CAPILLARY CELL: Neat

CONDENSED PHASE

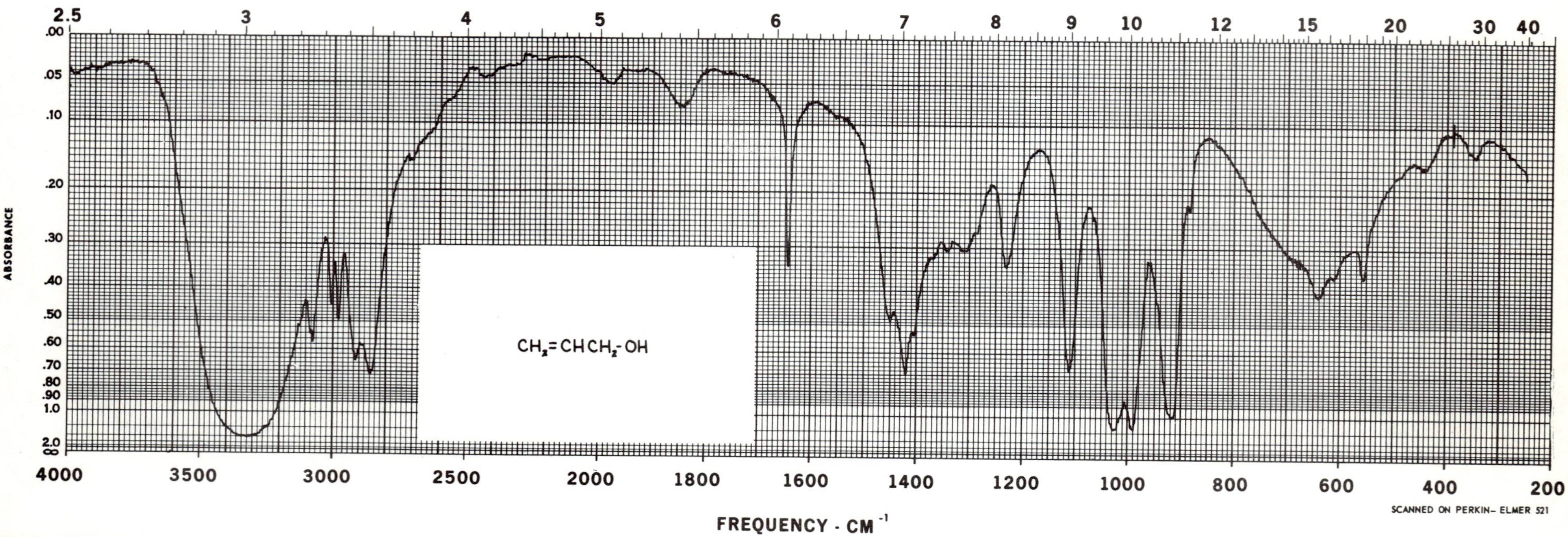

BOILING POINT:	97.2^{o}C/763 mm	INJECTION AMT.:	3.0 µl
VAPORIZATION TEMP.:	200°C	PRESSURE:	Atmospheric
CELL VOLUME:	7.5 cc	MAX./MIN.:	—
PATH LENGTH:	7.6 cm	SCALE EXPANSION:	0.96

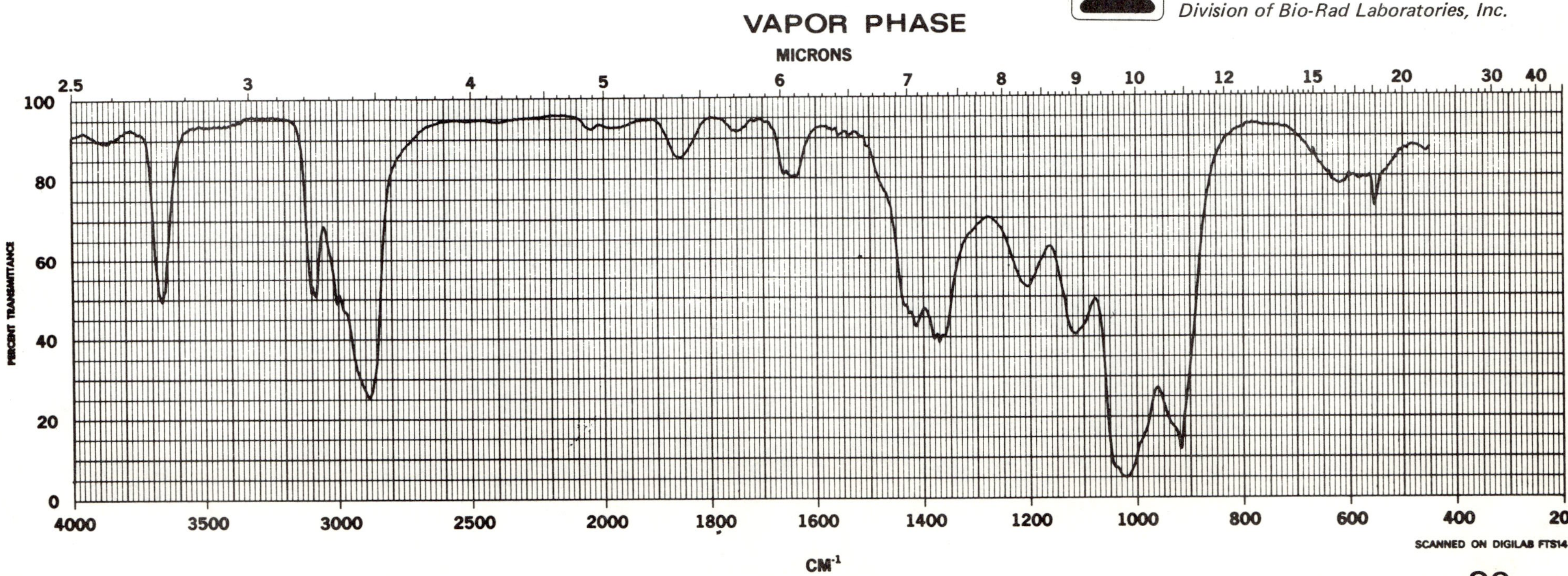

AMMONIA
AM–FOL

H_3N

MOLECULAR WEIGHT: 17.03

SOURCE OF SAMPLE: MCB Manufacturing Chemists
Norwood, Ohio

NH_3

CAS REGISTRY NO.: 7664-41-7

NIOSH NUMBER: BO08750

GAS CELL: 10 cm

A GAS AT 25°C

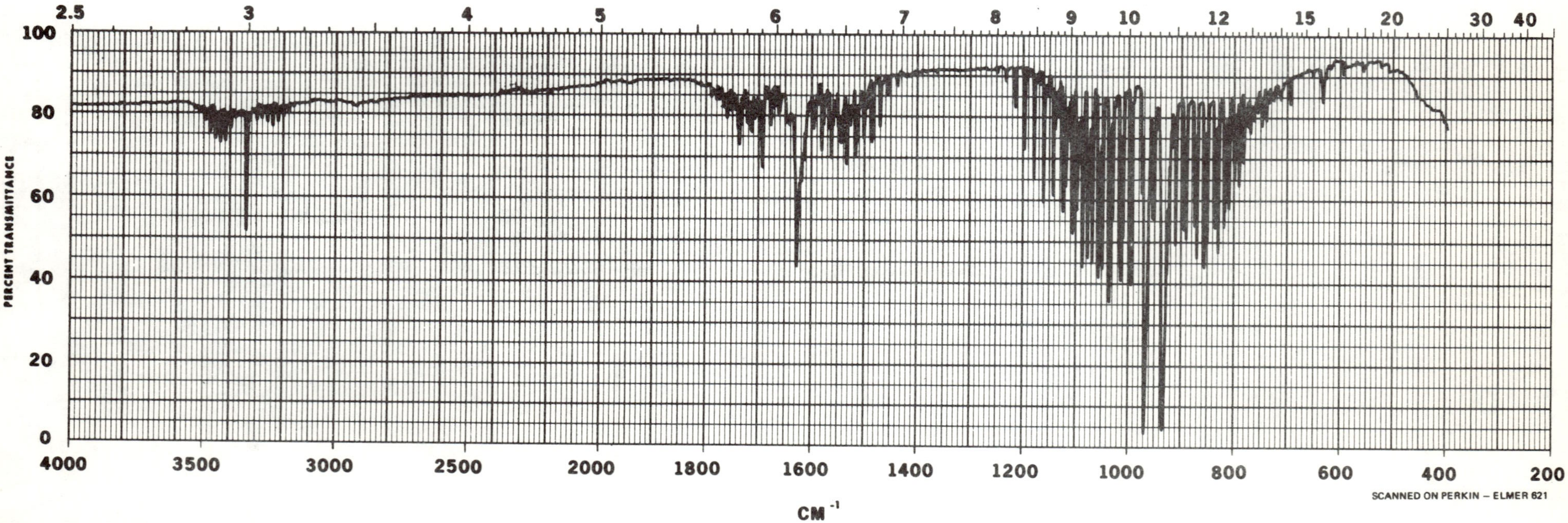

BOILING POINT:	–33.35°C	INJECTION AMT.:	250 µl
VAPORIZATION TEMP.:	200°C	PRESSURE:	Atmospheric
CELL VOLUME:	7.5 cc	MAX./MIN.:	104.3/42.6
PATH LENGTH:	7.6 cm	SCALE EXPANSION:	1.49

VAPOR PHASE

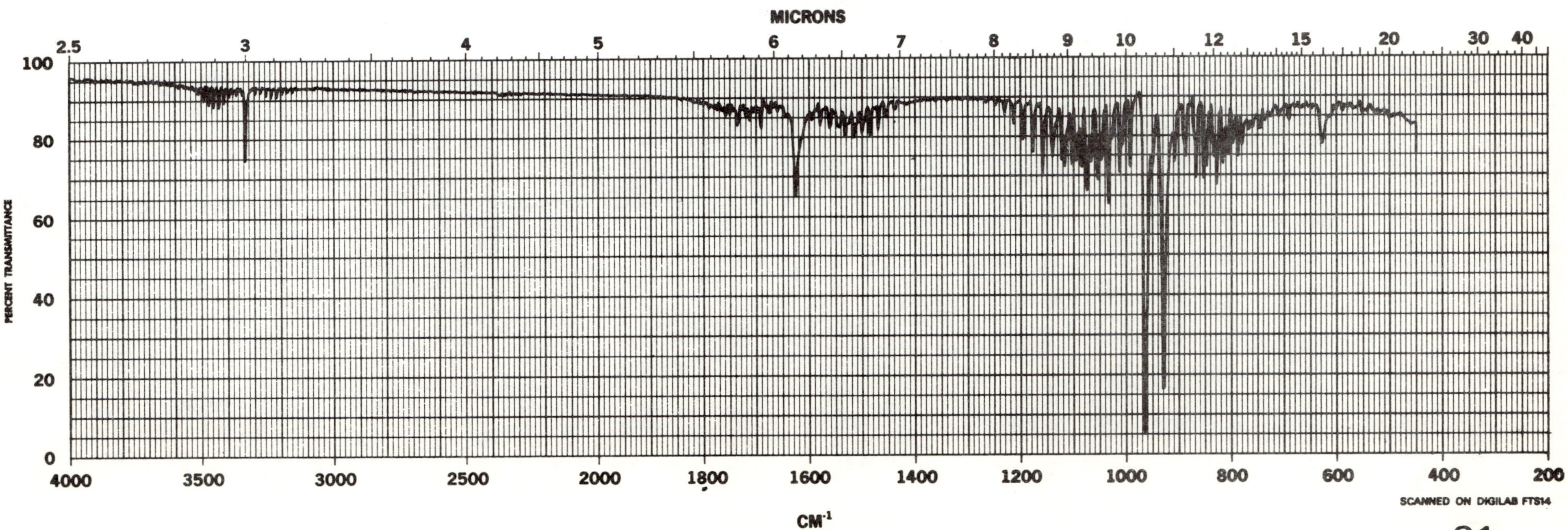

ANILINE
AMINOBENZENE
BENZENAMINE
BENZIDAM
AMINOPHEN

C_6H_7N

MOLECULAR WEIGHT: 93.13

SOURCE OF SAMPLE: Allied Chemical Corporation
New York, New York

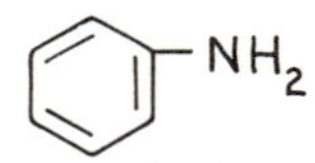

CAS REGISTRY NO.: 62-53-3

NIOSH NUMBER: BW66500

CAPILLARY CELL: Neat

CONDENSED PHASE

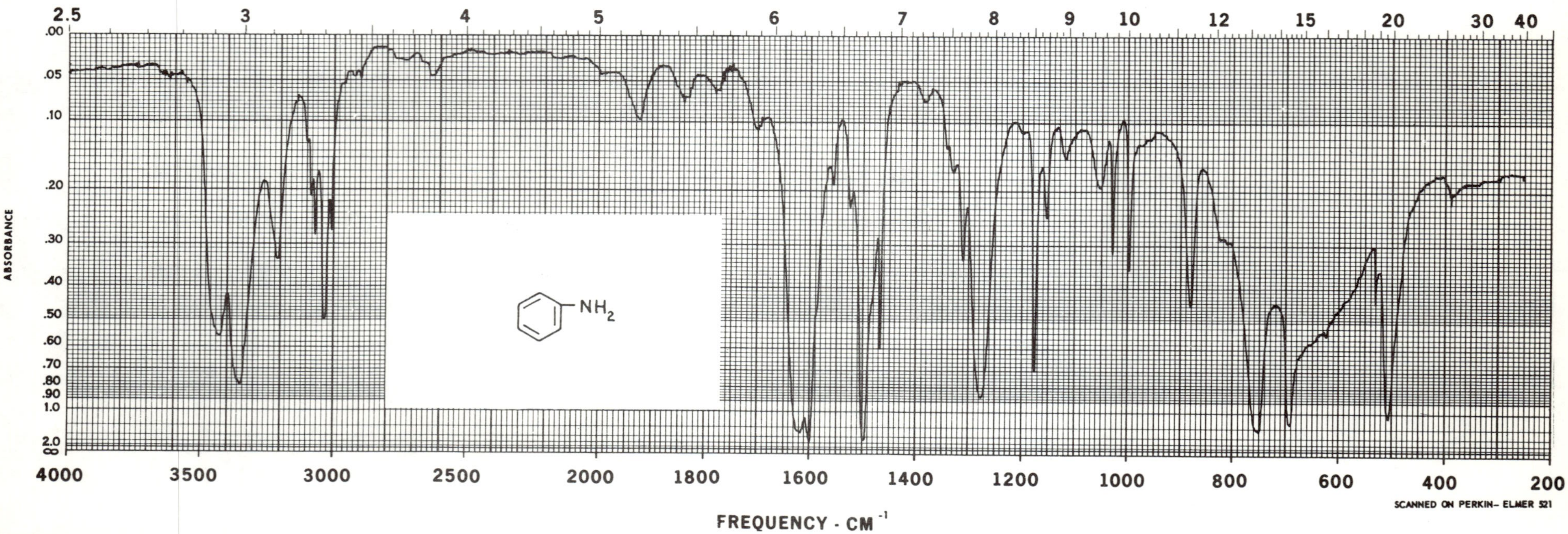

BOILING POINT:	184°C	INJECTION AMT.:	1.0 µl
VAPORIZATION TEMP.:	200°C	PRESSURE:	Atmospheric
CELL VOLUME:	7.5 cc	MAX./MIN.:	—
PATH LENGTH:	7.6 cm	SCALE EXPANSION:	1.02

VAPOR PHASE

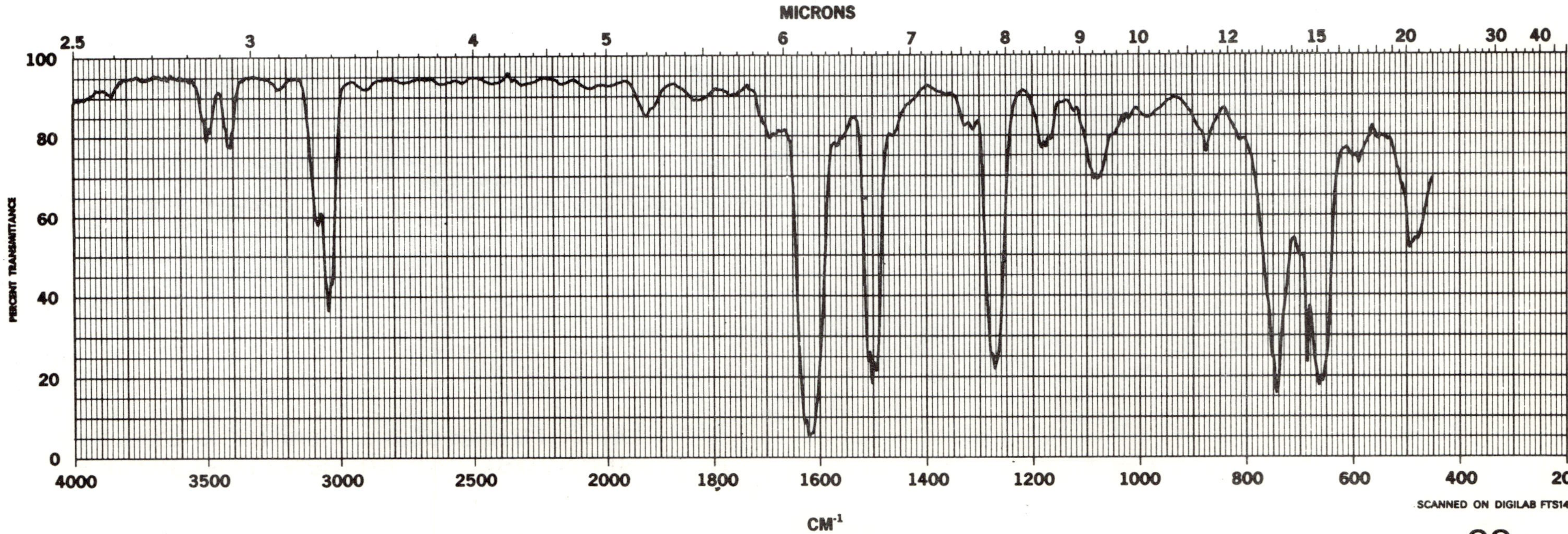

SCANNED ON DIGILAB FTS14

4,4′–METHYLENEBIS[2–CHLOROANILINE]
4,4′–METHYLENEBIS[2–CHLOROBENZENAMINE]
CYANASET
MOCA

$C_{13}H_{12}Cl_2N_2$

MOLECULAR WEIGHT:	267.16
SOURCE OF SAMPLE:	E. I. du Pont de Nemours and Company, Inc. Wilmington, Delaware
CAS REGISTRY NO.:	101-14-4
NIOSH NUMBER:	CY10500

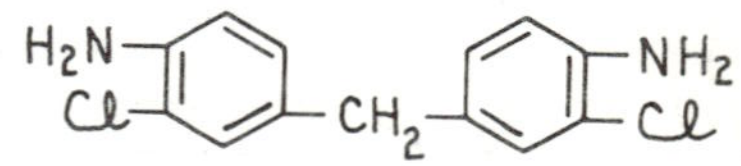

KBr Wafer

CONDENSED PHASE

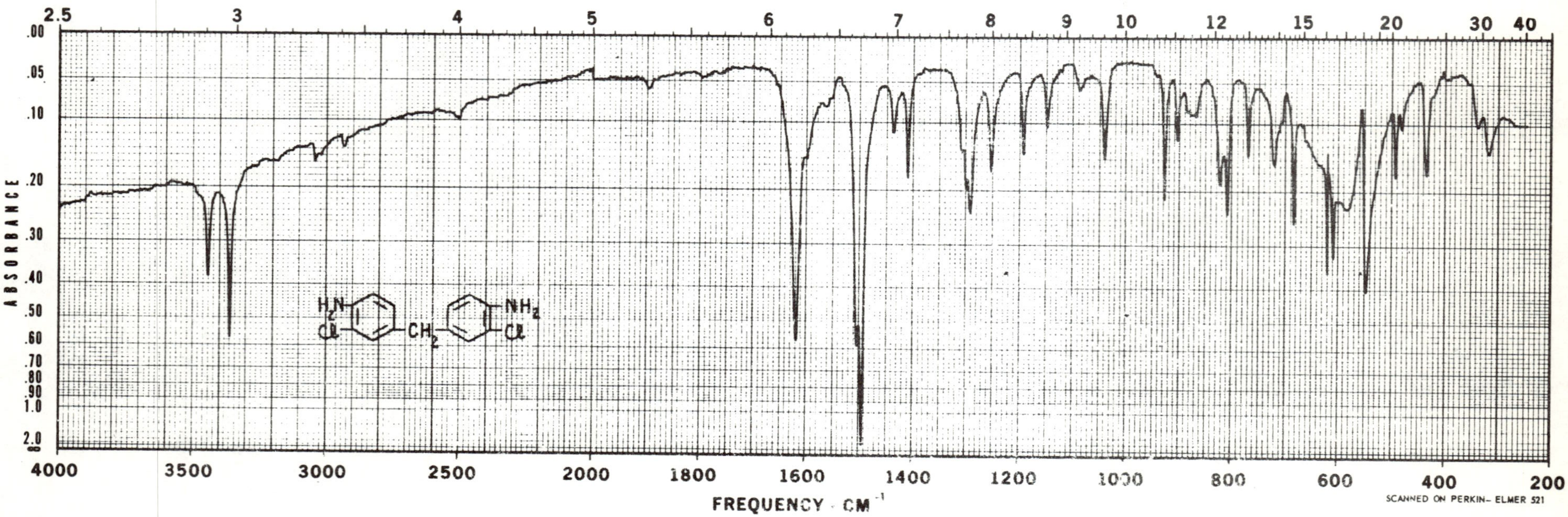

BOILING POINT:	–	INJECTION AMT.:	–
VAPORIZATION TEMP.:	280°C	PRESSURE:	Atmospheric
CELL VOLUME:	7.5 cc	MAX./MIN.:	99.6/39.0
PATH LENGTH:	7.6 cm	SCALE EXPANSION:	1.5

VAPOR PHASE

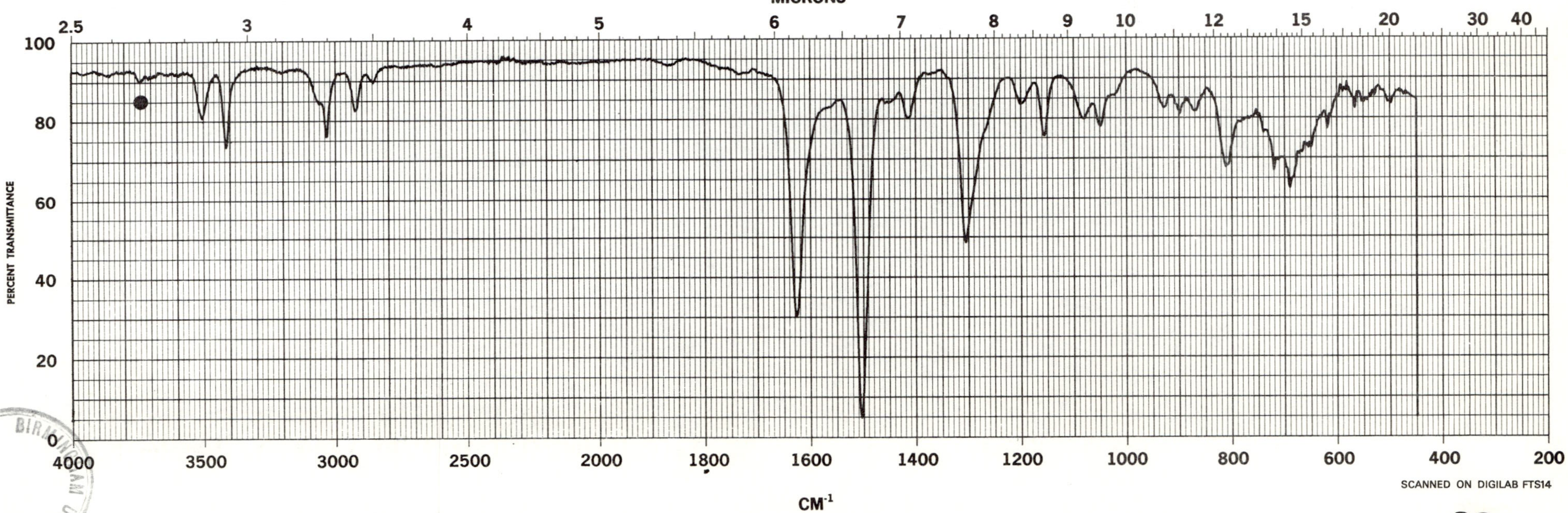

SCANNED ON DIGILAB FTS14

23

N–METHYL–N–NITROSOANILINE
N–METHYL–N–NITROSOBENZENAMINE
NMA

$C_7H_8N_2O$

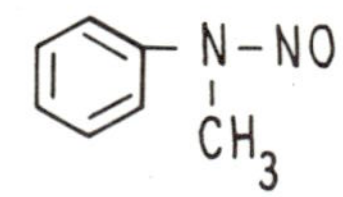

MOLECULAR WEIGHT: 136.15

SOURCE OF SAMPLE: Eastman Organic Chemicals
Rochester, New York

CAS REGISTRY NO.: 614-00-6

NIOSH NUMBER: BY57750

CAPILLARY CELL: Neat

CONDENSED PHASE

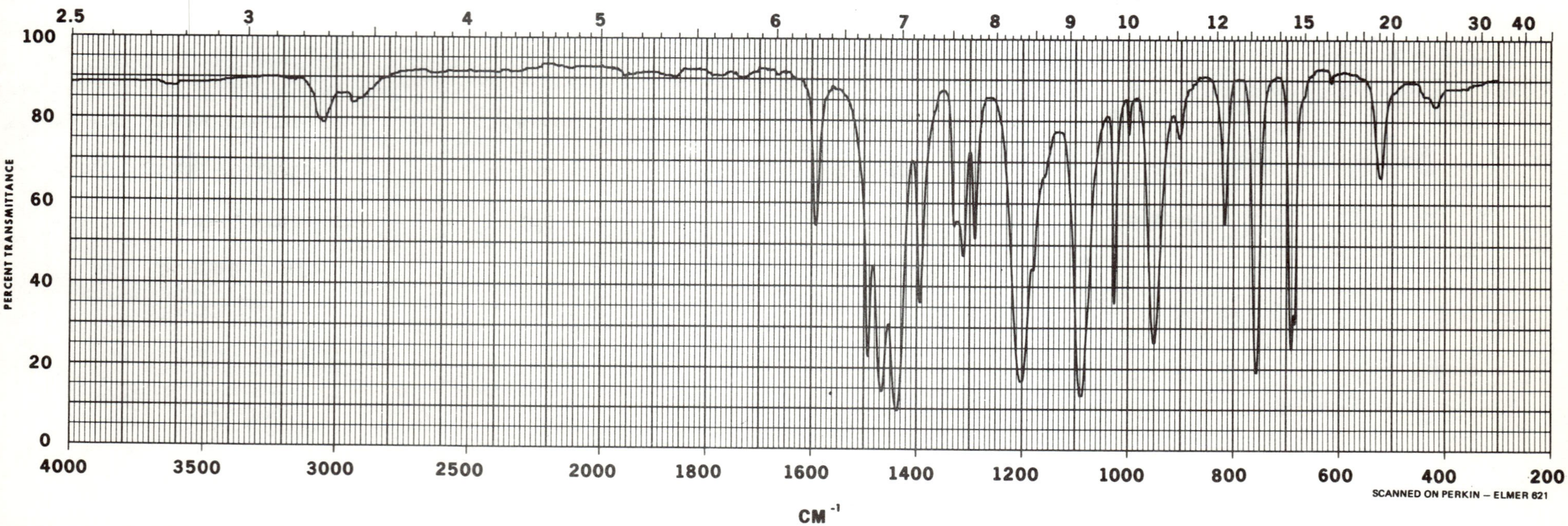

BOILING POINT:	225°C (dec.)	INJECTION AMT.:	0.8 µl
VAPORIZATION TEMP.:	200°C	PRESSURE:	Atmospheric
CELL VOLUME:	7.5 cc	MAX./MIN.:	—
PATH LENGTH:	7.6 cm	SCALE EXPANSION:	0.93

VAPOR PHASE

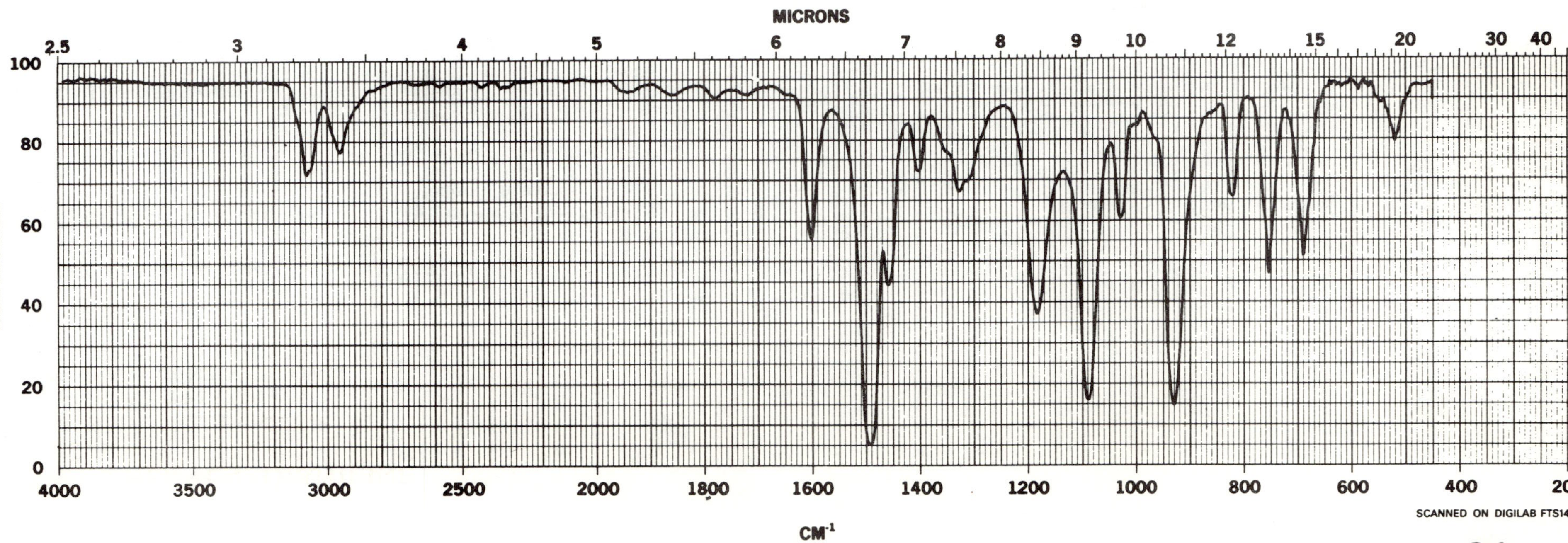

p–(PHENYLAZO)ANILINE
AAB
4--AMINOAZOBENZENE
ANILINE YELLOW
C. I. SOLVENT YELLOW 1

$C_{12}H_{11}N_3$

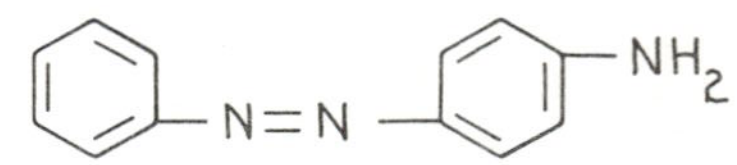

MOLECULAR WEIGHT: 197.24

SOURCE OF SAMPLE: University of Wisconsin, Madison, Wisconsin

CAS REGISTRY NO.: 60-09-3

NIOSH NUMBER: BY82250

KBr Wafer

CONDENSED PHASE

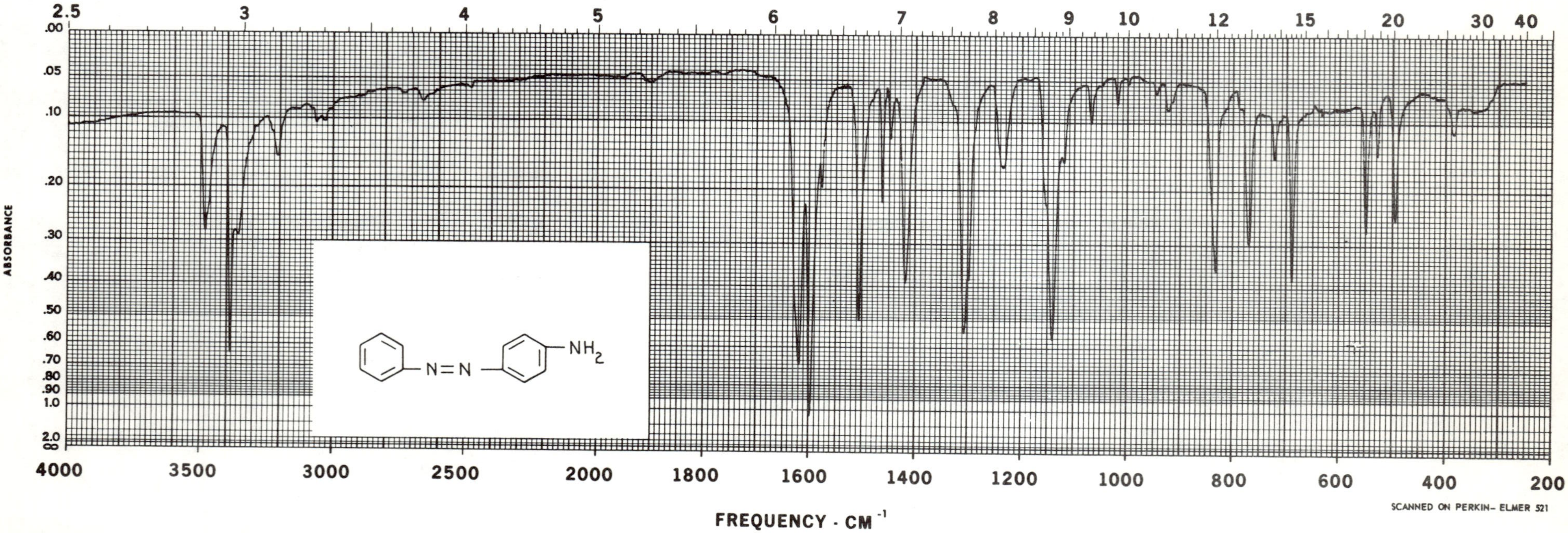

MELTING POINT:	124-124.5°C	INJECTION AMT.:	–
VAPORIZATION TEMP.:	280°C	PRESSURE:	Atmospheric
CELL VOLUME:	7.5 cc	MAX./MIN.:	90.0/3.4
PATH LENGTH:	7.6 cm	SCALE EXPANSION:	1.06

VAPOR PHASE

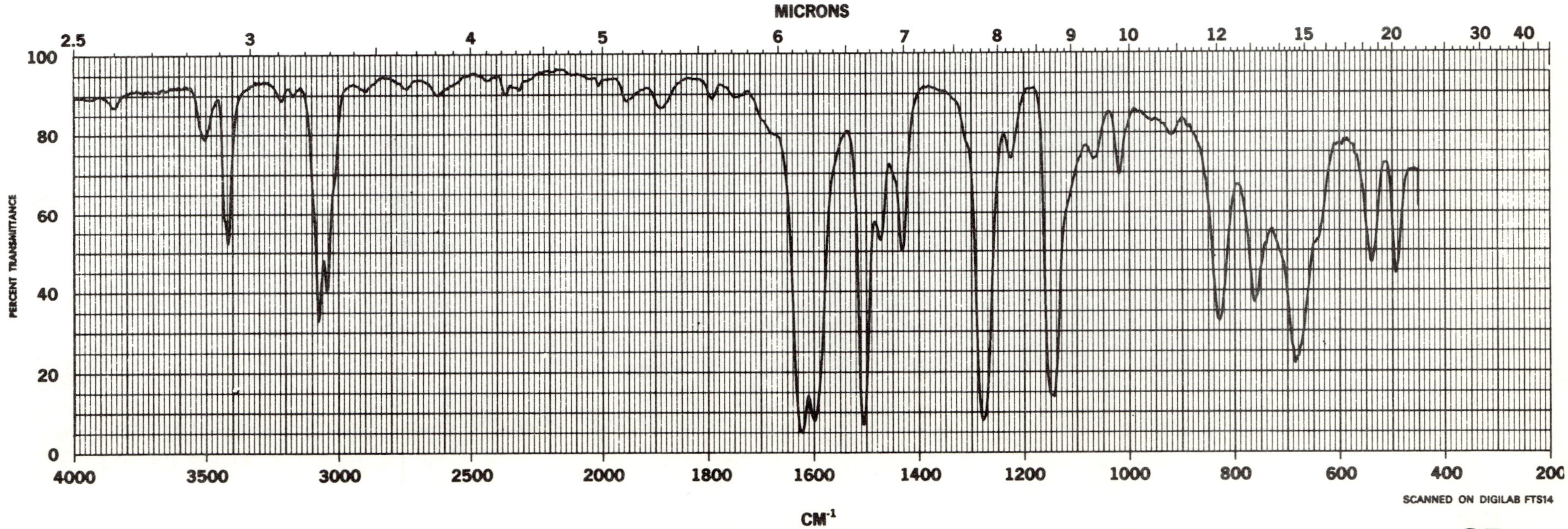

3,6–DICHLORO–o–ANISIC ACID
3,6–DICHLORO–2–METHOXYBENZOIC ACID
DICAMBA
BANLEN
BANVEL D

$C_8H_6Cl_2O_3$

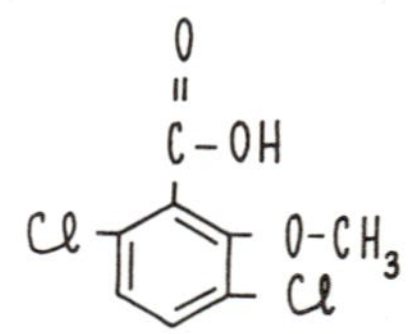

MOLECULAR WEIGHT:	221.04
SOURCE OF SAMPLE:	Velsicol Chemical Corporation Chicago, Illinois
CAS REGISTRY NO.:	1918-00-9
NIOSH NUMBER:	DG75250

KBr Wafer

CONDENSED PHASE

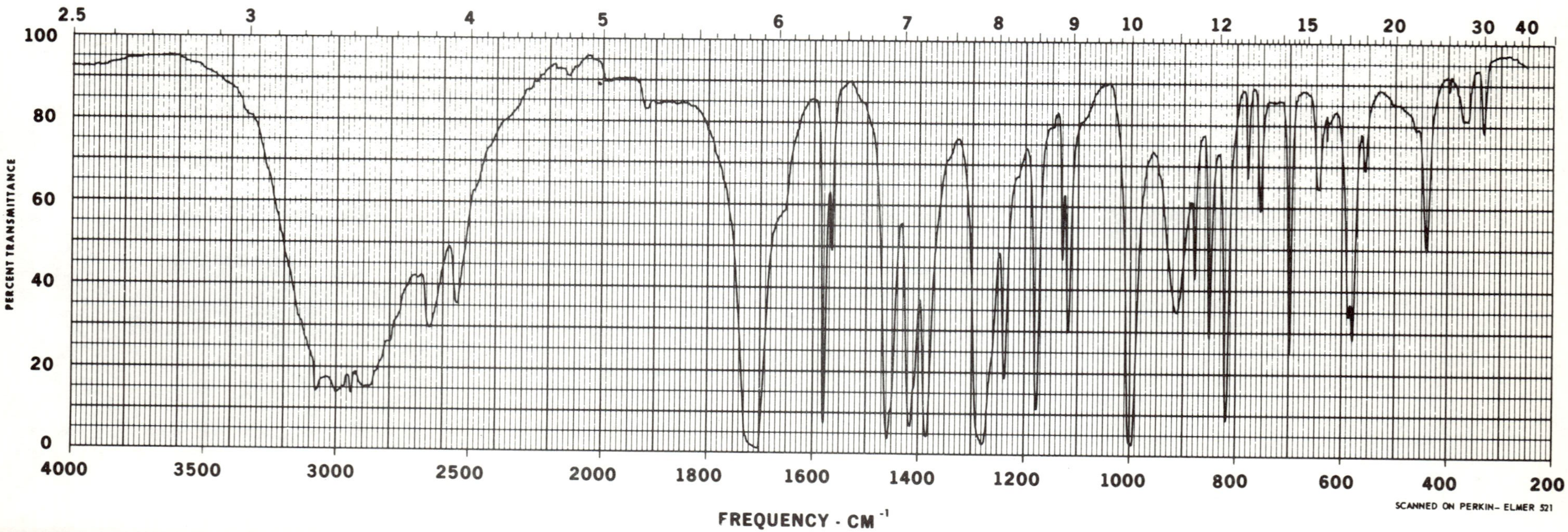

MELTING POINT:	114-116°C	INJECTION AMT.:	—
VAPORIZATION TEMP.:	240°C	PRESSURE:	Atmospheric
CELL VOLUME:	7.5 cc	MAX./MIN.:	99.0/9.4
PATH LENGTH:	7.6 cm	SCALE EXPANSION:	1.03

VAPOR PHASE

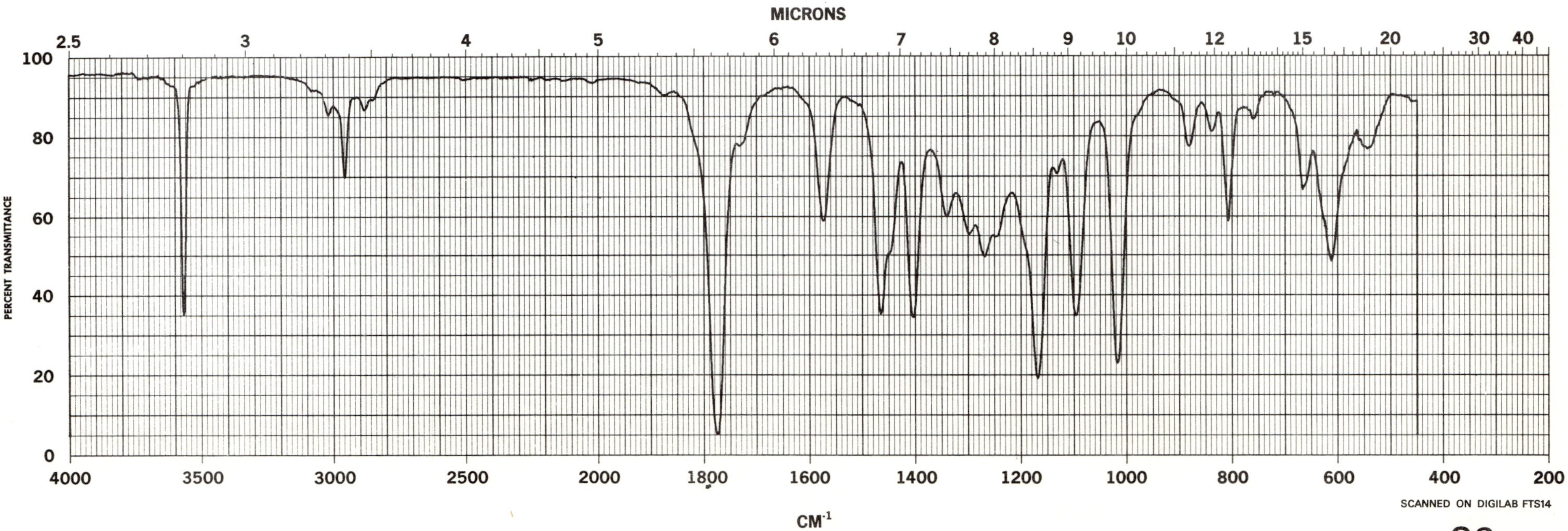

SCANNED ON DIGILAB FTS14

ANTHRACENE
GREEN OIL
PARANAPHTHALENE
TETRA OLIVE N2G

$C_{14}H_{10}$

MOLECULAR WEIGHT:	178.24
SOURCE OF SAMPLE:	Aldrich Chemical Company, Inc. Milwaukee, Wisconsin
CAS REGISTRY NO.:	120-12-7
NIOSH NUMBER:	CA93500

KBr Wafer

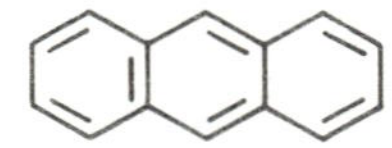

CONDENSED PHASE

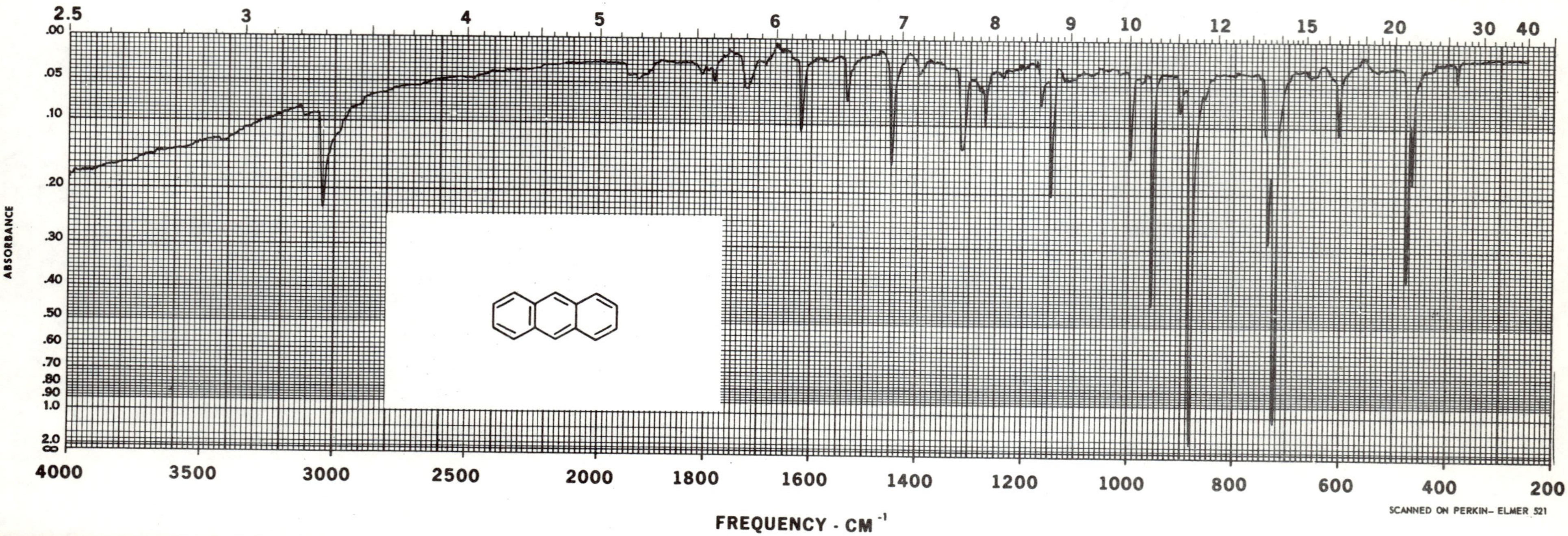

BOILING POINT:	340°C	INJECTION AMT.:	0.71 mg
VAPORIZATION TEMP.:	280°C	PRESSURE:	Atmospheric
CELL VOLUME:	7.5 cc	MAX./MIN.:	102.5/26.8
PATH LENGTH:	7.6 cm	SCALE EXPANSION:	1.22

VAPOR PHASE

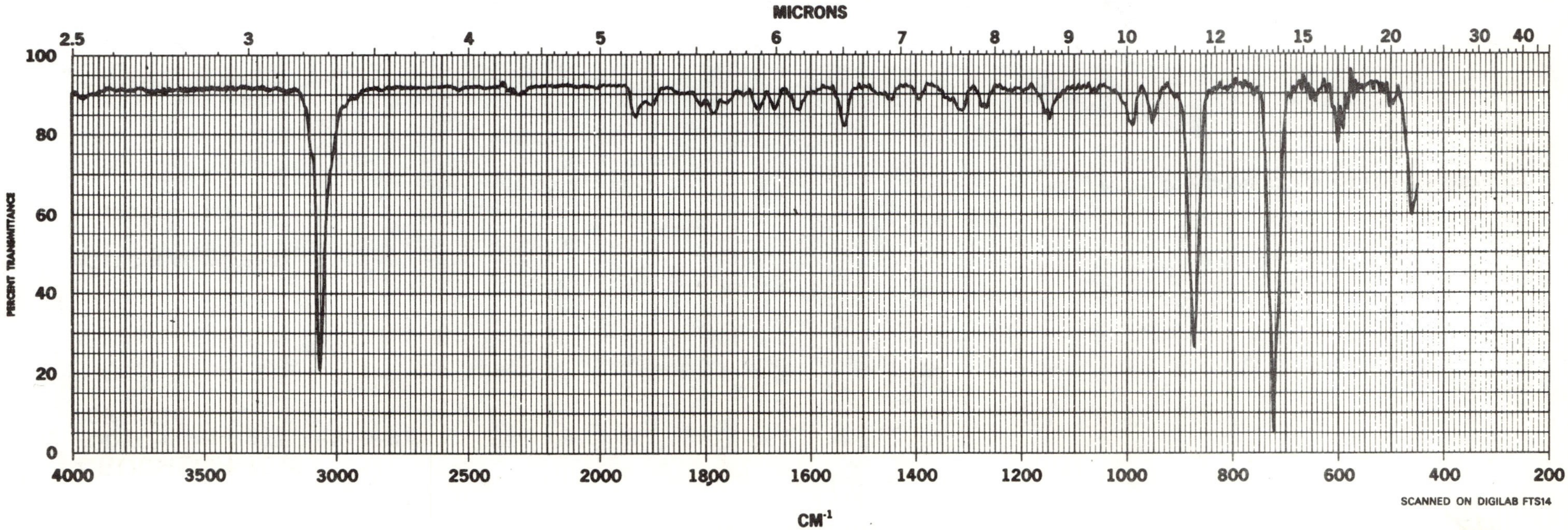

AROCHLOR 1016
POLYCHLORINATED BIPHENYL
PCB–1016

–

MOLECULAR WEIGHT: –

SOURCE OF SAMPLE: Chem Service Company
West Chester, Pennsylvania

* Refer to Appendix for GC/FT-IR Analysis

CAS REGISTRY NO.: –

NIOSH NUMBER: –

CAPILLARY CELL: Neat

CONDENSED PHASE

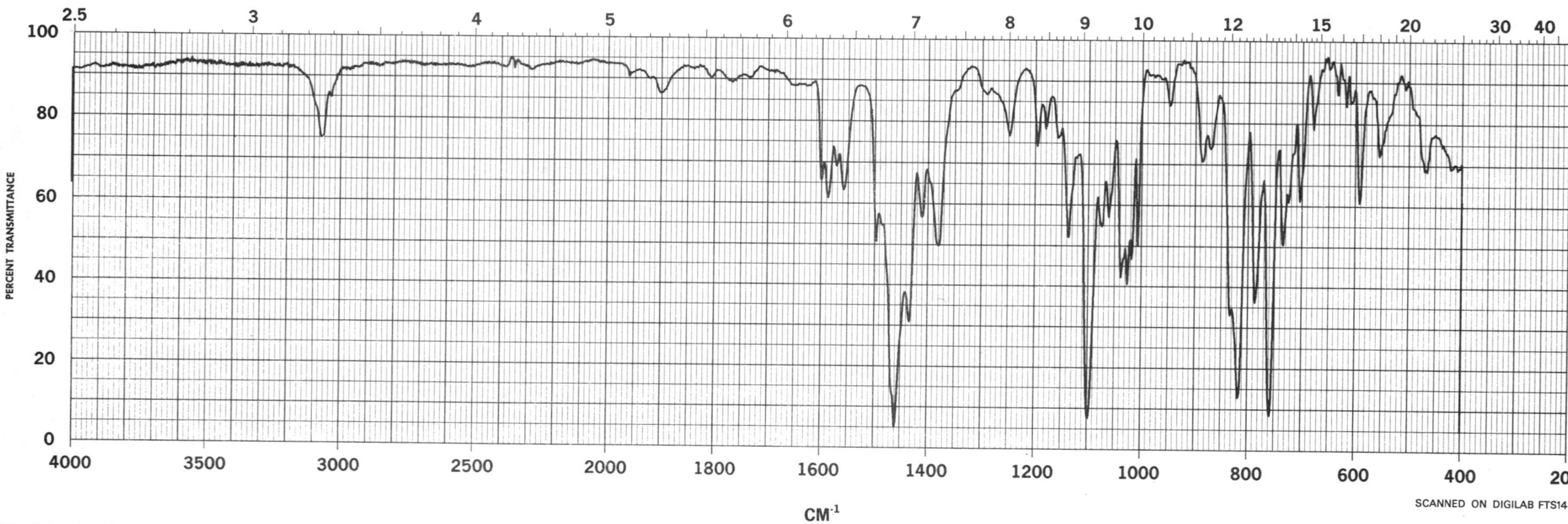

BOILING POINT:	–	INJECTION AMT.:	–
VAPORIZATION TEMP.:	280°C	PRESSURE:	Atmospheric
CELL VOLUME:	7.5 cc	MAX./MIN.:	9.9/10.0
PATH LENGTH:	7.6 cm	SCALE EXPANSION:	1.04

VAPOR PHASE

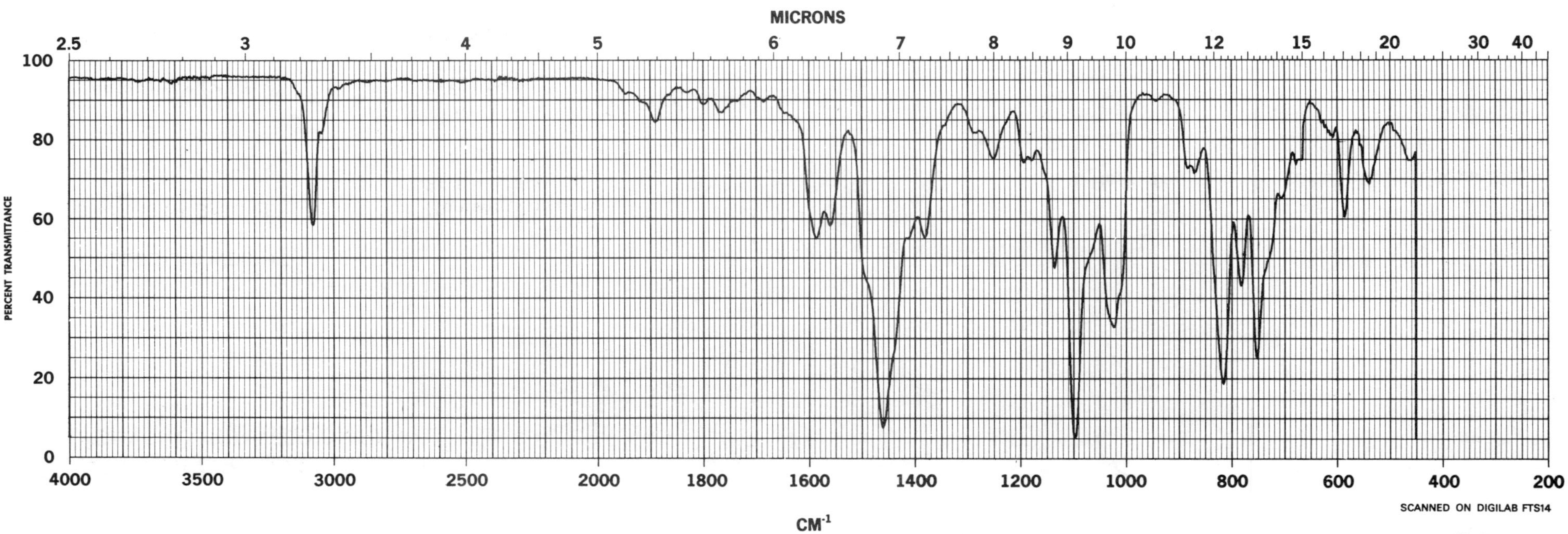

SCANNED ON DIGILAB FTS14

AROCHLOR 1221
POLYCHLORINATED BIPHENYL
PCB–1221

MOLECULAR WEIGHT: —

SOURCE OF SAMPLE: Monsanto Company
St. Louis, Missouri

* Refer to Appendix for GC/FT-IR Analysis

CAS REGISTRY NO.: 11104-28-2

NIOSH NUMBER: TQ13520

Between Salts

CONDENSED PHASE

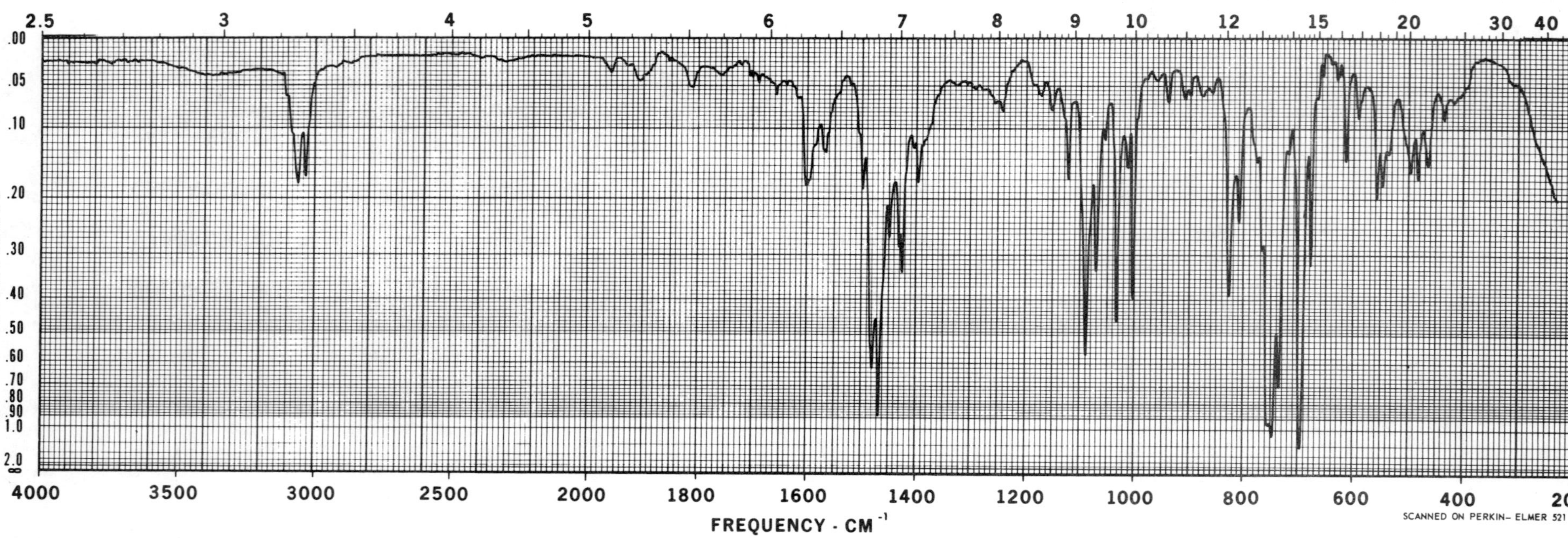

BOILING POINT:	275-320°C	INJECTION AMT.:	—
VAPORIZATION TEMP.:	200°C	PRESSURE:	Atmospheric
CELL VOLUME:	7.5 cc	MAX./MIN.:	100.7/41.3
PATH LENGTH:	7.6 cm	SCALE EXPANSION:	1.55

VAPOR PHASE

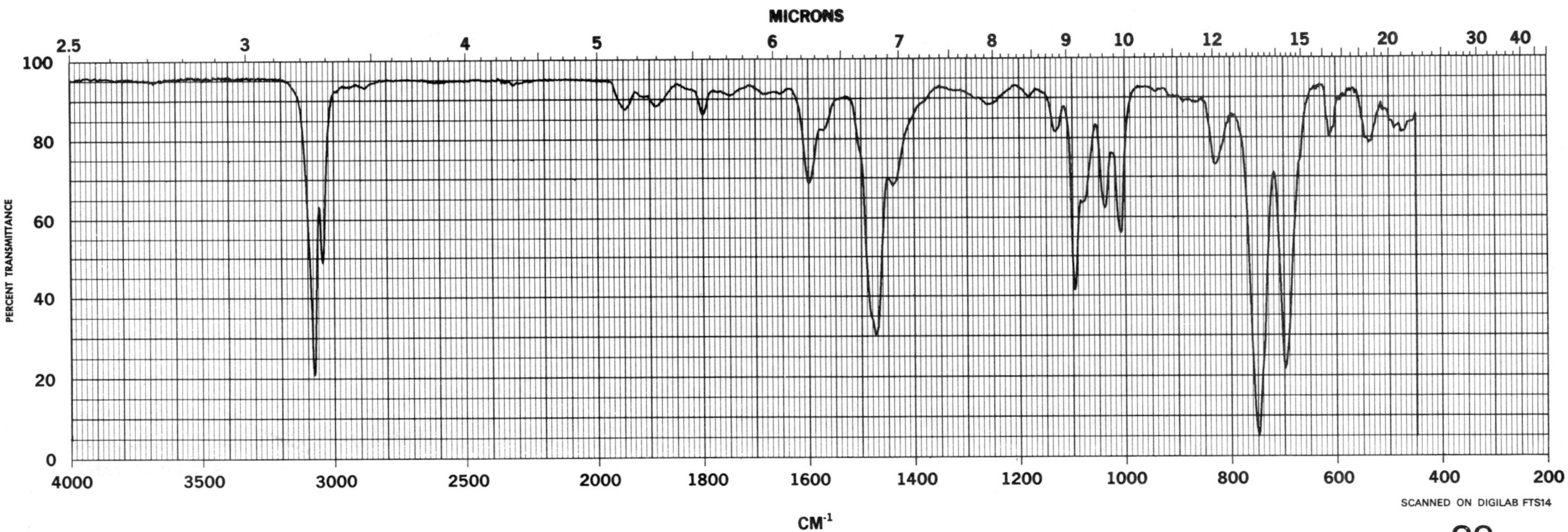

SCANNED ON DIGILAB FTS14

AROCHLOR 1232
POLYCHLORINATED BIPHENYL
PCB–1232

MOLECULAR WEIGHT: --

SOURCE OF SAMPLE: Monsanto Company
St. Louis, Missouri

* Refer to Appendix for GC/FT-IR Analysis

CAS REGISTRY NO.: 11141-16-5

NIOSH NUMBER: TQ13540

Between Salts

CONDENSED PHASE

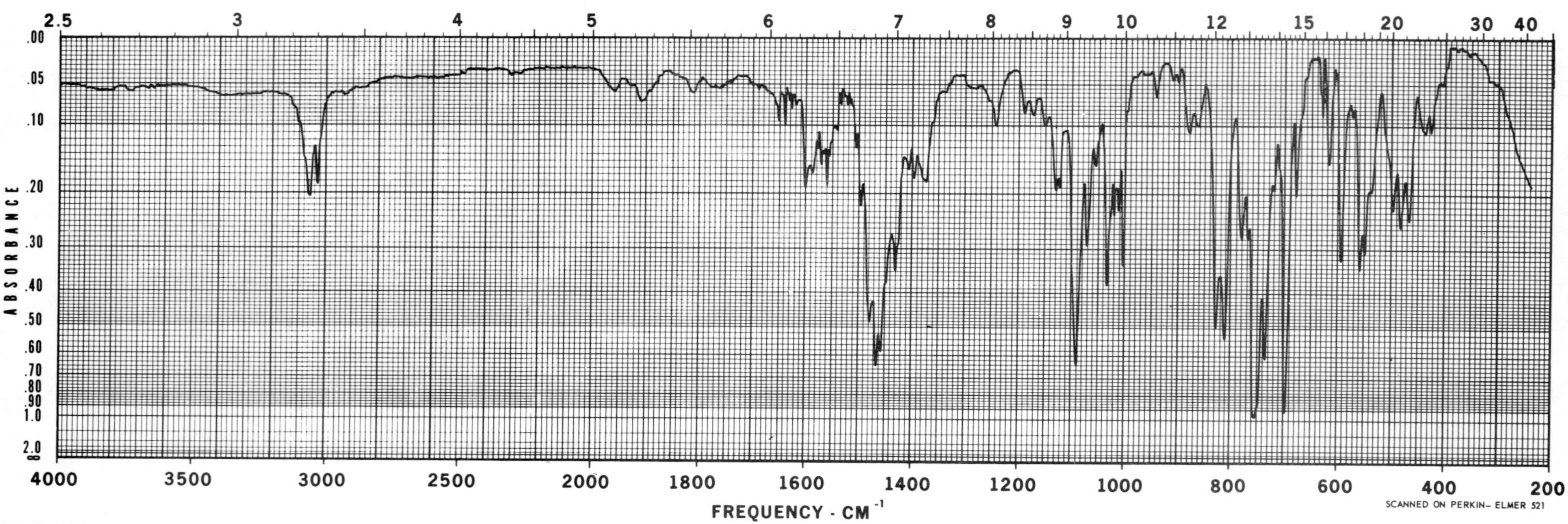

BOILING POINT:	290-325°C	INJECTION AMT.:	—
VAPORIZATION TEMP.:	200°C	PRESSURE:	Atmospheric
CELL VOLUME:	7.5 cc	MAX./MIN.:	100.9/48.9
PATH LENGTH:	7.6 cm	SCALE EXPANSION:	1.77

VAPOR PHASE

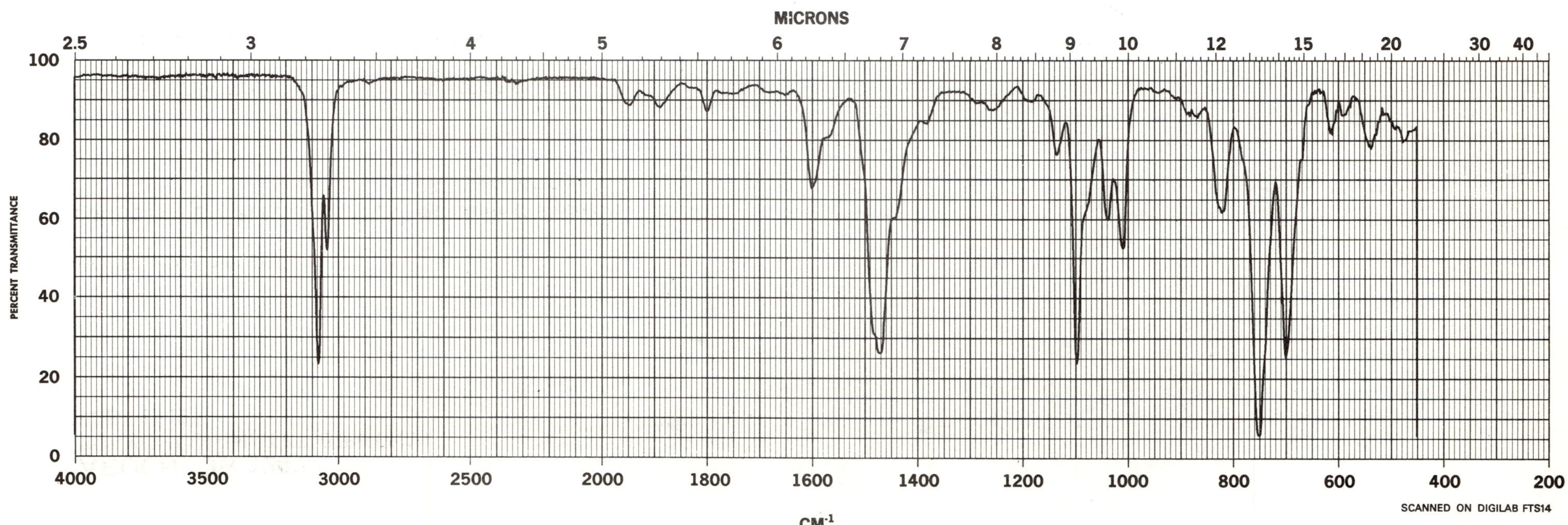

SCANNED ON DIGILAB FTS14

AROCHLOR 1242
POLYCHLORINATED BIPHENYL
PCB–1242

—

MOLECULAR WEIGHT:	--
SOURCE OF SAMPLE:	Monsanto Company St. Louis, Missouri
CAS REGISTRY NO.:	53469-21-9
NIOSH NUMBER:	TQ13560

Between Salts

* Refer to Appendix for GC/FT-IR Analysis

CONDENSED PHASE

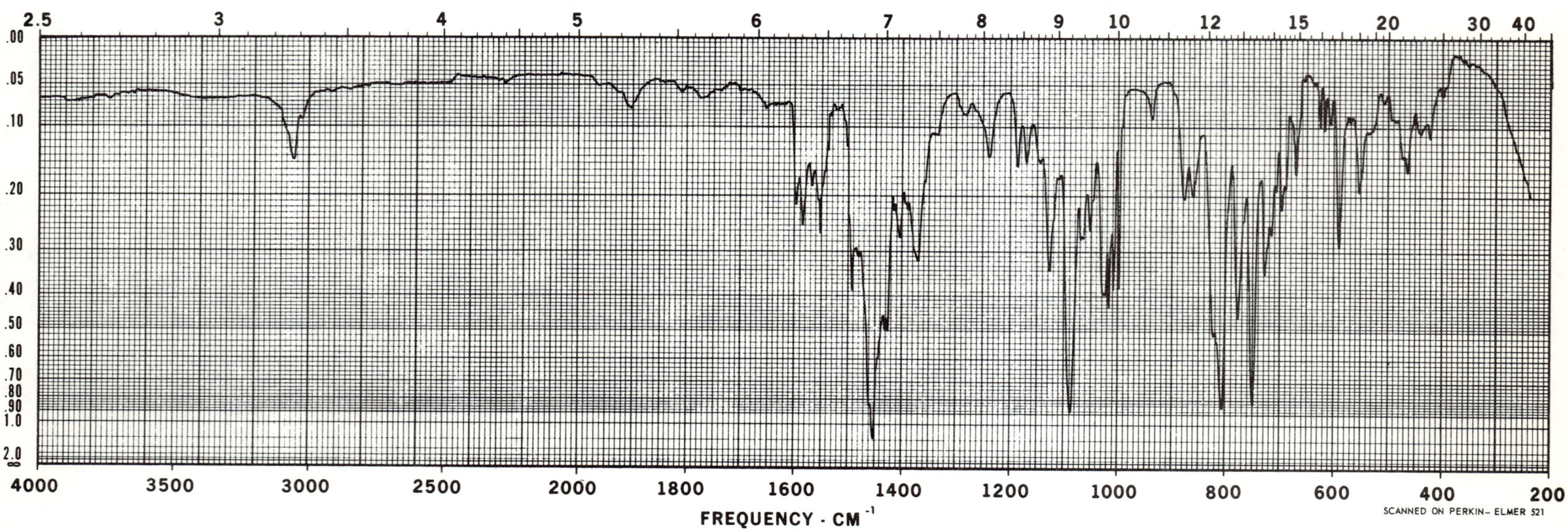

BOILING POINT:	325-366°C	INJECTION AMT.:	–
VAPORIZATION TEMP.:	240°C	PRESSURE:	Atmospheric
CELL VOLUME:	7.5 cc	MAX./MIN.:	100.1/42.4
PATH LENGTH:	7.6 cm	SCALE EXPANSION:	1.59

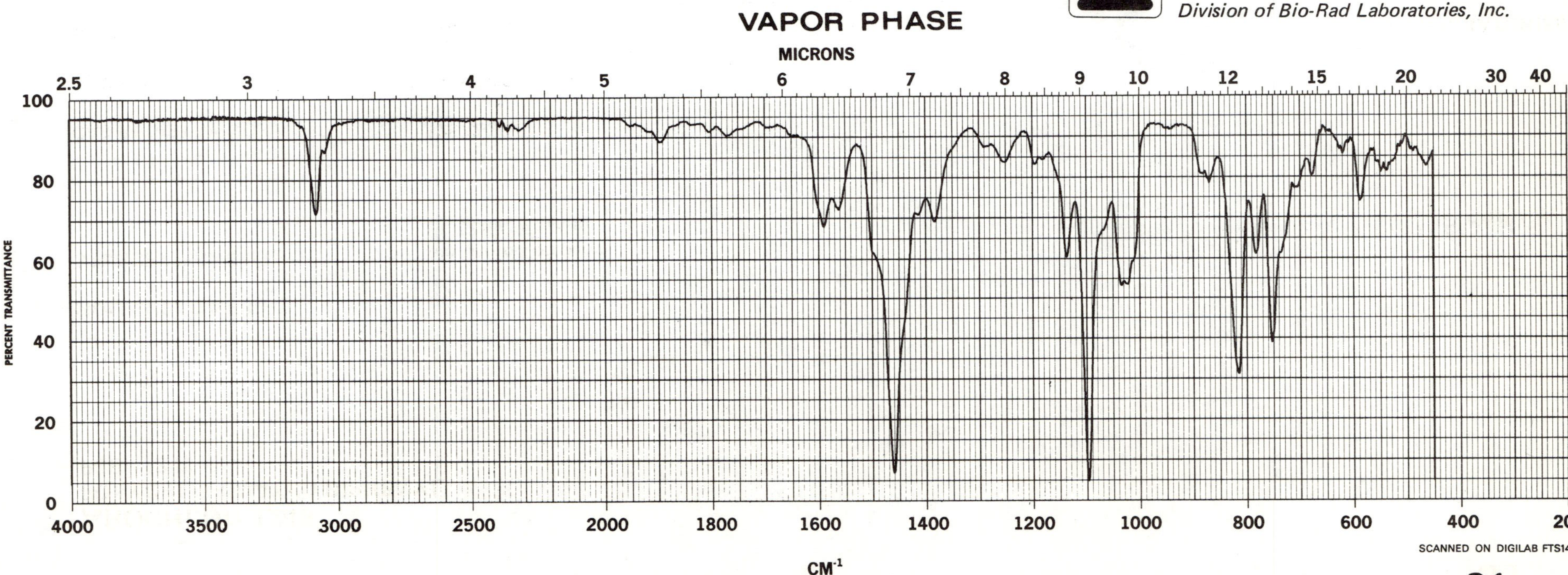

SCANNED ON DIGILAB FTS14

AROCHLOR 1248
POLYCHLORINATED BIPHENYL
PCB–1248

–

MOLECULAR WEIGHT: –

SOURCE OF SAMPLE: Monsanto Company
St. Louis, Missouri

* Refer to Appendix for GC/FT-IR Analysis

CAS REGISTRY NO.: 12672-29-6

NIOSH NUMBER: TQ13580

CAPILLARY CELL: Neat

CONDENSED PHASE

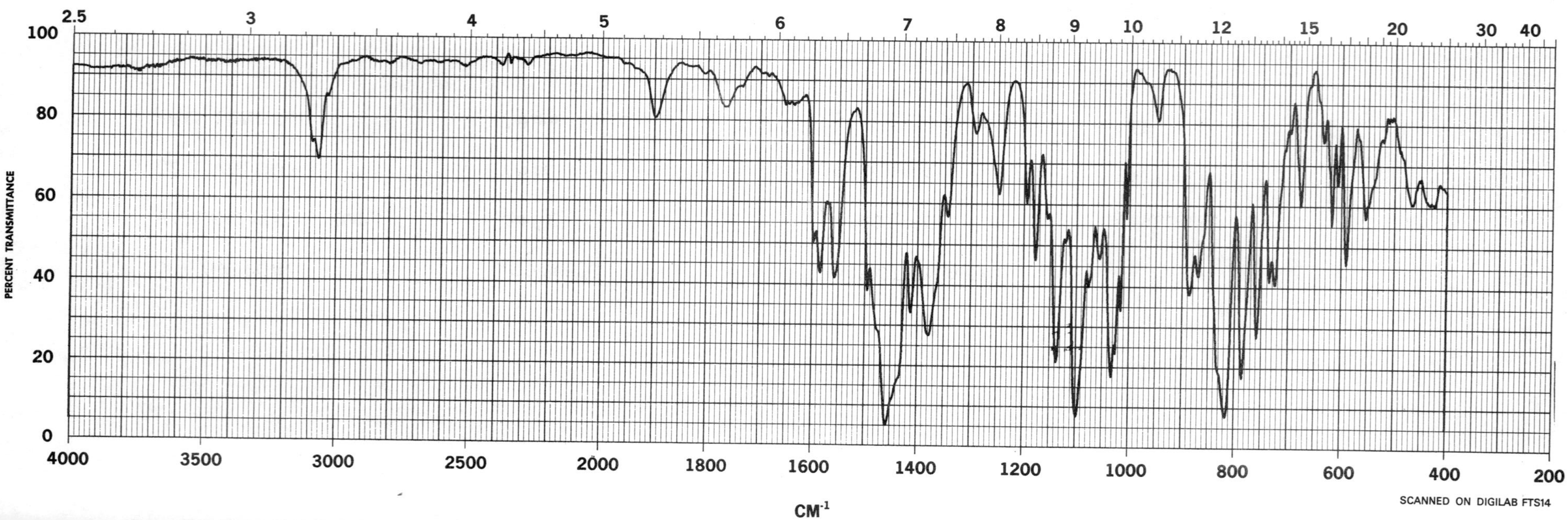

BOILING POINT:	—	INJECTION AMT.:	—
VAPORIZATION TEMP.:	280°C	PRESSURE:	Atmospheric
CELL VOLUME:	7.5 cc	MAX./MIN.:	106.3/16.0
PATH LENGTH:	7.6 cm	SCALE EXPANSION:	1.02

VAPOR PHASE

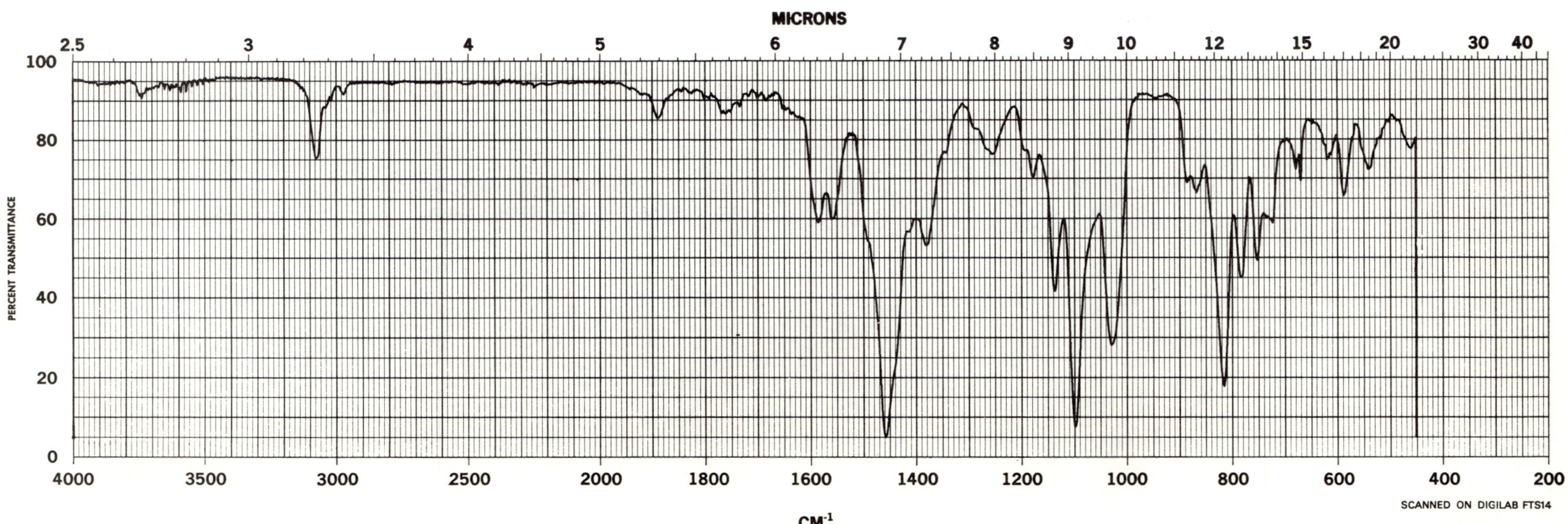

SCANNED ON DIGILAB FTS14

BENZENE
BENZOL
BENZOLE
CARBON OIL
PHENE

C_6H_6

MOLECULAR WEIGHT: 78.11

SOURCE OF SAMPLE: United States Testing Company, Inc.
Hoboken, New Jersey

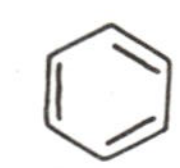

CAS REGISTRY NO.: 71-43-2

NIOSH NUMBER: CY14000

CAPILLARY CELL: Neat

CONDENSED PHASE

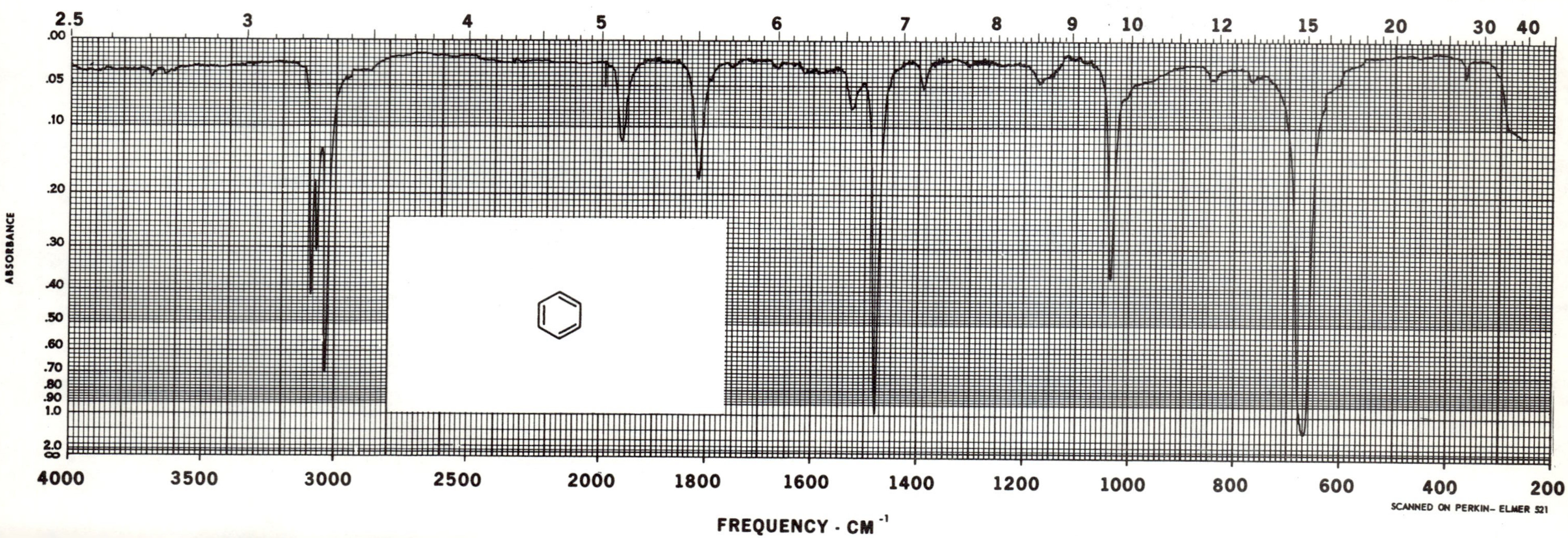

BOILING POINT:	80°C	INJECTION AMT.:	1.5 µl
VAPORIZATION TEMP.:	200°C	PRESSURE:	Atmospheric
CELL VOLUME:	7.5 cc	MAX./MIN.:	—
PATH LENGTH:	7.6 cm	SCALE EXPANSION:	0.94

VAPOR PHASE

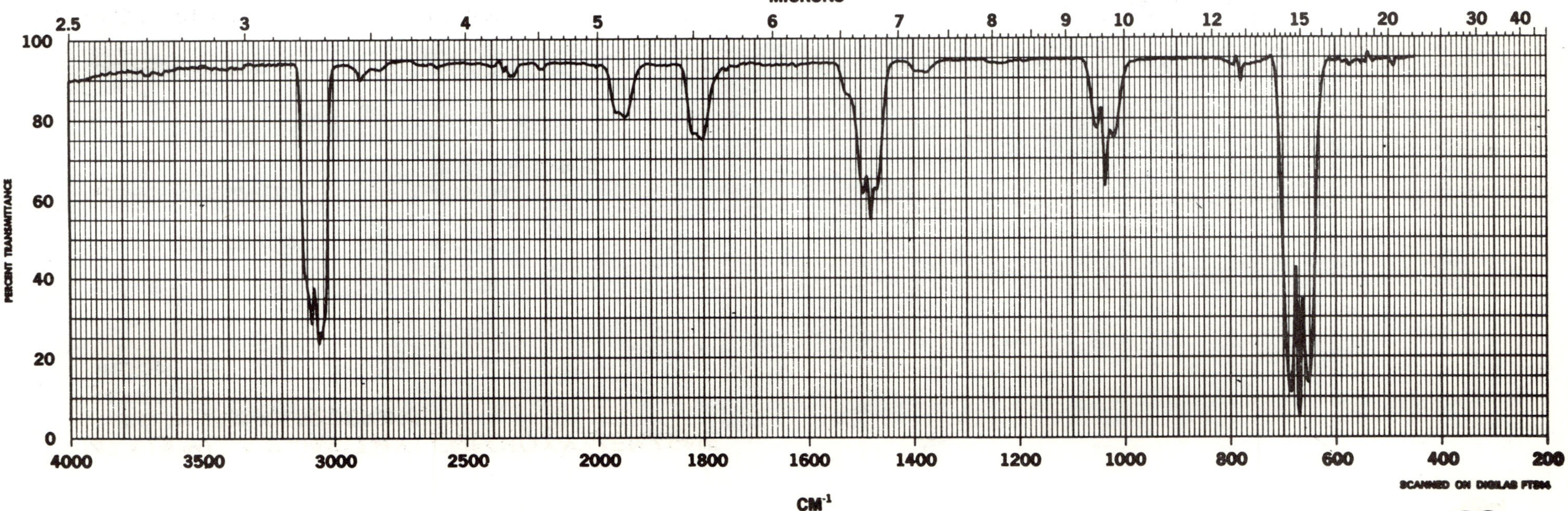

4–ALLYL–1,2–(METHYLENEDIOXY)BENZENE
5–ALLYL–1,3–BENZODIOXOLE
SAFROLE
SHIKIMOLE

$C_{10}H_{10}O_2$

MOLECULAR WEIGHT: 162.19

SOURCE OF SAMPLE: Fluka AG
Buchs, Switzerland

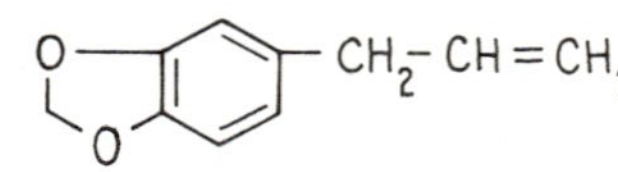

CAS REGISTRY NO.: 94-59-7

NIOSH NUMBER: CY28000

CAPILLARY CELL: Neat

CONDENSED PHASE

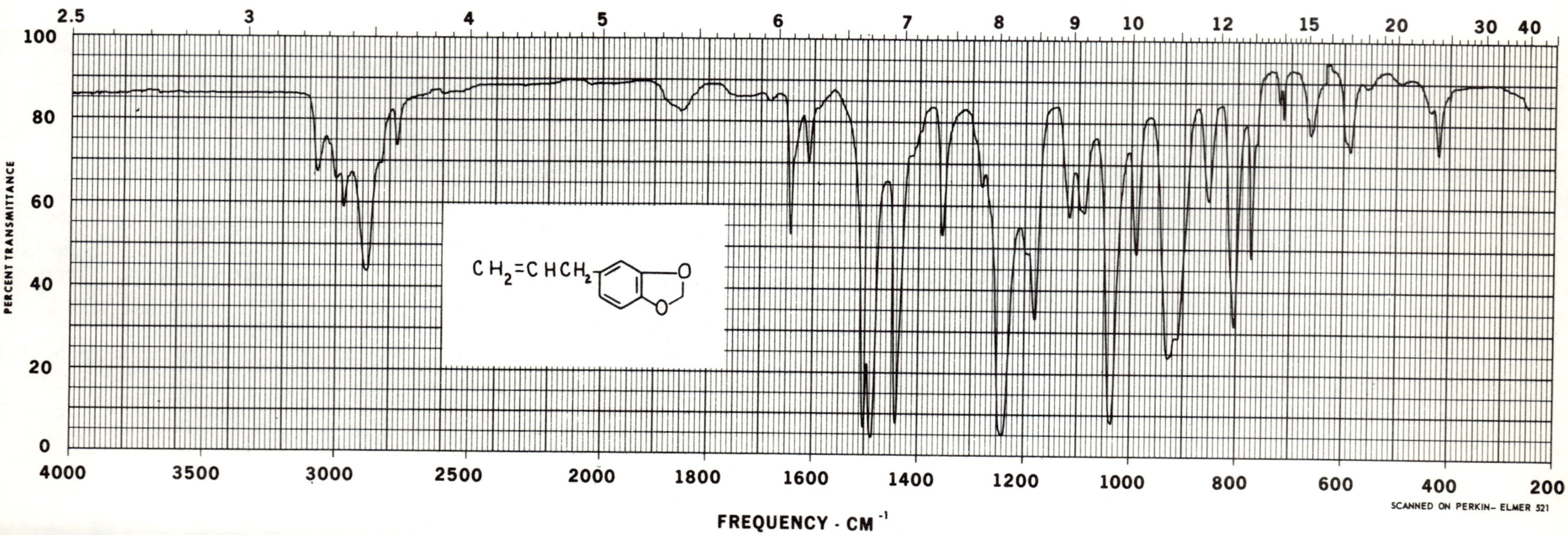

BOILING POINT:	232-234°C	INJECTION AMT.:	0.8 µl
VAPORIZATION TEMP.:	200°C	PRESSURE:	Atmospheric
CELL VOLUME:	7.5 cc	MAX./MIN.:	–
PATH LENGTH:	7.6 cm	SCALE EXPANSION:	1.08

VAPOR PHASE

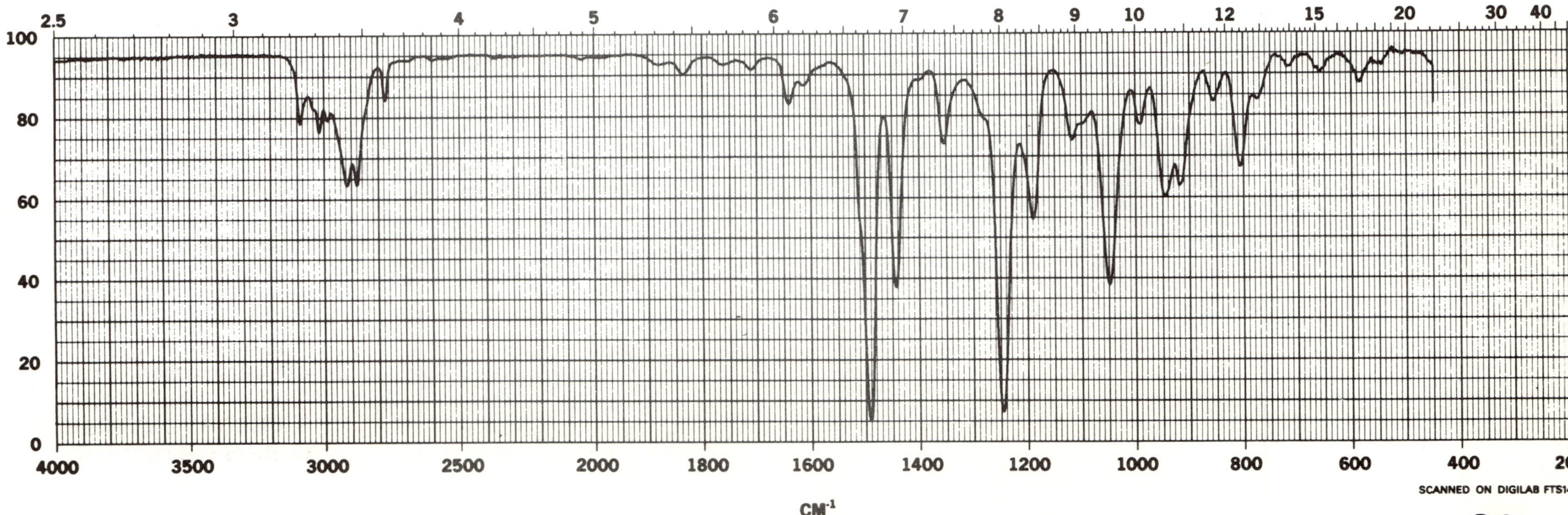

CHLOROBENZENE
BENZENE CHLORIDE
MCB
MONOCHLOROBENZENE
PHENYL CHLORIDE

C_6H_5Cl

MOLECULAR WEIGHT:	112.56
SOURCE OF SAMPLE:	Hooker Chemical Company Niagara Falls, New York
CAS REGISTRY NO.:	108-90-7
NIOSH NUMBER:	CZ01750
CAPILLARY CELL:	Neat

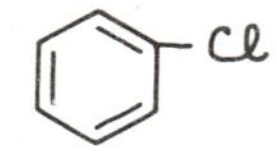

CONDENSED PHASE

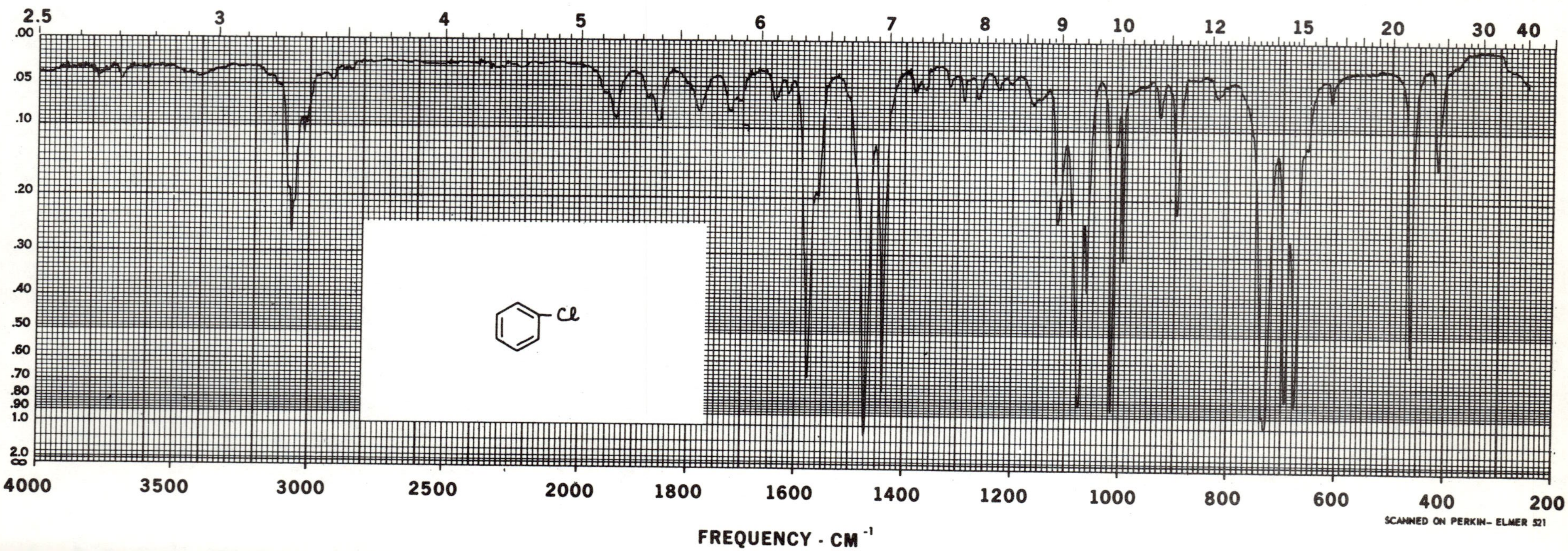

BOILING POINT:	$132^{o}C$	INJECTION AMT.:	1.0 µl
VAPORIZATION TEMP.:	$200^{o}C$	PRESSURE:	Atmospheric
CELL VOLUME:	7.5 cc	MAX./MIN.:	—
PATH LENGTH:	7.6 cm	SCALE EXPANSION:	1.04

VAPOR PHASE

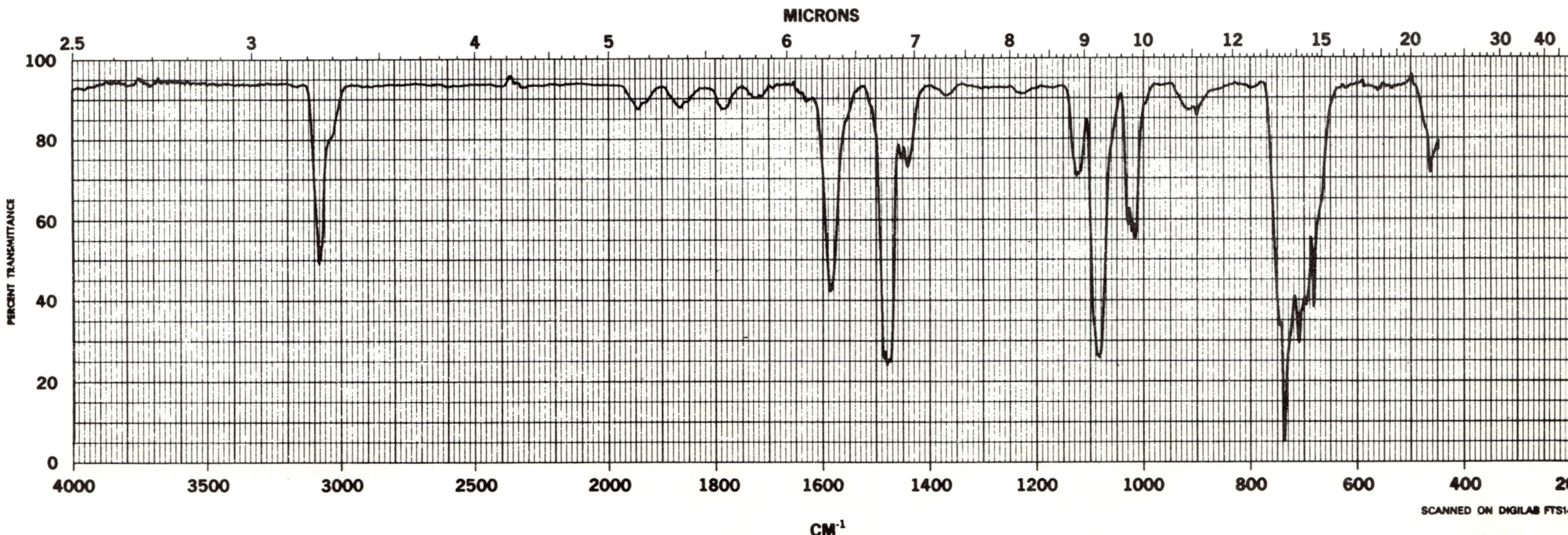

SCANNED ON DIGILAB FTS14

m–DICHLOROBENZENE
1,3–DICHLOROBENZENE

$C_6H_4Cl_2$

MOLECULAR WEIGHT: 147.00

SOURCE OF SAMPLE: Fluka AG, Buchs, Switzerland

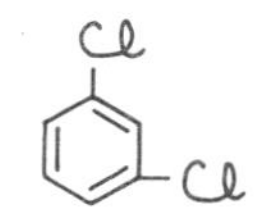

CAS REGISTRY NO.: 541-73-1

NIOSH NUMBER: –

CAPILLARY CELL: Neat

CONDENSED PHASE

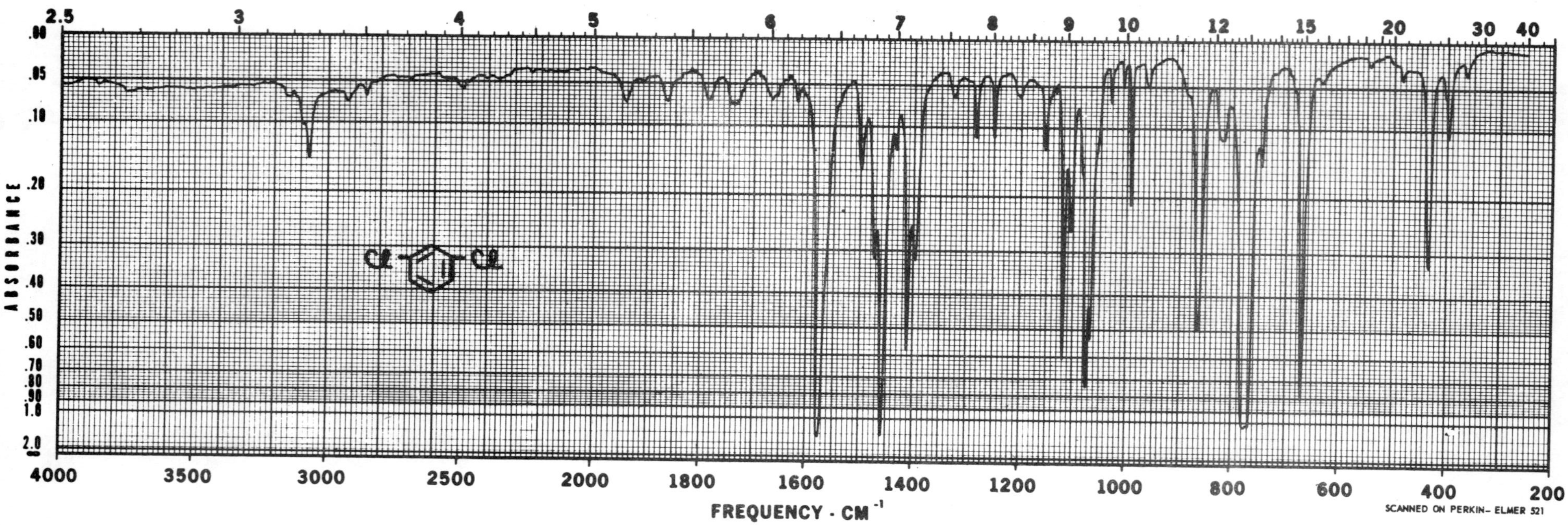

BOILING POINT:	172-173°C	INJECTION AMT.:	1.0 µl
VAPORIZATION TEMP.:	200°C	PRESSURE:	Atmospheric
CELL VOLUME:	7.5 cc	MAX./MIN.:	—
PATH LENGTH:	7.6 cm	SCALE EXPANSION:	0.94

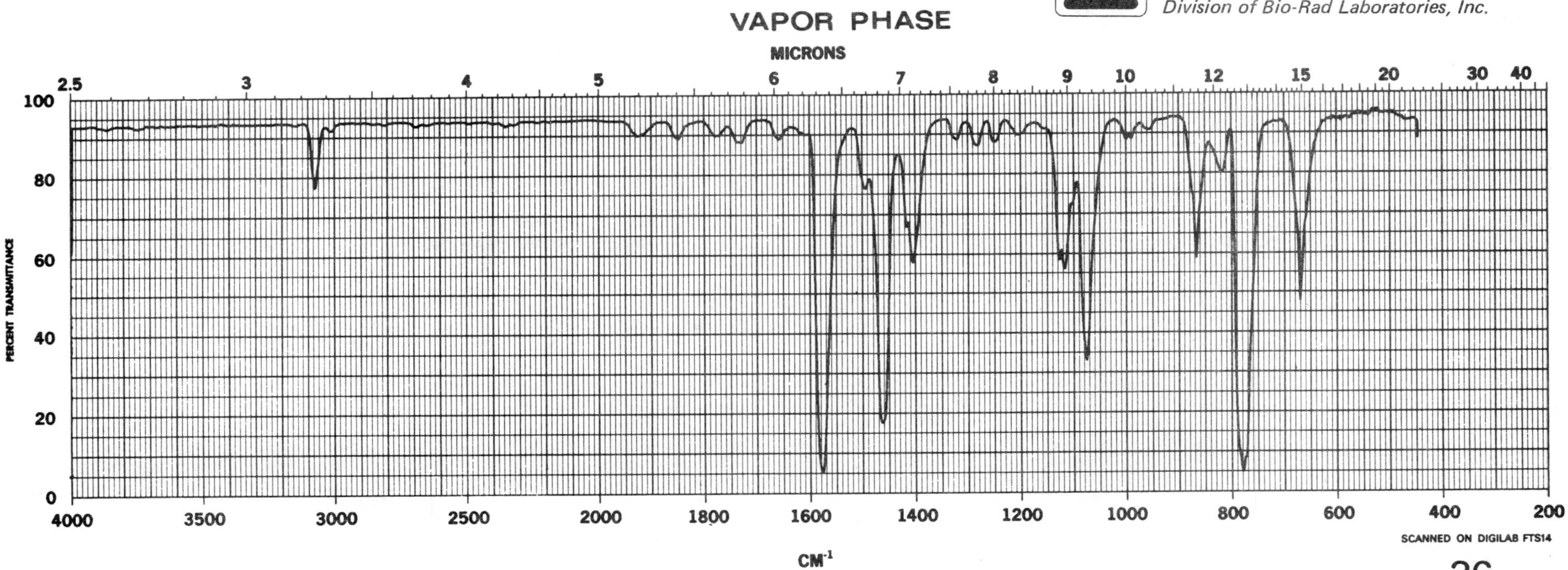

SCANNED ON DIGILAB FTS14

o–DICHLOROBENZENE

CHLOROBEN
1,2–DICHLOROBENZENE
DIZENE
ODCB

$C_6H_4Cl_2$

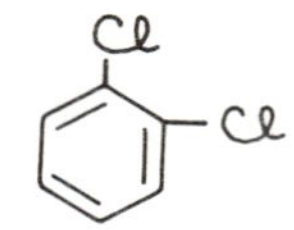

MOLECULAR WEIGHT:	147.00
SOURCE OF SAMPLE:	MCB Manufacturing Chemists Norwood, Ohio
CAS REGISTRY NO.:	95-50-1
NIOSH NUMBER:	CZ45000
CAPILLARY CELL:	Neat

CONDENSED PHASE

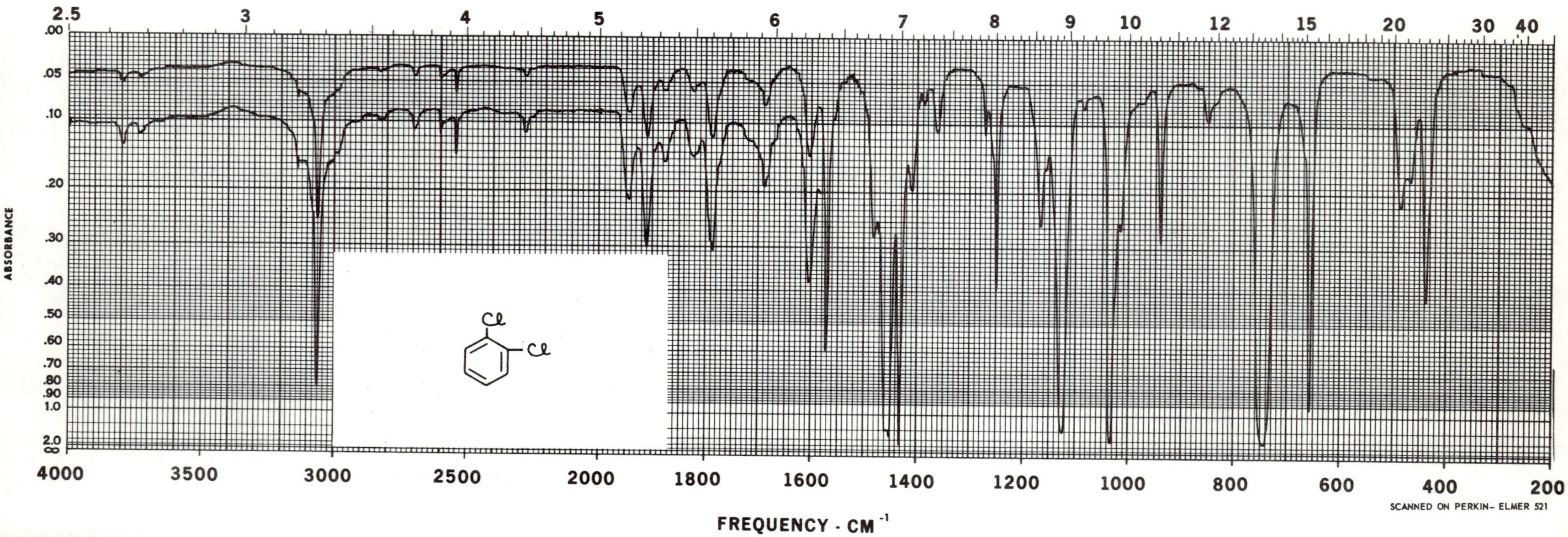

BOILING POINT:	179.5-180.5°C	INJECTION AMT.:	1.0 μl
VAPORIZATION TEMP.:	200°C	PRESSURE:	Atmospheric
CELL VOLUME:	7.5 cc	MAX./MIN.:	—
PATH LENGTH:	7.6 cm	SCALE EXPANSION:	0.99

VAPOR PHASE

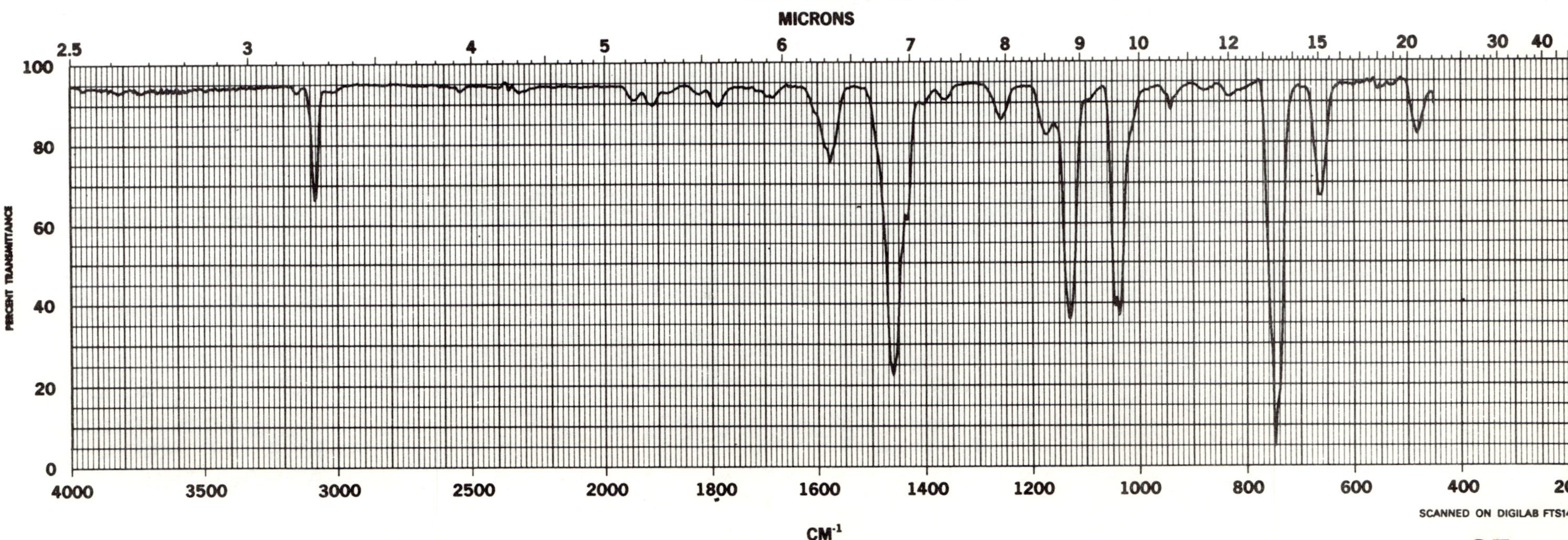

SCANNED ON DIGILAB FTS14

p–DICHLOROBENZENE

1,4–DICHLOROBENZENE
SANTOCHLOR 34
PARACIDE
PDCB

$C_6H_4Cl_2$

MOLECULAR WEIGHT:	147.00
SOURCE OF SAMPLE:	Aldrich Chemical Company, Inc. Milwaukee, Wisconsin
CAS REGISTRY NO.:	106-46-7
NIOSH NUMBER:	CZ45500
CAPILLARY CELL:	Melt

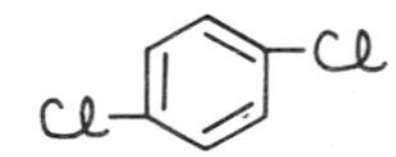

CONDENSED PHASE

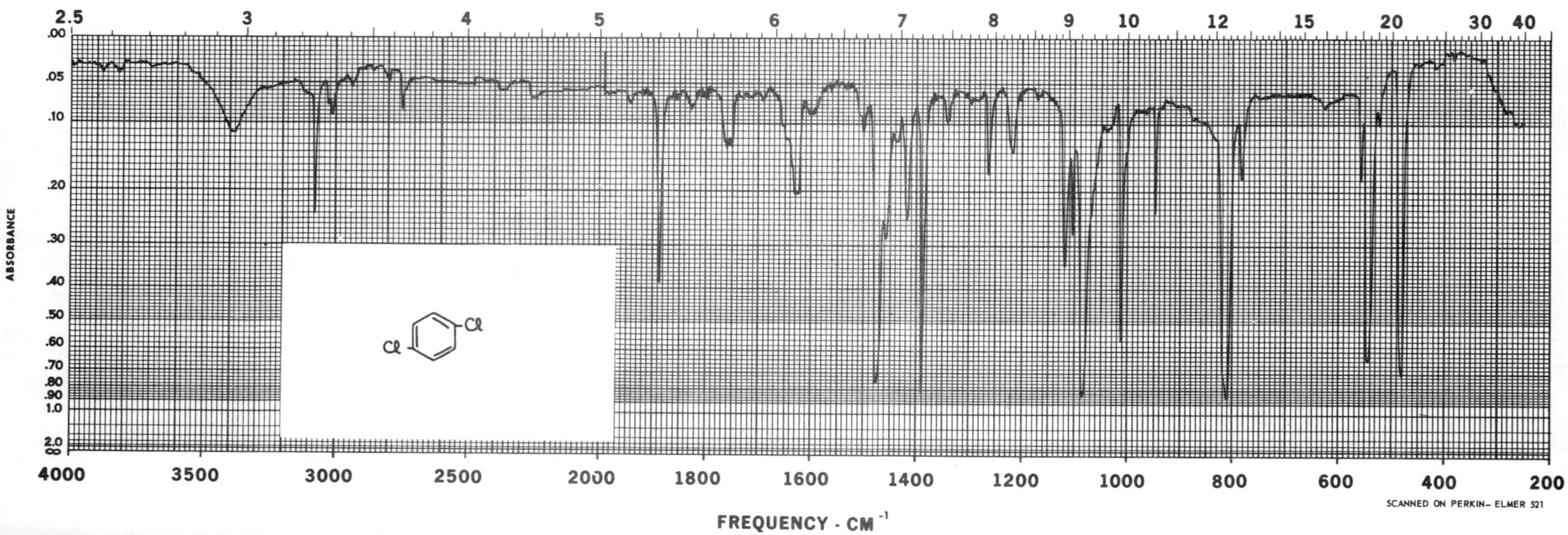

BOILING POINT:	173°C	INJECTION AMT.:	1.1 mg
VAPORIZATION TEMP.:	280°C	PRESSURE:	Atmospheric
CELL VOLUME:	7.5 cc	MAX./MIN.:	102.0/7.0
PATH LENGTH:	7.6 cm	SCALE EXPANSION:	0.97

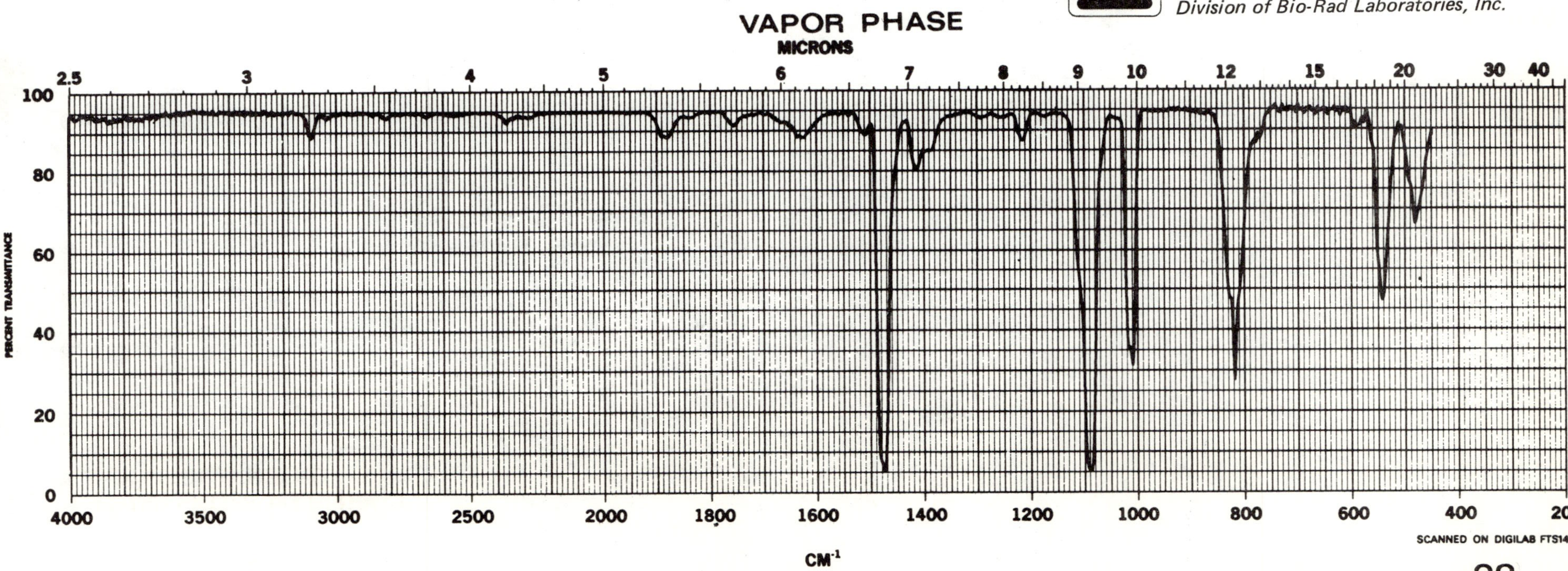

m–DINITROBENZENE
1,3–DINITROBENZENE

$C_6H_4N_2O_4$

MOLECULAR WEIGHT:	168.11
SOURCE OF SAMPLE:	Chem Service Company West Chester, Pennsylvania
CAS REGISTRY NO.:	99-65-0
NIOSH NUMBER:	CZ73500

KBr Wafer

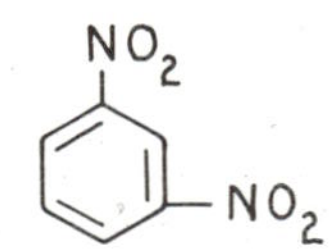

CONDENSED PHASE

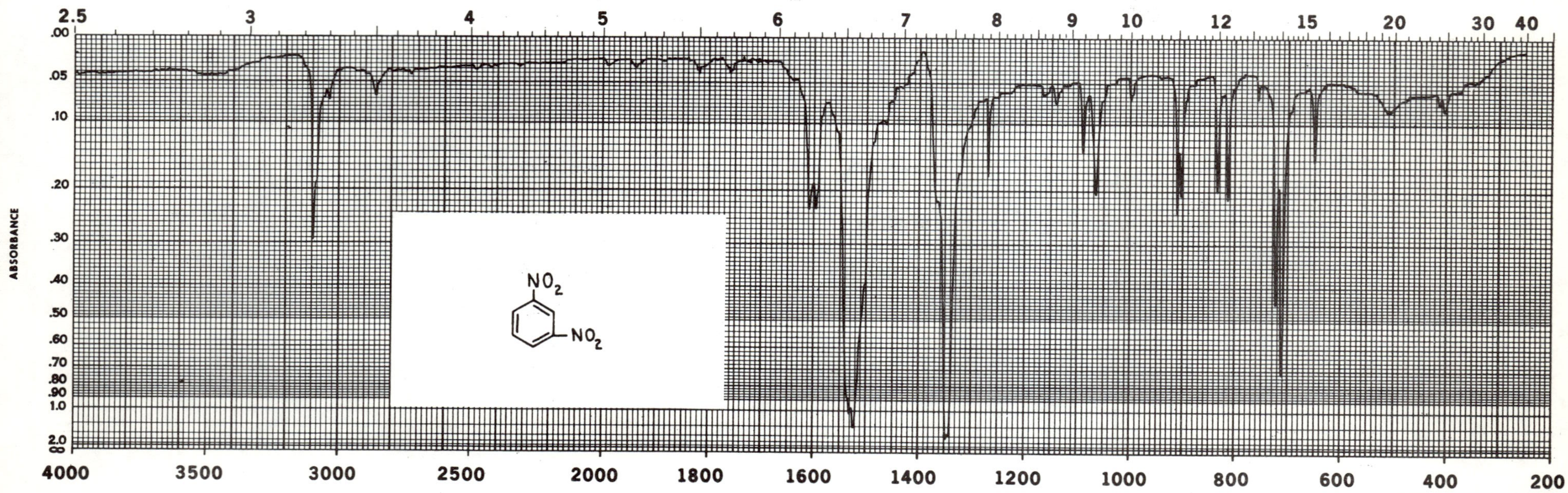

MELTING POINT:	117-118°C	INJECTION AMT.:	–
VAPORIZATION TEMP.:	280°C	PRESSURE:	Atmospheric
CELL VOLUME:	7.5 cc	MAX./MIN.:	102.6/13.7
PATH LENGTH:	7.6 cm	SCALE EXPANSION:	1.03

VAPOR PHASE

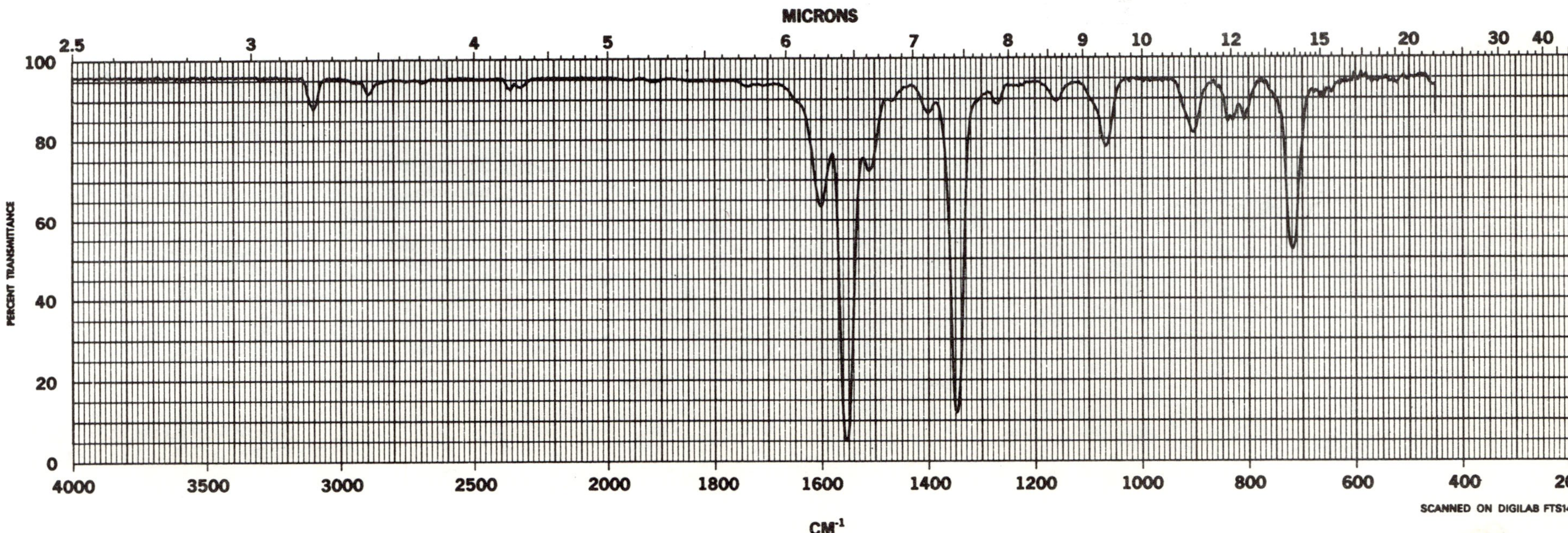

o–DINITROBENZENE
1,2–DINITROBENZENE

$C_6H_4N_2O_4$

MOLECULAR WEIGHT: 168.11

SOURCE OF SAMPLE: Calbiochem
Los Angeles, California

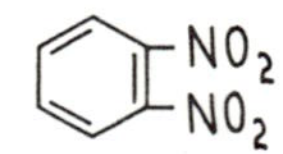

CAS REGISTRY NO.: 528-29-0

NIOSH NUMBER: --

KBr Wafer

CONDENSED PHASE

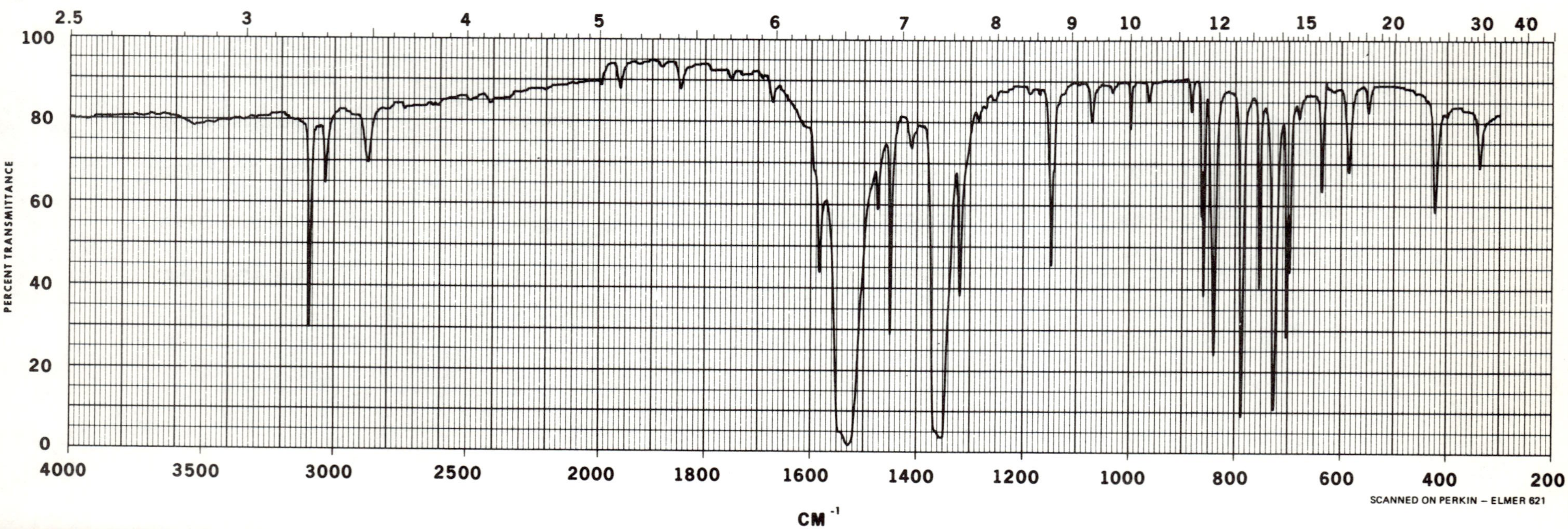

MELTING POINT:	117-118°C	INJECTION AMT.:	—
VAPORIZATION TEMP.:	280°C	PRESSURE:	Atmospheric
CELL VOLUME:	7.5 cc	MAX./MIN.:	98.6/14.2
PATH LENGTH:	7.6 cm	SCALE EXPANSION:	1.09

VAPOR PHASE

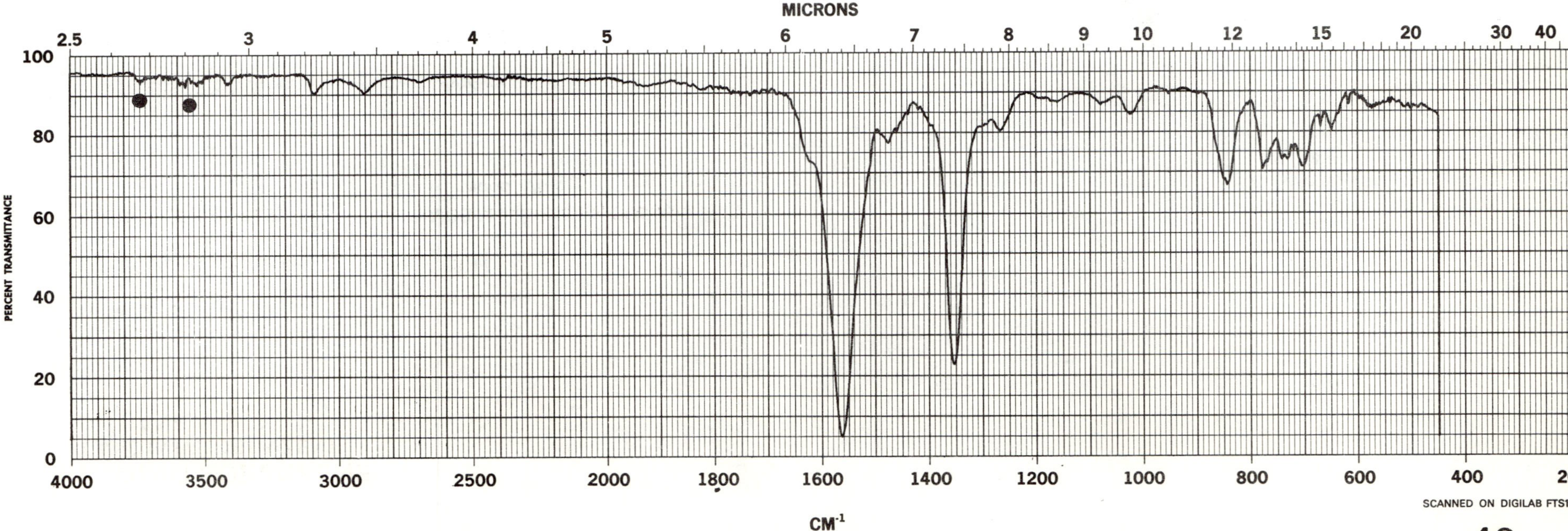

SCANNED ON DIGILAB FTS14

p–DINITROBENZENE
1,4–DINITROBENZENE
DITHANE A–4

$C_6H_4N_2O_4$

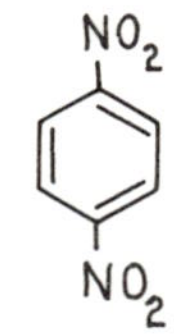

MOLECULAR WEIGHT: 168.11

SOURCE OF SAMPLE: MCB Manufacturing Chemists
Norwood, Ohio

CAS REGISTRY NO.: 100-25-4

NIOSH NUMBER: CZ75250

KBr Wafer

CONDENSED PHASE

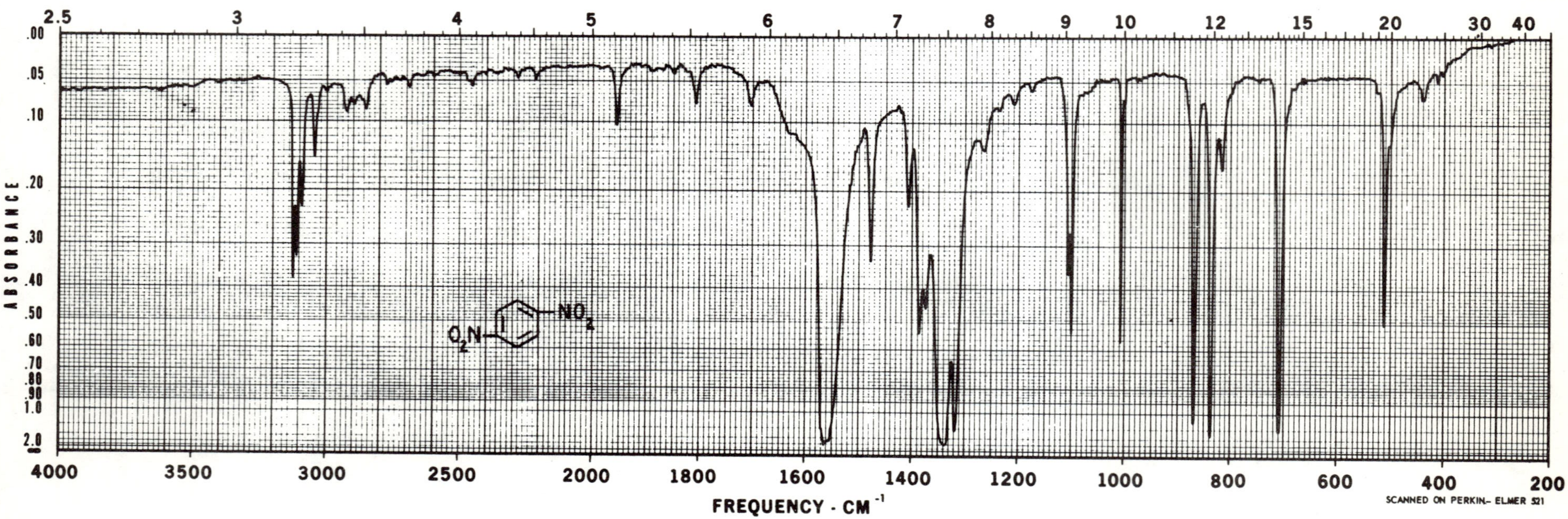

MELTING POINT:	170-171°C	INJECTION AMT.:	–
VAPORIZATION TEMP.:	240°C	PRESSURE:	Atmospheric
CELL VOLUME:	7.5 cc	MAX./MIN.:	101.7/6.0
PATH LENGTH:	7.6 cm	SCALE EXPANSION:	0.96

VAPOR PHASE

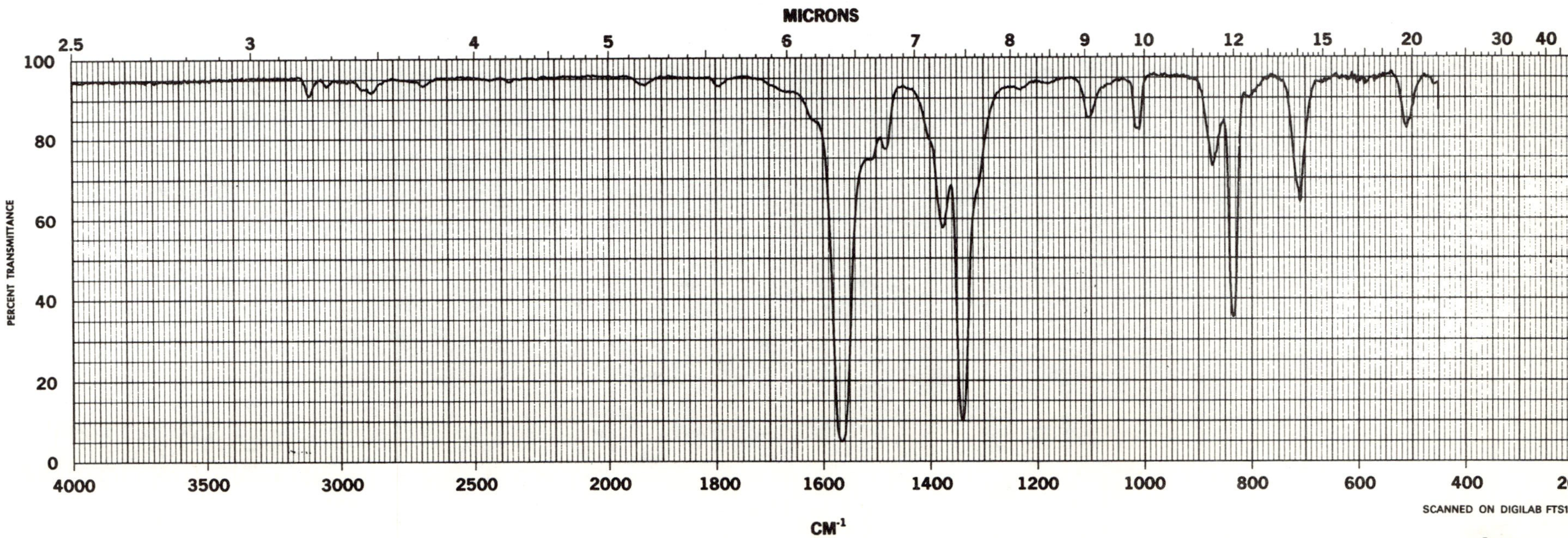

SCANNED ON DIGILAB FTS14

ETHYLBENZENE
PHENYLETHANE

C_8H_{10}

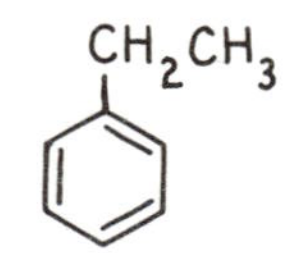

MOLECULAR WEIGHT:	106.17
SOURCE OF SAMPLE:	MCB Manufacturing Chemists Norwood, Ohio
CAS REGISTRY NO.:	100-41-4
NIOSH NUMBER:	DA07000
CAPILLARY CELL:	Neat

CONDENSED PHASE

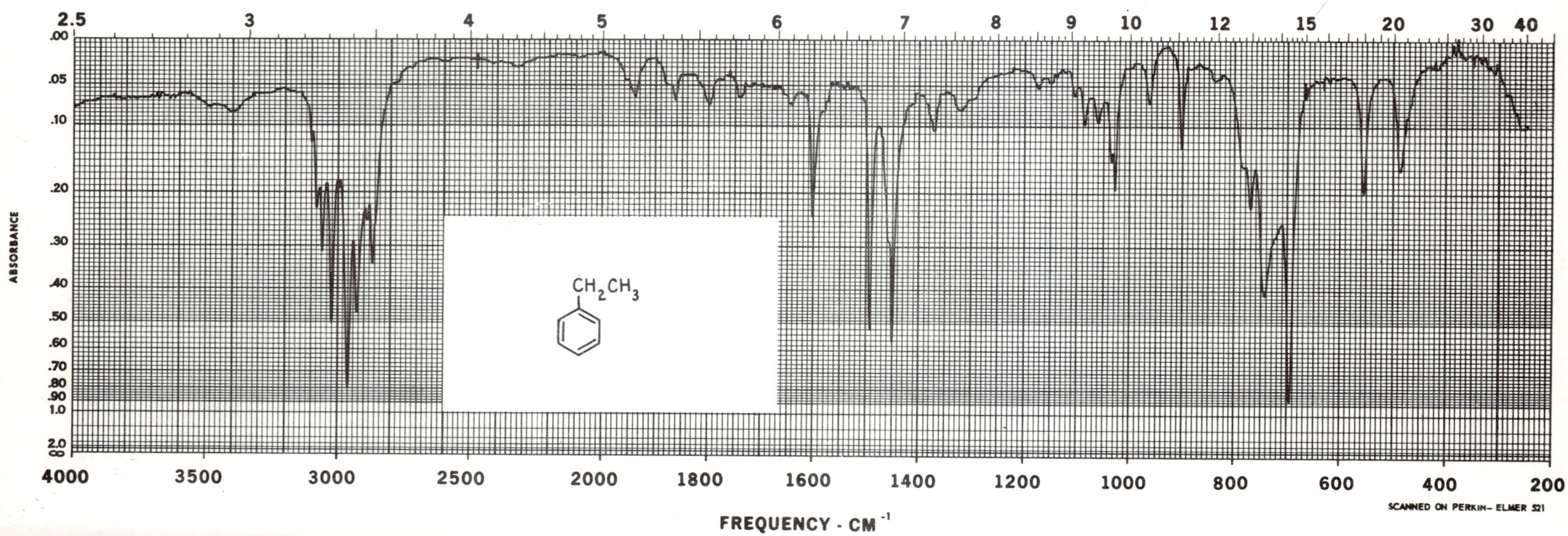

BOILING POINT:	134-136°C	INJECTION AMT.:	1.0 µl
VAPORIZATION TEMP.:	200°C	PRESSURE:	Atmospheric
CELL VOLUME:	7.5 cc	MAX./MIN.:	—
PATH LENGTH:	7.6 cm	SCALE EXPANSION:	1.11

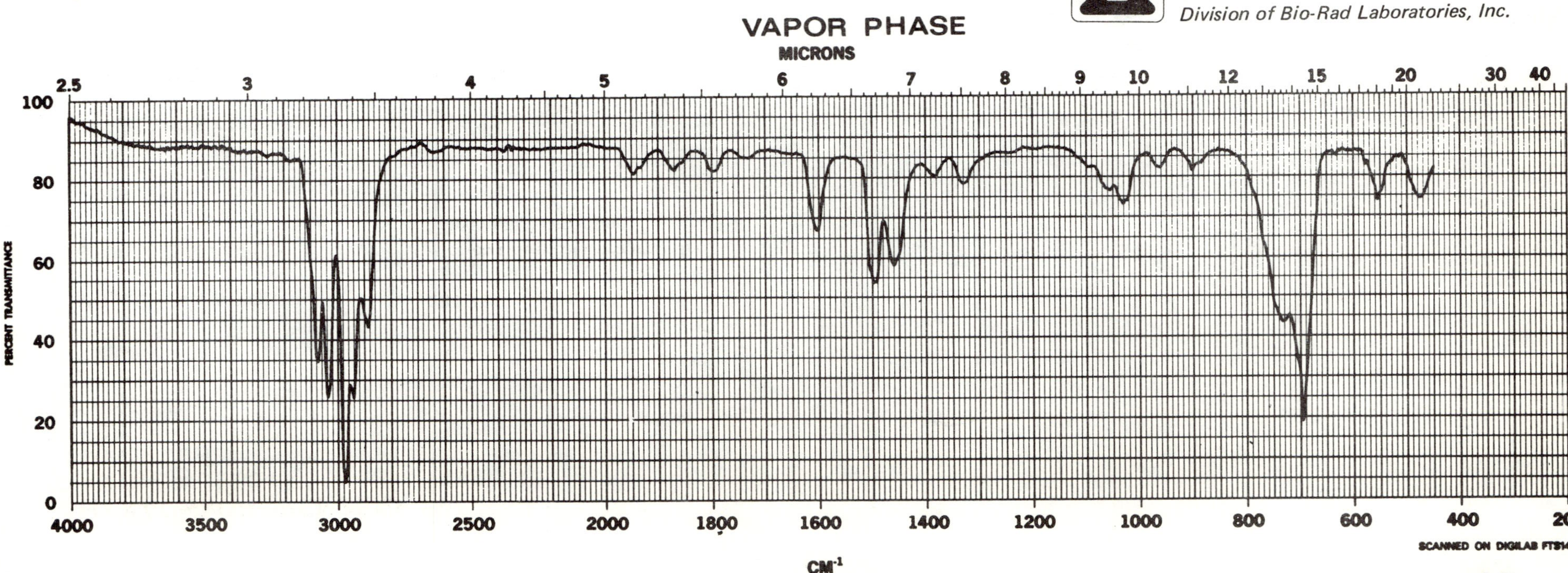

HEXACHLOROBENZENE
ANTI–CARIE 80
BUNT–CURE
HCB
SANOCIDE

C_6Cl_6

MOLECULAR WEIGHT: 284.73

SOURCE OF SAMPLE: Aldrich Chemical Company, Inc.
Milwaukee, Wisconsin

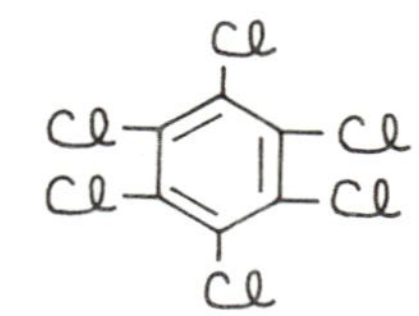

CAS REGISTRY NO.: 118-74-1

NIOSH NUMBER: DA29750

KBr Wafer

CONDENSED PHASE

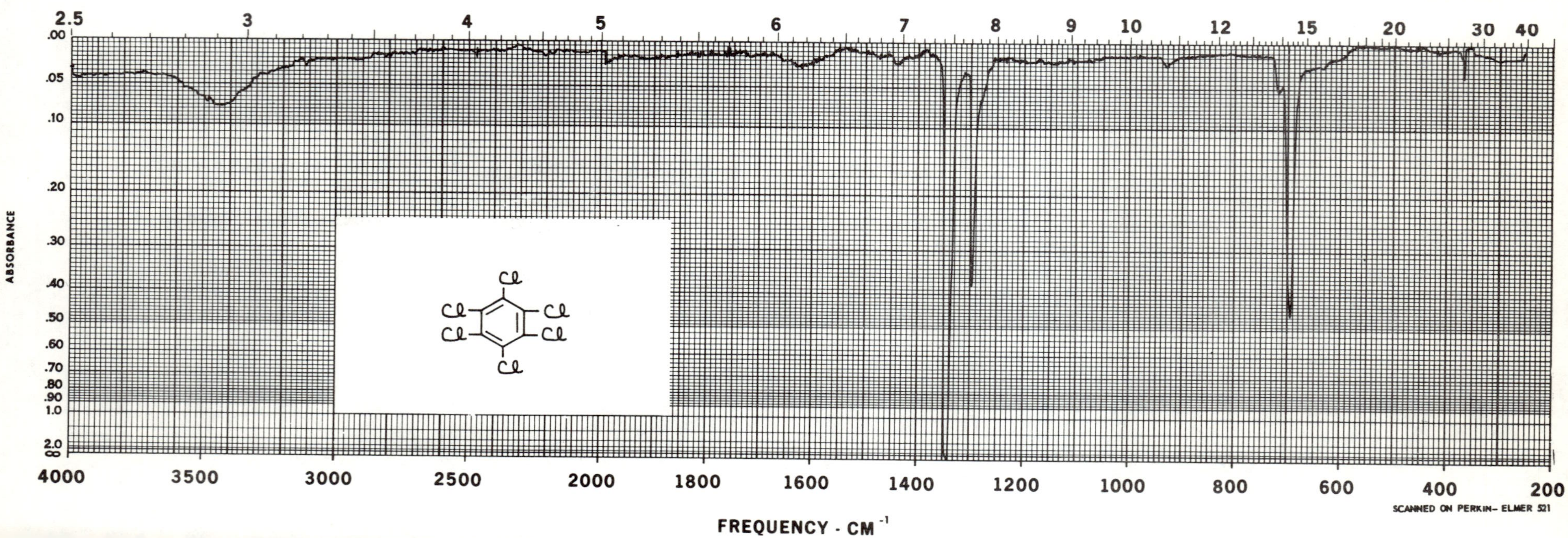

BOILING POINT:	$332^{o}C$	INJECTION AMT.:	0.82 mg
VAPORIZATION TEMP.:	$280^{o}C$	PRESSURE:	Atmospheric
CELL VOLUME:	7.5 cc	MAX./MIN.:	102.2/5.5
PATH LENGTH:	7.6 cm	SCALE EXPANSION:	0.95

VAPOR PHASE

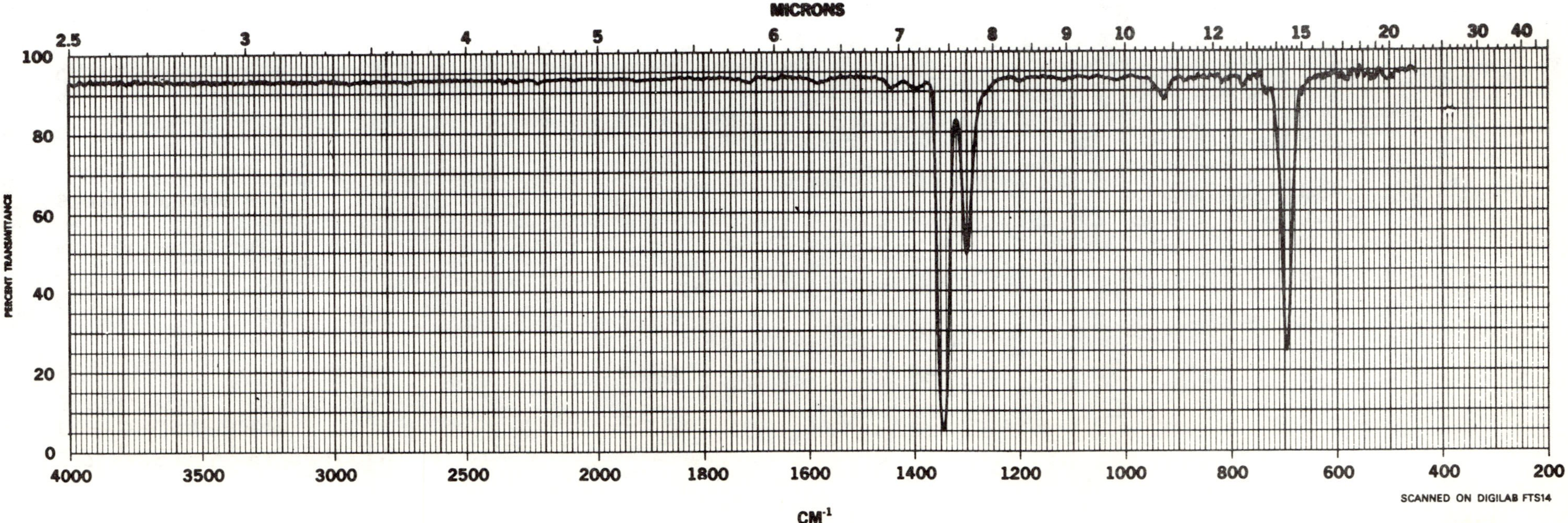

SCANNED ON DIGILAB FTS14

1,2–(METHYLENEDIOXY)–4–PROPYLBENZENE
DIHYDROSAFROLE
5–PROPYL–1,3–BENZODIOXOLE

$C_{10}H_{12}O_2$

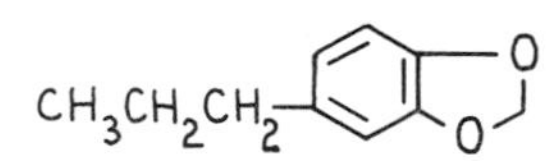

MOLECULAR WEIGHT:	164.21
SOURCE OF SAMPLE:	Aldrich Chemical Company, Inc. Milwaukee, Wisconsin
CAS REGISTRY NO.:	94-58-6
NIOSH NUMBER:	DA61250
CAPILLARY CELL:	Neat

CONDENSED PHASE

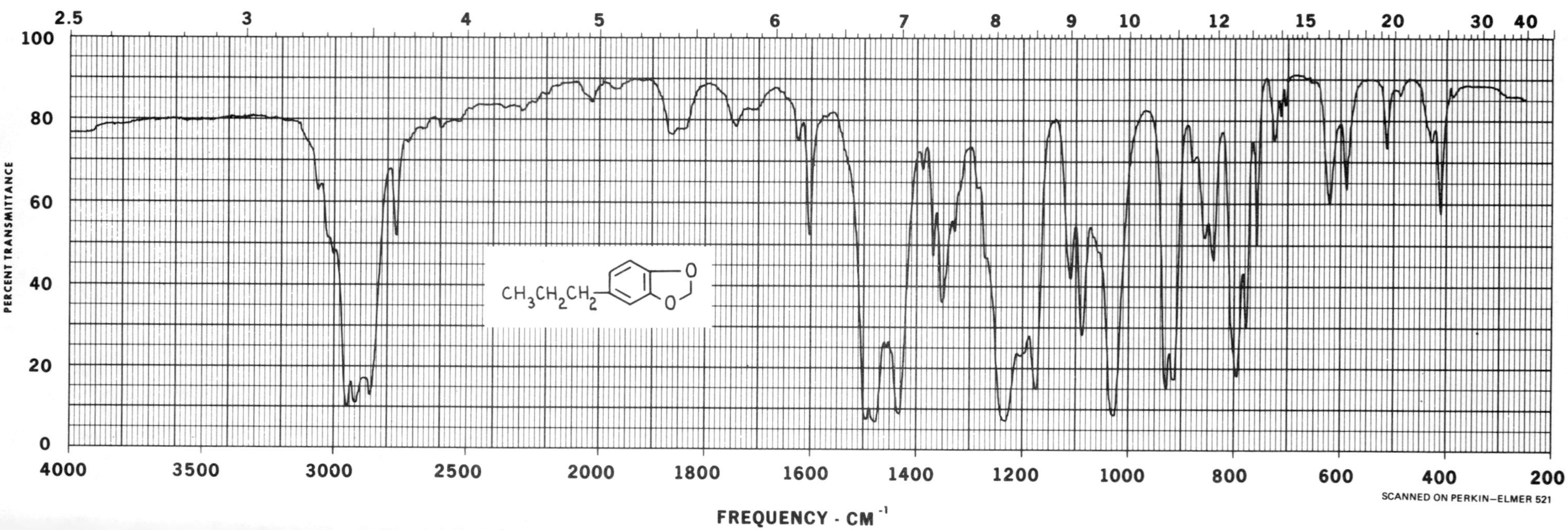

BOILING POINT:	–	INJECTION AMT.:	0.6 µl
VAPORIZATION TEMP.:	200°C	PRESSURE:	Atmospheric
CELL VOLUME:	7.5 cc	MAX./MIN.:	–
PATH LENGTH:	7.6 cm	SCALE EXPANSION:	1.06

VAPOR PHASE

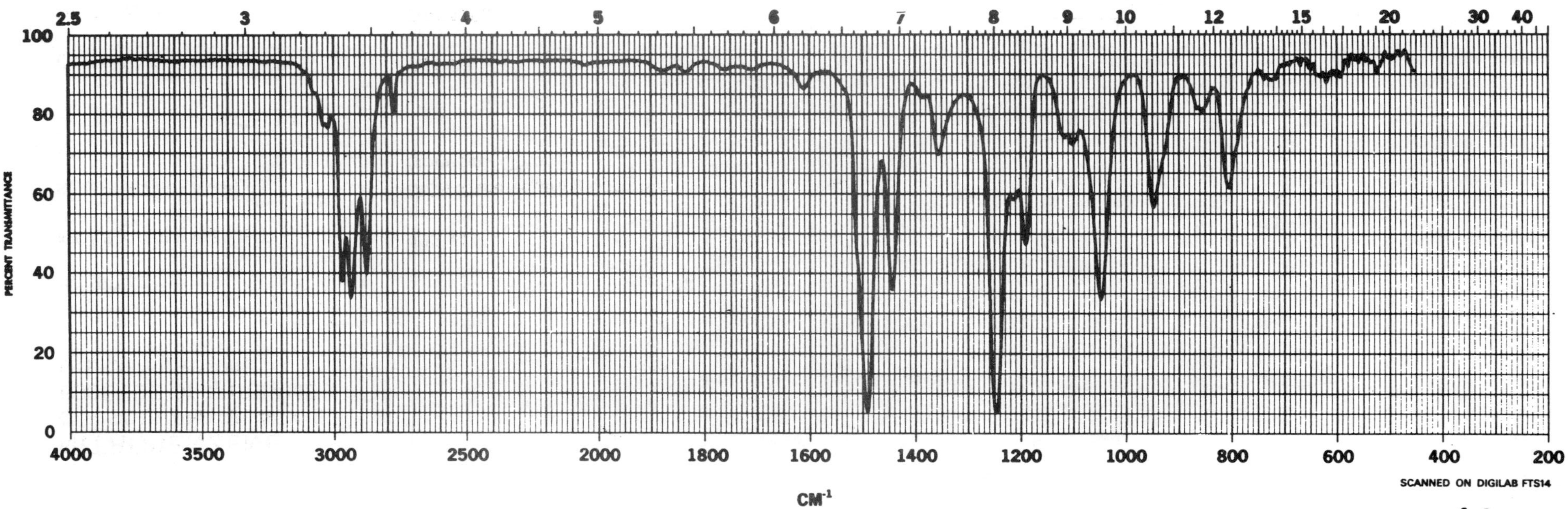

SCANNED ON DIGILAB FTS14

NITROBENZENE
C. I. SOLVENT BLACK 5
ESSENCE OF MIRBANE
OIL OF MIRBANE

$C_6H_5NO_2$

MOLECULAR WEIGHT: 123.11

SOURCE OF SAMPLE: MCB Manufacturing Chemists
Norwood, Ohio

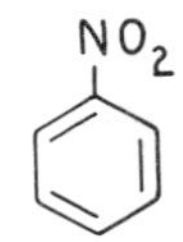

CAS REGISTRY NO.: 98-95-3

NIOSH NUMBER: DA64750

CAPILLARY CELL: Neat

CONDENSED PHASE

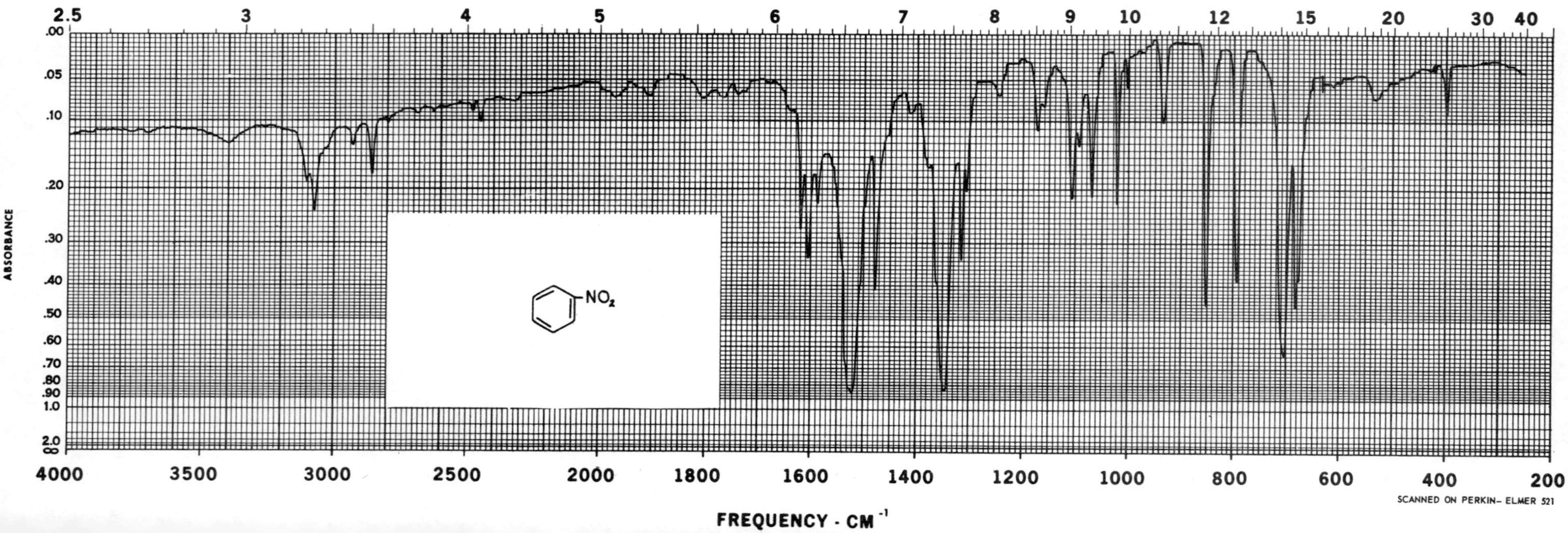

BOILING POINT:	210-211°C	INJECTION AMT.:	0.7 μl
VAPORIZATION TEMP.:	200°C	PRESSURE:	Atmospheric
CELL VOLUME:	7.5 cc	MAX./MIN.:	—
PATH LENGTH:	7.6 cm	SCALE EXPANSION:	0.92

VAPOR PHASE

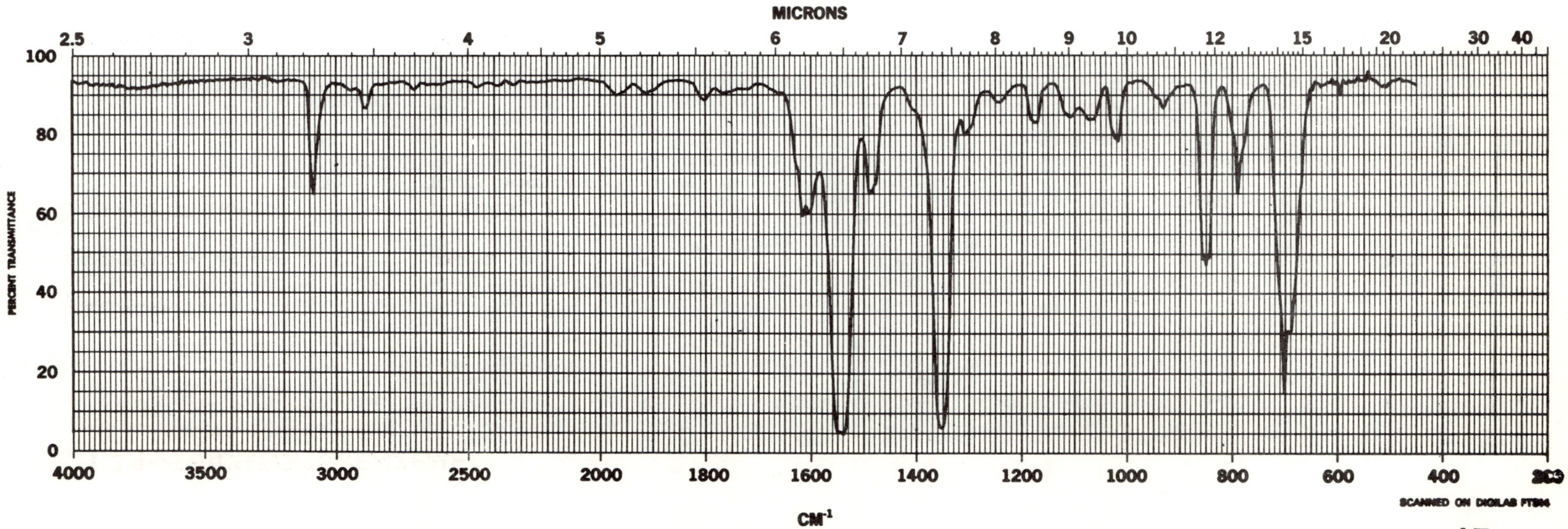

1,2,4–TRICHLOROBENZENE

$C_6H_3Cl_3$

MOLECULAR WEIGHT:	181.45
SOURCE OF SAMPLE:	Fluka AG Buchs, Switzerland
CAS REGISTRY NO.:	120-82-1
NIOSH NUMBER:	DC21000
CAPILLARY CELL:	Neat

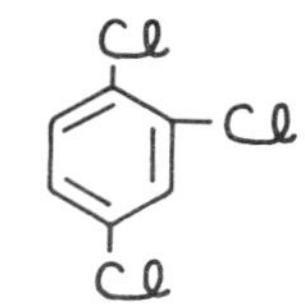

CONDENSED PHASE

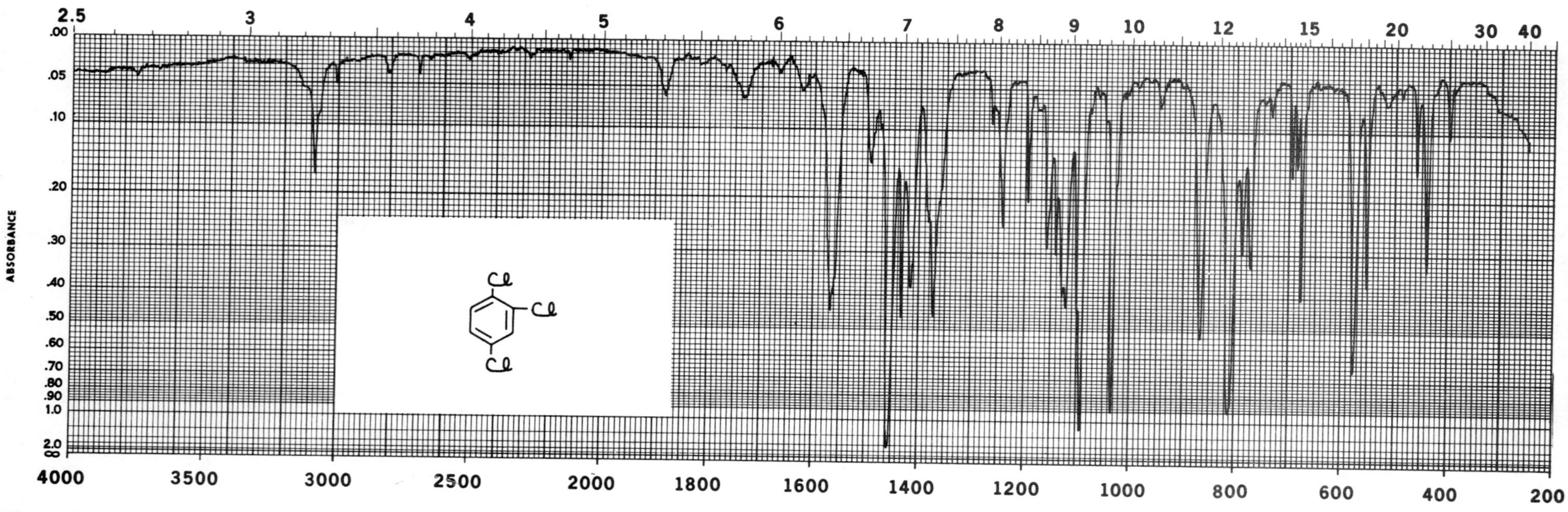

BOILING POINT:	214°C	INJECTION AMT.:	0.5 µl
VAPORIZATION TEMP.:	200°C	PRESSURE:	Atmospheric
CELL VOLUME:	7.5 cc	MAX./MIN.:	—
PATH LENGTH:	7.6 cm	SCALE EXPANSION:	1.10

VAPOR PHASE

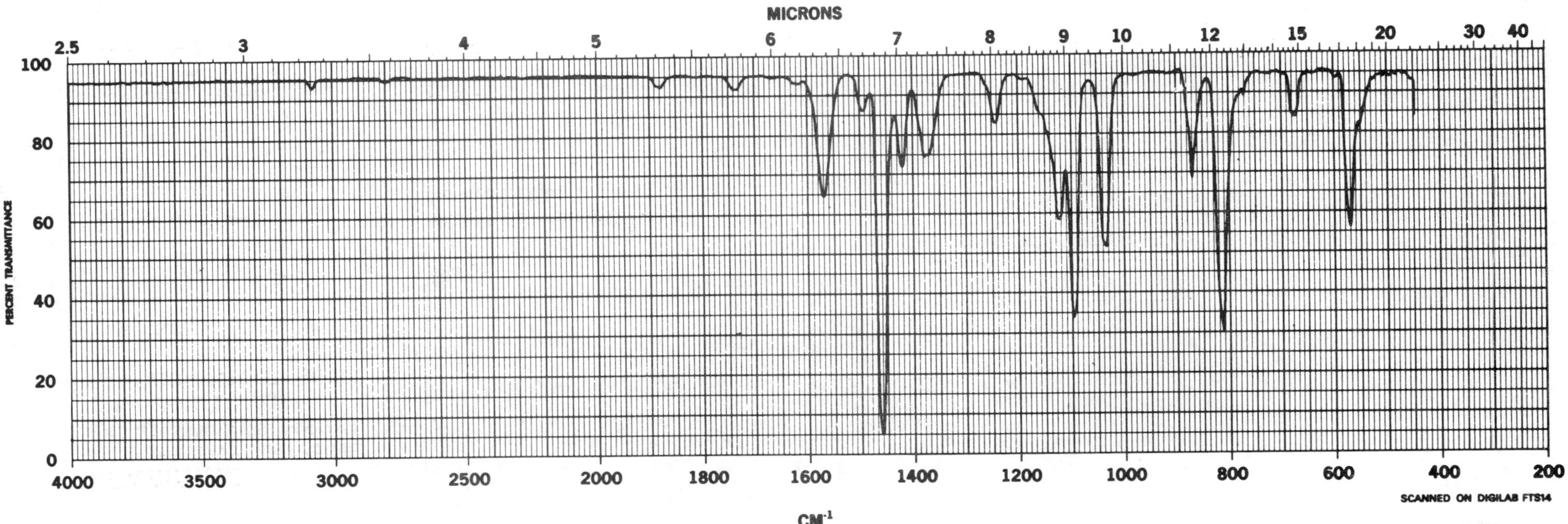

BENZIDINE
4,4′–BIPHENYLDIAMINE
4,4′–DIAMINOBIPHENYL
C. I. AZOIC DIAZO
FAST CORINTH BASE B

$C_{12}H_{12}N_2$

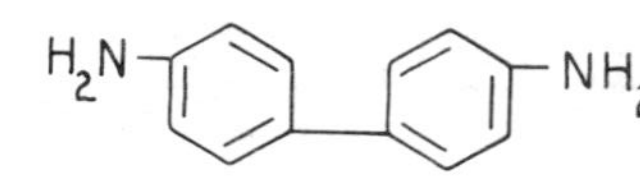

MOLECULAR WEIGHT: 184.24

SOURCE OF SAMPLE: MCB Manufacturing Chemists
Norwood, Ohio

CAS REGISTRY NO.: 92-87-5

NIOSH NUMBER: DC96250

KBr Wafer

CONDENSED PHASE

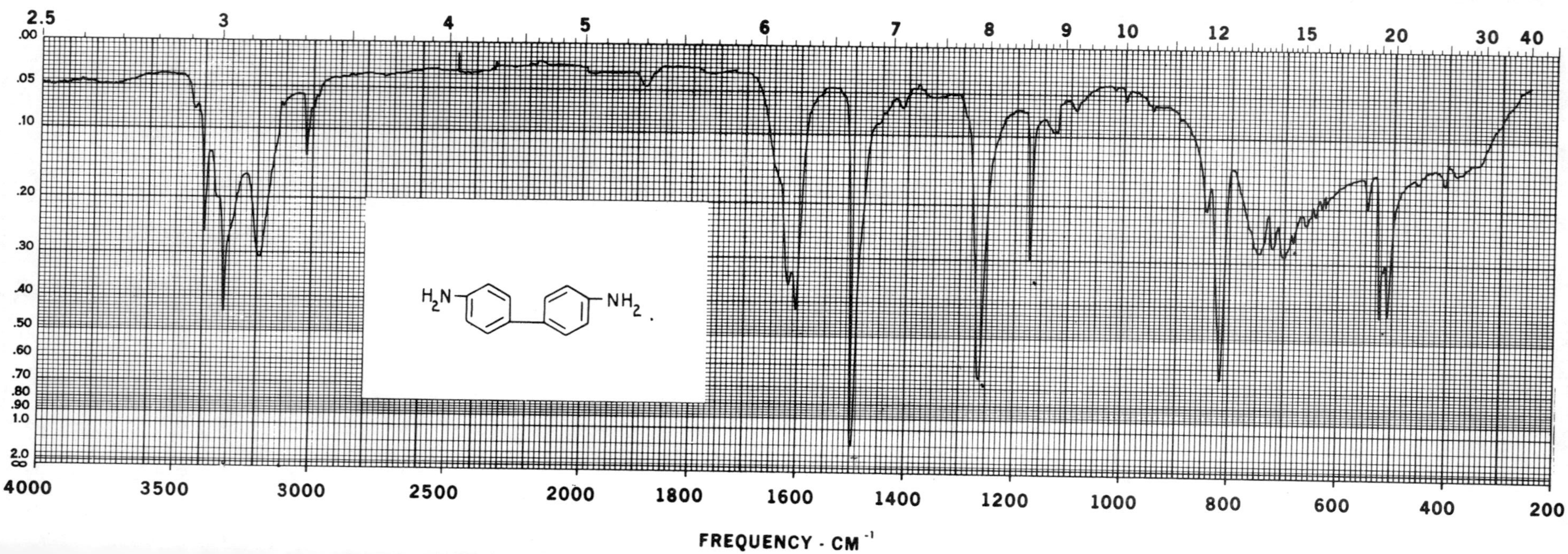

MELTING POINT:	127-129°C	INJECTION AMT.:	—
VAPORIZATION TEMP.:	280°C	PRESSURE:	Atmospheric
CELL VOLUME:	7.5 cc	MAX./MIN.:	99.6/29.2
PATH LENGTH:	7.6 cm	SCALE EXPANSION:	1.31

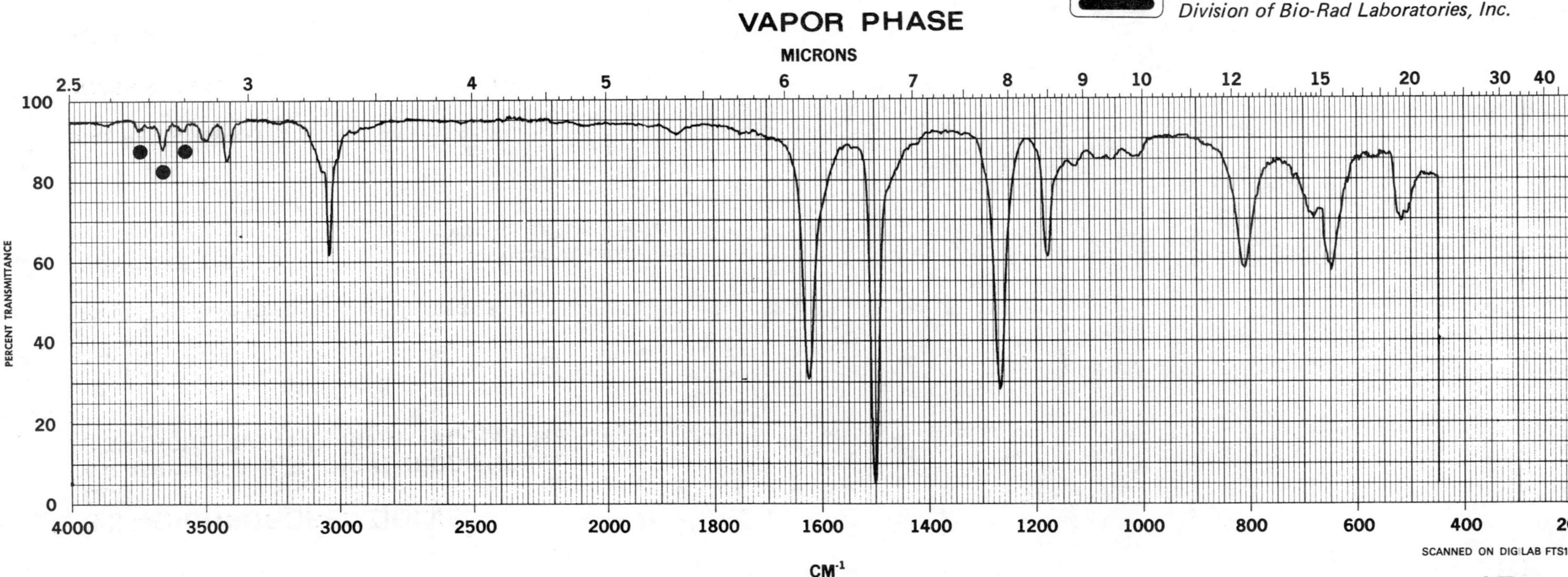

SCANNED ON DIGILAB FTS14

3,3′–DICHLOROBENZIDINE

DCB

4,4′–DIAMINO–3,3′–DICHLOROBIPHENYL

3,3′–DICHLORO–4,4′–BIPHENYLDIAMINE

$C_{12}H_{10}Cl_2N_2$

MOLECULAR WEIGHT:	253.13
SOURCE OF SAMPLE:	Sadtler Research Laboratories Philadelphia, Pennsylvania
CAS REGISTRY NO.:	91-94-1
NIOSH NUMBER:	DD05250

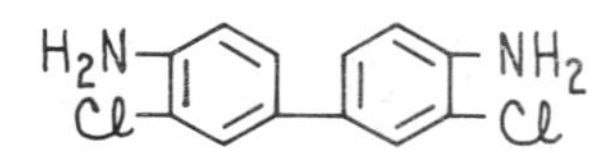

KBr Wafer

CONDENSED PHASE

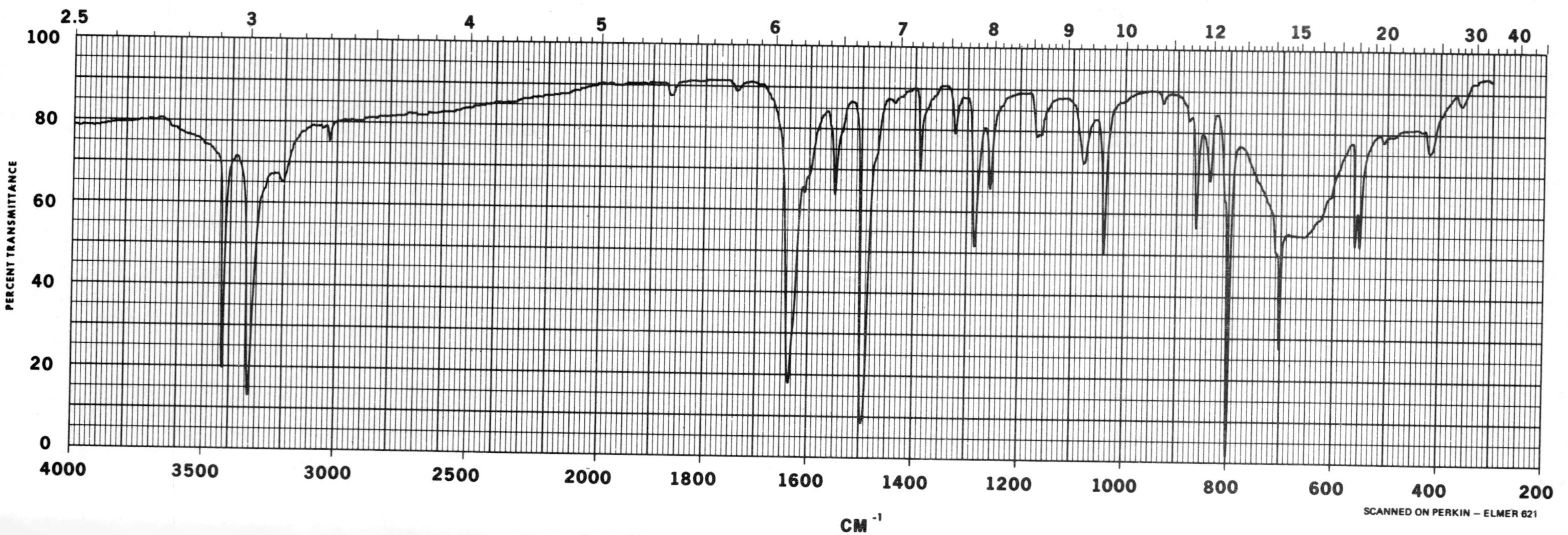

MELTING POINT:	132-133°C	INJECTION AMT.:	—
VAPORIZATION TEMP.:	280°C	PRESSURE:	Atmospheric
CELL VOLUME:	7.5 cc	MAX./MIN.:	98.7/36.9
PATH LENGTH:	7.6 cm	SCALE EXPANSION:	1.49

VAPOR PHASE

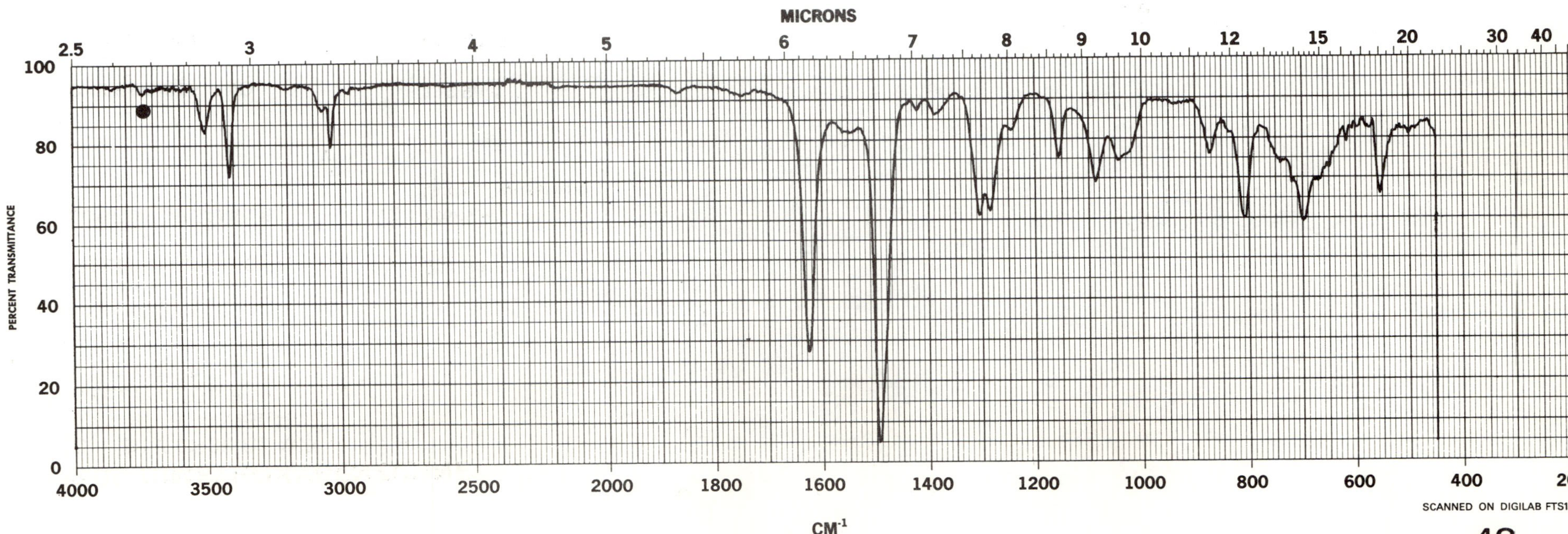

SCANNED ON DIGILAB FTS14

48

3,3'–DIMETHOXYBENZIDINE

C. I. DISPERSE BLACK 6
4,4'–BI–o–ANISIDINE
CELLITAZOL B
DIANISIDINE

$C_{14}H_{16}N_2O_2$

MOLECULAR WEIGHT: 244.30

SOURCE OF SAMPLE: Fluka AG, Buchs, Switzerland

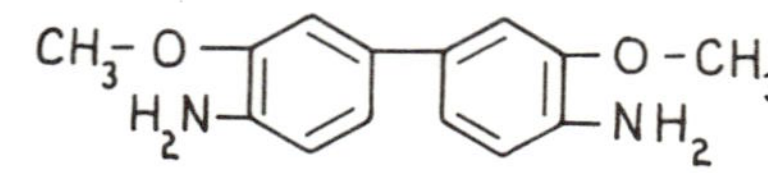

CAS REGISTRY NO.: 119-90-4

NIOSH NUMBER: DD08750

KBr Wafer

CONDENSED PHASE

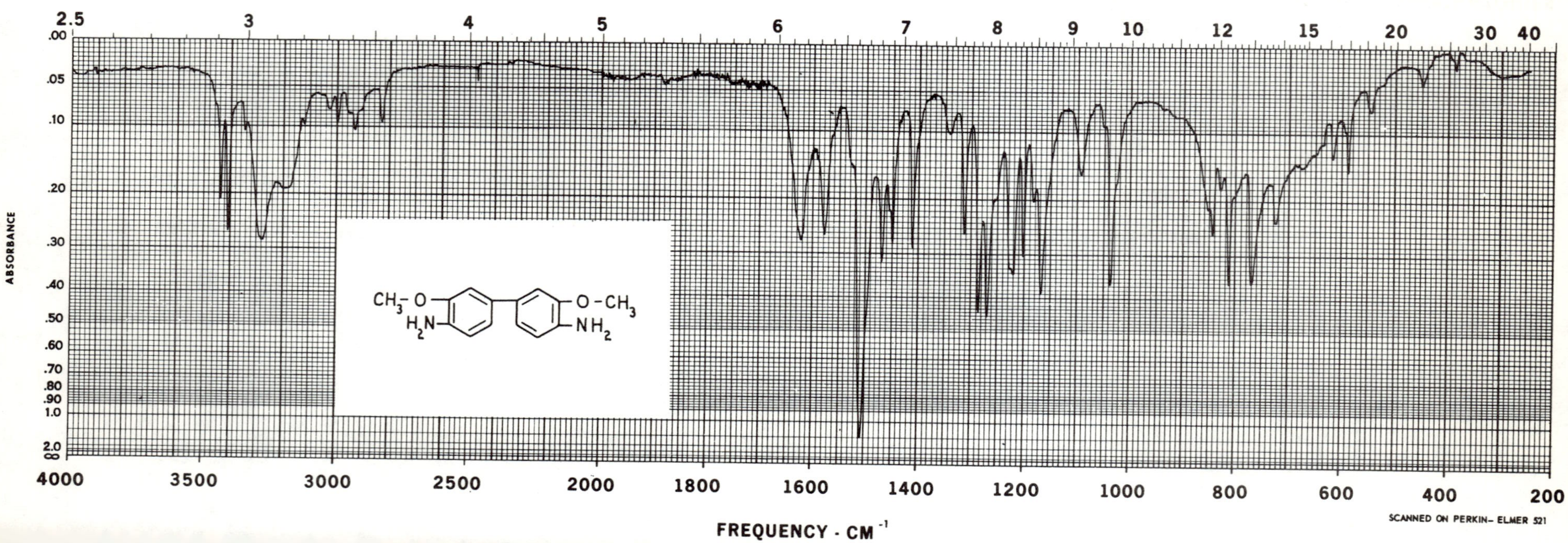

MELTING POINT:	137-139°C	INJECTION AMT.:	—
VAPORIZATION TEMP.:	280°C	PRESSURE:	Atmospheric
CELL VOLUME:	7.5 cc	MAX./MIN.:	98.6/30.8
PATH LENGTH:	7.6 cm	SCALE EXPANSION:	1.36

VAPOR PHASE

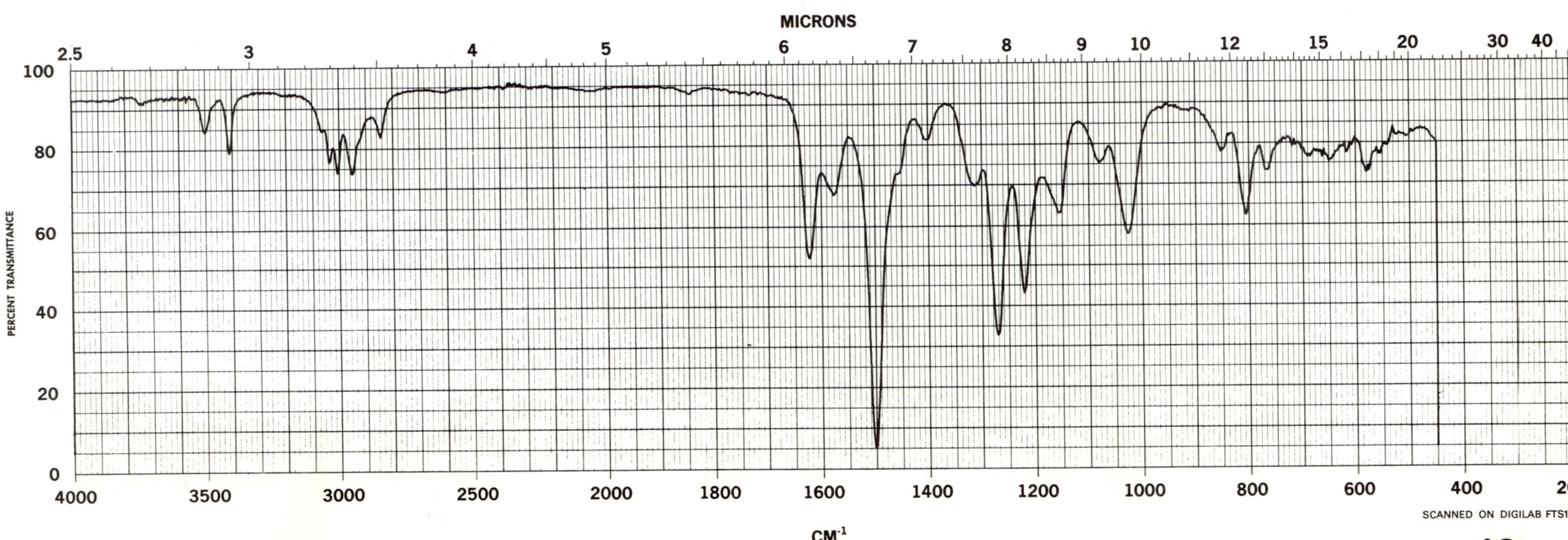

SCANNED ON DIGILAB FTS14

BENZOIC ACID
BENZENECARBOXYLIC ACID
RETARDER BA
RETARDEX
TENN—PLAS

$C_7H_6O_2$

MOLECULAR WEIGHT:	122.13
SOURCE OF SAMPLE:	Monsanto Company St. Louis, Missouri
CAS REGISTRY NO.:	65-85-0
NIOSH NUMBER:	DG08750

KBr Wafer

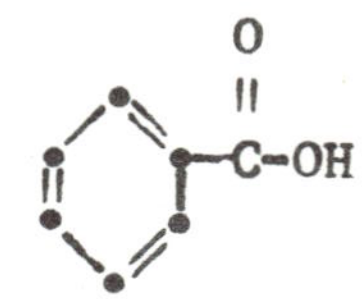

CONDENSED PHASE

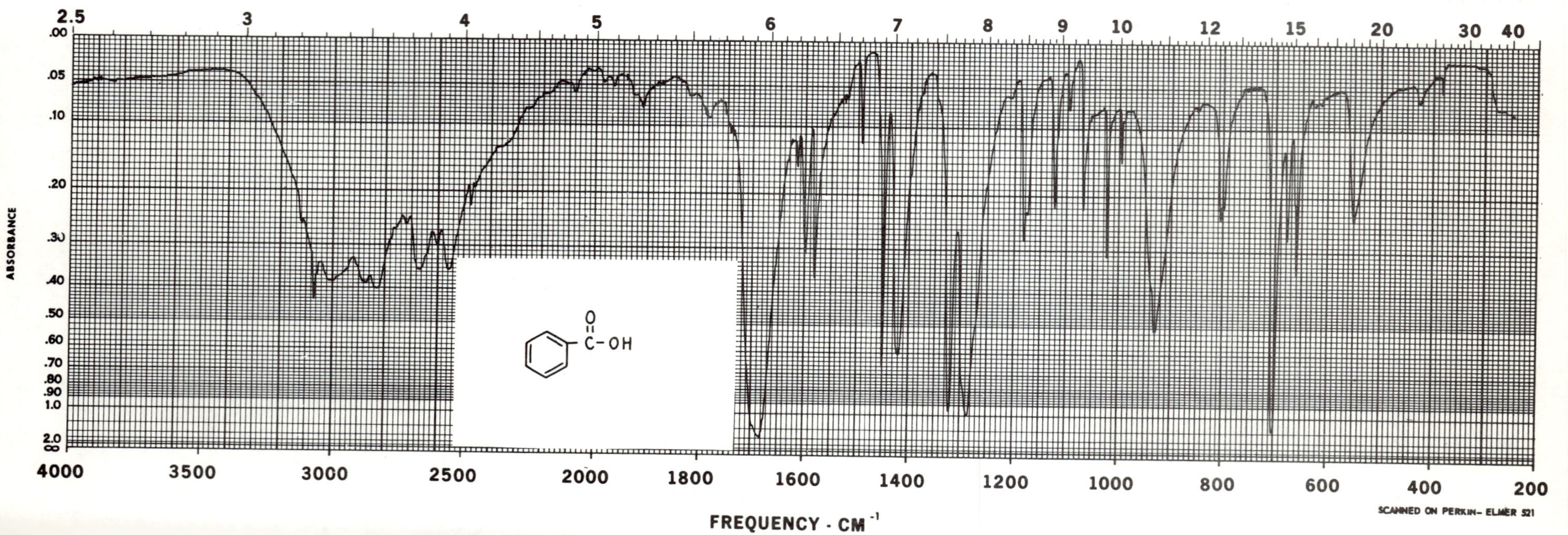

BOILING POINT:	132.5°C/10 mm	INJECTION AMT.:	0.68 mg
VAPORIZATION TEMP.:	280°C	PRESSURE:	Atmospheric
CELL VOLUME:	7.5 cc	MAX./MIN.:	—
PATH LENGTH:	7.6 cm	SCALE EXPANSION:	0.98

VAPOR PHASE

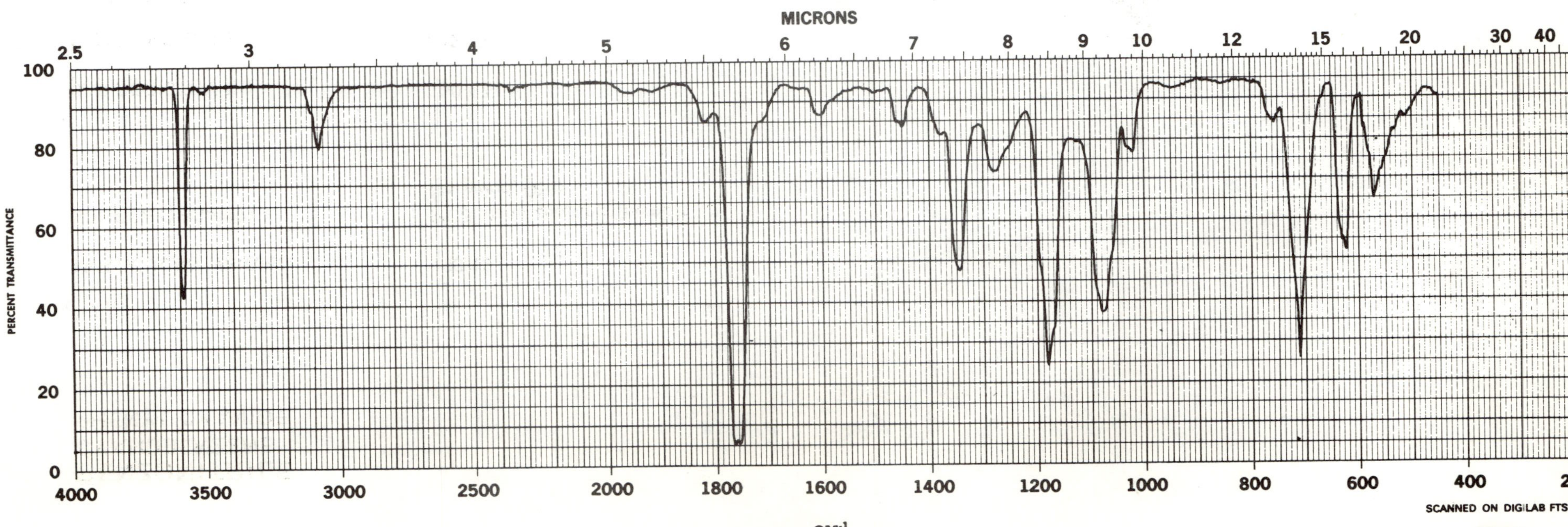

SCANNED ON DIGILAB FTS14

BENZONITRILE
CYANOBENZENE
PHENYL CYANIDE

C_7H_5N

MOLECULAR WEIGHT: 103.12

SOURCE OF SAMPLE: MCB Manufacturing Chemists
Norwood, Ohio

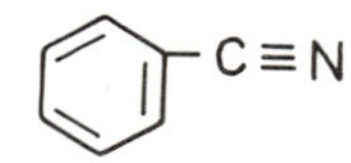

CAS REGISTRY NO.: 100-47-0

NIOSH NUMBER: DI24500

CAPILLARY CELL: Neat

CONDENSED PHASE

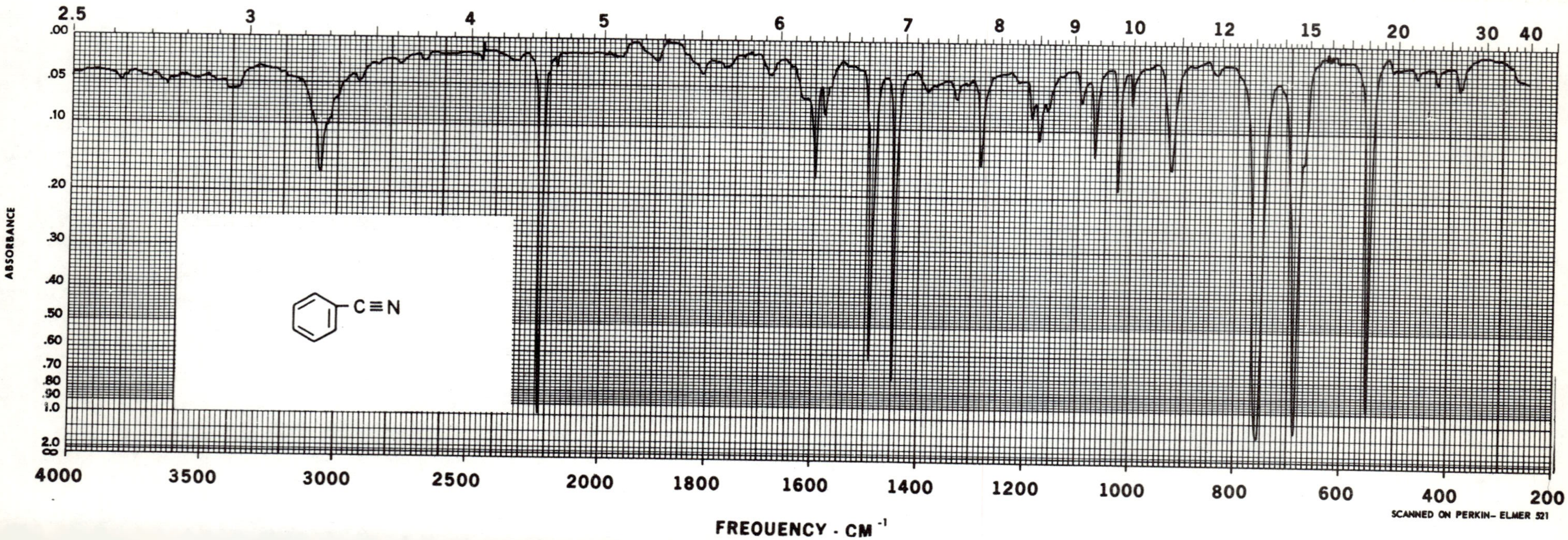

BOILING POINT:	188-189°C	INJECTION AMT.:	2.0 µl
VAPORIZATION TEMP.:	200°C	PRESSURE:	Atmospheric
CELL VOLUME:	7.5 cc	MAX./MIN.:	—
PATH LENGTH:	7.6 cm	SCALE EXPANSION:	1.08

VAPOR PHASE

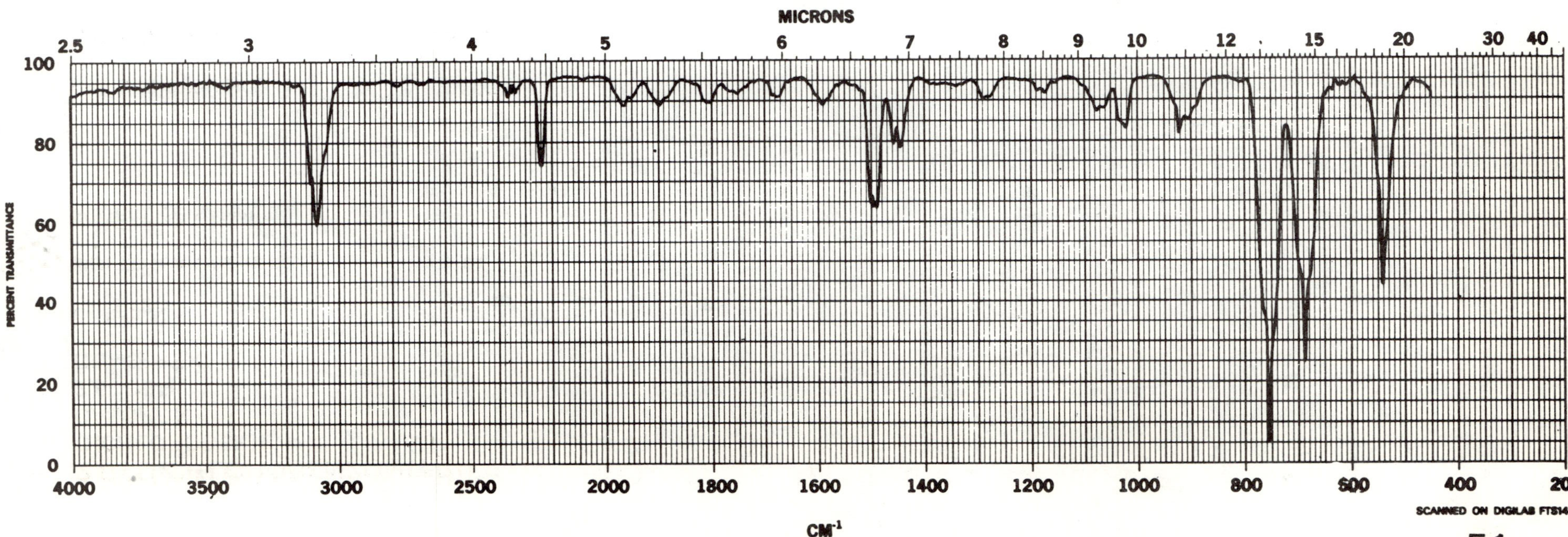

2,6–DICHLOROBENZONITRILE

DICHLOROBENIL
CASORON
2,6–DBN
DU–SPREX

$C_7H_3Cl_2N$

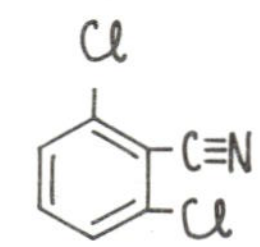

MOLECULAR WEIGHT:	172.01
SOURCE OF SAMPLE:	Fluka AG Buchs, Switzerland
CAS REGISTRY NO.:	1194-65-6
NIOSH NUMBER:	DI35000

KBr Wafer

CONDENSED PHASE

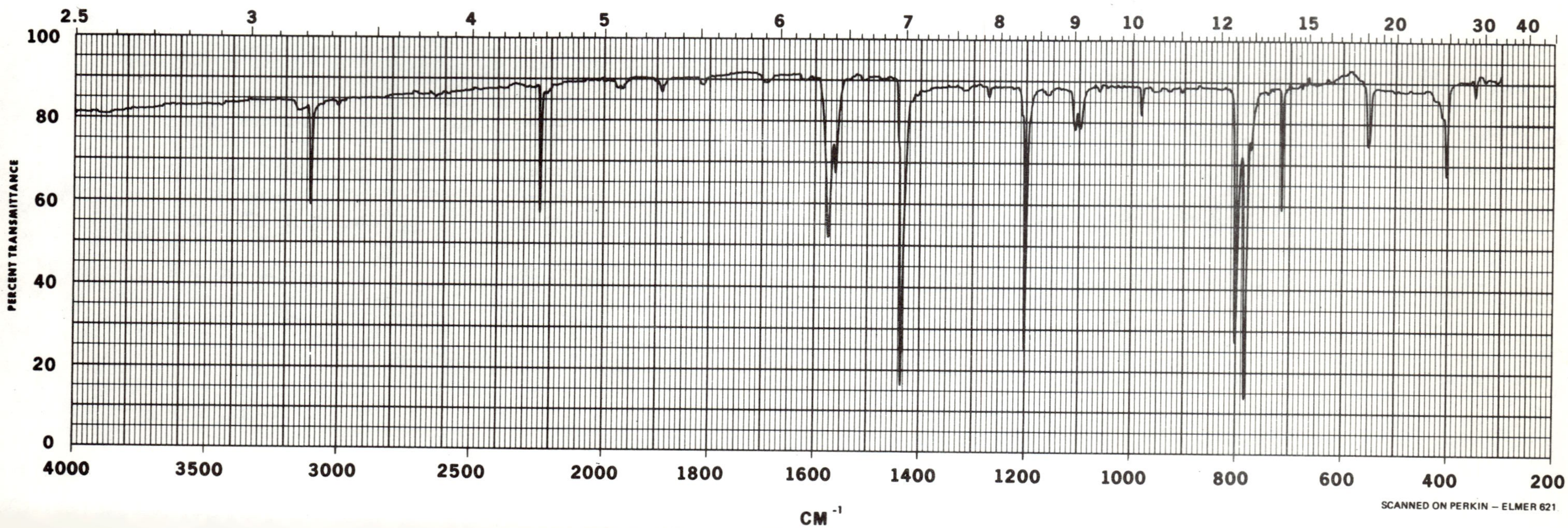

MELTING POINT:	141-143°C	INJECTION AMT.:	1.1 mg
VAPORIZATION TEMP.:	280°C	PRESSURE:	Atmospheric
CELL VOLUME:	7.5 cc	MAX./MIN.:	106.3/21.5
PATH LENGTH:	7.6 cm	SCALE EXPANSION:	1.08

VAPOR PHASE

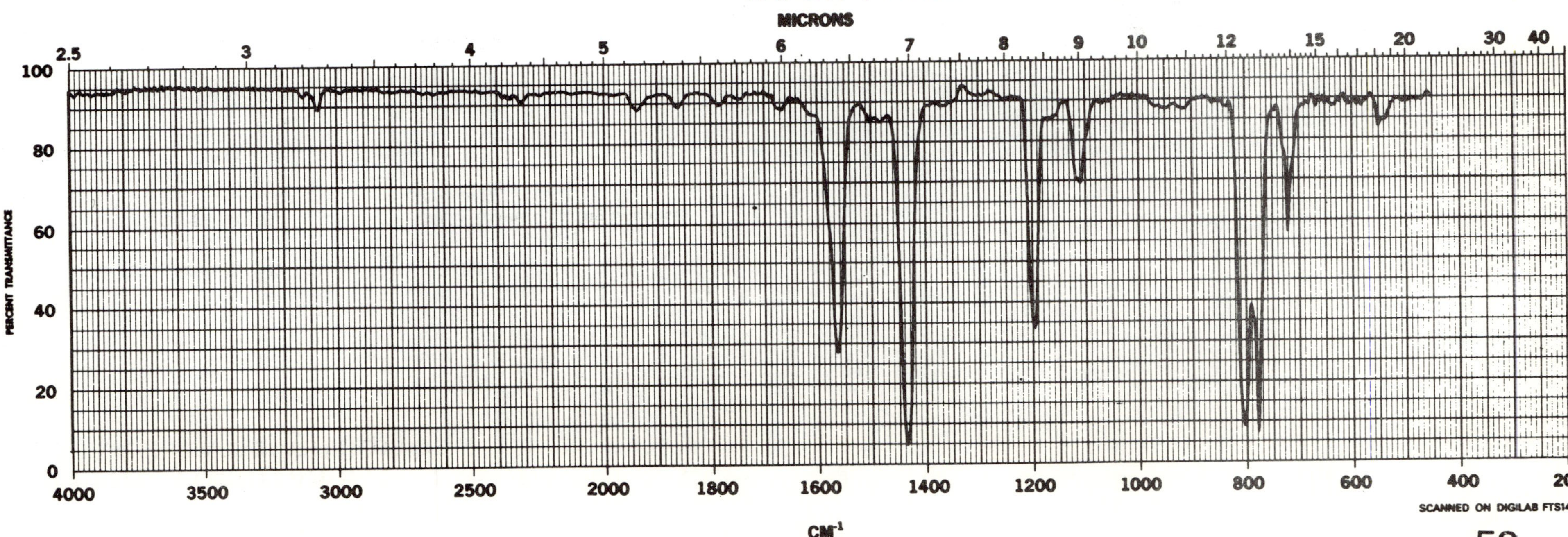

SCANNED ON DIGILAB FTS14

BENZOYL CHLORIDE
BENZENECARBONYL CHLORIDE

C_7H_5ClO

MOLECULAR WEIGHT: 140.57

SOURCE OF SAMPLE: MCB Manufacturing Chemists
Norwood, Ohio

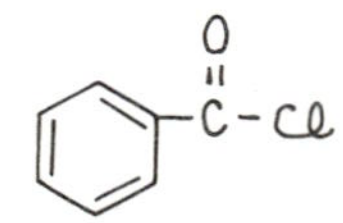

CAS REGISTRY NO.: 98-88-4

NIOSH NUMBER: DM66000

CAPILLARY CELL: Neat

CONDENSED PHASE

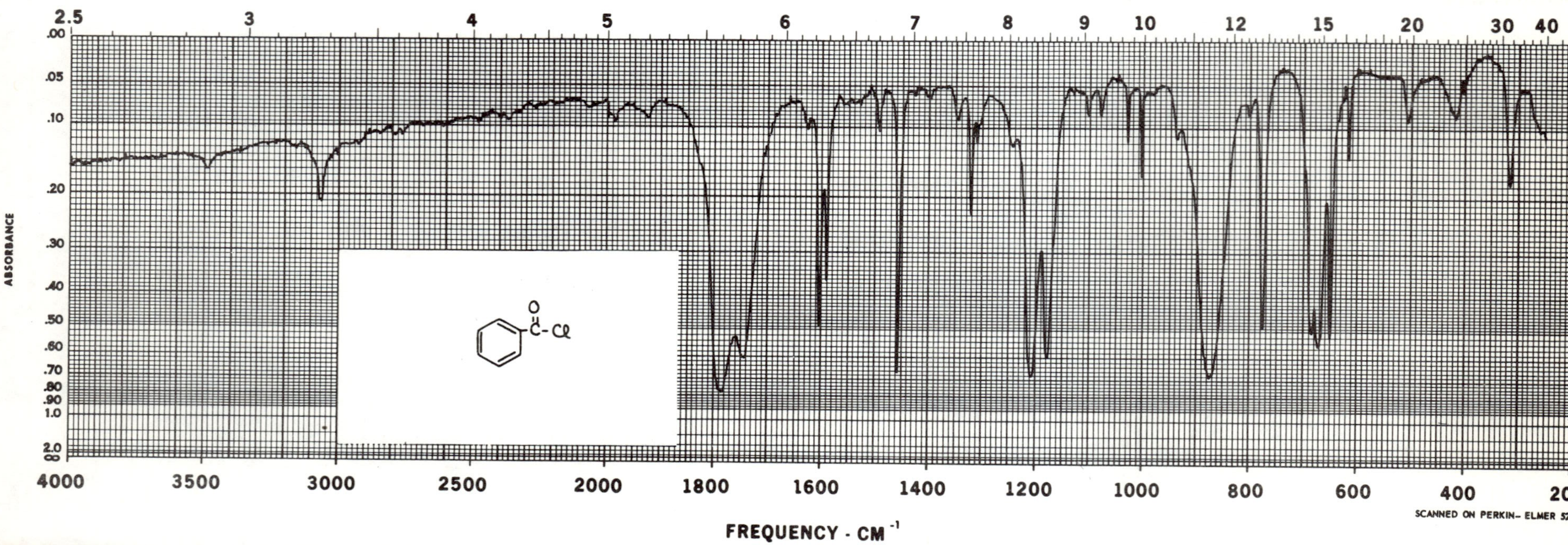

BOILING POINT:	70-74°C/11 mm	INJECTION AMT.:	–
VAPORIZATION TEMP.:	200°C	PRESSURE:	Atmospheric
CELL VOLUME:	7.5 cc	MAX./MIN.:	99.2/14.6
PATH LENGTH:	7.6 cm	SCALE EXPANSION:	1.09

VAPOR PHASE

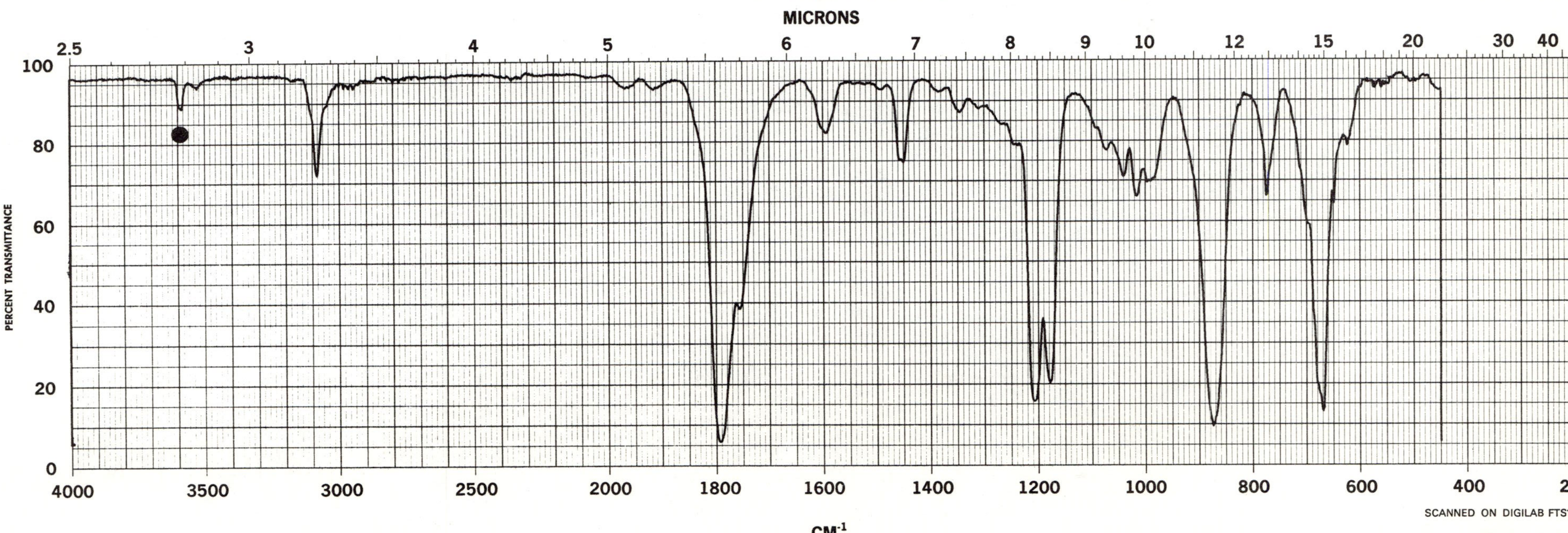

SCANNED ON DIGILAB FTS14

4–NITROBIPHENYL
PNB

$C_{12}H_9NO_2$

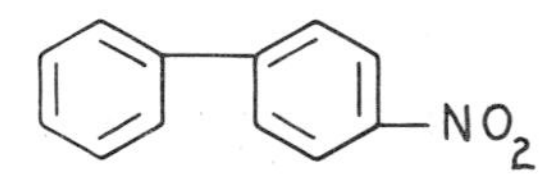

MOLECULAR WEIGHT: 199.21

SOURCE OF SAMPLE: MCB Manufacturing Chemists
Norwood, Ohio

CAS REGISTRY NO.: 92-93-3

NIOSH NUMBER: DV56000

KBr Wafer

CONDENSED PHASE

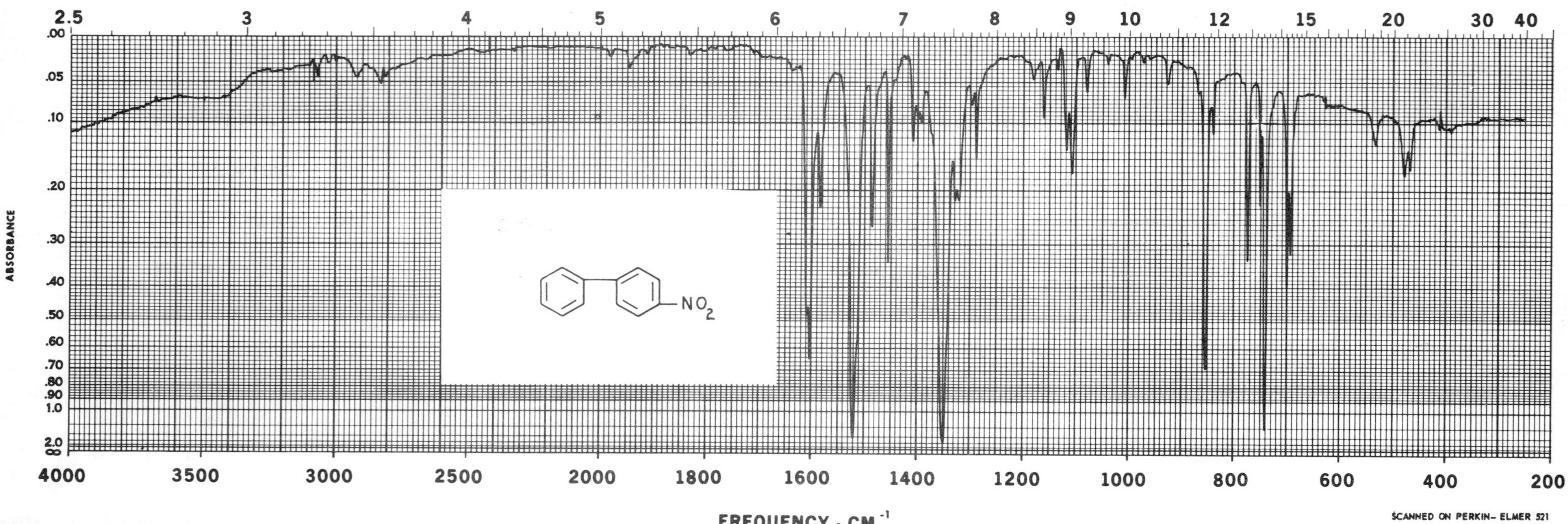

BOILING POINT:	340°C	INJECTION AMT.:	0.80 mg
VAPORIZATION TEMP.:	280°C	PRESSURE:	Atmospheric
CELL VOLUME:	7.5 cc	MAX./MIN.:	—
PATH LENGTH:	7.6 cm	SCALE EXPANSION:	1.11

VAPOR PHASE

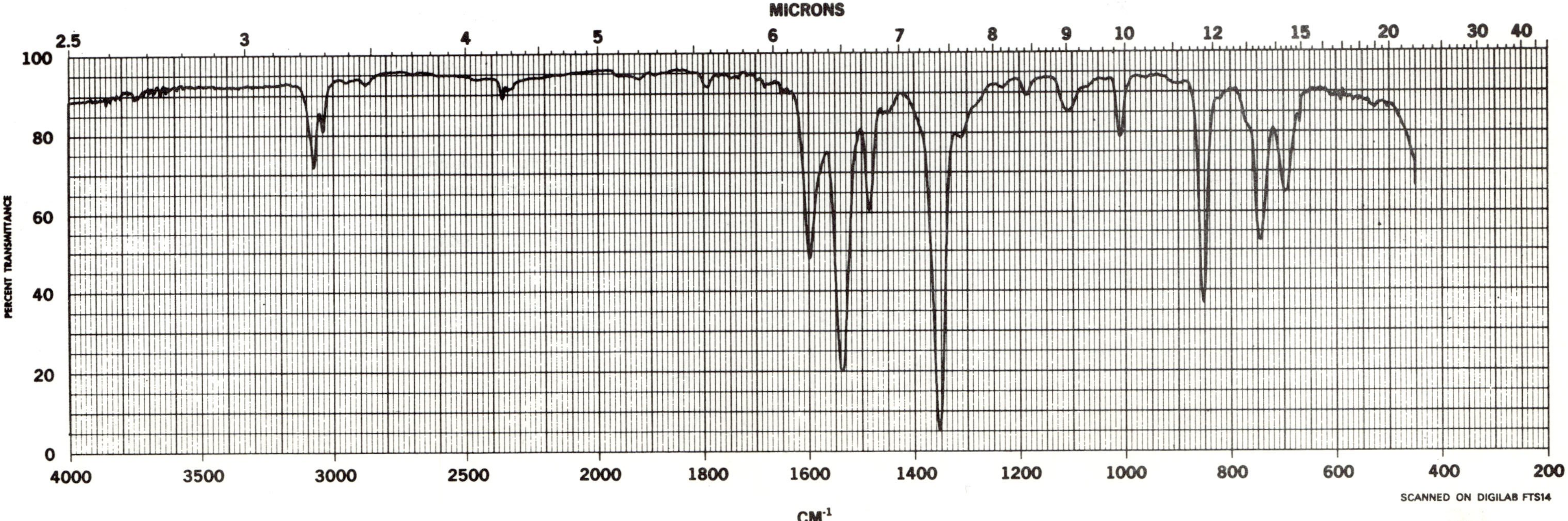

4–BIPHENYLAMINE
4–AMINOBIPHENYL

$C_{12}H_{11}N$

MOLECULAR WEIGHT: 169.23

SOURCE OF SAMPLE: Aldrich Chemical Company, Inc.
Milwaukee, Wisconsin

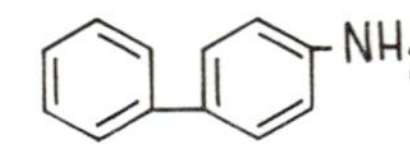

CAS REGISTRY NO.: 92-67-1

NIOSH NUMBER: DU89250

KBr Wafer

CONDENSED PHASE

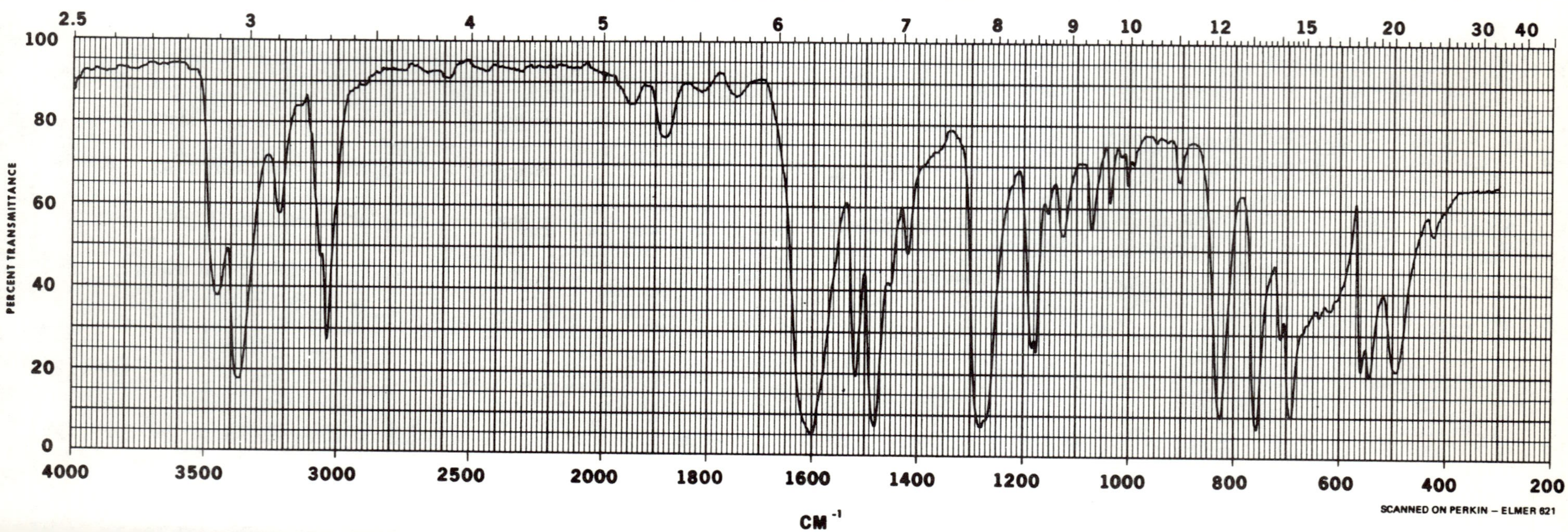

BOILING POINT:	191°C/15 mm	INJECTION AMT.:	0.93 mg
VAPORIZATION TEMP.:	280°C	PRESSURE:	Atmospheric
CELL VOLUME:	7.5 cc	MAX./MIN.:	88.7/22.4
PATH LENGTH:	7.6 cm	SCALE EXPANSION:	1.39

VAPOR PHASE

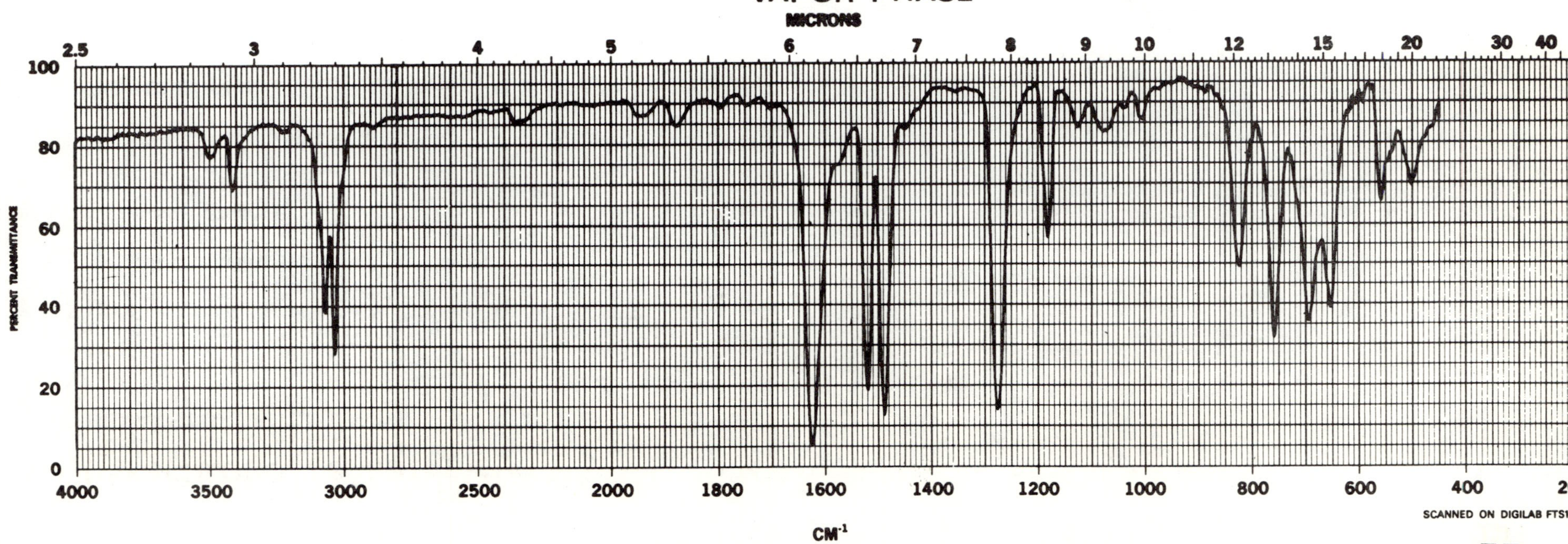

SCANNED ON DIGILAB FTS14

HEXACHLORO–1,3–BUTADIENE
HCBD
HEXACHLOROBUTADIENE

C_4Cl_6

MOLECULAR WEIGHT:	260.76
SOURCE OF SAMPLE:	Fluka AG Buchs, Switzerland
CAS REGISTRY NO.:	87-68-3
NIOSH NUMBER:	EJ07000
CAPILLARY CELL:	Neat

$Cl_2C=C(Cl)-C(Cl)=CCl_2$

CONDENSED PHASE

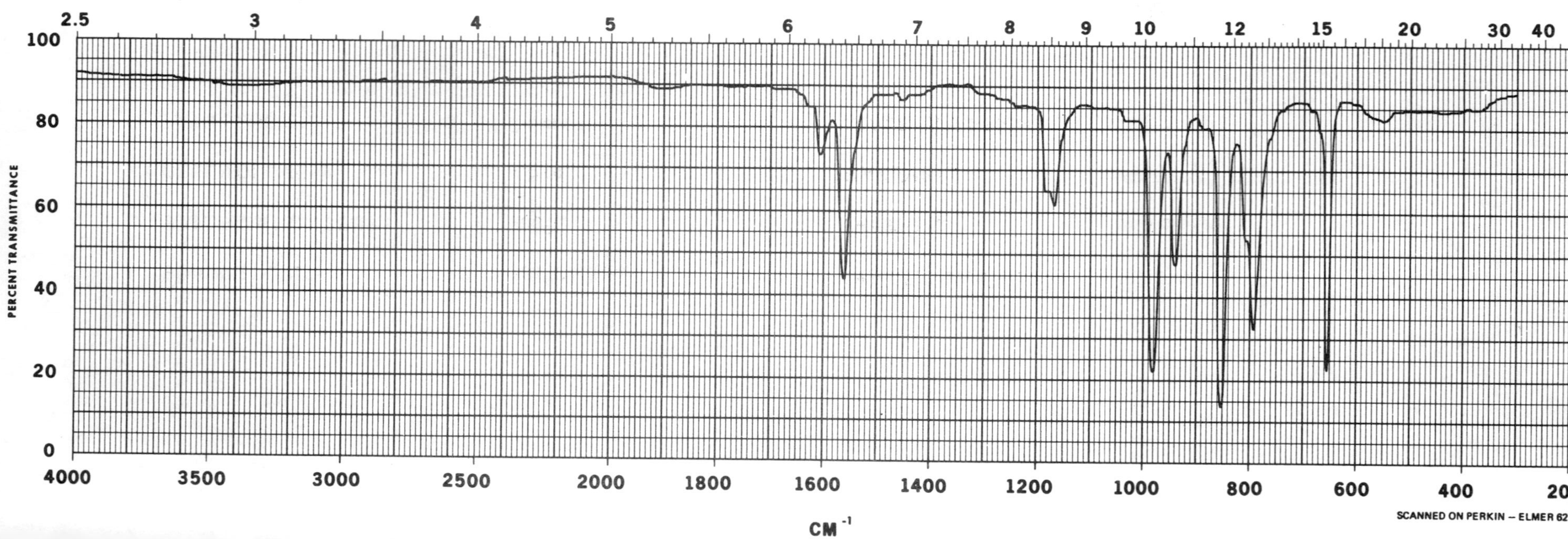

BOILING POINT:	210-220°C	INJECTION AMT.:	0.8 µl
VAPORIZATION TEMP.:	200°C	PRESSURE:	Atmospheric
CELL VOLUME:	7.5 cc	MAX./MIN.:	–
PATH LENGTH:	7.6 cm	SCALE EXPANSION:	1.03

VAPOR PHASE

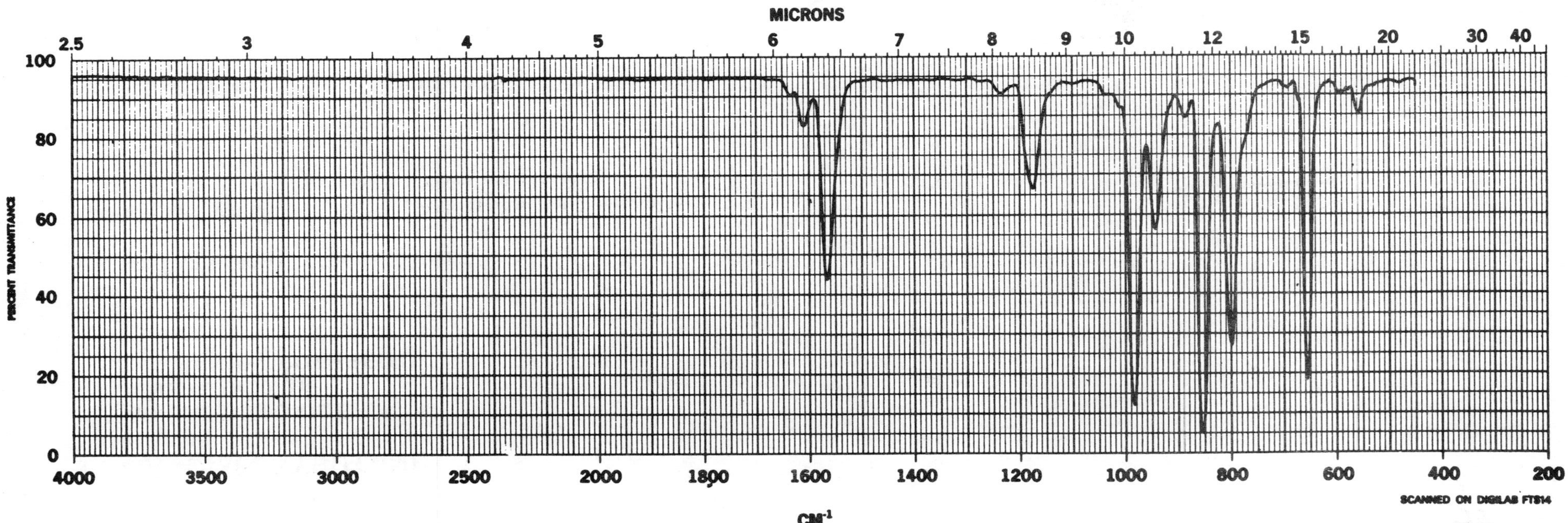

1,2:3,4–DIEPOXYBUTANE
BIOXIRANE
BUTADIENE DIEPOXIDE
BUTADIENE DIOXIDE
1,2:3,4–DIANHYDROTHREITOL

$C_4H_6O_2$

MOLECULAR WEIGHT: 86.09

SOURCE OF SAMPLE: Aldrich Chemical Company, Inc.
Milwaukee, Wisconsin

CAS REGISTRY NO.: 1464-53-5

NIOSH NUMBER: EJ82250

CAPILLARY CELL: Neat

CONDENSED PHASE

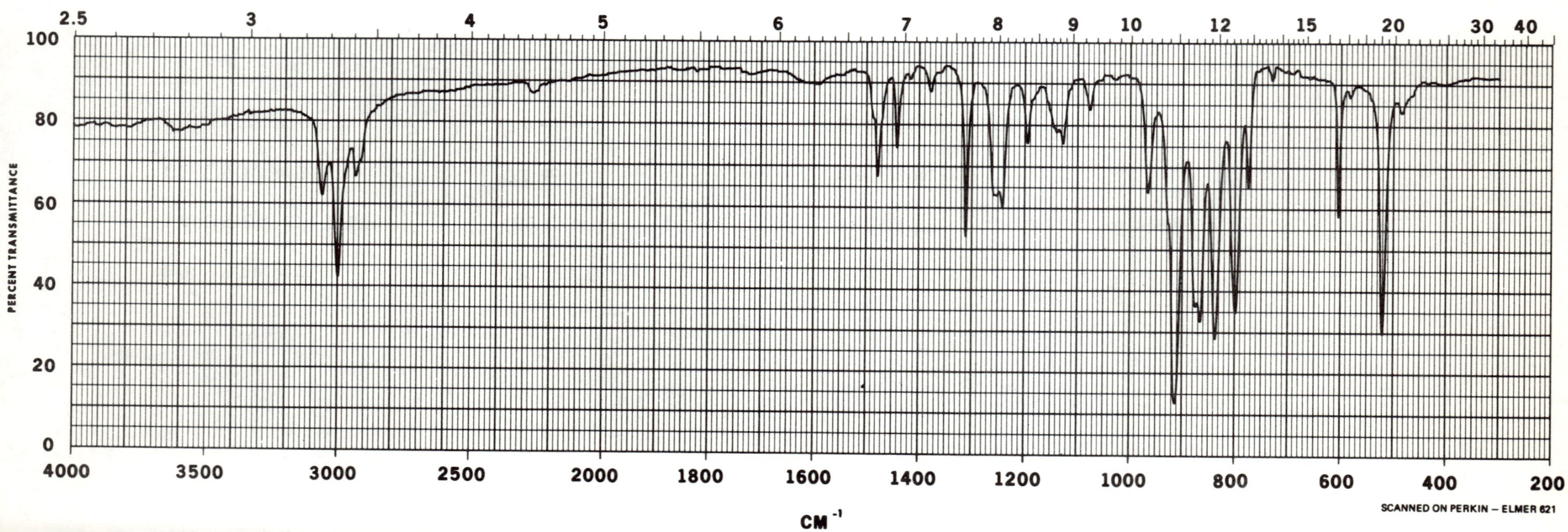

BOILING POINT:	138°C/767 mm	INJECTION AMT.:	1.0 µl
VAPORIZATION TEMP.:	200°C	PRESSURE:	Atmospheric
CELL VOLUME:	7.5 cc	MAX./MIN.:	—
PATH LENGTH:	7.6 cm	SCALE EXPANSION:	1.16

VAPOR PHASE

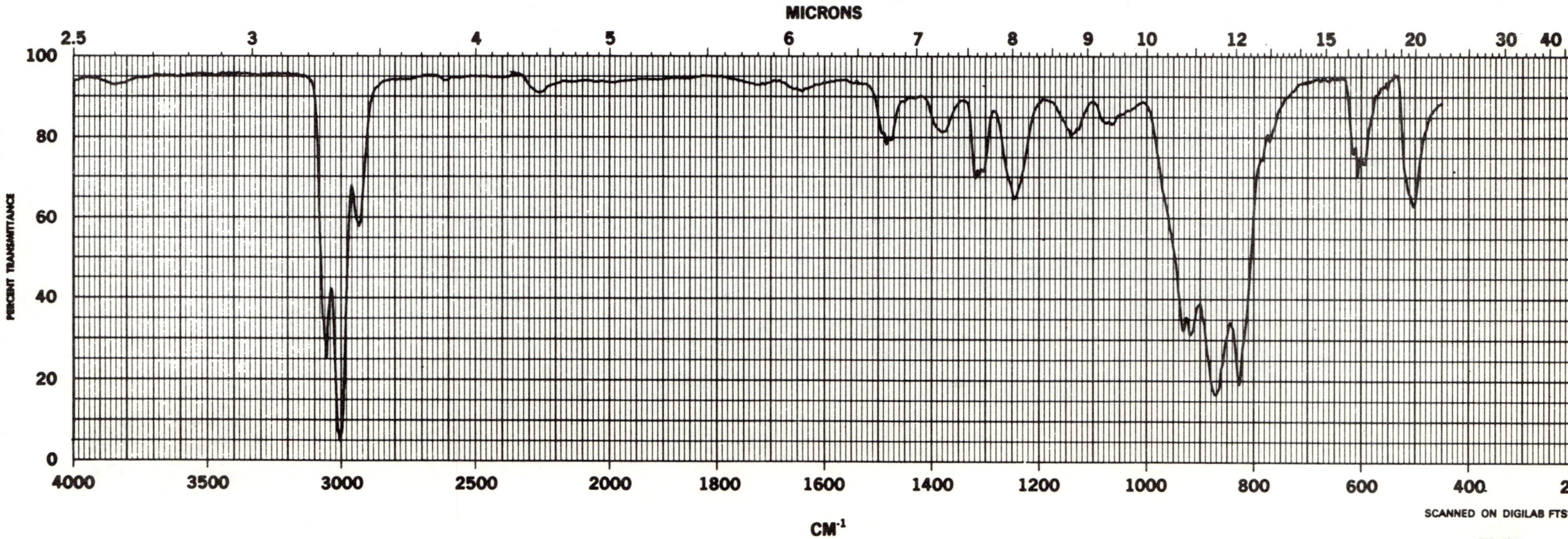

SCANNED ON DIGILAB FTS14

BUTYLAMINE
1–AMINOBUTANE
NORVALAMINE

$C_4H_{11}N$

MOLECULAR WEIGHT:	73.14
SOURCE OF SAMPLE:	MCB Manufacturing Chemists Norwood, Ohio
CAS REGISTRY NO.:	109-73-9
NIOSH NUMBER:	EO29750
0.025mm CsBr Cell:	Neat

$CH_3CH_2CH_2CH_2-NH_2$

CONDENSED PHASE

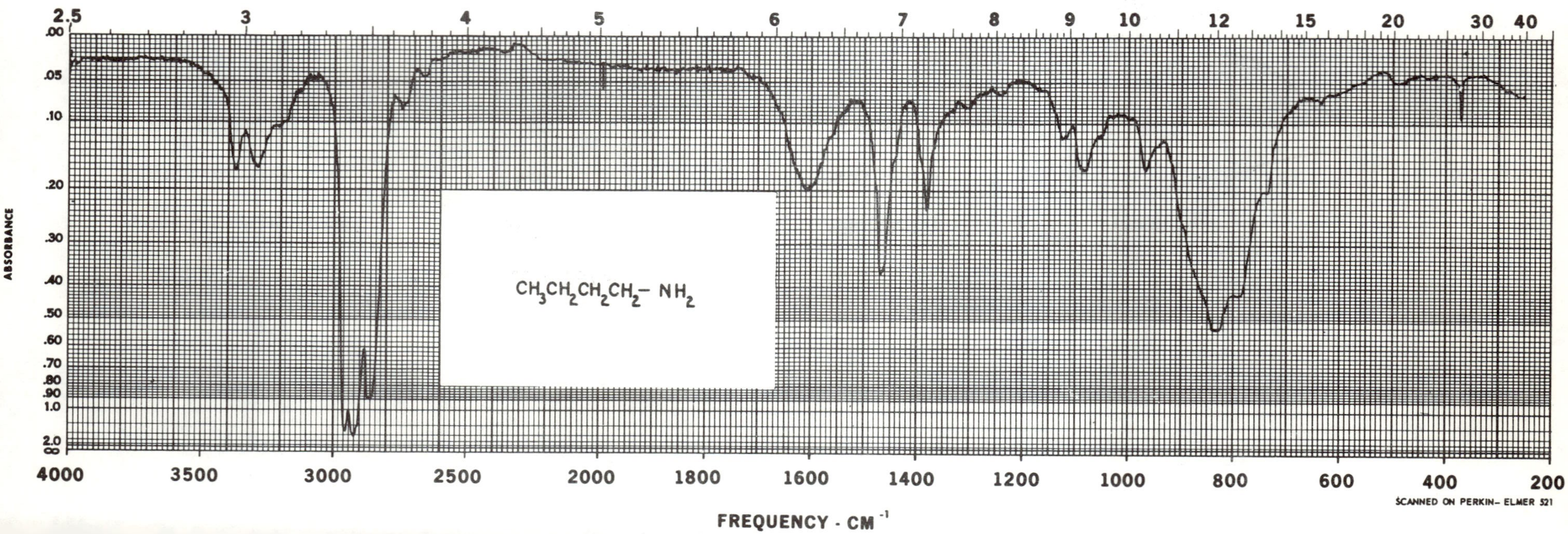

BOILING POINT:	76-78°C	INJECTION AMT.:	—
VAPORIZATION TEMP.:	200°C	PRESSURE:	Atmospheric
CELL VOLUME:	7.5 cc	MAX./MIN.:	99.0/5.8
PATH LENGTH:	7.6 cm	SCALE EXPANSION:	0.99

VAPOR PHASE

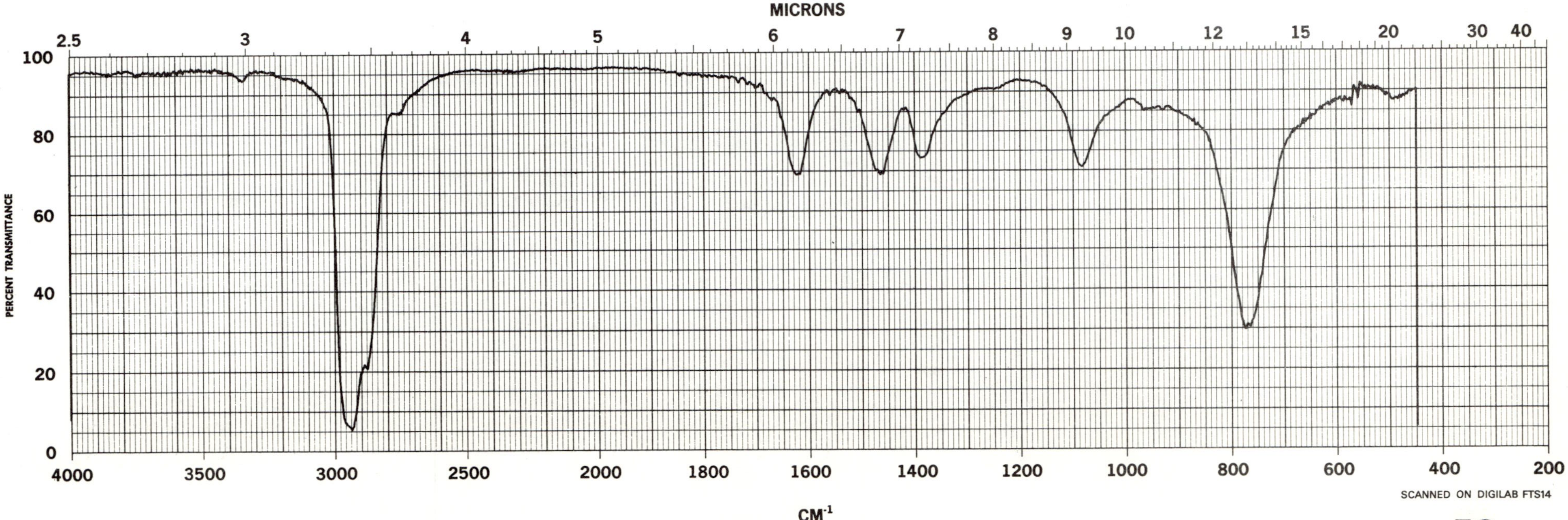

SCANNED ON DIGILAB FTS14

BUTYRIC ACID
BUTANOIC ACID
ETHYLACETIC ACID

$C_4H_8O_2$

MOLECULAR WEIGHT:	88.11
SOURCE OF SAMPLE:	Union Carbide Corporation Chemicals Division New York, New York
CAS REGISTRY NO.:	107-92-6
NIOSH NUMBER:	ES54250
CAPILLARY CELL:	Neat

$CH_3CH_2CH_2-C(=O)-OH$

CONDENSED PHASE

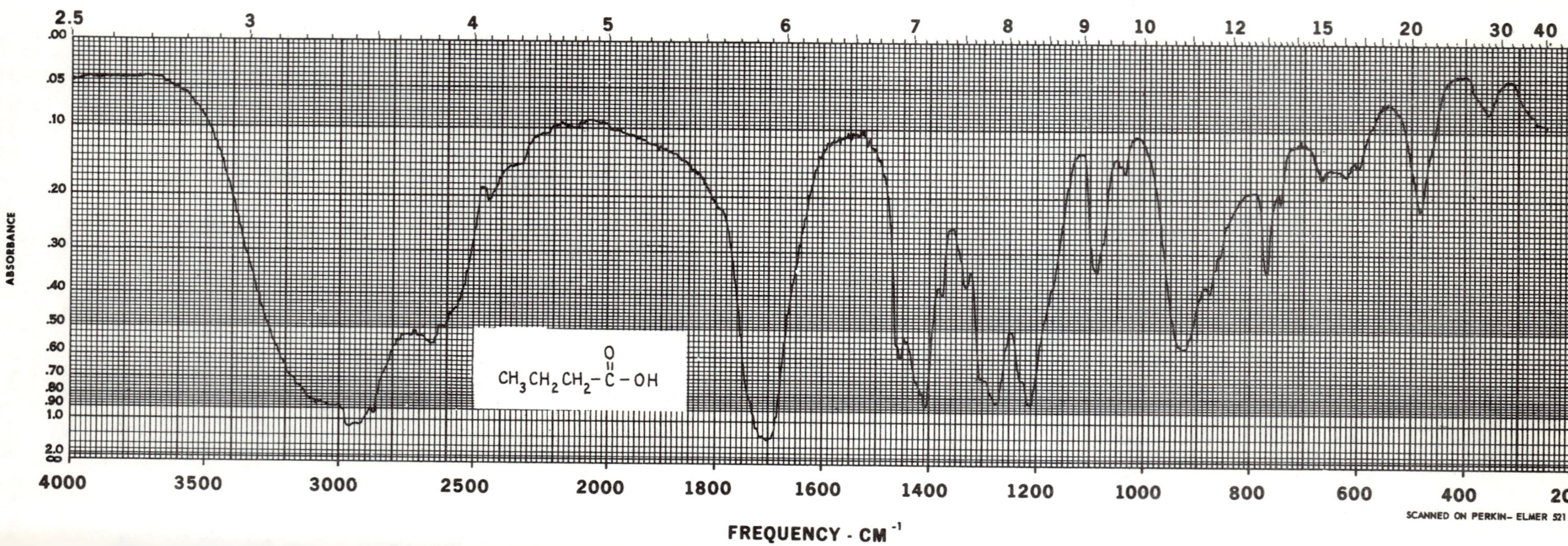

BOILING POINT:	162-163°C	INJECTION AMT.:	—
VAPORIZATION TEMP.:	200°C	PRESSURE:	Atmospheric
CELL VOLUME:	7.5 cc	MAX./MIN.:	99.5/12.0
PATH LENGTH:	7.6 cm	SCALE EXPANSION:	1.05

VAPOR PHASE

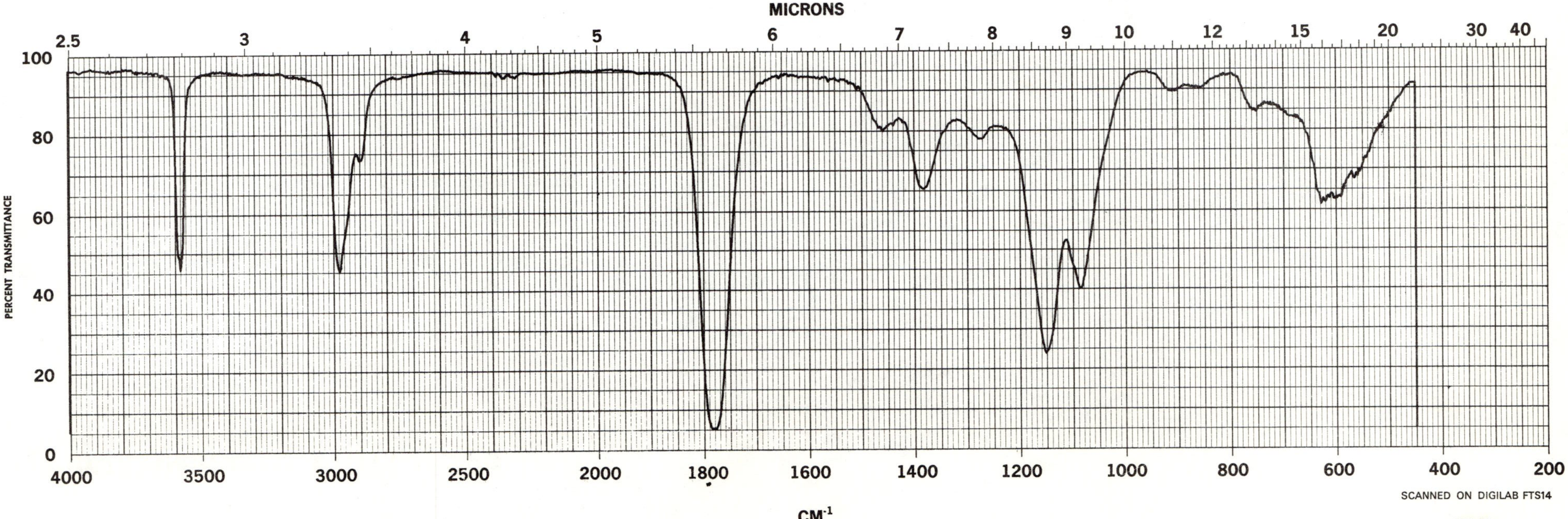

SCANNED ON DIGILAB FTS14

CARBAMIC ACID, ETHYL ESTER
ETHYL CARBAMATE
URETHANE
PRACARBAMIN

$C_3H_7NO_2$

MOLECULAR WEIGHT: 89.09

SOURCE OF SAMPLE: Fluka AG
Buchs, Switzerland

$H_2N-C(=O)-OCH_2CH_3$

CAS REGISTRY NO.: 51-97-6

NIOSH NUMBER: FA84000

CAPILLARY CELL: Melt

CONDENSED PHASE

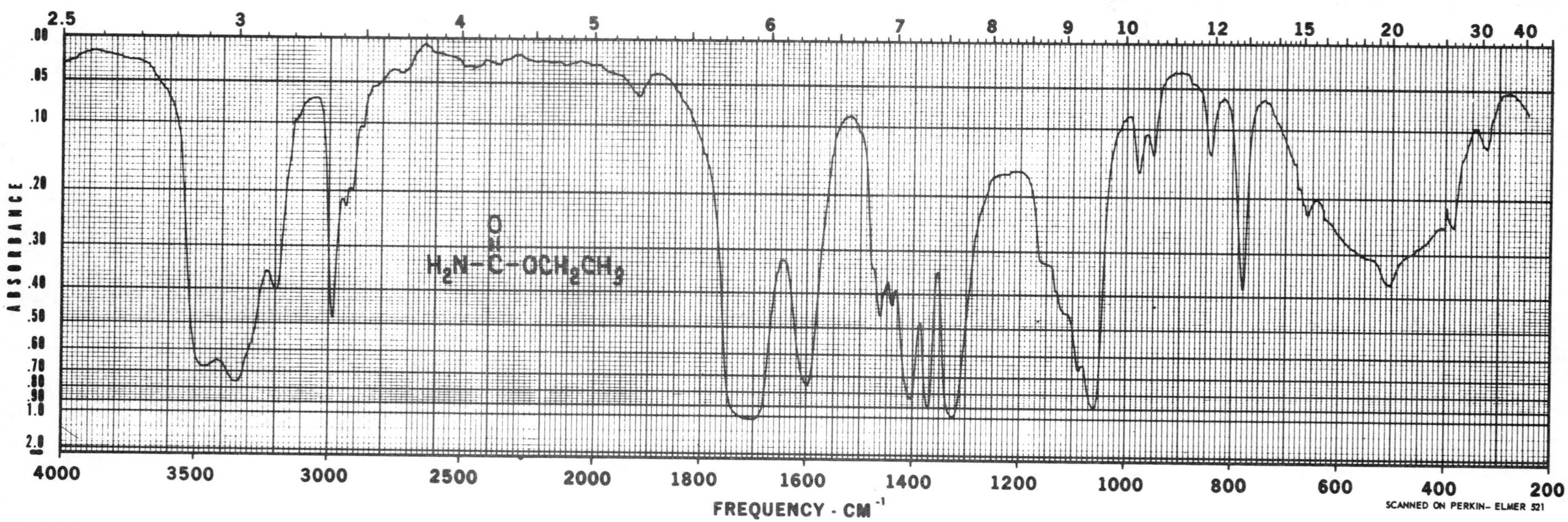

BOILING POINT:	182-184°C	INJECTION AMT.:	–
VAPORIZATION TEMP.:	160°C	PRESSURE:	Atmospheric
CELL VOLUME:	7.5 cc	MAX./MIN.:	101.2/7.7
PATH LENGTH:	7.6 cm	SCALE EXPANSION:	0.98

VAPOR PHASE

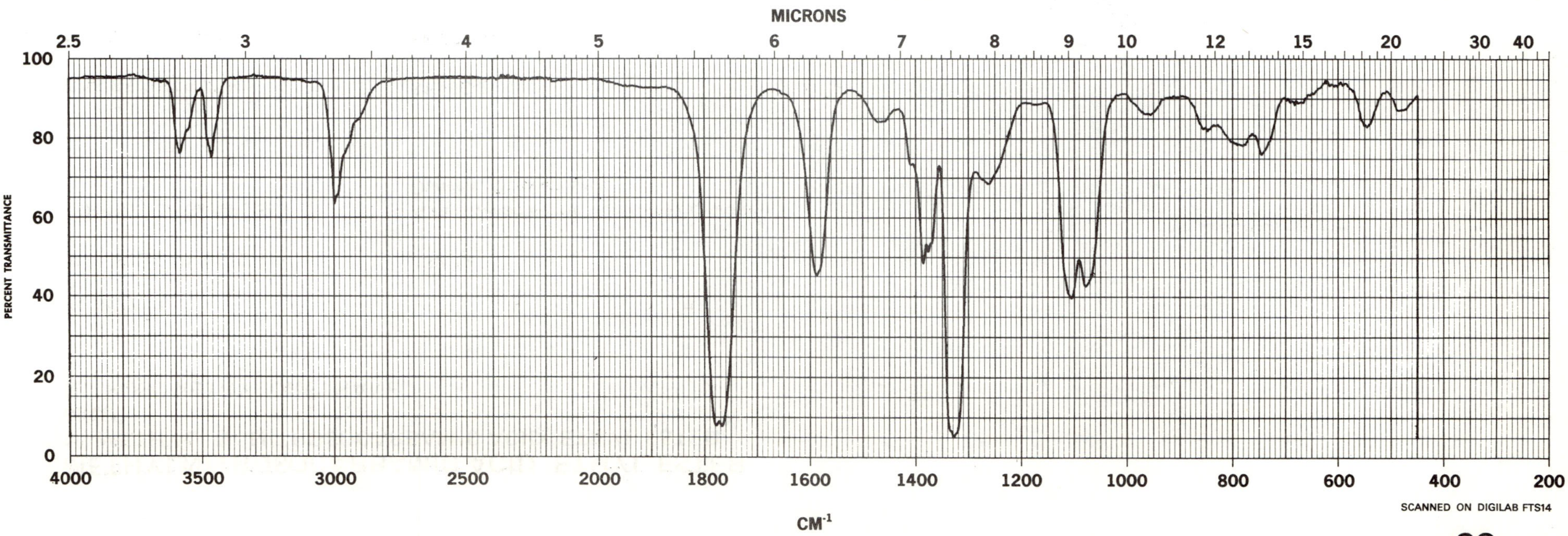

SCANNED ON DIGILAB FTS14

METHYLNITROSOCARBAMIC ACID, ETHYL ESTER
N–METHYL–N–NITROSOCARBAMIC ACID, ETHYL ESTER
MNU
NITROSOMETHYLURETHANE

$C_4H_8N_2O_3$

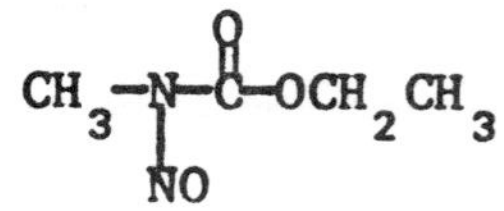

MOLECULAR WEIGHT:	132.12
SOURCE OF SAMPLE:	Alfa Division of the Ventron Corporation Danvers, Massachusetts
CAS REGISTRY NO.:	615-53-2
NIOSH NUMBER:	FC63000
CAPILLARY CELL:	Neat

CONDENSED PHASE

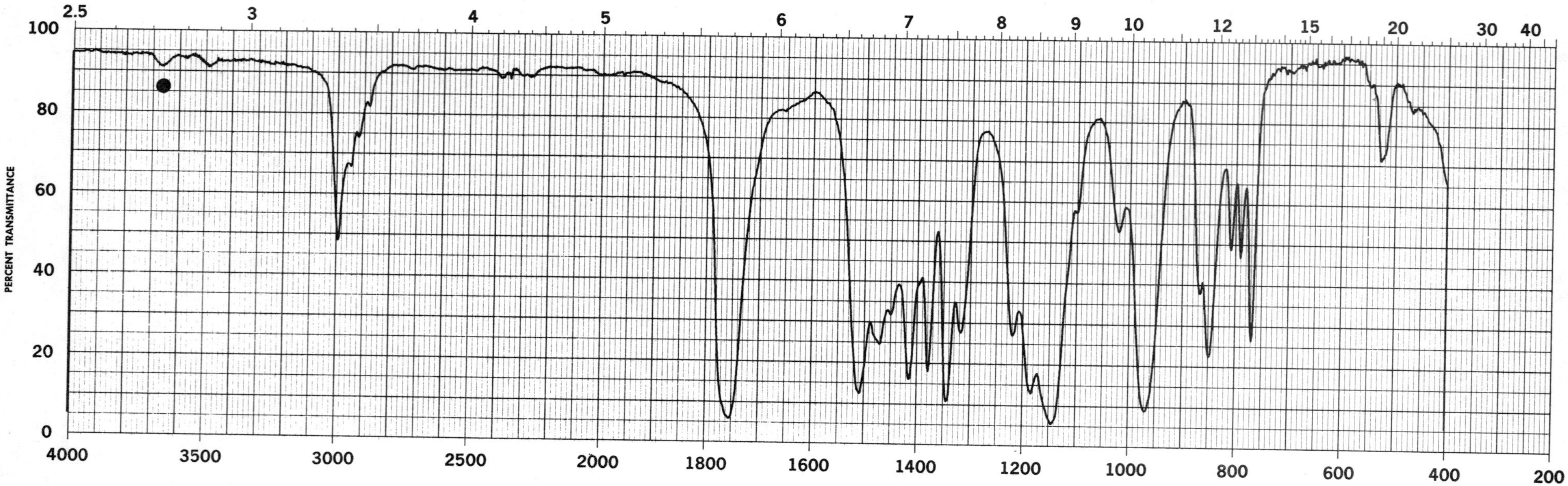

BOILING POINT:	–	INJECTION AMT.:	–
VAPORIZATION TEMP.:	70°C	PRESSURE:	Atmospheric
CELL VOLUME:	7.5 cc	MAX./MIN.:	103.2/13.9
PATH LENGTH:	7.6 cm	SCALE EXPANSION:	1.03

VAPOR PHASE

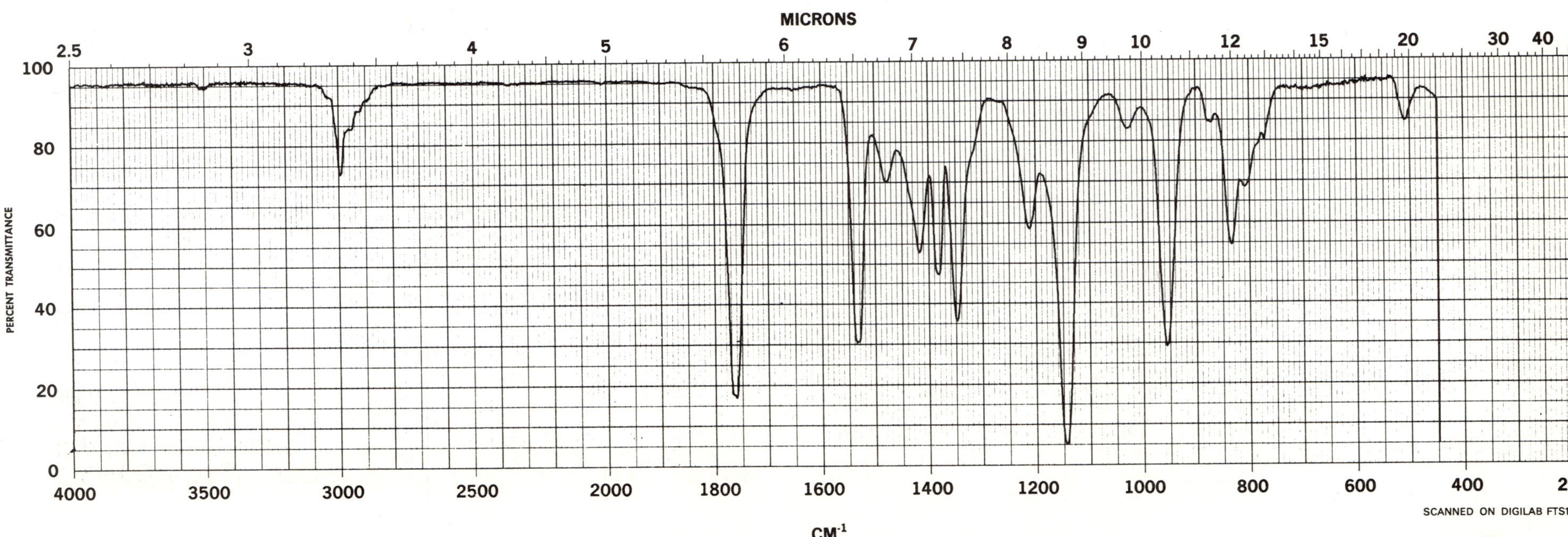

SCANNED ON DIGILAB FTS14

CARBON DISULFIDE
CARBON BISULFIDE
CARBON SULFIDE
WEEVILTOX

CS_2

S = C = S

MOLECULAR WEIGHT: 76.14

SOURCE OF SAMPLE: J. T. Baker Chemical Company
Phillipsburg, New Jersey

CAS REGISTRY NO.: 75-15-0

NIOSH NUMBER: FF66500

A: 2.5-15μ, 0.015 cm cell
15-40μ, Polyethylene cell
B: Capillary Cell: Neat

CONDENSED PHASE

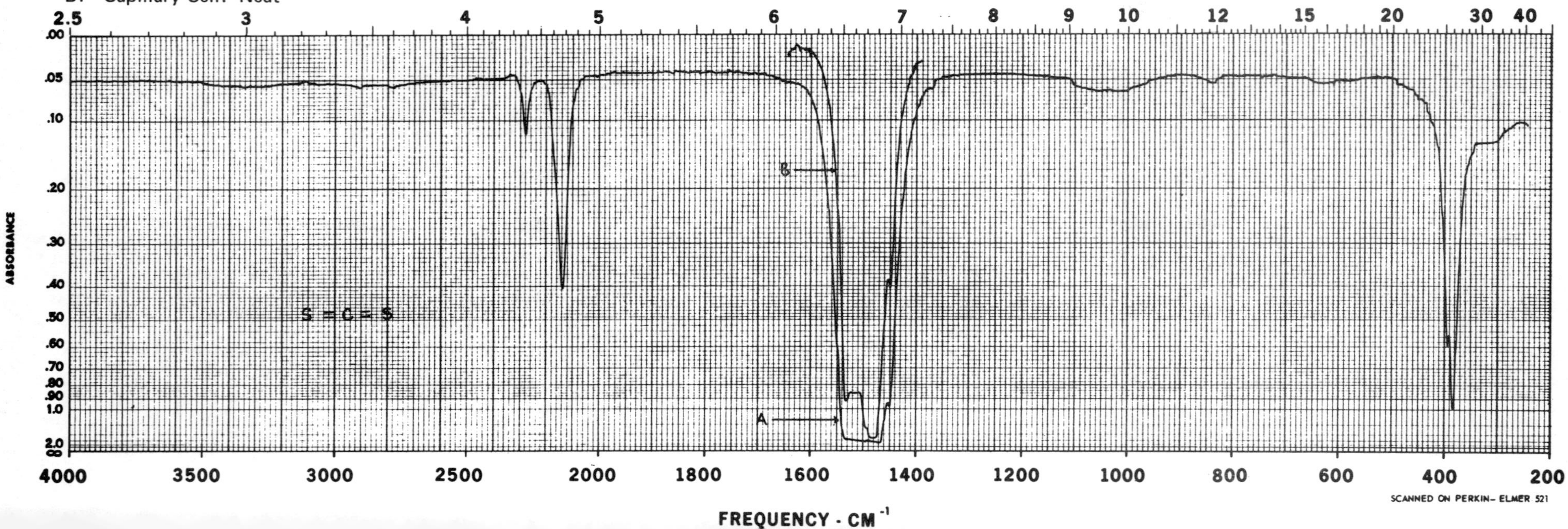

BOILING POINT:	$45^{o}C$	INJECTION AMT.:	0.1 μl
VAPORIZATION TEMP.:	$200^{o}C$	PRESSURE:	Atmospheric
CELL VOLUME:	7.5 cc	MAX./MIN.:	102.5/19.0
PATH LENGTH:	7.6 cm	SCALE EXPANSION:	0.91

VAPOR PHASE

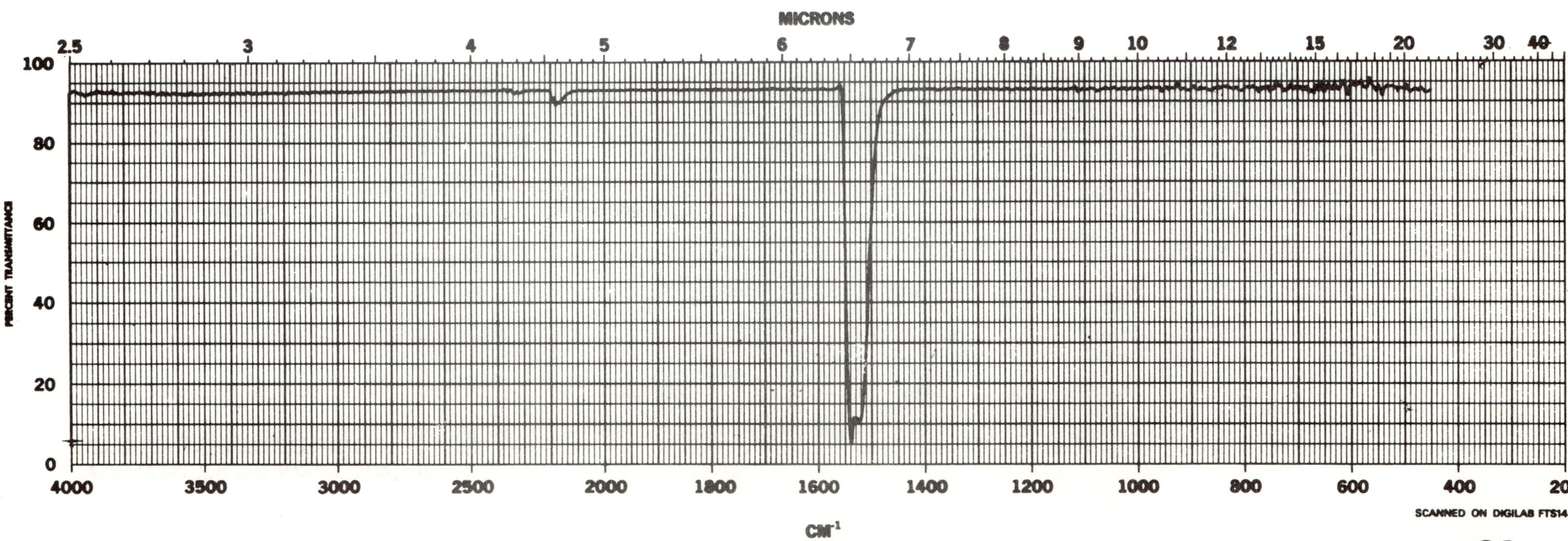

CARBON TETRACHLORIDE
TETRACHLOROMETHANE

CCl_4

MOLECULAR WEIGHT: 153.82

SOURCE OF SAMPLE: Aldrich Chemical Company, Inc.
Milwaukee, Wisconsin

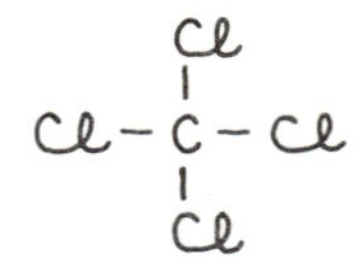

CAS REGISTRY NO.: 56-23-5

NIOSH NUMBER: FG49000

CAPILLARY CELL: Neat

CONDENSED PHASE

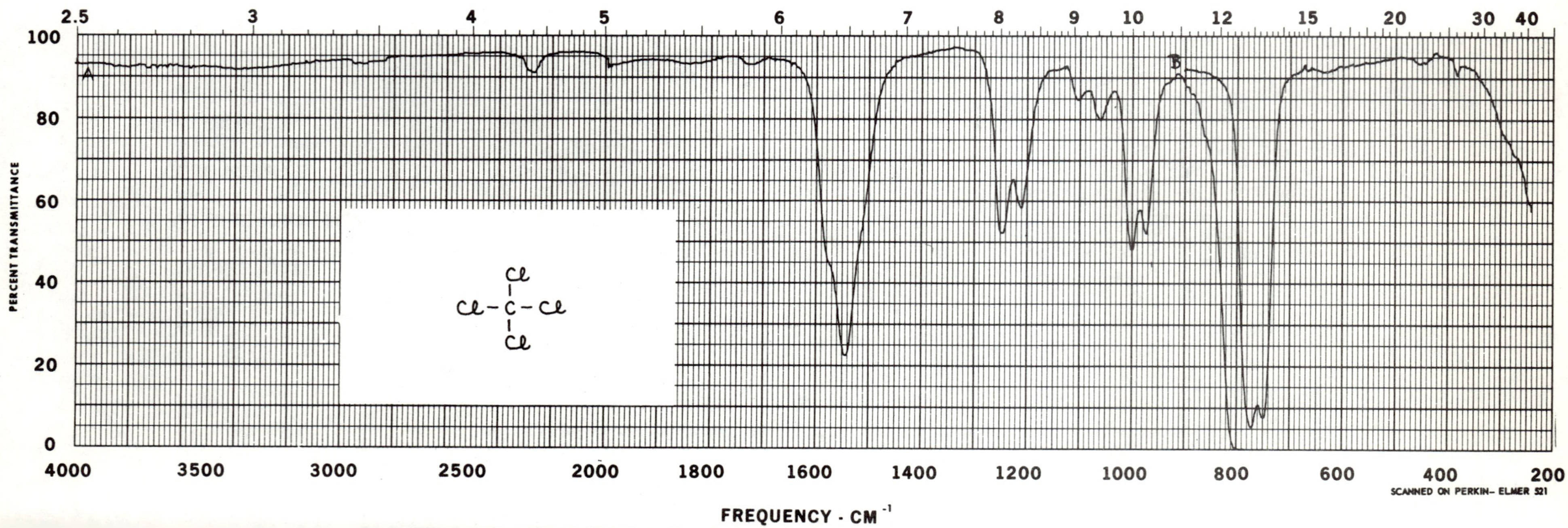

BOILING POINT:	77°C	INJECTION AMT.:	0.1 μl
VAPORIZATION TEMP.:	200°C	PRESSURE:	Atmospheric
CELL VOLUME:	7.5 cc	MAX./MIN.:	—
PATH LENGTH:	7.6 cm	SCALE EXPANSION:	0.92

VAPOR PHASE

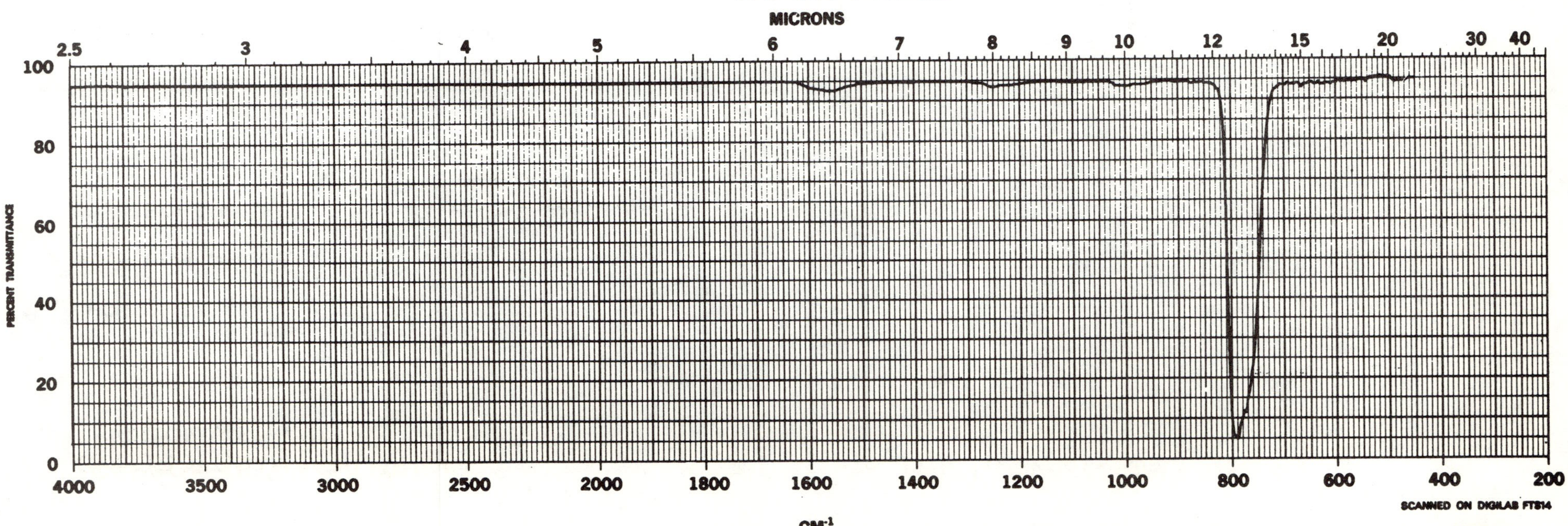

CHLOROFORM
TRICHLOROMETHANE

$CHCl_3$

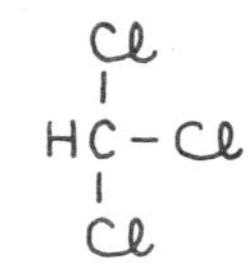

MOLECULAR WEIGHT: 119.38

SOURCE OF SAMPLE: Aldrich Chemical Company, Inc.
Milwaukee, Wisconsin

CAS REGISTRY NO.: 67-66-3

NIOSH NUMBER: FS91000

CAPILLARY CELL: Neat

CONDENSED PHASE

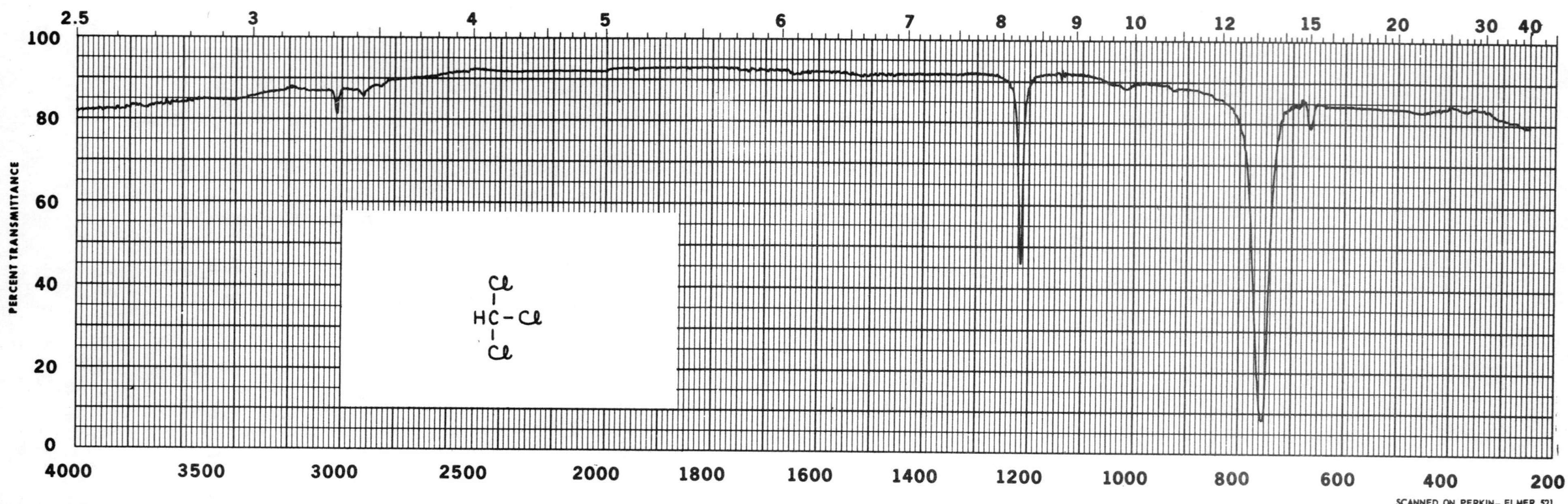

BOILING POINT:	60.5-61.5°C	INJECTION AMT.:	0.1 μl
VAPORIZATION TEMP.:	200°C	PRESSURE:	Atmospheric
CELL VOLUME:	7.5 cc	MAX./MIN.:	—
PATH LENGTH:	7.6 cm	SCALE EXPANSION:	0.91

VAPOR PHASE

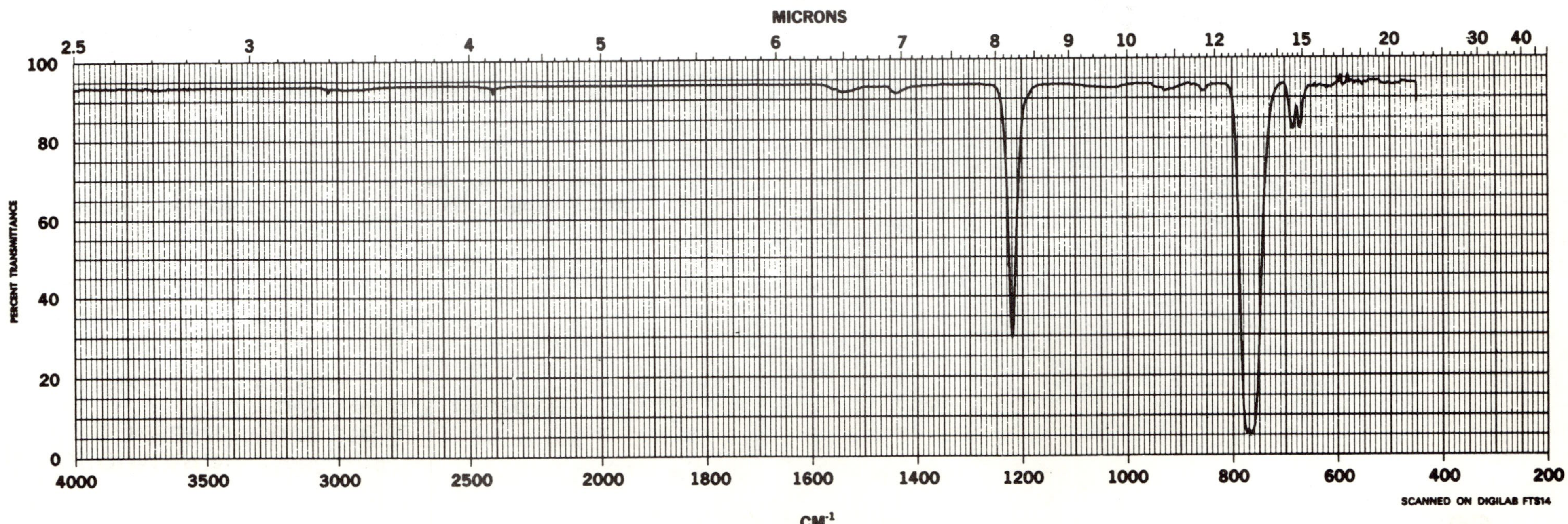

COUMARIN
2H–1–BENZOPYRAN–2–ONE
o–HYDROXYCINNAMIC ACID, δ–LACTONE
TONKA BEAN CAMPHOR

$C_9H_6O_2$

MOLECULAR WEIGHT:	146.15
SOURCE OF SAMPLE:	Dodge & Olcott Company New York, New York
CAS REGISTRY NO.:	91-64-5
NIOSH NUMBER:	GN42000
CAPILLARY CELL:	Melt

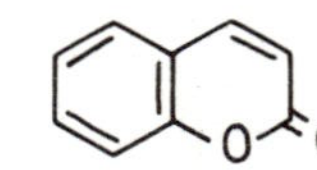

CONDENSED PHASE

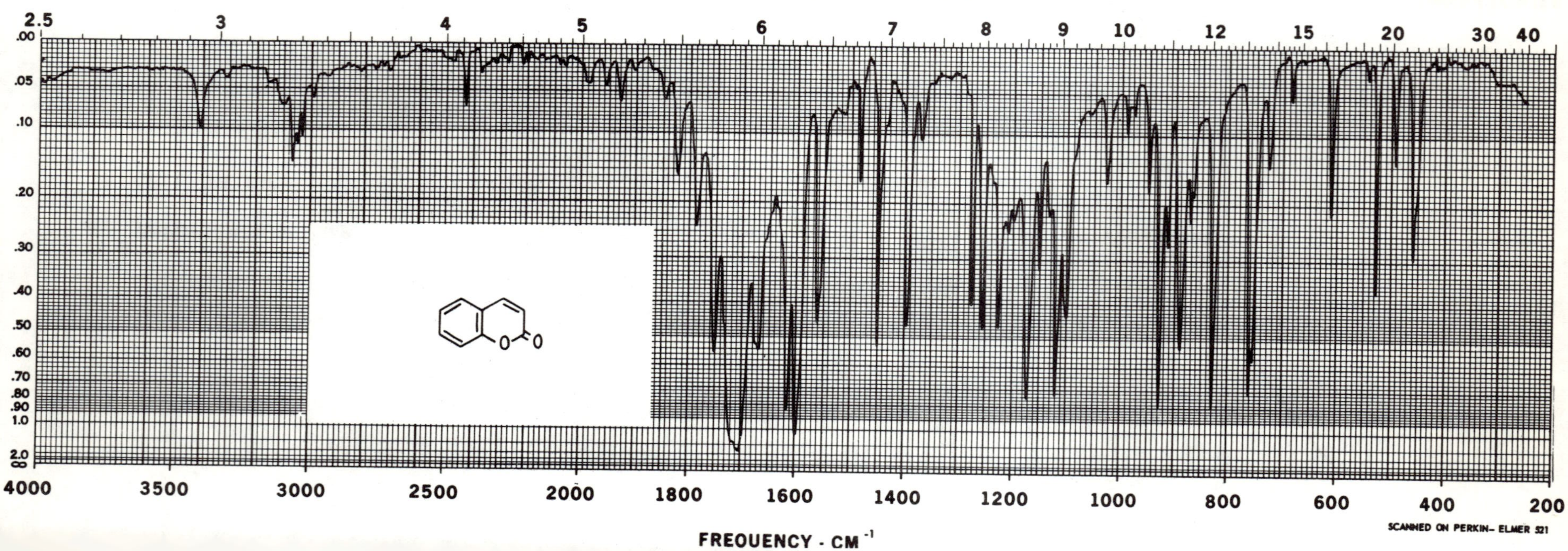

MELTING POINT:	68.8°C	INJECTION AMT.:	–
VAPORIZATION TEMP.:	280°C	PRESSURE:	Atmospheric
CELL VOLUME:	7.5 cc	MAX./MIN.:	101.7/10.4
PATH LENGTH:	7.6 cm	SCALE EXPANSION:	1.01

VAPOR PHASE

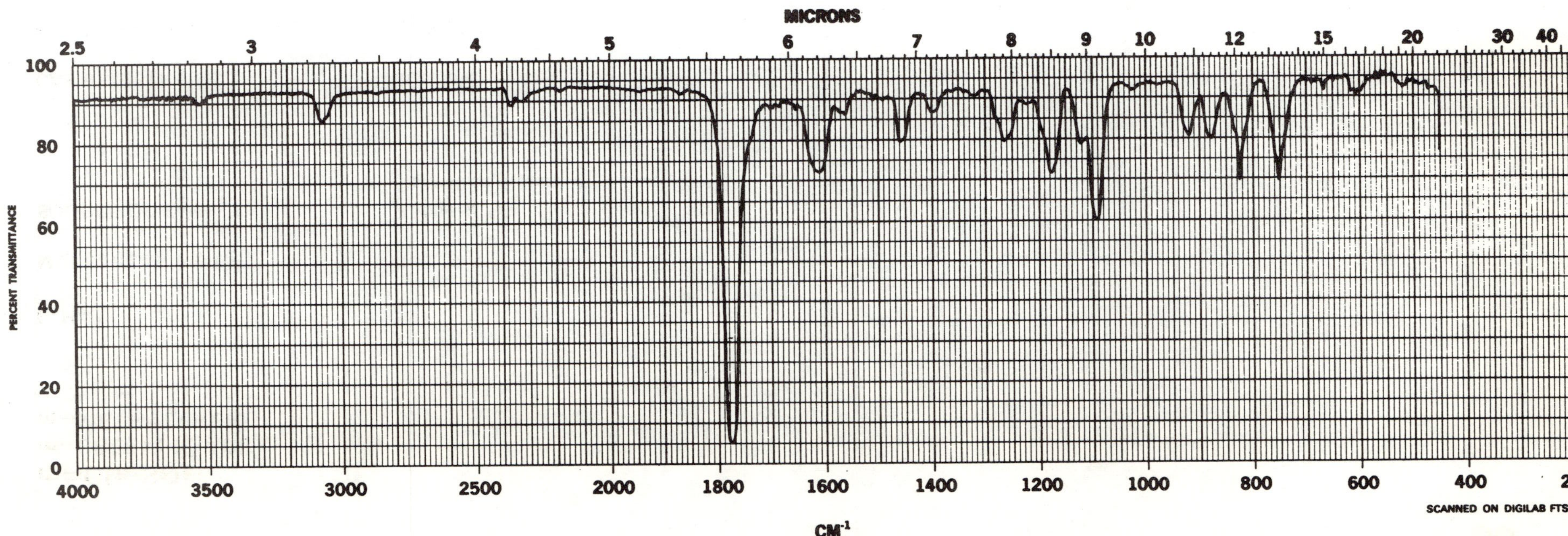

m—CRESOL
m—CRESYLIC ACID
m—METHYLPHENOL

C_7H_8O

MOLECULAR WEIGHT: 108.14

SOURCE OF SAMPLE: E. Merck AG
Darmstadt, Germany

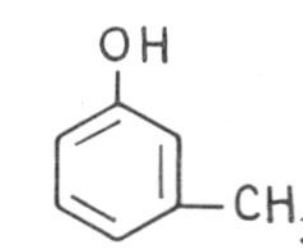

CAS REGISTRY NO.: 108-39-4

NIOSH NUMBER: GO61250

CAPILLARY CELL: Neat

CONDENSED PHASE

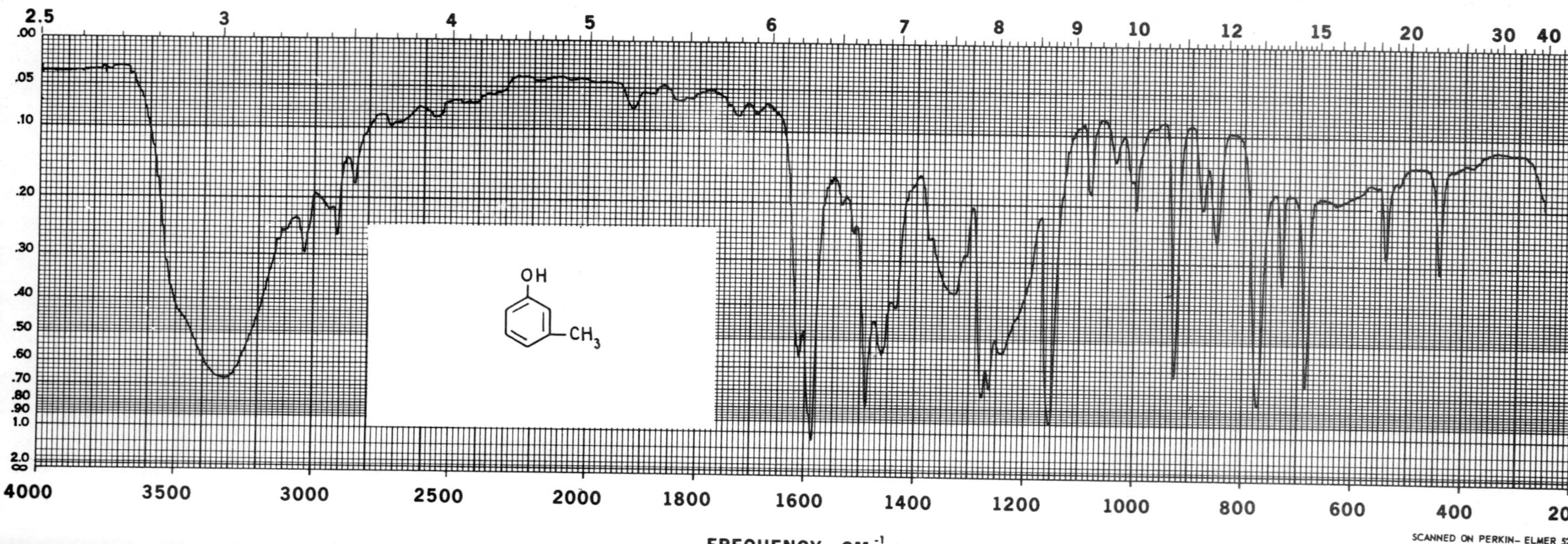

BOILING POINT:	203°C	INJECTION AMT.:	1.0 µl
VAPORIZATION TEMP.:	200°C	PRESSURE:	Atmospheric
CELL VOLUME:	7.5 cc	MAX./MIN.:	—
PATH LENGTH:	7.6 cm	SCALE EXPANSION:	1.04

VAPOR PHASE

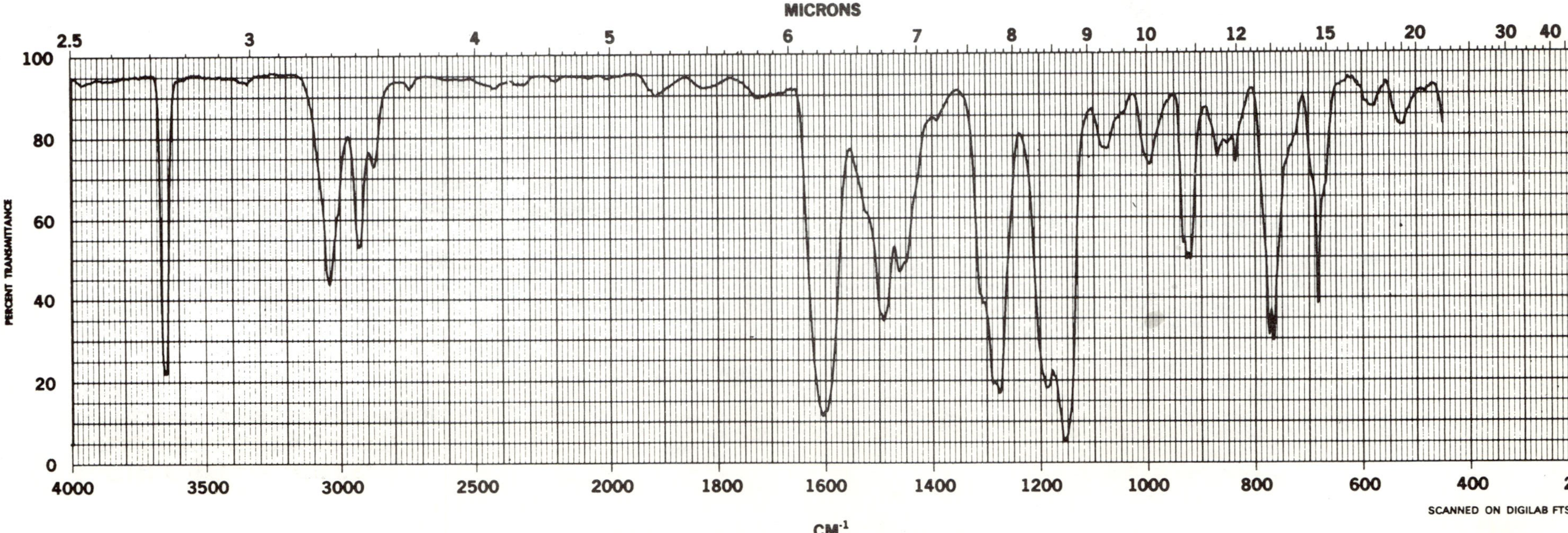

SCANNED ON DIGILAB FTS14

4–CHLORO–m–CRESOL
PARACHLOROMETA CRESOL
4–CHLORO–3–METHYLPHENOL
PAROL

C_7H_7ClO

MOLECULAR WEIGHT: 142.59

SOURCE OF SAMPLE: MCB Manufacturing Chemists
Norwood, Ohio

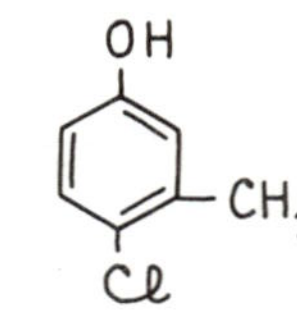

CAS REGISTRY NO.: 59-50-7

NIOSH NUMBER: GO71000

CAPILLARY CELL: Melt

CONDENSED PHASE

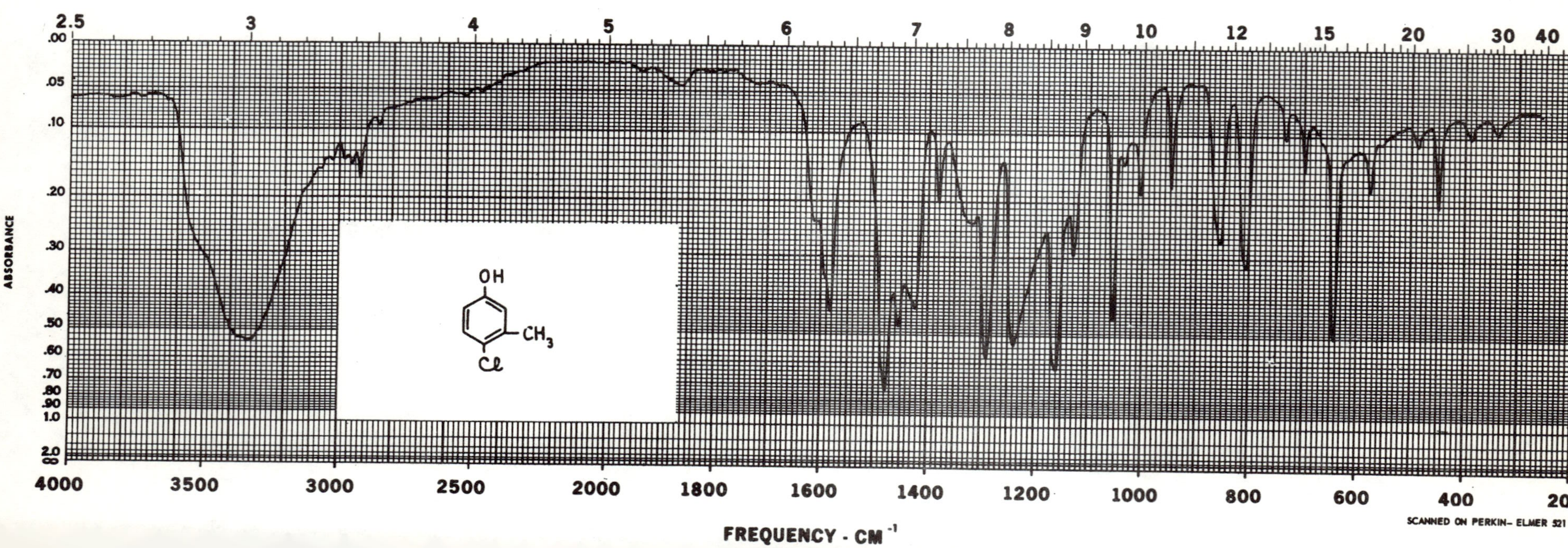

BOILING POINT:	235°C	INJECTION AMT.:	–
VAPORIZATION TEMP.:	240°C	PRESSURE:	Atmospheric
CELL VOLUME:	7.5 cc	MAX./MIN.:	101.8/11.7
PATH LENGTH:	7.6 cm	SCALE EXPANSION:	1.02

VAPOR PHASE

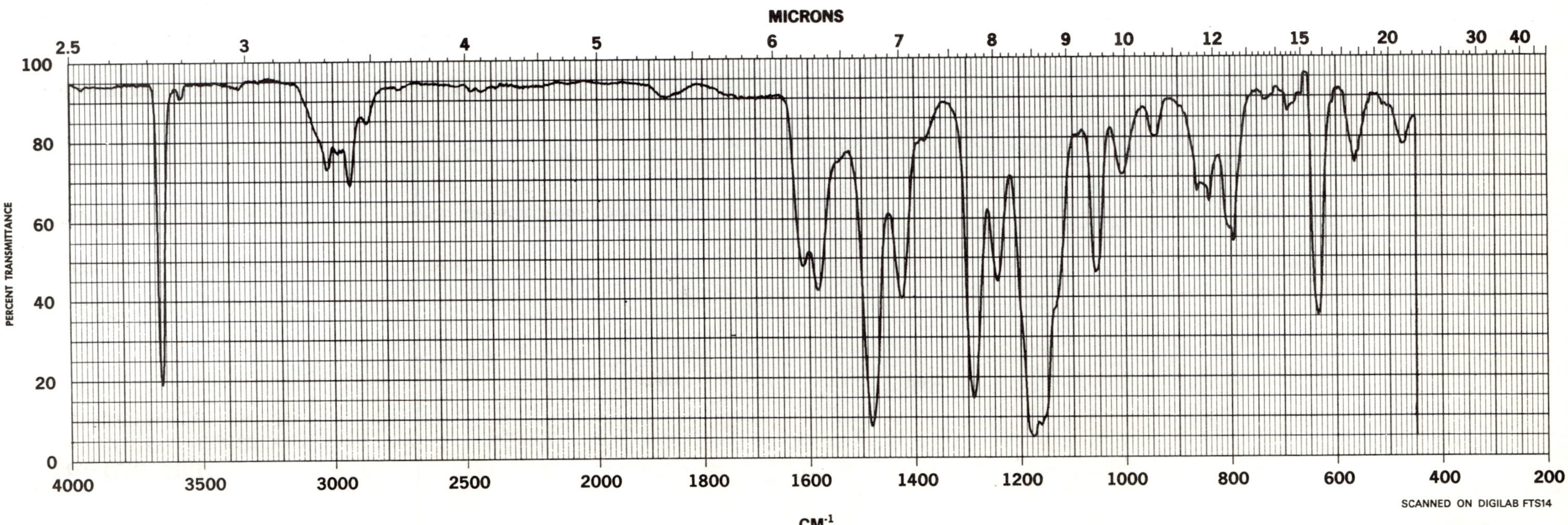

SCANNED ON DIGILAB FTS14

o—CRESOL
o—CRESYLIC ACID
o—METHYLPHENOL

C_7H_8O

MOLECULAR WEIGHT: 108.14

SOURCE OF SAMPLE: MCB Manufacturing Chemists
Norwood, Ohio

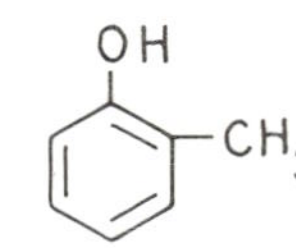

CAS REGISTRY NO.: 95-48-7

NIOSH NUMBER: GO63000

CAPILLARY CELL: Neat

CONDENSED PHASE

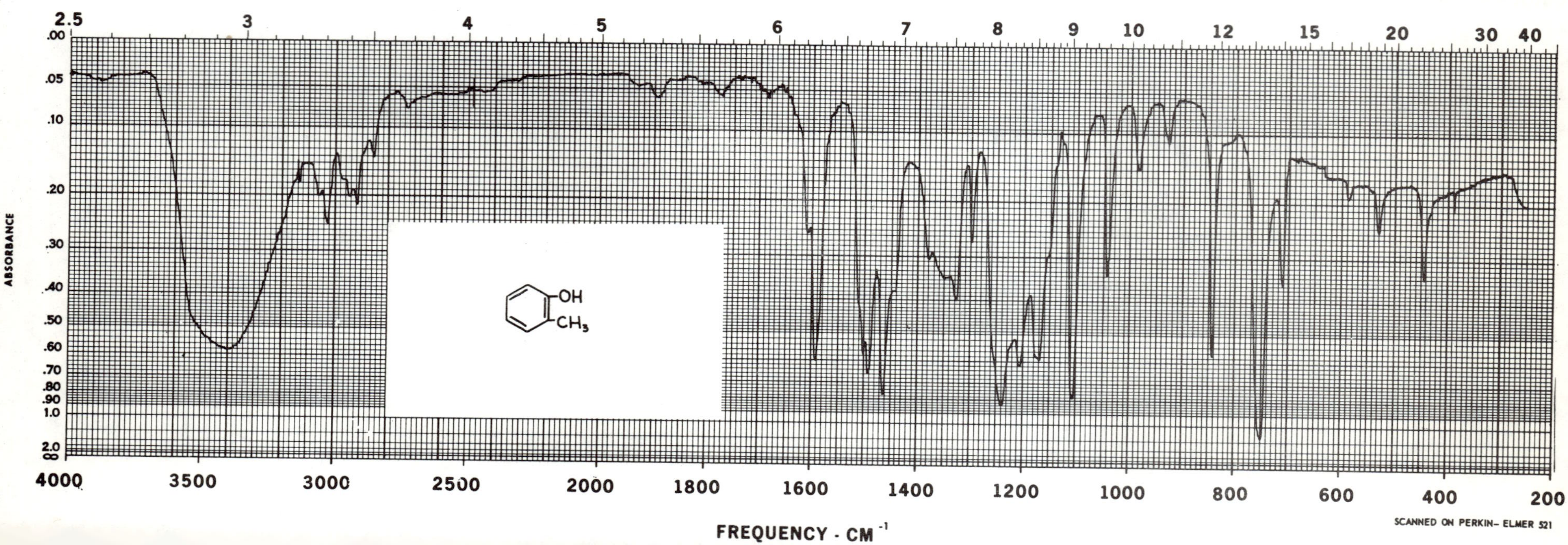

BOILING POINT:	190.8°C	INJECTION AMT.:	1.0 µl
VAPORIZATION TEMP.:	200°C	PRESSURE:	Atmospheric
CELL VOLUME:	7.5 cc	MAX./MIN.:	—
PATH LENGTH:	7.6 cm	SCALE EXPANSION:	1.05

VAPOR PHASE

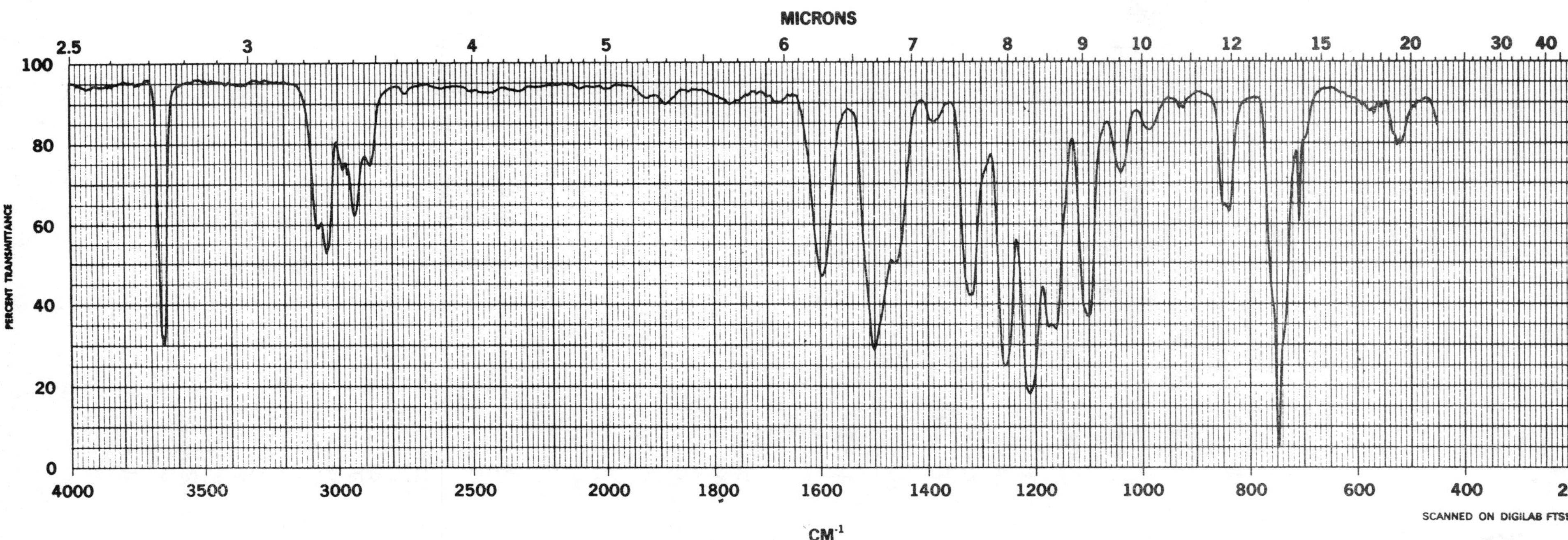

SCANNED ON DIGILAB FTS14

4,6–DINITRO–o–CRESOL
ANTINONIN
CAPSINE
DITROSOL
SANDOLIN A

$C_7H_6N_2O_5$

MOLECULAR WEIGHT: 198.14

SOURCE OF SAMPLE: Fluka AG, Buchs, Switzerland

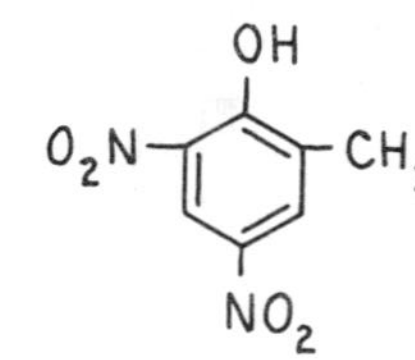

CAS REGISTRY NO.: 534-52-1

NIOSH NUMBER: GO96250

CAPILLARY CELL: Melt

CONDENSED PHASE

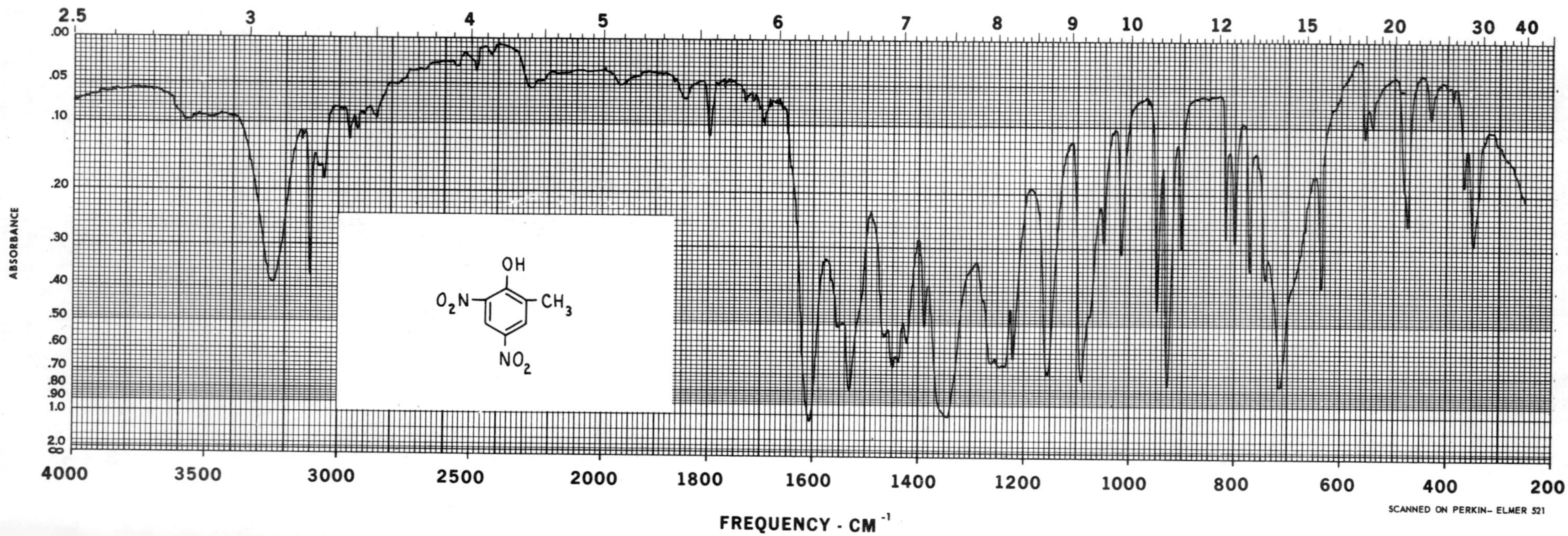

MELTING POINT:	84-86°C	INJECTION AMT.:	0.70 mg
VAPORIZATION TEMP.:	280°C	PRESSURE:	Atmospheric
CELL VOLUME:	7.5 cc	MAX./MIN.:	99.5/5.3
PATH LENGTH:	7.6 cm	SCALE EXPANSION:	0.98

VAPOR PHASE

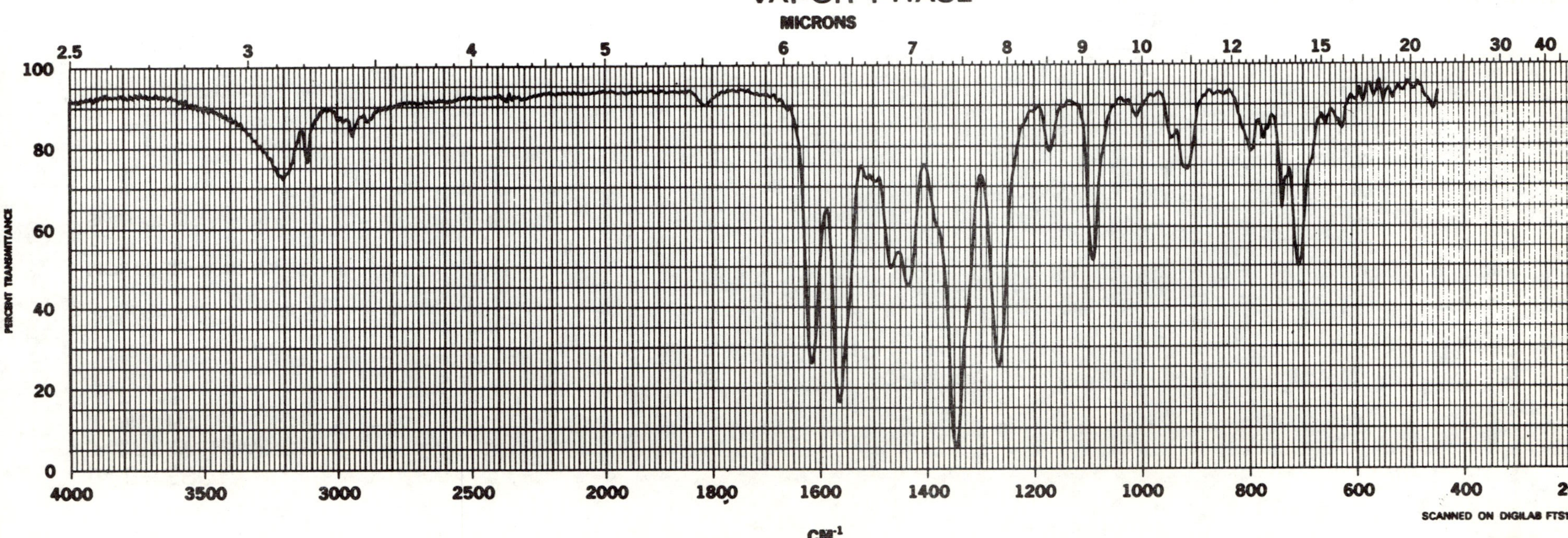

p–CRESOL
p–CRESYLIC ACID
p–METHYLPHENOL

C_7H_8O

MOLECULAR WEIGHT: 108.14

SOURCE OF SAMPLE: MCB Manufacturing Chemists
Norwood, Ohio

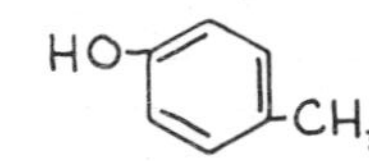

CAS REGISTRY NO.: 106-44-5

NIOSH NUMBER: GO64750

CAPILLARY CELL: Melt

CONDENSED PHASE

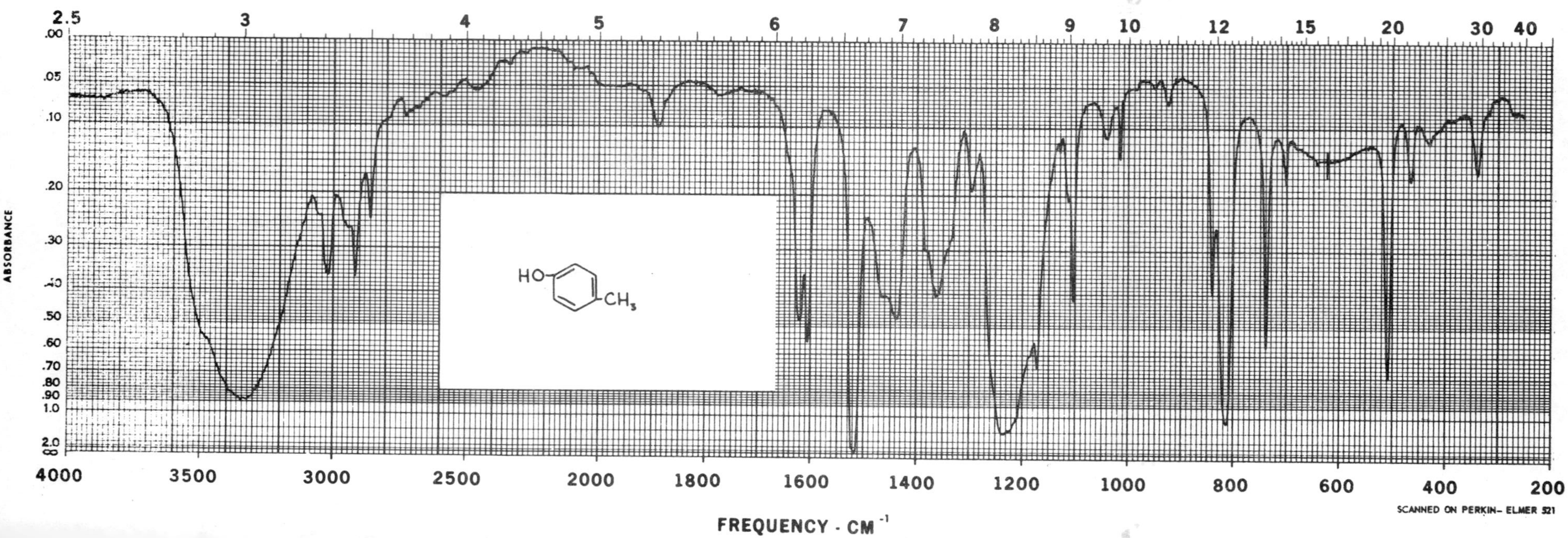

BOILING POINT:	202.5°C	INJECTION AMT.:	1.0 µl
VAPORIZATION TEMP.:	200°C	PRESSURE:	Atmospheric
CELL VOLUME:	7.5 cc	MAX./MIN.:	–
PATH LENGTH:	7.6 cm	SCALE EXPANSION:	0.94

VAPOR PHASE

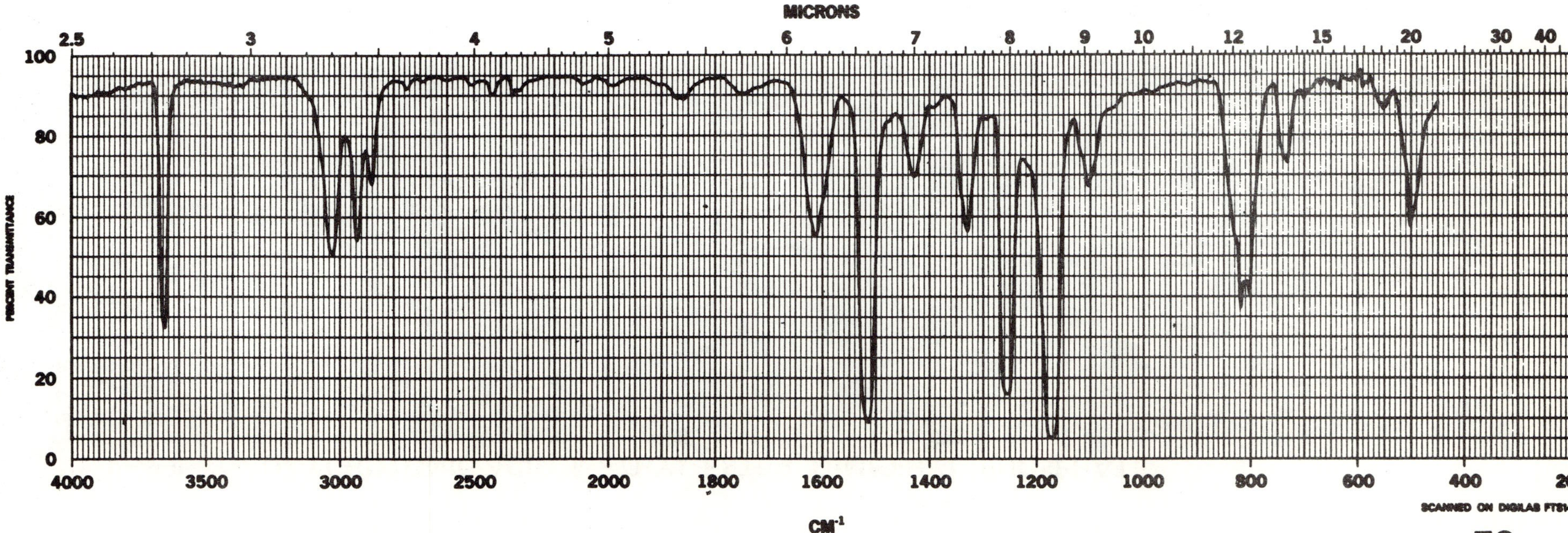

SCANNED ON DIGILAB FTS14

trans–3–HYDROXYCROTONIC ACID, METHYL ESTER, DIMETHYL PHOSPHATE
MENITE
MEVINPHOS
PHOSFENE
PHOSDRIN

$C_7H_{13}O_6P$

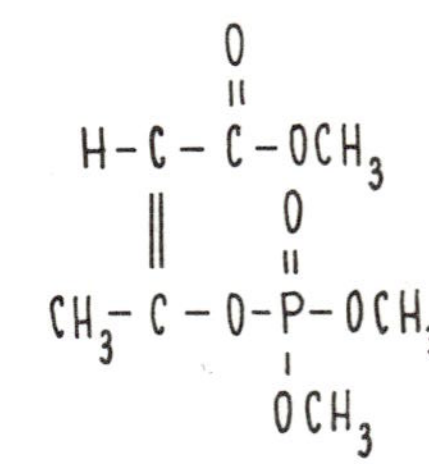

MOLECULAR WEIGHT: 224.15

SOURCE OF SAMPLE: Shell Chemical Company
Houston, Texas

CAS REGISTRY NO.: 298-01-1

NIOSH NUMBER: GQ52500

Between Salts

CONDENSED PHASE

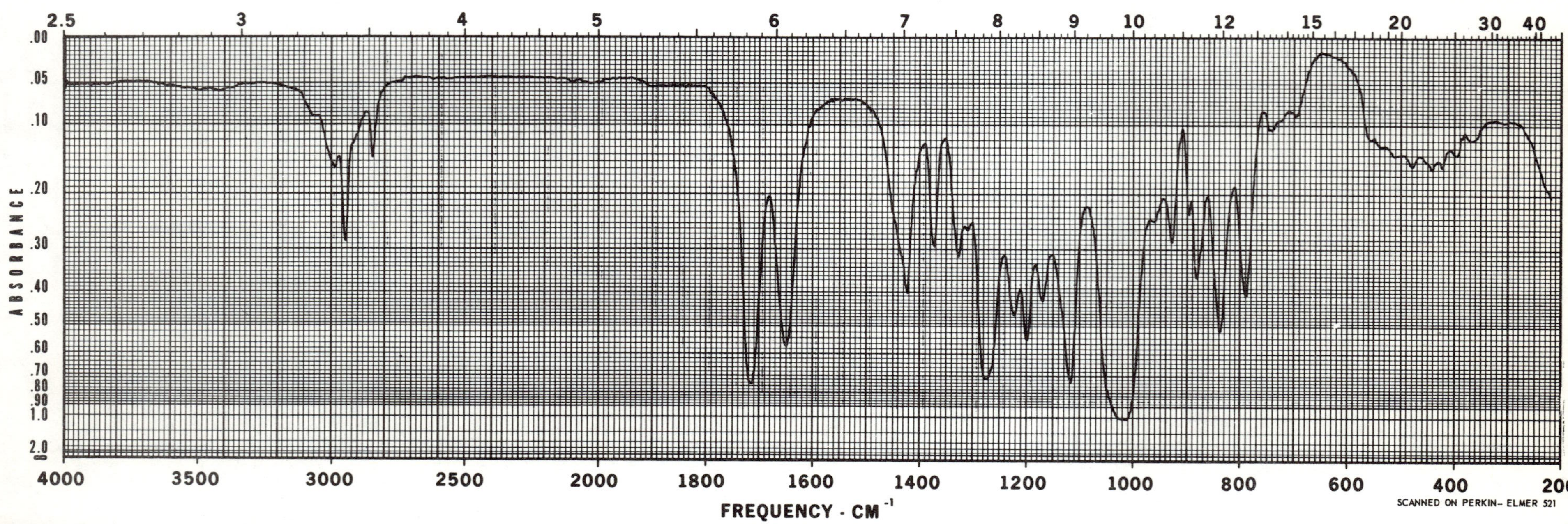

BOILING POINT:	—	INJECTION AMT.:	—
VAPORIZATION TEMP.:	200°C	PRESSURE:	Atmospheric
CELL VOLUME:	7.5 cc	MAX./MIN.:	100.5/8.7
PATH LENGTH:	7.6 cm	SCALE EXPANSION:	1.00

VAPOR PHASE

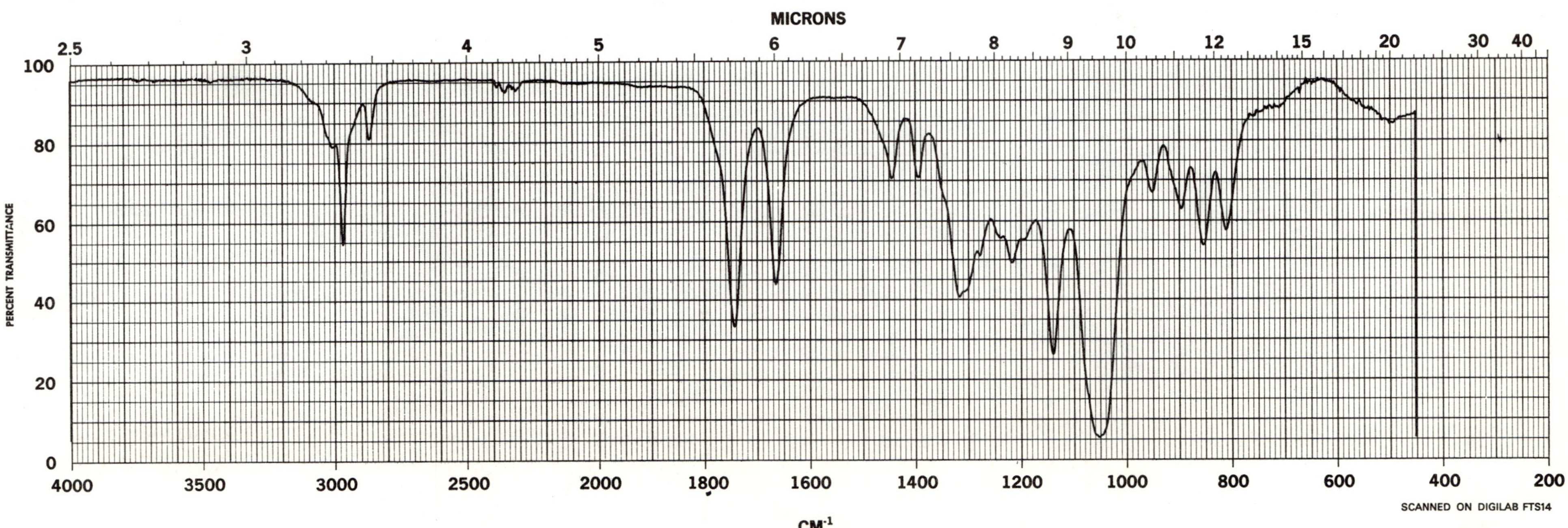

SCANNED ON DIGILAB FTS14

CYCLOHEXANE
HEXAHYDROBENZENE

C_6H_{12}

MOLECULAR WEIGHT: 84.16

SOURCE OF SAMPLE: MCB Manufacturing Chemists
Norwood, Ohio

CAS REGISTRY NO.: 110-82-7

NIOSH NUMBER: GU63000

CAPILLARY CELL: Neat

CONDENSED PHASE

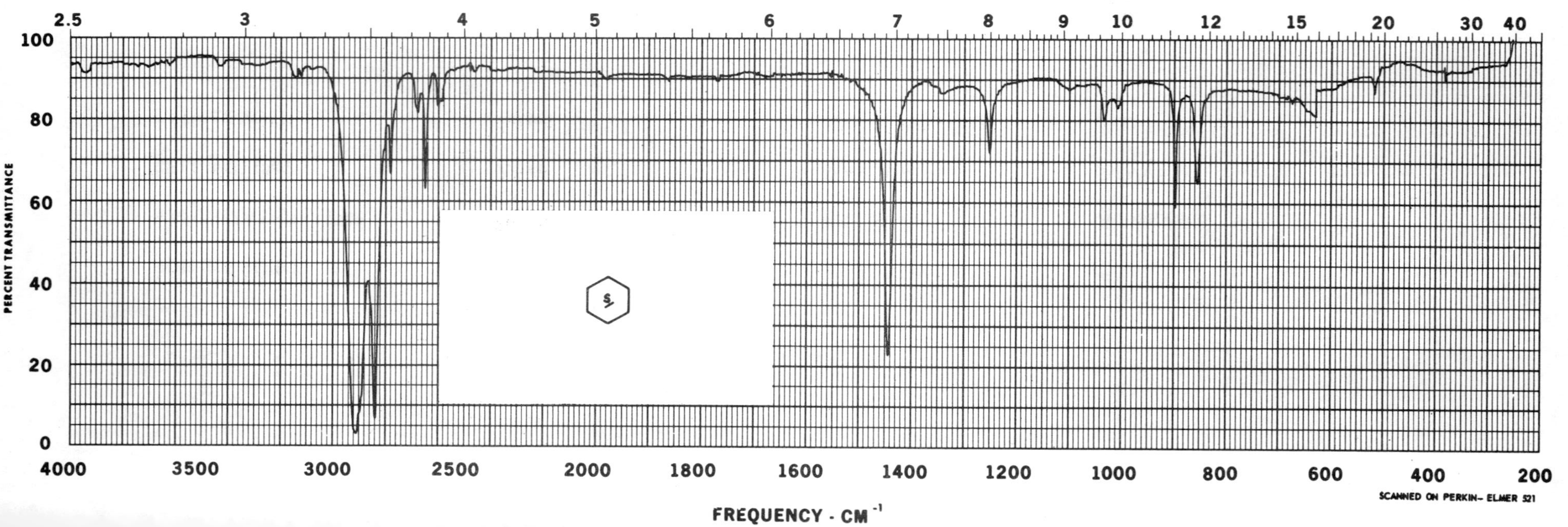

BOILING POINT:	80-81°C	INJECTION AMT.:	0.3 μl
VAPORIZATION TEMP.:	200°C	PRESSURE:	Atmospheric
CELL VOLUME:	7.5 cc	MAX./MIN.:	—
PATH LENGTH:	7.6 cm	SCALE EXPANSION:	0.89

VAPOR PHASE

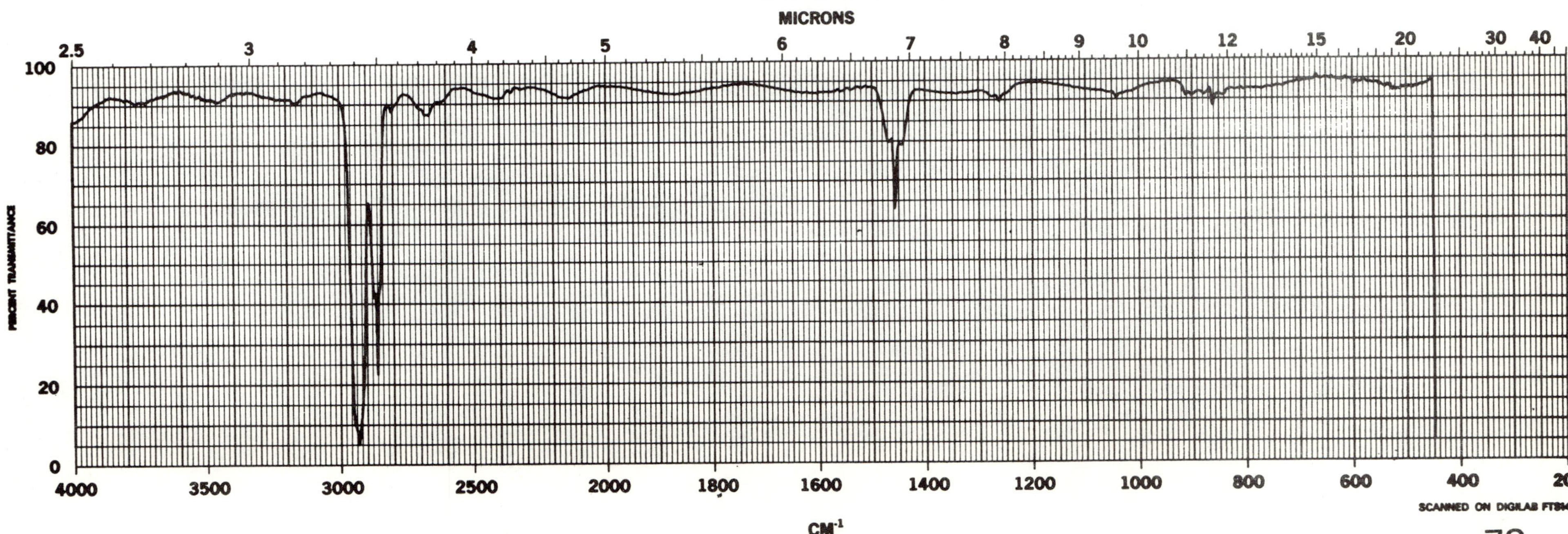

3,5,5–TRIMETHYL–2–CYCLOHEXEN–1–ONE
ISOPHORONE

$C_9H_{14}O$

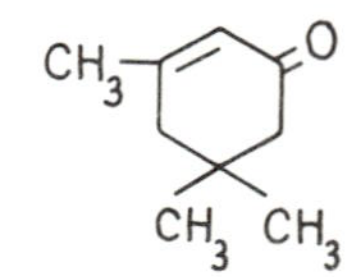

MOLECULAR WEIGHT: 138.21

SOURCE OF SAMPLE: Aldrich Chemical Company, Inc.
Milwaukee, Wisconsin

CAS REGISTRY NO.: 78-59-1

NIOSH NUMBER: GW77000

CAPILLARY CELL: Neat

CONDENSED PHASE

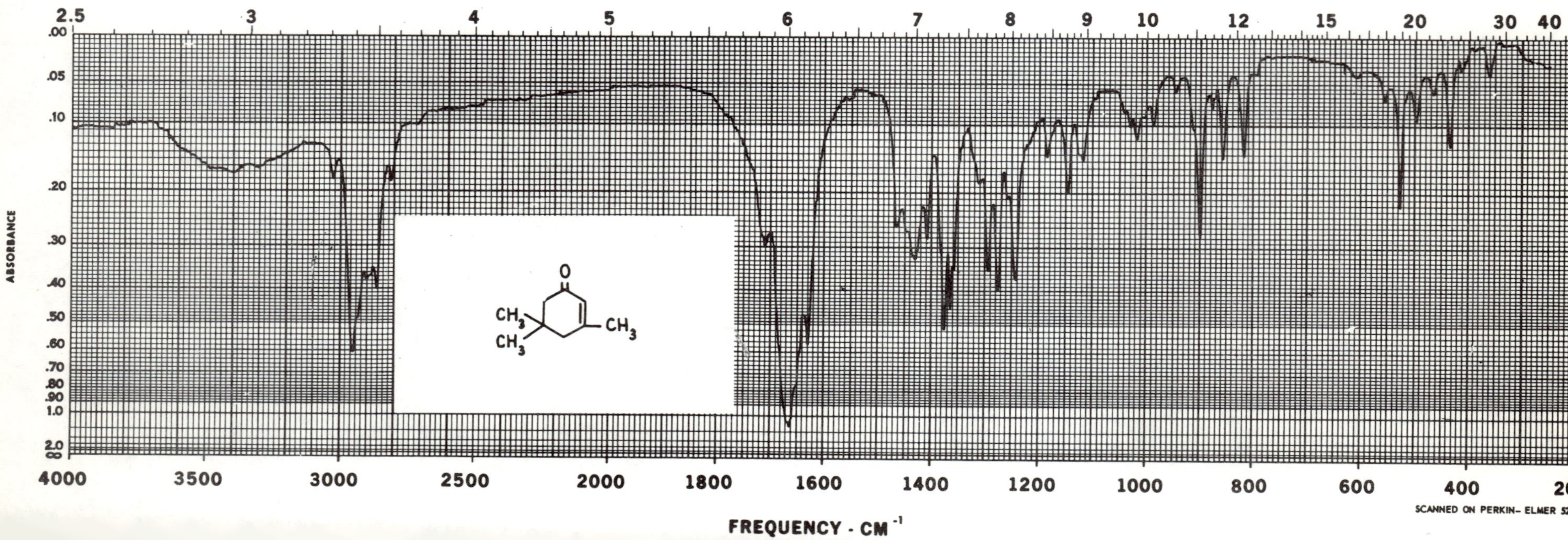

BOILING POINT:	213-214°C	INJECTION AMT.:	1.0 µl
VAPORIZATION TEMP.:	200°C	PRESSURE:	Atmospheric
CELL VOLUME:	7.5 cc	MAX./MIN.:	—
PATH LENGTH:	7.6 cm	SCALE EXPANSION:	0.94

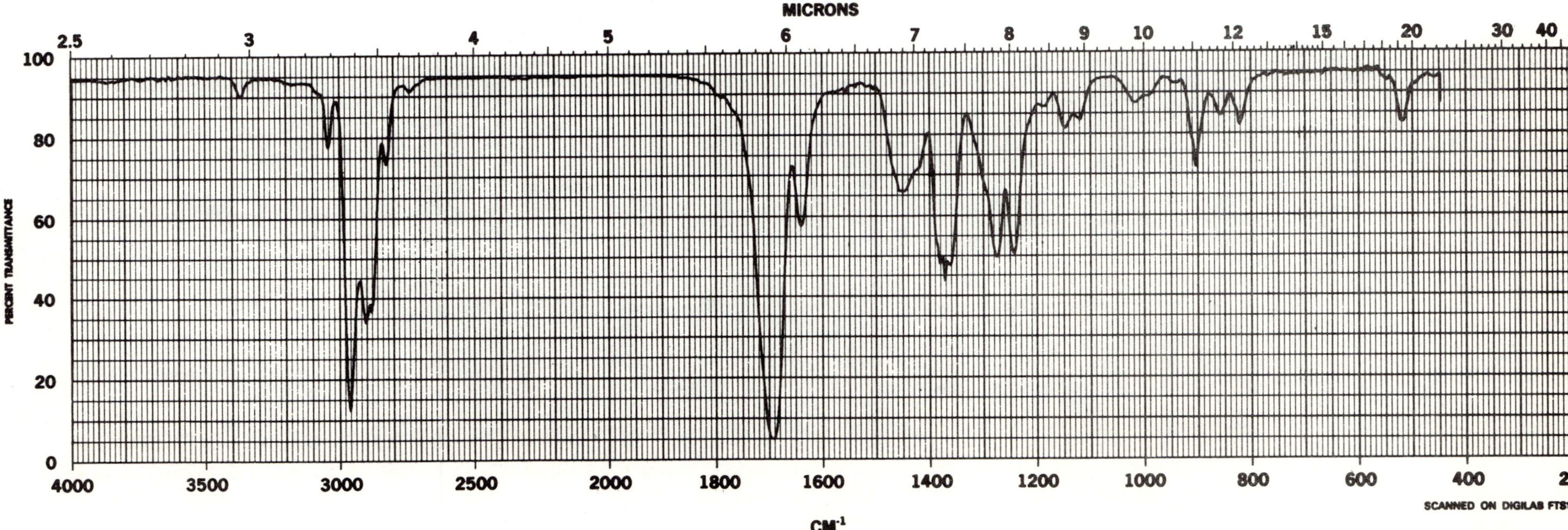

1,2,3,4,5,5–HEXACHLORO–1,3–CYCLOPENTADIENE

HEXACHLOROCYCLOPENTADIENE

C–56

C_5Cl_6

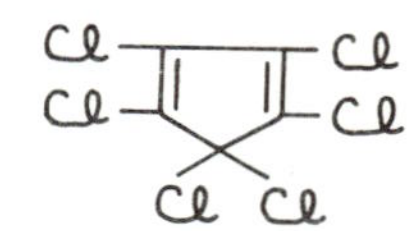

MOLECULAR WEIGHT: 272.77

SOURCE OF SAMPLE: Fluka AG
Buchs, Switzerland

CAS REGISTRY NO.: 77-47-4

NIOSH NUMBER: GY12250

CAPILLARY CELL: Neat

CONDENSED PHASE

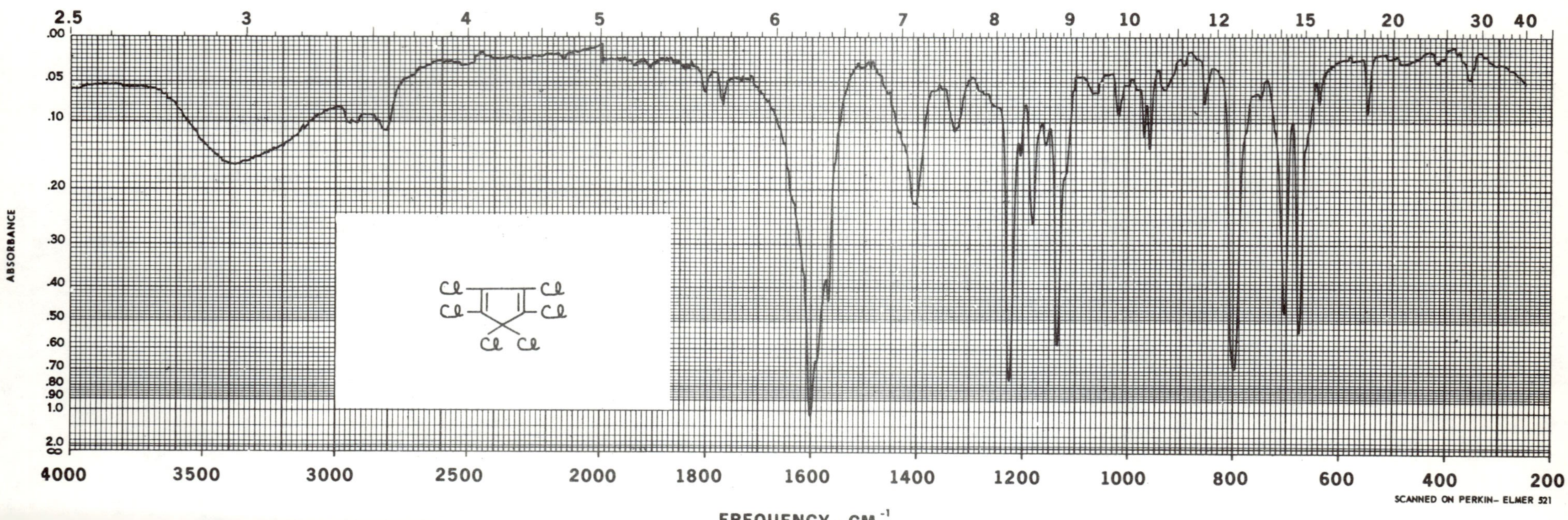

BOILING POINT:	239°C/753 mm	INJECTION AMT.:	0.8 µl
VAPORIZATION TEMP.:	200°C	PRESSURE:	Atmospheric
CELL VOLUME:	7.5 cc	MAX./MIN.:	—
PATH LENGTH:	7.6 cm	SCALE EXPANSION:	1.02

VAPOR PHASE

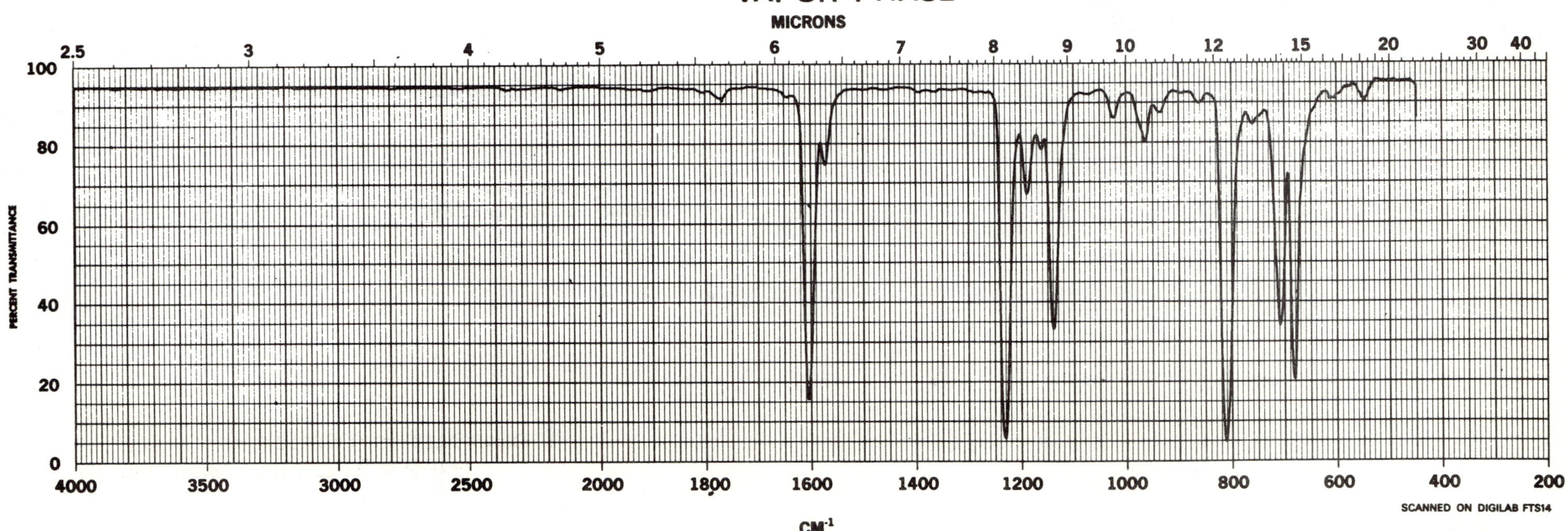

N–NITROSODIBUTYLAMINE
N–BUTYL–N–NITROSO–1–BUTANAMINE
DBN

$C_8H_{18}N_2O$

MOLECULAR WEIGHT: 158.25

SOURCE OF SAMPLE: Eastman Organic Chemicals
Rochester, New York

CAS REGISTRY NO.: 924-16-3

NIOSH NUMBER: EJ40250

CAPILLARY CELL: Neat

$CH_3CH_2CH_2CH_2-N(NO)-CH_2CH_2CH_2CH_3$

CONDENSED PHASE

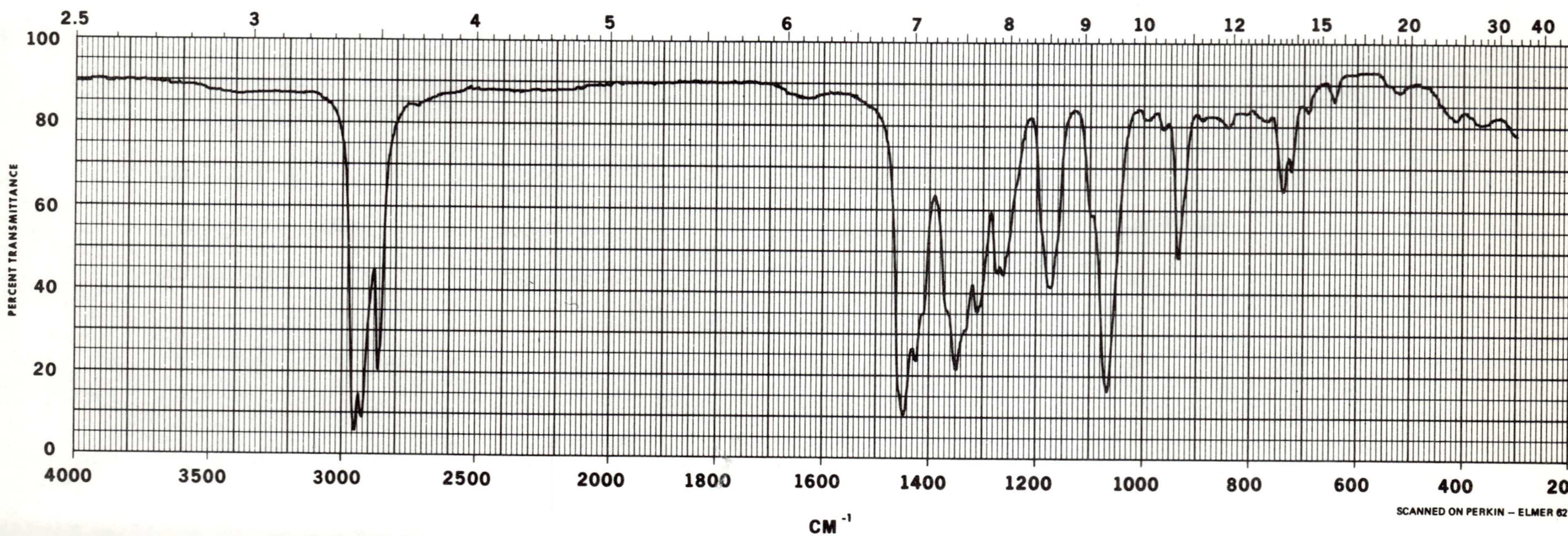

BOILING POINT:	104-105°C/8 mm	INJECTION AMT.:	—
VAPORIZATION TEMP.:	200°C	PRESSURE:	Atmospheric
CELL VOLUME:	7.5 cc	MAX./MIN.:	101.9/5.7
PATH LENGTH:	7.6 cm	SCALE EXPANSION:	0.96

VAPOR PHASE

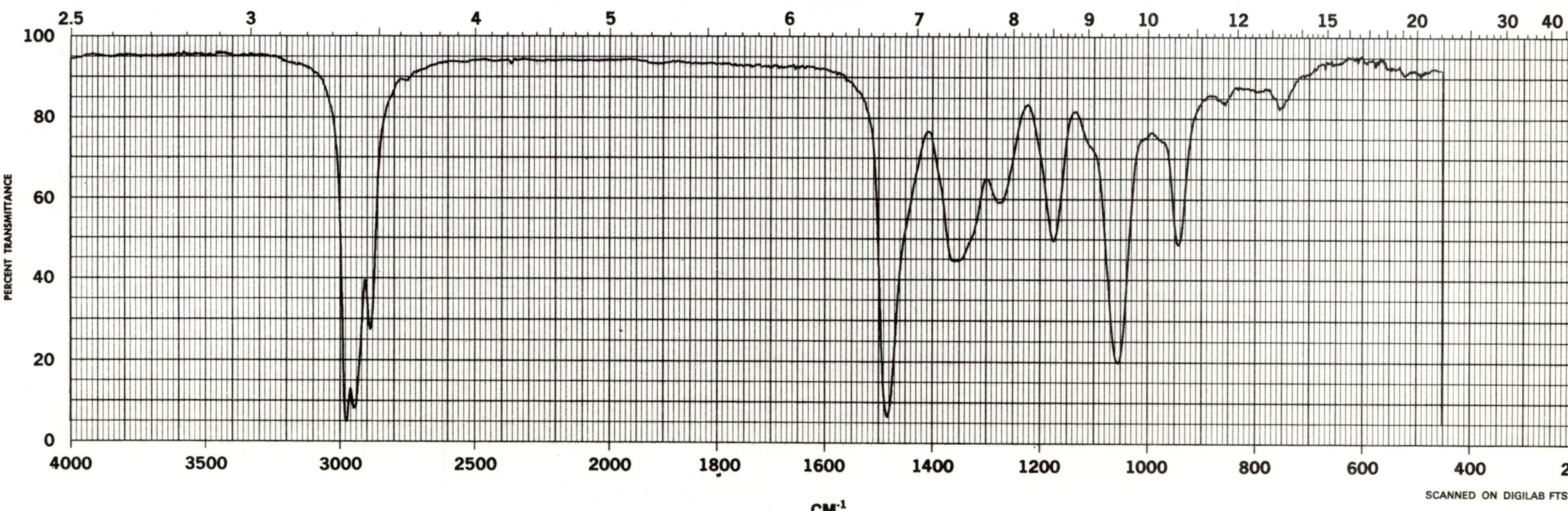

SCANNED ON DIGILAB FTS14

DIETHYLAMINE
DEN

$C_4H_{11}N$

$CH_3CH_2-NH-CH_2CH_3$

MOLECULAR WEIGHT: 73.14

SOURCE OF SAMPLE: Virginia Chemicals, Inc.
West Norfolk, Virginia

CAS REGISTRY NO.: 109-89-7

NIOSH NUMBER: HZ87500

CAPILLARY CELL: 0.015 mm

CONDENSED PHASE

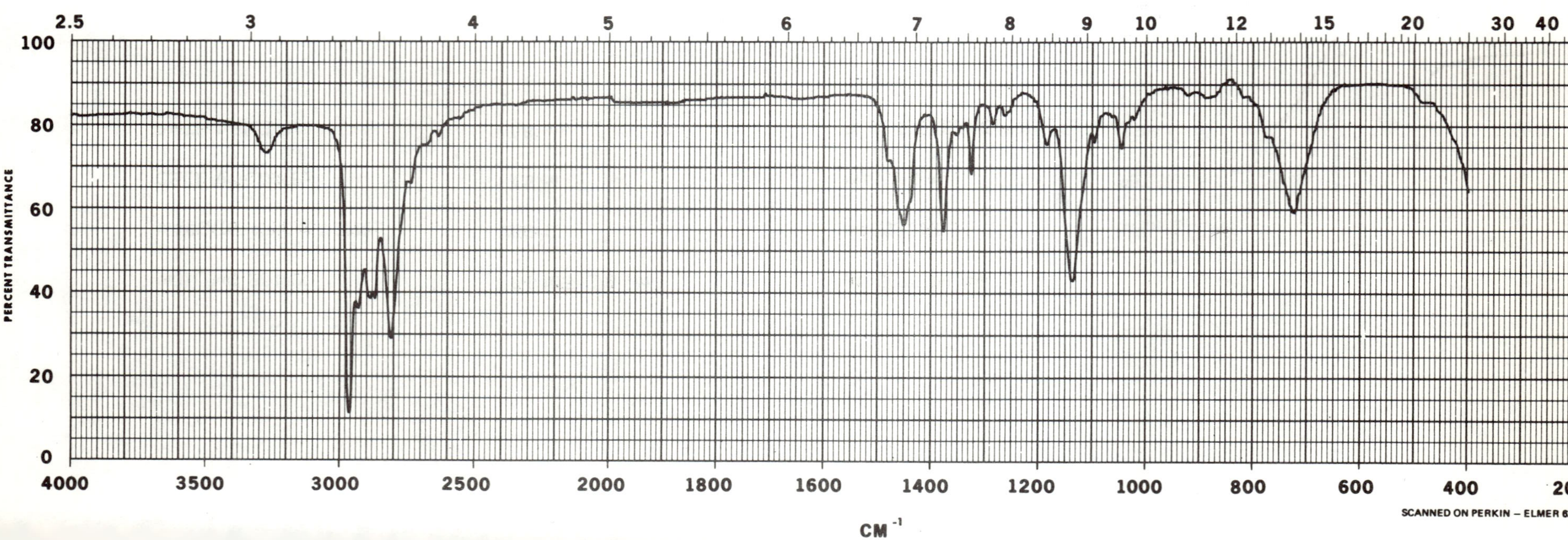

BOILING POINT:	55.9°C/760 mm	INJECTION AMT.:	1.0 μl
VAPORIZATION TEMP.:	200°C	PRESSURE:	Atmospheric
CELL VOLUME:	7.5 cc	MAX./MIN.:	—
PATH LENGTH:	7.6 cm	SCALE EXPANSION:	0.93

VAPOR PHASE

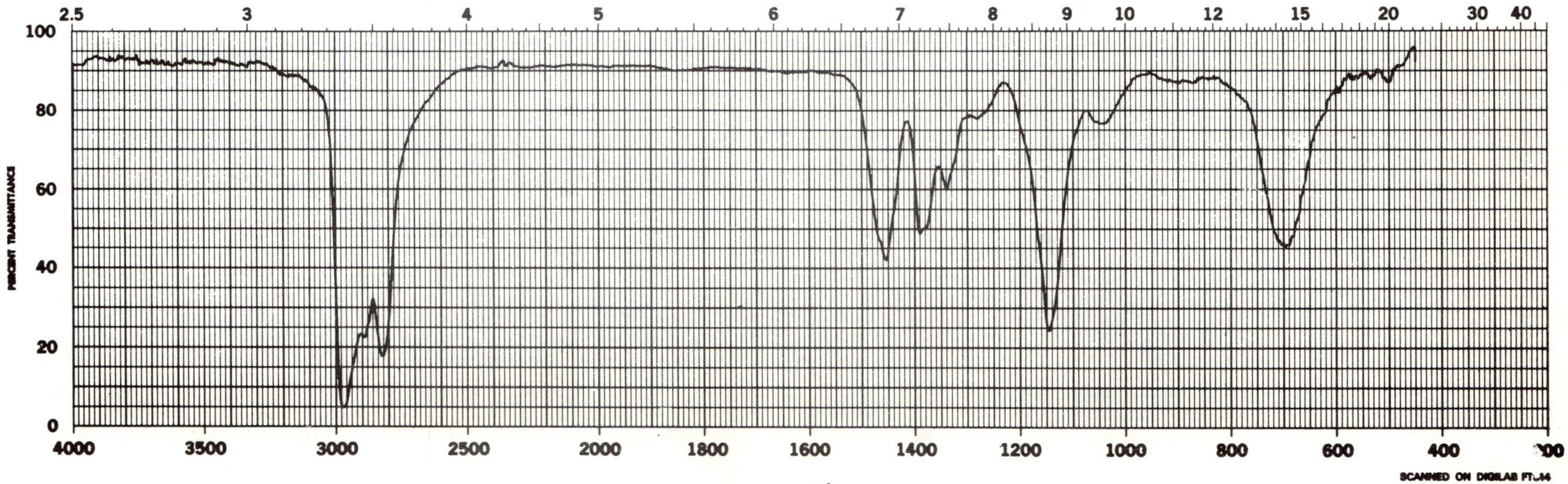

N—NITROSODIETHYLAMINE
DANA
DEN
DENA

$C_4H_{10}N_2O$

MOLECULAR WEIGHT: 102.14

SOURCE OF SAMPLE: The Ames Laboratories, Inc.
Milford, Connecticut

$CH_3CH_2-N(NO)-CH_2CH_3$

CAS REGISTRY NO.: 55-18-5

NIOSH NUMBER: IA35000

CAPILLARY CELL: Neat

CONDENSED PHASE

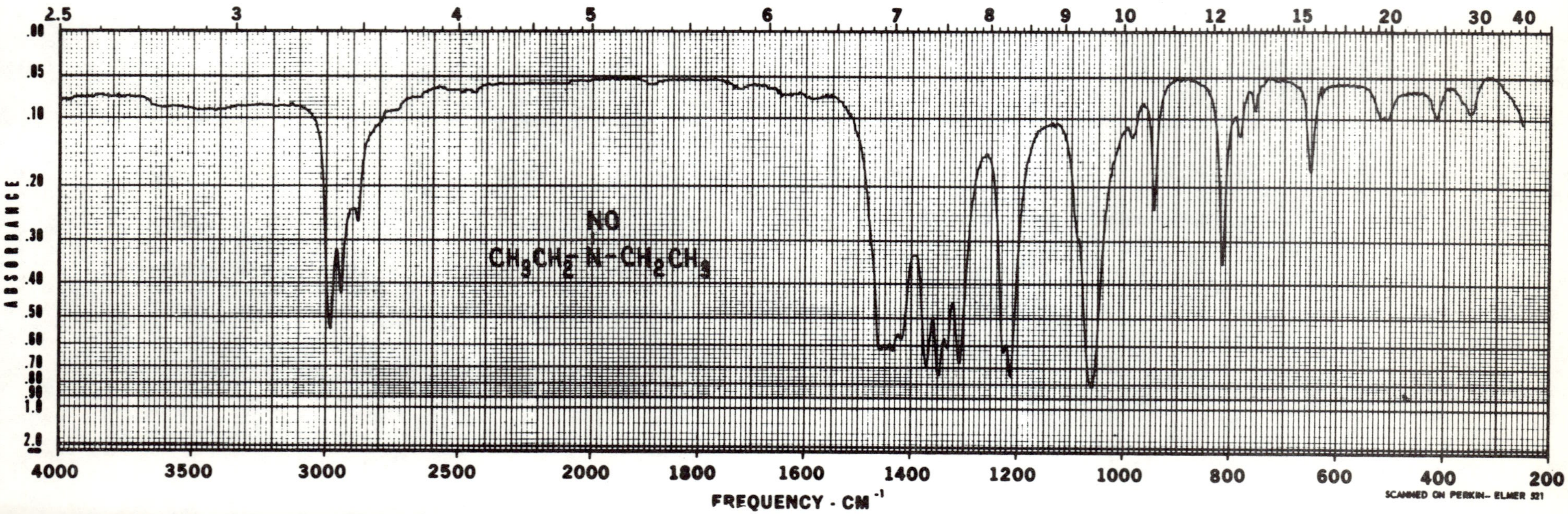

BOILING POINT:	176.9°C	INJECTION AMT.:	0.5 μl
VAPORIZATION TEMP.:	200°C	PRESSURE:	Atmospheric
CELL VOLUME:	7.5 cc	MAX./MIN.:	101.9/19.3
PATH LENGTH:	7.6 cm	SCALE EXPANSION:	1.11

VAPOR PHASE

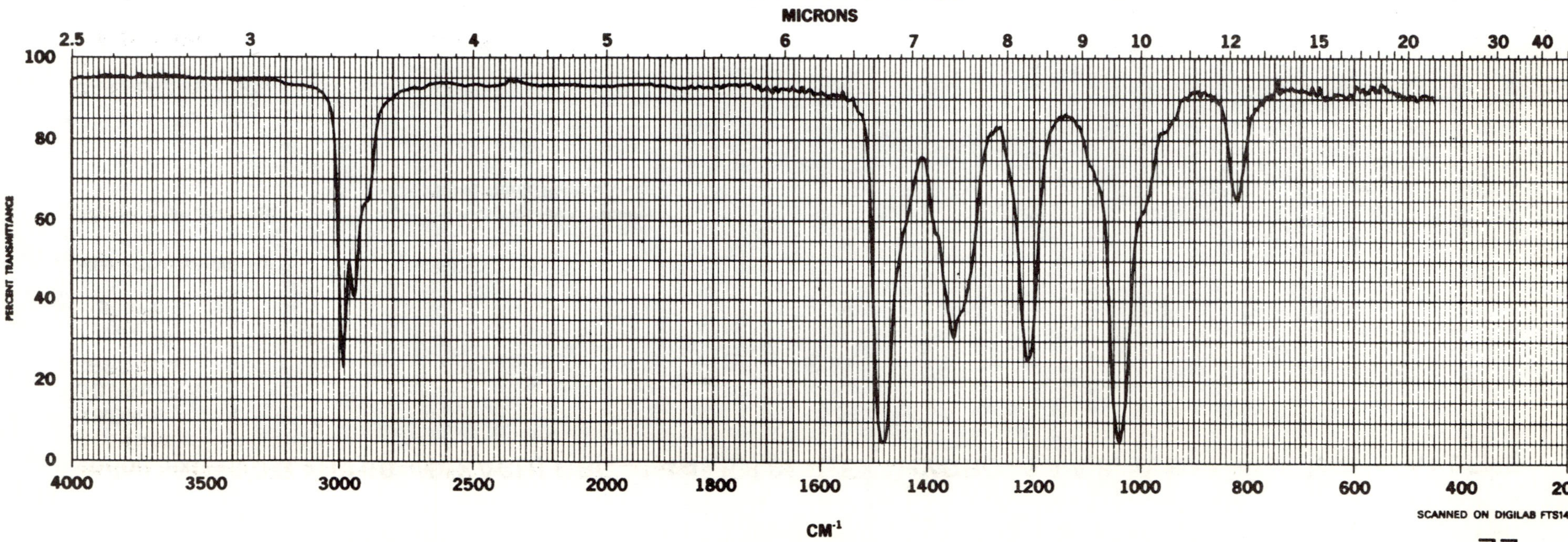

SCANNED ON DIGILAB FTS14

endo–,exo–1,2,3,4,10,10–HEXACHLORO–1,4,4a,5,8,8a–HEXAHYDRO–1,4:5,8–DIMETHANONAPHTHALENE
ALDRIN
HHDN
OCTALENE
SEEDRIN

$C_{12}H_8Cl_6$

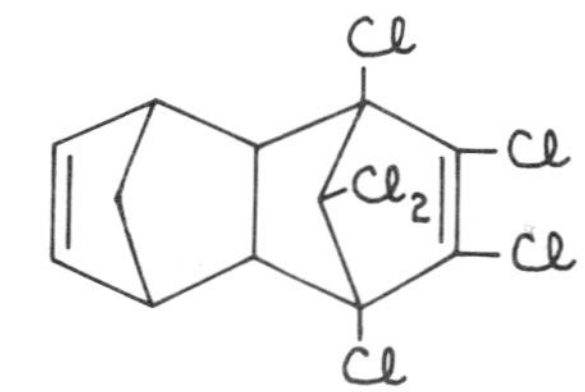

MOLECULAR WEIGHT: 364.92

SOURCE OF SAMPLE: Attapulgus Clay Company
Philadelphia, Pennsylvania

CAS REGISTRY NO.: 309-00-2

NIOSH NUMBER: IO21000

KBr Wafer

CONDENSED PHASE

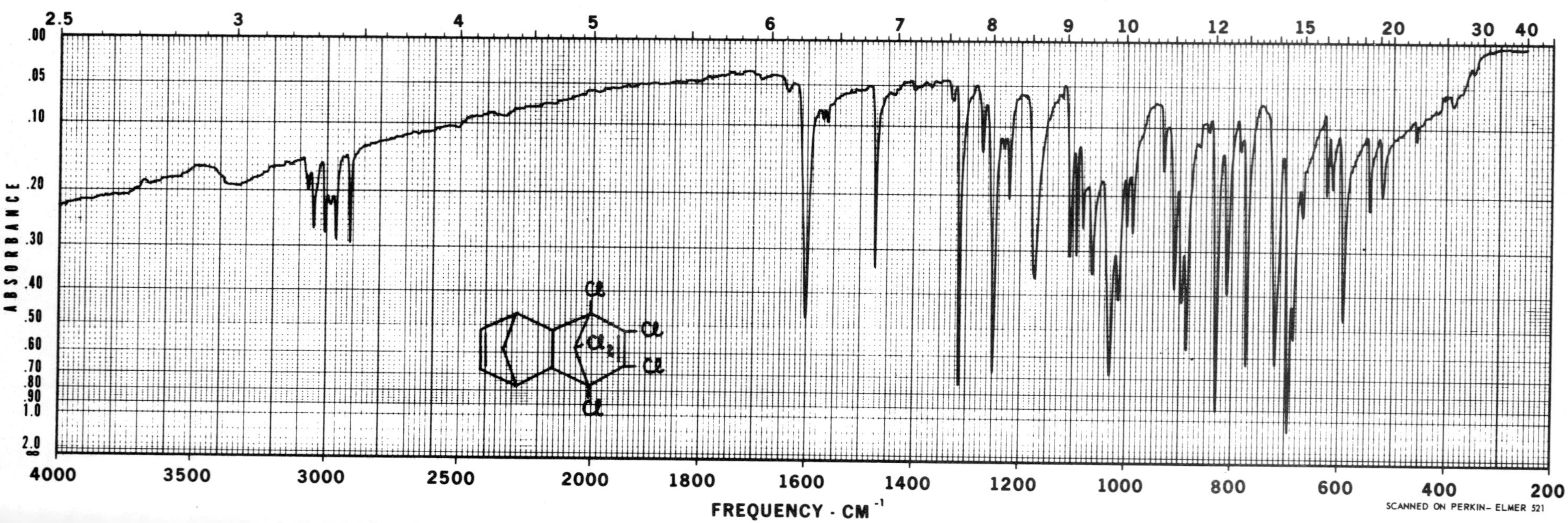

BOILING POINT:	—	INJECTION AMT.:	1.0 mg
VAPORIZATION TEMP.:	280°C	PRESSURE:	Atmospheric
CELL VOLUME:	7.5 cc	MAX./MIN.:	—
PATH LENGTH:	7.6 cm	SCALE EXPANSION:	1.33

VAPOR PHASE

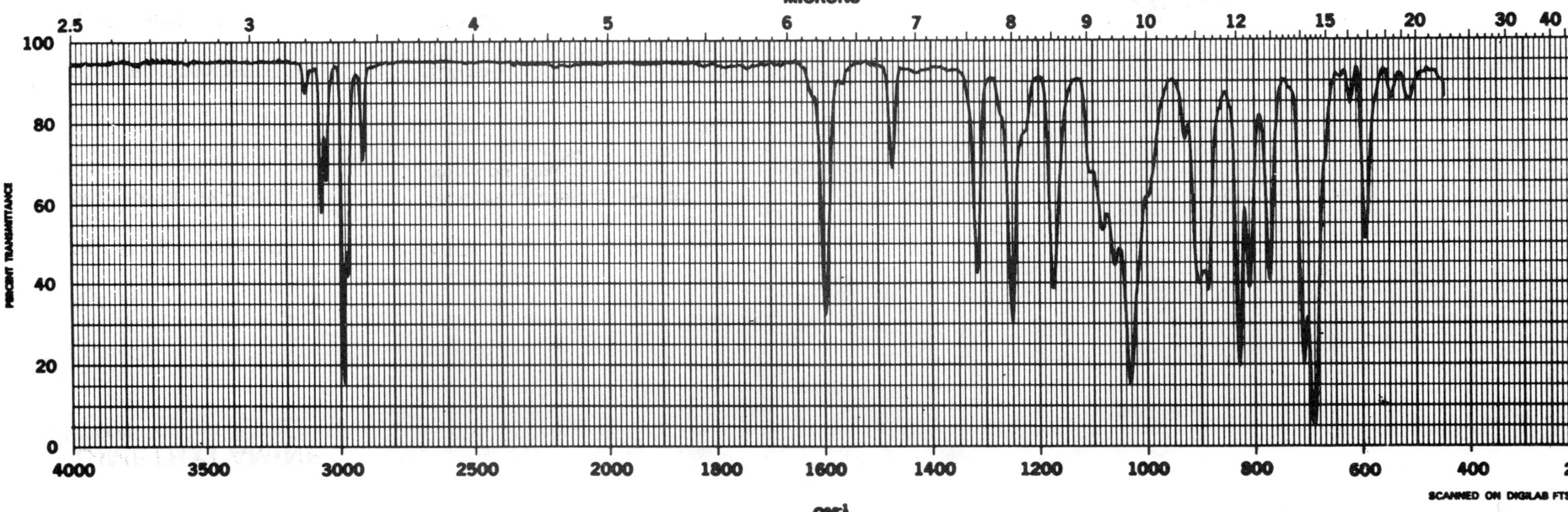

DIMETHYLAMINE
DMA

C_2H_7N

$CH_3-NH-CH_3$

MOLECULAR WEIGHT:	45.08
SOURCE OF SAMPLE:	MCB Manufacturing Chemists Norwood, Ohio
CAS REGISTRY NO.:	124-40-3
NIOSH NUMBER:	IP87500
Path Length:	10 cm
Complete Scan:	20 mm Hg

A GAS AT 25°C

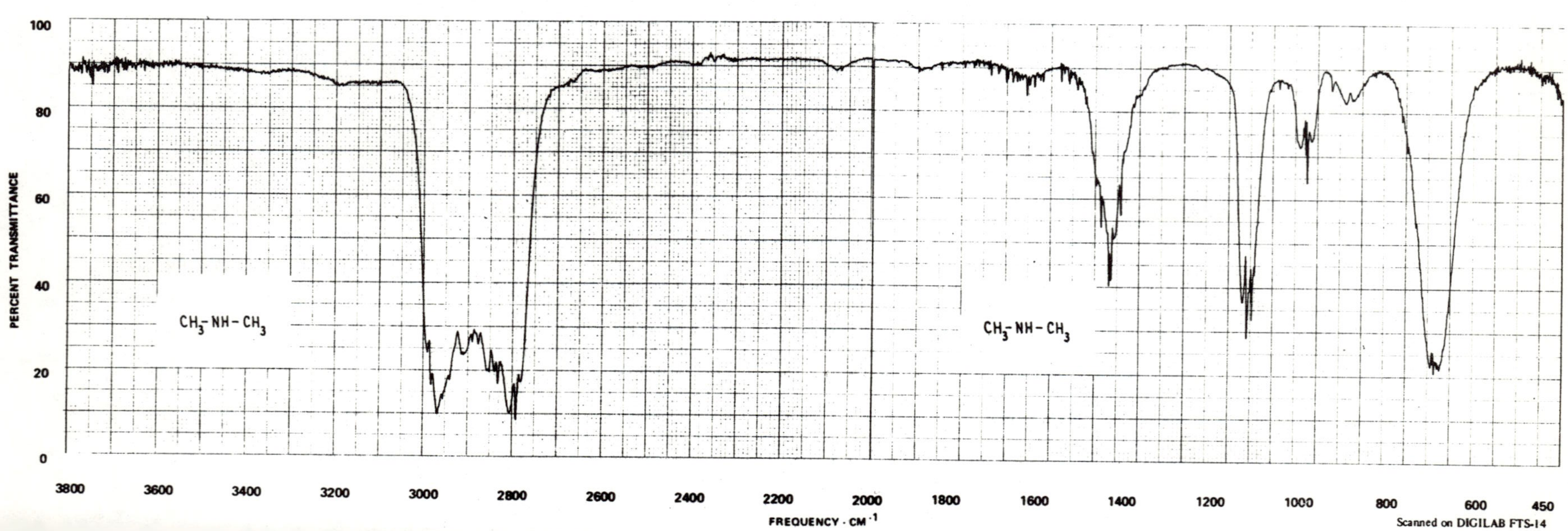

BOILING POINT:	6.9°C	INJECTION AMT.:	400 μl
VAPORIZATION TEMP.:	200°C	PRESSURE:	Atmospheric
CELL VOLUME:	7.5 cc	MAX./MIN.:	100.8/14.3
PATH LENGTH:	7.6 cm	SCALE EXPANSION:	1.06

VAPOR PHASE

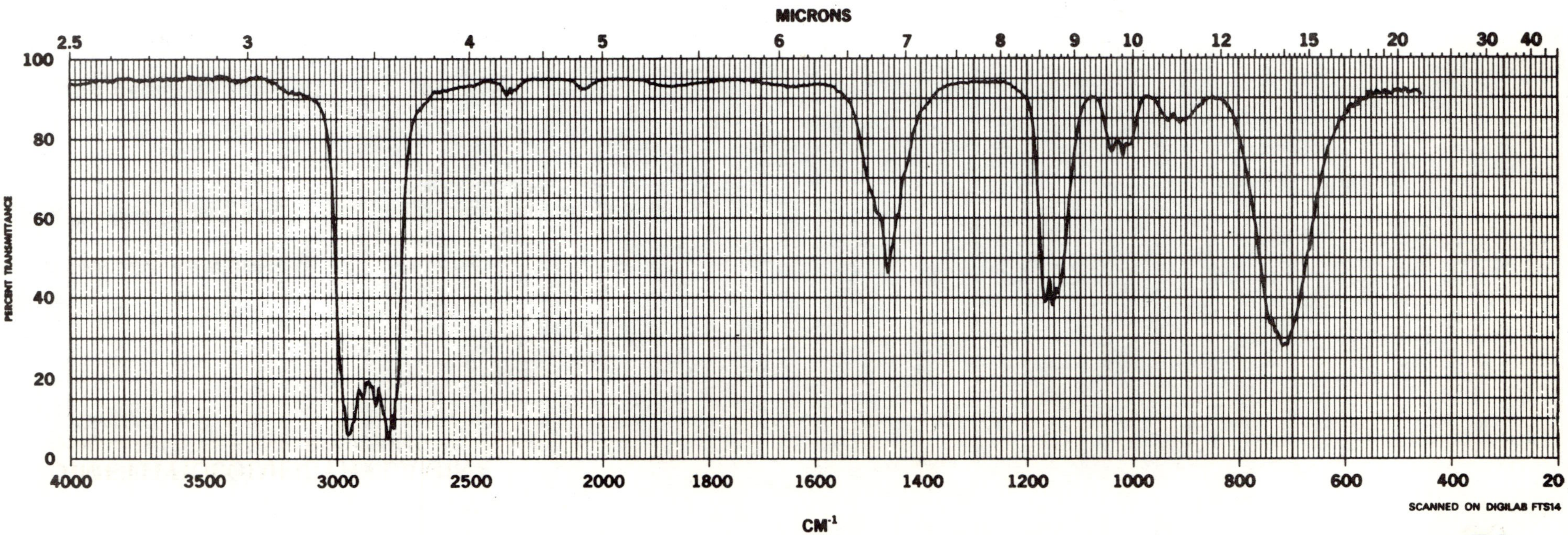

N—NITROSODIMETHYLAMINE
N,N—DIMETHYLNITROSAMINE
DMNA

$C_2H_6N_2O$

NO
|
CH_3 – N – CH_3

MOLECULAR WEIGHT: 74.08

SOURCE OF SAMPLE: Aldrich Chemical Company, Inc. Milwaukee, Wisconsin

CAS REGISTRY NO.: 62-75-9

NIOSH NUMBER: IQ05250

CAPILLARY CELL: Neat

CONDENSED PHASE

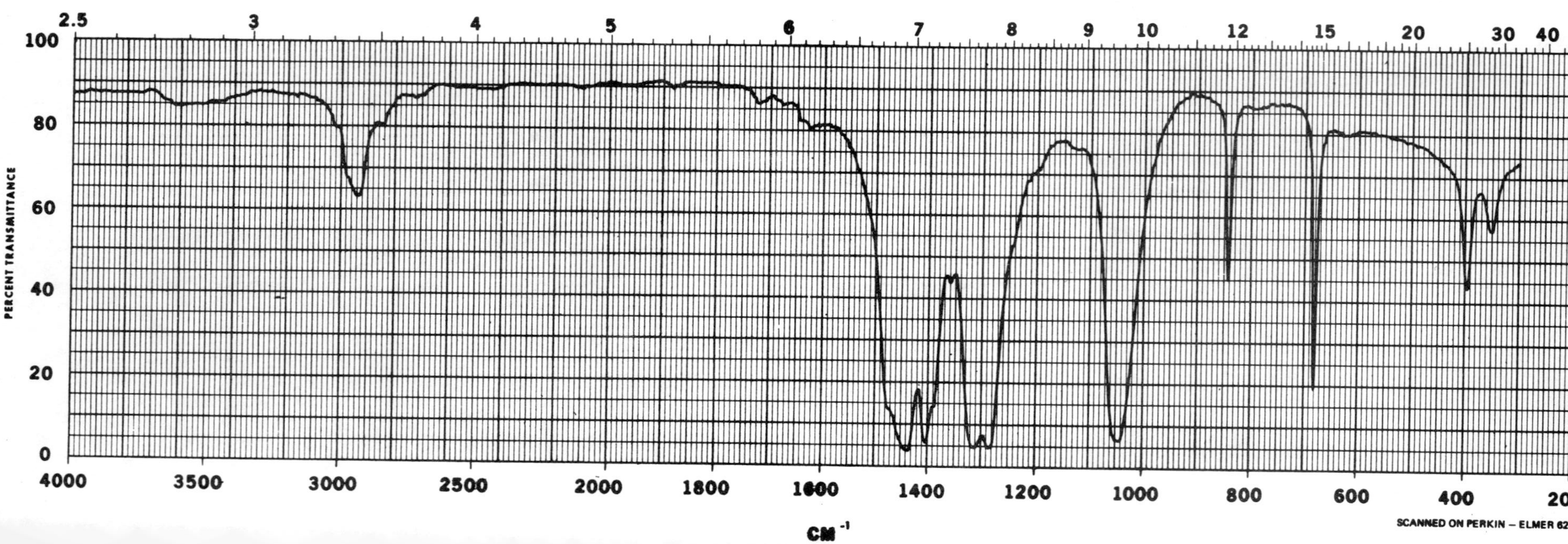

BOILING POINT:	$153^{o}C/774$ mm	INJECTION AMT.:	0.5 µl
VAPORIZATION TEMP.:	$200^{o}C$	PRESSURE:	Atmospheric
CELL VOLUME:	7.5 cc	MAX./MIN.:	100.1/3.1
PATH LENGTH:	7.6 cm	SCALE EXPANSION:	0.95

VAPOR PHASE

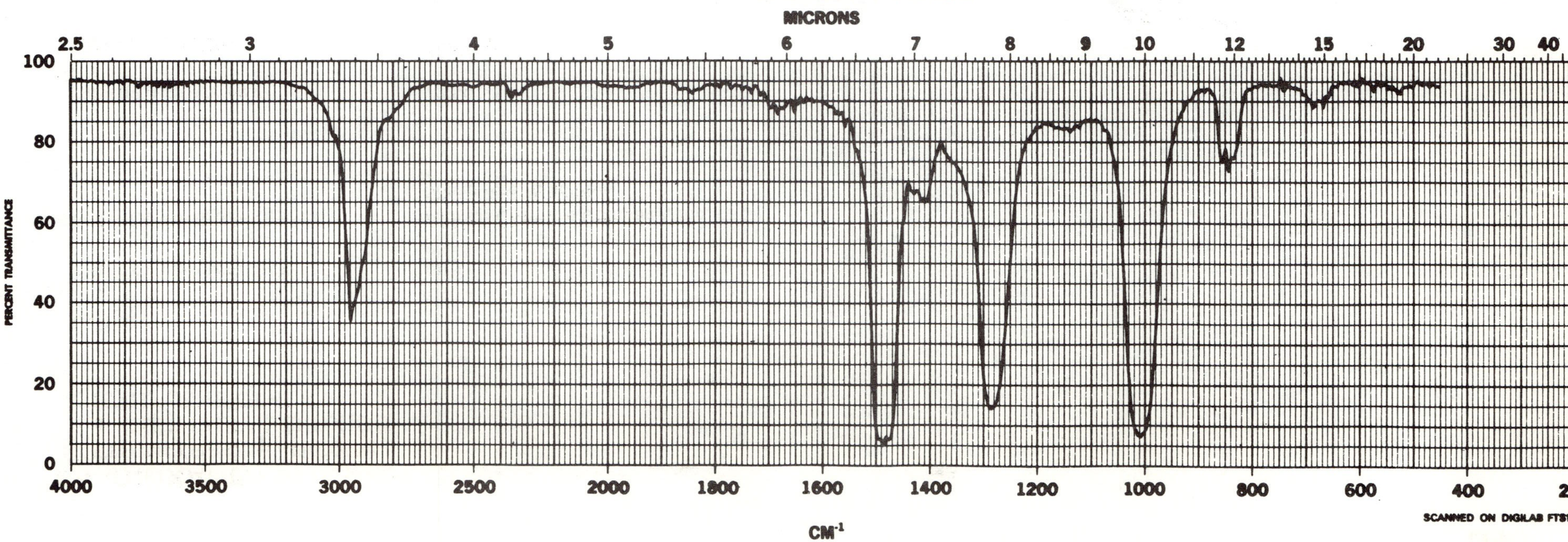

p–DIOXANE
TETRAHYDRO–p–DIOXIN
1,4–DIETHYLENE DIOXIDE
DIETHYLENE ETHER

$C_4H_8O_2$

MOLECULAR WEIGHT:	88.11
SOURCE OF SAMPLE:	Aldrich Chemical Company, Inc. Milwaukee, Wisconsin
CAS REGISTRY NO.:	123-91-1
NIOSH NUMBER:	JG82250
CAPILLARY CELL:	Neat

CONDENSED PHASE

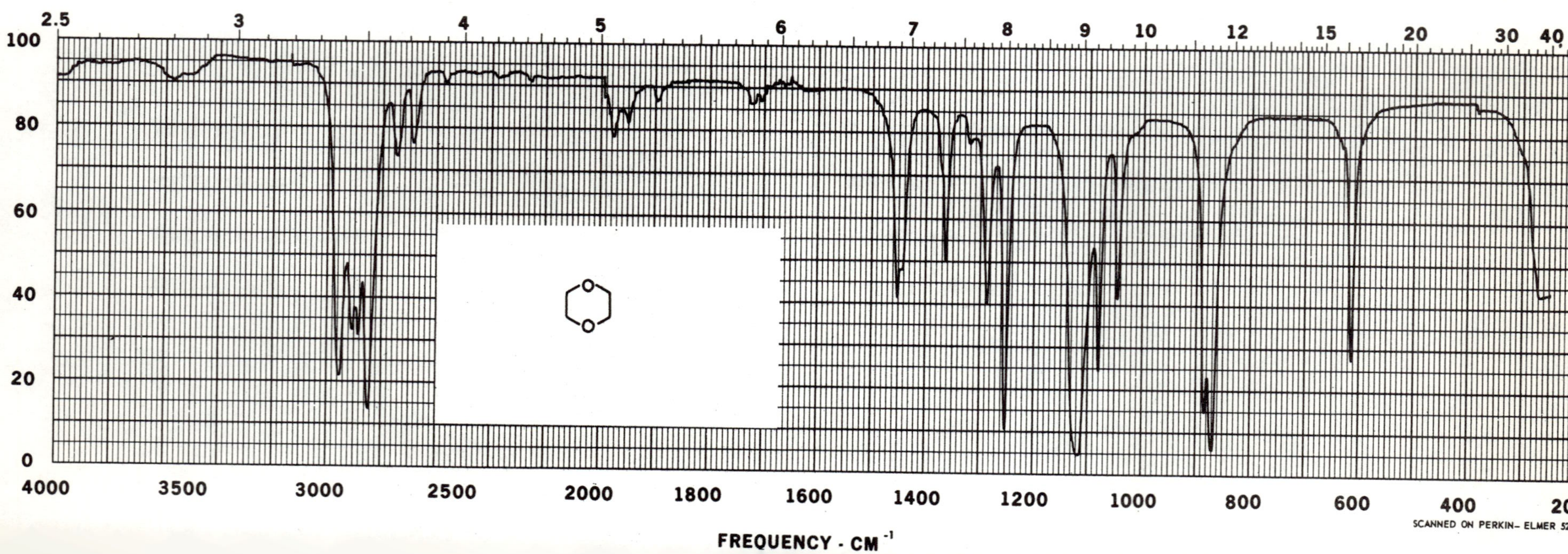

BOILING POINT:	101°C	INJECTION AMT.:	1.0 μl
VAPORIZATION TEMP.:	200°C	PRESSURE:	Atmospheric
CELL VOLUME:	7.5 cc	MAX./MIN.:	—
PATH LENGTH:	7.6 cm	SCALE EXPANSION:	0.92

VAPOR PHASE

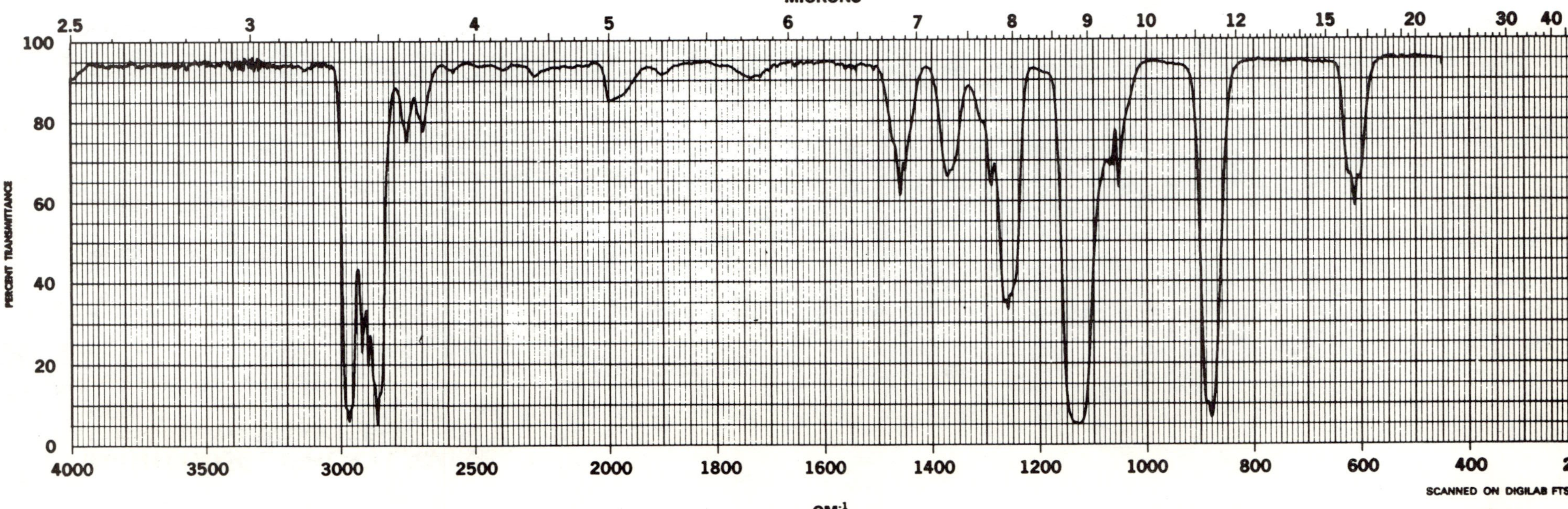

N—NITROSODIPROPYLAMINE
DIPROPYLNITROSAMINE
DPN

$C_6H_{14}N_2O$

MOLECULAR WEIGHT: 130.19

SOURCE OF SAMPLE: Eastman Organic Chemicals, Rochester, New York

$CH_3CH_2CH_2-N(NO)-CH_2CH_2CH_3$

CAS REGISTRY NO.: 621-64-7

NIOSH NUMBER: JL97000

CAPILLARY CELL: Neat

CONDENSED PHASE

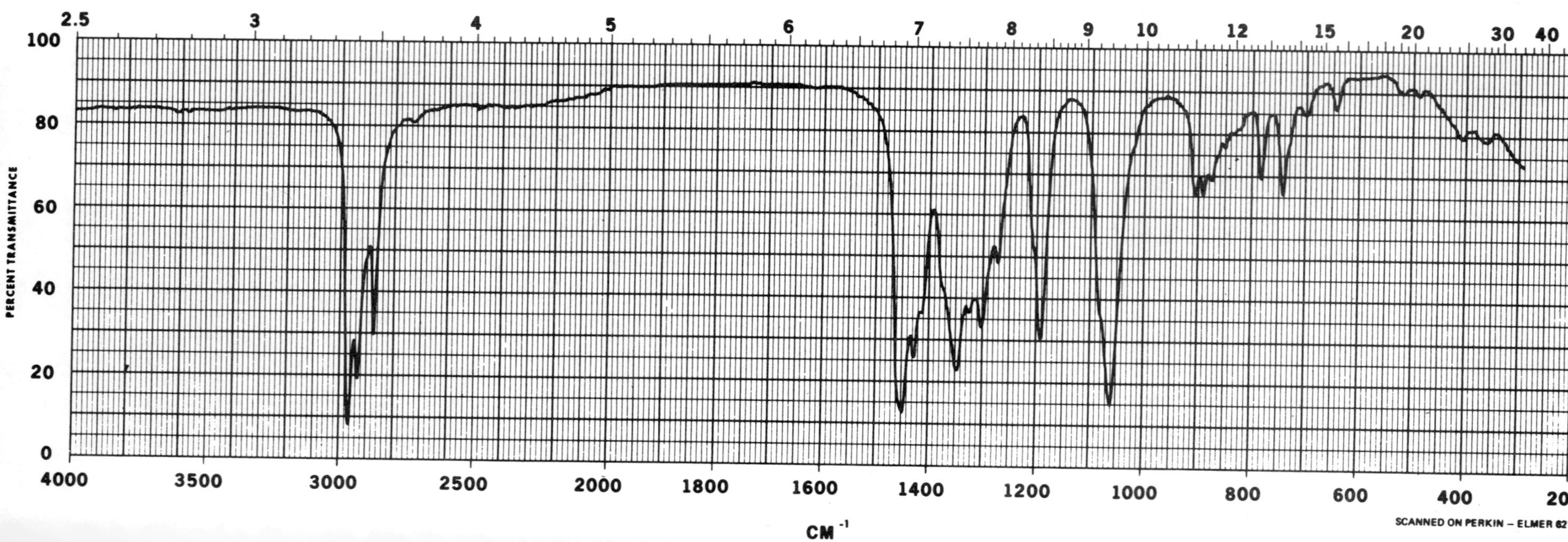

BOILING POINT:	–	INJECTION AMT.:	1.0 μl
VAPORIZATION TEMP.:	200°C	PRESSURE:	Atmospheric
CELL VOLUME:	7.5 cc	MAX./MIN.:	–
PATH LENGTH:	7.6 cm	SCALE EXPANSION:	0.99

VAPOR PHASE

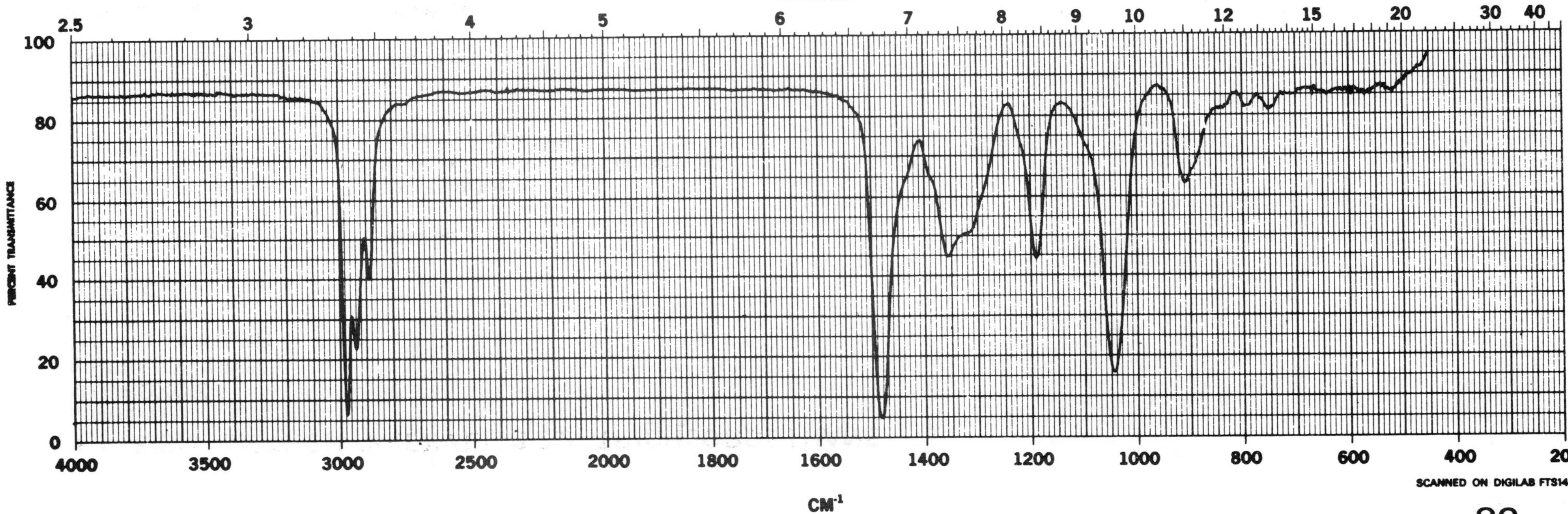

SCANNED ON DIGILAB FTS14

82

1,1–BIS(p–METHOXYPHENYL)–2,2,2–TRICHLOROETHANE

MARLATE
METHOXCIDE
METHOXY–DDT
METHOXYCHLOR

$C_{16}H_{15}Cl_3O_2$

MOLECULAR WEIGHT: 345.66

SOURCE OF SAMPLE: M. Hubacher
Ex-Lax Pharmaceutical Company, Inc.
Brooklyn, New York

CAS REGISTRY NO.: 72-43-5

NIOSH NUMBER: KJ36750

KBr Wafer

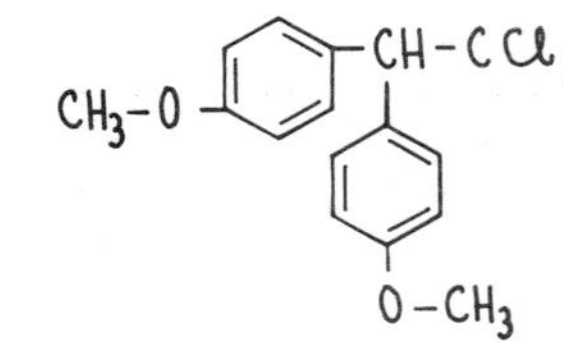

CONDENSED PHASE

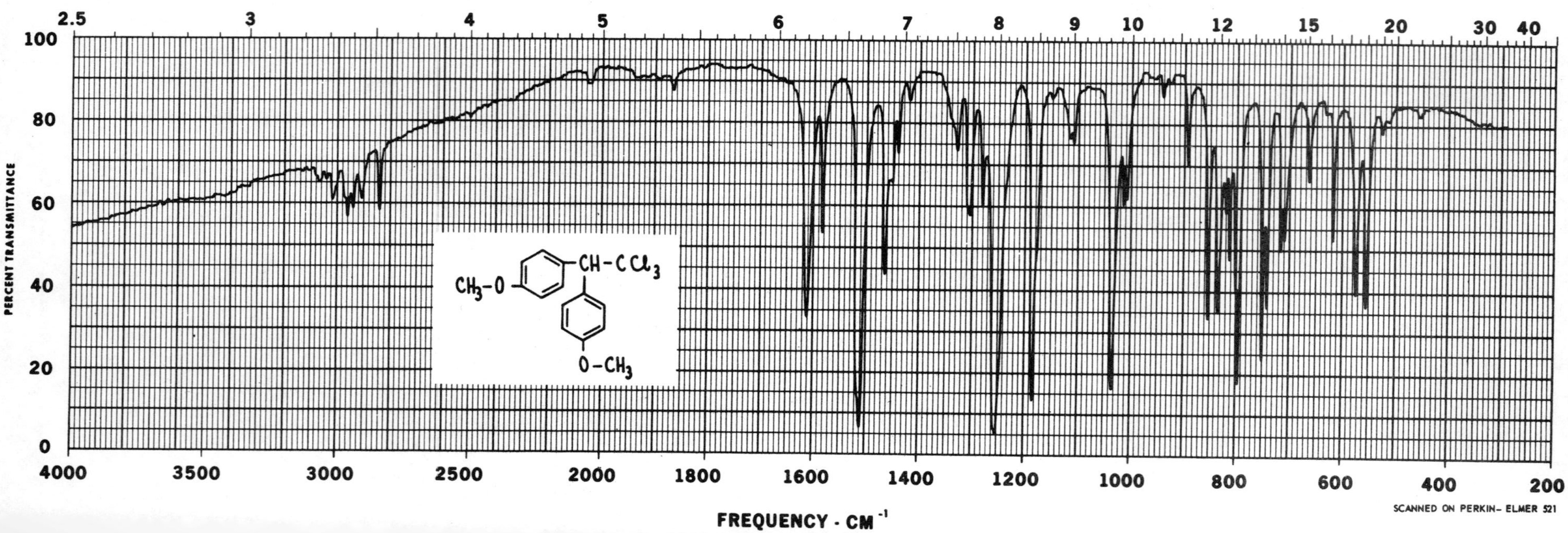

MELTING POINT:	87-88°C	INJECTION AMT.:	1.0 mg
VAPORIZATION TEMP.:	280°C	PRESSURE:	Atmospheric
CELL VOLUME:	7.5 cc	MAX./MIN.:	116.2/17.9
PATH LENGTH:	7.6 cm	SCALE EXPANSION:	0.94

VAPOR PHASE

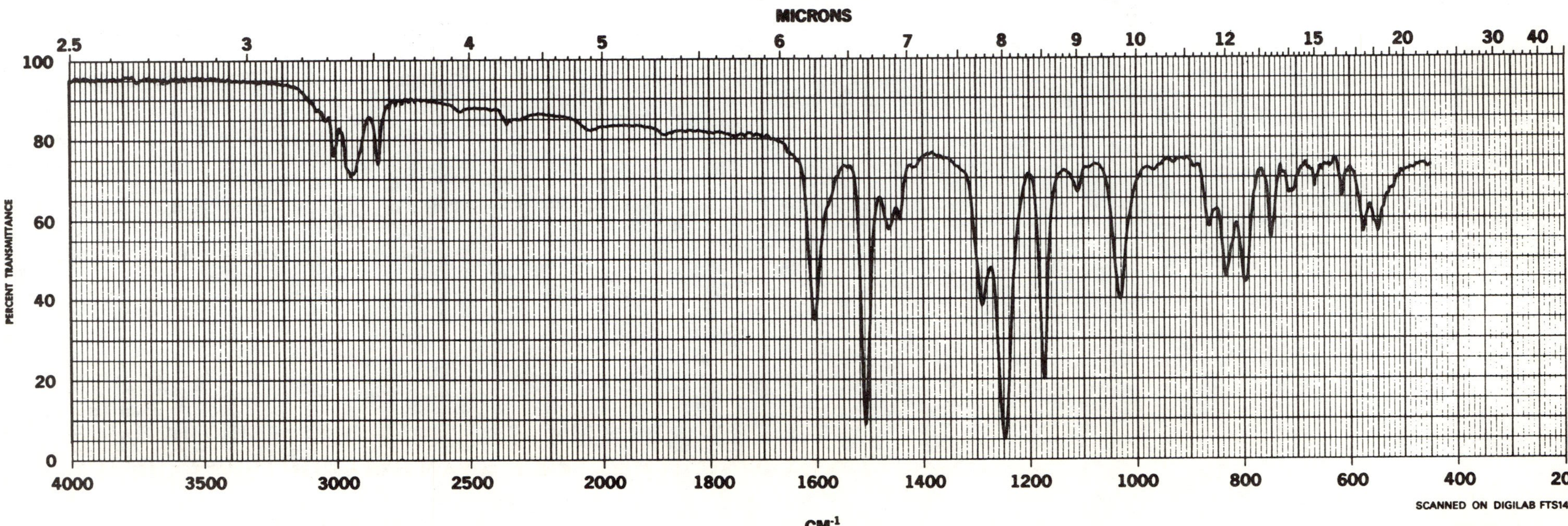

SCANNED ON DIGILAB FTS14

CHLOROETHANE
ETHYL CHLORIDE
KELENE
NARCOTILE

C_2H_5Cl

CH_3CH_2-Cl

MOLECULAR WEIGHT:	64.52
SOURCE OF SAMPLE:	MCB Manufacturing Chemists Norwood, Ohio
CAS REGISTRY NO.:	75-00-3
NIOSH NUMBER:	KH75250
GAS CELL:	10 cm

A GAS AT 25°C

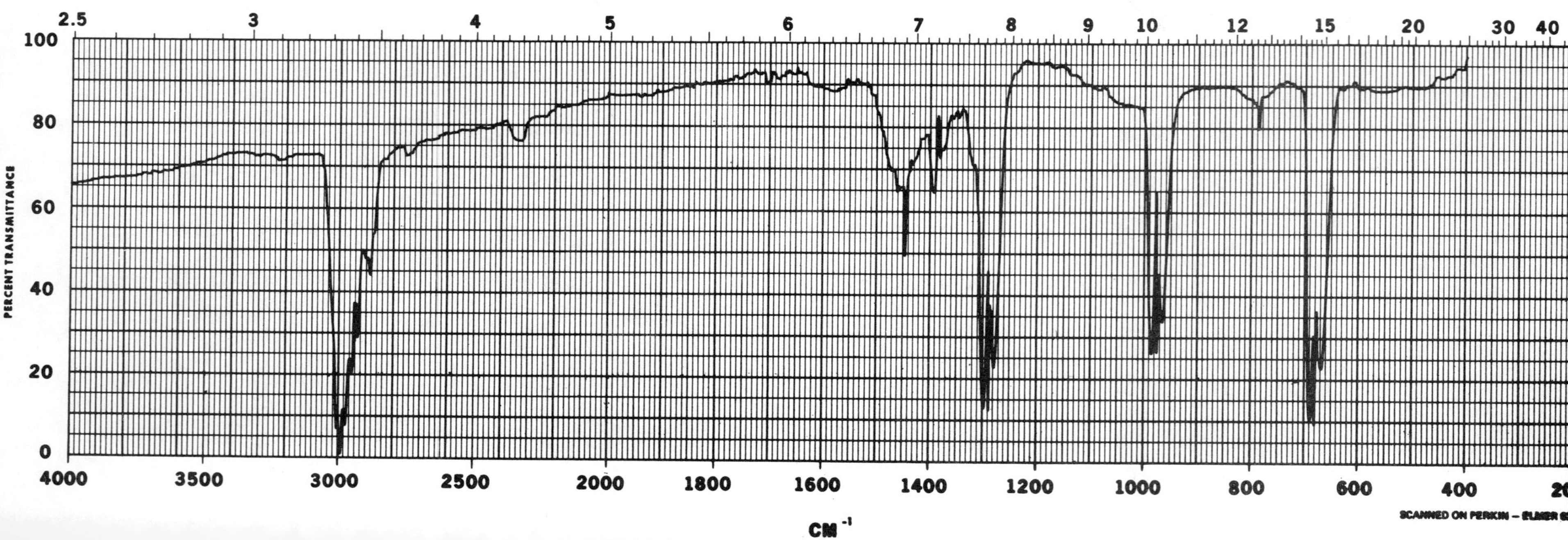

BOILING POINT:	12.27°C	INJECTION AMT.:	900 µl
VAPORIZATION TEMP.:	200°C	PRESSURE:	Atmospheric
CELL VOLUME:	7.5 cc	MAX./MIN.:	100.4/16.5
PATH LENGTH:	7.6 cm	SCALE EXPANSION:	1.10

VAPOR PHASE

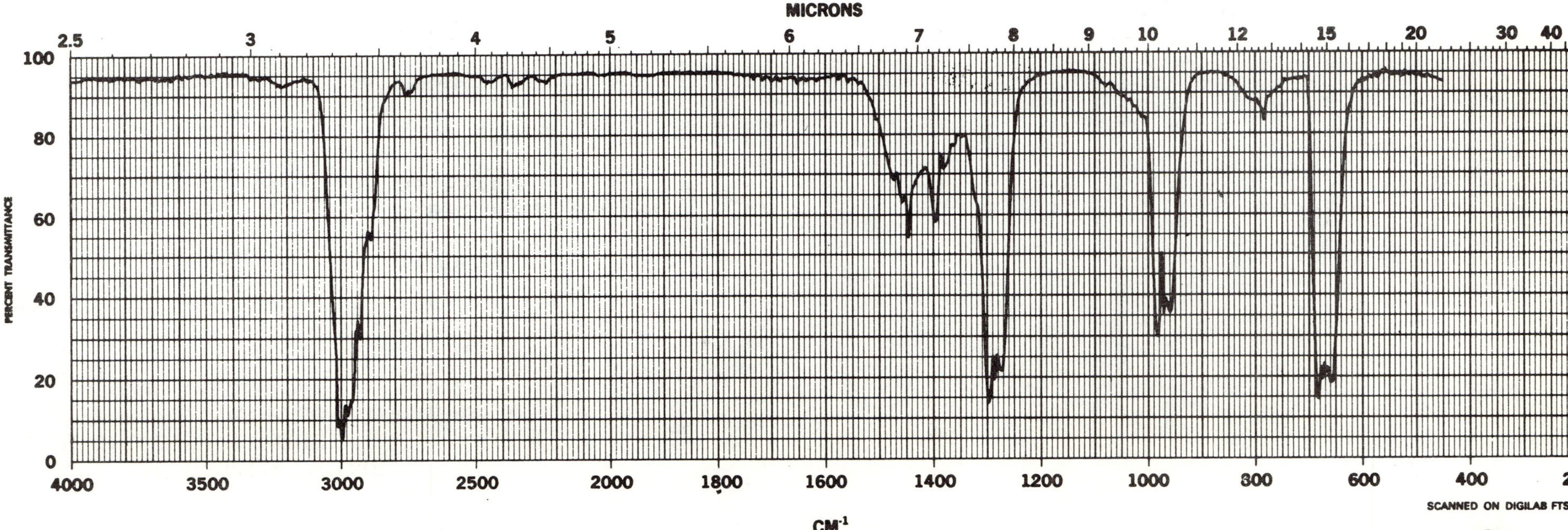

SCANNED ON DIGILAB FTS14

1,2–DIBROMOETHANE
ETHYLENE DIBROMIDE
DOWFUME 40/EDB/MC–2/W–8/W–85
GLYCOL DIBROMIDE

$C_2H_4Br_2$

MOLECULAR WEIGHT: 187.87

SOURCE OF SAMPLE: MCB Manufacturing Chemists
Norwood, Ohio

$Br-CH_2-CH_2-Br$

CAS REGISTRY NO.: 106-93-4

NIOSH NUMBER: KH92750

CAPILLARY CELL: Neat

CONDENSED PHASE

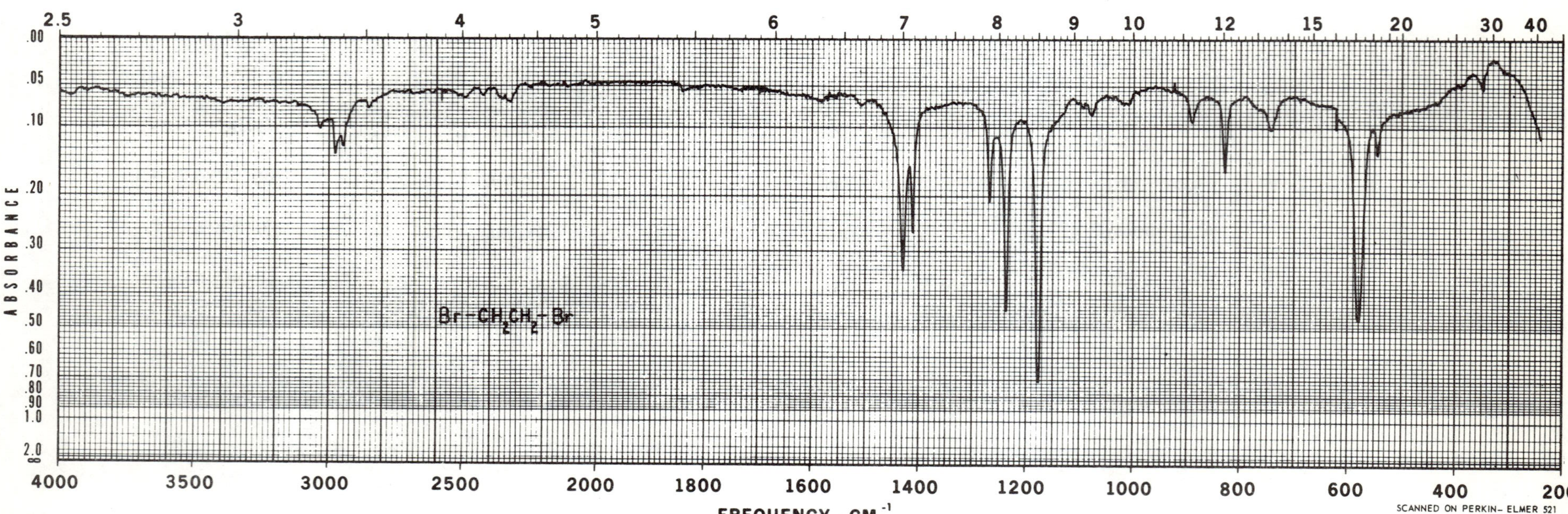

BOILING POINT:	131.5-132°C	INJECTION AMT.:	—
VAPORIZATION TEMP.:	160°C	PRESSURE:	Atmospheric
CELL VOLUME:	7.5 cc	MAX./MIN.:	108.7/11.0
PATH LENGTH:	7.6 cm	SCALE EXPANSION:	0.94

VAPOR PHASE

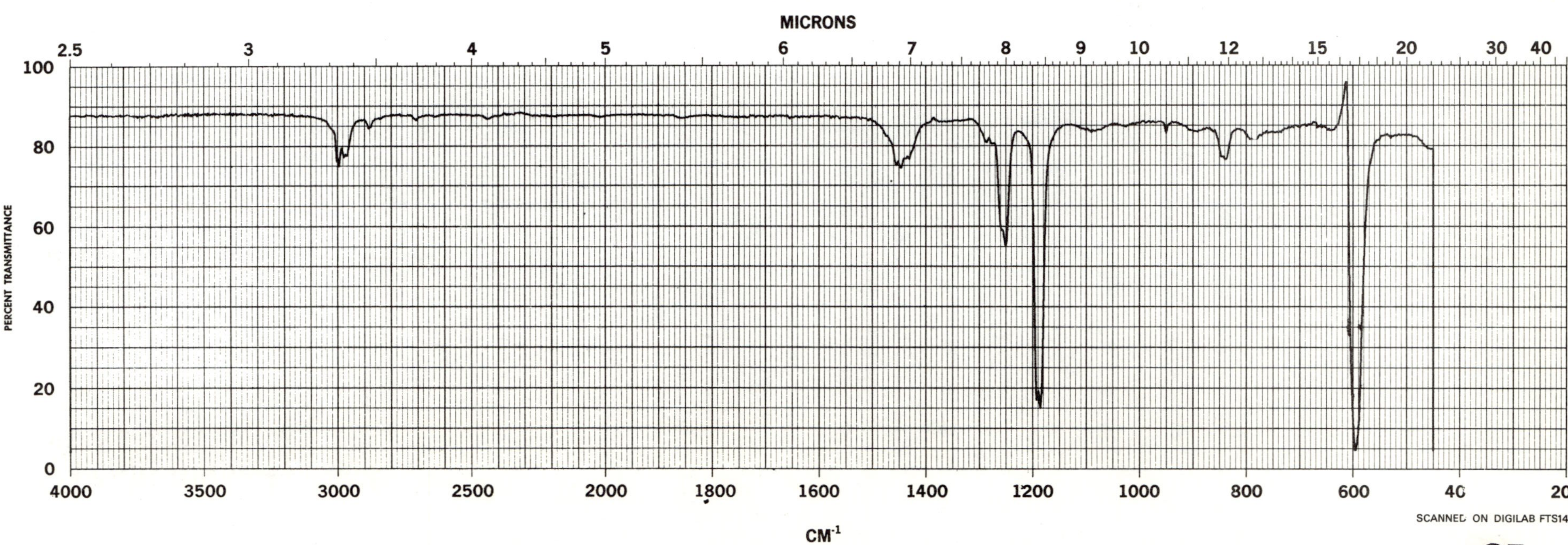

SCANNED ON DIGILAB FTS14

1,1–DICHLOROETHANE
ETHYLIDENE DICHLORIDE

$C_2H_4Cl_2$

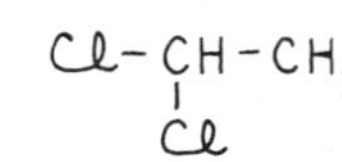

MOLECULAR WEIGHT: 98.96

SOURCE OF SAMPLE: Tokyo Kasei Kogyo Company Ltd.
Tokyo, Japan

CAS REGISTRY NO.: 75-34-3

NIOSH NUMBER: KI01750

CAPILLARY CELL: 0.015 mm

CONDENSED PHASE

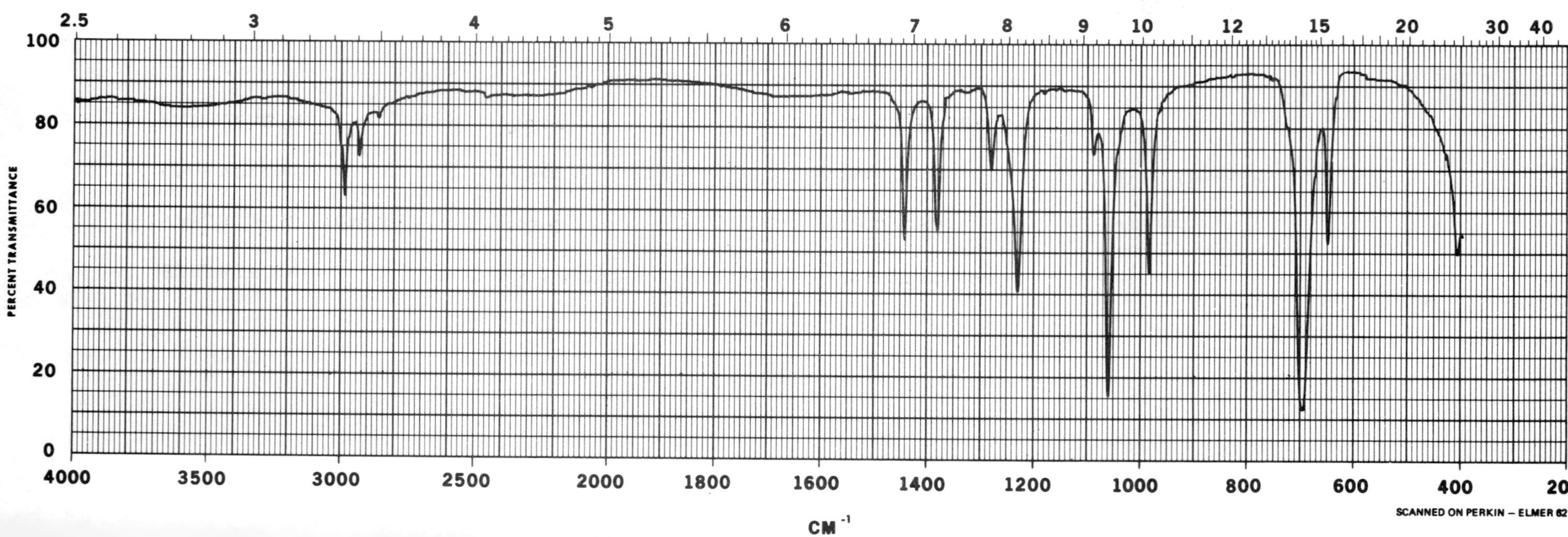

BOILING POINT:	57-58°C	INJECTION AMT.:	1.0 μl
VAPORIZATION TEMP.:	200°C	PRESSURE:	Atmospheric
CELL VOLUME:	7.5 cc	MAX./MIN.:	—
PATH LENGTH:	7.6 cm	SCALE EXPANSION:	0.96

VAPOR PHASE

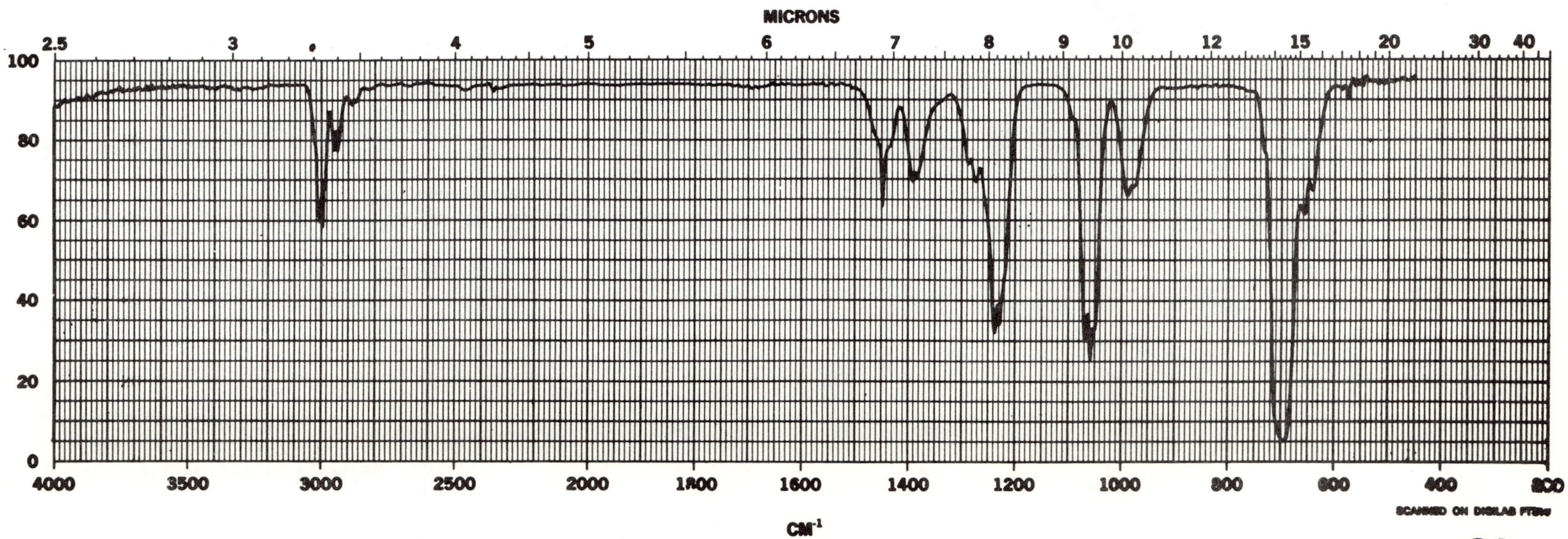

1,2–DICHLOROETHANE
ETHYLENE DICHLORIDE
GLYCOL DICHLORIDE

$C_2H_4Cl_2$

MOLECULAR WEIGHT: 98.96

SOURCE OF SAMPLE: Aldrich Chemical Company, Inc.
Milwaukee, Wisconsin

$Cl-CH_2CH_2-Cl$

CAS REGISTRY NO.: 107-06-2

NIOSH NUMBER: KI05250

CAPILLARY CELL: Neat

CONDENSED PHASE

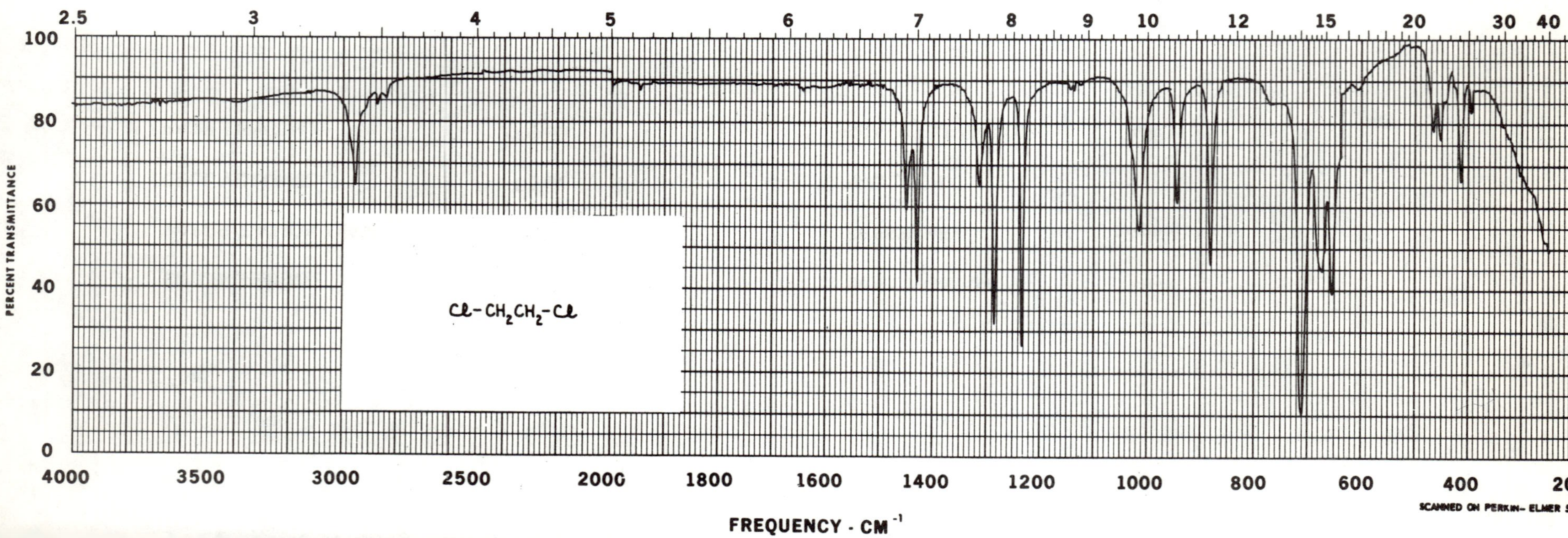

BOILING POINT:	81-85°C	INJECTION AMT.:	1.0 μl
VAPORIZATION TEMP.:	200°C	PRESSURE:	Atmospheric
CELL VOLUME:	7.5 cc	MAX./MIN.:	—
PATH LENGTH:	7.6 cm	SCALE EXPANSION:	0.93

VAPOR PHASE

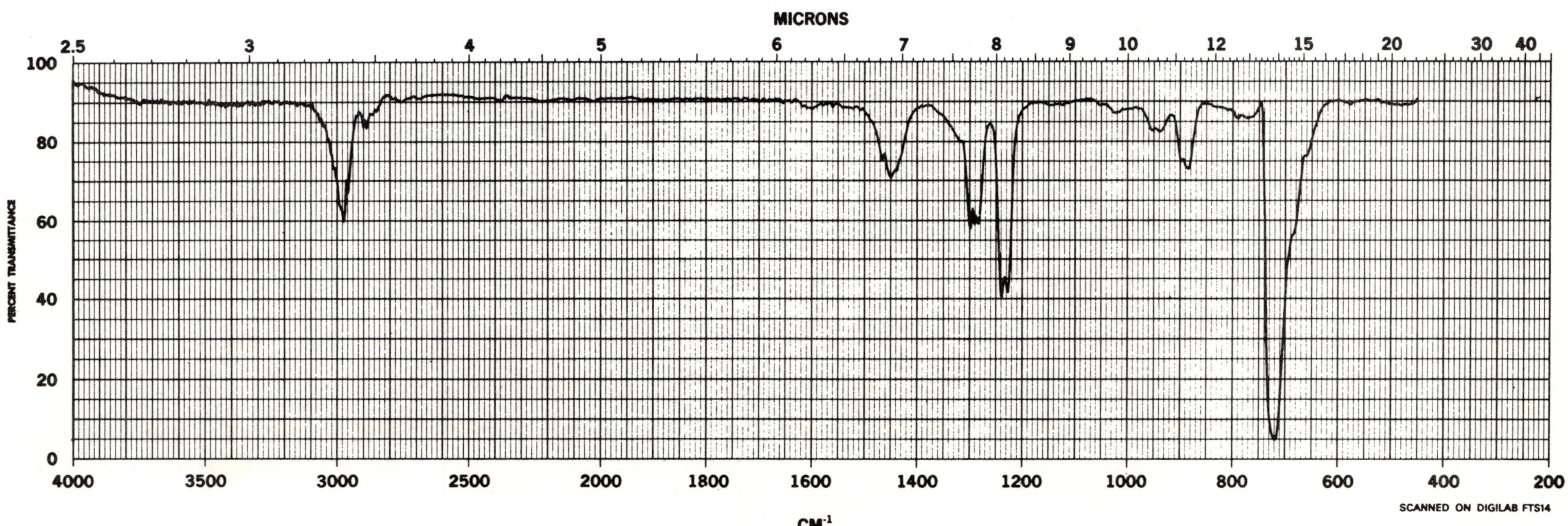

SCANNED ON DIGILAB FTS14

HEXACHLOROETHANE
ETHANE HEXACHLORIDE
PERCHLOROETHANE
PHENOHEP

C_2Cl_6

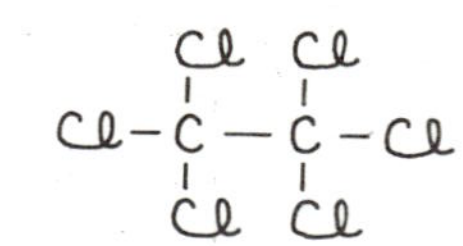

MOLECULAR WEIGHT: 236.74

SOURCE OF SAMPLE: Aldrich Chemical Company, Inc.
Milwaukee, Wisconsin

CAS REGISTRY NO.: 67-72-1

NIOSH NUMBER: KI40250

KBr Wafer

CONDENSED PHASE

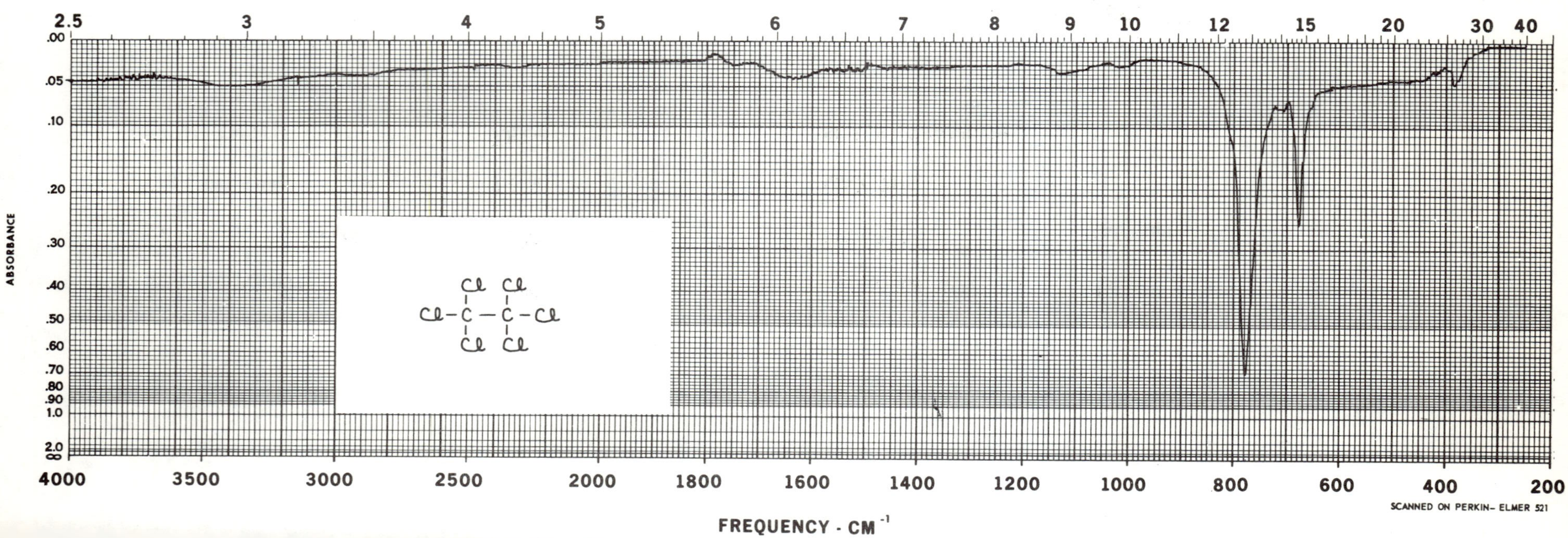

MELTING POINT:	185-186°C	INJECTION AMT.:	–
VAPORIZATION TEMP.:	240°C	PRESSURE:	Atmospheric
CELL VOLUME:	7.5 cc	MAX./MIN.:	99.8/9.1
PATH LENGTH:	7.6 cm	SCALE EXPANSION:	1.01

VAPOR PHASE

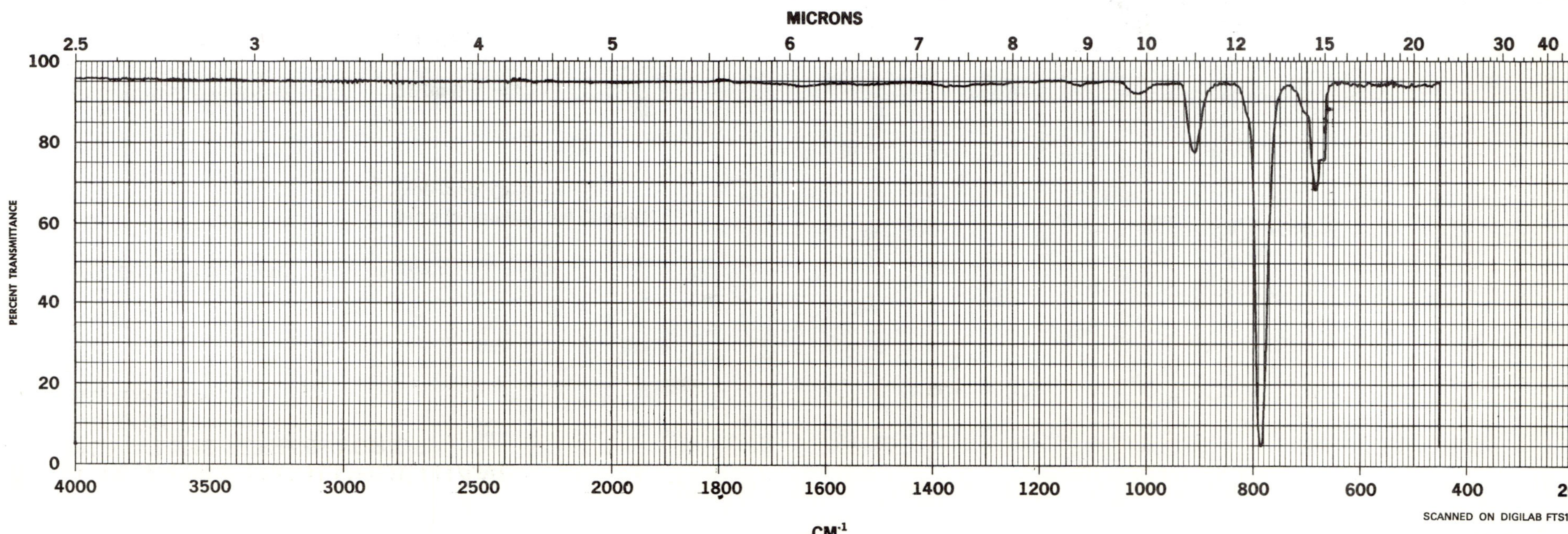

SCANNED ON DIGILAB FTS14

1,1,2,2–TETRACHLOROETHANE
ACETYLENE TETRACHLORIDE
CELLON
BONOFORM

$C_2H_2Cl_4$

MOLECULAR WEIGHT: 167.85

SOURCE OF SAMPLE: MCB Manufacturing Chemists
Norwood, Ohio

$Cl_2CHCHCl_2$

CAS REGISTRY NO.: 79-34-5

NIOSH NUMBER: KI85750

CAPILLARY CELL: Neat

CONDENSED PHASE

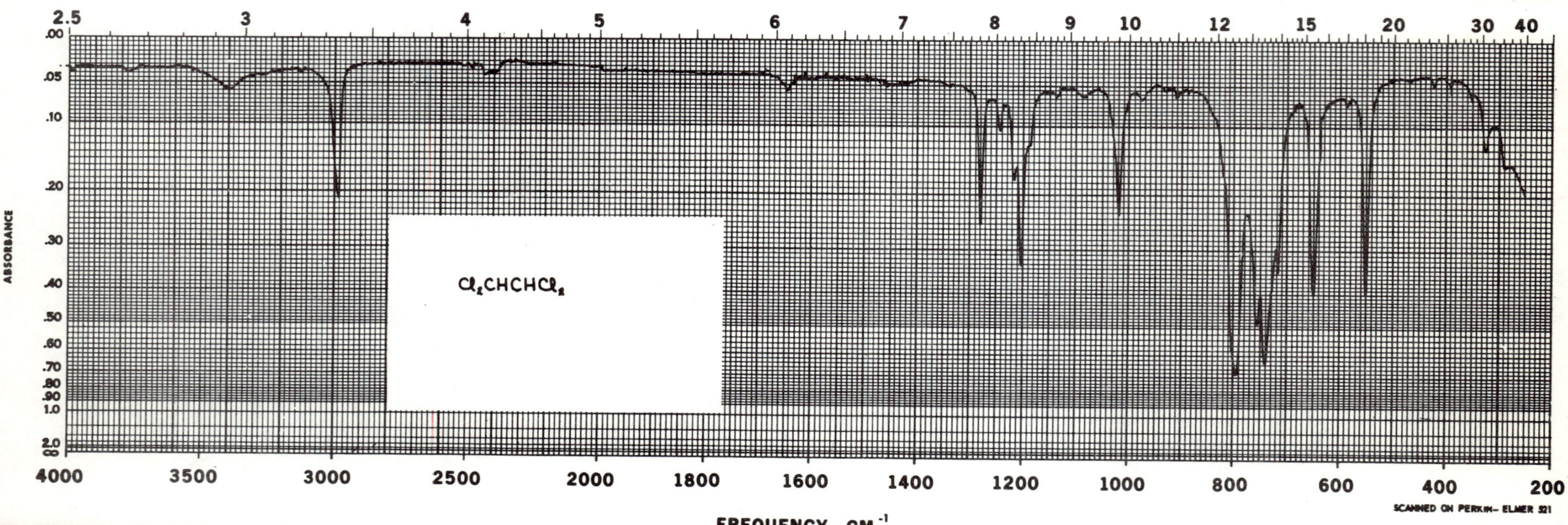

BOILING POINT:	144-146°C	INJECTION AMT.:	1.0 µl
VAPORIZATION TEMP.:	200°C	PRESSURE:	Atmospheric
CELL VOLUME:	7.5 cc	MAX./MIN.:	—
PATH LENGTH:	7.6 cm	SCALE EXPANSION:	1.03

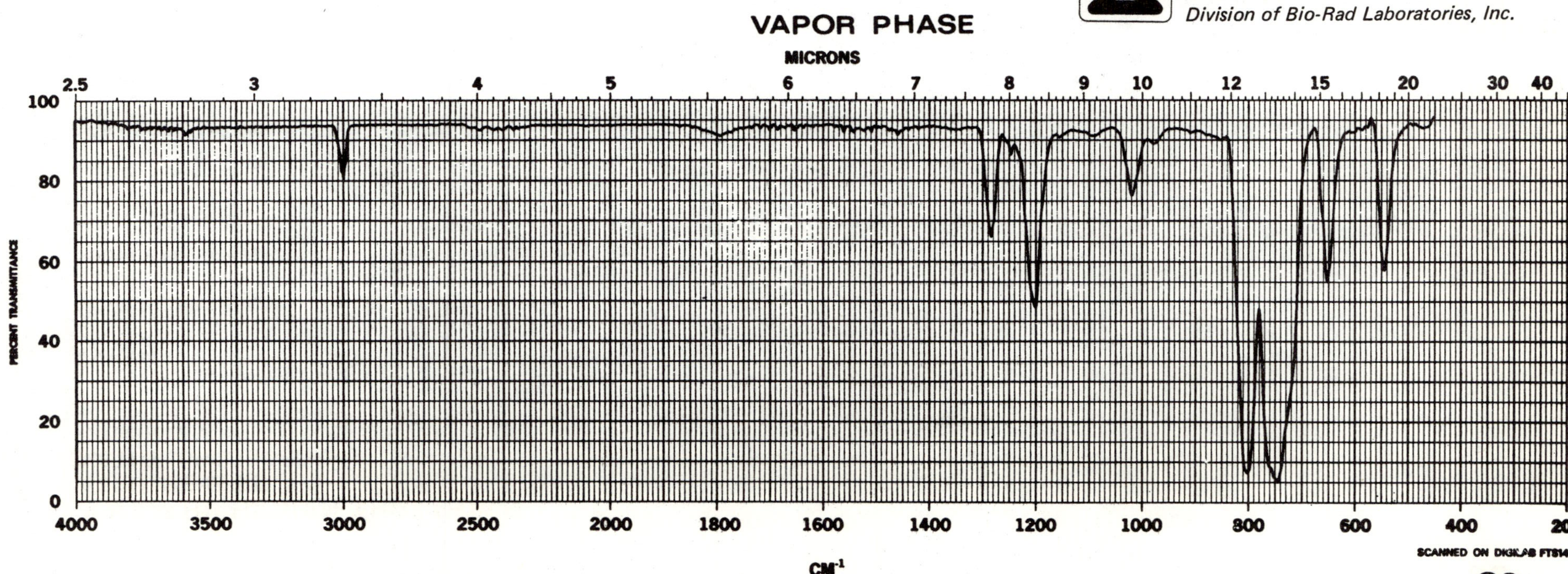

1,1,1–TRICHLOROETHANE
AEROTHENE TT
METHYLCHLOROFORM
CHLORTEN

$C_2H_3Cl_3$

MOLECULAR WEIGHT: 133.41

SOURCE OF SAMPLE: MCB Manufacturing Chemists
Norwood, Ohio

CAS REGISTRY NO.: 71-55-6

NIOSH NUMBER: KJ29750

CAPILLARY CELL: 0.015 mm

CH_3-CCl_3

CONDENSED PHASE

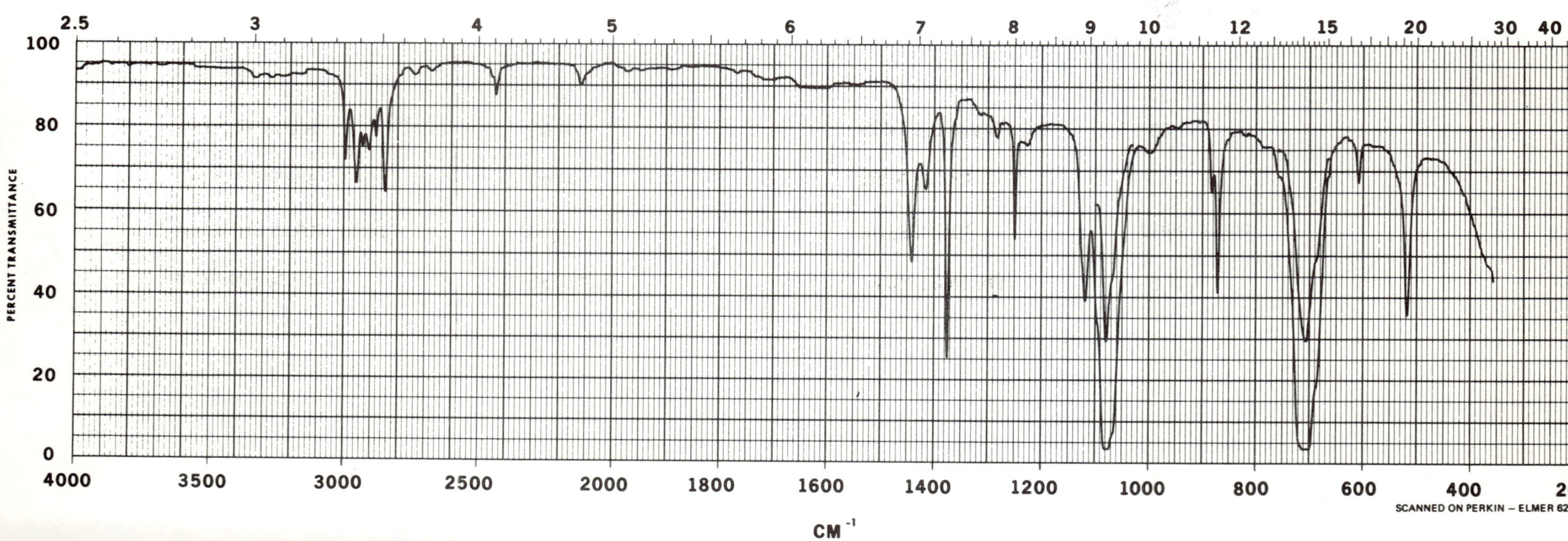

BOILING POINT:	74.1°C	INJECTION AMT.:	0.2 µl
VAPORIZATION TEMP.:	200°C	PRESSURE:	Atmospheric
CELL VOLUME:	7.5 cc	MAX./MIN.:	—
PATH LENGTH:	7.6 cm	SCALE EXPANSION:	0.97

VAPOR PHASE

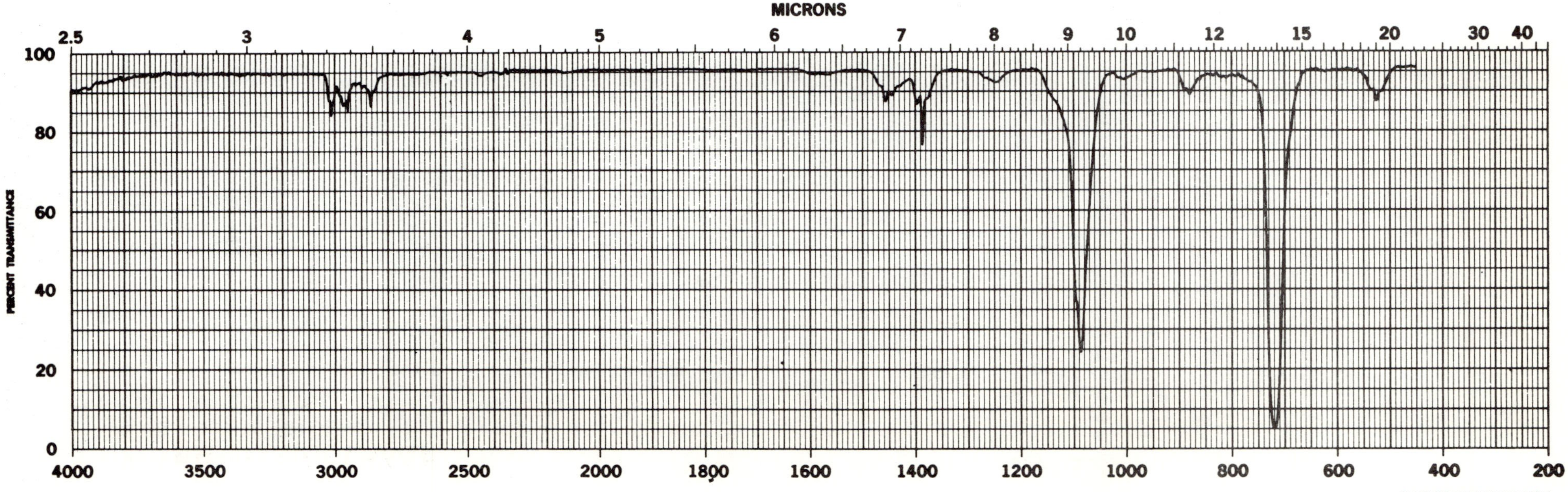

1,1,2–TRICHLOROETHANE
ETHANE TRICHLORIDE
VINYL TRICHLORIDE

$C_2H_3Cl_3$

MOLECULAR WEIGHT: 133.41

SOURCE OF SAMPLE: Fluka AG
Buchs, Switzerland

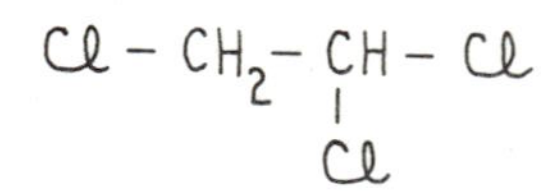

CAS REGISTRY NO.: 79-00-5

NIOSH NUMBER: KJ31500

CAPILLARY CELL: 0.015 mm

CONDENSED PHASE

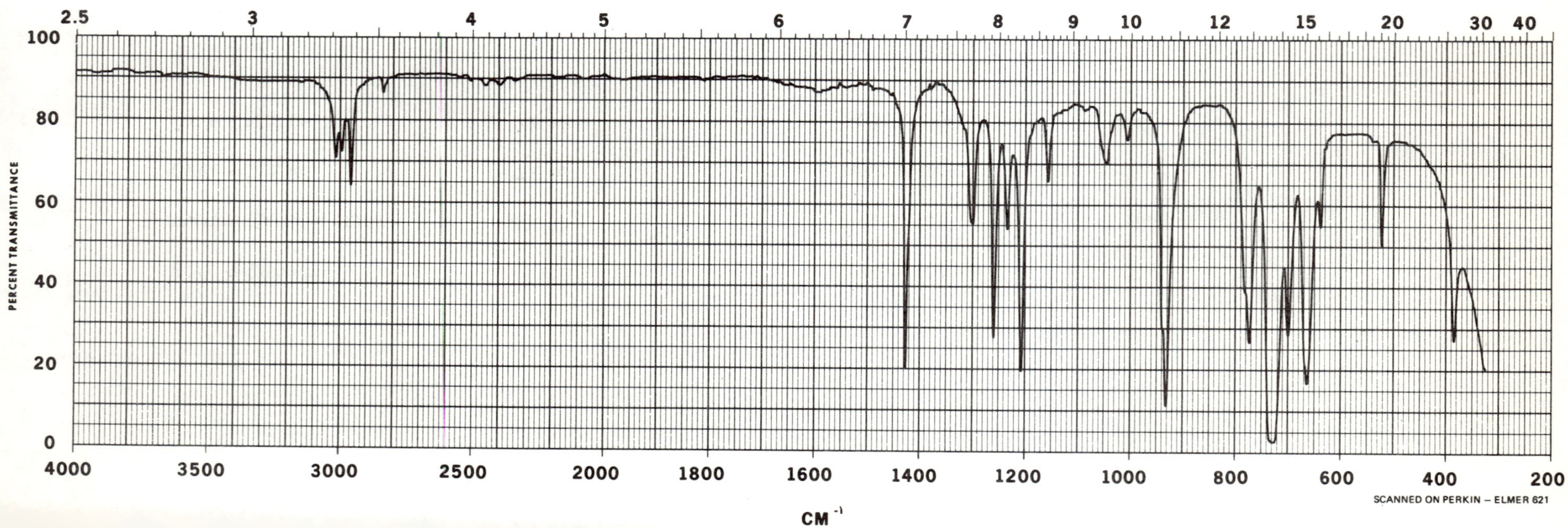

BOILING POINT:	110-115°C	INJECTION AMT.:	0.8 µl
VAPORIZATION TEMP.:	200°C	PRESSURE:	Atmospheric
CELL VOLUME:	7.5 cc	MAX./MIN.:	—
PATH LENGTH:	7.6 cm	SCALE EXPANSION:	0.95

VAPOR PHASE

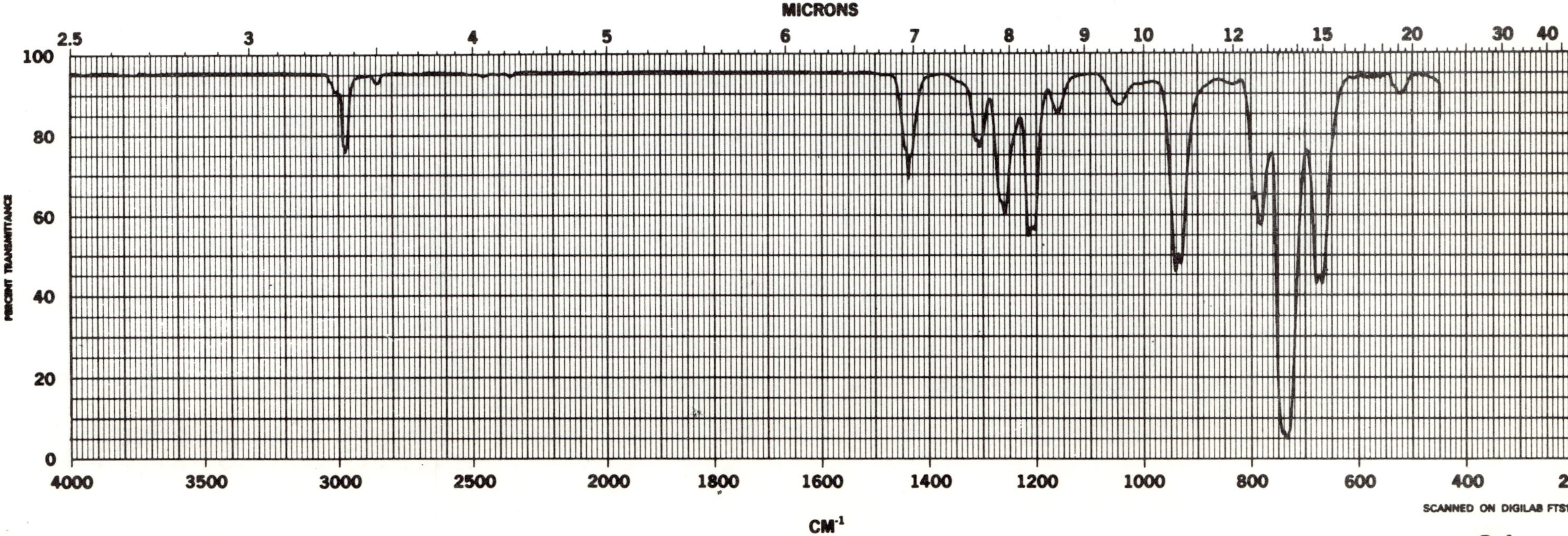

SCANNED ON DIGILAB FTS14

BIS(2–CHLOROETHYL) ETHER
CHLOREX
DCEE

$C_4H_8Cl_2O$

MOLECULAR WEIGHT:	143.01
SOURCE OF SAMPLE:	Fluka AG Buchs, Switzerland
CAS REGISTRY NO.:	111-44-4
NIOSH NUMBER:	KN08750
CAPILLARY CELL:	Neat

$Cl-CH_2CH_2-O-CH_2CH_2-Cl$

CONDENSED PHASE

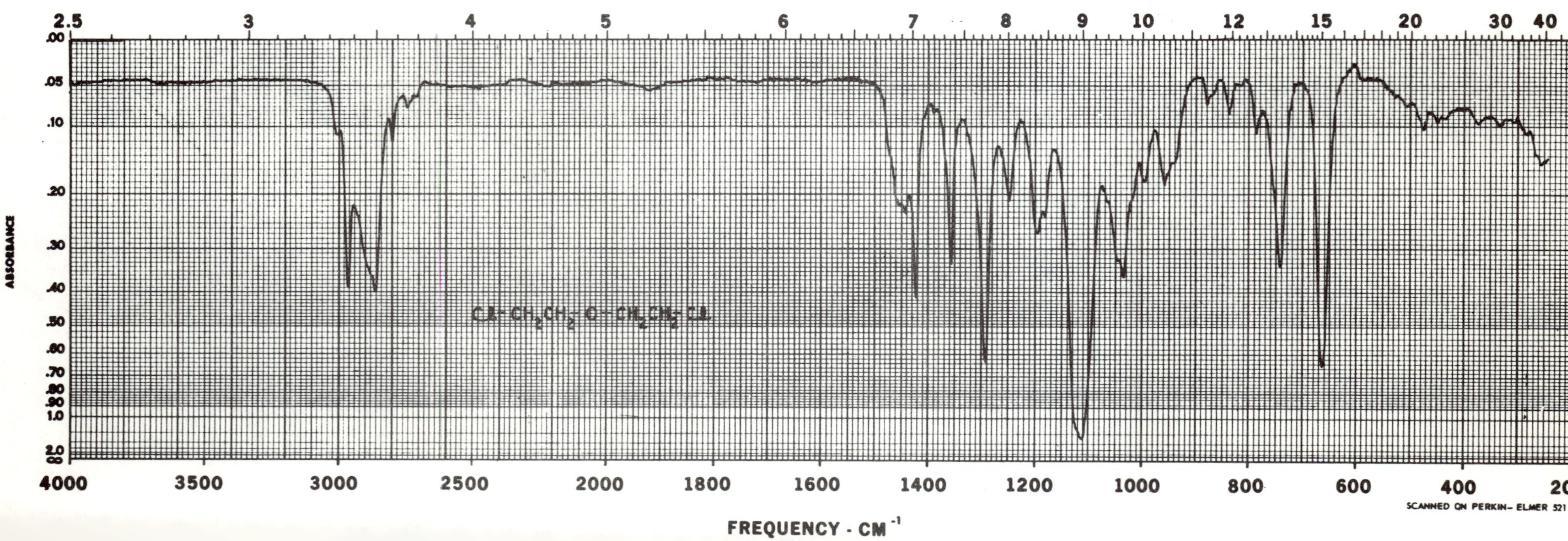

BOILING POINT:	178°C	INJECTION AMT.:	0.5 µl
VAPORIZATION TEMP.:	200°C	PRESSURE:	Atmospheric
CELL VOLUME:	7.5 cc	MAX./MIN.:	–
PATH LENGTH:	7.6 cm	SCALE EXPANSION:	1.14

VAPOR PHASE

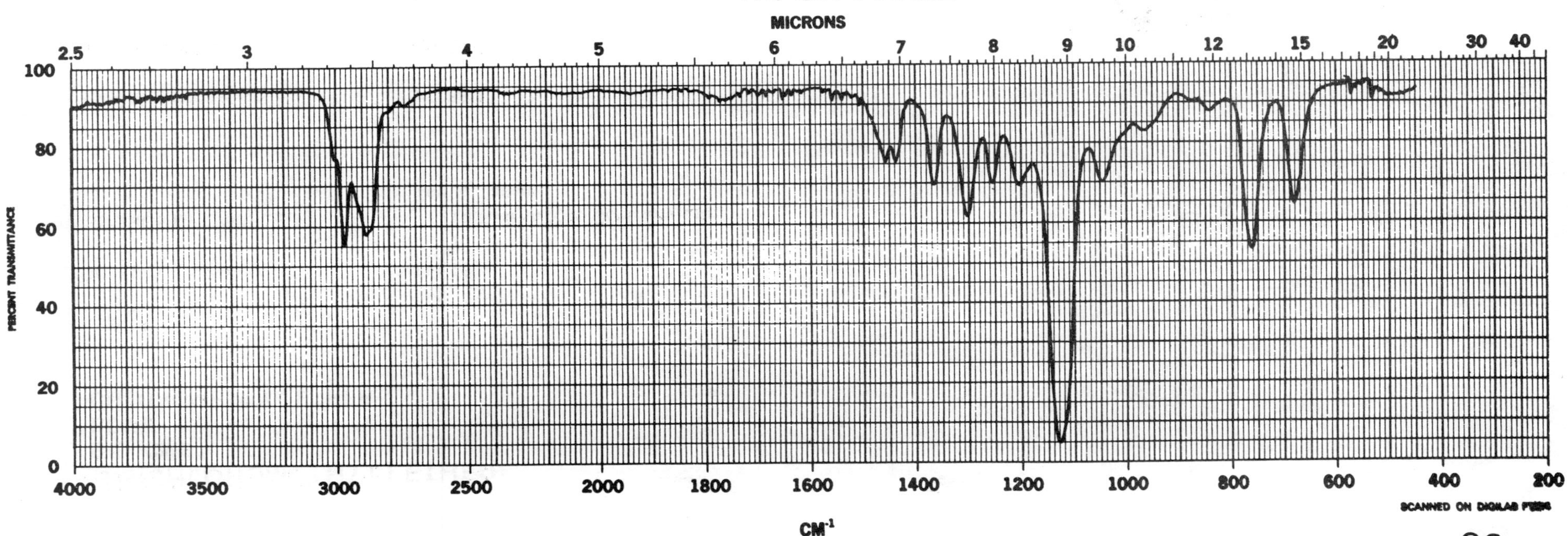

BIS(CHLOROMETHYL) ETHER

BIS–CME
BCME
CHLOROMETHYL ETHER

$C_2H_4Cl_2O$

MOLECULAR WEIGHT:	114.96
SOURCE OF SAMPLE:	Tokyo Kasei Kogyo Company Ltd. Tokyo, Japan
CAS REGISTRY NO.:	542-88-1
NIOSH NUMBER:	KN15750
CAPILLARY CELL:	0.015 mm

$Cl-CH_2-O-CH_2-Cl$

CONDENSED PHASE

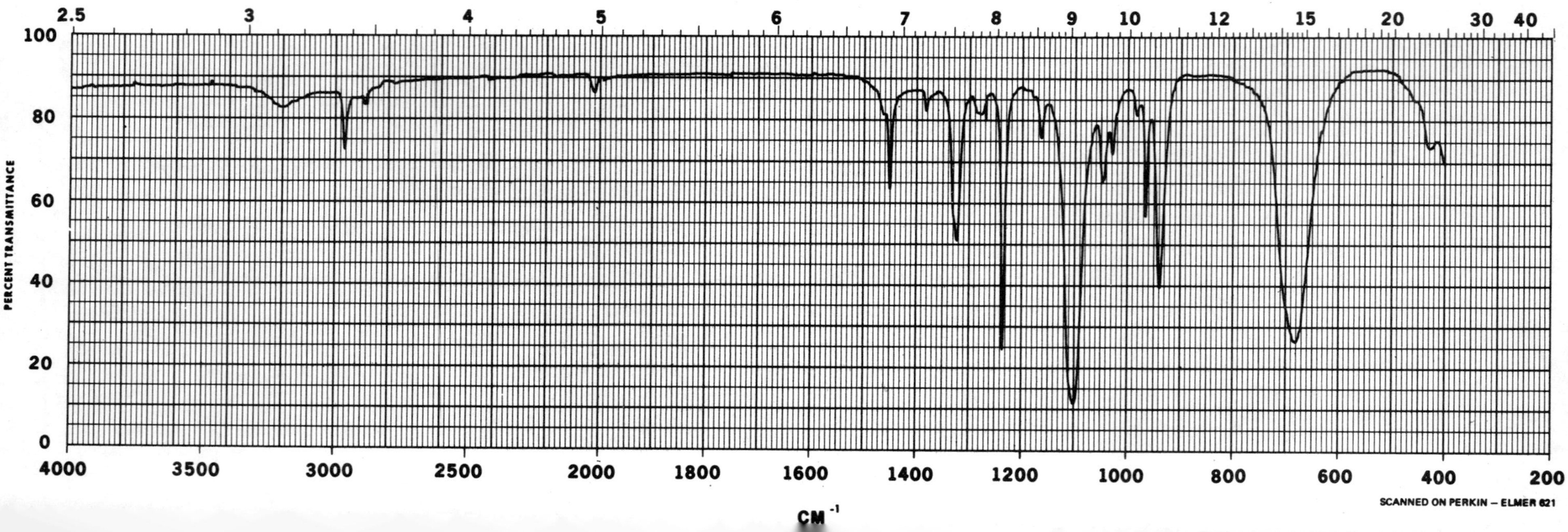

BOILING POINT:	58-60°C	INJECTION AMT.:	0.1 µl
VAPORIZATION TEMP.:	200°C	PRESSURE:	Atmospheric
CELL VOLUME:	7.5 cc	MAX./MIN.:	—
PATH LENGTH:	7.6 cm	SCALE EXPANSION:	0.93

VAPOR PHASE

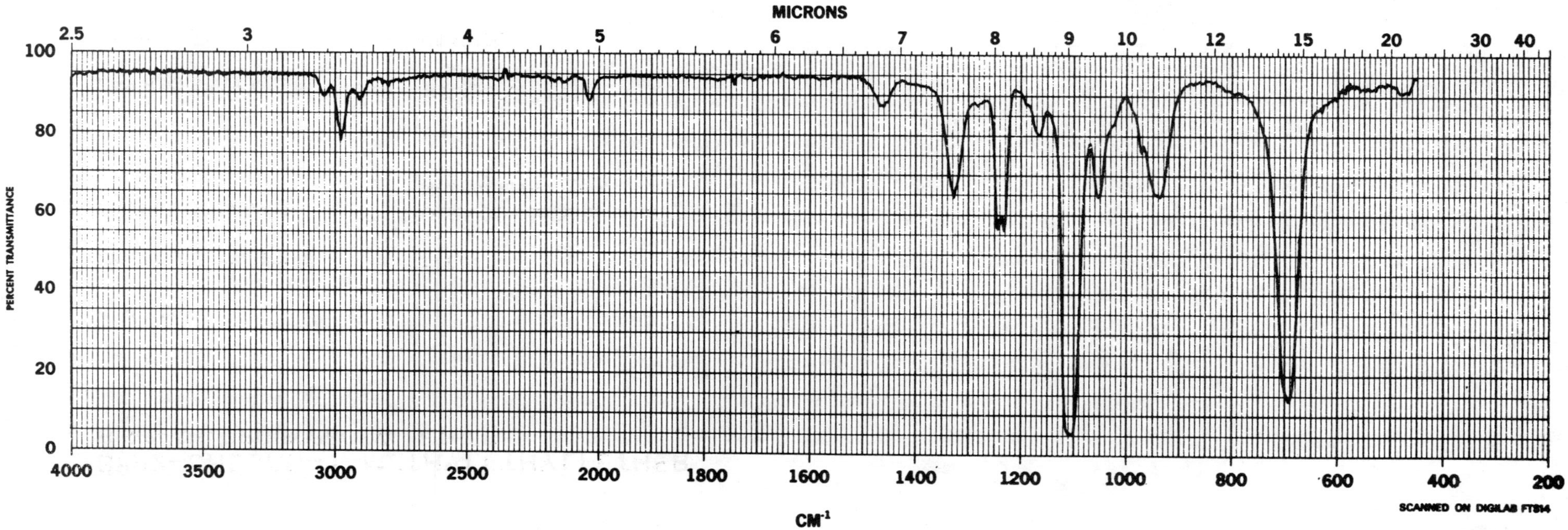

SCANNED ON DIGILAB FTS14

BIS(2–CHLORO–1–METHYLETHYL) ETHER
BIS(2–CHLOROISOPROPYL) ETHER

$C_6H_{12}Cl_2O$

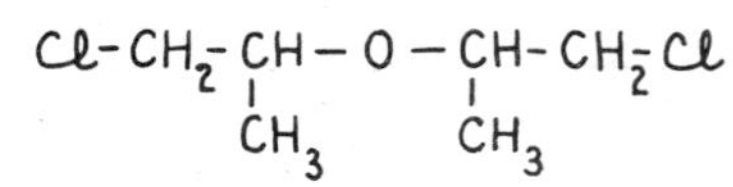

MOLECULAR WEIGHT:	171.07
SOURCE OF SAMPLE:	Union Carbide Corporation New York, New York
CAS REGISTRY NO.:	108-60-1
NIOSH NUMBER:	KN17500
CAPILLARY CELL:	Neat

CONDENSED PHASE

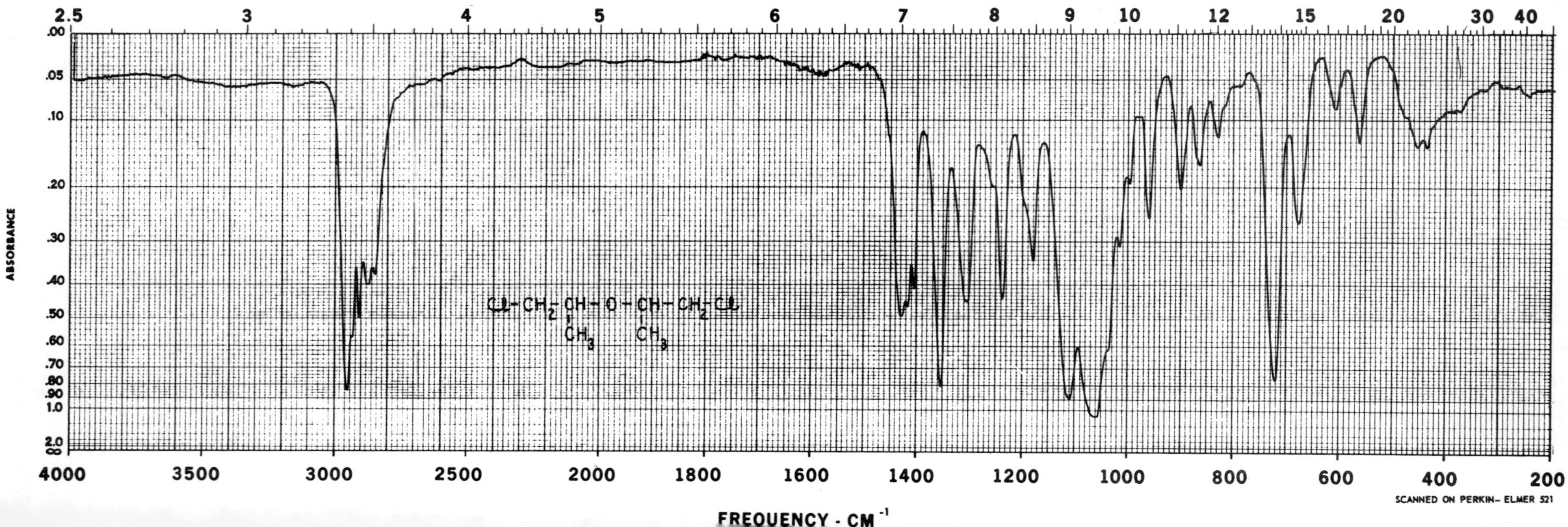

BOILING POINT:	187.1^{o}C	INJECTION AMT.:	—
VAPORIZATION TEMP.:	160°C	PRESSURE:	Atmospheric
CELL VOLUME:	7.5 cc	MAX./MIN.:	100.1/27.6
PATH LENGTH:	7.6 cm	SCALE EXPANSION:	1.27

VAPOR PHASE

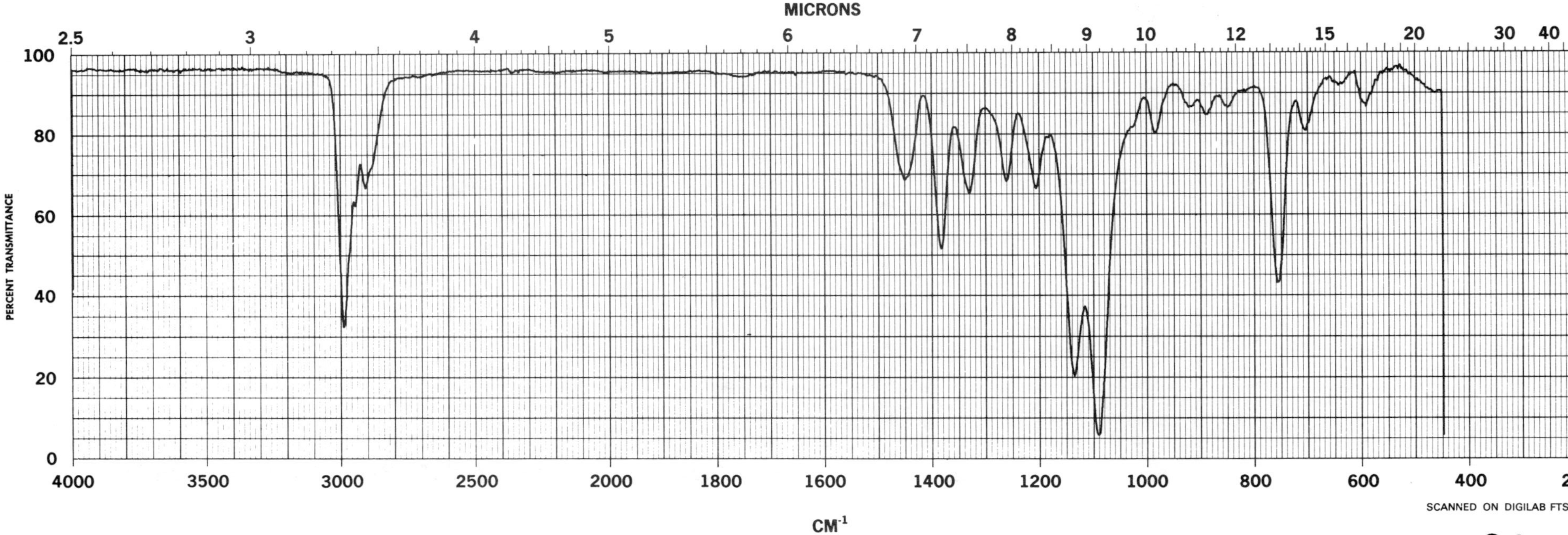

SCANNED ON DIGILAB FTS14

p–BROMOPHENYL PHENYL ETHER

$C_{12}H_9BrO$

MOLECULAR WEIGHT: 249.11

SOURCE OF SAMPLE: MCB Manufacturing Chemists
Norwood, Ohio

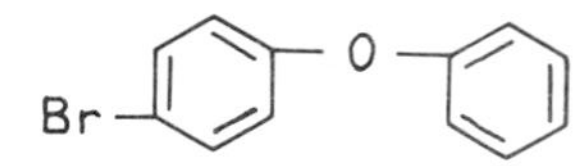

CAS REGISTRY NO.: 101-55-3

NIOSH NUMBER: –

CAPILLARY CELL: Neat

CONDENSED PHASE

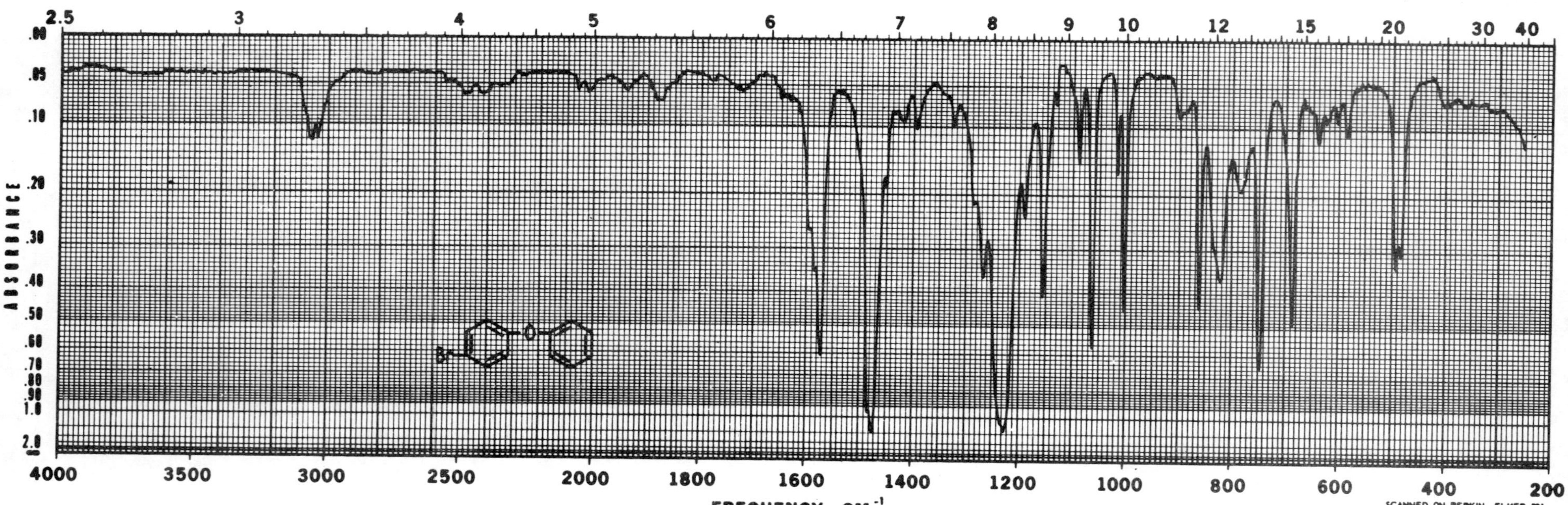

BOILING POINT:	142-145°C/5 mm	INJECTION AMT.:	0.5 μl
VAPORIZATION TEMP.:	240°C	PRESSURE:	Atmospheric
CELL VOLUME:	7.5 cc	MAX./MIN.:	—
PATH LENGTH:	7.6 cm	SCALE EXPANSION:	0.85

VAPOR PHASE

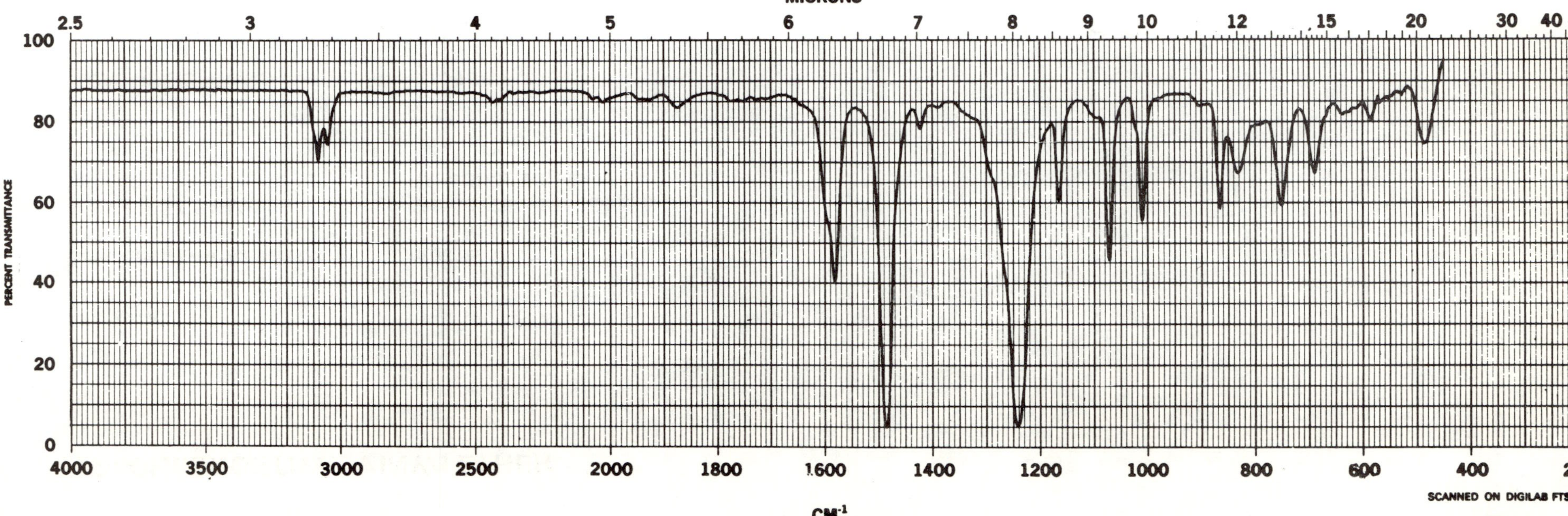

SCANNED ON DIGILAB FTS14

2—CHLOROETHYL VINYL ETHER

C_4H_7ClO

MOLECULAR WEIGHT: 106.55

SOURCE OF SAMPLE: Aldrich Chemical Company, Inc.
Milwaukee, Wisconsin

$CH_2=CH-O-CH_2CH_2Cl$

CAS REGISTRY NO.: 110-75-8

NIOSH NUMBER: KN63000

GAS CELL: 10 cm

A GAS AT 25°C

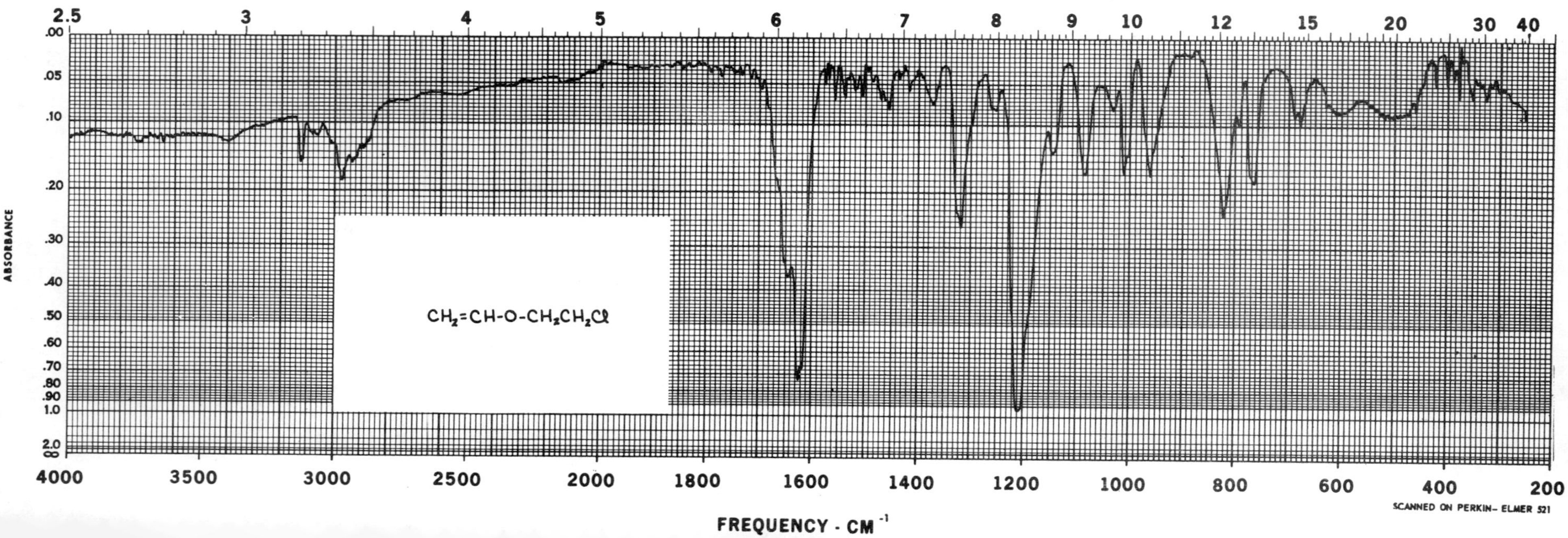

BOILING POINT:	109°C	INJECTION AMT.:	0.3 μl
VAPORIZATION TEMP.:	200°C	PRESSURE:	Atmospheric
CELL VOLUME:	7.5 cc	MAX./MIN.:	—
PATH LENGTH:	7.6 cm	SCALE EXPANSION:	1.15

VAPOR PHASE

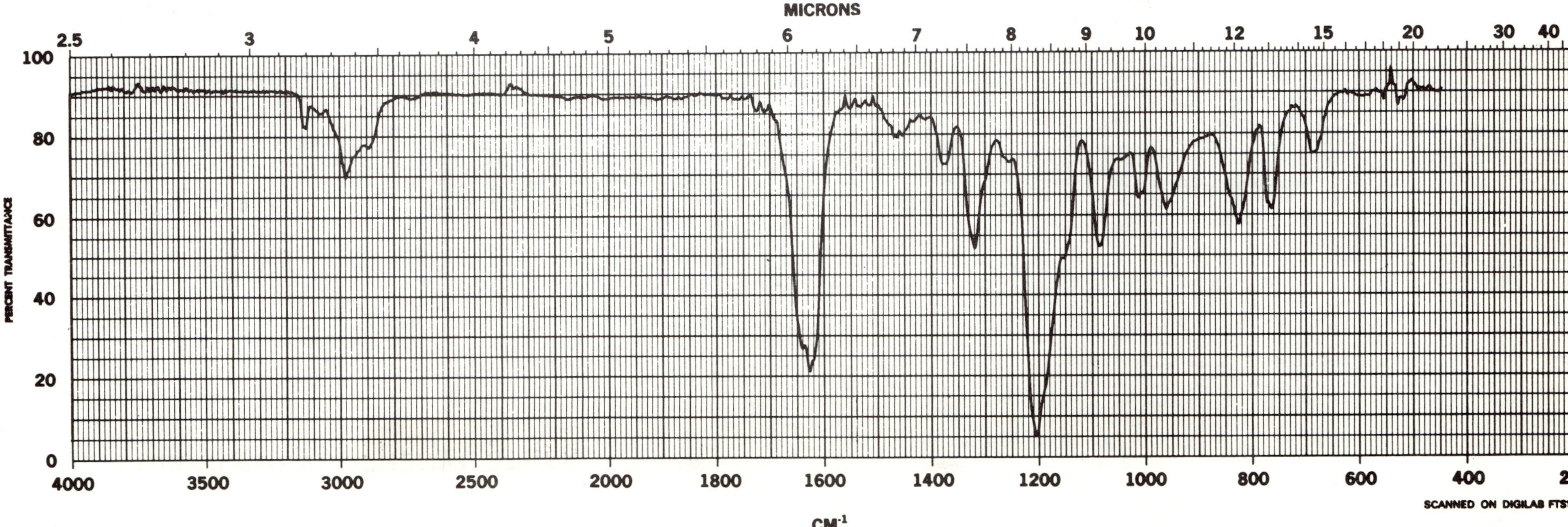

p–CHLOROPHENYL PHENYL ETHER

$C_{12}H_9ClO$

MOLECULAR WEIGHT: 204.66

SOURCE OF SAMPLE: Chem Service Company
West Chester, Pennsylvania

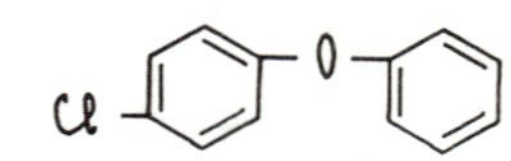

CAS REGISTRY NO.: 7005-72-3

NIOSH NUMBER: –

Between Salts

CONDENSED PHASE

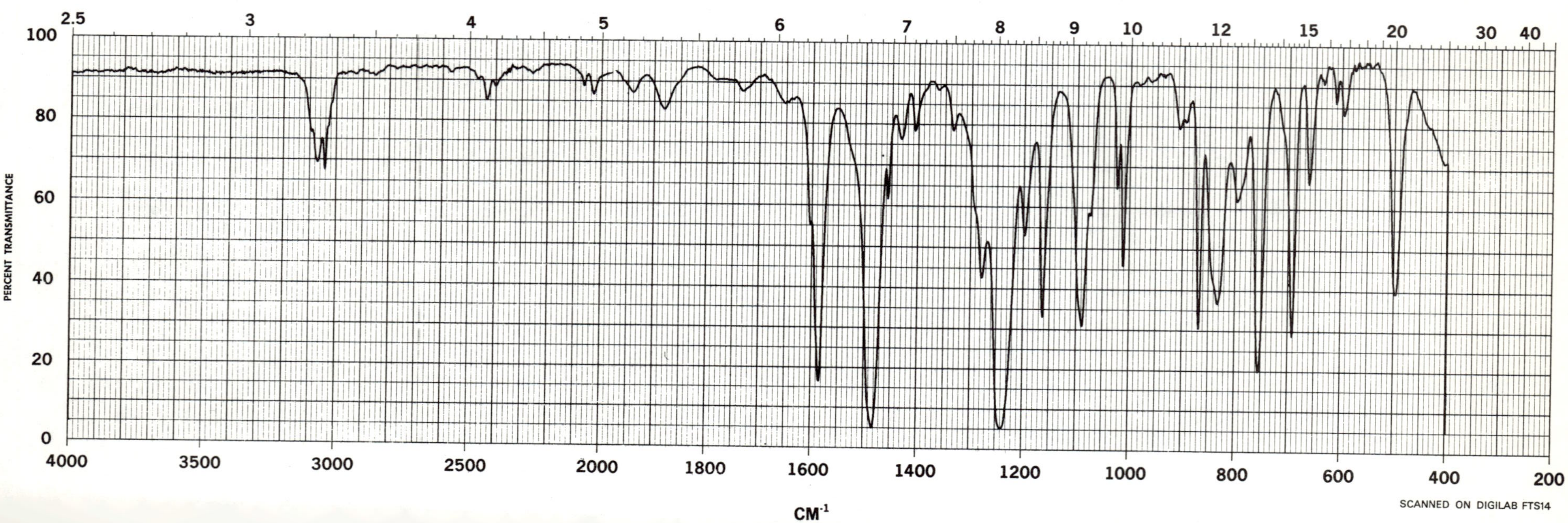

BOILING POINT:	—	INJECTION AMT.:	—
VAPORIZATION TEMP.:	200°C	PRESSURE:	Atmospheric
CELL VOLUME:	7.5 cc	MAX./MIN.:	90.1/24.9
PATH LENGTH:	7.6 cm	SCALE EXPANSION:	1.41

VAPOR PHASE

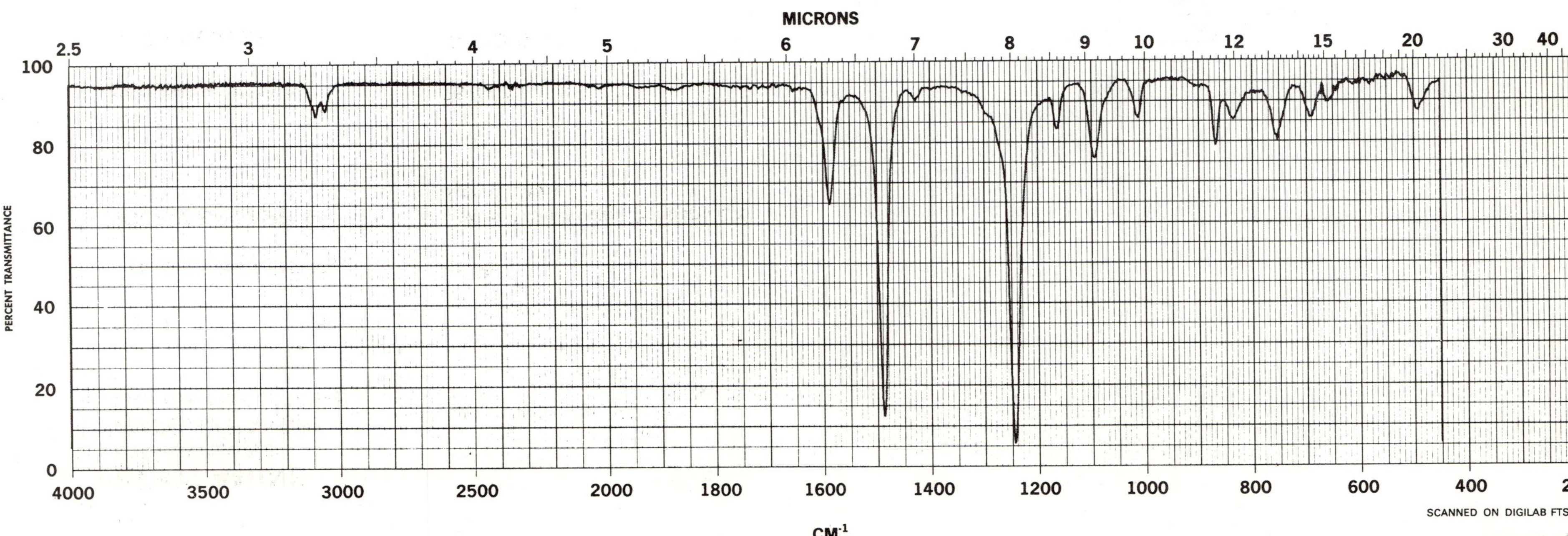

SCANNED ON DIGILAB FTS14

ETHYLAMINE
ETHANAMINE
AMINOETHANE
MONOETHYLAMINE

C_2H_7N

MOLECULAR WEIGHT:	45.08
SOURCE OF SAMPLE:	Tokyo Kasei Kogyo Company Ltd. Tokyo, Japan
CAS REGISTRY NO.:	75-04-7
NIOSH NUMBER:	KH21000
Path Length:	10 cm
Complete Scan:	25 mm Hg

$CH_3CH_2{-}NH_2$

A GAS AT 25°C

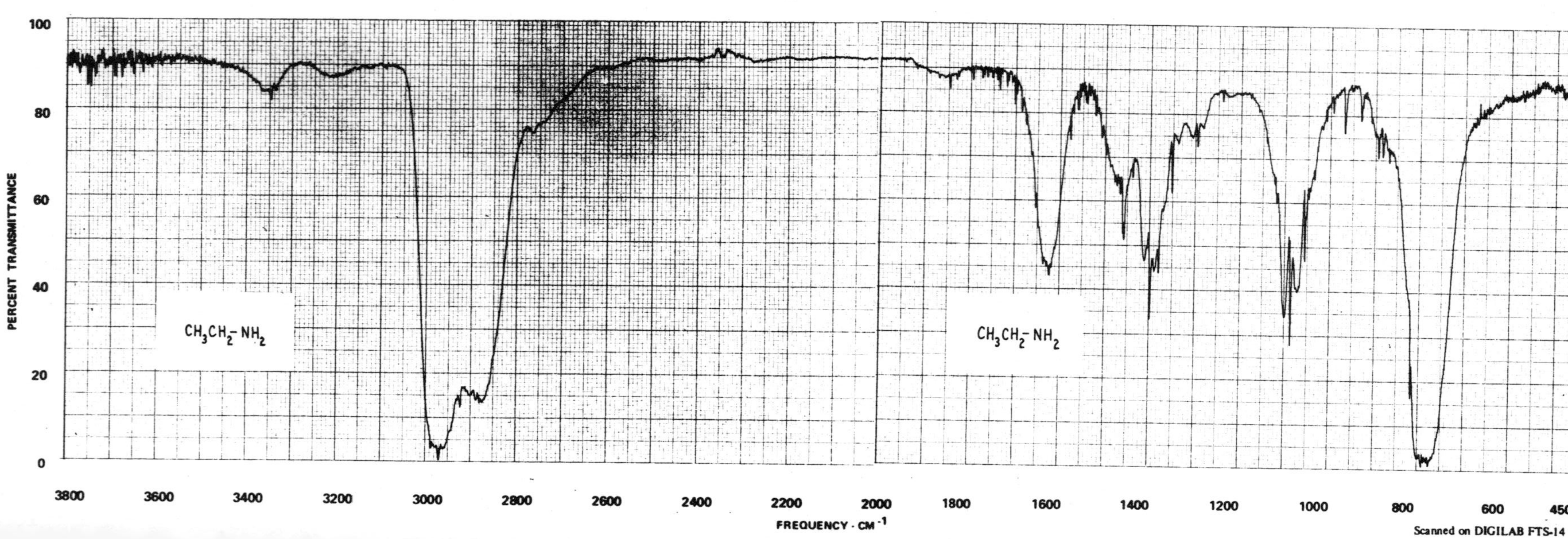

BOILING POINT:	16.6°C	INJECTION AMT.:	–
VAPORIZATION TEMP.:	200°C	PRESSURE:	Atmospheric
CELL VOLUME:	7.5 cc	MAX./MIN.:	100.2/3.2
PATH LENGTH:	7.6 cm	SCALE EXPANSION:	0.95

VAPOR PHASE

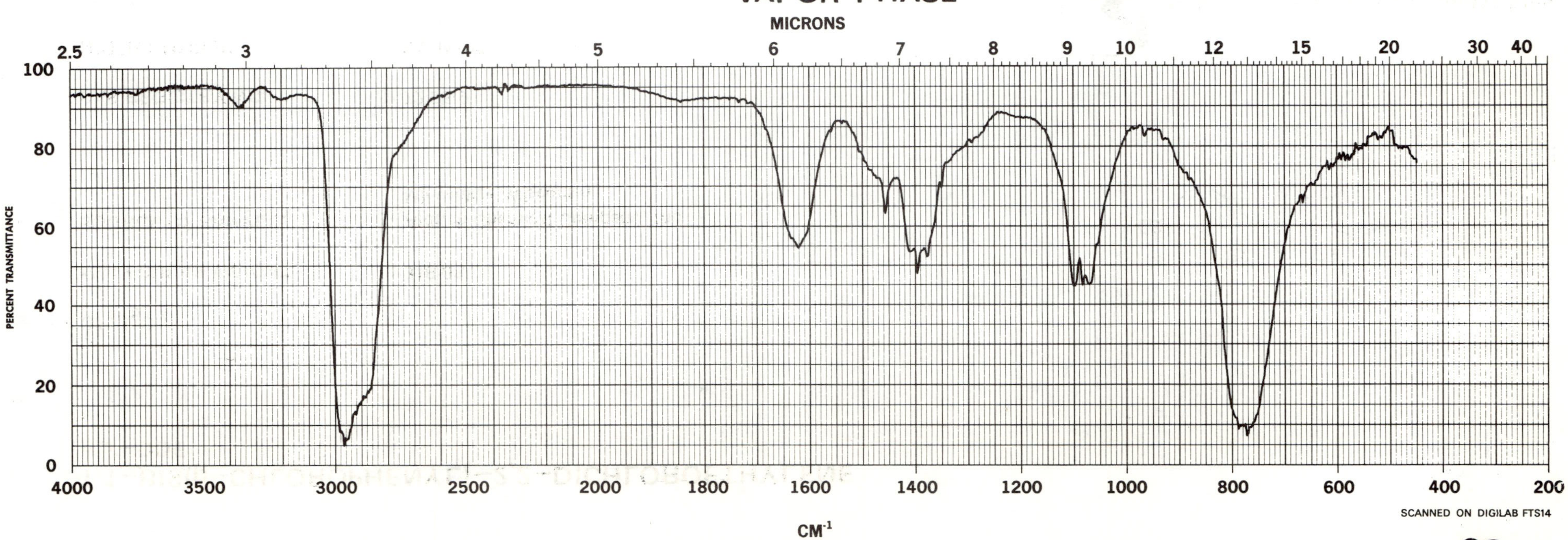

SCANNED ON DIGILAB FTS14

1,1–BIS(p–CHLOROPHENYL)–2,2–DICHLOROETHYLENE
p,p'–DDE

$C_{14}H_8Cl_4$

MOLECULAR WEIGHT: 318.03

SOURCE OF SAMPLE: Aldrich Chemical Company, Inc.
Milwaukee, Wisconsin

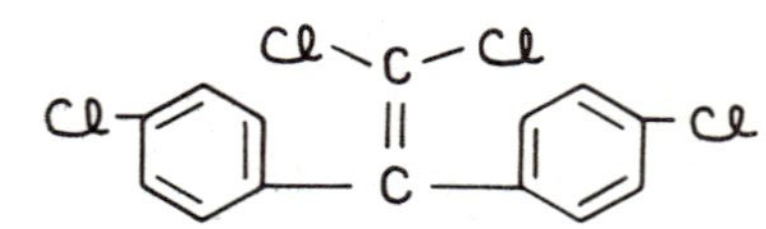

CAS REGISTRY NO.: 72-55-9

NIOSH NUMBER: KV94500

KBr Wafer

CONDENSED PHASE

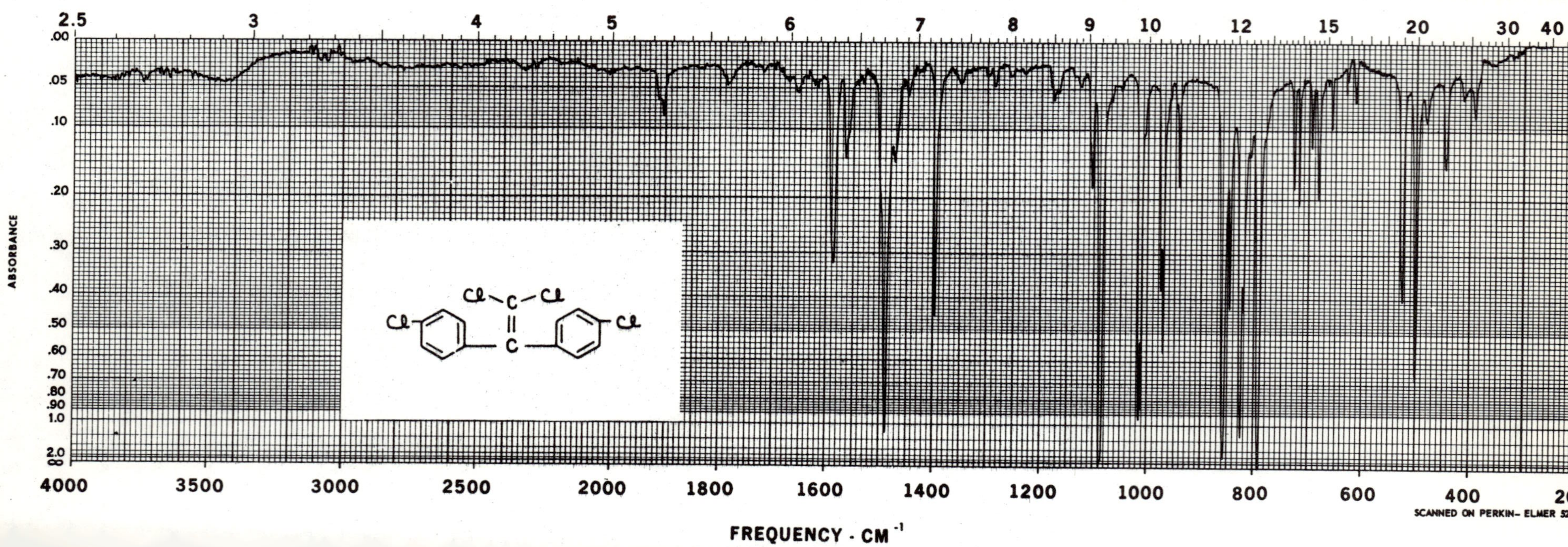

MELTING POINT:	88-90°C	INJECTION AMT.:	—
VAPORIZATION TEMP.:	280°C	PRESSURE:	Atmospheric
CELL VOLUME:	7.5 cc	MAX./MIN.:	103.2/6.7
PATH LENGTH:	7.6 cm	SCALE EXPANSION:	0.95

VAPOR PHASE

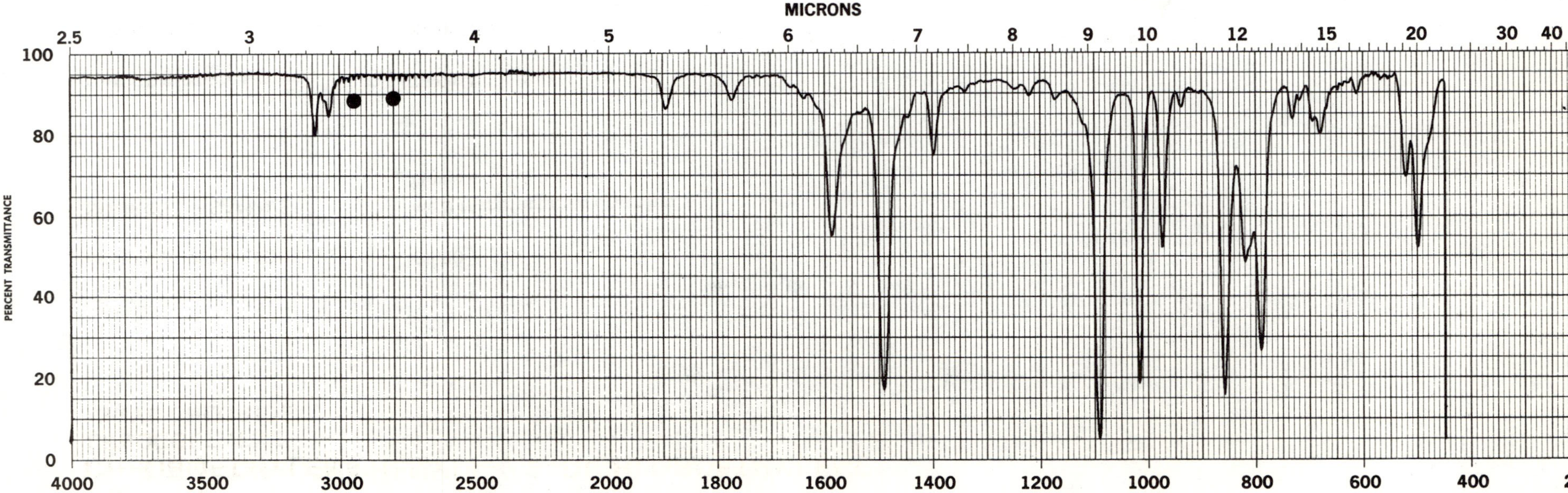

SCANNED ON DIGILAB FTS14

CHLOROETHYLENE
CHLOROETHENE
VINYL CHLORIDE

C_2H_3Cl

MOLECULAR WEIGHT: 62.50

SOURCE OF SAMPLE: MCB Manufacturing Chemists
Norwood, Ohio

$CH_2=CH-Cl$

CAS REGISTRY NO.: 75-01-4

NIOSH NUMBER: KU96250

GAS CELL: 10 cm
A = 350 mm vac.
B = 700 mm vac.

A GAS AT 25°C

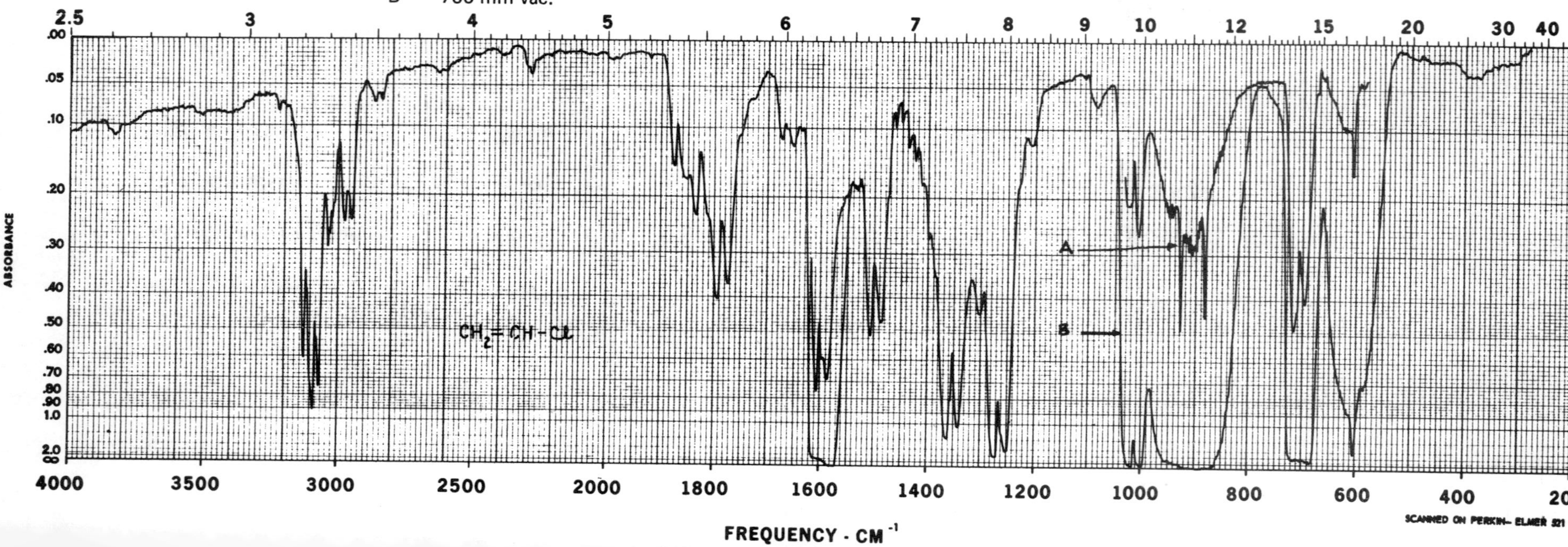

BOILING POINT:	−13.9°C	INJECTION AMT.:	400 μl
VAPORIZATION TEMP.:	200°C	PRESSURE:	Atmospheric
CELL VOLUME:	7.5 cc	MAX./MIN.:	99.8/54.2
PATH LENGTH:	7.6 cm	SCALE EXPANSION:	2.02

VAPOR PHASE

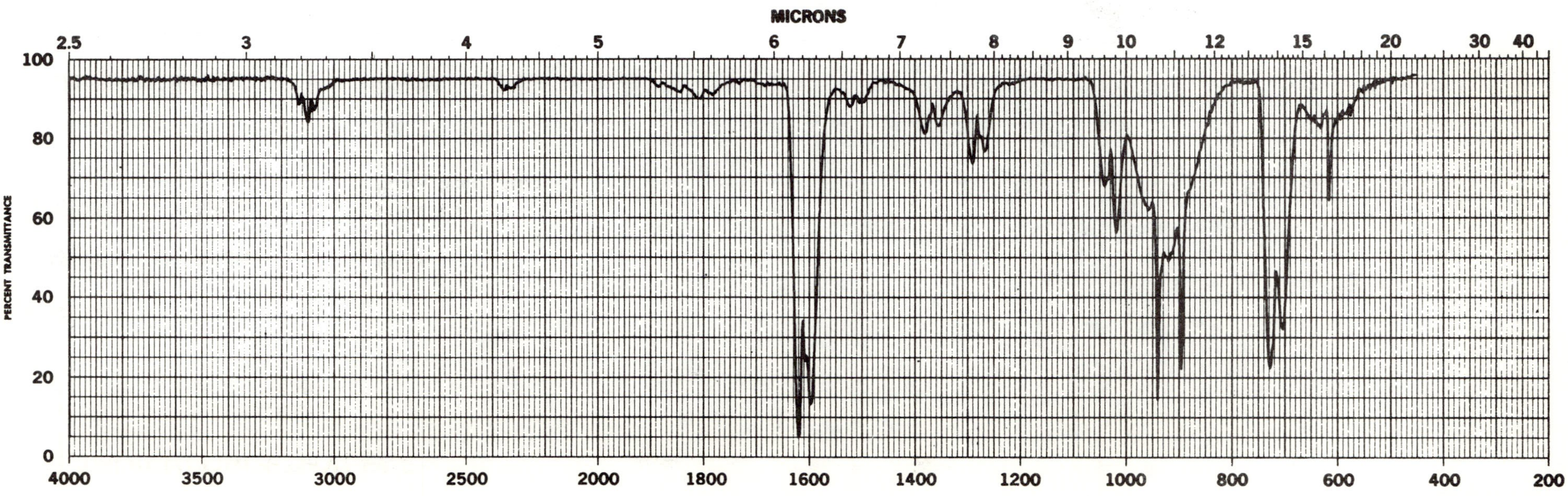

SCANNED ON DIGILAB FTS14

1,1–DICHLOROETHYLENE

1,1–DCE
VINYLIDENE CHLORIDE
1,1–DICHLOROETHENE

$C_2H_2Cl_2$

MOLECULAR WEIGHT: 96.94

SOURCE OF SAMPLE: Fluka AG
Buchs, Switzerland

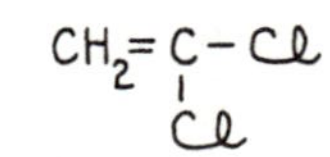

CAS REGISTRY NO.: 75-35-4

NIOSH NUMBER: KV92750

CAPILLARY CELL: 0.015 mm

CONDENSED PHASE

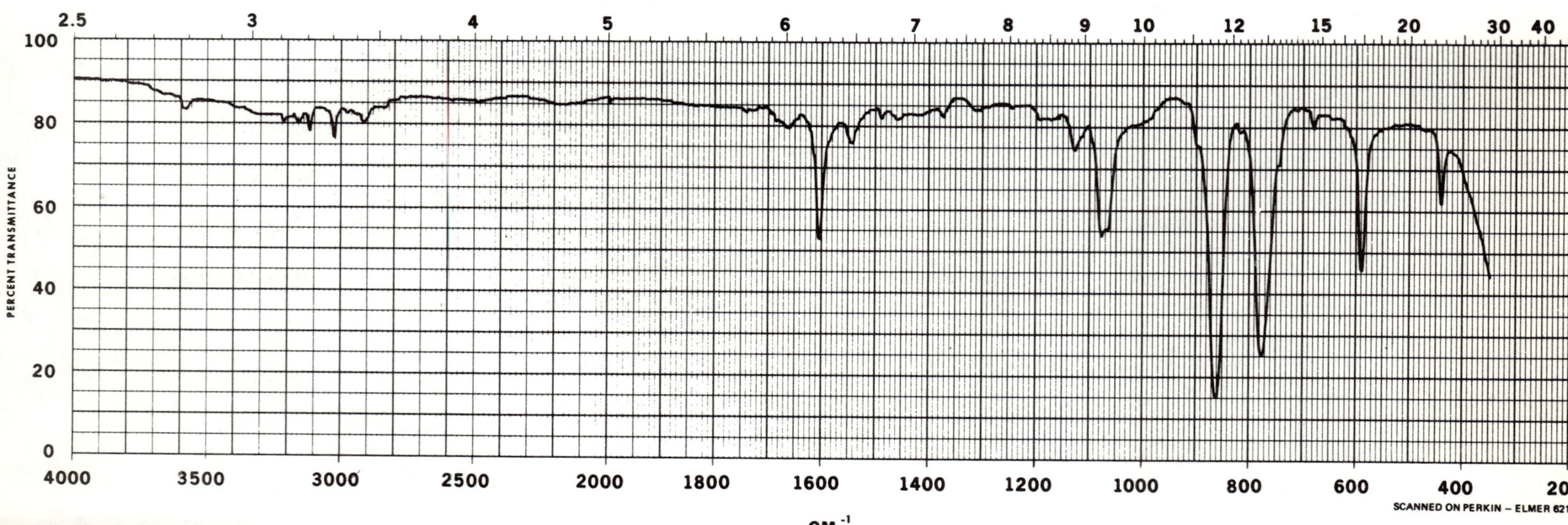

BOILING POINT:	30-32°C	INJECTION AMT.:	1.0 μl
VAPORIZATION TEMP.:	200°C	PRESSURE:	Atmospheric
CELL VOLUME:	7.5 cc	MAX./MIN.:	–
PATH LENGTH:	7.6 cm	SCALE EXPANSION:	0.95

VAPOR PHASE

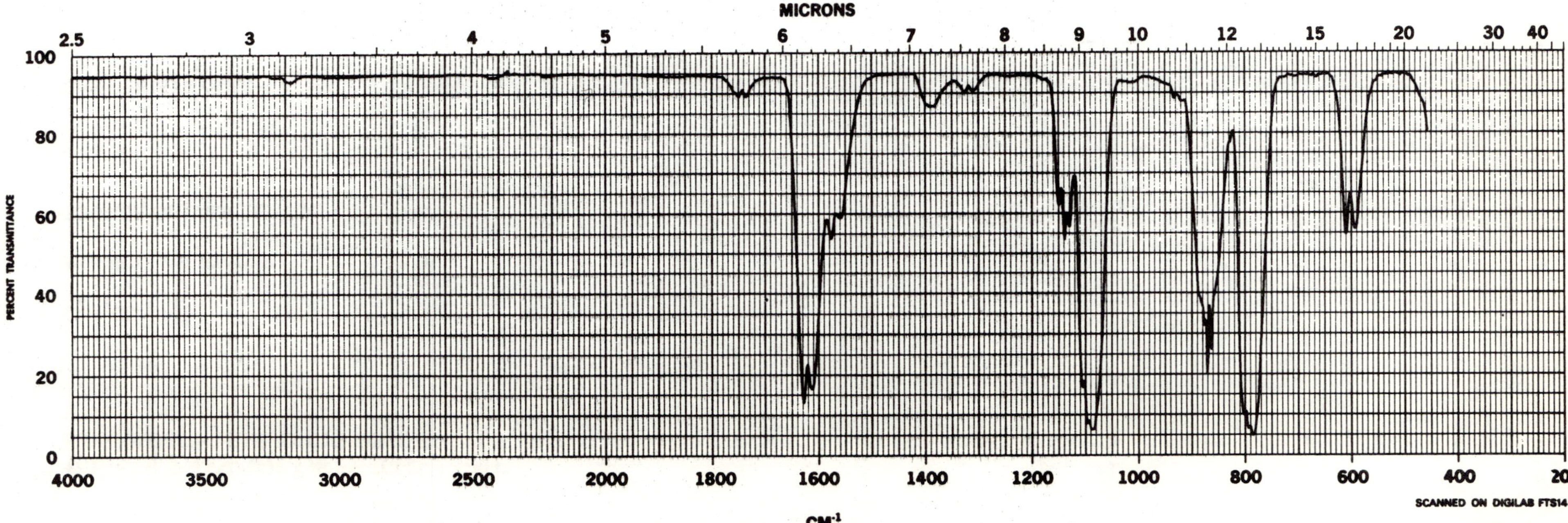

trans–1,2–DICHLOROETHYLENE
trans–ACETYLENE DICHLORIDE

$C_2H_2Cl_2$

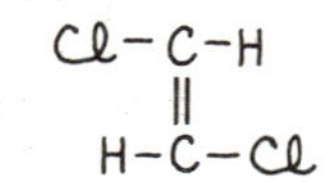

MOLECULAR WEIGHT:	96.94
SOURCE OF SAMPLE:	Fluka AG Buchs, Switzerland
CAS REGISTRY NO.:	156-60-5
NIOSH NUMBER:	KV94000
CAPILLARY CELL:	0.015 mm

CONDENSED PHASE

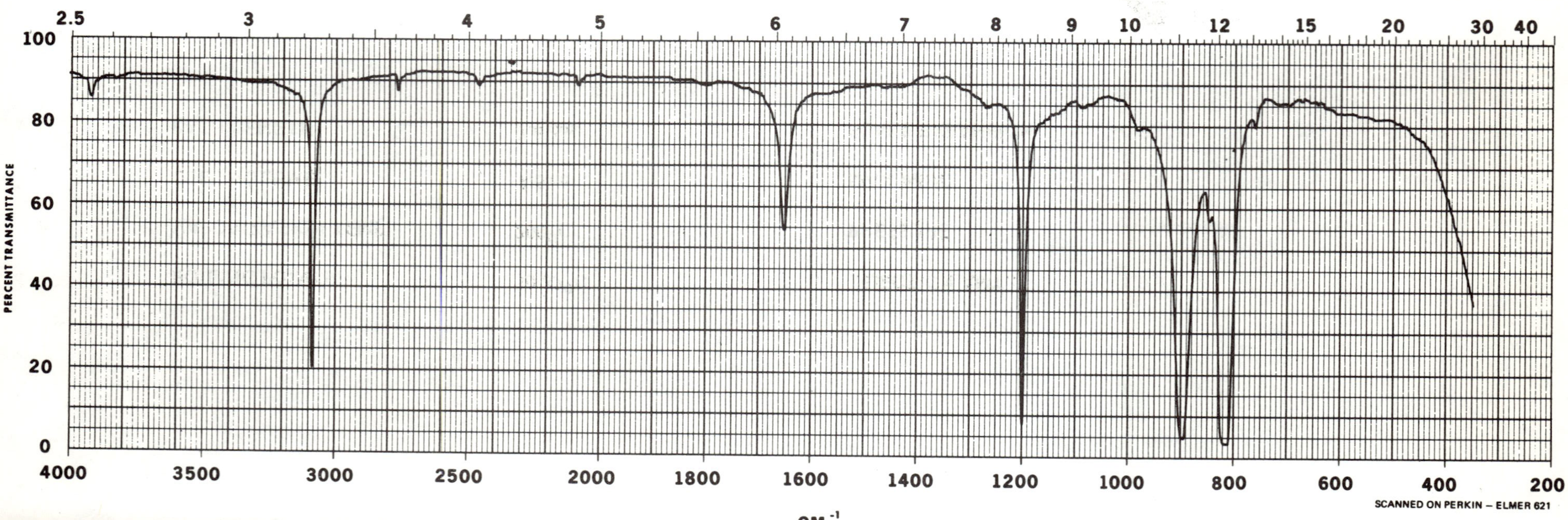

BOILING POINT:	48°C	INJECTION AMT.:	0.5 μl
VAPORIZATION TEMP.:	200°C	PRESSURE:	Atmospheric
CELL VOLUME:	7.5 cc	MAX./MIN.:	—
PATH LENGTH:	7.6 cm	SCALE EXPANSION:	0.91

VAPOR PHASE

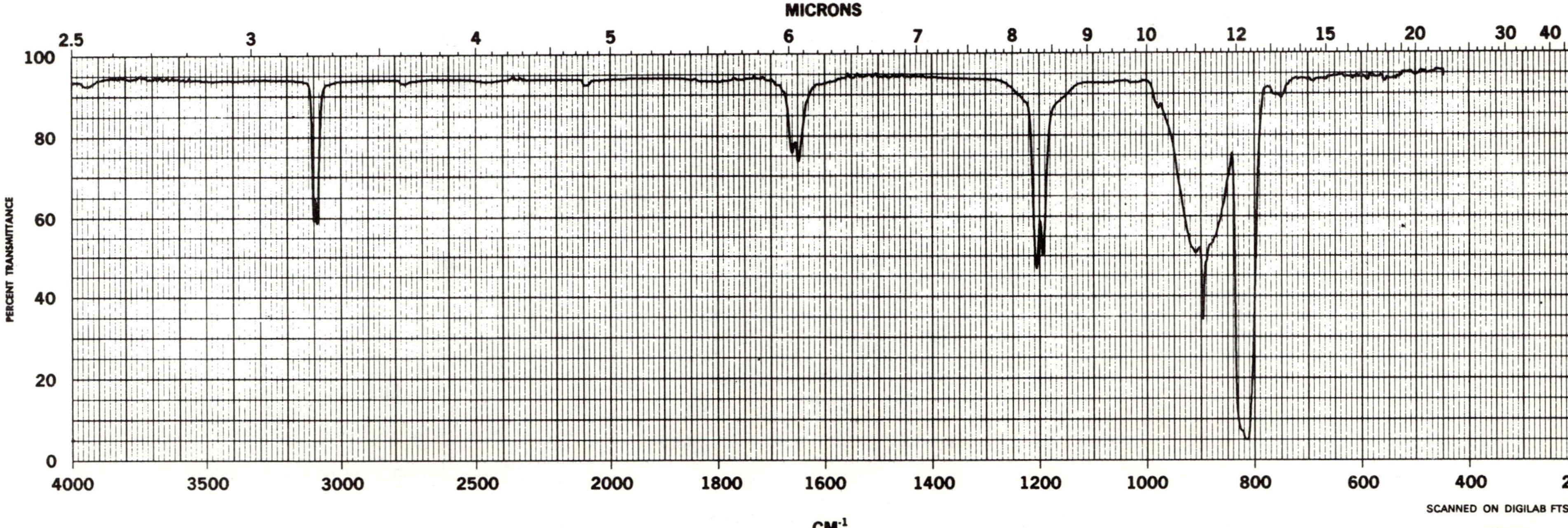

SCANNED ON DIGILAB FTS14

102

TETRACHLOROETHYLENE
PERCLENE
PERCHLOROETHYLENE
ETHYLENE TETRACHLORIDE
TETRACAP

C_2Cl_4

MOLECULAR WEIGHT: 165.83

SOURCE OF SAMPLE: MCB Manufacturing Chemists
Norwood, Ohio

$Cl_2C=CCl_2$

CAS REGISTRY NO.: 127-18-4

NIOSH NUMBER: KX38500

CAPILLARY CELL: Neat - Scale Expanded

CONDENSED PHASE

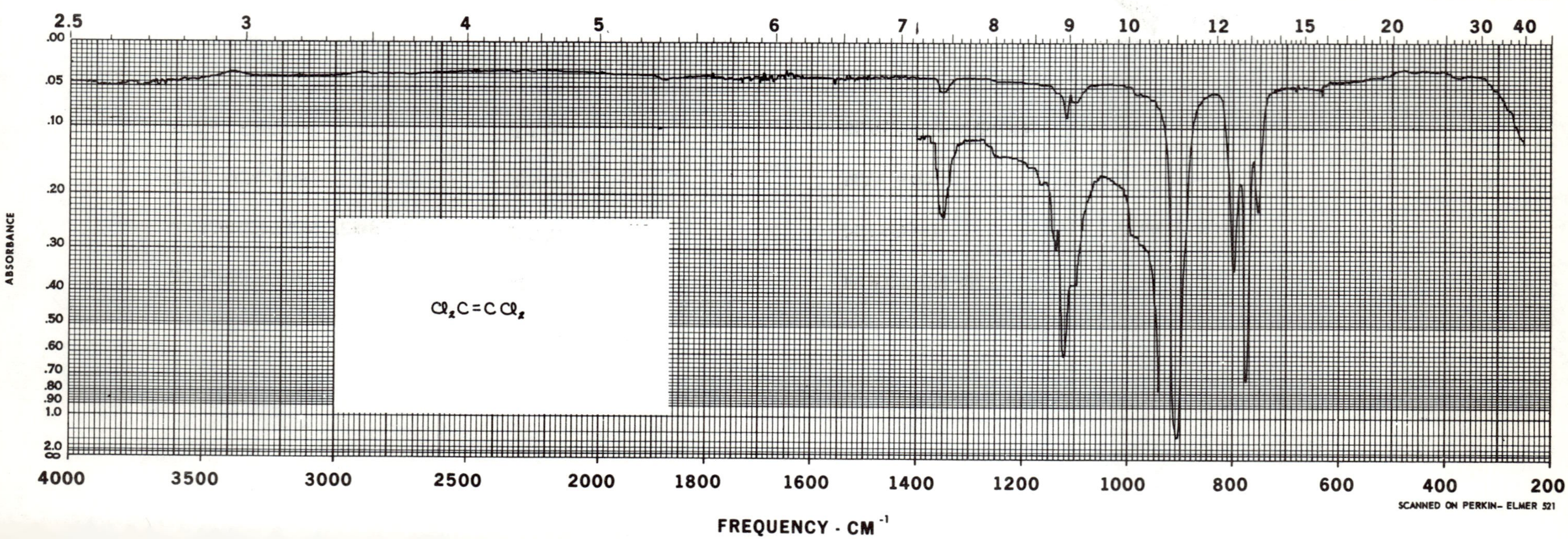

BOILING POINT:	121°C	INJECTION AMT.:	0.3 µl
VAPORIZATION TEMP.:	200°C	PRESSURE:	Atmospheric
CELL VOLUME:	7.5 cc	MAX./MIN.:	—
PATH LENGTH:	7.6 cm	SCALE EXPANSION:	0.97

VAPOR PHASE

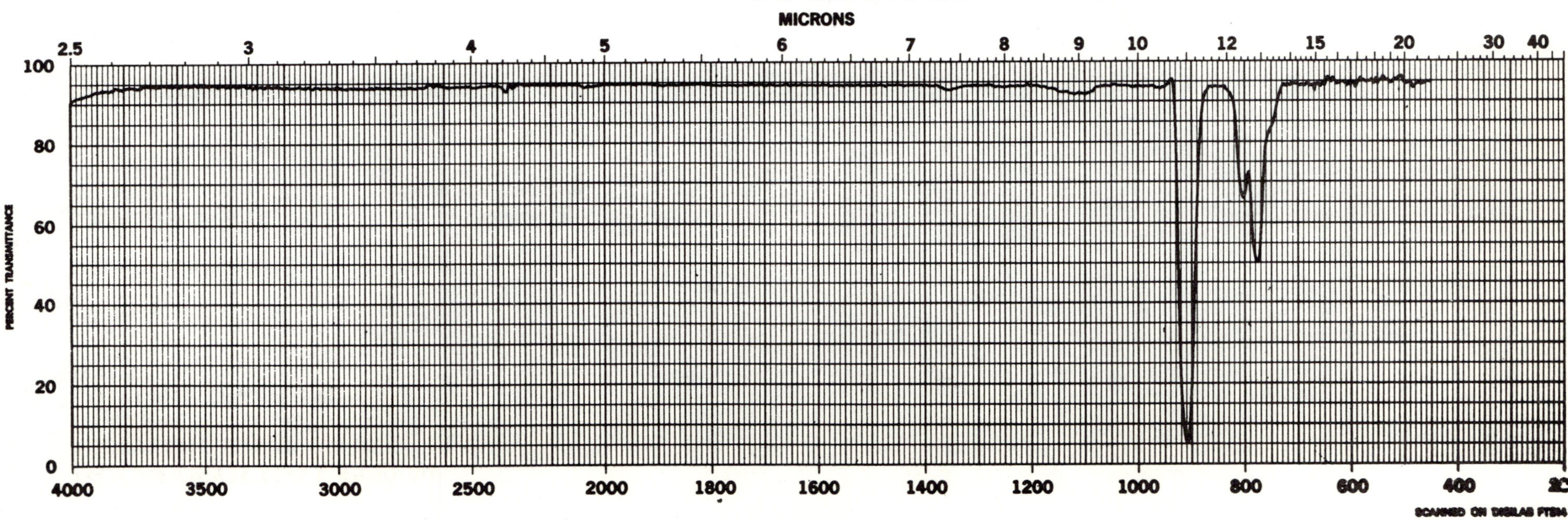

103

TRICHLOROETHYLENE
ACETYLENE TRICHLORIDE
TRICLENE
ETHYLENE TRICHLORIDE
TCE

C_2HCl_3

$Cl_2C=CHCl$

MOLECULAR WEIGHT:	131.39
SOURCE OF SAMPLE:	Aldrich Chemical Company, Inc. Milwaukee, Wisconsin
CAS REGISTRY NO.:	79-01-6
NIOSH NUMBER:	KX45500
CAPILLARY CELL:	Neat

CONDENSED PHASE

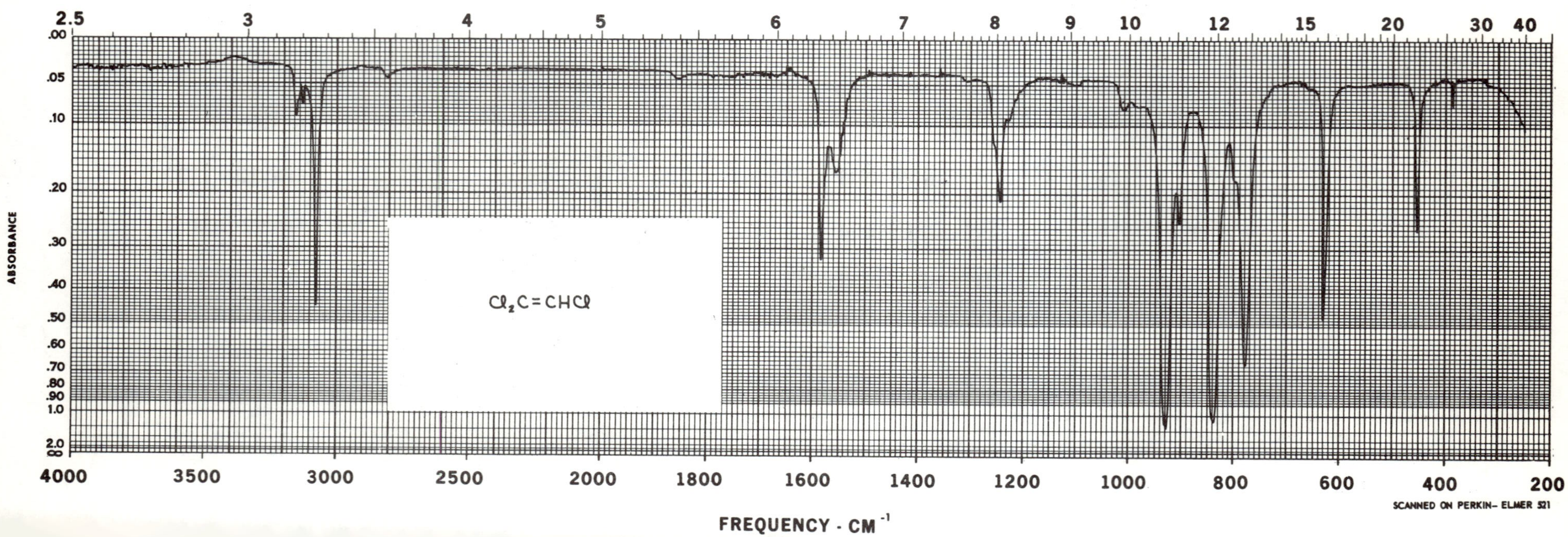

BOILING POINT:	87°C	INJECTION AMT.:	1.0 μl
VAPORIZATION TEMP.:	200°C	PRESSURE:	Atmospheric
CELL VOLUME:	7.5 cc	MAX./MIN.:	—
PATH LENGTH:	7.6 cm	SCALE EXPANSION:	0.95

VAPOR PHASE

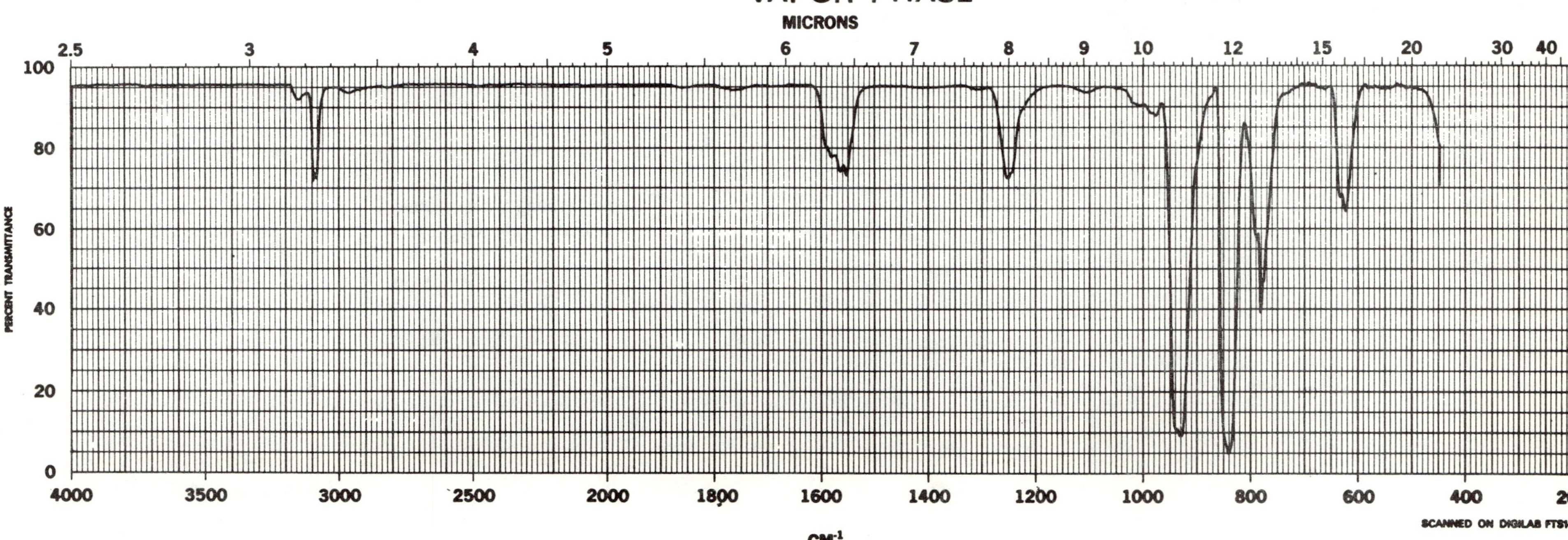

104

ETHYLENEDIAMINE
1,2–ETHANEDIAMINE

$C_2H_8N_2$

MOLECULAR WEIGHT:	60.10
SOURCE OF SAMPLE:	Chem Service Company West Chester, Pennsylvania
CAS REGISTRY NO.:	107-15-3
NIOSH NUMBER:	KH85750
CAPILLARY CELL:	Neat

$H_2N-CH_2CH_2-NH_2$

CONDENSED PHASE

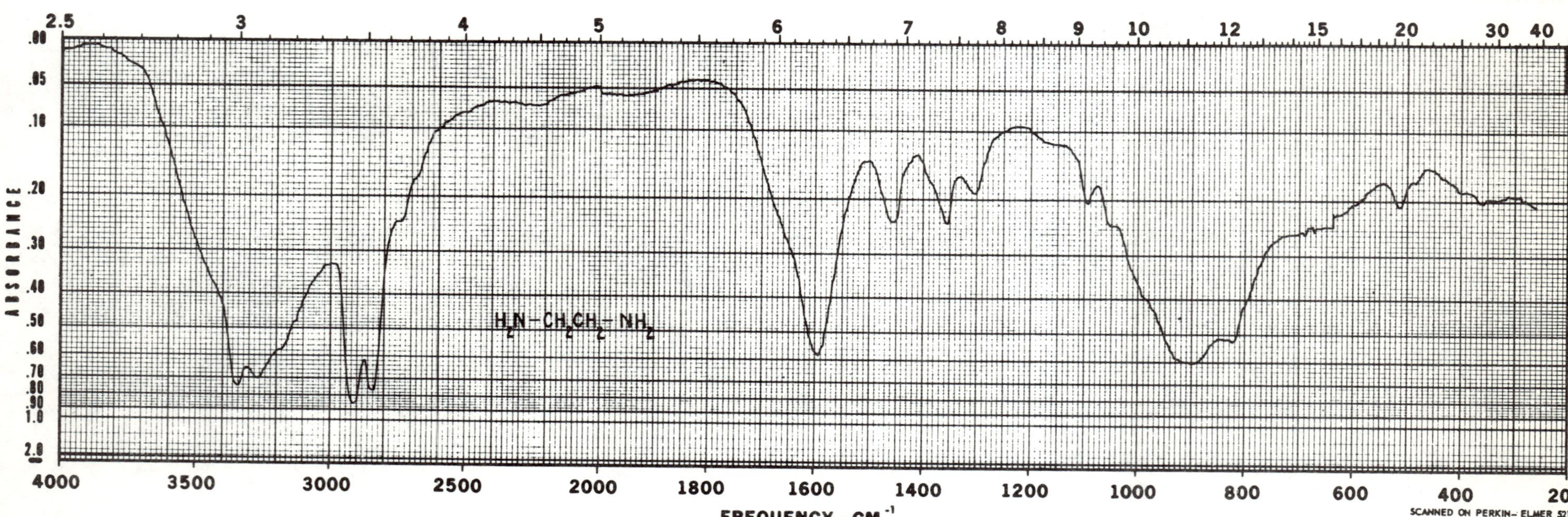

BOILING POINT:	117.2°C	INJECTION AMT.:	1.0 µl
VAPORIZATION TEMP.:	200°C	PRESSURE:	Atmospheric
CELL VOLUME:	7.5 cc	MAX./MIN.:	99.2/7.1
PATH LENGTH:	7.6 cm	SCALE EXPANSION:	1.00

VAPOR PHASE

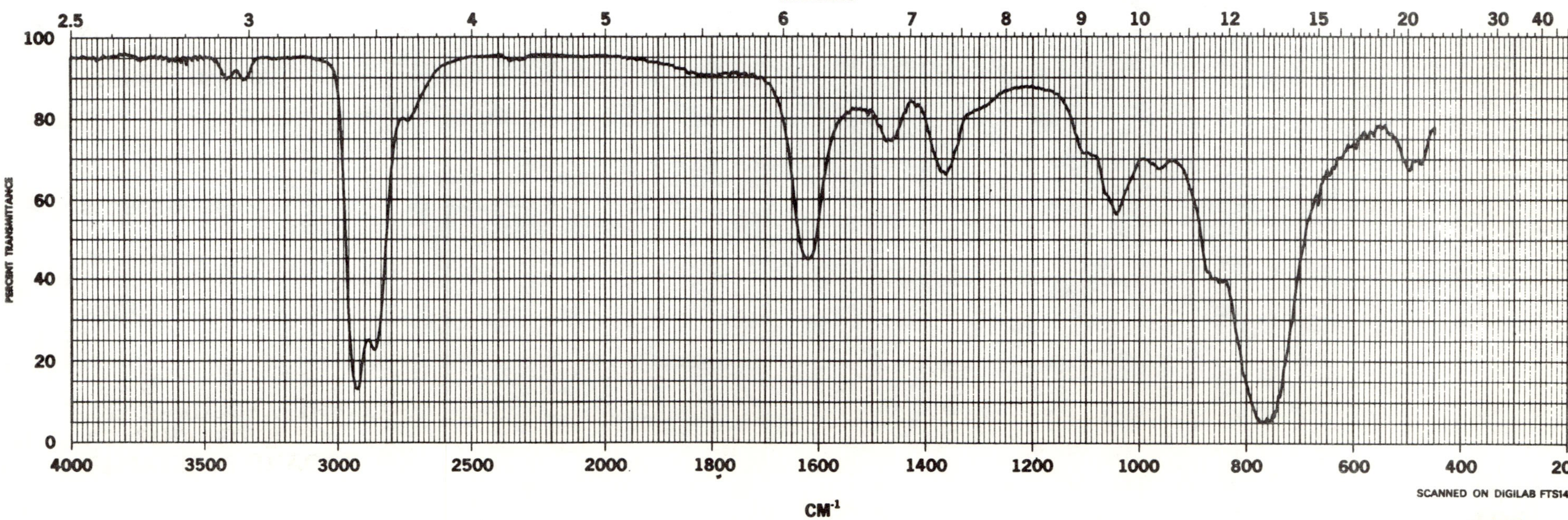

SCANNED ON DIGILAB FTS14

ETHYLENE GLYCOL
1,2--ETHANEDIOL
1,2–DIHYDROXYETHANE
TESCOL

$C_2H_6O_2$

MOLECULAR WEIGHT:	62.07
SOURCE OF SAMPLE:	Aldrich Chemical Company, Inc. Milwaukee, Wisconsin
CAS REGISTRY NO.:	107-21-1
NIOSH NUMBER:	KW29750
CAPILLARY CELL:	Neat

$HO-CH_2-CH_2-OH$

CONDENSED PHASE

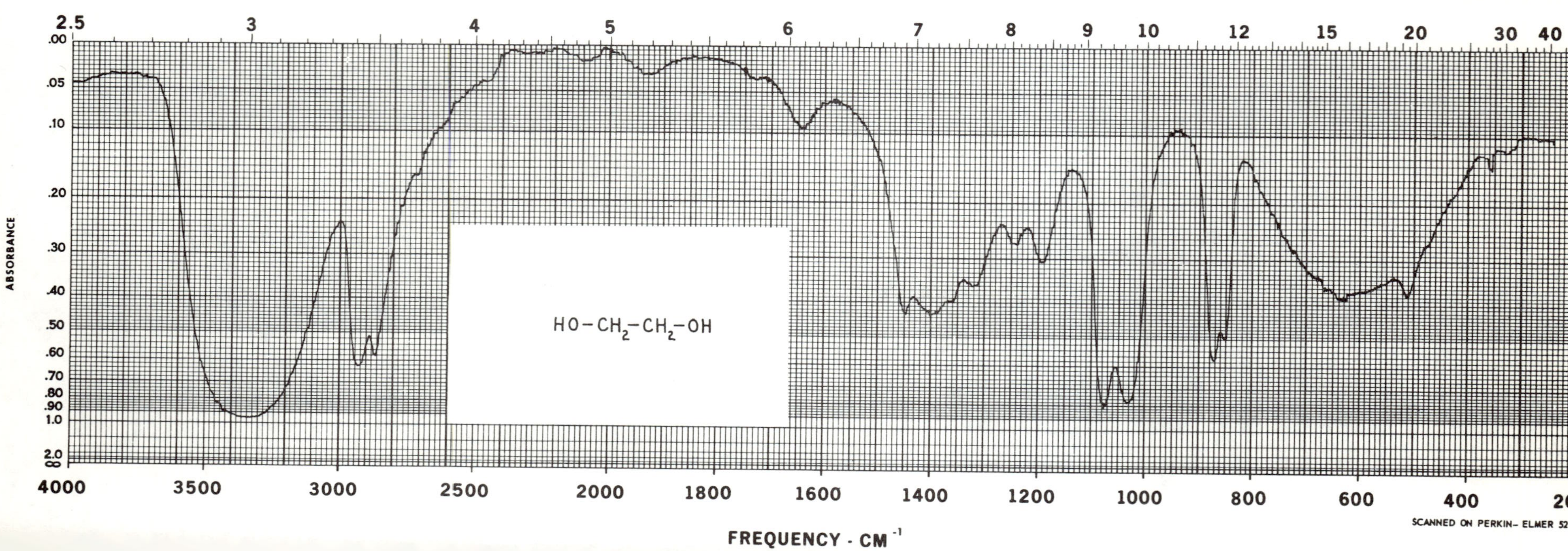

BOILING POINT:	196-198°C	INJECTION AMT.:	1.0 μl
VAPORIZATION TEMP.:	200°C	PRESSURE:	Atmospheric
CELL VOLUME:	7.5 cc	MAX./MIN.:	—
PATH LENGTH:	7.6 cm	SCALE EXPANSION:	1.01

VAPOR PHASE

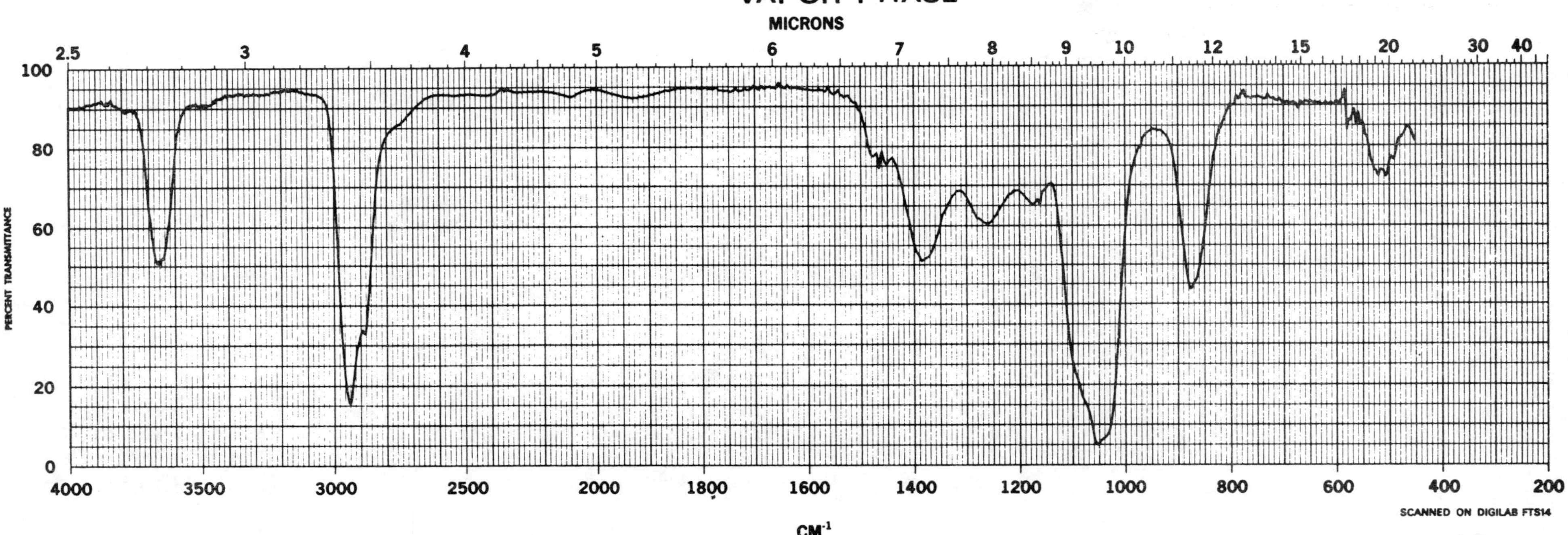

SCANNED ON DIGILAB FTS14

ETHYLENIMINE
AZACYCLOPROPANE
AZIRIDINE
DIMETHYLENIMINE

C_2H_5N

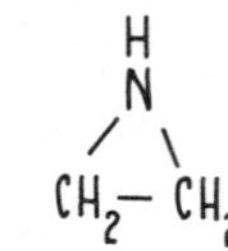

MOLECULAR WEIGHT:	43.07
SOURCE OF SAMPLE:	Chem Service Company West Chester, Pennsylvania
CAS REGISTRY NO.:	151-56-4
NIOSH NUMBER:	KX50750

Between Salts

CONDENSED PHASE

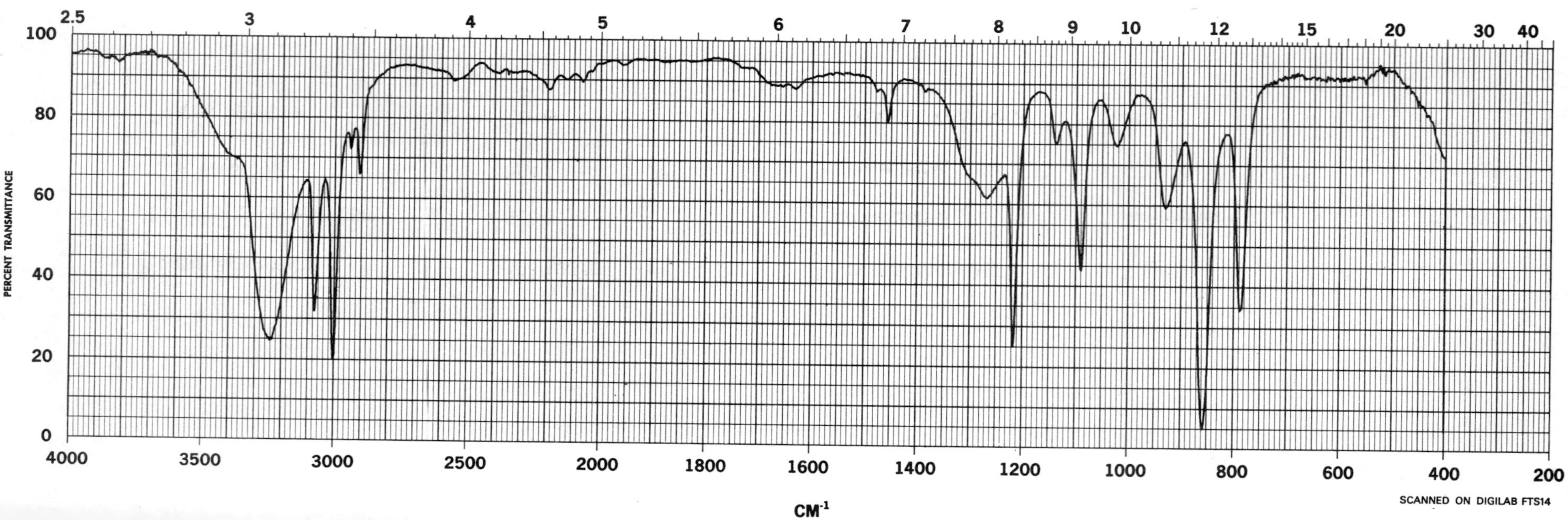

BOILING POINT:	56°C	INJECTION AMT.:	—
VAPORIZATION TEMP.:	70°C	PRESSURE:	Atmospheric
CELL VOLUME:	7.5 cc	MAX./MIN.:	96.8/25.8
PATH LENGTH:	7.6 cm	SCALE EXPANSION:	1.30

VAPOR PHASE

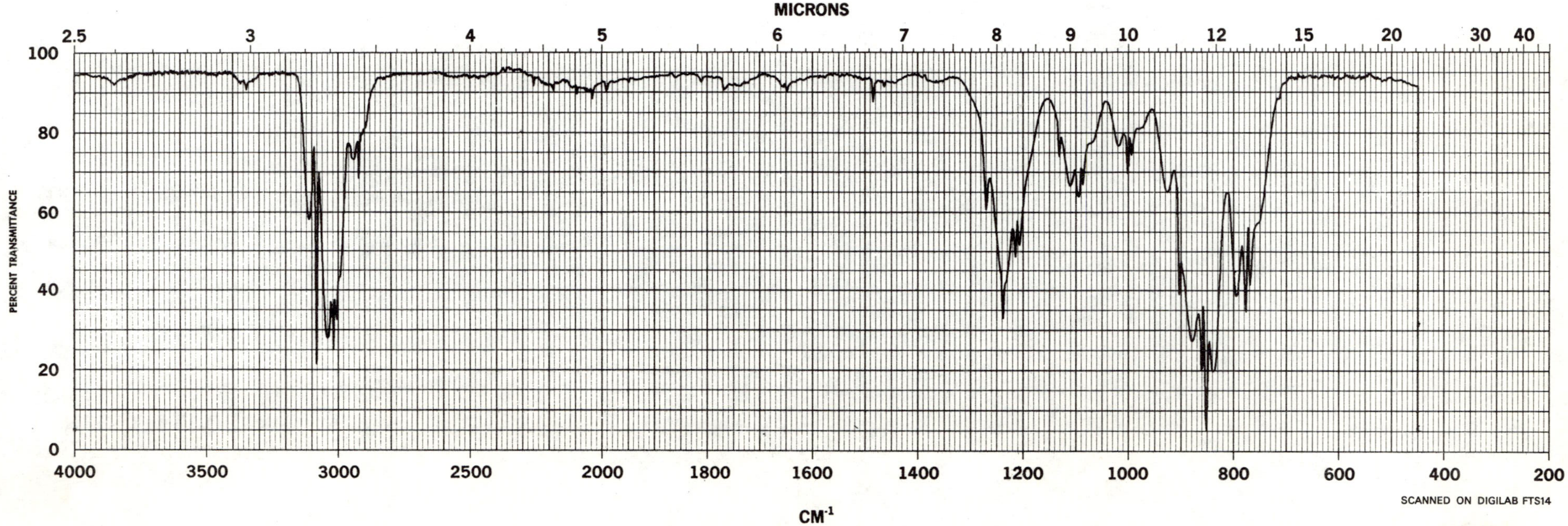

SCANNED ON DIGILAB FTS14

FLUORANTHENE
BENZO[jk] FLUORENE
IDRYL

$C_{16}H_{10}$

MOLECULAR WEIGHT: 202.26

SOURCE OF SAMPLE: James Hinton
Valparaiso, Florida

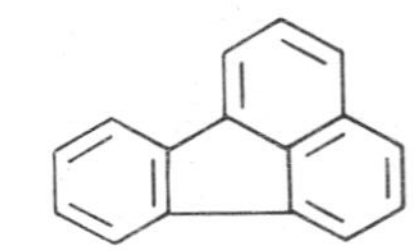

CAS REGISTRY NO.: 206-44-0

NIOSH NUMBER: LL40250

KBr Wafer

CONDENSED PHASE

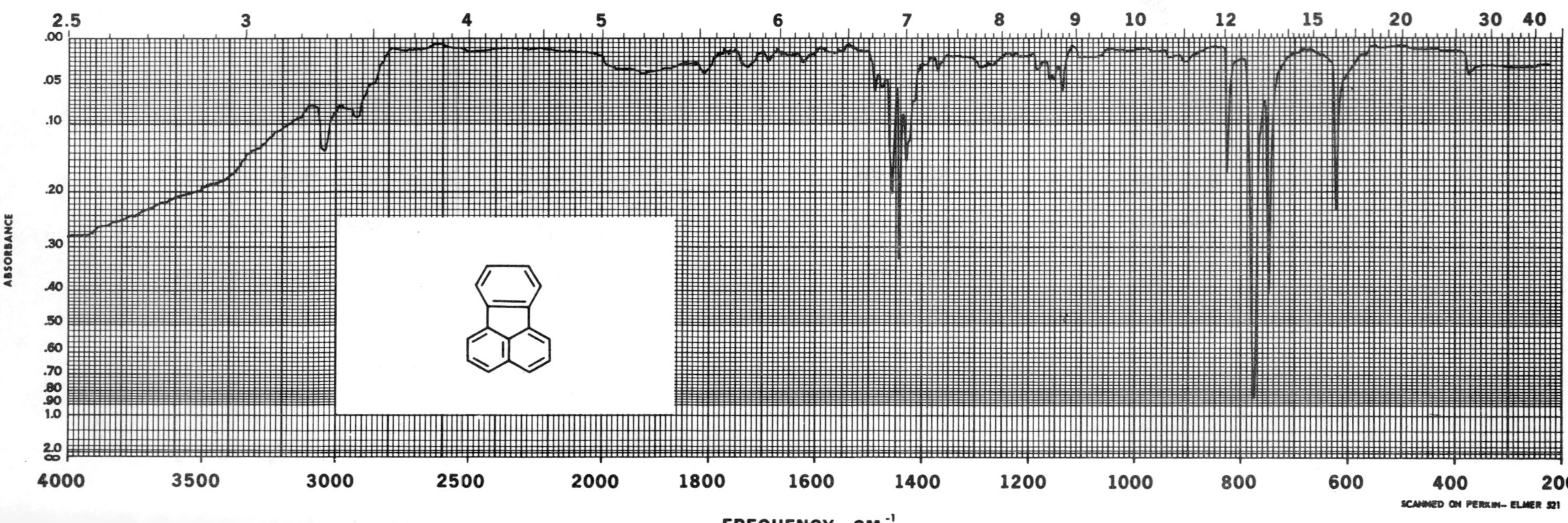

MELTING POINT:	111°C	INJECTION AMT.:	–
VAPORIZATION TEMP.:	280°C	PRESSURE:	Atmospheric
CELL VOLUME:	7.5 cc	MAX./MIN.:	90.3/7.4
PATH LENGTH:	7.6 cm	SCALE EXPANSION:	1.11

VAPOR PHASE

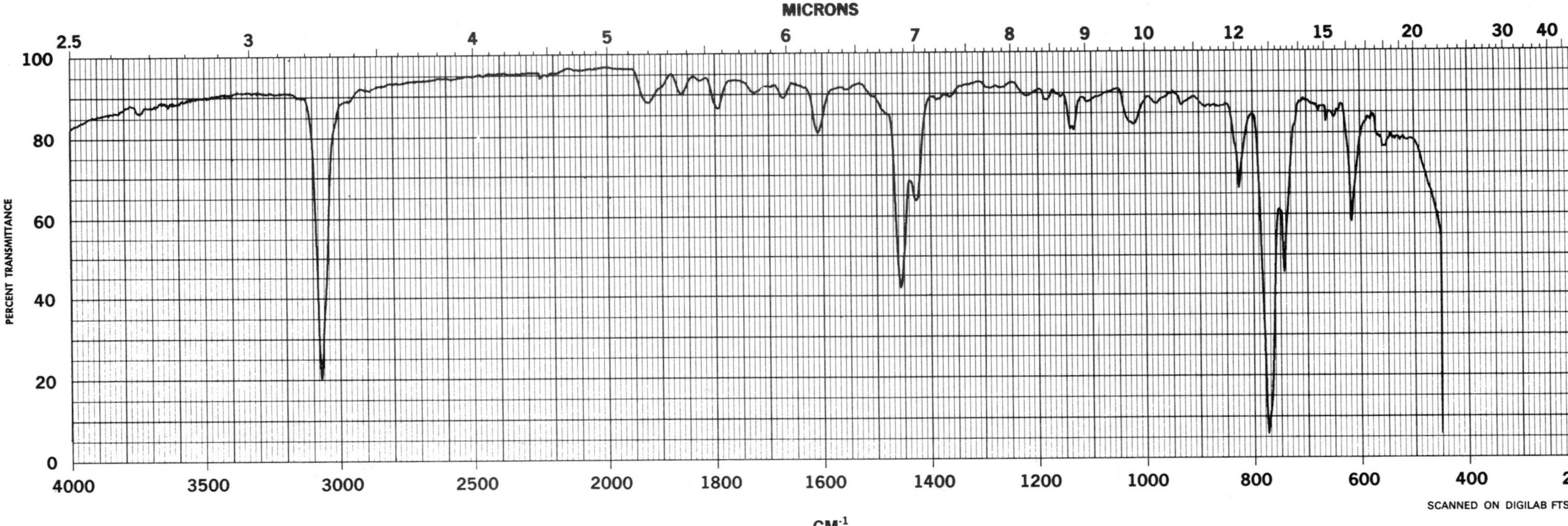

FLUOREN–2–AMINE
2–AMINOFLUORENE

$C_{13}H_{11}N$

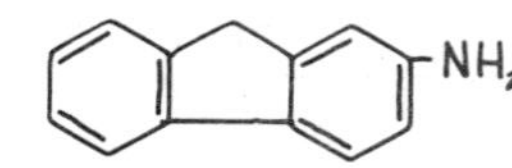

MOLECULAR WEIGHT: 181.24

SOURCE OF SAMPLE: Aldrich Chemical Company, Inc.
Milwaukee, Wisconsin

CAS REGISTRY NO.: 153-78-6

NIOSH NUMBER: LL50750

KBr Wafer

CONDENSED PHASE

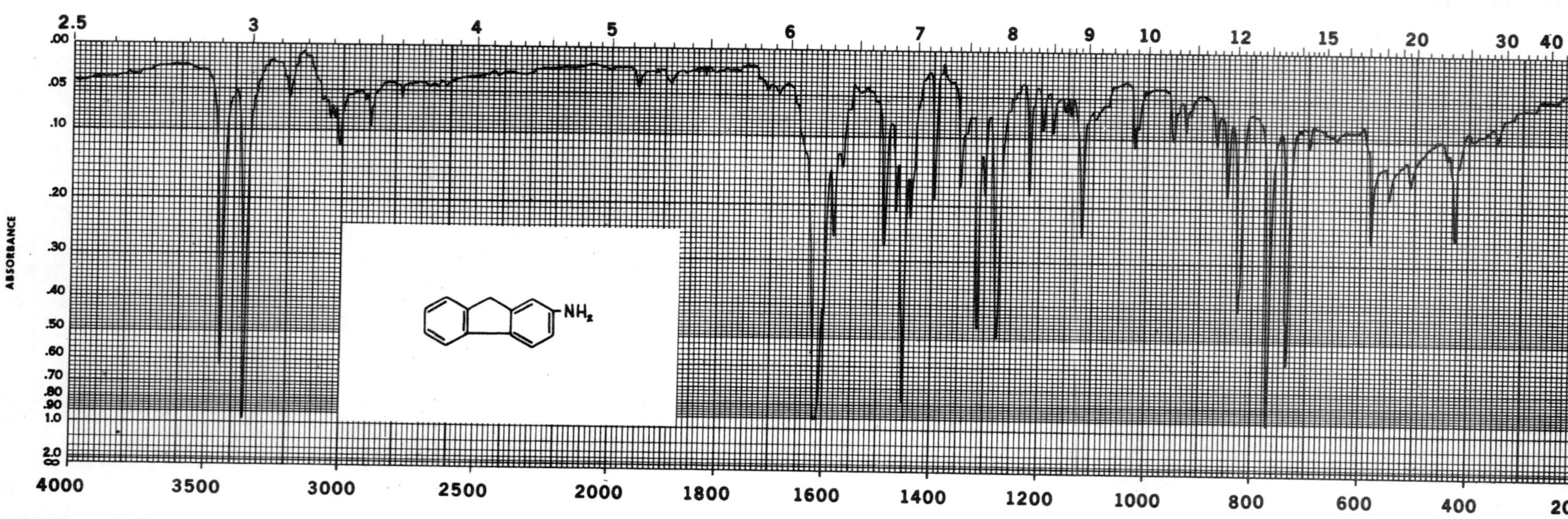

MELTING POINT:	125-127°C	INJECTION AMT.:	—
VAPORIZATION TEMP.:	280°C	PRESSURE:	Atmospheric
CELL VOLUME:	7.5 cc	MAX./MIN.:	102.7/34.7
PATH LENGTH:	7.6 cm	SCALE EXPANSION:	1.35

VAPOR PHASE

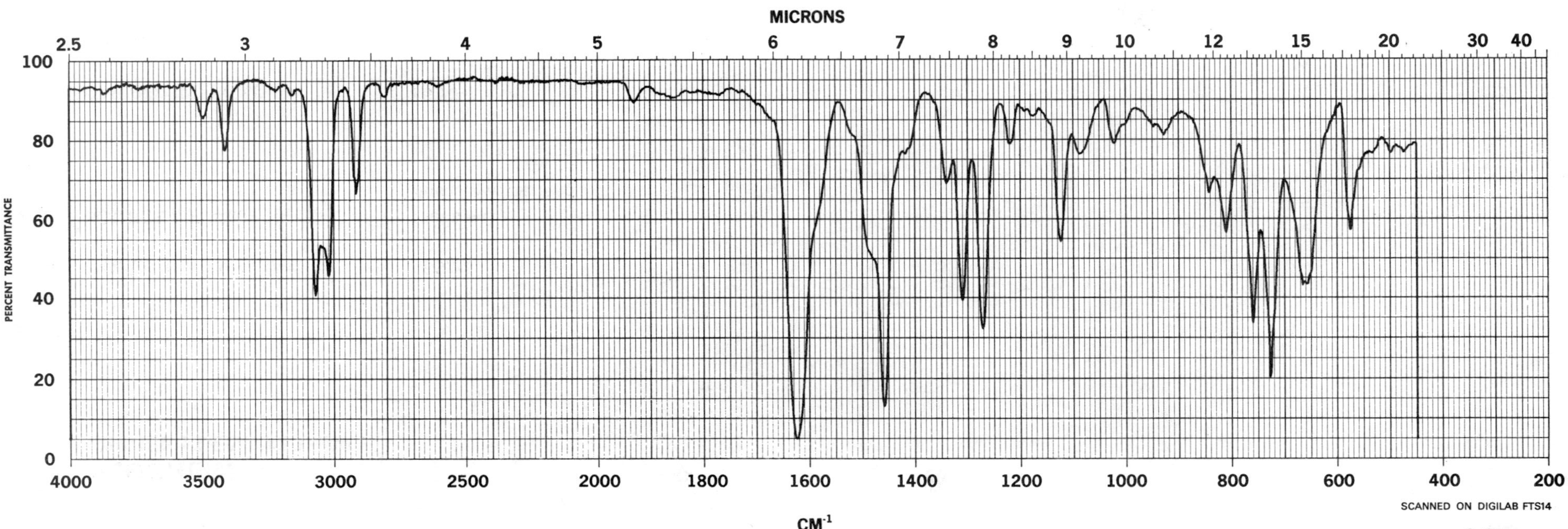

SCANNED ON DIGILAB FTS14

FLUORENE
DIPHENYLENEMETHANE

$C_{13}H_{10}$

MOLECULAR WEIGHT: 166.22

SOURCE OF SAMPLE: Fluka AG
Buchs, Switzerland

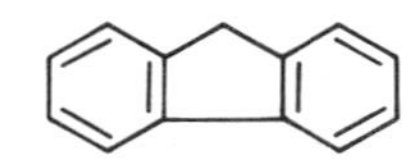

CAS REGISTRY NO.: 86-73-7

NIOSH NUMBER: –

KBr Wafer

CONDENSED PHASE

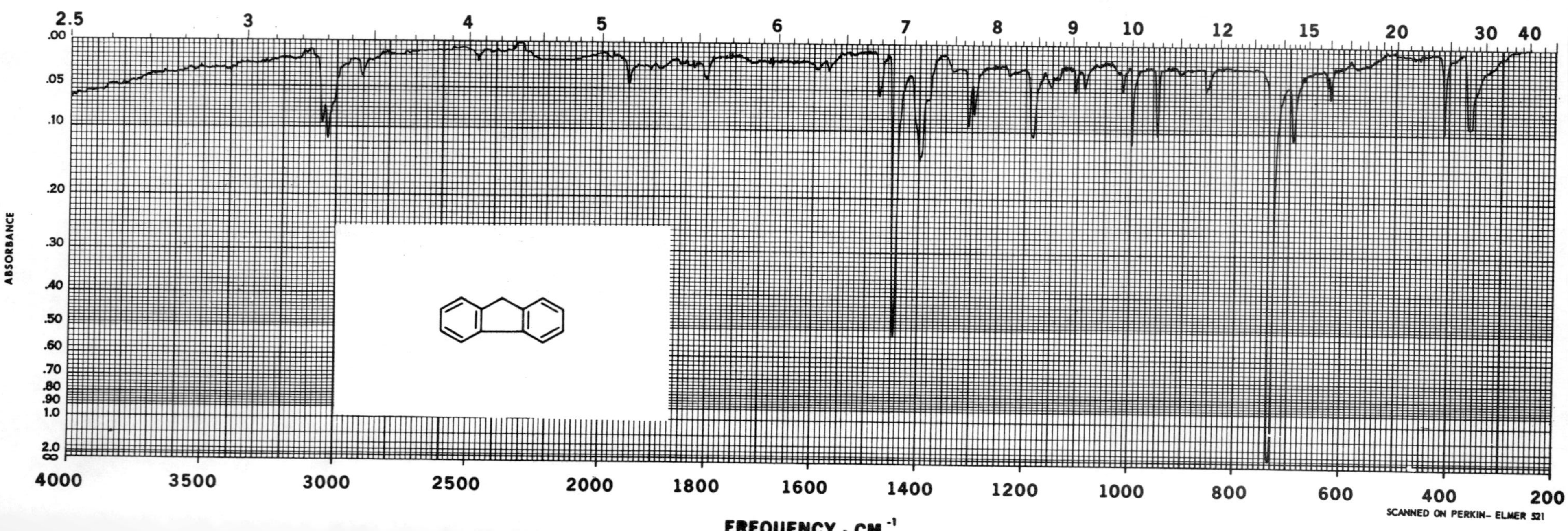

MELTING POINT:	113-115°C	INJECTION AMT.:	–
VAPORIZATION TEMP.:	280°C	PRESSURE:	Atmospheric
CELL VOLUME:	7.5 cc	MAX./MIN.:	100.7/21.3
PATH LENGTH:	7.6 cm	SCALE EXPANSION:	1.16

VAPOR PHASE

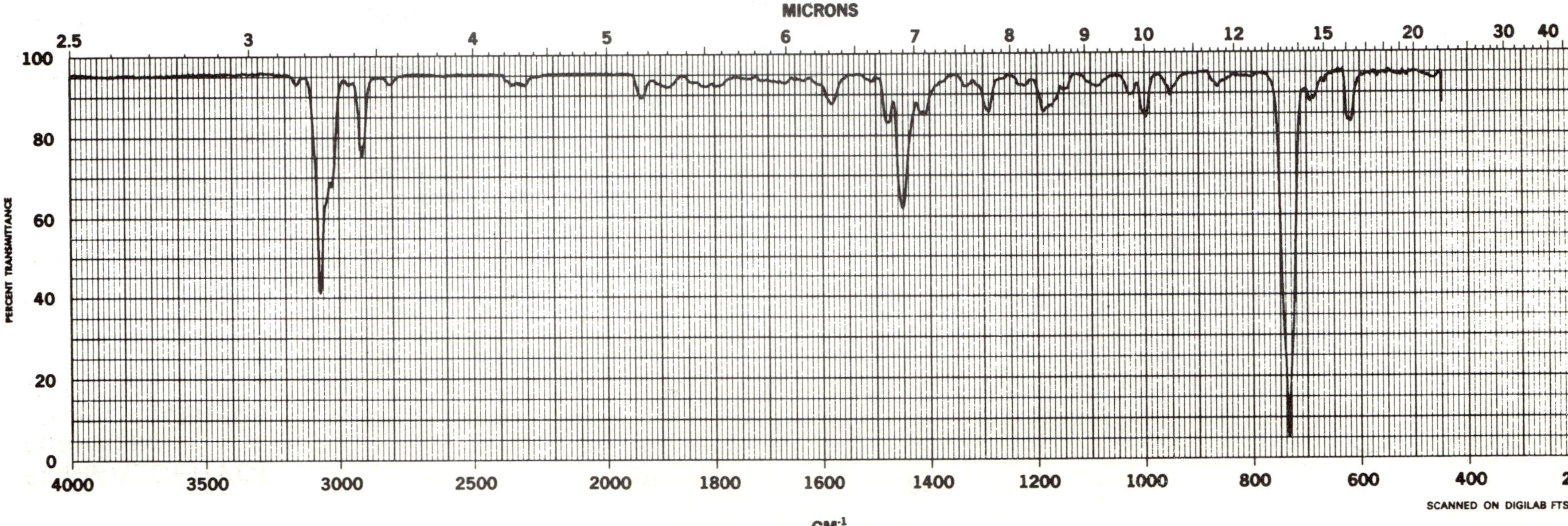

SCANNED ON DIGILAB FTS14

FORMALDEHYDE
FORMALIN
FORMIC ALDEHYDE
METHANAL

CH_2O

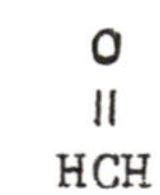

MOLECULAR WEIGHT:	30.02
SOURCE OF SAMPLE:	MCB Manufacturing Chemists Norwood, Ohio
CAS REGISTRY NO.:	50-00-0
NIOSH NUMBER:	LP89250

KBr Wafer

▲ indicates the presence of water

CONDENSED PHASE

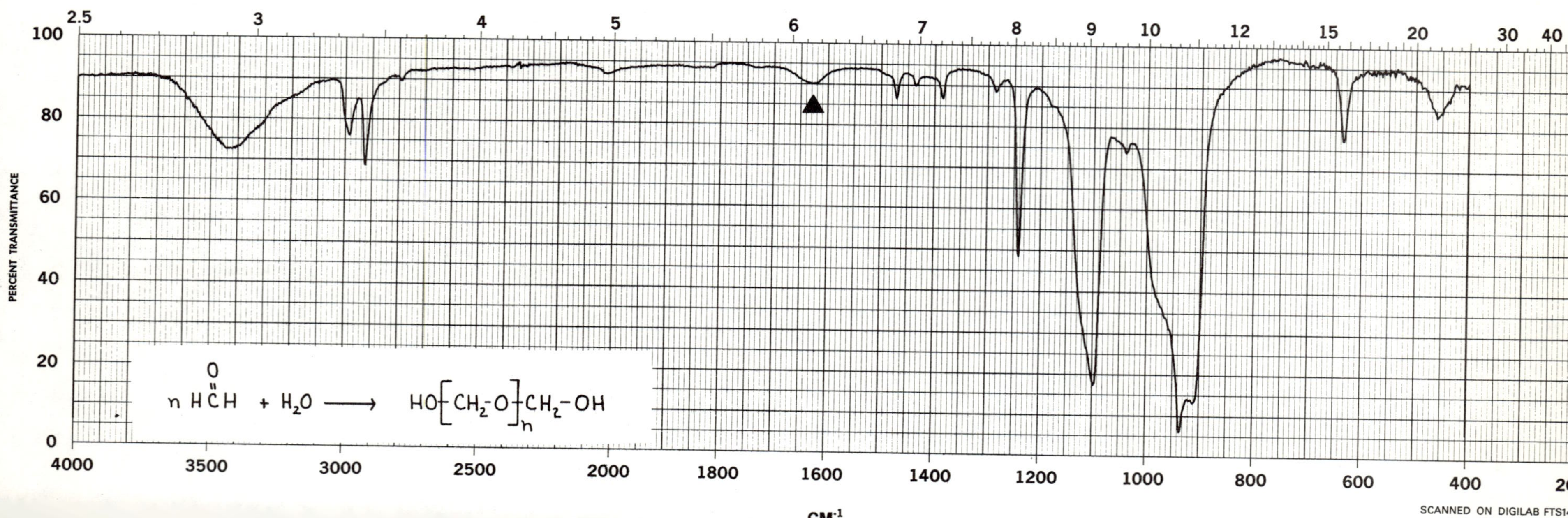

$$n\,HCHO + H_2O \longrightarrow HO{+}CH_2{-}O{+}_n CH_2{-}OH$$

BOILING POINT:	−21°C/760 mm	INJECTION AMT.:	—
VAPORIZATION TEMP.:	160°C	PRESSURE:	Atmospheric
CELL VOLUME:	7.5 cc	MAX./MIN.:	102.3/29.8
PATH LENGTH:	7.6 cm	SCALE EXPANSION:	1.27

VAPOR PHASE

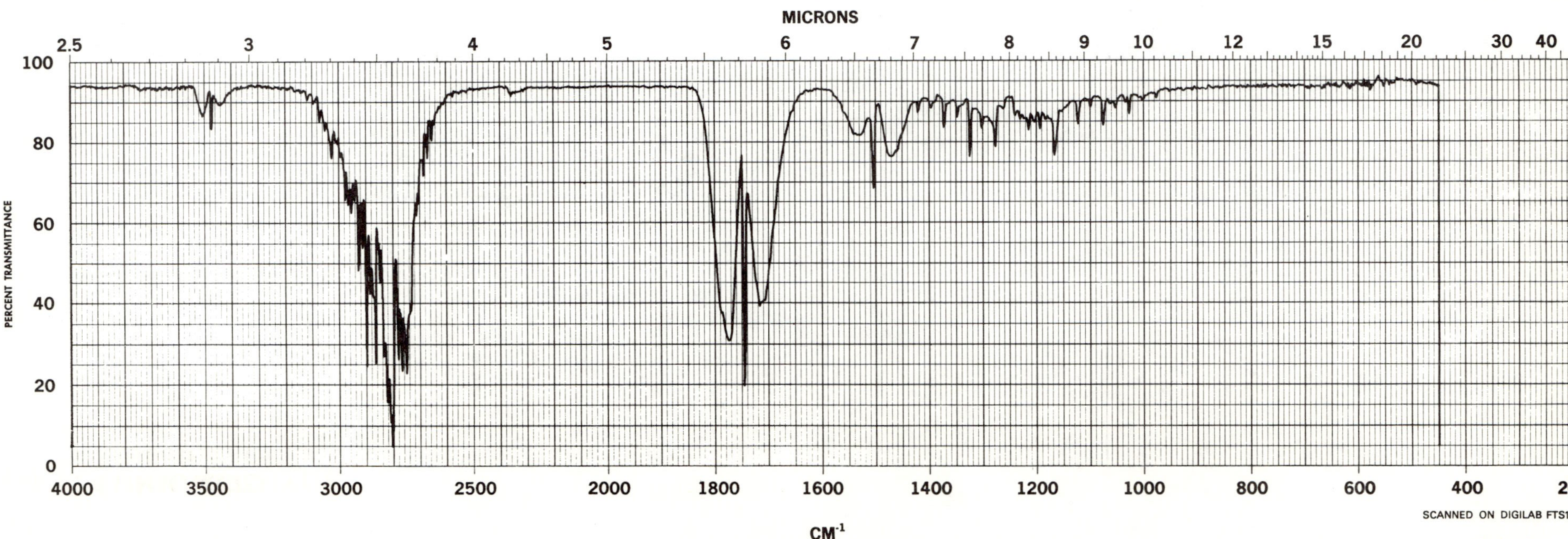

SCANNED ON DIGILAB FTS14

FORMIC ACID
METHANOIC ACID

CH_2O_2

MOLECULAR WEIGHT: 46.03

SOURCE OF SAMPLE: MCB Manufacturing Chemists
Norwood, Ohio

CAS REGISTRY NO.: 64-18-6

NIOSH NUMBER: LQ49000

CAPILLARY CELL: Neat

```
   O
   ‖
H-C-OH
```


CONDENSED PHASE

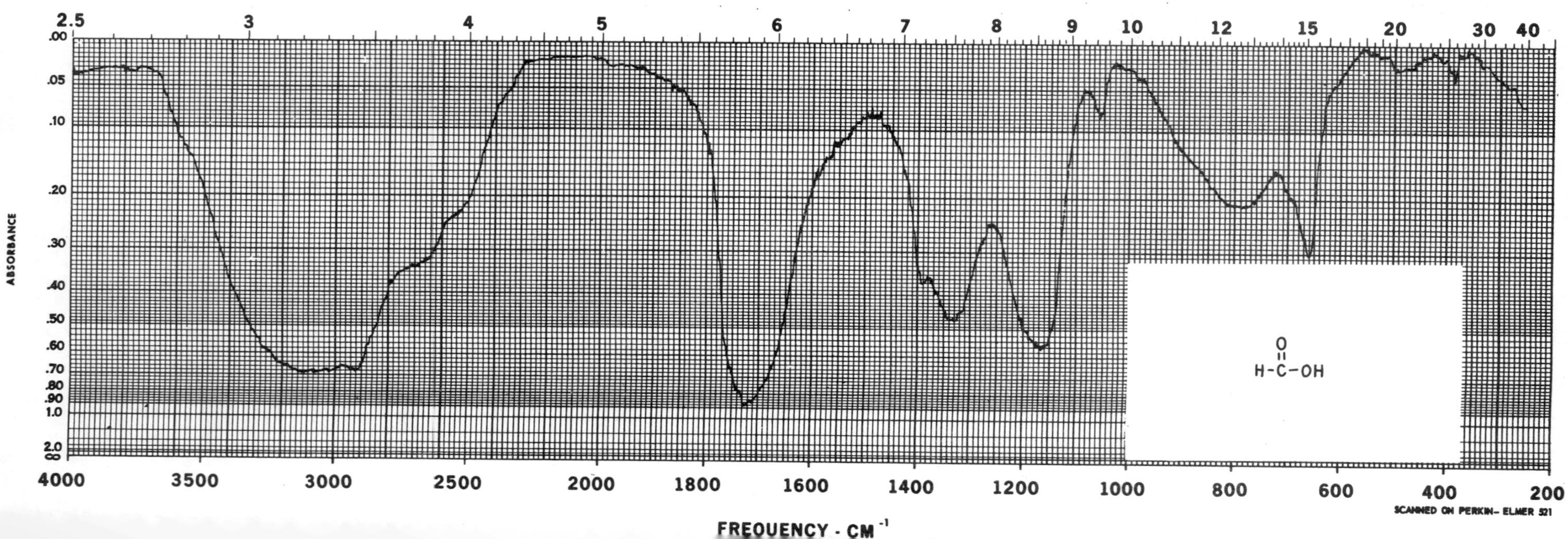

BOILING POINT:	100.8°C	INJECTION AMT.:	0.1 µl
VAPORIZATION TEMP.:	160°C	PRESSURE:	Atmospheric
CELL VOLUME:	7.5 cc	MAX./MIN.:	—
PATH LENGTH:	7.6 cm	SCALE EXPANSION:	0.97

VAPOR PHASE

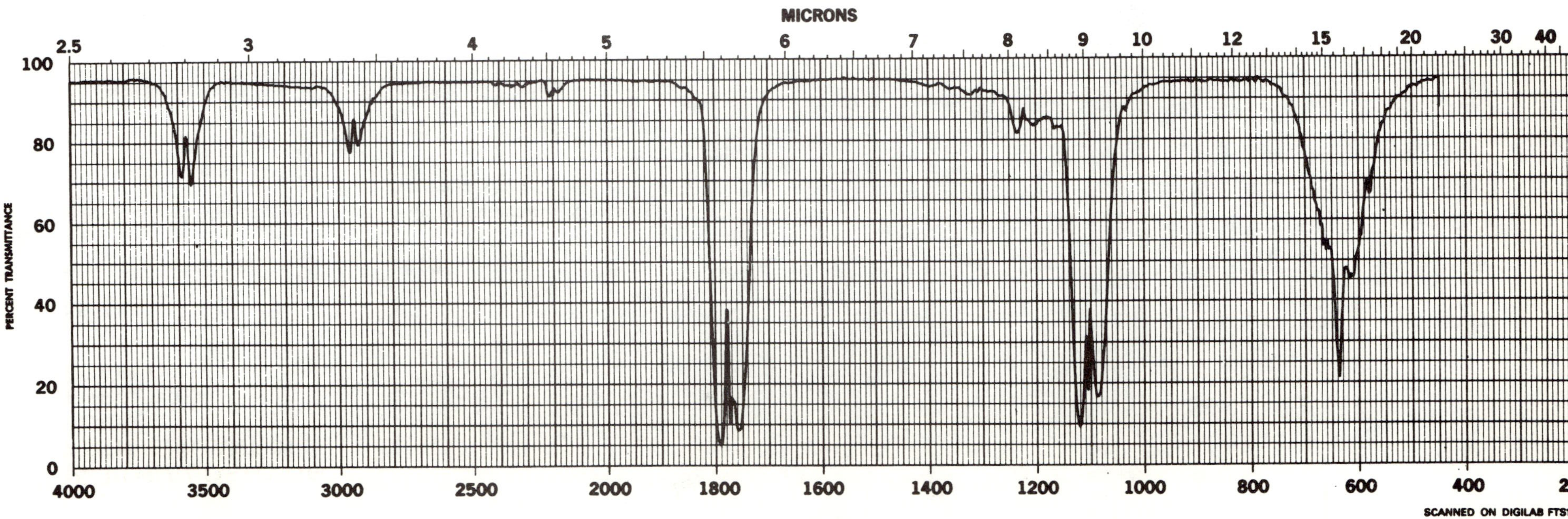

2–FURALDEHYDE
ARTIFICIAL ANT OIL
FURAL
FURFURAL
FURFUROLE

$C_5H_4O_2$

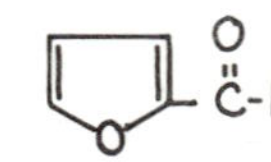

MOLECULAR WEIGHT: 96.09

SOURCE OF SAMPLE: MCB Manufacturing Chemists
Norwood, Ohio

CAS REGISTRY NO.: 98-01-1

NIOSH NUMBER: LT70000

CAPILLARY CELL: Neat

CONDENSED PHASE

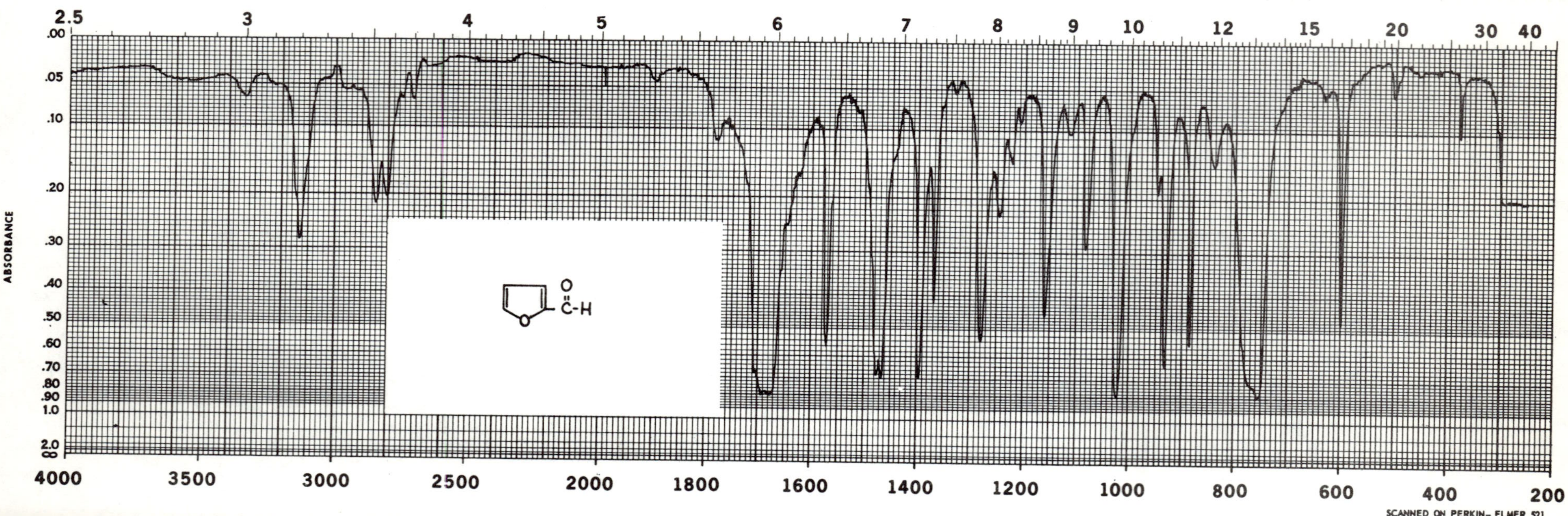

BOILING POINT:	41-42°C/5 mm	INJECTION AMT.:	0.5 μl
VAPORIZATION TEMP.:	200°C	PRESSURE:	Atmospheric
CELL VOLUME:	7.5 cc	MAX./MIN.:	—
PATH LENGTH:	7.6 cm	SCALE EXPANSION:	0.94

VAPOR PHASE

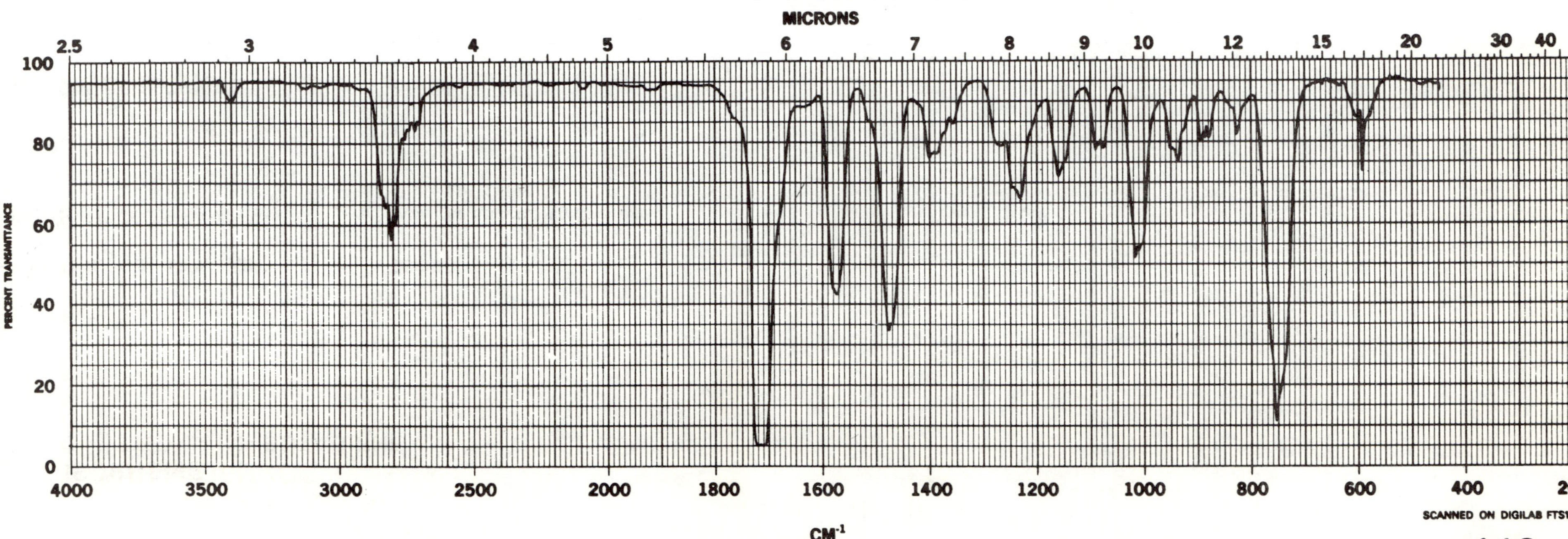

ISOPRENE
2–METHYL–1,3–BUTADIENE

C_5H_8

MOLECULAR WEIGHT: 68.12

SOURCE OF SAMPLE: MCB Manufacturing Chemists
Norwood, Ohio

$CH_2{=}C(CH_3){-}CH{=}CH_2$

CAS REGISTRY NO.: 78-79-5

NIOSH NUMBER: NT40370

CAPILLARY CELL: Neat

CONDENSED PHASE

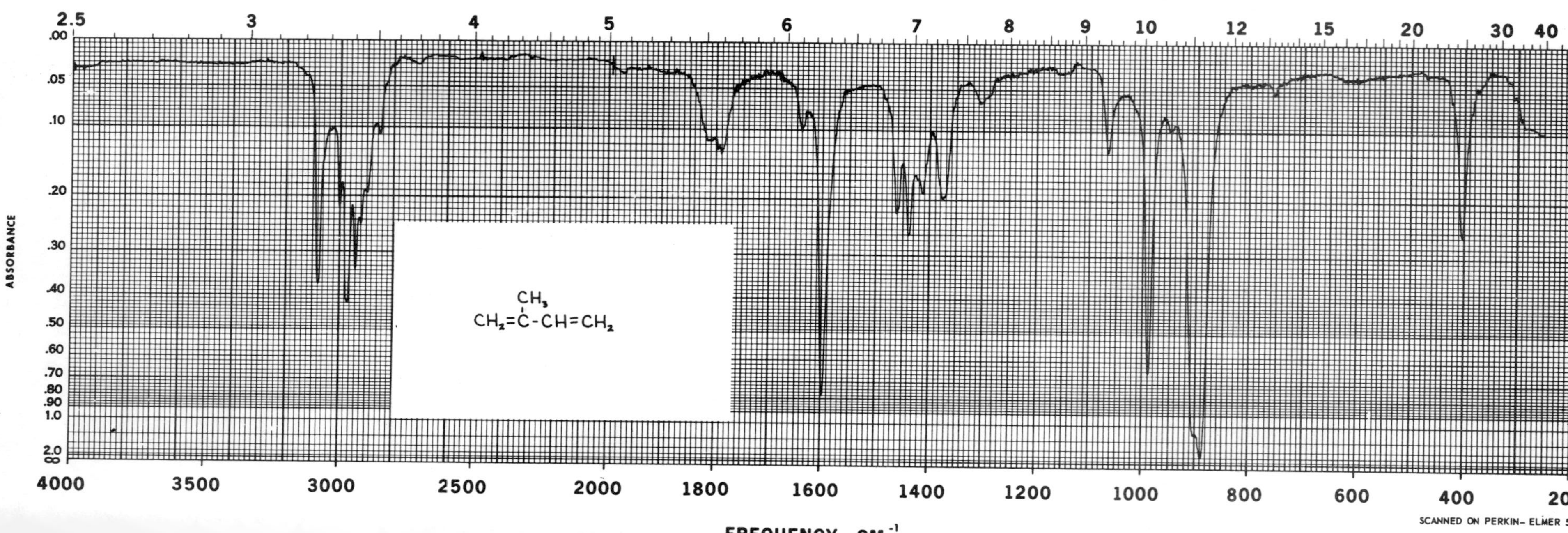

BOILING POINT:	34°C	INJECTION AMT.:	–
VAPORIZATION TEMP.:	160°C	PRESSURE:	Atmospheric
CELL VOLUME:	7.5 cc	MAX./MIN.:	104.0/12.3
PATH LENGTH:	7.6 cm	SCALE EXPANSION:	1.00

VAPOR PHASE

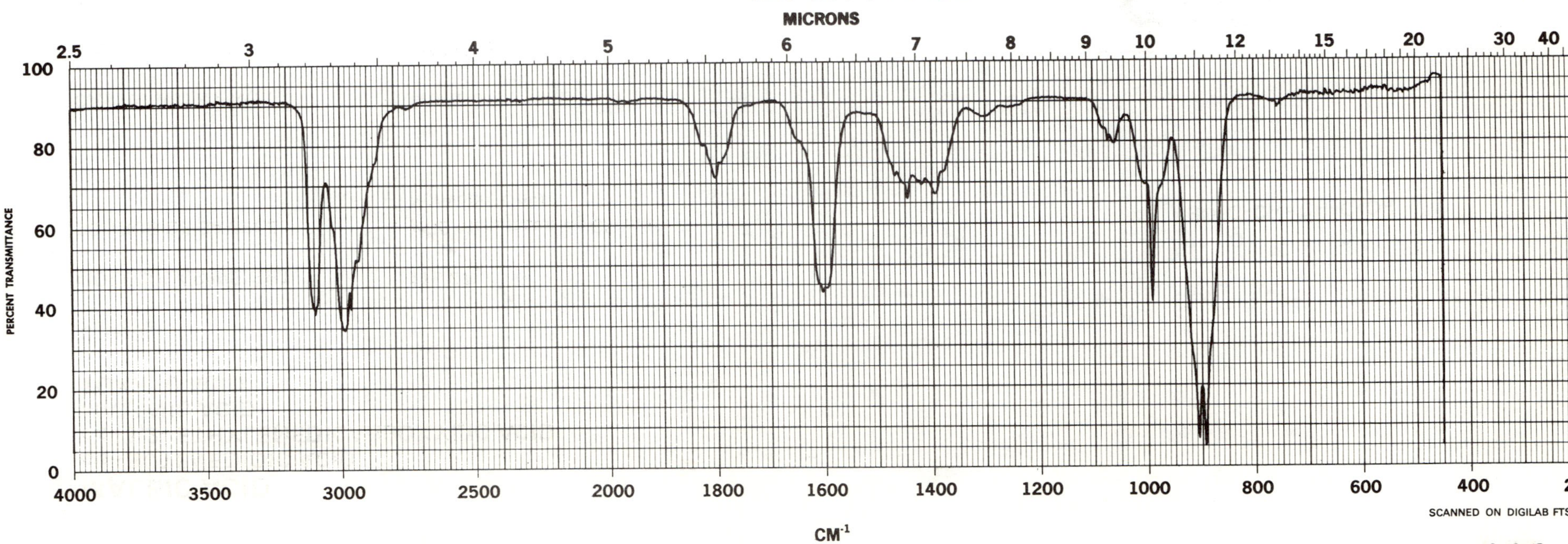

SCANNED ON DIGILAB FTS14

MALEIC ACID
cis–BUTENEDIOIC ACID
cis–1,2–ETHYLENEDICARBOXYLIC ACID
MALENIC ACID
TOXILIC ACID

$C_4H_4O_4$

MOLECULAR WEIGHT: 116.07

SOURCE OF SAMPLE: MCB Manufacturing Chemists
Norwood, Ohio

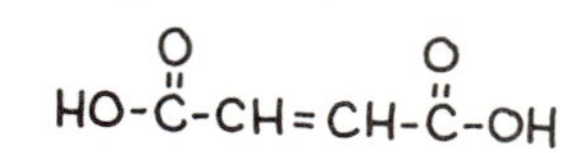

CAS REGISTRY NO.: 110-16-7

NIOSH NUMBER: OM96250

KBr Wafer

CONDENSED PHASE

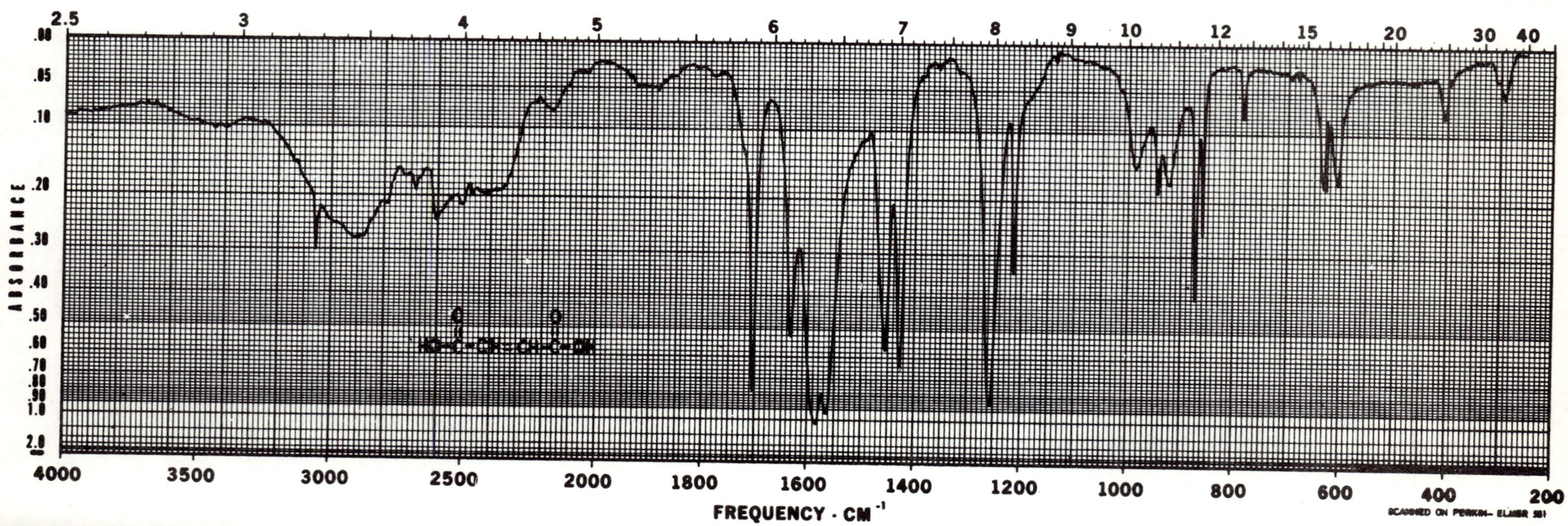

MELTING POINT:	139-141°C	INJECTION AMT.:	—
VAPORIZATION TEMP.:	160°C	PRESSURE:	Atmospheric
CELL VOLUME:	7.5 cc	MAX./MIN.:	92.2/5.7
PATH LENGTH:	7.6 cm	SCALE EXPANSION:	1.06

VAPOR PHASE

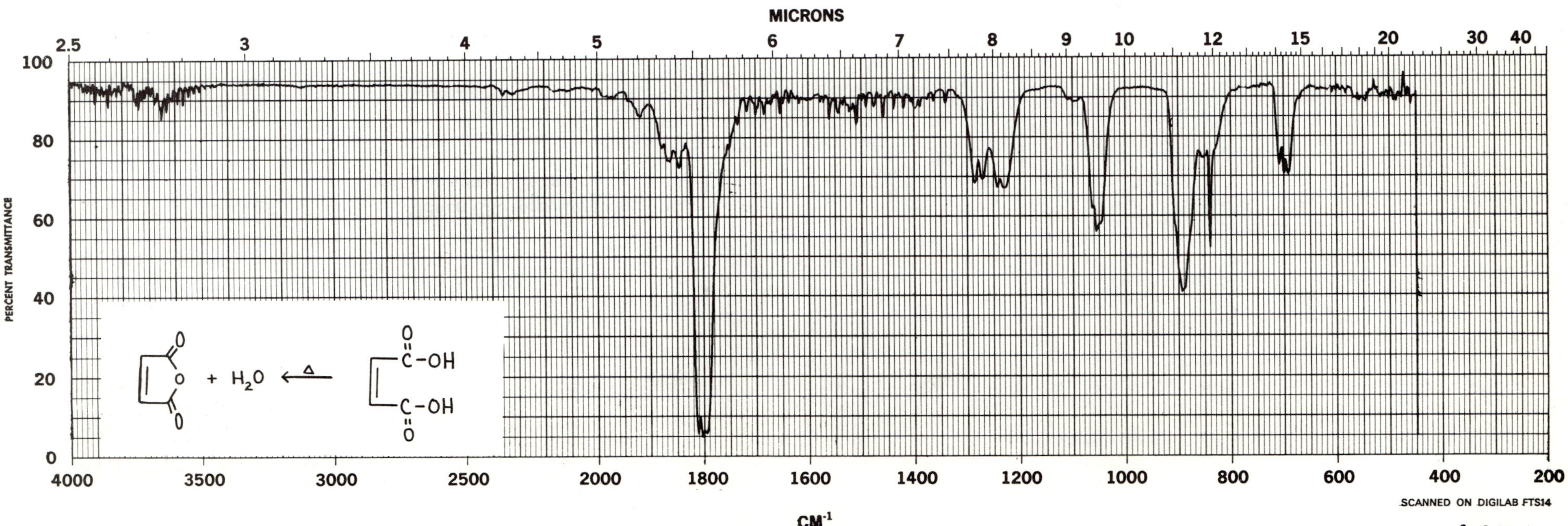

SCANNED ON DIGILAB FTS14

MALEIC ANHYDRIDE
2,5–FURANDIONE
LYTRON 810/820
TOXILIC ANHYDRIDE

$C_4H_2O_3$

MOLECULAR WEIGHT: 98.06

SOURCE OF SAMPLE: Aldrich Chemical Company, Inc.
Milwaukee, Wisconsin

CAS REGISTRY NO.: 108-31-6

NIOSH NUMBER: ON36750

CAPILLARY CELL: Melt

CONDENSED PHASE

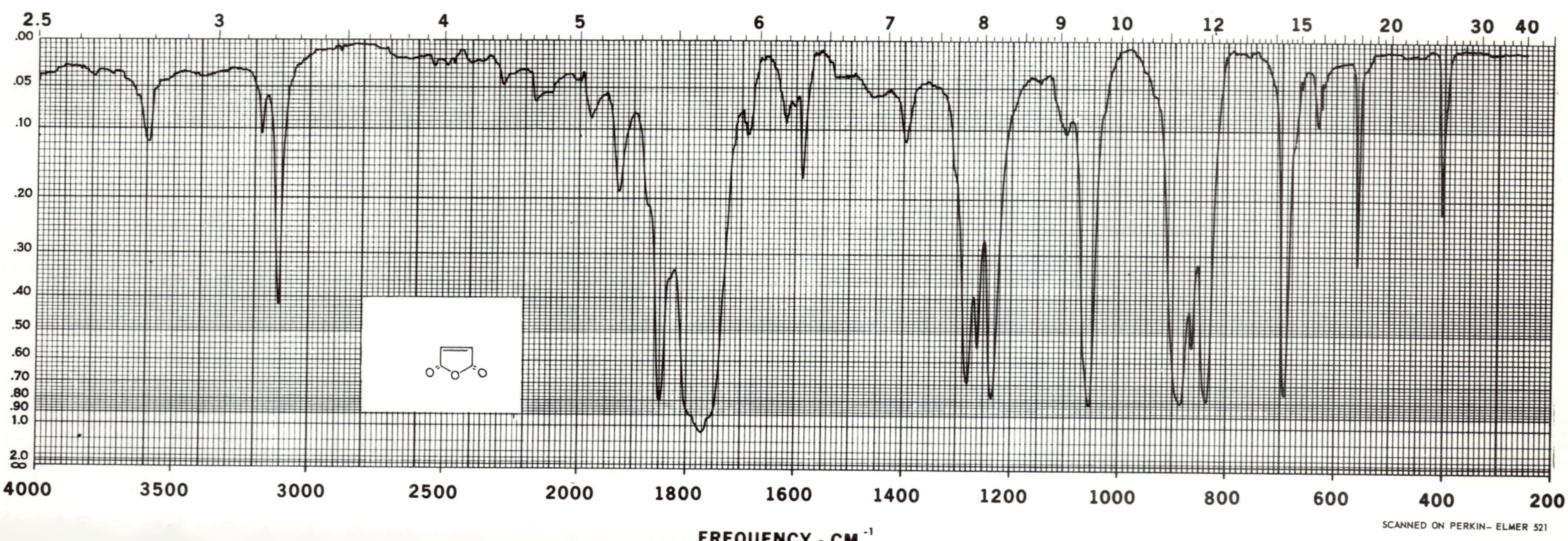

BOILING POINT:	199-200°C	INJECTION AMT.:	–
VAPORIZATION TEMP.:	240°C	PRESSURE:	Atmospheric
CELL VOLUME:	7.5 cc	MAX./MIN.:	102.7/23.7
PATH LENGTH:	7.6 cm	SCALE EXPANSION:	1.16

VAPOR PHASE

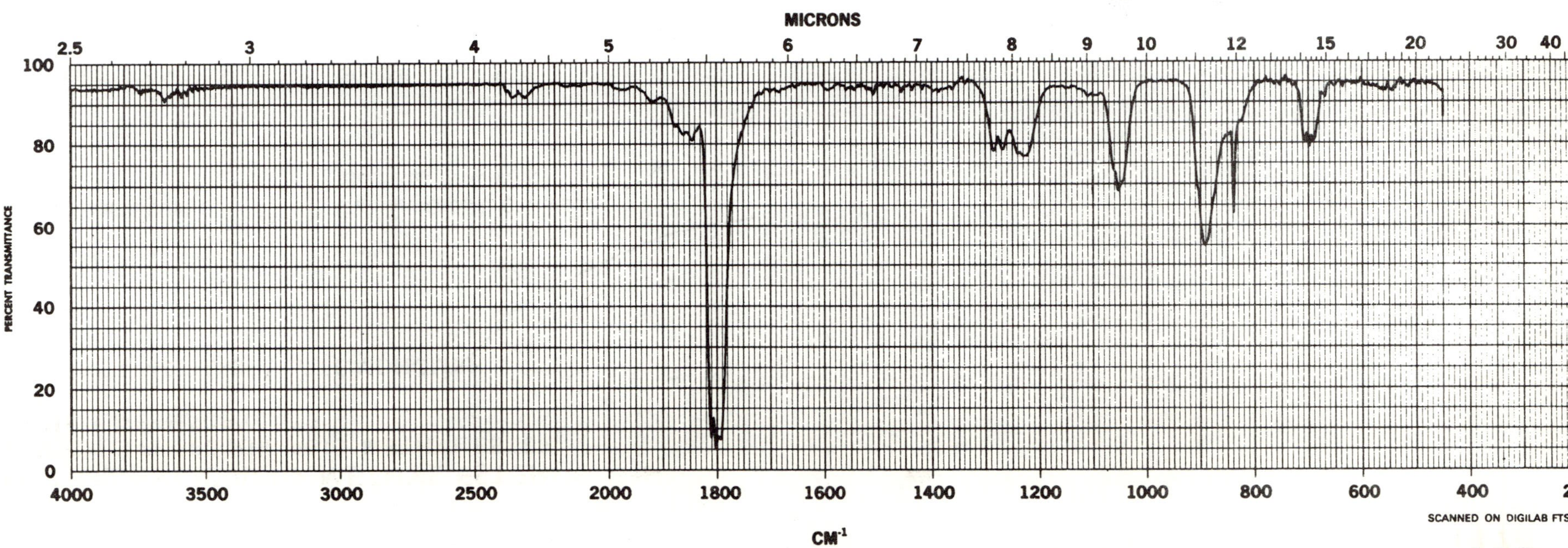

SCANNED ON DIGILAB FTS14

METHACRYLIC ACID, METHYL ESTER
2–METHYLPROPENOIC ACID, METHYL ESTER
METHYL METHACRYLATE
MME
MONOCITE METHACRYLATE MONOMER

$C_5H_8O_2$

MOLECULAR WEIGHT: 100.12

SOURCE OF SAMPLE: Eastman Organic Chemicals
Rochester, New York

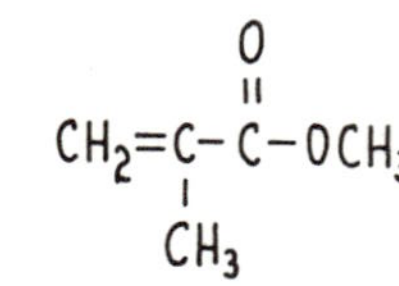

CAS REGISTRY NO.: 80-62-6

NIOSH NUMBER: OZ50750

CAPILLARY CELL: 0.015 mm

CONDENSED PHASE

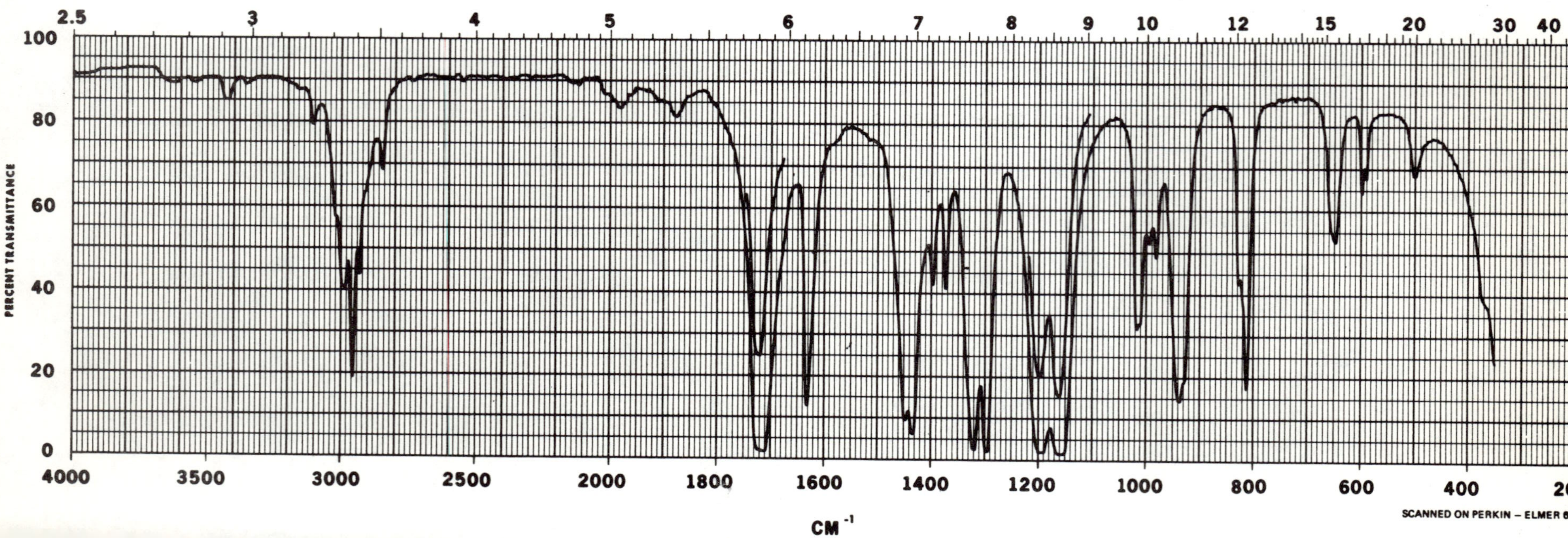

BOILING POINT:	100°C	INJECTION AMT.:	0.5 μl
VAPORIZATION TEMP.:	200°C	PRESSURE:	Atmospheric
CELL VOLUME:	7.5 cc	MAX./MIN.:	—
PATH LENGTH:	7.6 cm	SCALE EXPANSION:	0.93

VAPOR PHASE

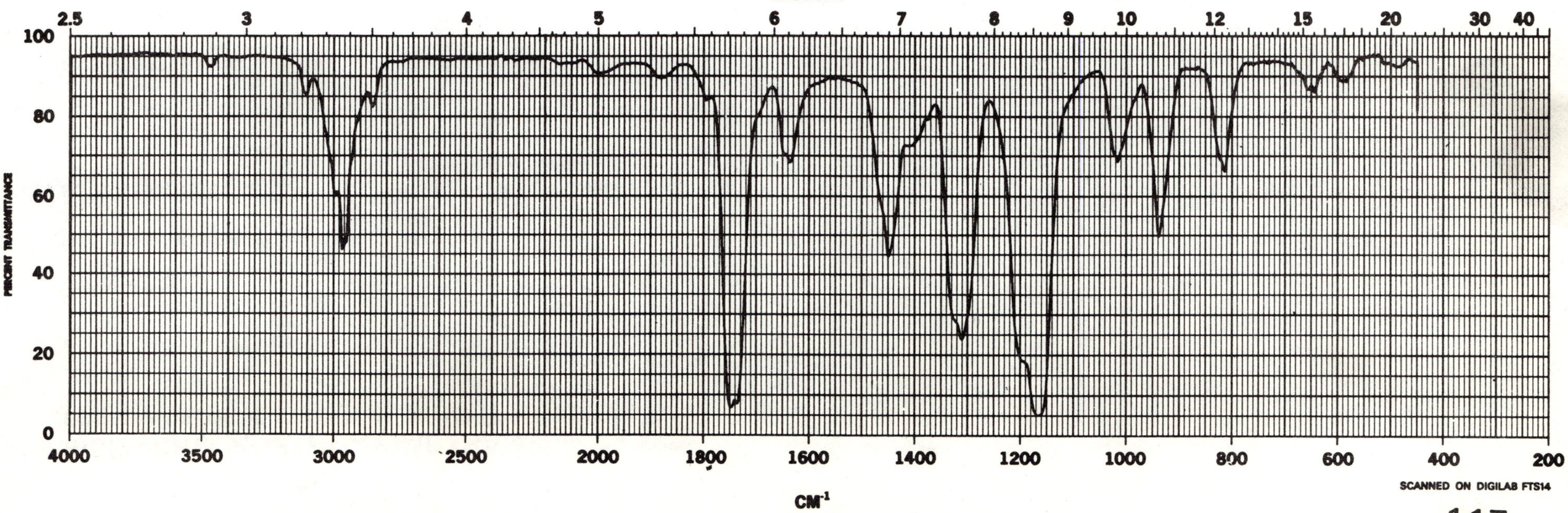

SCANNED ON DIGILAB FTS14

117

BIS(2–CHLOROETHOXY)METHANE
FORMALDEHYDE, BIS(2–CHLOROETHYL) ACETAL

$C_5H_{10}Cl_2O_2$

MOLECULAR WEIGHT: 173.04

SOURCE OF SAMPLE: Aldrich Chemical Company, Inc.
Milwaukee, Wisconsin

$Cl\text{-}CH_2CH_2\text{-}O-CH_2-O\text{-}CH_2CH_2\text{-}Cl$

CAS REGISTRY NO.: 111-91-1

NIOSH NUMBER: PA36750

CAPILLARY CELL: Neat

CONDENSED PHASE

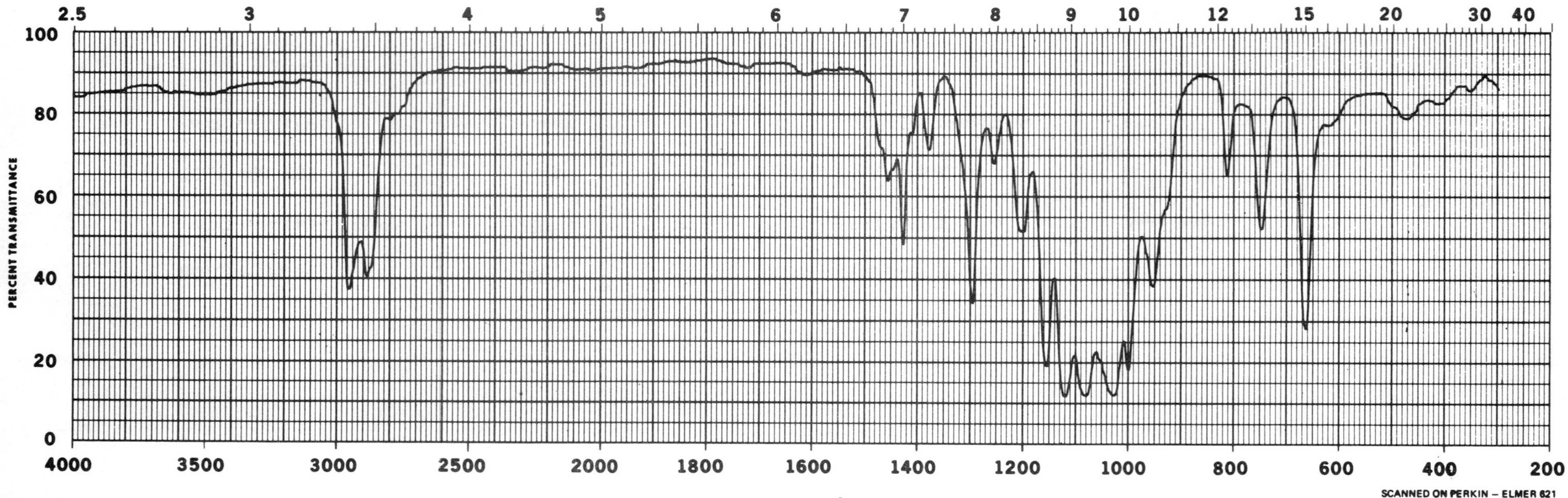

BOILING POINT:	218°C	INJECTION AMT.:	0.7 μl
VAPORIZATION TEMP.:	200°C	PRESSURE:	Atmospheric
CELL VOLUME:	7.5 cc	MAX./MIN.:	—
PATH LENGTH:	7.6 cm	SCALE EXPANSION:	0.98

VAPOR PHASE

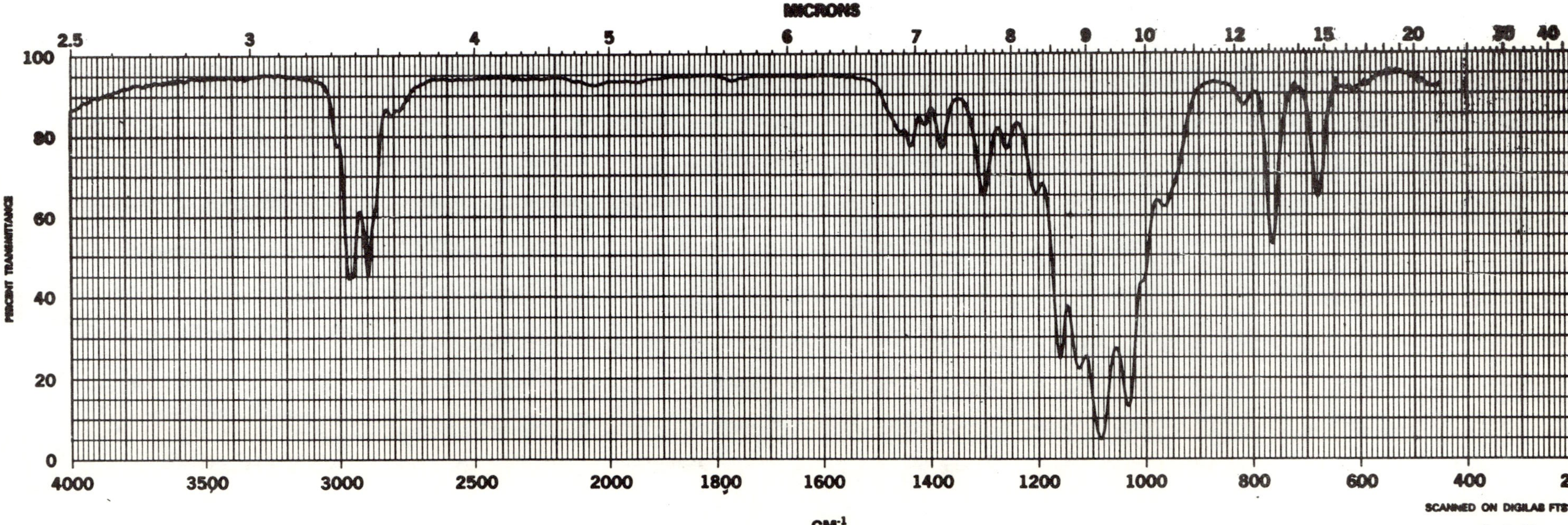

BROMOMETHANE
BROM-O-GAS
METHYL BROMIDE
METHOGAS
ROTOX

CH_3Br

Br − CH_3

MOLECULAR WEIGHT:	94.94
SOURCE OF SAMPLE:	MCB Manufacturing Chemists Norwood, Ohio
CAS REGISTRY NO.:	74-83-9
NIOSH NUMBER:	PA49000
GAS CELL:	10 cm A = 350 mm vac. B = 700 mm vac.

A GAS AT 25°C

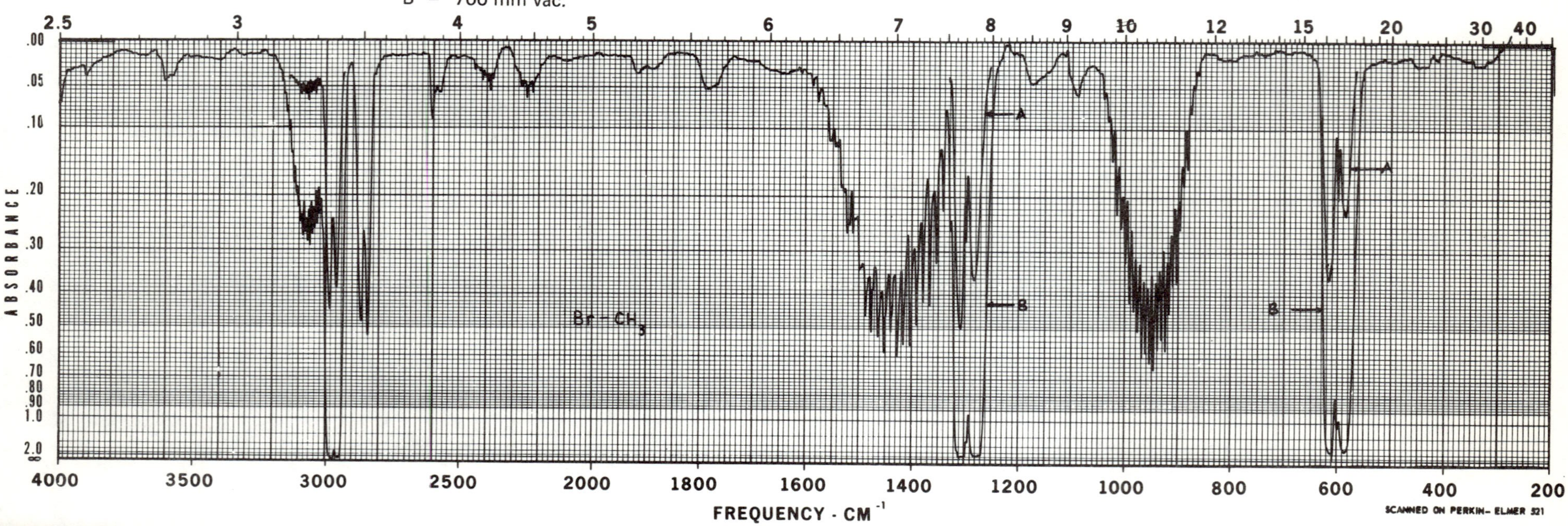

BOILING POINT:	3.6°C	INJECTION AMT.:	999 μl
VAPORIZATION TEMP.:	200°C	PRESSURE:	Atmospheric
CELL VOLUME:	7.5 cc	MAX./MIN.:	101.9/46.3
PATH LENGTH:	7.6 cm	SCALE EXPANSION:	0.94

VAPOR PHASE

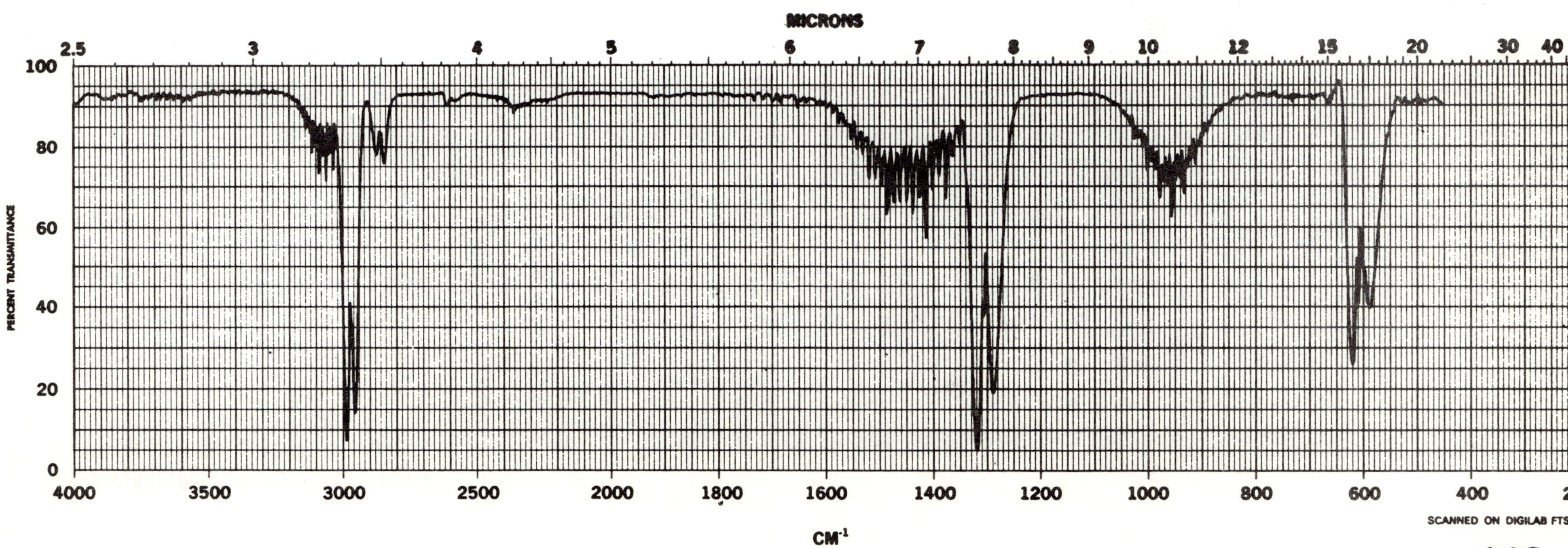

SCANNED ON DIGILAB FTS14

BROMODICHLOROMETHANE
DICHLOROBROMOMETHANE

$CHBrCl_2$

MOLECULAR WEIGHT: 163.83

SOURCE OF SAMPLE: Aldrich Chemical Company, Inc.
Milwaukee, Wisconsin

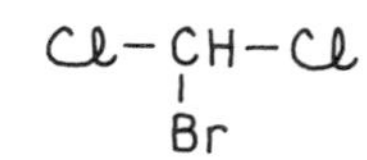

CAS REGISTRY NO.: 75-27-4

NIOSH NUMBER: –

CAPILLARY CELL: 0.015 mm

CONDENSED PHASE

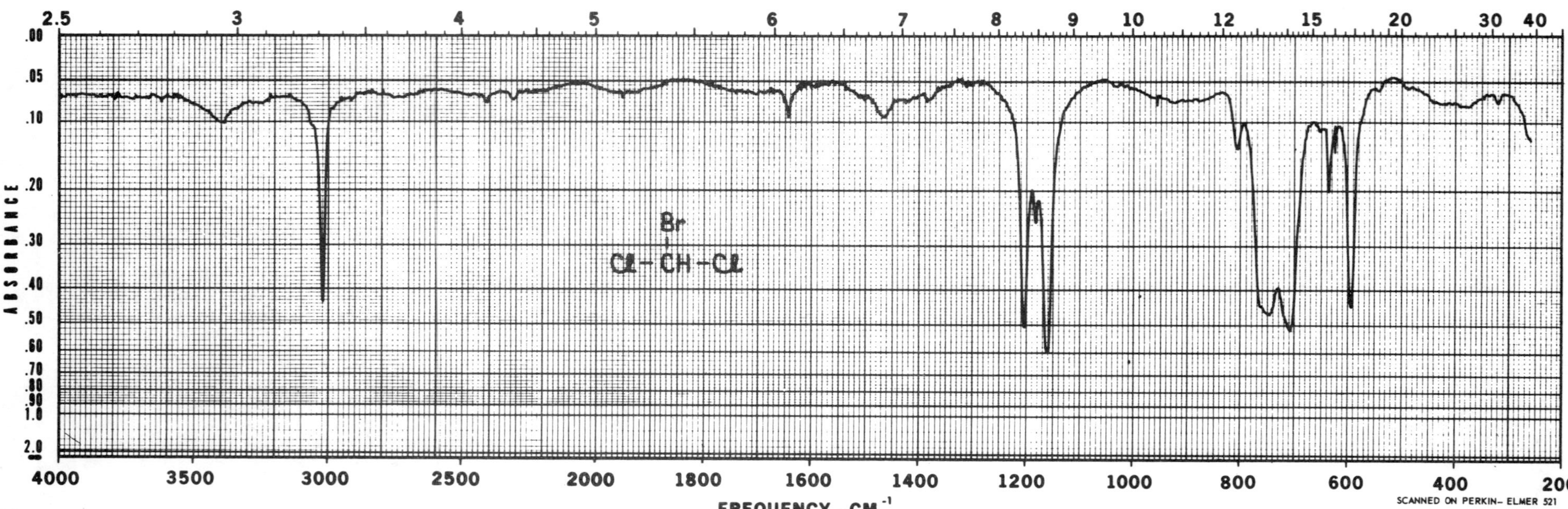

BOILING POINT:	$87^{o}C$	INJECTION AMT.:	–
VAPORIZATION TEMP.:	$200^{o}C$	PRESSURE:	Atmospheric
CELL VOLUME:	7.5 cc	MAX./MIN.:	99.4/12.6
PATH LENGTH:	7.6 cm	SCALE EXPANSION:	1.06

VAPOR PHASE

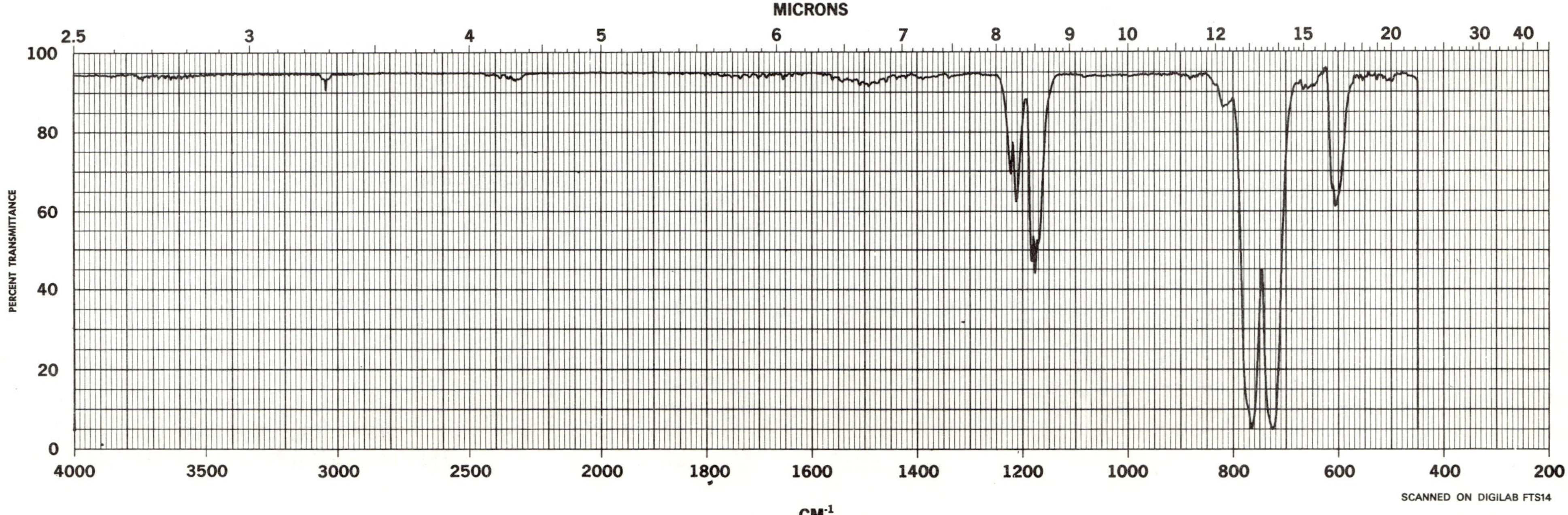

SCANNED ON DIGILAB FTS14

CHLORODIBROMOMETHANE
DIBROMOCHLOROMETHANE

$CHBr_2Cl$

MOLECULAR WEIGHT: 208.29

SOURCE OF SAMPLE: Fluka AG
Buchs, Switzerland

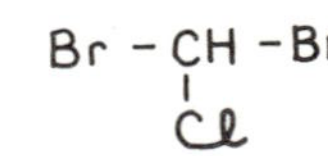

CAS REGISTRY NO.: 124-48-1

NIOSH NUMBER: -

CAPILLARY CELL: Neat

CONDENSED PHASE

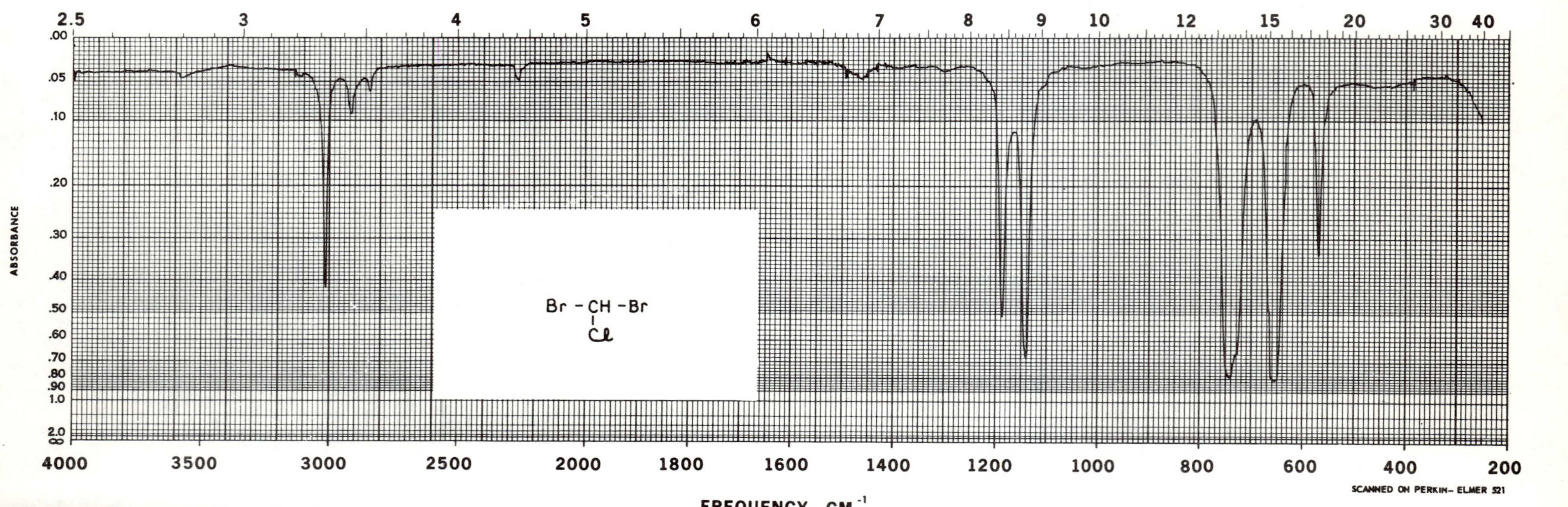

BOILING POINT:	118-120°C/748 mm	INJECTION AMT.:	0.5 µl
VAPORIZATION TEMP.:	200°C	PRESSURE:	Atmospheric
CELL VOLUME:	7.5 cc	MAX./MIN.:	—
PATH LENGTH:	7.6 cm	SCALE EXPANSION:	0.93

VAPOR PHASE

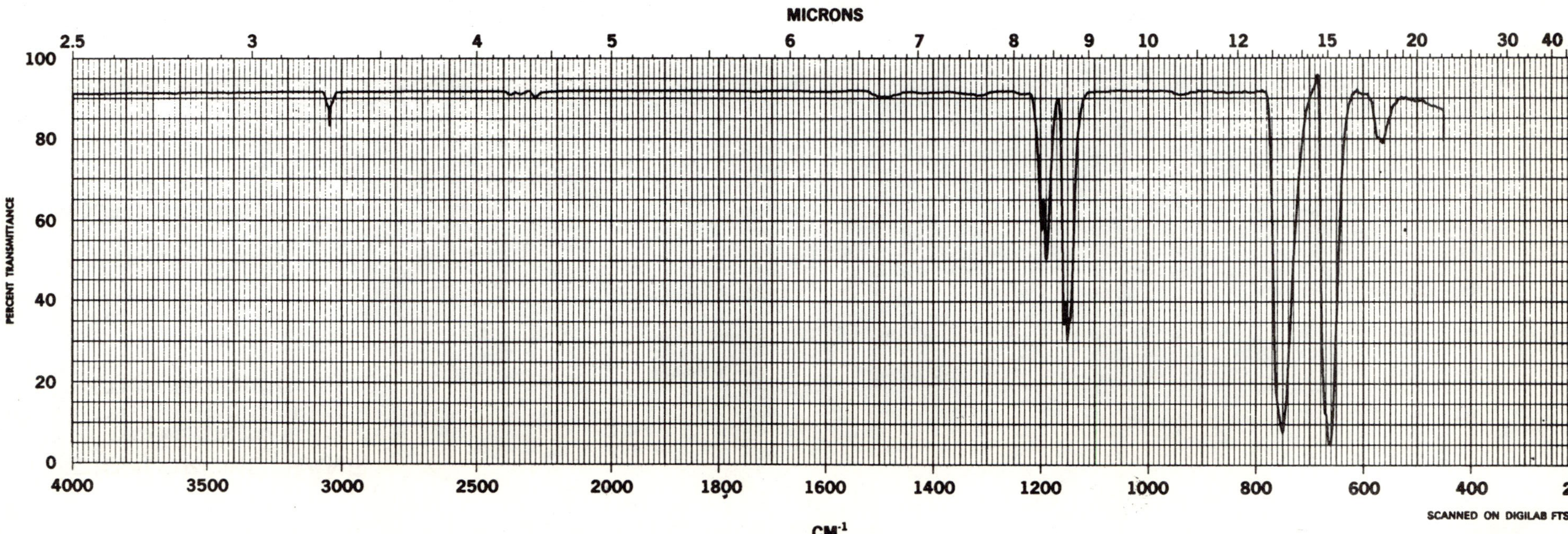

SCANNED ON DIGILAB FTS14

DICHLOROMETHANE
METHANE DICHLORIDE
METHYLENE CHLORIDE
METHYLENE DICHLORIDE

CH_2Cl_2

MOLECULAR WEIGHT: 84.93

SOURCE OF SAMPLE: Fluka AG
Buchs, Switzerland

CH_2Cl_2

CAS REGISTRY NO.: 75-09-2

NIOSH NUMBER: PA80500

CAPILLARY CELL: 0.015 mm

CONDENSED PHASE

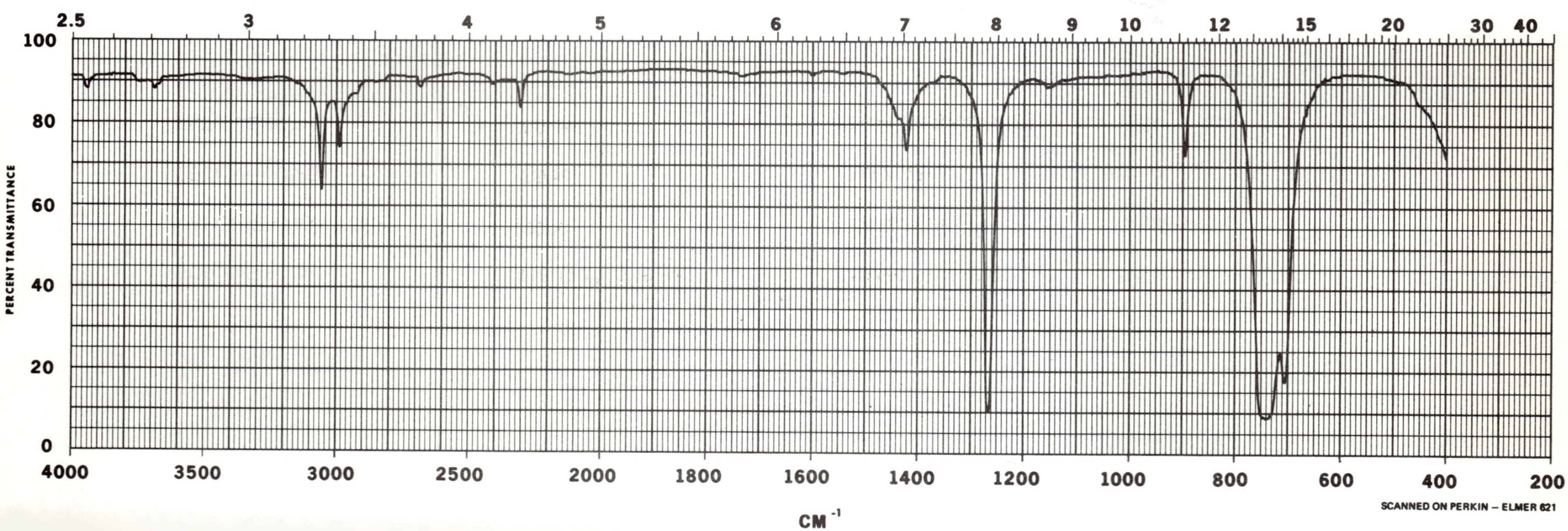

BOILING POINT:	39.5-40.5°C	INJECTION AMT.:	0.5 µl
VAPORIZATION TEMP.:	200°C	PRESSURE:	Atmospheric
CELL VOLUME:	7.5 cc	MAX./MIN.:	—
PATH LENGTH:	7.6 cm	SCALE EXPANSION:	0.92

VAPOR PHASE

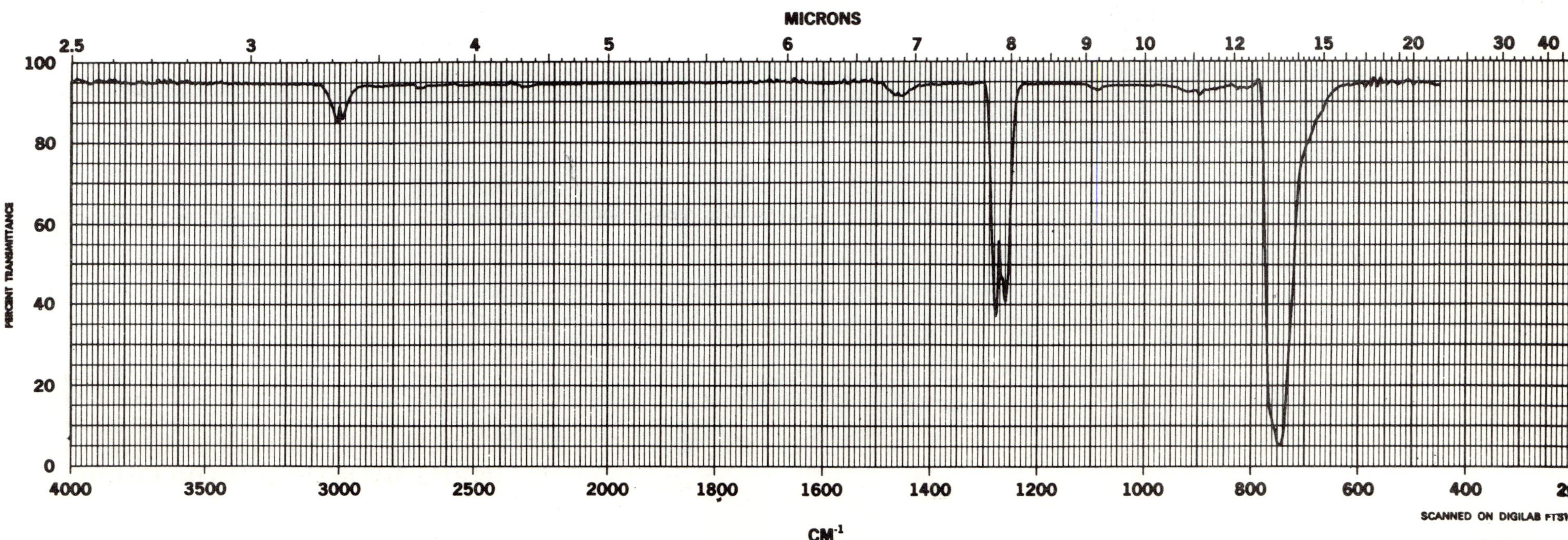

DICHLORODIFLUOROMETHANE

ALGOFRENE TYPE 2
ARCTON 6
FREON 12
GENETRON 12

CCl_2F_2

MOLECULAR WEIGHT:	120.91
SOURCE OF SAMPLE:	E. I. du Pont de Nemours and Company, Inc. Wilmington, Delaware
CAS REGISTRY NO.:	75-71-8
NIOSH NUMBER:	PA82000
GAS CELL:	10 cm @ 6.0 mm

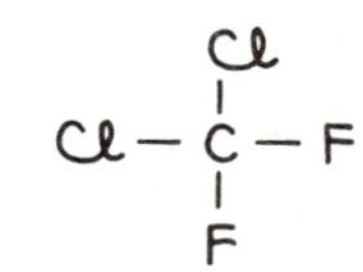

A GAS AT 25°C

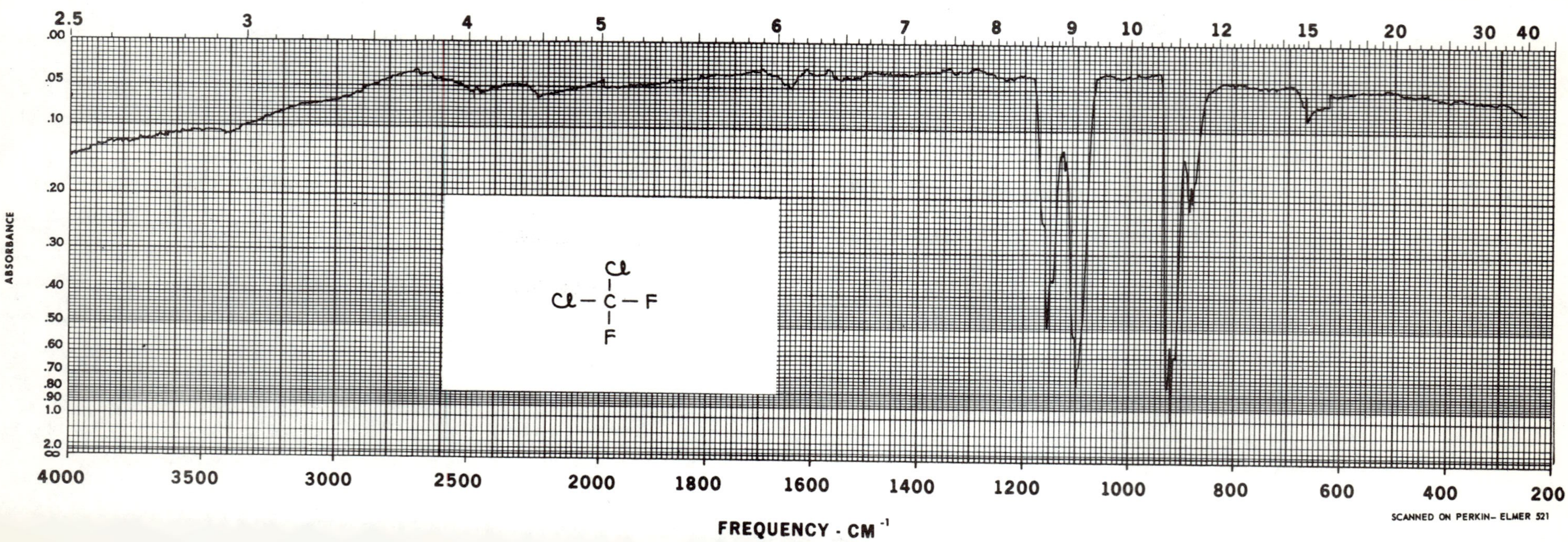

BOILING POINT:	−29.8°C	INJECTION AMT.:	100 μl
VAPORIZATION TEMP.:	200°C	PRESSURE:	Atmospheric
CELL VOLUME:	7.5 cc	MAX./MIN.:	100.5/6.6
PATH LENGTH:	7.6 cm	SCALE EXPANSION:	0.98

VAPOR PHASE

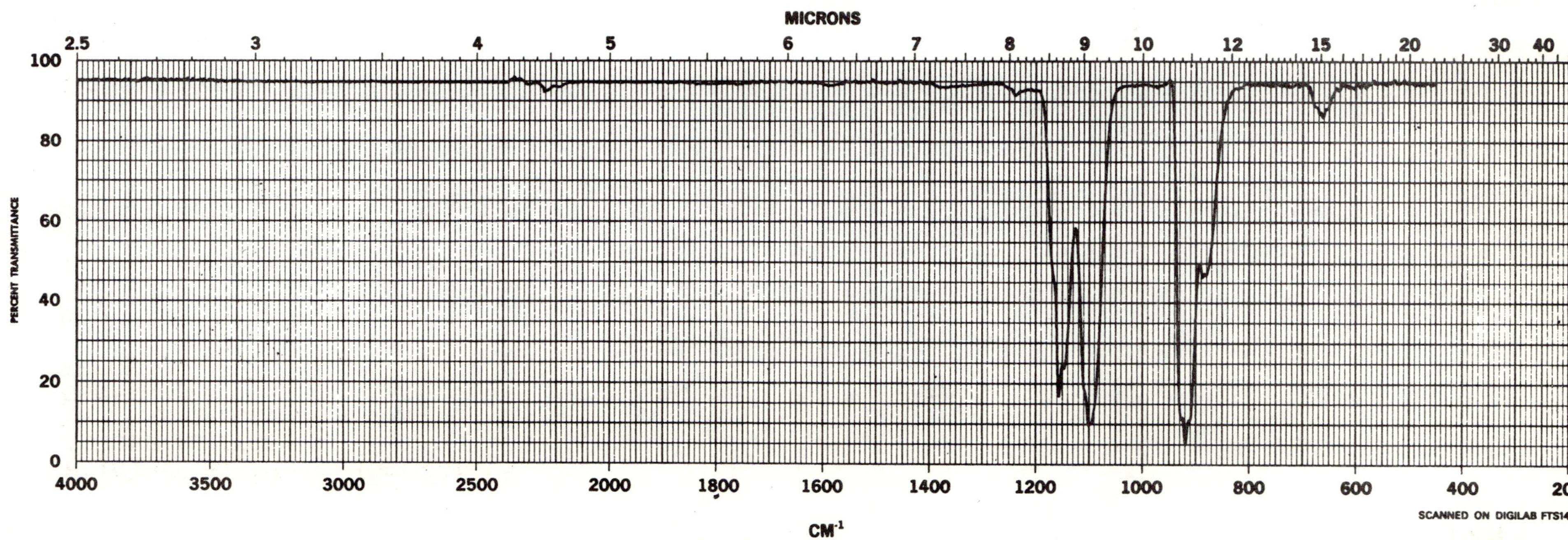

SCANNED ON DIGILAB FTS14

FLUOROTRICHLOROMETHANE
ALGOFRENE TYPE 1
TRICHLOROFLUOROMETHANE
GENETRON 11
FREON 11

CCl_3F

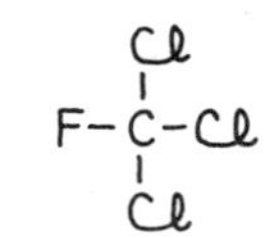

MOLECULAR WEIGHT:	137.37
SOURCE OF SAMPLE:	E. Merck AG Darmstadt, Germany
CAS REGISTRY NO.:	75-69-4
NIOSH NUMBER:	PB61250
GAS CELL:	10 cm

A GAS AT 25°C

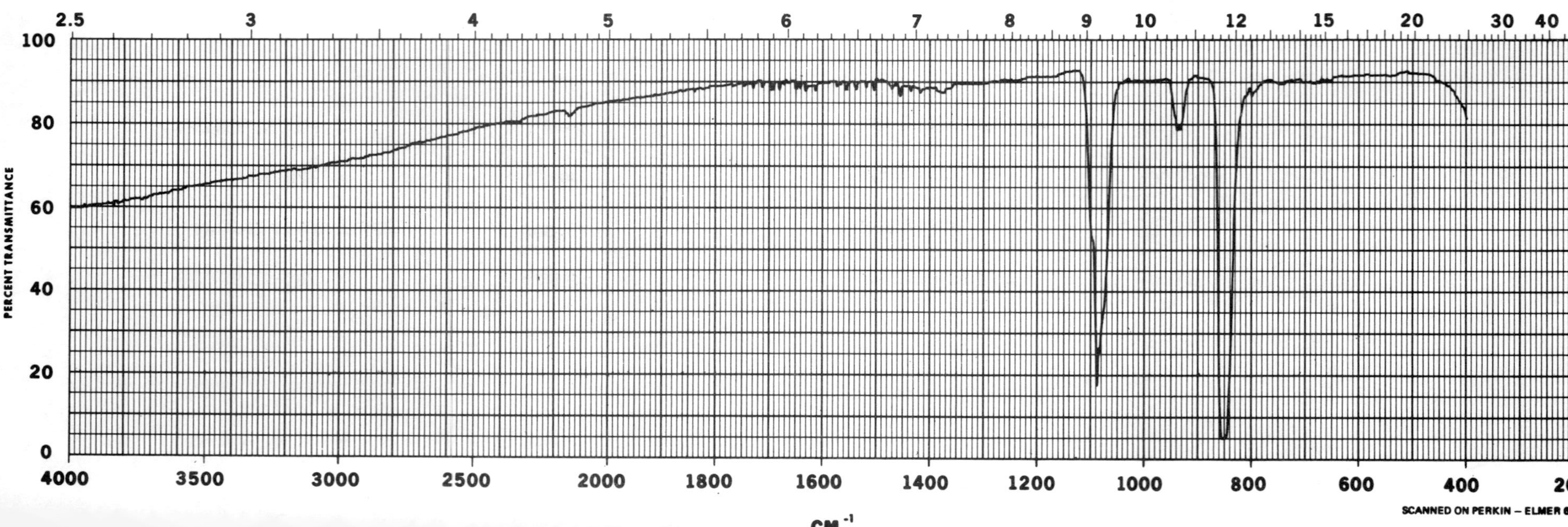

BOILING POINT:	—	INJECTION AMT.:	0.1 μl
VAPORIZATION TEMP.:	200°C	PRESSURE:	Atmospheric
CELL VOLUME:	7.5 cc	MAX./MIN.:	100.1/27.6
PATH LENGTH:	7.6 cm	SCALE EXPANSION:	1.25

VAPOR PHASE

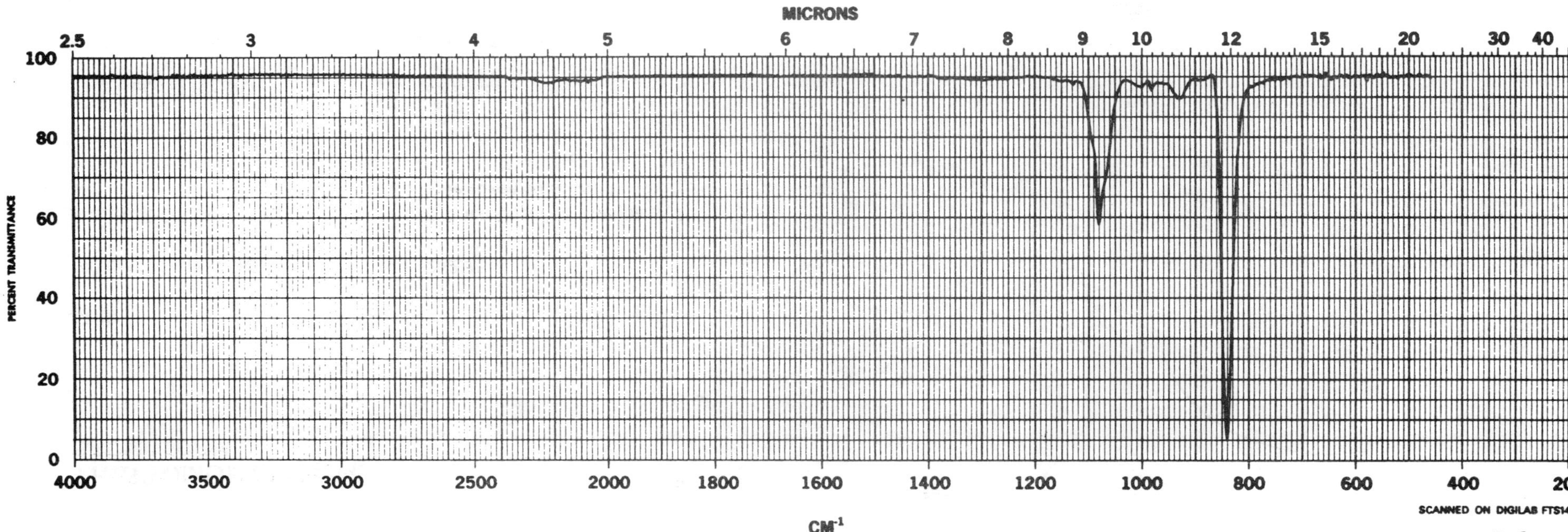

SCANNED ON DIGILAB FTS14

TRIBROMOMETHANE
BROMOFORM

$CHBr_3$

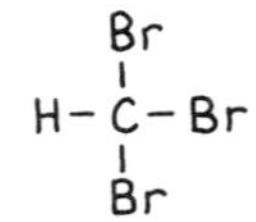

MOLECULAR WEIGHT:	252.75
SOURCE OF SAMPLE:	Aldrich Chemical Company, Inc. Milwaukee, Wisconsin
CAS REGISTRY NO.:	75-25-2
NIOSH NUMBER:	PB56000
CAPILLARY CELL:	Neat

CONDENSED PHASE

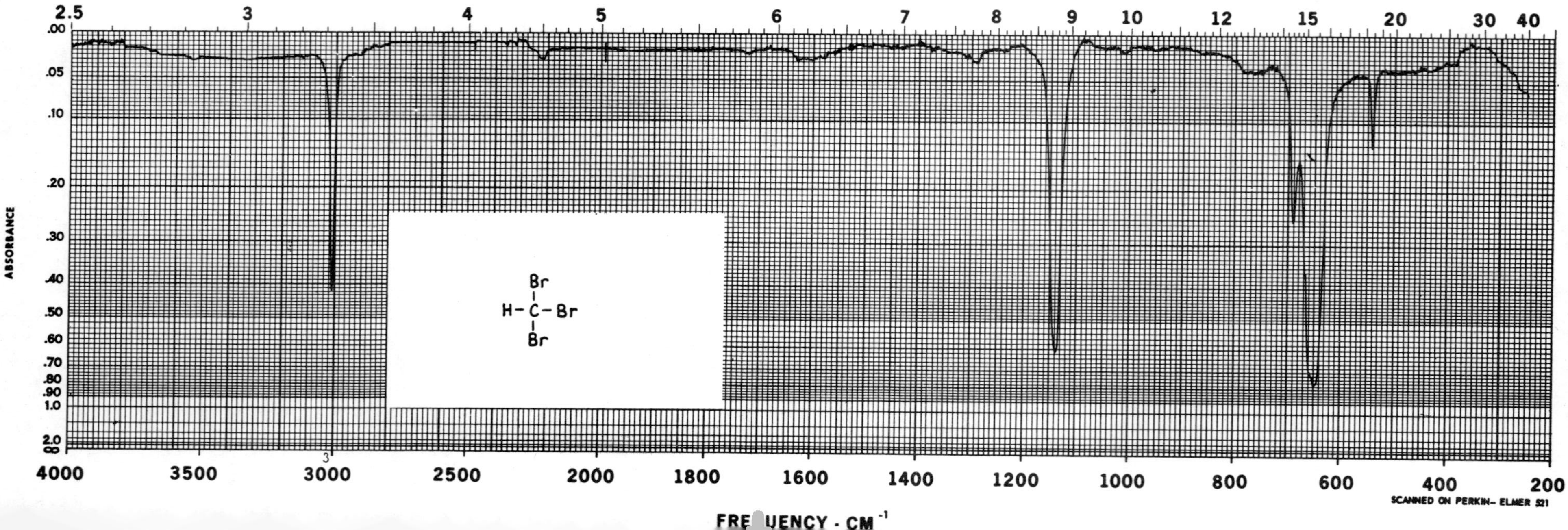

BOILING POINT:	146-150°C	INJECTION AMT.:	0.5 μl
VAPORIZATION TEMP.:	200°C	PRESSURE:	Atmospheric
CELL VOLUME:	7.5 cc	MAX./MIN.:	—
PATH LENGTH:	7.6 cm	SCALE EXPANSION:	0.91

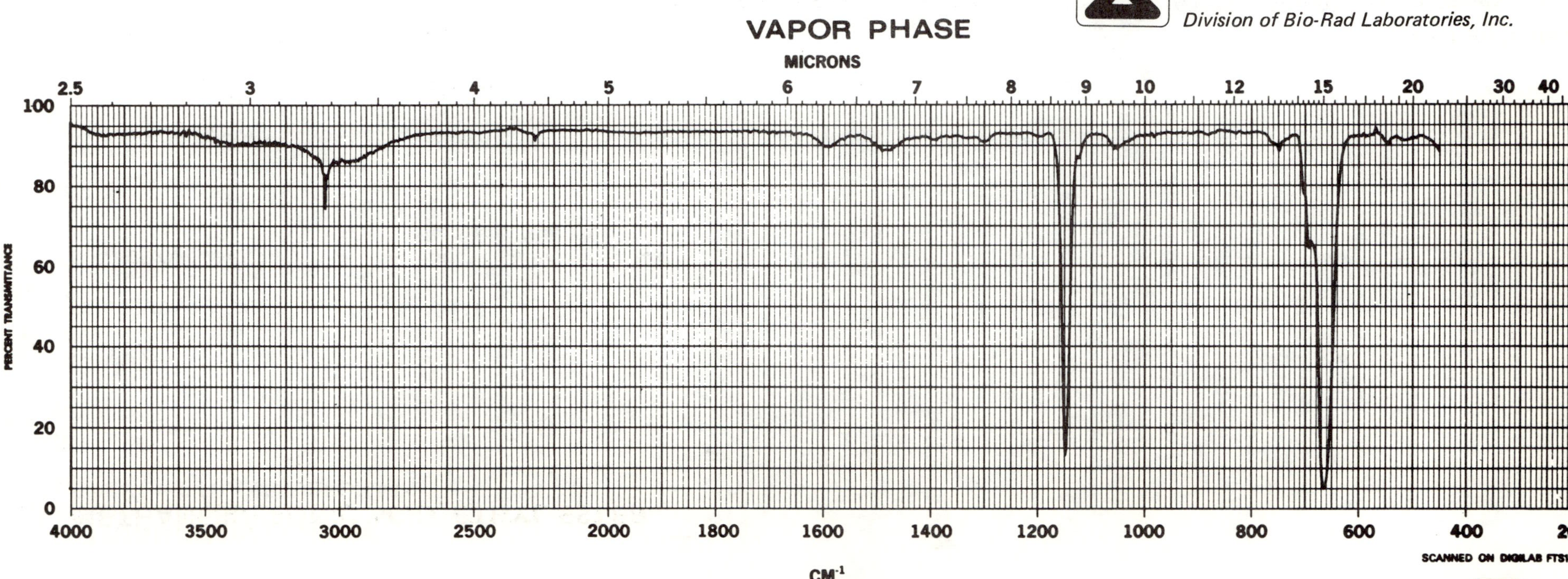

METHANESULFONIC ACID, ETHYL ESTER
ETHYL METHANESULFONATE

$C_3H_8O_3S$

MOLECULAR WEIGHT: 124.16

SOURCE OF SAMPLE: Eastman Organic Chemicals
Rochester, New York

$CH_3-SO_3-CH_2CH_3$

CAS REGISTRY NO.: 62-50-0

NIOSH NUMBER: PB21000

CAPILLARY CELL: Neat

CONDENSED PHASE

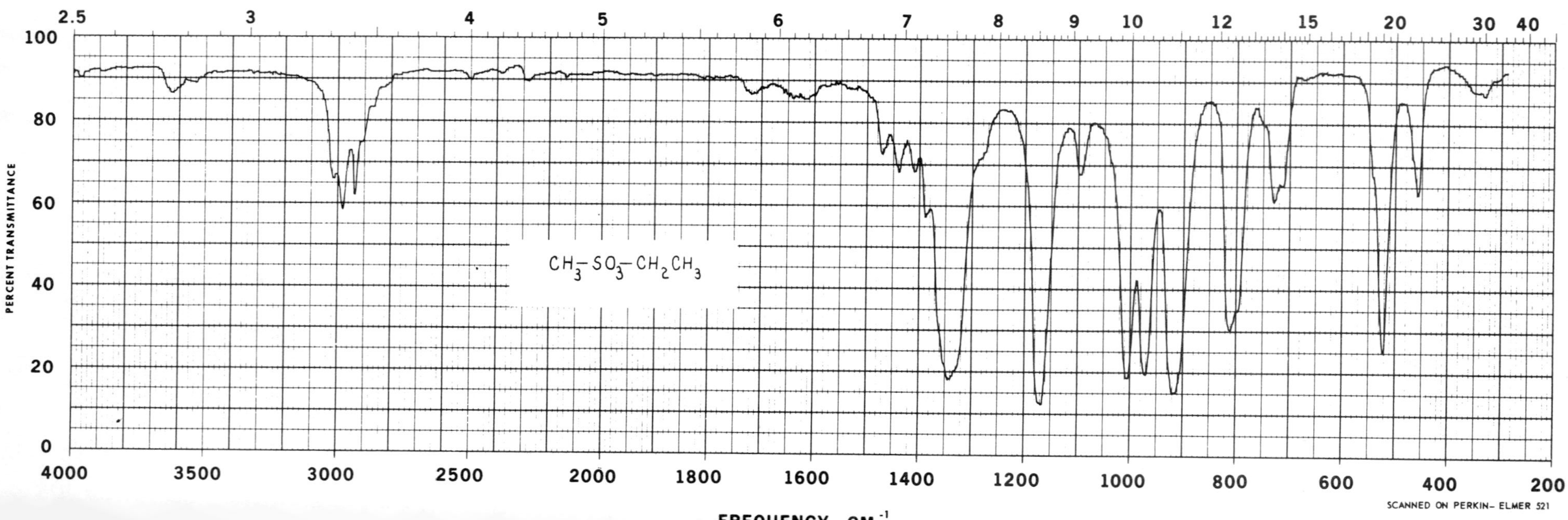

SCANNED ON PERKIN- ELMER 521

BOILING POINT:	85-86°C/10 mm	INJECTION AMT.:	0.5 μl
VAPORIZATION TEMP.:	240°C	PRESSURE:	Atmospheric
CELL VOLUME:	7.5 cc	MAX./MIN.:	100.6/8.4
PATH LENGTH:	7.6 cm	SCALE EXPANSION:	1.00

VAPOR PHASE

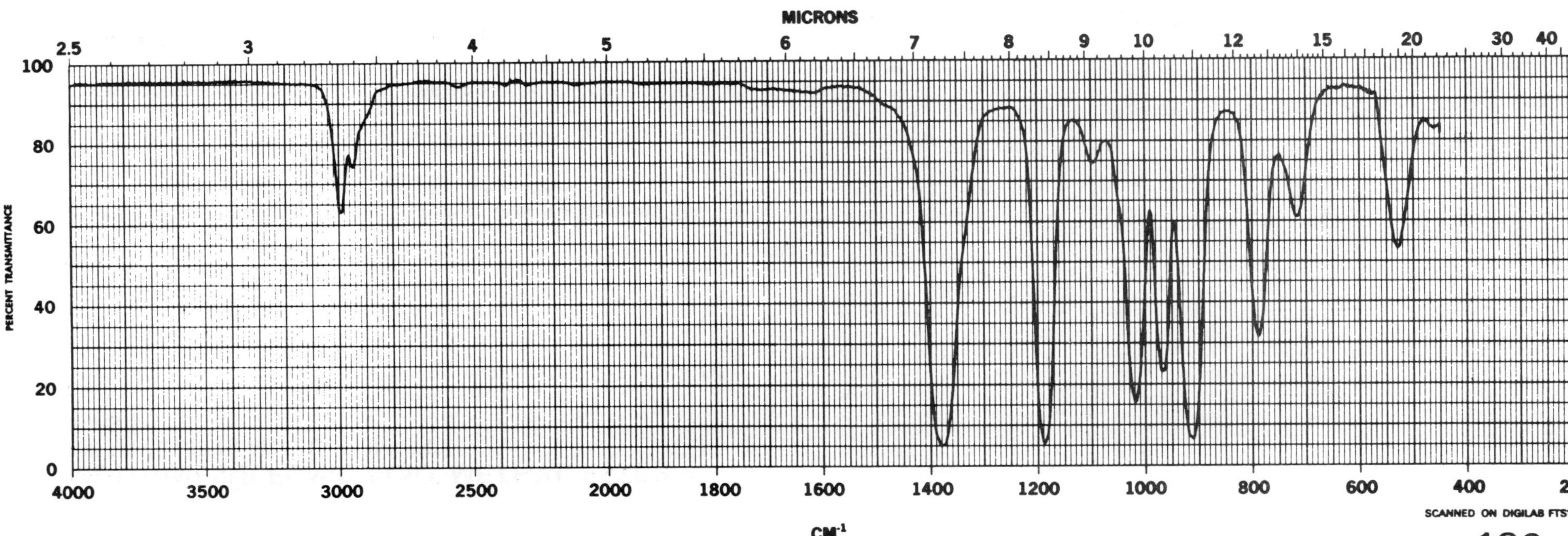

SCANNED ON DIGILAB FTS14

METHANESULFONIC ACID, METHYL ESTER
METHYL METHANESULFONATE
MMS

$C_2H_6O_3S$

MOLECULAR WEIGHT: 110.13

SOURCE OF SAMPLE: Aldrich Chemical Company, Inc.
Milwaukee, Wisconsin

$CH_3-SO_3-CH_3$

CAS REGISTRY NO.: 66-27-3

NIOSH NUMBER: PB26250

CAPILLARY CELL: Neat

CONDENSED PHASE

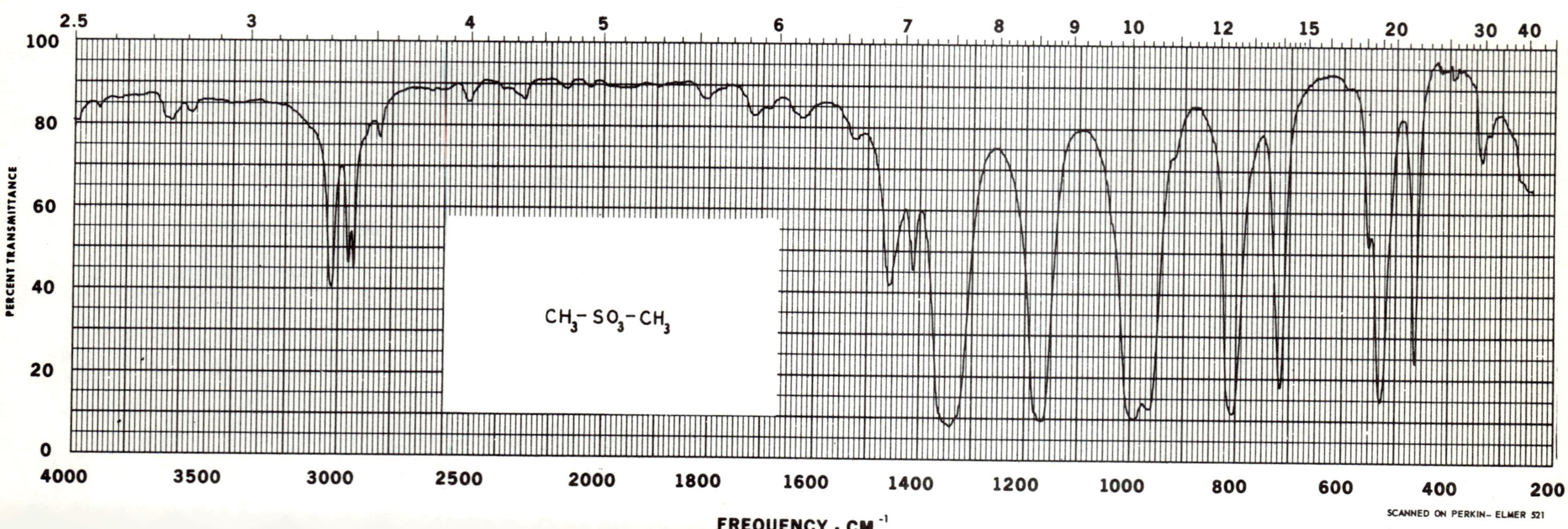

BOILING POINT:	202-203°C	INJECTION AMT.:	1.0 μl
VAPORIZATION TEMP.:	200°C	PRESSURE:	Atmospheric
CELL VOLUME:	7.5 cc	MAX./MIN.:	—
PATH LENGTH:	7.6 cm	SCALE EXPANSION:	0.93

VAPOR PHASE

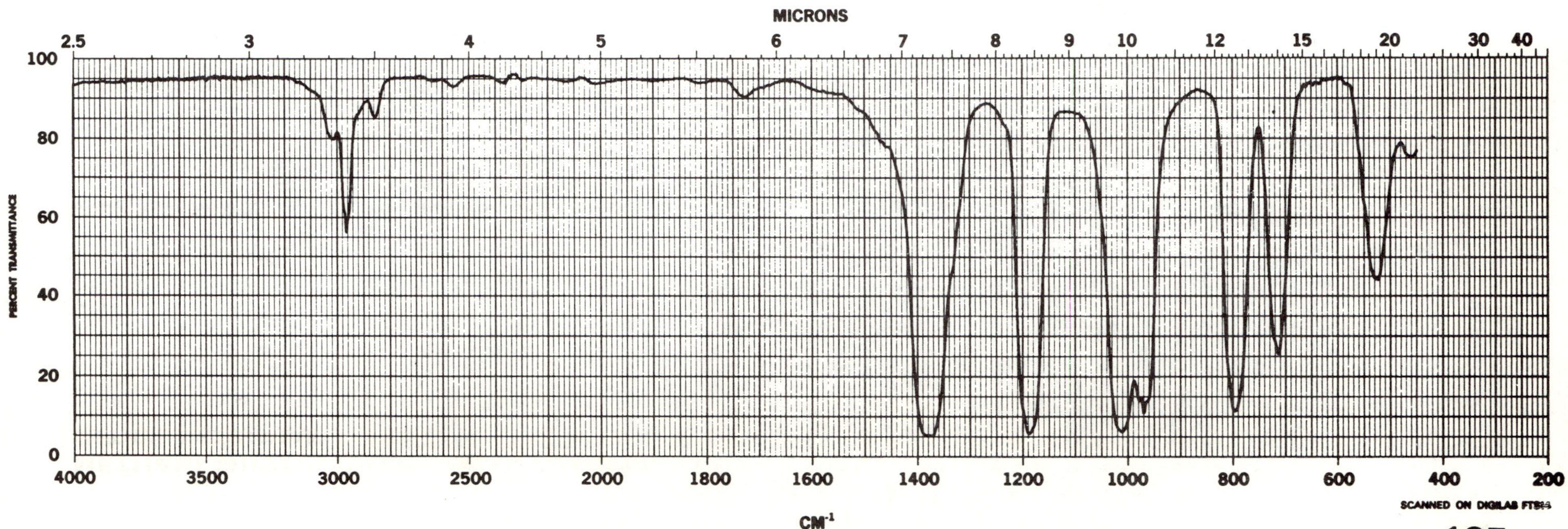

SCANNED ON DIGILAB FTS-14

METHANETHIOL
METHYL MERCAPTAN

CH_4S

MOLECULAR WEIGHT: 48.11

SOURCE OF SAMPLE: Matheson Gas Products
Division of Will Ross, Inc.
Lyndhurst, New Jersey

CH_3-SH

CAS REGISTRY NO.: 74-93-1

NIOSH NUMBER: PB43750

GAS CELL: 10 cm

A GAS AT 25°C

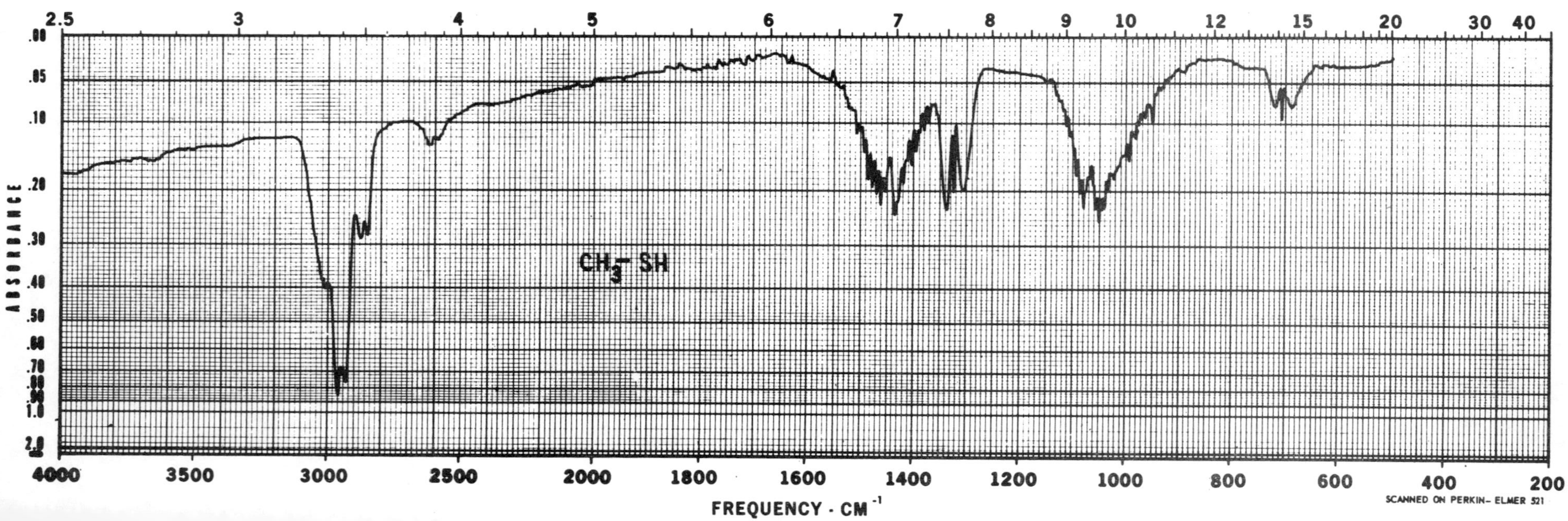

BOILING POINT:	5.96°C	INJECTION AMT.:	—
VAPORIZATION TEMP.:	23°C	PRESSURE:	Atmospheric
CELL VOLUME:	7.5 cc	MAX./MIN.:	100.5/61.2
PATH LENGTH:	7.6 cm	SCALE EXPANSION:	2.15

VAPOR PHASE

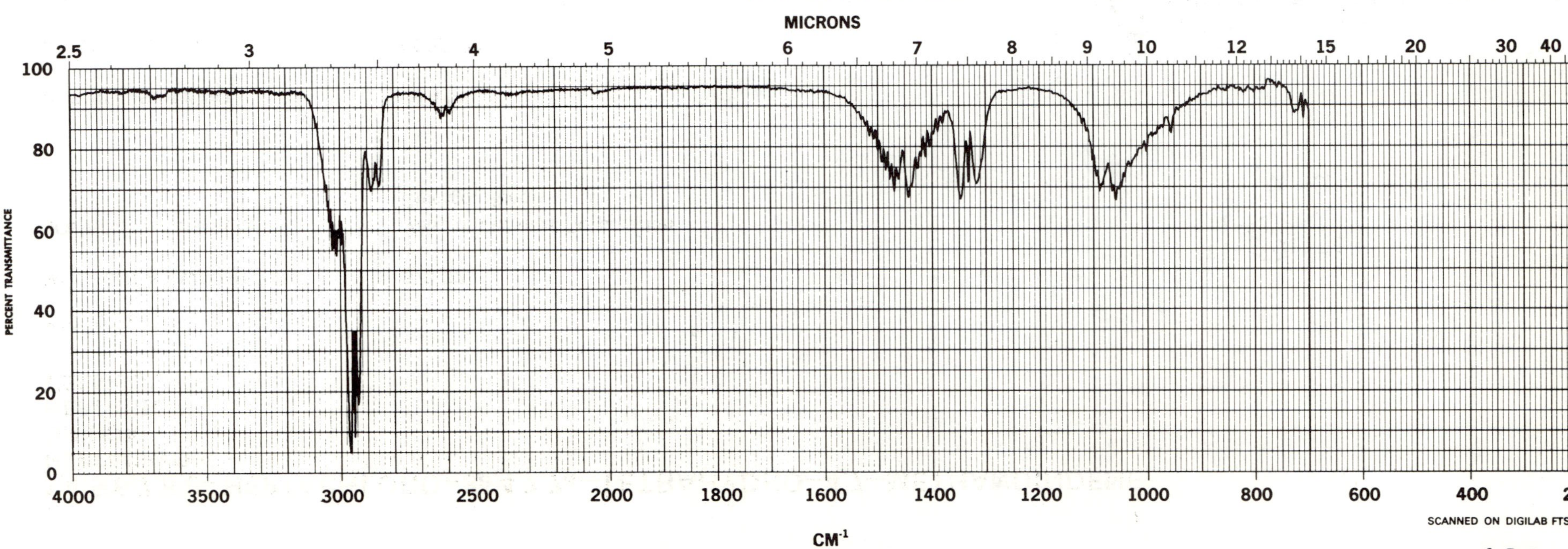

SCANNED ON DIGILAB FTS14

128

1,4,5,6,7,8,8–HEPTACHLORO–3a,4,7,7a–TETRAHYDRO–4,7–METHANOINDENE

HEPTACHLOR
DRINOX
HEPTAGRAN
VELSICOL 104

$C_{10}H_5Cl_7$

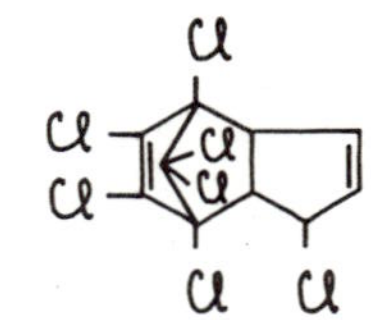

MOLECULAR WEIGHT:	373.32
SOURCE OF SAMPLE:	Velsicol Chemical Corporation Chicago, Illinois
CAS REGISTRY NO.:	76-44-8
NIOSH NUMBER:	PC07000

KBr Wafer

CONDENSED PHASE

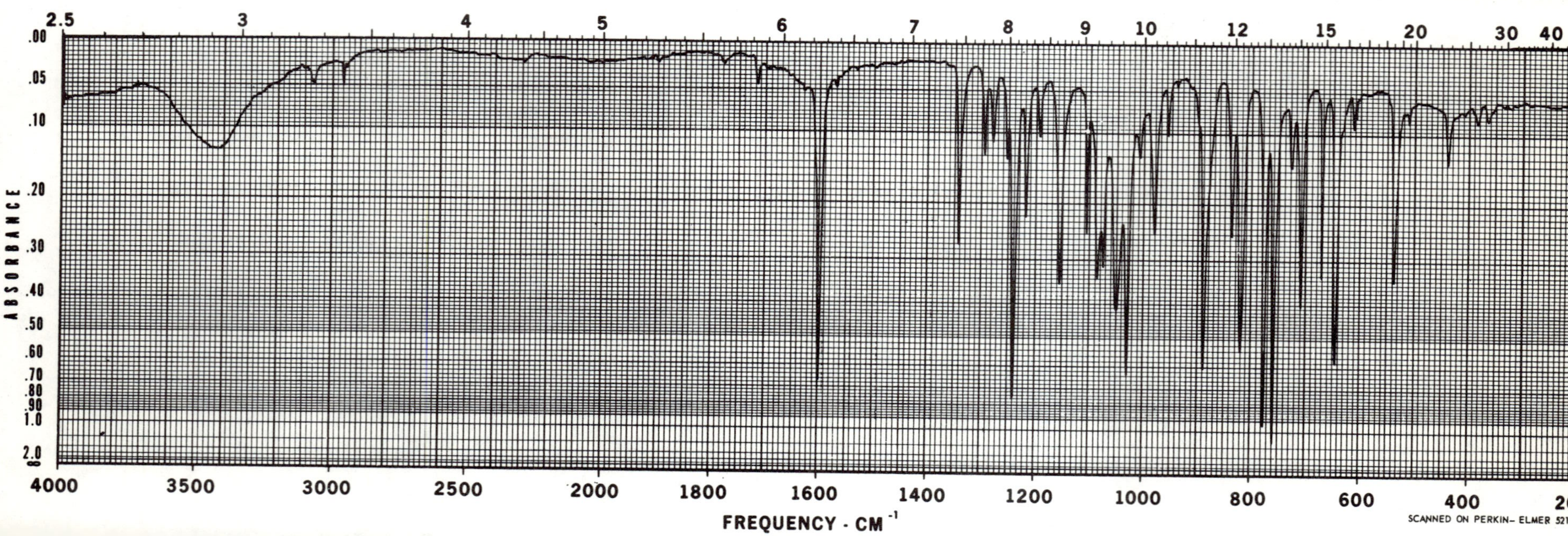

MELTING POINT:	95-96°C	INJECTION AMT.:	—
VAPORIZATION TEMP.:	200°C	PRESSURE:	Atmospheric
CELL VOLUME:	7.5 cc	MAX./MIN.:	100.1/50.7
PATH LENGTH:	7.6 cm	SCALE EXPANSION:	1.86

VAPOR PHASE

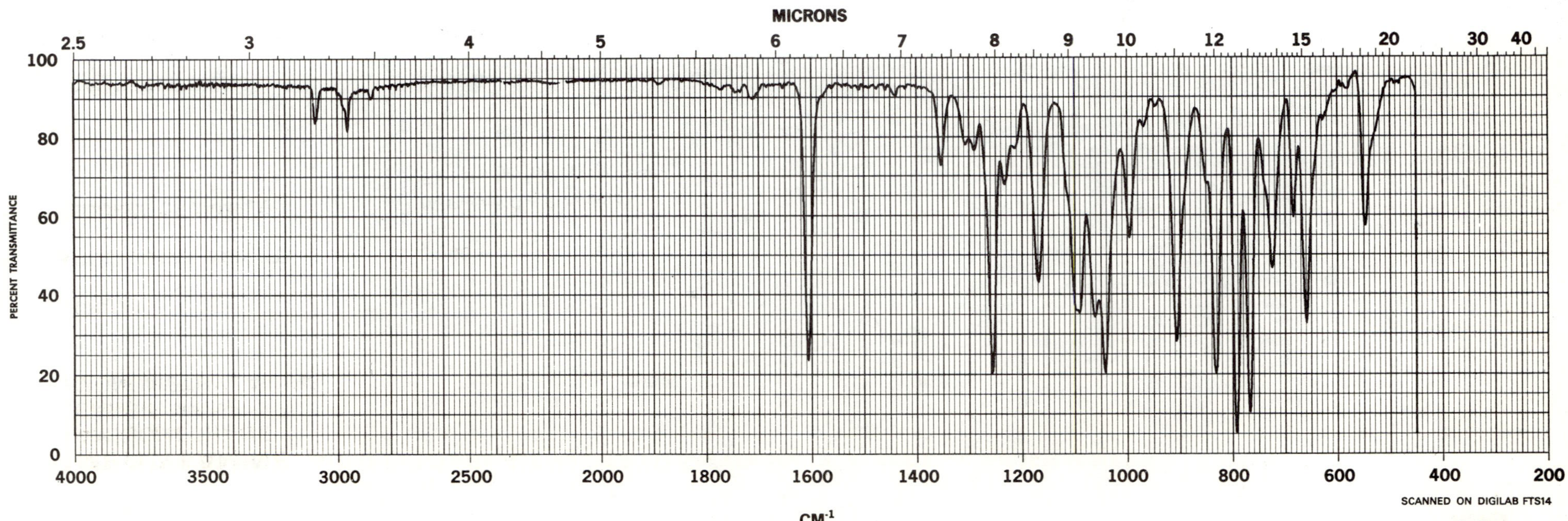

SCANNED ON DIGILAB FTS14

METHYLAMINE
AMINOMETHANE
CARBINAMINE
MONOMETHYLAMINE

CH_5N

$CH_3{-}NH_2$

MOLECULAR WEIGHT:	31.06
SOURCE OF SAMPLE:	Matheson Gas Products East Rutherford, New Jersey
CAS REGISTRY NO.:	74-89-5
NIOSH NUMBER:	PF63000
Path Length:	10 cm
Complete Scan:	20 mm Hg

A GAS AT 25°C

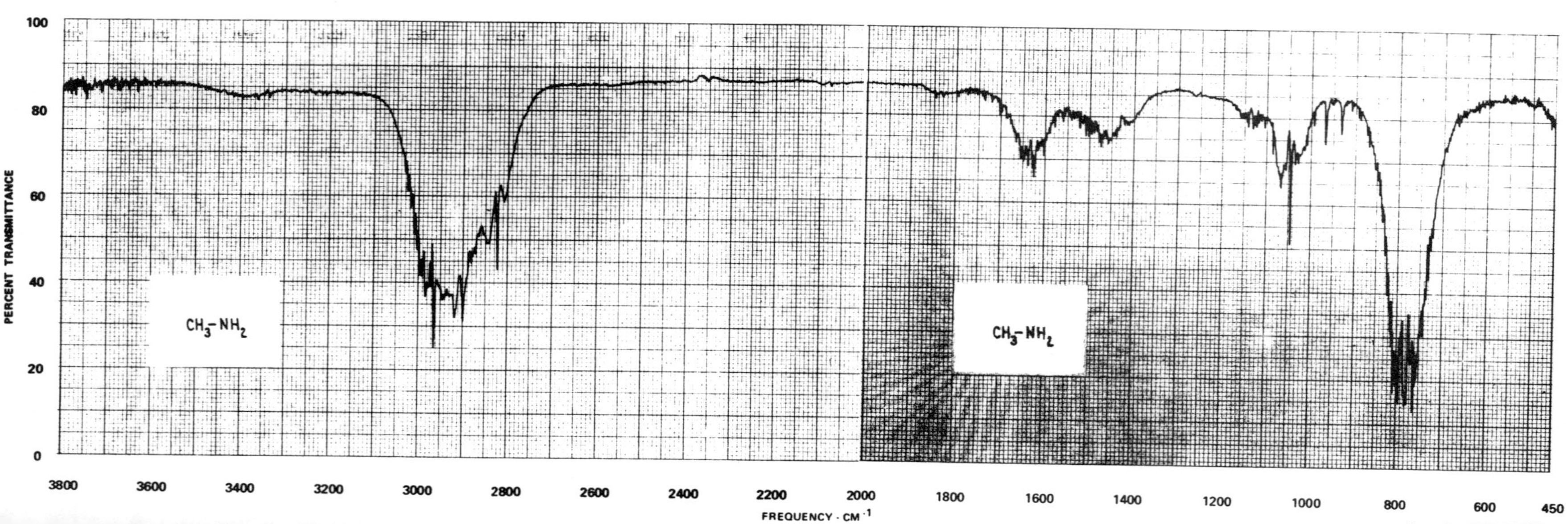

BOILING POINT:	$-6.3^{\circ}C$	INJECTION AMT.:	–
VAPORIZATION TEMP.:	$200^{\circ}C$	PRESSURE:	Atmospheric
CELL VOLUME:	7.5 cc	MAX./MIN.:	101.0/36.5
PATH LENGTH:	7.6 cm	SCALE EXPANSION:	1.43

VAPOR PHASE

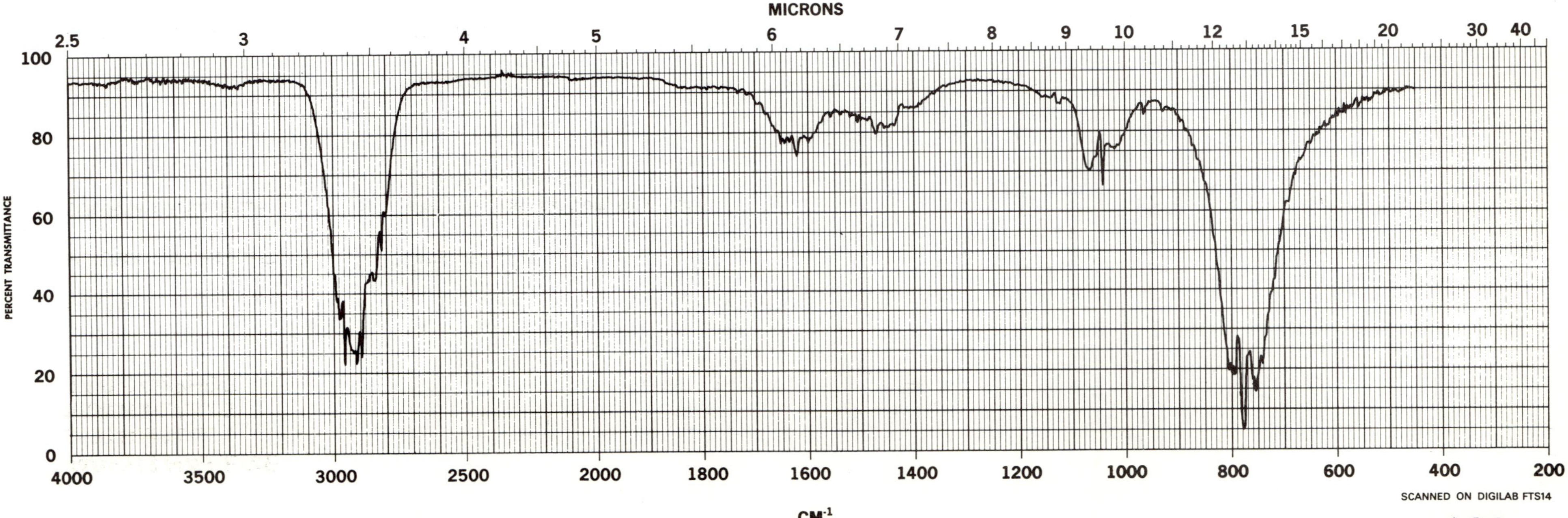

SCANNED ON DIGILAB FTS14

NAPHTHALENE
MOTH BALLS
MOTH FLAKES
TAR CAMPHOR

$C_{10}H_8$

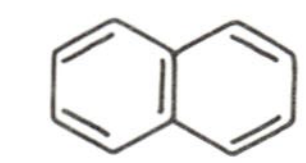

MOLECULAR WEIGHT:	128.18
SOURCE OF SAMPLE:	Aldrich Chemical Company, Inc. Milwaukee, Wisconsin
CAS REGISTRY NO.:	91-20-3
NIOSH NUMBER:	QJ05250
CAPILLARY CELL:	Melt

CONDENSED PHASE

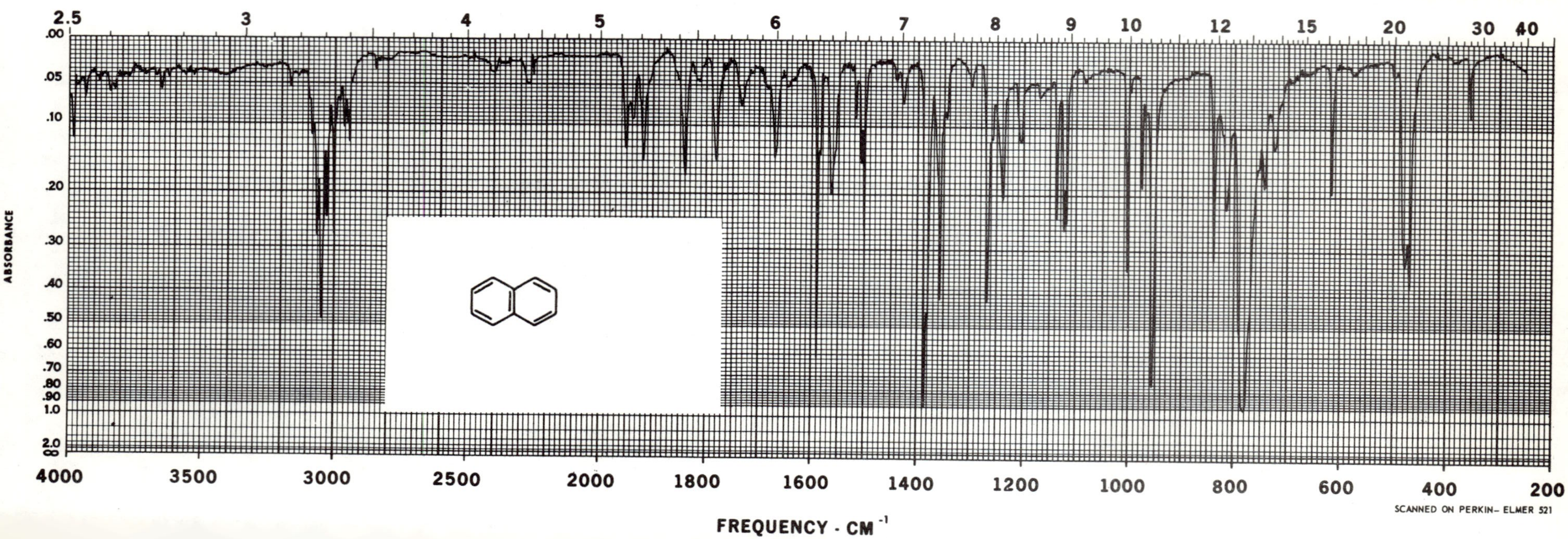

BOILING POINT:	218°C	INJECTION AMT.:	0.86 mg
VAPORIZATION TEMP.:	280°C	PRESSURE:	Atmospheric
CELL VOLUME:	7.5 cc	MAX./MIN.:	103.1/8.1
PATH LENGTH:	7.6 cm	SCALE EXPANSION:	0.97

VAPOR PHASE

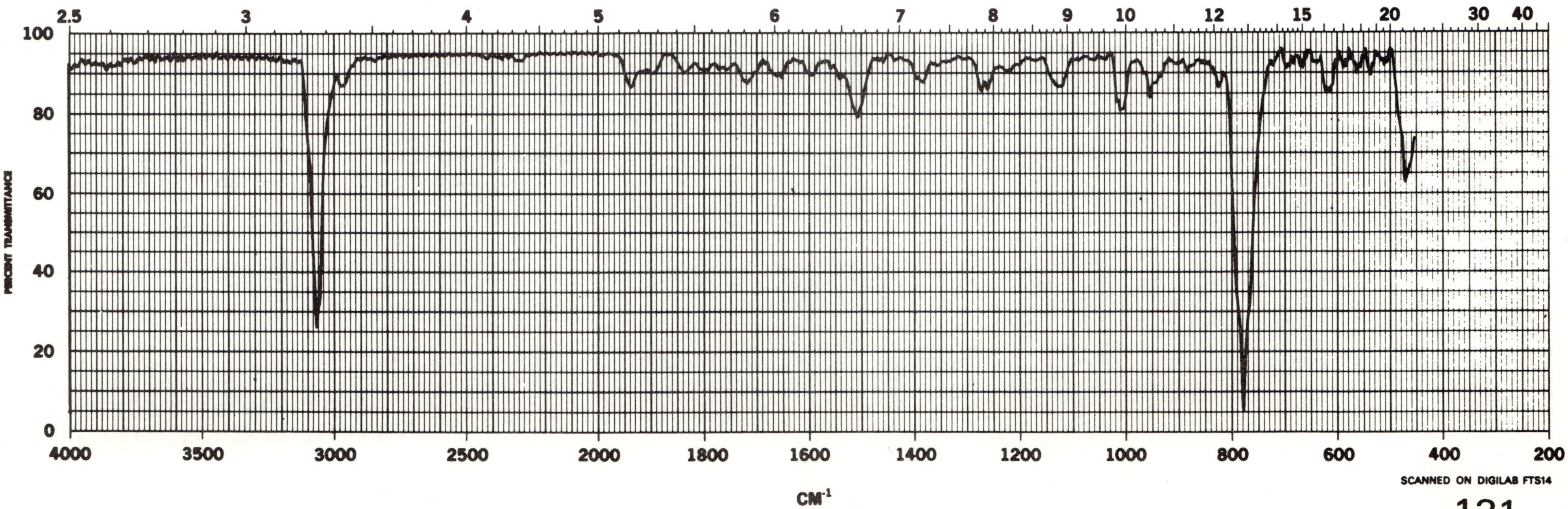

SCANNED ON DIGILAB FTS14

2–CHLORONAPHTHALENE
β–CHLORONAPHTHALENE

$C_{10}H_7Cl$

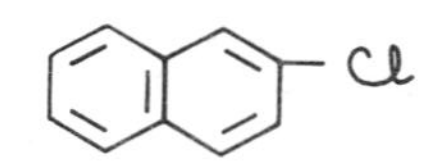

MOLECULAR WEIGHT:	162.62
SOURCE OF SAMPLE:	Fluka AG Buchs, Switzerland
CAS REGISTRY NO.:	91-58-7
NIOSH NUMBER:	QJ22750
CAPILLARY CELL:	Melt

CONDENSED PHASE

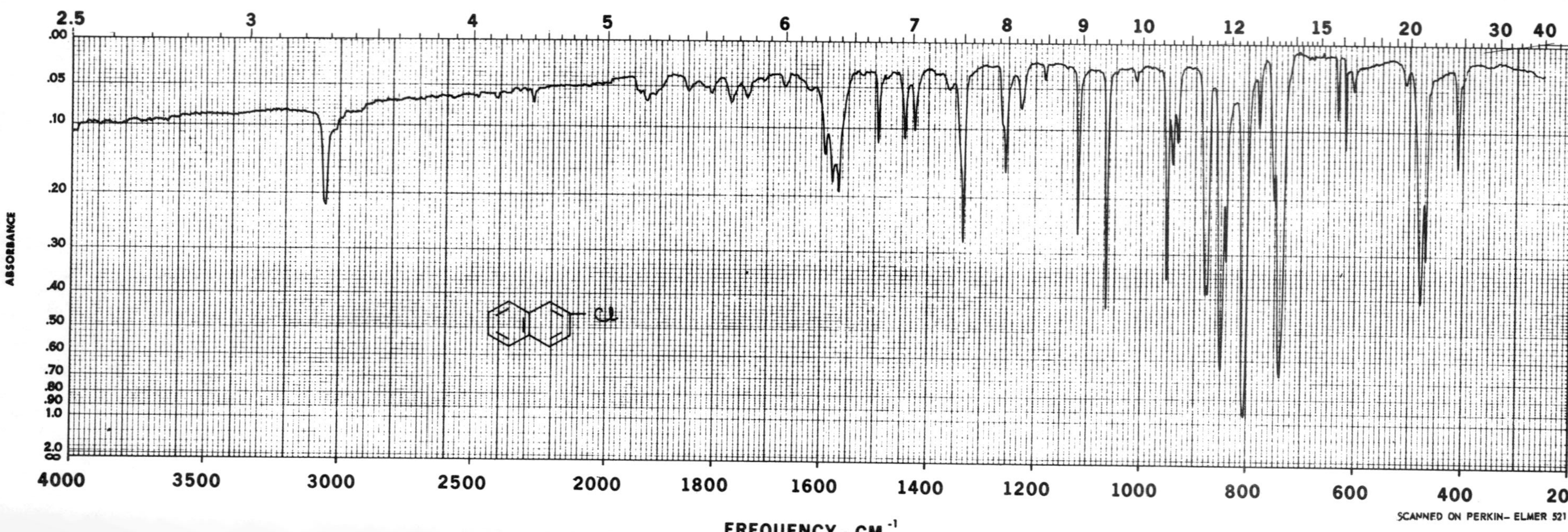

MELTING POINT:	57-58°C	INJECTION AMT.:	0.92 mg
VAPORIZATION TEMP.:	280°C	PRESSURE:	Atmospheric
CELL VOLUME:	7.5 cc	MAX./MIN.:	101.4/38.4
PATH LENGTH:	7.6 cm	SCALE EXPANSION:	1.46

VAPOR PHASE

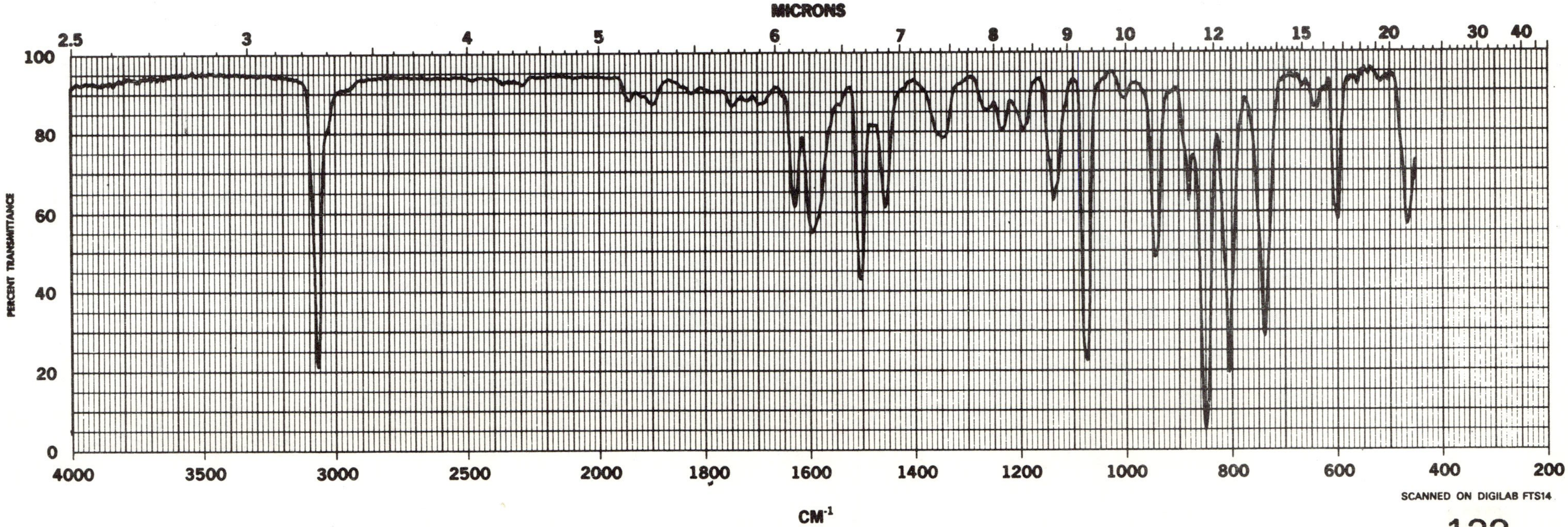

SCANNED ON DIGILAB FTS14

NAPHTHENIC ACID

MOLECULAR WEIGHT: --

SOURCE OF SAMPLE: Exxon Chemical Company
Houston, Texas

CAS REGISTRY NO.: 1338-24-5

NIOSH NUMBER: QK87500

Between Salts

CONDENSED PHASE

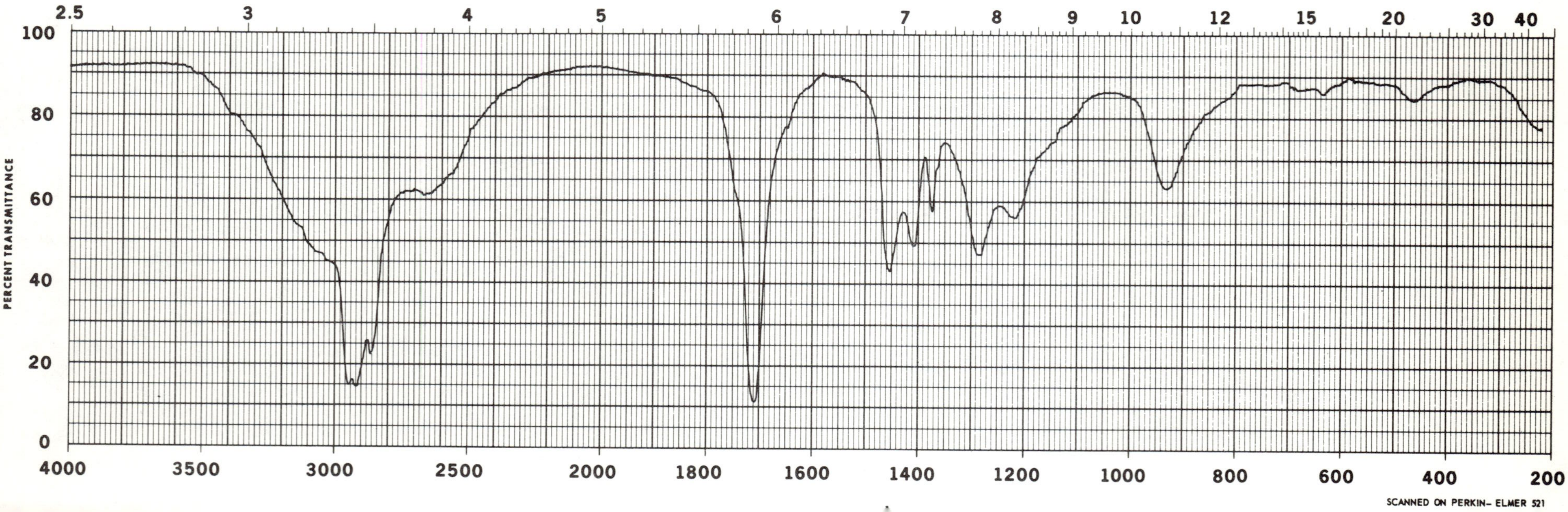

BOILING POINT:	—	INJECTION AMT.:	—
VAPORIZATION TEMP.:	160°C	PRESSURE:	Atmospheric
CELL VOLUME:	7.5 cc	MAX./MIN.:	102.3/39.6
PATH LENGTH:	7.6 cm	SCALE EXPANSION:	1.47

VAPOR PHASE

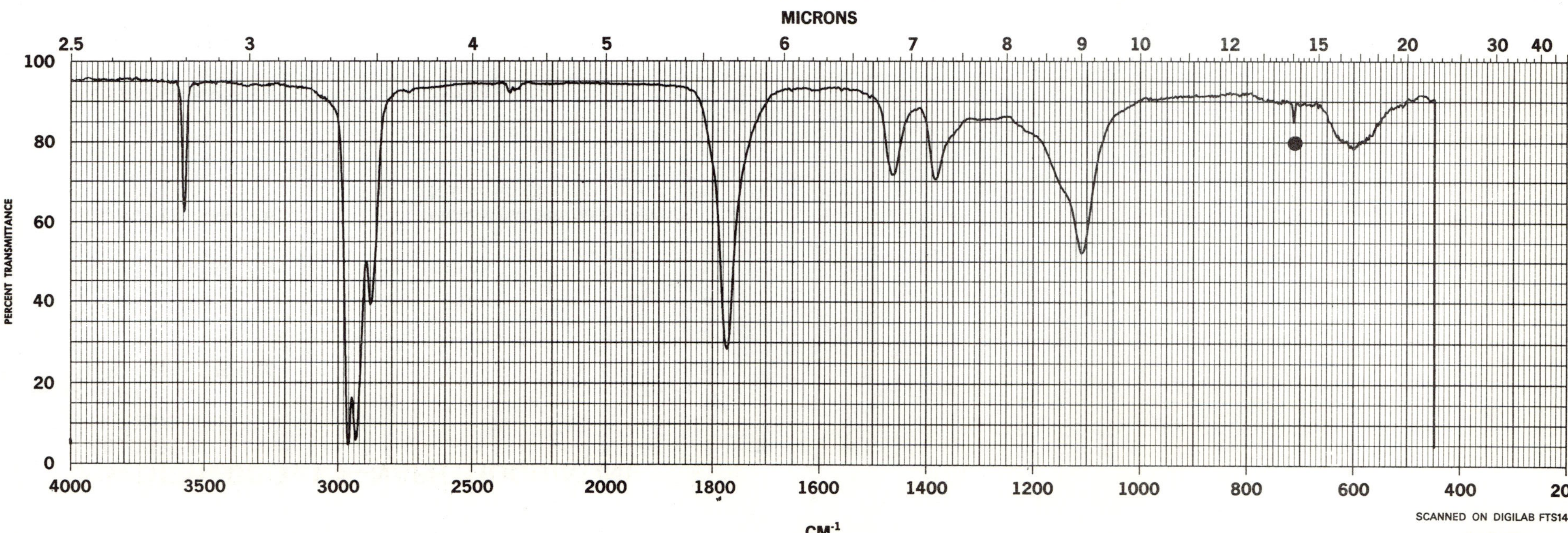

SCANNED ON DIGILAB FTS14

2,3–DICHLORO–1,4–NAPHTHOQUINONE
DICHLONE
PHYGON
SANQUINON
U. S. RUBBER 604

$C_{10}H_4Cl_2O_2$

MOLECULAR WEIGHT: 227.05

SOURCE OF SAMPLE: MCB Manufacturing Chemists
Norwood, Ohio

CAS REGISTRY NO.: 117-80-6

NIOSH NUMBER: QL75250

KBr Wafer

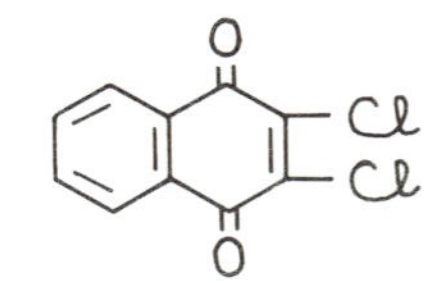

CONDENSED PHASE

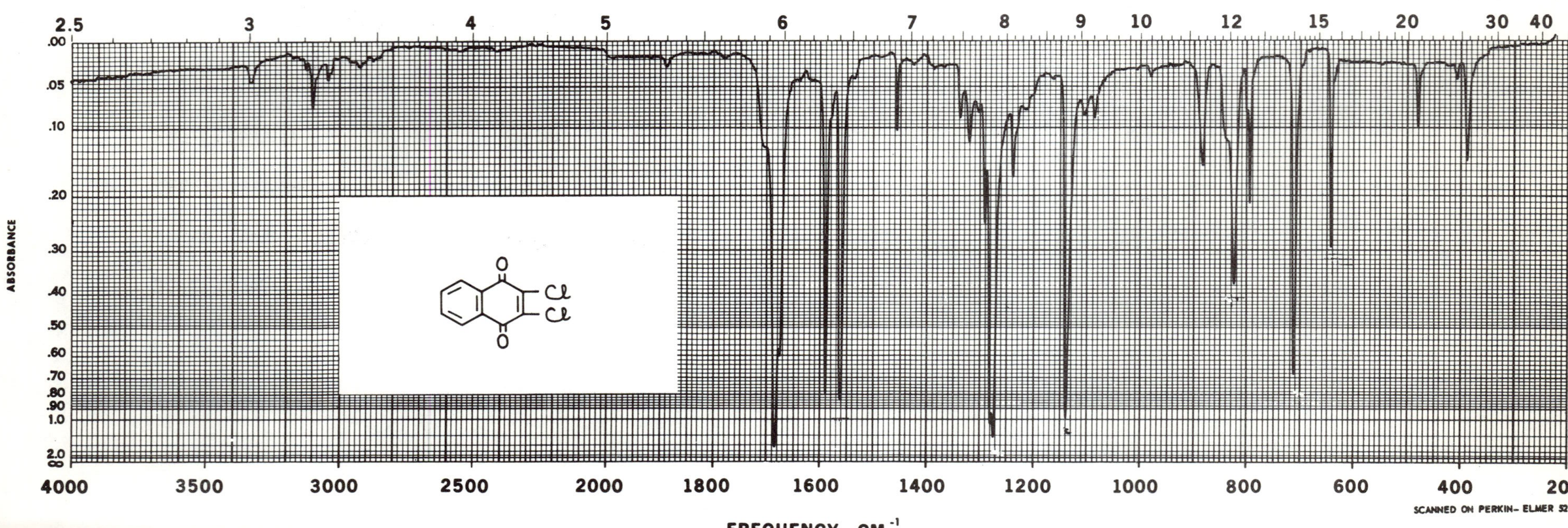

MELTING POINT:	195°C	INJECTION AMT.:	—
VAPORIZATION TEMP.:	280°C	PRESSURE:	Atmospheric
CELL VOLUME:	7.5 cc	MAX./MIN.:	104.8/4.3
PATH LENGTH:	7.6 cm	SCALE EXPANSION:	0.66

VAPOR PHASE

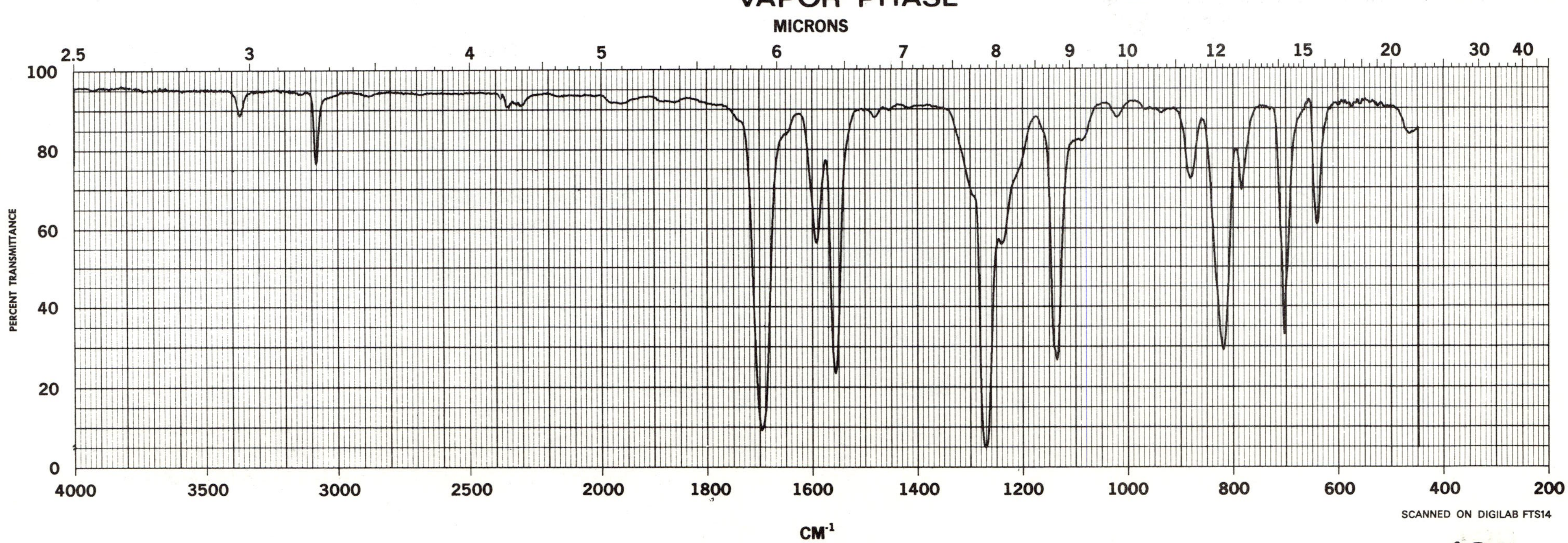

SCANNED ON DIGILAB FTS14

134

1—NAPHTHYLAMINE
α—NAPHTHYLAMINE
ANTIOXIDANT MB
FAST GARNET B

$C_{10}H_9N$

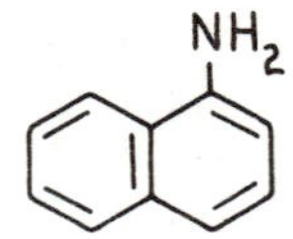

MOLECULAR WEIGHT: 143.19

SOURCE OF SAMPLE: J. T. Baker Chemical Company
Phillipsburg, New Jersey

CAS REGISTRY NO.: 134-32-7

NIOSH NUMBER: QM14000

CAPILLARY CELL: Melt

CONDENSED PHASE

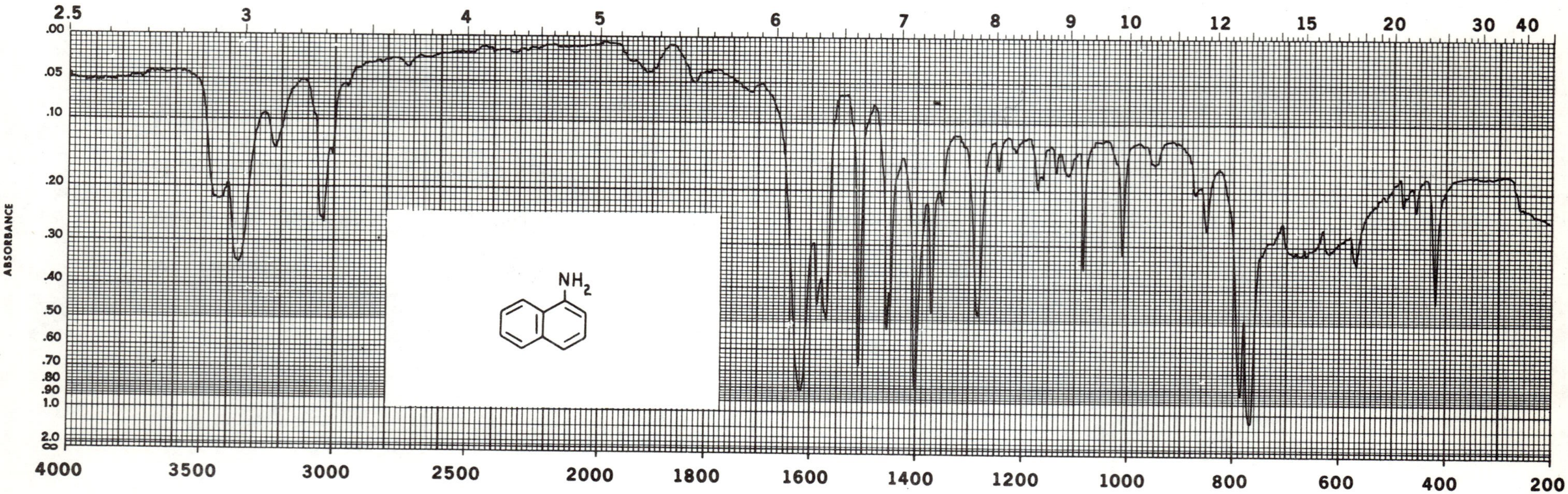

BOILING POINT:	301°C	INJECTION AMT.:	0.70 mg
VAPORIZATION TEMP.:	280°C	PRESSURE:	Atmospheric
CELL VOLUME:	7.5 cc	MAX./MIN.:	–
PATH LENGTH:	7.6 cm	SCALE EXPANSION:	1.30

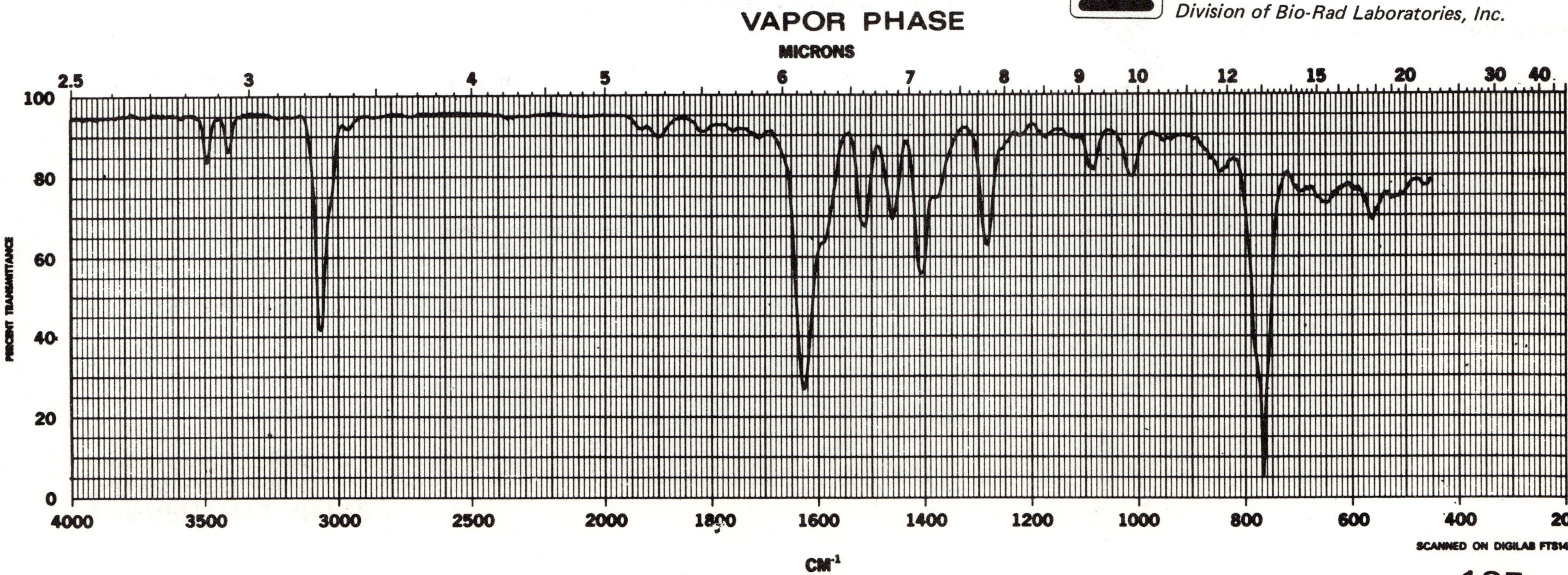

2–NAPHTHYLAMINE
β–NAPHTHYLAMINE
FAST SCARLET BASE B
USAF CB–22

$C_{10}H_9N$

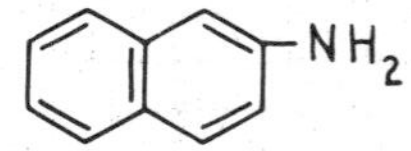

MOLECULAR WEIGHT: 143.19

SOURCE OF SAMPLE: Pfaltz and Bauer, Inc.
Stamford, Connecticut

CAS REGISTRY NO.: 91-59-8

NIOSH NUMBER: QM21000

KBr Wafer

CONDENSED PHASE

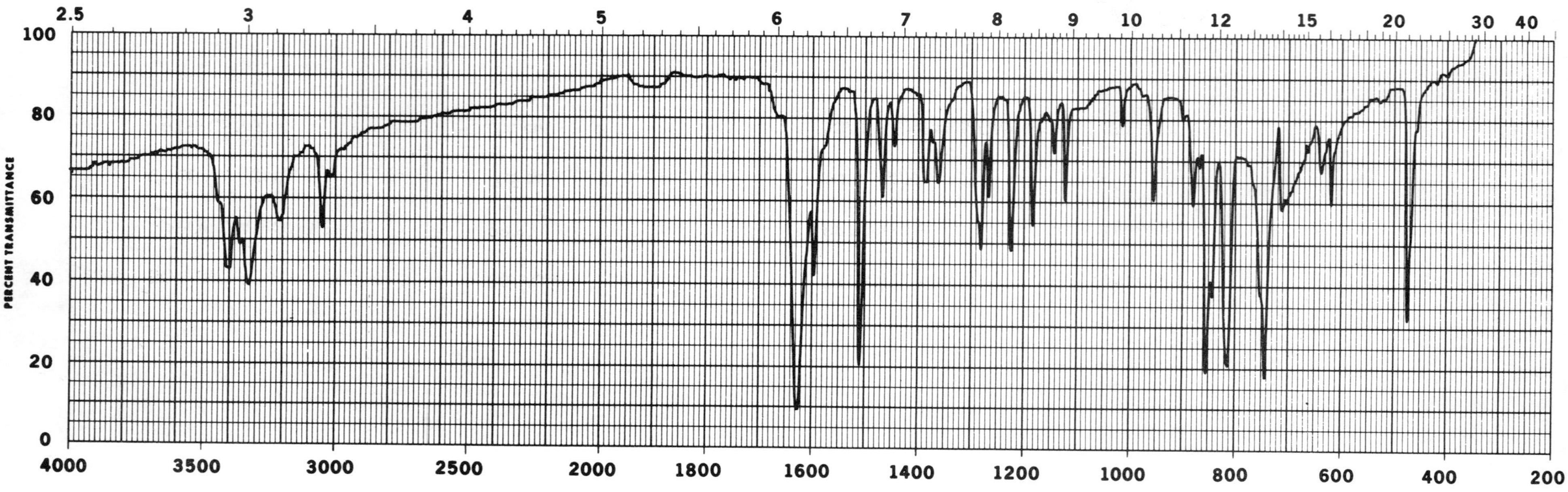

BOILING POINT:	306.1°C	INJECTION AMT.:	0.98 mg
VAPORIZATION TEMP.:	240°C	PRESSURE:	Atmospheric
CELL VOLUME:	7.5 cc	MAX./MIN.:	—
PATH LENGTH:	7.6 cm	SCALE EXPANSION:	1.17

VAPOR PHASE

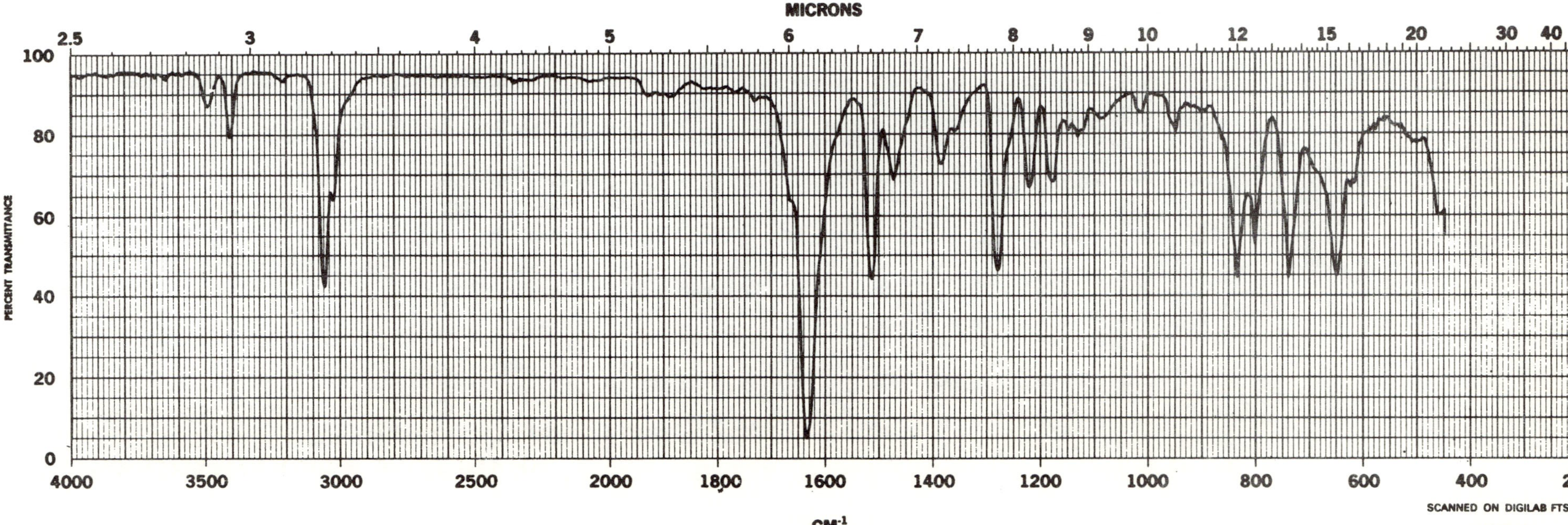

SCANNED ON DIGILAB FTS14

1,2–OXATHIOLANE, 2,2–DIOXIDE
3–HYDROXY–1–PROPANESULFONIC ACID, γ–SULTONE
PROPANESULTONE

$C_3H_6O_3S$

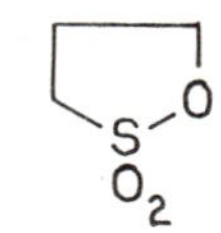

MOLECULAR WEIGHT:	122.14
SOURCE OF SAMPLE:	Aldrich Chemical Company, Inc. Milwaukee, Wisconsin
CAS REGISTRY NO.:	1120-71-4
NIOSH NUMBER:	RP54250
CAPILLARY CELL:	Melt

CONDENSED PHASE

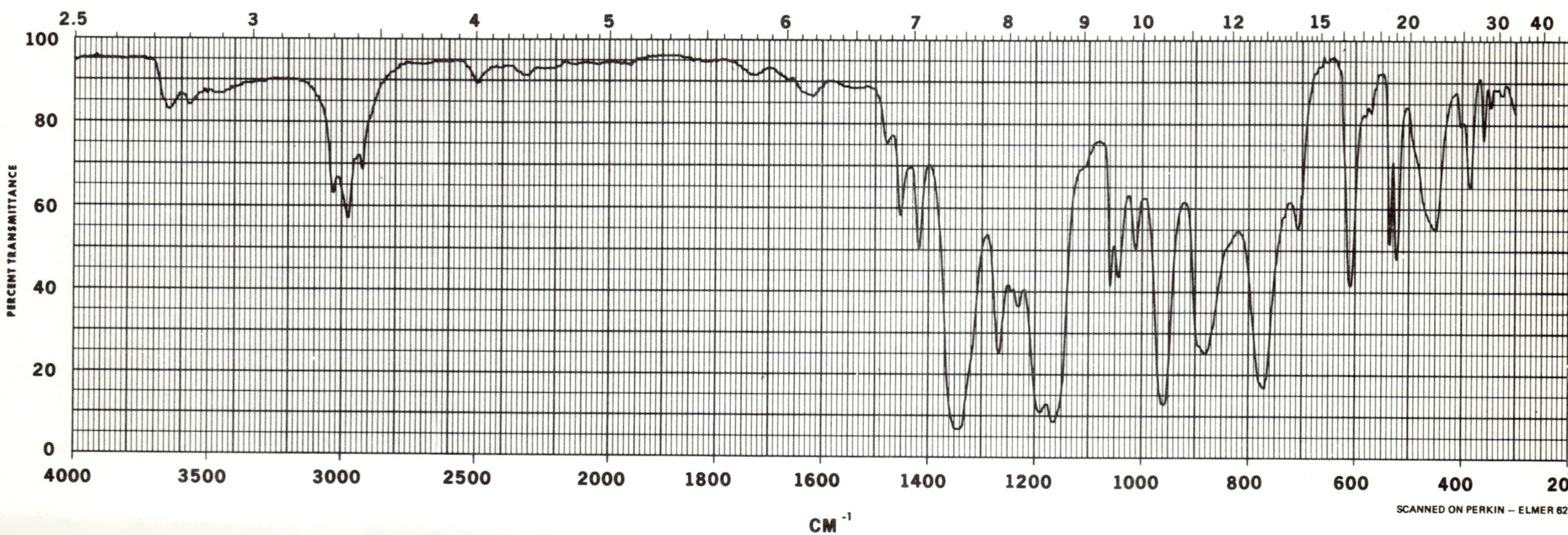

BOILING POINT:	180°C/30 mm	INJECTION AMT.:	0.7 μl
VAPORIZATION TEMP.:	280°C	PRESSURE:	Atmospheric
CELL VOLUME:	7.5 cc	MAX./MIN.:	–
PATH LENGTH:	7.6 cm	SCALE EXPANSION:	0.94

VAPOR PHASE

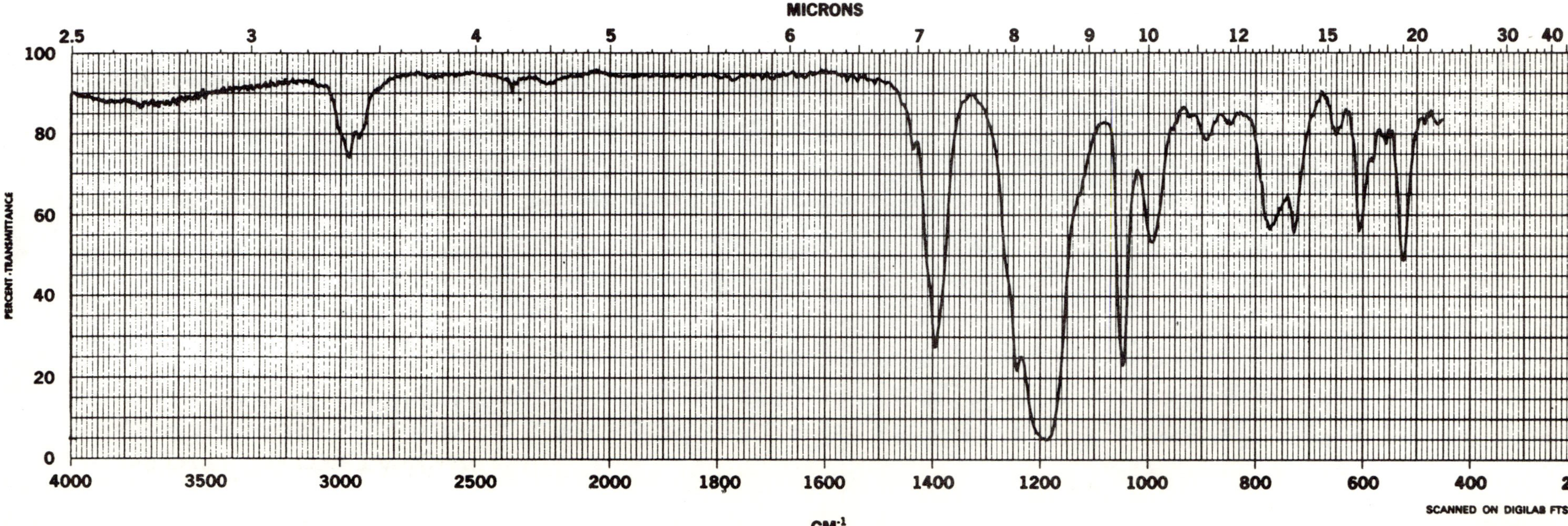

137

4–METHYL–2–OXETANONE
β–BUTYROLACTONE
3–HYDROXYBUTYRIC ACID, β–LACTONE

$C_4H_6O_2$

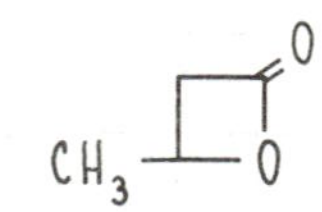

MOLECULAR WEIGHT:	86.09
SOURCE OF SAMPLE:	Eastman Chemical Products, Inc. Kingsport, Tennessee
CAS REGISTRY NO.:	3068-88-0
NIOSH NUMBER:	RQ80500
CAPILLARY CELL:	Neat

CONDENSED PHASE

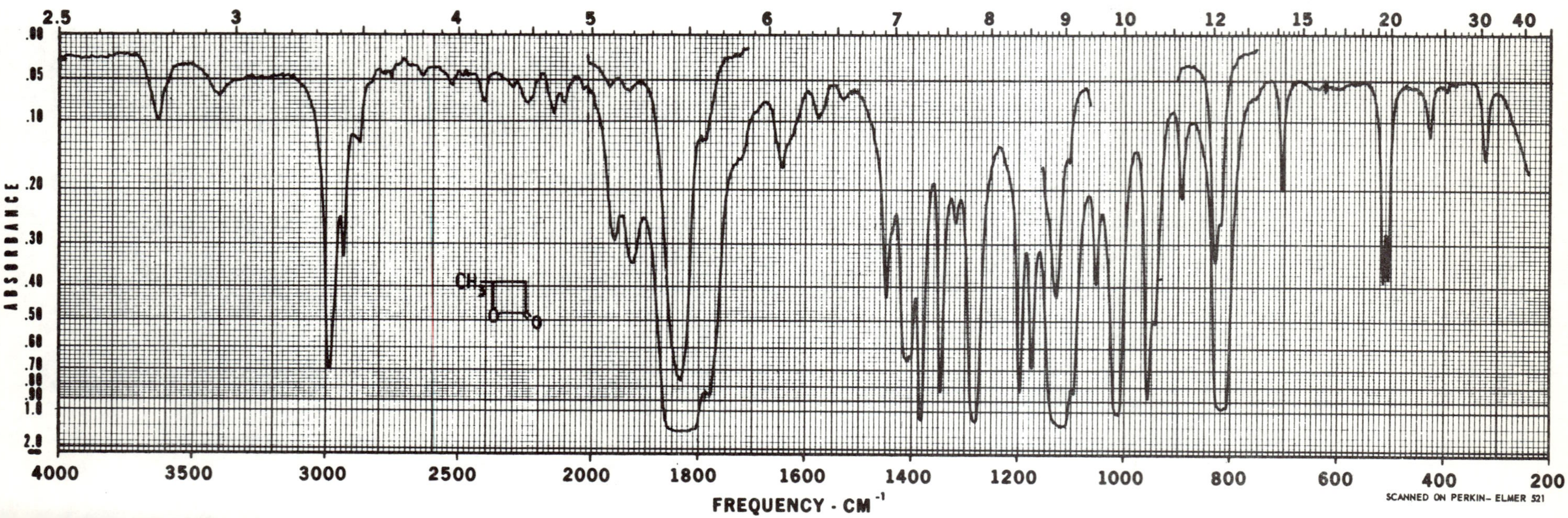

BOILING POINT:	71-73°C/29 mm	INJECTION AMT.:	—
VAPORIZATION TEMP.:	160°C	PRESSURE:	Atmospheric
CELL VOLUME:	7.5 cc	MAX./MIN.:	102.0/9.7
PATH LENGTH:	7.6 cm	SCALE EXPANSION:	1.00

VAPOR PHASE

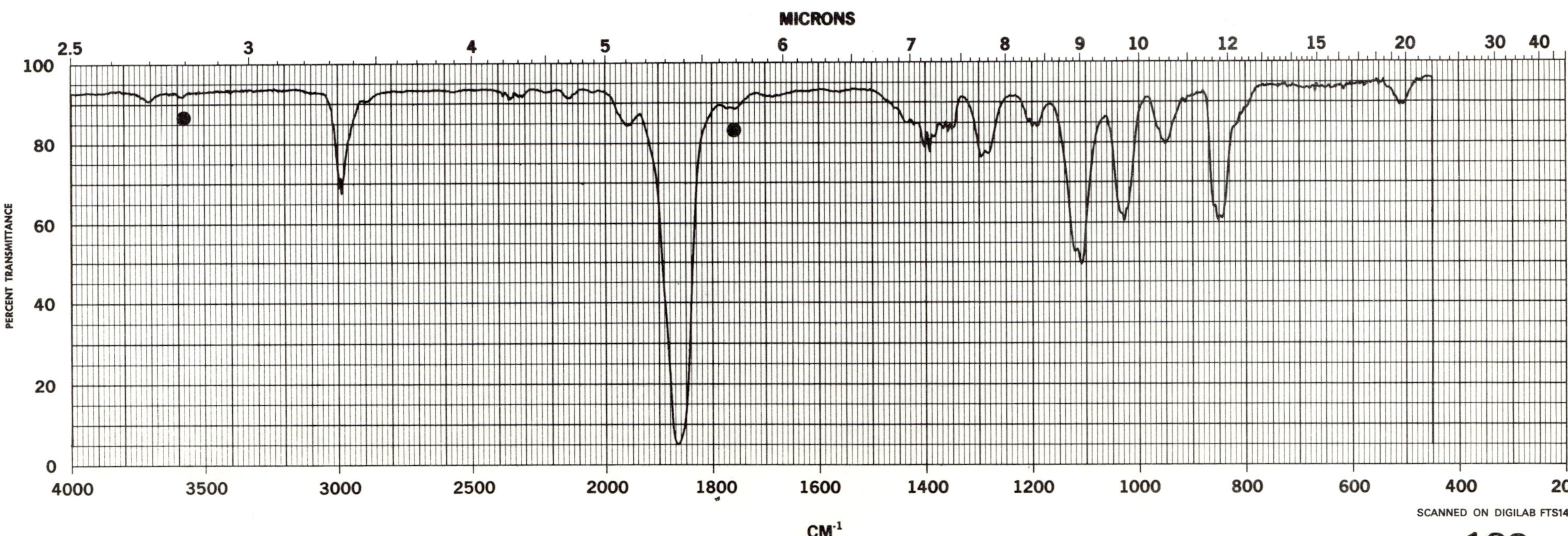

SCANNED ON DIGILAB FTS14

138

PHENANTHRENE

$C_{14}H_{10}$

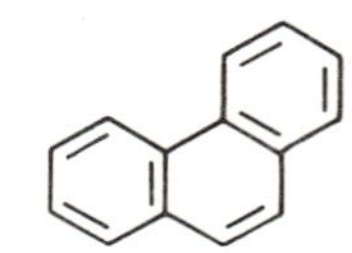

MOLECULAR WEIGHT: 178.24

SOURCE OF SAMPLE: MCB Manufacturing Chemists
Norwood, Ohio

CAS REGISTRY NO.: 85-01-8

NIOSH NUMBER: SF71750

KBr Wafer

CONDENSED PHASE

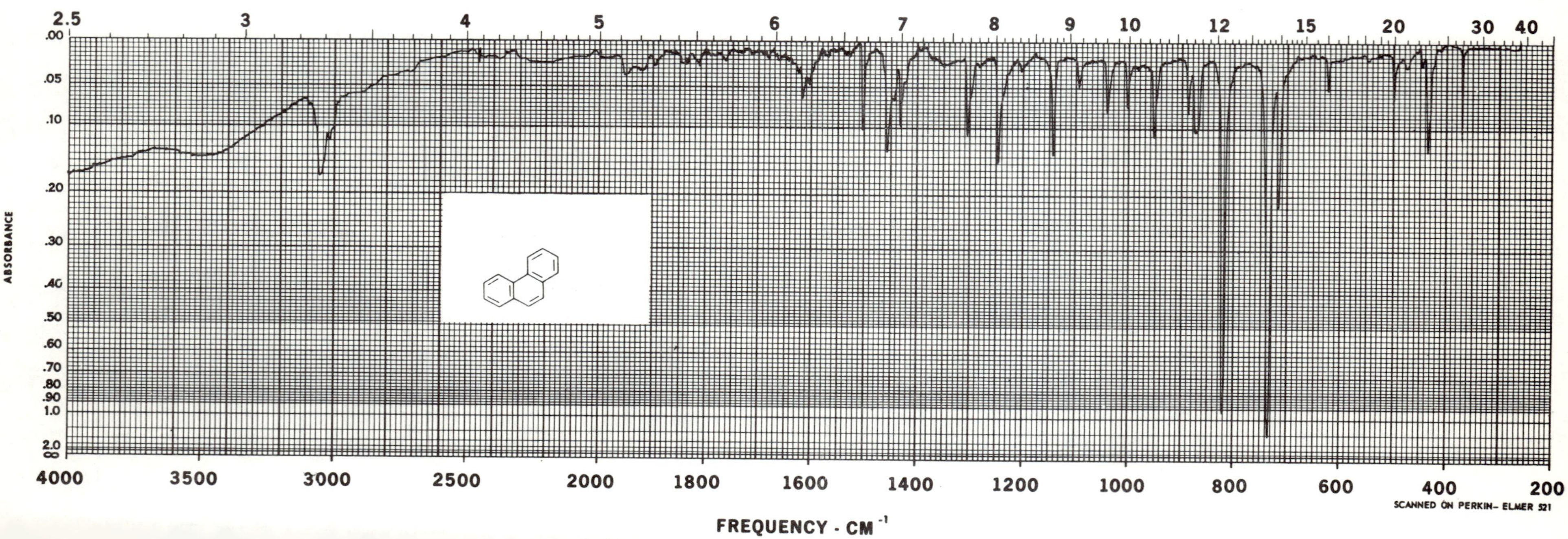

BOILING POINT:	336°C	INJECTION AMT.:	1.17 mg
VAPORIZATION TEMP.:	280°C	PRESSURE:	Atmospheric
CELL VOLUME:	7.5 cc	MAX./MIN.:	–
PATH LENGTH:	7.6 cm	SCALE EXPANSION:	1.02

VAPOR PHASE

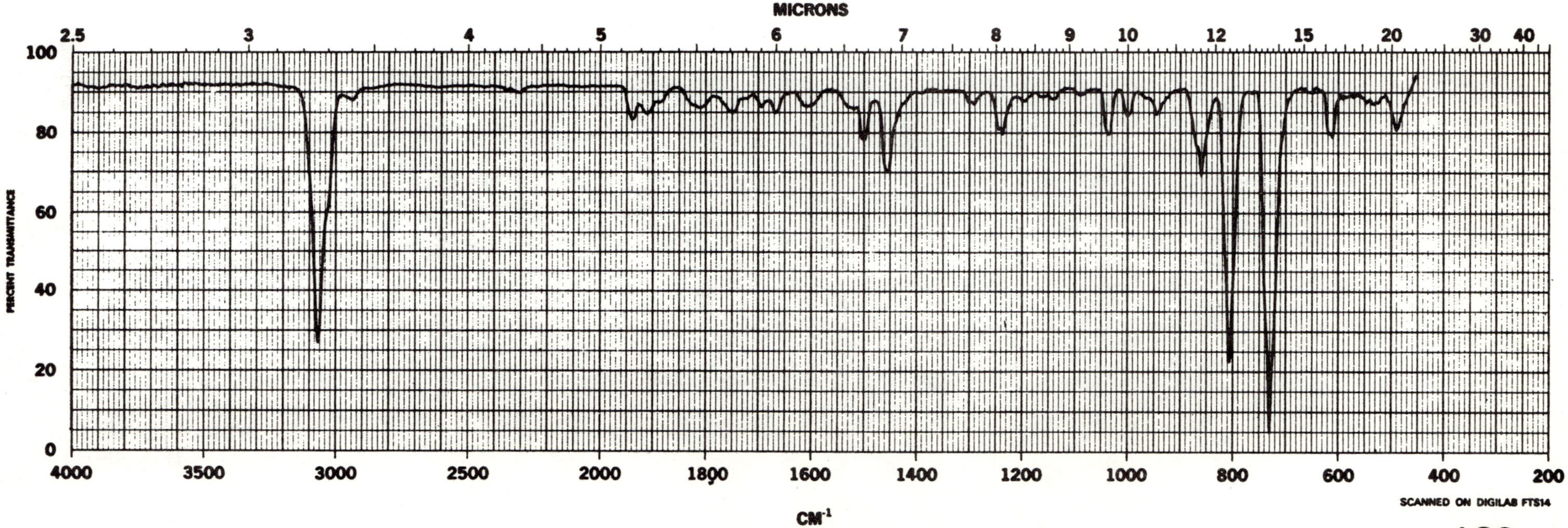

SCANNED ON DIGILAB FTS14

PHENOL
CARBOLIC ACID
HYDROXYBENZENE

C_6H_6O

MOLECULAR WEIGHT: 94.11

SOURCE OF SAMPLE: Aldrich Chemical Company, Inc.
Milwaukee, Wisconsin

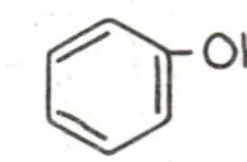

CAS REGISTRY NO.: 108-95-2

NIOSH NUMBER: SJ33250

CAPILLARY CELL: Melt

CONDENSED PHASE

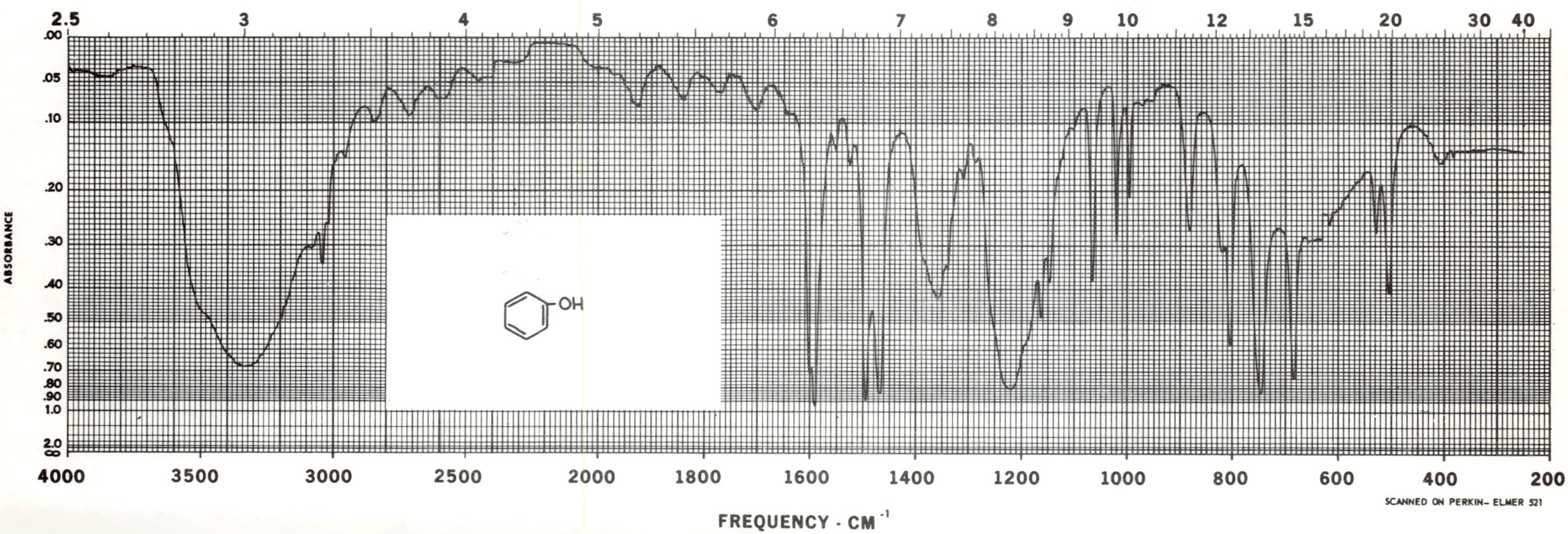

BOILING POINT:	181°C	INJECTION AMT.:	0.8 μl
VAPORIZATION TEMP.:	240°C	PRESSURE:	Atmospheric
CELL VOLUME:	7.5 cc	MAX./MIN.:	100.0/11.2
PATH LENGTH:	7.6 cm	SCALE EXPANSION:	1.04

VAPOR PHASE

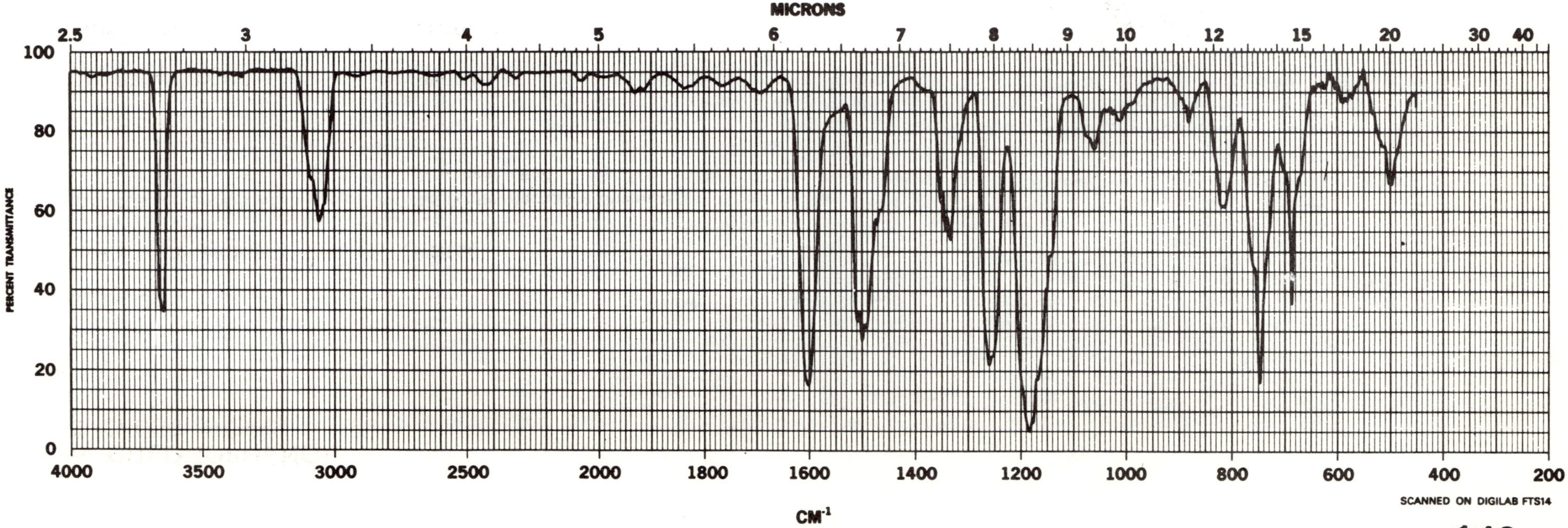

SCANNED ON DIGILAB FTS14

o–CHLOROPHENOL
2–CHLOROPHENOL

C_6H_5ClO

MOLECULAR WEIGHT: 128.56

SOURCE OF SAMPLE: MCB Manufacturing Chemists
Norwood, Ohio

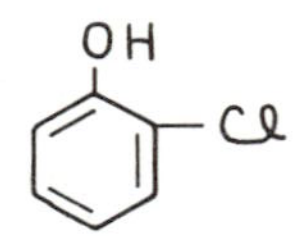

CAS REGISTRY NO.: 95-57-8

NIOSH NUMBER: SK26250

CAPILLARY CELL: Neat

CONDENSED PHASE

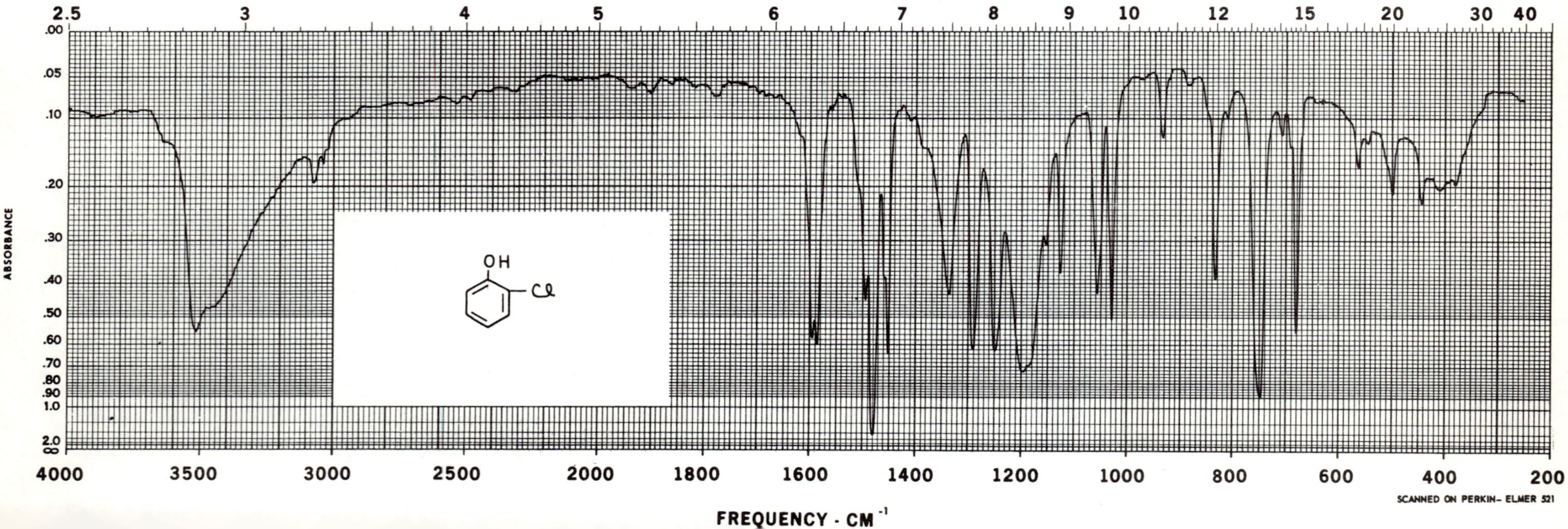

BOILING POINT:	176°C	INJECTION AMT.:	1.0 μl
VAPORIZATION TEMP.:	200°C	PRESSURE:	Atmospheric
CELL VOLUME:	7.5 cc	MAX./MIN.:	–
PATH LENGTH:	7.6 cm	SCALE EXPANSION:	0.96

VAPOR PHASE

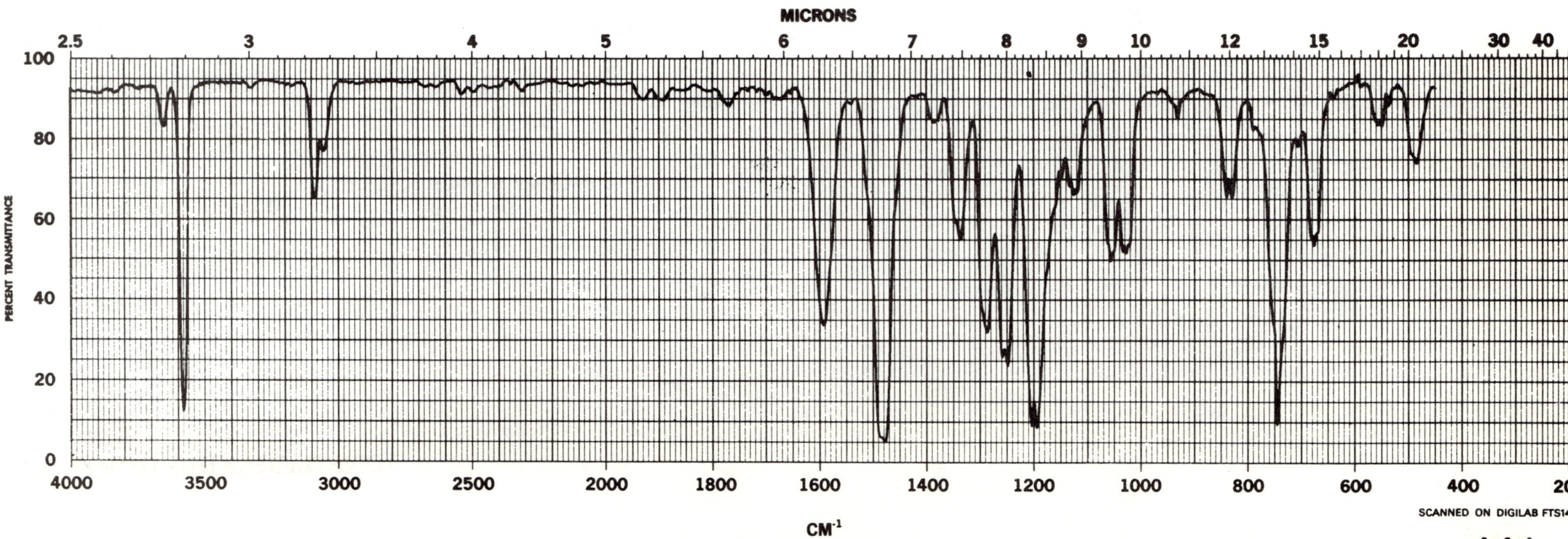

SCANNED ON DIGILAB FTS14

2,4—DICHLOROPHENOL
DCP

$C_6H_4Cl_2O$

MOLECULAR WEIGHT: 163.00

SOURCE OF SAMPLE: Fluka AG
Buchs, Switzerland

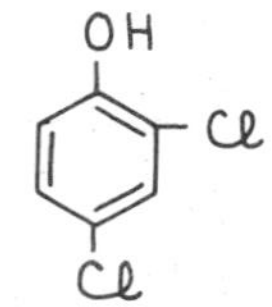

CAS REGISTRY NO.: 120-83-2

NIOSH NUMBER: SK85750

CAPILLARY CELL: Melt

CONDENSED PHASE

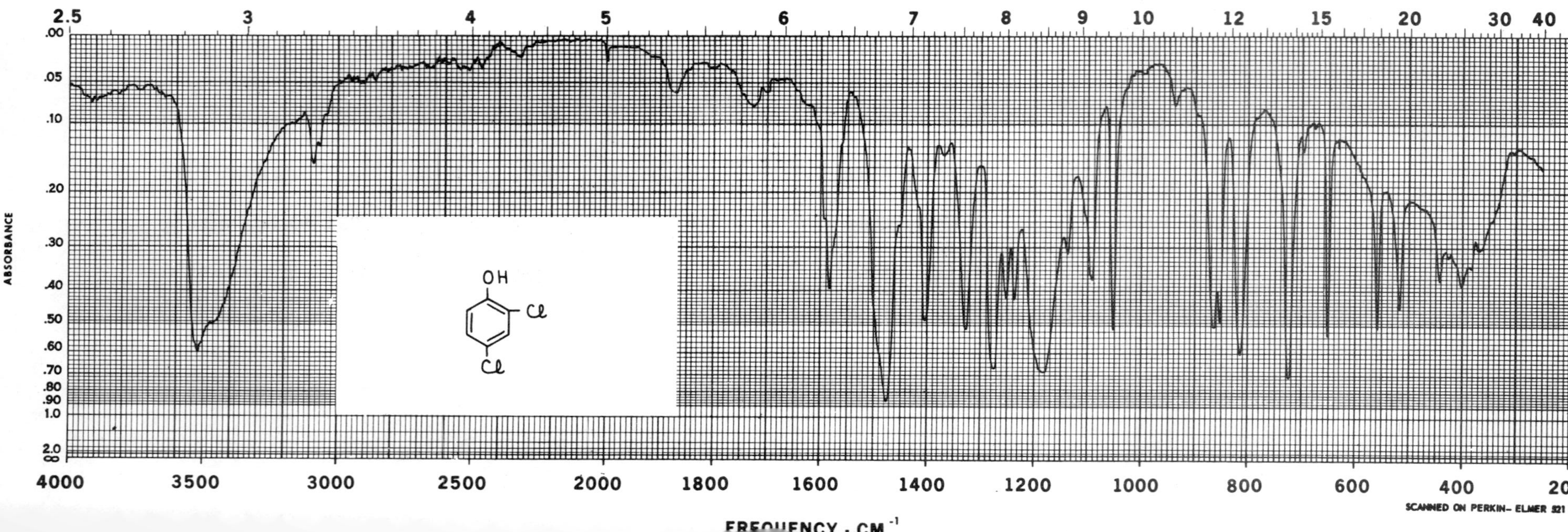

MELTING POINT:	37-40°C	INJECTION AMT.:	0.87 mg
VAPORIZATION TEMP.:	280°C	PRESSURE:	Atmospheric
CELL VOLUME:	7.5 cc	MAX./MIN.:	101.5/10.1
PATH LENGTH:	7.6 cm	SCALE EXPANSION:	1.01

VAPOR PHASE

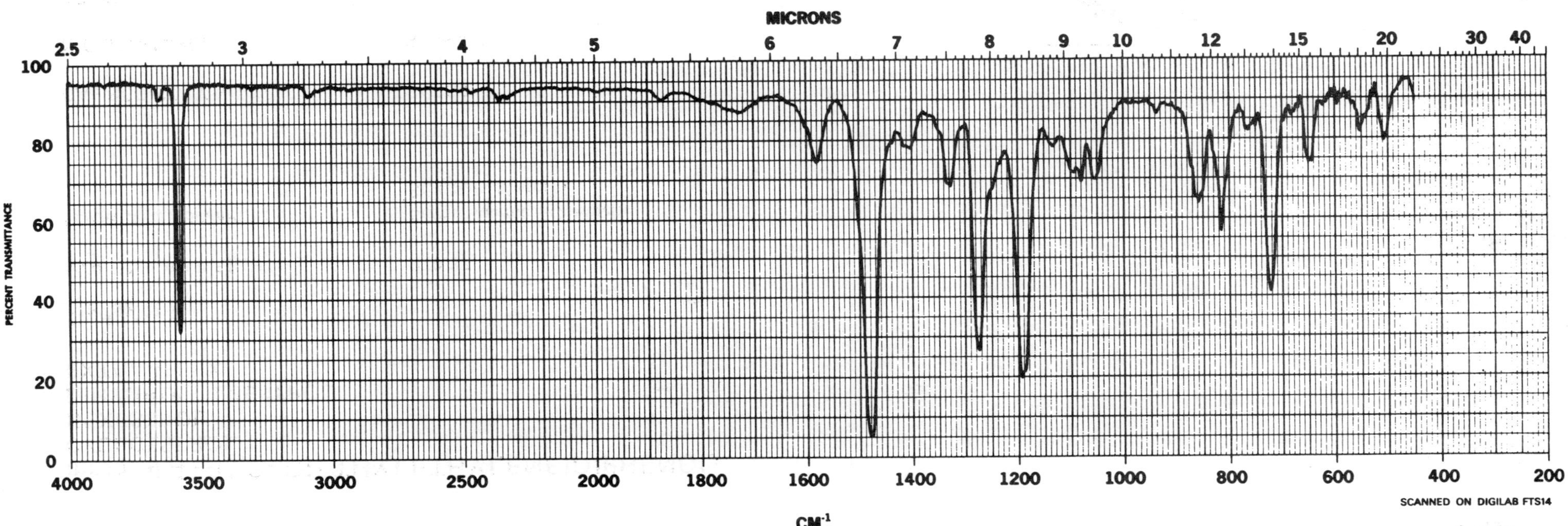

SCANNED ON DIGILAB FTS14

142

meso–4,4′–(1,2–DIETHYLETHYLENE)DIPHENOL
CYCLOESTROL
HEXESTROL
DIHYDROSTILBESTROL
EXTRA–PLEX

$C_{18}H_{22}O_2$

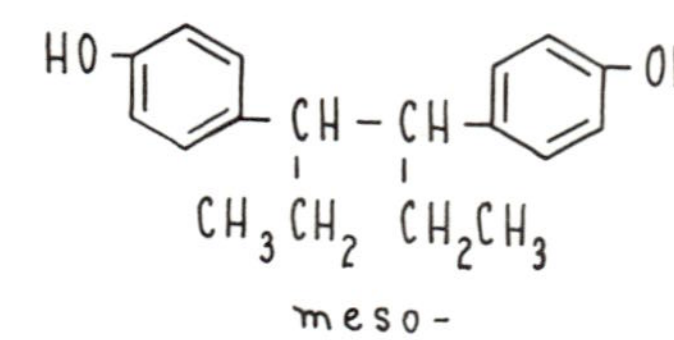

MOLECULAR WEIGHT: 270.37

SOURCE OF SAMPLE: Sigma Chemical Company
St. Louis, Missouri

CAS REGISTRY NO.: 84-16-2

NIOSH NUMBER: SL06000

KBr Wafer

CONDENSED PHASE

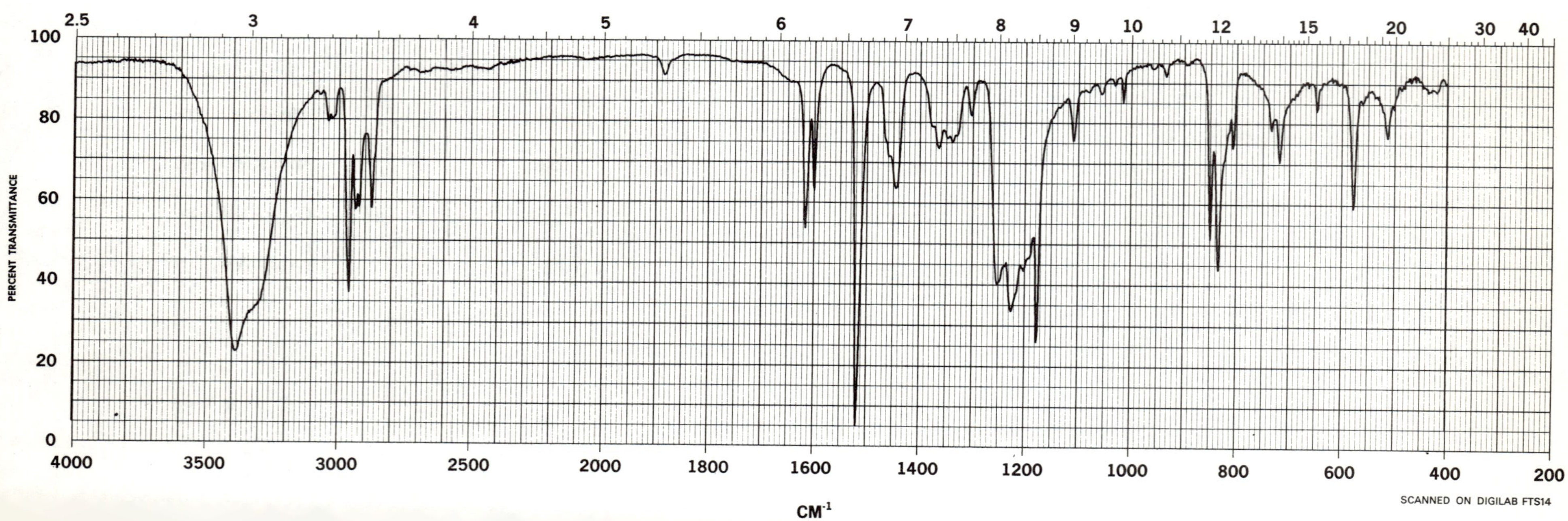

BOILING POINT:	—	INJECTION AMT.:	—
VAPORIZATION TEMP.:	$280^{O}C$	PRESSURE:	Atmospheric
CELL VOLUME:	7.5 cc	MAX./MIN.:	100.0/9.5
PATH LENGTH:	7.6 cm	SCALE EXPANSION:	1.02

VAPOR PHASE

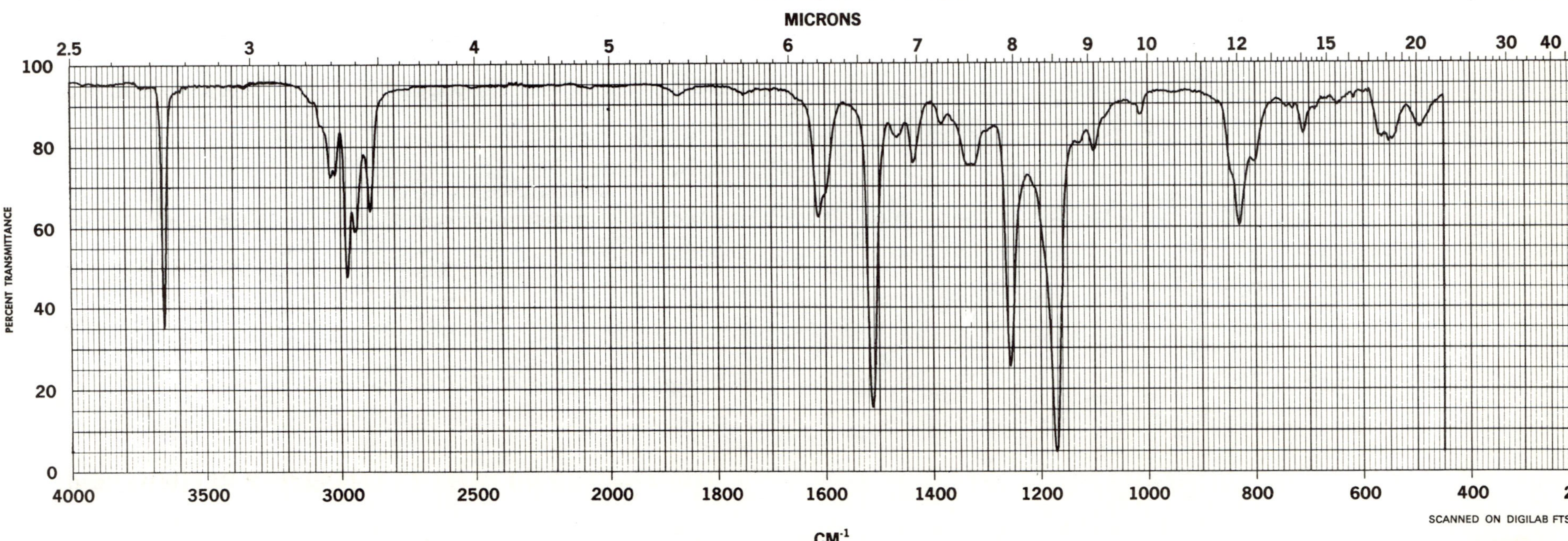

SCANNED ON DIGILAB FTS14

2,3–DINITROPHENOL

$C_6H_4N_2O_5$

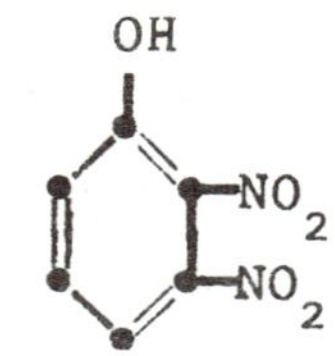

MOLECULAR WEIGHT: 184.11

SOURCE OF SAMPLE: Aldrich Chemical Company, Inc.
Milwaukee, Wisconsin

CAS REGISTRY NO.: 66-56-8

NIOSH NUMBER: SL27000

KBr Wafer

CONDENSED PHASE

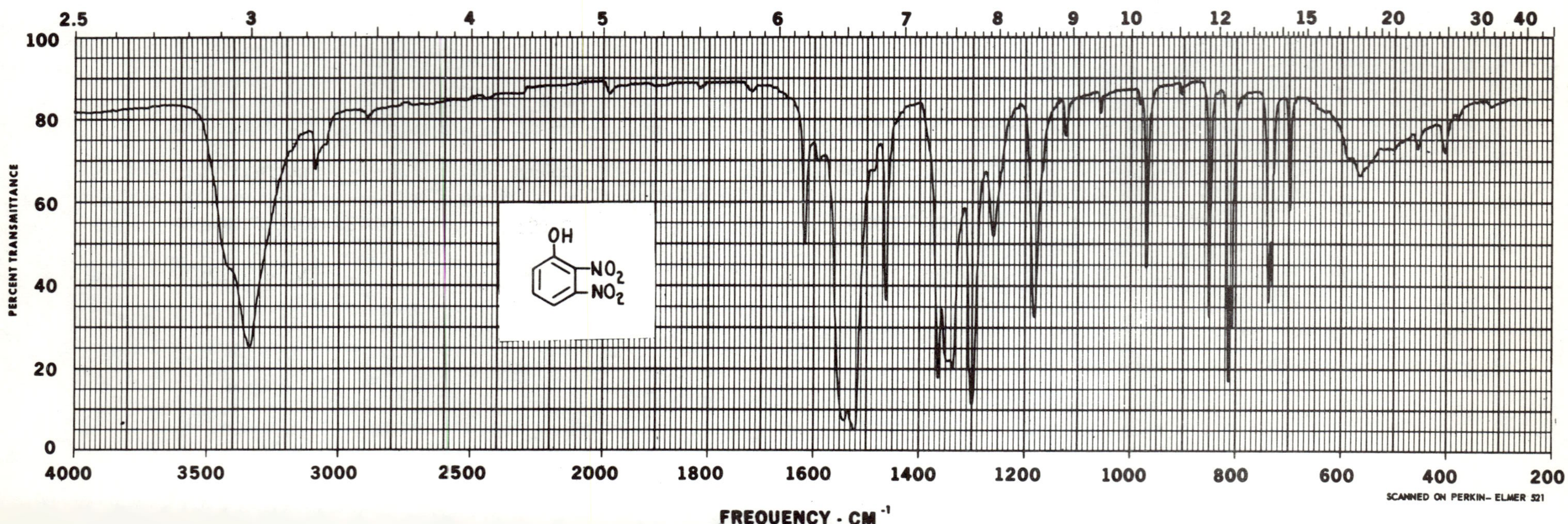

MELTING POINT:	$144^{o}C$	INJECTION AMT.:	0.80 mg
VAPORIZATION TEMP.:	$280^{o}C$	PRESSURE:	Atmospheric
CELL VOLUME:	7.5 cc	MAX./MIN.:	97.6/5.7
PATH LENGTH:	7.6 cm	SCALE EXPANSION:	1.00

VAPOR PHASE

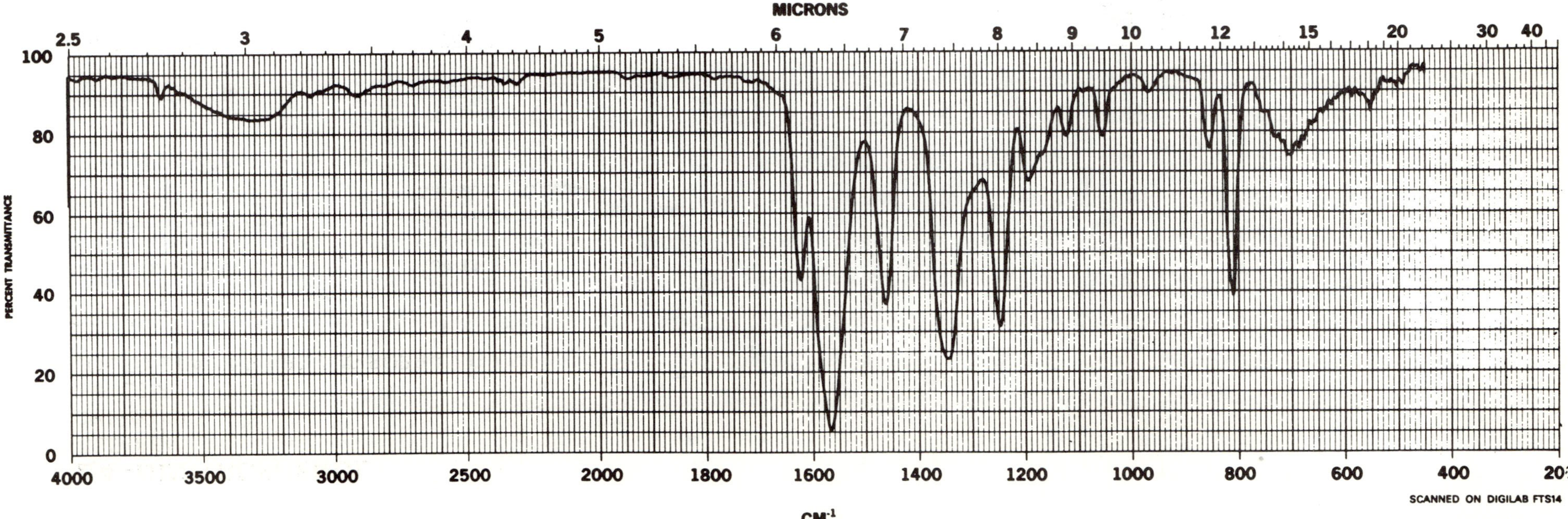

SCANNED ON DIGILAB FTS14

2,4–DINITROPHENOL
ALDIFEN
CHEMOX PE
2,4–DNP
SOLFO BLACK B / BB / 2B SUPRA / G / SB

$C_6H_4N_2O_5$

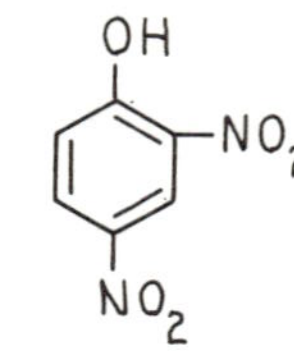

MOLECULAR WEIGHT: 184.11

SOURCE OF SAMPLE: Aldrich Chemical Company, Inc.
Milwaukee, Wisconsin

CAS REGISTRY NO.: 51-28-5

NIOSH NUMBER: SL28000

KBr Wafer

CONDENSED PHASE

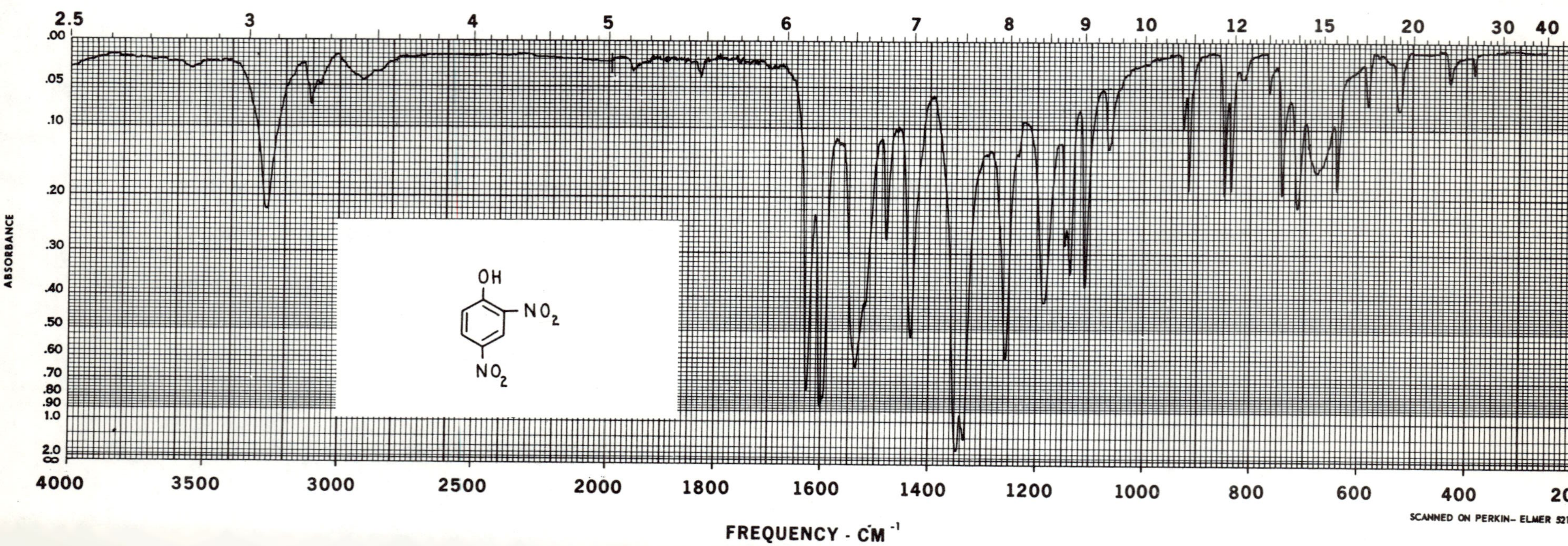

MELTING POINT:	106-108°C	INJECTION AMT.:	0.71 mg
VAPORIZATION TEMP.:	280°C	PRESSURE:	Atmospheric
CELL VOLUME:	7.5 cc	MAX./MIN.:	108.8/5.6
PATH LENGTH:	7.6 cm	SCALE EXPANSION:	0.89

VAPOR PHASE

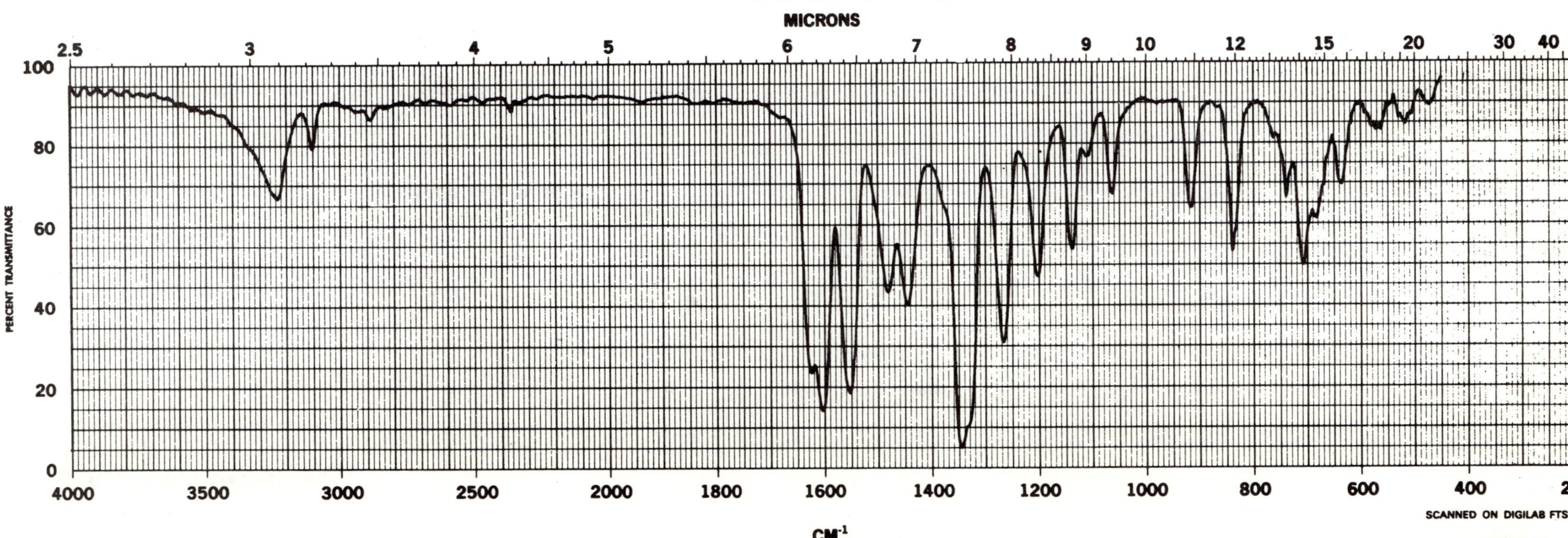

SCANNED ON DIGILAB FTS14

2,5–DINITROPHENOL
2,5–DNP
gamma–DINITROPHENOL

$C_6H_4N_2O_5$

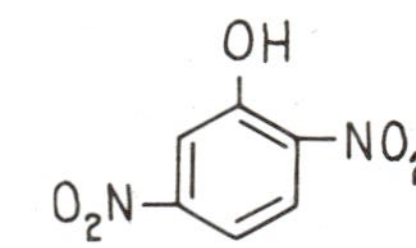

MOLECULAR WEIGHT: 184.11

SOURCE OF SAMPLE: Aldrich Chemical Company, Inc.
Milwaukee, Wisconsin

CAS REGISTRY NO.: 329-71-5

NIOSH NUMBER: SL29000

KBr Wafer

CONDENSED PHASE

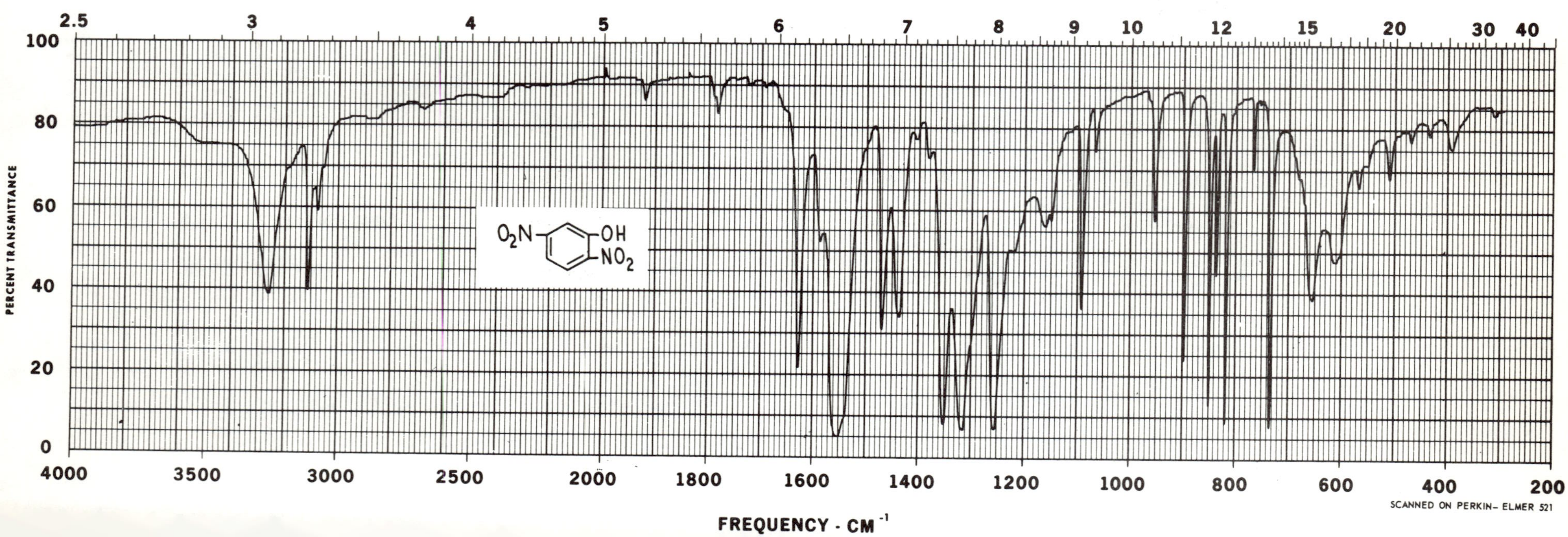

MELTING POINT:	106-109°C	INJECTION AMT.:	–
VAPORIZATION TEMP.:	240°C	PRESSURE:	Atmospheric
CELL VOLUME:	7.5 cc	MAX./MIN.:	97.1/2.2
PATH LENGTH:	7.6 cm	SCALE EXPANSION:	0.97

VAPOR PHASE

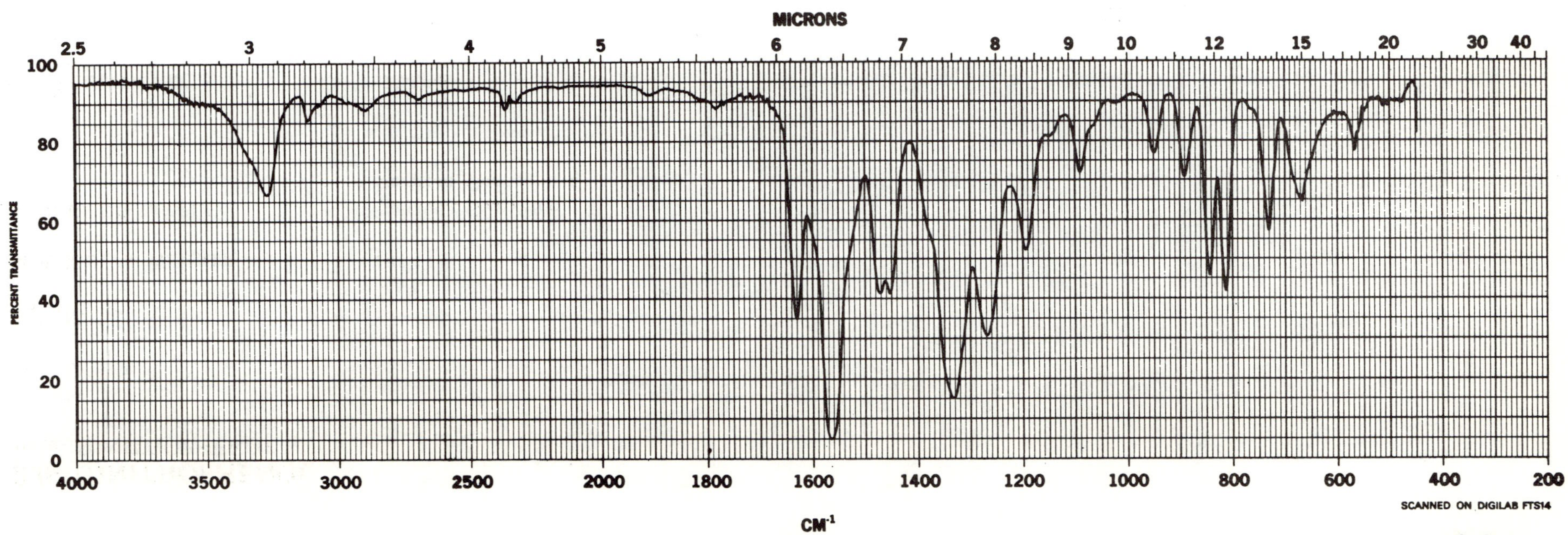

SCANNED ON DIGILAB FTS14

146

2,6–DINITROPHENOL
beta–DINITROPHENOL

$C_6H_4N_2O_5$

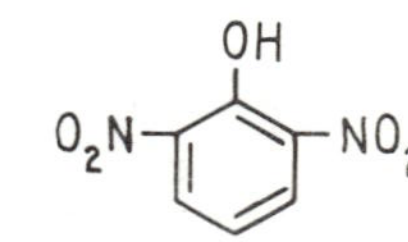

MOLECULAR WEIGHT: 184.11

SOURCE OF SAMPLE: Pfaltz and Bauer, Inc.
Stamford, Connecticut

CAS REGISTRY NO.: 573-56-8

NIOSH NUMBER: SL29750

KBr Wafer

CONDENSED PHASE

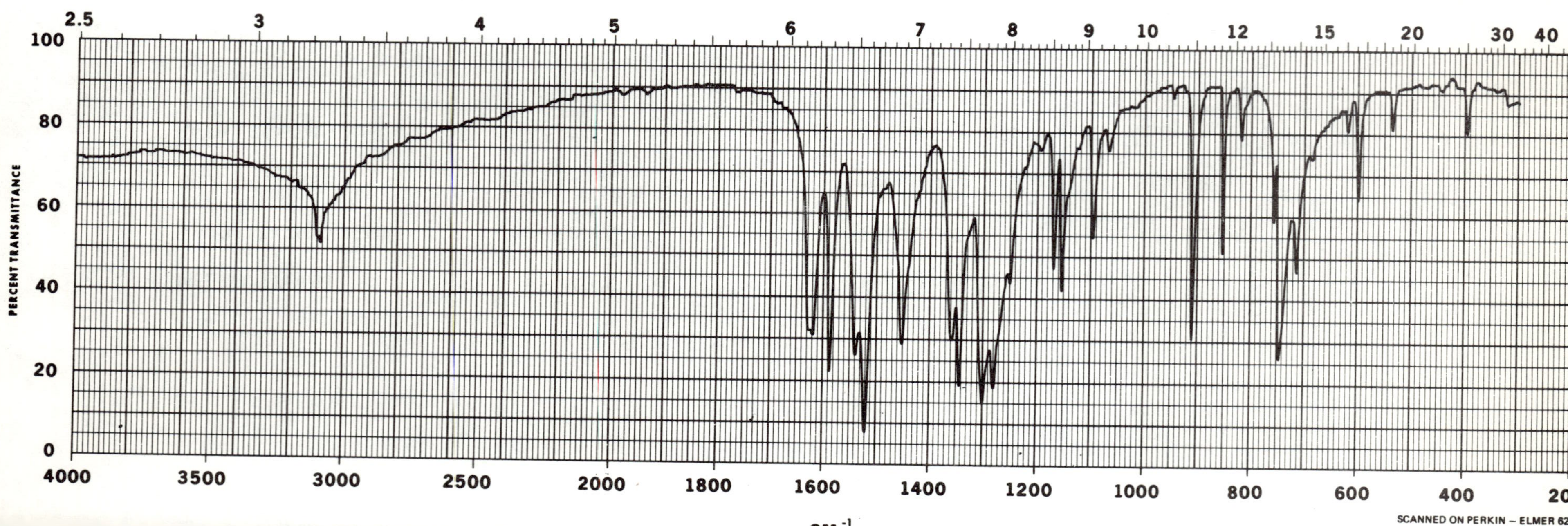

MELTING POINT:	63-64°C	INJECTION AMT.:	1.0 mg
VAPORIZATION TEMP.:	280°C	PRESSURE:	Atmospheric
CELL VOLUME:	7.5 cc	MAX./MIN.:	102.3/3.8
PATH LENGTH:	7.6 cm	SCALE EXPANSION:	0.93

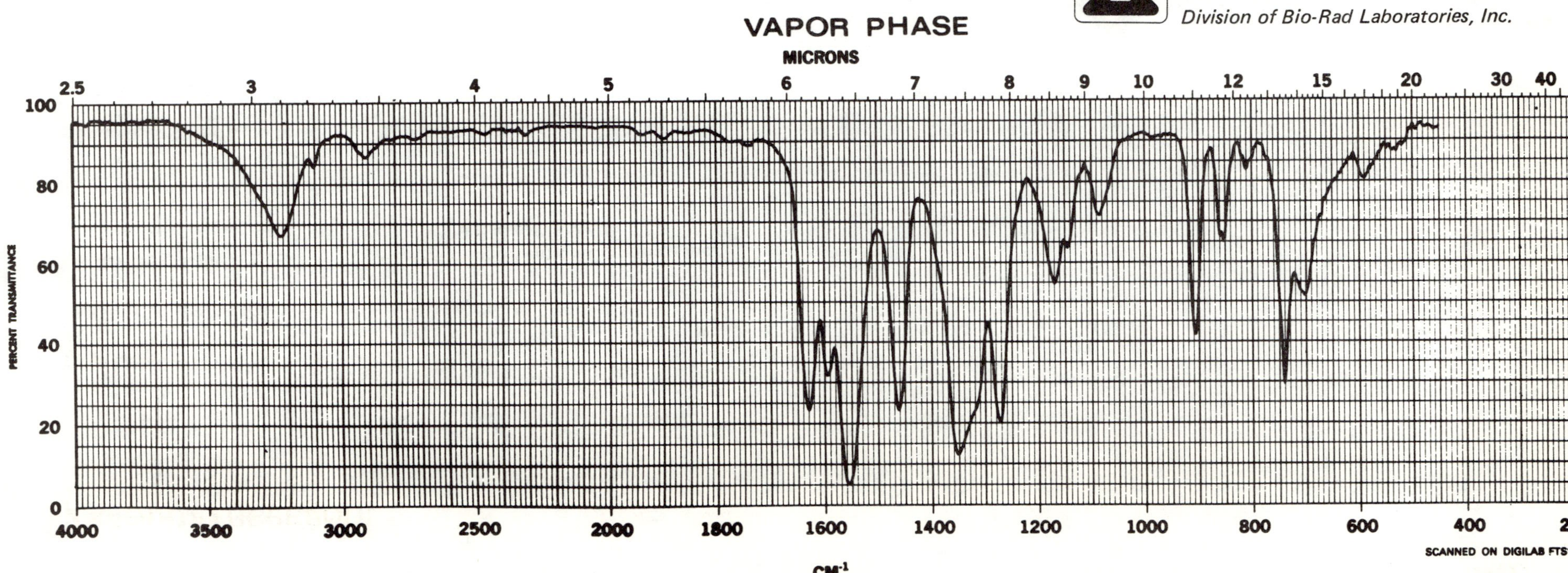

3,4–DINITROPHENOL

$C_6H_4N_2O_5$

MOLECULAR WEIGHT: 184.11

SOURCE OF SAMPLE: Aldrich Chemical Company, Inc.
Milwaukee, Wisconsin

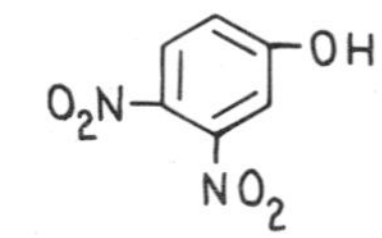

CAS REGISTRY NO.: 577-71-9

NIOSH NUMBER: SL30000

KBr Wafer

CONDENSED PHASE

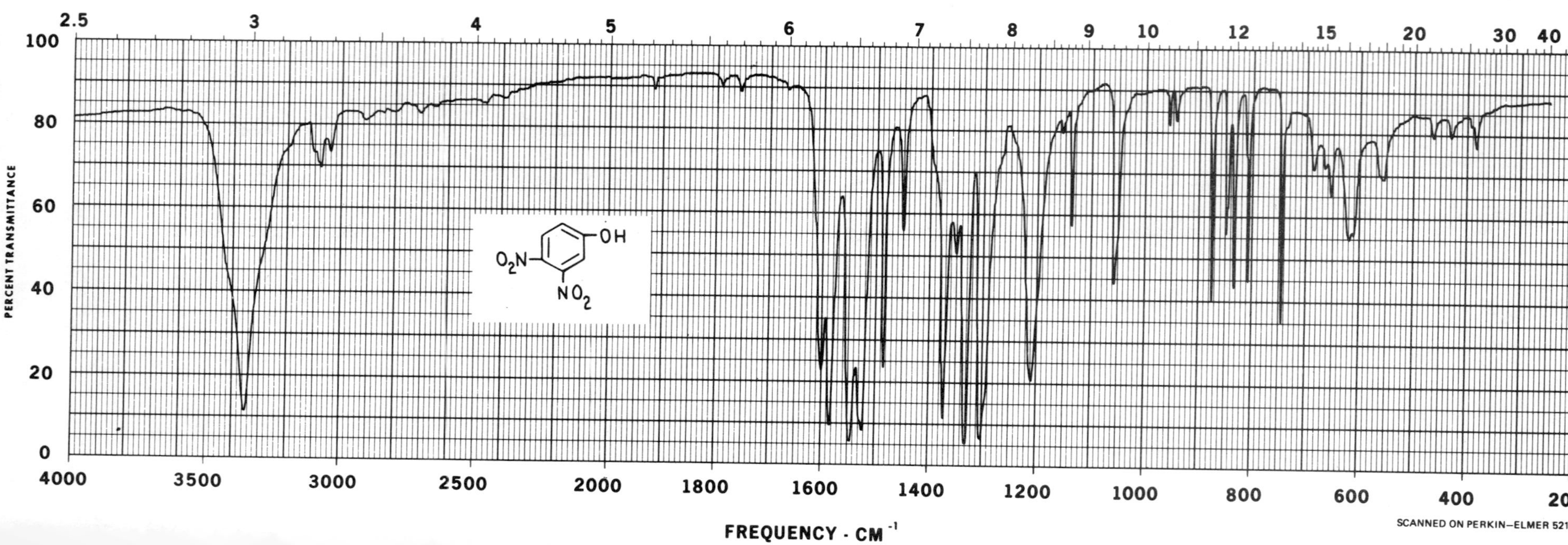

MELTING POINT:	134°C	INJECTION AMT.:	0.81 mg
VAPORIZATION TEMP.:	280°C	PRESSURE:	Atmospheric
CELL VOLUME:	7.5 cc	MAX./MIN.:	98.0/49.0
PATH LENGTH:	7.6 cm	SCALE EXPANSION:	1.88

VAPOR PHASE

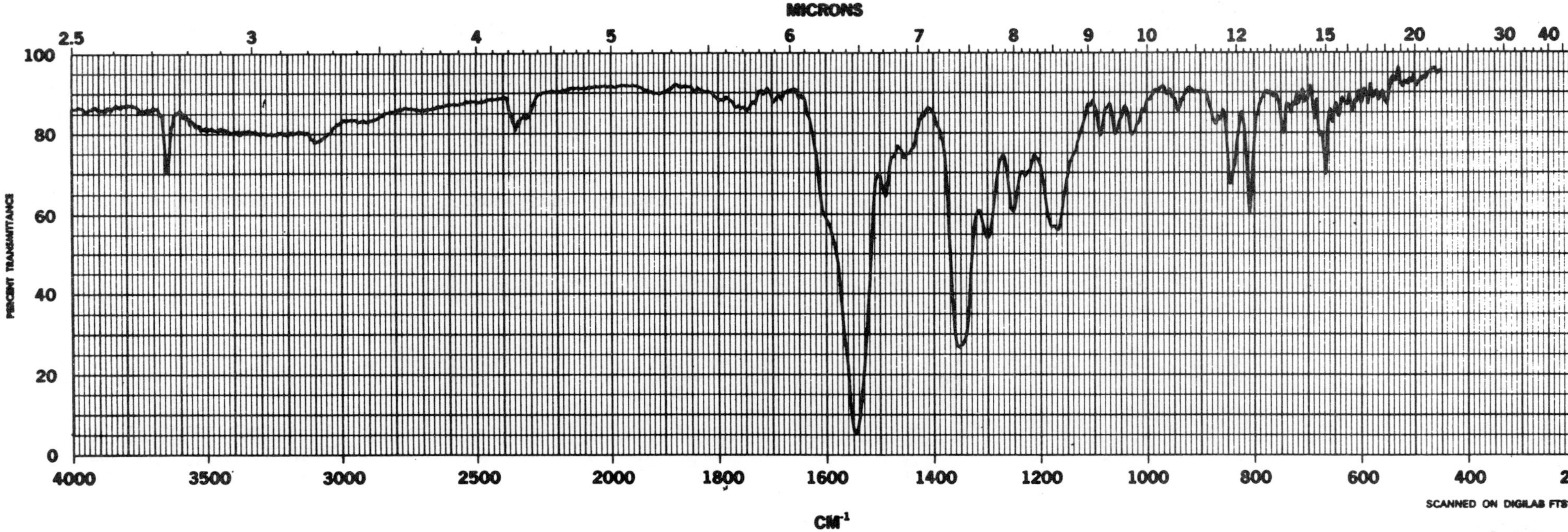

m—NITROPHENOL
3—NITROPHENOL

$C_6H_5NO_3$

MOLECULAR WEIGHT: 139.11

SOURCE OF SAMPLE: Eastman Organic Chemicals
Rochester, New York

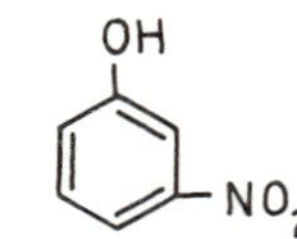

CAS REGISTRY NO.: 554-84-7

NIOSH NUMBER: SM19250

KBr Wafer

CONDENSED PHASE

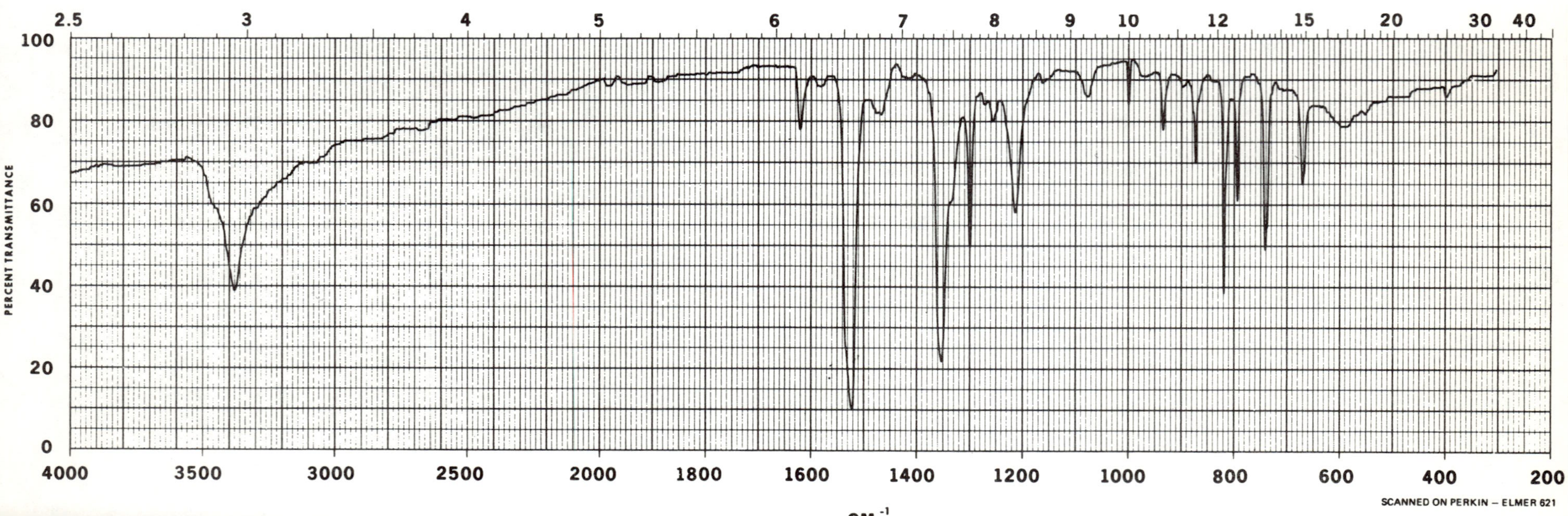

SCANNED ON PERKIN – ELMER 621

BOILING POINT:	216°C	INJECTION AMT.:	1.3 mg
VAPORIZATION TEMP.:	280°C	PRESSURE:	Atmospheric
CELL VOLUME:	7.5 cc	MAX./MIN.:	—
PATH LENGTH:	7.6 cm	SCALE EXPANSION:	0.91

VAPOR PHASE

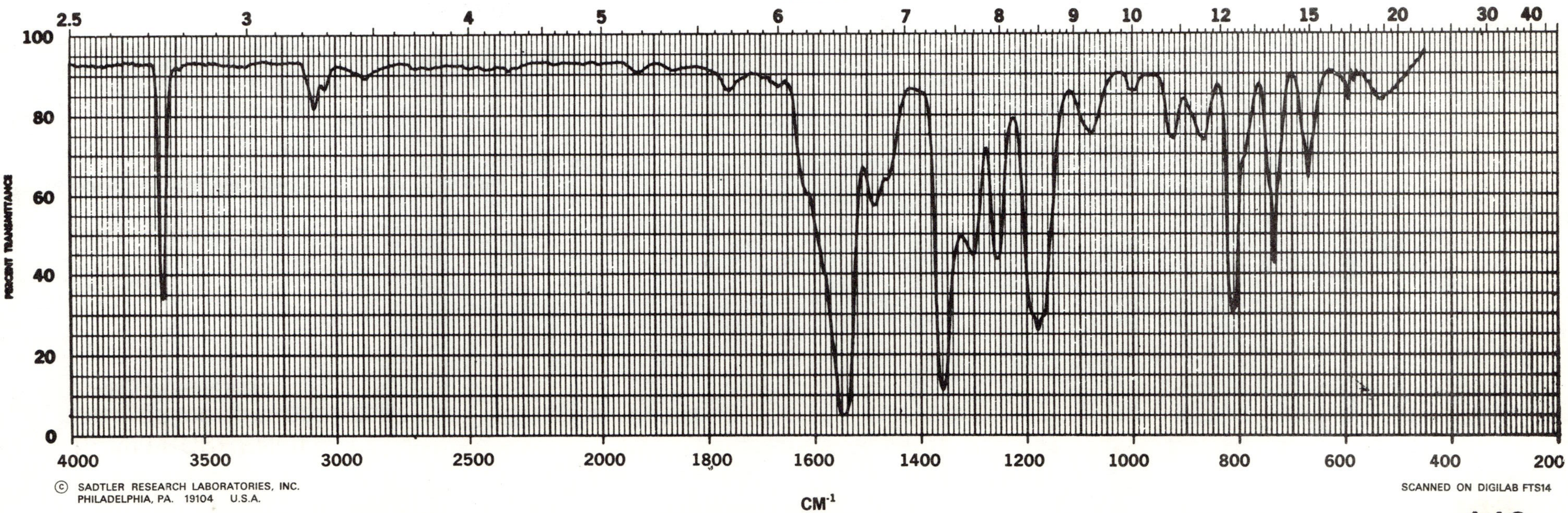

SCANNED ON DIGILAB FTS14

o—NITROPHENOL
2—HYDROXYNITROBENZENE
2—NITROPHENOL

$C_6H_5NO_3$

MOLECULAR WEIGHT: 139.11

SOURCE OF SAMPLE: Fluka AG
Buchs, Switzerland

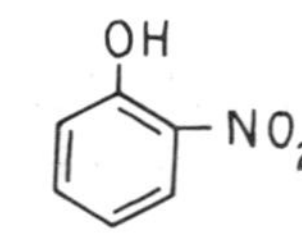

CAS REGISTRY NO.: 88-75-5

NIOSH NUMBER: SM21000

CAPILLARY CELL: Melt

CONDENSED PHASE

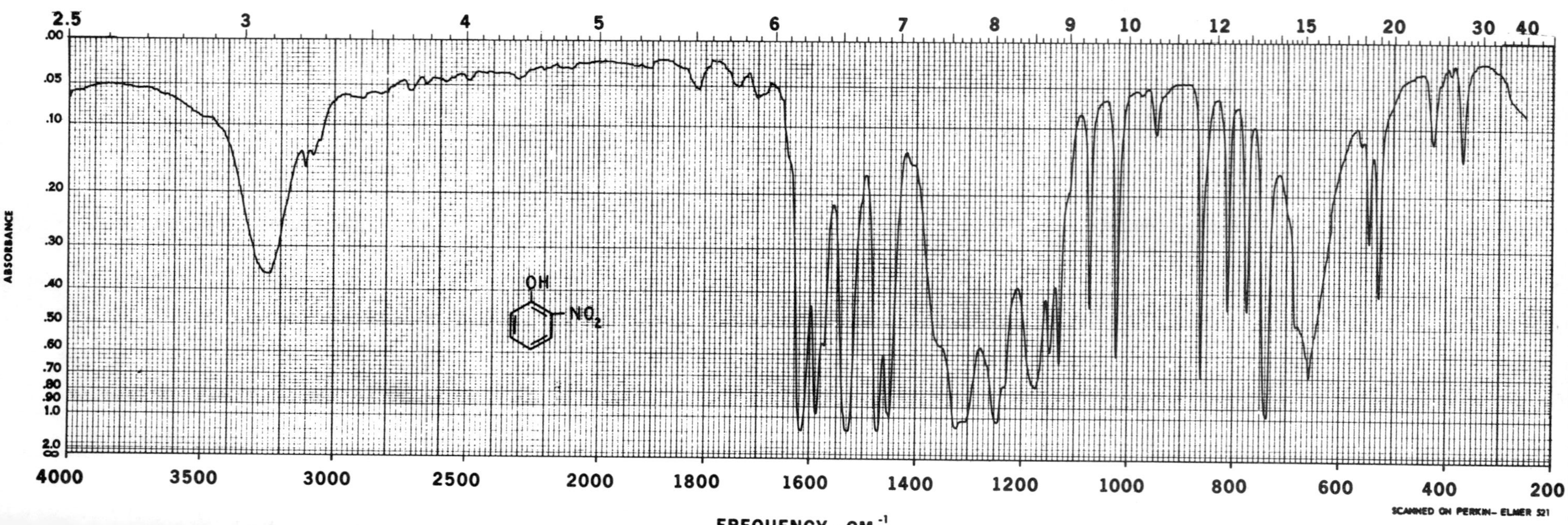

MELTING POINT:	45-46°C	INJECTION AMT.:	0.73 mg
VAPORIZATION TEMP.:	280°C	PRESSURE:	Atmospheric
CELL VOLUME:	7.5 cc	MAX./MIN.:	94.2/13.7
PATH LENGTH:	7.6 cm	SCALE EXPANSION:	1.14

VAPOR PHASE

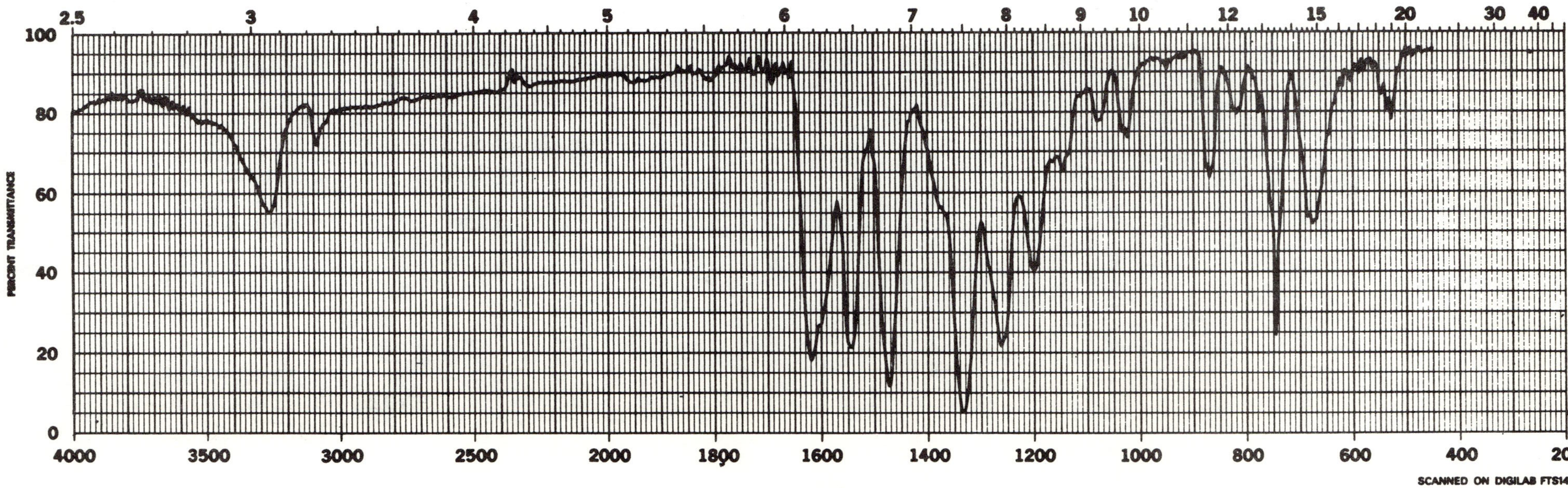

p—NITROPHENOL
4—HYDROXYNITROBENZENE
4—NITROPHENOL

$C_6H_5NO_3$

MOLECULAR WEIGHT: 139.11

SOURCE OF SAMPLE: Fluka AG
Buchs, Switzerland

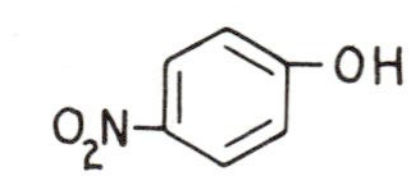

CAS REGISTRY NO.: 100-02-7

NIOSH NUMBER: SM22750

KBr Wafer

CONDENSED PHASE

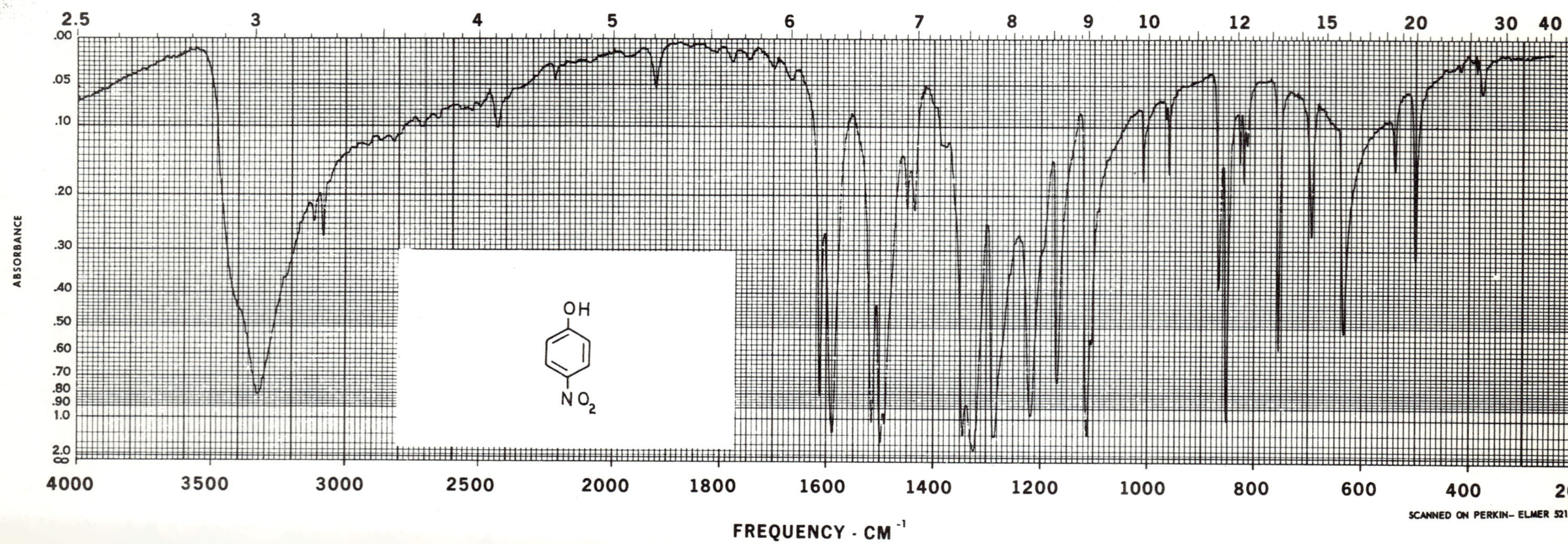

MELTING POINT:	111-113°C	INJECTION AMT.:	0.70 mg
VAPORIZATION TEMP.:	280°C	PRESSURE:	Atmospheric
CELL VOLUME:	7.5 cc	MAX./MIN.:	96.0/15.5
PATH LENGTH:	7.6 cm	SCALE EXPANSION:	1.14

VAPOR PHASE

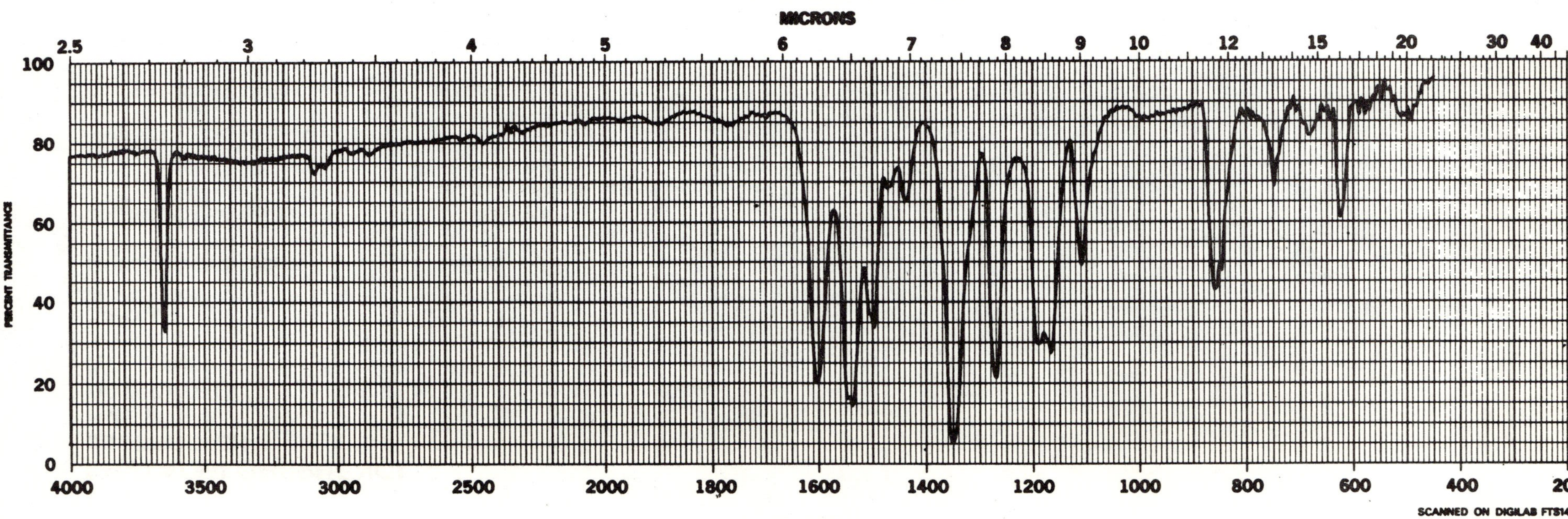

PENTACHLOROPHENOL
CHLOROPHEN
PCP
DOWICIDE G/T
WEEDONE

C_6HCl_5O

MOLECULAR WEIGHT: 266.34

SOURCE OF SAMPLE: J. Kaszubska
Szczecin Institute of Technology
Szczecin, Poland

CAS REGISTRY NO.: 87-86-5

NIOSH NUMBER: SM63000

KBr Wafer

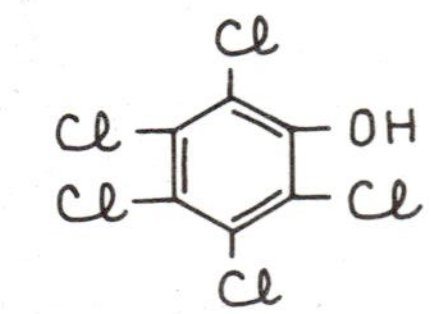

CONDENSED PHASE

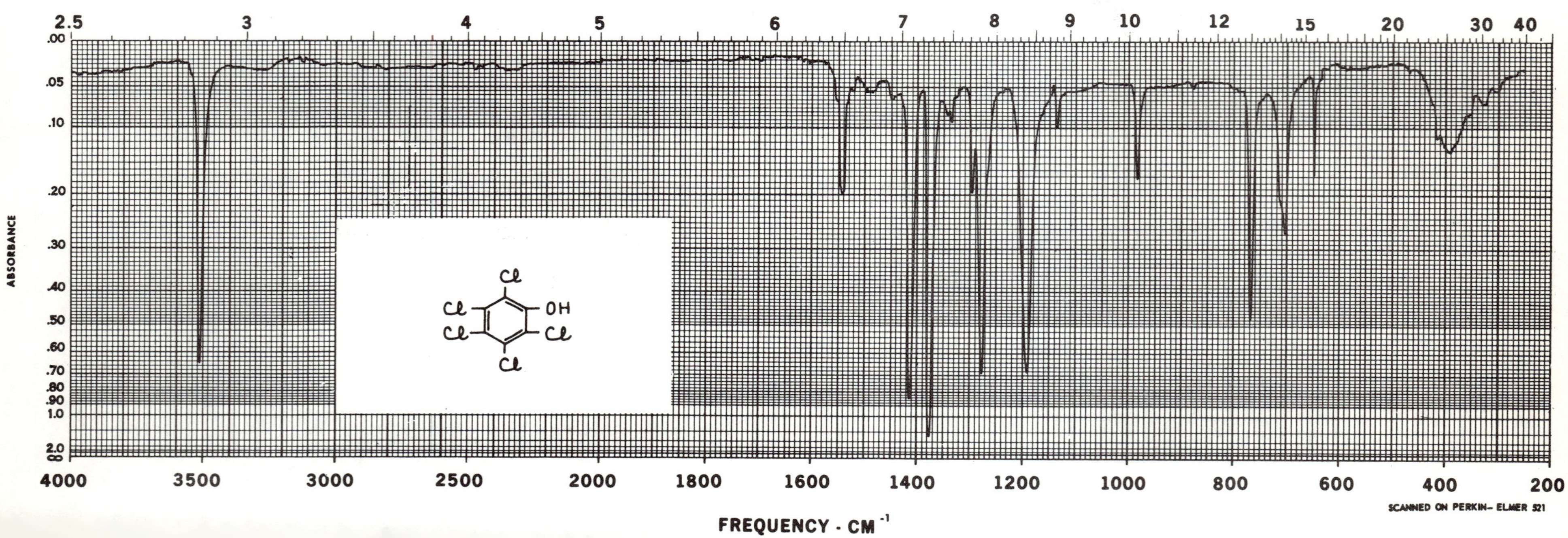

MELTING POINT:	186-187°C	INJECTION AMT.:	–
VAPORIZATION TEMP.:	280°C	PRESSURE:	Atmospheric
CELL VOLUME:	7.5 cc	MAX./MIN.:	101.5/7.5
PATH LENGTH:	7.6 cm	SCALE EXPANSION:	0.98

VAPOR PHASE

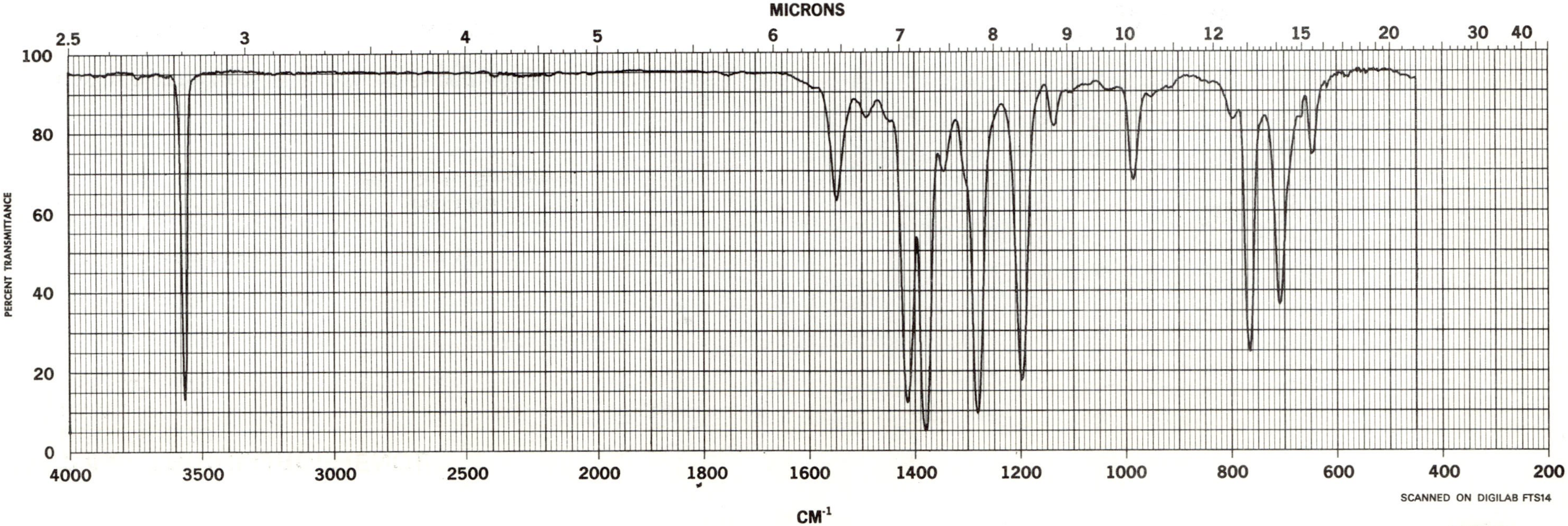

2,4,6–TRICHLOROPHENOL
DOWICIDE 2S
OMAL
PHENACHLOR

$C_6H_3Cl_3O$

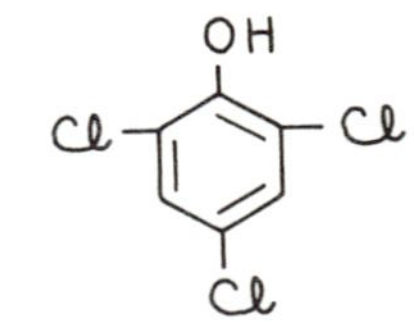

MOLECULAR WEIGHT:	197.45
SOURCE OF SAMPLE:	Fluka AG Buchs, Switzerland
CAS REGISTRY NO.:	88-06-2
NIOSH NUMBER:	SN15750
CAPILLARY CELL:	Melt

CONDENSED PHASE

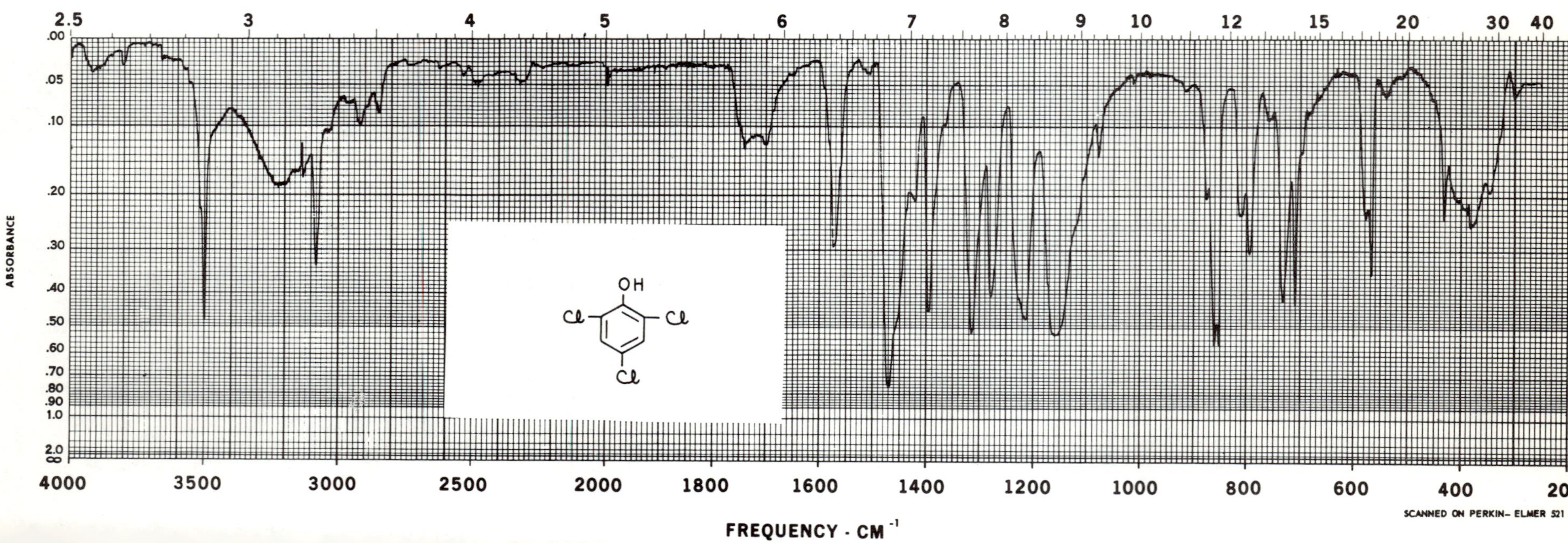

BOILING POINT:	246°C	INJECTION AMT.:	1.1 mg
VAPORIZATION TEMP.:	280°C	PRESSURE:	Atmospheric
CELL VOLUME:	7.5 cc	MAX./MIN.:	102.8/4.0
PATH LENGTH:	7.6 cm	SCALE EXPANSION:	0.93

VAPOR PHASE

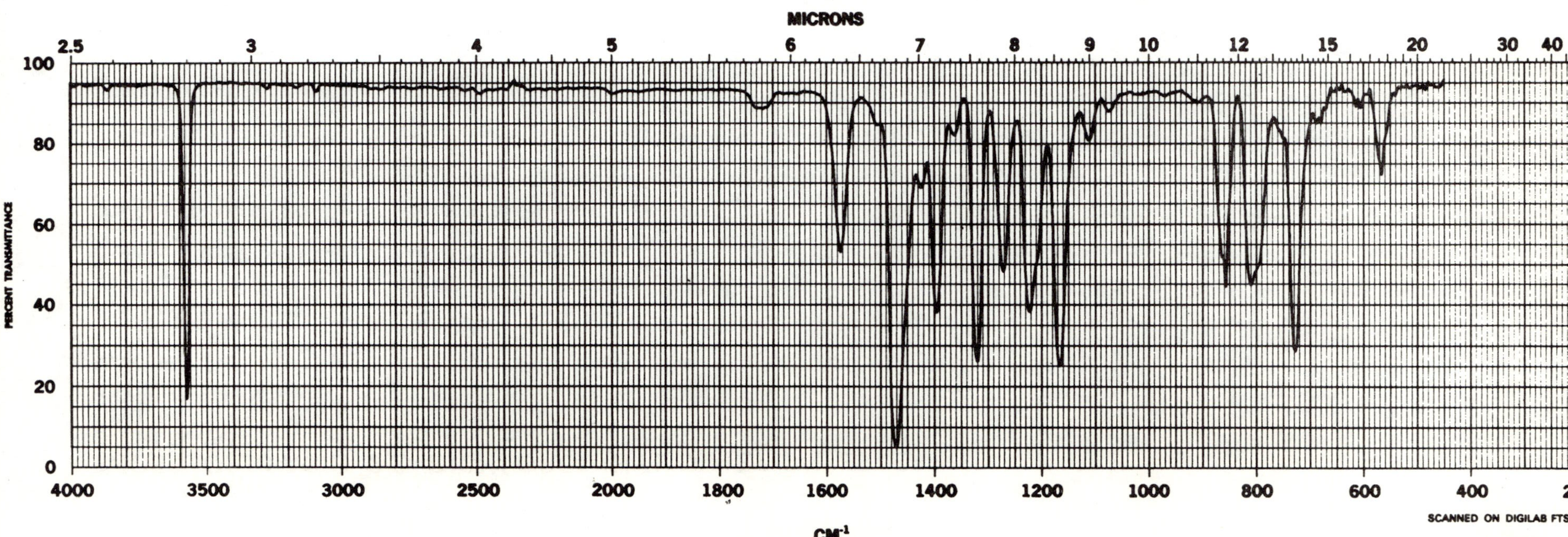

SCANNED ON DIGILAB FTS14

PHOSPHORIC ACID, 2,2–DICHLOROVINYL DIMETHYL ESTER
DICHLORVOS
DDVP
ATGARD V
VAPONA

$C_4H_7Cl_2O_4P$

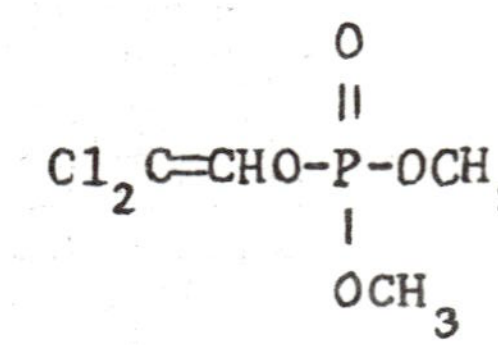

MOLECULAR WEIGHT: 220.98

SOURCE OF SAMPLE: S. B. Penick and Company
Agricultural and Farm Chemicals Division
Lyndhurst, New Jersey

CAS REGISTRY NO.: 62-73-7

NIOSH NUMBER: TC03500

Between Salts

CONDENSED PHASE

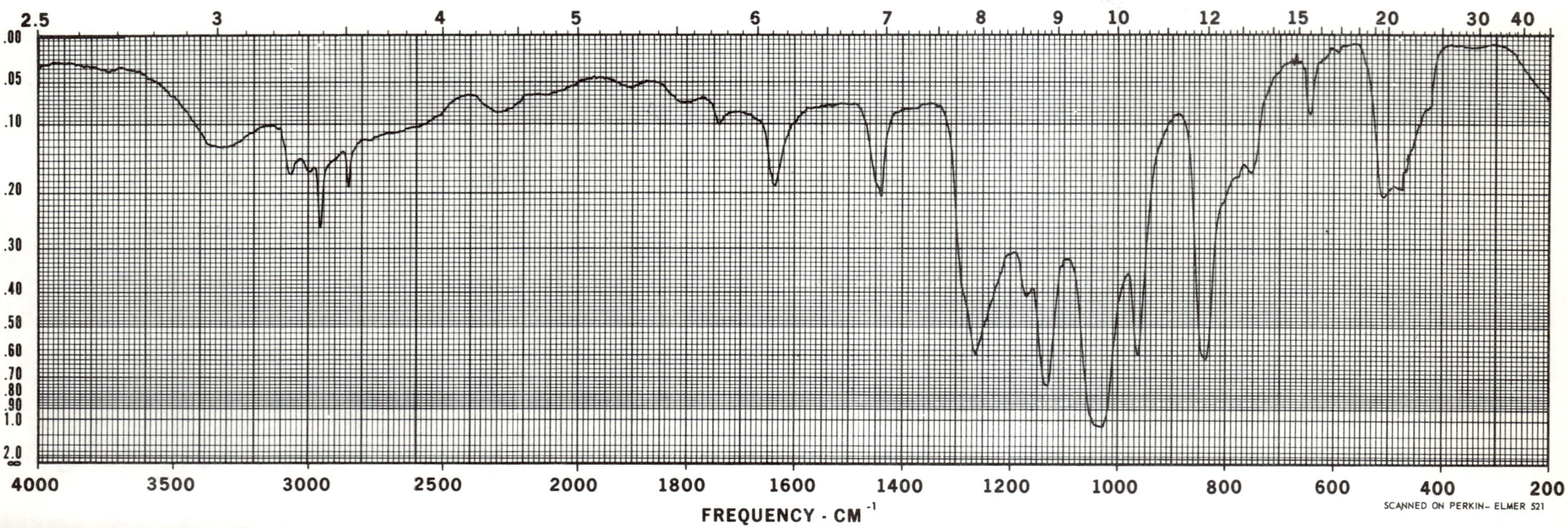

BOILING POINT:	87-88°C/0.7 mm	INJECTION AMT.:	–
VAPORIZATION TEMP.:	200°C	PRESSURE:	Atmospheric
CELL VOLUME:	7.5 cc	MAX./MIN.:	99.9/21.3
PATH LENGTH:	7.6 cm	SCALE EXPANSION:	1.17

VAPOR PHASE

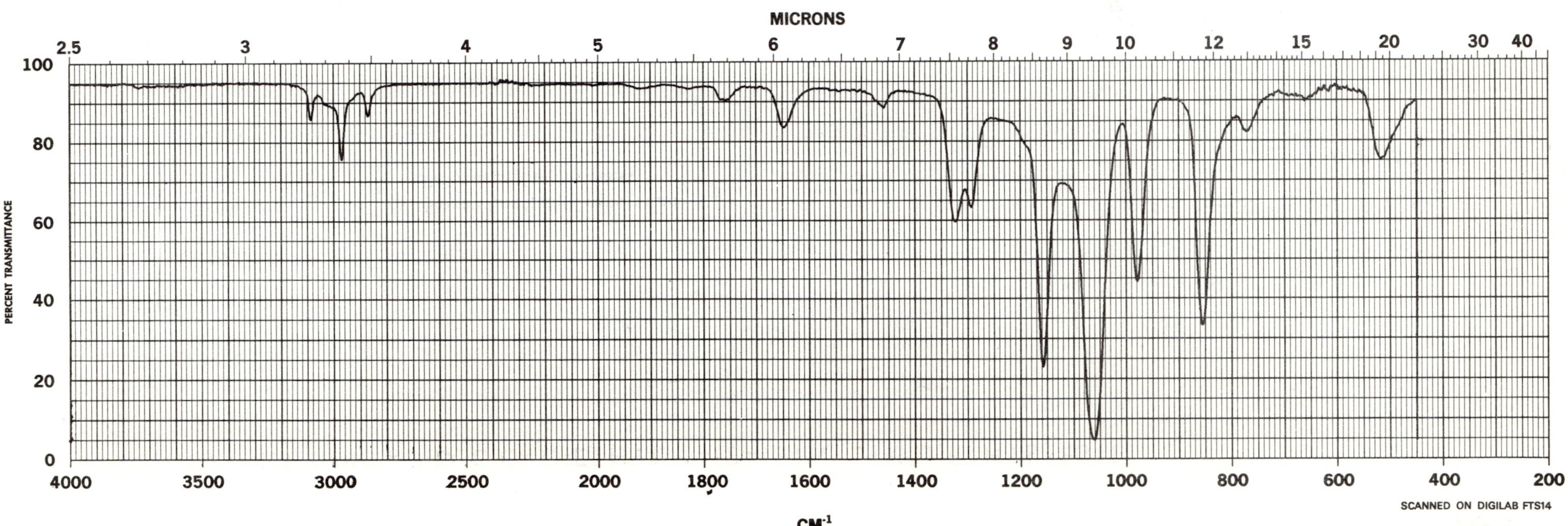

PHOSPHORODITHIOIC ACID, O,O–DIETHYL S–[2–(ETHYLTHIO)ETHYL] ESTER

DISULFOTON
DI–SYSTON
SOLVIREX
THIODEMETON

$C_8H_{19}O_2PS_3$

MOLECULAR WEIGHT:	274.41
SOURCE OF SAMPLE:	Mobay Chemical Corporation Pittsburgh, Pennsylvania
CAS REGISTRY NO.:	298-04-4
NIOSH NUMBER:	TD92750

$CH_3CH_2\text{-}S\text{-}CH_2CH_2S\text{-}P(=S)(OCH_2CH_3)\text{-}OCH_2CH_3$

Between Salts

CONDENSED PHASE

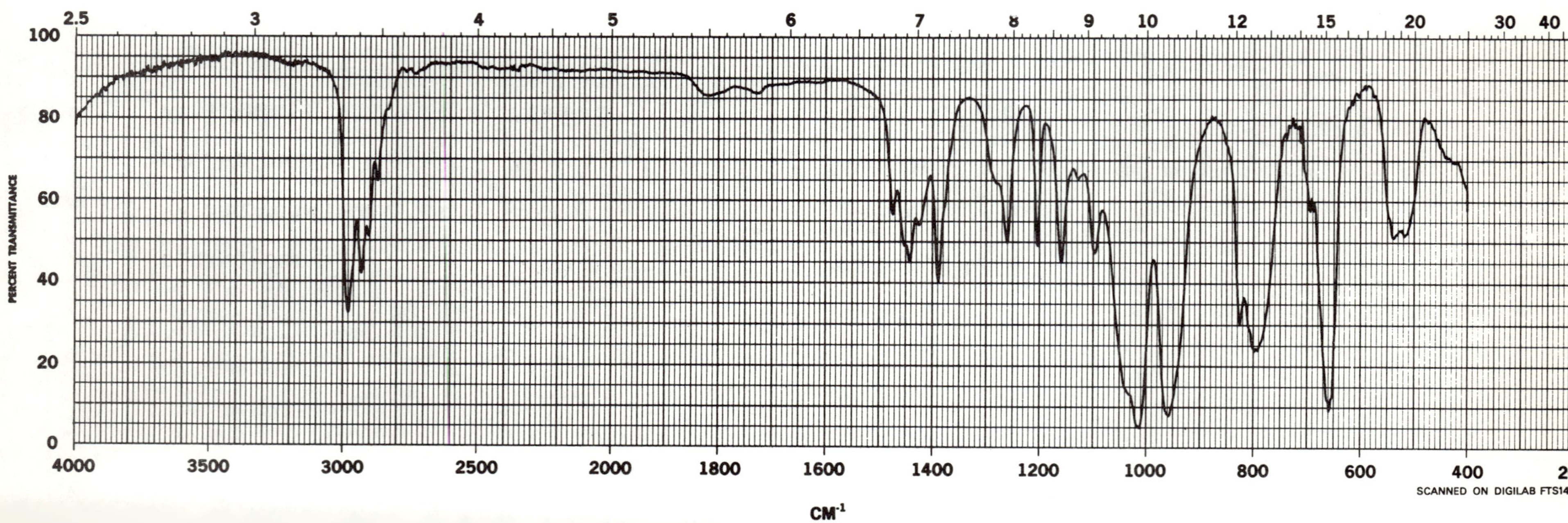

BOILING POINT:	—	INJECTION AMT.:	—
VAPORIZATION TEMP.:	200°C	PRESSURE:	Atmospheric
CELL VOLUME:	7.5 cc	MAX./MIN.:	99.5/12.6
PATH LENGTH:	7.6 cm	SCALE EXPANSION:	1.06

VAPOR PHASE

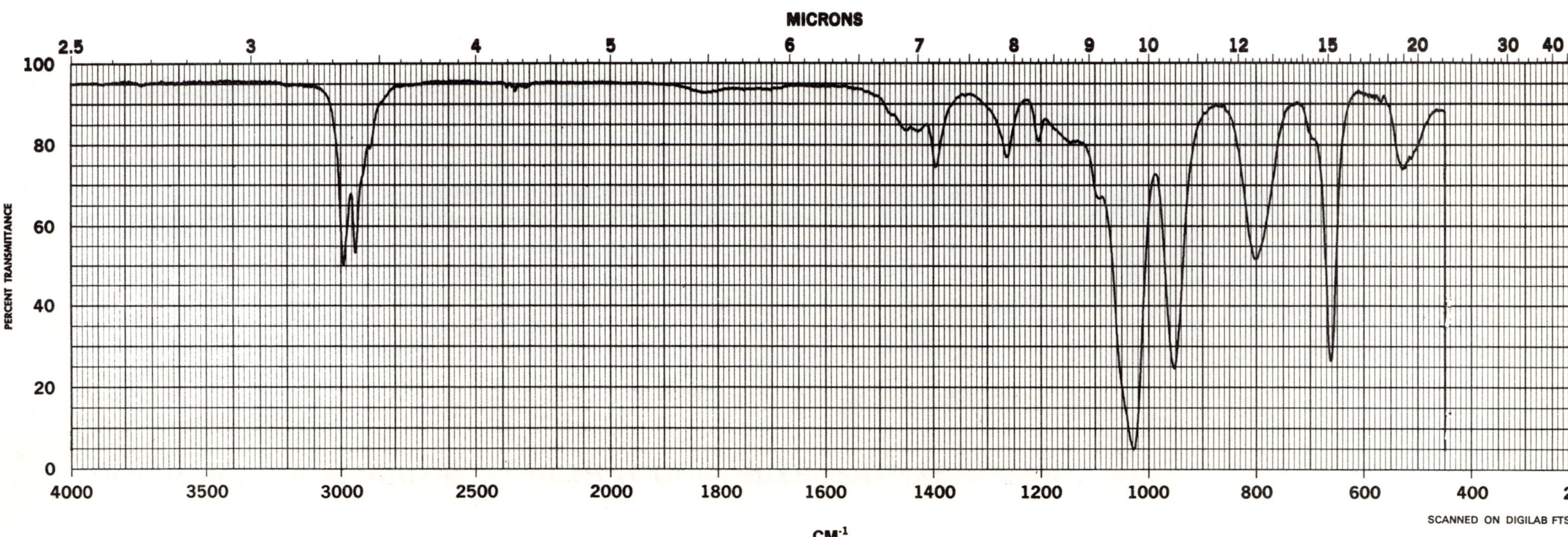

SCANNED ON DIGILAB FTS14

PHOSPHOROTHIOIC ACID, O,O–DIETHYL O–(p–NITROPHENYL) ESTER
ALKRON
ETHYL PARATHION
PARATHION
THIOPHOS

$C_{10}H_{14}NO_5PS$

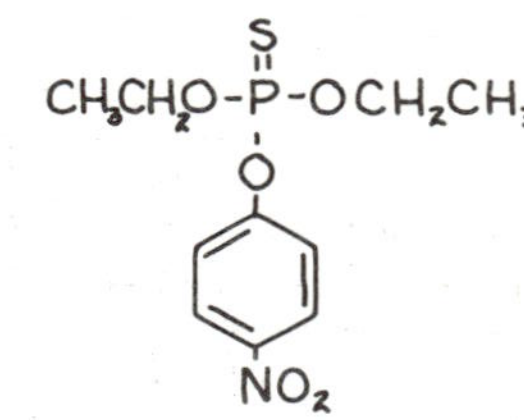

MOLECULAR WEIGHT: 291.26

SOURCE OF SAMPLE: American Potash and Chemical Corporation
Oklahoma City, Oklahoma

CAS REGISTRY NO.: 56-38-2

NIOSH NUMBER: TF45500

Between Salts

CONDENSED PHASE

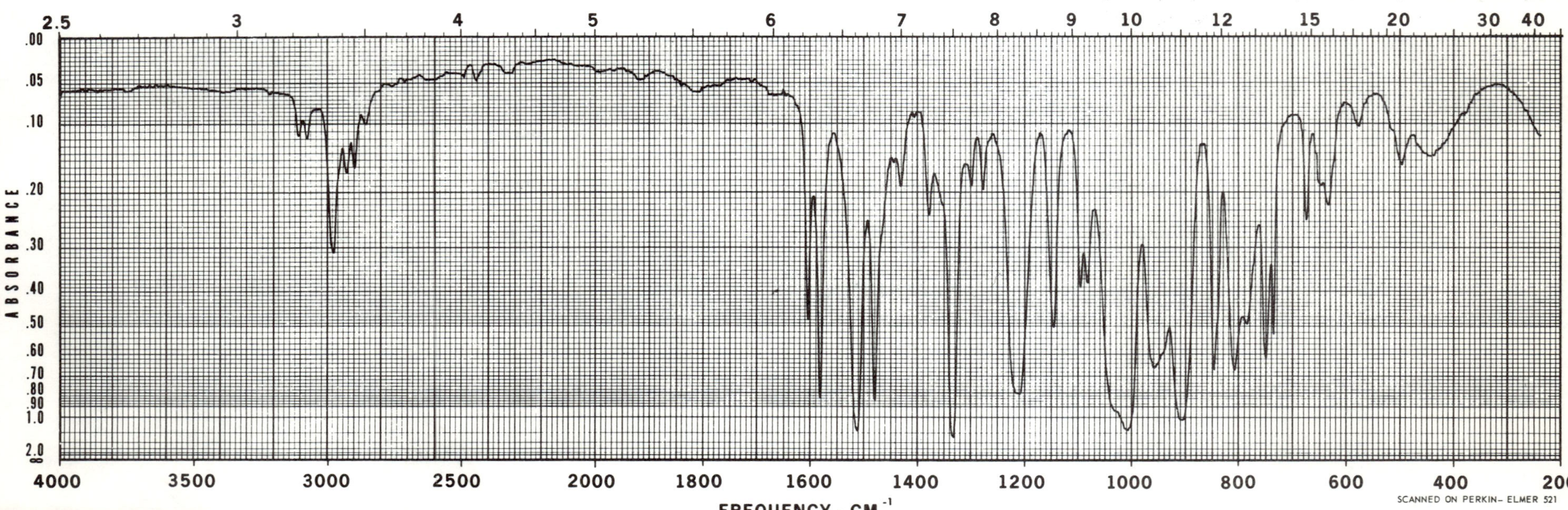

BOILING POINT:	375°C/760 mm	INJECTION AMT.:	—
VAPORIZATION TEMP.:	240°C	PRESSURE:	Atmospheric
CELL VOLUME:	7.5 cc	MAX./MIN.:	99.9/47.3
PATH LENGTH:	7.6 cm	SCALE EXPANSION:	1.75

VAPOR PHASE

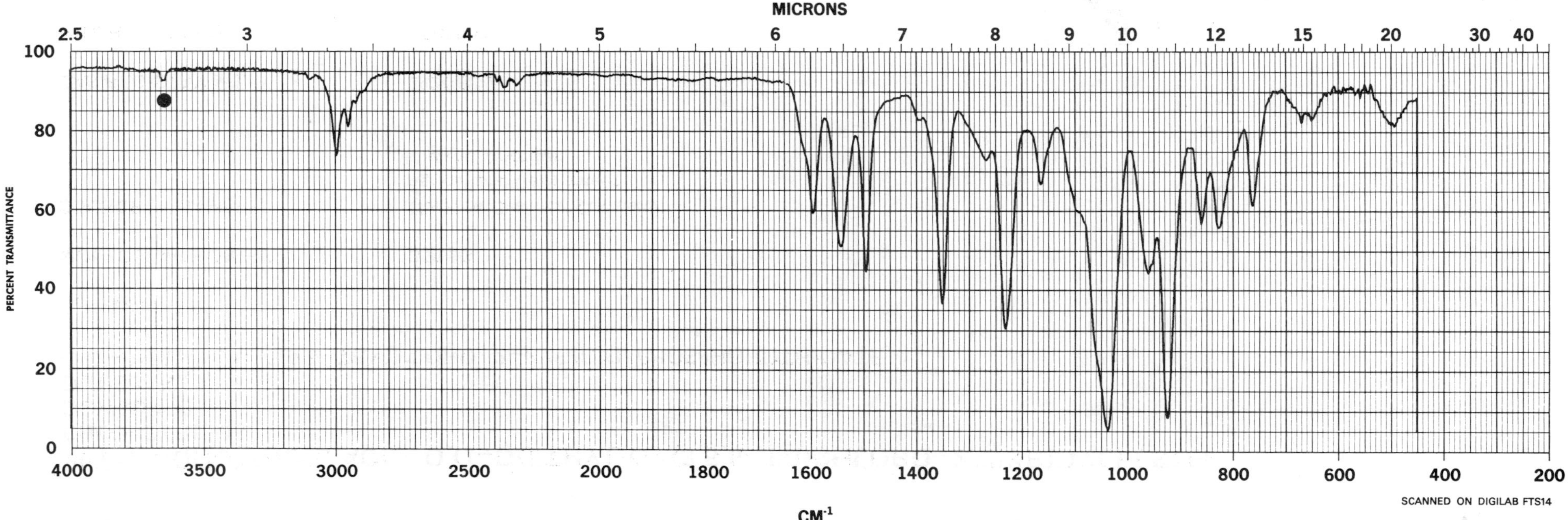

SCANNED ON DIGILAB FTS14

156

PHOSPHOROTHIOIC ACID, O,O–DIETHYL O–(3,5,6–TRICHLORO–2–PYRIDYL) ESTER
CHLORPYRIFOS
DOWCO 179
DURSBAN
O,O–DIETHYL O–3,5,6–TRICHLORO–2–PYRIDYL PHOSPHOROTHIOATE

$C_9H_{11}Cl_3NO_3PS$

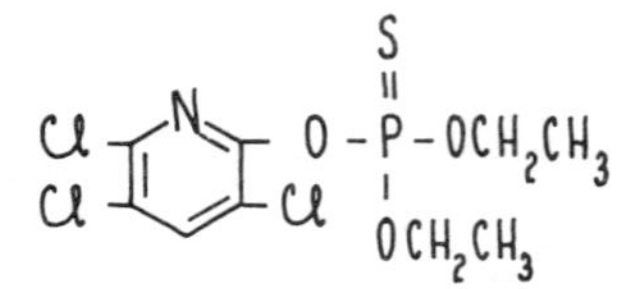

MOLECULAR WEIGHT: 350.59

SOURCE OF SAMPLE: The Dow Chemical Company
Agricultural Department
Midland, Michigan

CAS REGISTRY NO.: 2921-88-2

NIOSH NUMBER: TF63000

Film

CONDENSED PHASE

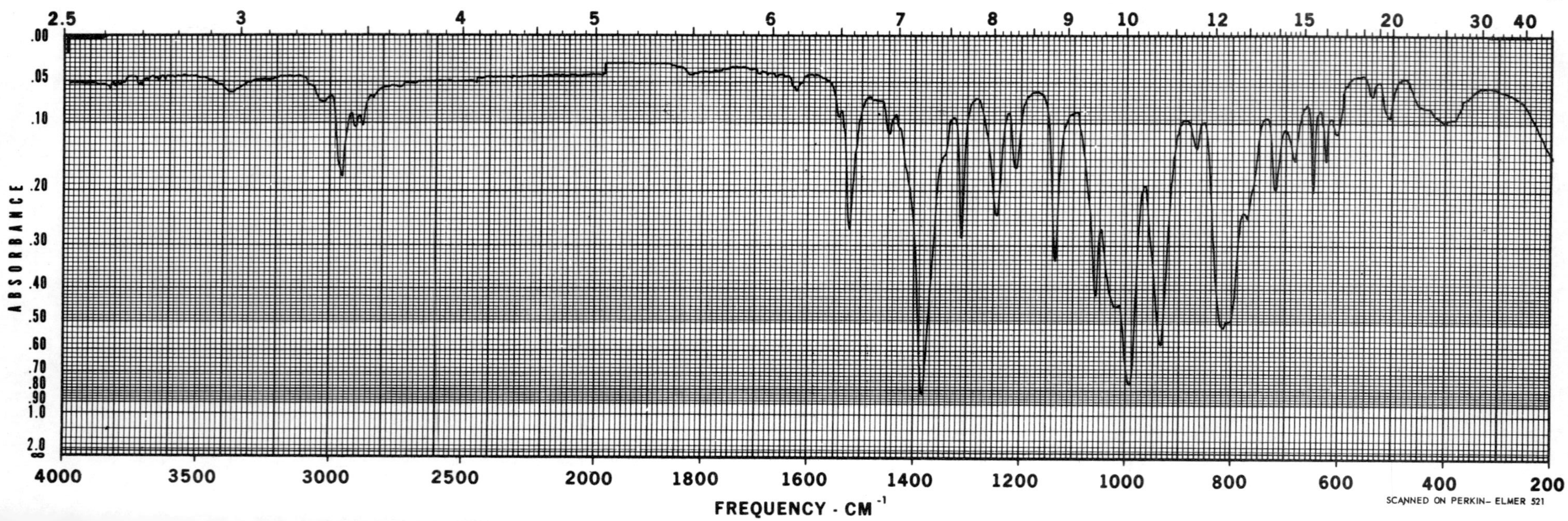

MELTING POINT:	41.5-43°C	INJECTION AMT.:	—
VAPORIZATION TEMP.:	200°C	PRESSURE:	Atmospheric
CELL VOLUME:	7.5 cc	MAX./MIN.:	89.8/21.6
PATH LENGTH:	7.6 cm	SCALE EXPANSION:	1.35

VAPOR PHASE

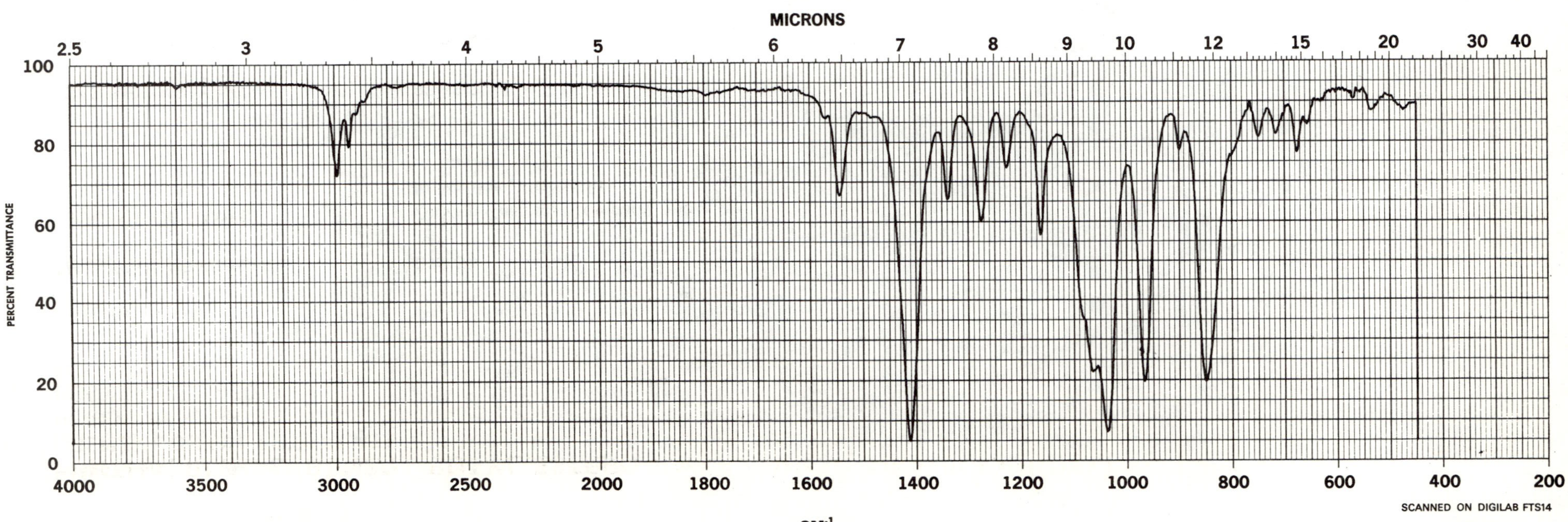

SCANNED ON DIGILAB FTS14

157

PHTHALIC ACID, BENZYL BUTYL ESTER
BENZYL BUTYL PHTHALATE
PALATINOL BB
SANTICIZER 160

$C_{19}H_{20}O_4$

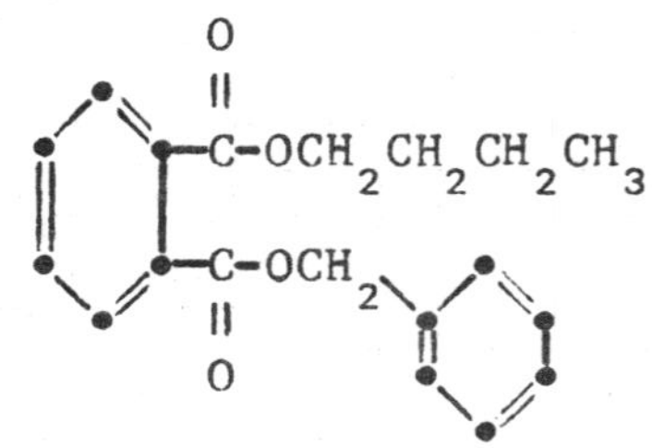

MOLECULAR WEIGHT: 312.37

SOURCE OF SAMPLE: Aldrich Chemical Company, Inc.
Milwaukee, Wisconsin

CAS REGISTRY NO.: 85-68-7

NIOSH NUMBER: TH99900

CAPILLARY CELL: Neat

CONDENSED PHASE

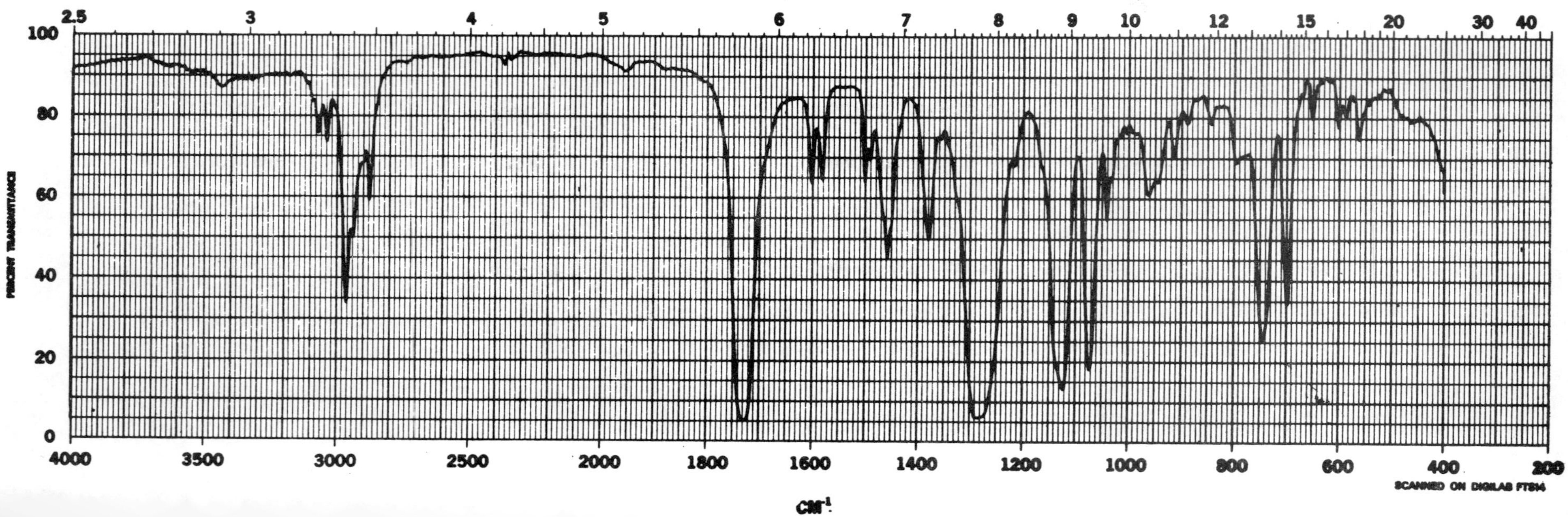

BOILING POINT:	–	INJECTION AMT.:	–
VAPORIZATION TEMP.:	280°C	PRESSURE:	Atmospheric
CELL VOLUME:	7.5 cc	MAX./MIN.:	100.5/28.0
PATH LENGTH:	7.6 cm	SCALE EXPANSION:	1.27

VAPOR PHASE

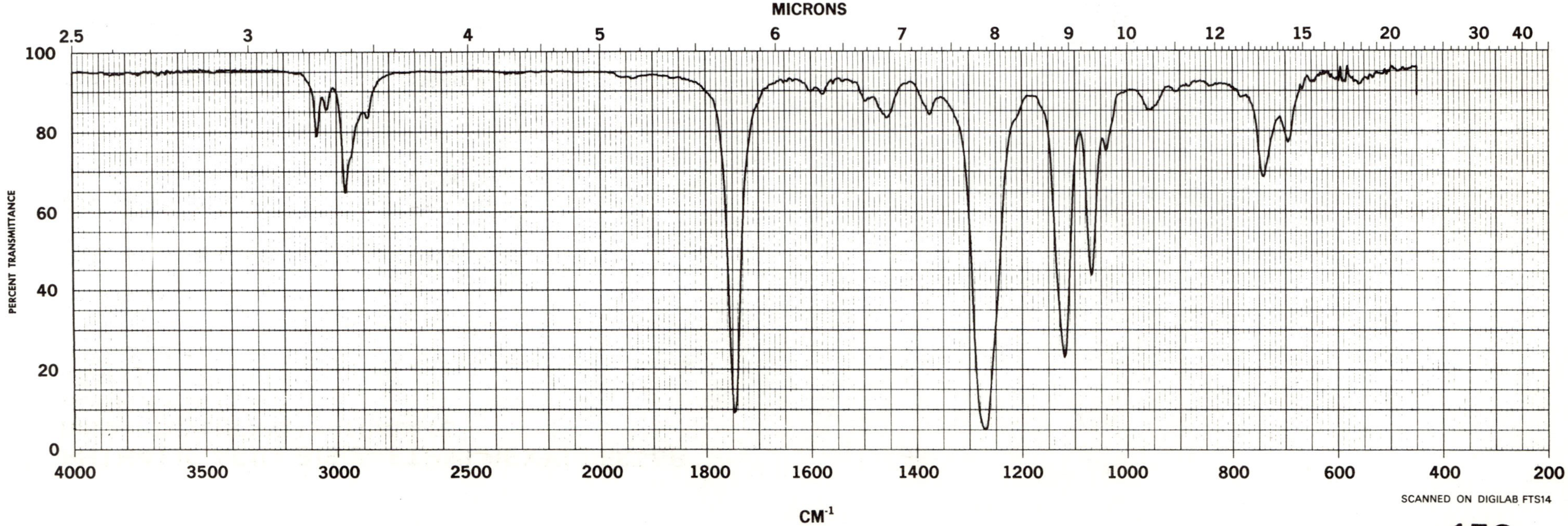

SCANNED ON DIGILAB FTS14

158

PHTHALIC ACID, BIS(2–ETHYLHEXYL) ESTER
BIS(2–ETHYLHEXYL) PHTHALATE
HOWFLEX DOP
STAFLEX DOP
WITCIZER 312

$C_{24}H_{38}O_4$

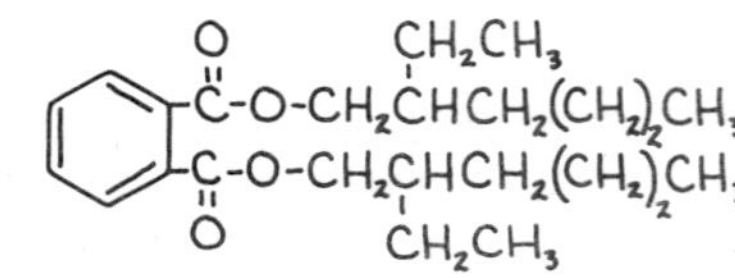

MOLECULAR WEIGHT: 390.57

SOURCE OF SAMPLE: Fluka AG
Buchs, Switzerland

CAS REGISTRY NO.: 117-81-7

NIOSH NUMBER: TI03500

CAPILLARY CELL: Neat

CONDENSED PHASE

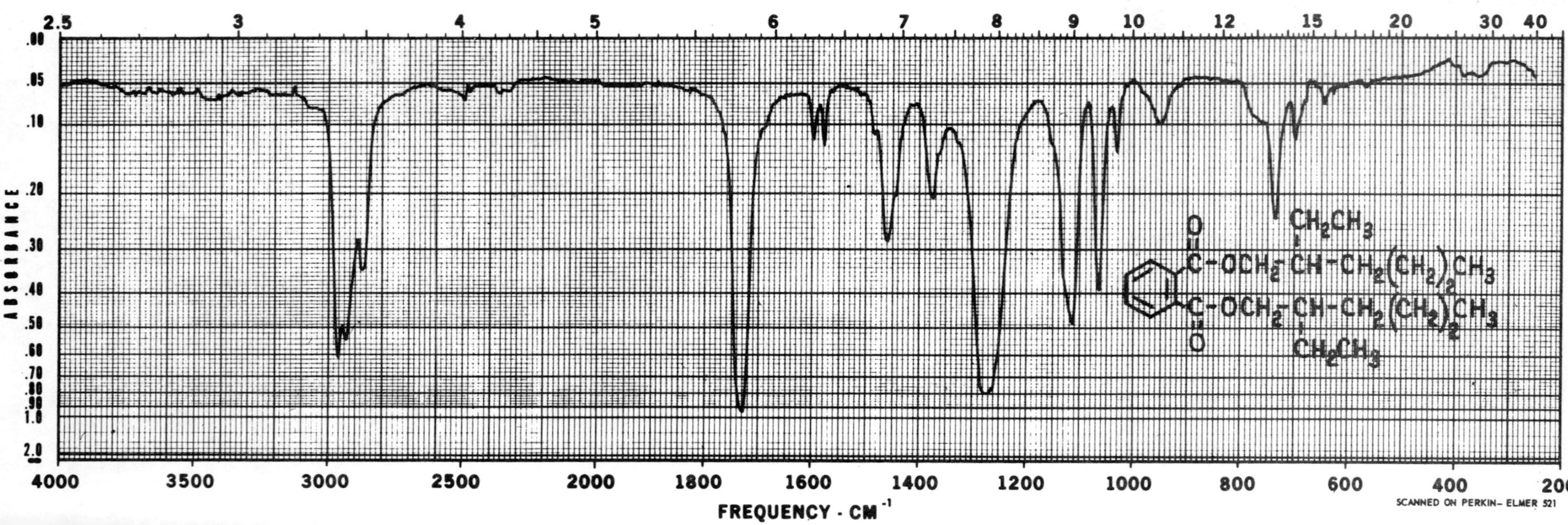

BOILING POINT:	222-230°C/4 mm	INJECTION AMT.:	—
VAPORIZATION TEMP.:	240°C	PRESSURE:	Atmospheric
CELL VOLUME:	7.5 cc	MAX./MIN.:	100.1/42.5
PATH LENGTH:	7.6 cm	SCALE EXPANSION:	1.60

VAPOR PHASE

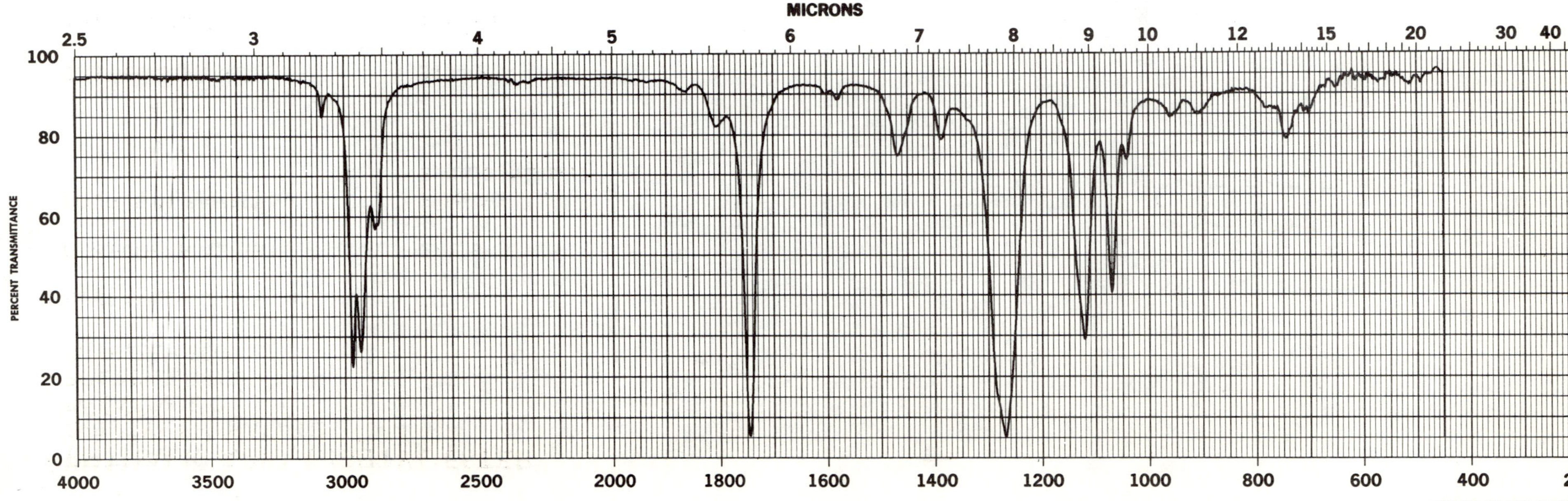

SCANNED ON DIGILAB FTS14

PHTHALIC ACID, DIBUTYL ESTER
DIBUTYL PHTHALATE
PALATINOL C
POLYCIZER DBP
WITCIZER 300

$C_{16}H_{22}O_4$

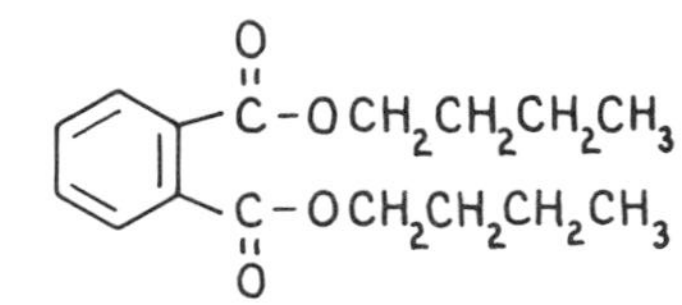

MOLECULAR WEIGHT: 278.35

SOURCE OF SAMPLE: Fluka AG
Buchs, Switzerland

CAS REGISTRY NO.: 84-74-2

NIOSH NUMBER: TI08750

CAPILLARY CELL: Neat

CONDENSED PHASE

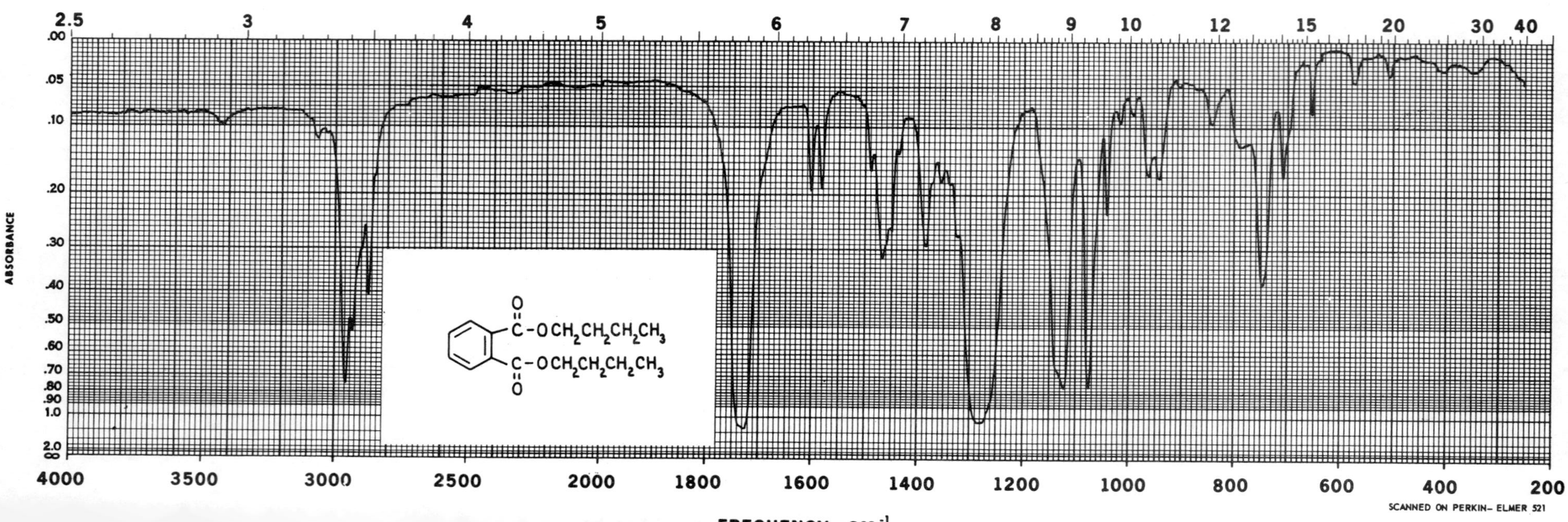

BOILING POINT:	340°C	INJECTION AMT.:	0.7 μl
VAPORIZATION TEMP.:	280°C	PRESSURE:	Atmospheric
CELL VOLUME:	7.5 cc	MAX./MIN.:	—
PATH LENGTH:	7.6 cm	SCALE EXPANSION:	1.02

VAPOR PHASE

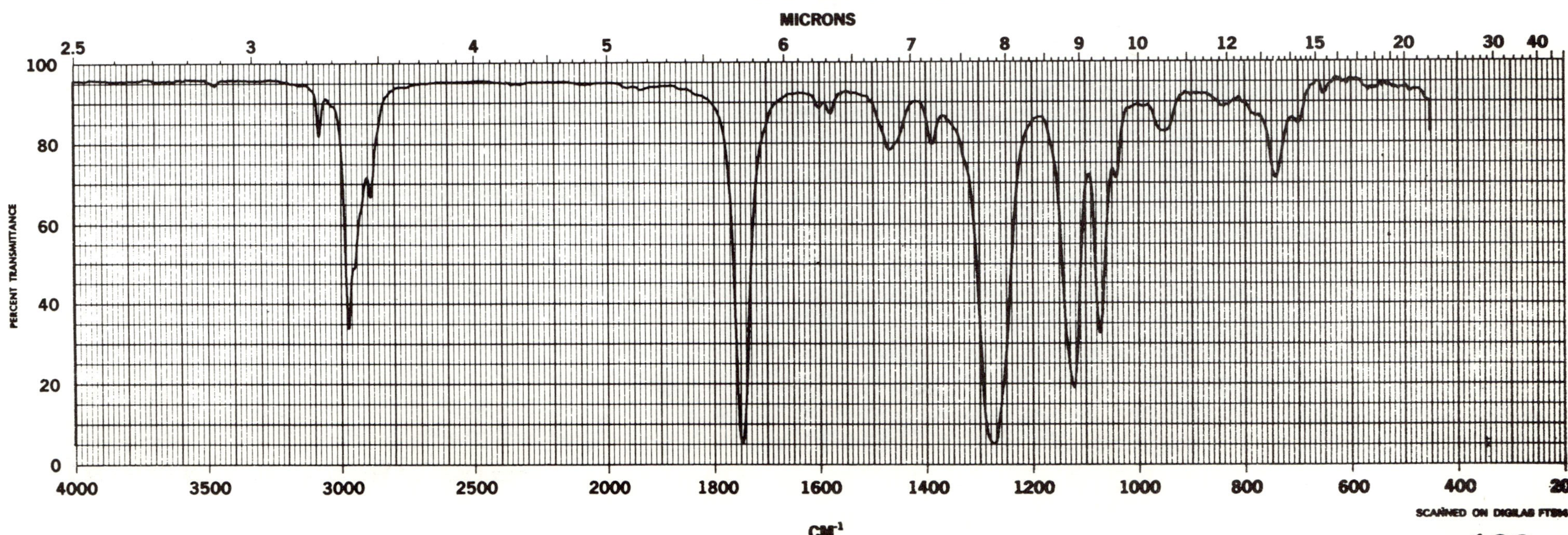

PHTHALIC ACID, DIETHYL ESTER
ANOZOL
DIETHYL PHTHALATE
PALATINOL A
SOLVANOL

$C_{12}H_{14}O_4$

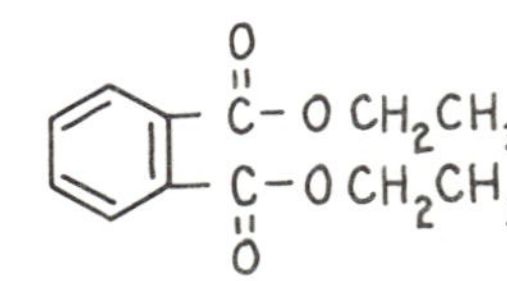

MOLECULAR WEIGHT:	222.24
SOURCE OF SAMPLE:	Eastman Organic Chemicals Rochester, New York
CAS REGISTRY NO.:	84-66-2
NIOSH NUMBER:	TI10500
CAPILLARY CELL:	Neat

CONDENSED PHASE

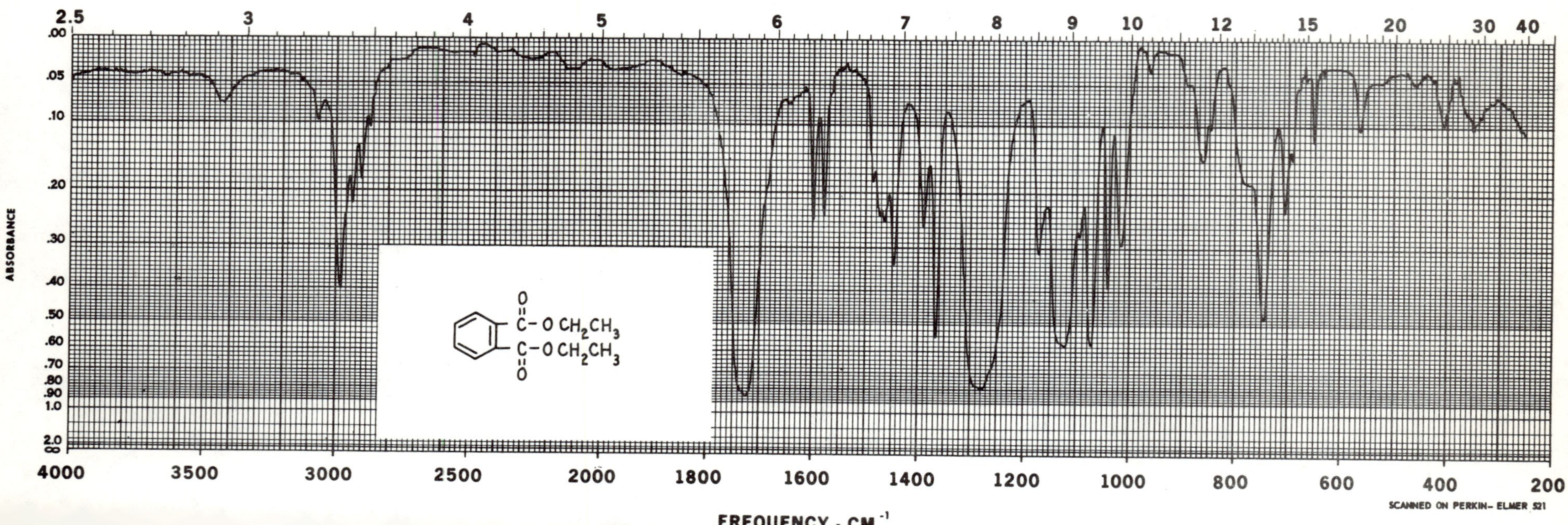

BOILING POINT:	298-299°C/735 mm	INJECTION AMT.:	0.5 μl
VAPORIZATION TEMP.:	280°C	PRESSURE:	Atmospheric
CELL VOLUME:	7.5 cc	MAX./MIN.:	—
PATH LENGTH:	7.6 cm	SCALE EXPANSION:	0.91

VAPOR PHASE

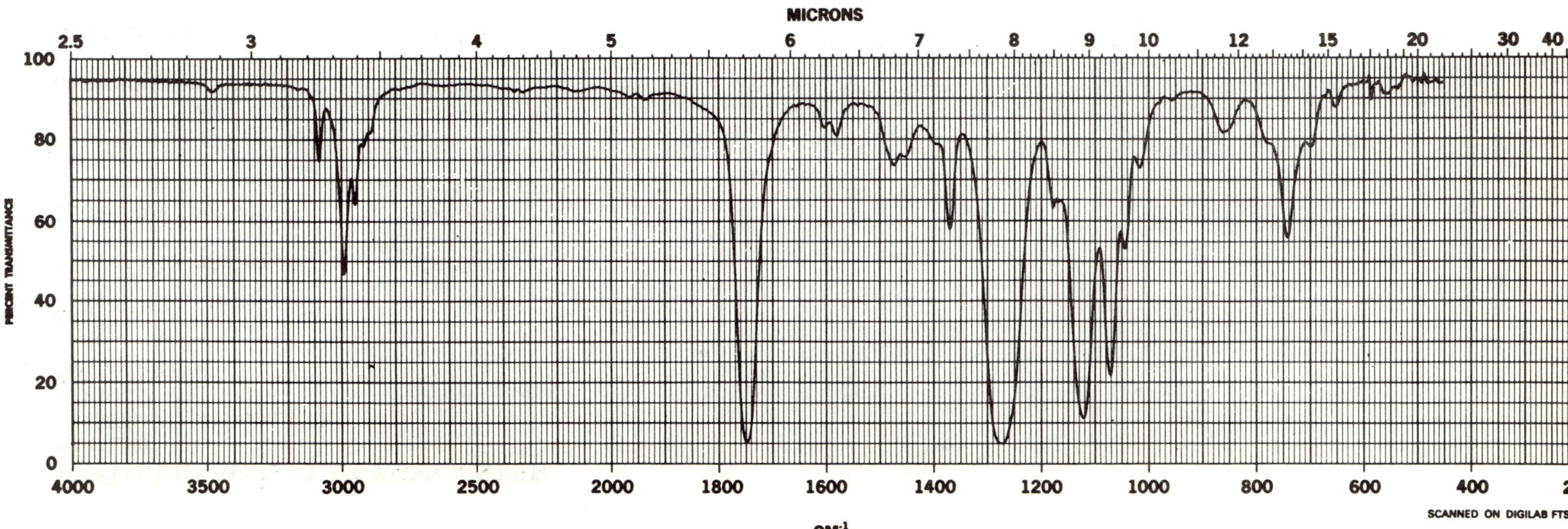

SCANNED ON DIGILAB FTS14

PHTHALIC ACID, DIMETHYL ESTER
AVOLIN
DIMETHYL PHTHALATE
MIPAX
SOLVAR ONE

$C_{10}H_{10}O_4$

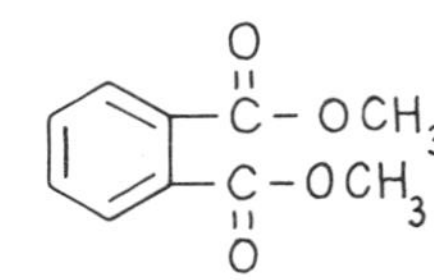

MOLECULAR WEIGHT: 194.19

SOURCE OF SAMPLE: Fluka AG, Buchs, Switzerland

CAS REGISTRY NO.: 131-11-3

NIOSH NUMBER: TI15750

CAPILLARY CELL: Neat

CONDENSED PHASE

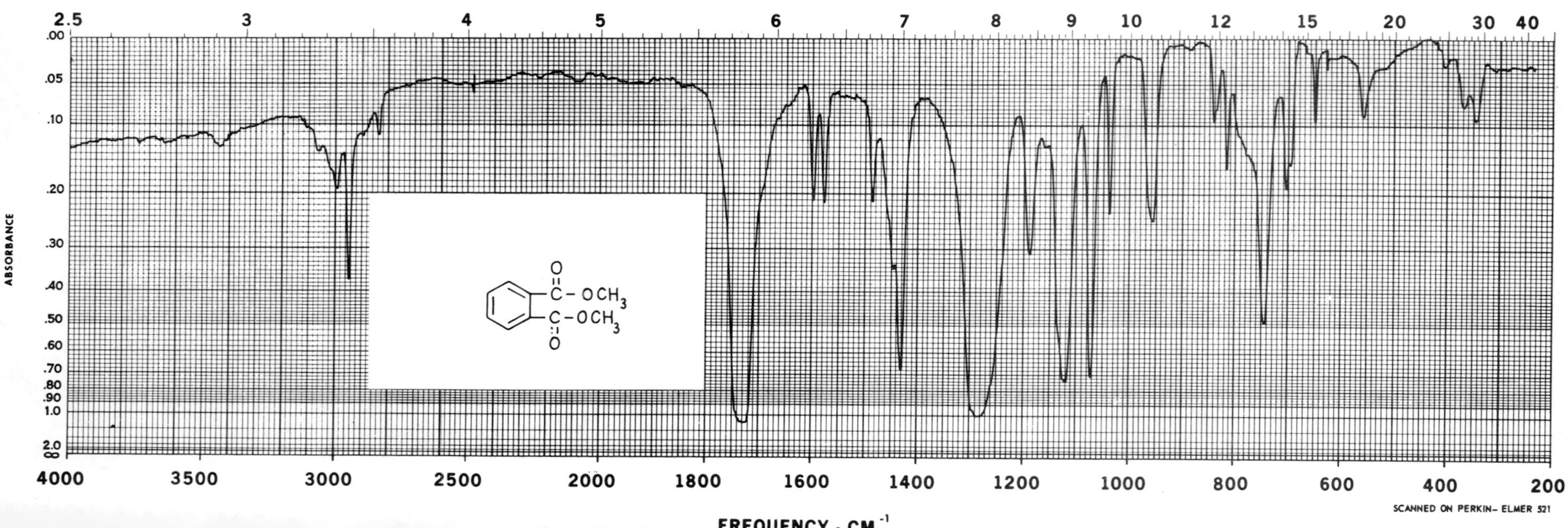

BOILING POINT:	151-152°C/12 mm	INJECTION AMT.:	0.5 µl
VAPORIZATION TEMP.:	240°C	PRESSURE:	Atmospheric
CELL VOLUME:	7.5 cc	MAX./MIN.:	—
PATH LENGTH:	7.6 cm	SCALE EXPANSION:	0.88

VAPOR PHASE

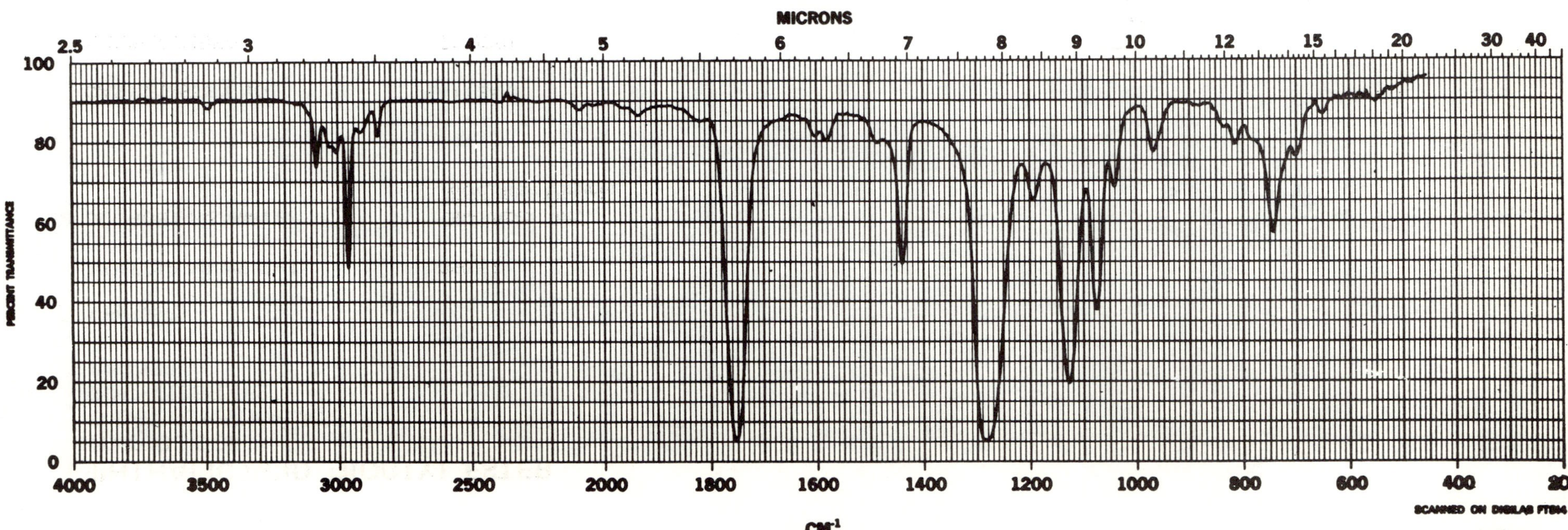

PHTHALIC ACID, DIOCTYL ESTER
HATCOL DOP
DIOCTYL PHTHALATE
POLYCIZER 162
PX–138

$C_{24}H_{38}O_4$

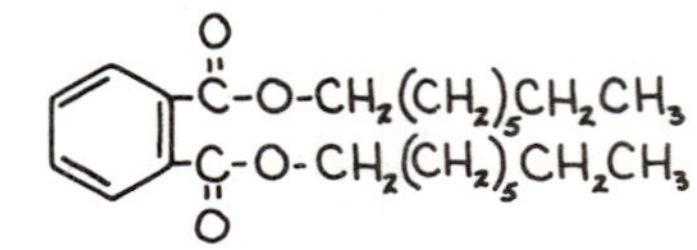

MOLECULAR WEIGHT: 390.57

SOURCE OF SAMPLE: Ohio-Apex, Inc.
Nitro, West Virginia

CAS REGISTRY NO.: 117-84-0

NIOSH NUMBER: TI19250

CAPILLARY CELL: Neat

CONDENSED PHASE

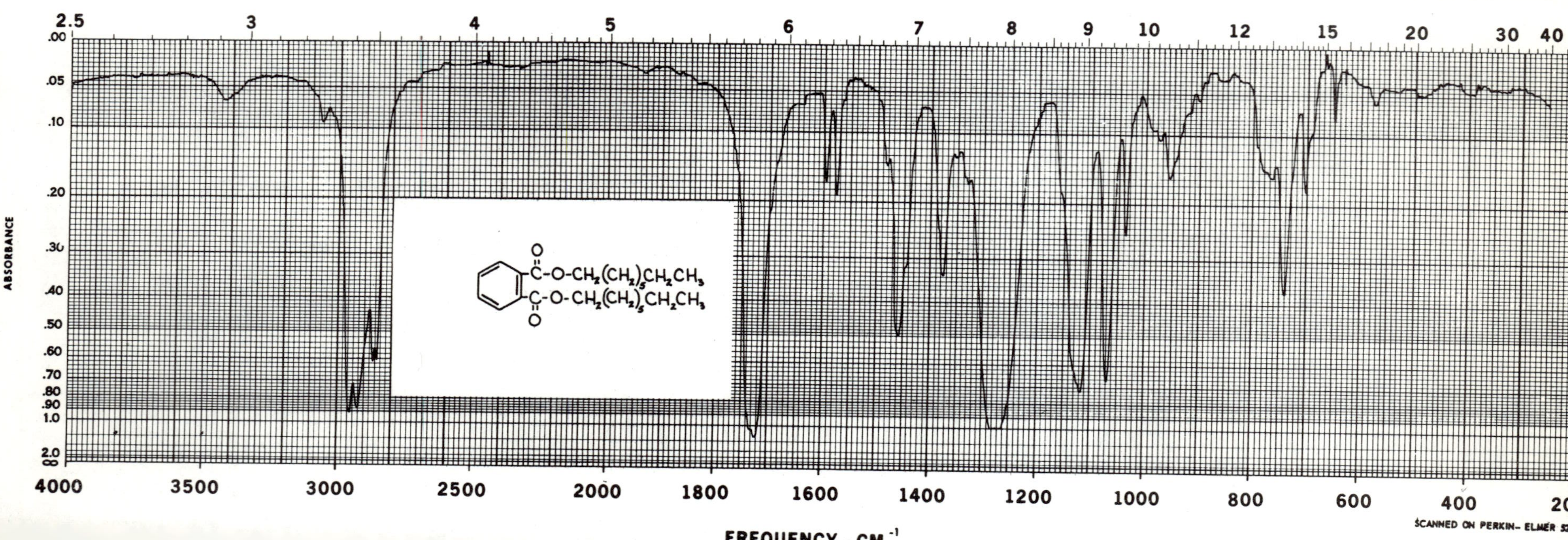

BOILING POINT:	–	INJECTION AMT.:	0.5 μl
VAPORIZATION TEMP.:	280°C	PRESSURE:	Atmospheric
CELL VOLUME:	7.5 cc	MAX./MIN.:	–
PATH LENGTH:	7.6 cm	SCALE EXPANSION:	1.30

VAPOR PHASE

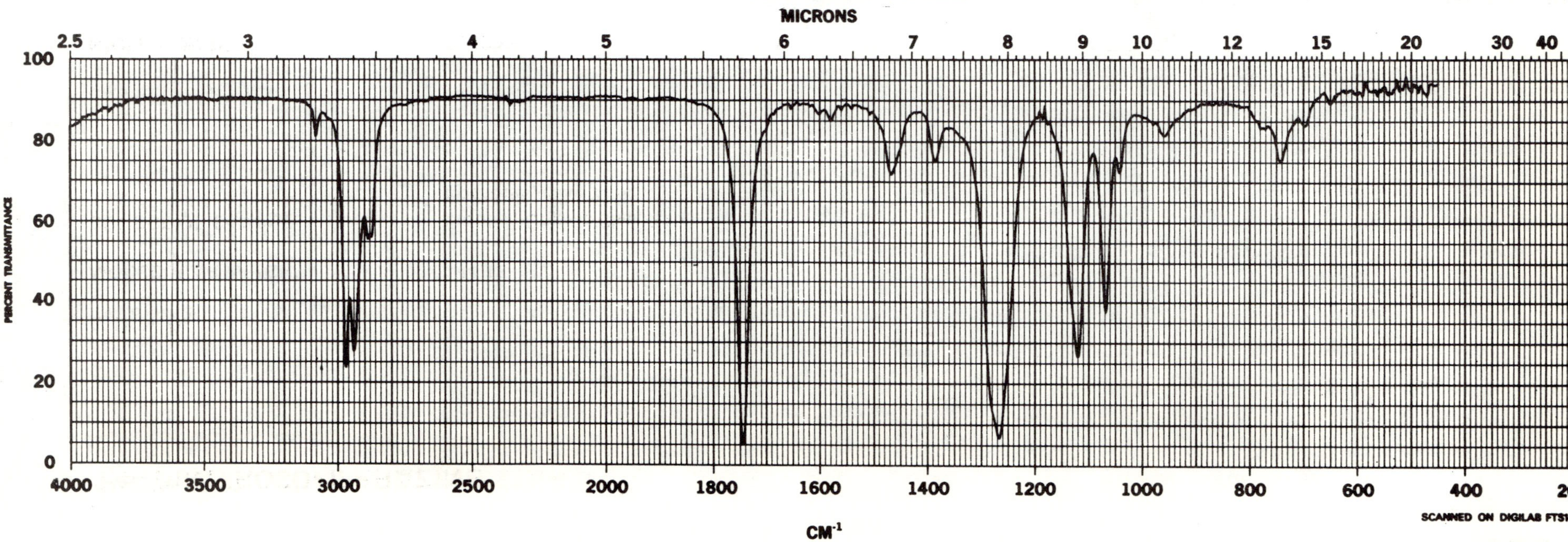

1,4–DINITROSOPIPERAZINE

$C_4H_8N_4O_2$

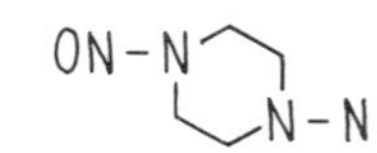

MOLECULAR WEIGHT: 144.13

SOURCE OF SAMPLE: Eastman Organic Chemicals
Rochester, New York

CAS REGISTRY NO.: 140-79-4

NIOSH NUMBER: TL63000

A = Mulled in Mineral Oil
B = Mulled in Perfluorinated Hydrocarbon

CONDENSED PHASE

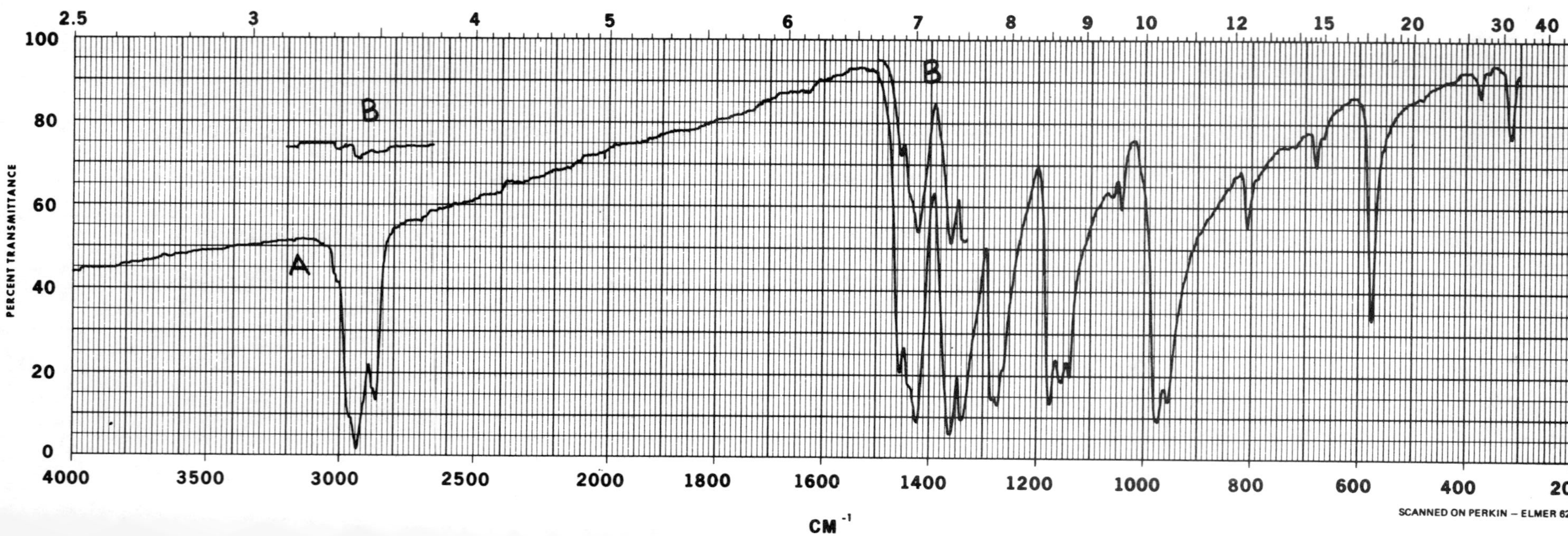

MELTING POINT:	158°C	INJECTION AMT.:	—
VAPORIZATION TEMP.:	200°C	PRESSURE:	Atmospheric
CELL VOLUME:	7.5 cc	MAX./MIN.:	94.4/19.4
PATH LENGTH:	7.6 cm	SCALE EXPANSION:	1.23

VAPOR PHASE

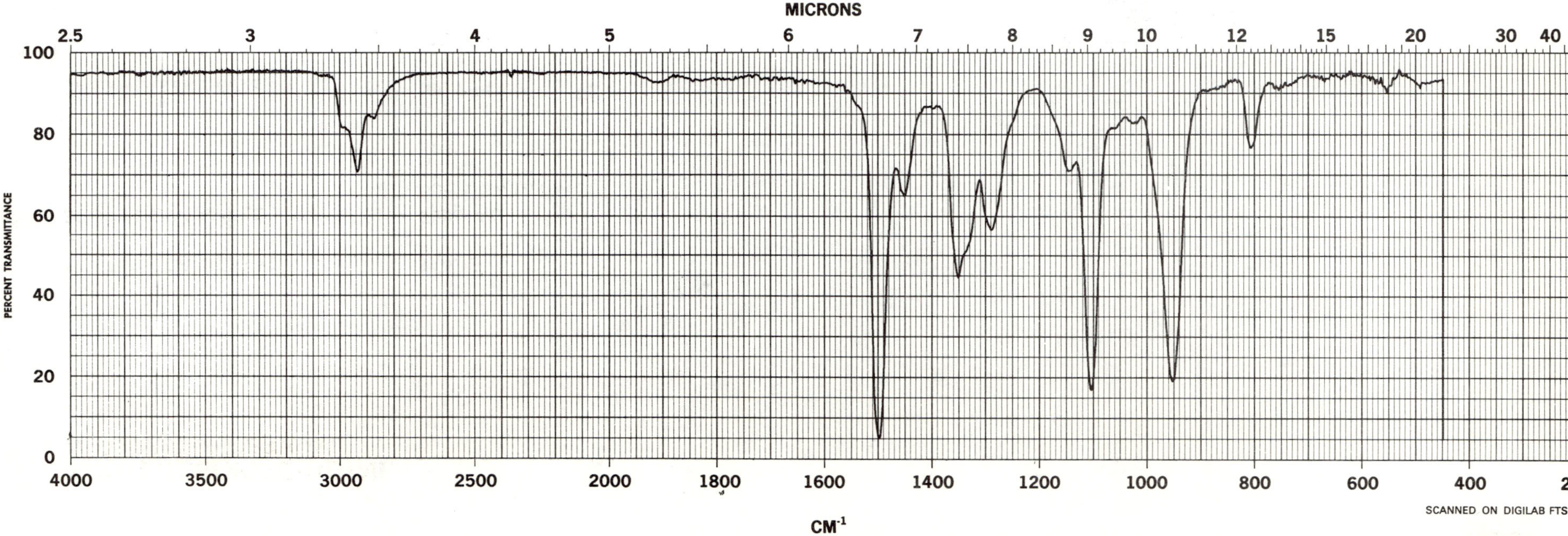

SCANNED ON DIGILAB FTS14

164

1—NITROSOPIPERIDINE
N—NITROSOPIPERIDINE

$C_5H_{10}N_2O$

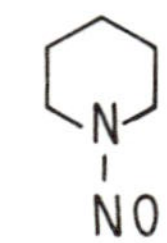

MOLECULAR WEIGHT:	114.15
SOURCE OF SAMPLE:	Fluka AG Buchs, Switzerland
CAS REGISTRY NO.:	100-75-4
NIOSH NUMBER:	TN21000
CAPILLARY CELL:	Neat

CONDENSED PHASE

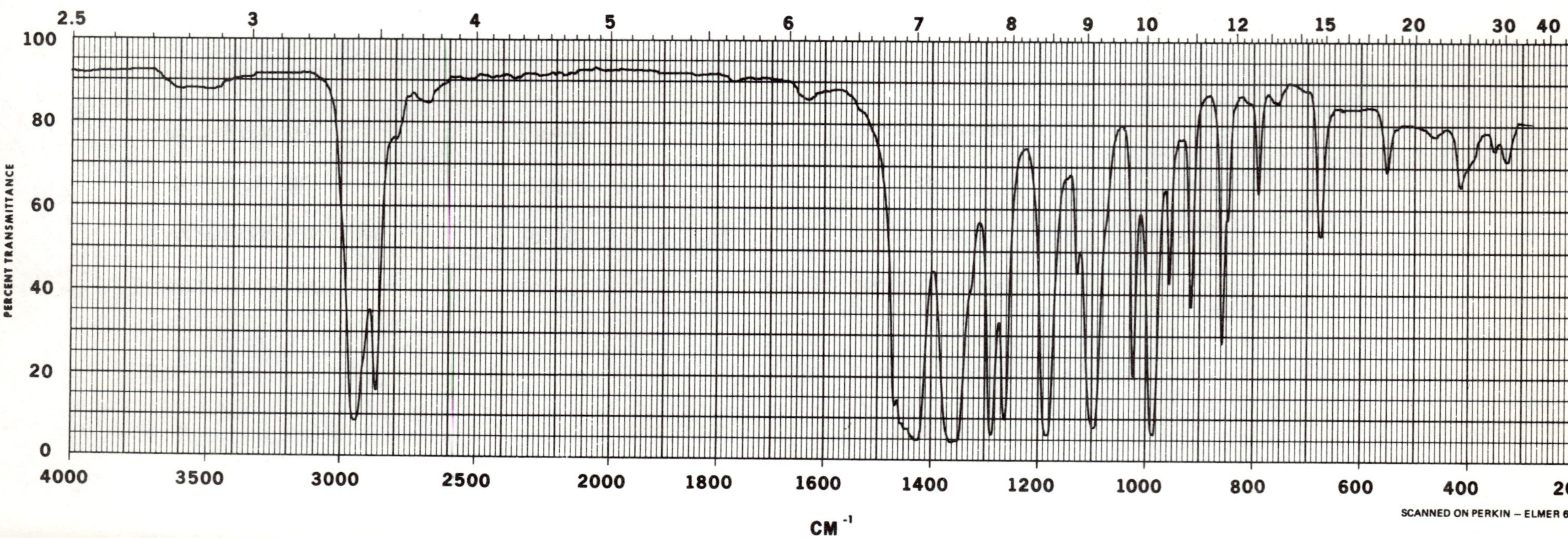

BOILING POINT:	216.5°C	INJECTION AMT.:	0.7 μl
VAPORIZATION TEMP.:	200°C	PRESSURE:	Atmospheric
CELL VOLUME:	7.5 cc	MAX./MIN.:	—
PATH LENGTH:	7.6 cm	SCALE EXPANSION:	1.01

VAPOR PHASE

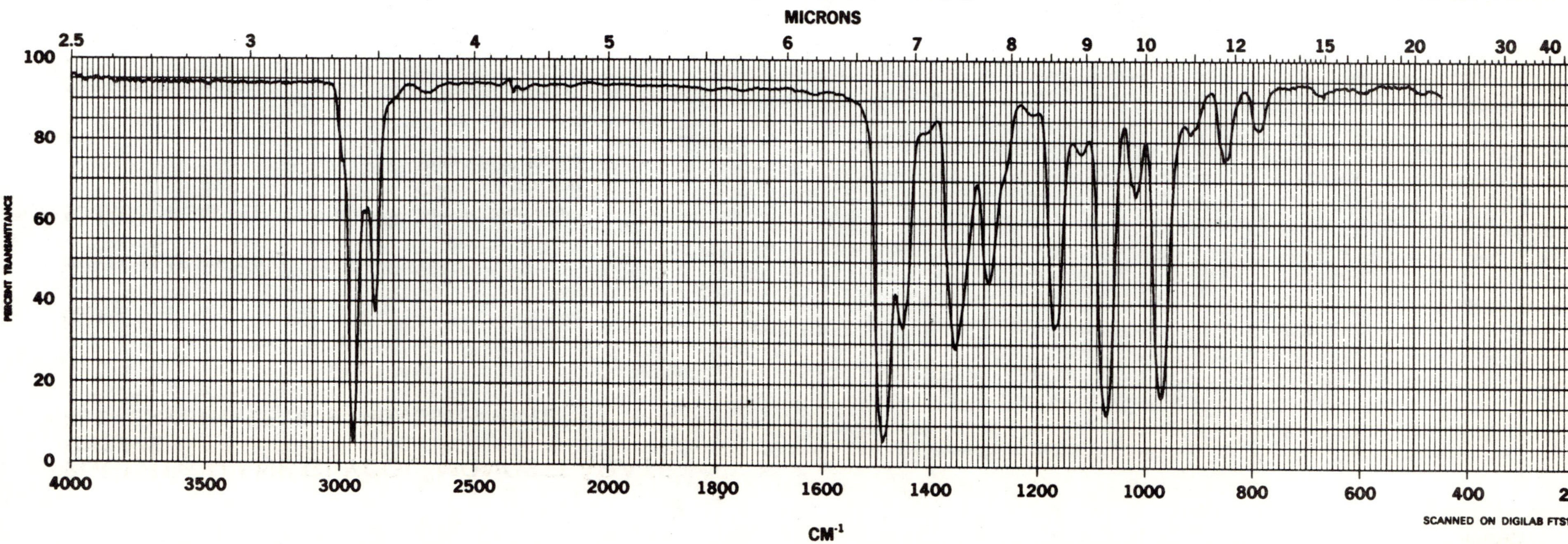

SCANNED ON DIGILAB FTS14

POLYETHYLENE GLYCOL, MONOSTEARATE
MYRJ 45
POLYOXYETHYLENE–8–MONOSTEARATE

$C_{99}H_{99}O_{99}$

MOLECULAR WEIGHT: --

SOURCE OF SAMPLE: Armak Company
Chemicals Division
Rutherford, New Jersey

$CH_3CH_2(CH_2)_{14}CH_2-\overset{O}{\overset{\|}{C}}-O(CH_2CH_2-O)_xH$

CAS REGISTRY NO.: --

NIOSH NUMBER: TQ54000

CAPILLARY CELL: Melt

CONDENSED PHASE

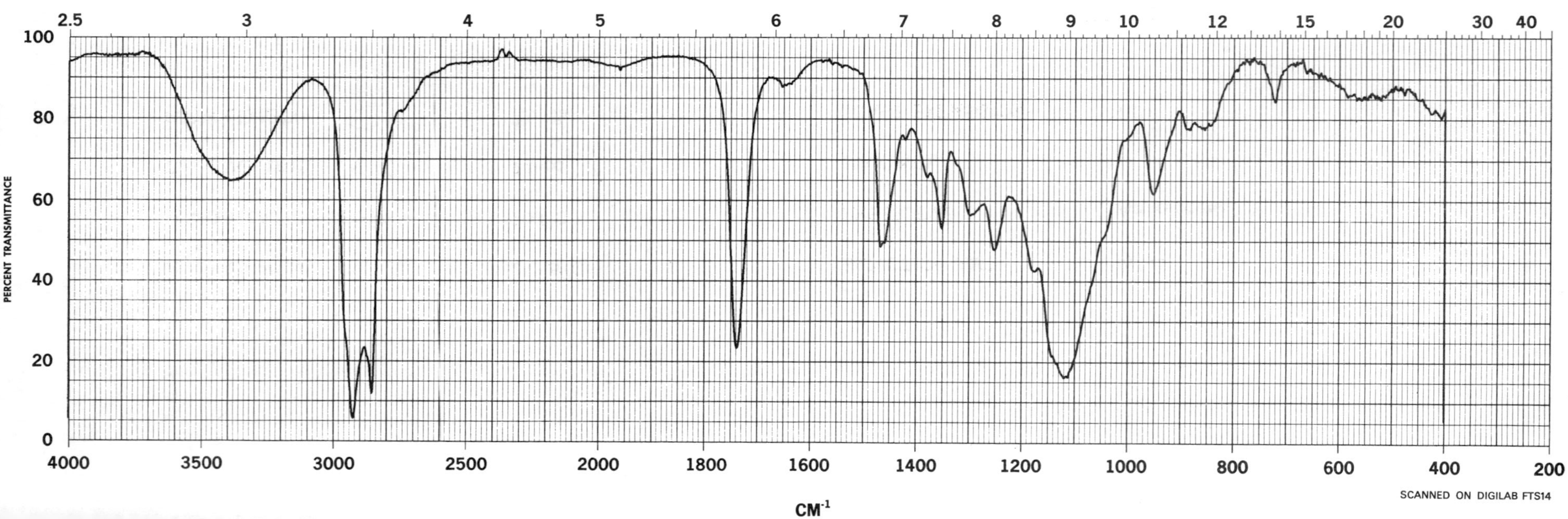

BOILING POINT:	—	INJECTION AMT.:	—
VAPORIZATION TEMP.:	280°C	PRESSURE:	Atmospheric
CELL VOLUME:	7.5 cc	MAX./MIN.:	108.4/36.6
PATH LENGTH:	7.6 cm	SCALE EXPANSION:	1.28

VAPOR PHASE

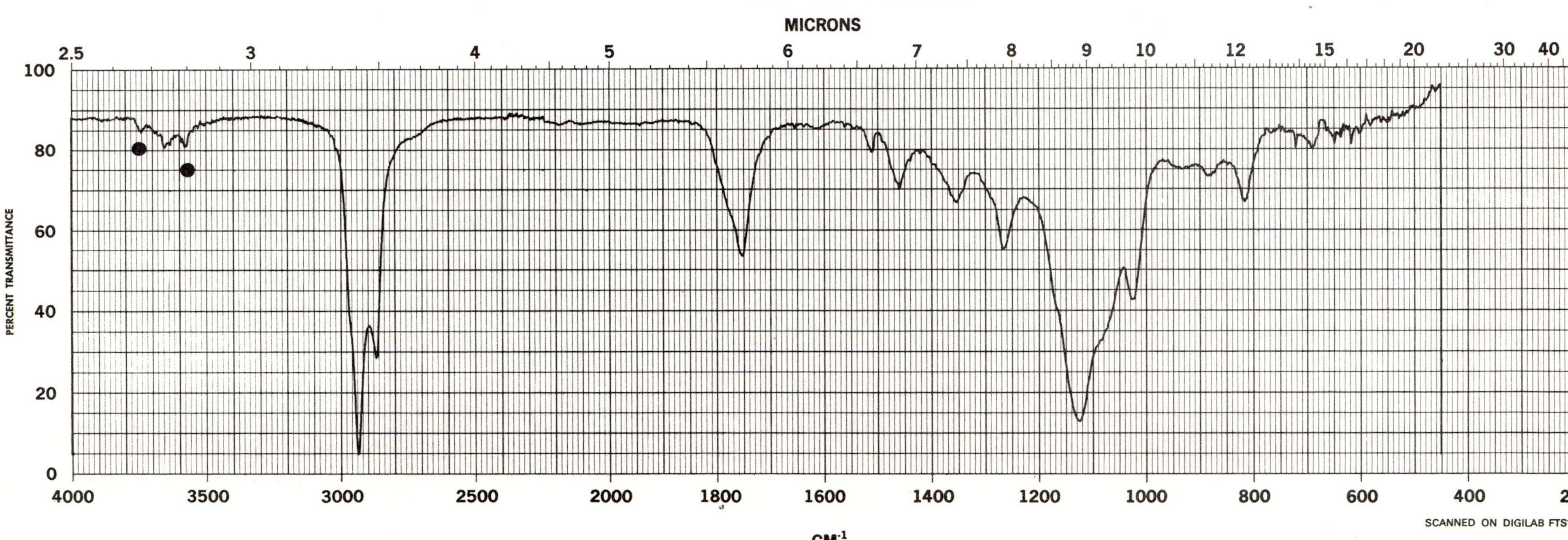

PROGESTERONE
PREGN–4–ENE–3,20–DIONE
LUTEOSTERONE
CORLUTIN
PROLIDON

$C_{21}H_{30}O_2$

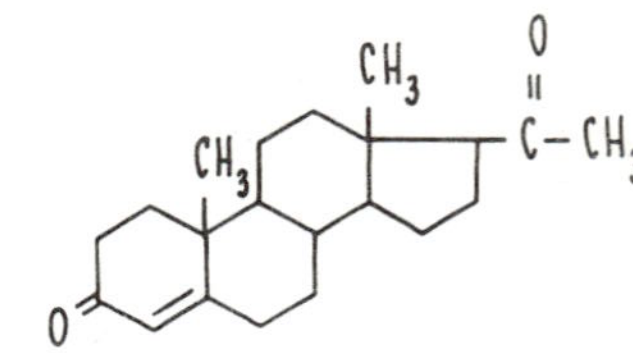

MOLECULAR WEIGHT: 314.47

SOURCE OF SAMPLE: Aldrich Chemical Company, Inc.
Milwaukee, Wisconsin

CAS REGISTRY NO.: 57-83-0

NIOSH NUMBER: TW01750

KBr Wafer

CONDENSED PHASE

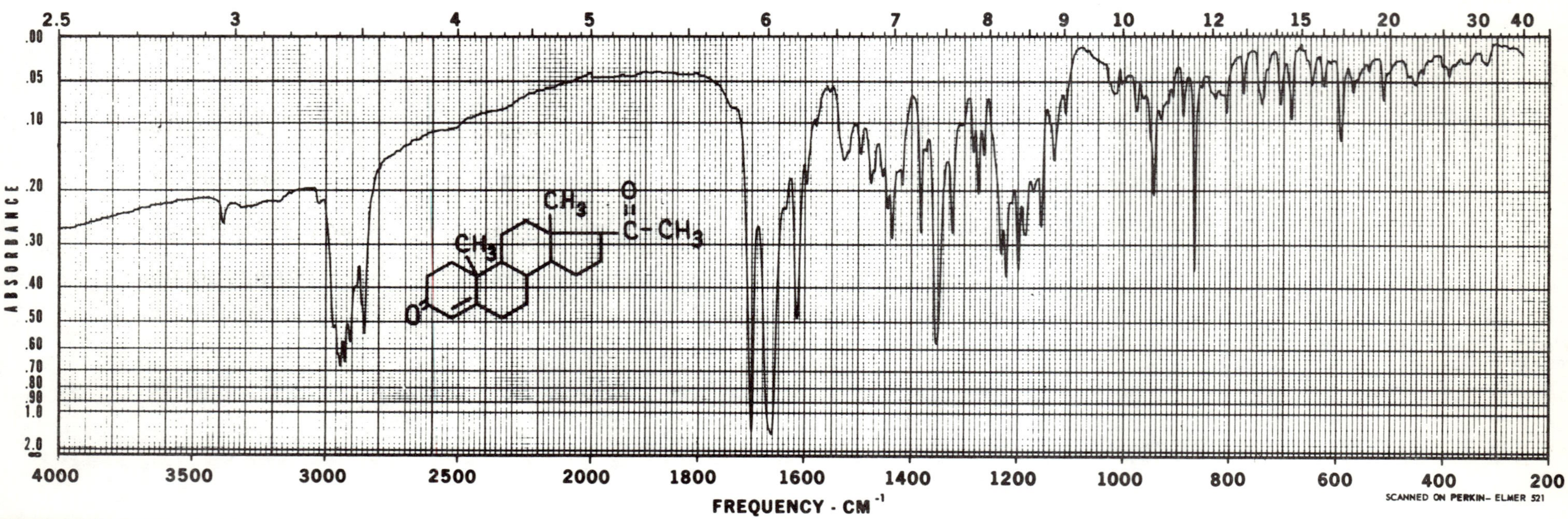

MELTING POINT:	129-131°C	INJECTION AMT.:	–
VAPORIZATION TEMP.:	280°C	PRESSURE:	Atmospheric
CELL VOLUME:	7.5 cc	MAX./MIN.:	102.0/49.8
PATH LENGTH:	7.6 cm	SCALE EXPANSION:	1.76

VAPOR PHASE

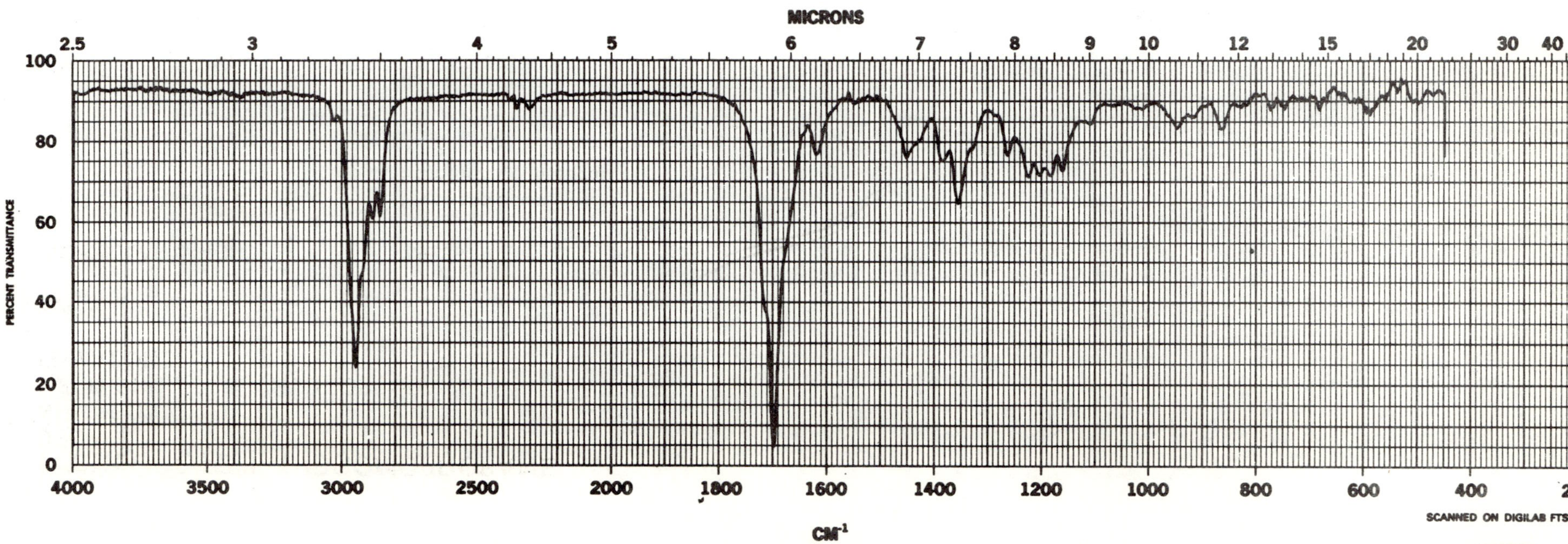

1–CHLORO–2,3–DIBROMOPROPANE
DBCP
NEMAGON
1,2–DIBROMO–3–CHLOROPROPANE
FUMAZONE

$C_3H_5Br_2Cl$

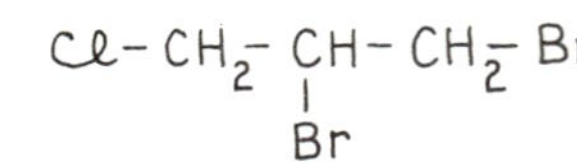

MOLECULAR WEIGHT: 236.35

SOURCE OF SAMPLE: Pfaltz and Bauer, Inc.
Stamford, Connecticut

CAS REGISTRY NO.: 96-12-8

NIOSH NUMBER: TX87500

CAPILLARY CELL: Neat

CONDENSED PHASE

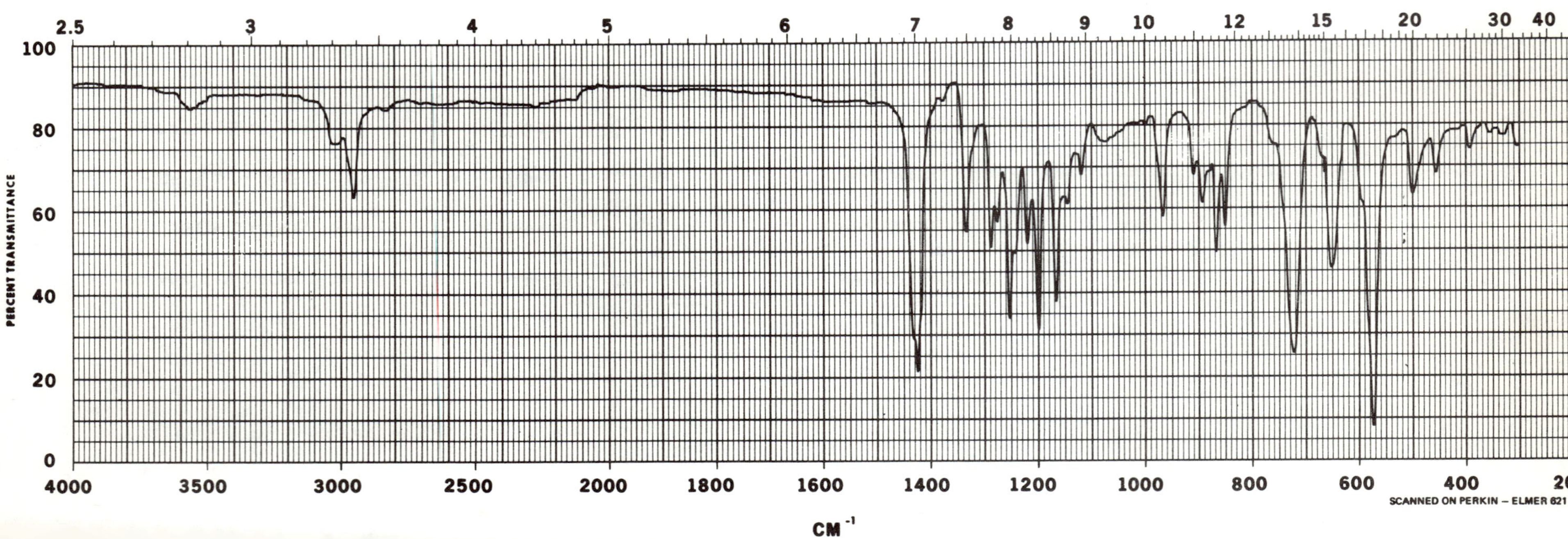

BOILING POINT:	—	INJECTION AMT.:	2.0 μl
VAPORIZATION TEMP.:	200°C	PRESSURE:	Atmospheric
CELL VOLUME:	7.5 cc	MAX./MIN.:	—
PATH LENGTH:	7.6 cm	SCALE EXPANSION:	1.09

VAPOR PHASE

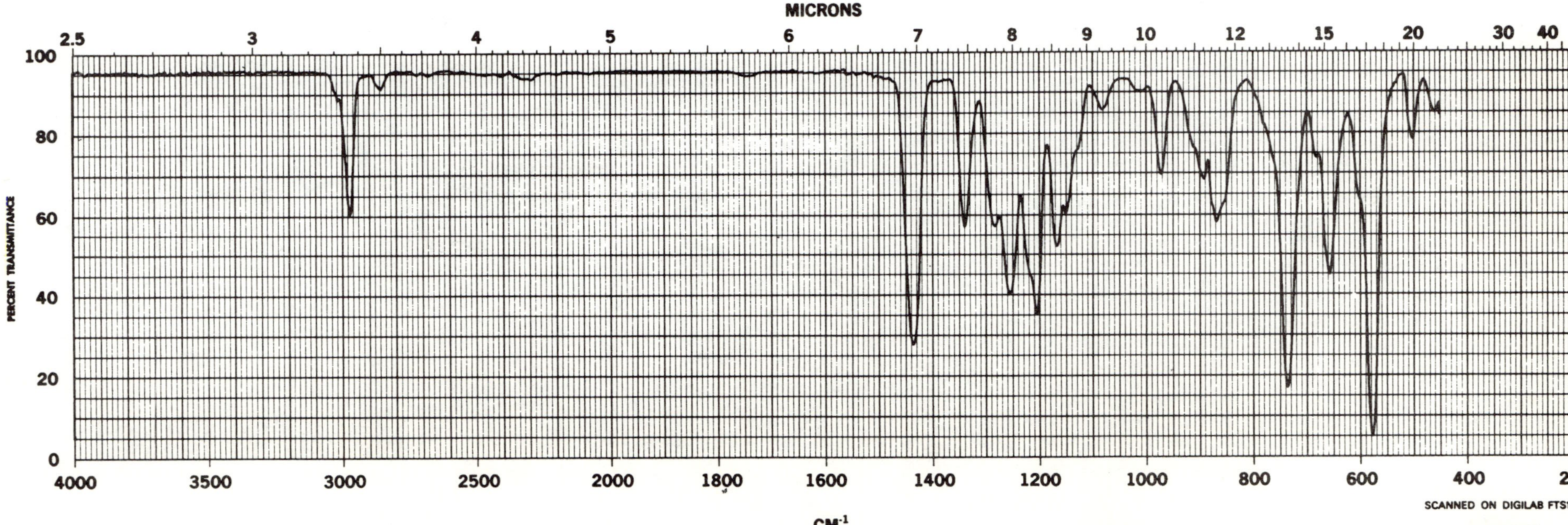

1,2–DICHLOROPROPANE
PROPYLENE DICHLORIDE

$C_3H_6Cl_2$

$CH_3-CH(Cl)-CH_2-Cl$

MOLECULAR WEIGHT:	112.99
SOURCE OF SAMPLE:	Aldrich Chemical Company, Inc. Milwaukee, Wisconsin
CAS REGISTRY NO.:	78-87-5
NIOSH NUMBER:	TX96250
CAPILLARY CELL:	Neat

CONDENSED PHASE

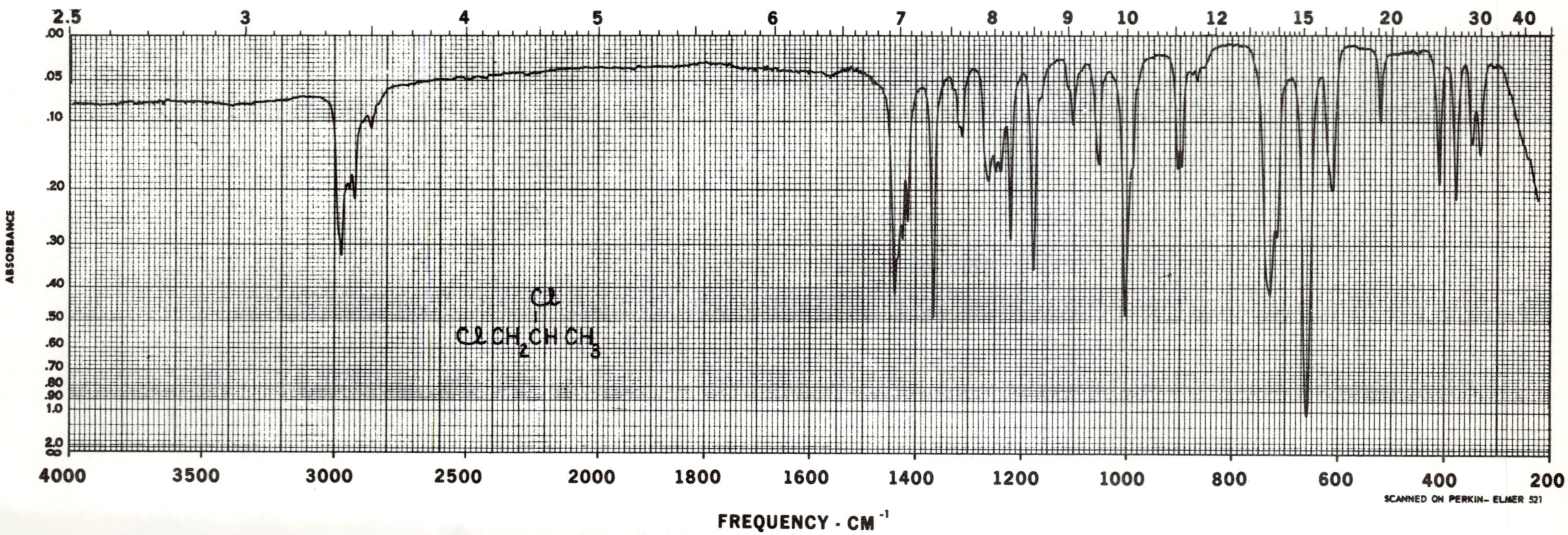

BOILING POINT:	95-96°C	INJECTION AMT.:	1.2 µl
VAPORIZATION TEMP.:	200°C	PRESSURE:	Atmospheric
CELL VOLUME:	7.5 cc	MAX./MIN.:	–
PATH LENGTH:	7.6 cm	SCALE EXPANSION:	1.40

VAPOR PHASE

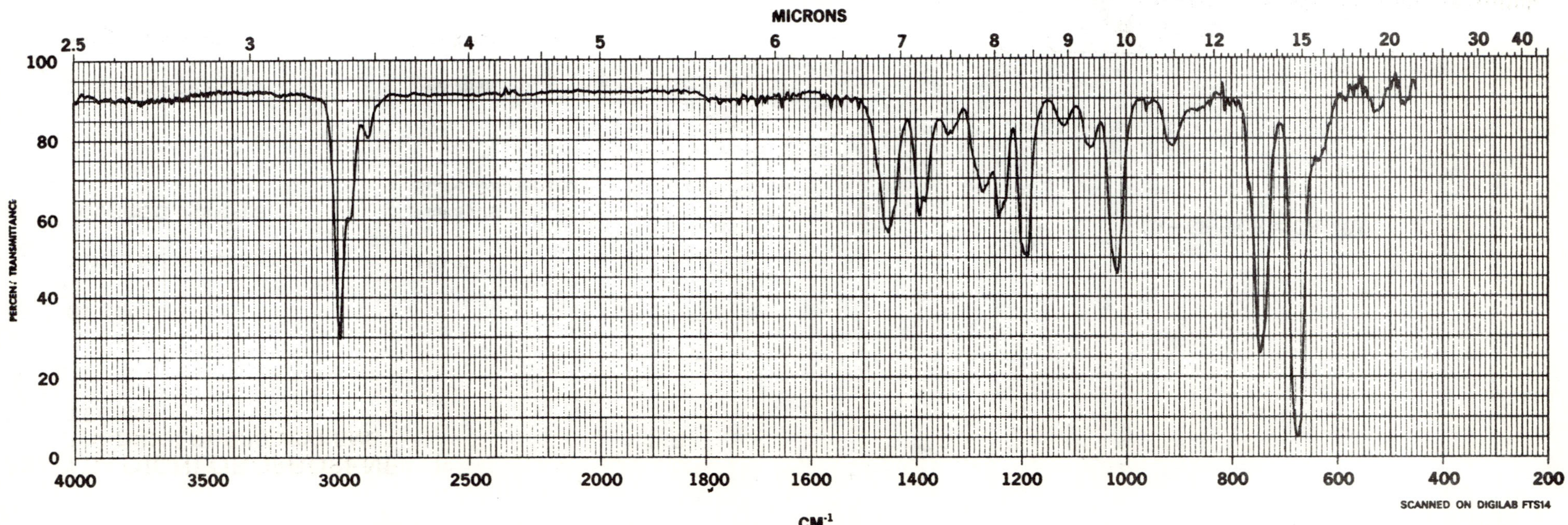

1,3–DICHLOROPROPENE
1,3–DICHLOROPROPYLENE
TELONE

$C_3H_4Cl_2$

MOLECULAR WEIGHT: 110.97

SOURCE OF SAMPLE: Fluka AG
Buchs, Switzerland

CAS REGISTRY NO.: 542-75-6

NIOSH NUMBER: UC83100

CAPILLARY CELL: 0.015 mm

$Cl-CH_2-CH=CH-Cl$

CONDENSED PHASE

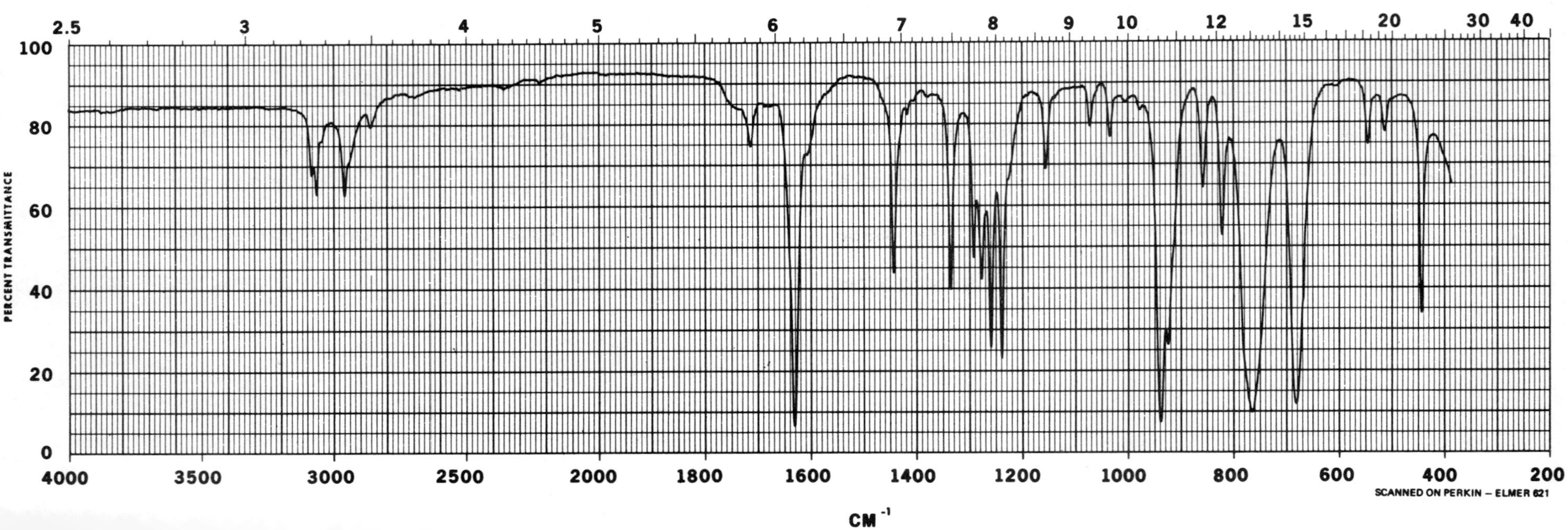

BOILING POINT:	97-112°C	INJECTION AMT.:	1.0 µl
VAPORIZATION TEMP.:	200°C	PRESSURE:	Atmospheric
CELL VOLUME:	7.5 cc	MAX./MIN.:	–
PATH LENGTH:	7.6 cm	SCALE EXPANSION:	1.20

VAPOR PHASE

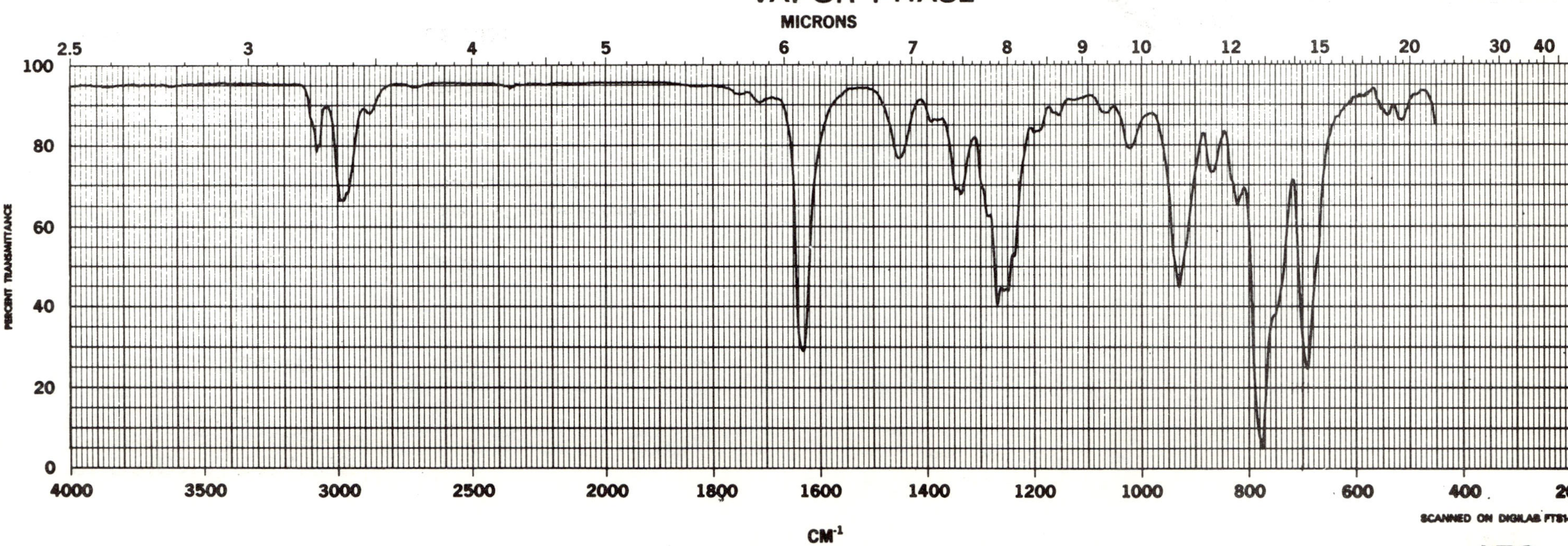

PROPIONALDEHYDE
METHYLACETALDEHYDE
PROPALDEHYDE
PROPANAL
PROPIONIC ALDEHYDE

C_3H_6O

$CH_3CH_2\text{-}\overset{O}{\overset{\|}{C}}\text{-}H$

MOLECULAR WEIGHT:	58.08
SOURCE OF SAMPLE:	Eastman Organic Chemicals Rochester, New York
CAS REGISTRY NO.:	123-38-6
NIOSH NUMBER:	UE03500
CAPILLARY CELL:	Neat

CONDENSED PHASE

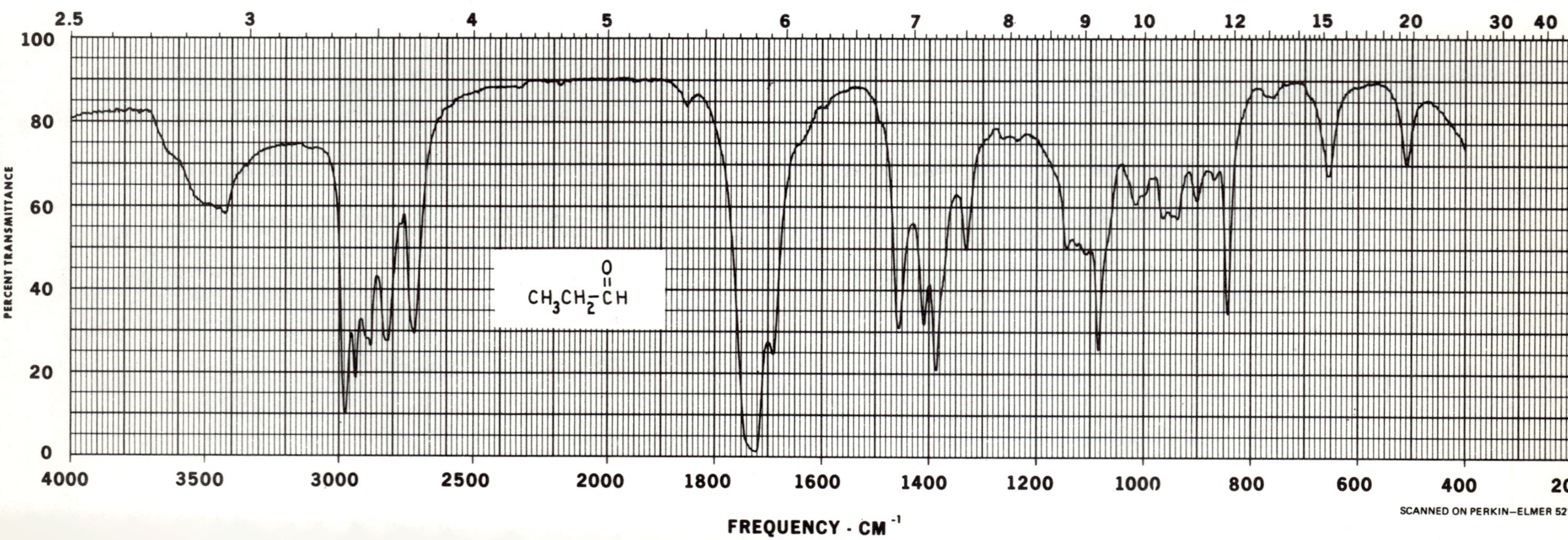

SCANNED ON PERKIN–ELMER 521

BOILING POINT:	48-50°C	INJECTION AMT.:	0.5 µl
VAPORIZATION TEMP.:	200°C	PRESSURE:	Atmospheric
CELL VOLUME:	7.5 cc	MAX./MIN.:	—
PATH LENGTH:	7.6 cm	SCALE EXPANSION:	1.17

VAPOR PHASE

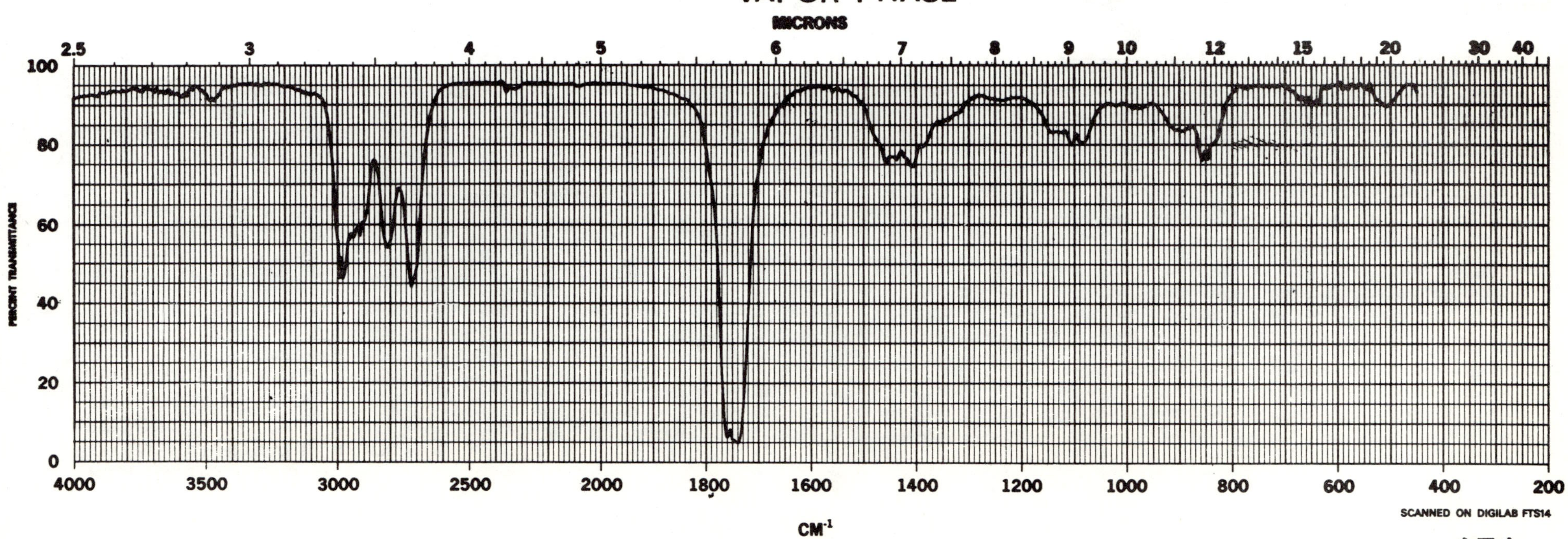

SCANNED ON DIGILAB FTS14

PROPIONIC ACID
ETHANECARBOXYLIC ACID
PROZOIN
ETHYLFORMIC ACID
PSEUDOACETIC ACID

$C_3H_6O_2$

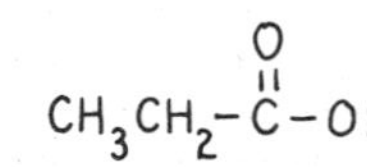

MOLECULAR WEIGHT: 74.08

SOURCE OF SAMPLE: E. I. du Pont de Nemours and Company, Inc.
Wilmington, Delaware

CAS REGISTRY NO.: 79-09-4

NIOSH NUMBER: UE59500

CAPILLARY CELL: Neat

CONDENSED PHASE

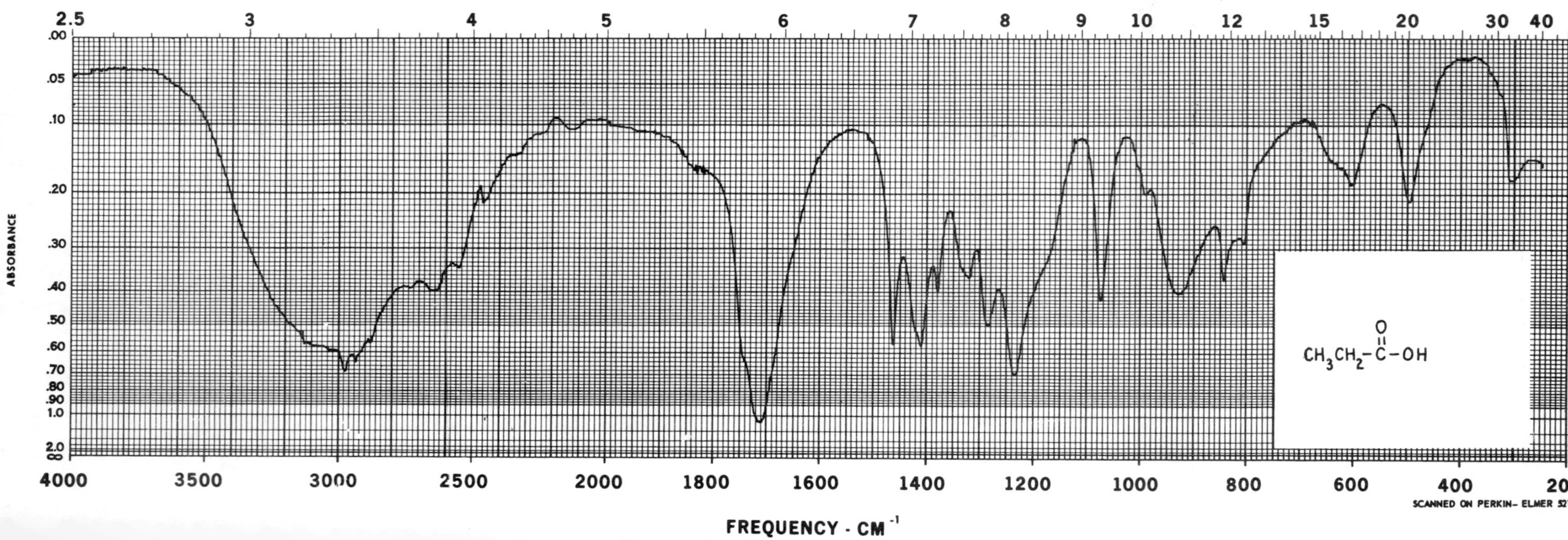

BOILING POINT:	140-141°C	INJECTION AMT.:	0.7 µl
VAPORIZATION TEMP.:	200°C	PRESSURE:	Atmospheric
CELL VOLUME:	7.5 cc	MAX./MIN.:	—
PATH LENGTH:	7.6 cm	SCALE EXPANSION:	0.93

VAPOR PHASE

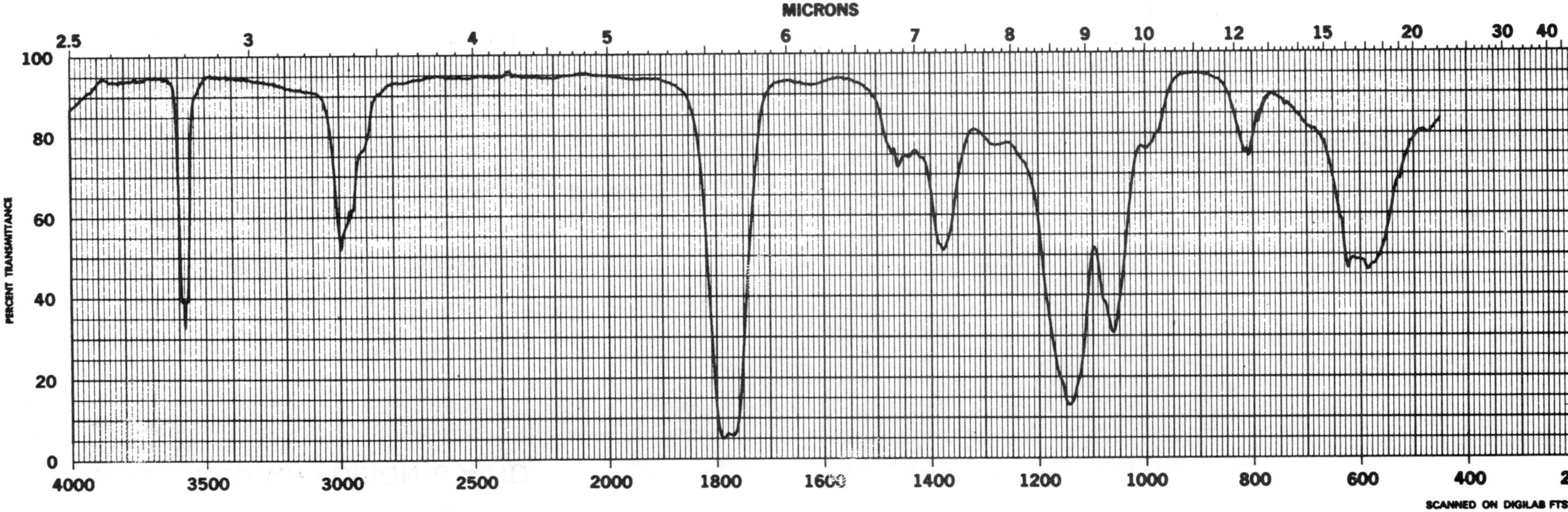

2,2--DICHLOROPROPIONIC ACID
DALAPON
DOWPON

$C_3H_4Cl_2O_2$

MOLECULAR WEIGHT: 142.97

SOURCE OF SAMPLE: Calbiochem
Los Angeles, California

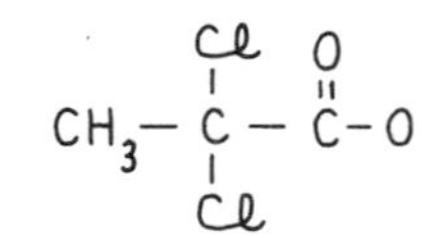

CAS REGISTRY NO.: 75-99-0

NIOSH NUMBER: UF06900

CAPILLARY CELL: Neat

CONDENSED PHASE

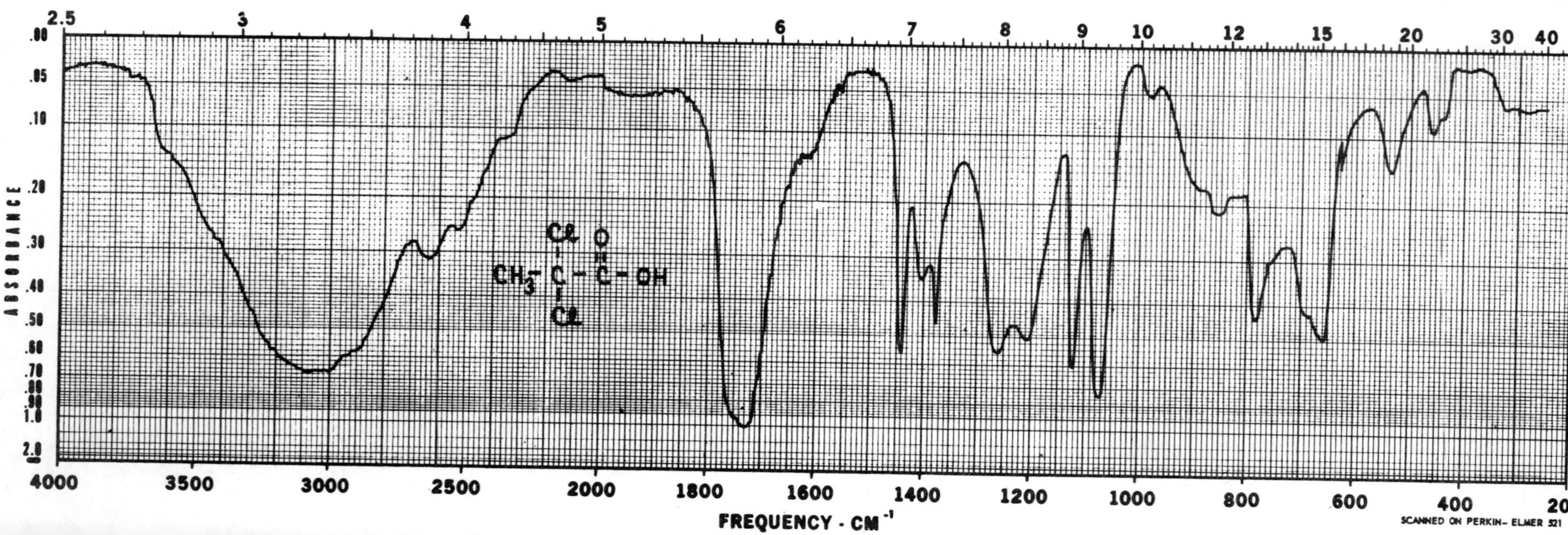

BOILING POINT:	98-100°C/18 mm	INJECTION AMT.:	–
VAPORIZATION TEMP.:	200°C	PRESSURE:	Atmospheric
CELL VOLUME:	7.5 cc	MAX./MIN.:	100.2/17.9
PATH LENGTH:	7.6 cm	SCALE EXPANSION:	1.12

VAPOR PHASE

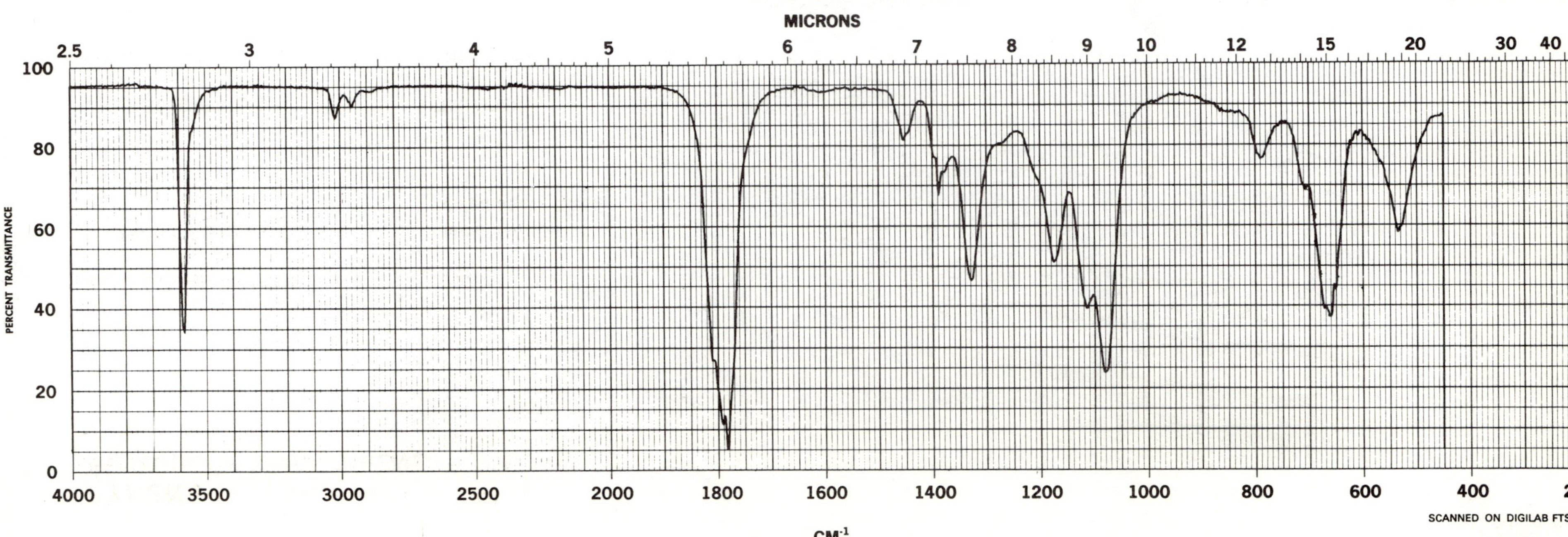

SCANNED ON DIGILAB FTS14

PYRENE
BENZO[def] PHENANTHRENE

$C_{16}H_{10}$

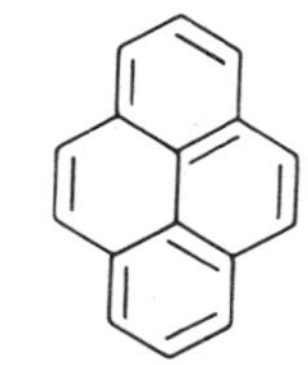

MOLECULAR WEIGHT:	202.26
SOURCE OF SAMPLE:	James Hinton Valparaiso, Florida
CAS REGISTRY NO.:	129-00-0
NIOSH NUMBER:	UR24500

KBr Wafer

CONDENSED PHASE

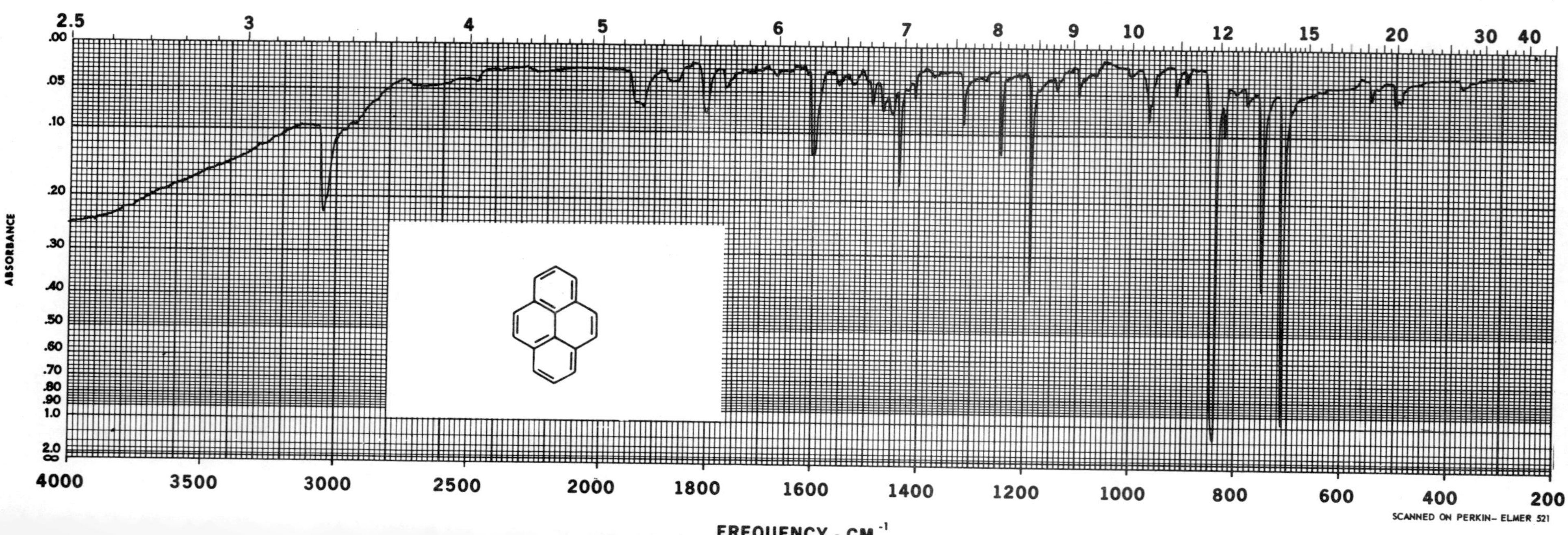

MELTING POINT:	156°C	INJECTION AMT.:	—
VAPORIZATION TEMP.:	280°C	PRESSURE:	Atmospheric
CELL VOLUME:	7.5 cc	MAX./MIN.:	99.1/23.2
PATH LENGTH:	7.6 cm	SCALE EXPANSION:	1.21

VAPOR PHASE

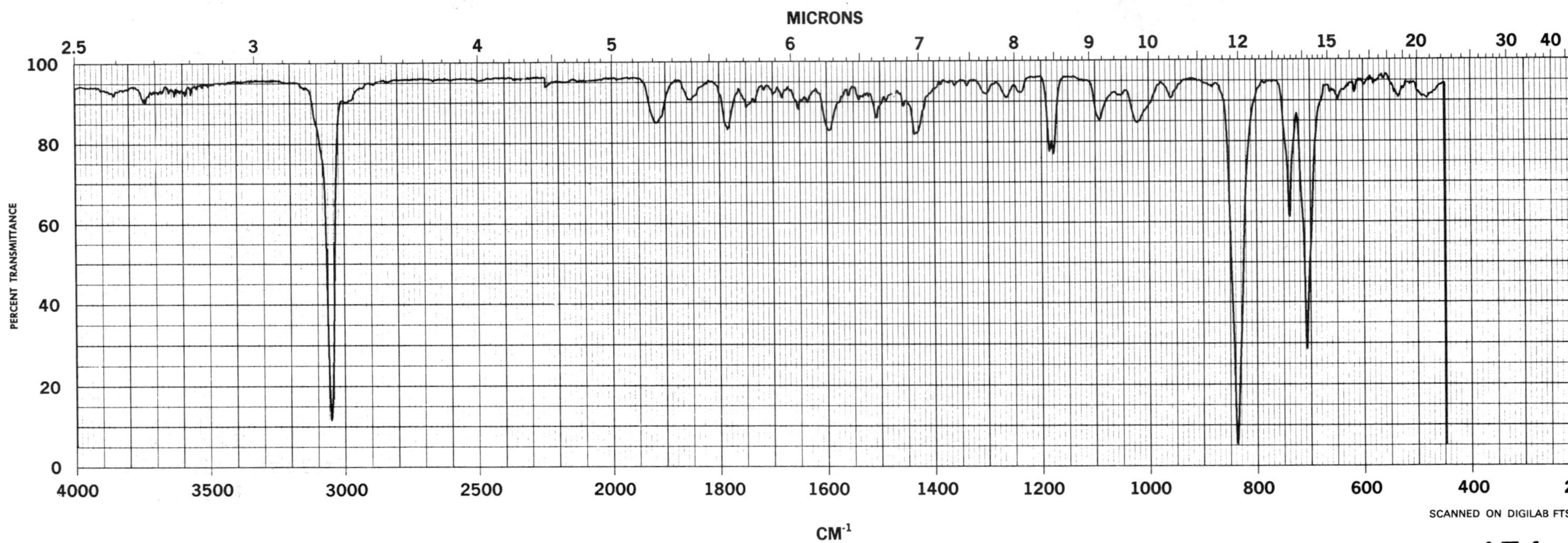

SCANNED ON DIGILAB FTS14

QUINOLINE
1–AZANAPHTHALENE
1–BENZAZINE
CHINOLINE
LEUCOLINE

C_9H_7N

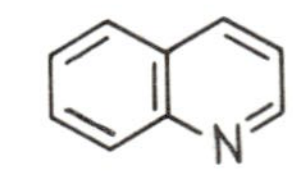

MOLECULAR WEIGHT:	129.16
SOURCE OF SAMPLE:	MCB Manufacturing Chemists Norwood, Ohio
CAS REGISTRY NO.:	91-22-5
NIOSH NUMBER:	VA92750
CAPILLARY CELL:	Neat

CONDENSED PHASE

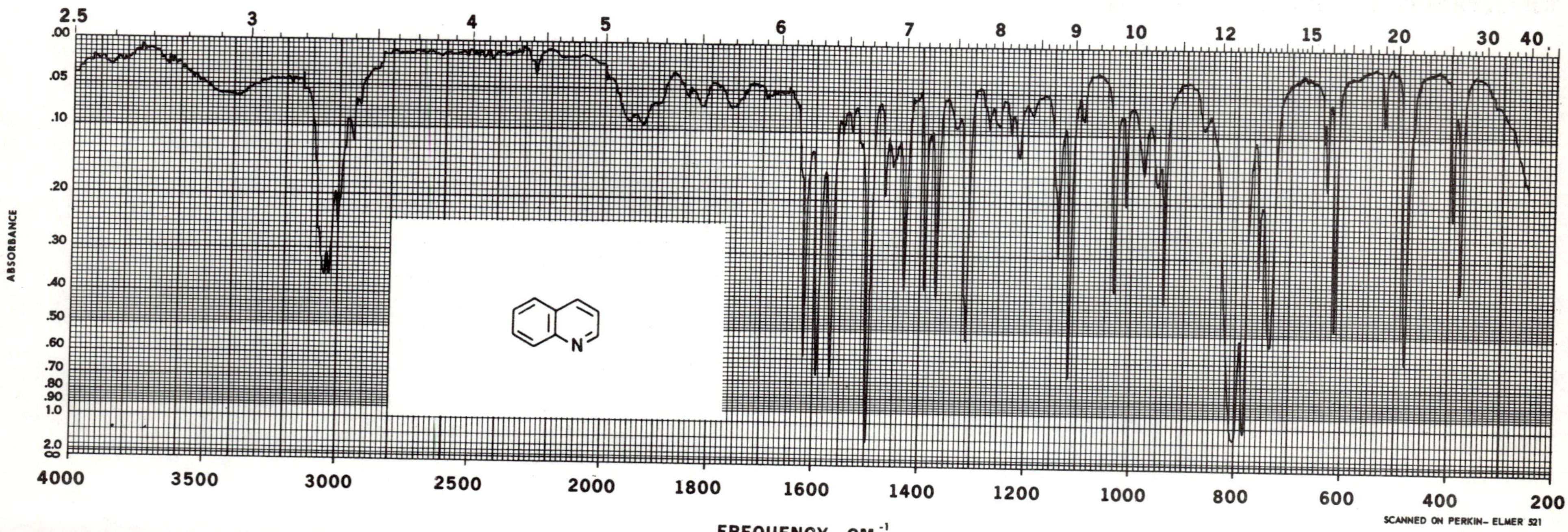

BOILING POINT:	237°C	INJECTION AMT.:	1.5 μl
VAPORIZATION TEMP.:	200°C	PRESSURE:	Atmospheric
CELL VOLUME:	7.5 cc	MAX./MIN.:	—
PATH LENGTH:	7.6 cm	SCALE EXPANSION:	1.00

VAPOR PHASE

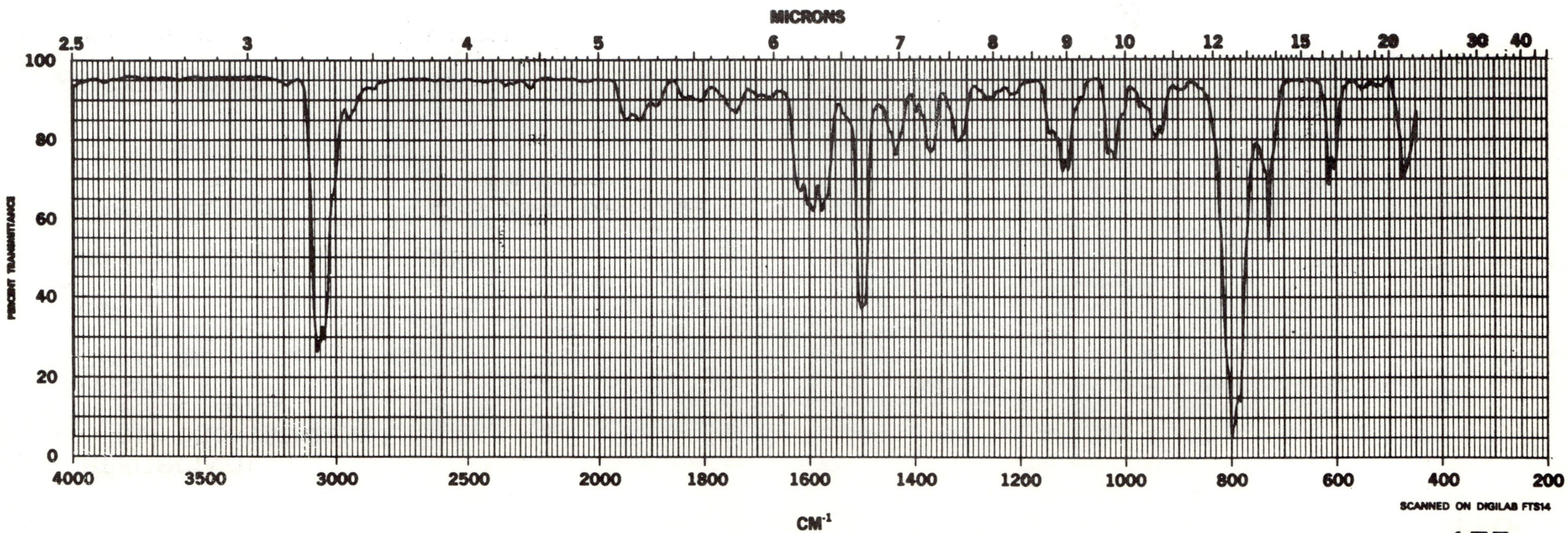

SCANNED ON DIGILAB FTS14

RESORCINOL
m–DIHYDROXYBENZENE
m–BENZENEDIOL
PELAGOL RS
RESORCINE

$C_6H_6O_2$

MOLECULAR WEIGHT: 110.11

SOURCE OF SAMPLE: Aldrich Chemical Company, Inc.
Milwaukee, Wisconsin

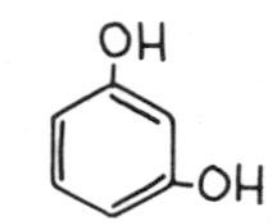

CAS REGISTRY NO.: 108-46-3

NIOSH NUMBER: VG96250

KBr Wafer

CONDENSED PHASE

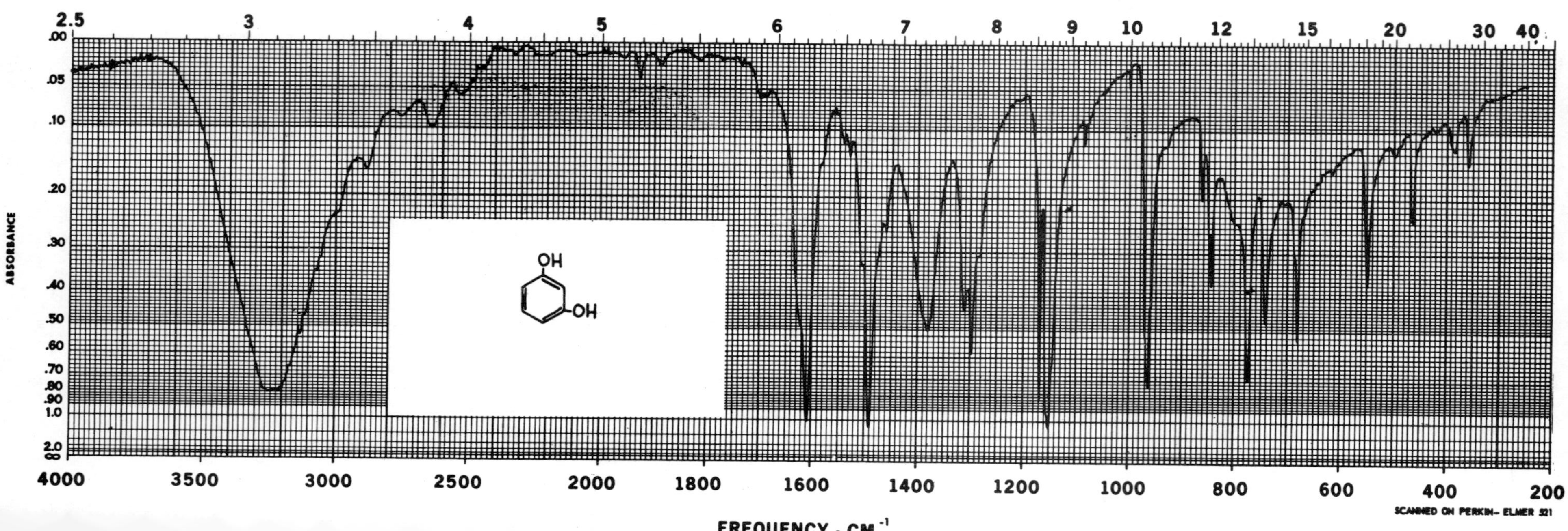

BOILING POINT:	178°C/16 mm	INJECTION AMT.:	1.13 mg
VAPORIZATION TEMP.:	280°C	PRESSURE:	Atmospheric
CELL VOLUME:	7.5 cc	MAX./MIN.:	100.8/19.3
PATH LENGTH:	7.6 cm	SCALE EXPANSION:	1.13

VAPOR PHASE

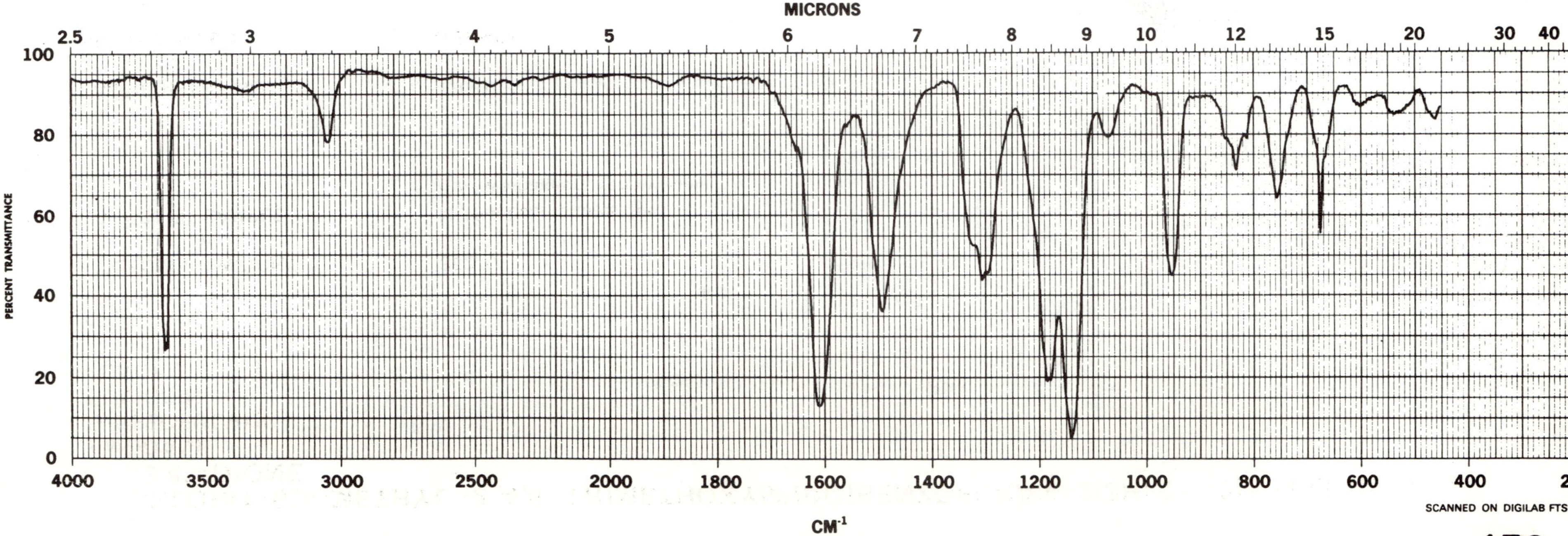

SCANNED ON DIGILAB FTS14

7–CHLORO–6′β–METHYL–2′,4,6–TRIMETHOXYSPIRO[BENZOFURAN–2(3H),1′–[2] CYCLOHEXENE]–3,4′–DIONE

AMUDANE
GRISEOFULVIN
FULVICIN
PONCYL

$C_{17}H_{17}ClO_6$

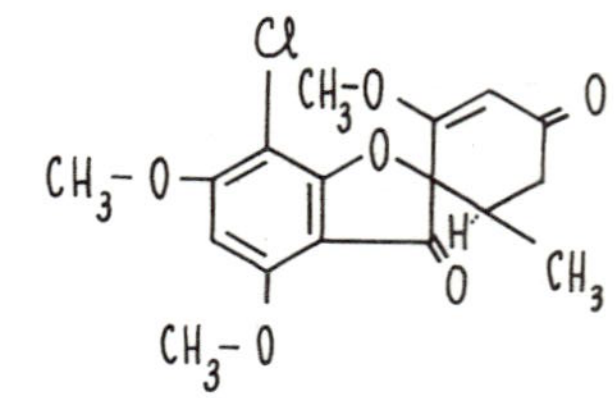

MOLECULAR WEIGHT: 352.77

SOURCE OF SAMPLE: McNeil Laboratories, Inc.
Fort Washington, Pennsylvania

CAS REGISTRY NO.: 126-07-8

NIOSH NUMBER: WG98000

KBr Wafer

CONDENSED PHASE

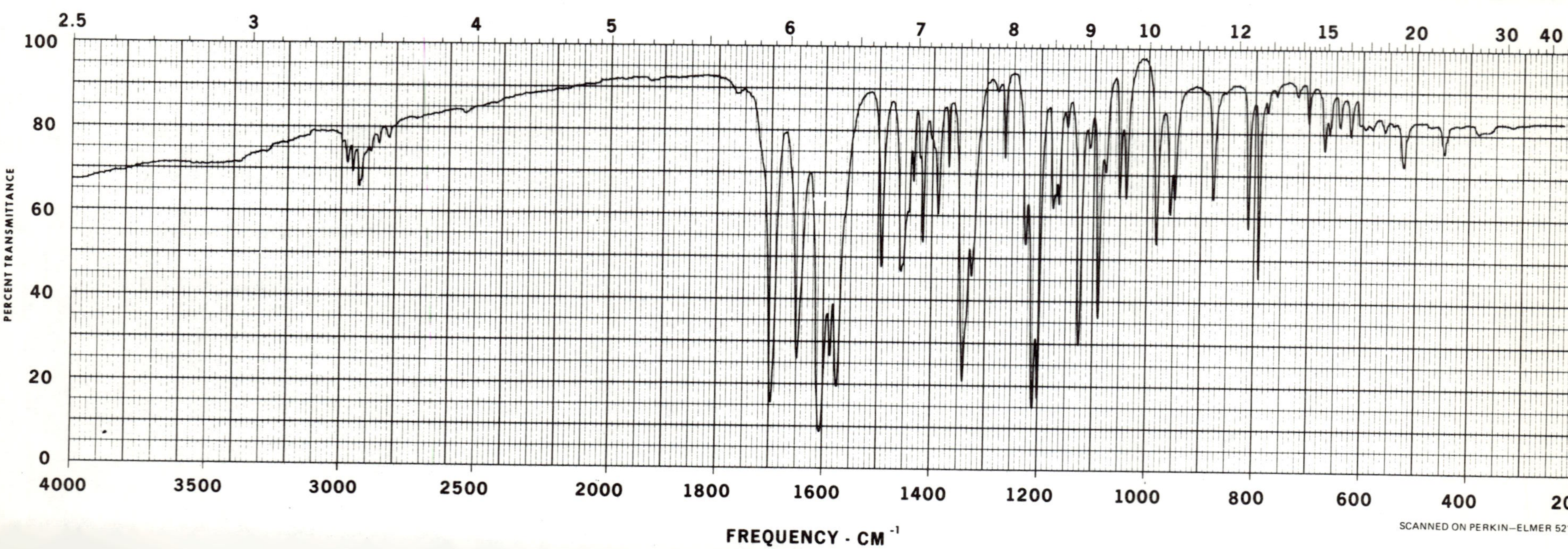

MELTING POINT:	220°C	INJECTION AMT.:	—
VAPORIZATION TEMP.:	280°C	PRESSURE:	Atmospheric
CELL VOLUME:	7.5 cc	MAX./MIN.:	97.1/61.8
PATH LENGTH:	7.6 cm	SCALE EXPANSION:	2.61

VAPOR PHASE

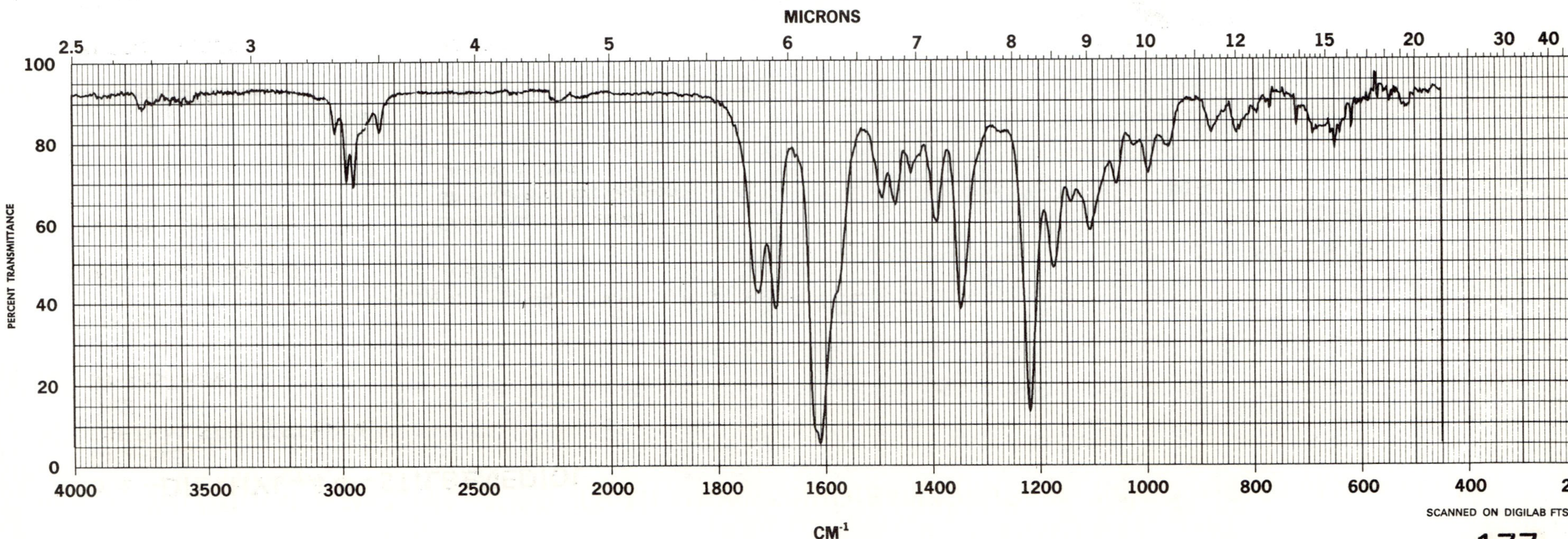

SCANNED ON DIGILAB FTS14

177

α,α'–DIETHYL–4,4'–STILBENEDIOL
BIO–DES
DIETHYLSTILBESTEROL
DES
SYNESTRIN

$C_{18}H_{20}O_2$

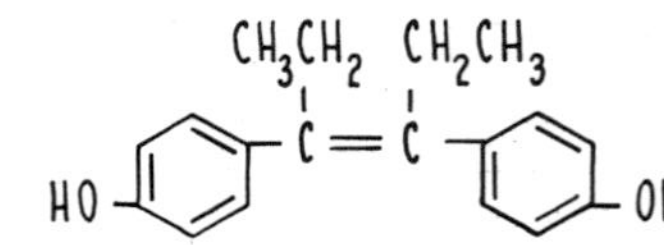

MOLECULAR WEIGHT:	268.36
SOURCE OF SAMPLE:	Chemetron Corporation Chemical Products Division Chicago, Illinois
CAS REGISTRY NO.:	56-53-1
NIOSH NUMBER:	WJ56000

KBr Wafer

CONDENSED PHASE

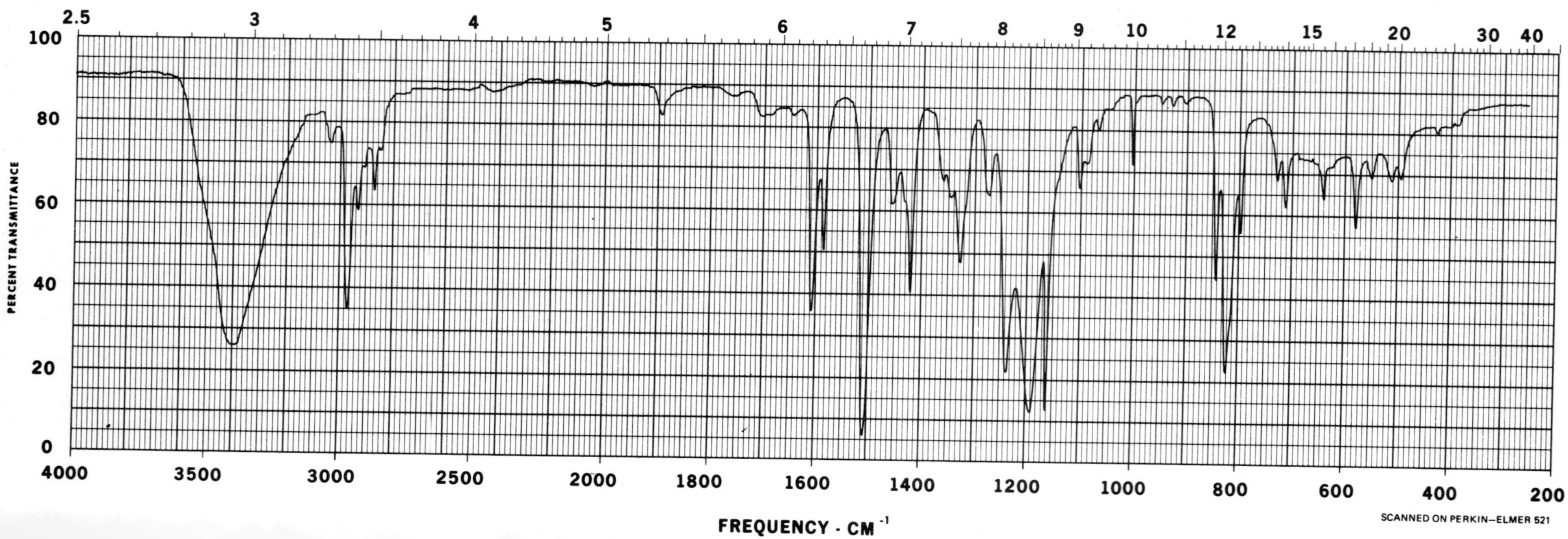

MELTING POINT:	169-171°C	INJECTION AMT.:	—
VAPORIZATION TEMP.:	280°C	PRESSURE:	Atmospheric
CELL VOLUME:	7.5 cc	MAX./MIN.:	105.5/21.4
PATH LENGTH:	7.6 cm	SCALE EXPANSION:	1.03

VAPOR PHASE

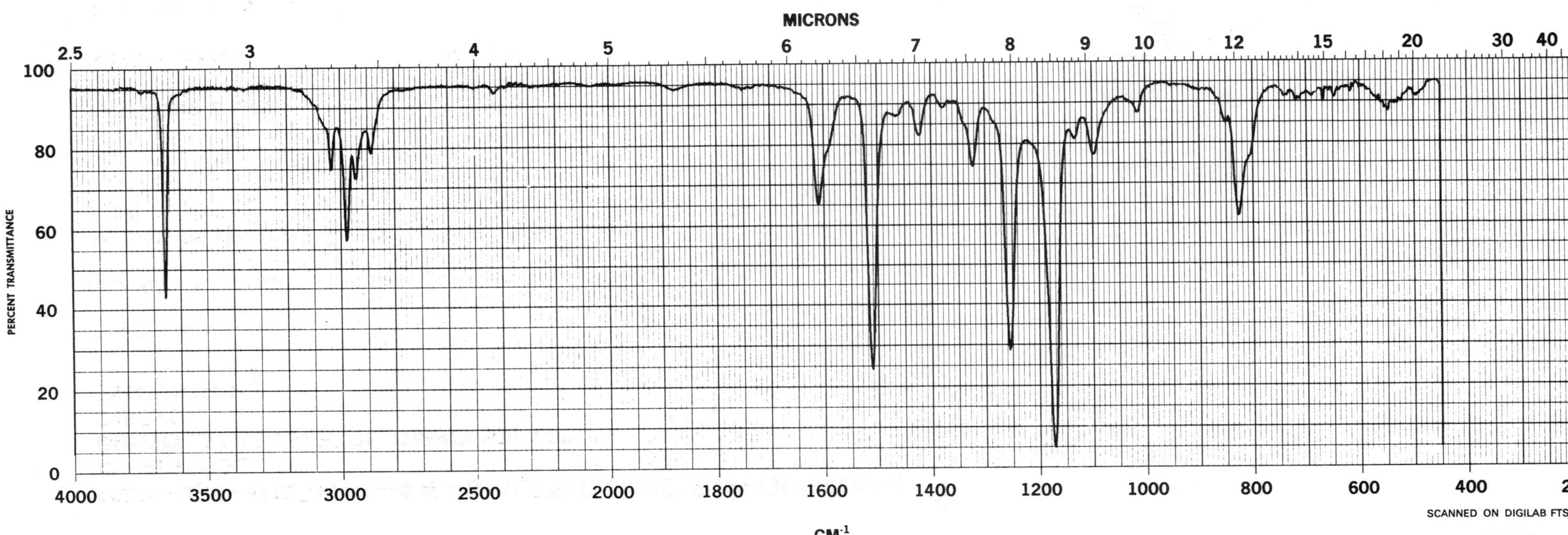

SCANNED ON DIGILAB FTS14

178

trans–*a,a'*–DIETHYL–4,4'–STILBENEDIOL, DIPROPIONATE
CLINESTROL
DIETHYLSTILBESTEROL, DIPROPIONATE
DIBESTIL
SYNTESTRINE

$C_{24}H_{28}O_4$

MOLECULAR WEIGHT: 380.49

SOURCE OF SAMPLE: Chem Service Company
West Chester, Pennsylvania

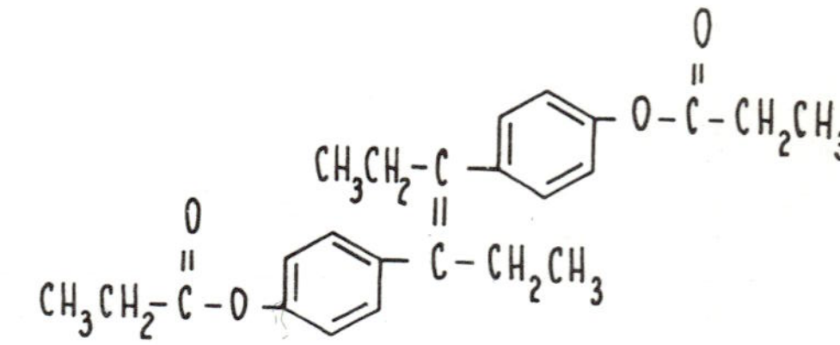

CAS REGISTRY NO.: 130-80-3

NIOSH NUMBER: WJ57750

Film (Cast from $CHCl_3$)

CONDENSED PHASE

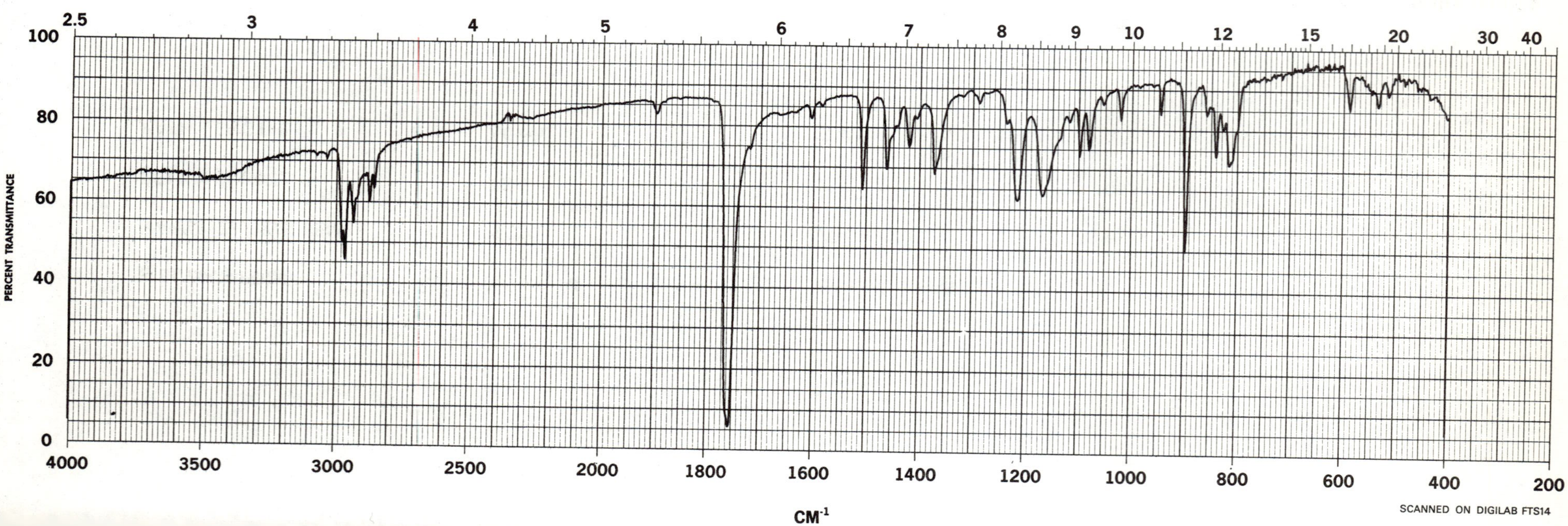

BOILING POINT:	—	INJECTION AMT.:	—
VAPORIZATION TEMP.:	280°C	PRESSURE:	Atmospheric
CELL VOLUME:	7.5 cc	MAX./MIN.:	102.1/18.2
PATH LENGTH:	7.6 cm	SCALE EXPANSION:	1.20

VAPOR PHASE

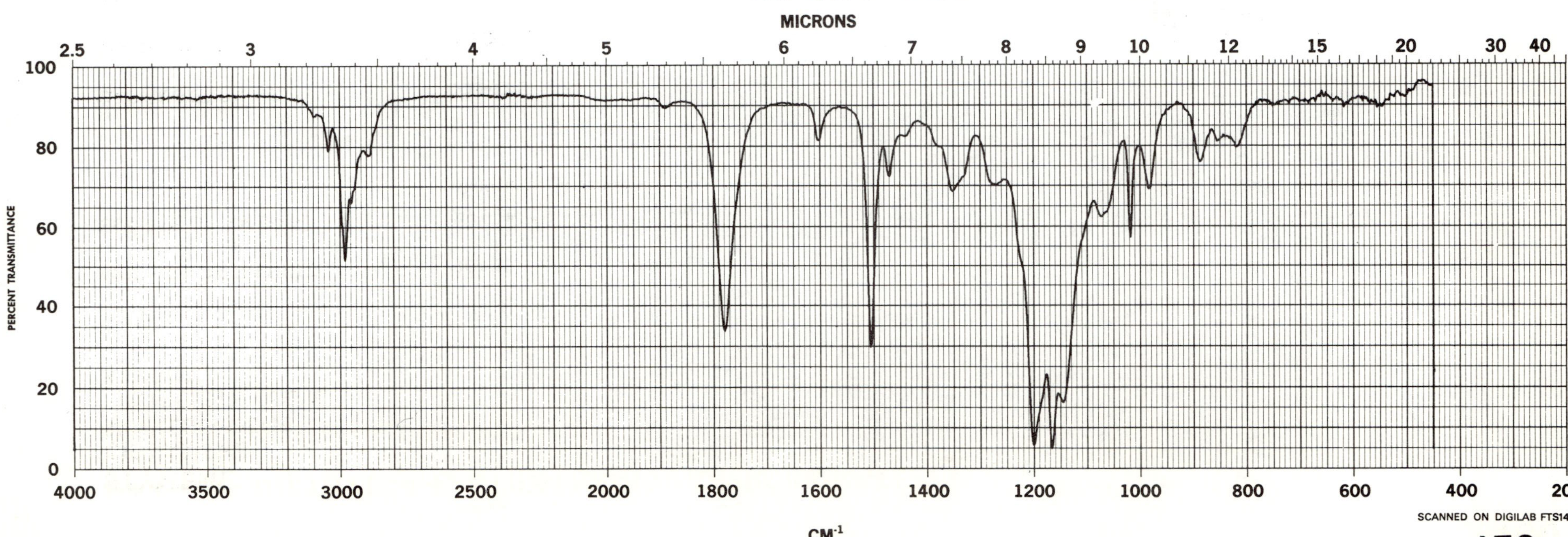

SCANNED ON DIGILAB FTS14

STYRENE
CINNAMENE
VINYLBENZENE
PHENYLETHYLENE
STYRON

C_8H_8

MOLECULAR WEIGHT:	104.15
SOURCE OF SAMPLE:	Aldrich Chemical Company, Inc. Milwaukee, Wisconsin
CAS REGISTRY NO.:	100-42-5
NIOSH NUMBER:	WL36750
CAPILLARY CELL:	Neat

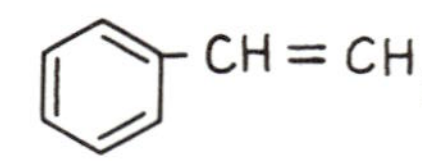

CONDENSED PHASE

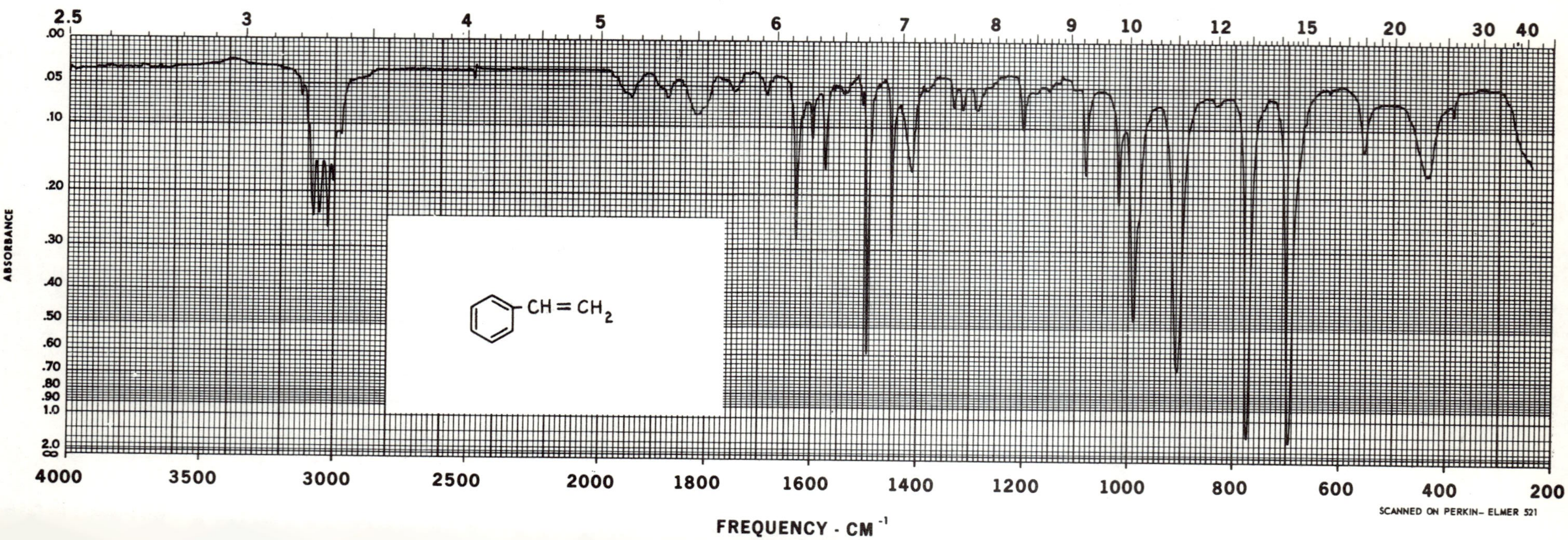

BOILING POINT:	145-146°C	INJECTION AMT.:	—
VAPORIZATION TEMP.:	200°C	PRESSURE:	Atmospheric
CELL VOLUME:	7.5 cc	MAX./MIN.:	99.9/1.20
PATH LENGTH:	7.6 cm	SCALE EXPANSION:	0.93

VAPOR PHASE

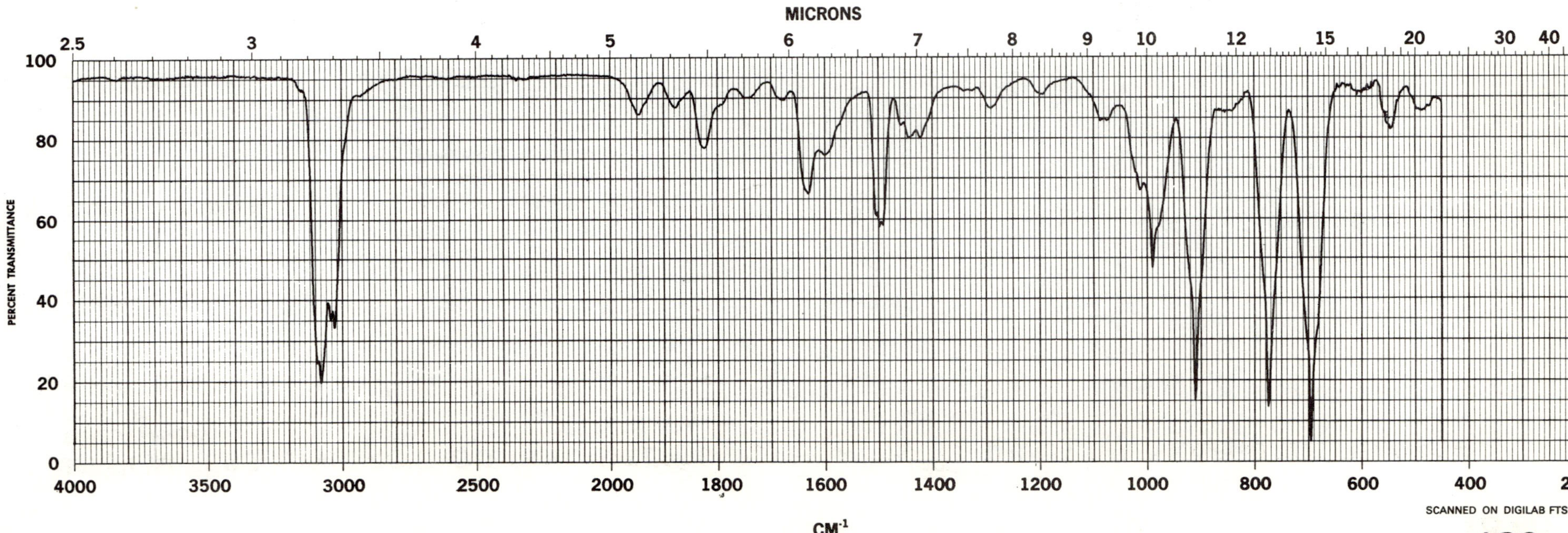

SCANNED ON DIGILAB FTS14

MERCAPTOSUCCINIC ACID, DIETHYL ESTER, S–ESTER WITH O,O–DIMETHYLPHOSPHORODITHIOATE

MALATHION
CYTHION
KARBOFOS
ZITHIOL

$C_{10}H_{19}O_6PS_2$

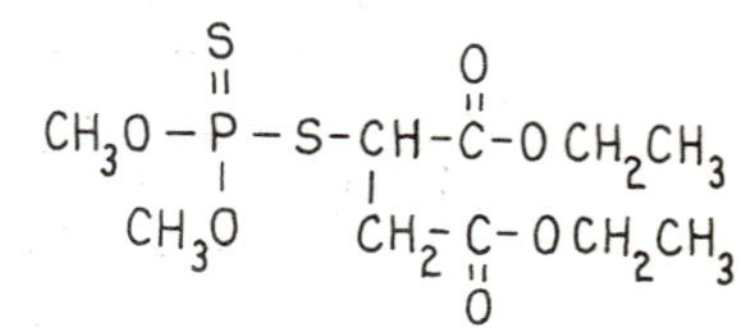

MOLECULAR WEIGHT: 330.36

SOURCE OF SAMPLE: American Cyanamid Company
Agricultural Division
Princeton, New Jersey

CAS REGISTRY NO.: 121-75-5

NIOSH NUMBER: WM84000

Between Salts

CONDENSED PHASE

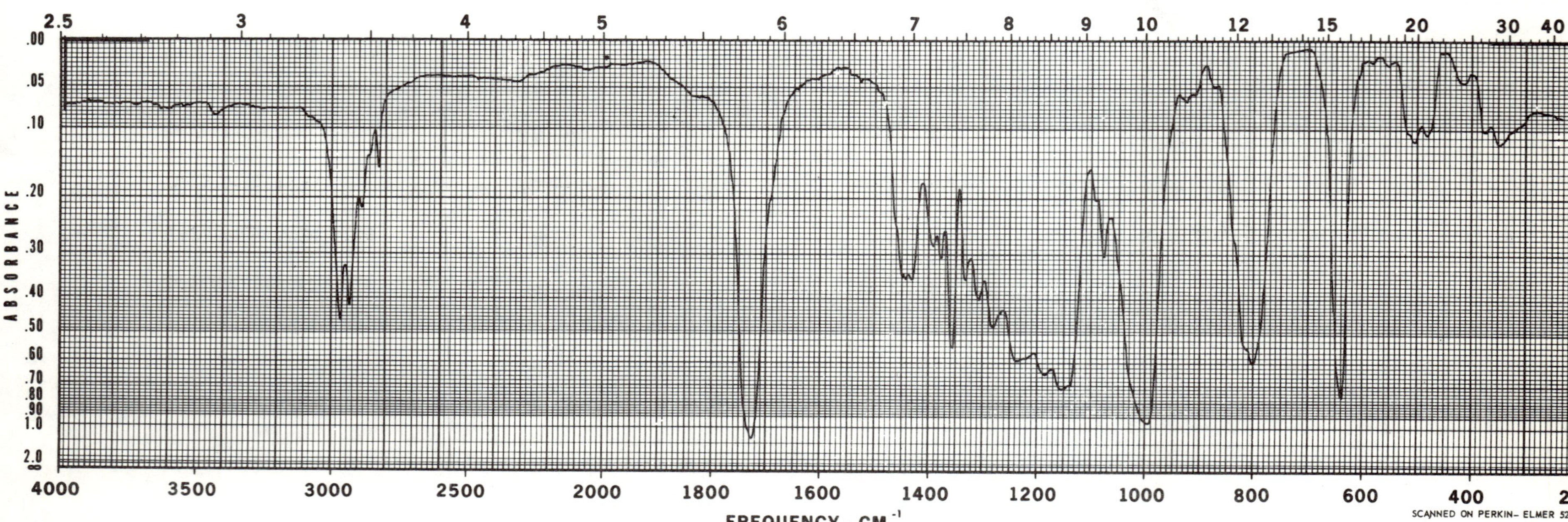

BOILING POINT:	156-157°C/0.7 mm	INJECTION AMT.:	–
VAPORIZATION TEMP.:	200°C	PRESSURE:	Atmospheric
CELL VOLUME:	7.5 cc	MAX./MIN.:	100.2/41.7
PATH LENGTH:	7.6 cm	SCALE EXPANSION:	1.57

VAPOR PHASE

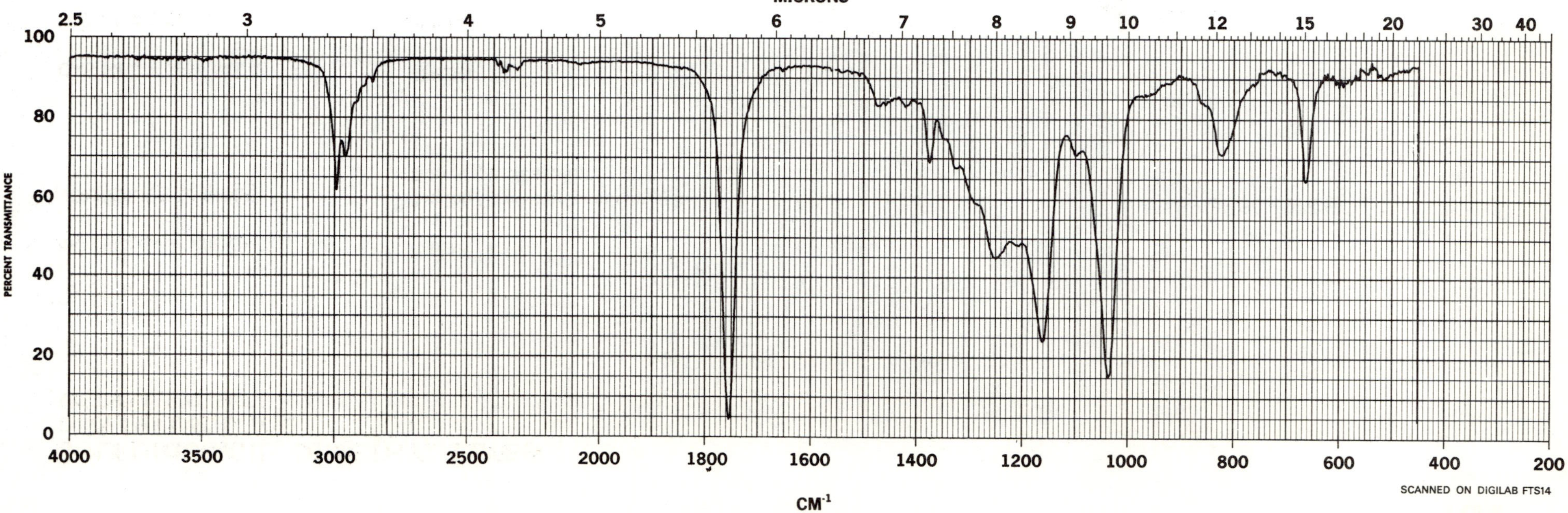

SCANNED ON DIGILAB FTS14

SULFURIC ACID, DIMETHYL ESTER
DIMETHYL SULFATE
METHYL SULFATE

$C_2H_6O_4S$

MOLECULAR WEIGHT: 126.13

SOURCE OF SAMPLE: Aldrich Chemical Company, Inc.
Milwaukee, Wisconsin

$CH_3-O-SO_2-O-CH_3$

CAS REGISTRY NO.: 77-78-1

NIOSH NUMBER: WS82250

Between Salts

CONDENSED PHASE

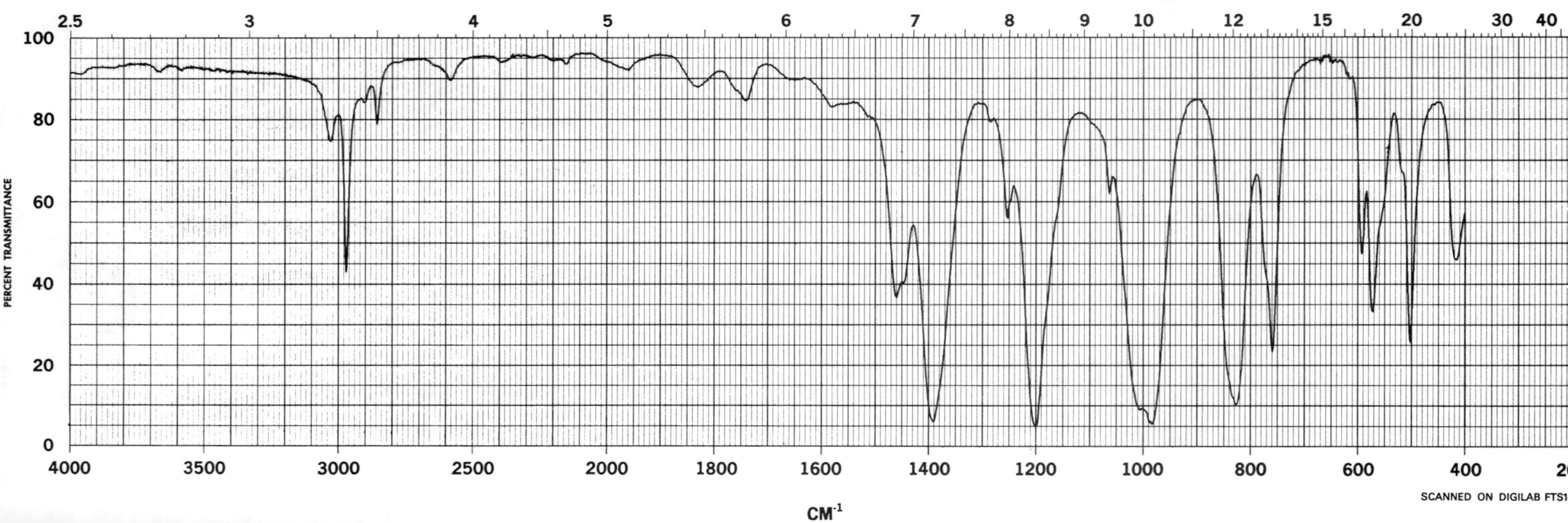

BOILING POINT:	188°C	INJECTION AMT.:	0.5 µl
VAPORIZATION TEMP.:	200°C	PRESSURE:	Atmospheric
CELL VOLUME:	7.5 cc	MAX./MIN.:	—
PATH LENGTH:	7.6 cm	SCALE EXPANSION:	0.94

VAPOR PHASE

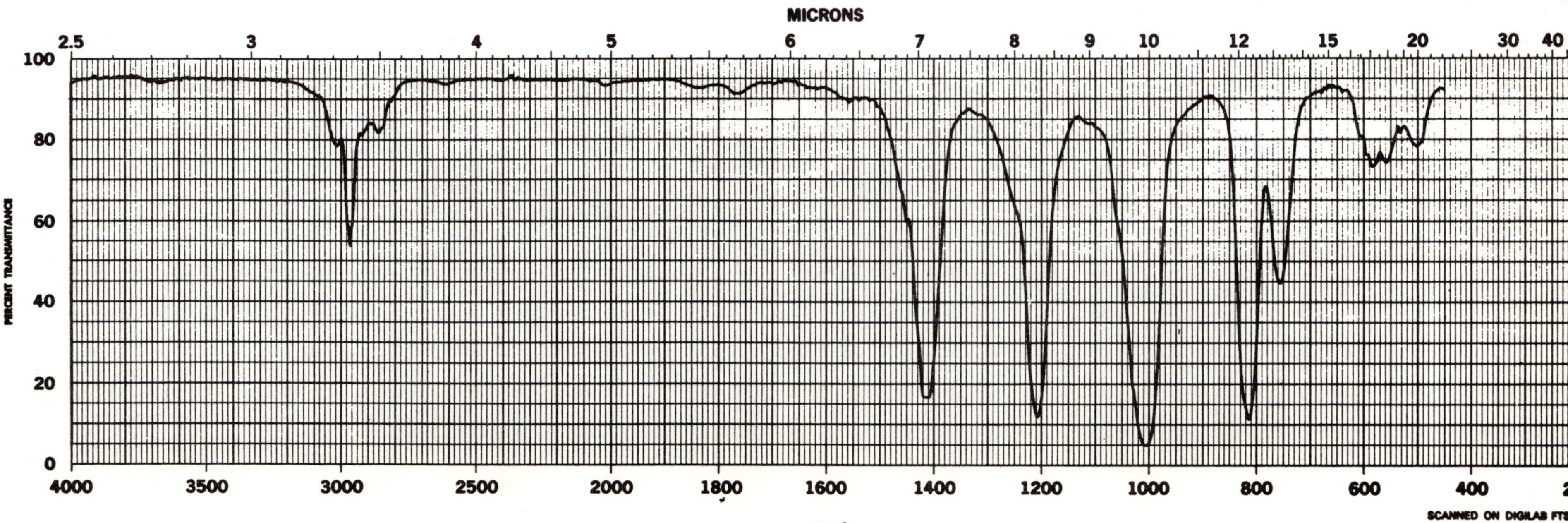

TOLUENE
METHYLBENZENE
TOLUOL
METHACIDE
PHENYLMETHANE

C_7H_8

MOLECULAR WEIGHT:	92.14
SOURCE OF SAMPLE:	MCB Manufacturing Chemists Norwood, Ohio
CAS REGISTRY NO.:	108-88-3
NIOSH NUMBER:	XS52500
CAPILLARY CELL:	Neat

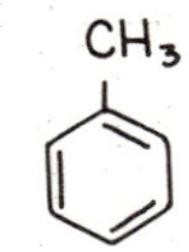

CONDENSED PHASE

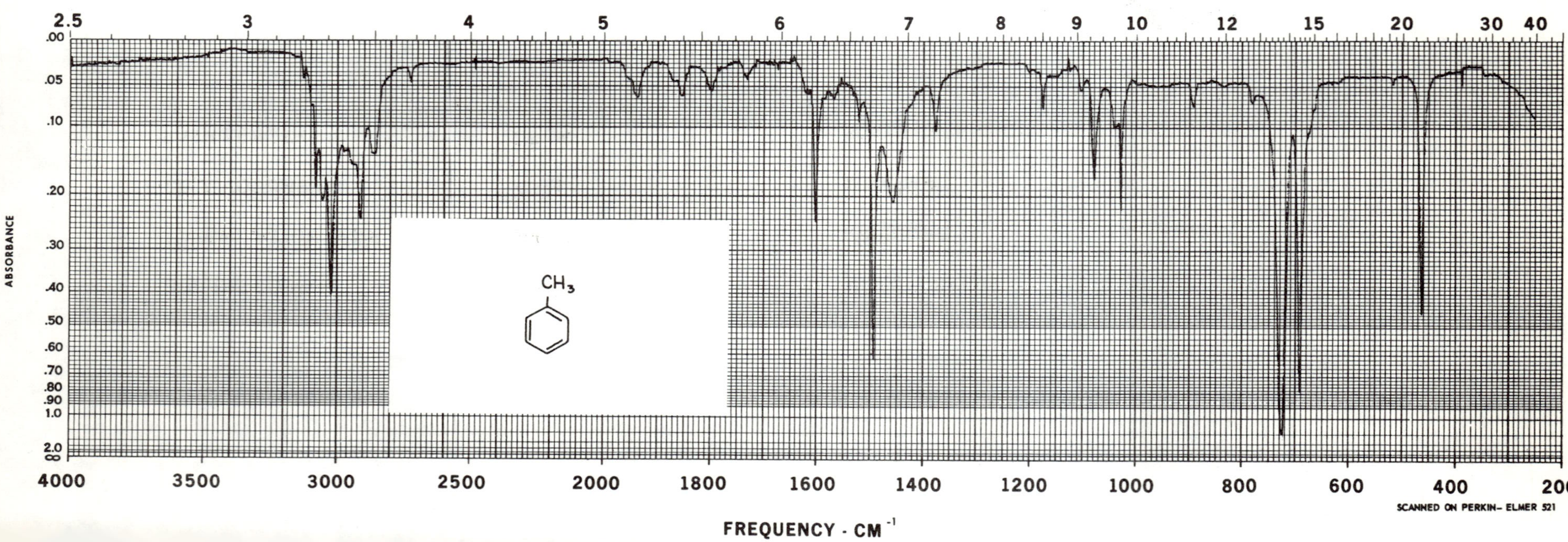

BOILING POINT:	110.6°C	INJECTION AMT.:	1.0 µl
VAPORIZATION TEMP.:	200°C	PRESSURE:	Atmospheric
CELL VOLUME:	7.5 cc	MAX./MIN.:	—
PATH LENGTH:	7.6 cm	SCALE EXPANSION:	1.05

VAPOR PHASE

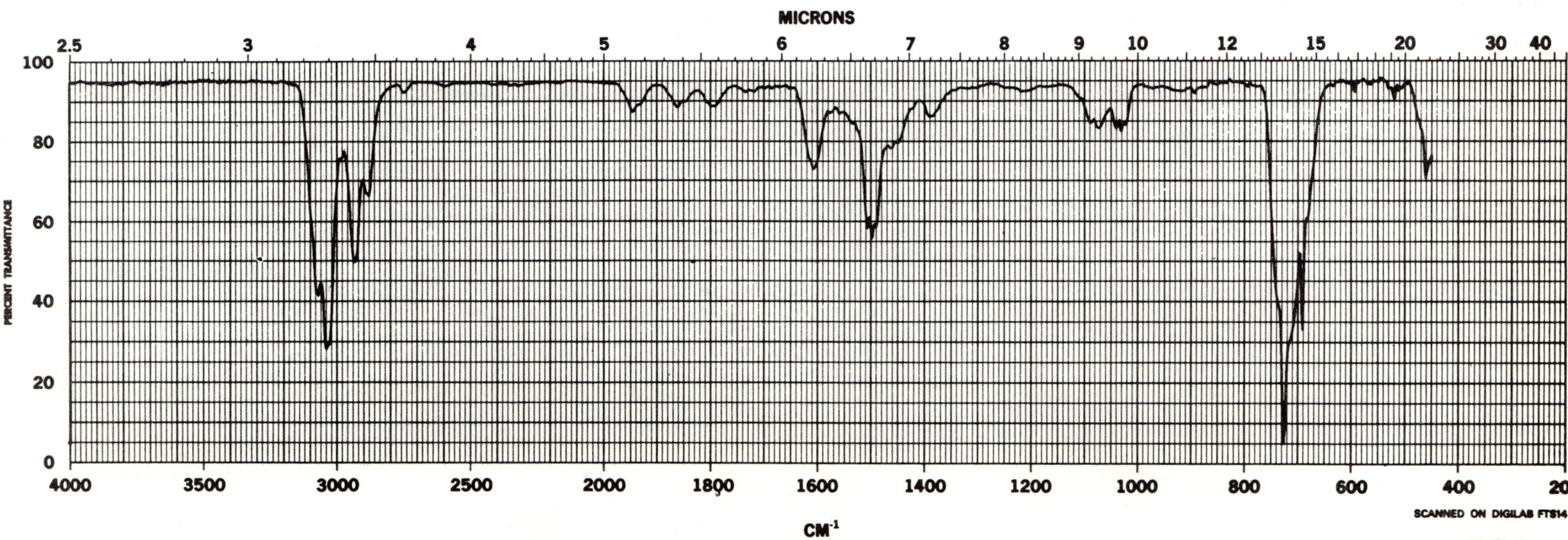

a—CHLOROTOLUENE
BENZYL CHLORIDE

C_7H_7Cl

MOLECULAR WEIGHT: 126.59

SOURCE OF SAMPLE: Aldrich Chemical Company, Inc.
Milwaukee, Wisconsin

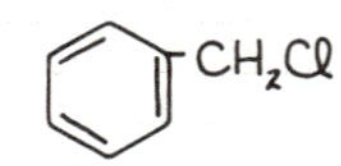

CAS REGISTRY NO.: 100-44-7

NIOSH NUMBER: XS89250

CAPILLARY CELL: Neat

CONDENSED PHASE

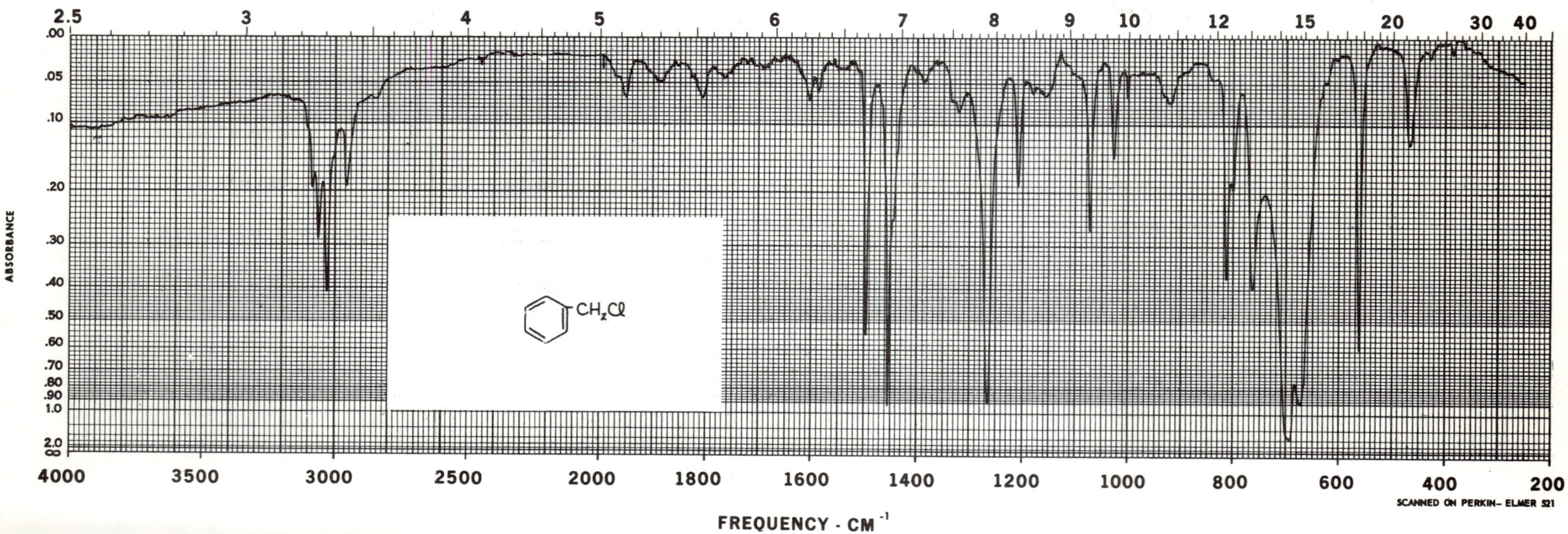

BOILING POINT:	177-181°C	INJECTION AMT.:	1.0 μl
VAPORIZATION TEMP.:	200°C	PRESSURE:	Atmospheric
CELL VOLUME:	7.5 cc	MAX./MIN.:	—
PATH LENGTH:	7.6 cm	SCALE EXPANSION:	1.03

VAPOR PHASE

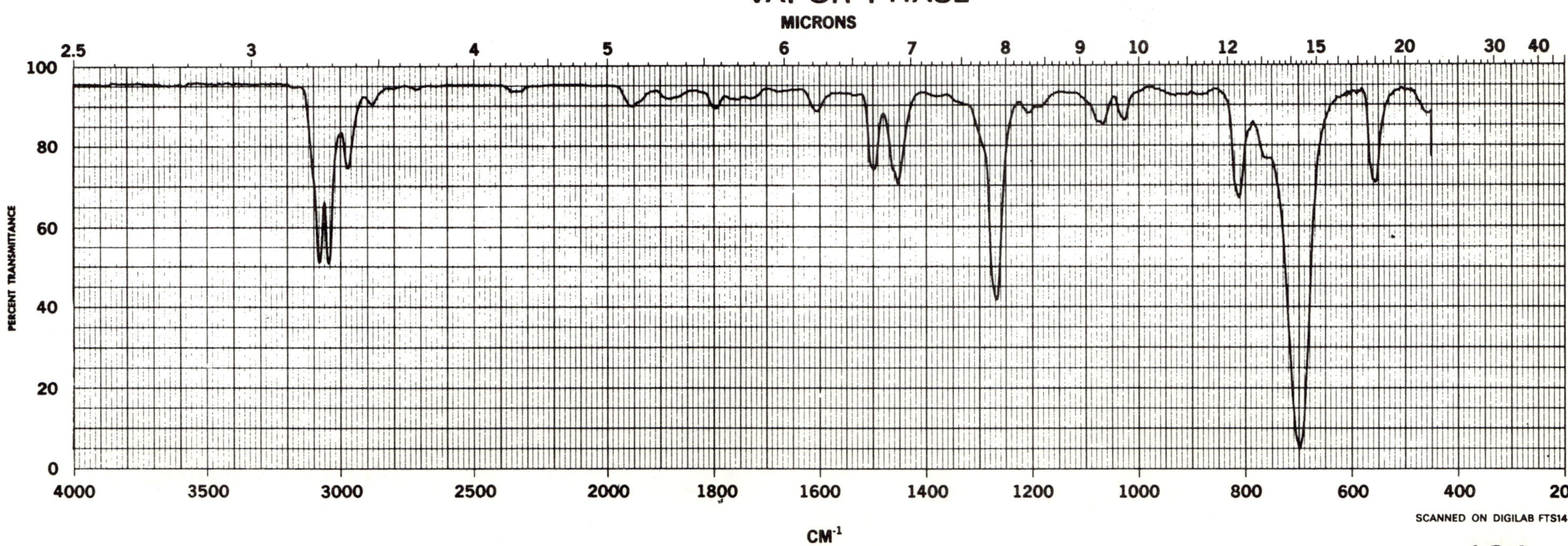

2,4–DINITROTOLUENE
2,4–DNT

$C_7H_6N_2O_4$

MOLECULAR WEIGHT: 182.14

SOURCE OF SAMPLE: MCB Manufacturing Chemists
Norwood, Ohio

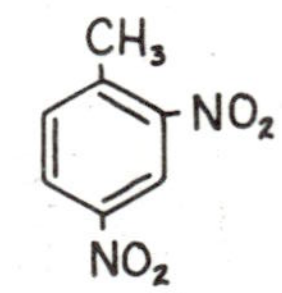

CAS REGISTRY NO.: 121-14-2

NIOSH NUMBER: XT15750

CAPILLARY CELL: Melt

CONDENSED PHASE

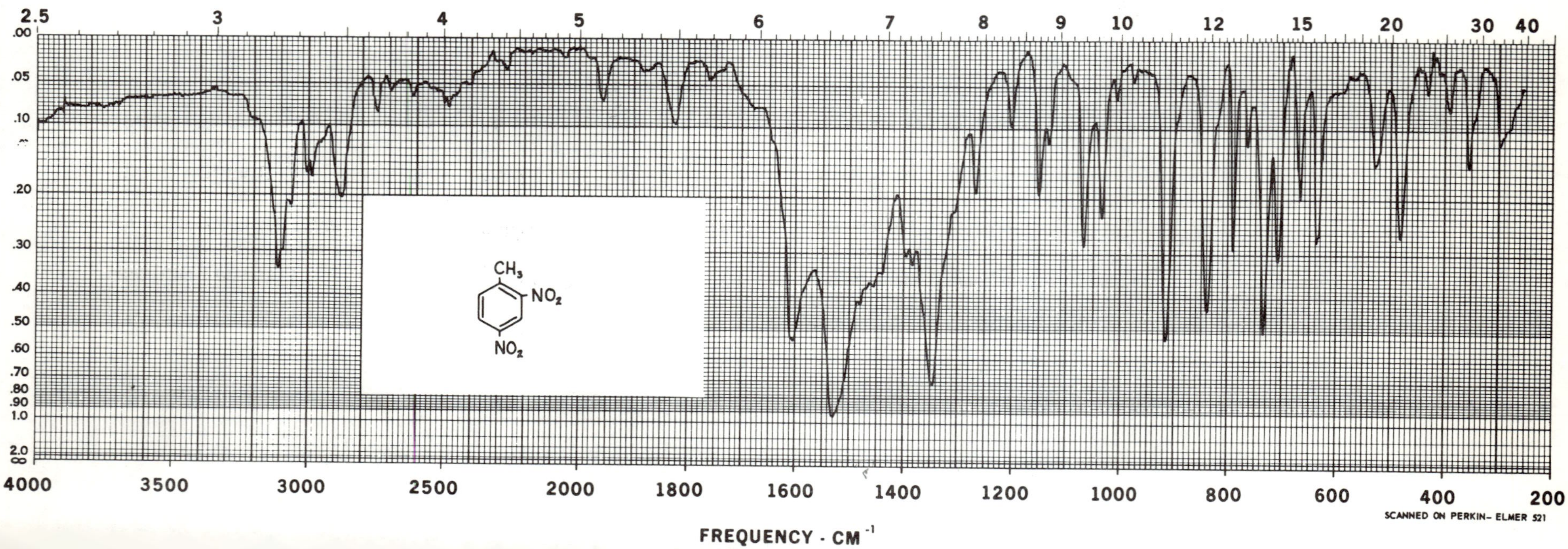

BOILING POINT:	300°C	INJECTION AMT.:	1.0 mg
VAPORIZATION TEMP.:	280°C	PRESSURE:	Atmospheric
CELL VOLUME:	7.5 cc	MAX./MIN.:	—
PATH LENGTH:	7.6 cm	SCALE EXPANSION:	0.91

VAPOR PHASE

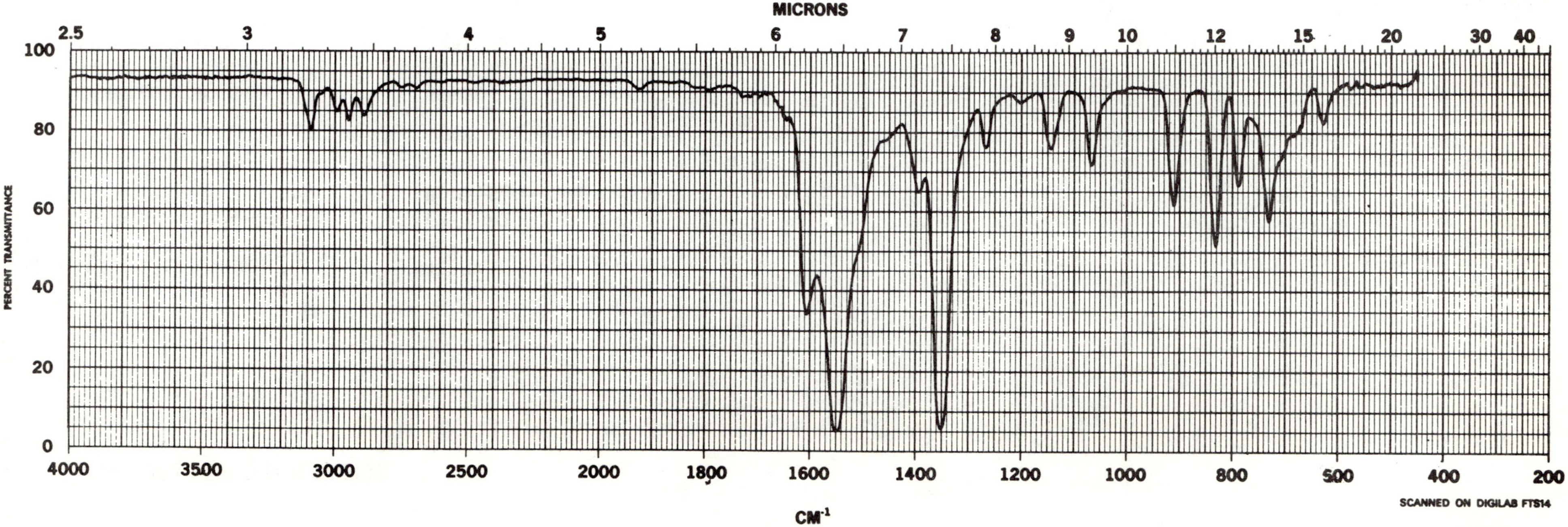

2,6–DINITROTOLUENE

$C_7H_6N_2O_4$

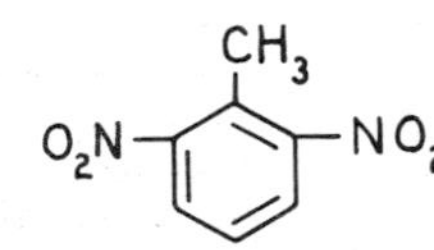

MOLECULAR WEIGHT:	182.14
SOURCE OF SAMPLE:	Fluka AG Buchs, Switzerland
CAS REGISTRY NO.:	606-20-2
NIOSH NUMBER:	XT19250
CAPILLARY CELL:	Melt

CONDENSED PHASE

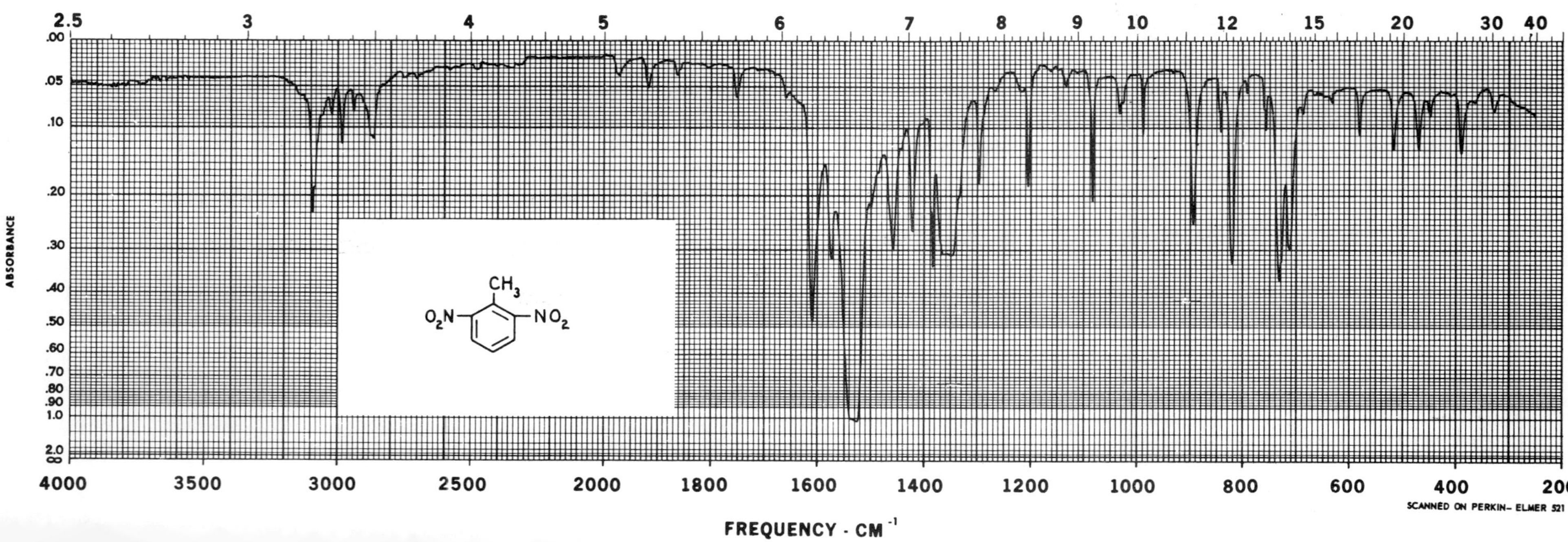

MELTING POINT:	61-64°C	INJECTION AMT.:	1.0 mg
VAPORIZATION TEMP.:	280°C	PRESSURE:	Atmospheric
CELL VOLUME:	7.5 cc	MAX./MIN.:	—
PATH LENGTH:	7.6 cm	SCALE EXPANSION:	0.91

VAPOR PHASE

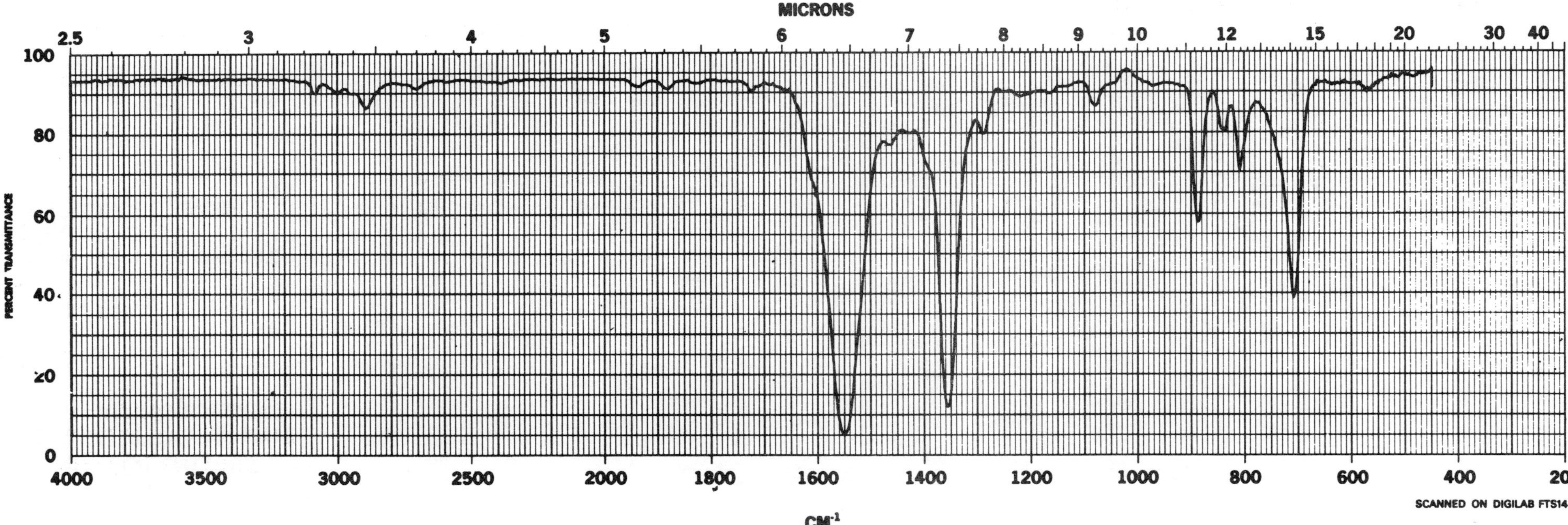

SCANNED ON DIGILAB FTS14

o—TOLUIDINE
o—METHYLANILINE
o—AMINOTOLUENE

C_7H_9N

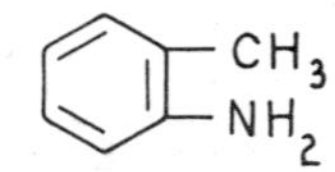

MOLECULAR WEIGHT: 107.16

SOURCE OF SAMPLE: Eastman Organic Chemicals
Rochester, New York

CAS REGISTRY NO.: 95-53-4

NIOSH NUMBER: XU29750

CAPILLARY CELL: Neat

CONDENSED PHASE

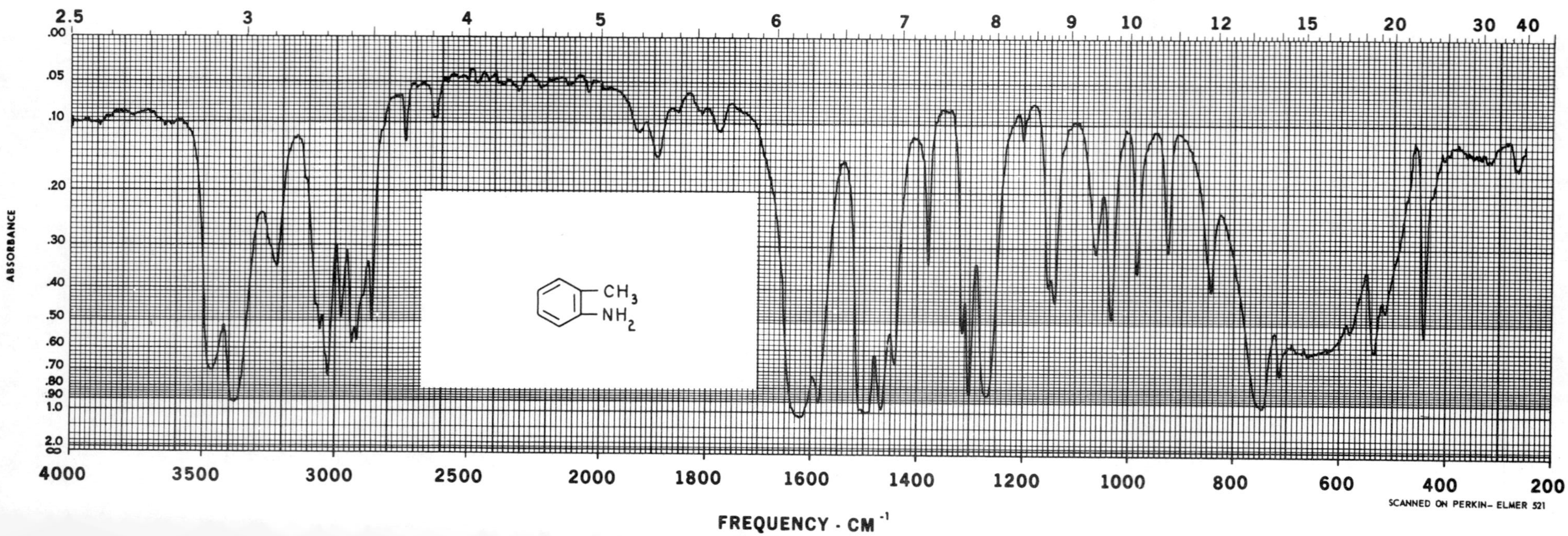

BOILING POINT:	199-200°C	INJECTION AMT.:	1.0 µl
VAPORIZATION TEMP.:	200°C	PRESSURE:	Atmospheric
CELL VOLUME:	7.5 cc	MAX./MIN.:	—
PATH LENGTH:	7.6 cm	SCALE EXPANSION:	1.06

VAPOR PHASE

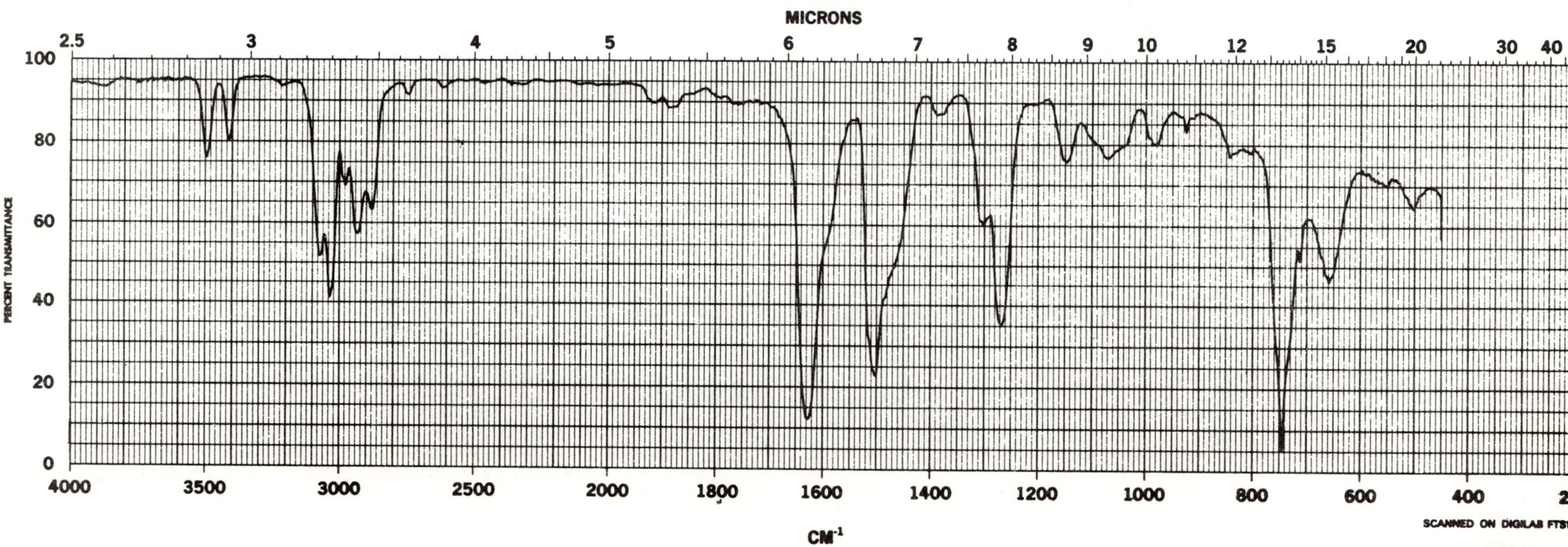

3,3–DIMETHYL–1–PHENYLTRIAZENE
PDMT

$C_8H_{11}N_3$

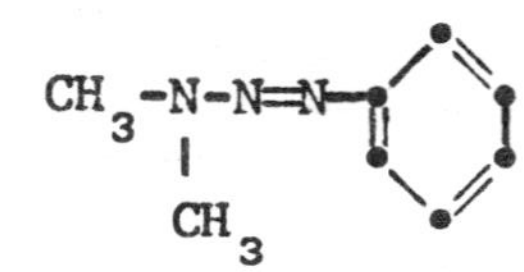

MOLECULAR WEIGHT: 149.20

SOURCE OF SAMPLE: Aldrich Chemical Company, Inc.
Milwaukee, Wisconsin

CAS REGISTRY NO.: 7227-91-0

NIOSH NUMBER: XY21000

CAPILLARY CELL: Neat

CONDENSED PHASE

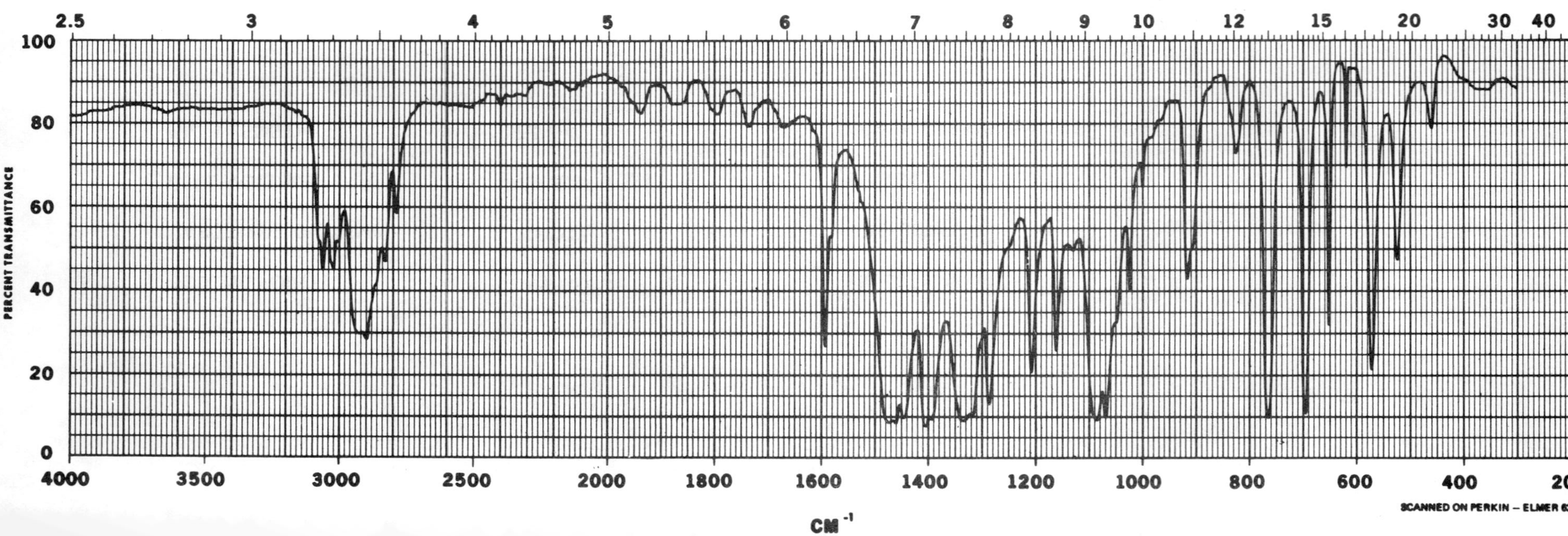

BOILING POINT:	–	INJECTION AMT.:	1.0 µl
VAPORIZATION TEMP.:	200°C	PRESSURE:	Atmospheric
CELL VOLUME:	7.5 cc	MAX./MIN.:	–
PATH LENGTH:	7.6 cm	SCALE EXPANSION:	0.95

VAPOR PHASE

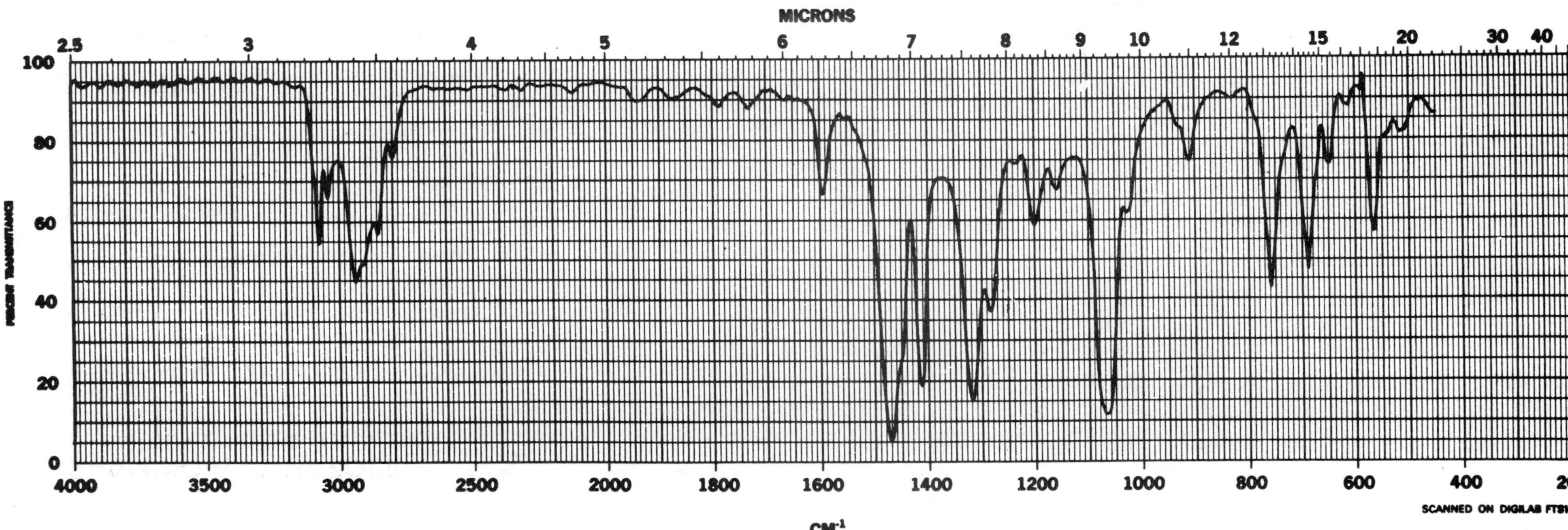

3–AMINO–s–TRIAZOLE
3–AMINO–1H–1,2,4–TRIAZOLE
AMITROLE
AMIZOL
WEEDOCLOR

$C_2H_4N_4$

MOLECULAR WEIGHT: 84.08

SOURCE OF SAMPLE: Aldrich Chemical Company, Inc.
Milwaukee, Wisconsin

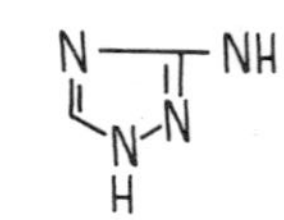

CAS REGISTRY NO.: 61-82-5

NIOSH NUMBER: XZ38500

KBr Wafer

CONDENSED PHASE

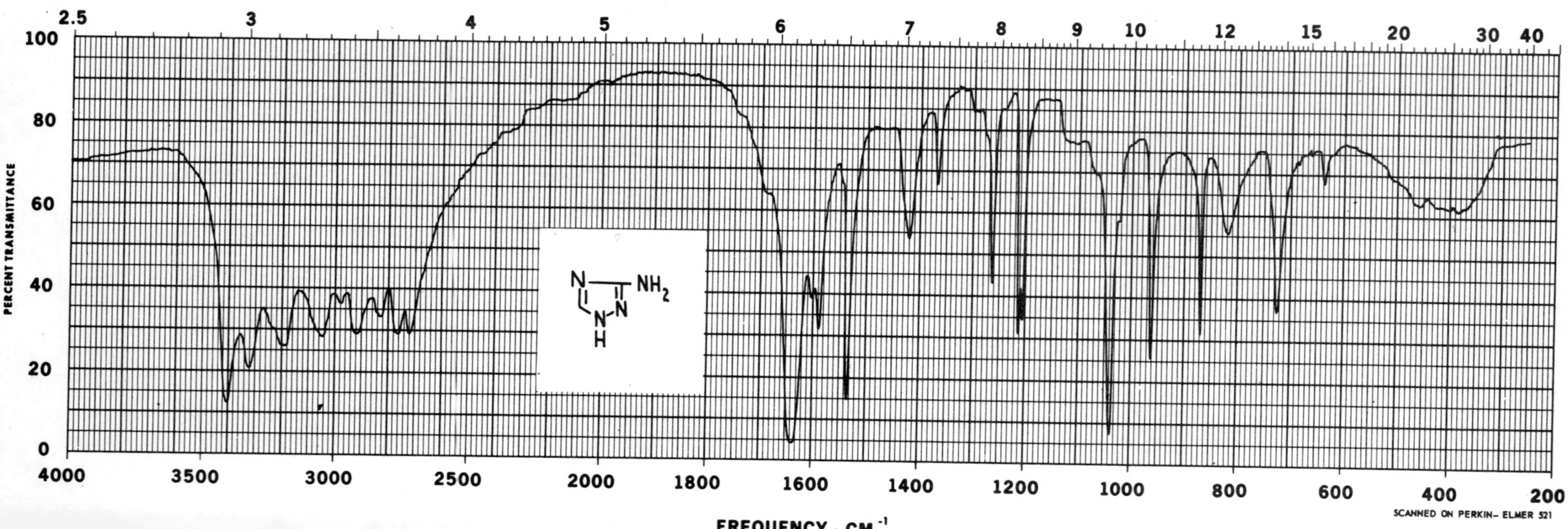

MELTING POINT:	153-156°C	INJECTION AMT.:	1.2 mg
VAPORIZATION TEMP.:	280°C	PRESSURE:	Atmospheric
CELL VOLUME:	7.5 cc	MAX./MIN.:	92.4/5.9
PATH LENGTH:	7.6 cm	SCALE EXPANSION:	1.06

VAPOR PHASE

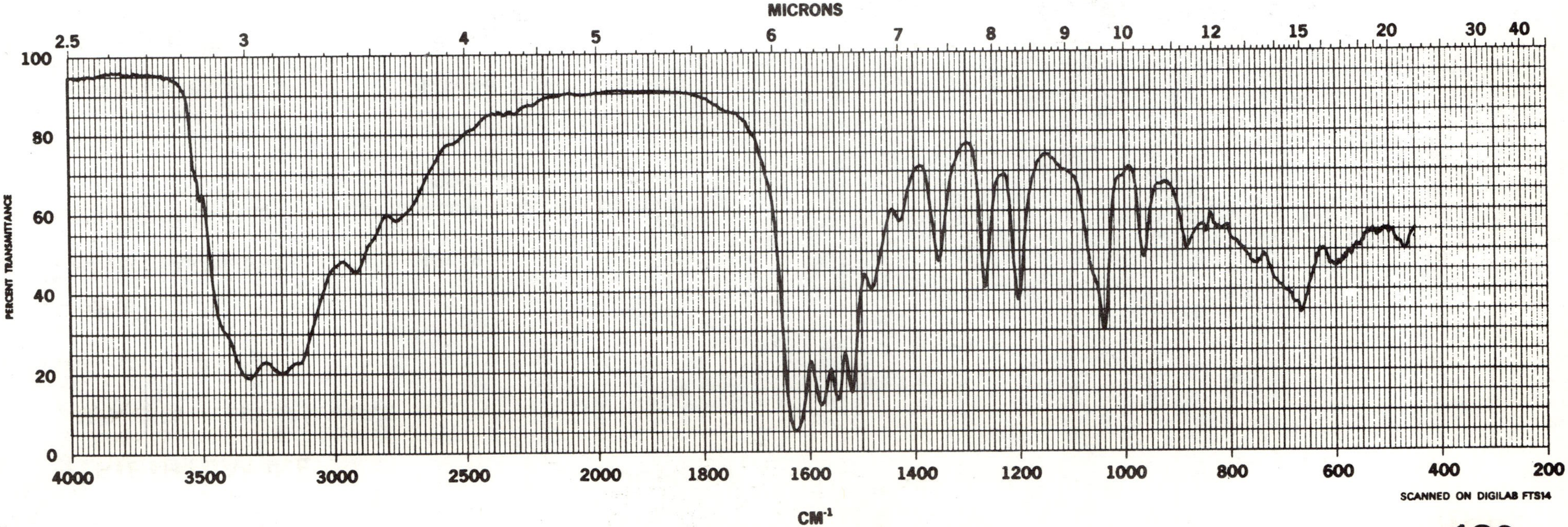

TRIETHYLAMINE
TEN
N,N–DIETHYLETHANAMINE

$C_6H_{15}N$

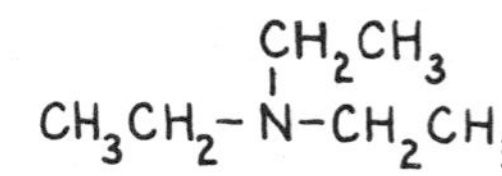

MOLECULAR WEIGHT: 101.19

SOURCE OF SAMPLE: MCB Manufacturing Chemists
Norwood, Ohio

CAS REGISTRY NO.: 121-44-8

NIOSH NUMBER: YE01750

CAPILLARY CELL: Neat

CONDENSED PHASE

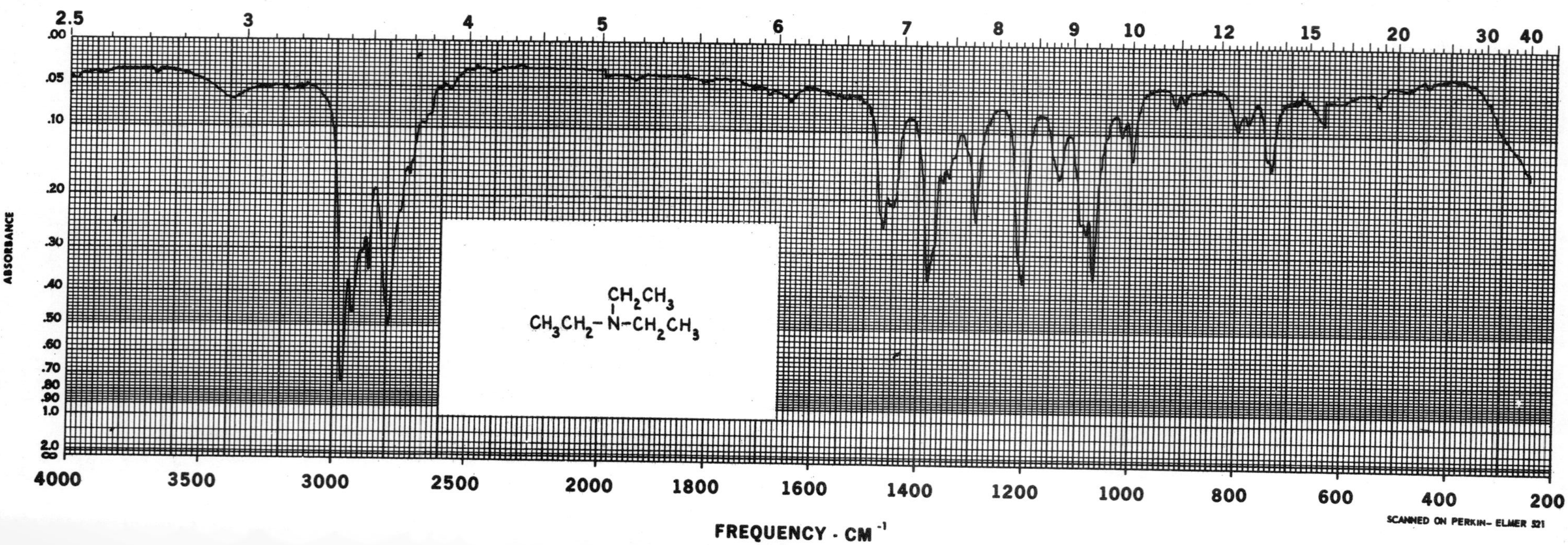

BOILING POINT:	90°C	INJECTION AMT.:	1.0 μl
VAPORIZATION TEMP.:	200°C	PRESSURE:	Atmospheric
CELL VOLUME:	7.5 cc	MAX./MIN.:	—
PATH LENGTH:	7.6 cm	SCALE EXPANSION:	0.92

VAPOR PHASE

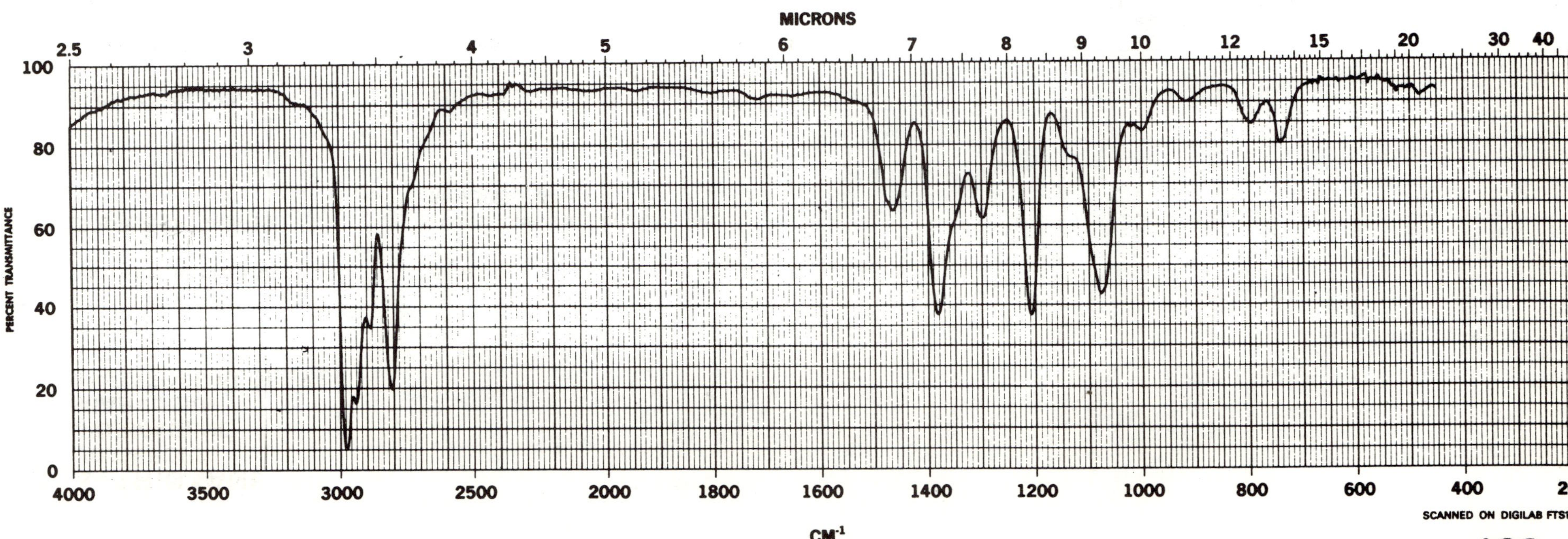

190

s—TRIOXANE
ALDEFORM
FORMAGENE
MARVOSAN
PARAFORMALDEHYDE

$C_3H_6O_3$

MOLECULAR WEIGHT:	90.08
SOURCE OF SAMPLE:	Aldrich Chemical Company, Inc. Milwaukee, Wisconsin
CAS REGISTRY NO.:	110-88-3
NIOSH NUMBER:	YM14000

KBr Wafer

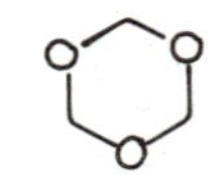

CONDENSED PHASE

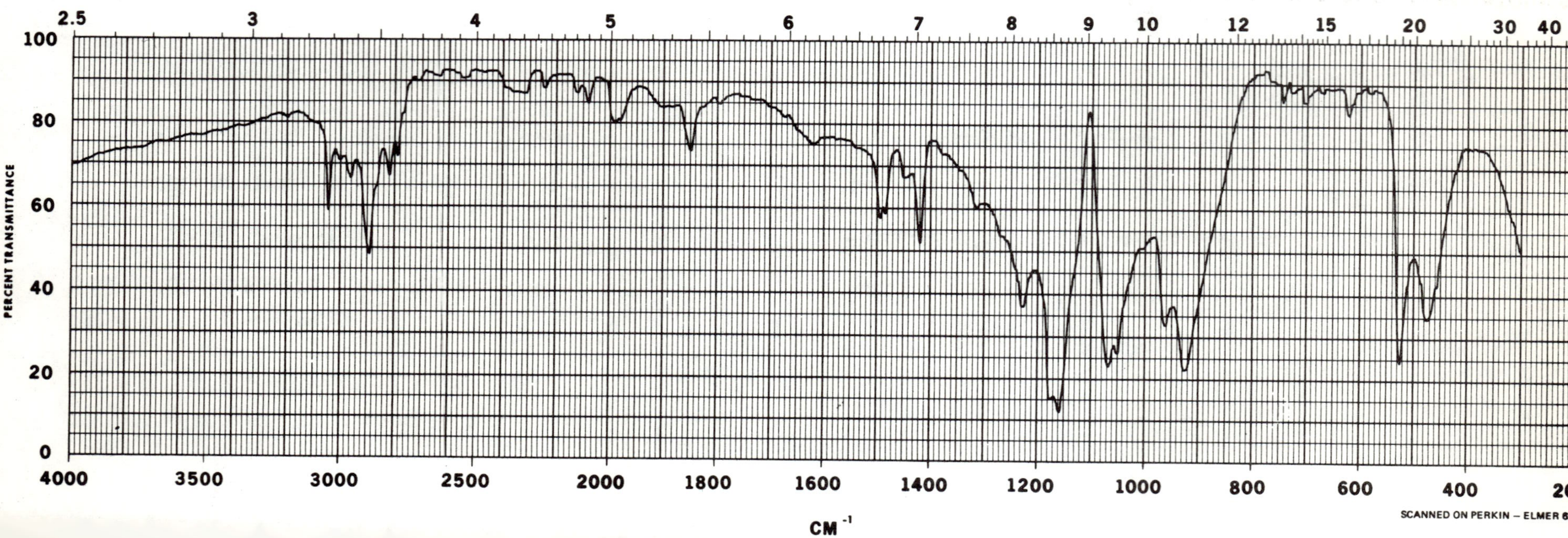

MELTING POINT:	61-62.5°C	INJECTION AMT.:	—
VAPORIZATION TEMP.:	100°C	PRESSURE:	Atmospheric
CELL VOLUME:	7.5 cc	MAX./MIN.:	98.8/7.3
PATH LENGTH:	7.6 cm	SCALE EXPANSION:	1.01

VAPOR PHASE

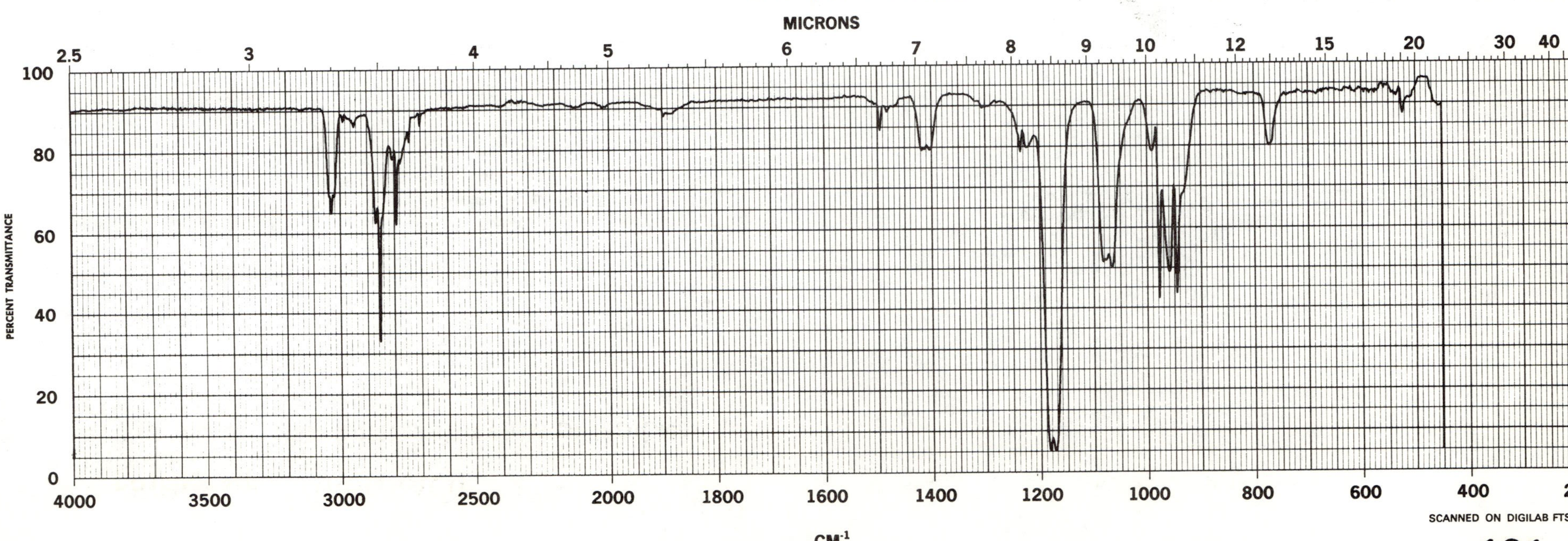

SCANNED ON DIGILAB FTS14

6–PROPYL–2–THIOURACIL
PROCASIL
6–PROPYL–2–THIO–2,4(1H,3H)–PYRIMIDINEDIONE
PTU

$C_7H_{10}N_2OS$

MOLECULAR WEIGHT: 170.23

SOURCE OF SAMPLE: Fluka AG
Buchs, Switzerland

CAS REGISTRY NO.: 51-52-5

NIOSH NUMBER: YR14000

KBr Wafer

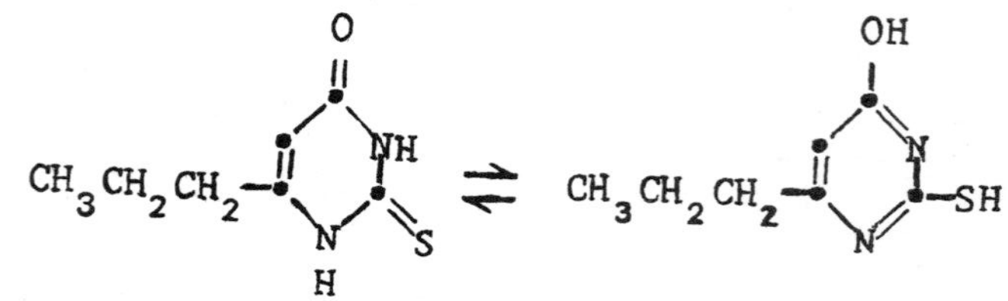

CONDENSED PHASE

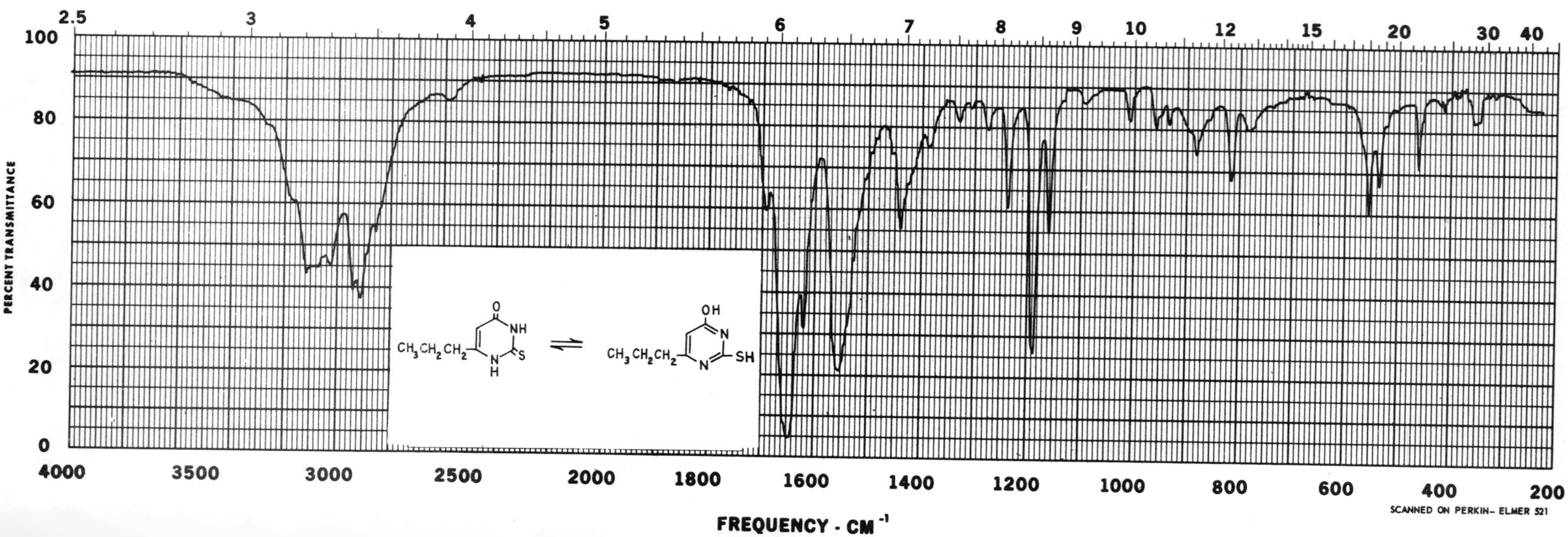

MELTING POINT:	219°C	INJECTION AMT.:	–
VAPORIZATION TEMP.:	280°C	PRESSURE:	Atmospheric
CELL VOLUME:	7.5 cc	MAX./MIN.:	102.0/38.0
PATH LENGTH:	7.6 cm	SCALE EXPANSION:	1.44

VAPOR PHASE

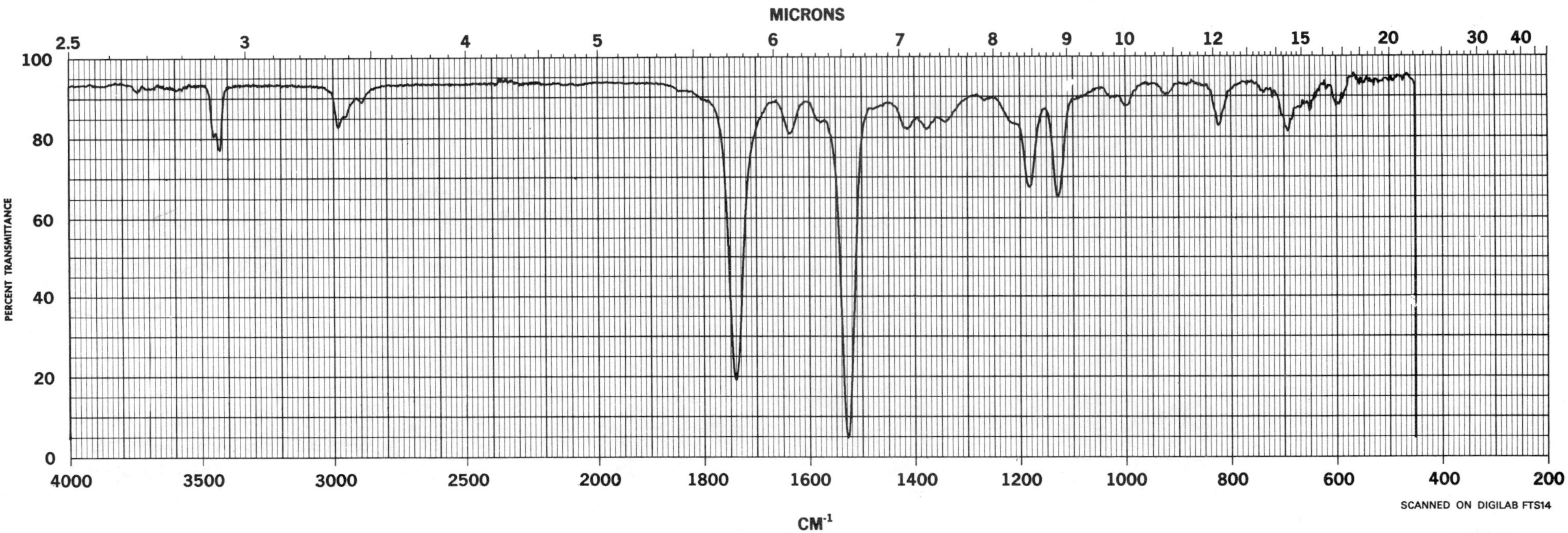

SCANNED ON DIGILAB FTS14

192

m–XYLENE
1,3–DIMETHYLBENZENE
m–XYLOL

C_8H_{10}

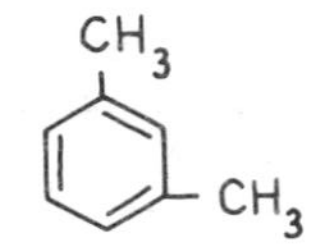

MOLECULAR WEIGHT: 106.17

SOURCE OF SAMPLE: MCB Manufacturing Chemists
Norwood, Ohio

CAS REGISTRY NO.: 108-38-3

NIOSH NUMBER: ZE22750

CAPILLARY CELL: Neat

CONDENSED PHASE

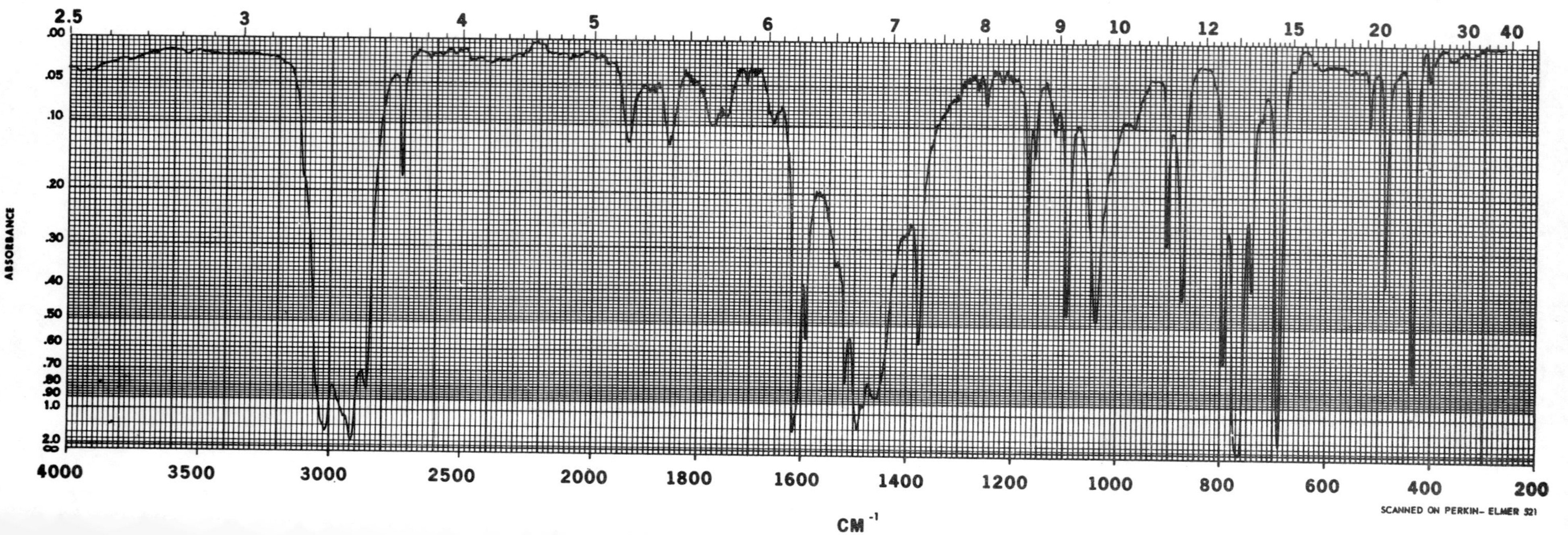

BOILING POINT:	138.5-139°C	INJECTION AMT.:	1.5 µl
VAPORIZATION TEMP.:	200°C	PRESSURE:	Atmospheric
CELL VOLUME:	7.5 cc	MAX./MIN.:	—
PATH LENGTH:	7.6 cm	SCALE EXPANSION:	1.11

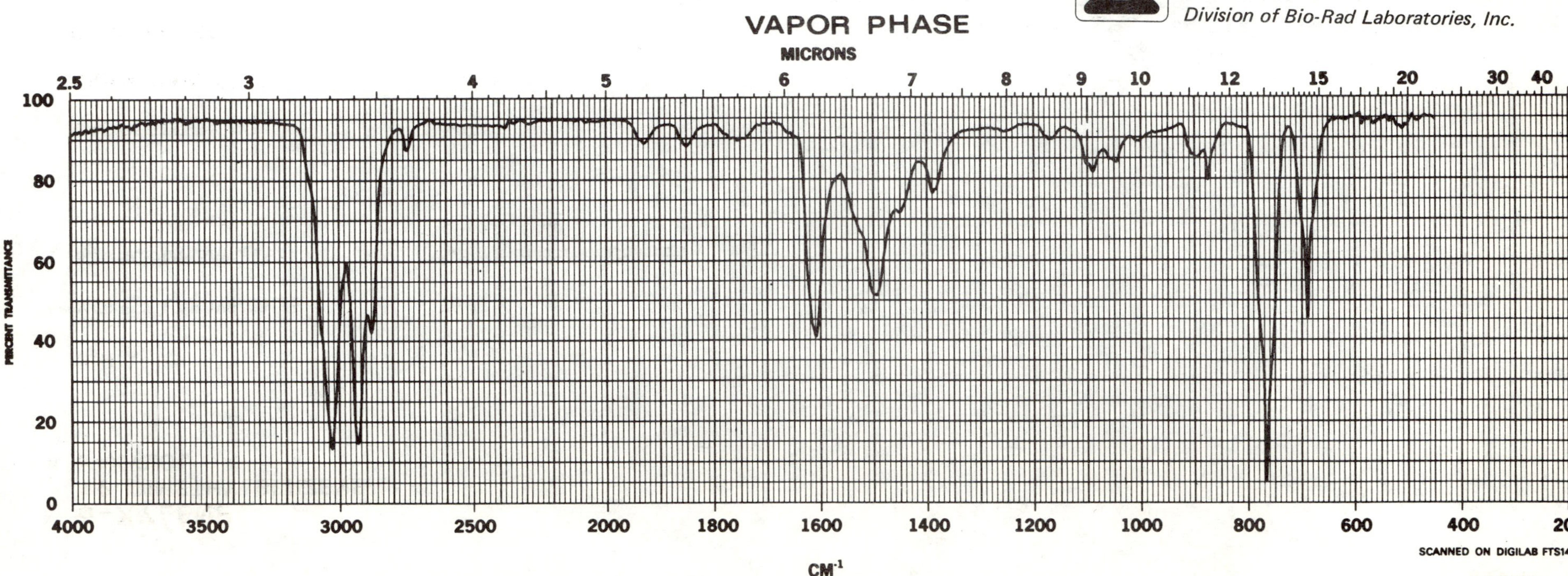

o–XYLENE
1,2–DIMETHYLBENZENE
o–XYLOL

C_8H_{10}

MOLECULAR WEIGHT:	106.17
SOURCE OF SAMPLE:	MCB Manufacturing Chemists Norwood, Ohio
CAS REGISTRY NO.:	95-47-6
NIOSH NUMBER:	ZE24500
CAPILLARY CELL:	Neat

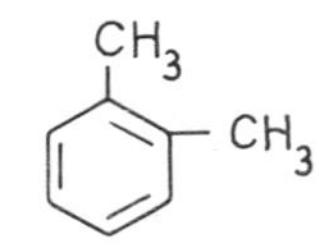

CONDENSED PHASE

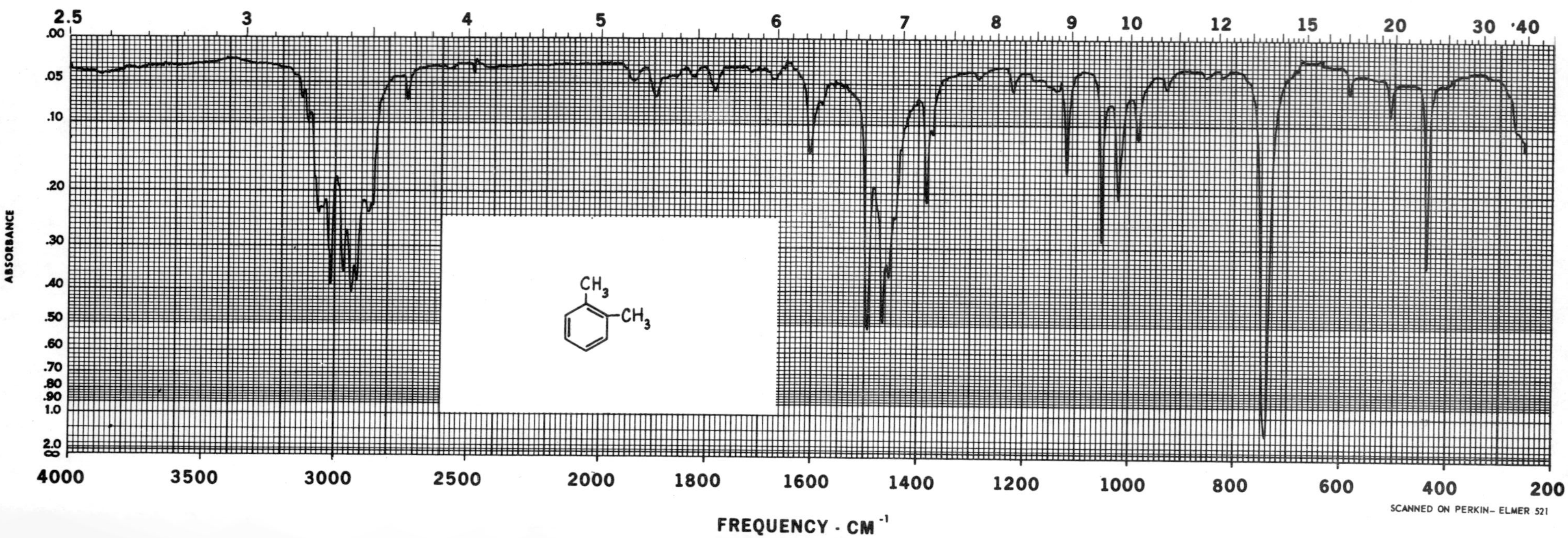

BOILING POINT:	143.5-144.5^{o}C	INJECTION AMT.:	1.0 μl
VAPORIZATION TEMP.:	200°C	PRESSURE:	Atmospheric
CELL VOLUME:	7.5 cc	MAX./MIN.:	—
PATH LENGTH:	7.6 cm	SCALE EXPANSION:	1.02

VAPOR PHASE

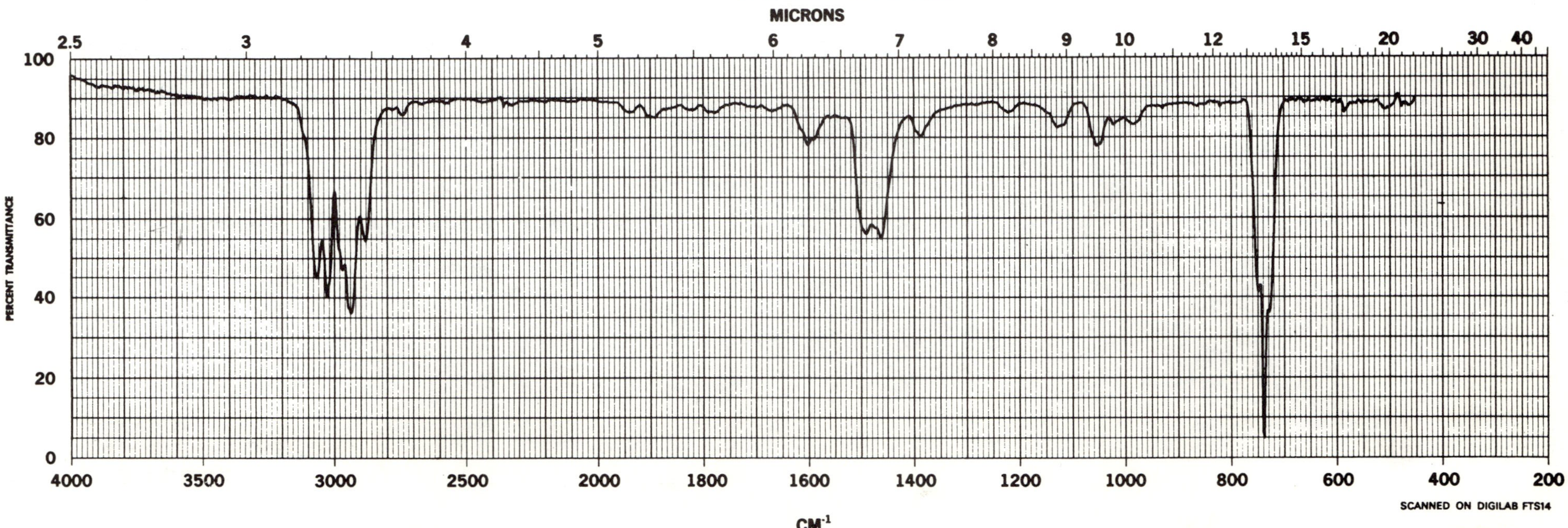

194

p–XYLENE
1,4–DIMETHYLBENZENE
p–XYLOL

C_8H_{10}

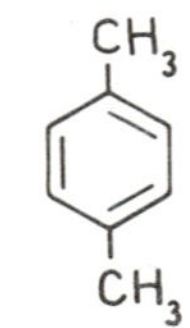

MOLECULAR WEIGHT: 106.17

SOURCE OF SAMPLE: MCB Manufacturing Chemists
Norwood, Ohio

CAS REGISTRY NO.: 106-42-3

NIOSH NUMBER: ZE26250

CAPILLARY CELL: Neat

CONDENSED PHASE

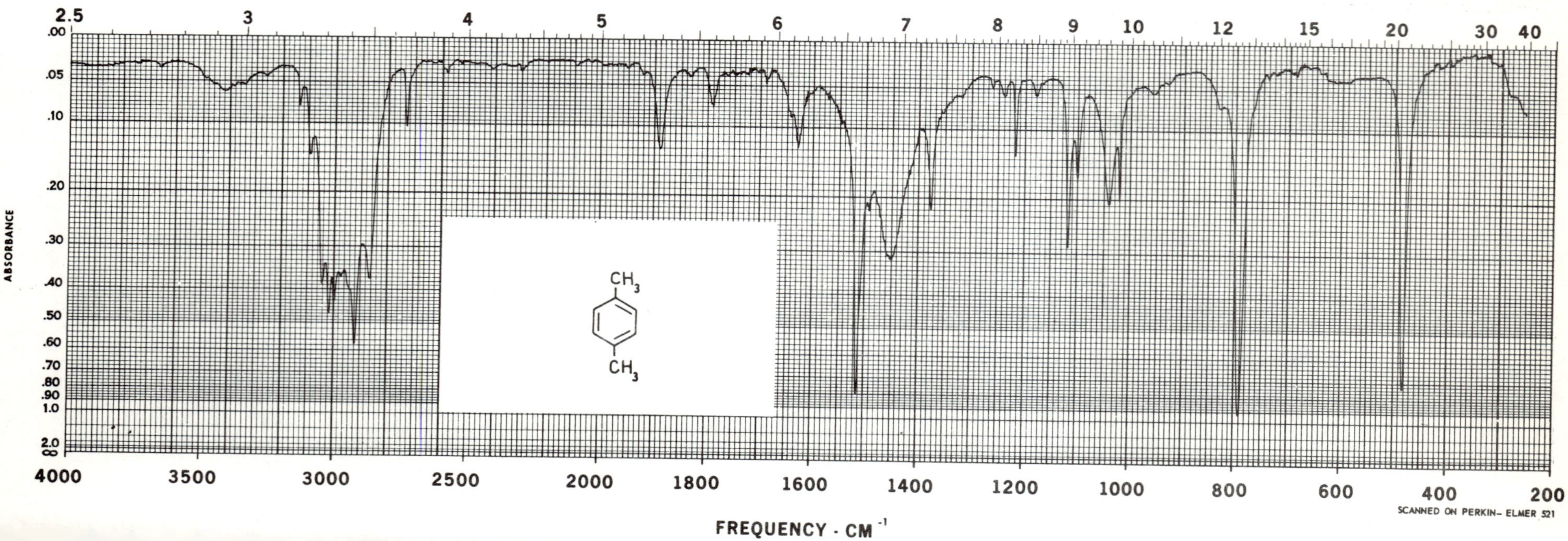

BOILING POINT:	138°C	INJECTION AMT.:	1.0 µl
VAPORIZATION TEMP.:	200°C	PRESSURE:	Atmospheric
CELL VOLUME:	7.5 cc	MAX./MIN.:	—
PATH LENGTH:	7.6 cm	SCALE EXPANSION:	1.21

VAPOR PHASE

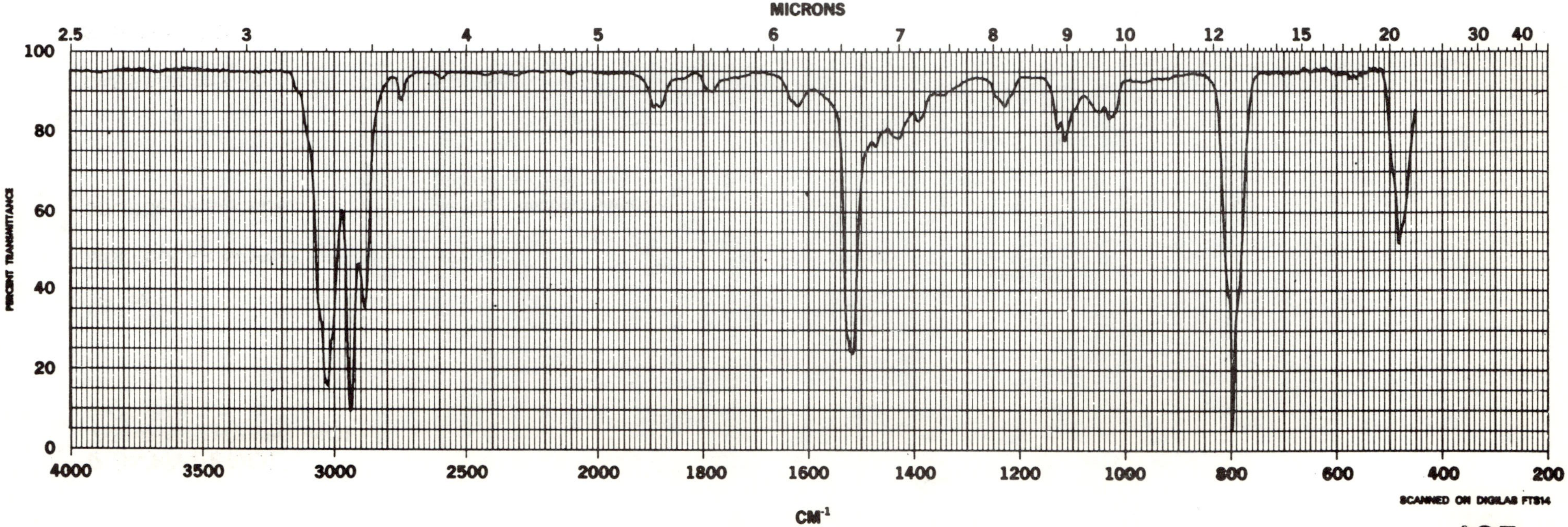

2,3–XYLENOL
2,3–DIMETHYLPHENOL

$C_8H_{10}O$

MOLECULAR WEIGHT:	122.17
SOURCE OF SAMPLE:	Aldrich Chemical Company, Inc. Milwaukee, Wisconsin
CAS REGISTRY NO.:	526-75-0
NIOSH NUMBER:	--
CAPILLARY CELL:	Melt

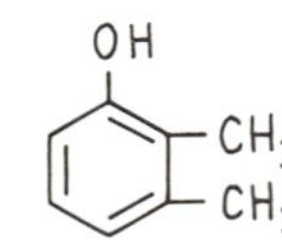

CONDENSED PHASE

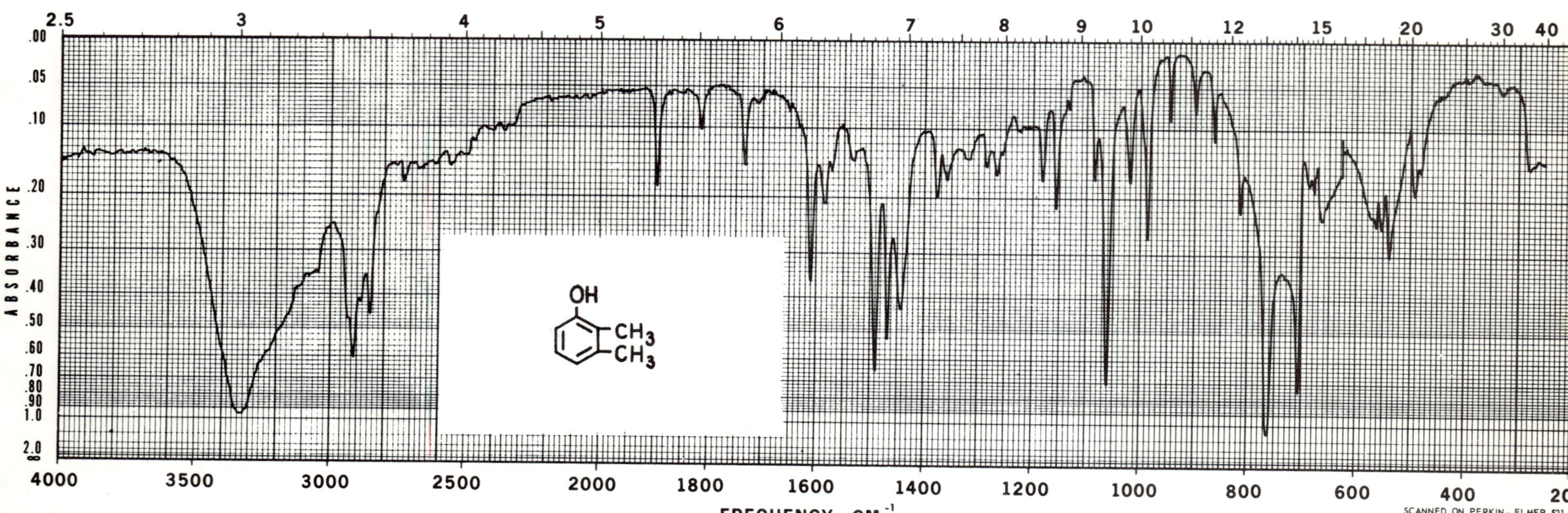

BOILING POINT:	217°C	INJECTION AMT.:	1.2 mg
VAPORIZATION TEMP.:	280°C	PRESSURE:	Atmospheric
CELL VOLUME:	7.5 cc	MAX./MIN.:	—
PATH LENGTH:	7.6 cm	SCALE EXPANSION:	1.69

VAPOR PHASE

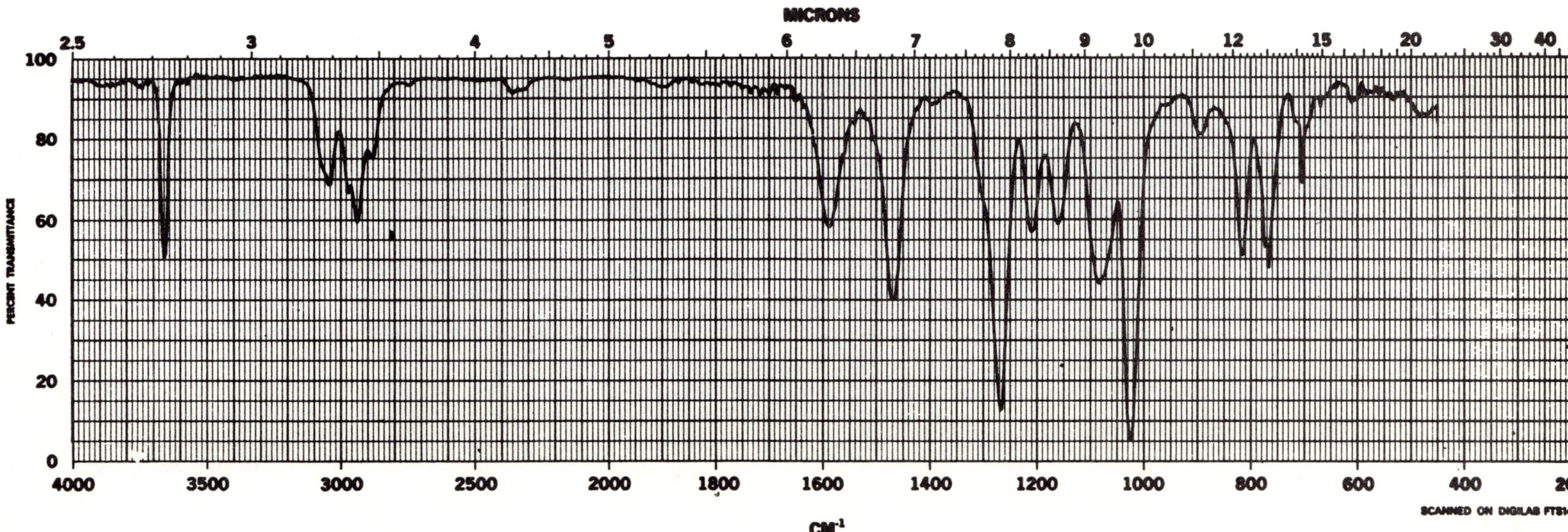

2,4–XYLENOL
2,4–DIMETHYLPHENOL

$C_8H_{10}O$

MOLECULAR WEIGHT: 122.17

SOURCE OF SAMPLE: Aldrich Chemical Company, Inc.
Milwaukee, Wisconsin

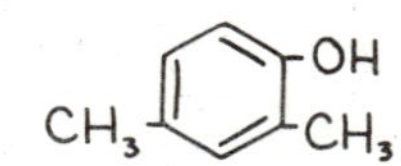

CAS REGISTRY NO.: 105-67-9

NIOSH NUMBER: ZE56000

CAPILLARY CELL: Neat

CONDENSED PHASE

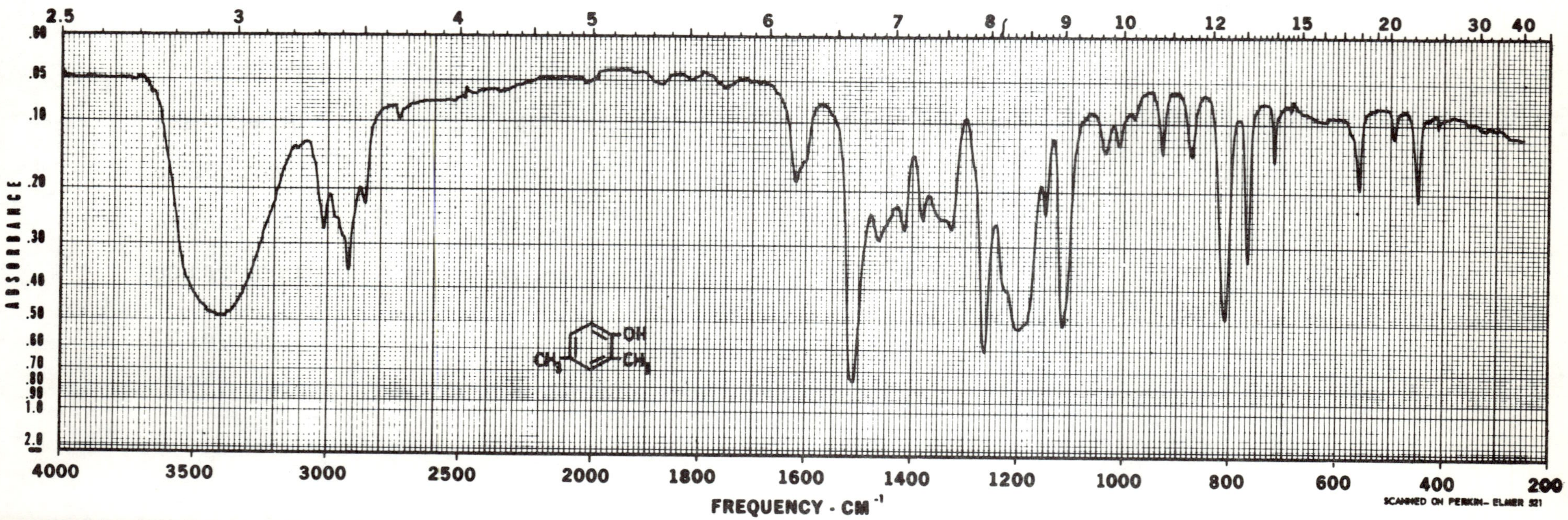

BOILING POINT:	212°C/766 mm	INJECTION AMT.:	1.0 µl
VAPORIZATION TEMP.:	240°C	PRESSURE:	Atmospheric
CELL VOLUME:	7.5 cc	MAX./MIN.:	100.0/14.5
PATH LENGTH:	7.6 cm	SCALE EXPANSION:	1.08

VAPOR PHASE

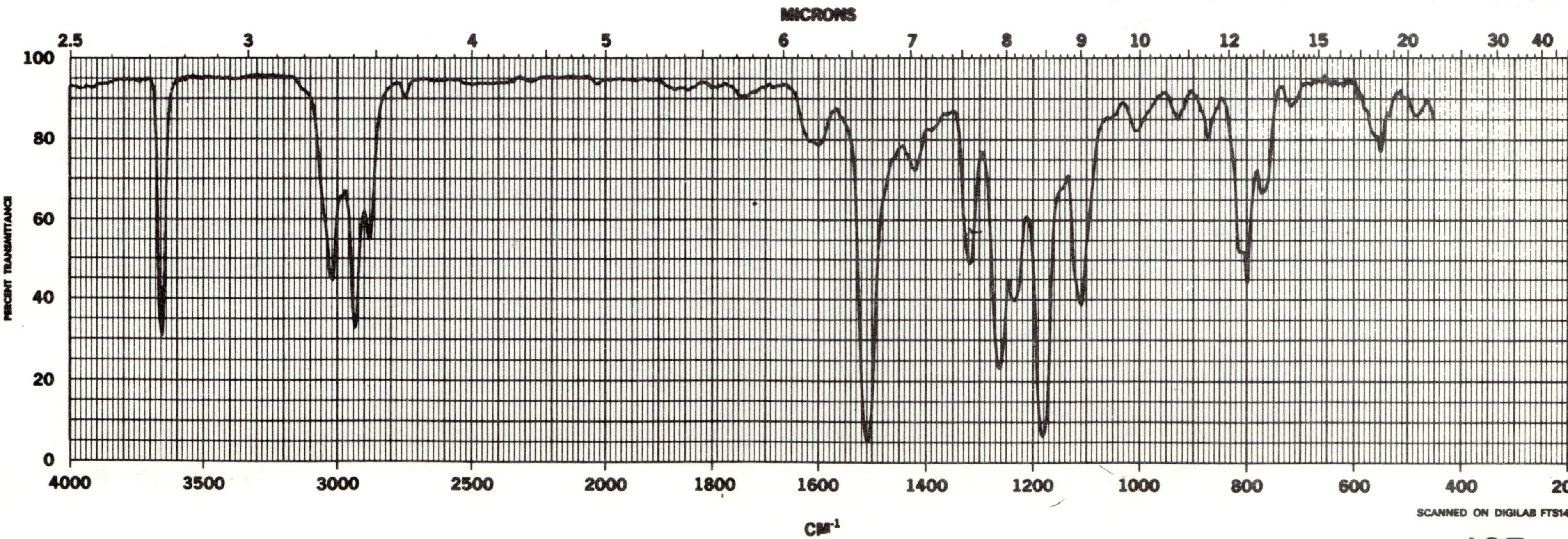

SCANNED ON DIGILAB FTS14

2,5–XYLENOL
2,5–DIMETHYLPHENOL
2,5–DMP

$C_8H_{10}O$

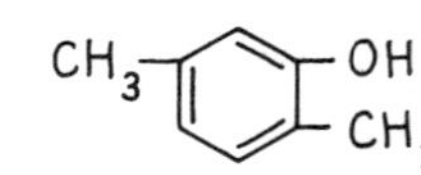

MOLECULAR WEIGHT:	122.17
SOURCE OF SAMPLE:	MCB Manufacturing Chemists Norwood, Ohio
CAS REGISTRY NO.:	95-87-4
NIOSH NUMBER:	ZE57750
CAPILLARY CELL:	Melt

CONDENSED PHASE

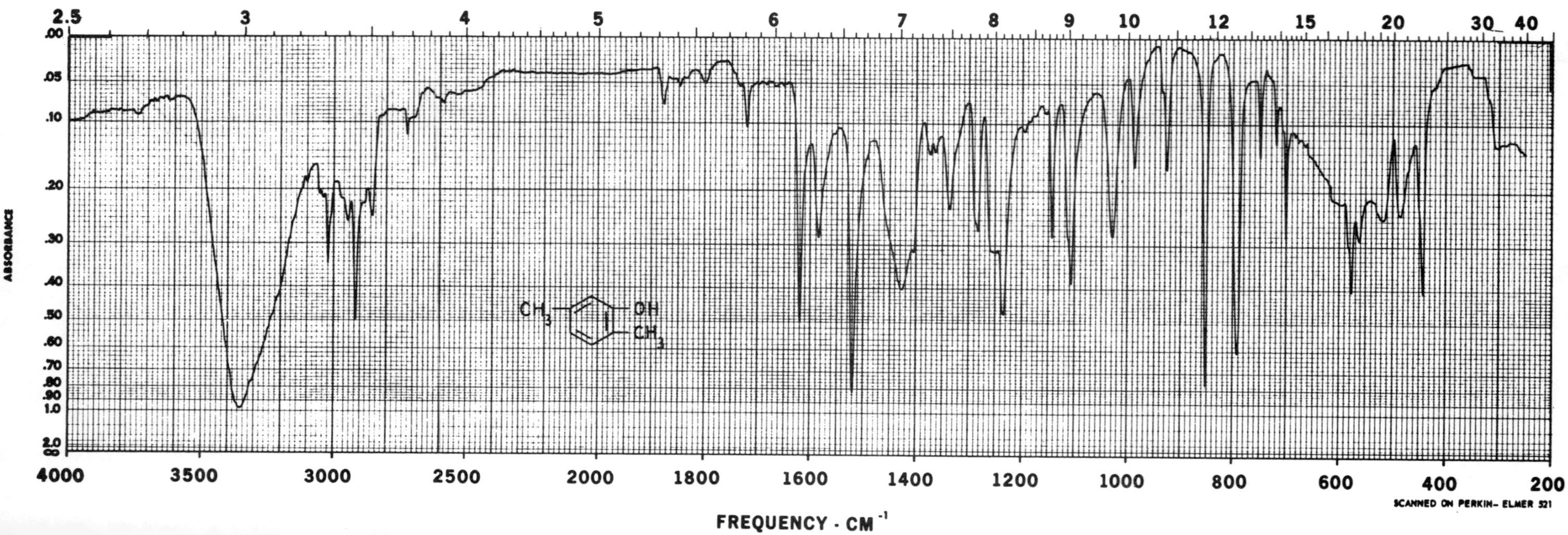

BOILING POINT:	211°C	INJECTION AMT.:	–
VAPORIZATION TEMP.:	280°C	PRESSURE:	Atmospheric
CELL VOLUME:	7.5 cc	MAX./MIN.:	102.1/18.6
PATH LENGTH:	7.6 cm	SCALE EXPANSION:	1.10

VAPOR PHASE

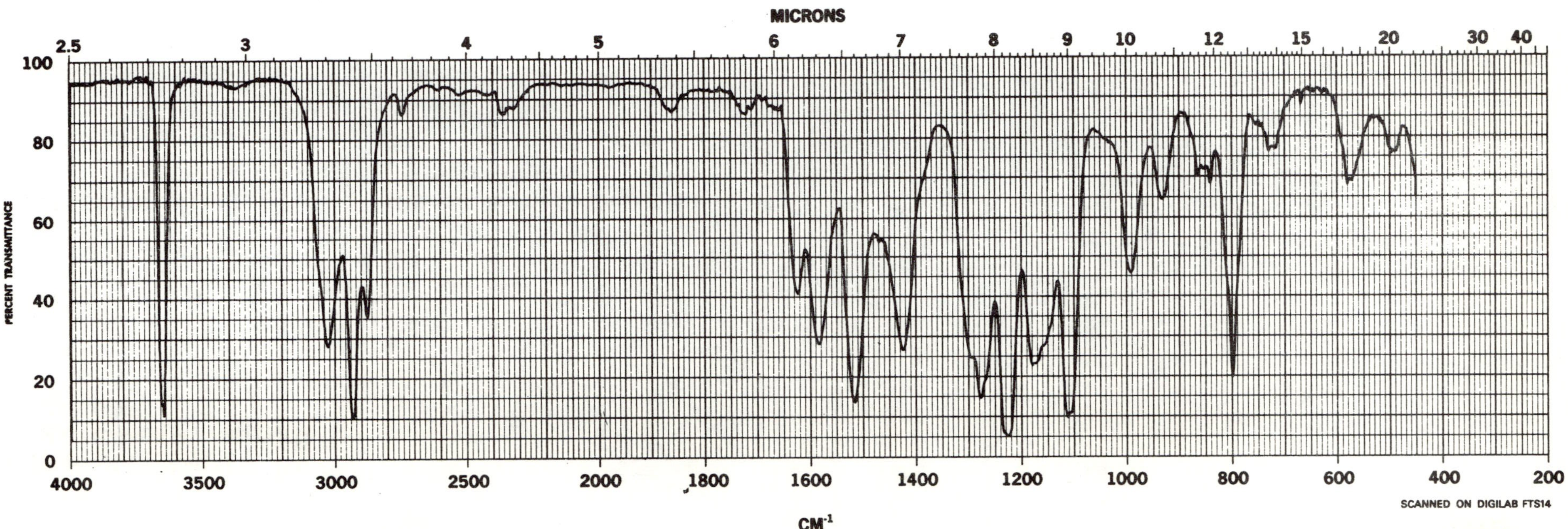

SCANNED ON DIGILAB FTS14

2,6–XYLENOL
2,6–DIMETHYLPHENOL
2,6–DMP

$C_8H_{10}O$

MOLECULAR WEIGHT: 122.17

SOURCE OF SAMPLE: Aldrich Chemical Company, Inc.
Milwaukee, Wisconsin

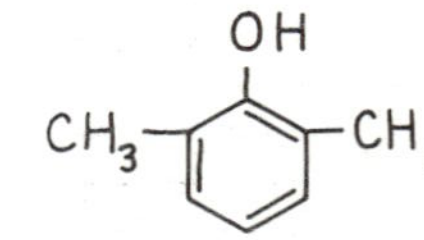

CAS REGISTRY NO.: 576-26-1

NIOSH NUMBER: ZE61250

CAPILLARY CELL: Melt

CONDENSED PHASE

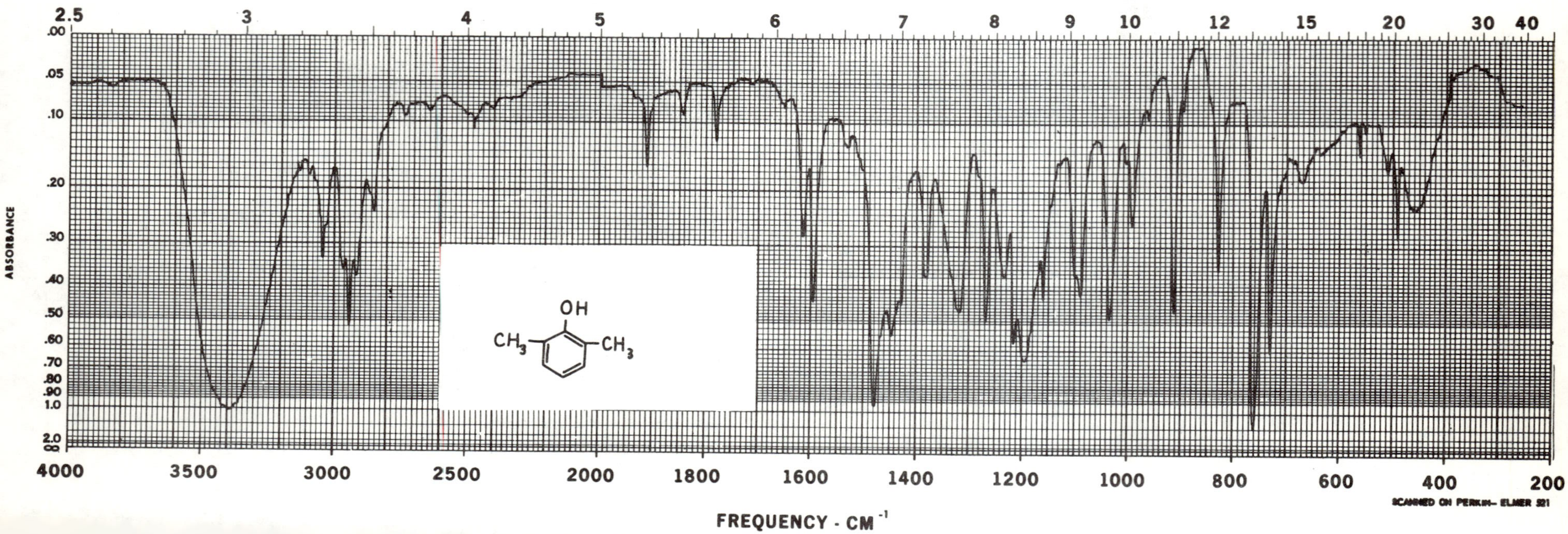

BOILING POINT:	$203^{o}C$	INJECTION AMT.:	0.87 mg
VAPORIZATION TEMP.:	$240^{o}C$	PRESSURE:	Atmospheric
CELL VOLUME:	7.5 cc	MAX./MIN.:	—
PATH LENGTH:	7.6 cm	SCALE EXPANSION:	1.08

VAPOR PHASE

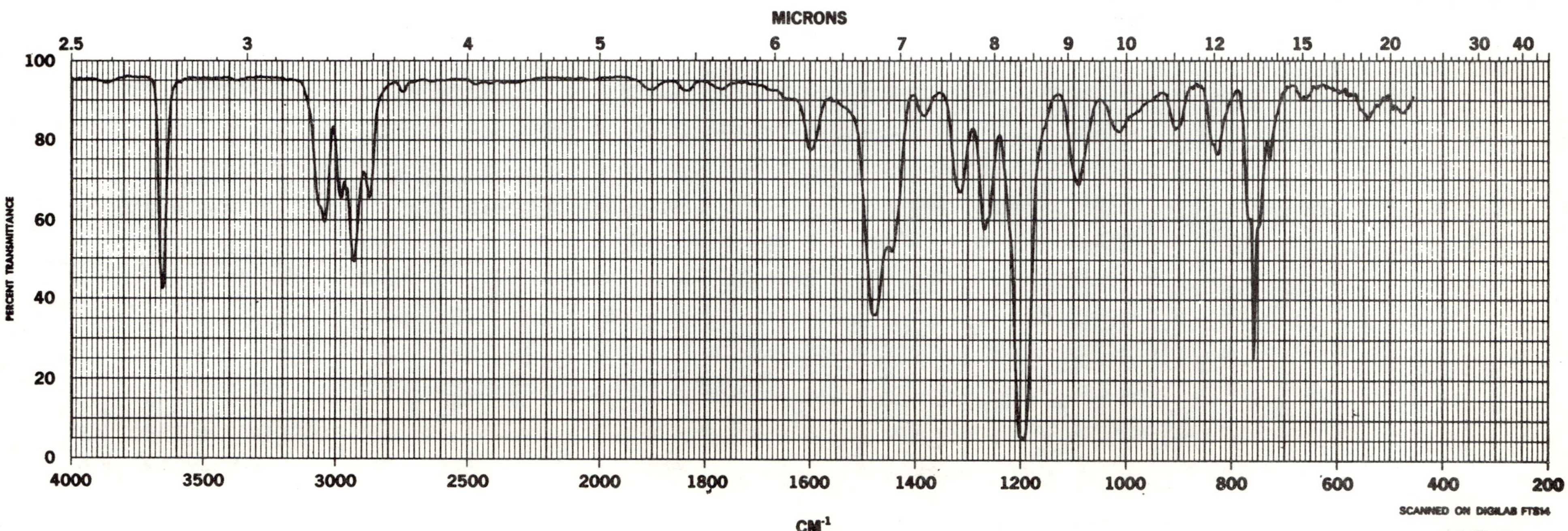

3,4–XYLENOL
3,4–DIMETHYLPHENOL
3,4–DMP

$C_8H_{10}O$

MOLECULAR WEIGHT: 122.17

SOURCE OF SAMPLE: MCB Manufacturing Chemists
Norwood, Ohio

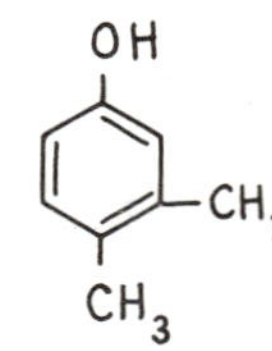

CAS REGISTRY NO.: 95-65-8

NIOSH NUMBER: ZE63000

CAPILLARY CELL: Melt

CONDENSED PHASE

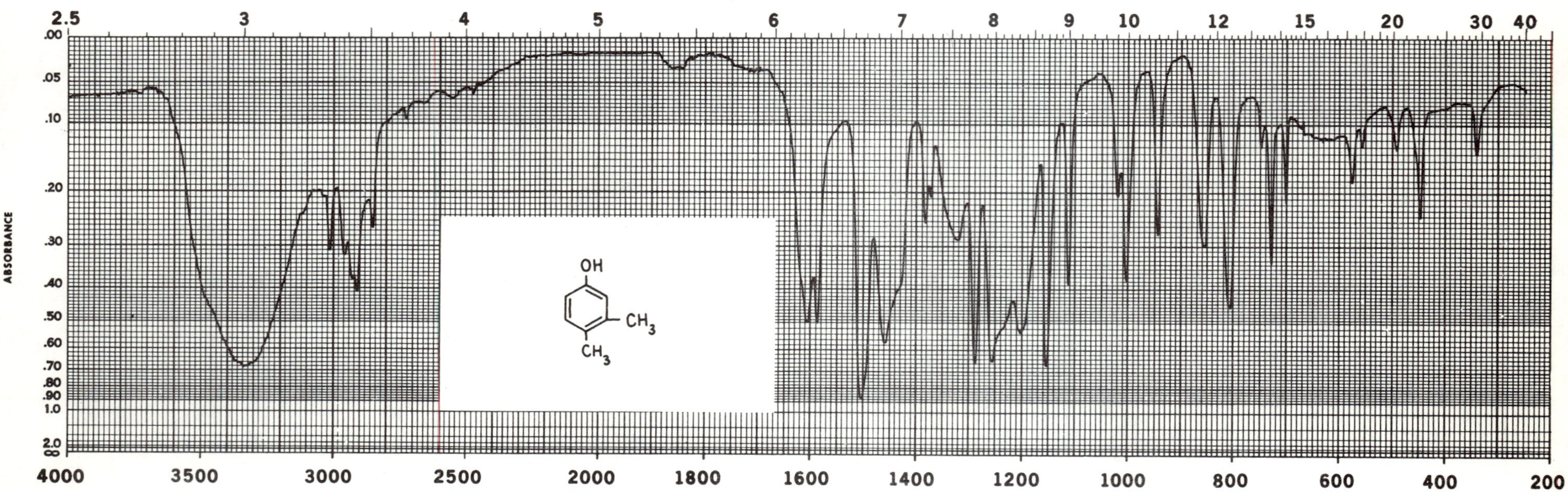

MELTING POINT:	62.5°C	INJECTION AMT.:	–
VAPORIZATION TEMP.:	280°C	PRESSURE:	Atmospheric
CELL VOLUME:	7.5 cc	MAX./MIN.:	99.7/4.2
PATH LENGTH:	7.6 cm	SCALE EXPANSION:	0.96

VAPOR PHASE

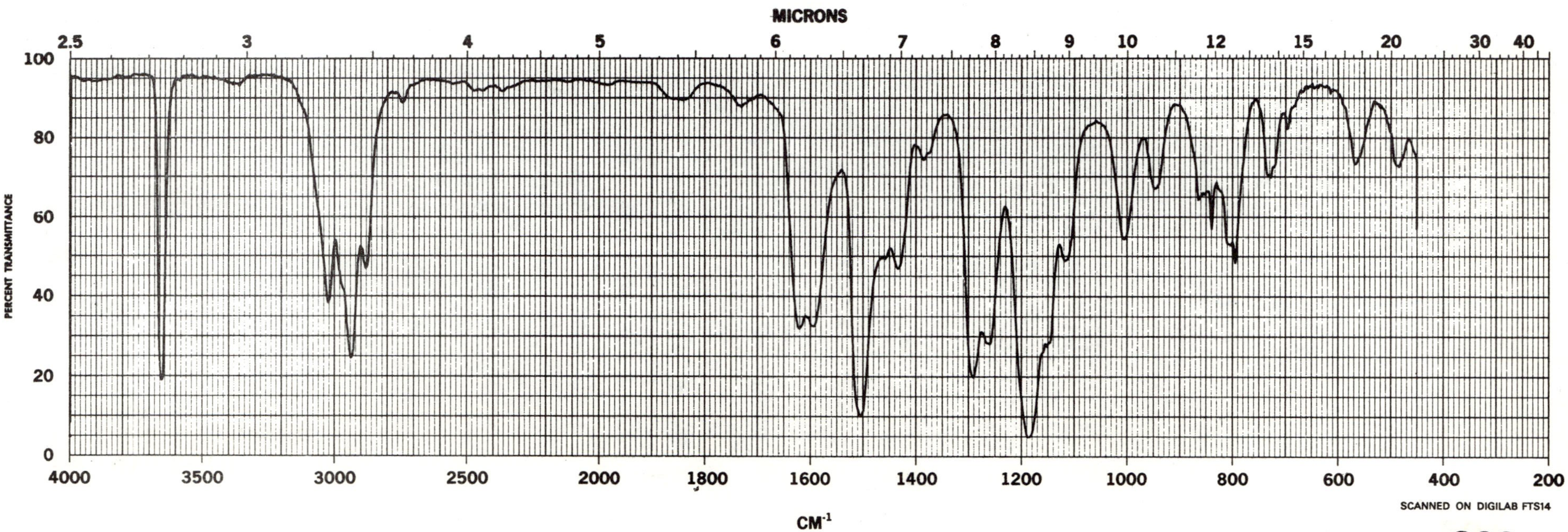

200

3,5–XYLENOL
3,5–DIMETHYLPHENOL
3,5–DMP

$C_8H_{10}O$

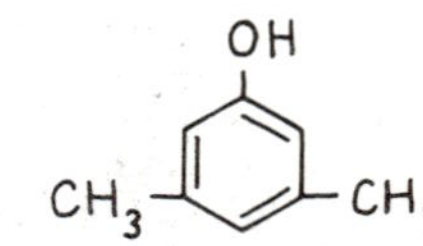

MOLECULAR WEIGHT:	122.17
SOURCE OF SAMPLE:	Aldrich Chemical Company, Inc. Milwaukee, Wisconsin
CAS REGISTRY NO.:	108-68-9
NIOSH NUMBER:	ZE64750
CAPILLARY CELL:	Melt

CONDENSED PHASE

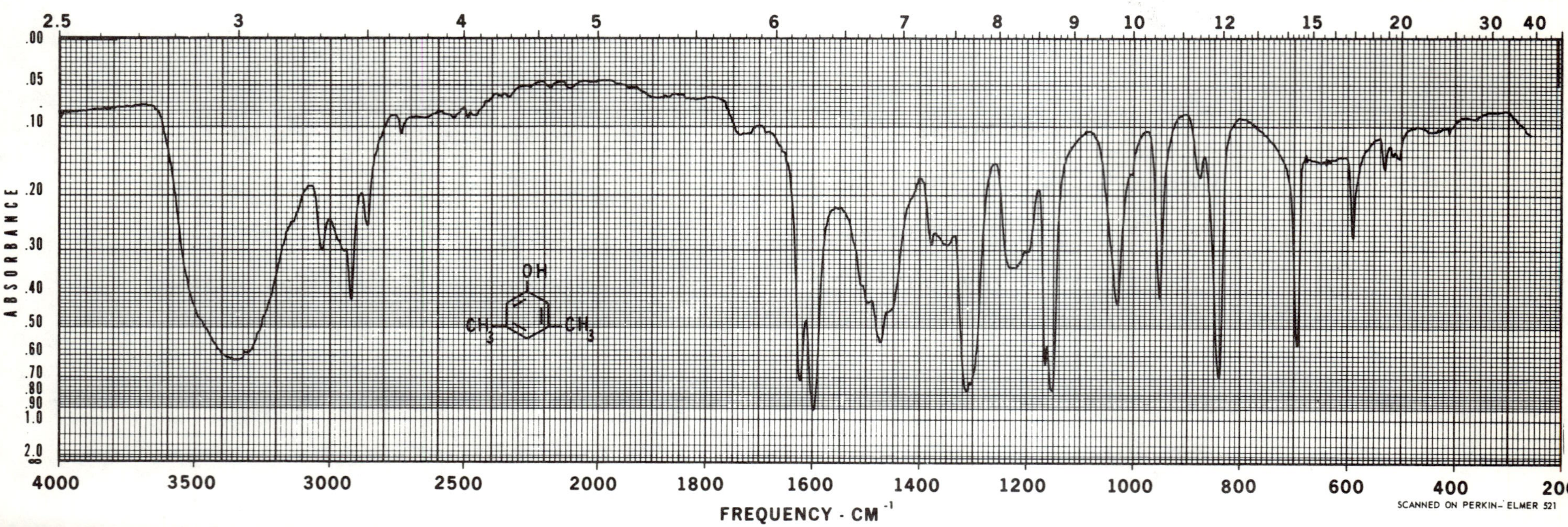

MELTING POINT:	68°C	INJECTION AMT.:	—
VAPORIZATION TEMP.:	240°C	PRESSURE:	Atmospheric
CELL VOLUME:	7.5 cc	MAX./MIN.:	96.7/14.4
PATH LENGTH:	7.6 cm	SCALE EXPANSION:	1.12

VAPOR PHASE

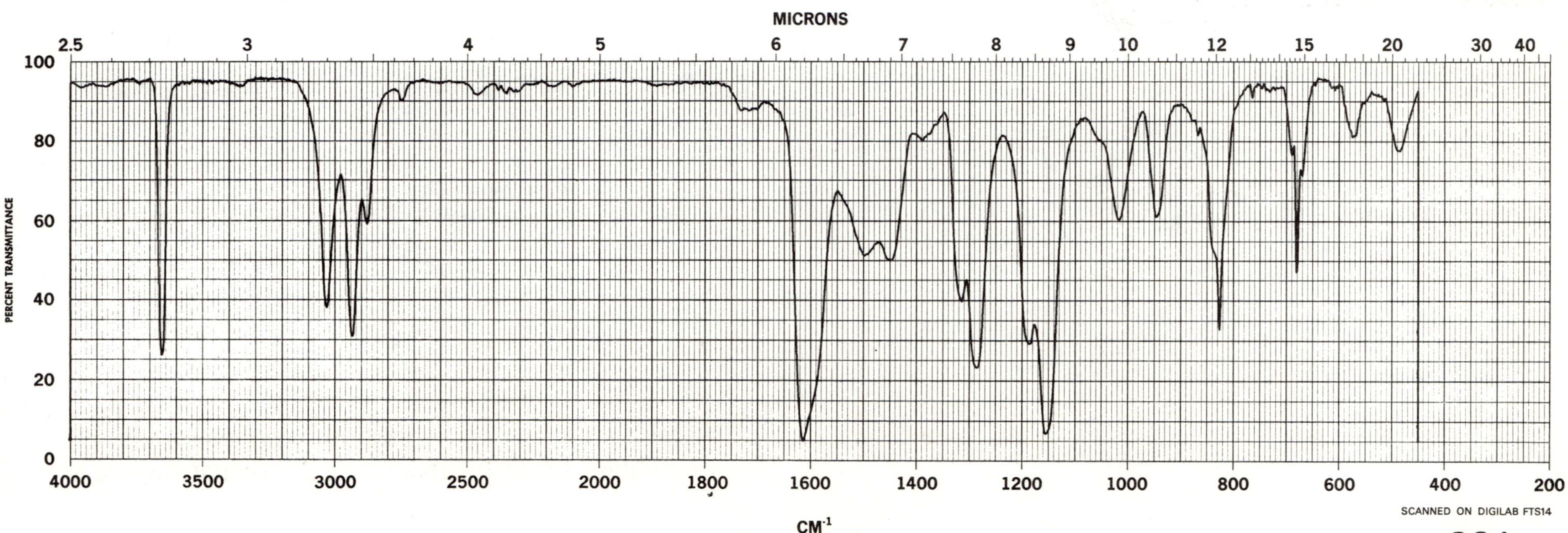

SCANNED ON DIGILAB FTS14

N,N′–FLUOREN–2,7–YLENEBISACETAMIDE
2,7–DIACETAMIDOFLUORENE
2,7–FAA

$C_{17}H_{16}N_2O_2$

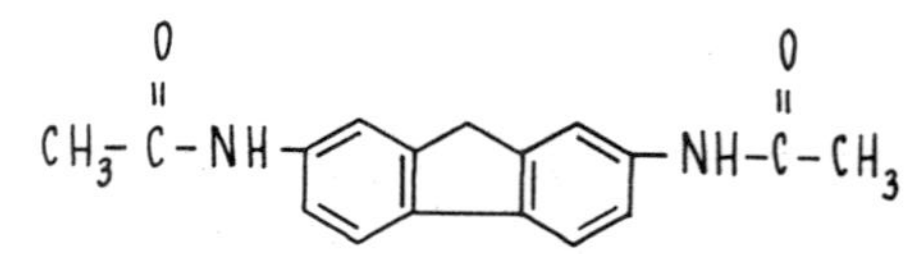

MOLECULAR WEIGHT:	280.33
SOURCE OF SAMPLE:	University of Florida Gainesville, Florida
CAS REGISTRY NO.:	304-28-9
NIOSH NUMBER:	AC07000

KBr Wafer

CONDENSED PHASE

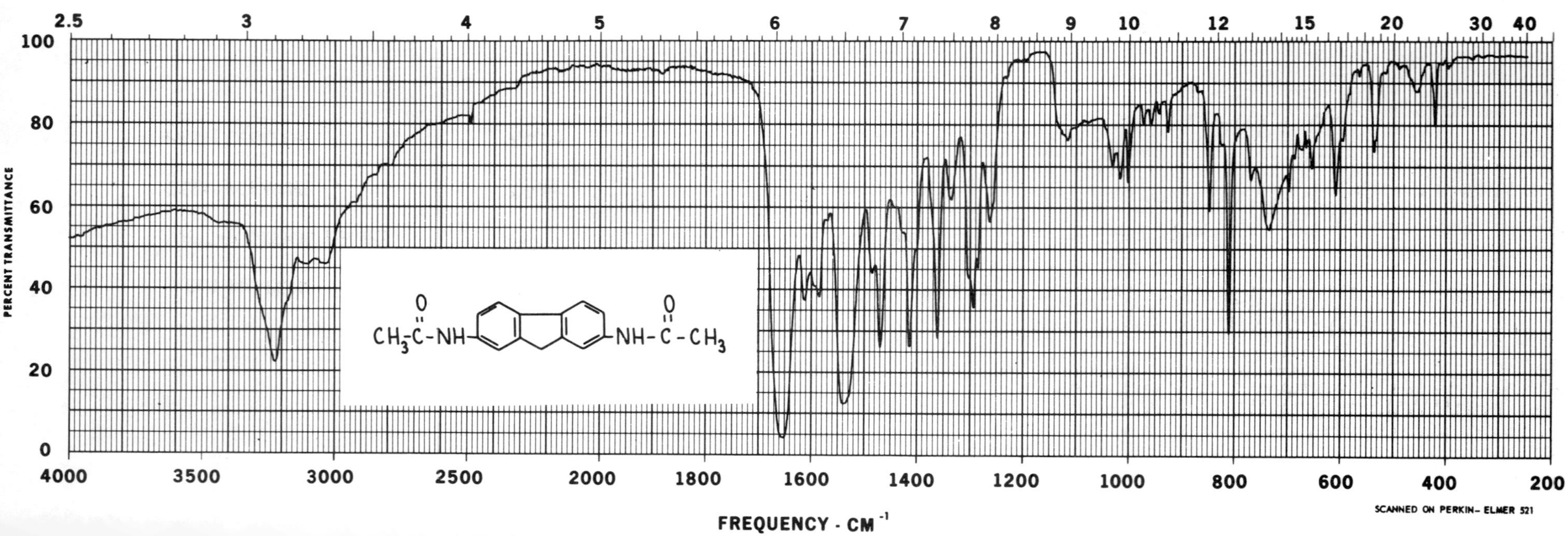

N–2–PHENANTHRYLACETAMIDE
2–ACETAMIDOPHENANTHRENE

$C_{16}H_{13}NO$

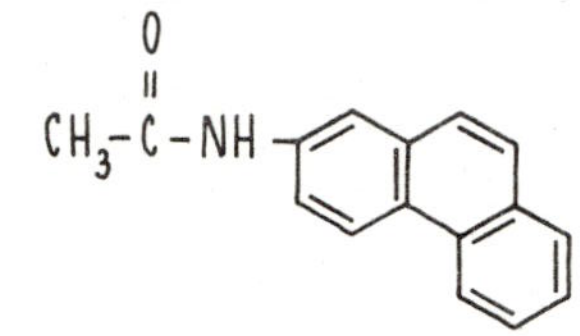

MOLECULAR WEIGHT:	235.29
SOURCE OF SAMPLE:	University of Florida Gainesville, Florida
CAS REGISTRY NO.:	4120-77-8
NIOSH NUMBER:	AC71750

KBr Wafer

CONDENSED PHASE

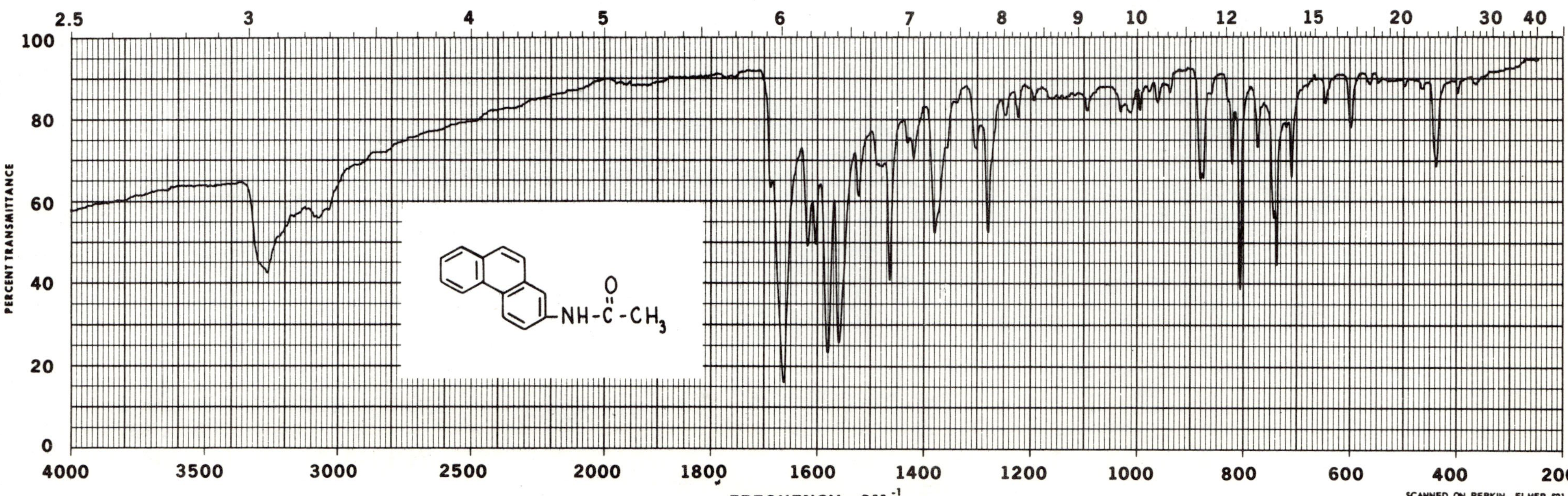

4'–PHENYLACETANILIDE
4–ACETAMIDOBIPHENYL
N–4–BIPHENYLYLACETAMIDE

$C_{14}H_{13}NO$

MOLECULAR WEIGHT: 211.27

SOURCE OF SAMPLE: S. Laham
Department of National Health & Welfare
Ottawa, Canada

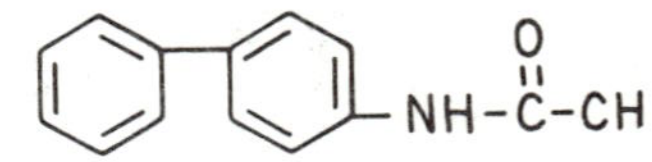

CAS REGISTRY NO.: 4075-79-0

NIOSH NUMBER: AE61250

KBr Wafer

CONDENSED PHASE

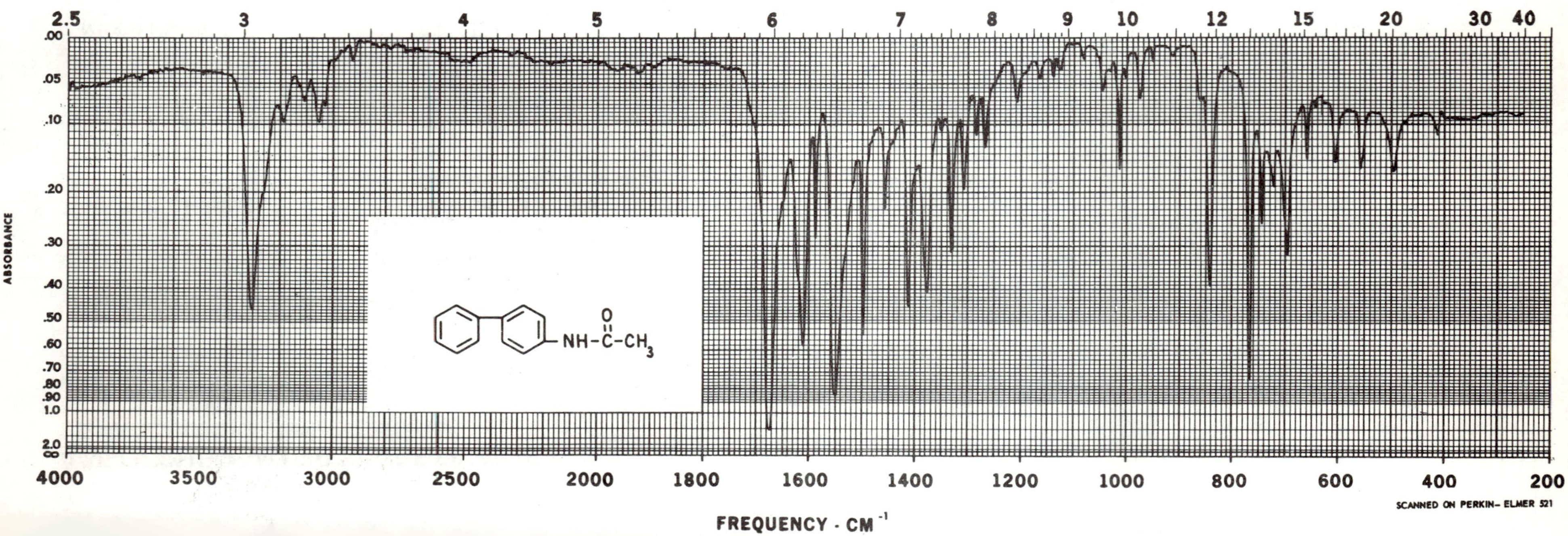

ACETIC ACID, LEAD (2+) SALT
LEAD ACETATE
LEAD (II) ACETATE
LEAD DIACETATE
SALT OF SATURN

$C_4H_6O_4Pb$

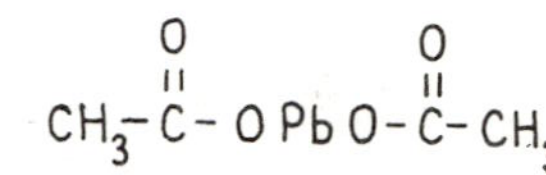

MOLECULAR WEIGHT: 325.28

SOURCE OF SAMPLE: Chem Service Company
West Chester, Pennsylvania

CAS REGISTRY NO.: 301-04-2

NIOSH NUMBER: AI52500

4000 - 1350 cm^{-1} = Mulled in Perfluorinated Hydrocarbon
1350 - 400 cm^{-1} = Mulled in Mineral Oil

CONDENSED PHASE

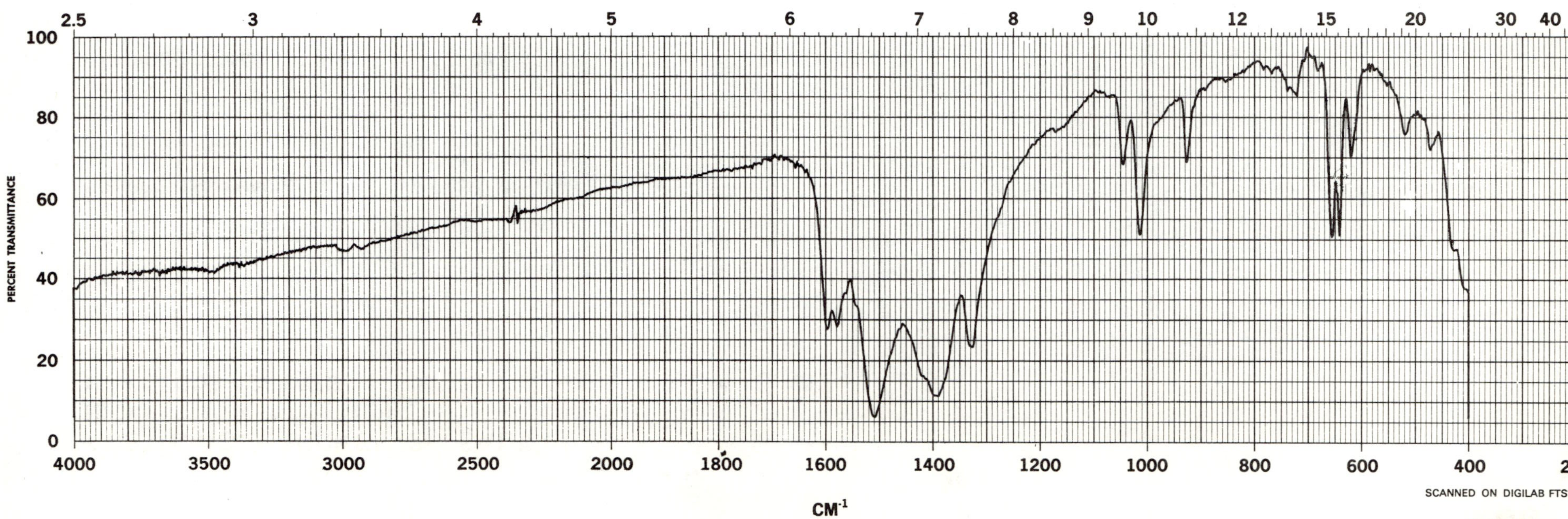

SCANNED ON DIGILAB FTS14

ACETIC ACID, NICKEL (2+) SALT
NICKEL (II) ACETATE (1:2)
NICKELOUS ACETATE

$C_4H_6NiO_4$

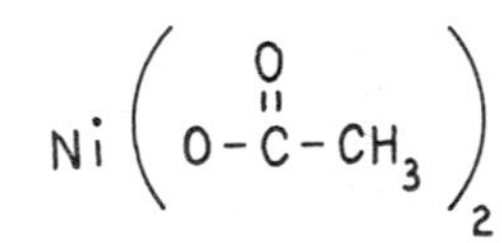

MOLECULAR WEIGHT: 176.80

SOURCE OF SAMPLE: Rocky Mountain Research, Inc.
Denver, Colorado

CAS REGISTRY NO.: 373-02-4

NIOSH NUMBER: QR61250

KBr Wafer

CONDENSED PHASE

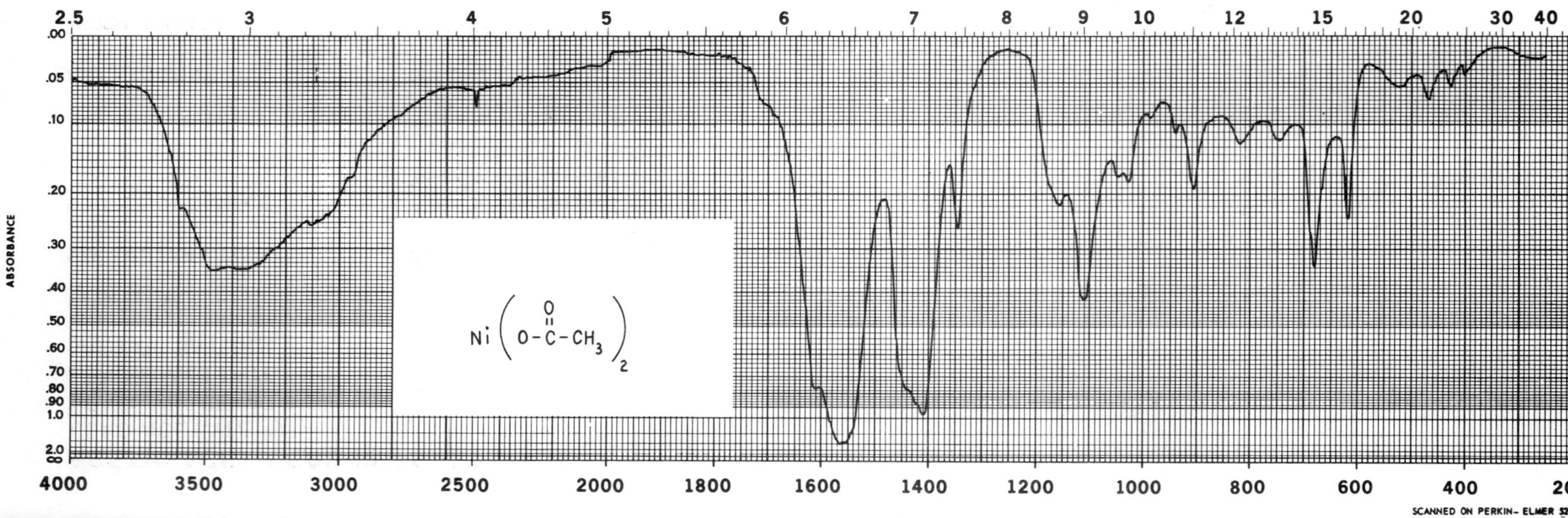

(ETHYLENEDINITRILO)TETRAACETIC ACID
ETHYLENEDIAMINETETRAACETIC ACID
EDTA
VERSENE
NULLAPON B ACID

$C_{10}H_{16}N_2O_8$

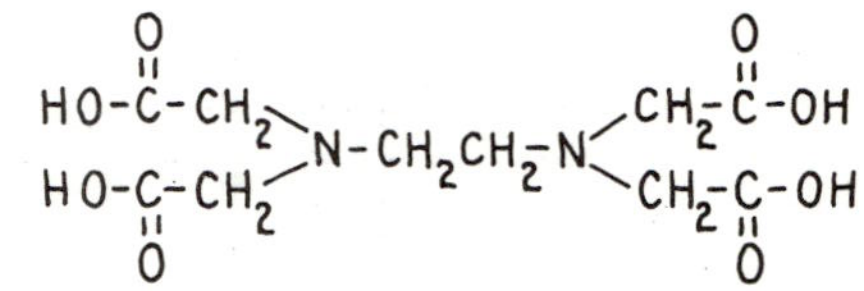

MOLECULAR WEIGHT: 292.25

SOURCE OF SAMPLE: MCB Manufacturing Chemists
Norwood, Ohio

CAS REGISTRY NO.: 60-00-4

NIOSH NUMBER: AH40250

KBr Wafer

CONDENSED PHASE

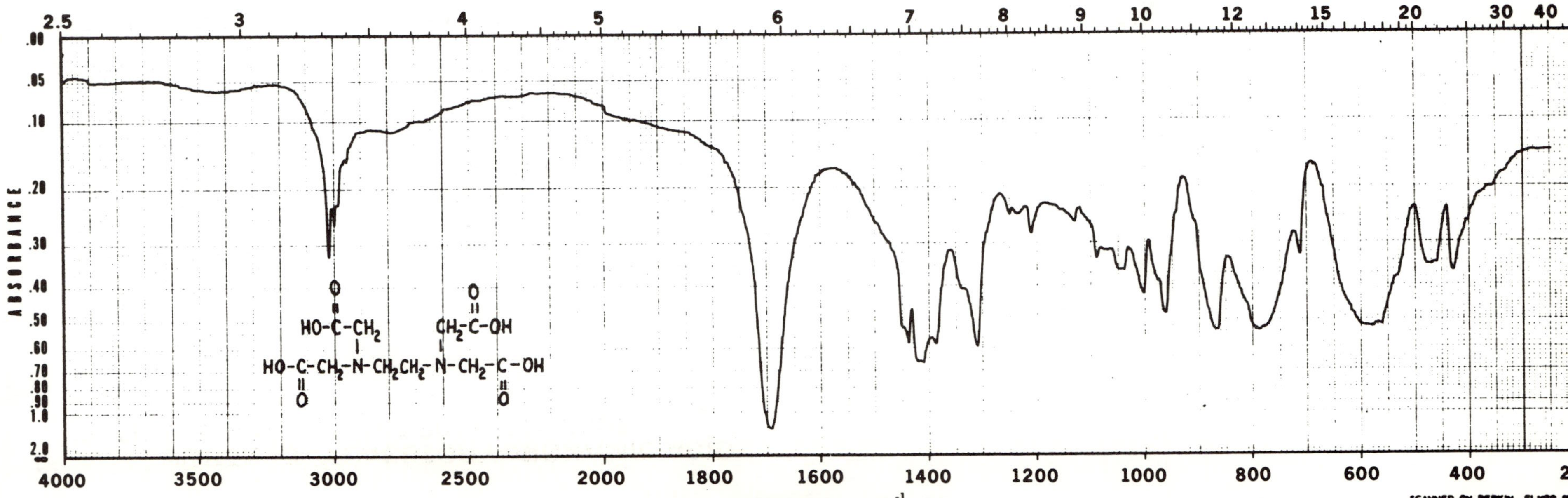

N–FLUOREN–3–YLACETOHYDROXAMIC ACID
N–FLUOREN–3–YL–N–HYDROXYACETAMIDE

$C_{15}H_{13}NO_2$

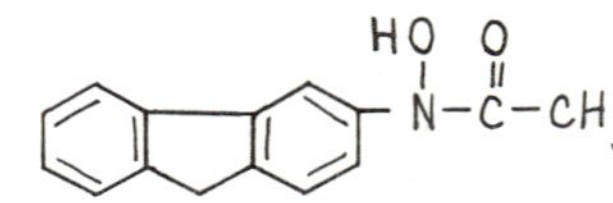

MOLECULAR WEIGHT: 239.28

SOURCE OF SAMPLE: H. Gutmann
Veterans Administration Hospital
Minneapolis, Minnesota

CAS REGISTRY NO.: 22225-32-7

NIOSH NUMBER: AK87500

KBr Wafer

CONDENSED PHASE

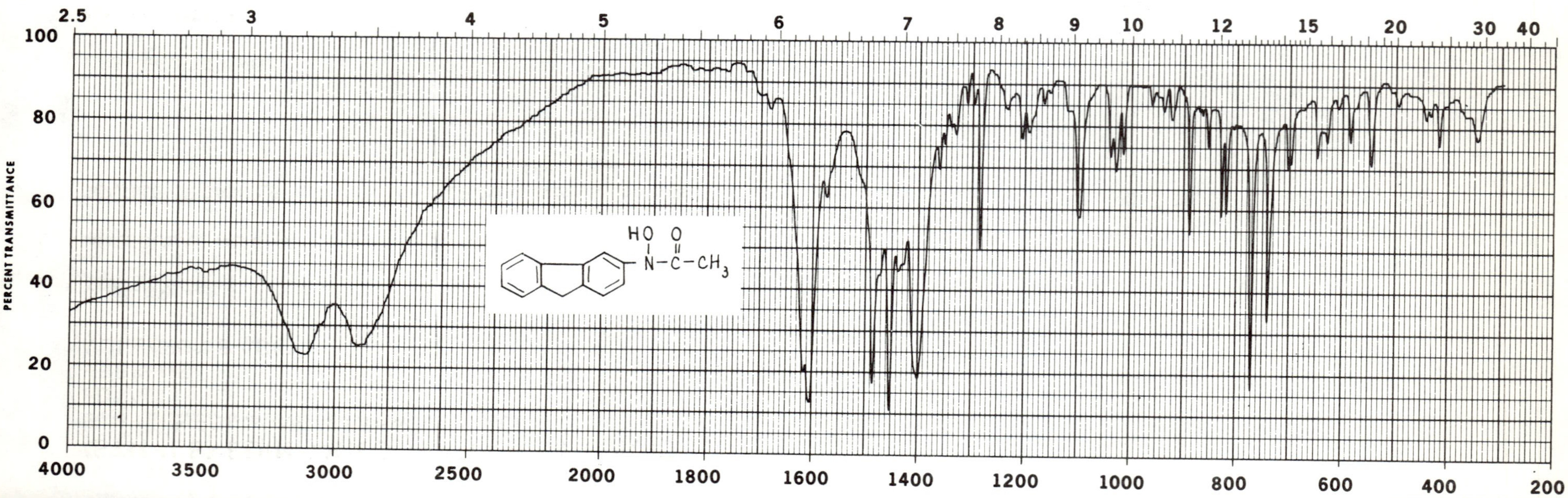

L–3– { p–[BIS(2–CHLOROETHYL)AMINO] PHENYL } ALANINE
ALKERAN
MELPHALAN
NSC–8806
L–PAM

$C_{13}H_{18}Cl_2N_2O_2$

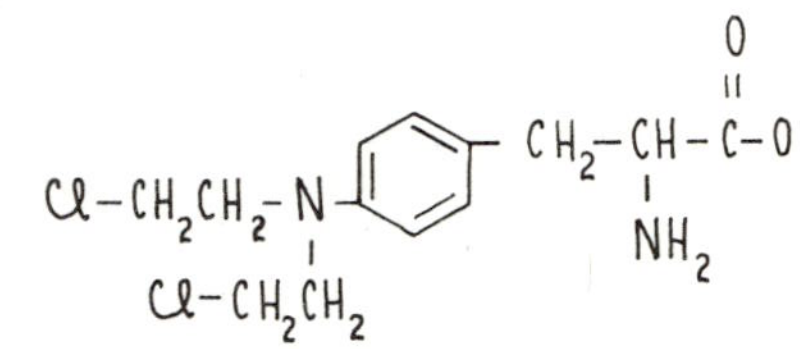

MOLECULAR WEIGHT: 305.21

SOURCE OF SAMPLE: Burroughs Wellcome and Company, Inc.
Research Triangle Park, North Carolina

CAS REGISTRY NO.: 148-82-3

NIOSH NUMBER: AY36750

KBr Wafer

CONDENSED PHASE

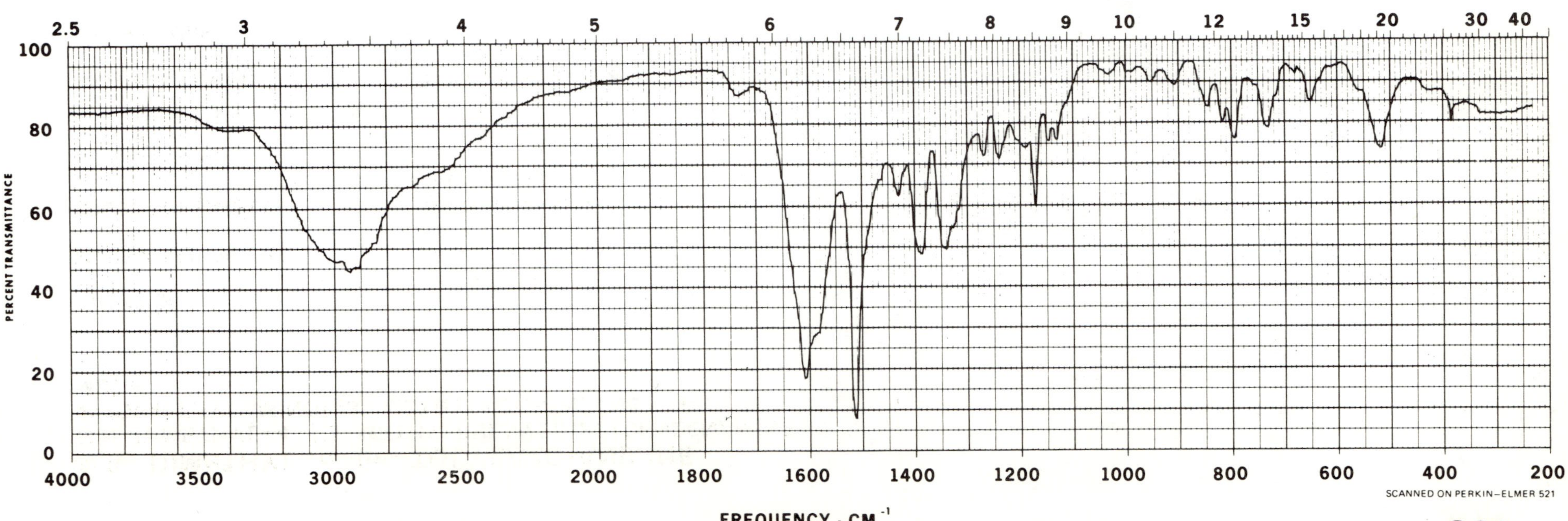

SCANNED ON PERKIN–ELMER 521

N,N–DIMETHYL–p–(m–TOLYLAZO)ANILINE
N,N–DIMETHYL–4–[(3–METHYLPHENYL)AZO] BENZENAMINE
4–(DIMETHYLAMINO)–3'–METHYLAZOBENZENE
MDAB

$C_{15}H_{17}N_3$

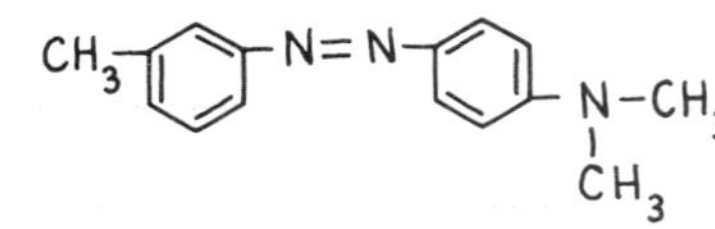

MOLECULAR WEIGHT: 239.32

SOURCE OF SAMPLE: University of Wisconsin
Madison, Wisconsin

CAS REGISTRY NO.: 55-80-1

NIOSH NUMBER: –

KBr Wafer

CONDENSED PHASE

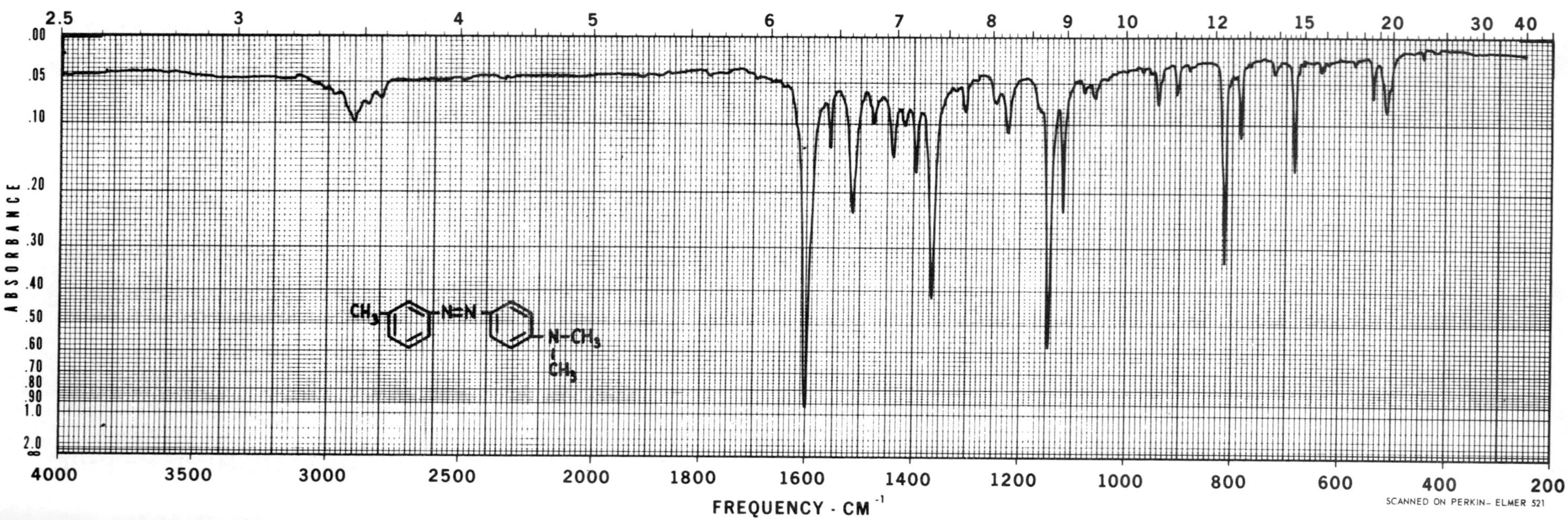

N–METHYL–p–(PHENYLAZO)ANILINE
N–METHYL–4–AMINOAZOBENZENE

$C_{13}H_{13}N_3$

MOLECULAR WEIGHT: 211.27

SOURCE OF SAMPLE: University of Wisconsin
Milwaukee, Wisconsin

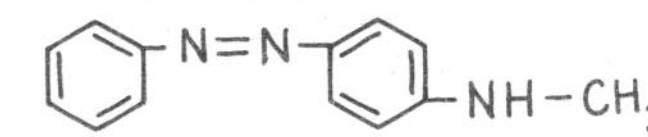

CAS REGISTRY NO.: 621-90-9

NIOSH NUMBER: BY59500

Film

CONDENSED PHASE

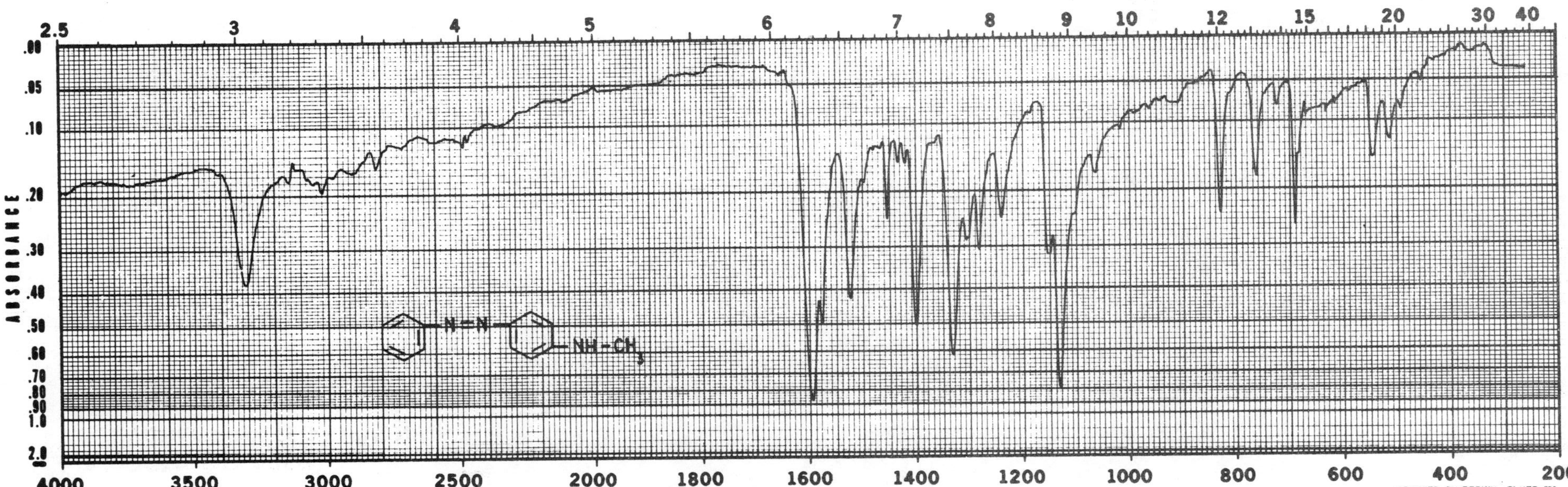

AROCHLOR 1260
POLYCHLORINATED BIPHENYL
PCB–1260

–

MOLECULAR WEIGHT:	–
SOURCE OF SAMPLE:	Monsanto Company St. Louis, Missouri
CAS REGISTRY NO.:	11096-82-5
NIOSH NUMBER:	TQ13620

Film

* Refer to Appendix for GC/FT-IR Analysis

CONDENSED PHASE

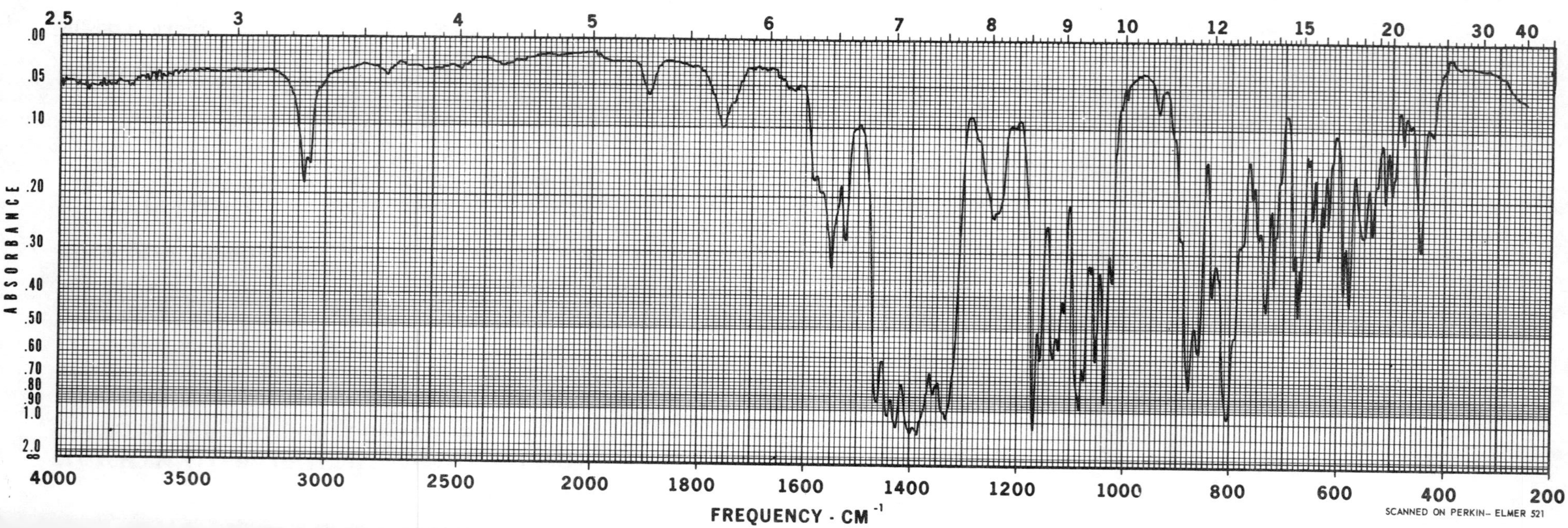

ARSENIC ACID, CALCIUM SALT (2:3)
CALCIUM ARSENATE
CHIP–CAL
PENCAL
TRICALCIUM ARSENATE

$As_2Ca_3O_8$

$Ca_3(AsO_4)_2$

MOLECULAR WEIGHT:	398.08
SOURCE OF SAMPLE:	Alfa Division of the Ventron Corporation Danvers, Massachusetts
CAS REGISTRY NO.:	10103-62-5
NIOSH NUMBER:	CG08300

4000 - 1350 cm^{-1} = Mulled in Perfluorinated Hydrocarbon
1350 - 400 cm^{-1} = Mulled in Mineral Oil

CONDENSED PHASE

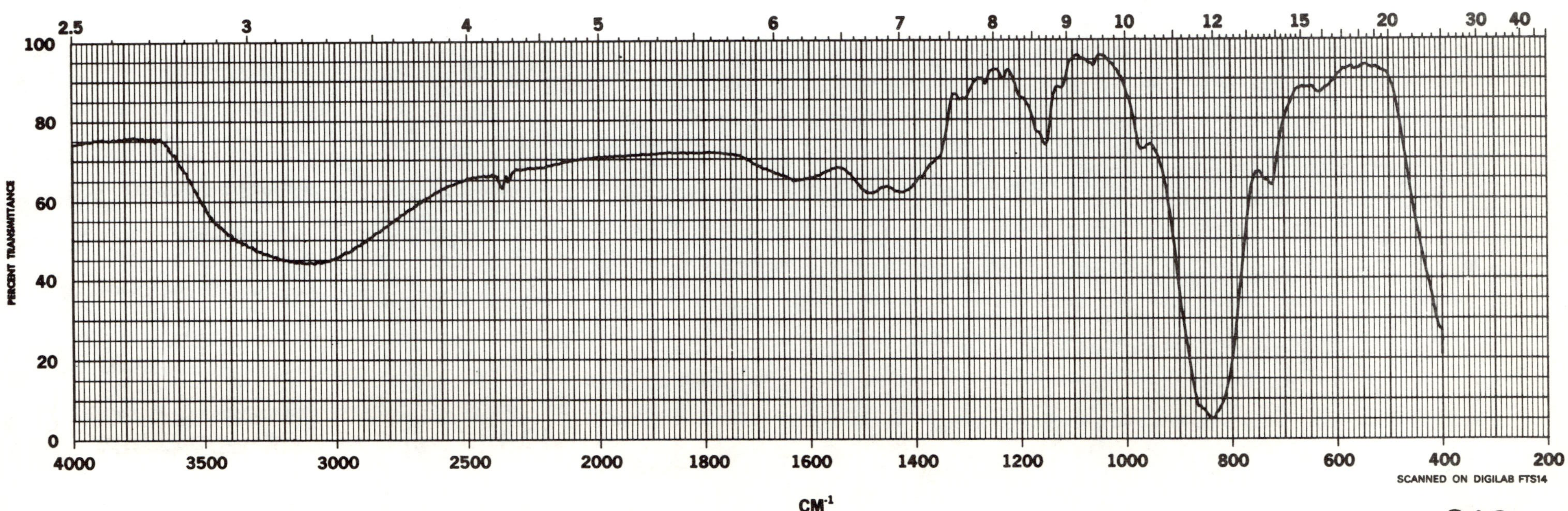

ARSENIC ACID, LEAD (2+) SALT (1:1)
LEAD ARSENATE
SCHULTENITE
GYPSINE
SOPRABEL

$HAsO_4Pb$

MOLECULAR WEIGHT:	347.12
SOURCE OF SAMPLE:	Chem Service Company West Chester, Pennsylvania
CAS REGISTRY NO.:	7784-40-9
NIOSH NUMBER:	CG09800

$PbHAsO_4$

4000 - 1350 cm^{-1} = Mulled in Perfluorinated Hydrocarbon
1350 - 400 cm^{-1} = Mulled in Mineral Oil

CONDENSED PHASE

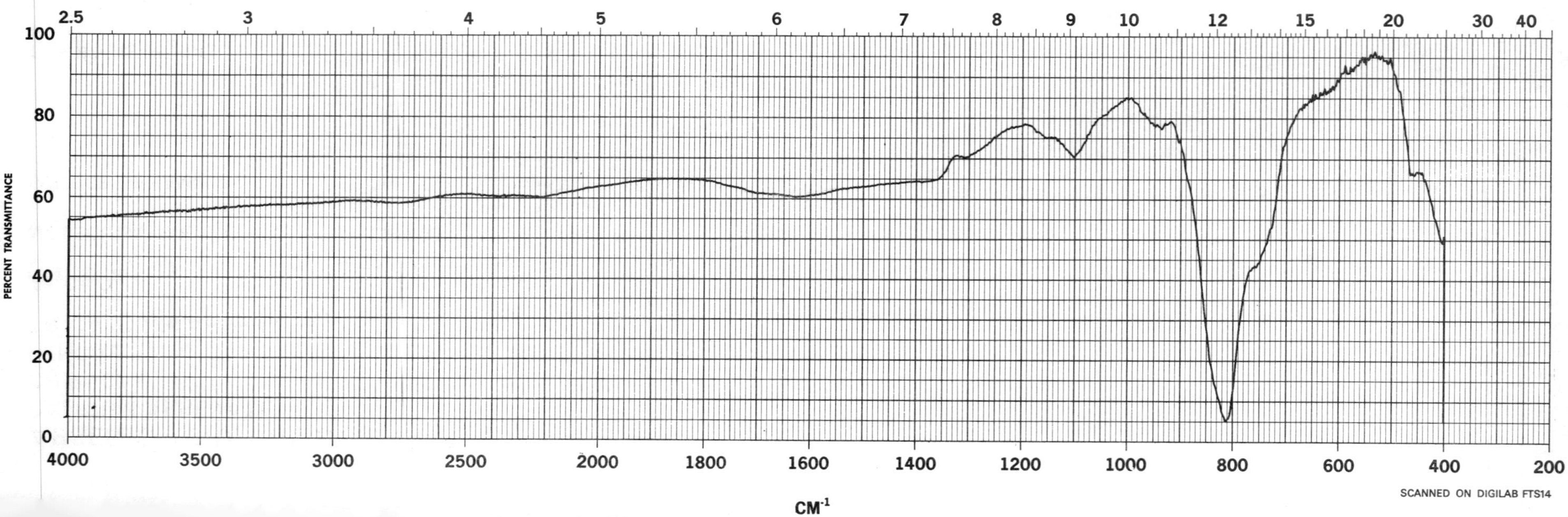

ARSENIC ACID, MONOPOTASSIUM SALT
POTASSIUM ACID ARSENATE
POTASSIUM ARSENATE
POTASSIUM DIHYDROGEN ARSENATE
MACQUER'S SALT

H_2AsKO_4

MOLECULAR WEIGHT: 180.04

SOURCE OF SAMPLE: MCB Manufacturing Chemists
Norwood, Ohio

KH_2AsO_4

CAS REGISTRY NO.: 7784-41-0

NIOSH NUMBER: CG11000

KBr Wafer

CONDENSED PHASE

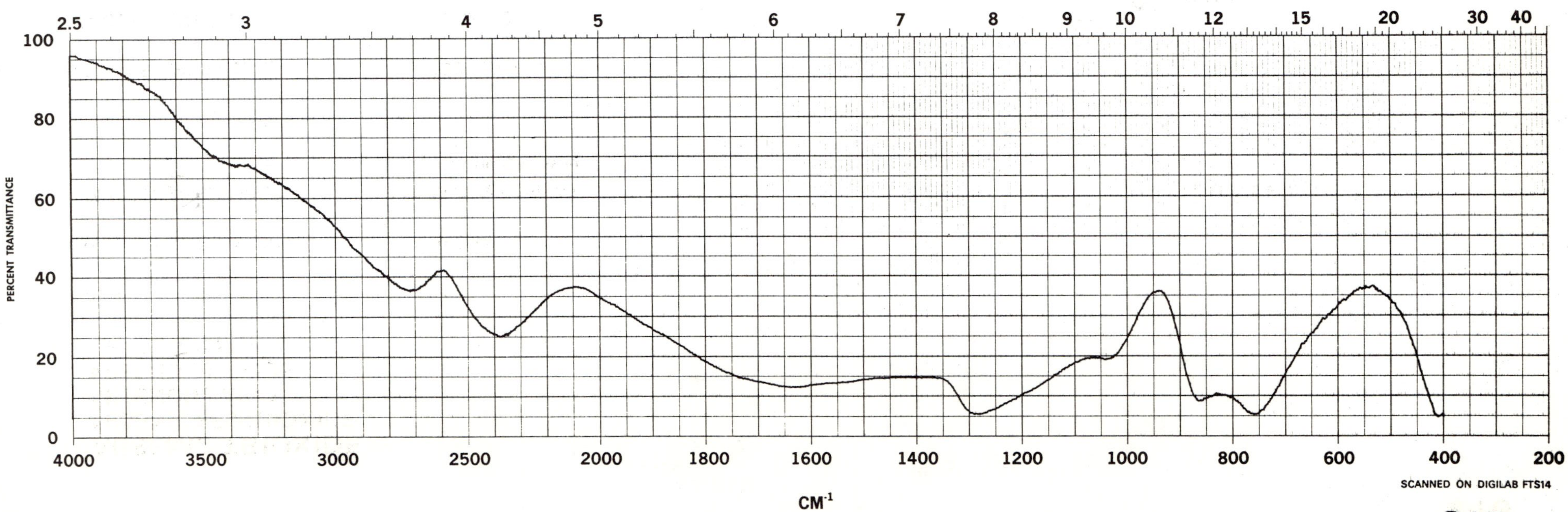

SCANNED ON DIGILAB FTS14

ARSENIC TRIOXIDE
ARSENIC SESQUIOXIDE
ARSENOUS ACID
ARSENOUS OXIDE
WHITE ARSENIC

As_2O_3

MOLECULAR WEIGHT: 197.84

SOURCE OF SAMPLE: Mallinckrodt, Inc.
St. Louis, Missouri

As_2O_3

CAS REGISTRY NO.: 1327-53-3

NIOSH NUMBER: CG33250

KBr Wafer

CONDENSED PHASE

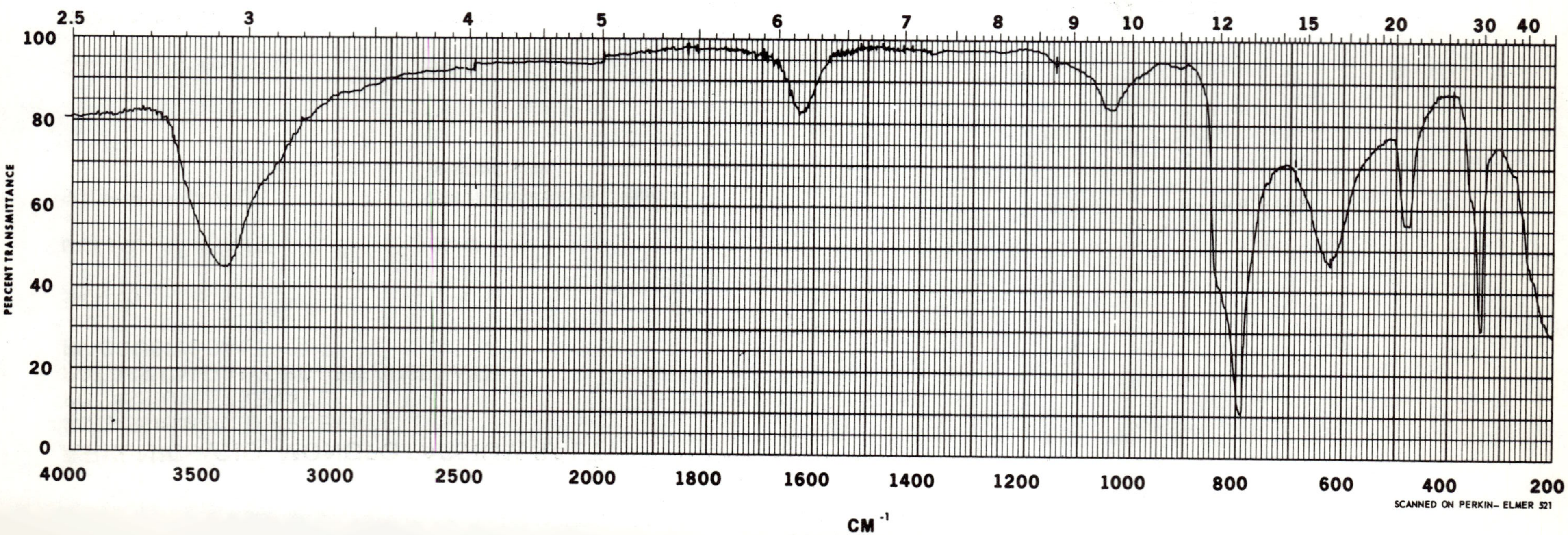

ASBESTOS
AMIANTHUS
AMOSITE
NATIVE CALCIUM–MAGNESIUM SILICATE
ACTINOLITE

–

MOLECULAR WEIGHT: –

SOURCE OF SAMPLE: Asbestos Corporation Ltd.
Quebec, Canada

CAS REGISTRY NO.: 1332-21-4

NIOSH NUMBER: CI64750

KBr Wafer

CONDENSED PHASE

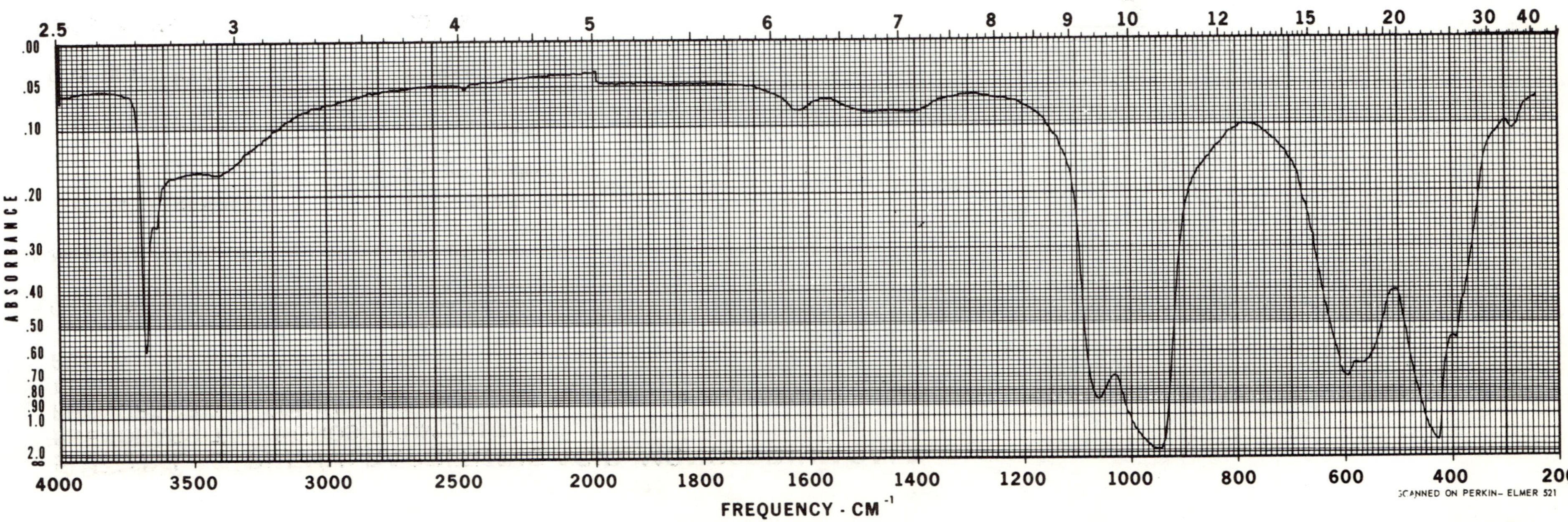

5–ETHYL–5–PHENYLBARBITURIC ACID, MONOSODIUM SALT
PHENOBARBITAL, MONOSODIUM SALT
PHENOBARBITAL SODIUM
LUMINAL SODIUM
PBS

$C_{12}H_{11}N_2NaO_3$

MOLECULAR WEIGHT: 254.22

SOURCE OF SAMPLE: E. Merck AG
Darmstadt, Germany

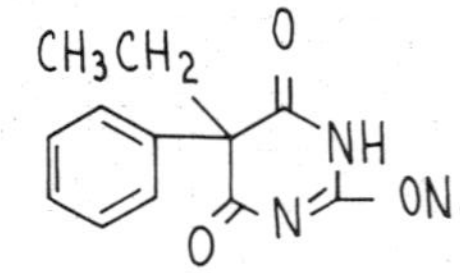

CAS REGISTRY NO.: 57-30-7

NIOSH NUMBER: CQ70000

KBr Wafer

CONDENSED PHASE

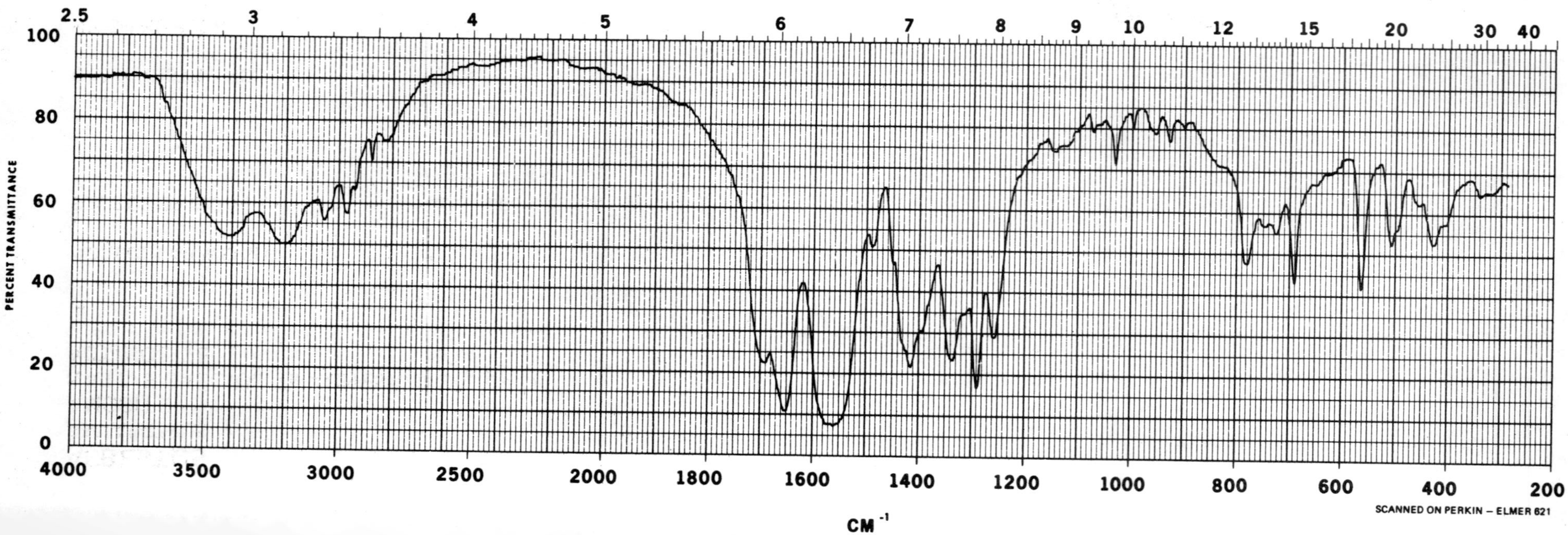

BENZ[e] ACEPHENANTHRYLENE
3,4–BENZOFLUORANTHENE
BENZO(b)FLUORANTHENE

$C_{20}H_{12}$

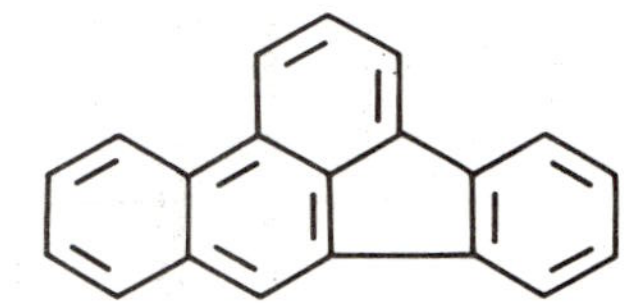

MOLECULAR WEIGHT: 252.32

SOURCE OF SAMPLE: Aldrich Chemical Company, Inc.
Milwaukee, Wisconsin

CAS REGISTRY NO.: 205-99-2

NIOSH NUMBER: CU14000

KBr Wafer

CONDENSED PHASE

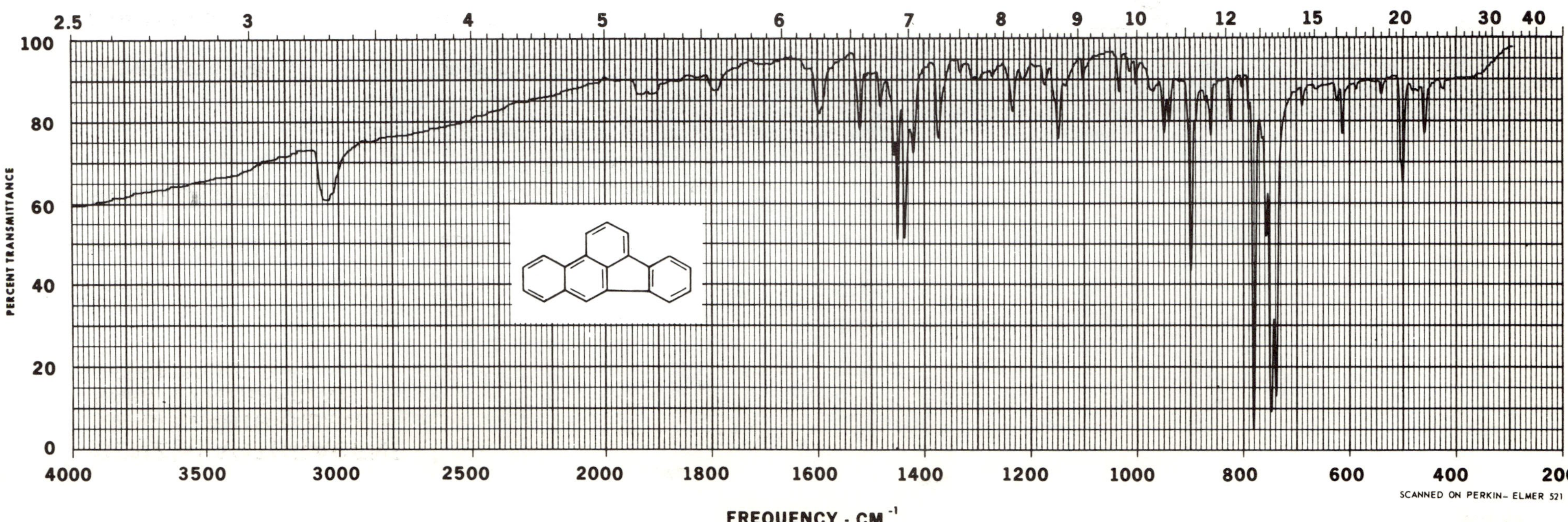

BENZ[a] ANTHRACENE
1,2–BENZANTHRACENE
NAPHTHANTHRACENE
TETRAPHENE

$C_{18}H_{12}$

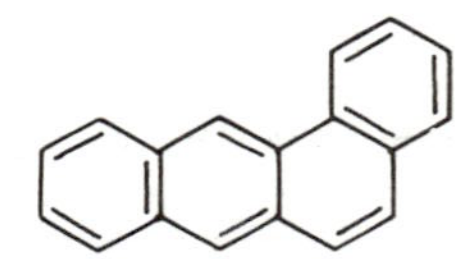

MOLECULAR WEIGHT: 228.30

SOURCE OF SAMPLE: Aldrich Chemical Company, Inc.
Milwaukee, Wisconsin

CAS REGISTRY NO.: 56-55-3

NIOSH NUMBER: CV92750

KBr Wafer

CONDENSED PHASE

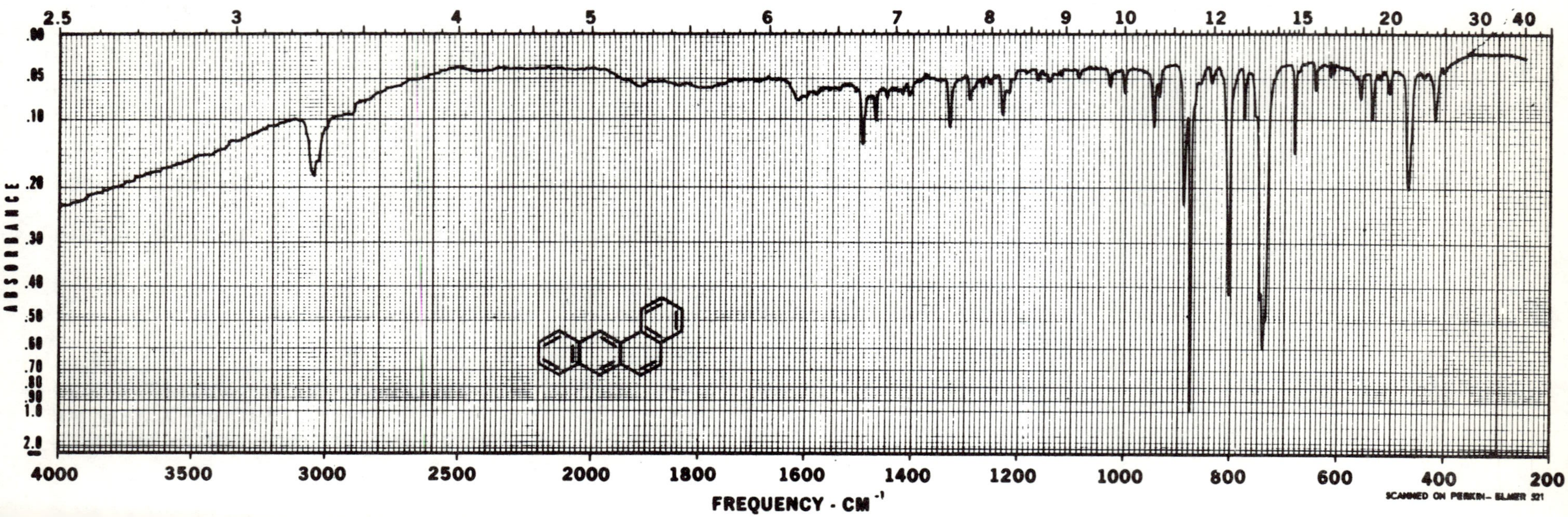

7,12–DIMETHYLBENZ[a] ANTHRACENE
DBA
9,10–DIMETHYL–1,2–BENZANTHRACENE
DMBA

$C_{20}H_{16}$

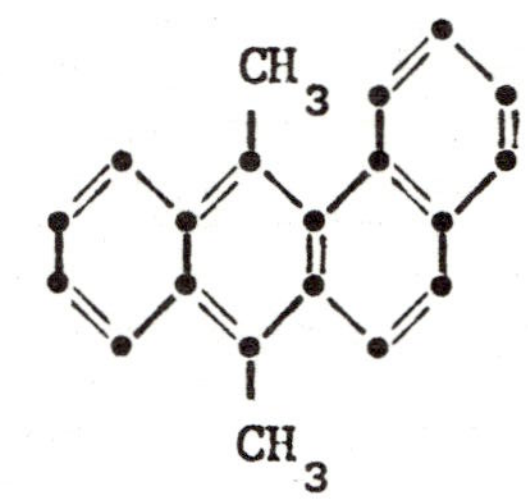

MOLECULAR WEIGHT: 256.35

SOURCE OF SAMPLE: Tokyo Kasei Kogyo Company Ltd.
Tokyo, Japan

CAS REGISTRY NO.: 57-97-6

NIOSH NUMBER: CW38500

KBr Wafer

CONDENSED PHASE

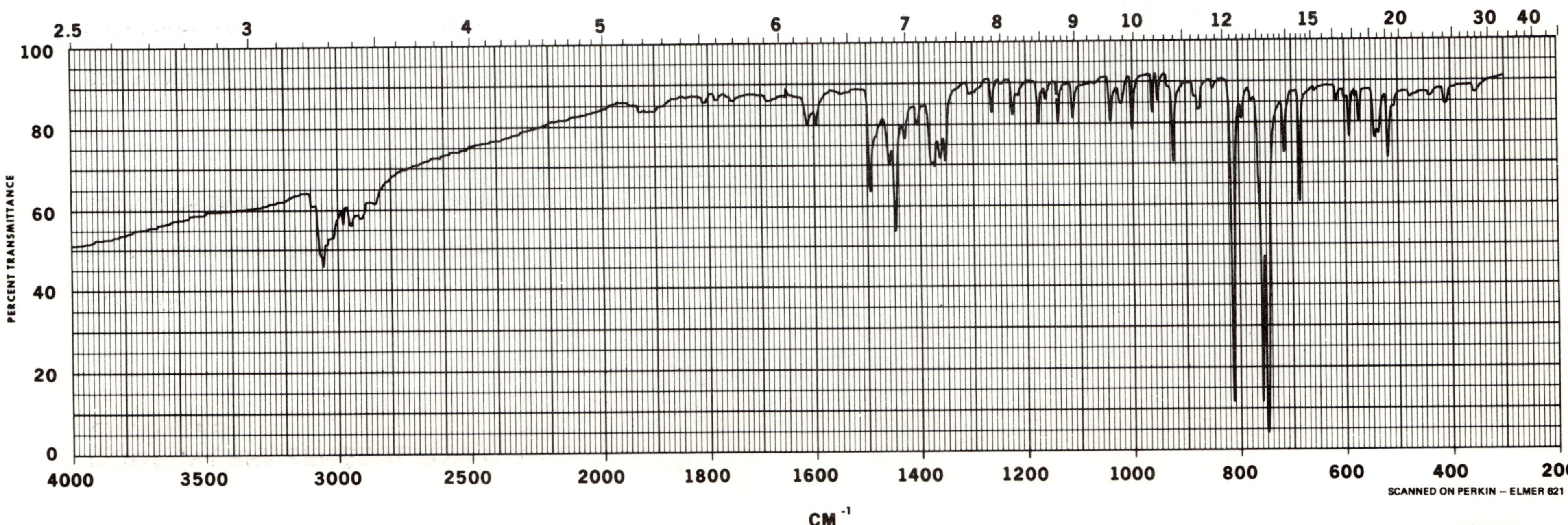

DODECYLBENZENESULFONIC ACID
LAKEWAY SA

$C_{18}H_{30}O_{3}S$

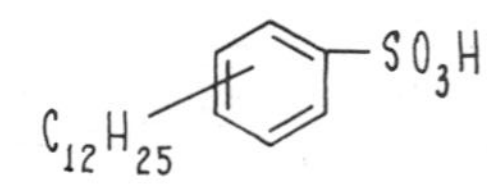

MOLECULAR WEIGHT: 326.50

SOURCE OF SAMPLE: Tanatex Chemical Company
Division of Sybron Corporation
Lyndhurst, New Jersey

CAS REGISTRY NO.: 27176-87-0

NIOSH NUMBER: DB64750

Film

CONDENSED PHASE

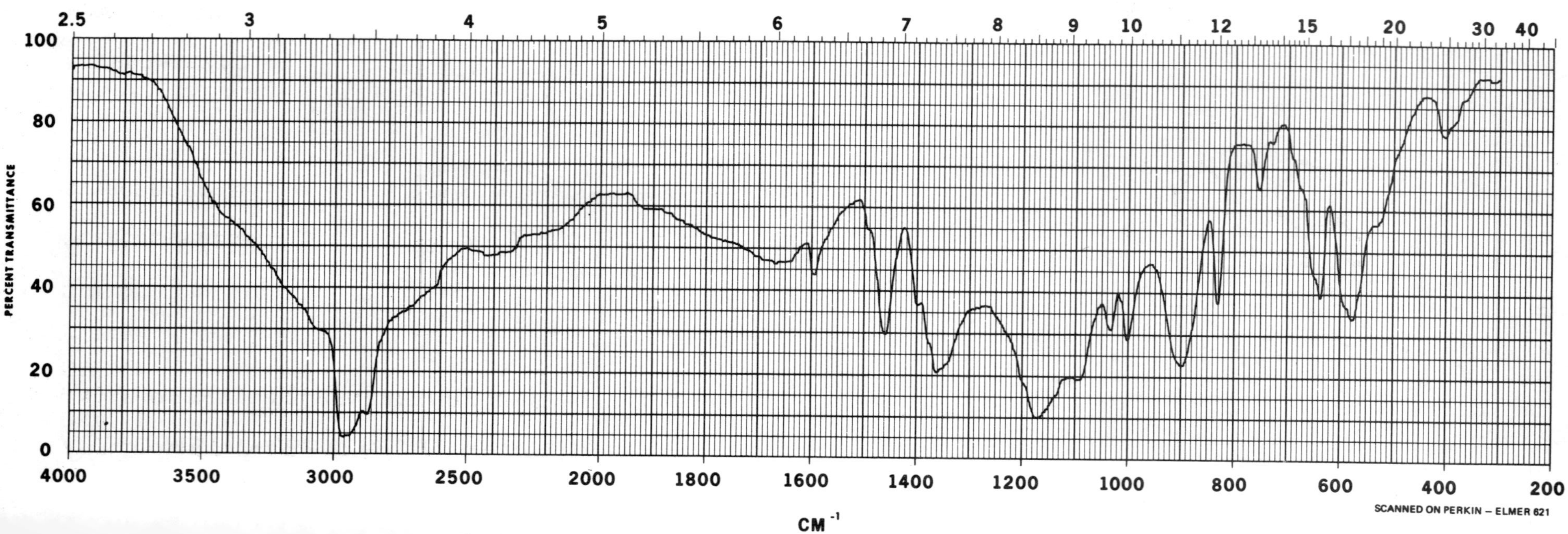

DODECYLBENZENESULFONIC ACID, CALCIUM SALT (2:1)

$C_{36}H_{58}CaO_6S_2$

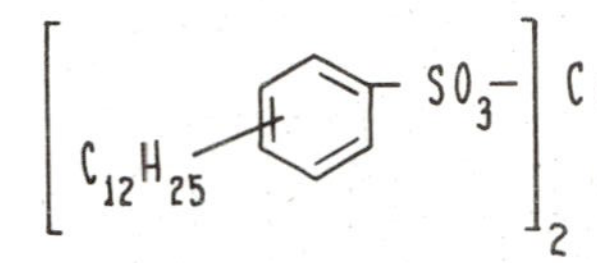

MOLECULAR WEIGHT: 691.07

SOURCE OF SAMPLE: ICI Organics, Inc.
Stamford, Connecticut

CAS REGISTRY NO.: –

NIOSH NUMBER: –

Film

CONDENSED PHASE

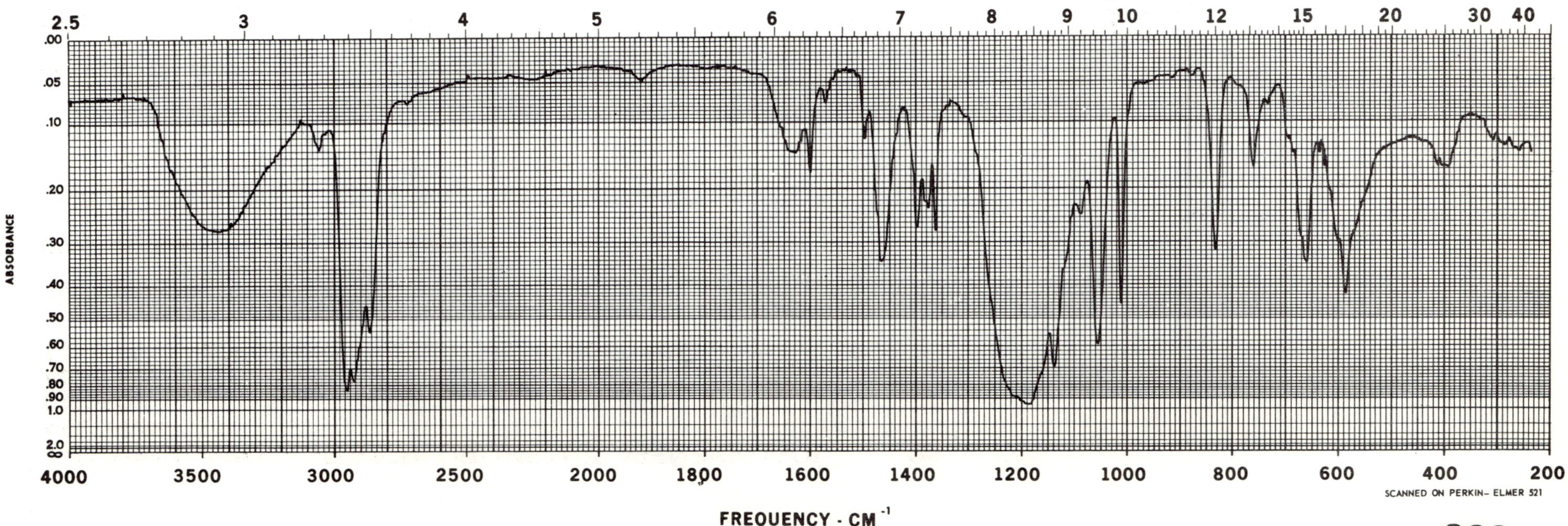

4,4′–DICHLORO–α–(TRICHLOROMETHYL)BENZHYDROL
KELTHANE
DICOFOL
ACARIN
1,1–BIS(p–CHLOROPHENYL)–2,2,2–TRICHLOROETHANOL

$C_{14}H_9Cl_5O$

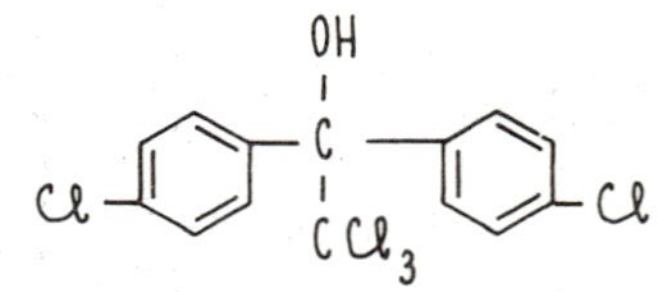

MOLECULAR WEIGHT:	370.49
SOURCE OF SAMPLE:	Industria Prodotti Chimici S.p.A. Italy
CAS REGISTRY NO.:	115-32-2
NIOSH NUMBER:	DC84000
CAPILLARY CELL:	Melt

CONDENSED PHASE

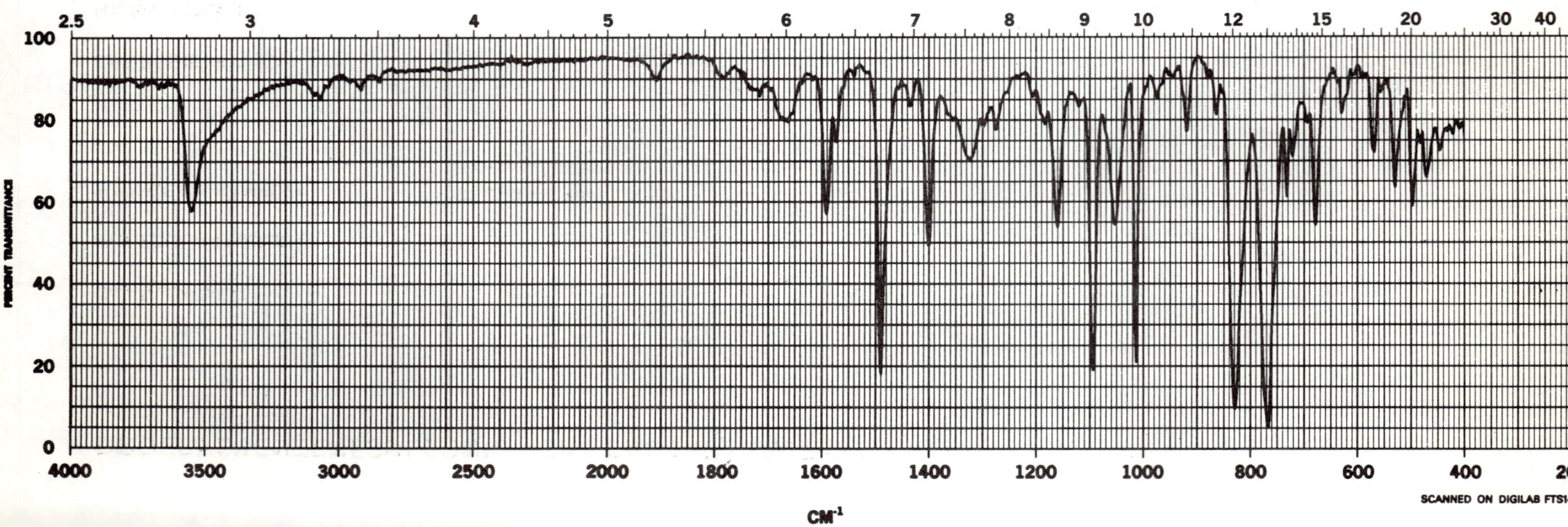

BENZIDINE, SULFATE (1:1)
4,4'–DIAMINOBIPHENYL, SULFATE (1:1)
4,4'–BIPHENYLDIAMINE, SULFATE (1:1)

$C_{12}H_{12}N_2 \cdot H_2SO_4$

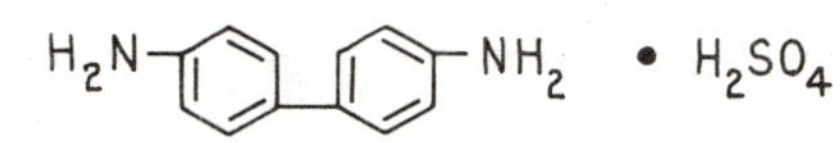

MOLECULAR WEIGHT: 282.32

SOURCE OF SAMPLE: Fluka AG
Buchs, Switzerland

CAS REGISTRY NO.: 531-86-2

NIOSH NUMBER: DD15750

KBr Wafer

CONDENSED PHASE

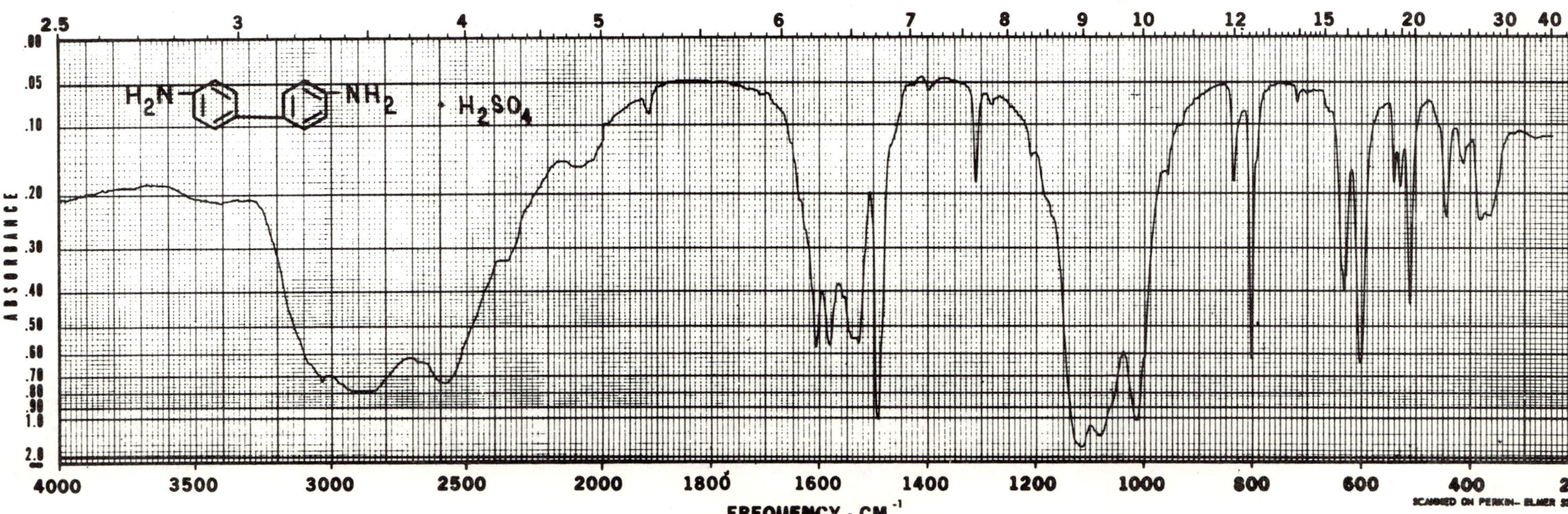

3,3'–DIMETHOXYBENZIDINE, DIHYDROCHLORIDE
4,4'–BI–o–ANISIDINE, DIHYDROCHLORIDE
C. I. DISPERSE BLACK 6, DIHYDROCHLORIDE

$C_{14}H_{16}N_2O_2 \cdot 2HCl$

MOLECULAR WEIGHT: 317.22

SOURCE OF SAMPLE: Eastman Organic Chemicals
Rochester, New York

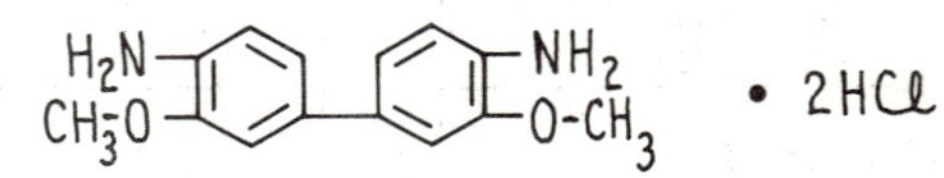

CAS REGISTRY NO.: 20325-40-4

NIOSH NUMBER: DD10500

KBr Wafer

CONDENSED PHASE

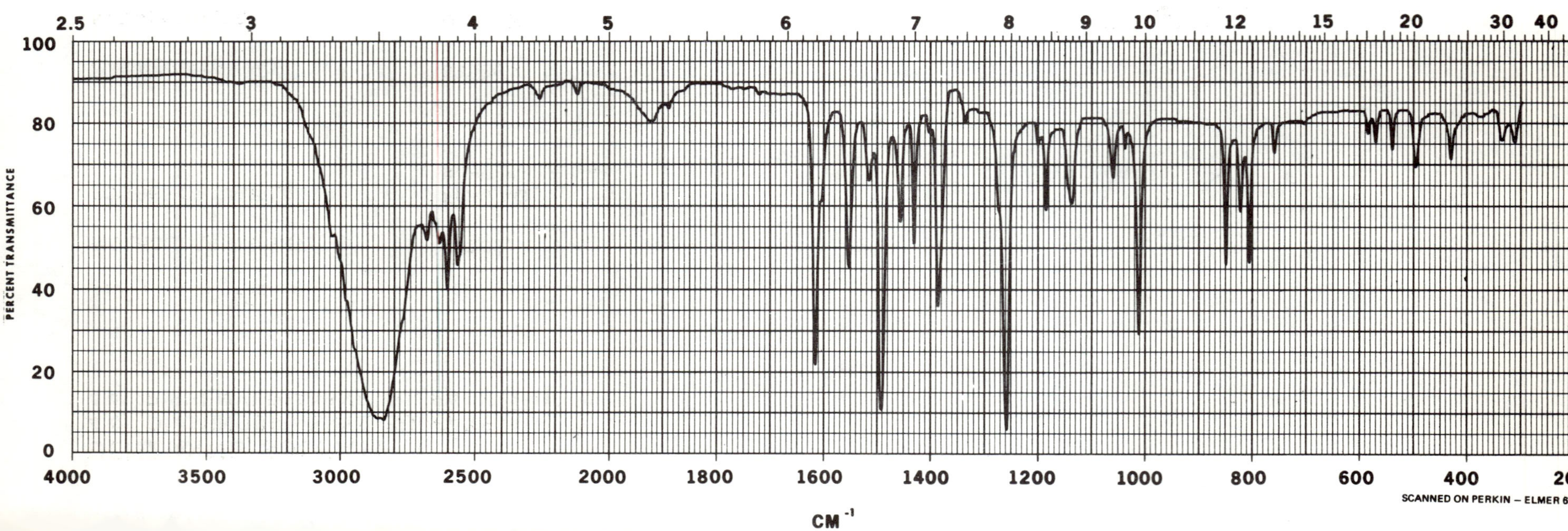

1,2–BENZISOTHIAZOLIN–3–ONE, 1,1–DIOXIDE
1,2–BENZISOTHIAZOL–3–OL, 1,1–DIOXIDE
1,2–BENZISOTHIAZOL–3(2H)–ONE, 1,1–DIOXIDE
SACCHARIN
o–BENZOSULFIMIDE

$C_7H_5NO_3S$

MOLECULAR WEIGHT: 183.19

SOURCE OF SAMPLE: MCB Manufacturing Chemists
Norwood, Ohio

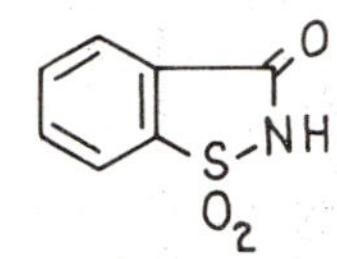

CAS REGISTRY NO.: 81-07-2

NIOSH NUMBER: DE42000

KBr Wafer

CONDENSED PHASE

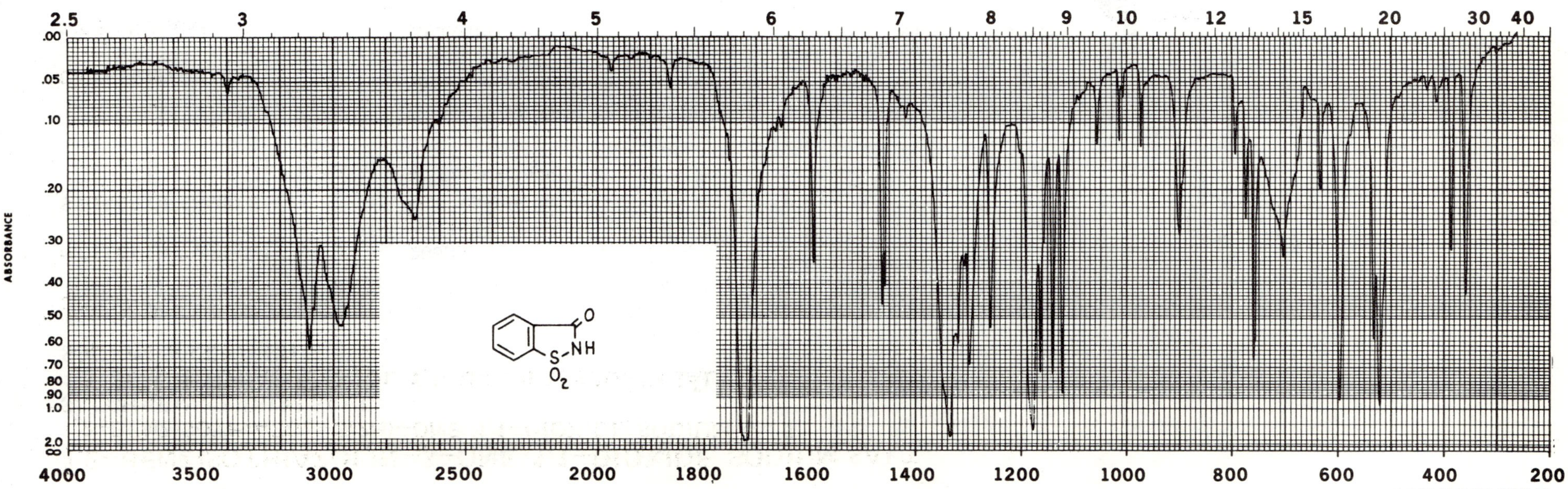

1,2–BENZISOTHIAZOLIN–3–ONE, 1,1–DIOXIDE, SODIUM SALT

1,2–BENZISOTHIAZOL–3(2H)–ONE, 1,1–DIOXIDE, SODIUM SALT
SACCHARIN, SODIUM SALT
1,2–BENZISOTHIAZOL–3–OL, 1,1–DIOXIDE, SODIUM SALT
SODIUM SACCHARIN

$C_7H_4NNaO_3S$

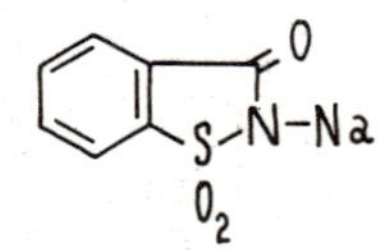

MOLECULAR WEIGHT: 205.17

SOURCE OF SAMPLE: Mallinckrodt, Inc.
St. Louis, Missouri

CAS REGISTRY NO.: 128-44-9

NIOSH NUMBER: DE45500

KBr Wafer

CONDENSED PHASE

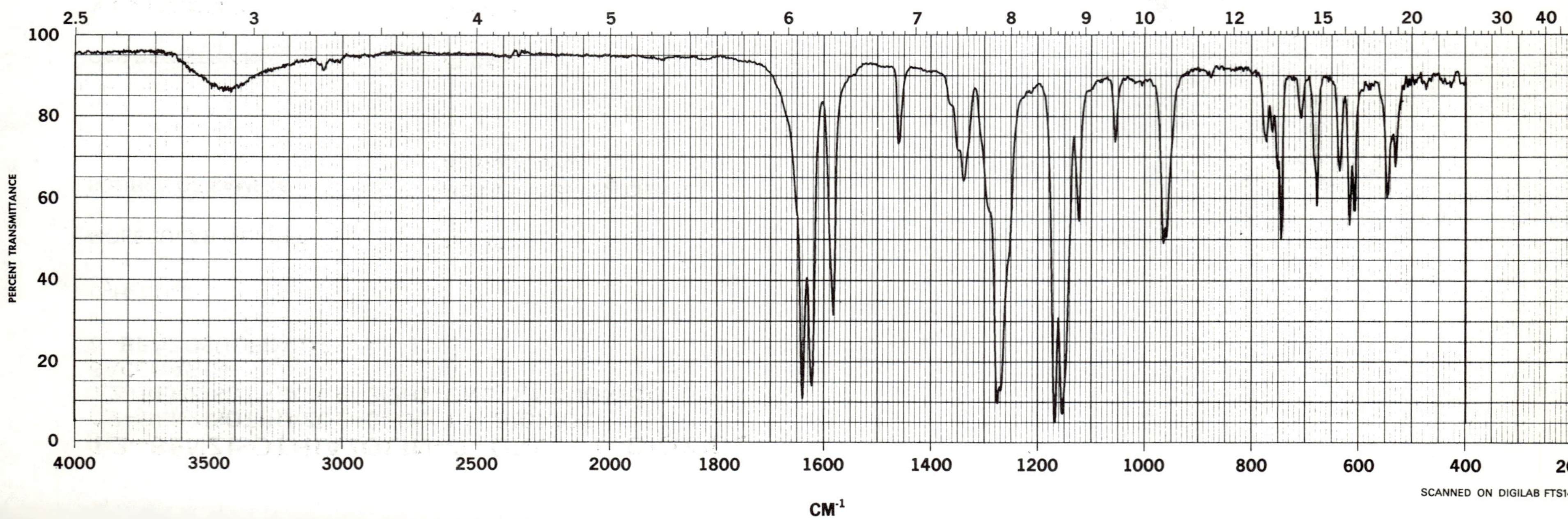

BENZO[k] FLUORANTHENE
11,12–BENZOFLUORANTHENE

$C_{20}H_{12}$

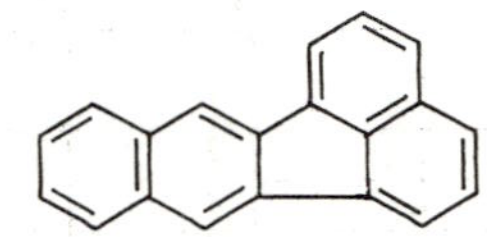

MOLECULAR WEIGHT: 252.32

SOURCE OF SAMPLE: Chem Service Company
West Chester, Pennsylvania

CAS REGISTRY NO.: 207-08-9

NIOSH NUMBER: DF63500

KBr Wafer

▲ broad absorption impurity

CONDENSED PHASE

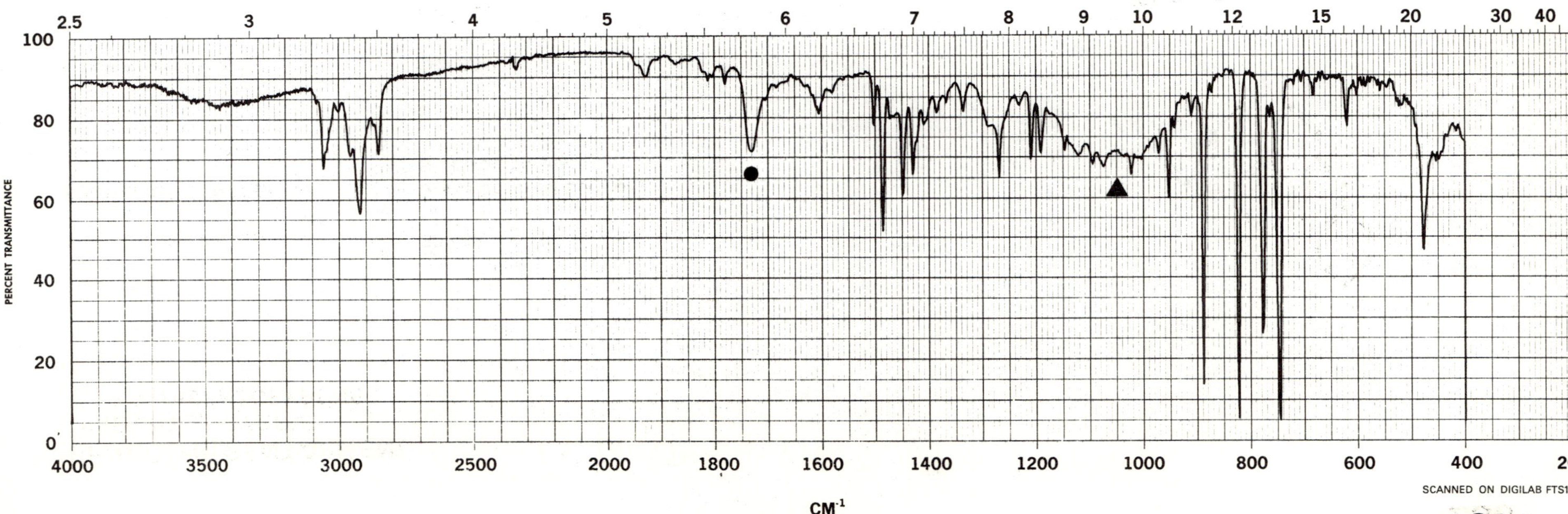

SCANNED ON DIGILAB FTS14

BENZO[rst] PENTAPHENE
DIBENZO[a,i] PYRENE
1,2:7,8–DIBENZOPYRENE

$C_{24}H_{14}$

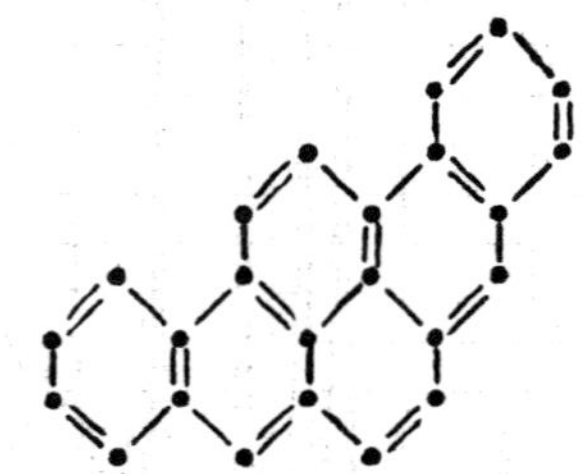

MOLECULAR WEIGHT: 302.38

SOURCE OF SAMPLE: Sigma Chemical Company
St. Louis, Missouri

CAS REGISTRY NO.: 189-55-9

NIOSH NUMBER: DI57750

KBr Wafer

▲ broad absorption impurity

CONDENSED PHASE

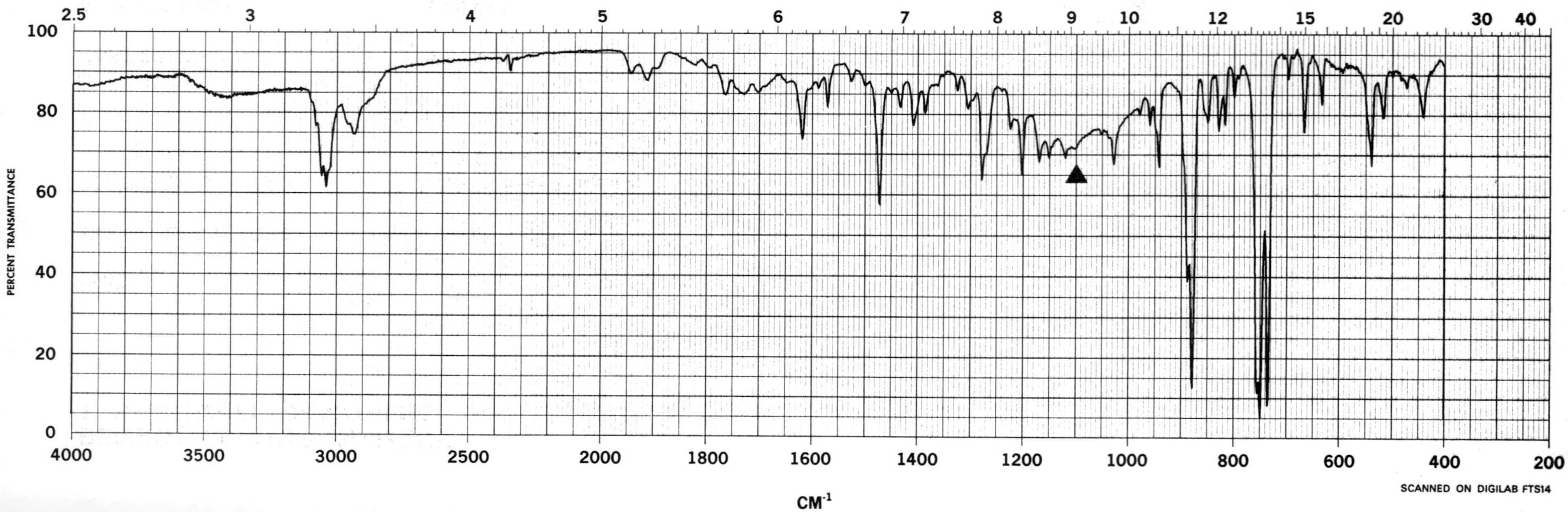

BENZO[ghi] PERYLENE
1,12–BENZOPERYLENE

$C_{22}H_{12}$

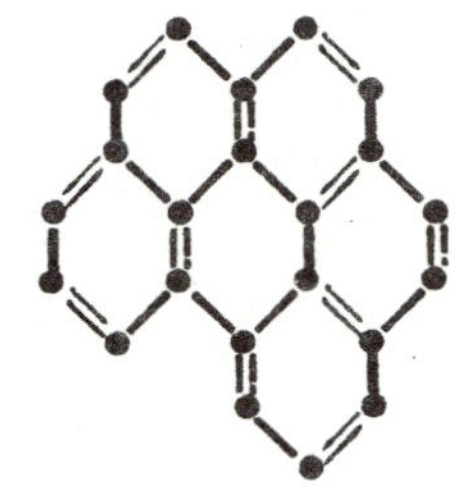

MOLECULAR WEIGHT: 276.34

SOURCE OF SAMPLE: Aldrich Chemical Company, Inc.
Milwaukee, Wisconsin

CAS REGISTRY NO.: 191-24-2

NIOSH NUMBER: –

KBr Wafer

CONDENSED PHASE

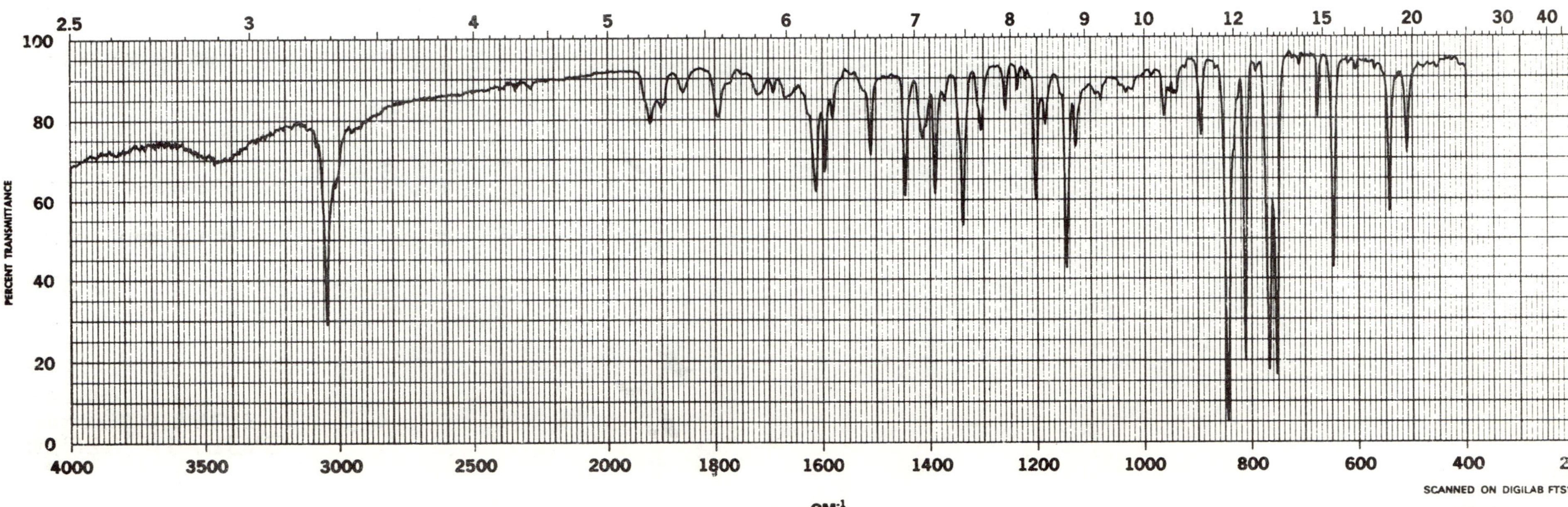

SCANNED ON DIGILAB FTS14

BENZO[a]PYRENE
3,4–BENZOPYRENE
3,4–BP

$C_{20}H_{12}$

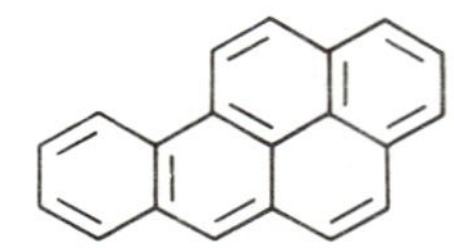

MOLECULAR WEIGHT:	252.32
SOURCE OF SAMPLE:	Fluka AG Buchs, Switzerland
CAS REGISTRY NO.:	50-32-8
NIOSH NUMBER:	DJ36750

KBr Wafer

CONDENSED PHASE

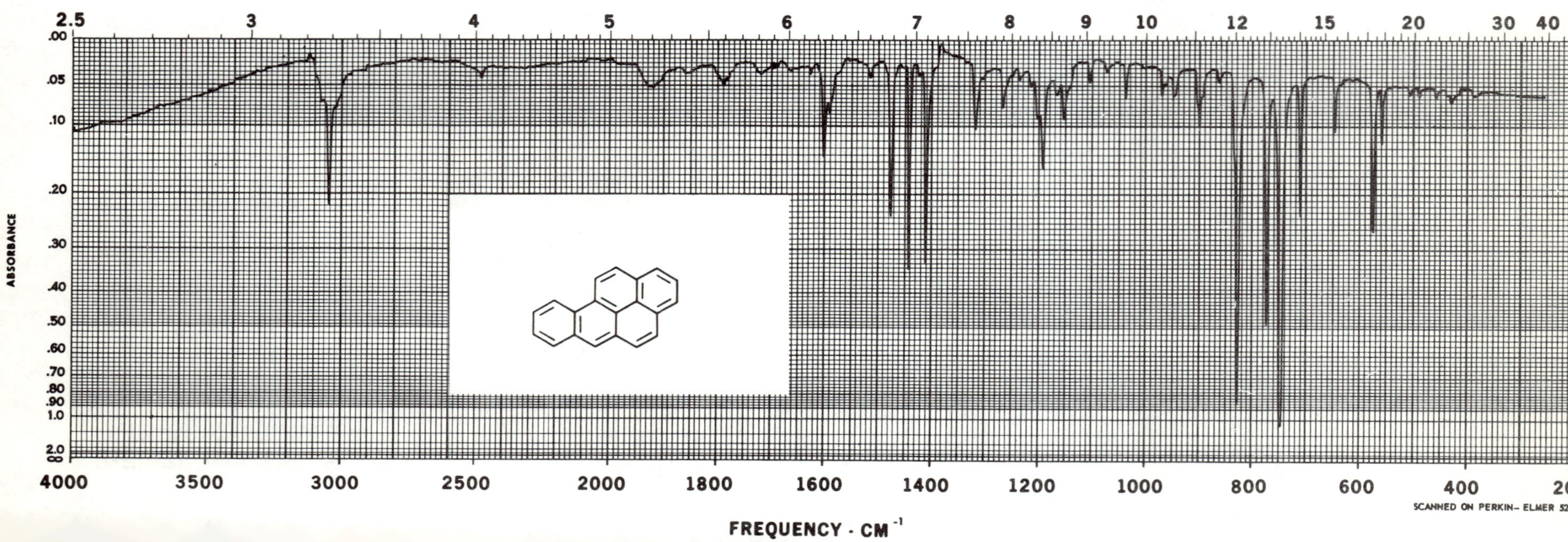

BENZO[b] TRIPHENYLENE
DIBENZ[a,c] ANTHRACENE
1,2:3,4–DIBENZANTHRACENE

$C_{22}H_{14}$

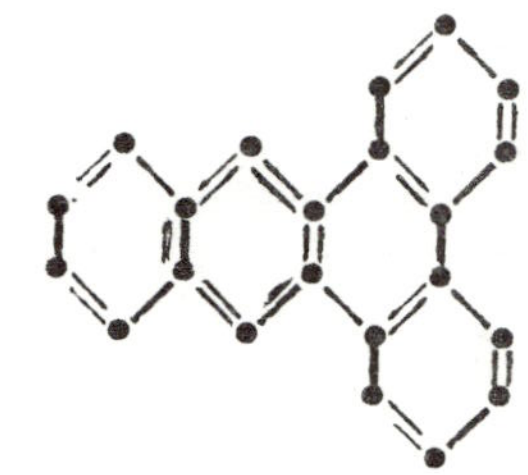

MOLECULAR WEIGHT: 278.36

SOURCE OF SAMPLE: Research Organic/Inorganic Chemical Corporation
Belleville, New Jersey

CAS REGISTRY NO.: 215-58-7

NIOSH NUMBER: DM19250

KBr Wafer

CONDENSED PHASE

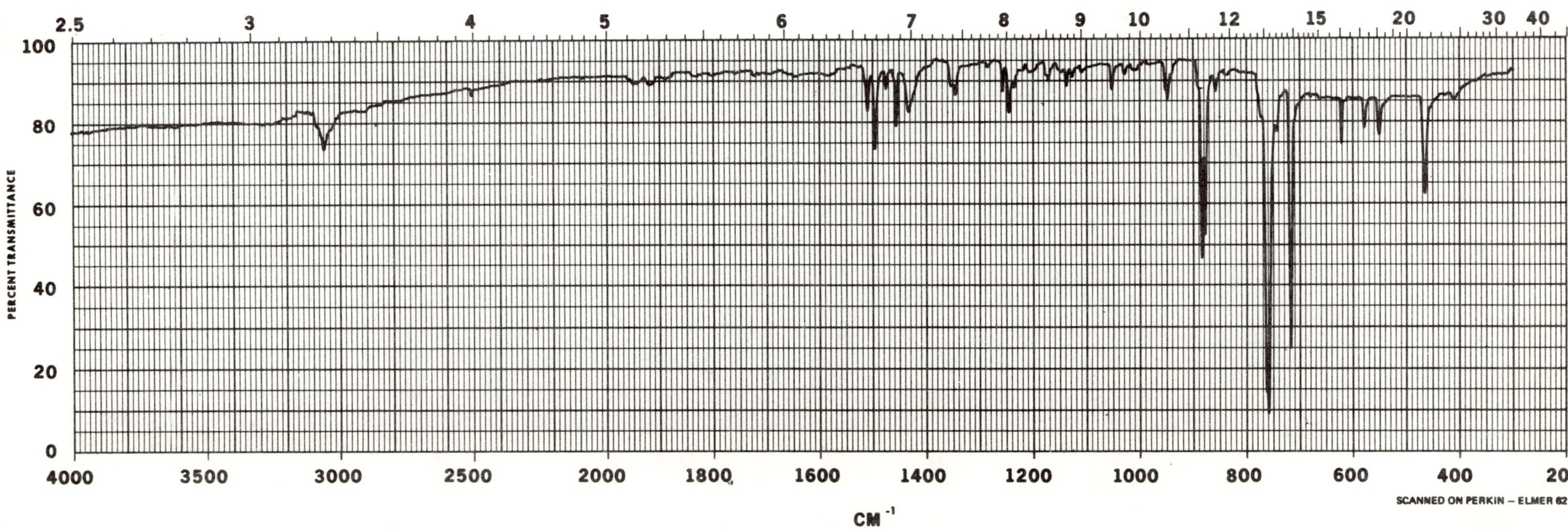

BERYLLIUM OXIDE
BERYLLIA

BeO

MOLECULAR WEIGHT: 25.01

SOURCE OF SAMPLE: Fluka AG
Buchs, Switzerland

Be O

CAS REGISTRY NO.: 1304-56-9

NIOSH NUMBER: DS40250

Mulled in Mineral Oil

CONDENSED PHASE

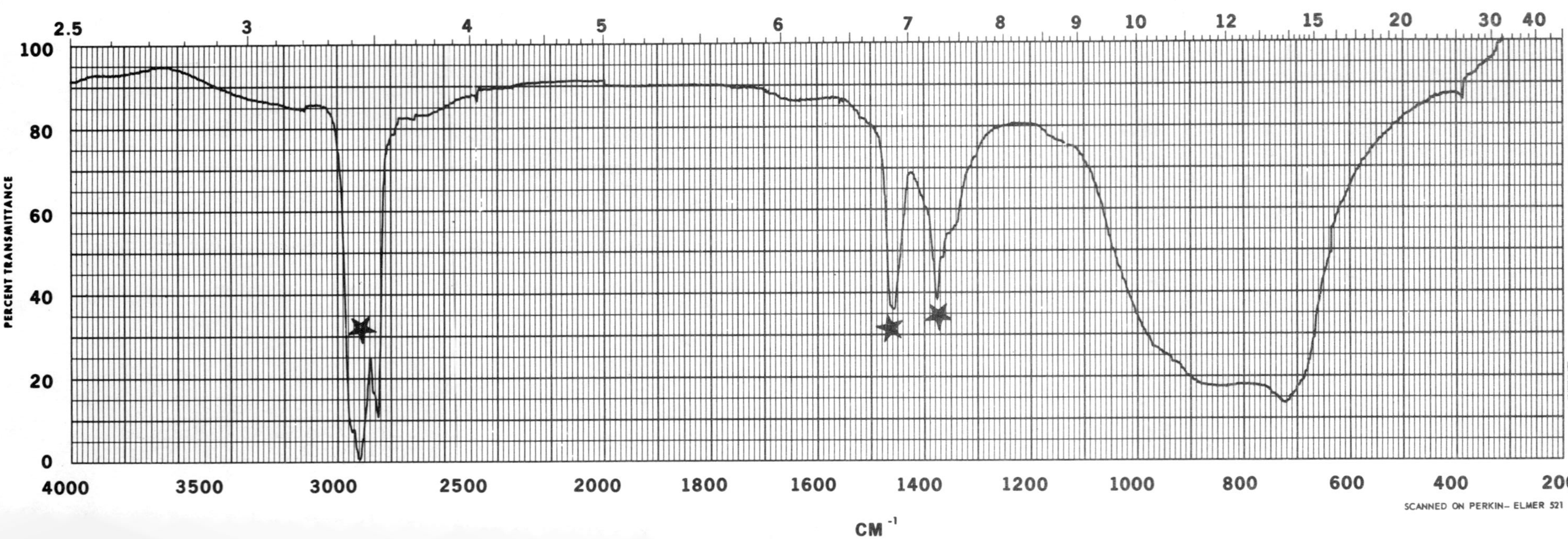

DL–2–AMINO–4–(ETHYLTHIO)BUTYRIC ACID
DL–ETHIONINE
ETH
S–ETHYL–DL–HOMOCYSTEINE

$C_6H_{13}NO_2S$

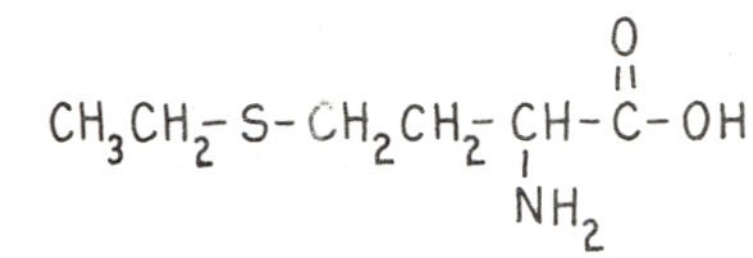

MOLECULAR WEIGHT: 163.24

SOURCE OF SAMPLE: Fluka AG
Buchs, Switzerland

CAS REGISTRY NO.: 67-21-0

NIOSH NUMBER: ES70000

KBr Wafer

CONDENSED PHASE

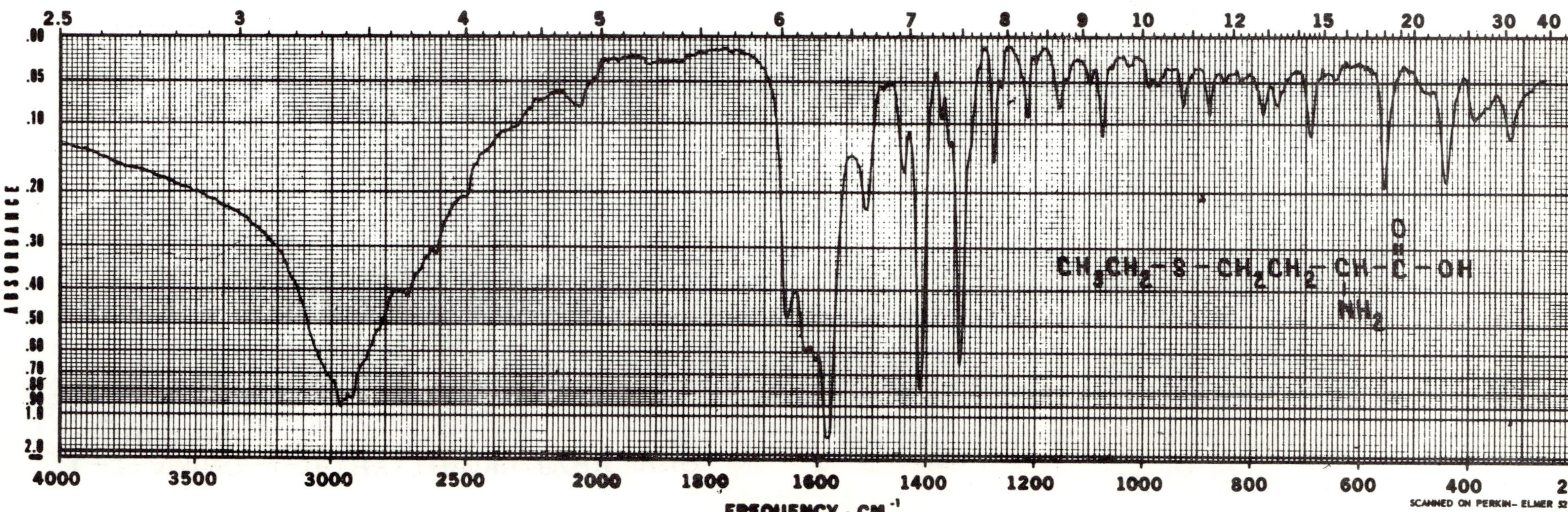

L–2–AMINO–4–(ETHYLTHIO)BUTYRIC ACID
L–ETHIONINE
S–ETHYL–L–HOMOCYSTEINE

$C_6H_{13}NO_2S$

MOLECULAR WEIGHT: 163.24

SOURCE OF SAMPLE: Fluka AG
Buchs, Switzerland

$CH_3CH_2-S-CH_2CH_2-CH(NH_2)-C(=O)-OH$

CAS REGISTRY NO.: 13073-35-3

NIOSH NUMBER: –

KBr Wafer

CONDENSED PHASE

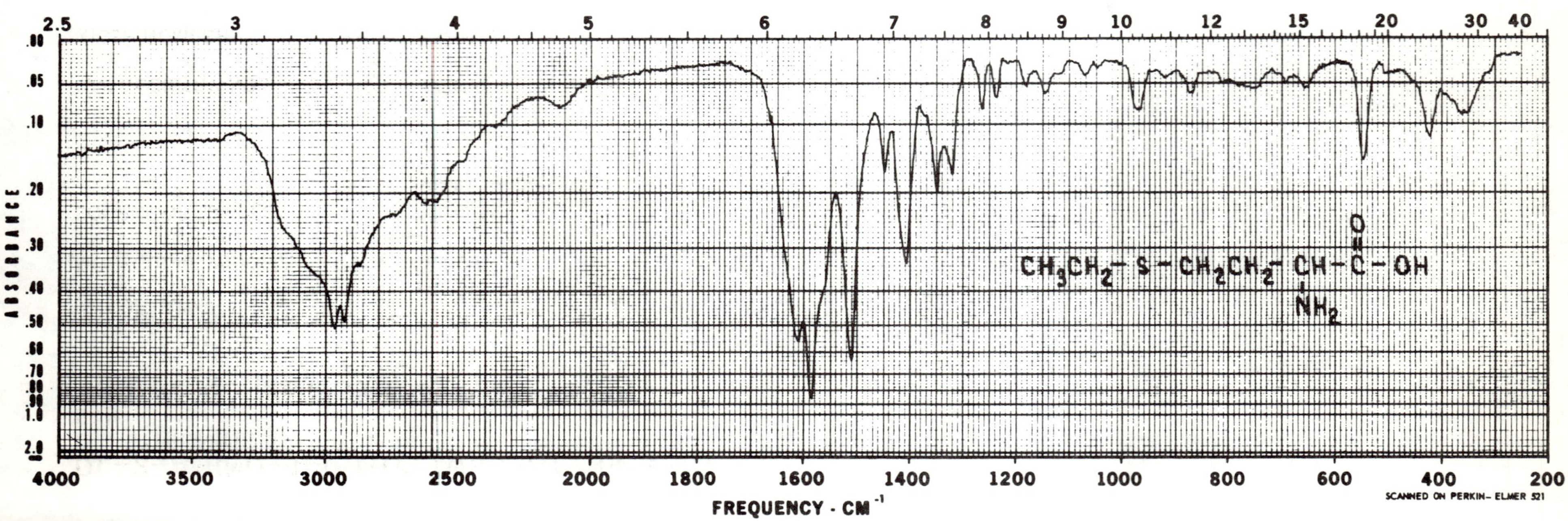

4– { p–[BIS(2–CHLOROETHYL)AMINO] PHENYL } BUTYRIC ACID

CHLORAMBUCIL
LEUKERAN
ELCORIL
NSC–3088

$C_{14}H_{19}Cl_2NO_2$

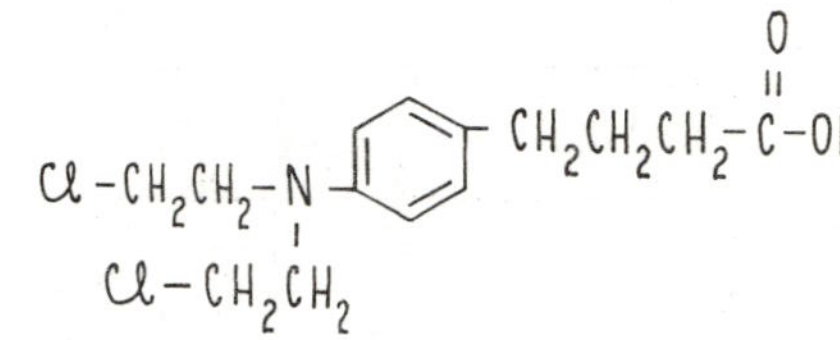

MOLECULAR WEIGHT: 304.22

SOURCE OF SAMPLE: Burroughs Wellcome and Company, Inc.
Research Triangle Park, North Carolina

CAS REGISTRY NO.: 305-03-3

NIOSH NUMBER: ES75250

KBr Wafer

CONDENSED PHASE

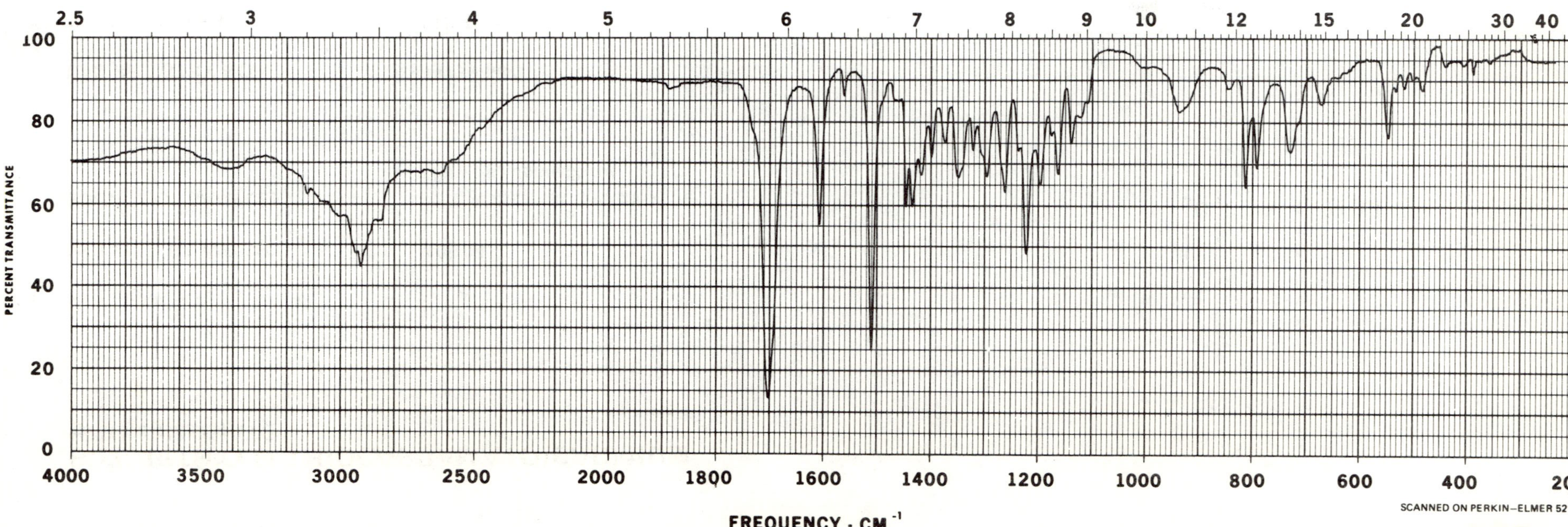

BIS(DIETHYLDITHIOCARBAMATO)CADMIUM
ETHYL CADMATE

$C_{10}H_{20}CdN_2S_4$

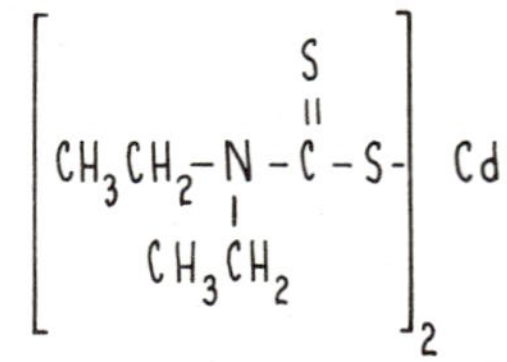

MOLECULAR WEIGHT: 408.94

SOURCE OF SAMPLE: R. T. Vanderbilt Company, Inc.
Norwalk, Connecticut

CAS REGISTRY NO.: 14239-68-0

NIOSH NUMBER: EU98500

CAPILLARY CELL: Melt

CONDENSED PHASE

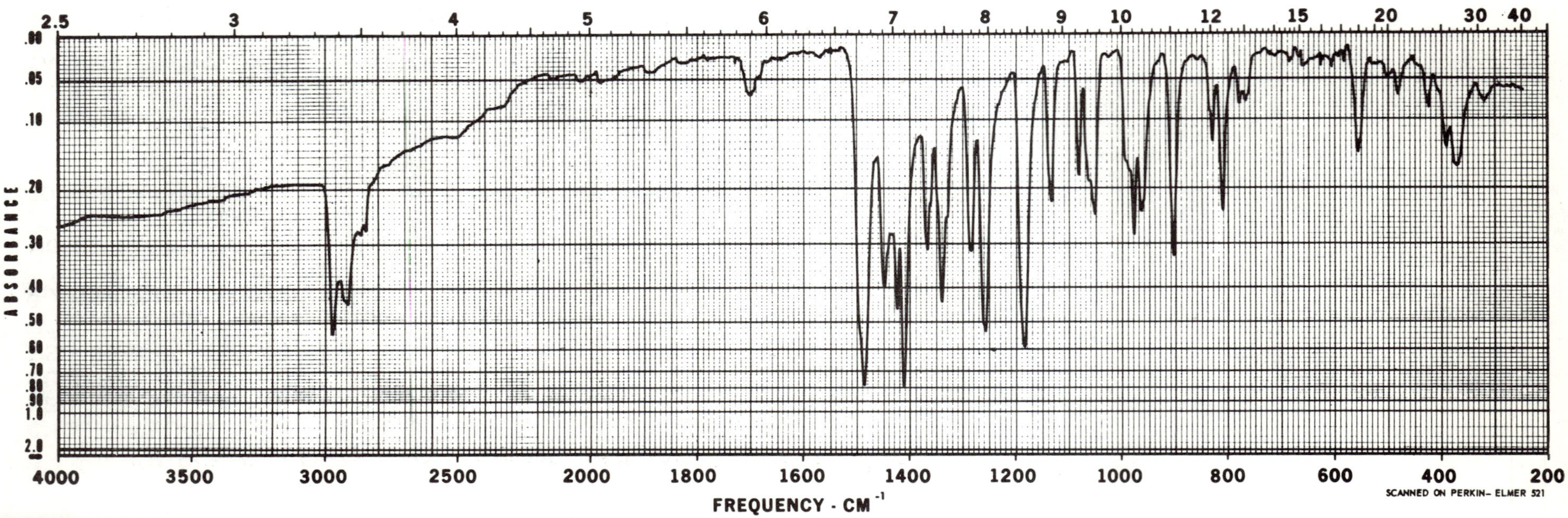

CADMIUM CHLORIDE, DIHYDRATE
CADDY
VI–CAD

$CdCl_2 \cdot 2H_2O$

MOLECULAR WEIGHT: 203.34

SOURCE OF SAMPLE: Sadtler Research Laboratories
Philadelphia, Pennsylvania

$CdCl_2 \cdot 2H_2O$

CAS REGISTRY NO.: 34330-64-8

NIOSH NUMBER: -

4000 - 1350 cm^{-1} = Mulled in Perfluorinated Hydrocarbon
1350 - 400 cm^{-1} = Mulled in Mineral Oil

CONDENSED PHASE

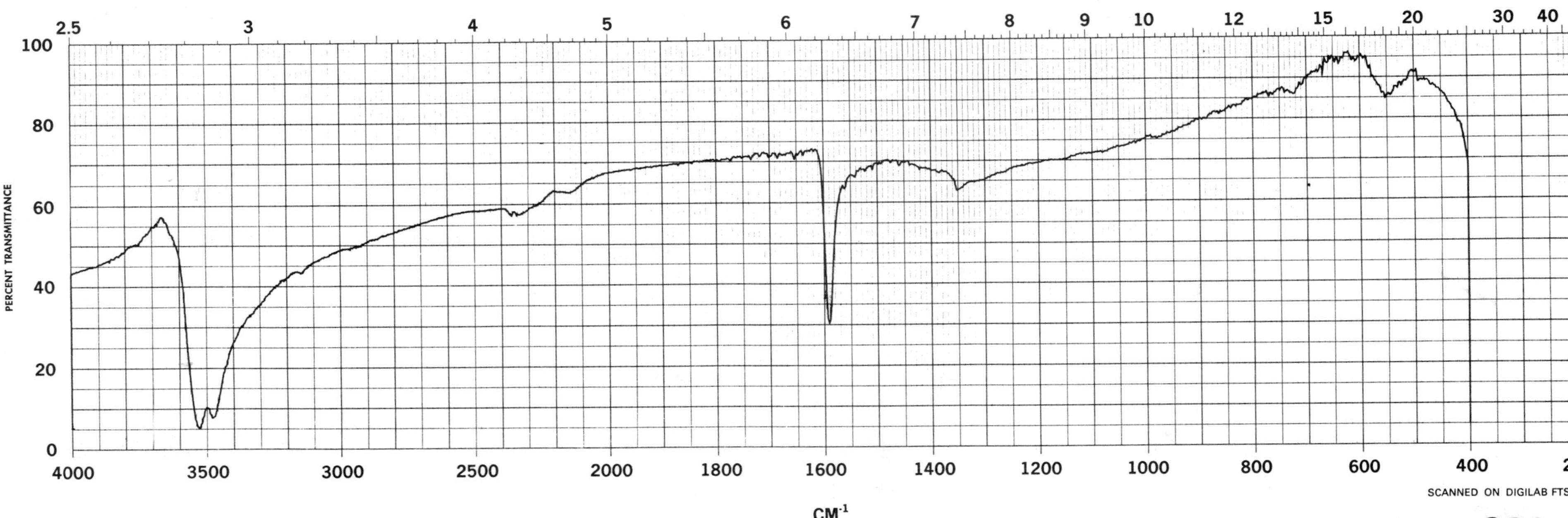

SCANNED ON DIGILAB FTS14

CADMIUM OXIDE

CdO

MOLECULAR WEIGHT: 128.40

SOURCE OF SAMPLE: Aldrich Chemical Company, Inc.
Milwaukee, Wisconsin

Cd O

CAS REGISTRY NO.: 1306-19-0

NIOSH NUMBER: EV19250

4000 - 1350 cm^{-1} = Mulled in Perfluorinated Hydrocarbon
1350 - 400 cm^{-1} = Mulled in Mineral Oil

CONDENSED PHASE

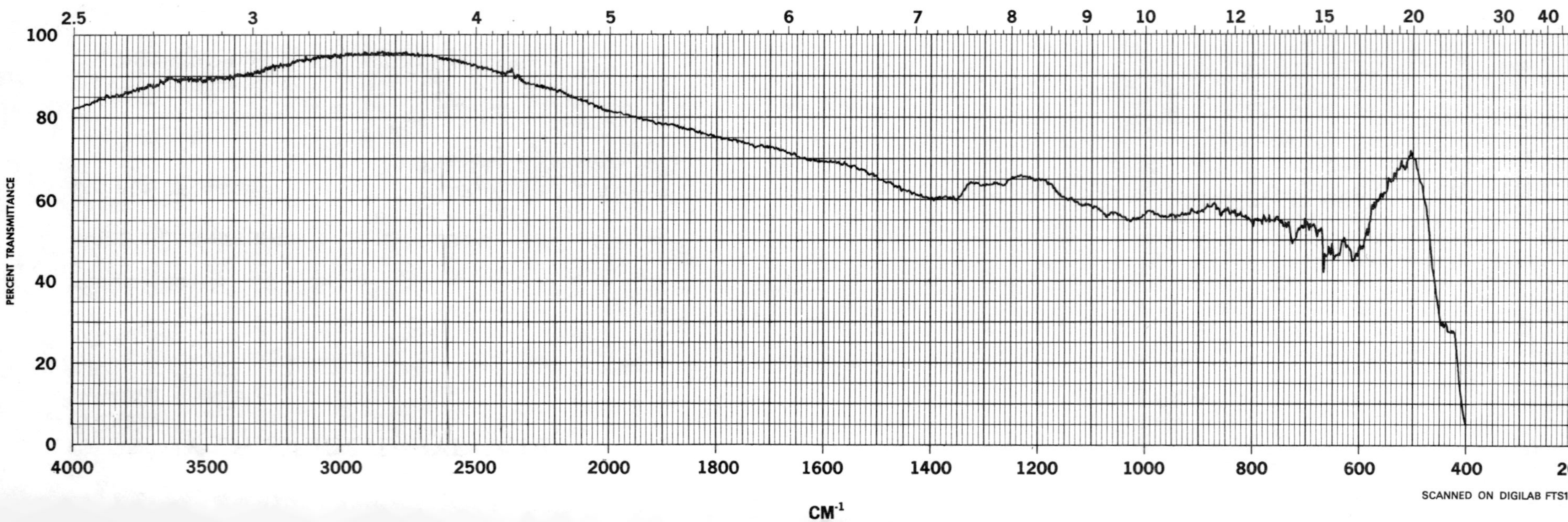

SCANNED ON DIGILAB FTS14

METHYLCARBAMIC ACID, 4–(DIMETHYLAMINO)–3,5–XYLYL ESTER
DOWCO 139
MEXACARBATE
4–(DIMETHYLAMINO)–3,5–XYLENOL, METHYLCARBAMATE
ZECTRANE

$C_{12}H_{18}N_2O_2$

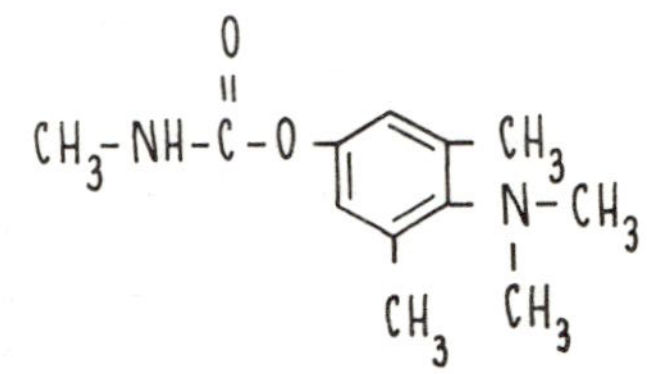

MOLECULAR WEIGHT:	222.29
SOURCE OF SAMPLE:	The Dow Chemical Company Midland, Michigan
CAS REGISTRY NO.:	315-18-4
NIOSH NUMBER:	FC07000
CAPILLARY CELL:	Melt

CONDENSED PHASE

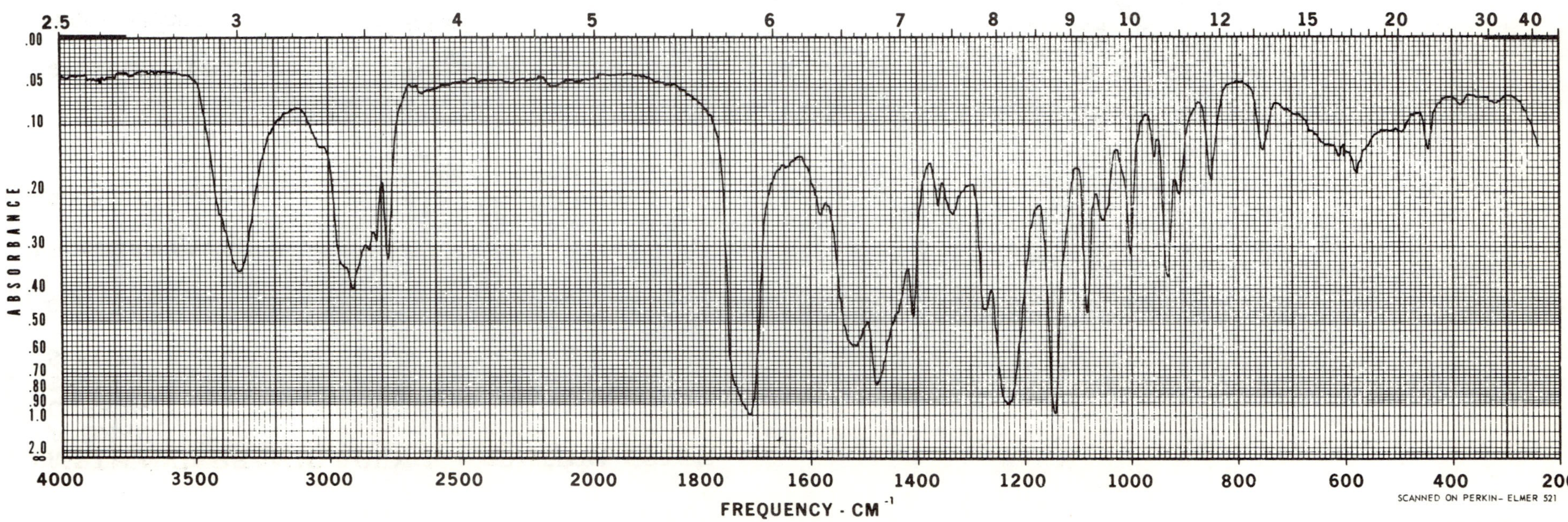

METHYLCARBAMIC ACID, 1–NAPHTHYL ESTER
ATOXAN
CARBARYL
SEVIN
SEPTENE

$C_{12}H_{11}NO_2$

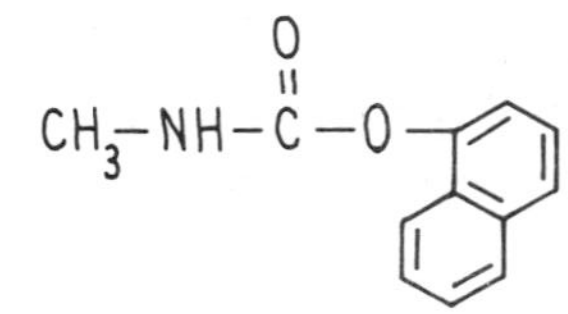

MOLECULAR WEIGHT:	201.23
SOURCE OF SAMPLE:	Frinton Laboratories, Inc. Vineland, New Jersey
CAS REGISTRY NO.:	63-25-2
NIOSH NUMBER:	FC59500

KBr Wafer

CONDENSED PHASE

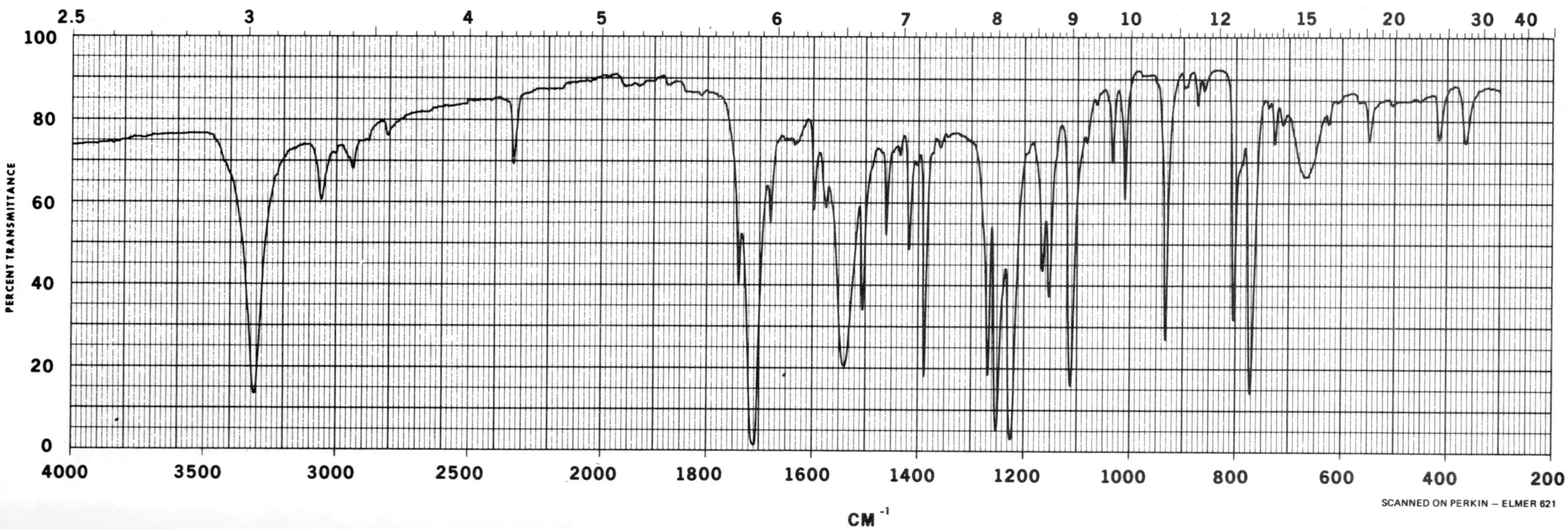

SCANNED ON PERKIN – ELMER 621

3–METHYLCHOLANTHRENE
1,2–DIHYDRO–3–METHYLBENZ[j] ACEANTHRYLENE
MC

$C_{21}H_{16}$

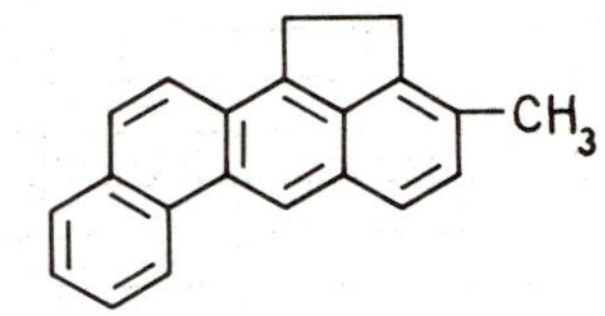

MOLECULAR WEIGHT: 268.36

SOURCE OF SAMPLE: Sigma Chemical Company
St. Louis, Missouri

CAS REGISTRY NO.: 56-49-5

NIOSH NUMBER: FZ36750

KBr Wafer

CONDENSED PHASE

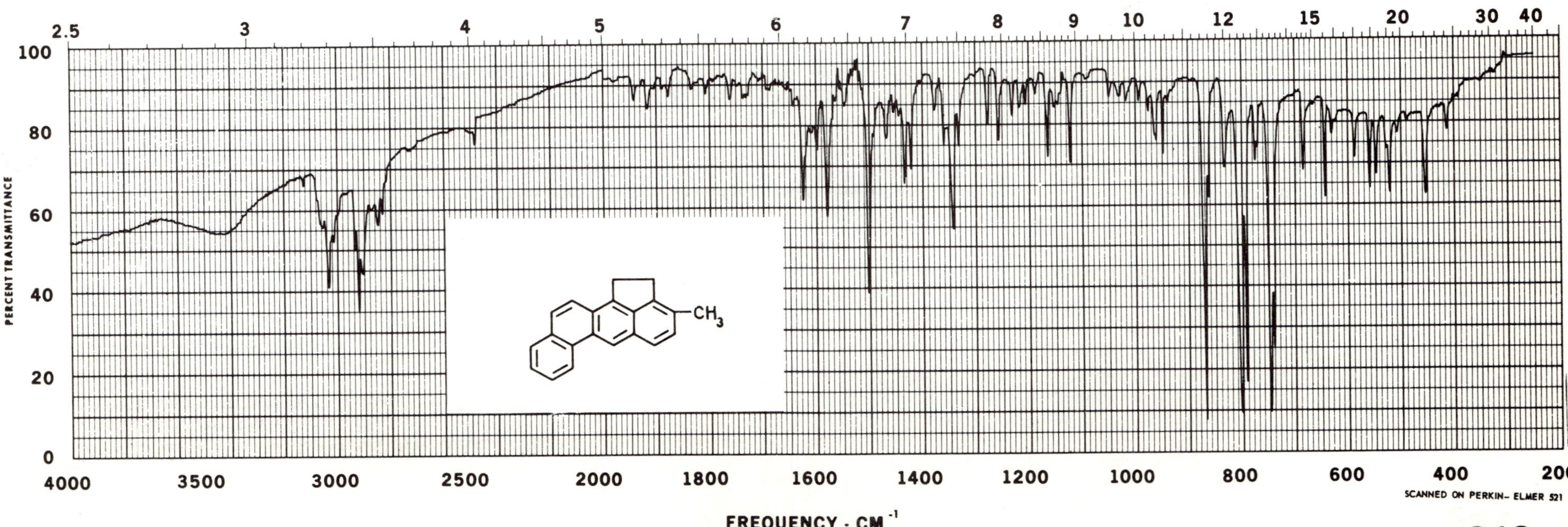

CHROMIC ACID, CALCIUM SALT (1:1)
CALCIUM CHROMATE (VI)
CALCIUM CHROME YELLOW
GELBIN
YELLOW ULTRAMARINE

$CaCrO_4$

MOLECULAR WEIGHT: 156.07

SOURCE OF SAMPLE: Gallard-Schlesinger Chemical Manufacturing Corporation
Carle Place, New York

$CaCrO_4$

CAS REGISTRY NO.: 13765-19-0

NIOSH NUMBER: GB27500

KBr Wafer

CONDENSED PHASE

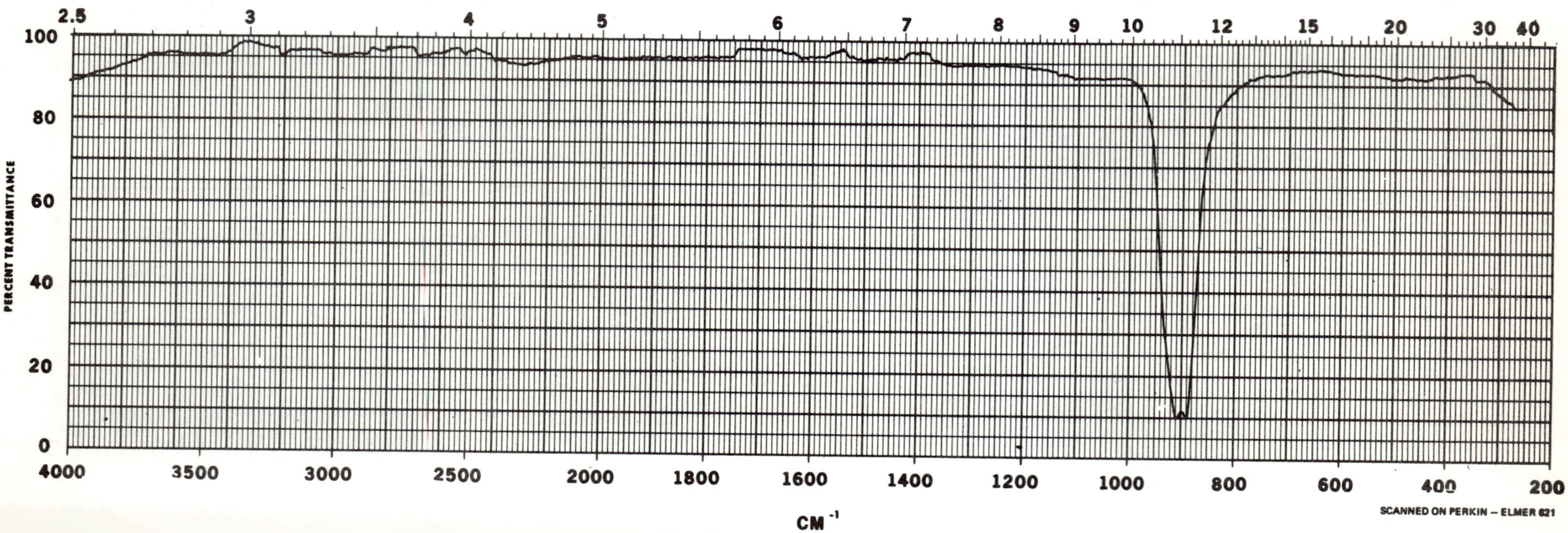

CHROMIC ACID, DIPOTASSIUM SALT
POTASSIUM CHROMATE (VI)
NEUTRAL POTASSIUM CHROMATE
TARAPACAITE

CrK_2O_4

MOLECULAR WEIGHT: 194.20

SOURCE OF SAMPLE: Mallinckrodt, Inc.
St. Louis, Missouri

CAS REGISTRY NO.: 7789-00-6

NIOSH NUMBER: GB29400

KBr Wafer

K_2CrO_4

CONDENSED PHASE

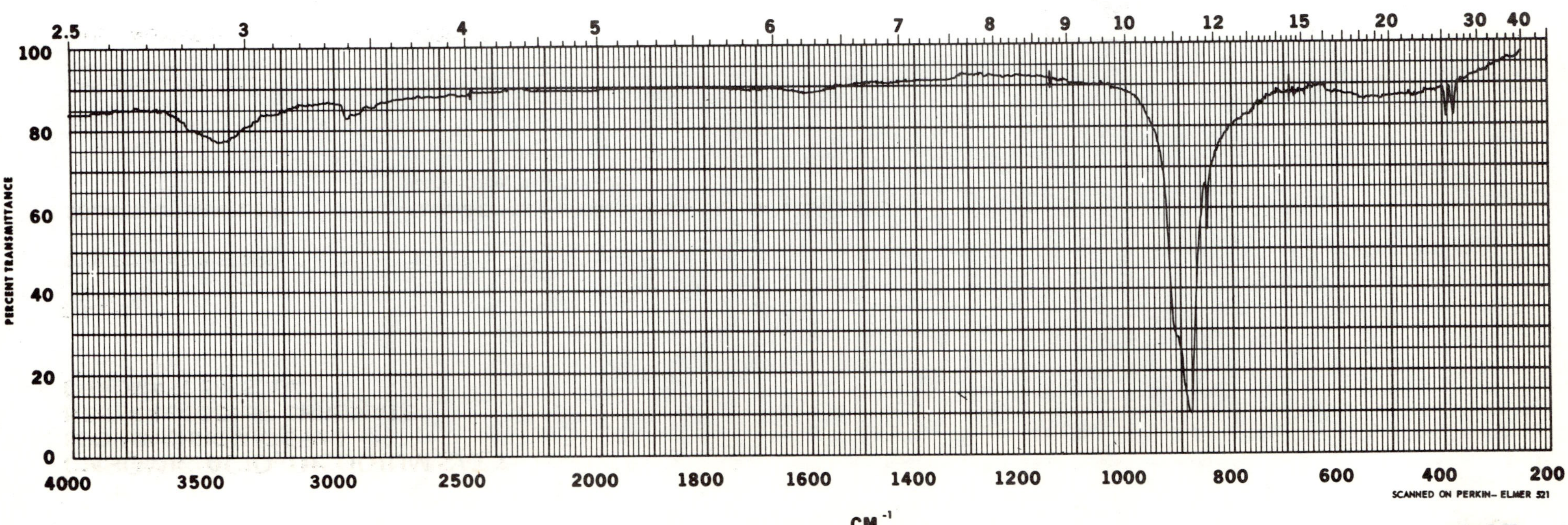

CHROMIC ACID, DISODIUM SALT
SODIUM CHROMATE (VI)
NEUTRAL SODIUM CHROMATE
CHROMATE OF SODA

$CrNa_2O_4$

MOLECULAR WEIGHT: 161.97

SOURCE OF SAMPLE: Gallard-Schlesinger Chemical Manufacturing Corporation
Carle Place, New York

Na_2CrO_4

CAS REGISTRY NO.: 7775-11-3

NIOSH NUMBER: GB29550

KBr Wafer

CONDENSED PHASE

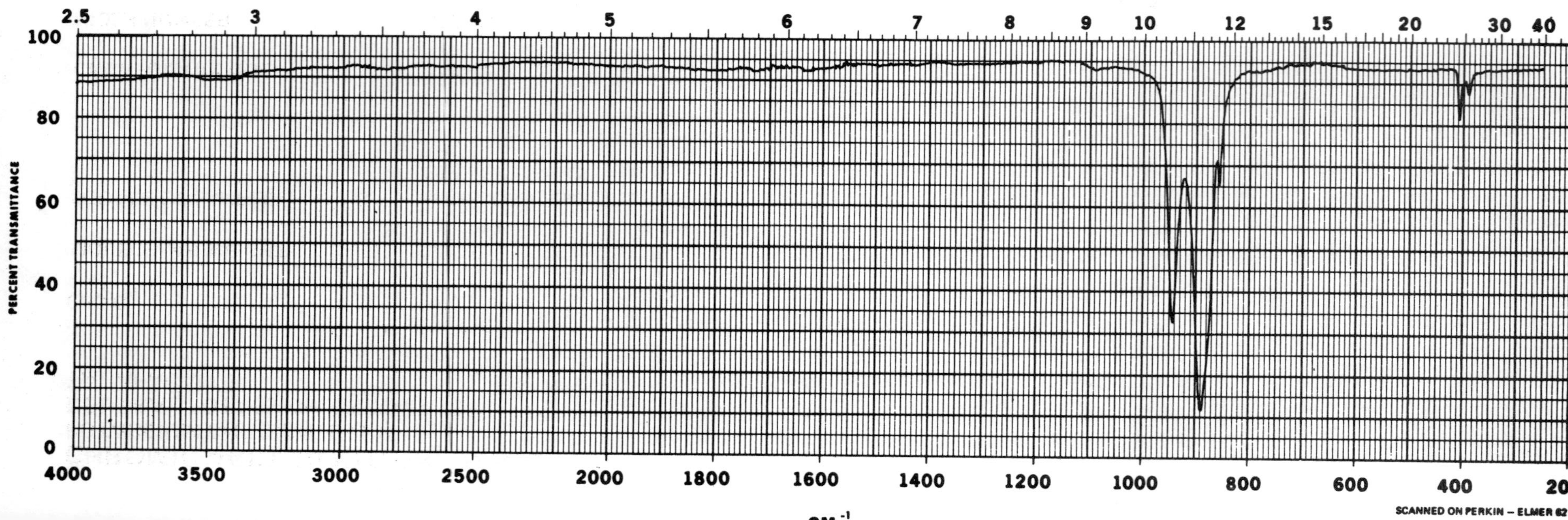

CHROMIC ACID, LEAD (2+) SALT (1:1)
LEAD CHROMATE (VI)
CHROME YELLOW
COLOGNE YELLOW
KING'S YELLOW

CrO_4Pb

MOLECULAR WEIGHT: 323.18

SOURCE OF SAMPLE: Chem Service Company
West Chester, Pennsylvania

$PbCrO_4$

CAS REGISTRY NO.: 7758-97-6

NIOSH NUMBER: GB29750

Mulled in Mineral Oil

CONDENSED PHASE

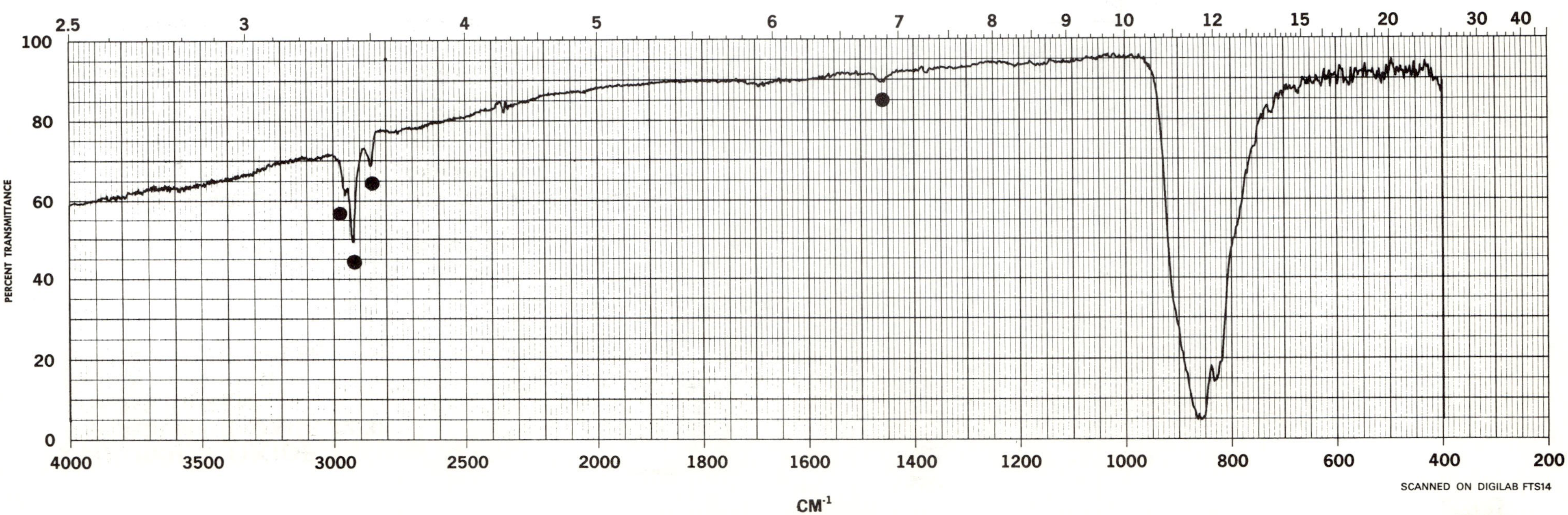

SCANNED ON DIGILAB FTS14

CHROMIUM OXIDE
CHROMIUM (III) OXIDE (2:3)
CHROME GREEN
GREEN CINNABAR
ULTRAMARINE GREEN

Cr_2O_3

MOLECULAR WEIGHT: 151.99

SOURCE OF SAMPLE: MCB Manufacturing Chemists
Norwood, Ohio

Cr_2O_3

CAS REGISTRY NO.: 1308-04-9

NIOSH NUMBER: GB64750

Mulled in Mineral Oil

CONDENSED PHASE

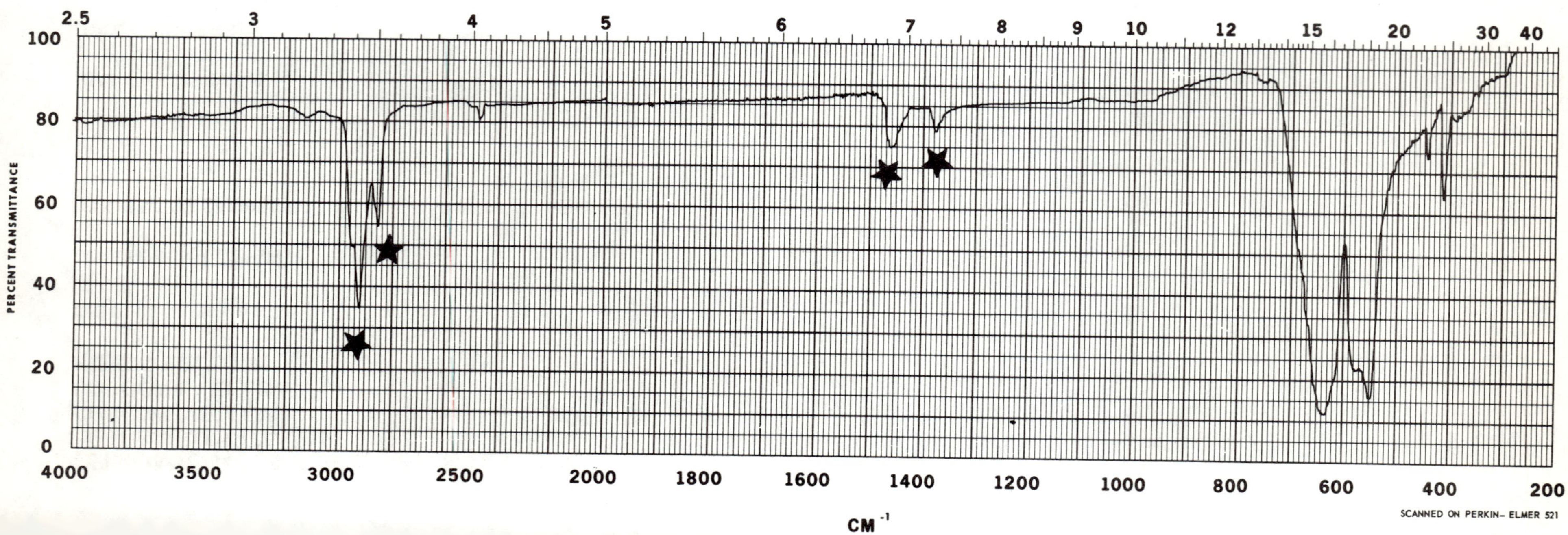

CHROMIUM OXIDE
CHROMIUM (VI) OXIDE (1:3)
CHROMIUM TRIOXIDE
CHROMIC ACID
CHROMIC ANHYDRIDE

CrO_3

MOLECULAR WEIGHT: 99.99

SOURCE OF SAMPLE: Mallinckrodt, Inc.
St. Louis, Missouri

CAS REGISTRY NO.: 1333-82-0

NIOSH NUMBER: GB66500

KBr Wafer

$Cr O_3$

CONDENSED PHASE

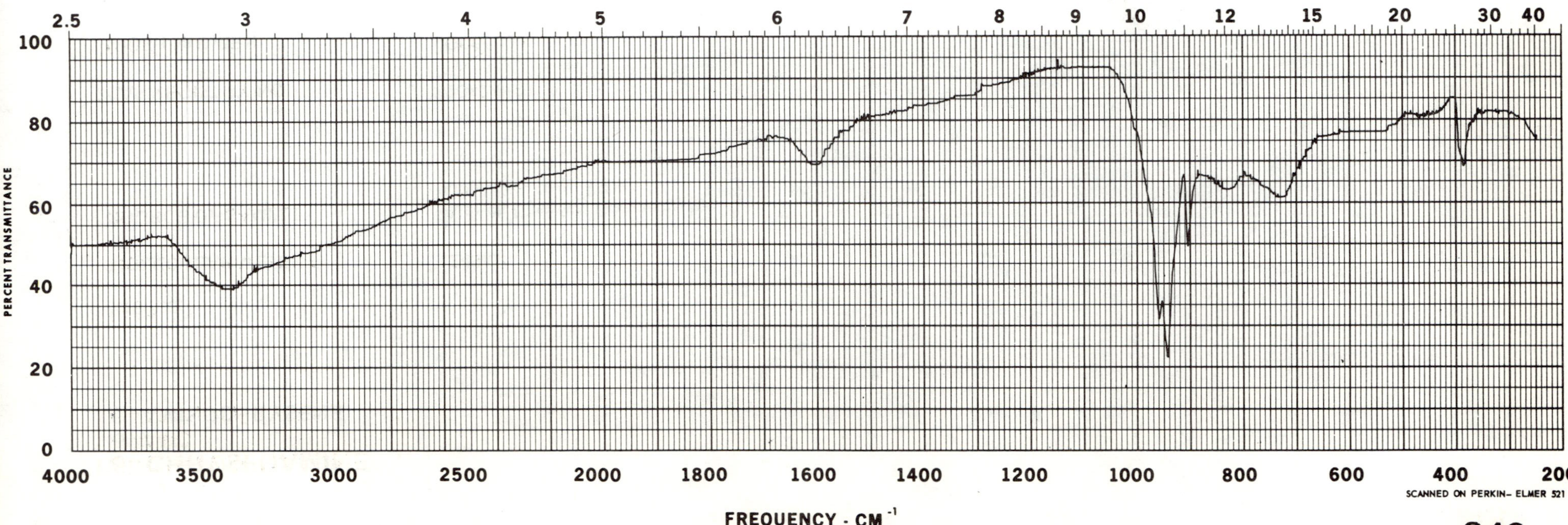

SCANNED ON PERKIN- ELMER 521

6–CHRYSENAMINE
6–AMC
6–AMINOCHRYSENE

$C_{18}H_{13}N$

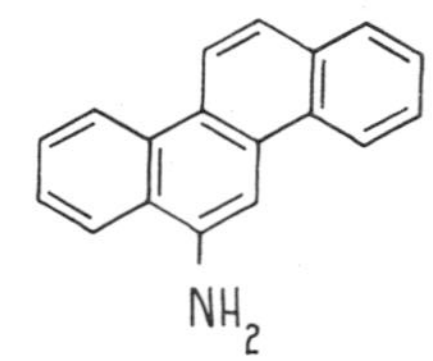

MOLECULAR WEIGHT: 243.31

SOURCE OF SAMPLE: Aldrich Chemical Company, Inc.
Milwaukee, Wisconsin

CAS REGISTRY NO.: 2642-98-0

NIOSH NUMBER: GC05000

KBr Wafer

CONDENSED PHASE

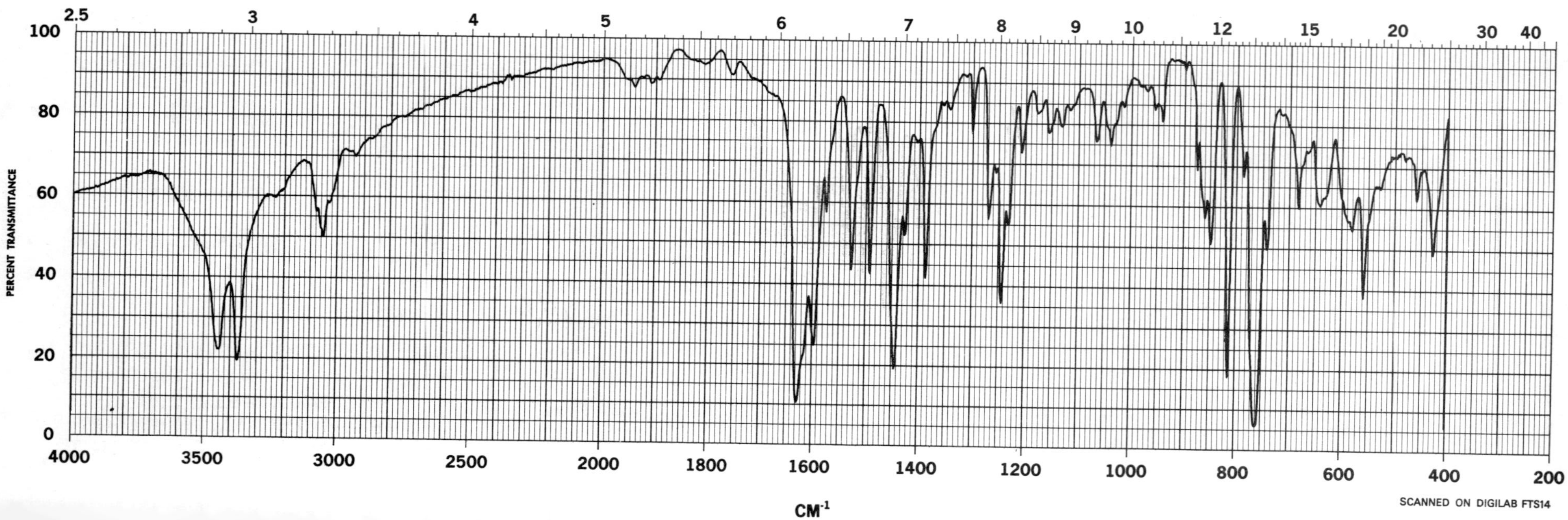

CHRYSENE
1,2–BENZOPHENANTHRENE
BENZO[a] PHENANTHRENE
1,2:5,6–DIBENZONAPHTHALENE

$C_{18}H_{12}$

MOLECULAR WEIGHT: 228.30

SOURCE OF SAMPLE: James Hinton
Valparaiso, Florida

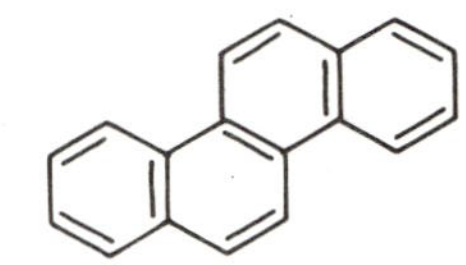

CAS REGISTRY NO.: 218-01-9

NIOSH NUMBER: GC07000

KBr Wafer

CONDENSED PHASE

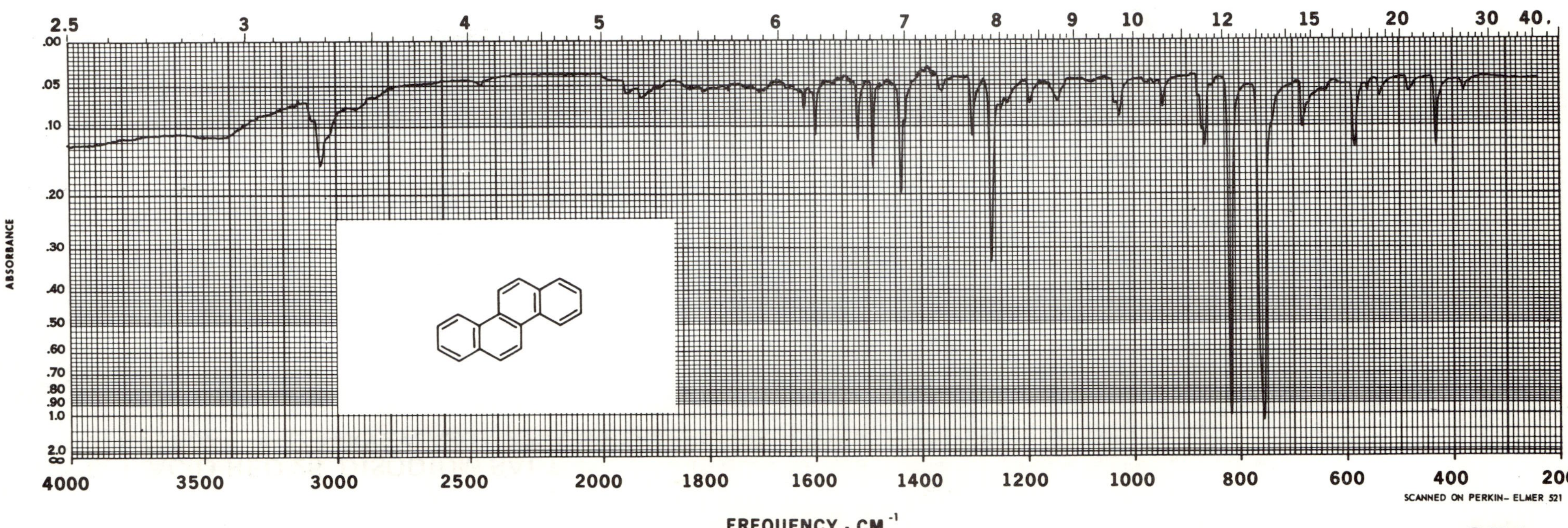

C. I. ACID RED 26, DISODIUM SALT
3–HYDROXY–4–(2,4–XYLYLAZO)–2,7–NAPHTHALENEDISULFONIC ACID, DISODIUM SALT

$C_{18}H_{14}N_2Na_2O_7S_2$

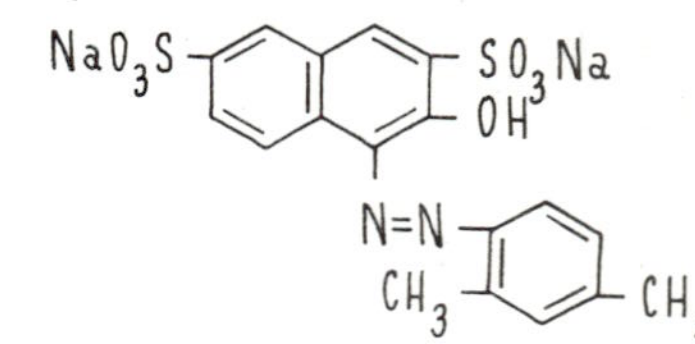

MOLECULAR WEIGHT: 480.43

SOURCE OF SAMPLE: Allied Chemical Corporation
New York, New York

CAS REGISTRY NO.: 3761-53-3

NIOSH NUMBER: QJ68250

KBr Wafer

CONDENSED PHASE

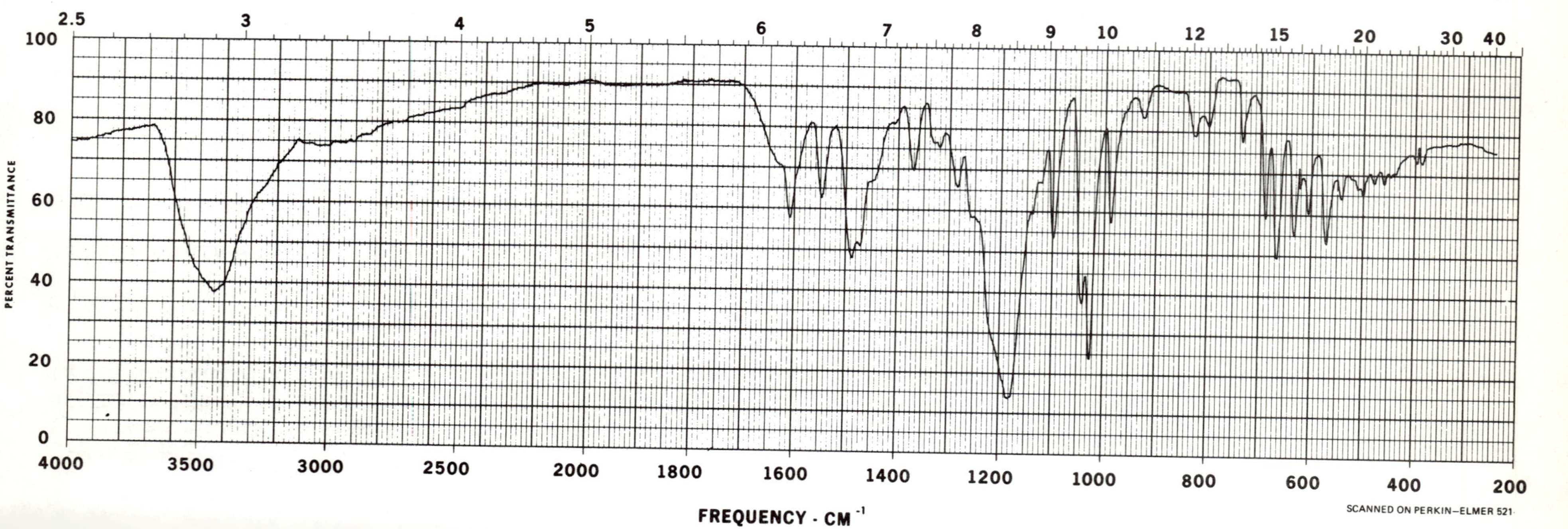

SCANNED ON PERKIN–ELMER 521

C. I. BASIC RED 9, MONOHYDROCHLORIDE
PARAFUCHSINE
a–(p–AMINOPHENYL)–*a*–(4–IMINO–2,5–CYCLOHEXADIEN–1–YLIDENE)–p–TOLUIDINE, MONOHYDROCHLORIDE
PARAROSANILINE
PARAMAGENTA

$C_{19}H_{17}N_3 \cdot HCl$

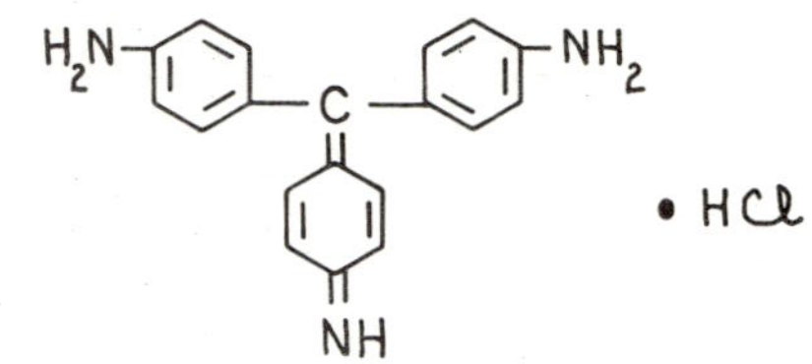

MOLECULAR WEIGHT: 323.82

SOURCE OF SAMPLE: Eastman Organic Chemicals
Rochester, New York

CAS REGISTRY NO.: 569-61-9

NIOSH NUMBER: XU33000

KBr Wafer

CONDENSED PHASE

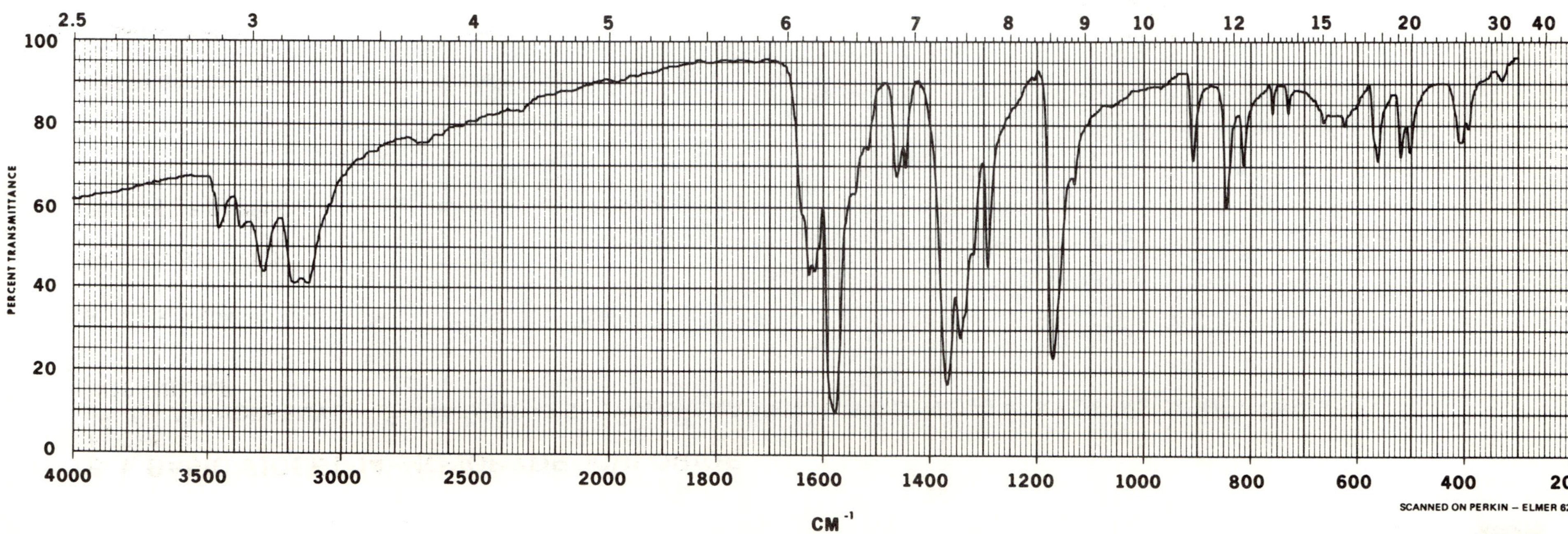

SCANNED ON PERKIN – ELMER 621

C. I. BASIC VIOLET 14, MONOHYDROCHLORIDE

FUCHSINE
a^4–(p–AMINOPHENYL)–a^4–(4–IMINO–2,5–CYCLOHEXADIEN–1–YLIDENE)–2,4–XYLIDINE, MONOHYDROCHLORIDE
ROSANILINE

$C_{20}H_{19}N_3 \cdot HCl$

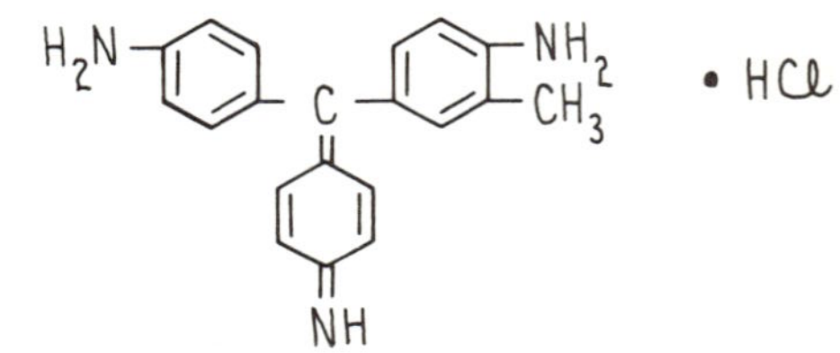

MOLECULAR WEIGHT:	337.86
SOURCE OF SAMPLE:	Eastman Organic Chemicals Rochester, New York
CAS REGISTRY NO.:	632-99-5
NIOSH NUMBER:	ZE97400

KBr Wafer

CONDENSED PHASE

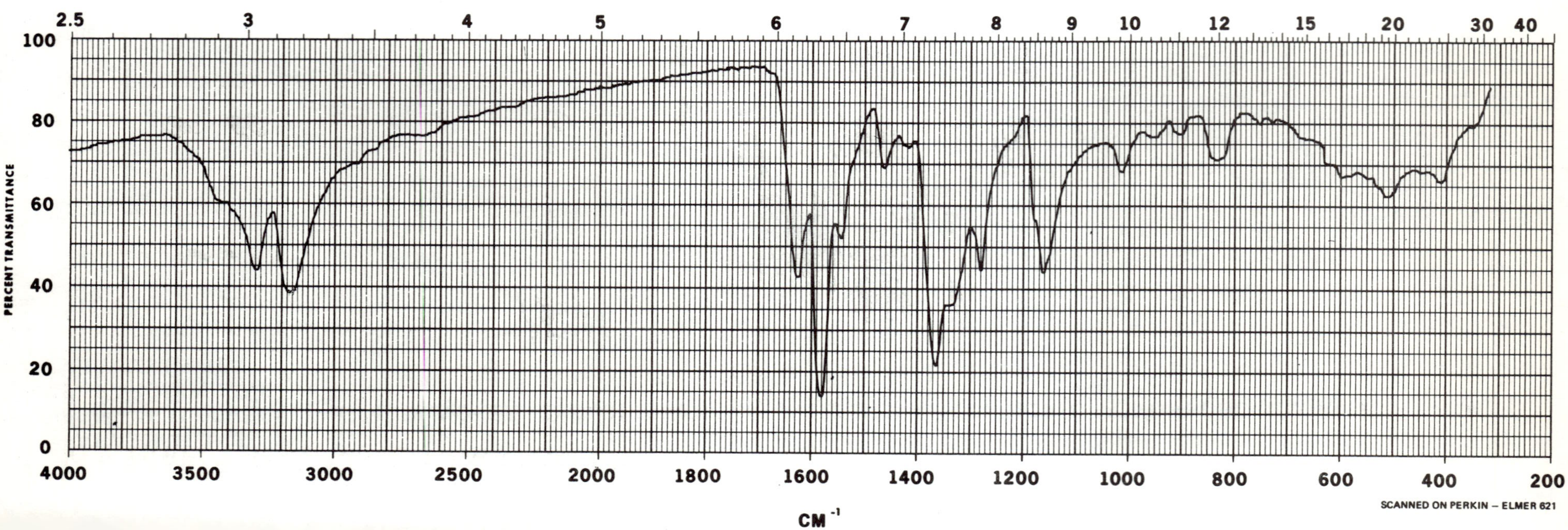

C. I. DIRECT BLUE 14, TETRASODIUM SALT
AMANIL SKY BLUE
BENZAMINE BLUE
3,3'–[(3,3'–DIMETHYL–4,4'–BIPHENYLYLENE)BIS(AZO)] BIS[5–AMINO–4–HYDROXY–2,7–NAPHTHALENEDISULFONIC ACID], TETRASODIUM SALT
DIRECT BLUE 3B
$C_{34}H_{28}N_6Na_4O_{14}S_4$

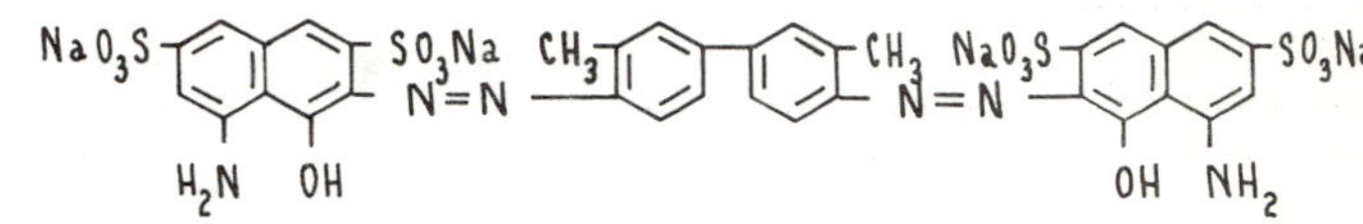

MOLECULAR WEIGHT: 960.82

SOURCE OF SAMPLE: Fluka AG
Buchs, Switzerland

CAS REGISTRY NO.: 72-57-1

NIOSH NUMBER: QJ64750

KBr Wafer

CONDENSED PHASE

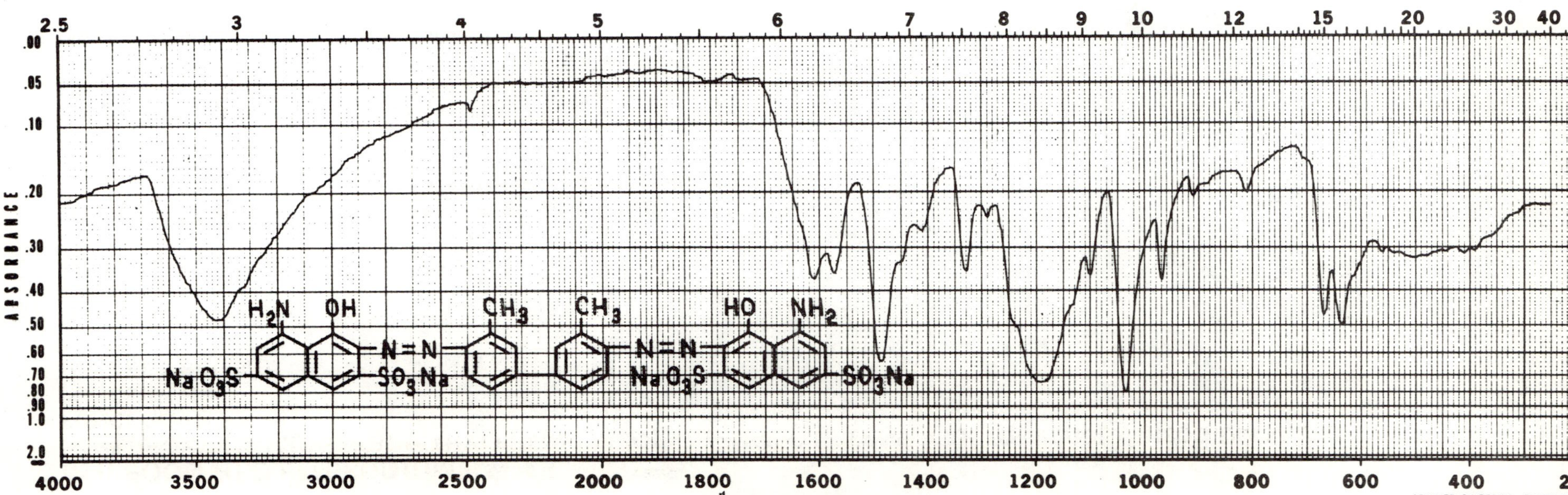

C. I. FOOD RED 6, DISODIUM SALT
3–HYDROXY–4–[(2,4,5–TRIMETHYLPHENYL)AZO] –2,7–NAPHTHALENEDISULFONIC ACID, DISODIUM SALT
PONCEAU 3R
PONCEAU RN
USACERT RED NO. 1

$C_{19}H_{16}N_2Na_2O_7S_2$

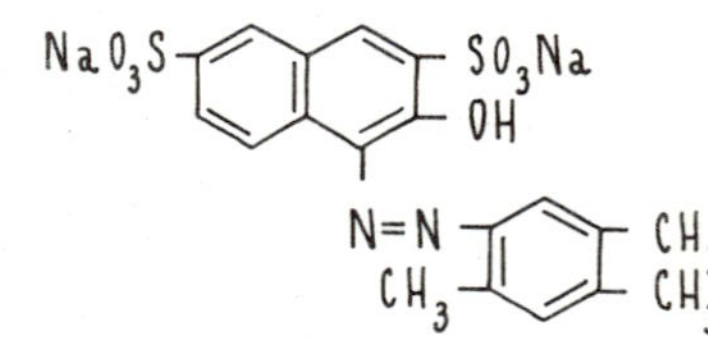

MOLECULAR WEIGHT: 494.46

SOURCE OF SAMPLE: American Cyanamid Company
Dyes and Textiles Department
Bound Brook, New Jersey

CAS REGISTRY NO.: 3564-09-8

NIOSH NUMBER: QJ66500

KBr Wafer

CONDENSED PHASE

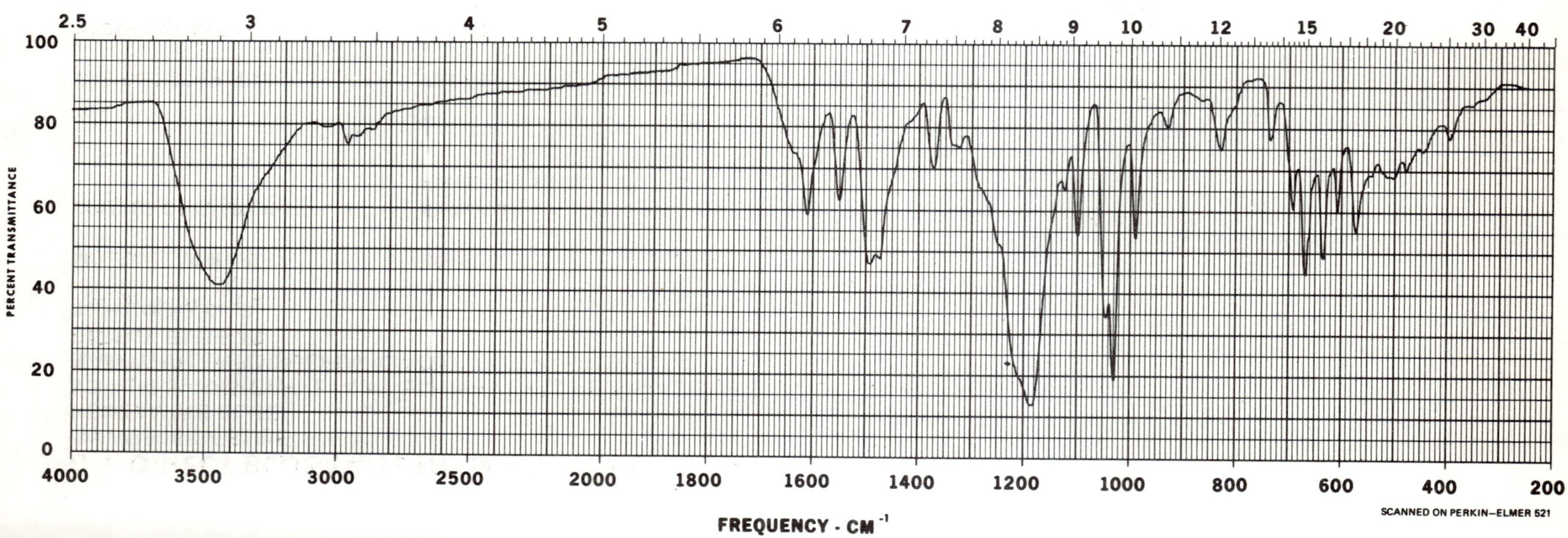

C. I. SOLVENT ORANGE 2
1–(o–TOLYLAZO)–2–NAPHTHOL
FAT ORANGE II
OIL ORANGE OPEL

$C_{17}H_{14}N_2O$

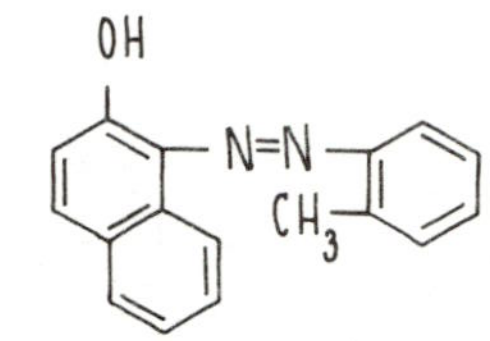

MOLECULAR WEIGHT: 262.31

SOURCE OF SAMPLE: Allied Chemical Corporation
Specialty Chemicals Division
Morristown, New Jersey

CAS REGISTRY NO.: 2646-17-5

NIOSH NUMBER: QL54250

KBr Wafer

CONDENSED PHASE

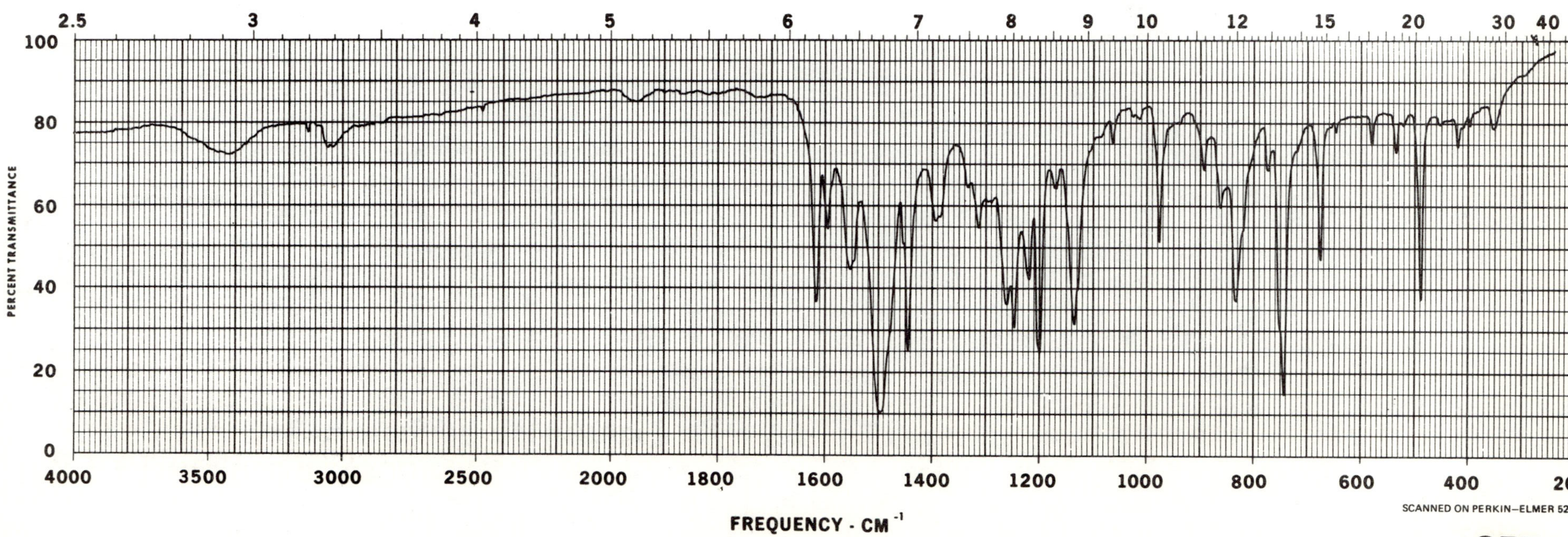

SCANNED ON PERKIN–ELMER 521

C. I. SOLVENT RED 80
1–[(2,5–DIMETHOXYPHENYL)AZO] –2–NAPHTHOL
CITRUS RED 2

$C_{18}H_{16}N_2O_3$

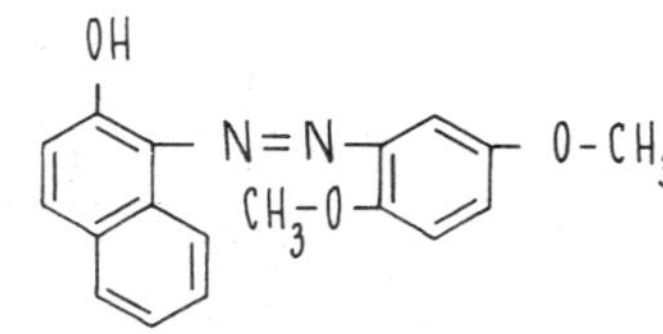

MOLECULAR WEIGHT: 308.34

SOURCE OF SAMPLE: Allied Chemical Corporation
Specialty Chemicals Department
Morristown, New Jersey

CAS REGISTRY NO.: 6358-53-8

NIOSH NUMBER: QL36750

KBr Wafer

CONDENSED PHASE

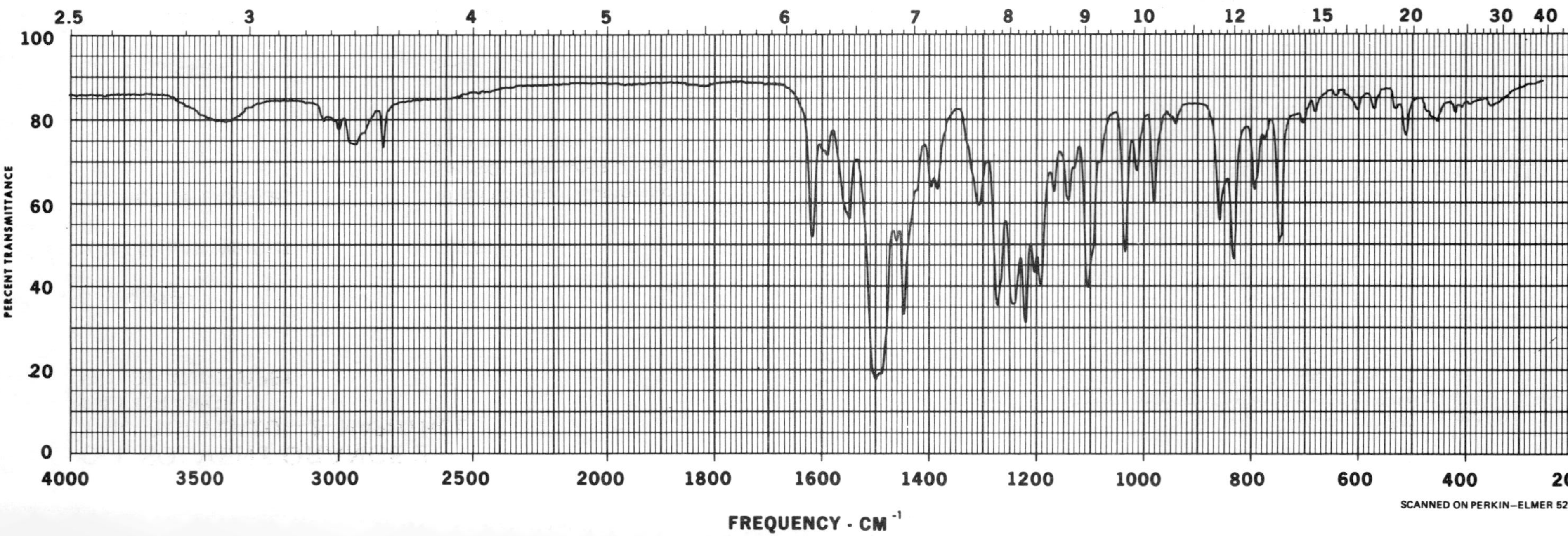

C. I. SOLVENT YELLOW 2

N,N–DIMETHYL–p–(PHENYLAZO)ANILINE
BUTTER YELLOW
METHYL YELLOW
p–(DIMETHYLAMINO)AZOBENZENE

$C_{14}H_{15}N_3$

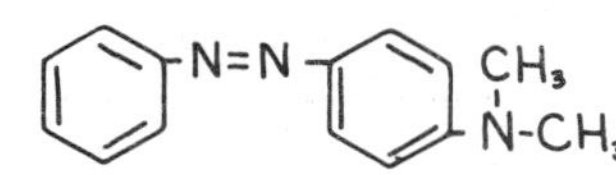

MOLECULAR WEIGHT:	225.30
SOURCE OF SAMPLE:	Eastman Organic Chemicals Rochester, New York
CAS REGISTRY NO.:	60-11-7
NIOSH NUMBER:	BX73500

KBr Wafer

CONDENSED PHASE

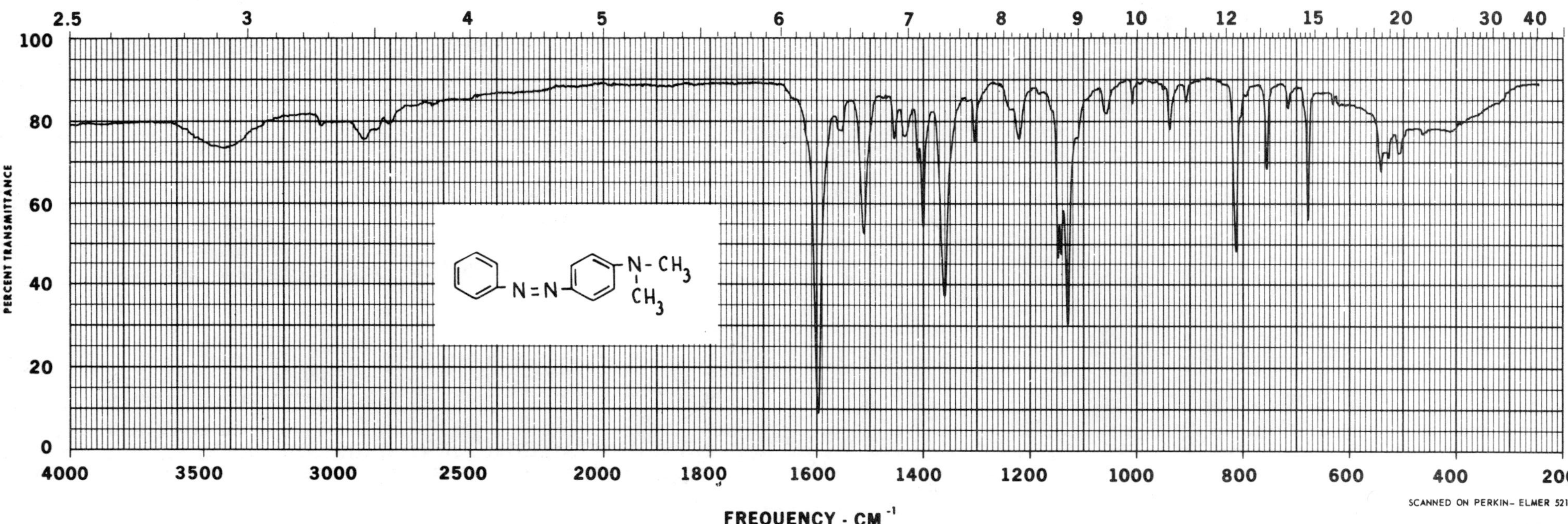

C. I. SOLVENT YELLOW 34
4,4′–IMIDOCARBONYLBIS[N,N–DIMETHYLANILINE]
AURAMINE
WAXOLINE YELLOW 0
YELLOW PYOCTANINE

$C_{17}H_{21}N_3$

MOLECULAR WEIGHT: 267.38

SOURCE OF SAMPLE: L & R Dyestuffs Corporation, Clifton, New Jersey

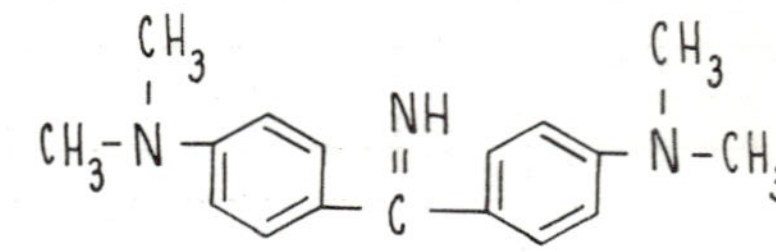

CAS REGISTRY NO.: 492-80-8

NIOSH NUMBER: BY35000

KBr Wafer

CONDENSED PHASE

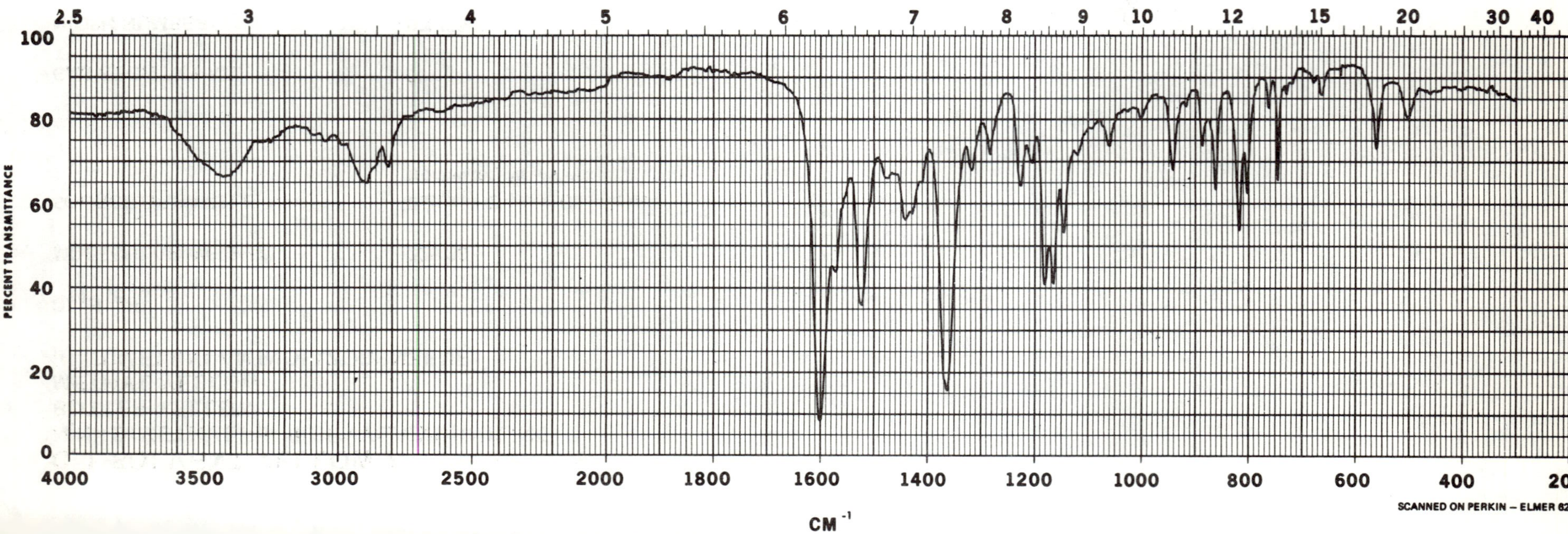

BIS(ACETATO)HEXAMETAARSENITOTETRACOPPER
CUPRIC ACETOARSENITE
(ACETATO)TRIMETAARSENITODICOPPER
COPPER ACETATE ARSENITE
PARIS GREEN

$C_4H_6As_6Cu_4O_{16}$

$Cu(C_2H_3O_2)_2 \cdot 3Cu(AsO_2)_2$

MOLECULAR WEIGHT: 1013.77

SOURCE OF SAMPLE: Chipman Chemical Company, Inc.
New Brunswick, New Jersey

CAS REGISTRY NO.: 1299-88-3

NIOSH NUMBER: GL64750

KBr Wafer

CONDENSED PHASE

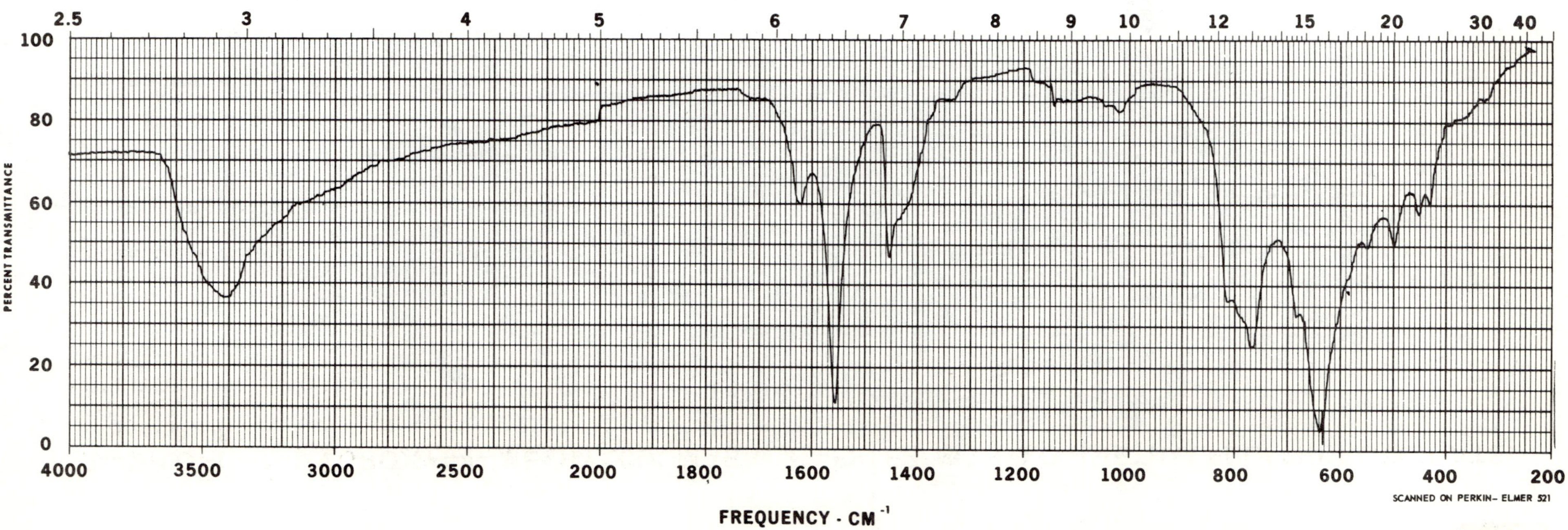

3–CHLORO–7–HYDROXY–4–METHYLCOUMARIN, O–ESTER WITH O,O–DIETHYLPHOSPHOROTHIOATE

COUMAPHOS
CO–RAL
RESITOX
SUNTOL

$C_{14}H_{16}ClO_5PS$

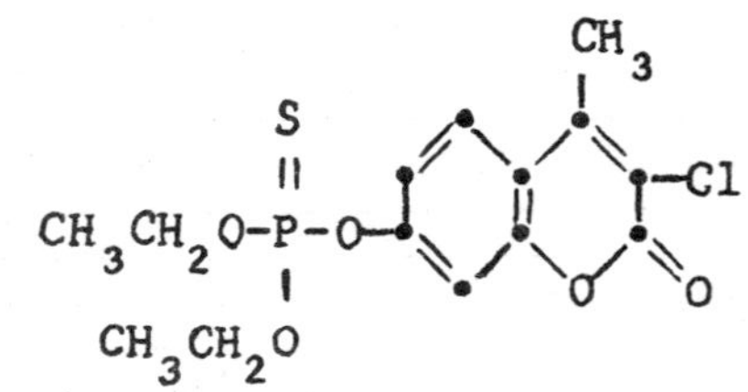

MOLECULAR WEIGHT: 362.77

SOURCE OF SAMPLE: Pesticides Repository
U. S. Public Health Service
Perrine, Florida

CAS REGISTRY NO.: 56-72-4

NIOSH NUMBER: GN63000

KBr Wafer

CONDENSED PHASE

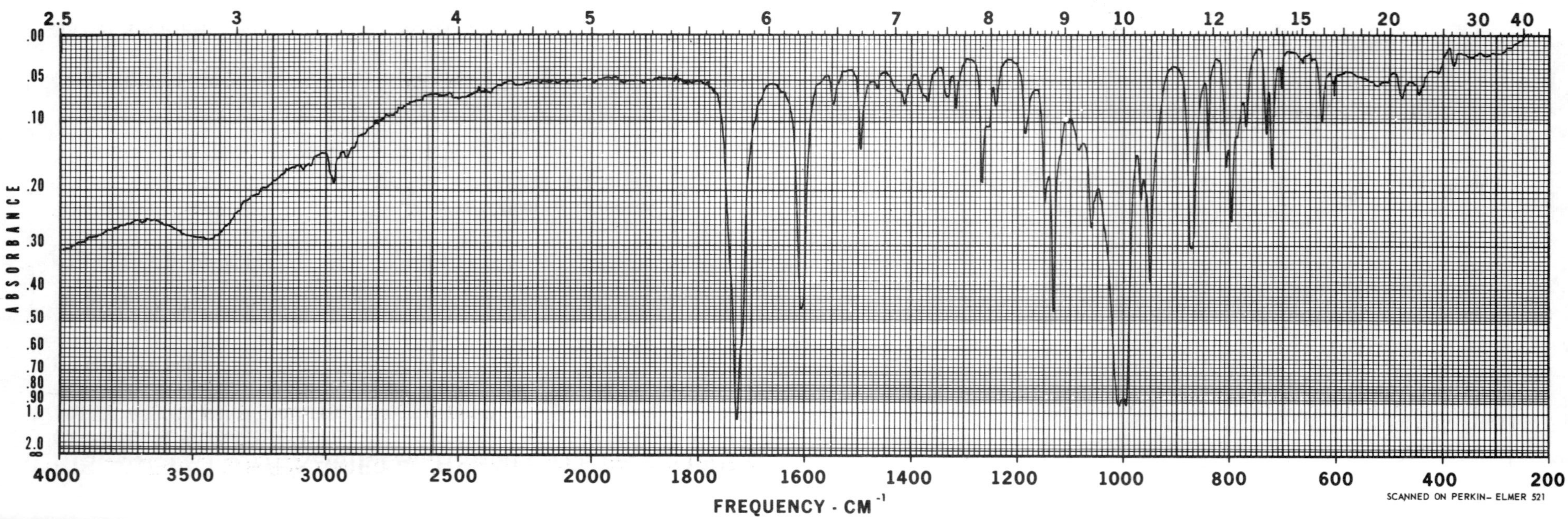

CRESOL
CRESYLIC ACID

C_7H_8O

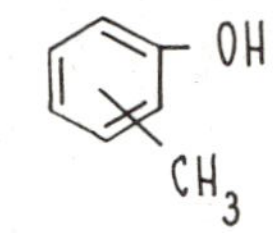

MOLECULAR WEIGHT:	108.14
SOURCE OF SAMPLE:	Allied Chemical Corporation New York, New York
CAS REGISTRY NO.:	1319-77-3
NIOSH NUMBER:	GO59500

Between Salts

CONDENSED PHASE

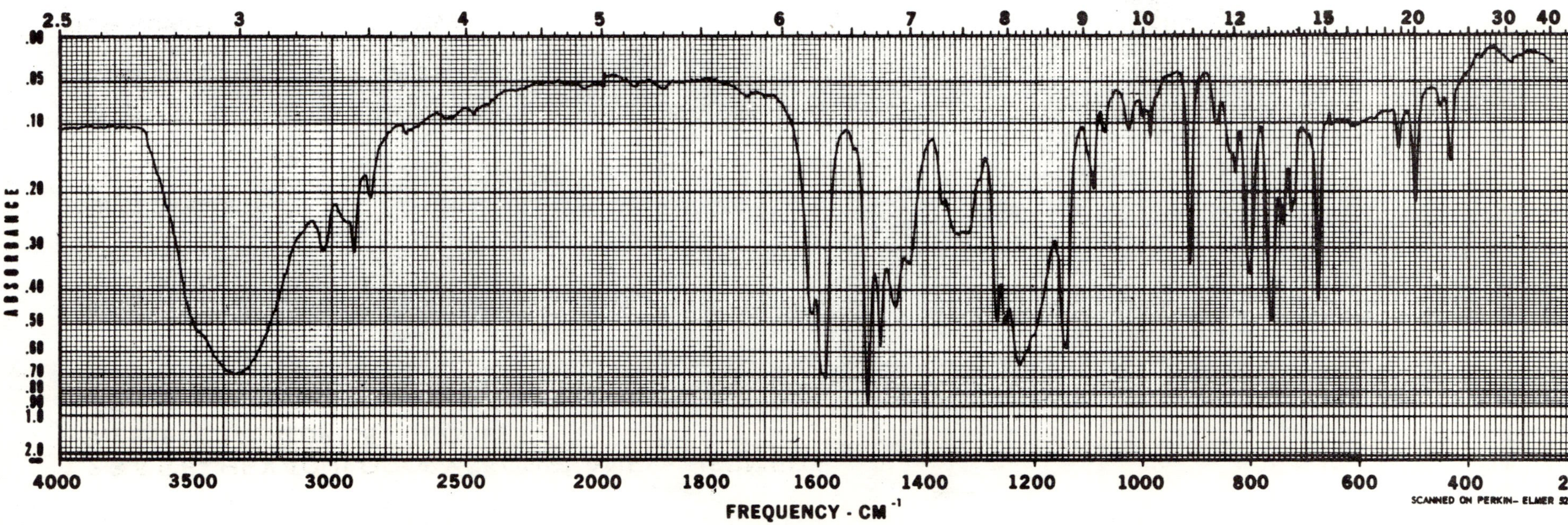

1,2,3,4,5,6–HEXACHLOROCYCLOHEXANE, *a*–ISOMER
BENZENE HEXACHLORIDE, *a*–ISOMER
***a*–BHC**
***a*–LINDANE**

$C_6H_6Cl_6$

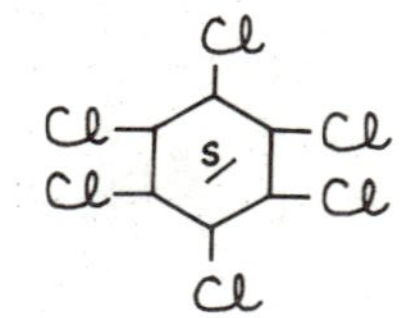

MOLECULAR WEIGHT: 290.83

SOURCE OF SAMPLE: MCB Manufacturing Chemists
Norwood, Ohio

CAS REGISTRY NO.: 319-84-6

NIOSH NUMBER: GV35000

KBr Wafer

CONDENSED PHASE

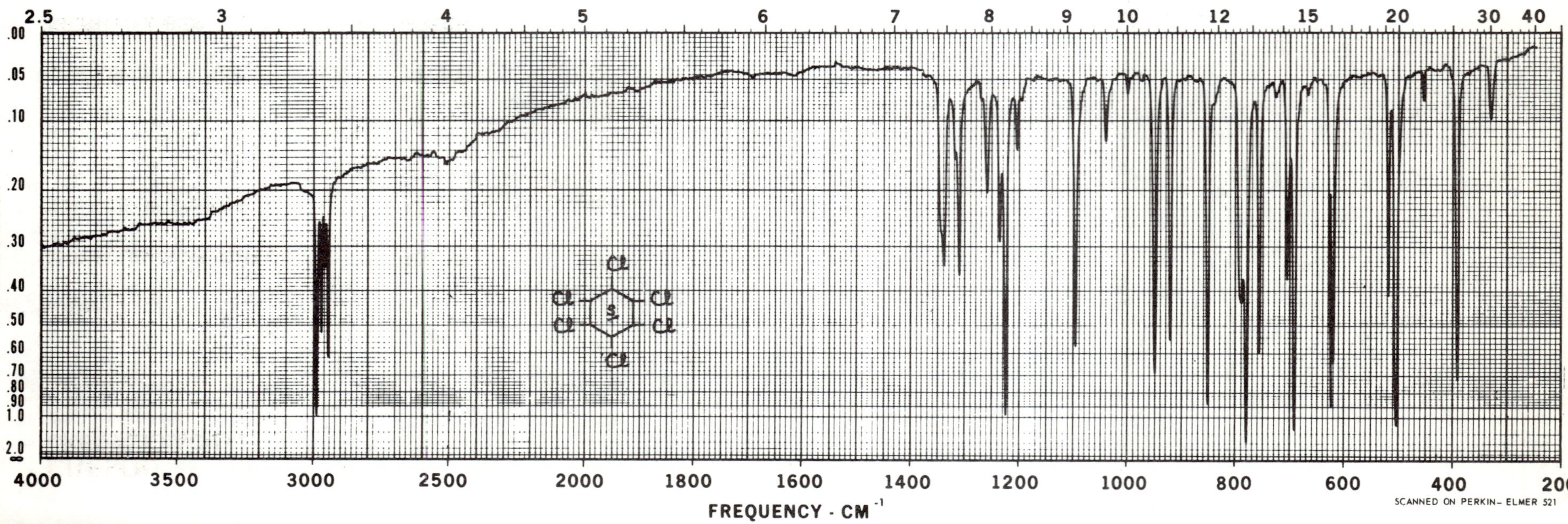

1,2,3,4,5,6–HEXACHLOROCYCLOHEXANE, β–ISOMER
BENZENE HEXACHLORIDE, β–ISOMER
β–BHC
β–LINDANE

$C_6H_6Cl_6$

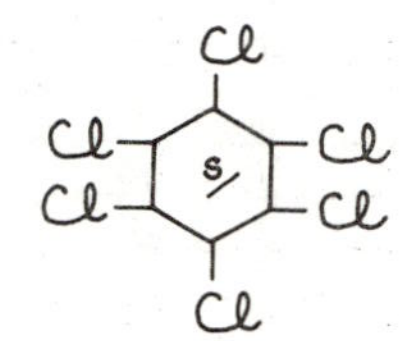

MOLECULAR WEIGHT: 290.83

SOURCE OF SAMPLE: Pennwalt Corporation
King of Prussia, Pennsylvania

CAS REGISTRY NO.: 391-85-7

NIOSH NUMBER: GV43750

KBr Wafer

CONDENSED PHASE

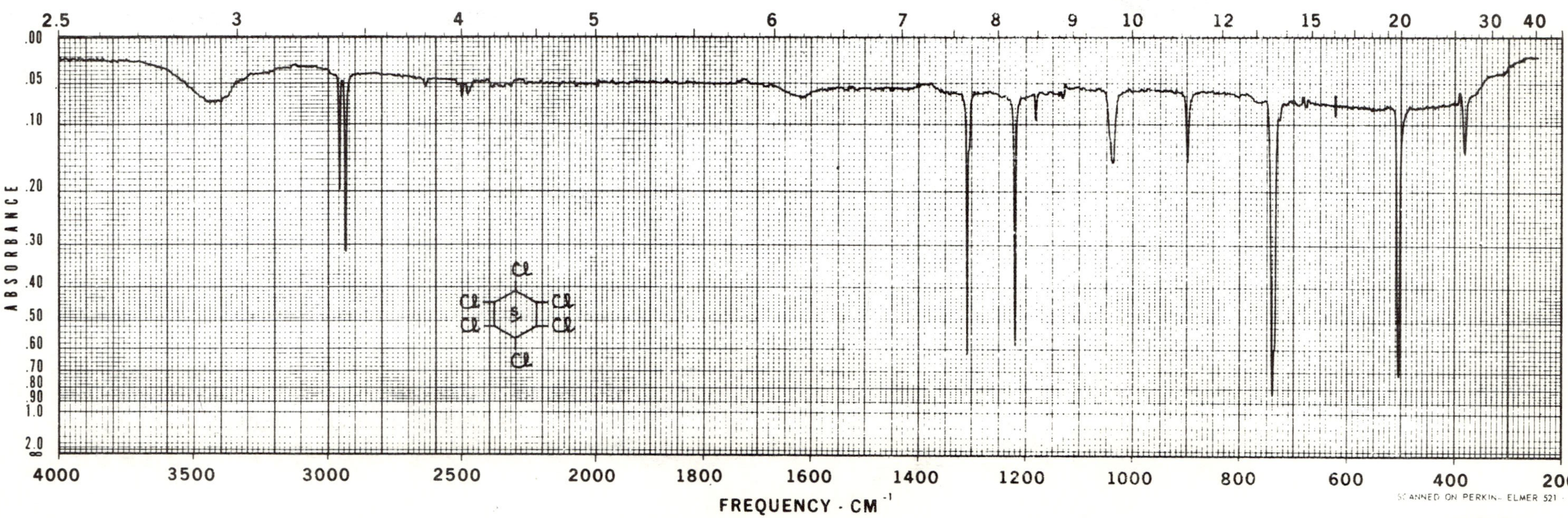

1,2,3,4,5,6–HEXACHLOROCYCLOHEXANE, δ–ISOMER
BENZENE HEXACHLORIDE, δ–ISOMER
δ–BHC
δ–LINDANE

$C_6H_6Cl_6$

MOLECULAR WEIGHT: 290.83

SOURCE OF SAMPLE: Aldrich Chemical Company, Inc.
Milwaukee, Wisconsin

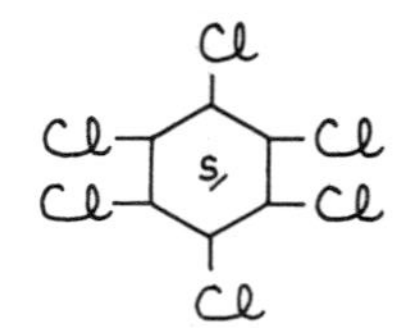

CAS REGISTRY NO.: 319-86-8

NIOSH NUMBER: GV45500

KBr Wafer

CONDENSED PHASE

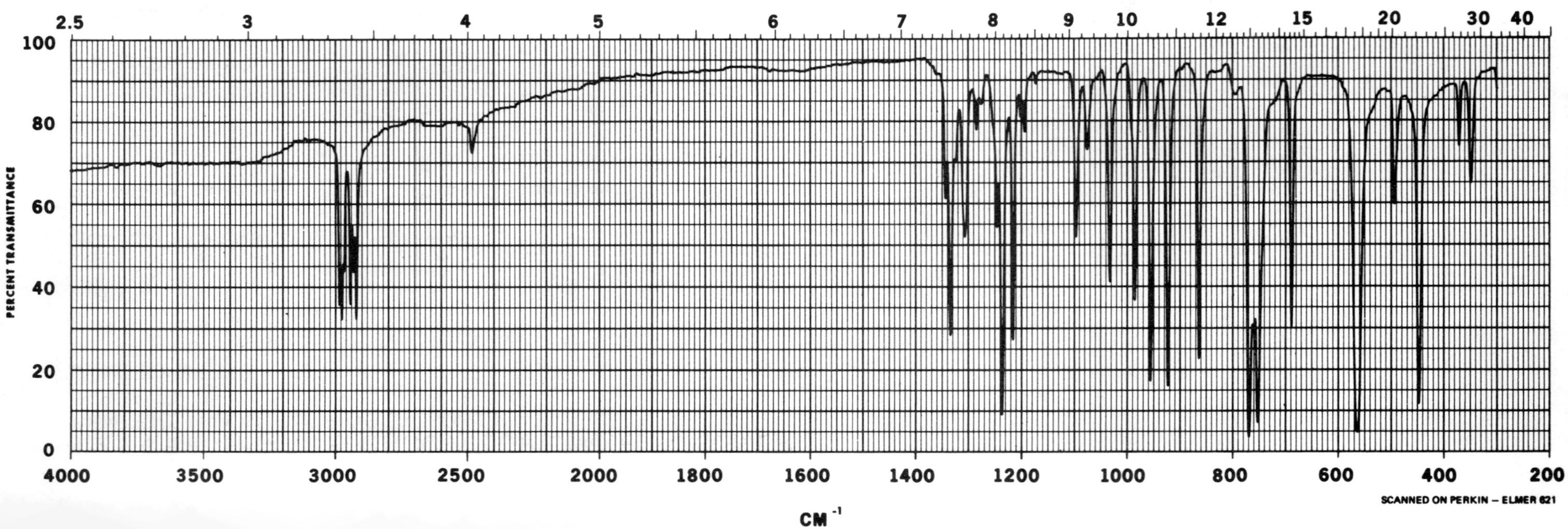

1,2,3,4,5,6–HEXACHLOROCYCLOHEXANE, γ–ISOMER
AFICIDE
BENZENE HEXACHLORIDE, γ–ISOMER
LINDANE
KWELL

$C_6H_6Cl_6$

MOLECULAR WEIGHT: 290.83

SOURCE OF SAMPLE: James Hinton, Valparaiso, Florida

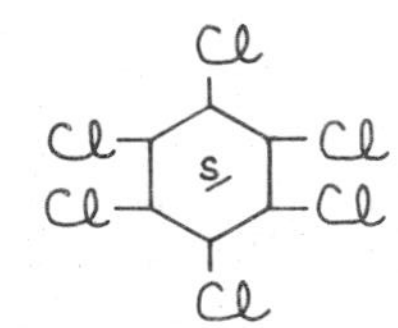

CAS REGISTRY NO.: 58-89-9

NIOSH NUMBER: GV49000

KBr Wafer

CONDENSED PHASE

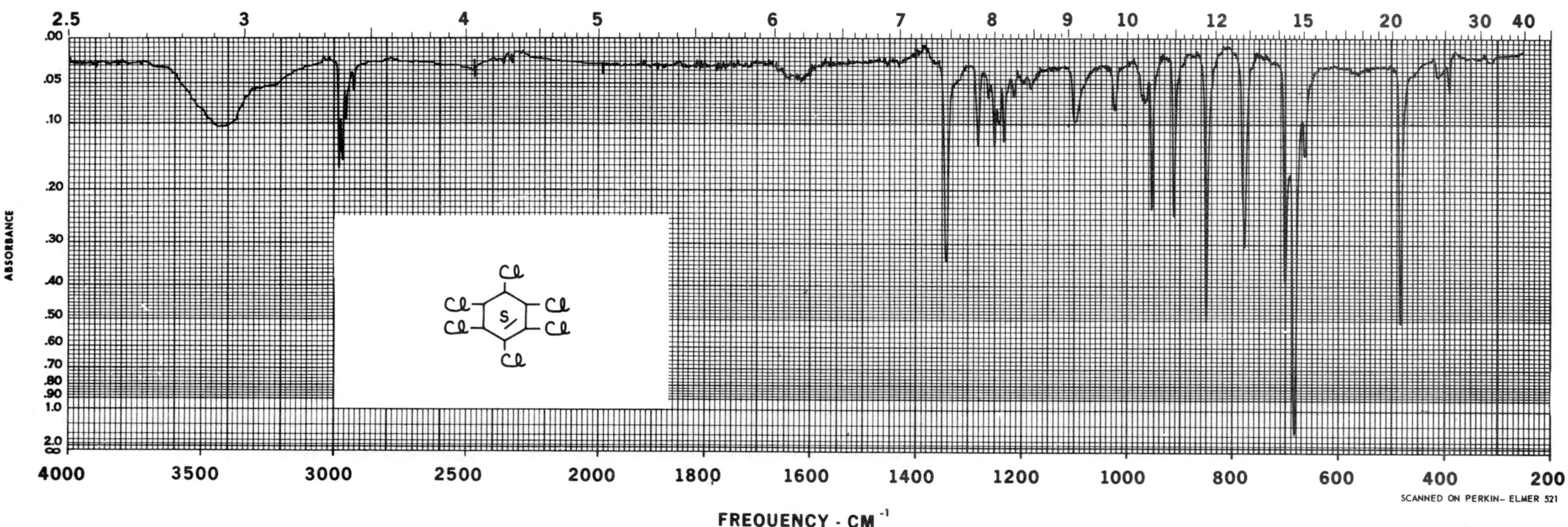

1,2,3,4,5,6–HEXACHLOROCYCLOHEXANE (MIXED ISOMERS)
BENZENE HEXACHLORIDE (MIXED ISOMERS)

$C_6H_6Cl_6$

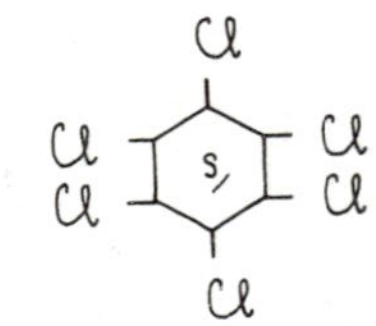

MOLECULAR WEIGHT:	290.83
SOURCE OF SAMPLE:	Diamond Shamrock Corporation Cleveland, Ohio
CAS REGISTRY NO.:	60291-32-9
NIOSH NUMBER:	GV49500

KBr Wafer

CONDENSED PHASE

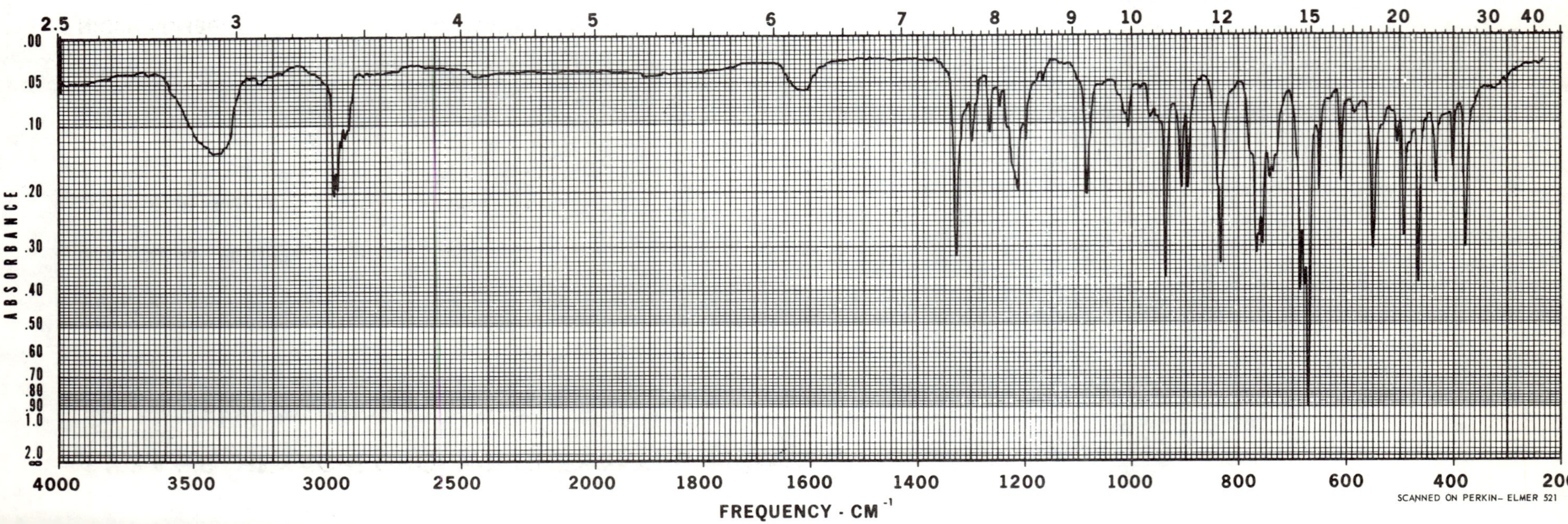

N–[(TRICHLOROMETHYL)THIO] –4–CYCLOHEXENE–1,2–DICARBOXIMIDE
CAPTAN
ORTHOCIDE
AMERCIDE
VANGUARD K

$C_9H_8Cl_3NO_2S$

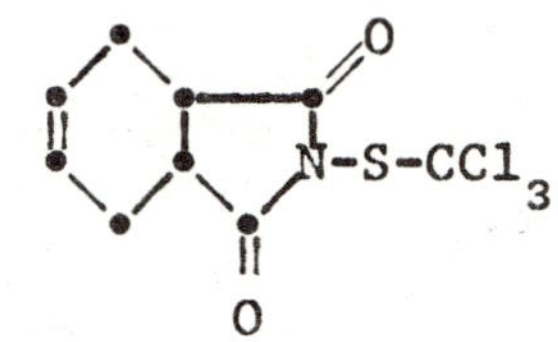

MOLECULAR WEIGHT: 300.59

SOURCE OF SAMPLE: Ortho Pharmaceutical Corporation
Raritan, New Jersey

CAS REGISTRY NO.: 133-06-2

NIOSH NUMBER: GW50750

KBr Wafer

CONDENSED PHASE

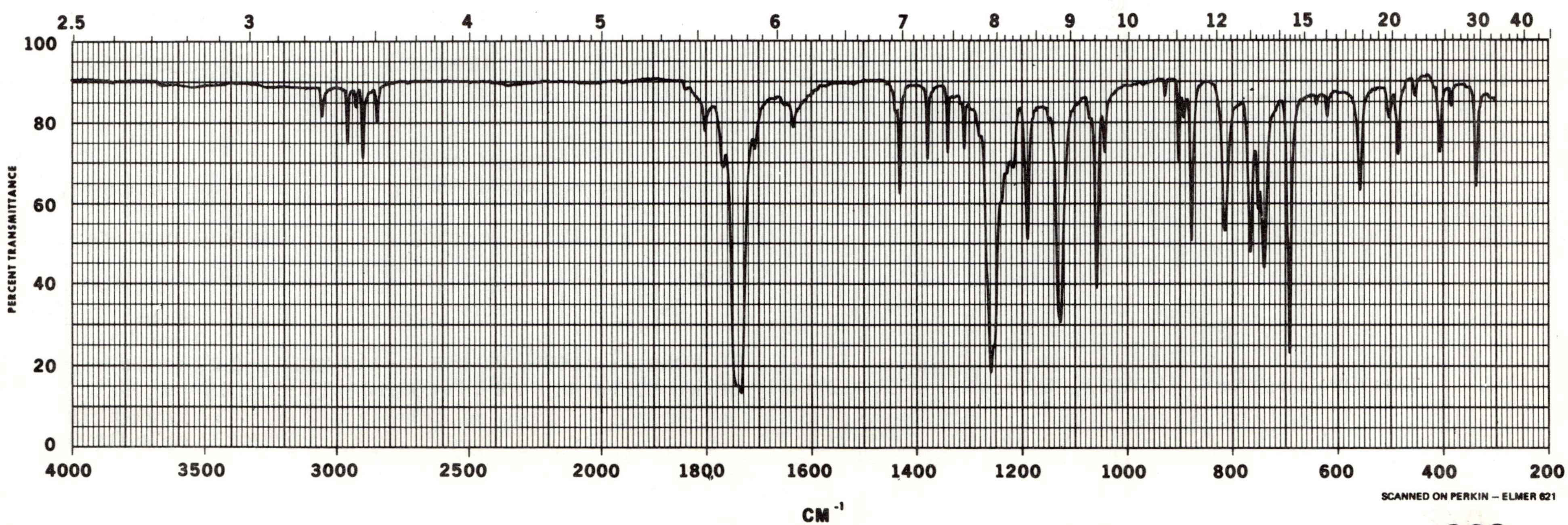

DAUNOMYCIN
RUBOMYCIN C
(1S,3S)–3–ACETYL–6,11–DIOXO–1,2,3,4,6,11–HEXAHYDRO–10–METHOXY–3,5,12–TRIHYDROXY–1–NAPHTHACENYL 3–AMINO–2,3,6–TRIDEOXY–α–L–lyxo–HEXOPYRANOSIDE
RUBIDOMYCIN
CERUBIDIN
$C_{27}H_{29}NO_{10}$

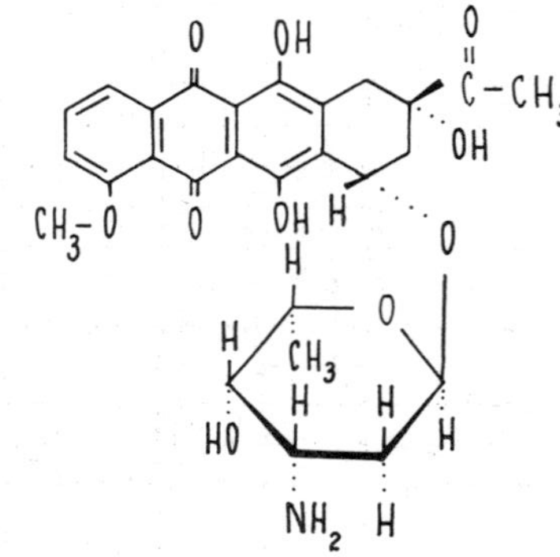

MOLECULAR WEIGHT: 527.53

SOURCE OF SAMPLE: Sigma Chemical Company
St. Louis, Missouri

CAS REGISTRY NO.: 20830-81-3

NIOSH NUMBER: HB78750

KBr Wafer

CONDENSED PHASE

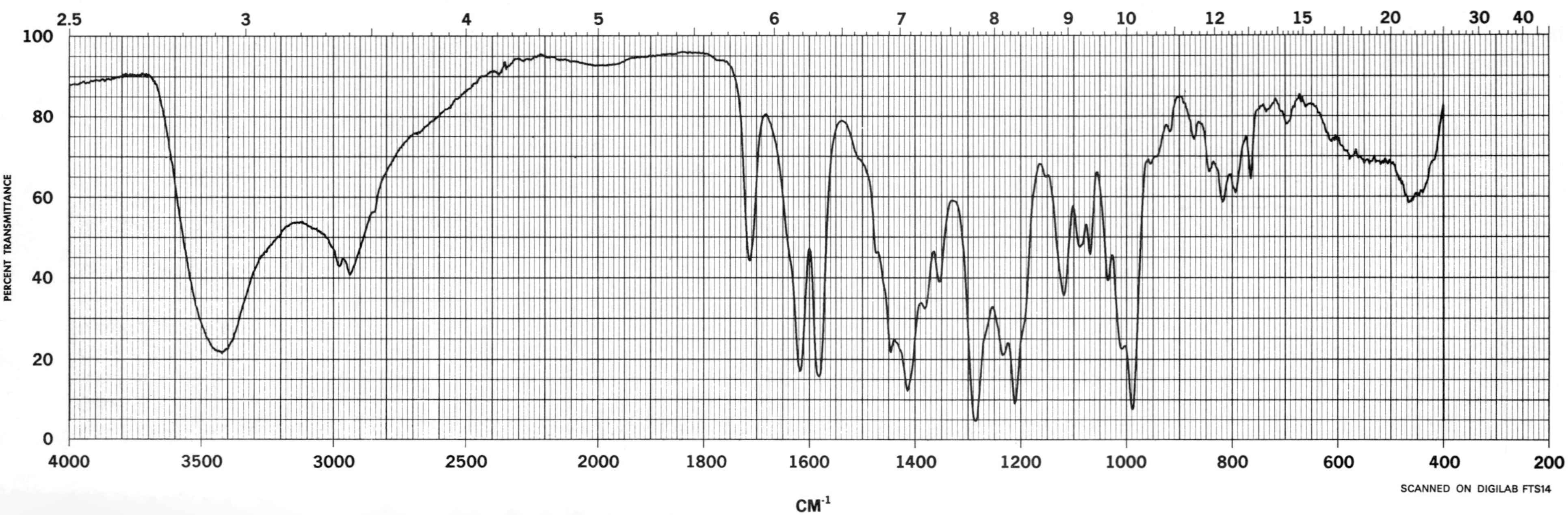

DIBENZ[a,h] ACRIDINE
1,2:5,6–DIBENZACRIDINE

$C_{21}H_{13}N$

MOLECULAR WEIGHT: 279.34

SOURCE OF SAMPLE: Aldrich Chemical Company, Inc.
Milwaukee, Wisconsin

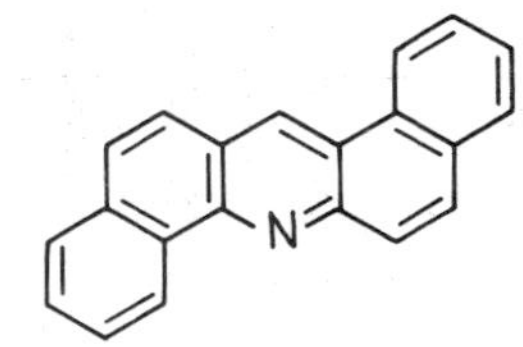

CAS REGISTRY NO.: 226-36-8

NIOSH NUMBER: HN08750

KBr Wafer

CONDENSED PHASE

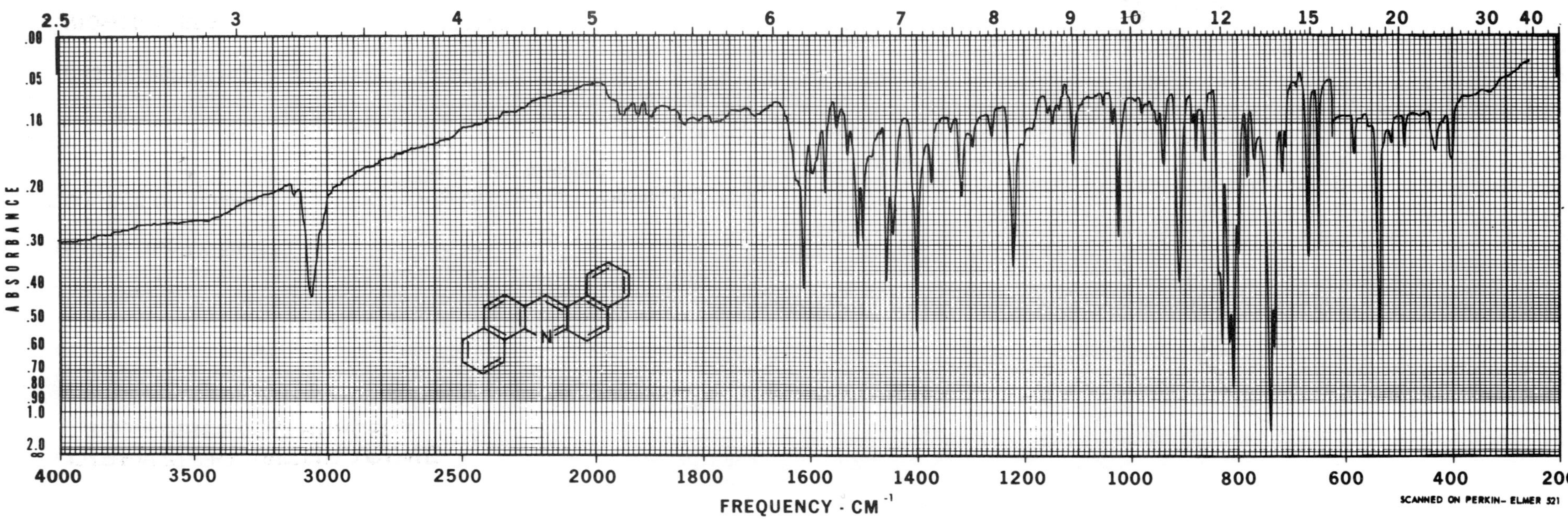

DIBENZ[a,h] ANTHRACENE
1,2:5,6–DIBENZANTHRACENE

$C_{22}H_{14}$

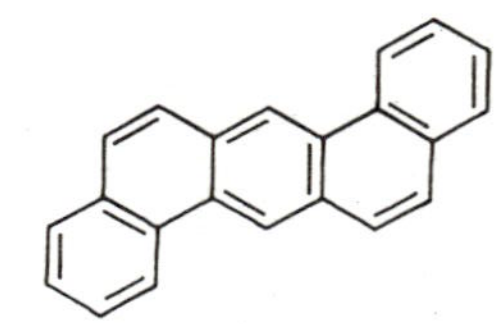

MOLECULAR WEIGHT: 278.36

SOURCE OF SAMPLE: Aldrich Chemical Company, Inc.
Milwaukee, Wisconsin

CAS REGISTRY NO.: 53-70-3

NIOSH NUMBER: HN26250

KBr Wafer

CONDENSED PHASE

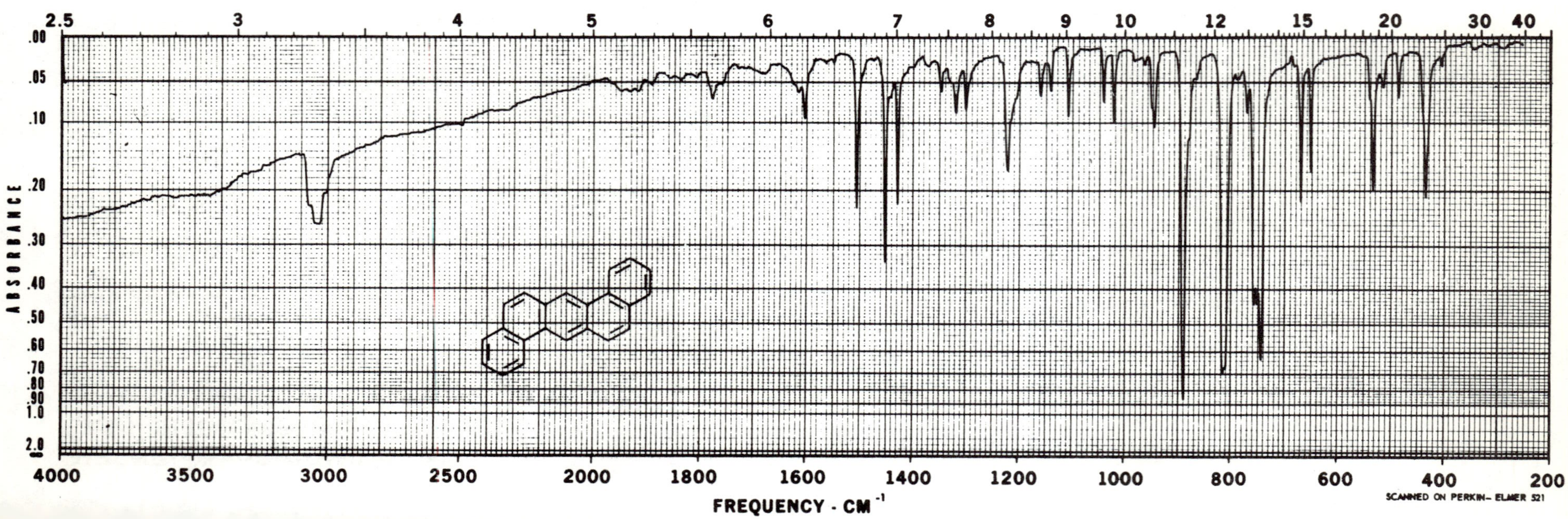

7H–DIBENZO[c,g] CARBAZOLE
3,4:5,6–DIBENZOCARBAZOLE

$C_{20}H_{13}N$

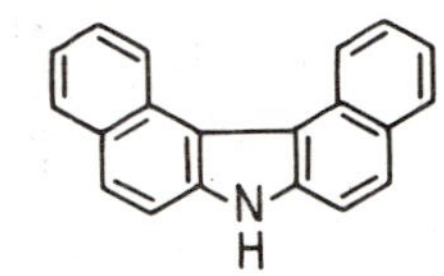

MOLECULAR WEIGHT: 267.33

SOURCE OF SAMPLE: Aldrich Chemical Company, Inc.
Milwaukee, Wisconsin

CAS REGISTRY NO.: 194-59-2

NIOSH NUMBER: HO56000

KBr Wafer

CONDENSED PHASE

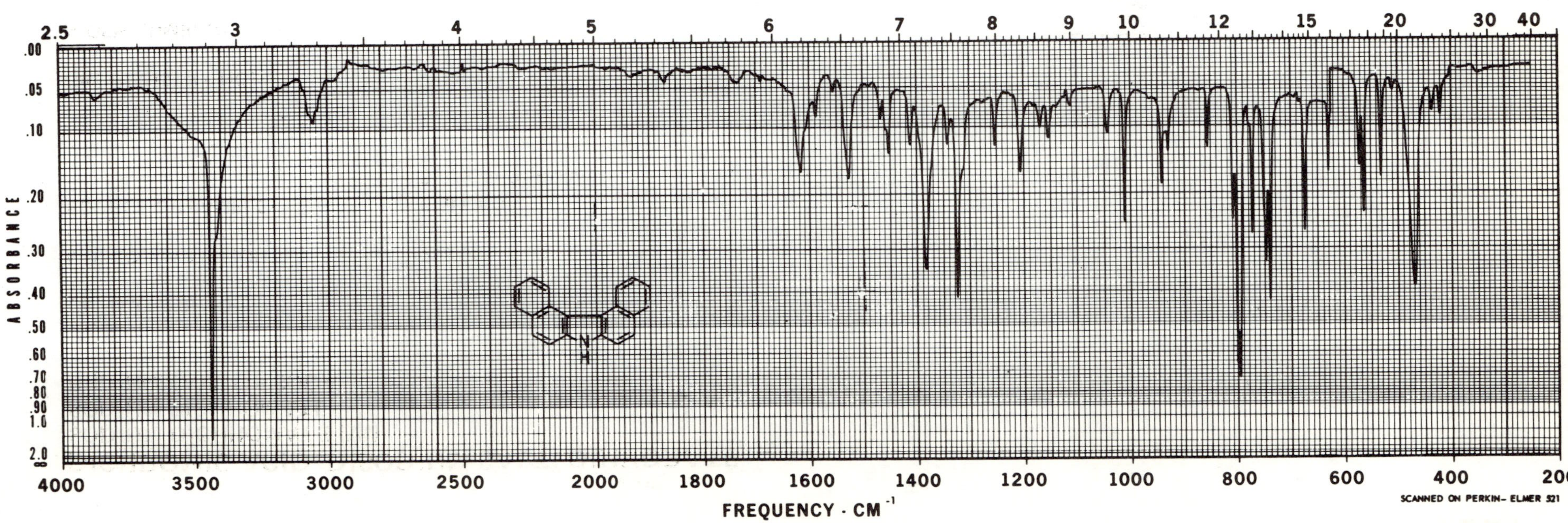

DICHROMIC ACID, DISODIUM SALT, DIHYDRATE
SODIUM DICHROMATE (VI)
SODIUM BICHROMATE
BICHROMATE OF SODA

$Cr_2Na_2O_7 \cdot 2H_2O$

$Na_2Cr_2O_7 \cdot 2H_2O$

MOLECULAR WEIGHT: 298.00

SOURCE OF SAMPLE: Fluka AG
Buchs, Switzerland

CAS REGISTRY NO.: 7789-12-0

NIOSH NUMBER: –

4000 - 1350 cm^{-1} = Mulled in Perfluorinated Hydrocarbon
1350 - 400 cm^{-1} = Mulled in Mineral Oil

CONDENSED PHASE

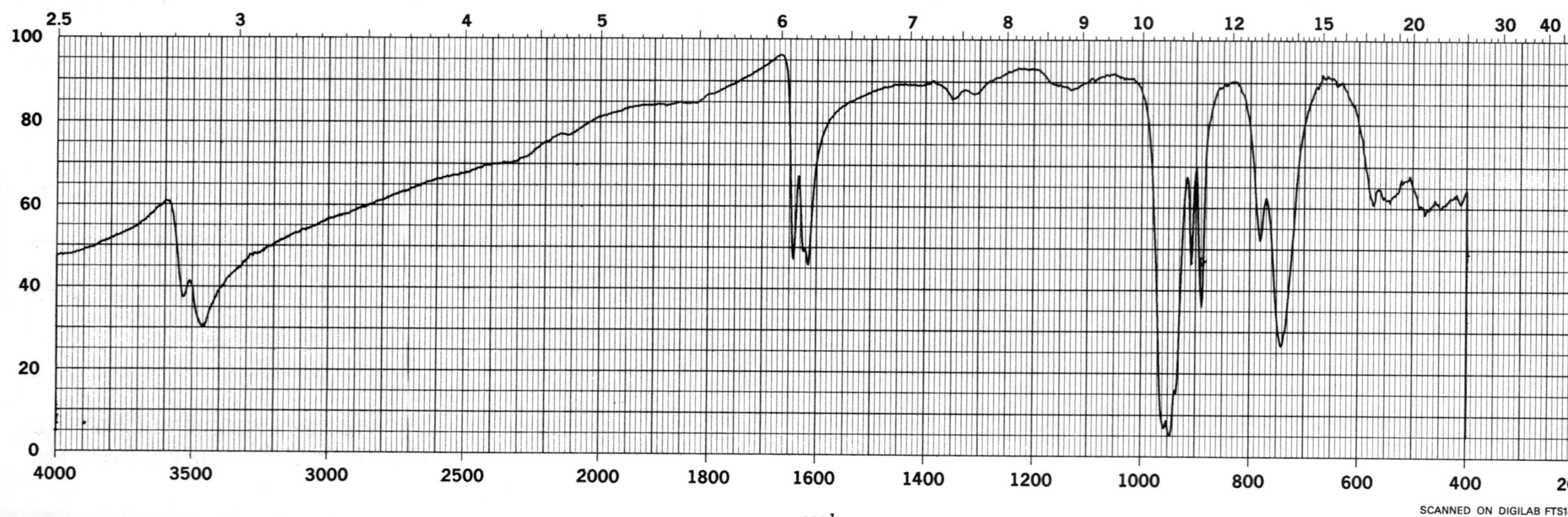

endo–,endo–6,7–EPOXY–1,2,3,4,10,10–HEXACHLORO–1,4,4a,5,6,7,8,8a–OCTAHYDRO–1,4:5,8–DIMETHANONAPHTHALENE
ENDRIN
HEXADRIN
MENDRIN

$C_{12}H_8Cl_6O$

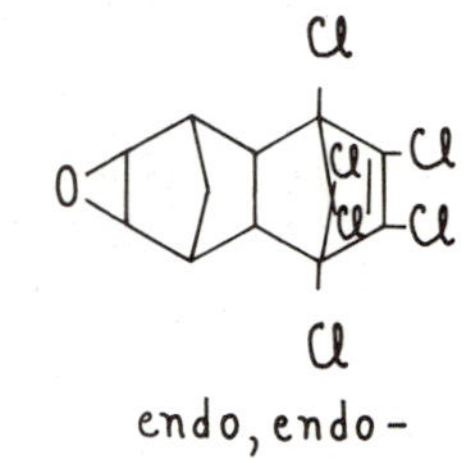

MOLECULAR WEIGHT: 380.91

SOURCE OF SAMPLE: Tokyo Kasei Kogyo Company Ltd.
Tokyo, Japan

CAS REGISTRY NO.: 72-20-8

NIOSH NUMBER: IO15750

KBr Wafer

CONDENSED PHASE

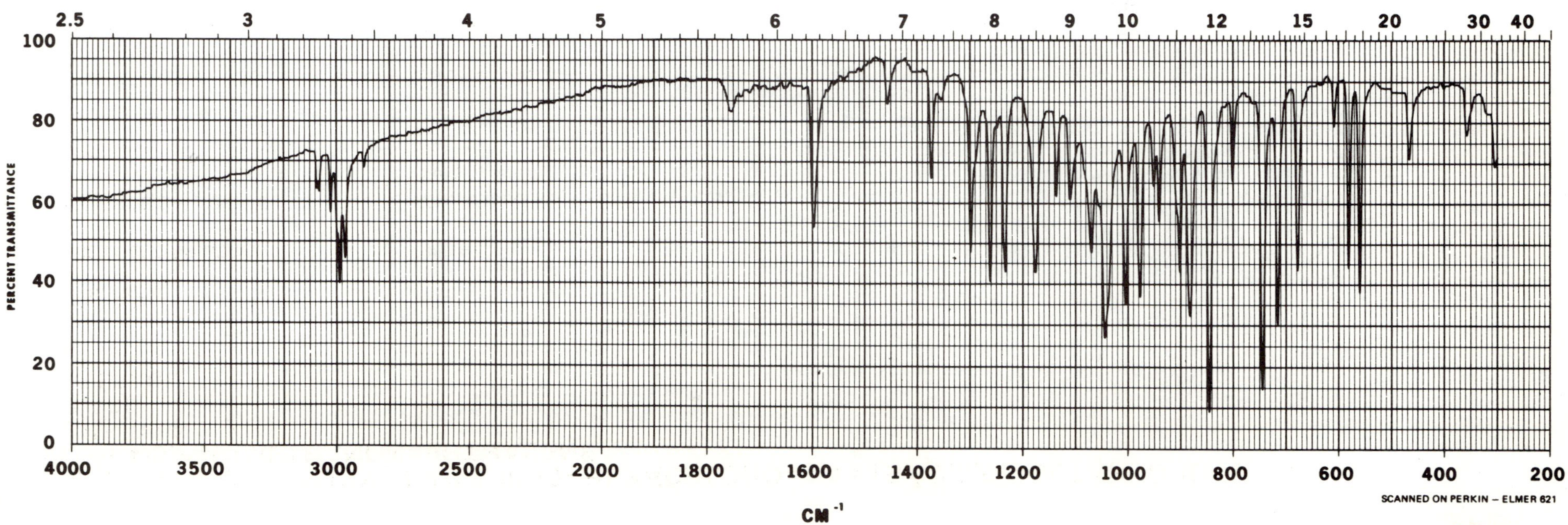

SCANNED ON PERKIN – ELMER 621

endo–,exo–6,7–EPOXY–1,2,3,4,10,10–HEXACHLORO–1,4,4a,5,6,7,8,8a–OCTAHYDRO–1,4:5,8–DIMETHANONAPHTHALENE
DIELDRIN
HEOD
ILLOXOL

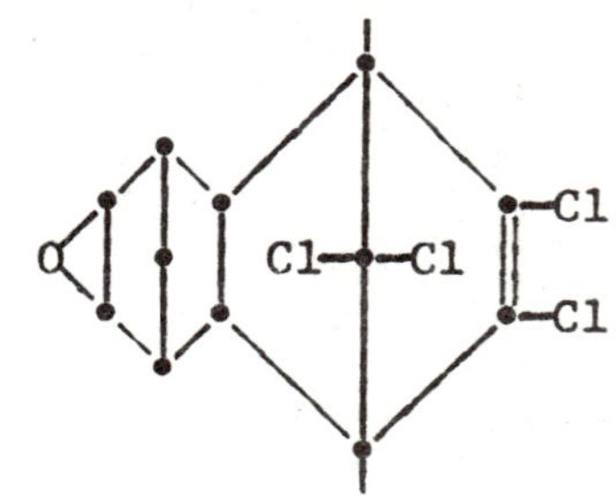

$C_{12}H_8Cl_6O$

MOLECULAR WEIGHT: 380.91

SOURCE OF SAMPLE: Tokyo Kasei Kogyo Company Ltd.
Tokyo, Japan

CAS REGISTRY NO.: 60-57-1

NIOSH NUMBER: IO17500

KBr Wafer

CONDENSED PHASE

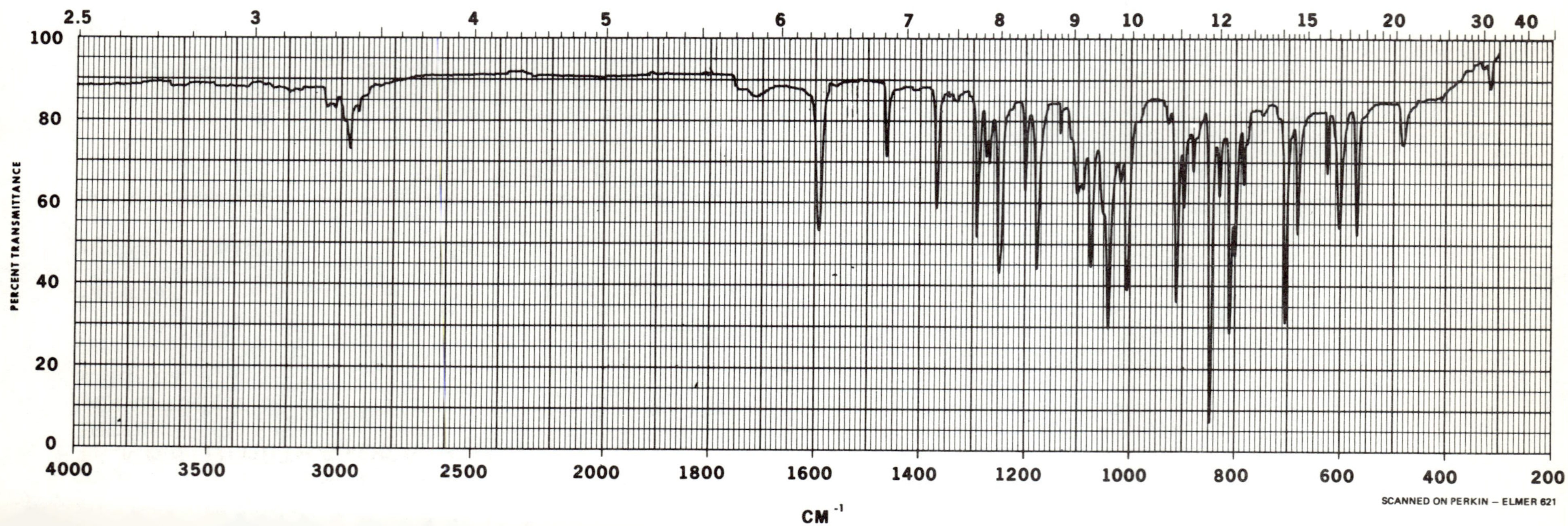

N–NITROSODIPHENYLAMINE
REDAX
VULCATARD
WILTROL N
RETARDER 2N

$C_{12}H_{10}N_2O$

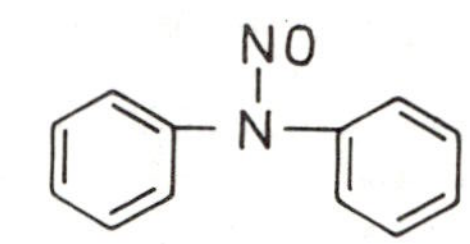

MOLECULAR WEIGHT:	198.23
SOURCE OF SAMPLE:	A. Gennaro Philadelphia College of Pharmacy and Science Philadelphia, Pennsylvania
CAS REGISTRY NO.:	86-30-6
NIOSH NUMBER:	JJ98000
CAPILLARY CELL:	Melt

CONDENSED PHASE

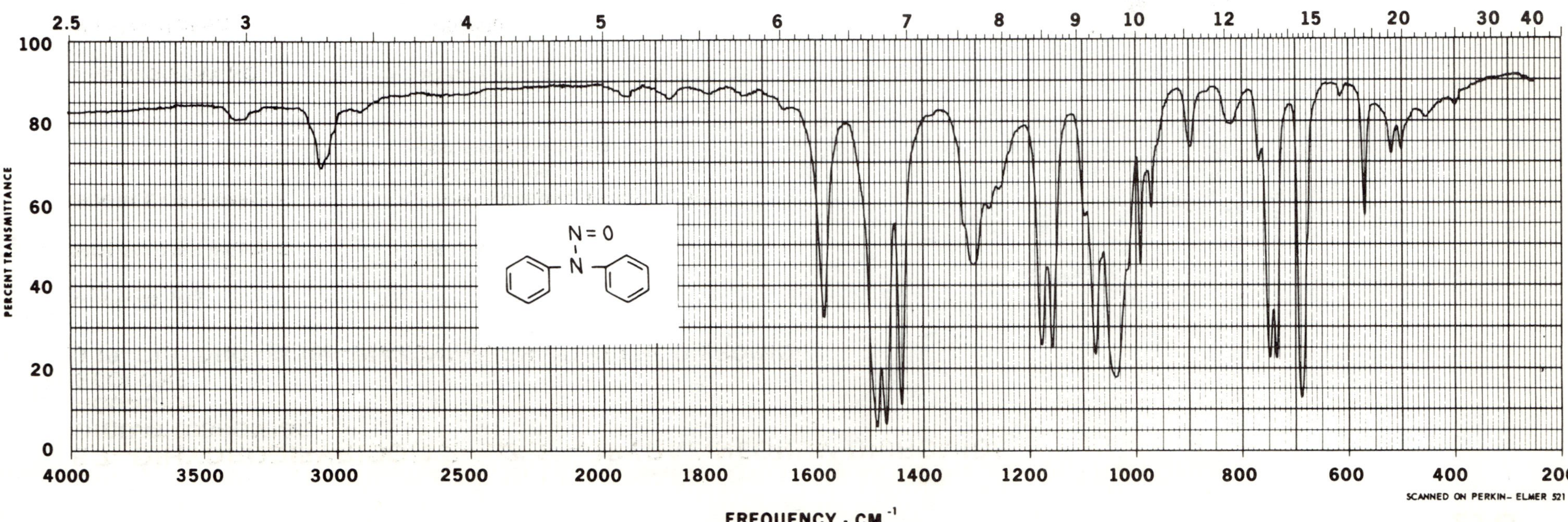

6,7–DIHYDRODIPYRIDO[1,2–a:2′,1′–c] PYRAZINEDIIUM DIBROMIDE
DIQUAT DIBROMIDE
AQUACIDE
REGLONE
9,10–DIHYDRO–8a,10a–DIAZONIAPHENANTHRENE DIBROMIDE

$C_{12}H_{12}Br_2N_2$

MOLECULAR WEIGHT: 344.06

SOURCE OF SAMPLE: Chevron Chemical Company, Inc.
Ortho Division
Richmond, California

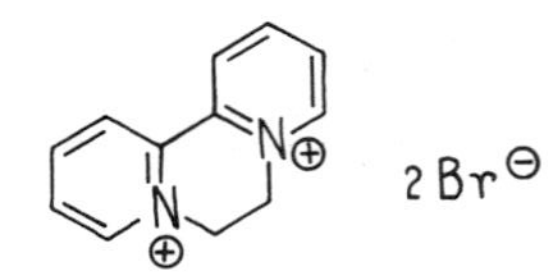

CAS REGISTRY NO.: 85-00-7

NIOSH NUMBER: JM56900

KBr Wafer

CONDENSED PHASE

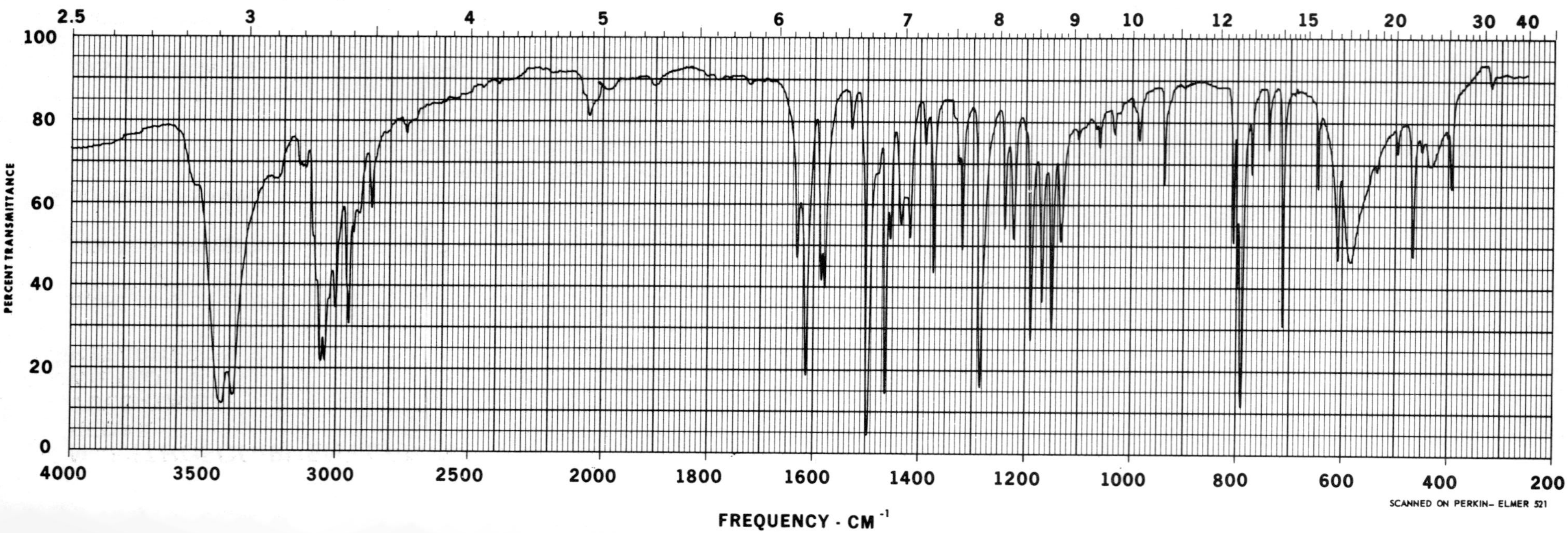

2,2a,3,3,4,7—HEXACHLORODECAHYDRO—1,2,4—METHENOCYCLOPENTA[cd]PENTALENE—5—CARBOXALDEHYDE
ENDRIN ALDEHYDE

$C_{12}H_8Cl_6O$

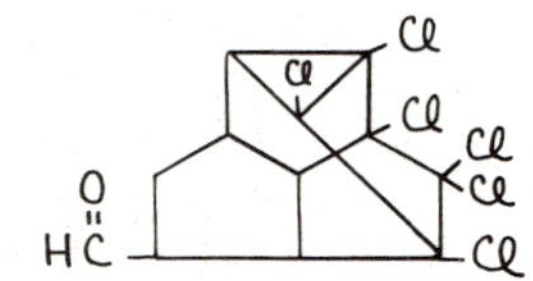

MOLECULAR WEIGHT: 380.91

SOURCE OF SAMPLE: Chem Service Company
West Chester, Pennsylvania

CAS REGISTRY NO.: 7421-93-4

NIOSH NUMBER: -

KBr Wafer

CONDENSED PHASE

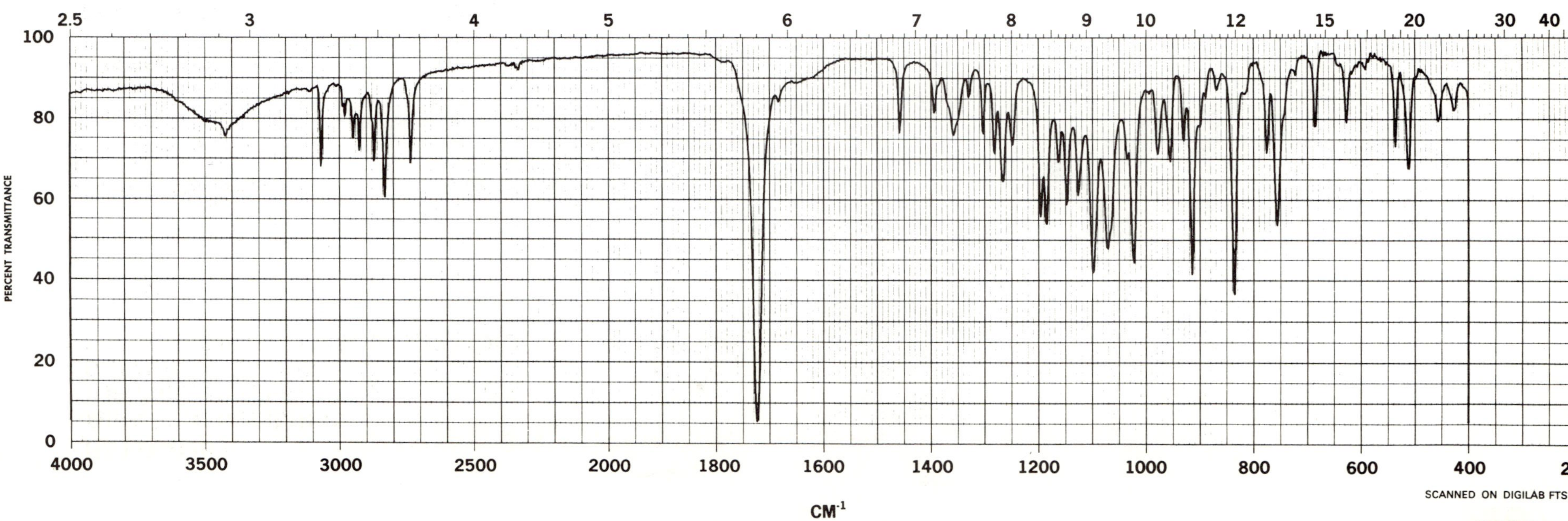

SCANNED ON DIGILAB FTS14

ESTRADIOL
DIHYDROFOLLICULIN
DIHYDROTHEELIN
DIHYDROXYESTRIN
ESTRA–1,3,5(10)–TRIENE–3,17β–DIOL

$C_{18}H_{24}O_2$

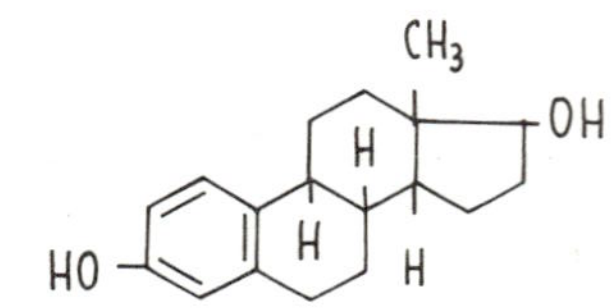

MOLECULAR WEIGHT: 272.39

SOURCE OF SAMPLE: Schering Corporation
Bloomfield, New Jersey

CAS REGISTRY NO.: 50-28-2

NIOSH NUMBER: KG29750

KBr Wafer

CONDENSED PHASE

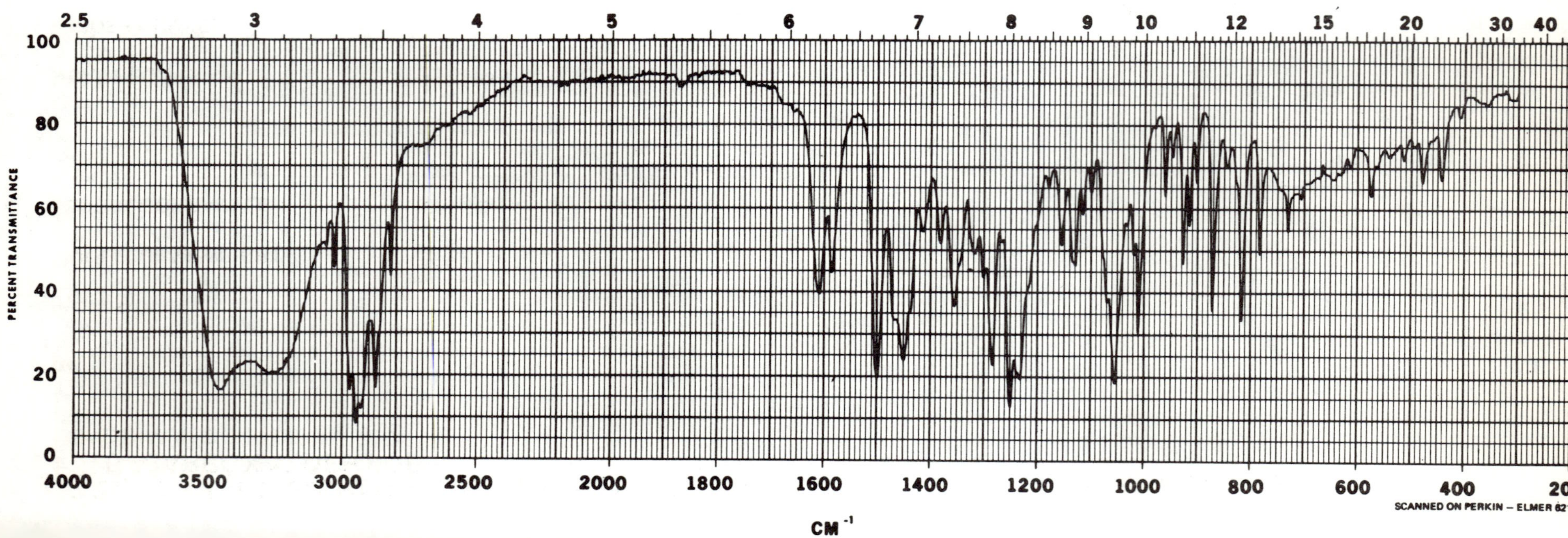

ESTRADIOL, 3–BENZOATE
DIHYDROFOLLICULIN BENZOATE
ESTRA–1,3,5(10)–TRIENE–3,17β–DIOL, 3–BENZOATE
17β–ESTRADIOL, 3–BENZOATE
PROGYNON BENZOATE

$C_{25}H_{28}O_3$

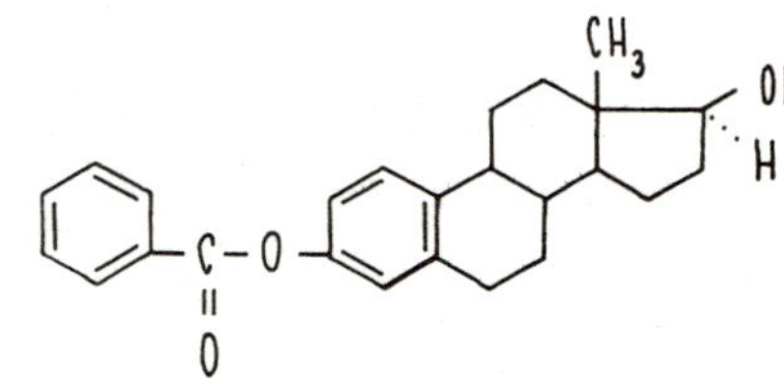

MOLECULAR WEIGHT: 376.50

SOURCE OF SAMPLE: Sigma Chemical Company
St. Louis, Missouri

CAS REGISTRY NO.: 50-50-0

NIOSH NUMBER: KG40500

Film (Cast from $CHCl_3$)

CONDENSED PHASE

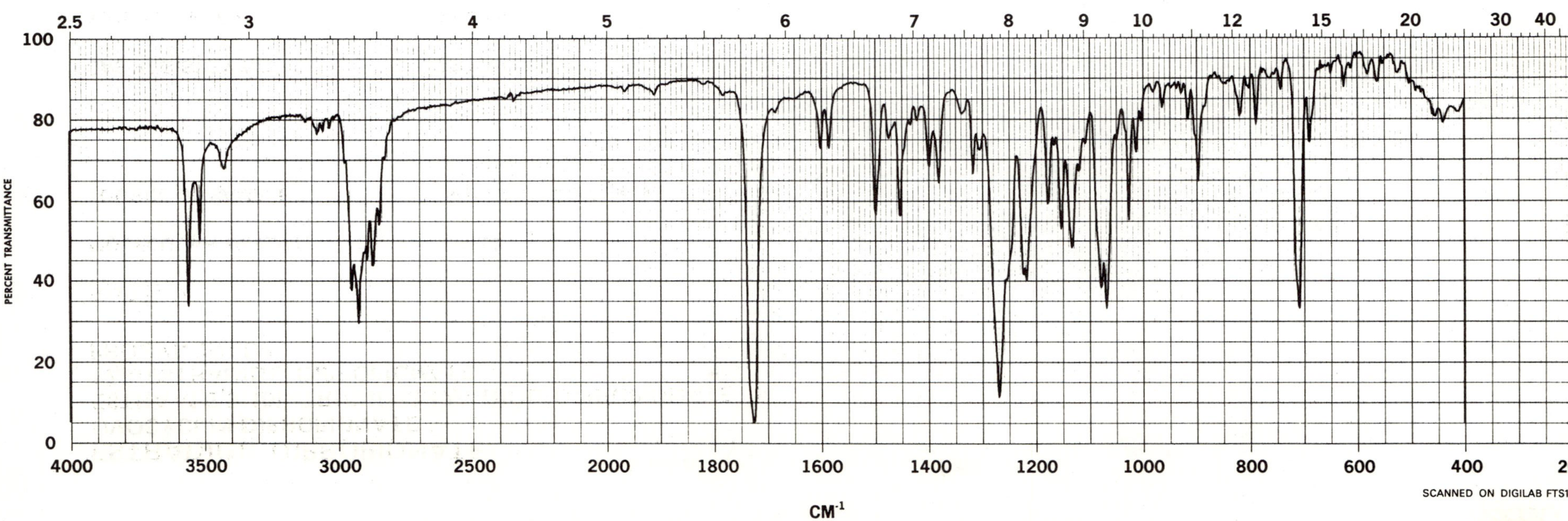

SCANNED ON DIGILAB FTS14

ESTRADIOL, DIPROPIONATE
OVOCYCLIN DIPROPIONATE
ESTRA–1,3,5(10)–TRIENE–3,17β–DIOL, DIPROPIONATE
17β–ESTRADIOL, DIPROPIONATE
PROGYNON–DP

$C_{24}H_{32}O_4$

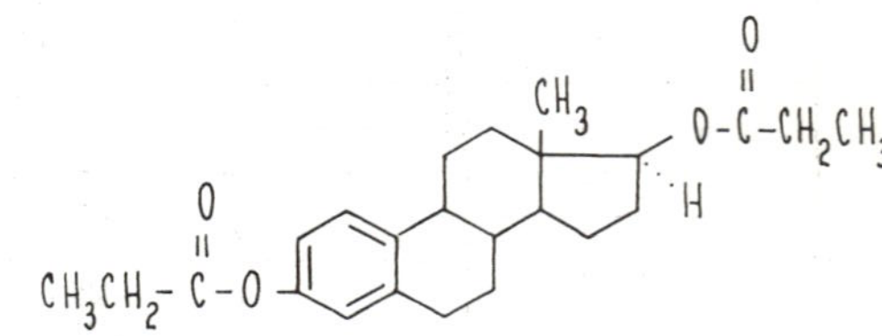

MOLECULAR WEIGHT: 384.52

SOURCE OF SAMPLE: Sigma Chemical Company
St. Louis, Missouri

CAS REGISTRY NO.: 113-38-2

NIOSH NUMBER: KG47250

Film (Cast from $CHCl_3$)

CONDENSED PHASE

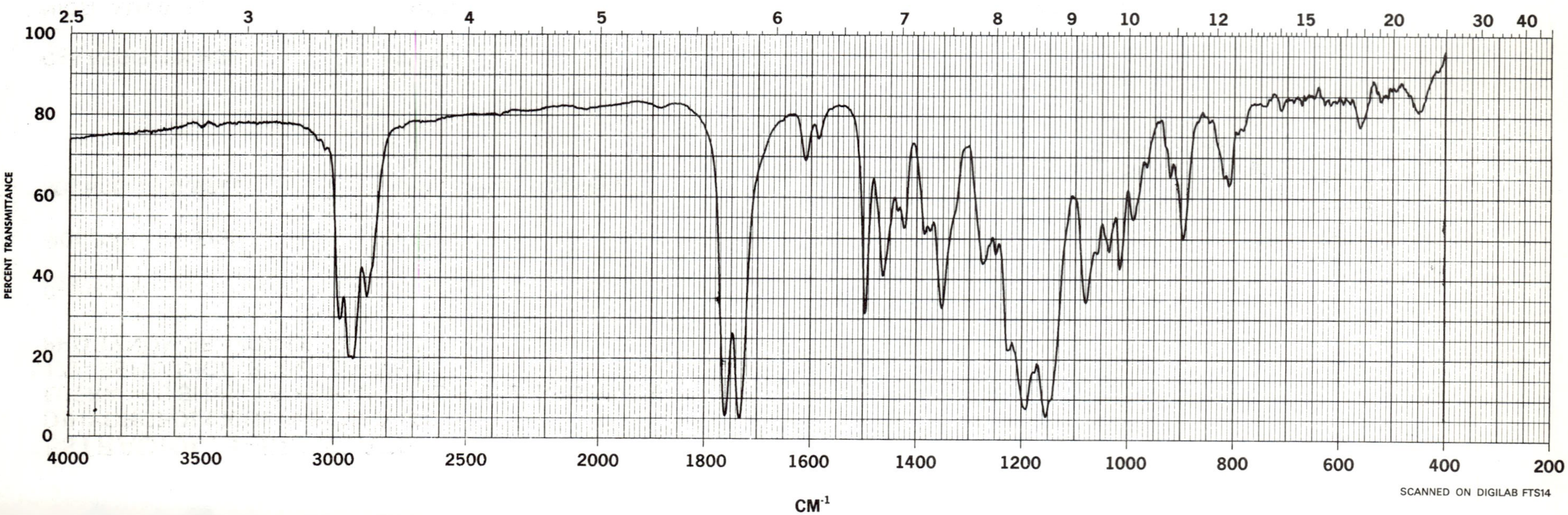

17α–ESTRADIOL
ESTRA–1,3,5(10)–TRIENE–3,17α–DIOL

$C_{18}H_{24}O_2$

MOLECULAR WEIGHT: 272.39

SOURCE OF SAMPLE: Sigma Chemical Company
St. Louis, Missouri

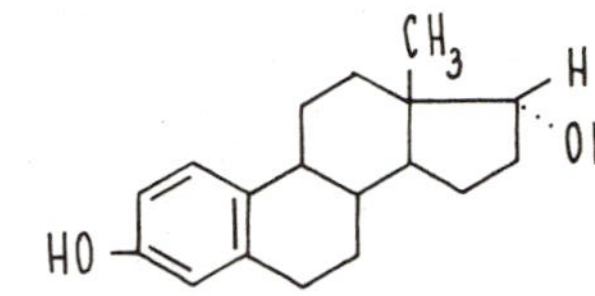

CAS REGISTRY NO.: 57-91-0

NIOSH NUMBER: KG37500

KBr Wafer

CONDENSED PHASE

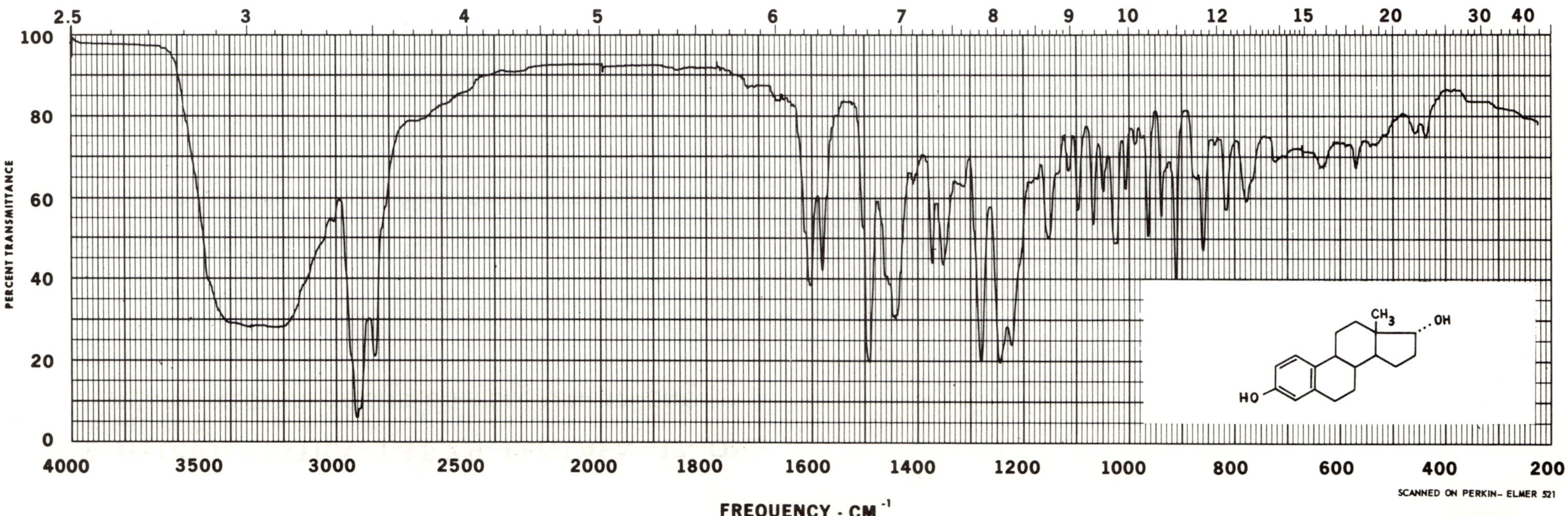

3–HYDROXYESTRA–1,3,5,7,9–PENTAEN–17–ONE
EQUILENIN
3–HYDROXYESTRA–1,3,5(10),6,8–PENTAEN–17–ONE

$C_{18}H_{18}O_2$

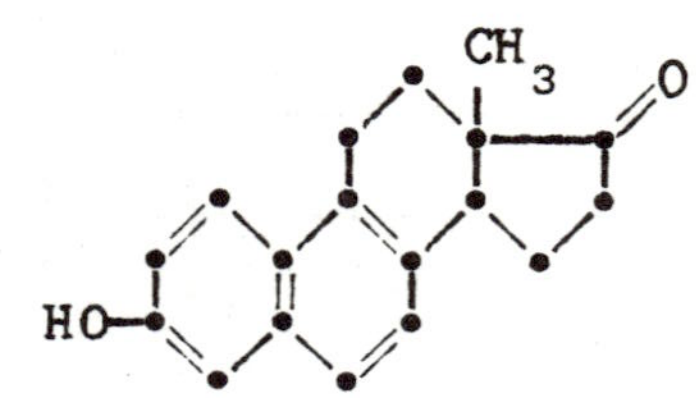

MOLECULAR WEIGHT: 266.34

SOURCE OF SAMPLE: Sigma Chemical Company
St. Louis, Missouri

CAS REGISTRY NO.: 517-09-9

NIOSH NUMBER: KG61250

KBr Wafer

CONDENSED PHASE

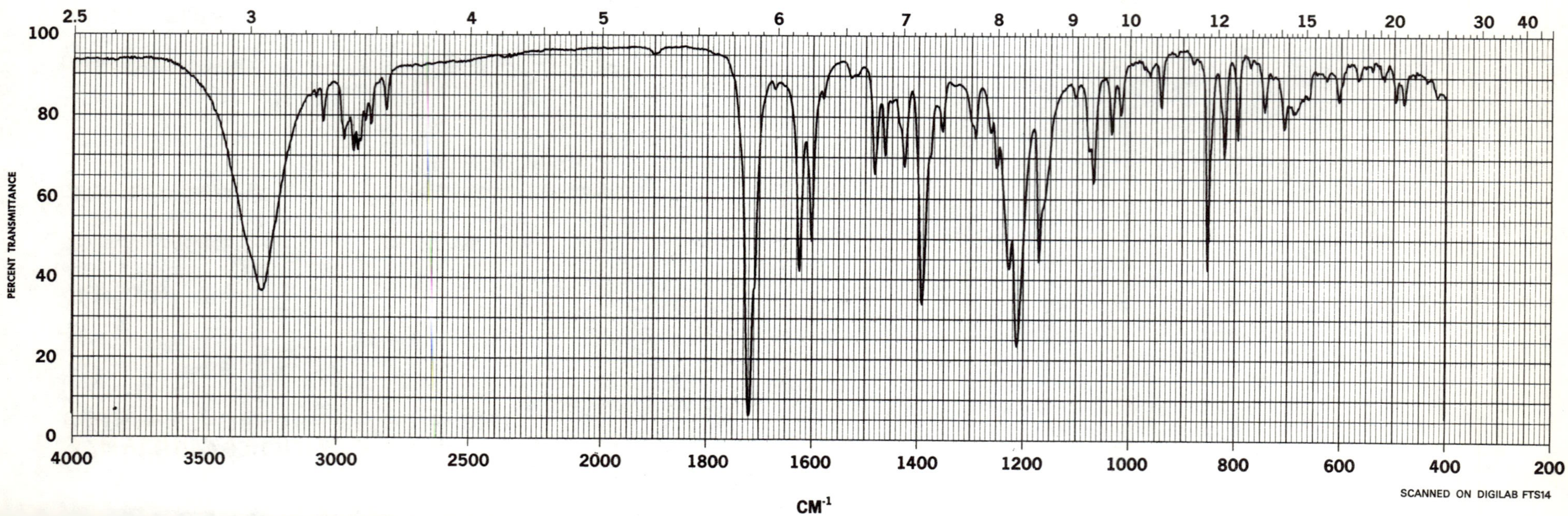

3–HYDROXYESTRA–1,3,5(10),7–TETRAEN–17–ONE
EQUILIN

$C_{18}H_{20}O_2$

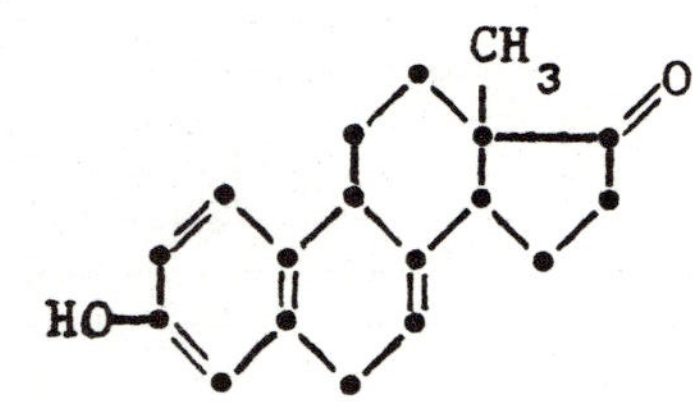

MOLECULAR WEIGHT: 268.36

SOURCE OF SAMPLE: Sigma Chemical Company
St. Louis, Missouri

CAS REGISTRY NO.: 474-86-2

NIOSH NUMBER: KG66500

KBr Wafer

CONDENSED PHASE

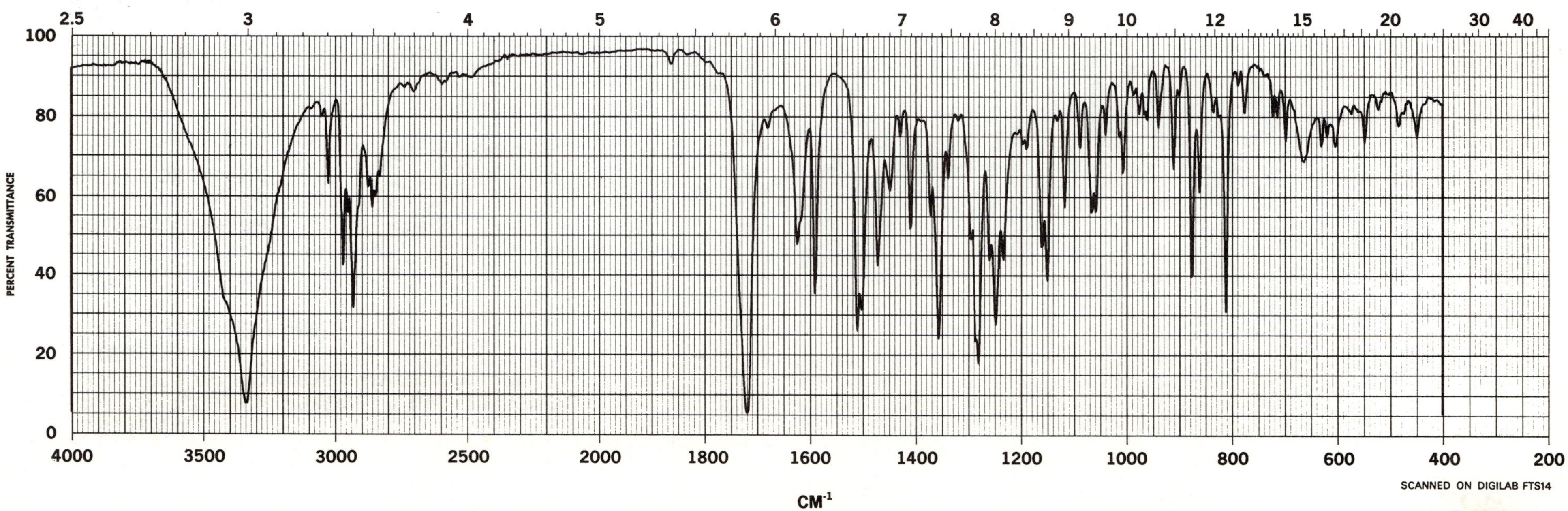

SCANNED ON DIGILAB FTS14

ESTRIOL
ESTRA–1,3,5(10)–TRIENE–3,16α,17β–TRIOL
FOLLICULAR HORMONE HYDRATE
THEELOL
TRIHYDROXYESTRIN

$C_{18}H_{24}O_3$

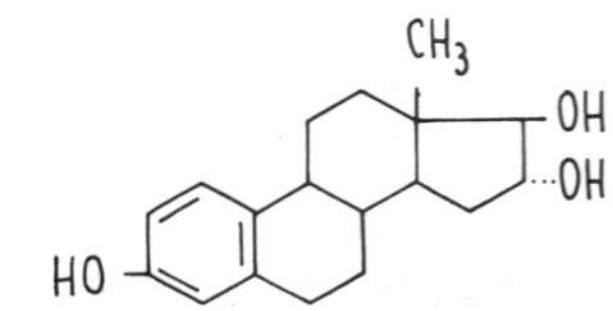

MOLECULAR WEIGHT: 288.39

SOURCE OF SAMPLE: Parke, Davis and Company
Detroit, Michigan

CAS REGISTRY NO.: 50-27-1

NIOSH NUMBER: KG82250

KBr Wafer

CONDENSED PHASE

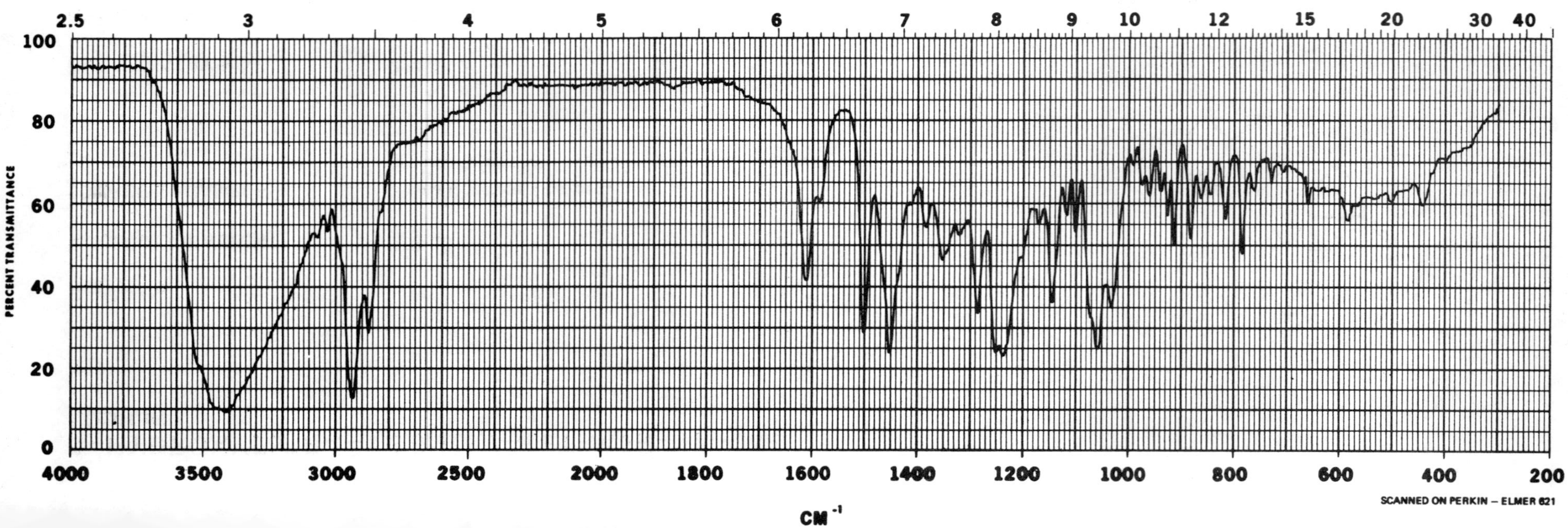

ESTRONE
FOLLICULIN
KETOHYDROXYESTRIN
THEELIN
3–HYDROXYESTRA–1,3,5(10)–TRIEN–17–ONE
MENFORMON
$C_{18}H_{22}O_2$

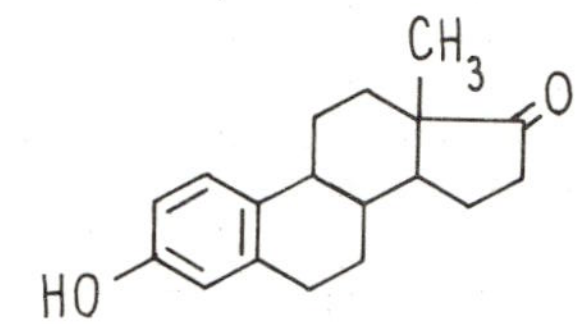

MOLECULAR WEIGHT: 270.37

SOURCE OF SAMPLE: Sadtler Research Laboratories
Philadelphia, Pennsylvania

CAS REGISTRY NO.: 53-16-7

NIOSH NUMBER: KG85750

KBr Wafer

CONDENSED PHASE

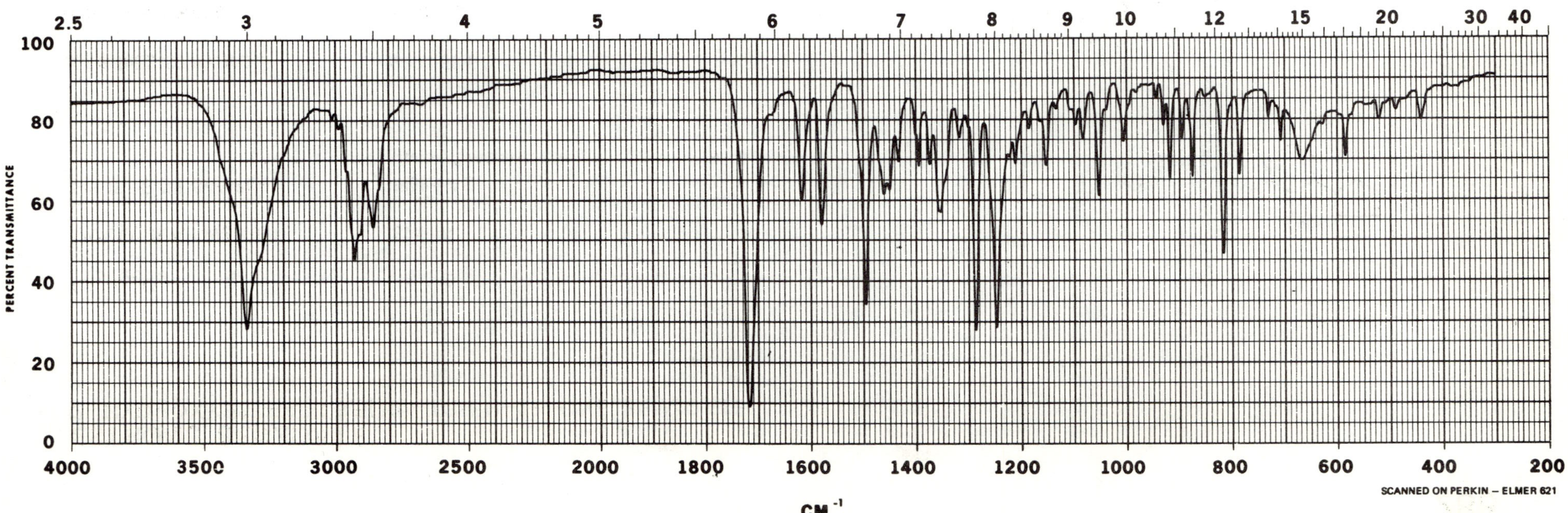

SCANNED ON PERKIN – ELMER 621

ESTRONE, BENZOATE
3–HYDROXYESTRA–1,3,5(10)–TRIEN–17–ONE, BENZOATE

$C_{25}H_{26}O_3$

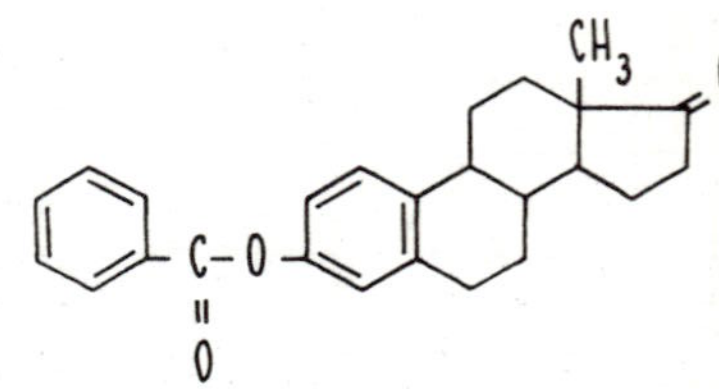

MOLECULAR WEIGHT:	374.48
SOURCE OF SAMPLE:	Fluka AG Buchs, Switzerland
CAS REGISTRY NO.:	2393-53-5
NIOSH NUMBER:	KG87500

Film (Cast from $CHCl_3$)

CONDENSED PHASE

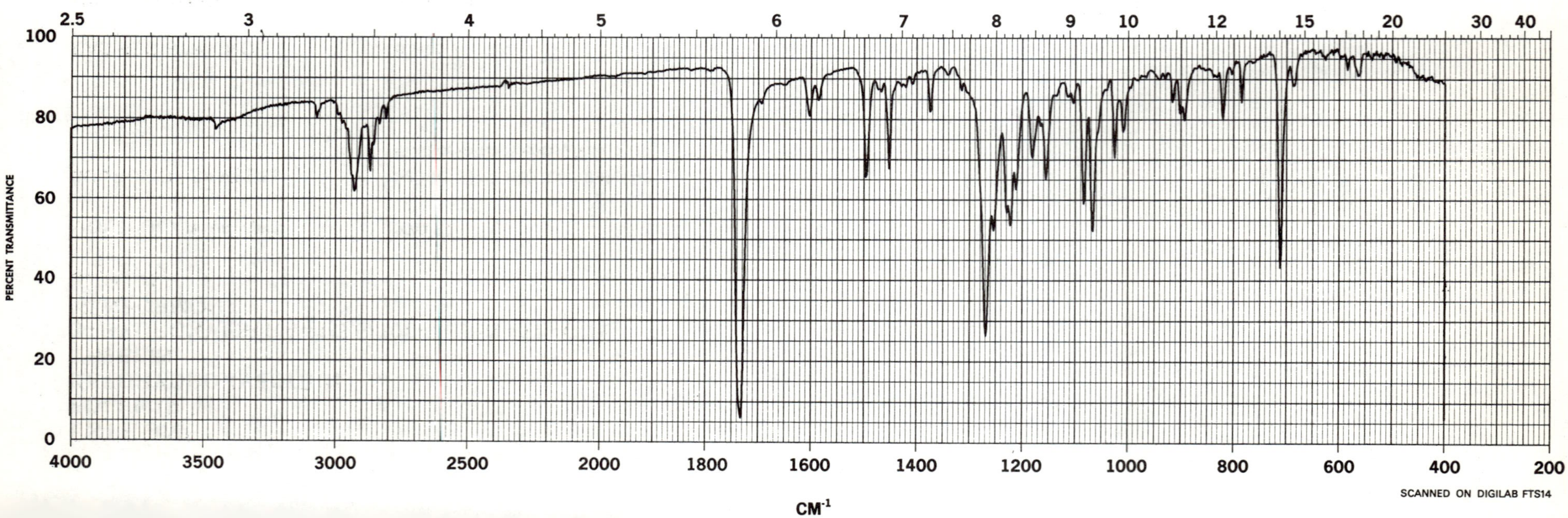

1,1–BIS(p–CHLOROPHENYL)–2,2–DICHLOROETHANE
p,p'–DDD
RHOTHANE D–3
TDE
DILENE

$C_{14}H_{10}Cl_4$

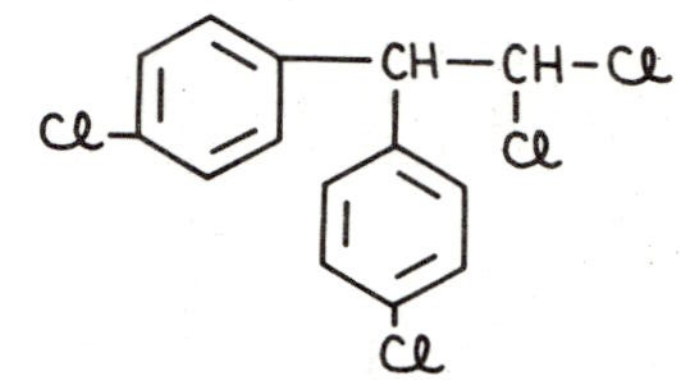

MOLECULAR WEIGHT:	320.05
SOURCE OF SAMPLE:	Rohm & Haas Company Philadelphia, Pennsylvania
CAS REGISTRY NO.:	72-54-8
NIOSH NUMBER:	KI07000

A = Mulled in Mineral Oil
B = Mulled in Perfluorinated Hydrocarbon

CONDENSED PHASE

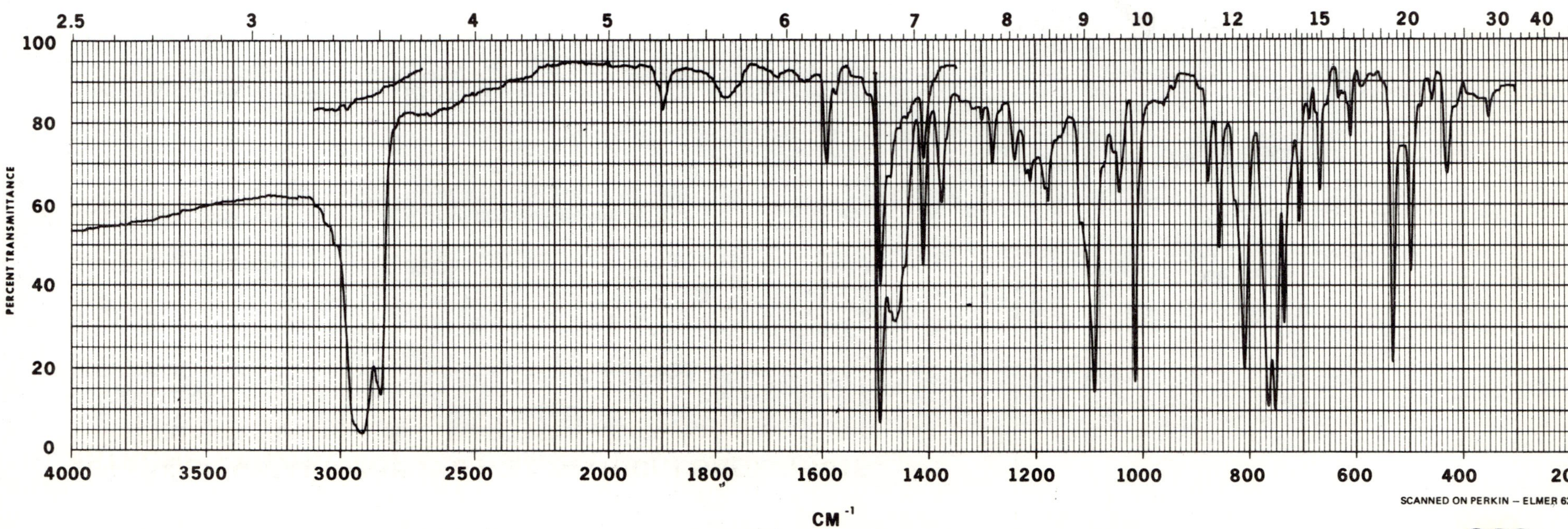

SCANNED ON PERKIN – ELMER 621

1,1–BIS(p–CHLOROPHENYL)–2,2,2–TRICHLOROETHANE
ANOFEX
p,p′–DDT
DICOPHANE
ZERDANE

$C_{14}H_9Cl_5$

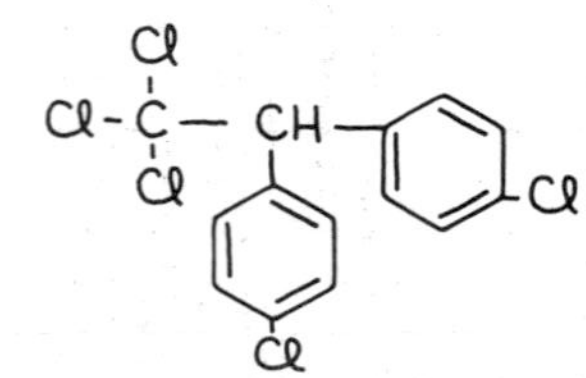

MOLECULAR WEIGHT: 354.49

SOURCE OF SAMPLE: Aldrich Chemical Company, Inc.
Milwaukee, Wisconsin

CAS REGISTRY NO.: 50-29-3

NIOSH NUMBER: KJ33250

Film (Cast from $CHCl_3$)

CONDENSED PHASE

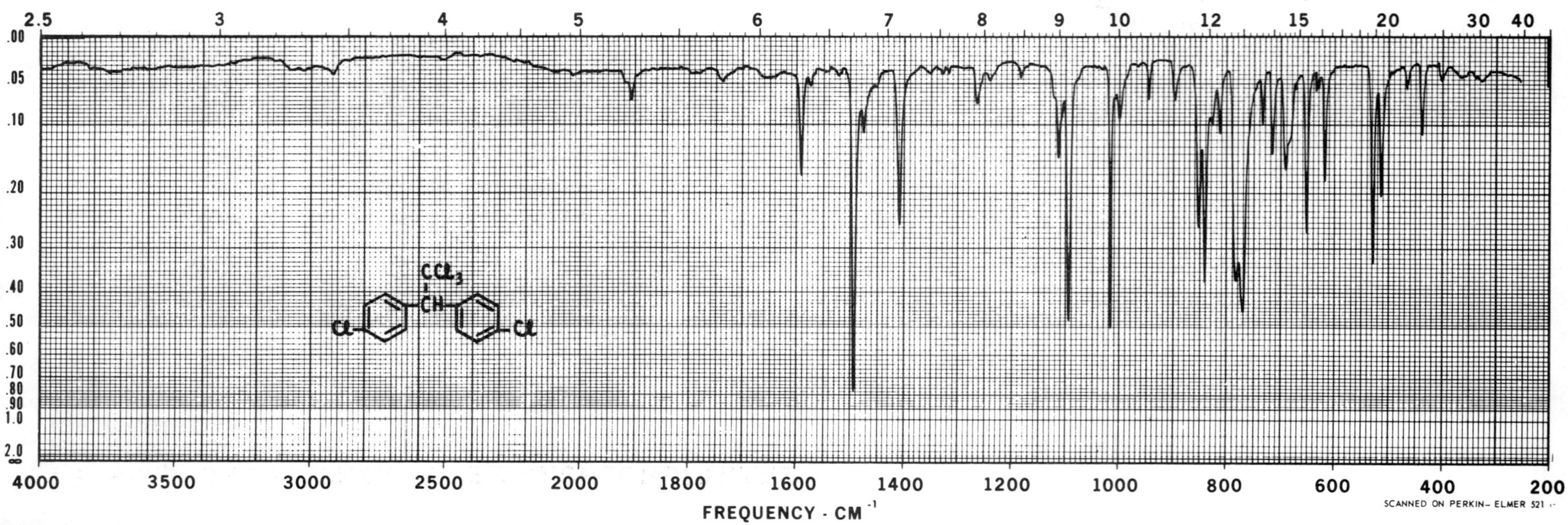

FLUORENE–2,7–DIAMINE
2,7–DIAMINOFLUORENE

$C_{13}H_{12}N_2$

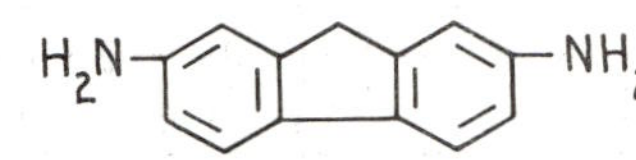

MOLECULAR WEIGHT: 196.25

SOURCE OF SAMPLE: Calbiochem
Los Angeles, California

CAS REGISTRY NO.: 525-64-4

NIOSH NUMBER: LL69800

KBr Wafer

CONDENSED PHASE

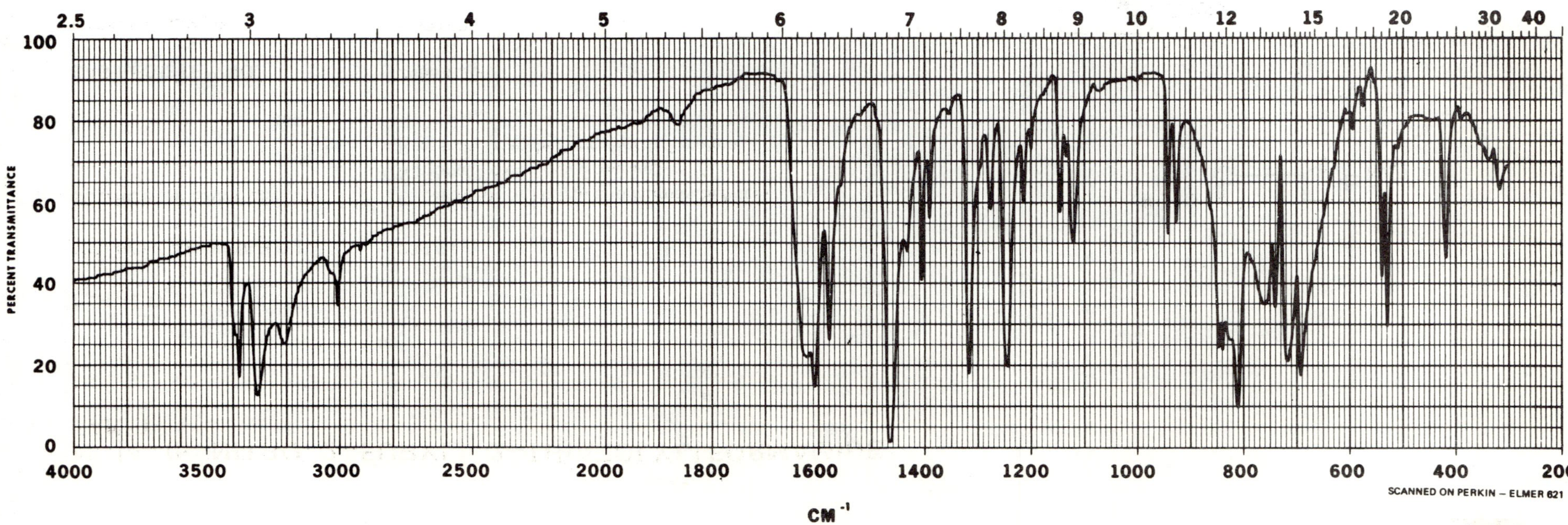

N–[4–(5–NITRO–2–FURYL)–2–THIAZOLYL] FORMAMIDE
FANFT

$C_8H_5N_3O_4S$

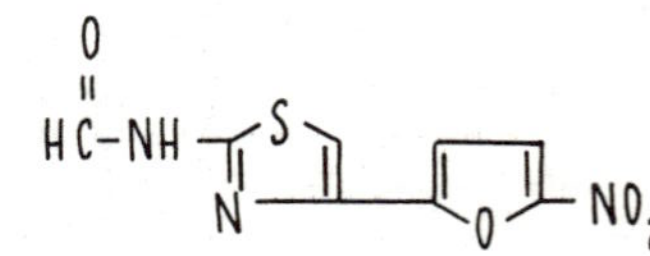

MOLECULAR WEIGHT: 239.21

SOURCE OF SAMPLE: Department of Health and Human Services
National Toxicology Program
Research Triangle Park, North Carolina

CAS REGISTRY NO.: 24554-26-5

NIOSH NUMBER: LQ31500

KBr Wafer

CONDENSED PHASE

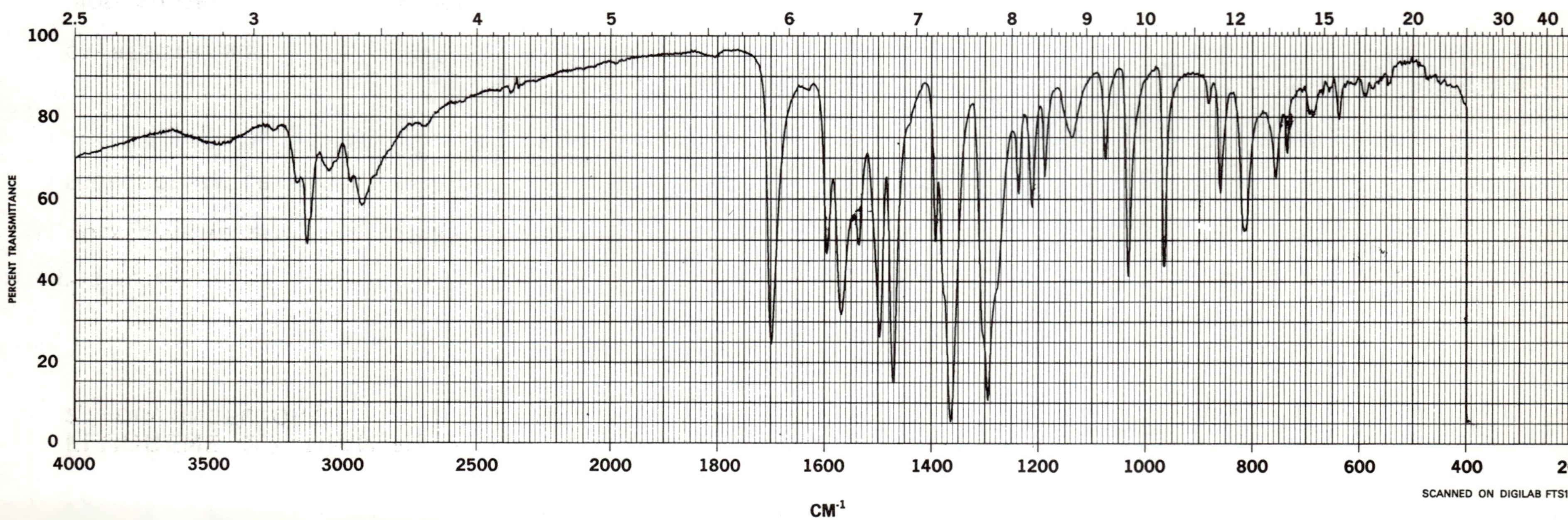

FUMARIC ACID

trans–BUTENEDIOIC ACID
trans–1,2–ETHYLENEDICARBOXYLIC ACID
trans–1,2–ETHENEDICARBOXYLIC ACID

$C_4H_4O_4$

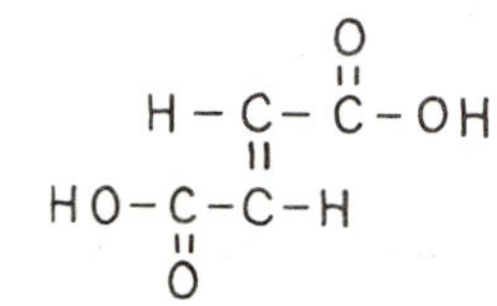

MOLECULAR WEIGHT:	116.07
SOURCE OF SAMPLE:	A. Gennaro Philadelphia College of Pharmacy and Science Philadelphia, Pennsylvania
CAS REGISTRY NO.:	110-17-8
NIOSH NUMBER:	LS96250

KBr Wafer

CONDENSED PHASE

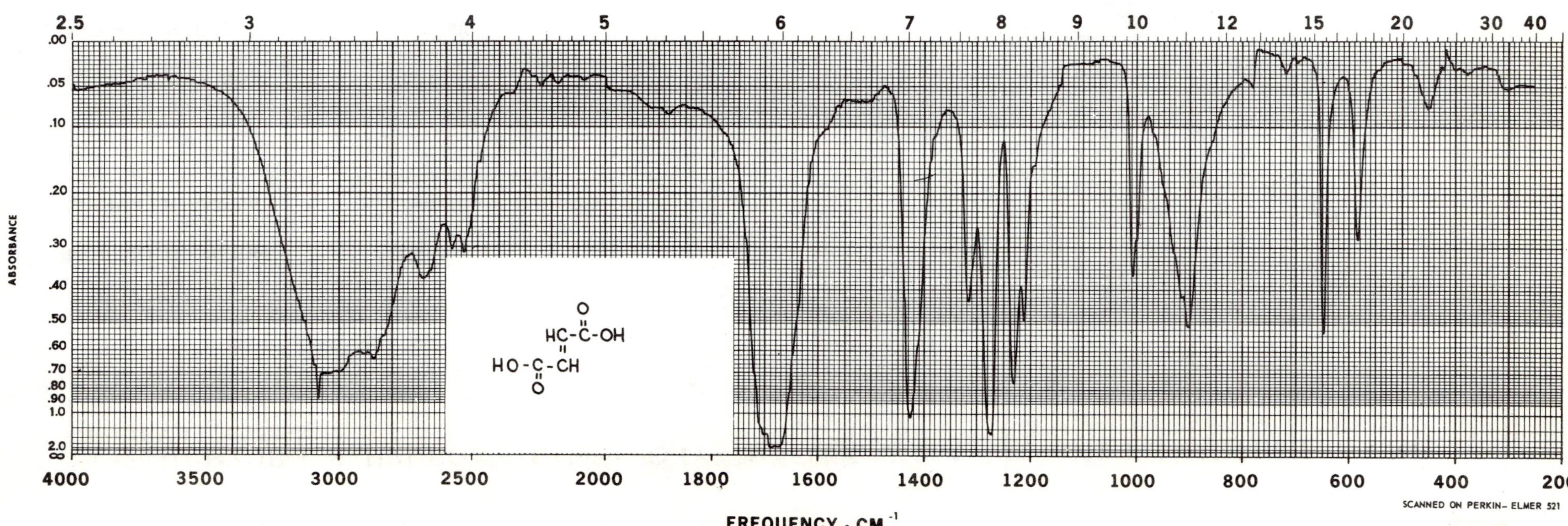

3a,12c–DIHYDRO–8–HYDROXY–6–METHOXY–7H–FURO[3′,2′:4,5] FURO[2,3-c] XANTHEN–7–ONE
STERIGMATOCYSTIN

$C_{18}H_{10}O_6$

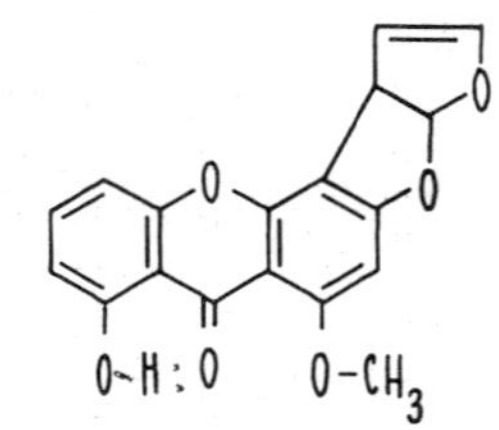

MOLECULAR WEIGHT: 322.28

SOURCE OF SAMPLE: Sigma Chemical Company
St. Louis, Missouri

CAS REGISTRY NO.: 10048-13-2

NIOSH NUMBER: LV17500

Film (Cast from $CHCl_3$)

CONDENSED PHASE

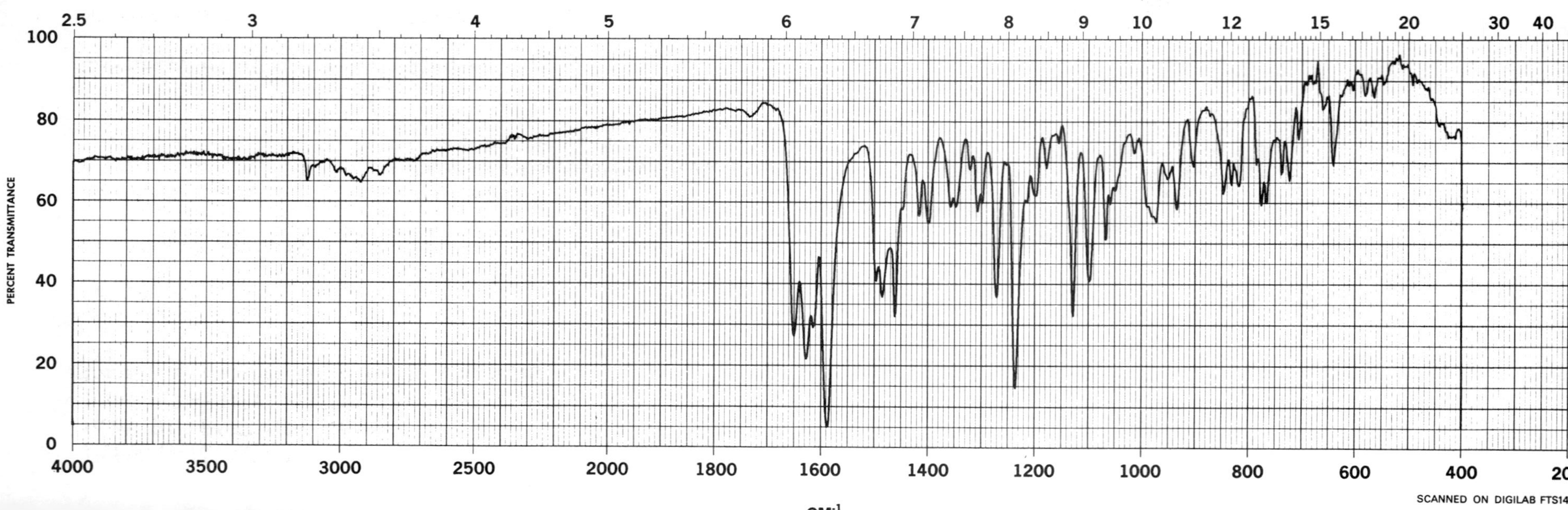

2–DEOXY–2–(3–METHYL–3–NITROSOUREIDO)–D–GLUCOPYRANOSE
STR
STREPTOZOTOCIN

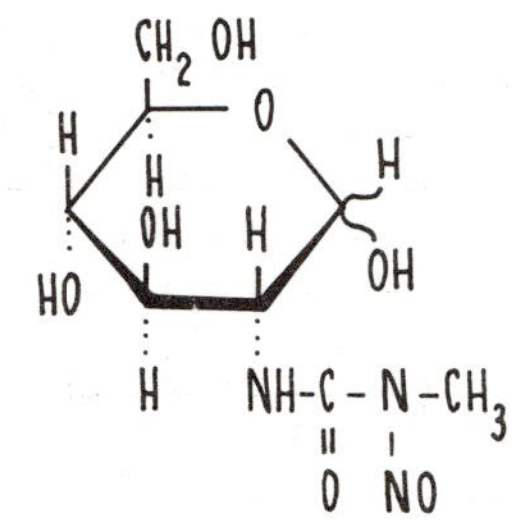

$C_8H_{15}N_3O_7$

MOLECULAR WEIGHT: 265.22

SOURCE OF SAMPLE: Sigma Chemical Company
St. Louis, Missouri

CAS REGISTRY NO.: 18883-66-4

NIOSH NUMBER: LZ57750

KBr Wafer

CONDENSED PHASE

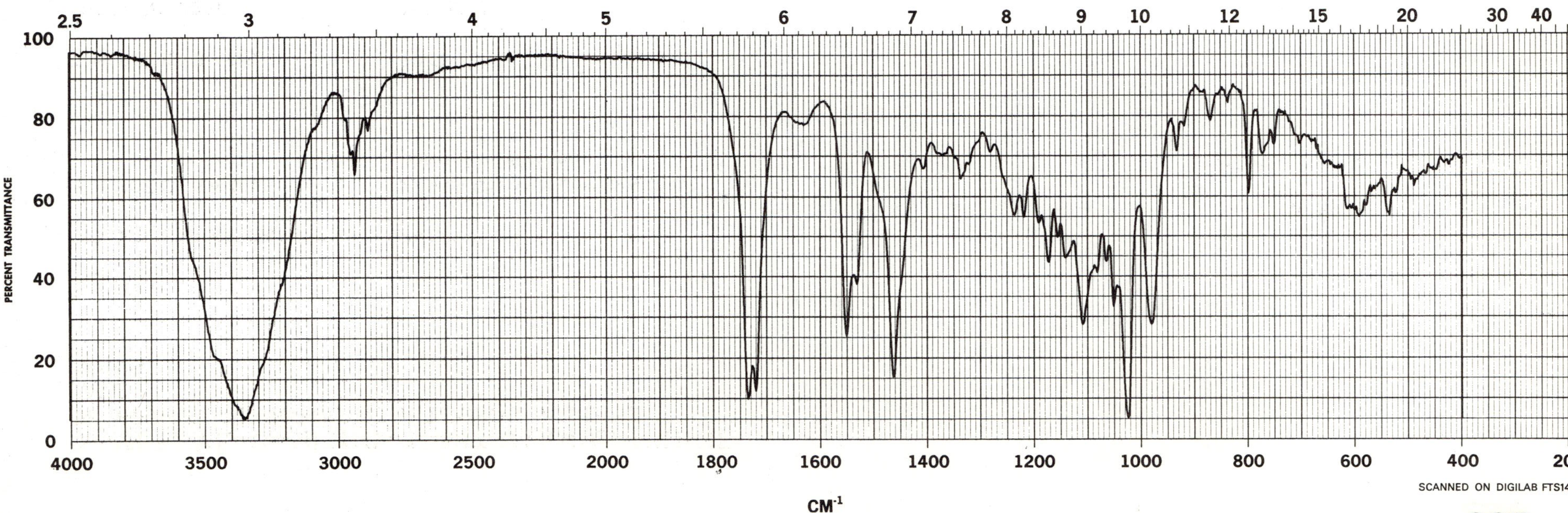

SCANNED ON DIGILAB FTS14

1—METHYL—3—NITRO—1—NITROSOGUANIDINE
N—METHYL—N'—NITRO—N—NITROSOGUANIDINE
MNG
MNNG

$C_2H_5N_5O_3$

MOLECULAR WEIGHT: 147.09

SOURCE OF SAMPLE: Aldrich Chemical Company, Inc.
Milwaukee, Wisconsin

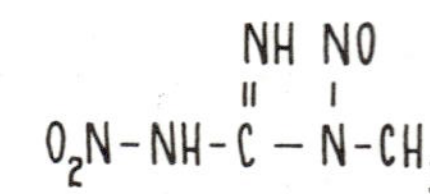

CAS REGISTRY NO.: 70-25-7

NIOSH NUMBER: MF42000

KBr Wafer

CONDENSED PHASE

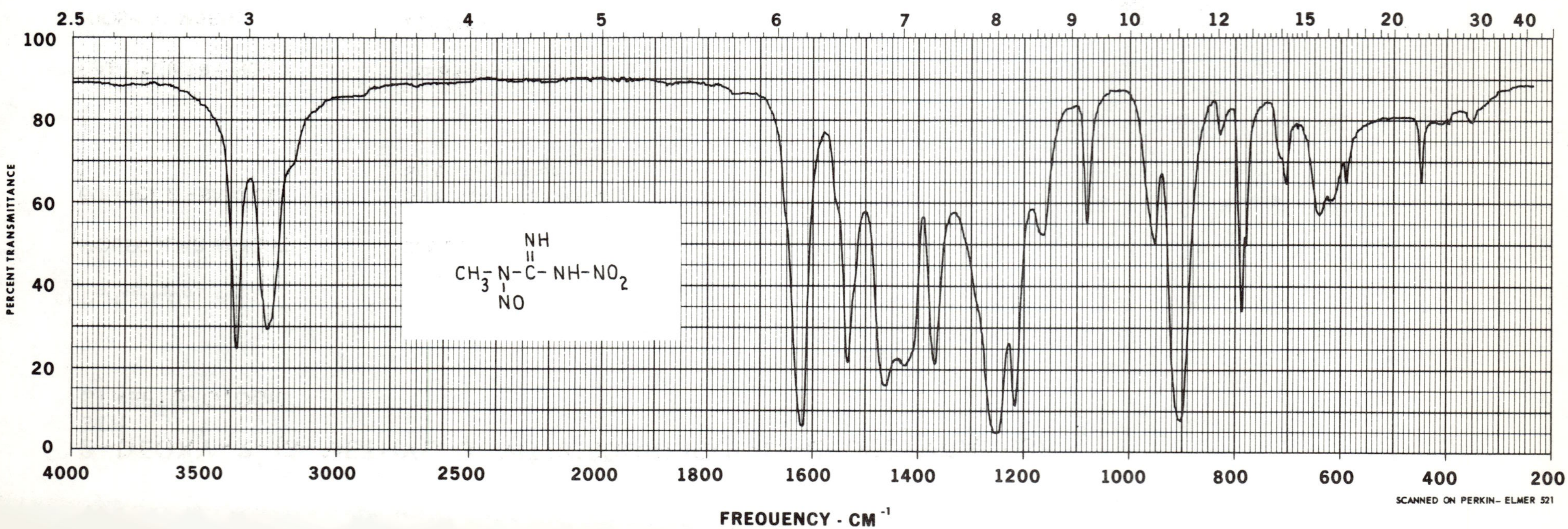

HEMATITE
BLOOD STONE
IRON ORE
RED IRON ORE

Fe_2O_3

MOLECULAR WEIGHT: 159.69

SOURCE OF SAMPLE: Aldrich Chemical Company, Inc.
Milwaukee, Wisconsin

$\alpha - Fe_2O_3$

CAS REGISTRY NO.: 1317-60-8

NIOSH NUMBER: MH76000

Mulled in Mineral Oil

CONDENSED PHASE

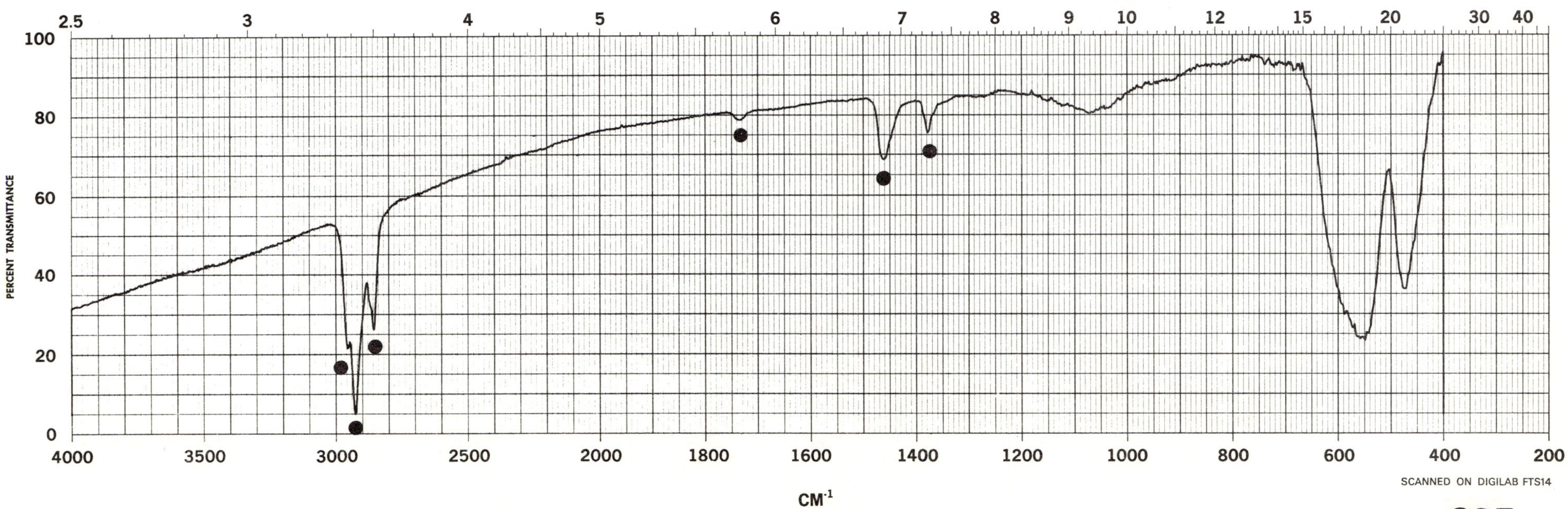

SCANNED ON DIGILAB FTS14

HYDRAZINE
DIAMIDE
DIAMINE

H_4N_2

H_2N-NH_2

MOLECULAR WEIGHT: 32.05

SOURCE OF SAMPLE: Aldrich Chemical Company, Inc.
Milwaukee, Wisconsin

CAS REGISTRY NO.: 302-01-2

NIOSH NUMBER: MU71750

Between Salts

CONDENSED PHASE

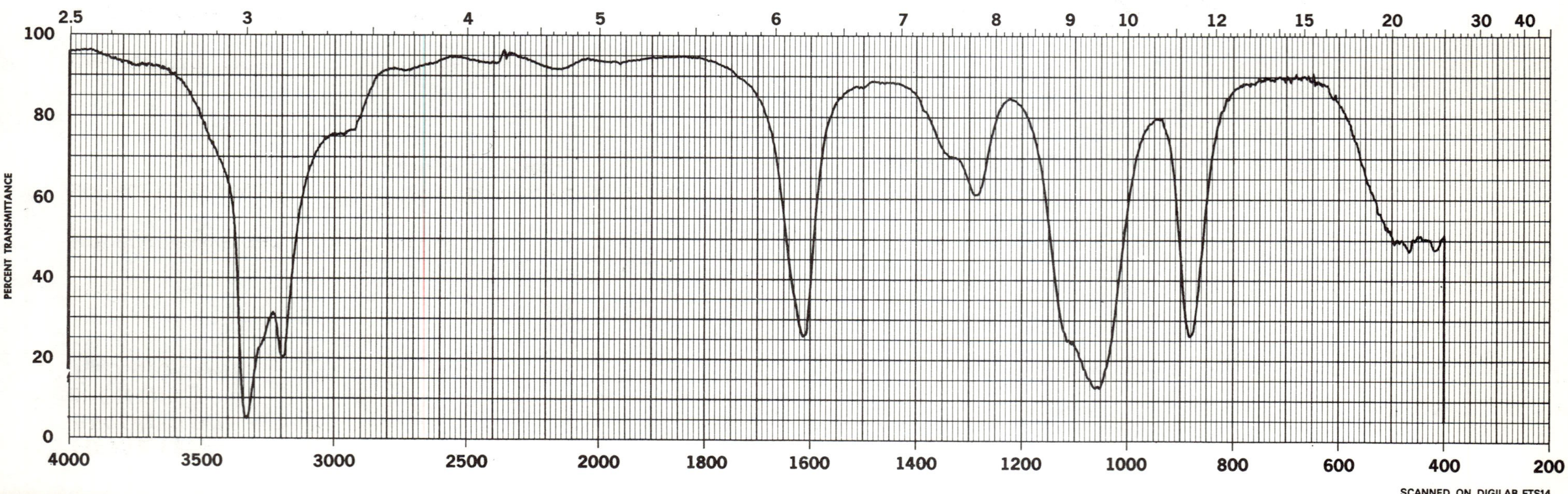

HYDRAZINE, MONOHYDRATE
HYDRAZINE HYDRATE

$H_4N_2 \cdot H_2O$

$H_2N-NH_2 \bullet H_2O$

MOLECULAR WEIGHT: 50.06

SOURCE OF SAMPLE: Fairmount Chemical Company
Newark, New Jersey

CAS REGISTRY NO.: 7803-57-8

NIOSH NUMBER: MV80500

CAPILLARY CELL: Neat

CONDENSED PHASE

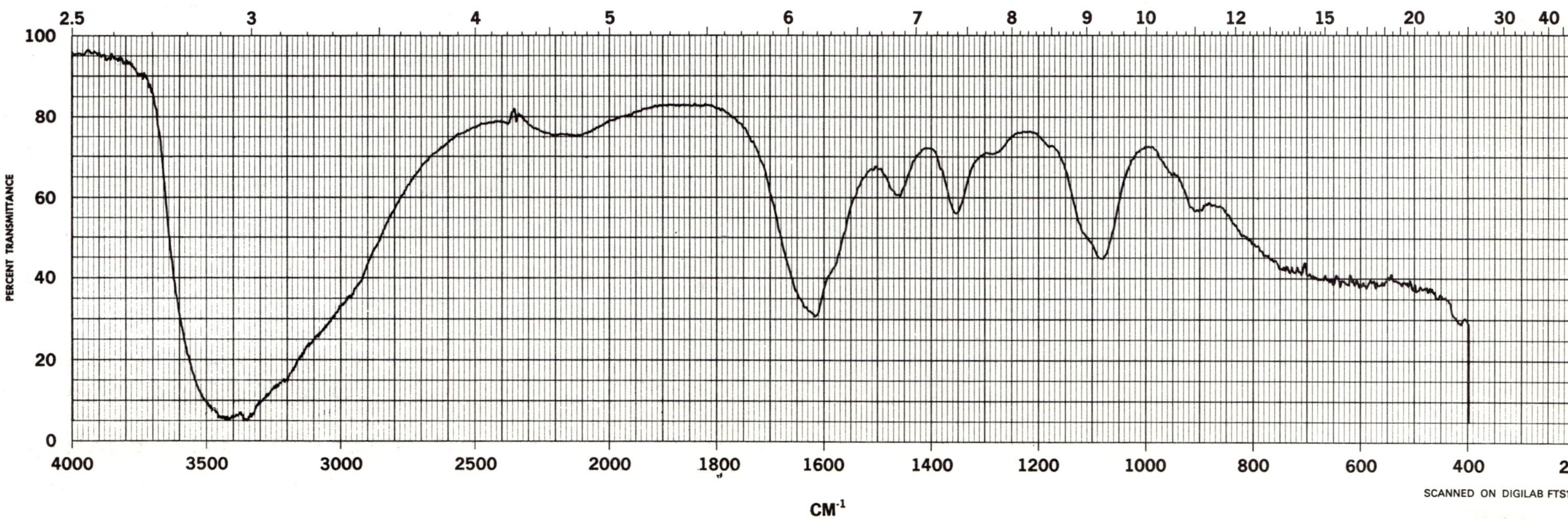

SCANNED ON DIGILAB FTS14

299

HYDRAZINE, SULFATE (1:1)

HS
HYDRAZINIUM SULFATE
HYDRAZONIUM SULFATE

$H_4N_2 \cdot H_2SO_4$

MOLECULAR WEIGHT: 130.12

SOURCE OF SAMPLE: E. Merck AG, Darmstadt, Germany

CAS REGISTRY NO.: 10034-93-2

NIOSH NUMBER: MV96250

KBr Wafer

$H_2N-NH_2 \bullet H_2SO_4$

CONDENSED PHASE

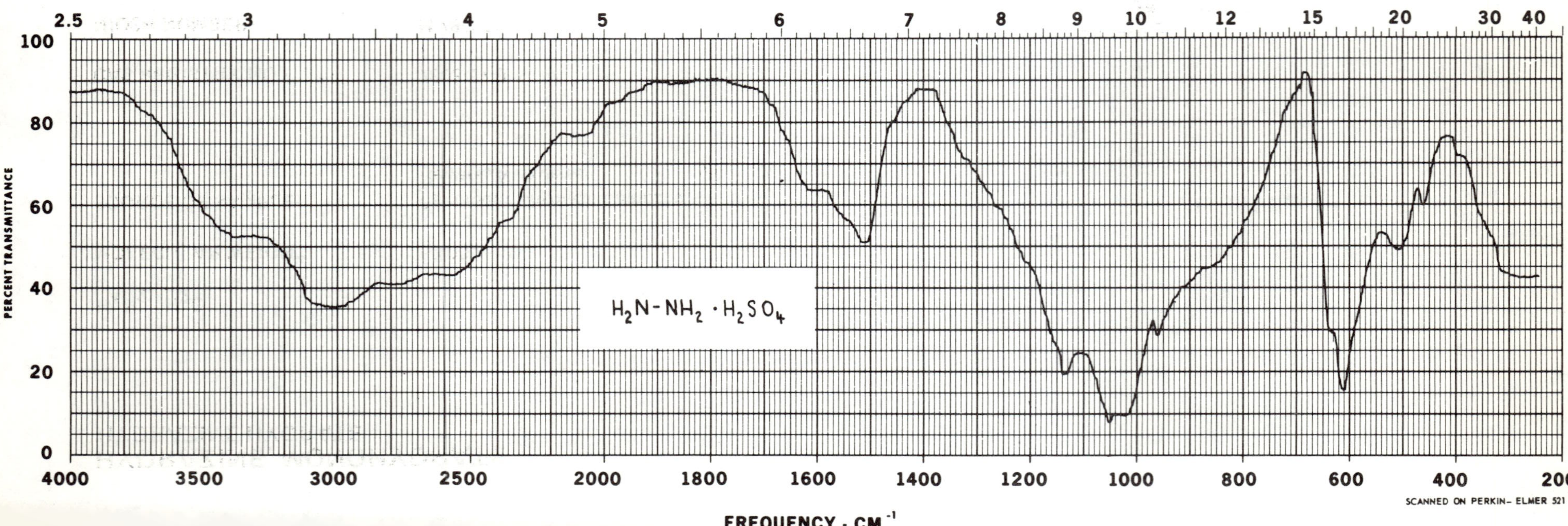

1—ACETYL—2—PICOLINOYLHYDRAZINE

$C_8H_9N_3O_2$

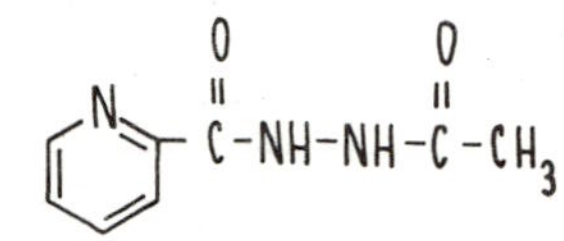

MOLECULAR WEIGHT: 179.18

SOURCE OF SAMPLE: Department of Health and Human Services
National Toxicology Program
Research Triangle Park, North Carolina

CAS REGISTRY NO.: 1743-31-7

NIOSH NUMBER: MU77000

KBr Wafer

CONDENSED PHASE

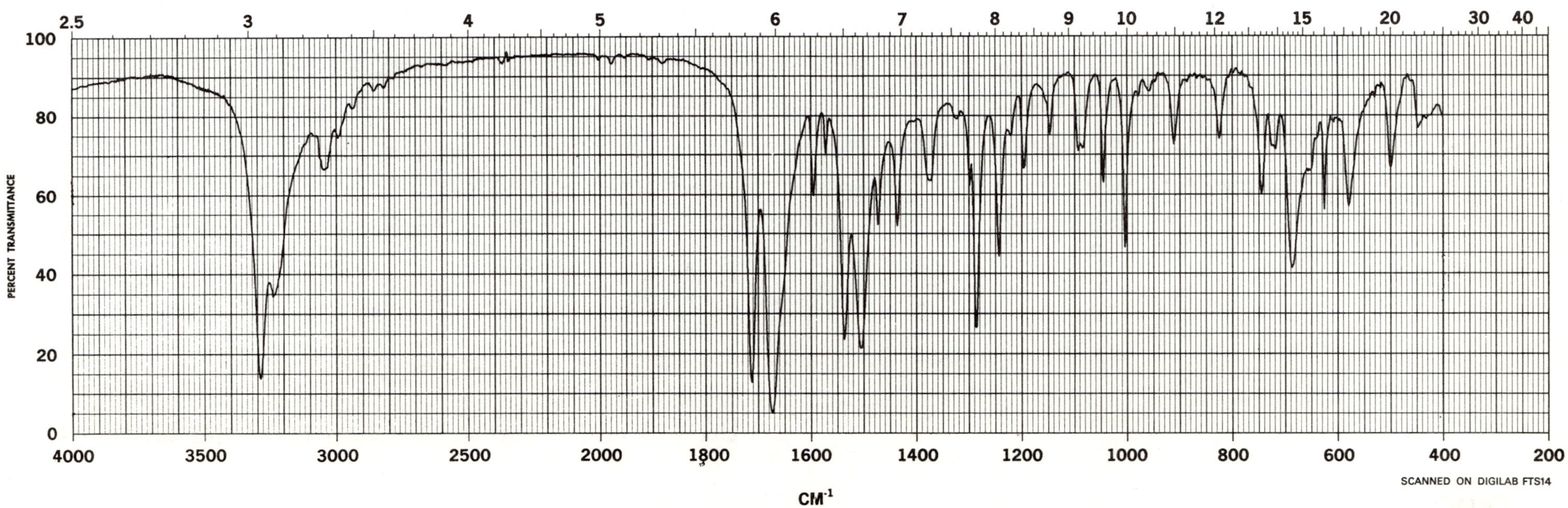

SCANNED ON DIGILAB FTS14

1,1–DIMETHYLHYDRAZINE, MONOHYDROCHLORIDE

$C_2H_8N_2 \cdot HCl$

$H_2N-N(CH_3)-CH_3 \cdot HCl$

MOLECULAR WEIGHT: 96.56

SOURCE OF SAMPLE: Tokyo Kasei Kogyo Company Ltd.
Tokyo, Japan

CAS REGISTRY NO.: 593-82-8

NIOSH NUMBER: MV29000

A = Mulled in Mineral Oil
B = Mulled in Perfluorinated Hydrocarbon

CONDENSED PHASE

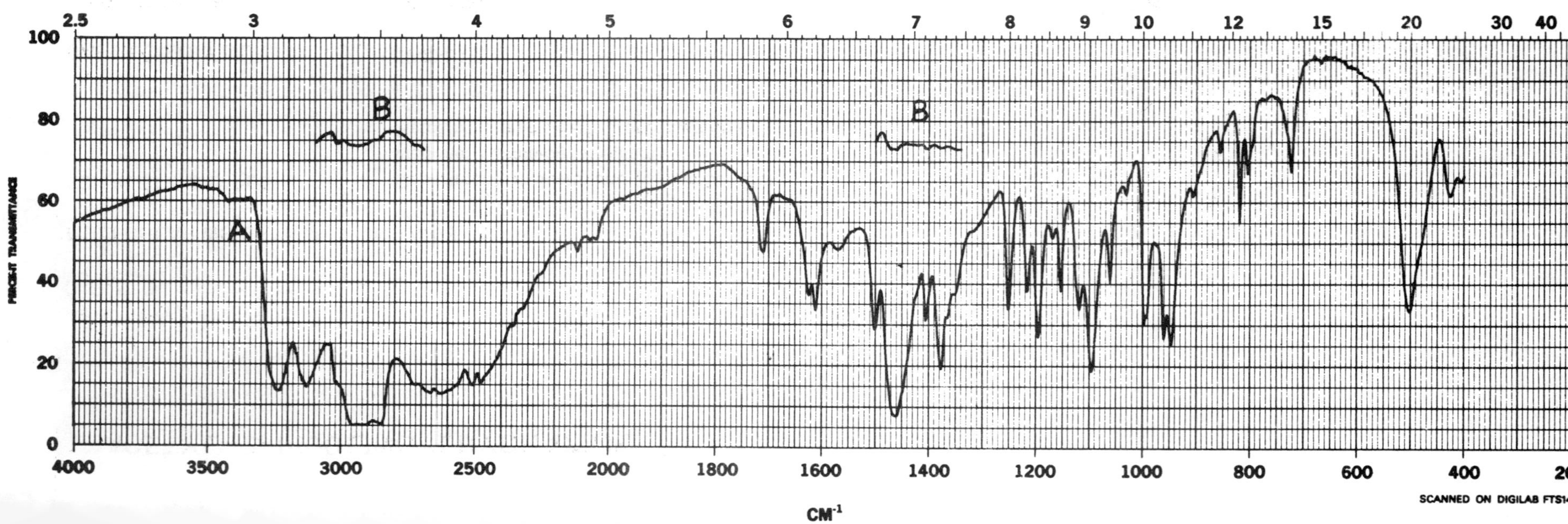

1,2–DIMETHYLHYDRAZINE, DIHYDROCHLORIDE
N,N'–DIMETHYLHYDRAZINE, DIHYDROCHLORIDE
sym–DIMETHYLHYDRAZINE, DIHYDROCHLORIDE

$C_2H_8N_2 \cdot 2HCl$

MOLECULAR WEIGHT: 133.02

SOURCE OF SAMPLE: Aldrich Chemical Company, Inc.
Milwaukee, Wisconsin

$CH_3-NH-NH-CH_3 \quad \bullet 2HCl$

CAS REGISTRY NO.: 306-37-6

NIOSH NUMBER: MV28000

KBr Wafer

CONDENSED PHASE

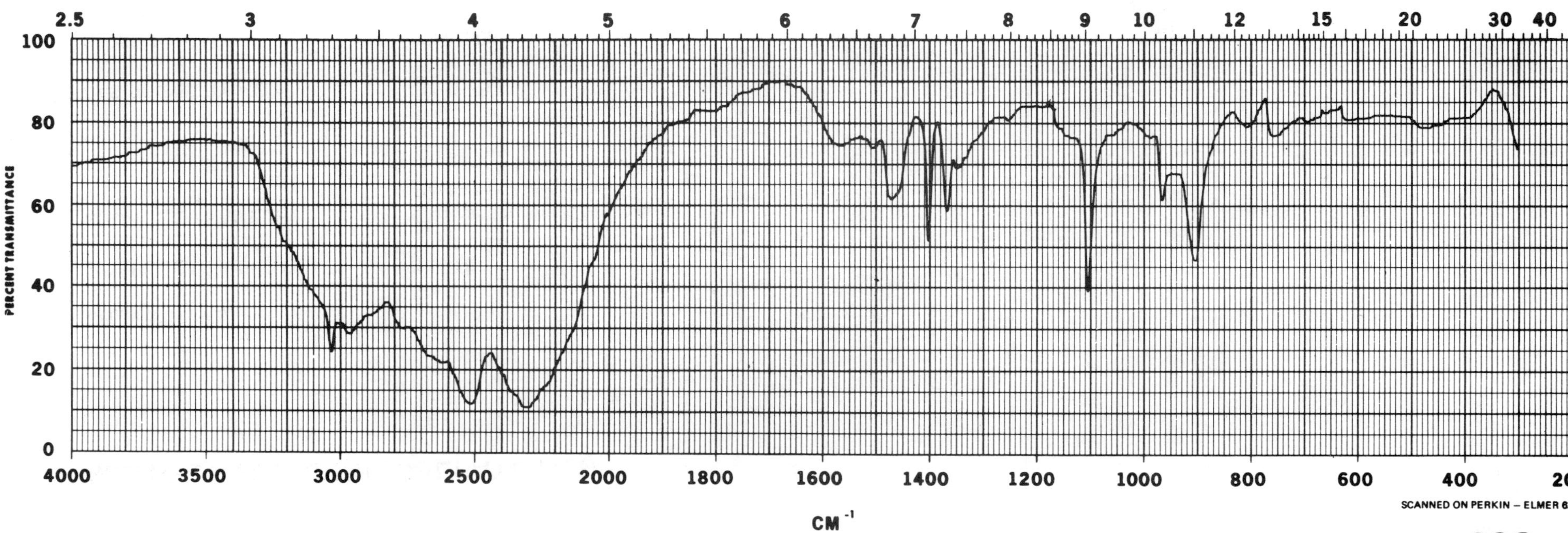

SCANNED ON PERKIN – ELMER 621

METHYLHYDRAZINE, SULFATE (1:1)

$CH_6N_2 \cdot H_2SO_4$

MOLECULAR WEIGHT: 144.15

SOURCE OF SAMPLE: Eastman Organic Chemicals
Rochester, New York

CH_3-NH-NH_2 · H_2SO_4

CAS REGISTRY NO.: 302-15-8

NIOSH NUMBER: MV77500

KBr Wafer

CONDENSED PHASE

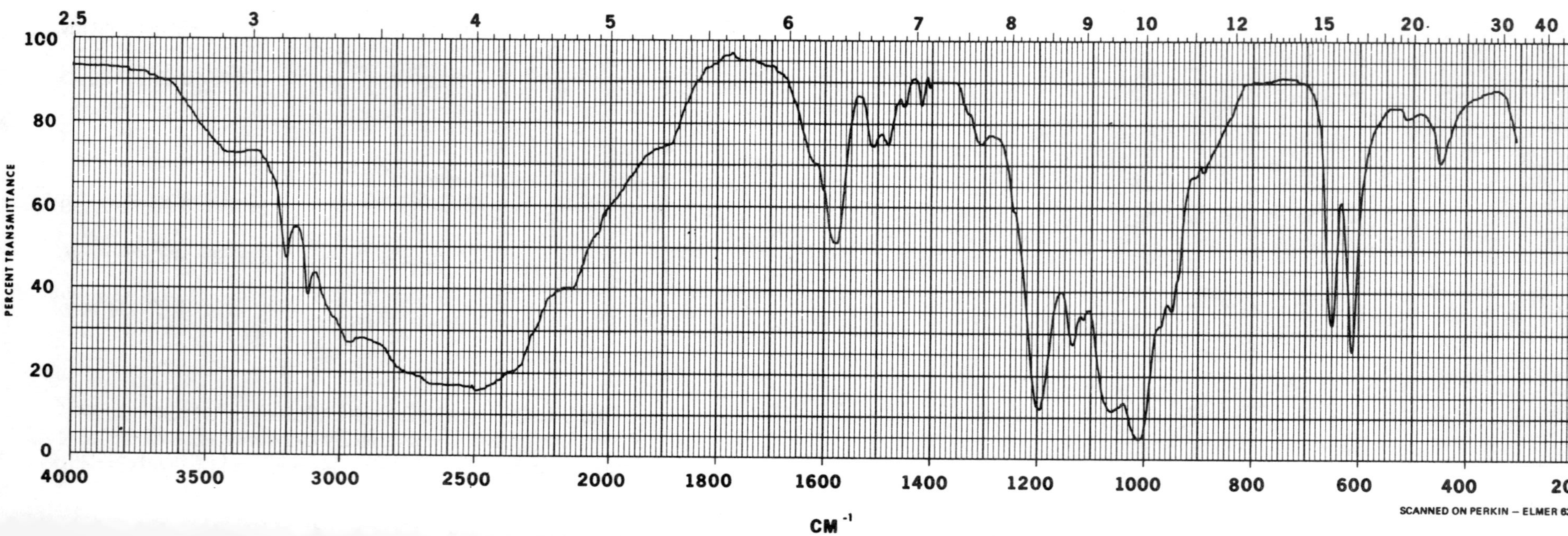

PHENYLHYDRAZINE, MONOHYDROCHLORIDE

$C_6H_8N_2 \cdot HCl$

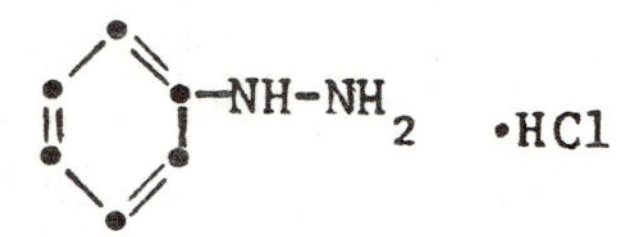

MOLECULAR WEIGHT: 144.61

SOURCE OF SAMPLE: Merck and Company, Inc.
Rahway, New Jersey

CAS REGISTRY NO.: 59-88-1

NIOSH NUMBER: MV90000

KBr Wafer

CONDENSED PHASE

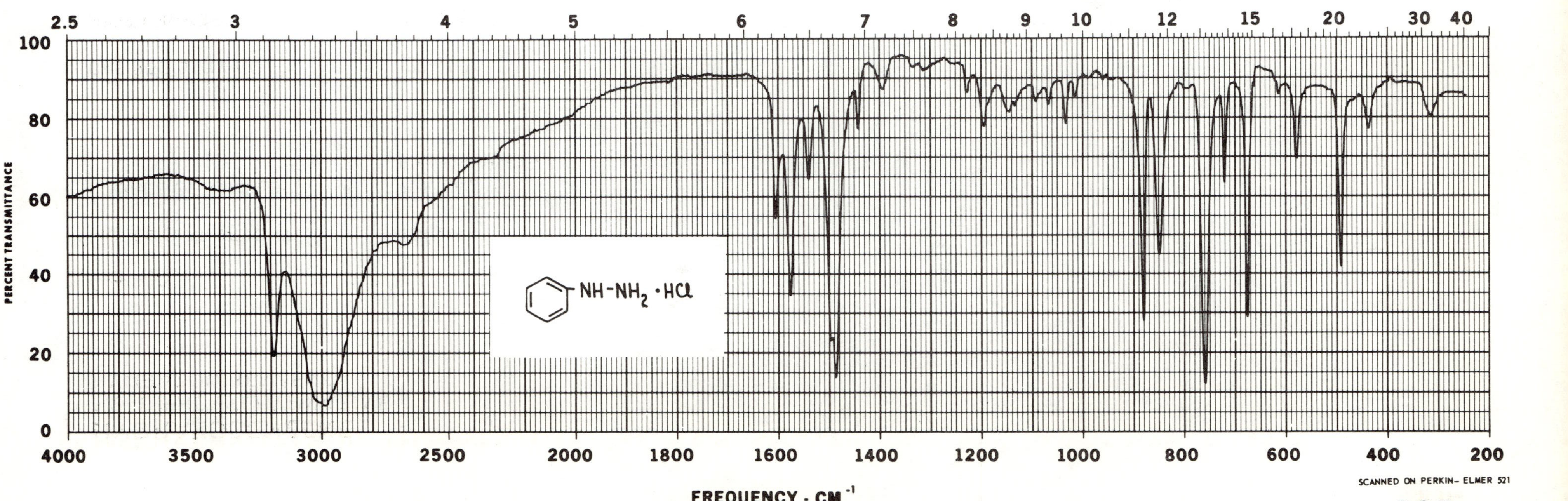

HYDRAZOBENZENE
1,2–DIPHENYLHYDRAZINE
HYDRAZODIBENZENE

$C_{12}H_{12}N_2$

MOLECULAR WEIGHT: 184.24

SOURCE OF SAMPLE: Fluka AG
Buchs, Switzerland

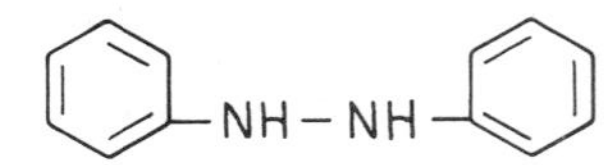

CAS REGISTRY NO.: 122-66-7

NIOSH NUMBER: MW26250

KBr Wafer

CONDENSED PHASE

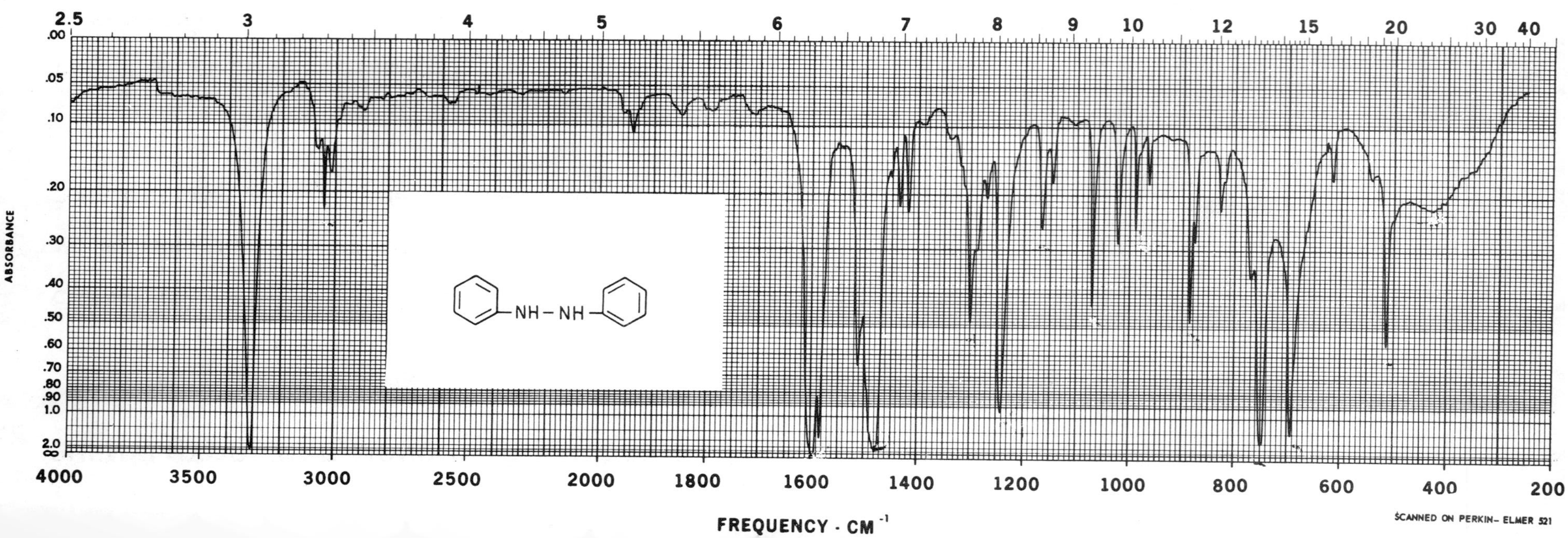

ISONICOTINIC ACID, HYDRAZIDE
AMIDON
ISONIAZID
4–PYRIDINECARBOXYLIC ACID, HYDRAZIDE
ZINADON

$C_6H_7N_3O$

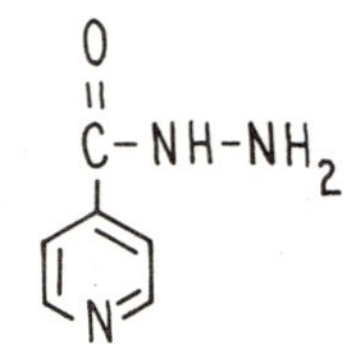

MOLECULAR WEIGHT: 137.14

SOURCE OF SAMPLE: Eastman Organic Chemicals
Rochester, New York

CAS REGISTRY NO.: 54-85-3

NIOSH NUMBER: NS17500

KBr Wafer

CONDENSED PHASE

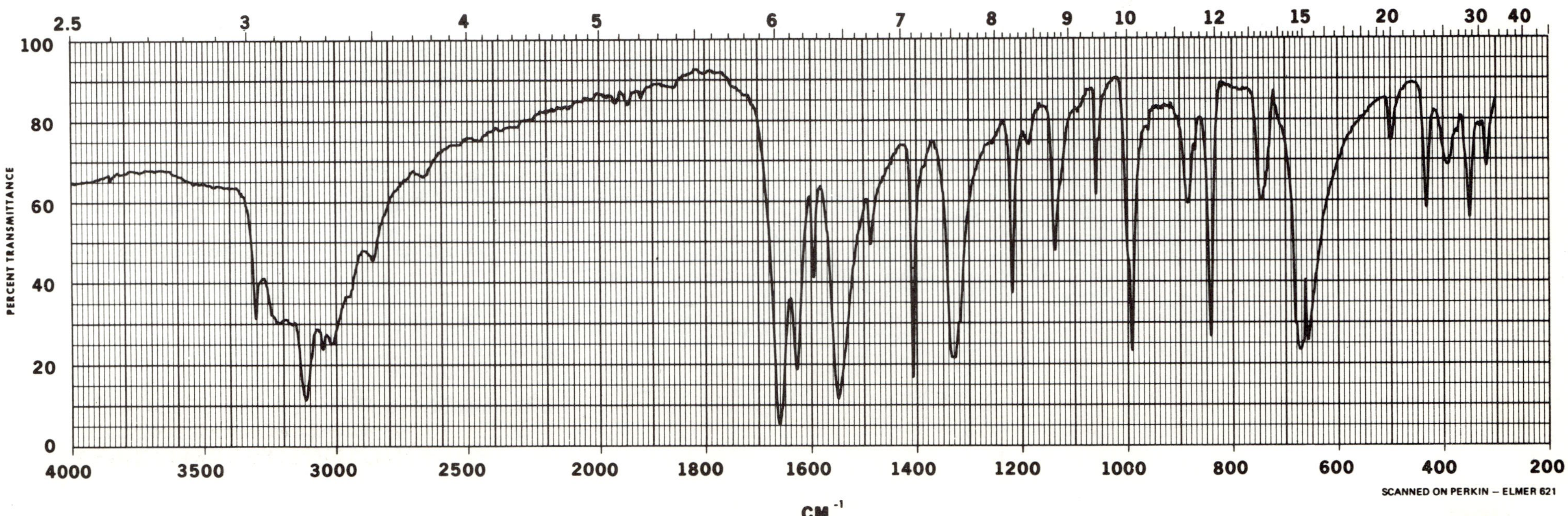

SCANNED ON PERKIN – ELMER 621

2–METHYLLACTONITRILE
ACETONE CYANOHYDRIN
2–HYDROXY–2–METHYLPROPIONITRILE

C_4H_7NO

MOLECULAR WEIGHT: 85.11

SOURCE OF SAMPLE: Eastman Organic Chemicals
Rochester, New York

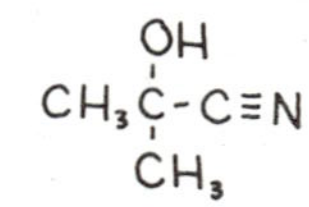

CAS REGISTRY NO.: 75-86-5

NIOSH NUMBER: OD92750

CAPILLARY CELL: Neat

CONDENSED PHASE

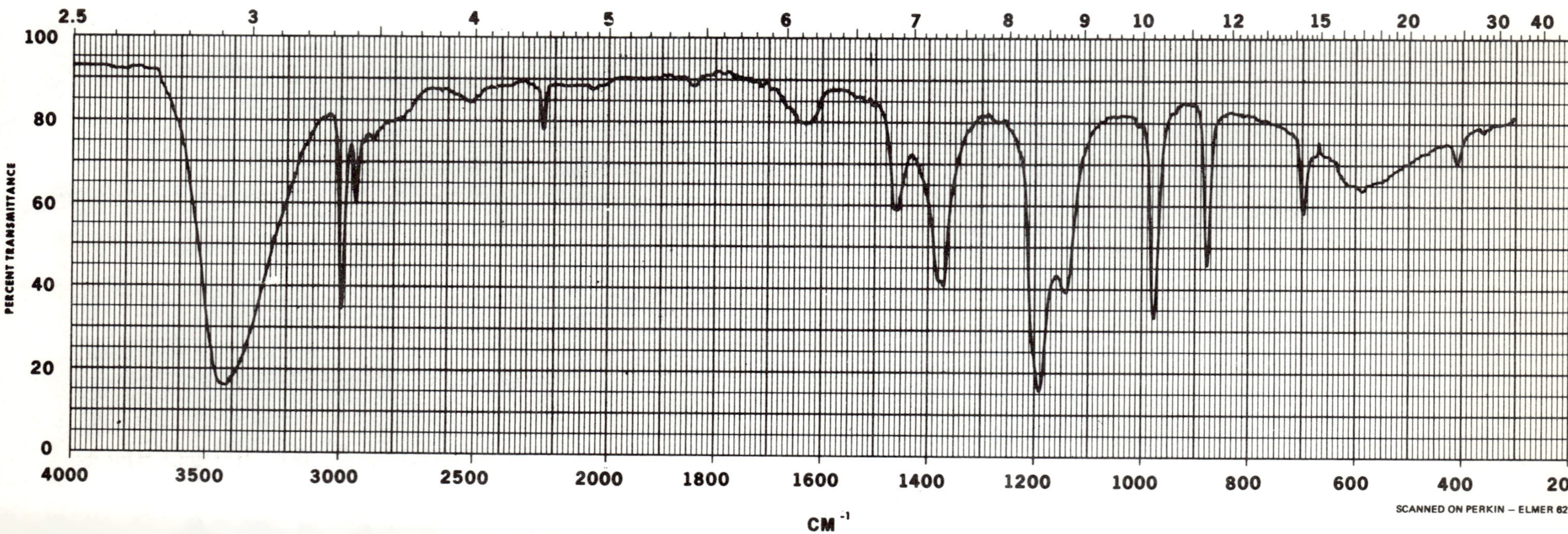

SCANNED ON PERKIN – ELMER 621

LASIOCARPINE

2–METHYL–2–BUTENOIC ACID, 7– { [2,3–DIHYDROXY–2–(1–METHOXYETHYL)–3–METHYL–1–OXOBUTOXY] METHYL } –2,3,5,7a–TETRAHYDRO–1H–PYRROLIZIN–1–YL ESTER

$C_{21}H_{33}NO_7$

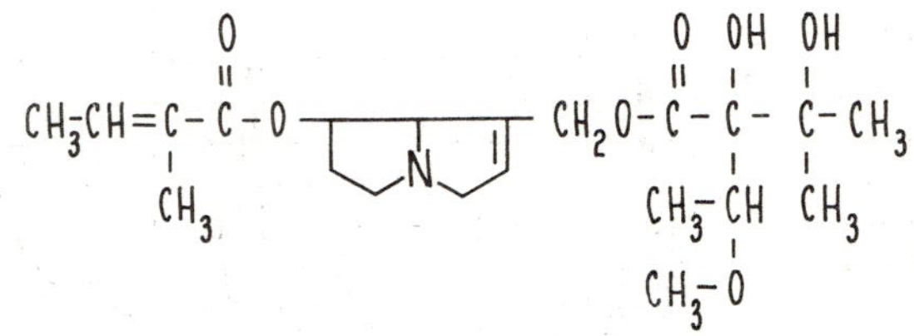

MOLECULAR WEIGHT: 411.55

SOURCE OF SAMPLE: Department of Health and Human Services
National Toxicology Program
Research Triangle Park, North Carolina

CAS REGISTRY NO.: 303-34-4

NIOSH NUMBER: OE78750

KBr Wafer

CONDENSED PHASE

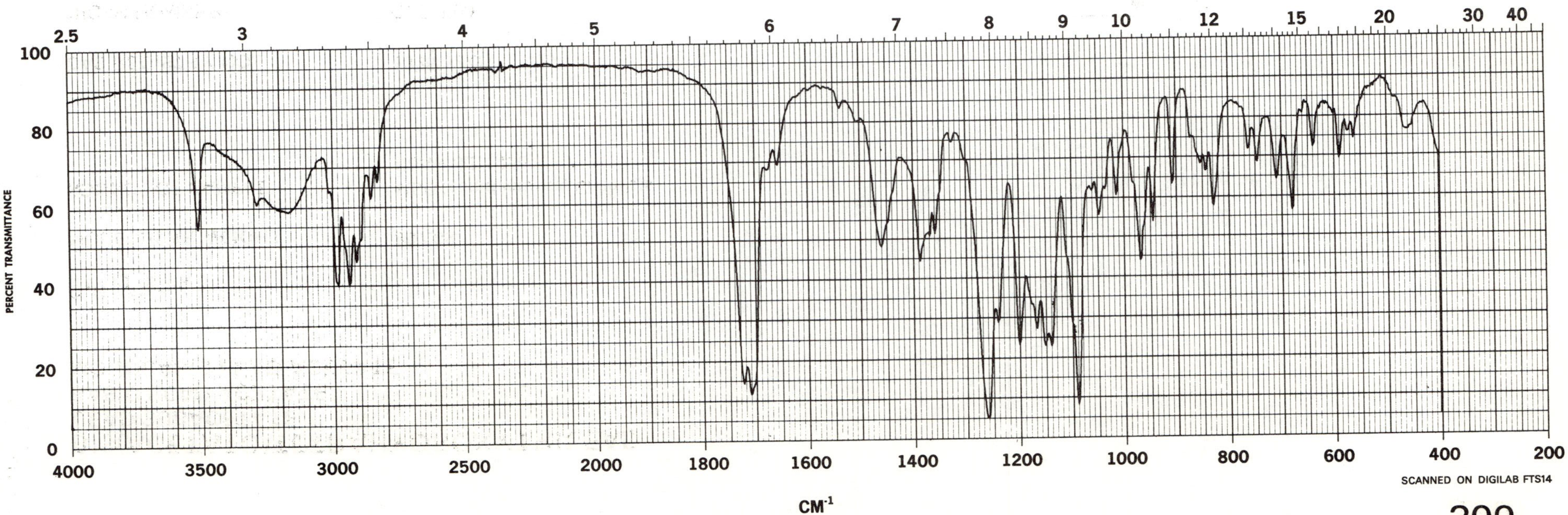

SCANNED ON DIGILAB FTS14

BIS(ACETATO)TETRAHYDROXYTRILEAD
LEAD SUBACETATE
BASIC LEAD ACETATE
MONOBASIC LEAD ACETATE

$C_4H_{10}O_8Pb_3$

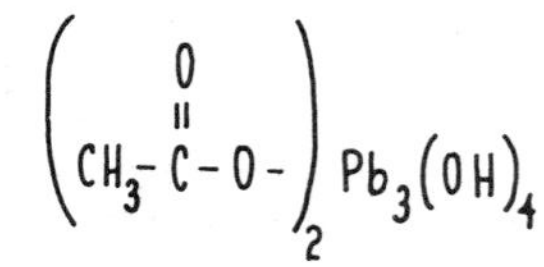

MOLECULAR WEIGHT: 807.69

SOURCE OF SAMPLE: MCB Manufacturing Chemists
Norwood, Ohio

CAS REGISTRY NO.: 1335-32-6

NIOSH NUMBER: OF87500

4000 - 1350 cm^{-1} = Mulled in Perfluorinated Hydrocarbon
1350 - 400 cm^{-1} = Mulled in Mineral Oil

CONDENSED PHASE

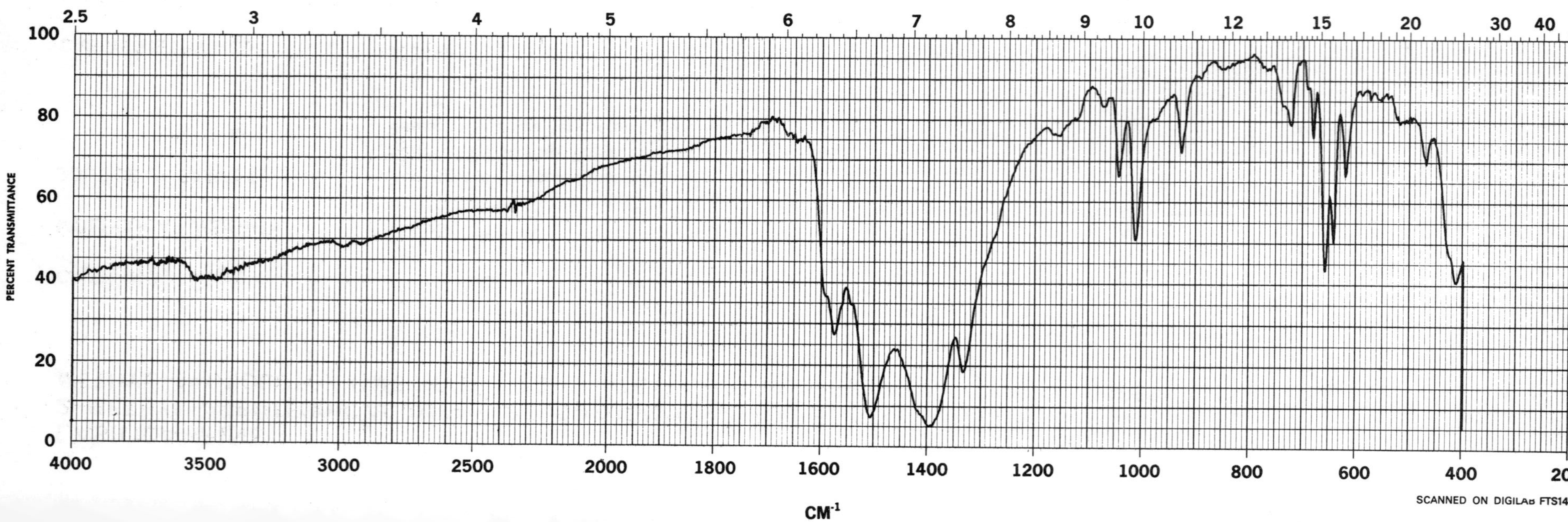

METHACRYLIC ACID, SODIUM SALT
SODIUM METHACRYLATE
2–METHYLACRYLIC ACID, SODIUM SALT

$C_4H_5NaO_2$

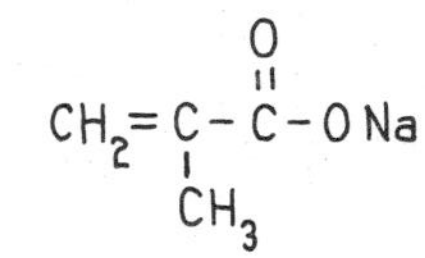

MOLECULAR WEIGHT: 108.07

SOURCE OF SAMPLE: Eastman Organic Chemicals
Rochester, New York

CAS REGISTRY NO.: –

NIOSH NUMBER: –

KBr Wafer

CONDENSED PHASE

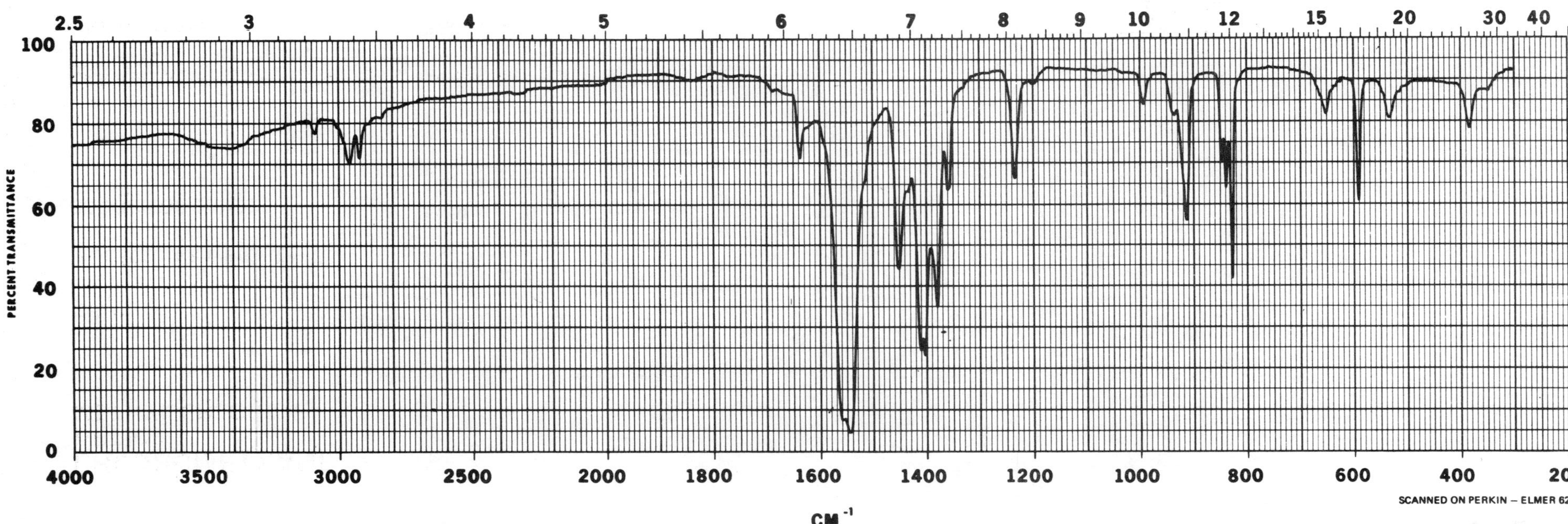

CHLOROMETHANE
ARTIC
METHYL CHLORIDE

CH_3Cl

MOLECULAR WEIGHT: 50.49

SOURCE OF SAMPLE: MCB Manufacturing Chemists
Norwood, Ohio

CAS REGISTRY NO.: 74-87-3

NIOSH NUMBER: PA63000

GAS CELL: 10 cm

CH_3-Cl

A GAS AT 25°C

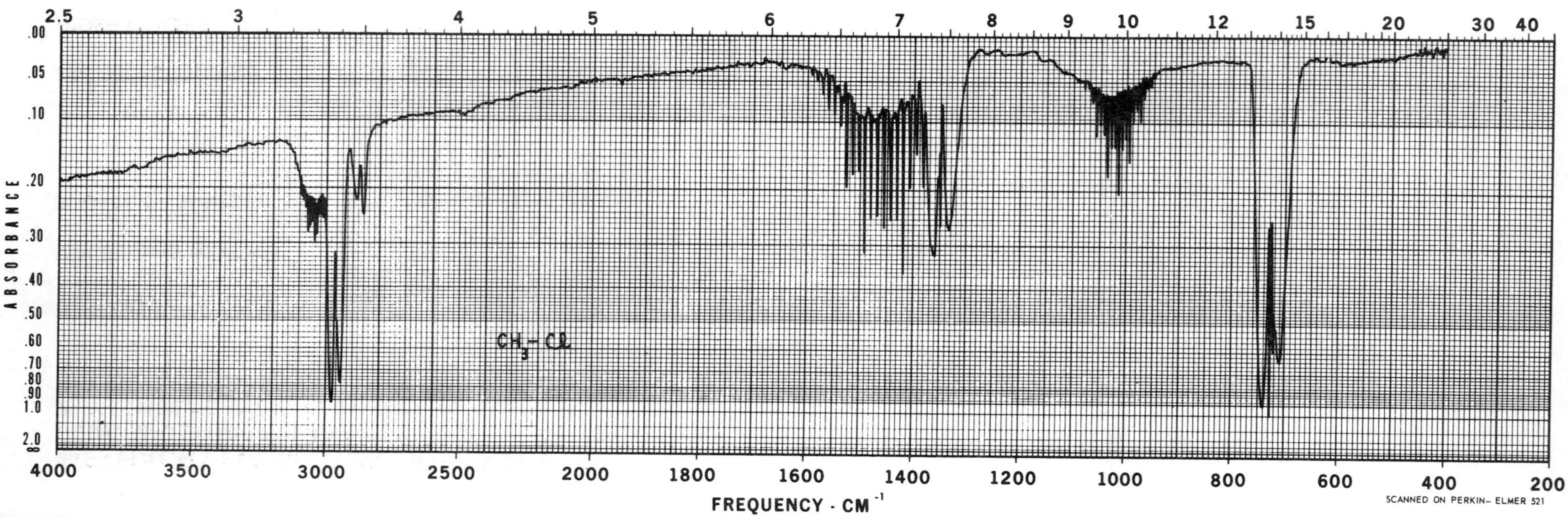

2,3–EPOXY–1,4,5,6,7,8,8–HEPTACHLORO–3a,4,7,7a–TETRAHYDRO–4,7–METHANOINDAN
HEPTACHLOR EPOXIDE
ENT 25,584
VELSICOL 53-CS-17

$C_{10}H_5Cl_7O$

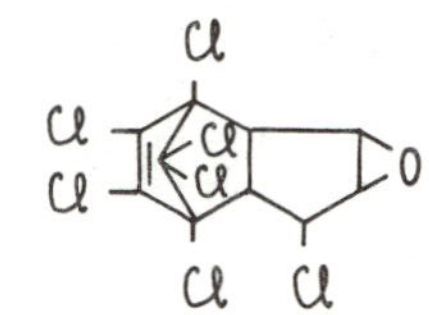

MOLECULAR WEIGHT:	389.32
SOURCE OF SAMPLE:	Pesticides Repository U. S. Public Health Service Perrine, Florida
CAS REGISTRY NO.:	1024-57-3
NIOSH NUMBER:	PB94500

Film

CONDENSED PHASE

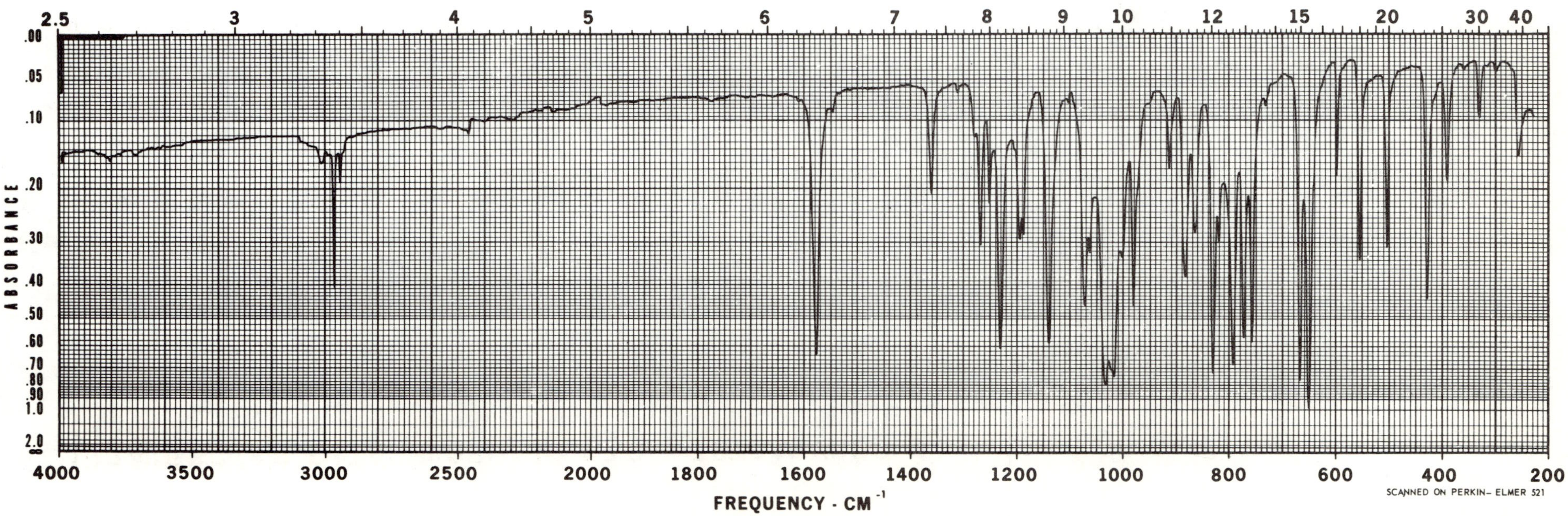

1α,2,4β,5,6,7β,8,8–OCTACHLORO–3aα,4,7,7aα–TETRAHYDRO–4,7–METHANOINDAN
CHLORDAN (MIXTURE OF ISOMERS)

* Refer to Appendix for GC/FT-IR Analysis

$C_{10}H_8Cl_8$

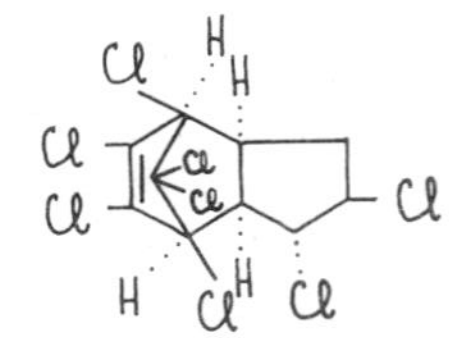

MOLECULAR WEIGHT: 411.80

SOURCE OF SAMPLE: Velsicol Chemical Corporation
Chicago, Illinois

CAS REGISTRY NO.: –

NIOSH NUMBER: –

Between Salts

CONDENSED PHASE

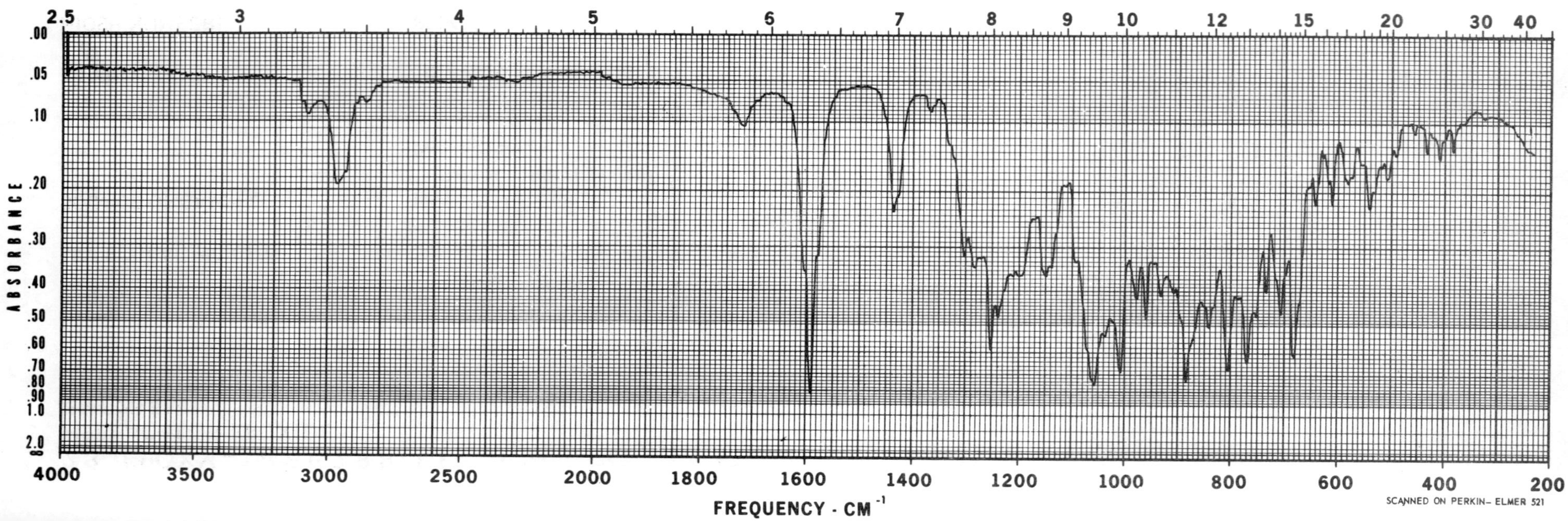

1,1a,2,2,3,3a,4,5,5,5a,5b,6–DODECACHLOROOCTAHYDRO–1,3,4–METHENO–1H–CYCLOBUTA[cd] PENTALENE
BICHLORENDO
DECHLORANE
MIREX
HEXACHLOROCYCLOPENTADIENE, DIMER

$C_{10}Cl_{12}$

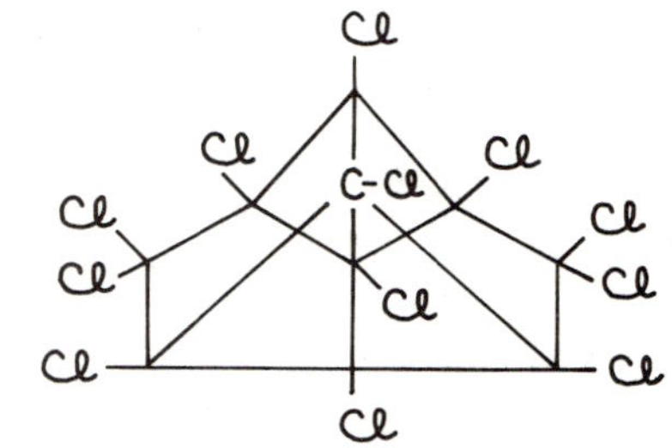

MOLECULAR WEIGHT: 545.55

SOURCE OF SAMPLE: Aldrich Chemical Company, Inc.
Milwaukee, Wisconsin

CAS REGISTRY NO.: 2385-85-5

NIOSH NUMBER: PC82250

KBr Wafer

CONDENSED PHASE

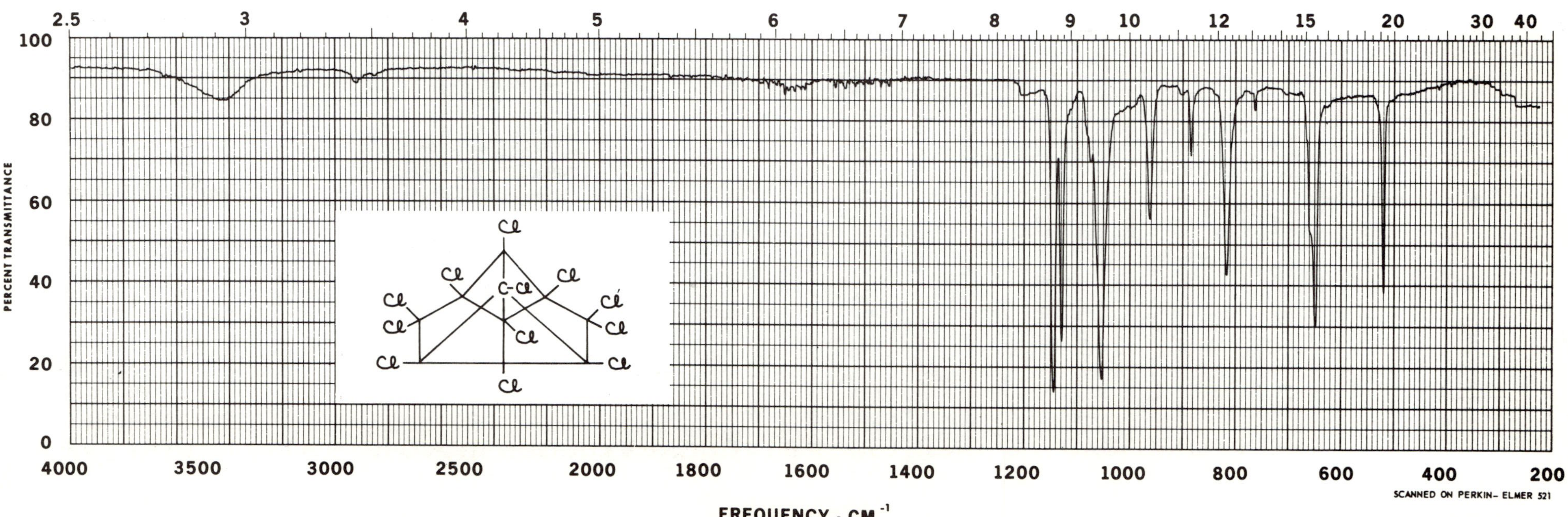

1,1a,3,3a,4,5,5,5a,5b,6–DECACHLOROOCTAHYDRO–1,3,4–METHENO–2H–CYCLOBUTA[cd] PENTALEN–2–ONE
CHLORDECONE
KEPONE
MEREX

$C_{10}Cl_{10}O$

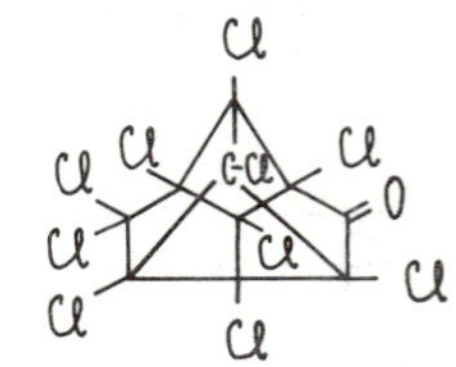

MOLECULAR WEIGHT:	490.64
SOURCE OF SAMPLE:	Allied Chemical Corporation Agricultural Division Houston, Texas
CAS REGISTRY NO.:	143-50-0
NIOSH NUMBER:	PC85750

KBr Wafer

CONDENSED PHASE

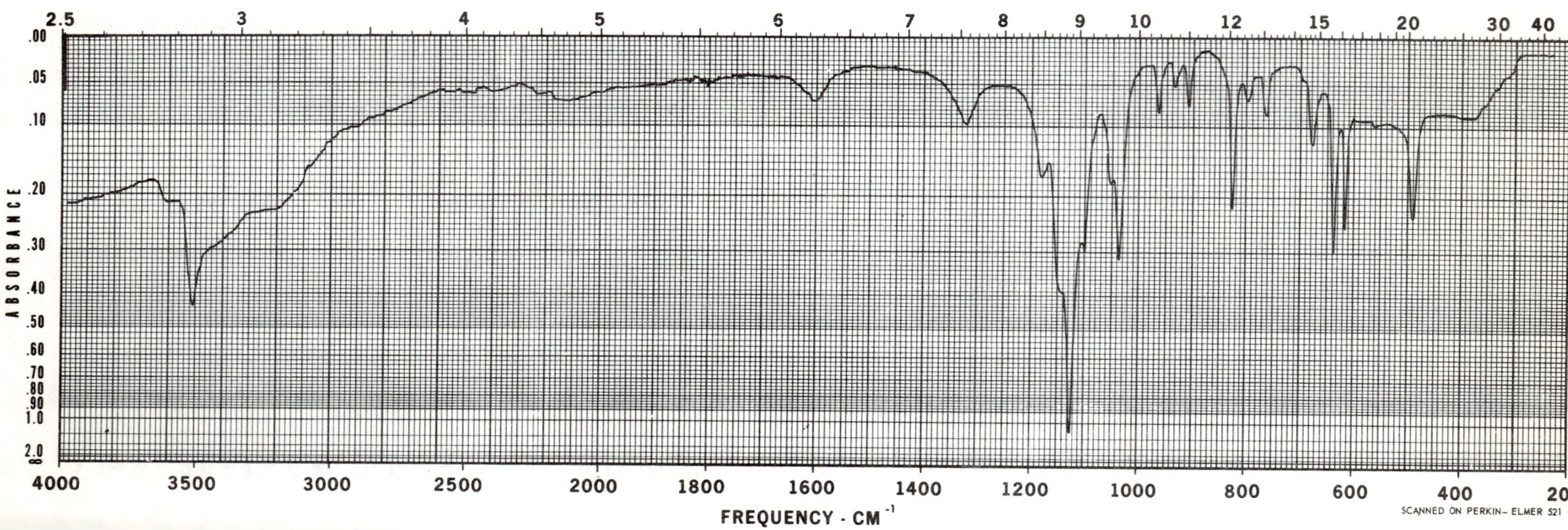

4–(ETHYLSULFONYL)–1–NAPHTHALENESULFONAMIDE
ENS

$C_{12}H_{13}NO_4S_2$

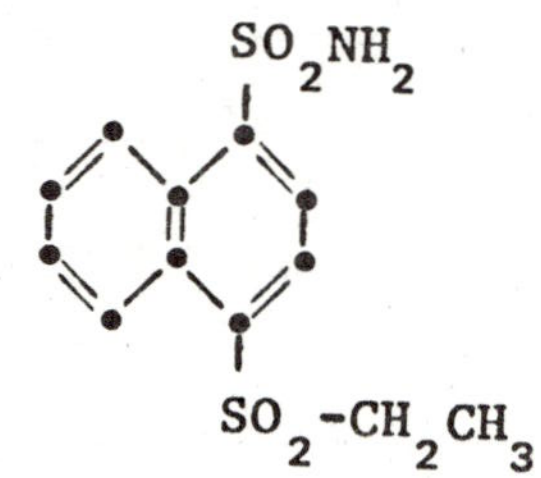

MOLECULAR WEIGHT: 299.37

SOURCE OF SAMPLE: T. W. Whaley and G. H. Daub
University of California
Los Alamos, New Mexico

CAS REGISTRY NO.: 842-00-2

NIOSH NUMBER: WR10500

KBr Wafer

CONDENSED PHASE

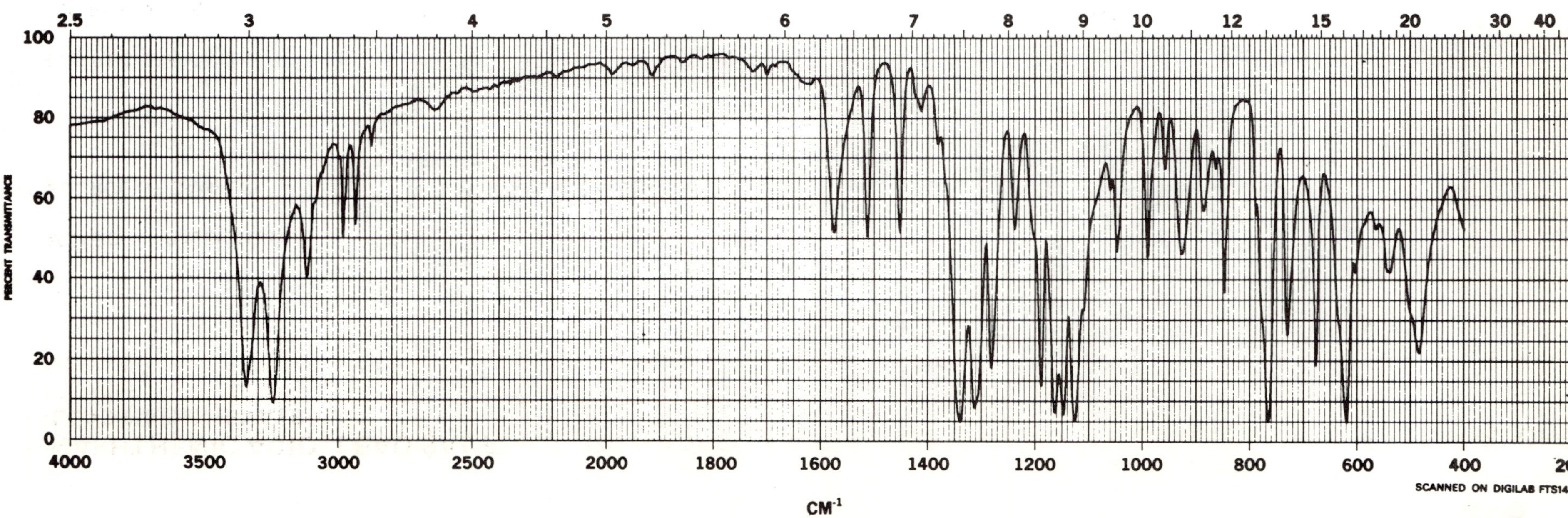

NAPHTHENIC ACID, LEAD SALT
LEAD NAPHTHENATE

–

MOLECULAR WEIGHT: –

SOURCE OF SAMPLE: A. B. M. Industrial Products Ltd.
Cheshire, England

CAS REGISTRY NO.: –

NIOSH NUMBER: OG20250

Between Salts

CONDENSED PHASE

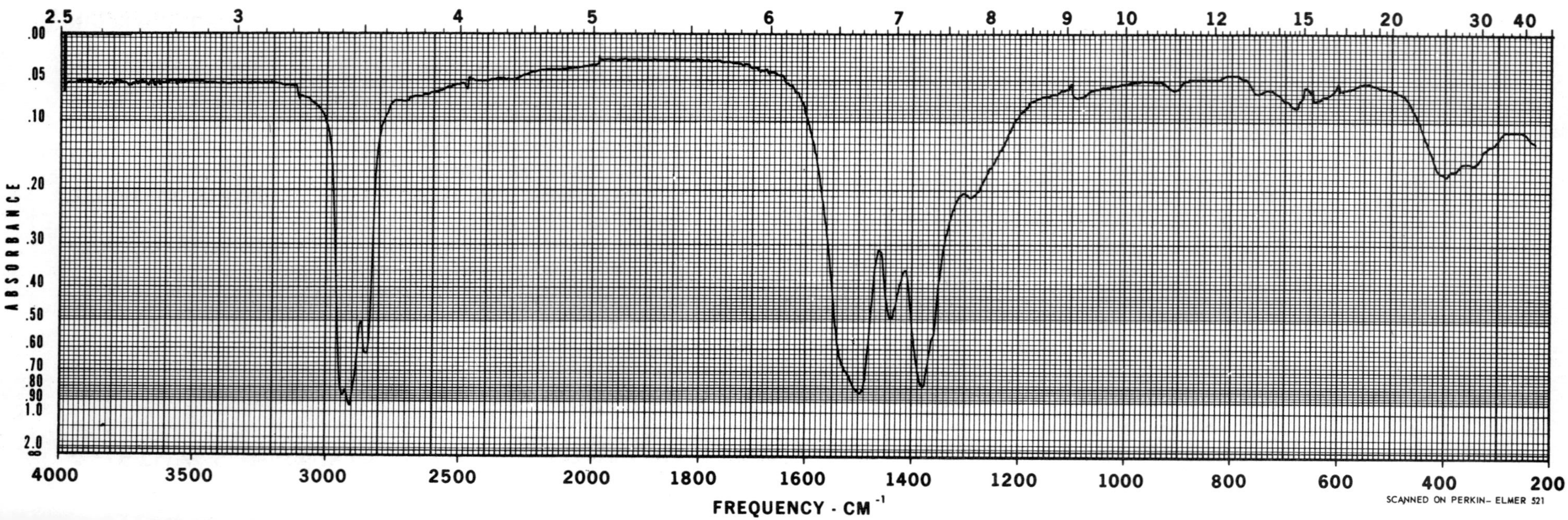

BIS(DIBUTYLDITHIOCARBAMATO)NICKEL
DIBUTYLDITHIOCARBAMIC ACID, NICKEL SALT
VANGUARD N
NBC

$C_{18}H_{36}N_2NiS_4$

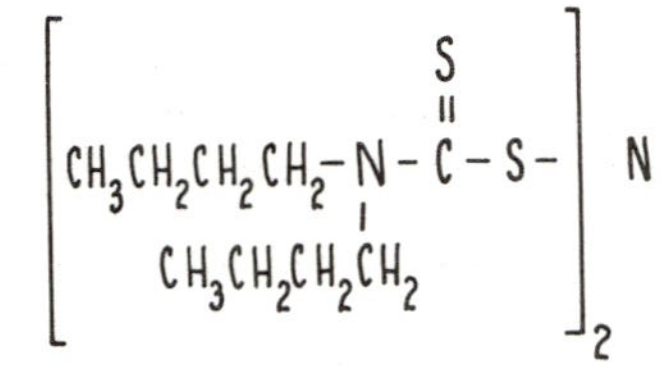

MOLECULAR WEIGHT:	467.47
SOURCE OF SAMPLE:	E. I. du Pont de Nemours and Company, Inc. Wilmington, Delaware
CAS REGISTRY NO.:	13927-77-0
NIOSH NUMBER:	QR61400

KBr Wafer

CONDENSED PHASE

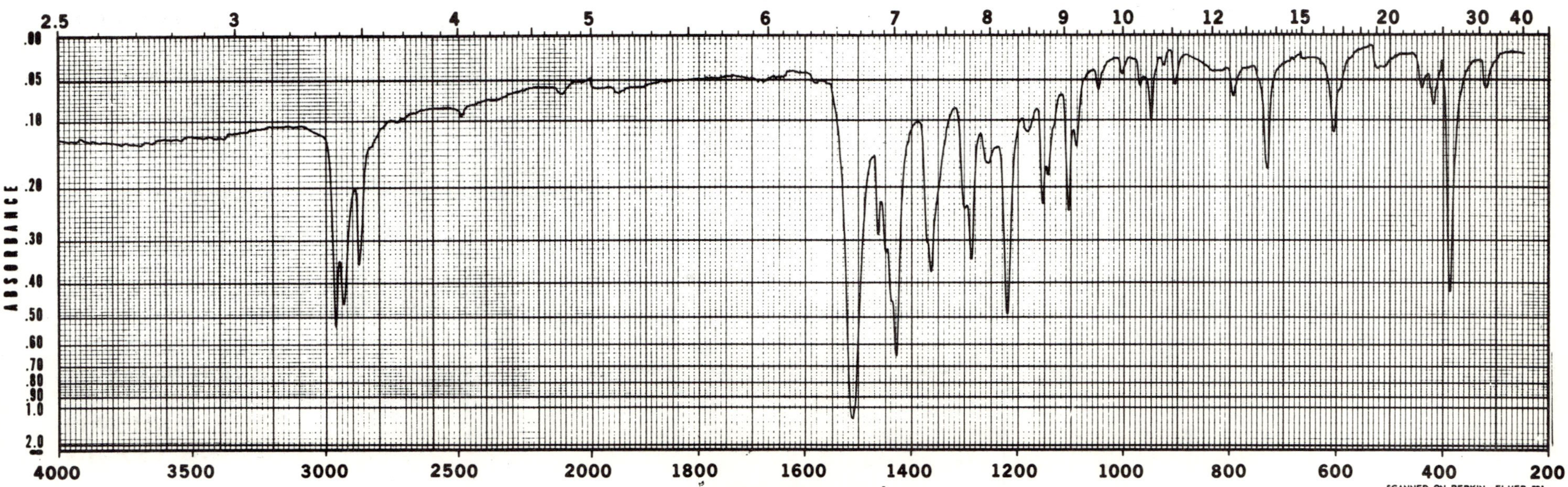

NICKEL OXIDE
BUNSENITE
NICKEL (II) OXIDE (1:1)
NICKEL MONOXIDE
NICKELOUS OXIDE

NiO

MOLECULAR WEIGHT:	74.71
SOURCE OF SAMPLE:	Aldrich Chemical Company, Inc. Milwaukee, Wisconsin
CAS REGISTRY NO.:	1313-99-1
NIOSH NUMBER:	QR84000

Mulled in Mineral Oil

NiO

CONDENSED PHASE

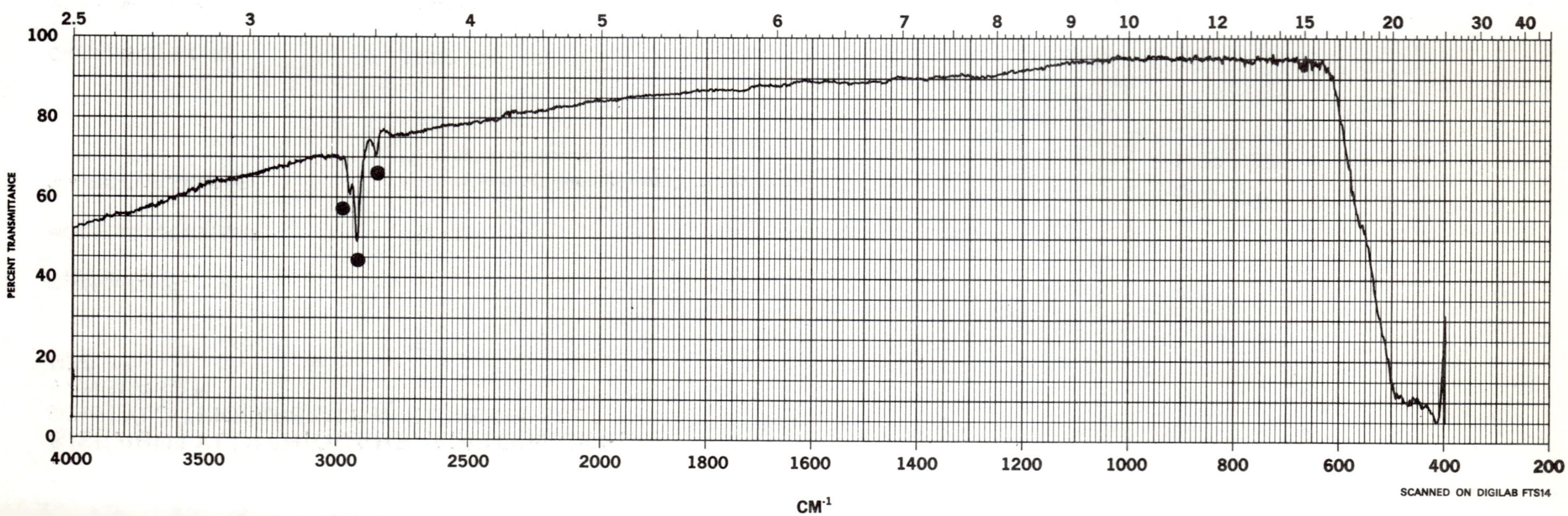

1,4,5,6,7,7–HEXACHLORO–5–NORBORNENE–2,3–DIMETHANOL, CYCLIC SULFATE
ENDOSULFAN SULFATE

$C_9H_6Cl_6O_4S$

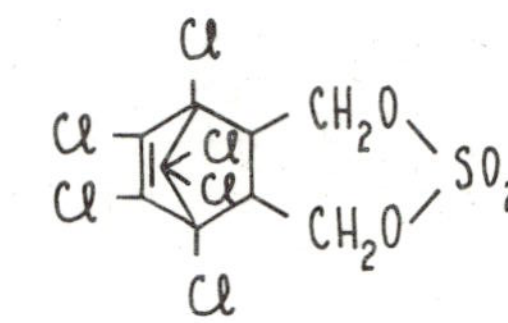

MOLECULAR WEIGHT: 422.93

SOURCE OF SAMPLE: U. S. EPA Repository
Research Triangle Park
North Carolina

CAS REGISTRY NO.: 1031-07-8

NIOSH NUMBER: –

KBr Wafer

CONDENSED PHASE

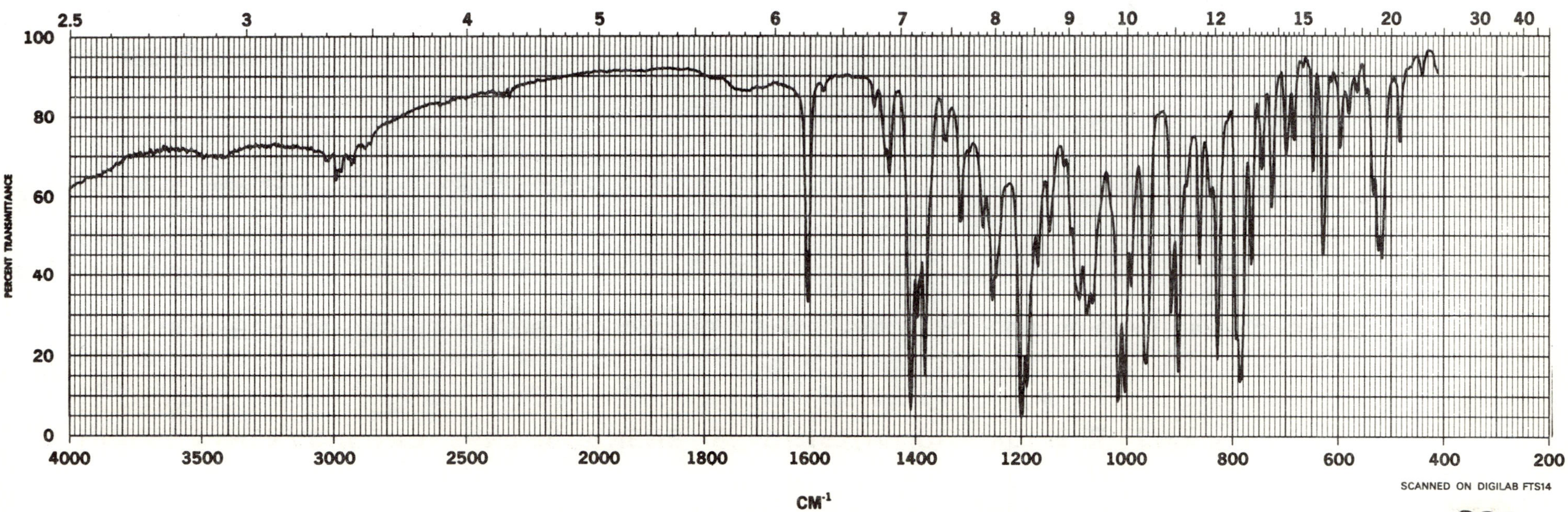

1,4,5,6,7,7–HEXACHLORO–5–NORBORNENE–2,3–DIMETHANOL, CYCLIC SULFITE

ENDOSULFAN
THIODAN
MALIX
6,7,8,9,10,10–HEXACHLORO–1,5,5a,6,9,9a–HEXAHYDRO–6,9–METHANO–2,4,3–BENZODIOXATHIEPIN, 3–OXIDE

$C_9H_6Cl_6O_3S$

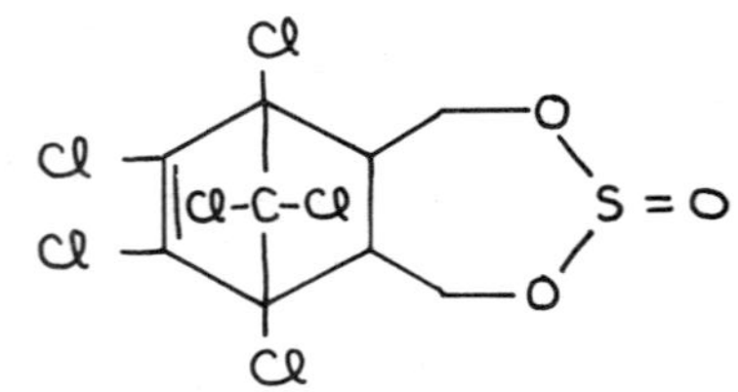

MOLECULAR WEIGHT: 406.93

SOURCE OF SAMPLE: FMC Corporation
Niagara Chemical Division
Middleport, New York

CAS REGISTRY NO.: 115-29-7

NIOSH NUMBER: RB92750

KBr Wafer

CONDENSED PHASE

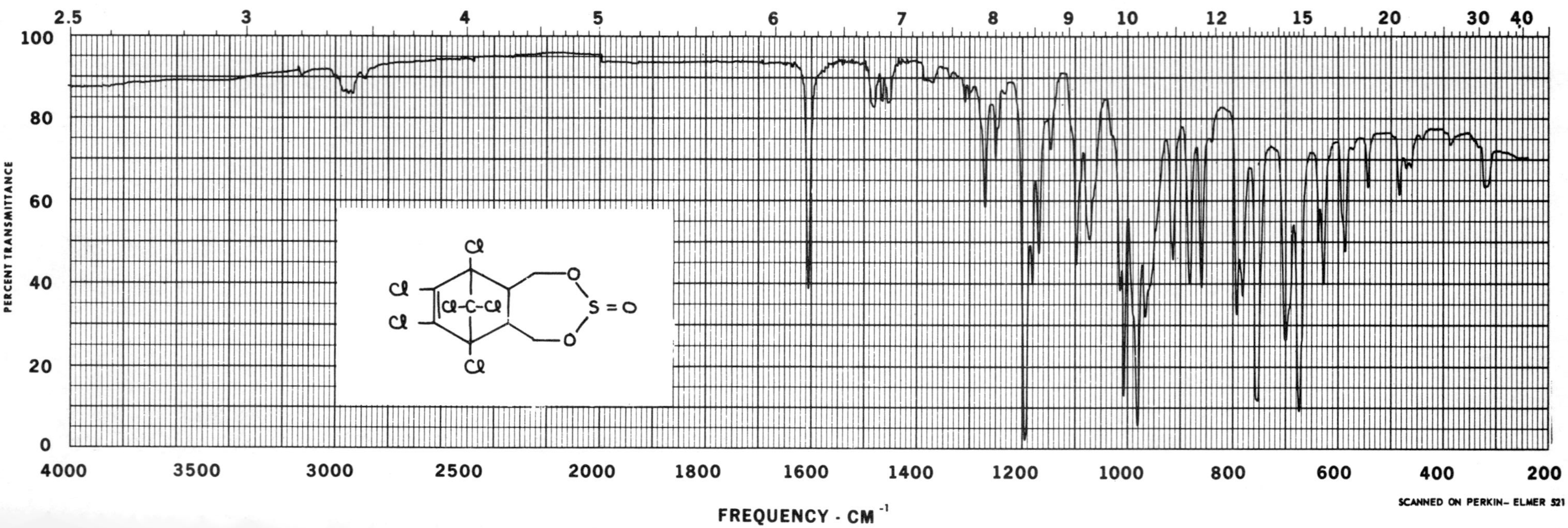

19–NOR–17α–PREGNA–1,3,5(10)–TRIEN–20–YNE–3,17–DIOL
17–ETHYNYLESTRADIOL
ETHINYL ESTRADIOL
ESTINYL
NOVESTROL

$C_{20}H_{24}O_2$

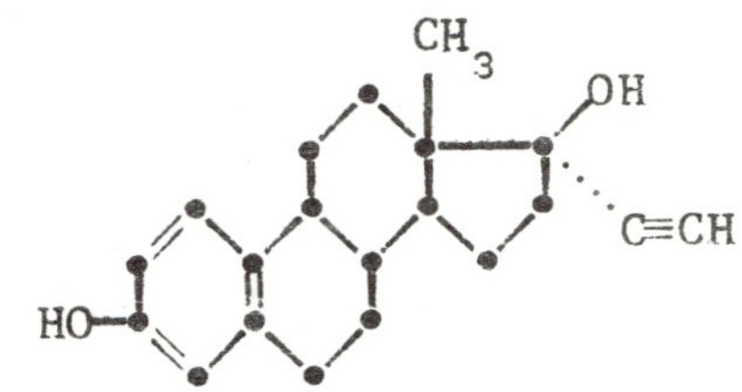

MOLECULAR WEIGHT:	296.41
SOURCE OF SAMPLE:	E. Merck AG Darmstadt, Germany
CAS REGISTRY NO.:	57-63-6
NIOSH NUMBER:	RC89250

KBr Wafer

CONDENSED PHASE

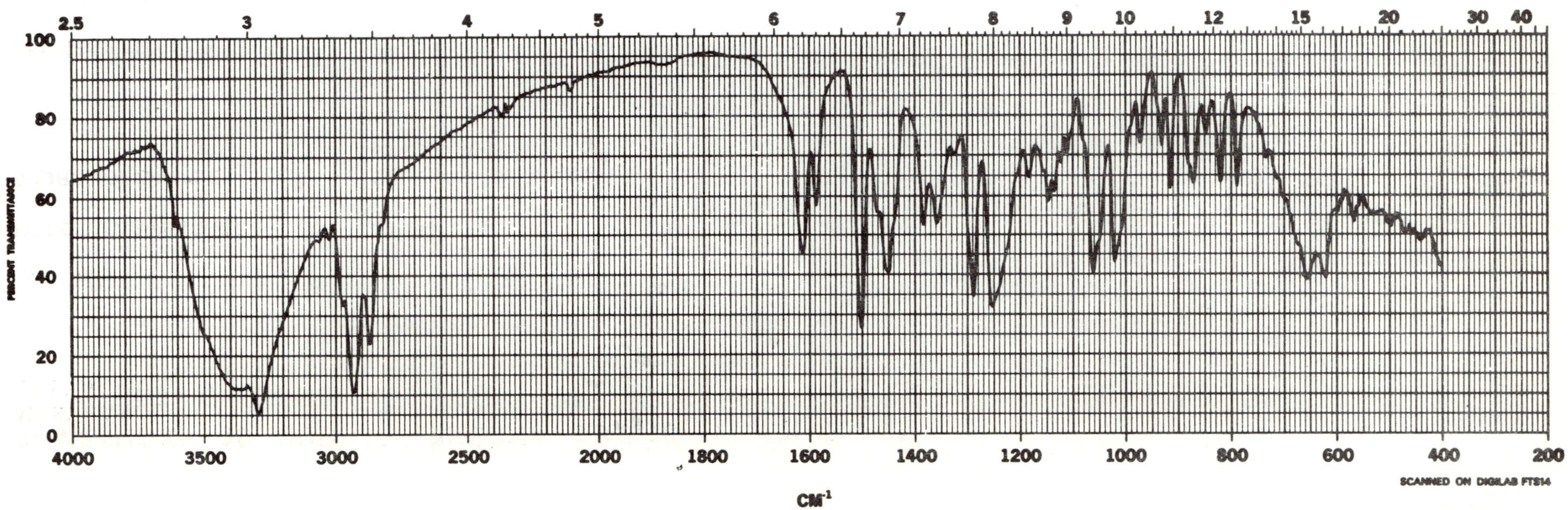

3–METHOXY–19–NOR–17*a*–PREGNA–1,3,5(10)–TRIEN–20–YN–17–OL
MESTRANOL
DELTA–MVE

$C_{21}H_{26}O_2$

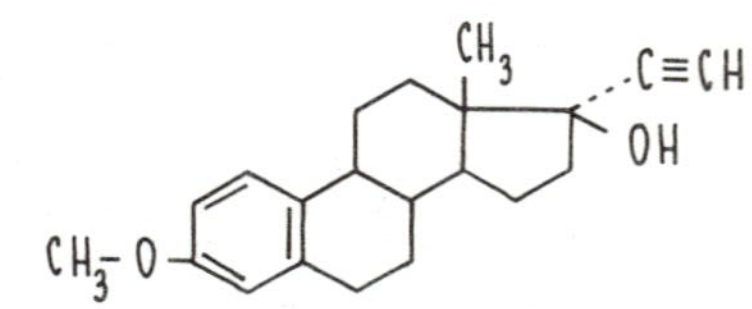

MOLECULAR WEIGHT:	310.44
SOURCE OF SAMPLE:	Sigma Chemical Company St. Louis, Missouri
CAS REGISTRY NO.:	72-33-3
NIOSH NUMBER:	RC89600

KBr Wafer

CONDENSED PHASE

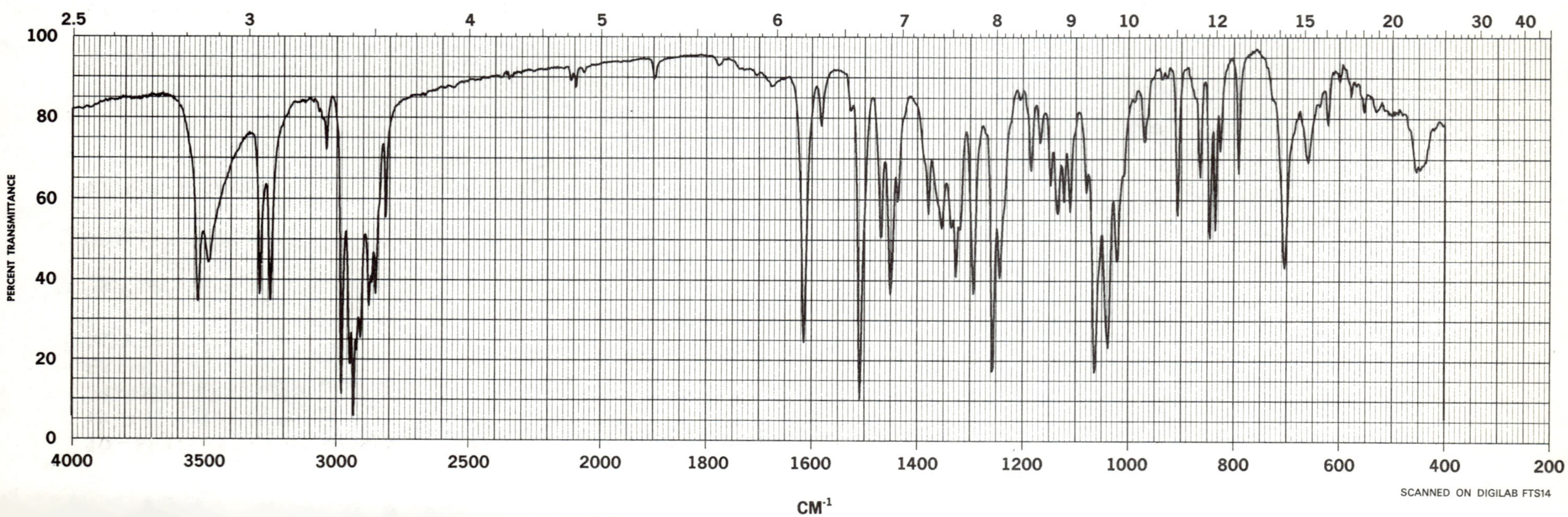

17–HYDROXY–19–NOR–17α–PREGN–4–EN–20–YN–3–ONE
NORETHINDRONE
NORETHISTERONE
NORLUTIN

$C_{20}H_{26}O_2$

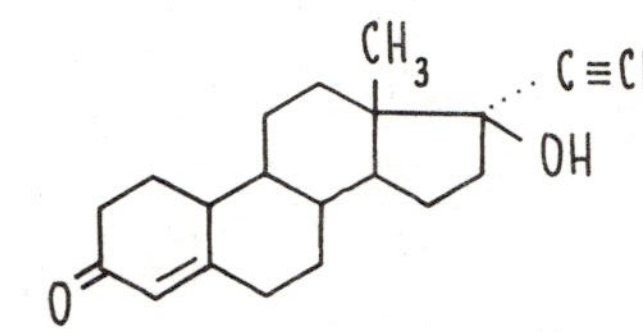

MOLECULAR WEIGHT: 298.43

SOURCE OF SAMPLE: Parke, Davis and Company
Detroit, Michigan

CAS REGISTRY NO.: 68-22-4

NIOSH NUMBER: RC89750

KBr Wafer

CONDENSED PHASE

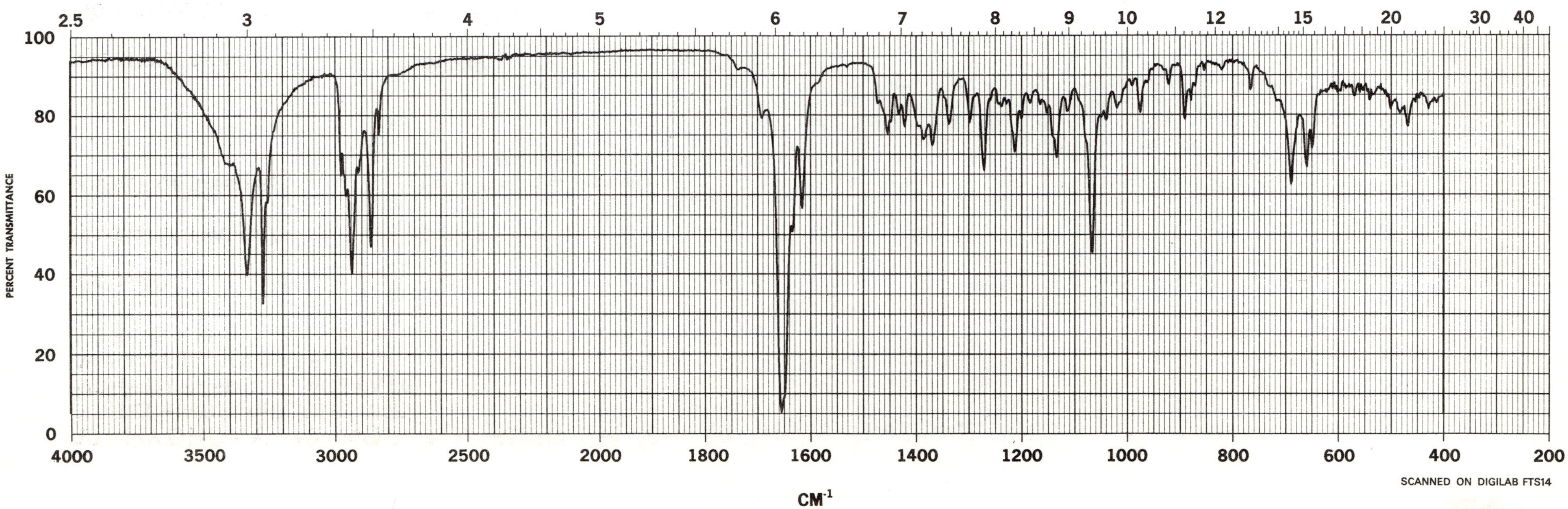

SCANNED ON DIGILAB FTS14

17–HYDROXY–19–NOR–17α–PREGN–4–EN–20–YN–3–ONE, ACETATE
NORETHINDRONE, ACETATE
NORETHISTERONE, ACETATE
NORLUTIN, ACETATE
NORLUTATE

$C_{22}H_{28}O_3$

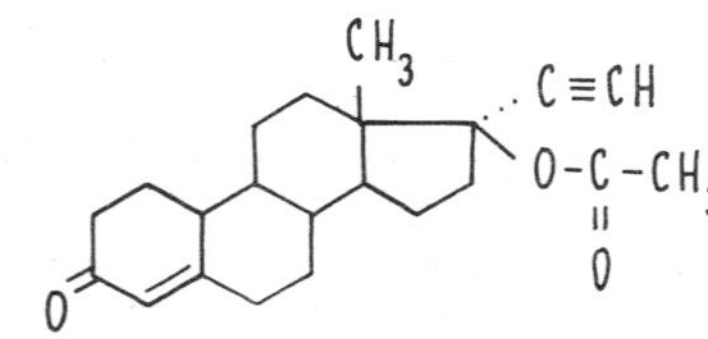

MOLECULAR WEIGHT: 340.47

SOURCE OF SAMPLE: Parke, Davis and Company
Detroit, Michigan

CAS REGISTRY NO.: 51-98-9

NIOSH NUMBER: RC89650

KBr Wafer

CONDENSED PHASE

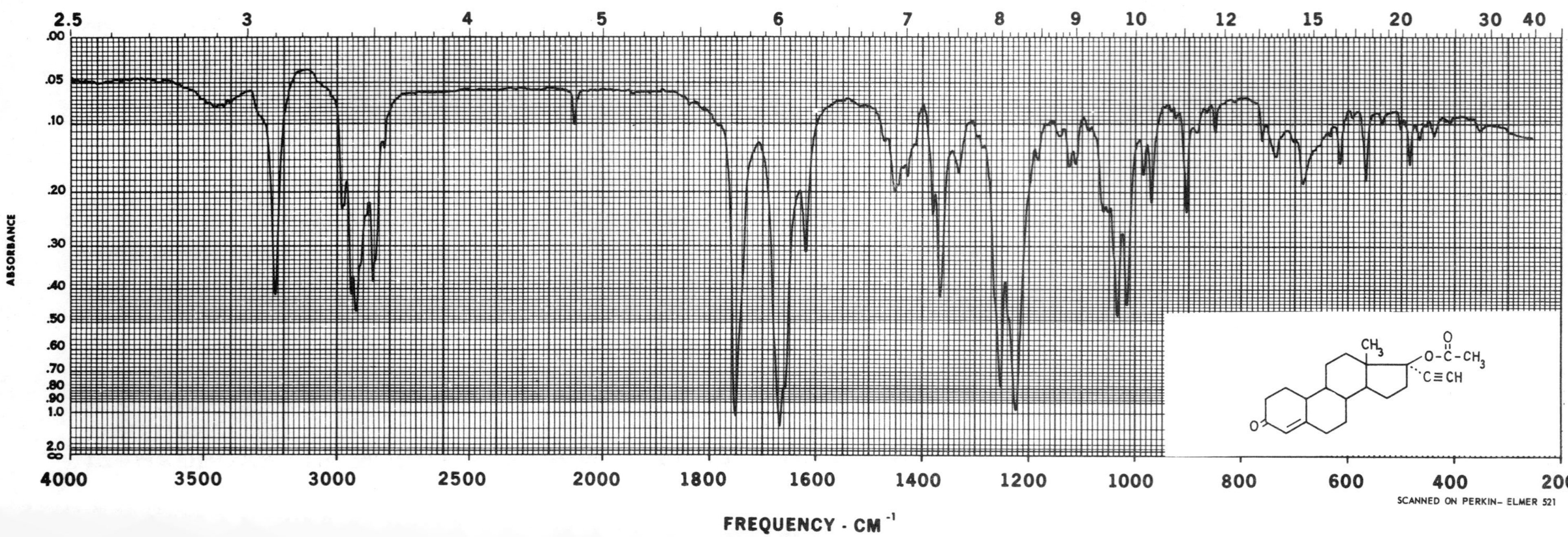

17–HYDROXY–19–NOR–17α–PREGN–5(10)–EN–20–YN–3–ONE
NORETHYNODREL
NORETHINYNODREL

$C_{20}H_{26}O_2$

MOLECULAR WEIGHT: 298.43

SOURCE OF SAMPLE: Sigma Chemical Company
St. Louis, Missouri

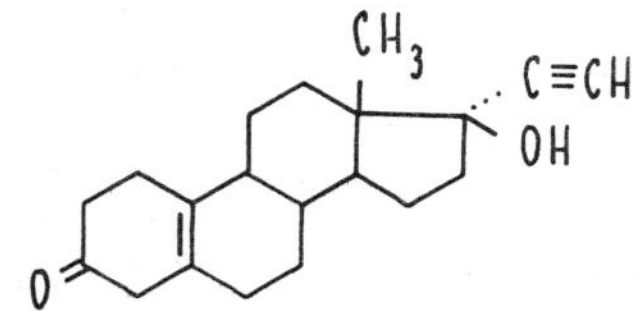

CAS REGISTRY NO.: 68-23-5

NIOSH NUMBER: RC89800

KBr Wafer

CONDENSED PHASE

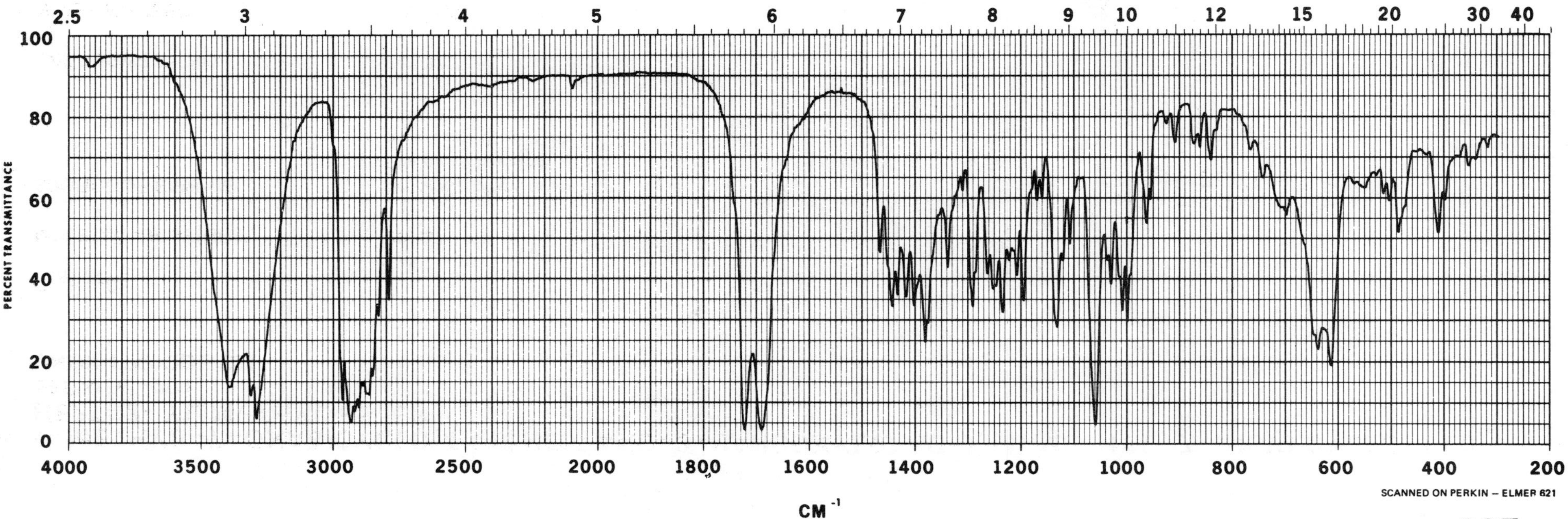

SCANNED ON PERKIN – ELMER 621

4–METHYL–7–OXABICYCLO[4.1.0] HEPTANE–3–CARBOXYLIC ACID, (4–METHYL–7–OXABICYCLO[4.1.0]-HEPT–3–YL)METHYL ESTER
EPOXIDE–201
UNOX EPOXIDE 201

$C_{16}H_{24}O_4$

MOLECULAR WEIGHT: 280.37

SOURCE OF SAMPLE: Union Carbide Corporation
Chemicals Division
New York, New York

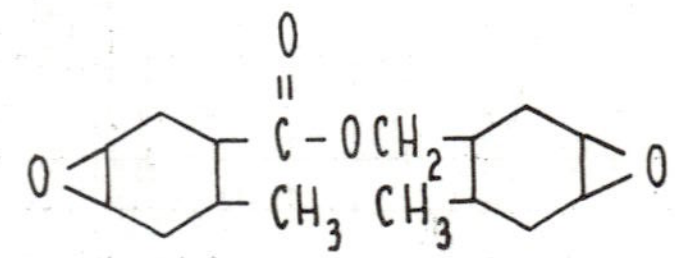

CAS REGISTRY NO.: 141-37-7

NIOSH NUMBER: RN77000

CAPILLARY CELL: Neat

CONDENSED PHASE

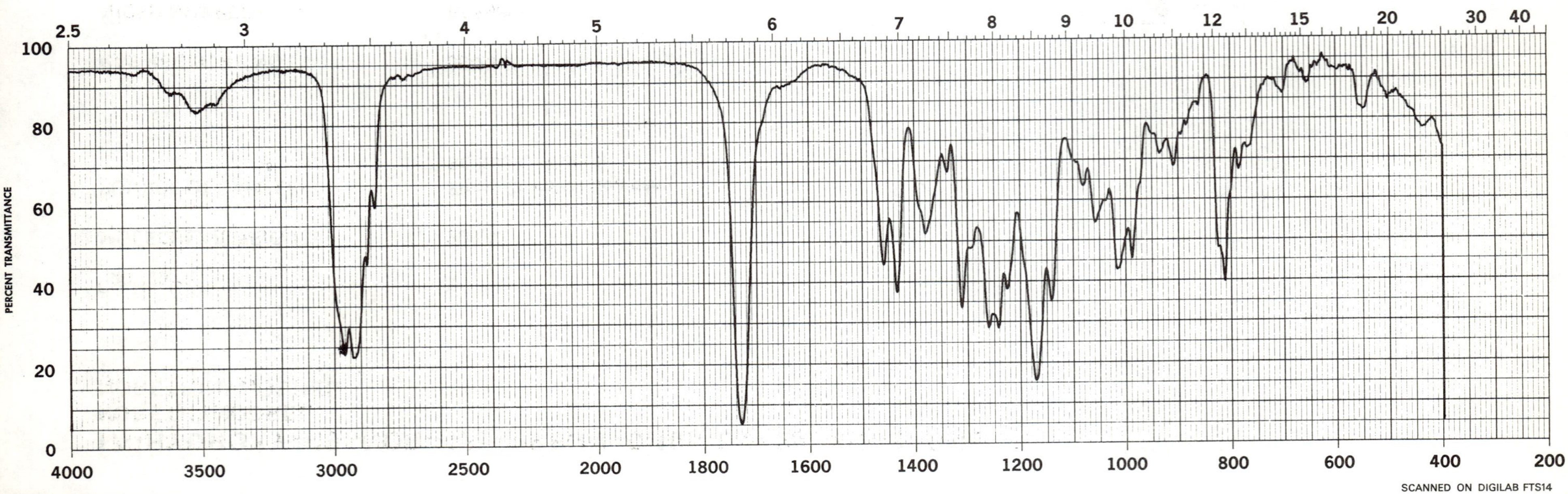

2–[BIS(2–CHLOROETHYL)AMINO] TETRAHYDRO–2H–1,3,2–OXAZAPHOSPHORINE, 2–OXIDE

ASTA
CYTOXAN
ENDOXAN
GENOXAL

$C_7H_{15}Cl_2N_2O_2P \cdot H_2O$

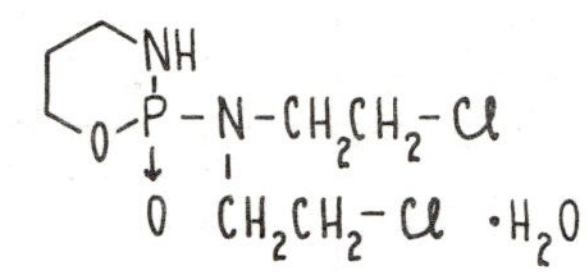

MOLECULAR WEIGHT: 279.13

SOURCE OF SAMPLE: Mead Johnson and Company, Evansville, Indiana

CAS REGISTRY NO.: 50-18-0

NIOSH NUMBER: RP59500

CAPILLARY CELL: Melt

CONDENSED PHASE

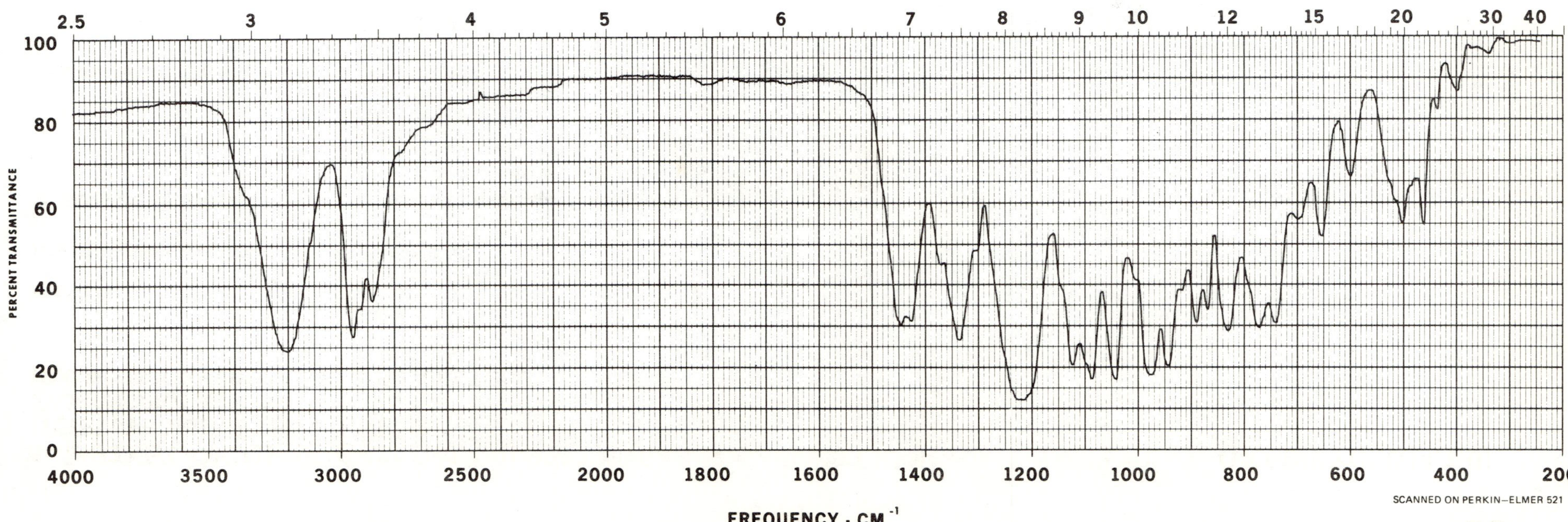

SCANNED ON PERKIN–ELMER 521

2–OXETANONE
BETAPRONE
β–PROPIOLACTONE
β–HYDROXYPROPIONIC ACID, LACTONE

$C_3H_4O_2$

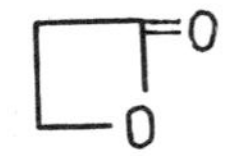

MOLECULAR WEIGHT: 72.06

SOURCE OF SAMPLE: Eastman Organic Chemicals
Rochester, New York

CAS REGISTRY NO.: 57-57-8

NIOSH NUMBER: RQ73500

CAPILLARY CELL: 0.015 mm

CONDENSED PHASE

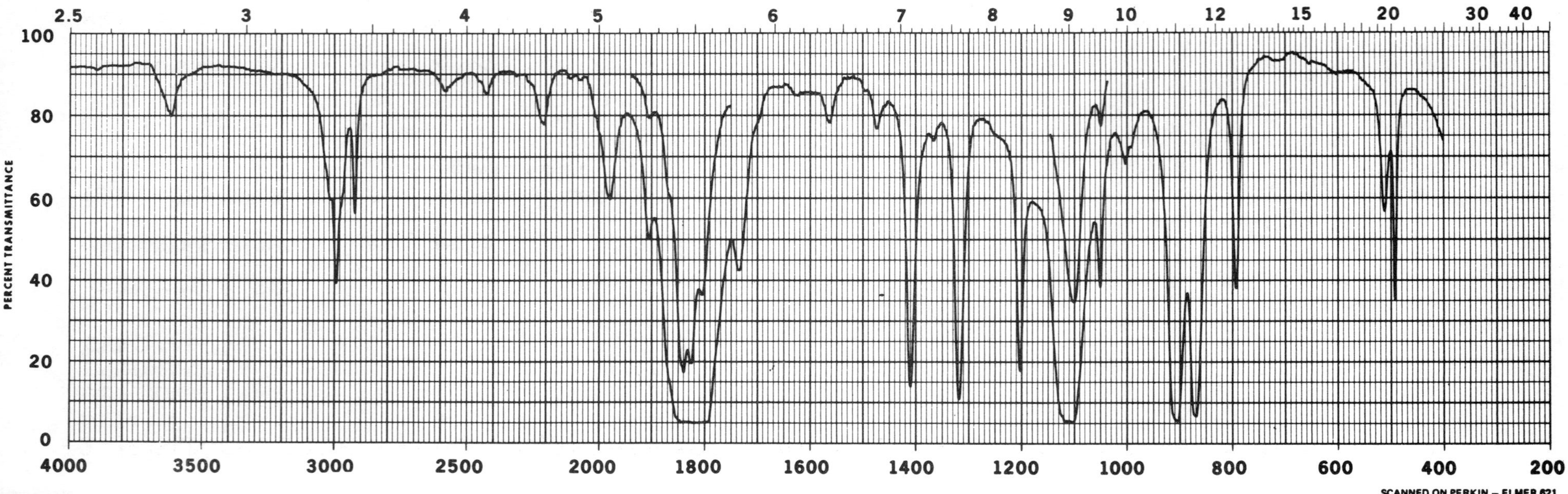

(1–HYDROXY–2,2,2–TRICHLOROETHYL)PHOSPHONIC ACID, DIMETHYL ESTER

CHLOROPHOS
DANEX
TRICHLOROFON
TRINEX

$C_4H_8Cl_3O_4P$

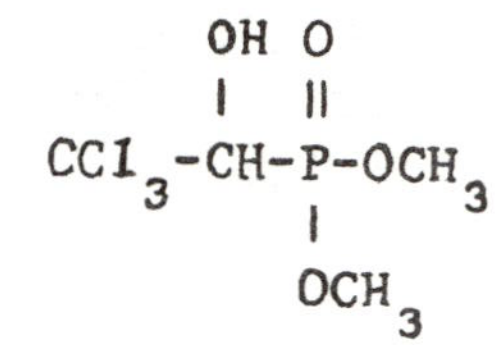

MOLECULAR WEIGHT: 257.44

SOURCE OF SAMPLE: Chemagro Corporation
Kansas City, Missouri

CAS REGISTRY NO.: 52-68-6

NIOSH NUMBER: TA07000

Film

CONDENSED PHASE

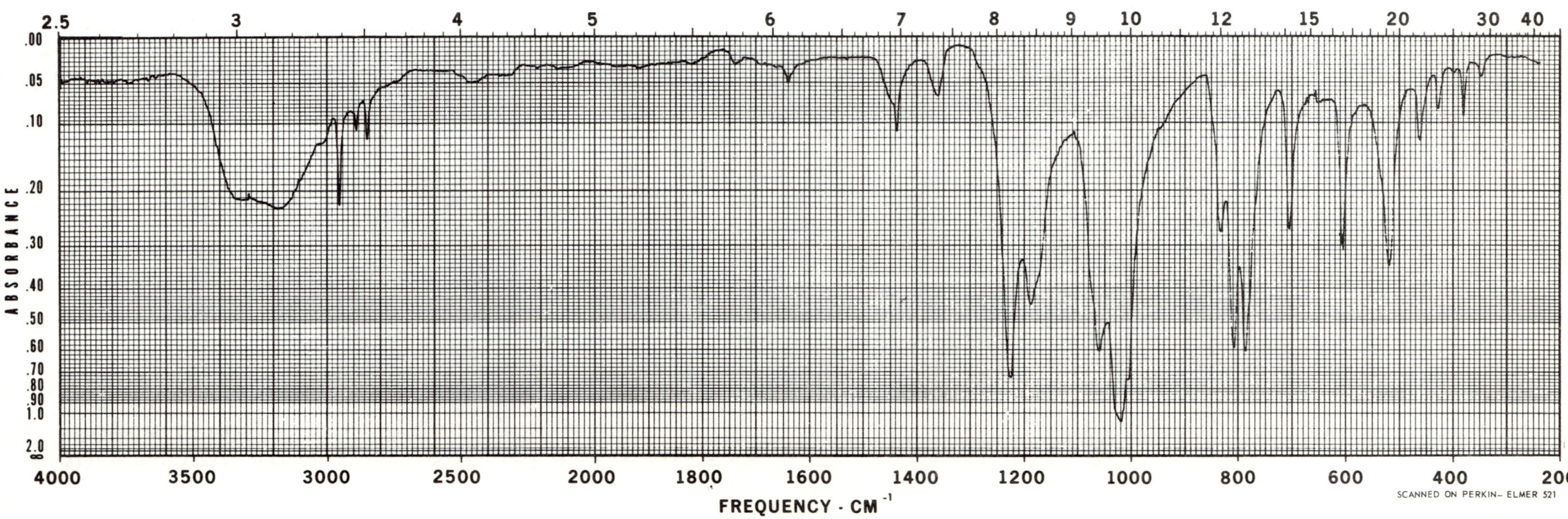

SCANNED ON PERKIN– ELMER 521

PHOSPHORIC ACID, 1,2–DIBROMO–2,2–DICHLOROETHYL DIMETHYL ESTER
BROMCHLOPHOS
BROMEX
DIBROM
NALED

$C_4H_7Br_2Cl_2O_4P$

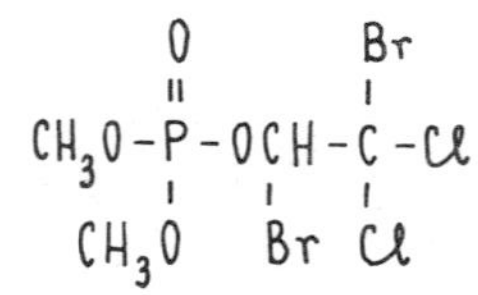

MOLECULAR WEIGHT:	380.80
SOURCE OF SAMPLE:	Chevron Chemical Company Ortho Division Richmond, California
CAS REGISTRY NO.:	300-76-5
NIOSH NUMBER:	TB94500

Between Salts

CONDENSED PHASE

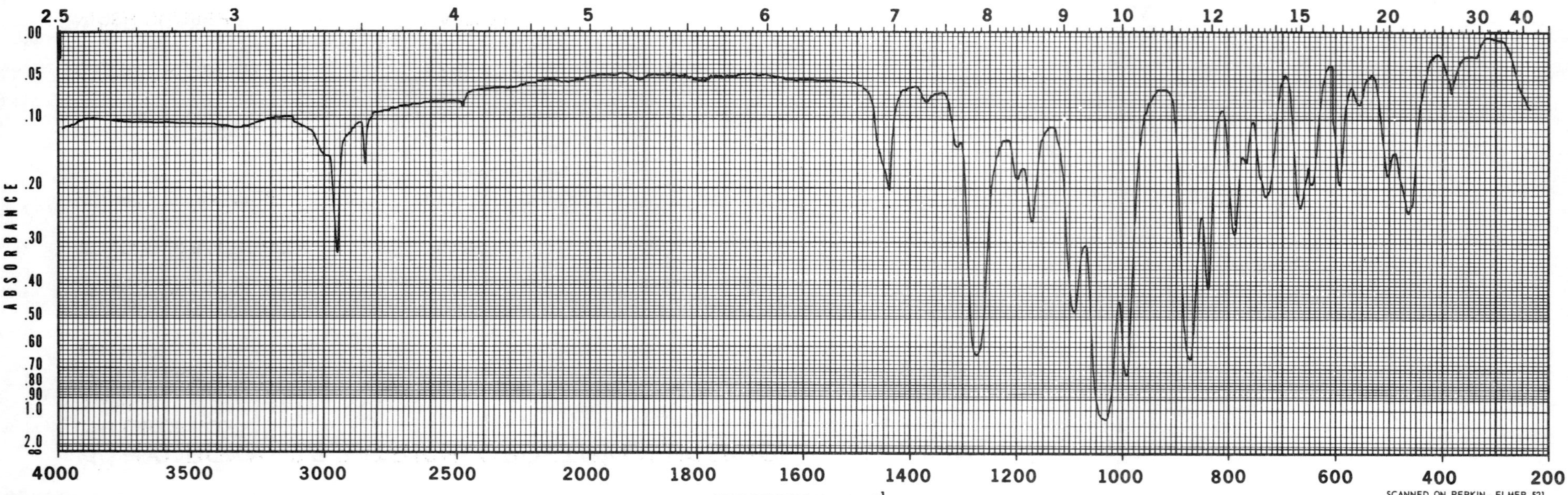

PHOSPHORIC ACID, LEAD (2+) SALT (2:3)
LEAD ORTHOPHOSPHATE
LEAD (II) PHOSPHATE (3:2)
TRILEAD PHOSPHATE

$O_8P_2Pb_3$

$Pb_3(PO_4)_2$

MOLECULAR WEIGHT:	811.51
SOURCE OF SAMPLE:	MCB Manufacturing Chemists Norwood, Ohio
CAS REGISTRY NO.:	7446-27-7
NIOSH NUMBER:	OG36750

KBr Wafer

CONDENSED PHASE

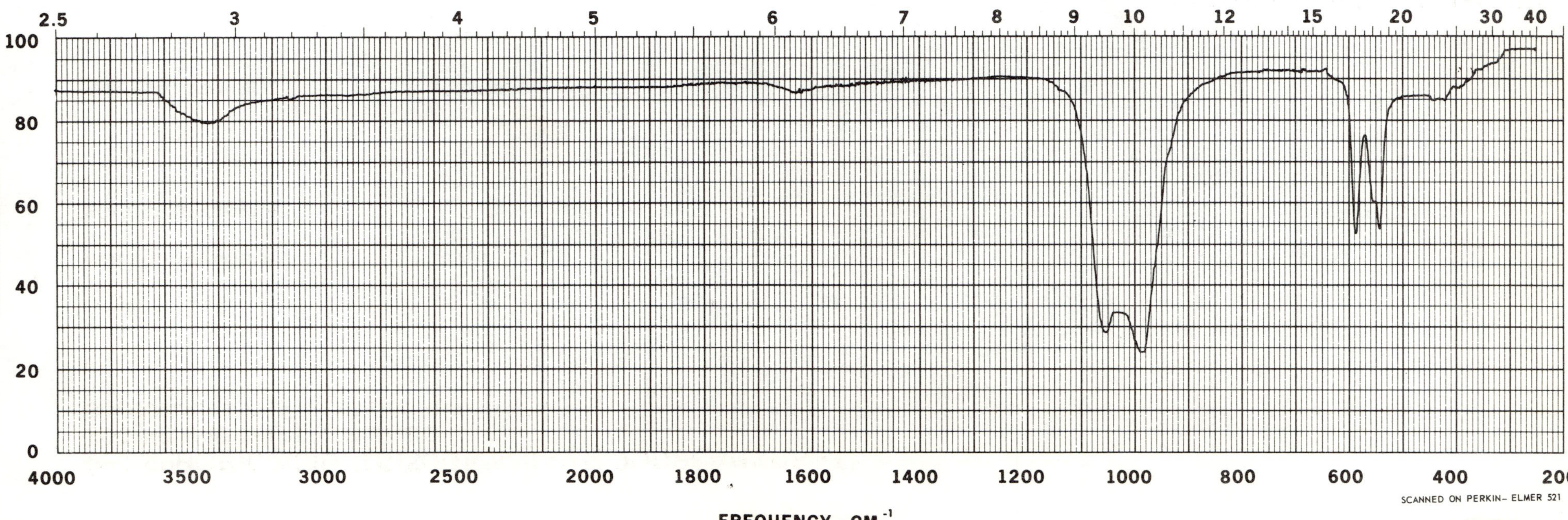

SCANNED ON PERKIN- ELMER 521

PHOSPHORIC ACID, TRISODIUM SALT, DODECAHYDRATE
SODIUM PHOSPHATE TRIBASIC
SODIUM PHOSPHATE, DODECAHYDRATE

$Na_3O_4P \cdot 12H_2O$

MOLECULAR WEIGHT: 380.12

SOURCE OF SAMPLE: MCB Manufacturing Chemists
Norwood, Ohio

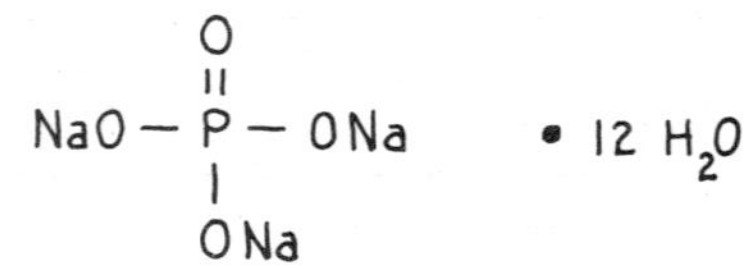

CAS REGISTRY NO.: 10101-89-0

NIOSH NUMBER: WD59500

KBr Wafer

CONDENSED PHASE

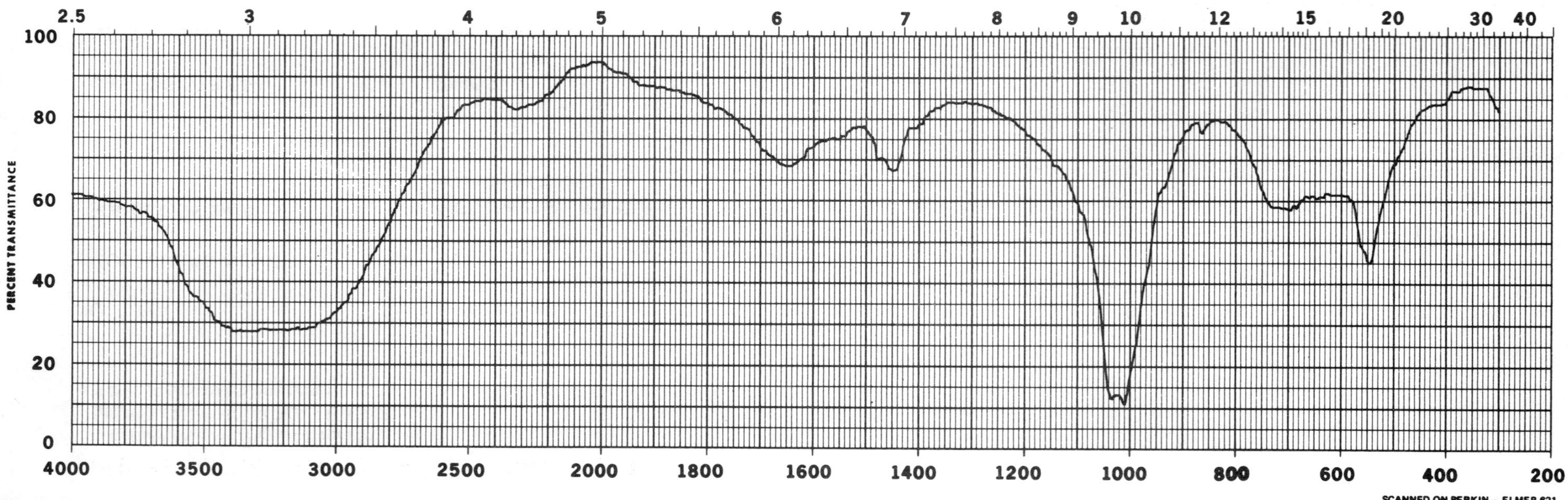

PHOSPHORODITHIOIC ACID, O,O–DIMETHYL ESTER, S–ESTER WITH 3–(MERCAPTOMETHYL)–1,2,3–BENZOTRIAZIN–4(3H)–ONE
GUTHION
AZINOPHOS METHYL

$C_{10}H_{12}N_3O_3PS_2$

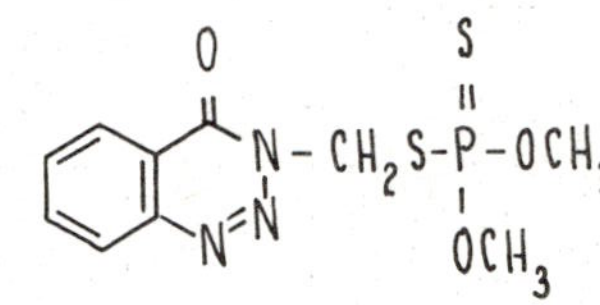

MOLECULAR WEIGHT: 317.33

SOURCE OF SAMPLE: Chemagro Corporation
Kansas City, Missouri

CAS REGISTRY NO.: 86-50-0

NIOSH NUMBER: TE19250

Film

CONDENSED PHASE

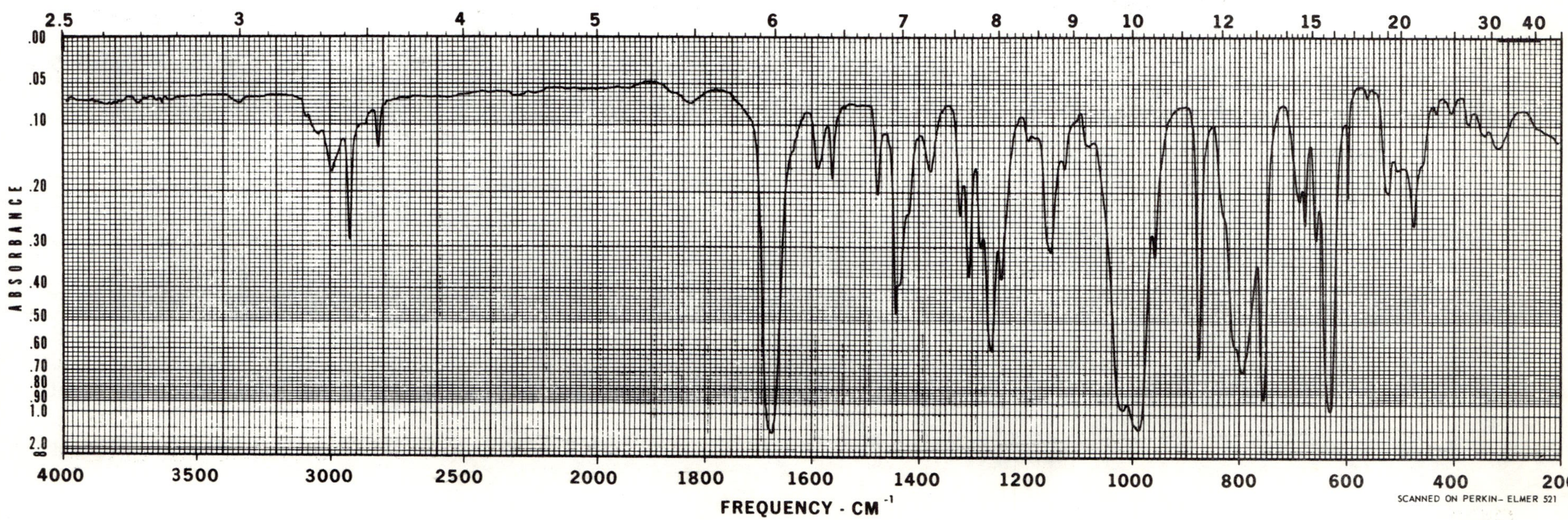

PHOSPHORODITHIOIC ACID, S,S'–METHYLENE O,O,O',O'–TETRAETHYL ESTER

ETHION
BLADAN
PHOSPHOTOX E
RODOCID

$C_9H_{22}O_4P_2S_4$

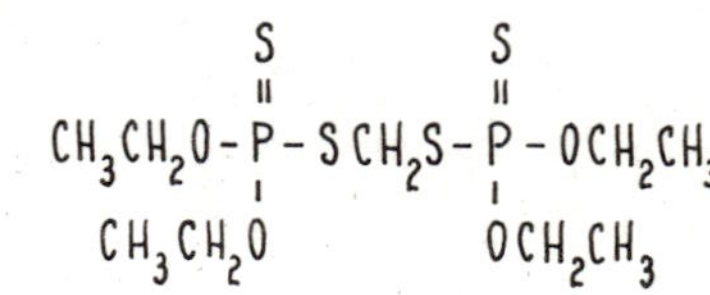

MOLECULAR WEIGHT: 384.48

SOURCE OF SAMPLE: FMC Corporation
Niagara Chemical Division
Middleport, New York

CAS REGISTRY NO.: 563-12-2

NIOSH NUMBER: TE45500

Between Salts

CONDENSED PHASE

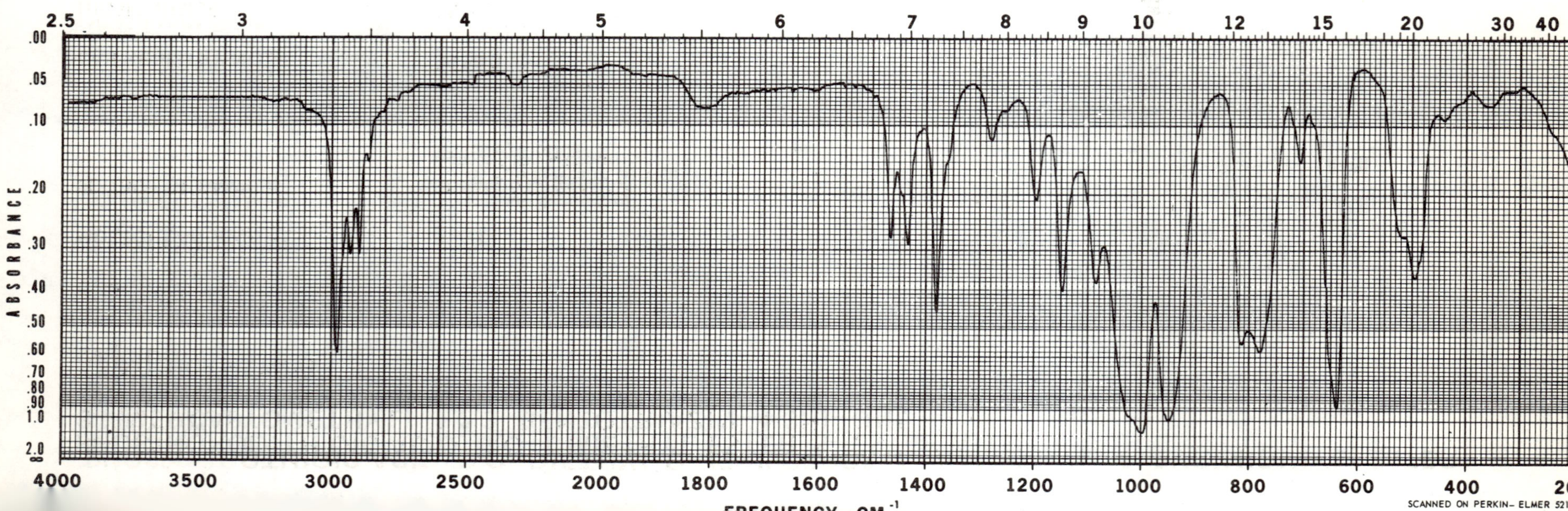

PHOSPHOROTHIOIC ACID, O,O–DIETHYL O–(2–ISOPROPYL–6–METHYL–4–PYRIMIDINYL) ESTER
DIAZINON
ALFA–TOX
GARDENTOX
NUCIDOL

$C_{12}H_{21}N_2O_3PS$

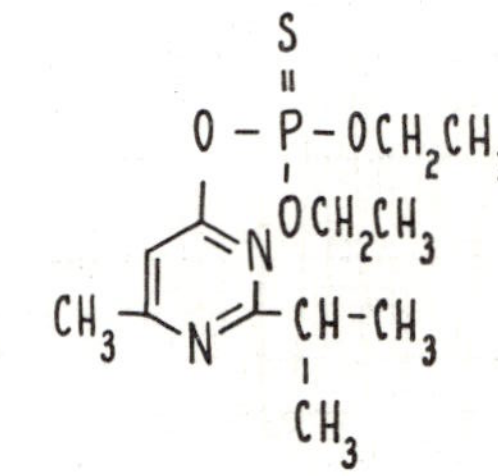

MOLECULAR WEIGHT:	304.35
SOURCE OF SAMPLE:	Pesticides Repository U. S. Public Health Service Perrine, Florida
CAS REGISTRY NO.:	333-41-5
NIOSH NUMBER:	TF33250

Between Salts

CONDENSED PHASE

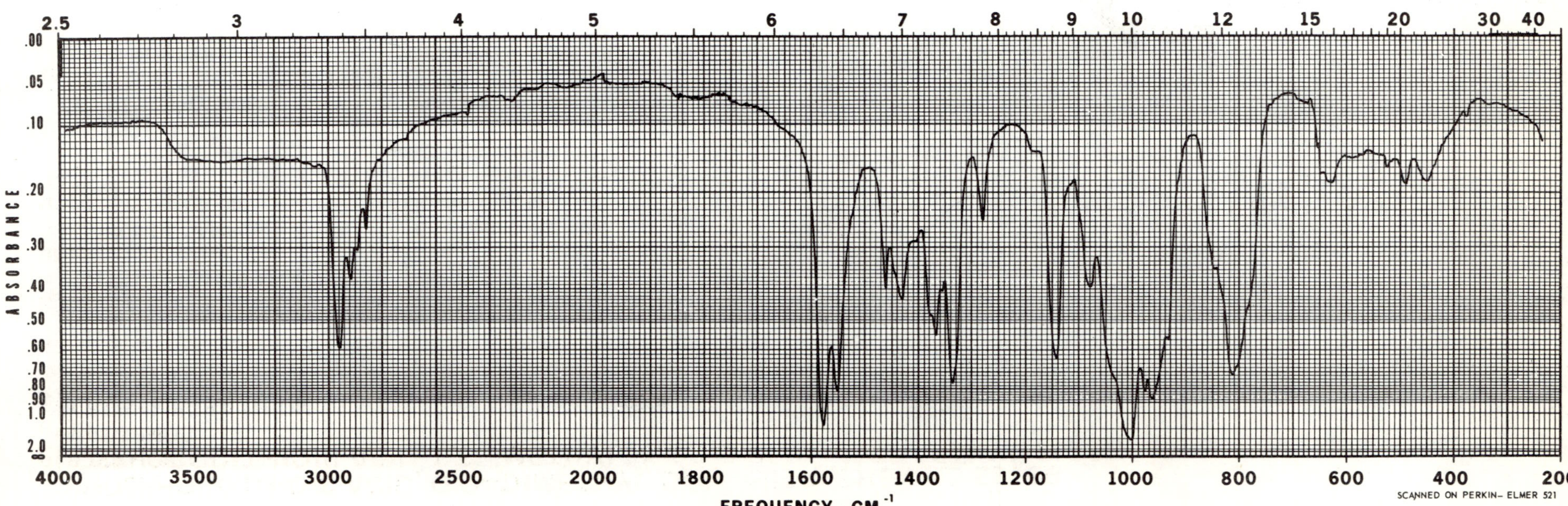

SCANNED ON PERKIN– ELMER 521

PHOSPHOROTHIOIC ACID, O,O–DIMETHYL O–(p–NITROPHENYL) ESTER
AZOPHOS
METAPHOS
METHYL PARATHION
NITROX 80

$C_8H_{10}NO_5PS$

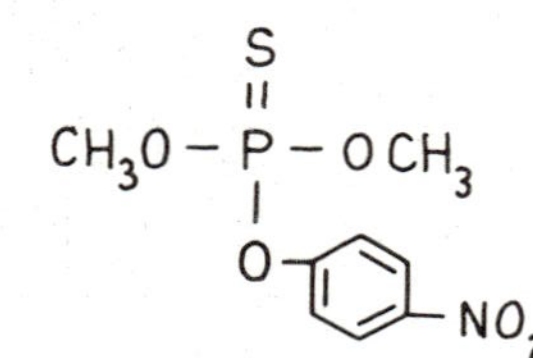

MOLECULAR WEIGHT: 263.21

SOURCE OF SAMPLE: Chemagro Corporation
Kansas City, Missouri

CAS REGISTRY NO.: 298-00-0

NIOSH NUMBER: TG01750

Between Salts

CONDENSED PHASE

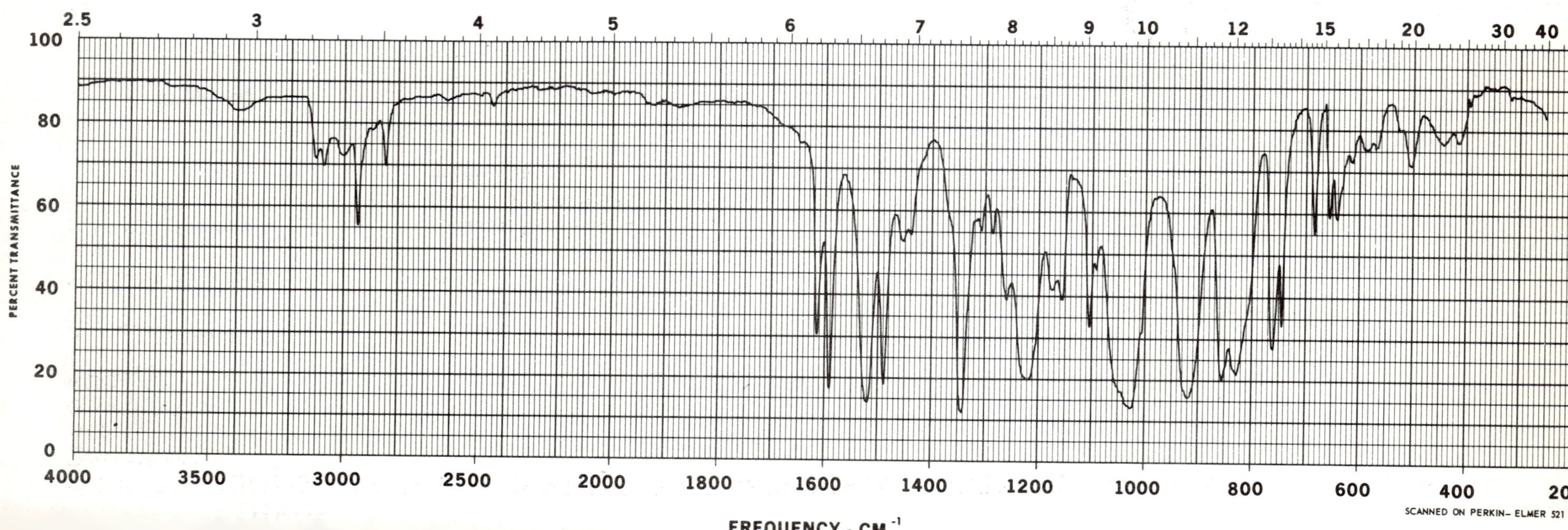

2,6–DIAMINO–3–(PHENYLAZO)PYRIDINE, MONOHYDROCHLORIDE
PHENAZOPYRIDINE HYDROCHLORIDE
AZODYNE
PYRIDIUM
MALLOPHENE

$C_{11}H_{11}N_5 \cdot HCl$

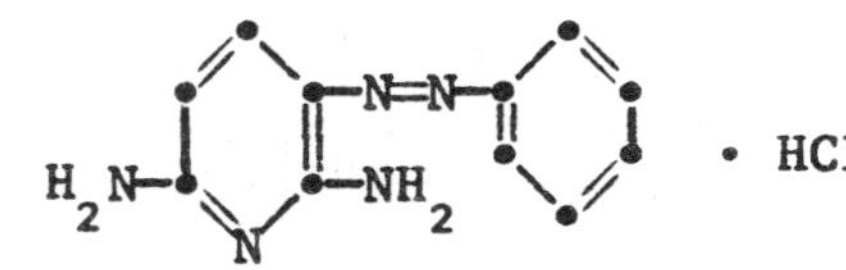

MOLECULAR WEIGHT: 249.71

SOURCE OF SAMPLE: Aldrich Chemical Company, Inc. Milwaukee, Wisconsin

CAS REGISTRY NO.: 136-40-3

NIOSH NUMBER: US78750

KBr Wafer

CONDENSED PHASE

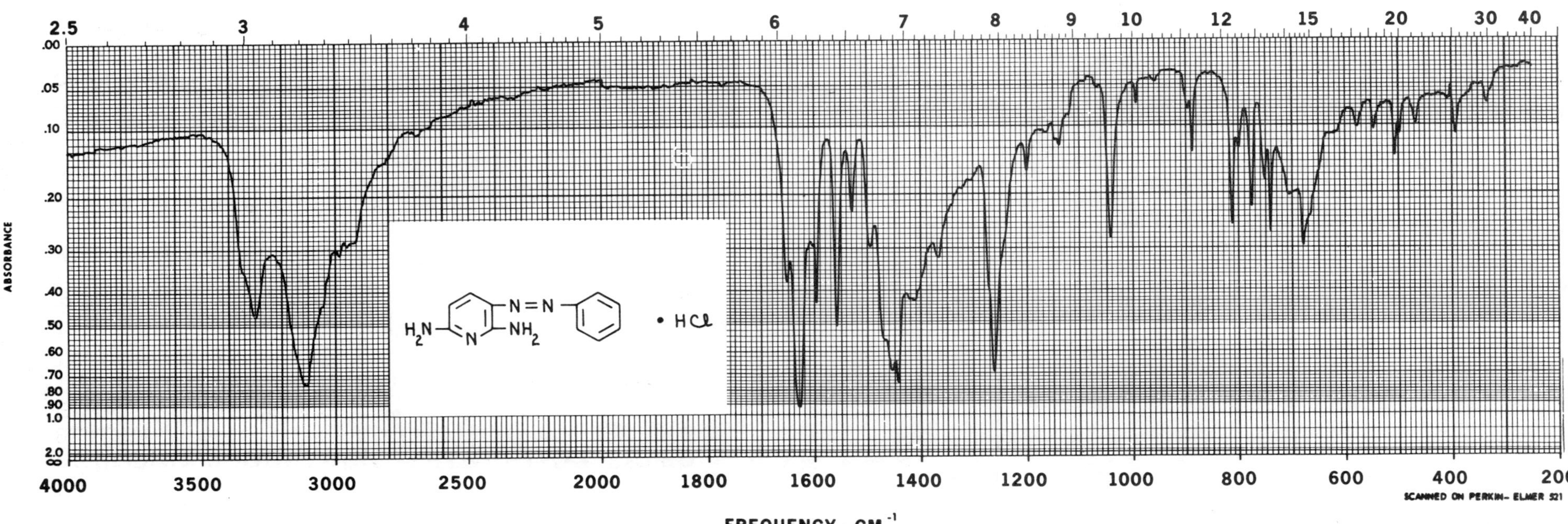

PYROPHOSPHORIC ACID, TETRAETHYL ESTER
BLADEX
ETHYL PYROPHOSPHATE
TETRAETHYLPYROPHOSPHATE
TETRON

$C_8H_{20}O_7P_2$

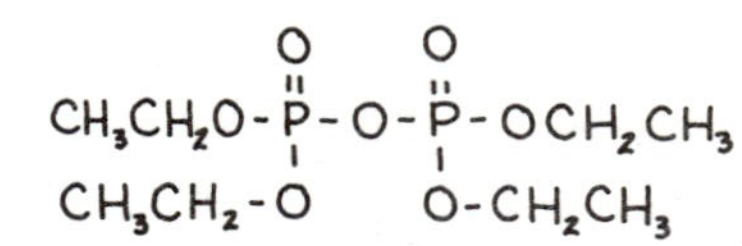

MOLECULAR WEIGHT: 290.19

SOURCE OF SAMPLE: Stauffer Chemical Company
Agricultural Chemical Division
Westport, Connecticut

CAS REGISTRY NO.: 107-49-3

NIOSH NUMBER: UX68250

CAPILLARY CELL: Neat

CONDENSED PHASE

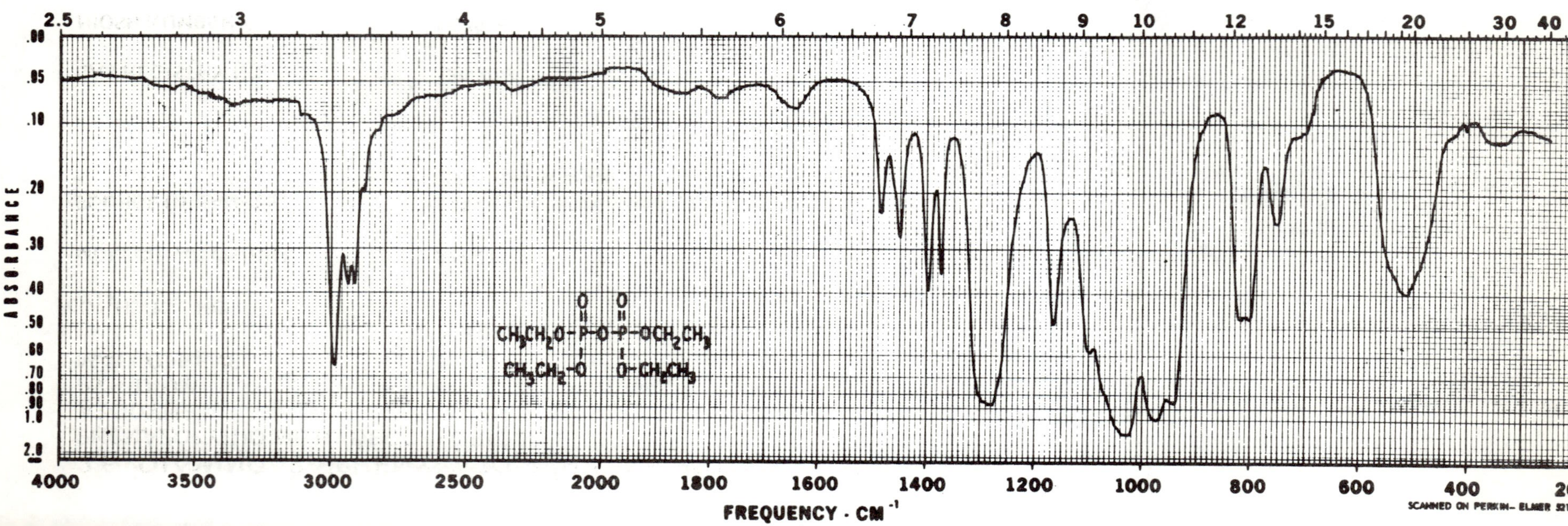

4–NITROQUINOLINE, 1–OXIDE
4–NQO

$C_9H_6N_2O_3$

MOLECULAR WEIGHT: 190.16

SOURCE OF SAMPLE: Aldrich Chemical Company, Inc.
Milwaukee, Wisconsin

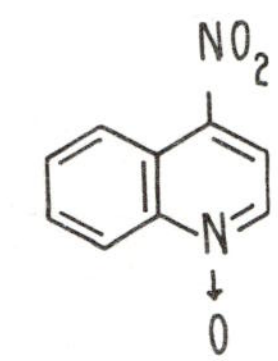

CAS REGISTRY NO.: 56-57-5

NIOSH NUMBER: VC21000

KBr Wafer

CONDENSED PHASE

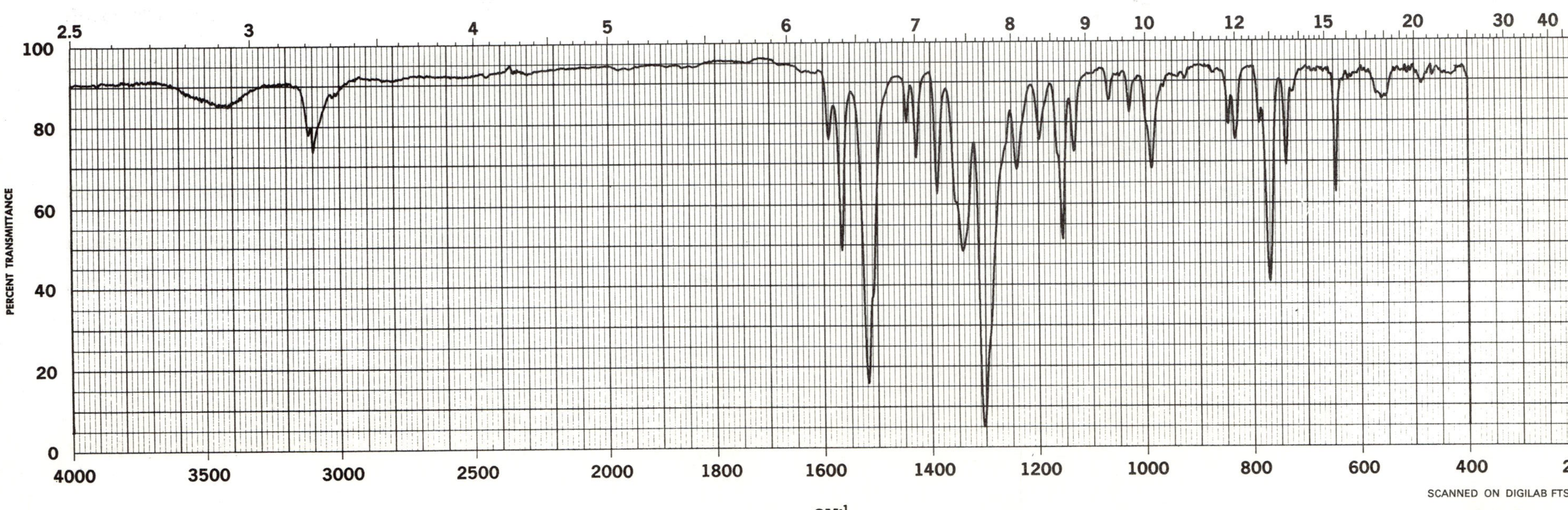

SCANNED ON DIGILAB FTS14

4–NITROSTILBENE

$C_{14}H_{11}NO_2$

MOLECULAR WEIGHT: 225.25

SOURCE OF SAMPLE: Maybridge Chemical Company Ltd. North Cornwall, England

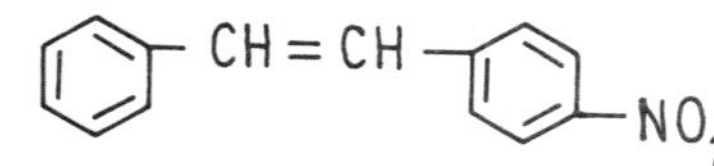

CAS REGISTRY NO.: 4003-94-5

NIOSH NUMBER: WJ69500

KBr Wafer

CONDENSED PHASE

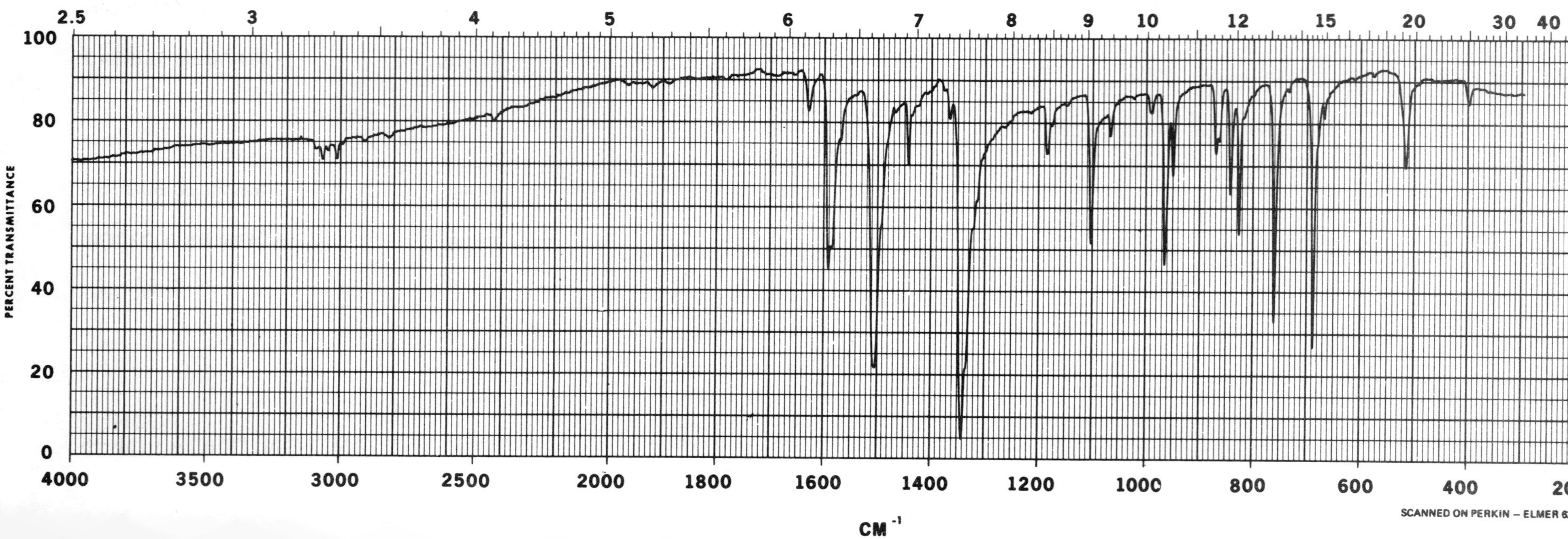

STRYCHNINE
2,4a,5,5a,7,8,15,15a,15b,15c–DECAHYDRO–4,6–METHANO–6H,14H–INDOLO[3,2,1–ij] OXEPINO[2,3,4–de] –
PYRROLO[2,3–h] QUINOLIN–14–ONE
KWIK–KIL
MOUSE–TOX
RO–DEX

$C_{21}H_{22}N_2O_2$

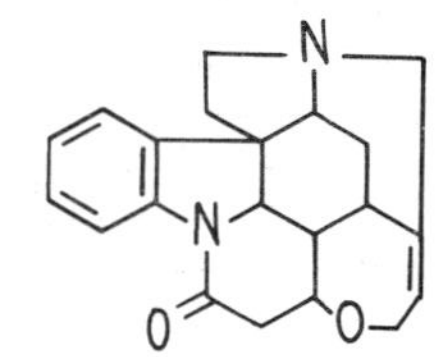

MOLECULAR WEIGHT:	334.42
SOURCE OF SAMPLE:	E. Merck AG Darmstadt, Germany
CAS REGISTRY NO.:	57-24-9
NIOSH NUMBER:	WL22750

KBr Wafer

CONDENSED PHASE

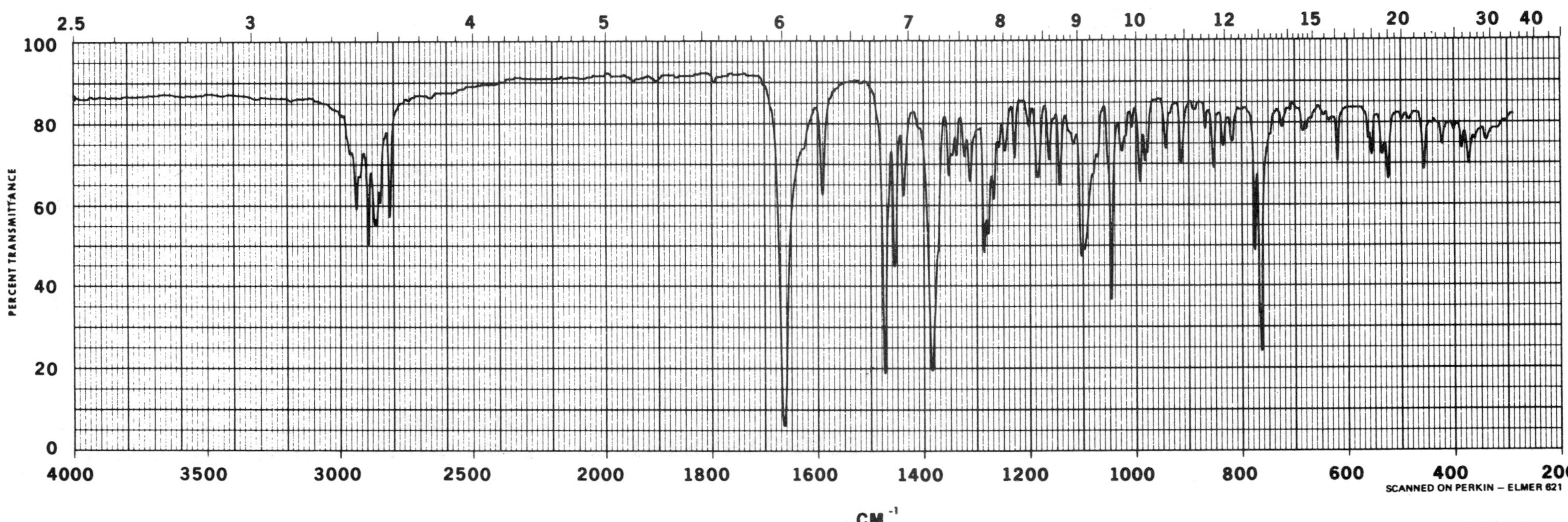

SULFURIC ACID, BERYLLIUM SALT (1:1), TETRAHYDRATE
BERYLLIUM SULFATE, TETRAHYDRATE (1:1:4)

$BeO_4S \cdot 4H_2O$

MOLECULAR WEIGHT: 177.14

SOURCE OF SAMPLE: Research Organic/Inorganic Chemical Corporation
Belleville, New Jersey

$BeSO_4 \cdot 4H_2O$

CAS REGISTRY NO.: 7787-56-6

NIOSH NUMBER: DS50000

KBr Wafer

CONDENSED PHASE

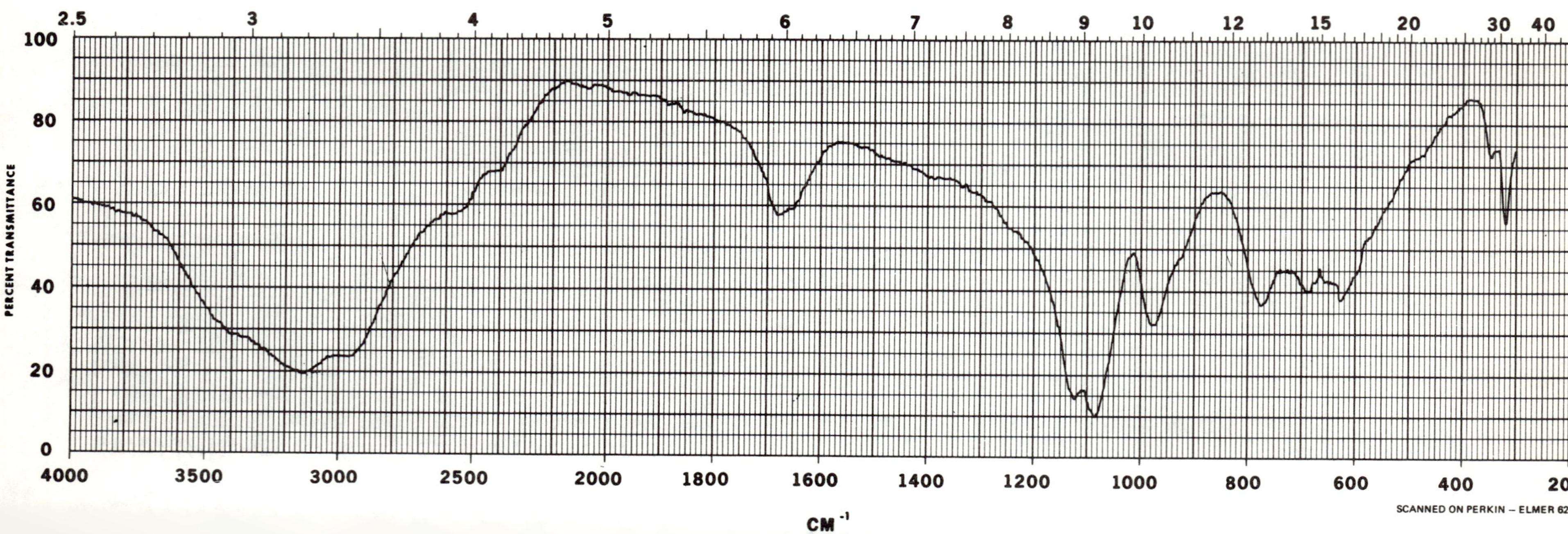

SULFURIC ACID, CADMIUM SALT (1:1)
CADMIUM SULFATE (1:1)

CdO_4S

MOLECULAR WEIGHT: 208.46

SOURCE OF SAMPLE: Mallinckrodt, Inc.
St. Louis, Missouri

$CdSO_4$

CAS REGISTRY NO.: 10124-36-4

NIOSH NUMBER: EV27000

KBr Wafer

CONDENSED PHASE

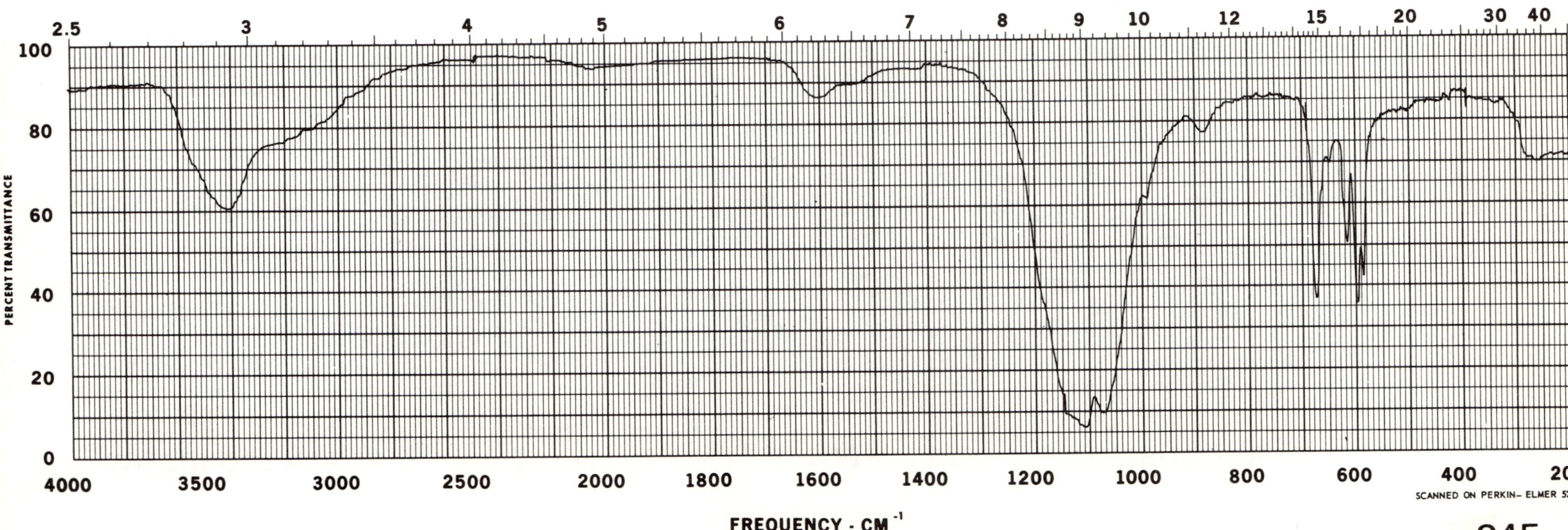

SCANNED ON PERKIN-ELMER 521

SULFUROUS ACID, 2–(p–tert–BUTYLPHENOXY)–1–METHYLETHYL 2–CHLOROETHYL ESTER
ARAMITE
ARATRON
CES
ORTHO–MITE

$C_{15}H_{23}ClO_4S$

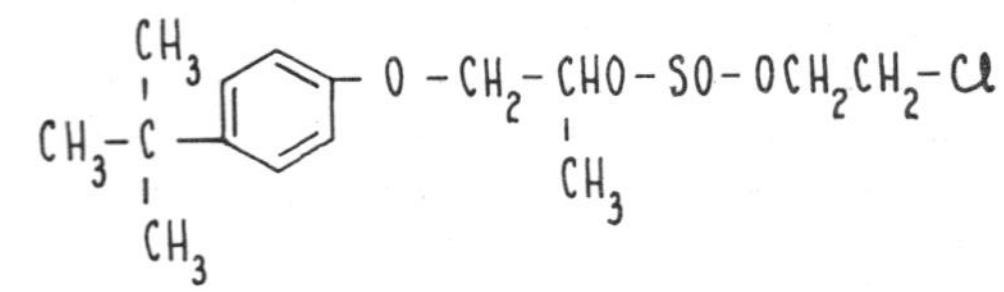

MOLECULAR WEIGHT:	334.87
SOURCE OF SAMPLE:	Uniroyal, Inc. Chemical Division Naugatuck, Connecticut
CAS REGISTRY NO.:	140-57-8
NIOSH NUMBER:	WT29750

KBr Wafer

CONDENSED PHASE

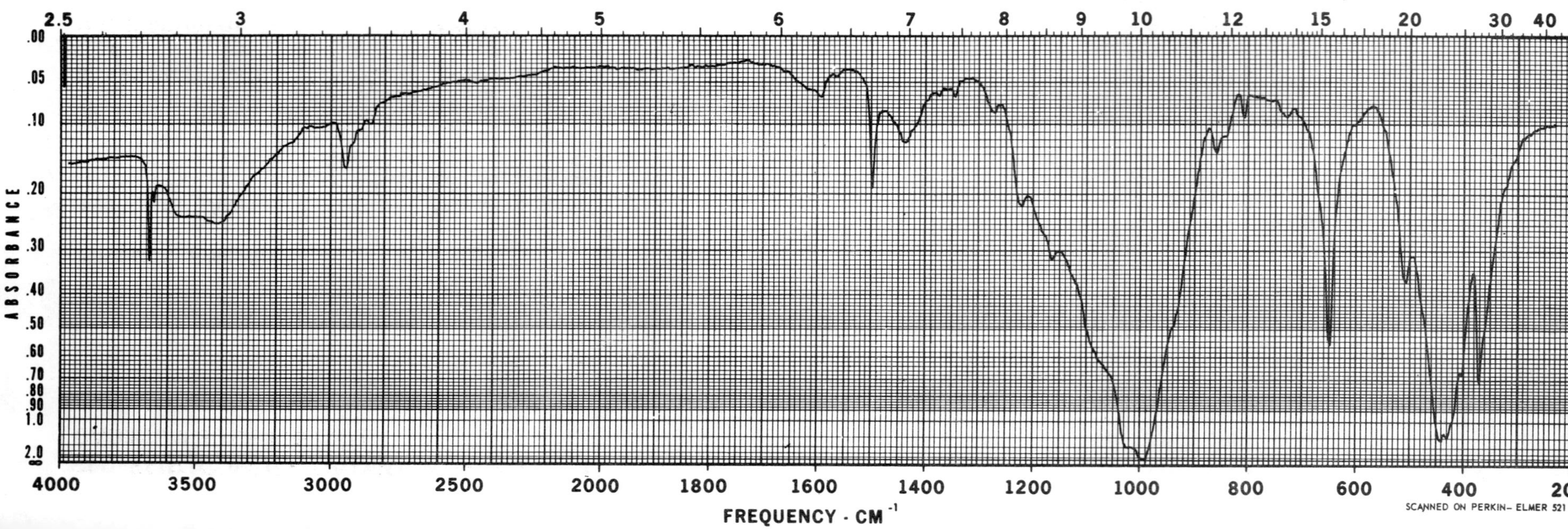

TANNIC ACID
TANNIN
GALLOTANNIN
GALLOTANNIC ACID
GLYCERITE

$C_{27}H_{24}O_{18}$

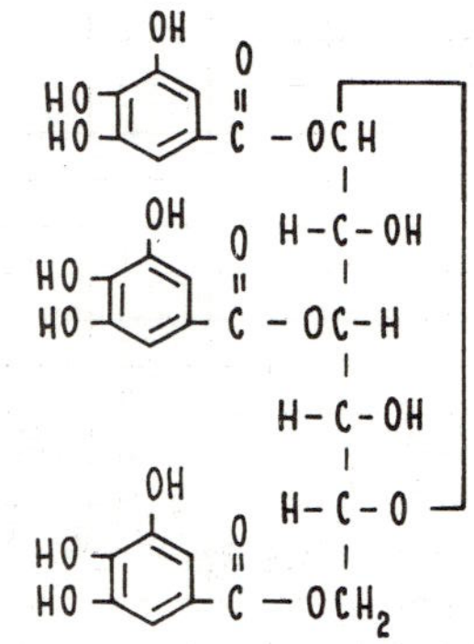

MOLECULAR WEIGHT:	636.48
SOURCE OF SAMPLE:	Mallinckrodt, Inc. St. Louis, Missouri
CAS REGISTRY NO.:	1401-55-4
NIOSH NUMBER:	WW50750

Film (Cast from Acetone)

CONDENSED PHASE

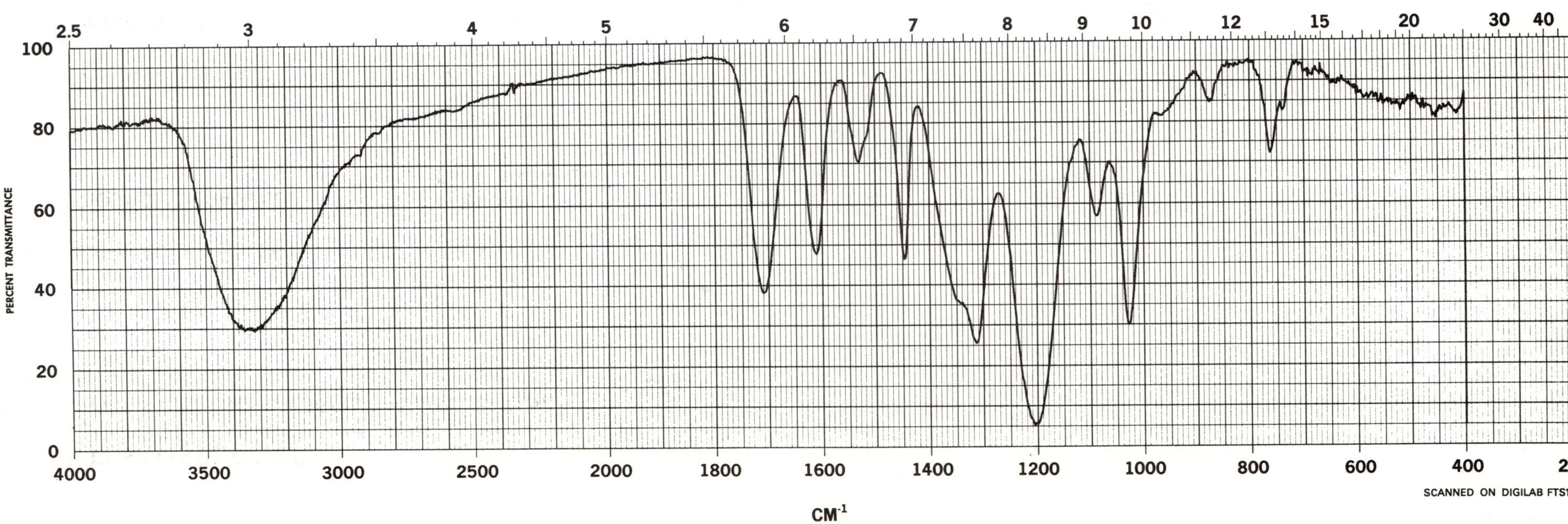

SCANNED ON DIGILAB FTS14

TESTOSTERONE
ANDROLIN
17β–HYDROXYANDROST–4–EN–3–ONE
MERTESTATE
TESLEN

$C_{19}H_{28}O_2$

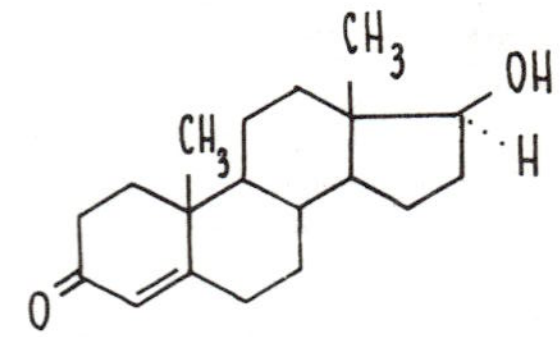

MOLECULAR WEIGHT: 288.43

SOURCE OF SAMPLE: Fluka AG
Buchs, Switzerland

CAS REGISTRY NO.: 58-22-0

NIOSH NUMBER: XA30300

KBr Wafer

CONDENSED PHASE

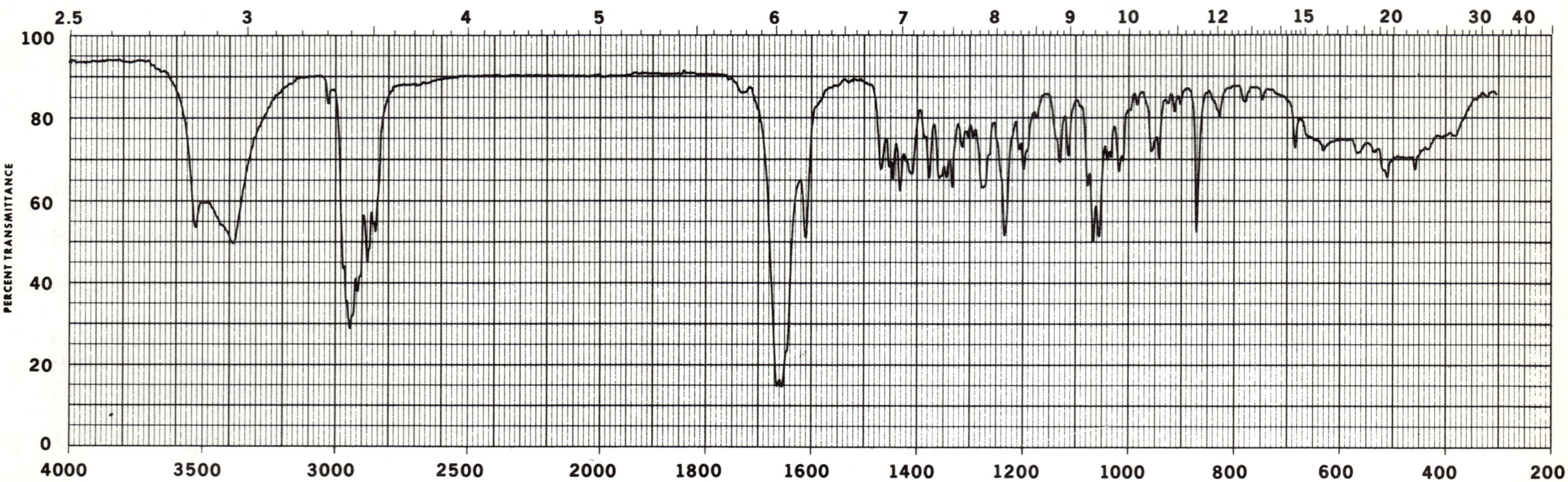

TESTOSTERONE, PROPIONATE
AGOVIRIN
17β–HYDROXYANDROST–4–EN–3–ONE, PROPIONATE
HORMOTESTON
ORETON

$C_{22}H_{32}O_3$

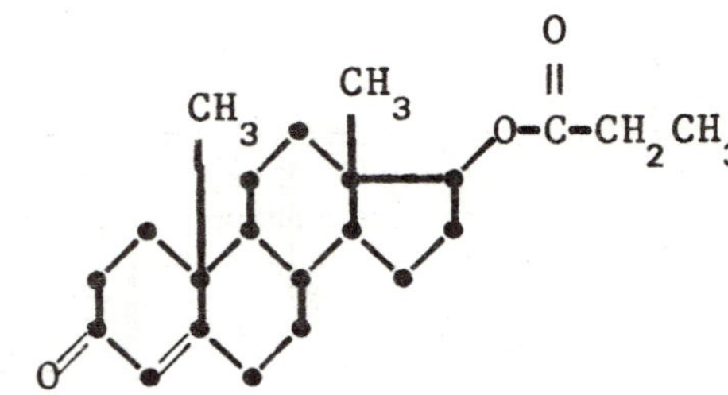

MOLECULAR WEIGHT: 344.50

SOURCE OF SAMPLE: Fluka AG
Buchs, Switzerland

CAS REGISTRY NO.: 57-85-2

NIOSH NUMBER: XA31150

KBr Wafer

CONDENSED PHASE

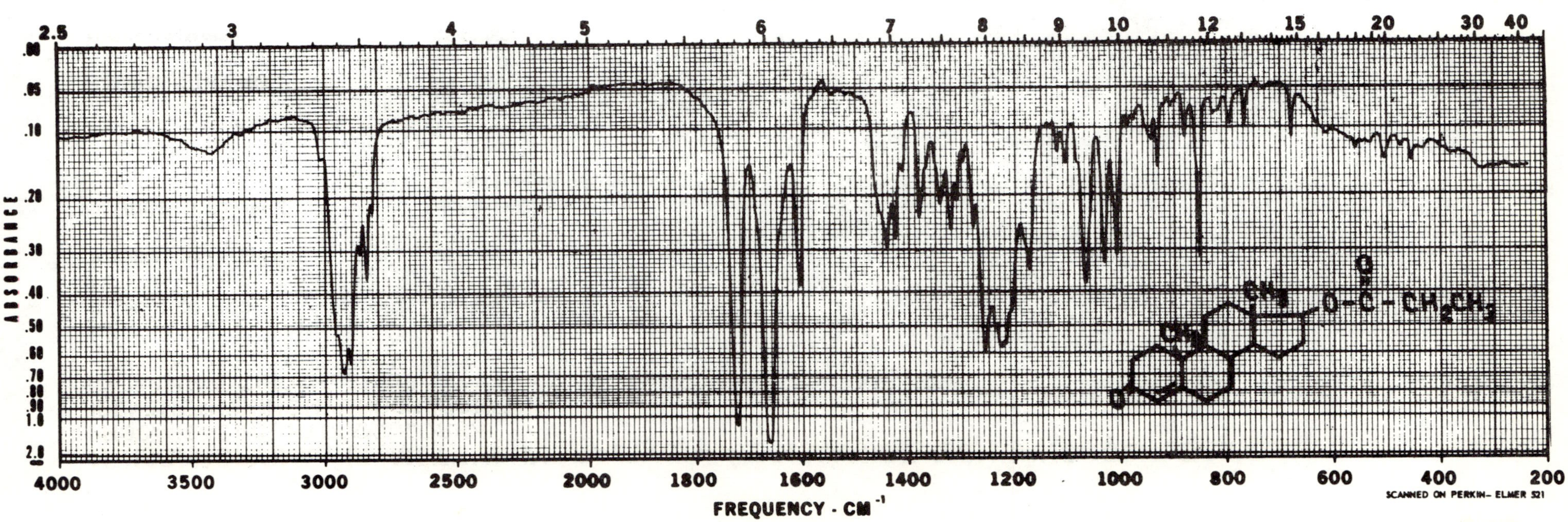

N—ISOPROPYL—α—(2—METHYLHYDRAZINO)—p—TOLUAMIDE, MONOHYDROCHLORIDE
PROCARBAZINE, HYDROCHLORIDE
IBENZMETHYZINE HYDROCHLORIDE
NATULAN HYDROCHLORIDE

$C_{12}H_{19}N_3O \cdot HCl$

MOLECULAR WEIGHT: 257.77

SOURCE OF SAMPLE: Hoffman-LaRoche, Inc.
Roche Laboratories Division
Nutley, New Jersey

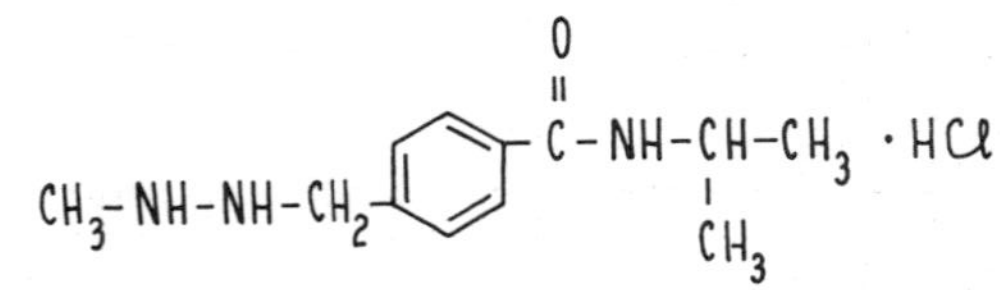

CAS REGISTRY NO.: 366-70-1

NIOSH NUMBER: XS47250

KBr Wafer

CONDENSED PHASE

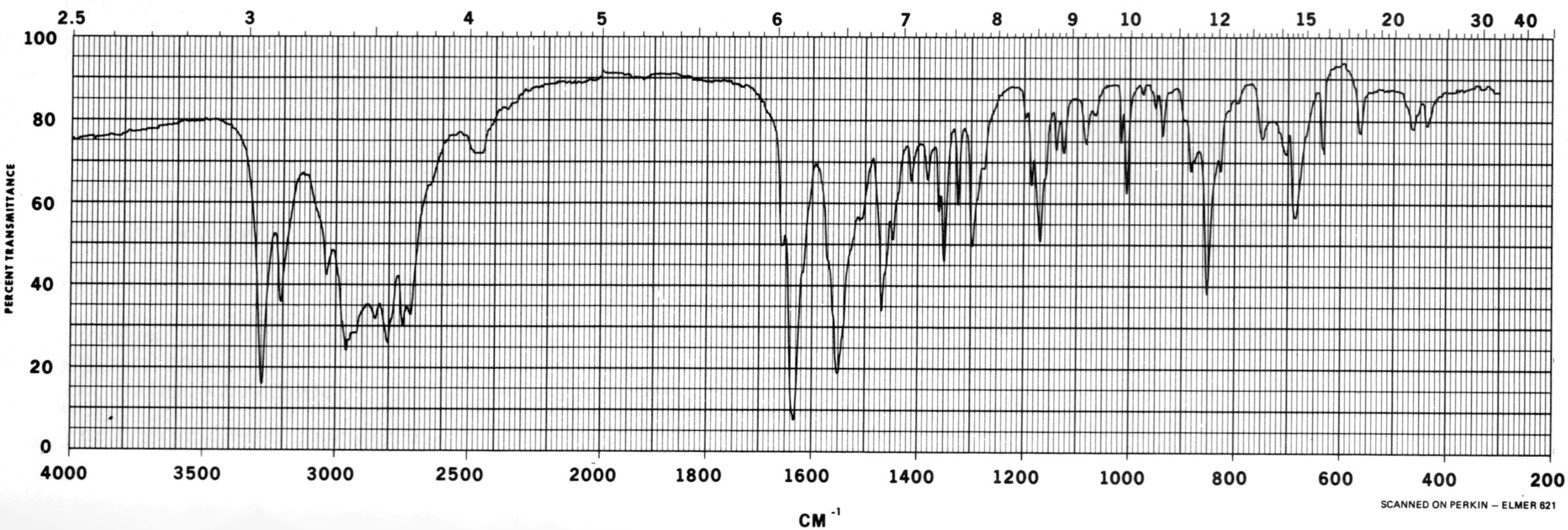

TOXAPHENE
CHLORINATED CAMPHENE
ALLTOX
PENPHENE
POLYCHLORCAMPHENE

$C_{10}H_{10}Cl_8$

MOLECULAR WEIGHT: 413.82

SOURCE OF SAMPLE: Hercules Inc.
Agricultural Chemicals
Wilmington, Delaware

CAS REGISTRY NO.: 8001-35-2

NIOSH NUMBER: XW52500

Film

CONDENSED PHASE

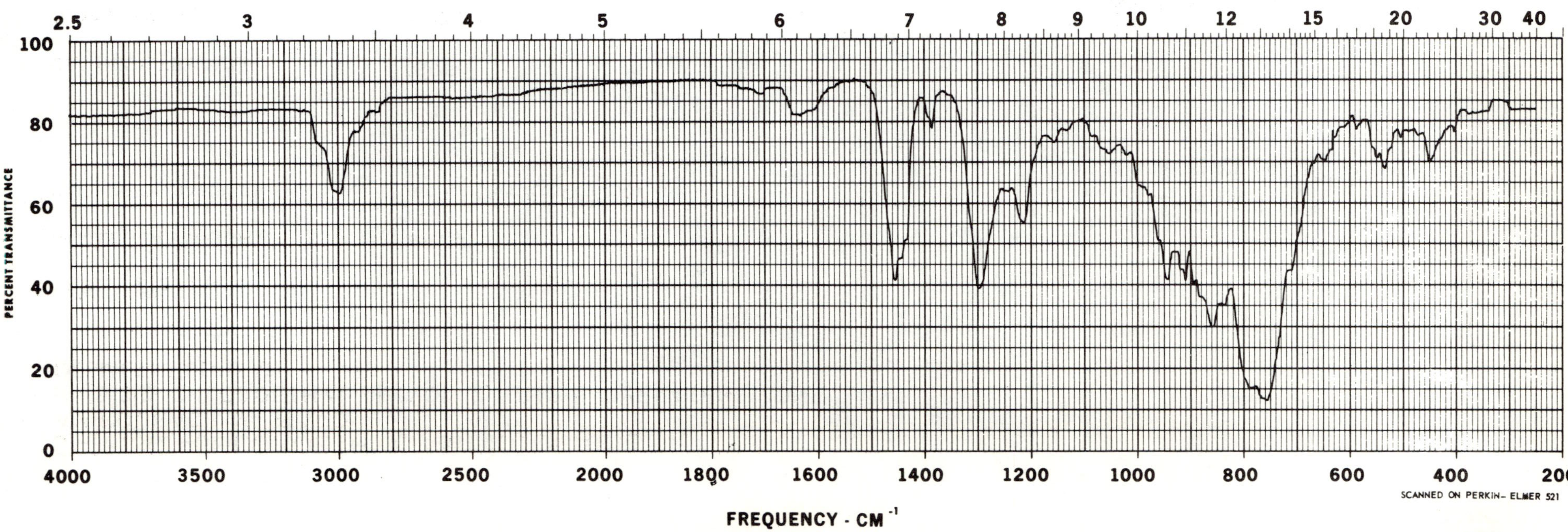

5–[BIS(2–CHLOROETHYL)AMINO] URACIL
AMINOURACIL MUSTARD
5–[BIS(2–CHLOROETHYL)AMINO] –2,4(1H,3H)–PYRIMIDINEDIONE
DESMETHYLDOPAN

$C_8H_{11}Cl_2N_3O_2$

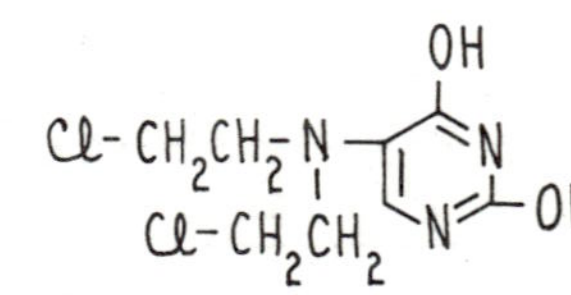

MOLECULAR WEIGHT:	252.10
SOURCE OF SAMPLE:	Aldrich Chemical Company, Inc. Milwaukee, Wisconsin
CAS REGISTRY NO.:	66-75-1
NIOSH NUMBER:	YQ89250

KBr Wafer

CONDENSED PHASE

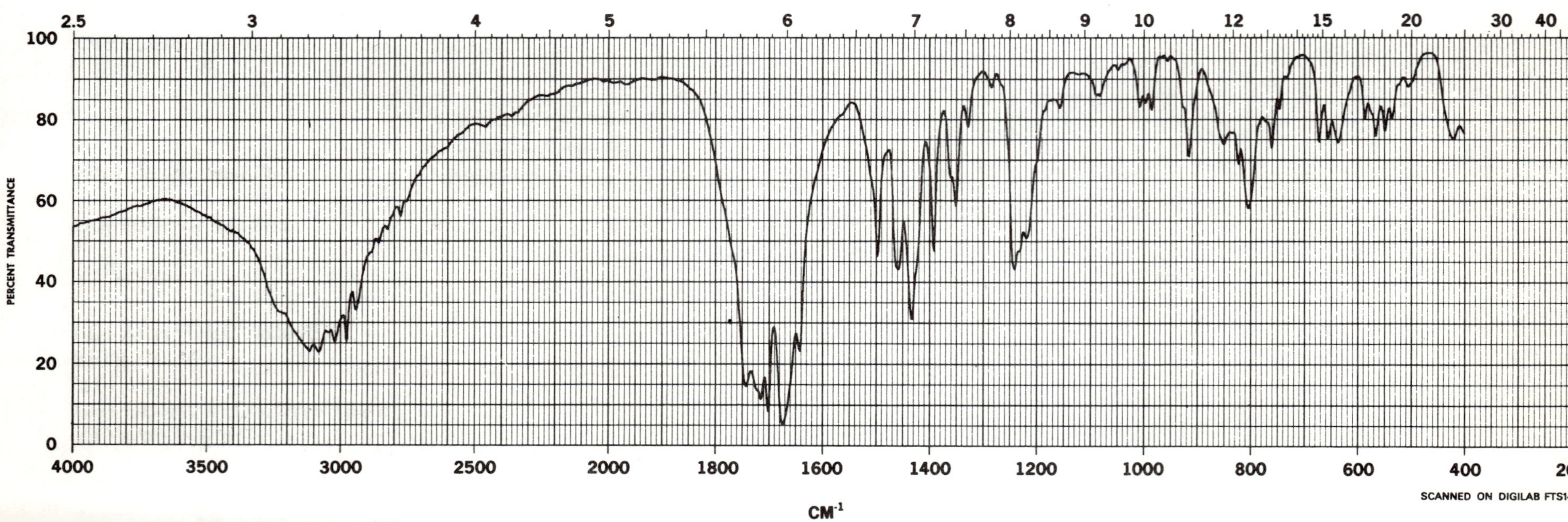

6–METHYL–2–THIOURACIL
ALKIRON
METACIL
6–METHYL–2–THIO–2,4(1H,3H)–PYRIMIDINEDIONE
MURACIL

$C_5H_6N_2OS$

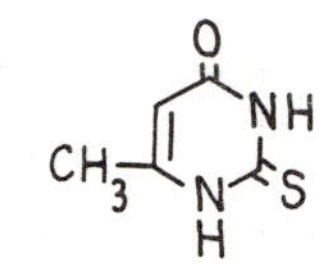

MOLECULAR WEIGHT: 142.18

SOURCE OF SAMPLE: Philadelphia College of Pharmacy and Science
Philadelphia, Pennsylvania

CAS REGISTRY NO.: 56-04-2

NIOSH NUMBER: YR08750

KBr Wafer

CONDENSED PHASE

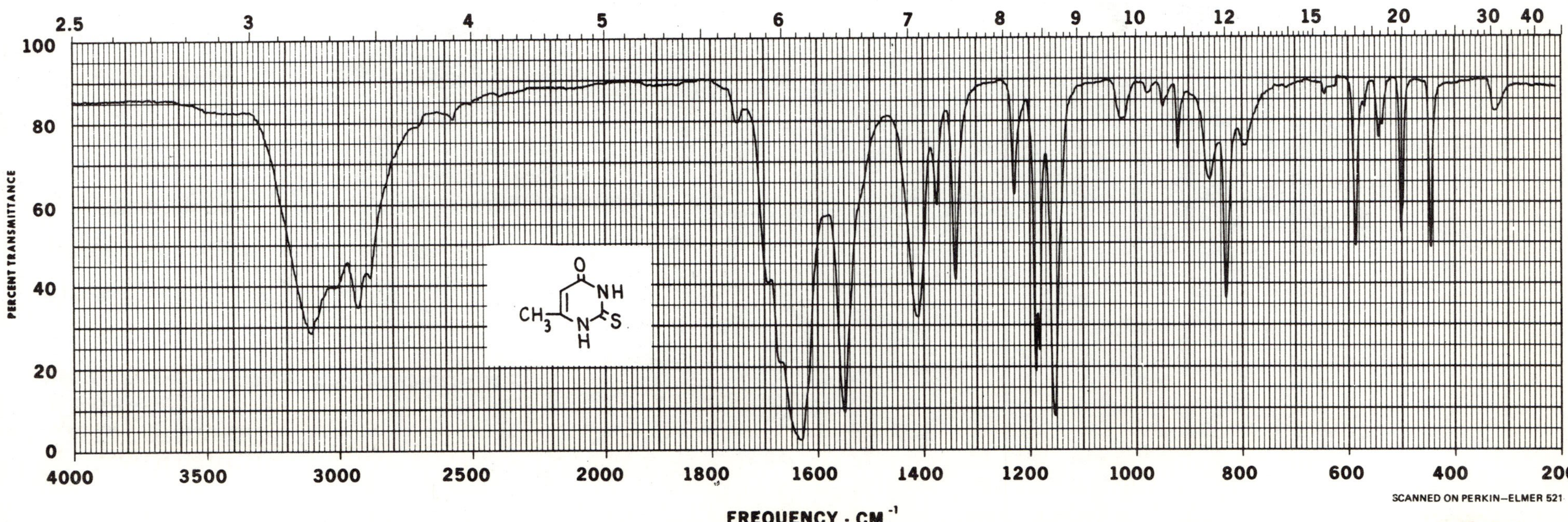

2–THIOURACIL
ANTAGOTHYROID
DERACIL
2–THIO–2,4(1H,3H)–PYRIMIDINEDIONE

$C_4H_4N_2OS$

MOLECULAR WEIGHT: 128.15

SOURCE OF SAMPLE: Eastman Organic Chemicals
Rochester, New York

CAS REGISTRY NO.: 141-90-2

NIOSH NUMBER: YR15750

KBr Wafer

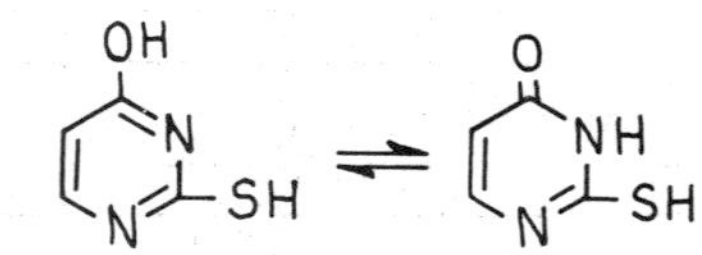

CONDENSED PHASE

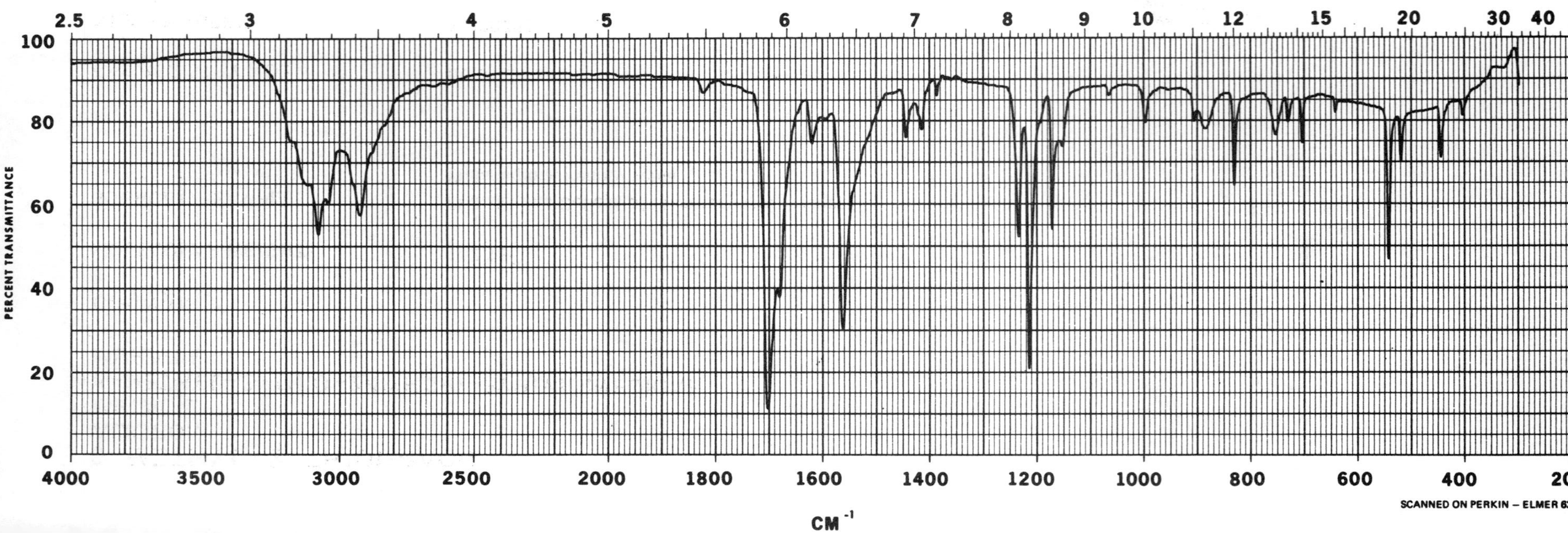

BIS(ACETATO)DIOXOURANIUM
URANYL ACETATE
URANIUM OXYACETATE

$C_4H_6O_6U$

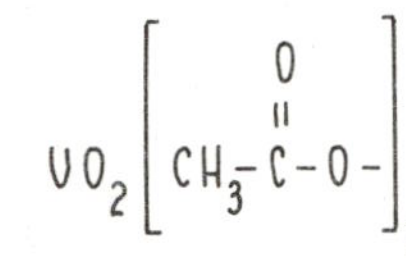

MOLECULAR WEIGHT: 388.12

SOURCE OF SAMPLE: Chem Service Company
West Chester, Pennsylvania

CAS REGISTRY NO.: 541-09-3

NIOSH NUMBER: YR36750

4000 - 1350 cm^{-1} = Mulled in Perfluorinated Hydrocarbon
1350 - 400 cm^{-1} = Mulled in Mineral Oil

CONDENSED PHASE

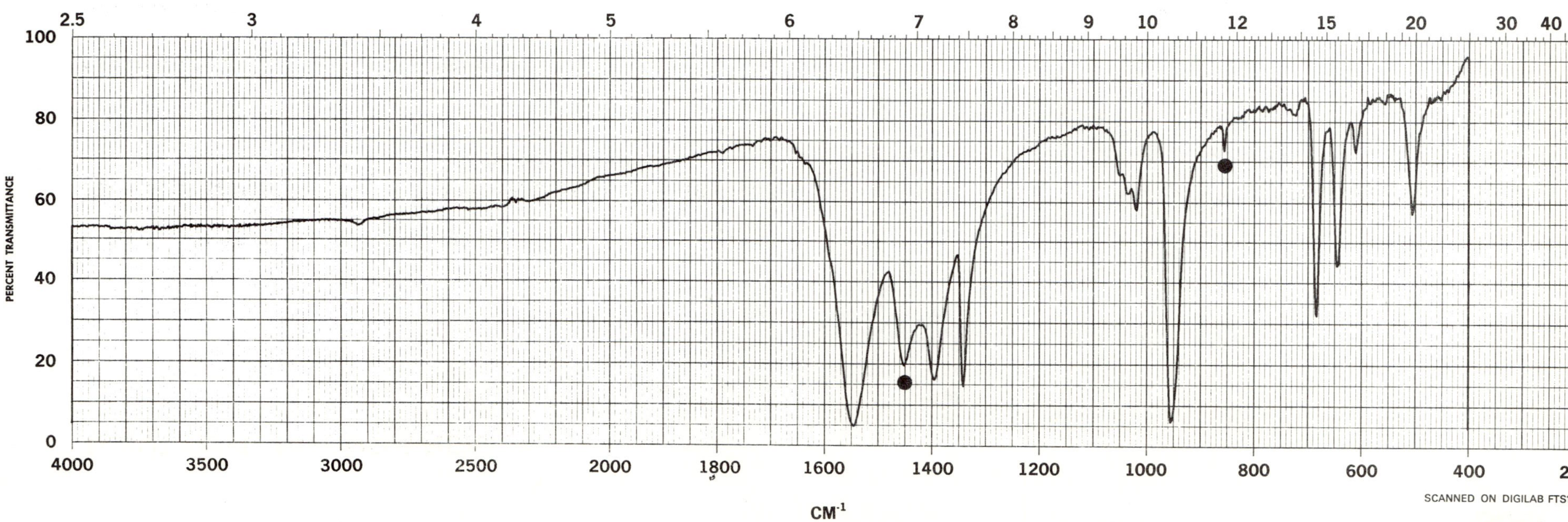

SCANNED ON DIGILAB FTS14

1—BUTYL—1—NITROSOUREA

$C_5H_{11}N_3O_2$

MOLECULAR WEIGHT: 145.16

SOURCE OF SAMPLE: Alfa Division of the Ventron Corporation
Danvers, Massachusetts

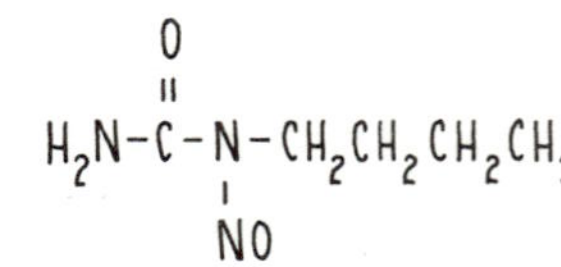

CAS REGISTRY NO.: 869-01-2

NIOSH NUMBER: YS38500

Film

CONDENSED PHASE

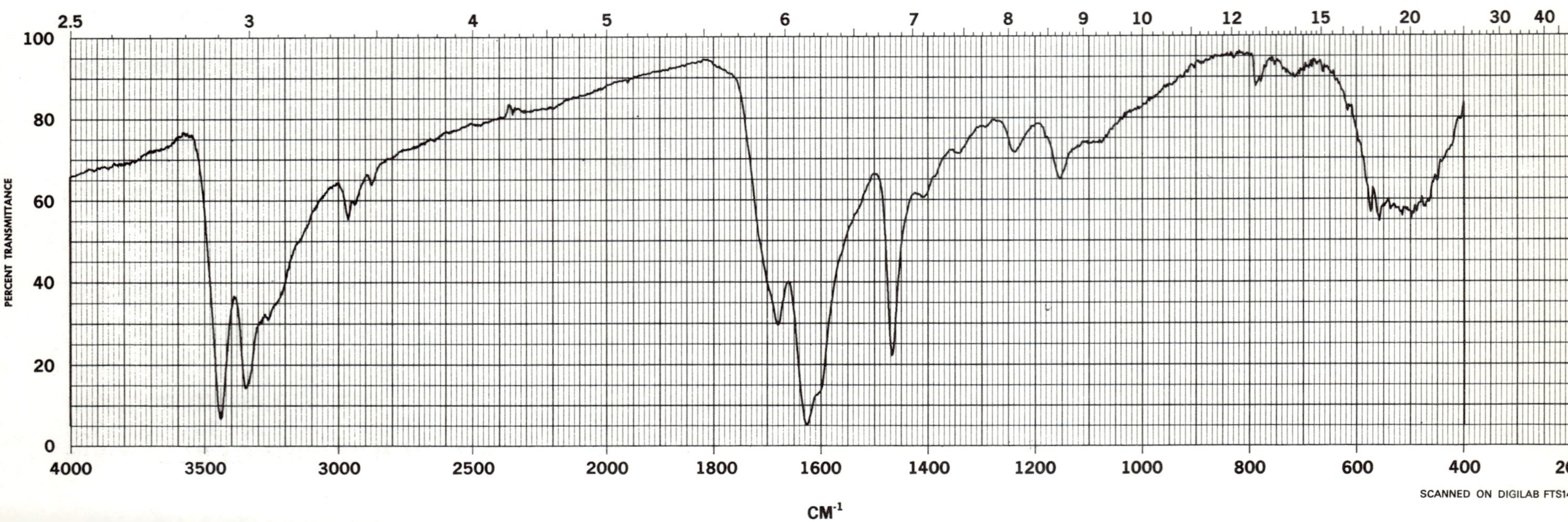

3–(3,4–DICHLOROPHENYL)–1,1–DIMETHYLUREA

DIURON
DICHLORFENIDIM
DIUREX
KARMEX

$C_9H_{10}Cl_2N_2O$

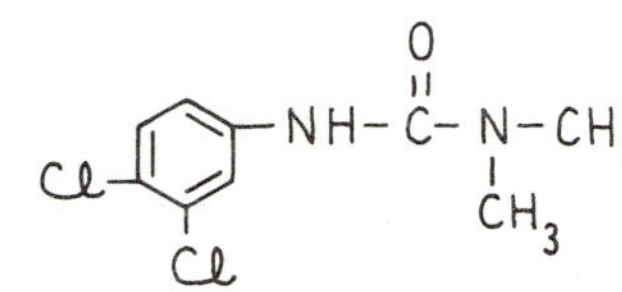

MOLECULAR WEIGHT: 233.10

SOURCE OF SAMPLE: Sadtler Research Laboratories
Philadelphia, Pennsylvania

CAS REGISTRY NO.: 330-54-1

NIOSH NUMBER: YS89250

KBr Wafer

CONDENSED PHASE

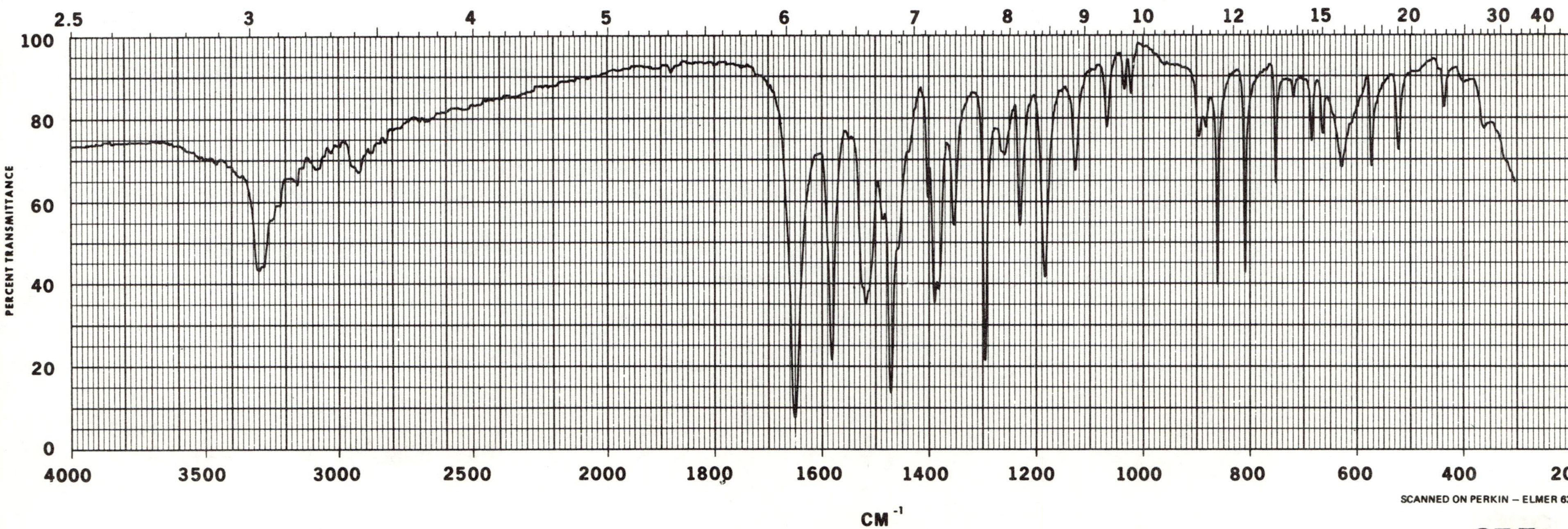

1–ETHYL–1–NITROSOUREA
ENU

$C_3H_7N_3O_2$

MOLECULAR WEIGHT: 117.11

SOURCE OF SAMPLE: Fluka AG
Buchs, Switzerland

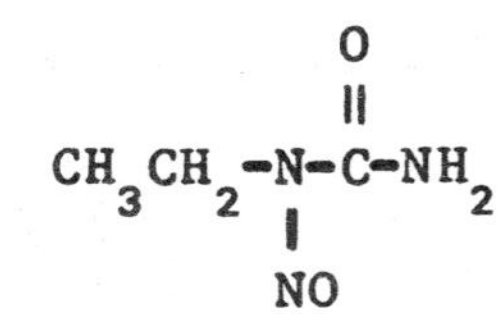

CAS REGISTRY NO.: 759-73-9

NIOSH NUMBER: YT31500

Split Mull

CONDENSED PHASE

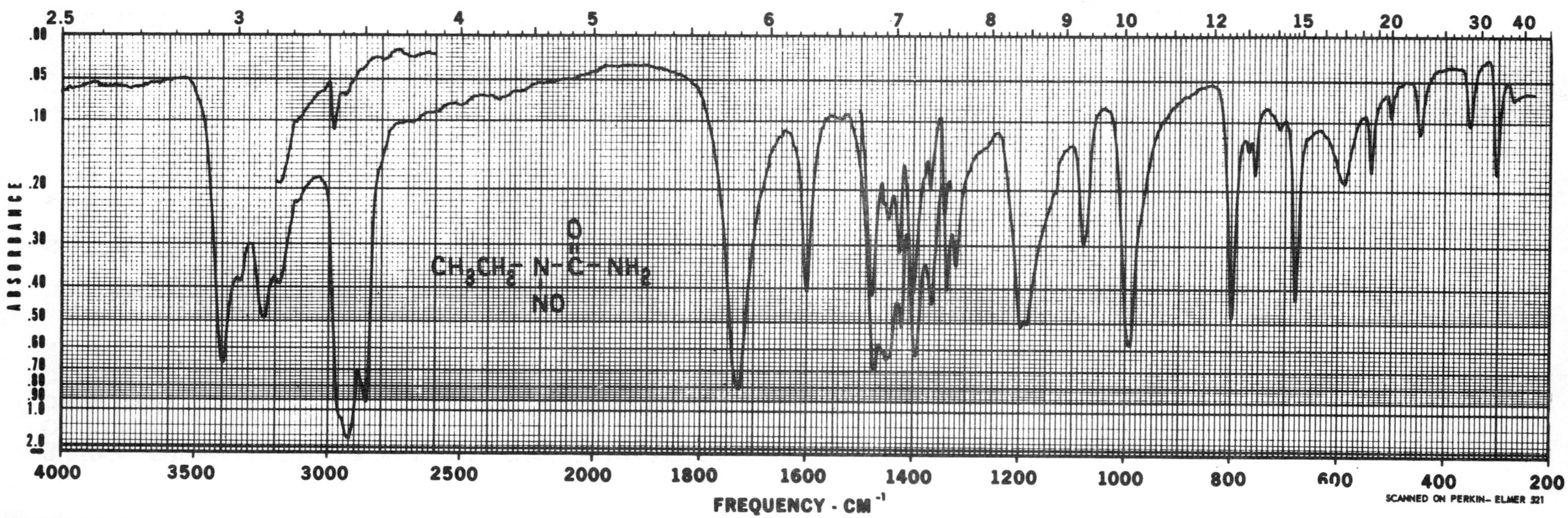

THIOUREA
THIOCARBAMIDE
THU

CH_4N_2S

MOLECULAR WEIGHT: 76.12

SOURCE OF SAMPLE: MCB Manufacturing Chemists
Norwood, Ohio

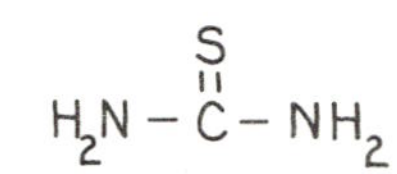

CAS REGISTRY NO.: 62-56-6

NIOSH NUMBER: YU28000

KBr Wafer

CONDENSED PHASE

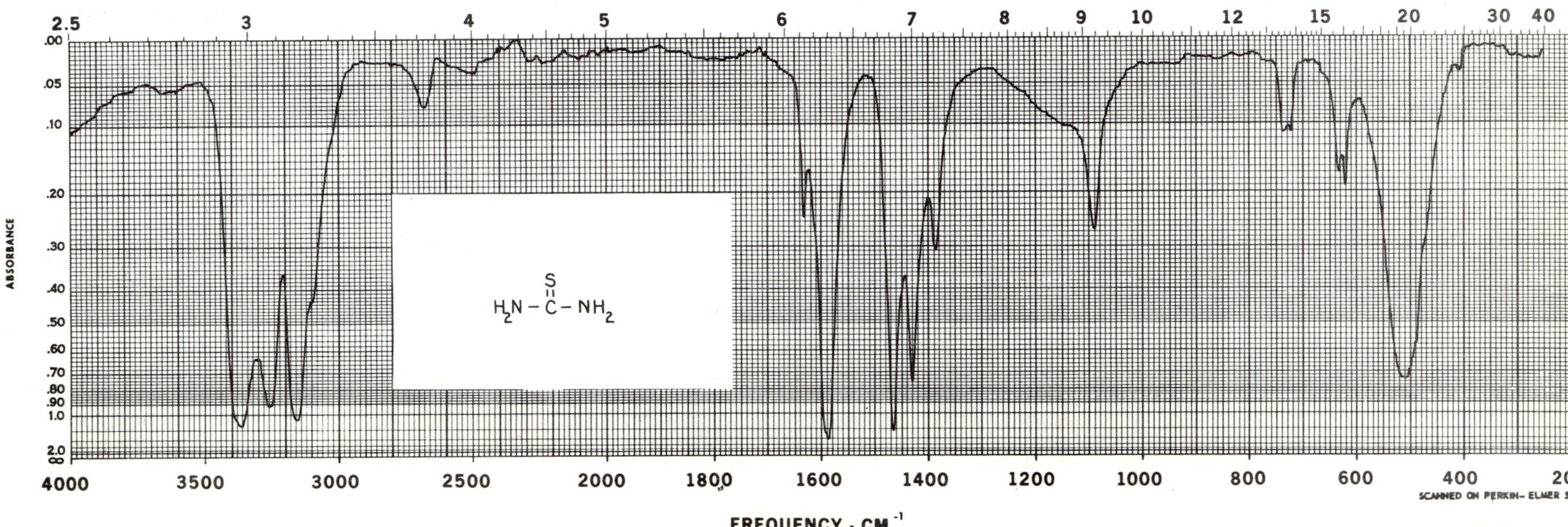

XYLENE
XYLOL
DIMETHYLBENZENE

C_8H_{10}

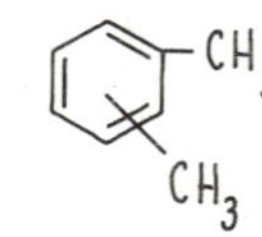

MOLECULAR WEIGHT:	106.17
SOURCE OF SAMPLE:	Allied Chemical Corporation New York, New York
CAS REGISTRY NO.:	1330-20-7
NIOSH NUMBER:	ZE21000

Between Salts

CONDENSED PHASE

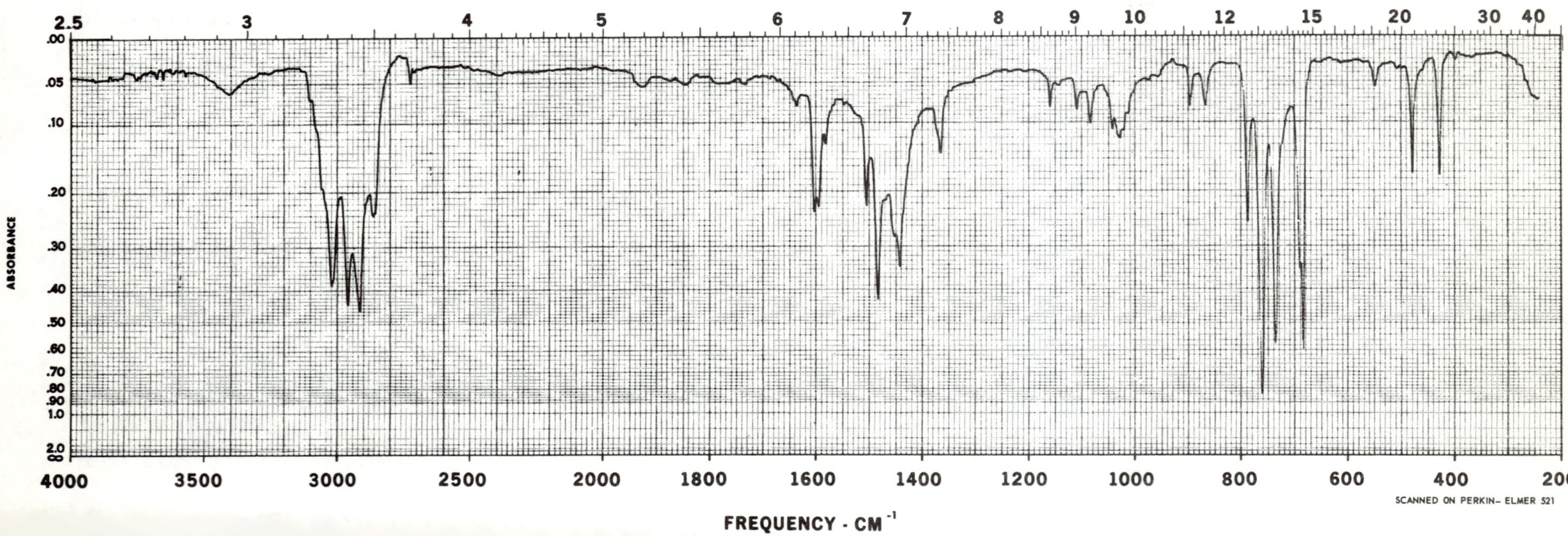

PHOSPHORIC ACID, BERYLLIUM SALT (2:3)
BERYLLIUM PHOSPHATE

$Be_3O_8P_2$

$Be_3(PO_4)_2$

MOLECULAR WEIGHT: 216.98

SOURCE OF SAMPLE: Amend Drug and Chemical Company, Inc.
New York, New York

CAS REGISTRY NO.: —

NIOSH NUMBER: —

KBr Wafer

CONDENSED PHASE

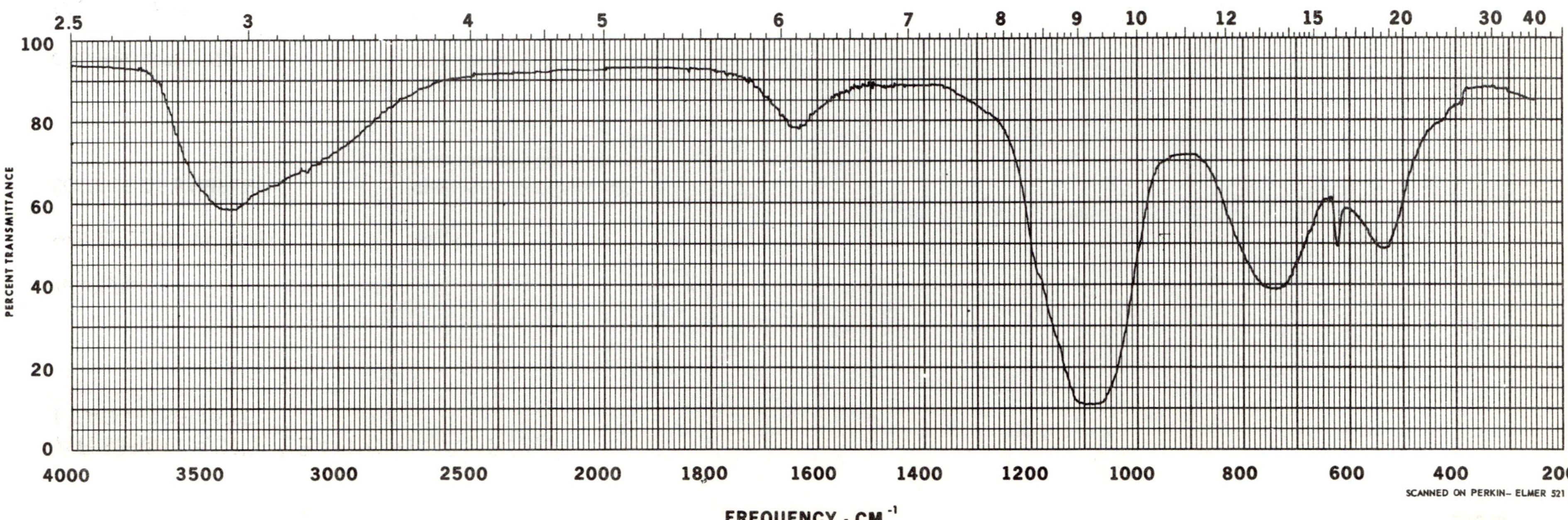

ARSENIC ACID, LEAD (2+) SALT (2:3)
LEAD ARSENATE

$As_2O_8Pb_3$

MOLECULAR WEIGHT: 899.41

SOURCE OF SAMPLE: Chem Service Company
West Chester, Pennsylvania

$Pb_3(AsO_4)_2$

CAS REGISTRY NO.: –

NIOSH NUMBER: –

Split Mull

CONDENSED PHASE

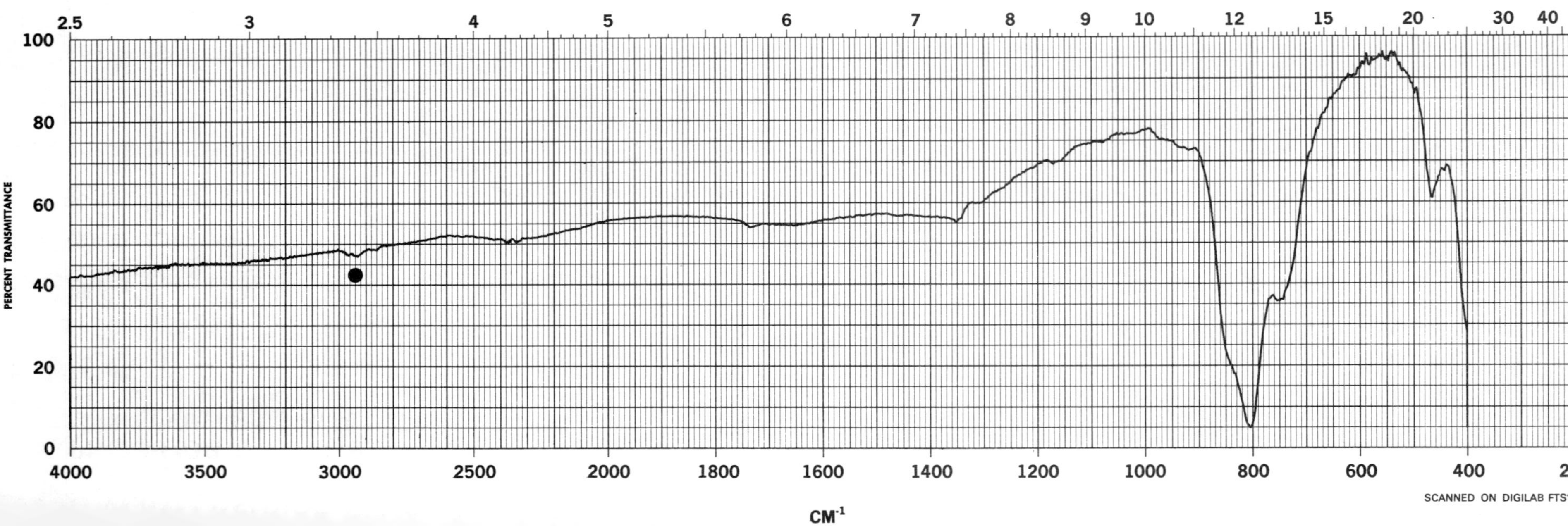

SCANNED ON DIGILAB FTS14

GLYCIDALDEHYDE
EPIHYDRINALDEHYDE
2,3–EPOXY–1–PROPANAL
2,3–EPOXYPROPIONALDEHYDE

$C_3H_4O_2$

MOLECULAR WEIGHT:	72.06	CAS REGISTRY NO.:	765-34-4
SOURCE OF SAMPLE:	Shell Chemical Company Houston, Texas	NIOSH NUMBER:	MB31500
BOILING POINT:	112-113°C		
VAPORIZATION TEMP.:	200°C	PRESSURE:	Atmospheric
CELL VOLUME:	7.5 cc	MAX./MIN.:	—
PATH LENGTH:	7.6 cm	SCALE EXPANSION:	0.95

VAPOR PHASE

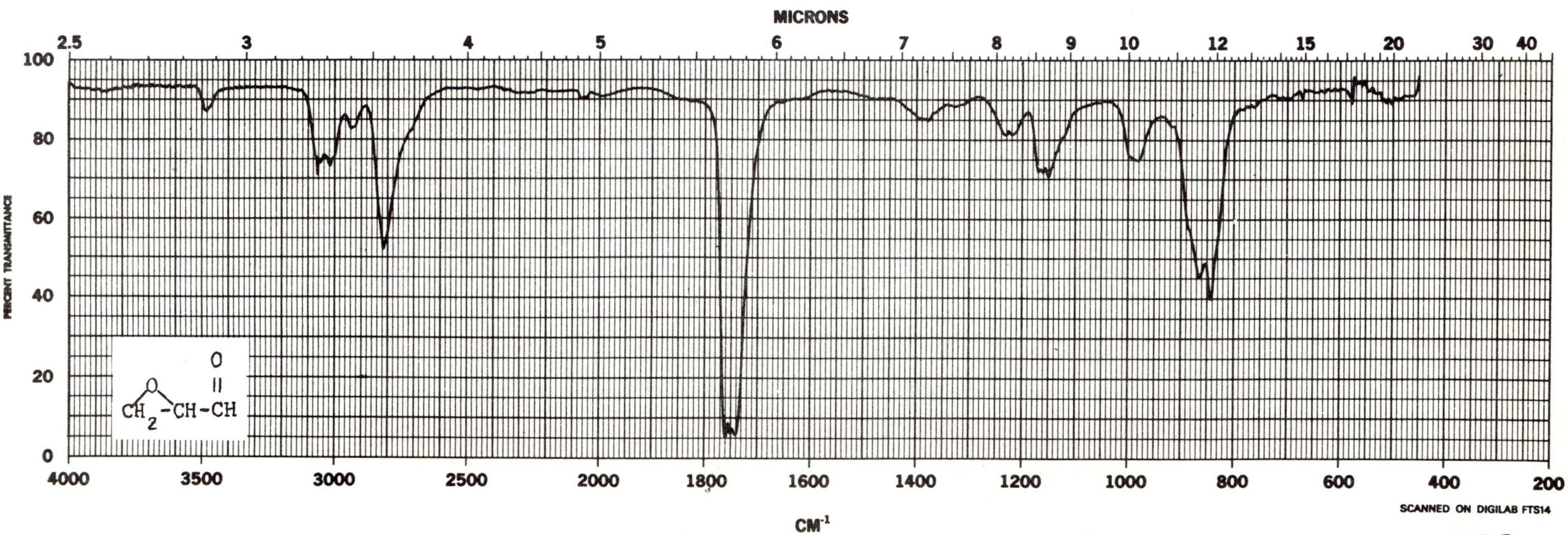

SCANNED ON DIGILAB FTS14

THE
INFRARED SPECTRA
HANDBOOK OF

PRIORITY POLLUTANTS AND TOXIC CHEMICALS

APPENDIX

INTRODUCTION

The purpose of this Appendix is to provide a Gas Chromatography/Fourier Transform Infrared Spectroscopy (GC/FT-IR) analysis of certain chemical mixtures whose infrared spectra appear in this handbook. These materials are generally identified by generic, common or trade names and consist of isomers or families of chemical compounds rather than pure materials. These mixtures are generally analyzed by GC/FT-IR or GC/MS to identify the individual components. While a vapor phase infrared reference spectrum of the mixture is presented in the main part of the handbook for those compounds that could be vaporized in the optical cell, we selected six of the PCBs and chlordan to show more detailed analysis. No attempt was made to identify each component isolated during the GC/FT-IR analysis of these seven materials, however two of the spectra of the pure components were searched against Sadtler's vapor phase infrared reference spectra library to demonstrate the versatility of GC/FT-IR with spectral search. This handbook is intended to serve as a desk reference for both condensed phase and vapor phase spectra of published polluting chemicals. Much more detailed chemical analyses are generally required for positive identification of pure components and this handbook is designed as a starting point for that analysis.

The Appendix format consists of a detailed GC/FT-IR analysis for each of the seven materials. For each material GC operating conditions, the gas chromatogram for FID detection and a table of peak values with corresponding retention times precede the spectra.

GAS CHROMATOGRAPHY/FT-IR DATA

The analyses of the six PCBs and chlordan were completed using a Digilab Model FTS-20 spectrometer with a GC/C$^{T.M.}$ capillary accessory (Ref. 1). A narrow band MCT detector was used for measuring each spectrum, 0.6 seconds per spectrum. The optical cell was a gold-coated lightpipe with KBr windows. Transfer lines were 0.25 mm fused silica attached to the lightpipe to minimize dead volume. The gas chromatograph was a Hewlett Packard, Model 5880 (Ref. 2). A J & W, SE-30, 30 meter, 0.32 mm (I.D.), fused silica column was used (Ref. 3), and the exit end of the lightpipe was connected to the FID detector with minimum loss in chromatography resolution.

OPERATING CONDITIONS

The preparation and injected amount of the samples were identical for all of the seven substances. A saturated solution of the compound in diethyl ether was prepared and a 50 μl aliquot of this solution was diluted to 1 ml. A 2 μl portion of the diluted solution was injected into the gas chromatograph for analysis using manual splitless injection technique. In each case the temperature of the lightpipe was maintained at 270°C, the injector at 260°C and the FID detector at 290°C. The carrier gas was helium and the column, J & W, SE-30, fused silica, 30 m long and 0.32 mm in diameter (I.D.) The flow rate was approximately 1.5 ml/min and the inlet pressure was maintained at 60 psi. Other pertinent data such as temperature programming appears with the chromatogram at the beginning of the component spectra for each material.

GC/FT-IR WITH SPECTRAL SEARCH

Automatic library searches of unknown mass spectra have been used for some time to assist the analyst in identifying unknown compounds. Only within the last three years has a vapor phase infrared reference spectra collection become large enough to allow automatic library searching in GC/FT-IR analyses. To demonstrate how spectral searching has been incorporated into GC/FT-IR, two component spectra were selected from the Arochlor 1232 analysis and searched against Sadtler's Vap-ir T.M. data base. The Vap-ir data base consists of infrared reference spectra for pure materials measured in the vapor state above ambient temperature. These reference spectra are stored directly on the magnetic disk of the FT-IR instrument and they are in a format compatible with Digilab's FTS-GC/S T.M. software (Ref. 1). The GC/S software automatically searches the GC peaks selected from a reconstructed chromatogram after the infrared spectral data has been collected.

The infrared spectra for GC peaks 1 and 3 for Arochlor 1232 are shown on page 23 of the Appendix. As soon as the chromatogram was completed these spectra were automatically searched against the Vap-ir library. The search results for peak 1 and peak 3 are given in Figure 1 and Figure 2 respectively.

The search results for each peak show the ten best matches from a library of 6800 reference spectra. The order of best matches is given as the HIT number with the best match as 1. The compound number is the numerical identification number for the compound in the library (i.e. number 1 HIT is compound 2165), and the index is the degree of match for the unknown with the library spectrum. A perfect match would give an index of zero.

The Chemical Abstracts registry number is also given for each compound in the search report along with the compound name.

Figures 3 and 4 show the reference spectra for Biphenyl (No. 2716) and 4-Chlorobiphenyl (No. 2595) from Sadtler's Vap-ir reference library. Comparison of these reference spectra with the spectra of peaks 1 and 3, respectively, for the Arochlor 1232 analysis on page 23 of the Appendix will confirm the search results. In addition, a comparison of all of the spectra obtained from the GC/FT-IR analysis with the spectrum of the Arochlor 1232 mixture (Handbook Spectrum 30) shows that the mixture is the sum of the individual component spectra. The reference spectrum for the mixture may be used to identify the material by chemical class, but detailed GC/FT-IR analysis with search is needed for identification of the pure components.

Ref. 1: Digilab, Cambridge, Massachusetts

Ref. 2: Hewlett Packard, Avondale, Pennsylvania

Ref. 3: J & W Scientific, Inc., Rancho Cordova, California

FIGURE 1 SEARCH RESULTS FOR AROCHLOR1232, GC (1); PEAK 1

LIBNAME=VAPOLIB68
STP=3200.00
ENP= 700.00

HIT	CMPD	INDEX			COMPOUND NAME
1	2716	0.0616	02716	92-52-4	BIPHENYL
2	3777	0.3010	03777	831-91-4	SULFIDE, BENZYL PHENYL,
3	6764	0.3463	06764	86-29-3	ACETONITRILE, DIPHENYL–,
4	3045	0.3766	03045	13865-49-1	SULFIDE, PHENETHYL PHENYL,
5	6714	0.4023	06714	91-00-9	METHYLAMINE, 1,1–DIPHENYL–,
6	5288	0.4338	05288	519-73-3	METHANE, TRIPHENYL–,
7	154	0.4358	00154	101-81-5	METHANE, DIPHENYL–,
8	2165	0.4577	02165	84-15-1	BENZENE, O–DIPHENYL–,
9	3211	0.4590	03211	90-99-3	METHANE, CHLORODIPHENYL–,
10	2109	0.4692	02109	882-33-7	PHENYL DISULFIDE

FIGURE 2 SEARCH RESULTS FOR AROCHLOR1232, GC (3); PEAK 3

LIBNAME=VAPOLIB68
STP=3200.00
ENP= 700.00

HIT	CMPD	INDEX			COMPOUND NAME
1	2595	0.0928	02595	2051-62-9	BIPHENYL, 4–CHLORO–,
2	4627	0.5248	04627	53830-54-9	SULFIDE, BENZYL P–CHLOROPHENYL,
3	2204	0.6162	02204	2050-68-2	BIPHENYL, 4,4PR–DICHLORO–,
4	3039	0.6325	03039	134-83-8	METHANE, CHLORO/P–CHLOROPHENYL/–PHENYL–,
5	5416	0.6726	05416	92-94-4	BENZENE, P–DIPHENYL–,
6	698	0.6855	00698	4350-41-8	PYRINE, 2–/P–CHLOROBENZYL/–,
7	6009	0.6918	06009	1591-31-7	BIPHENYL, 4–IODO–,
8	2653	0.7013	02653	106-46-7	BENZENE, P–DICHLORO–,
9	4458	0.7061	04458	21650-66-8	AZOXYBENZENE, 4,4PR–DICHLORO–,TRANS–,
10	5281	0.7167	05281	623-03-0	BENZONITRILE, P–CHLORO–,

2716V BIPHENYL

FIGURE 3

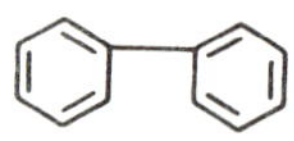

$C_{12}H_{10}$ **Mol. Wt.** 154.21

Boiling Point:	255°C	**IR:**	163 K
Vaporization Temp.:	280°C	**Injection Amount:**	.70 mg
Cell Volume:	7.5 cc	**Pressure:**	atmospheric
Path Length:	7.6 cm	**Max./Min.:**	99.6/37.6
Source of Sample:		**Scale Expansion:**	1.48

Aldrich Chemical Company, Inc.,
Milwaukee, Wisconsin

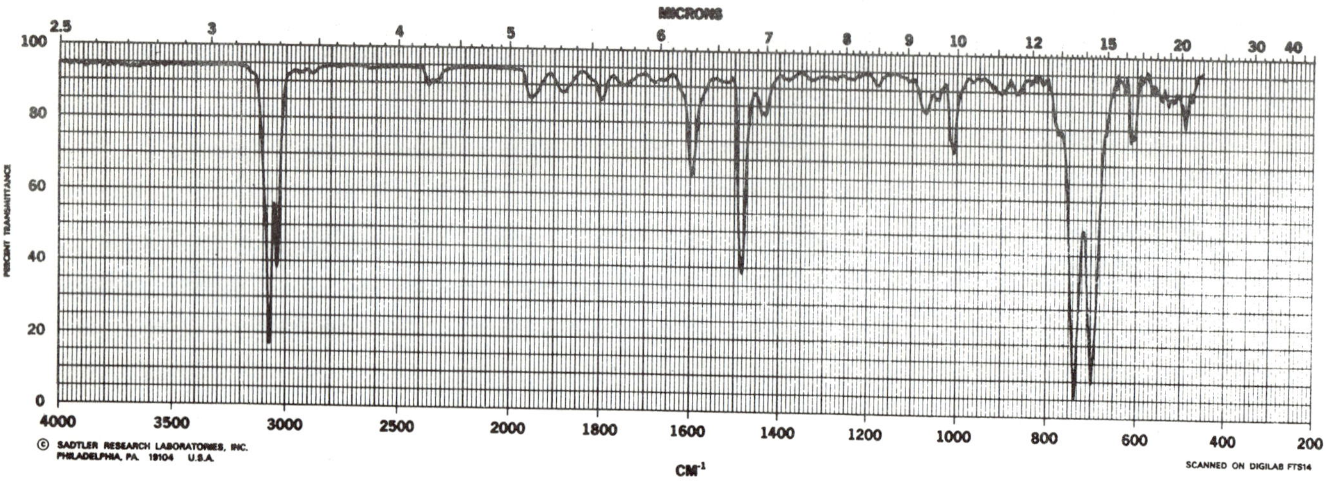

2595V 4-CHLOROBIPHENYL

FIGURE 4

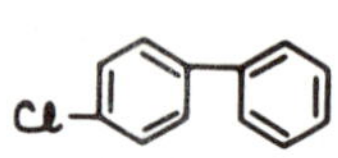

$C_{12}H_9Cl$ **Mol. Wt.** 188.66

Boiling Point:	282°C	**IR:**	28367 K
Vaporization Temp.:	280°C	**Injection Amount:**	.88 mg
Cell Volume:	7.5 cc	**Pressure:**	atmospheric
Path Length:	7.6 cm	**Max./Min.:**	100.7/27.8
Source of Sample:		**Scale Expansion:**	1.26

Federal Phosphorous Company,
Anniaton, Alabama

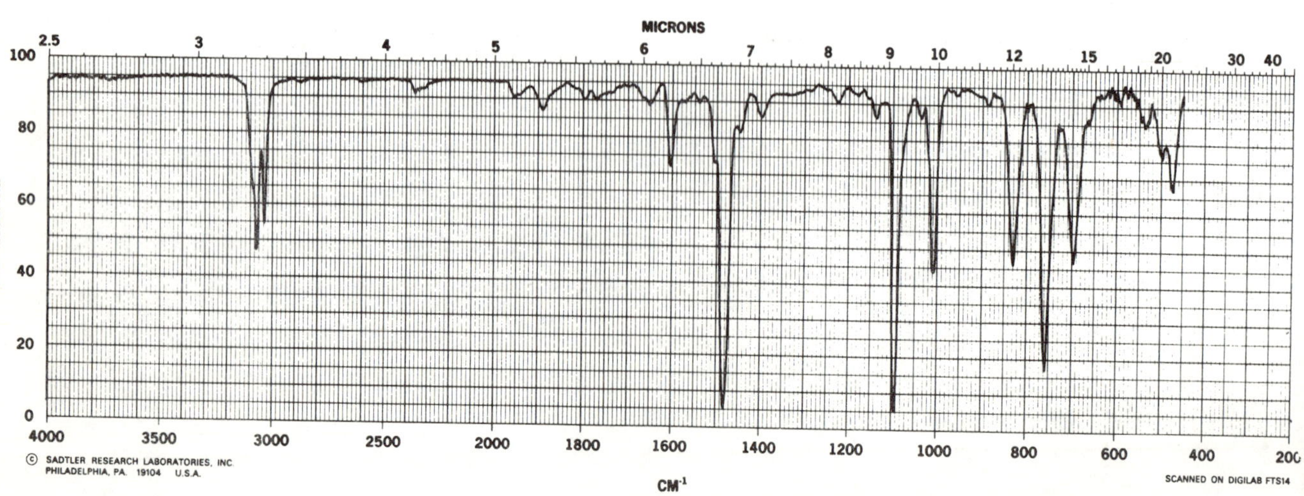

TABLE OF CONTENTS

AROCHLOR 1016

GAS CHROMATOGRAPH DATA

TYPE = HP 5880 G.C.
MANUAL SPLITLESS INJECTION

SAMPLE PREPARATION = 50 μl of a saturated solution of the compound in diethyl ether was diluted to 1 ml.

INJECTED AMT. = 2 μl of the diluted solution

DET. TEMP. = 290°C

INJ. TEMP. = 260°C

CARRIER GAS = He

FLOW RATE = ∿ 1.5 ml./min.

PRESSURE = 60 PSI

DETECTOR = FID

COLUMN = J & W, SE–30, FUSED SILICA
30 m. LONG, 0.32 mm. DIA.

FILM THICKNESS = 0.25 microns

OVEN PROGRAM RATES 1 AND 2:

INITAL TEMP. = 35°C

INITAL TIME = 0.40 min.

LEVEL 1

PRGM RATE = 25.00°C/min.

FINAL TEMP. = 120°C

FINAL TIME = 0.00 min.

LEVEL 2

PROGRAM RATE = 4.00°C/min.

FINAL TEMP. = 210°C

FINAL TIME = 20.00 min.

PEAK	RETENTION TIME	PEAK	RETENTION TIME
1	12.80	15	21.32
2	14.43	16	21.42
3	15.28	17	21.71
4	16.31	18	22.06
5	16.74	19	22.24
6	17.07	20	22.40
7	17.70	21	22.81
8	18.84	22	22.96
9	18.93	23	23.29
10	19.14	24	23.49
11	19.52	25	24.16
12	20.28	26	24.29
13	20.73	27	24.44
14	21.12		

IR DATA

DIGILAB FTS 20 E WITH DIGILAB GC/C[T.M.] ACCESSORY

PURGE = N_2

DETECTOR = LIQUID N_2 COOLED MCT (NARROW BAND)

OPTICAL CELL = GOLD COATED LIGHTPIPE WITH KBr WINDOWS

[illegible]

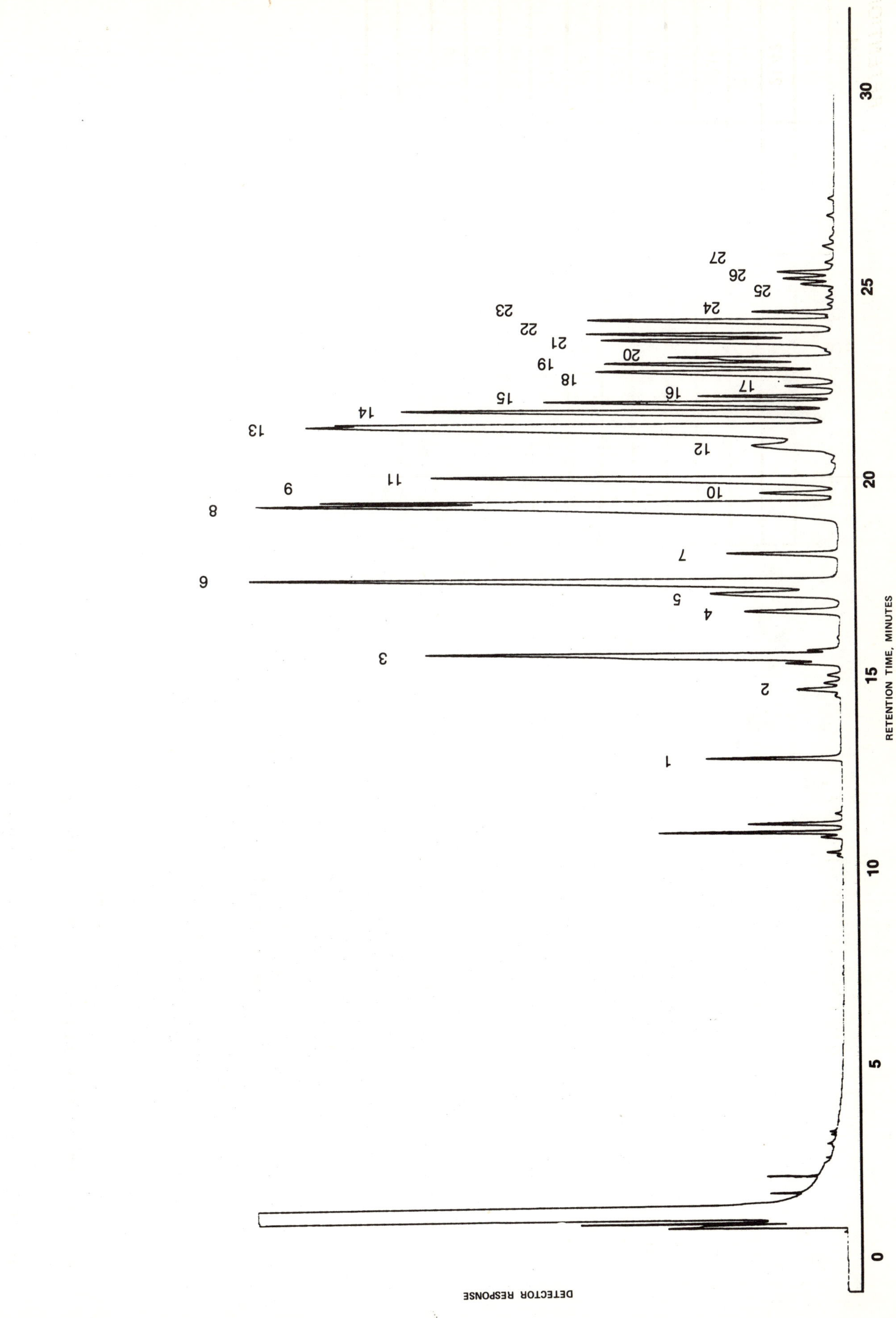

AROCHLOR 1016

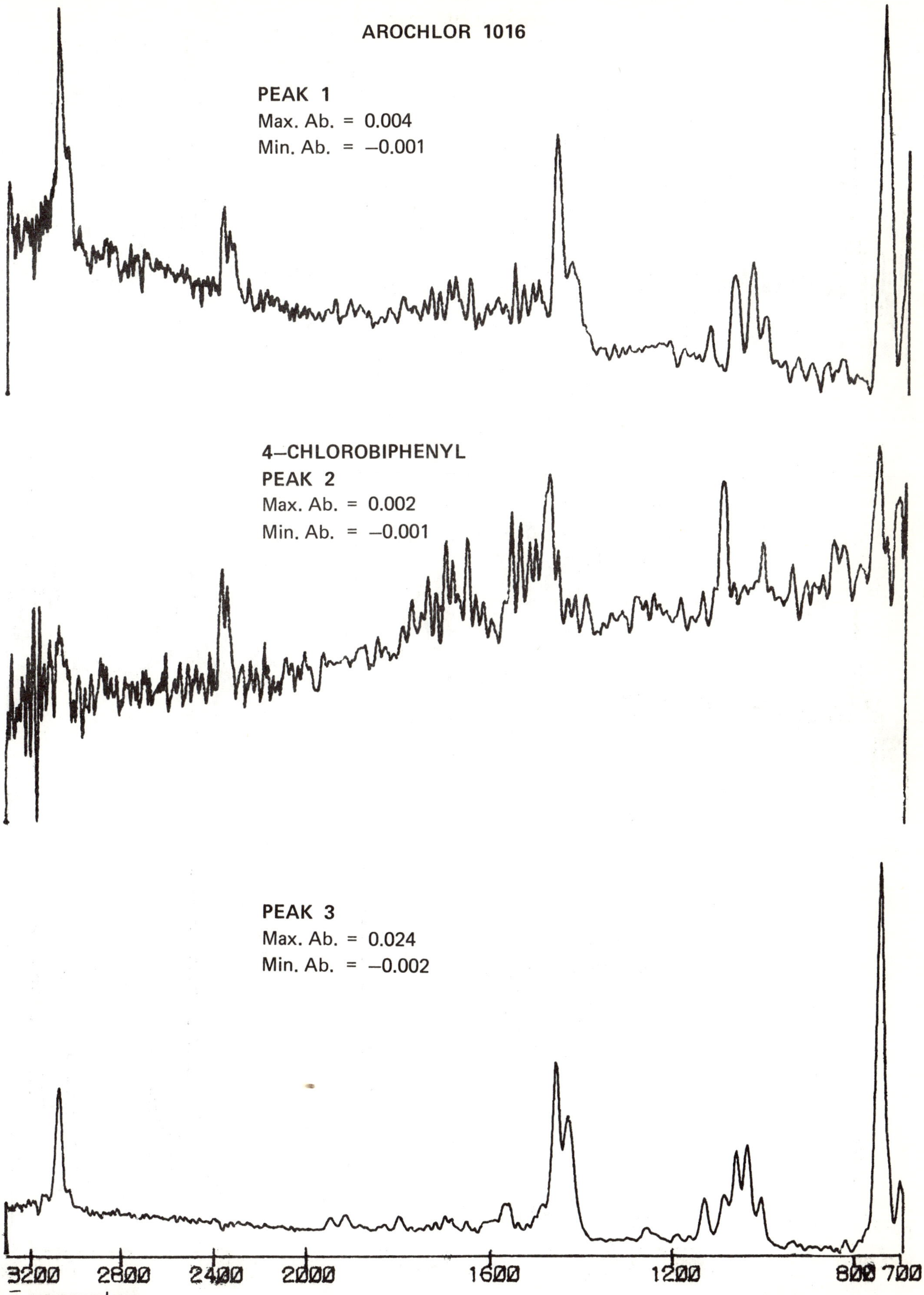
AROCHLOR 1016
PEAK 1
Max. Ab. = 0.004
Min. Ab. = –0.001
4–CHLOROBIPHENYL
PEAK 2
Max. Ab. = 0.002
Min. Ab. = –0.001
PEAK 3
Max. Ab. = 0.024
Min. Ab. = –0.002
3200
2800
2400
2000
1600
1200
800
700
$\bar{\nu}$, wavenumbers

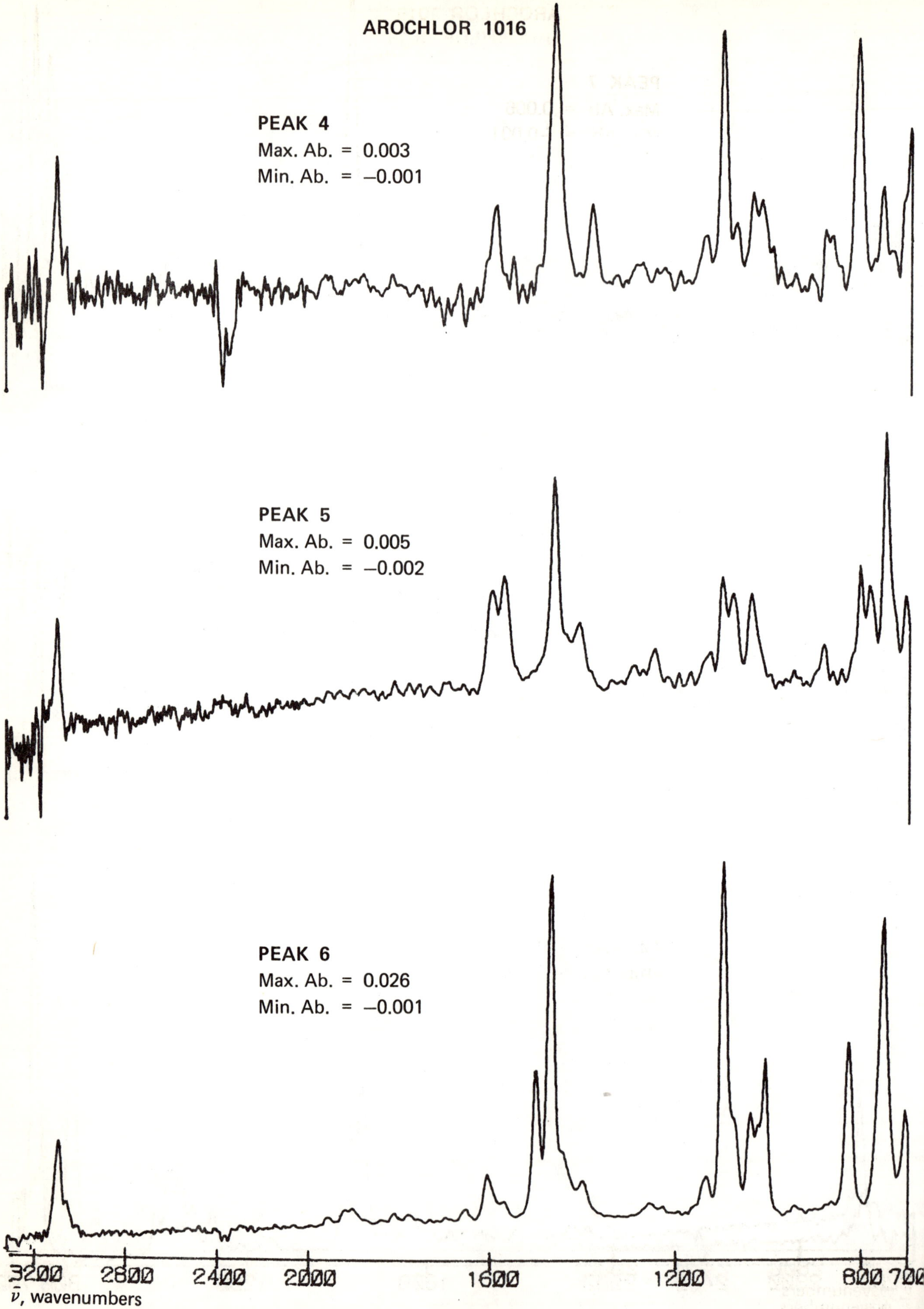

AROCHLOR 1016
PEAK 4
Max. Ab. = 0.003
Min. Ab. = –0.001
PEAK 5
Max. Ab. = 0.005
Min. Ab. = –0.002
PEAK 6
Max. Ab. = 0.026
Min. Ab. = –0.001
3200
2800
2400
2000
1600
1200
800
700
$\bar{\nu}$, wavenumbers

AROCHLOR 1016

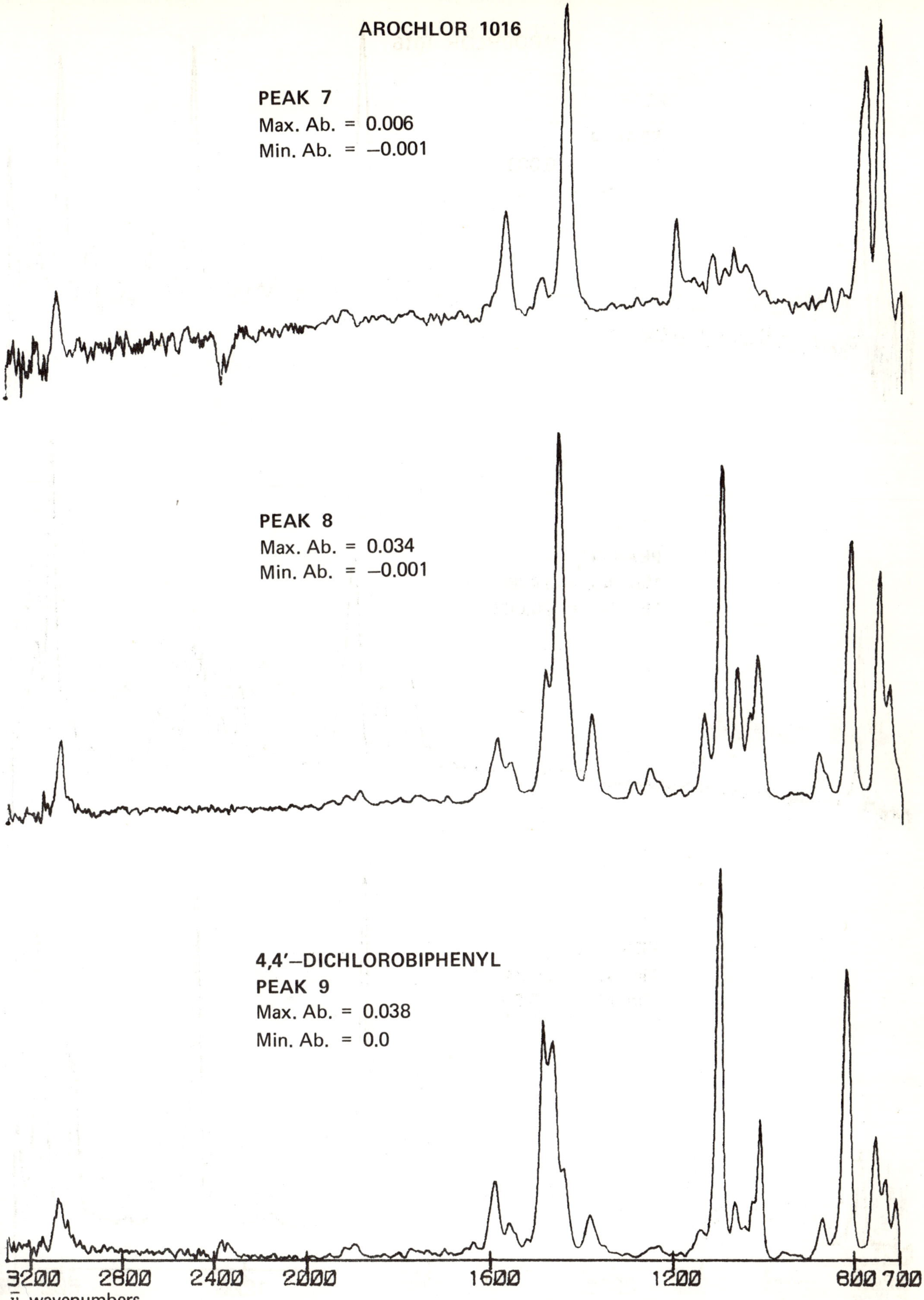

AROCHLOR 1016

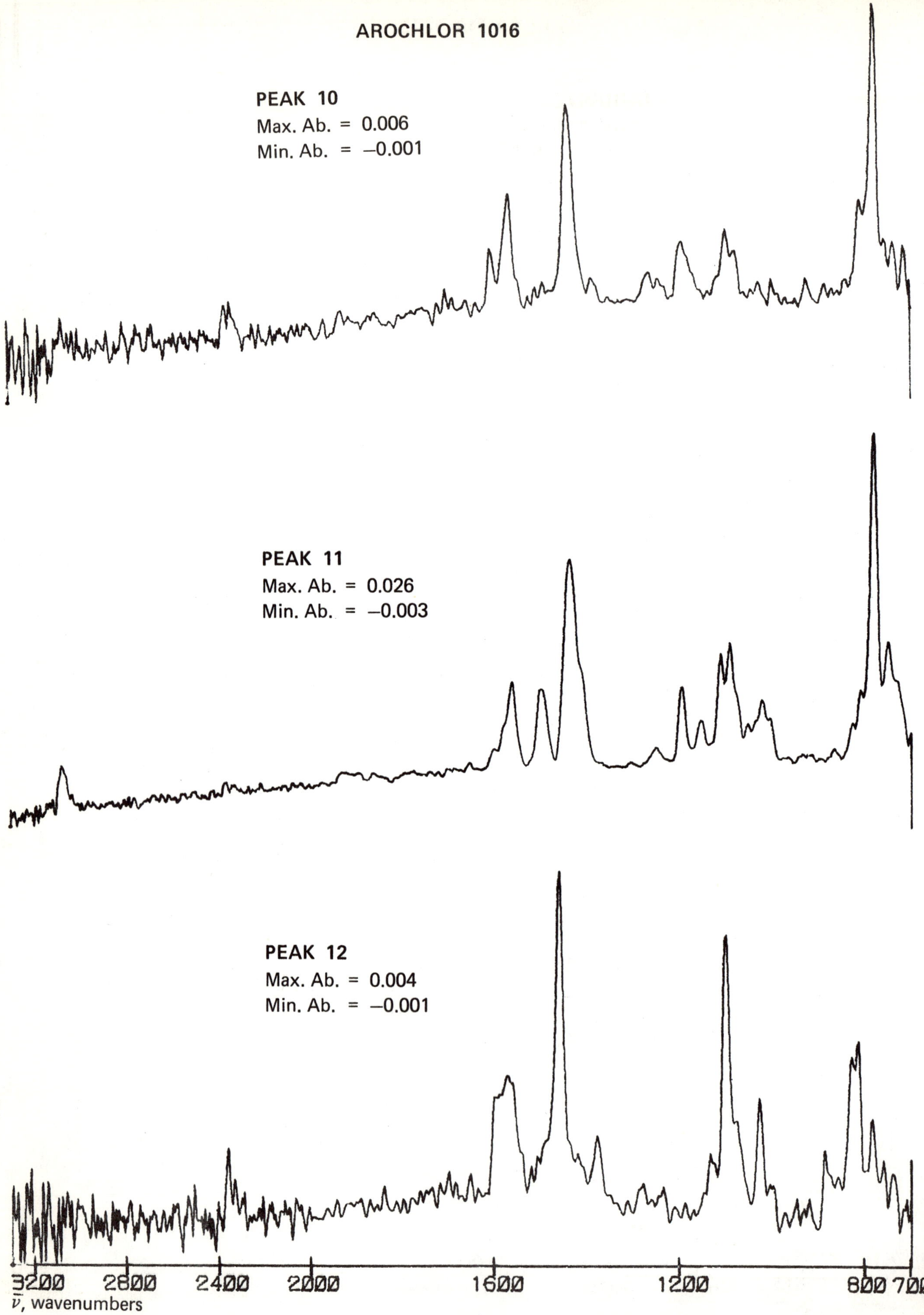

AROCHLOR 1016

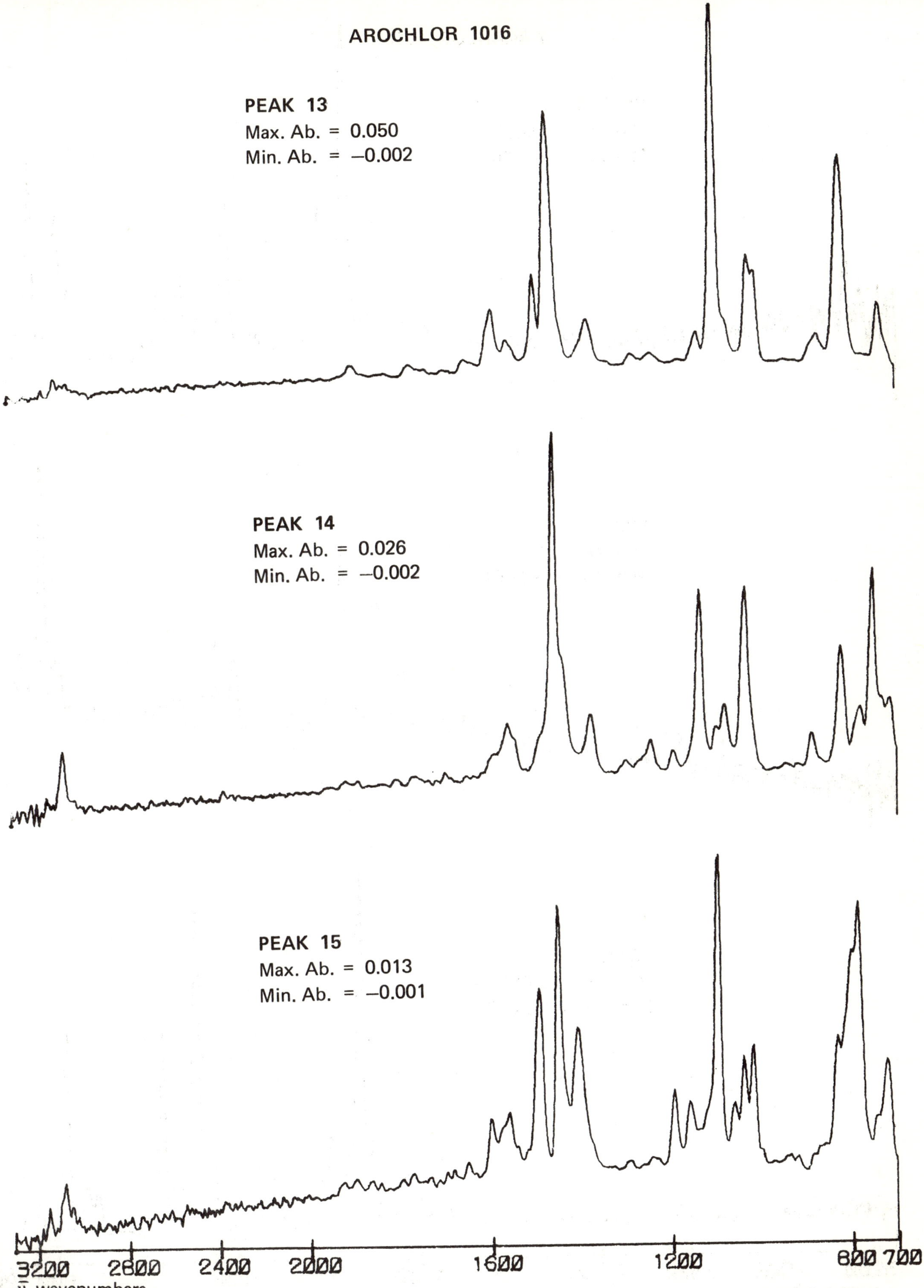

AROCHLOR 1016

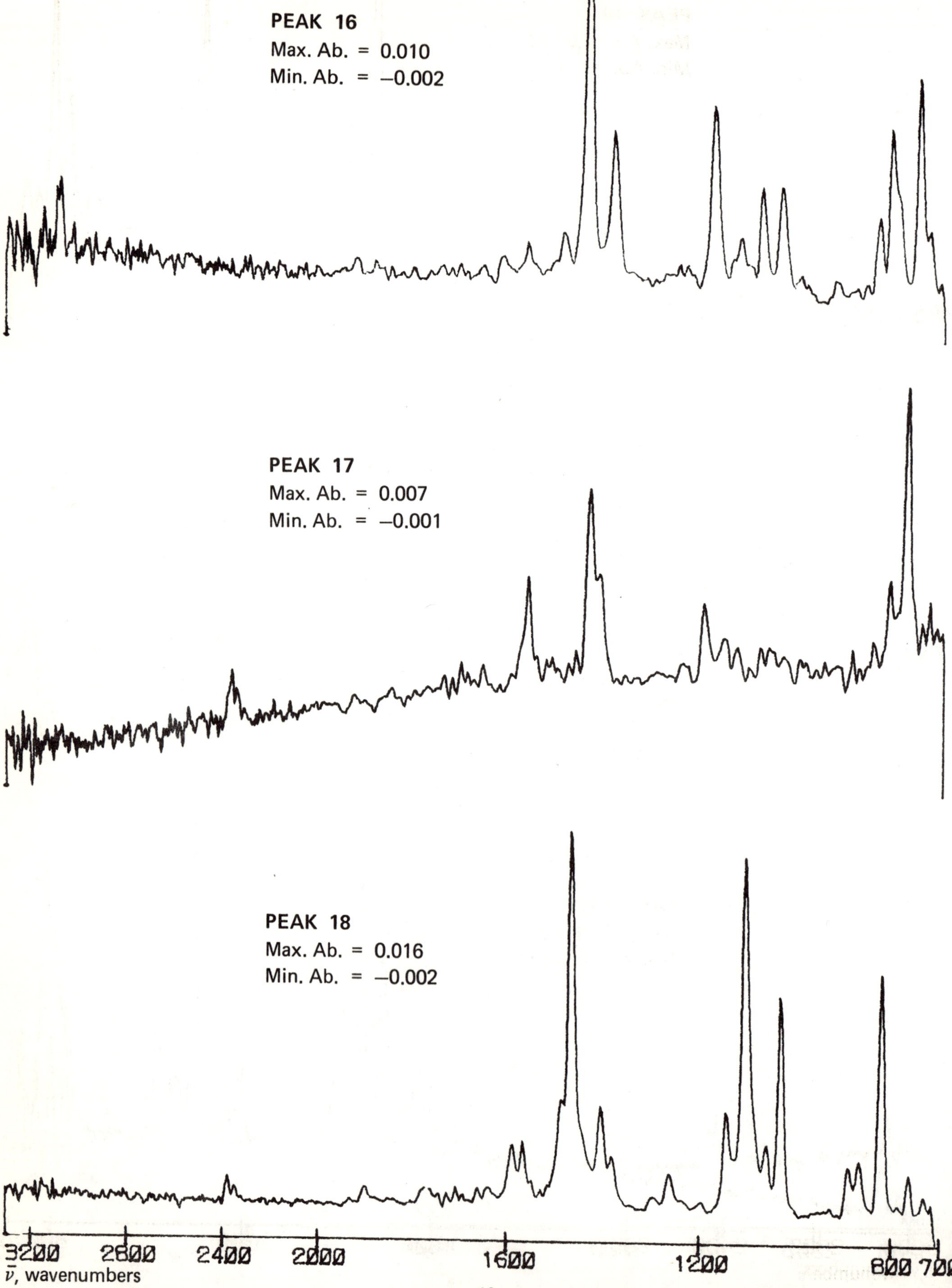

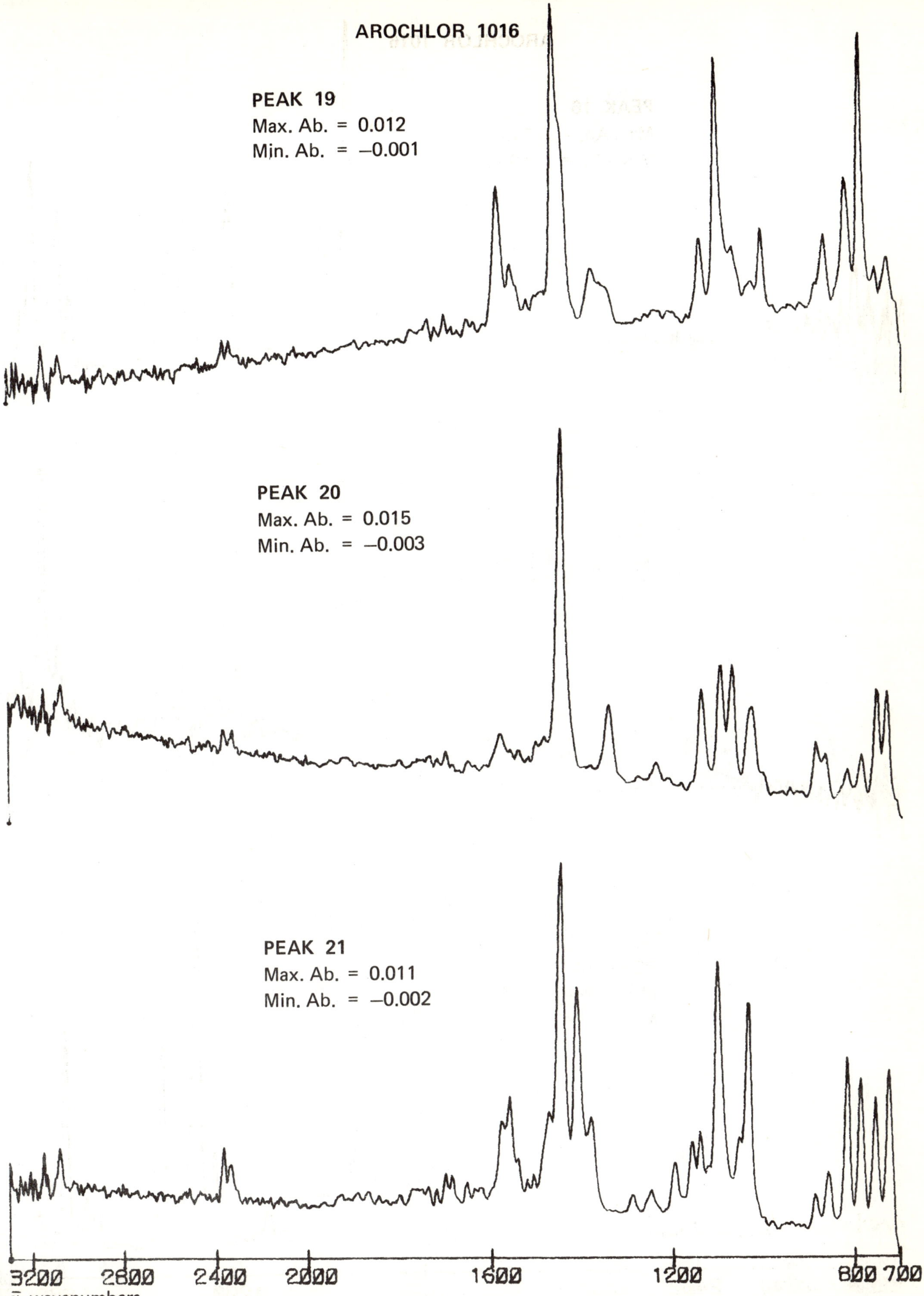

ARCHLOR 1016
PEAK 19
Max. Ab. = 0.012
Min. Ab. = –0.001
PEAK 20
Max. Ab. = 0.015
Min. Ab. = –0.003
PEAK 21
Max. Ab. = 0.011
Min. Ab. = –0.002
3200
2800
2400
2000
1600
1200
800
700
$\bar{\nu}$, wavenumbers

AROCHLOR 1016

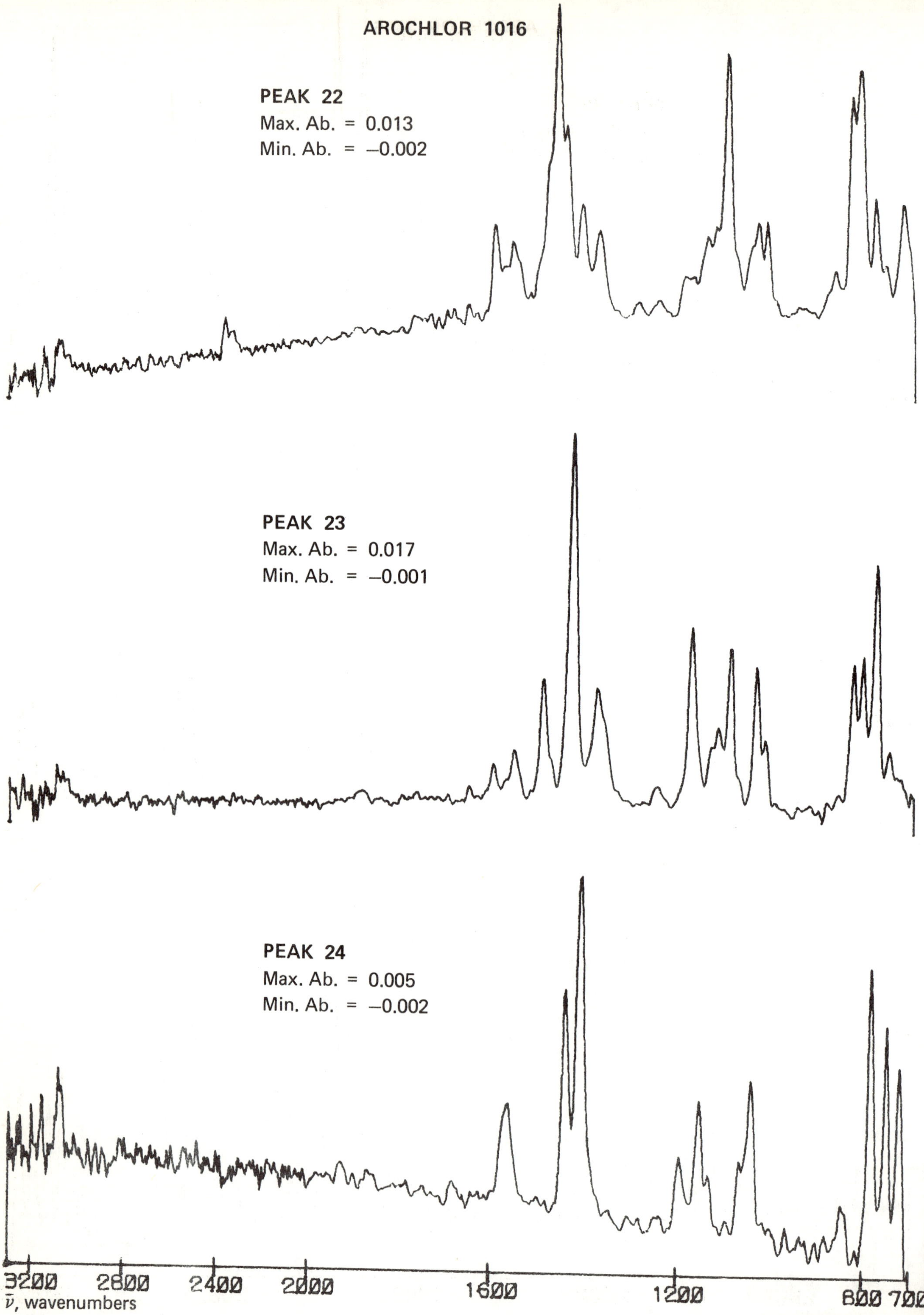

AROCHLOR 1016

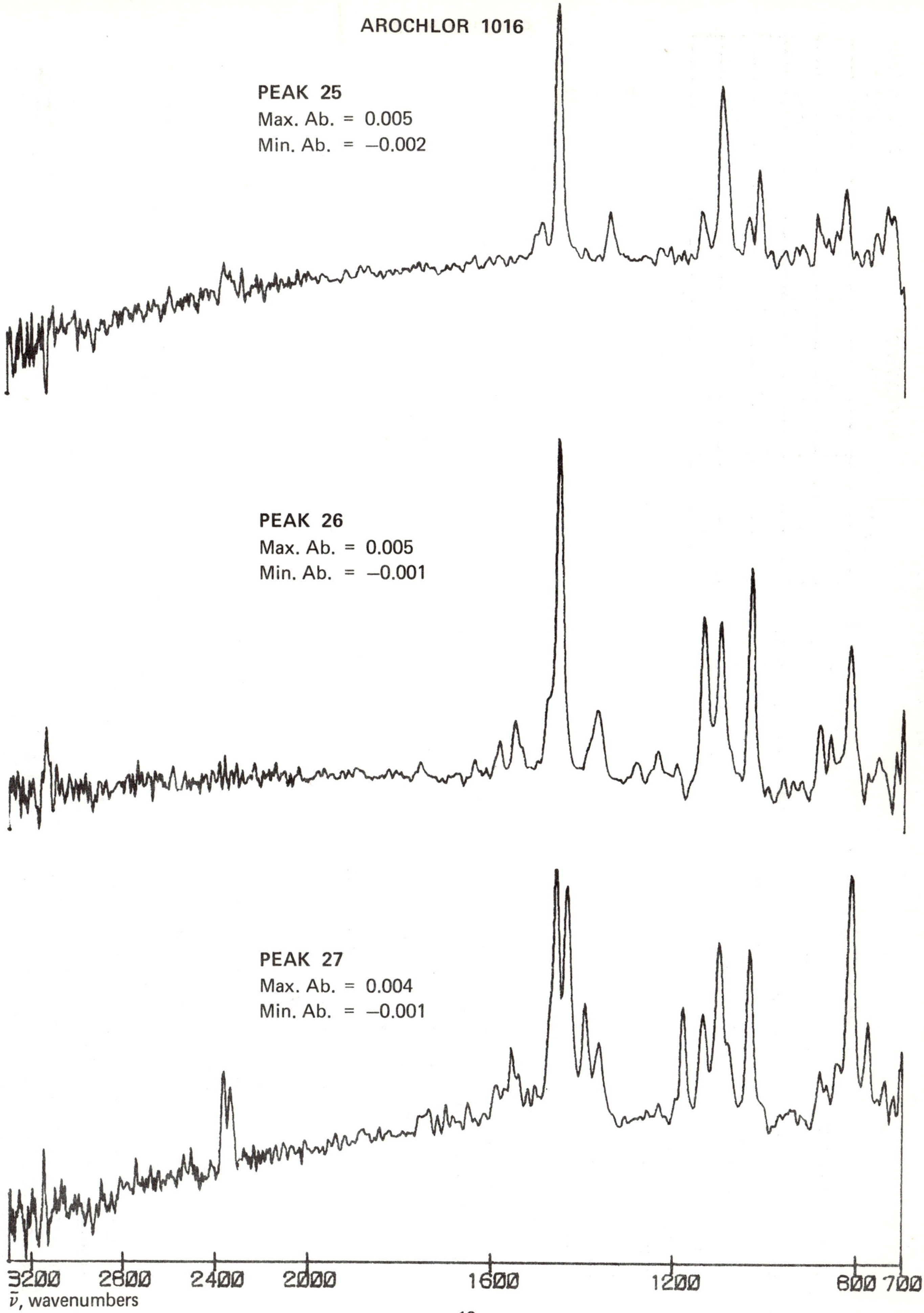

AROCHLOR 1221

PEAK	RETENTION TIME	PEAK	RETENTION TIME
1	10.55	6	16.29
2	12.89	7	16.67
3	14.28	8	16.94
4	14.48	9	18.90
5	15.20		

GAS CHROMATOGRAPH DATA

TYPE = HP 5880 G.C.
MANUAL SPLITLESS INJECTION

SAMPLE PREPARATION = 50 μl of a saturated solution of the compound in diethyl ether was diluted to 1 ml.

INJECTED AMT. = 2 μl of the diluted solution

DET. TEMP. = 290°C

INJ. TEMP. = 260°C

CARRIER GAS = He

FLOW RATE = ∿ 1.5 ml./min.

PRESSURE = 60 PSI

DETECTOR = FID

COLUMN = J & W, SE–30, FUSED SILICA
30 m. LONG, 0.32 mm. DIA.

FILM THICKNESS = 0.25 microns

OVEN PROGRAM RATES 1 AND 2:

INITAL TEMP. = 35°C

INITAL TIME = 0.40 min.

LEVEL 1

PRGM RATE = 25.00°C/min.

FINAL TEMP. = 120°C

FINAL TIME = 0.00 min.

LEVEL 2

PROGRAM RATE = 4.00°C/min.

FINAL TEMP. = 210°C

FINAL TIME = 20.00 min.

IR DATA

DIGILAB FTS 20 E WITH DIGILAB GC/C$^{T.M.}$ ACCESSORY

PURGE = N_2

DETECTOR = LIQUID N_2 COOLED MCT (NARROW BAND)

OPTICAL CELL = GOLD COATED LIGHTPIPE WITH KBr WINDOWS

LIGHTPIPE TEMP. = 270°C

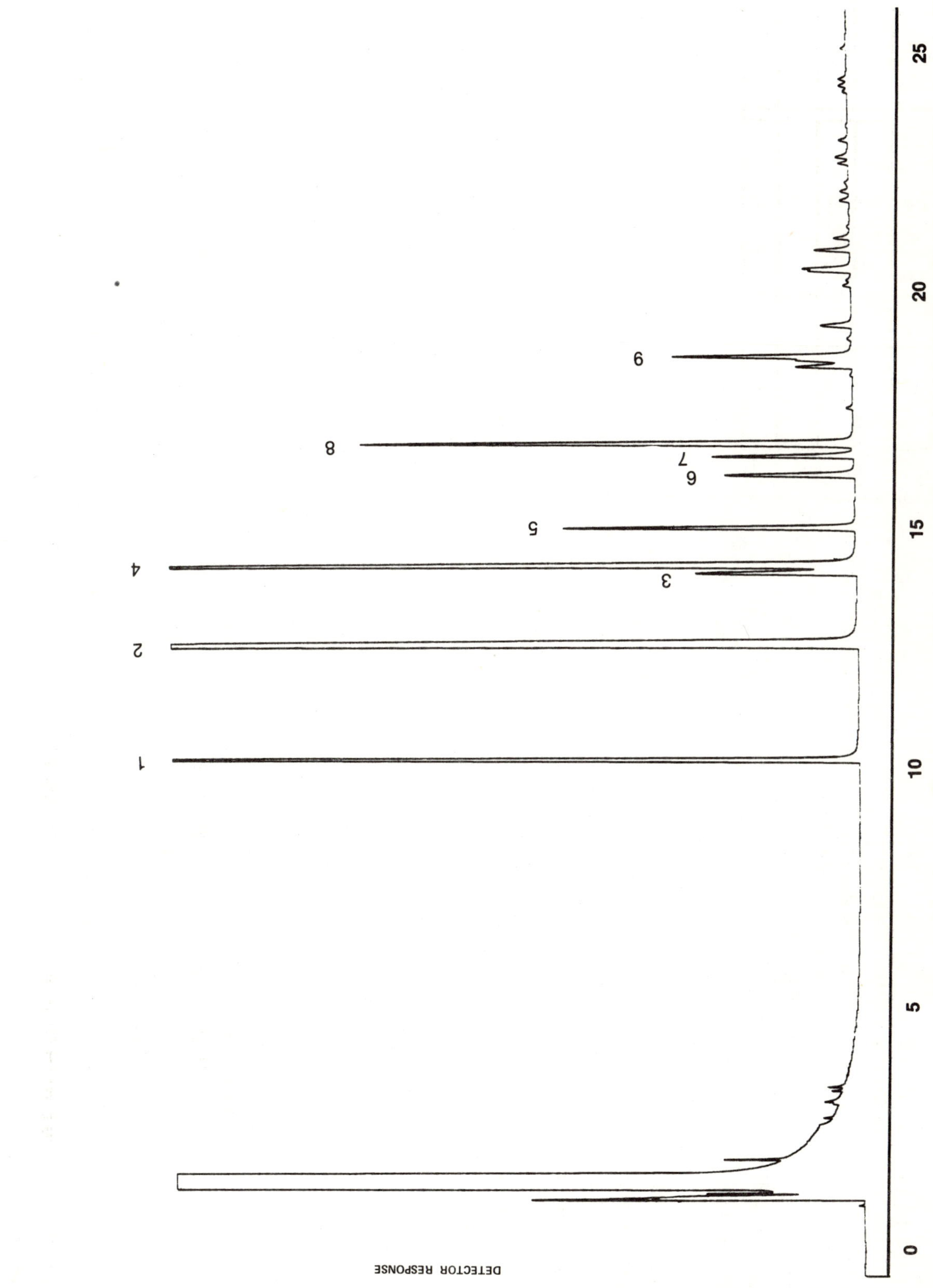

AROCHLOR 1221

AROCHLOR 1221

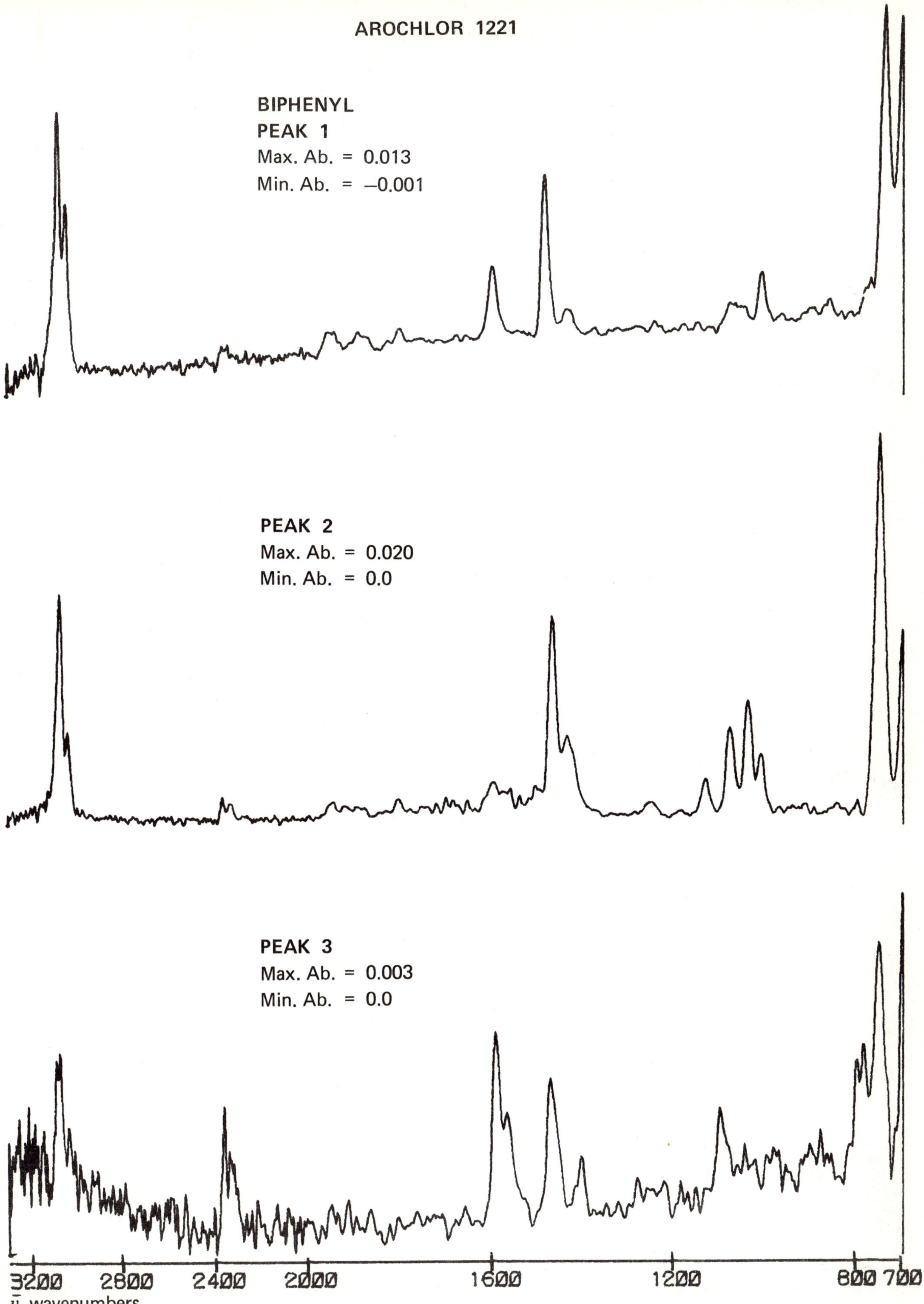

AROCHLOR 1221

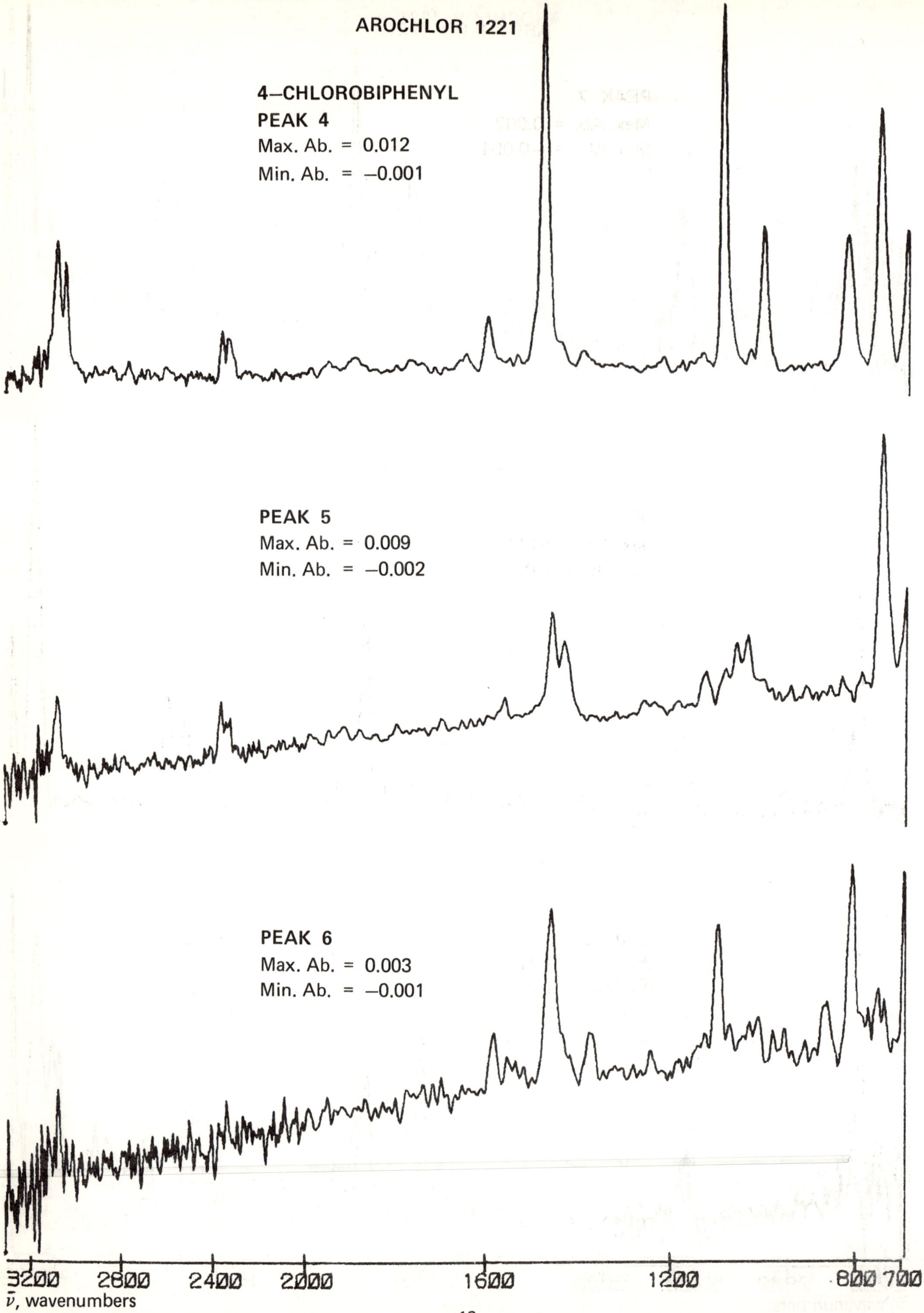

AROCHLOR 1221

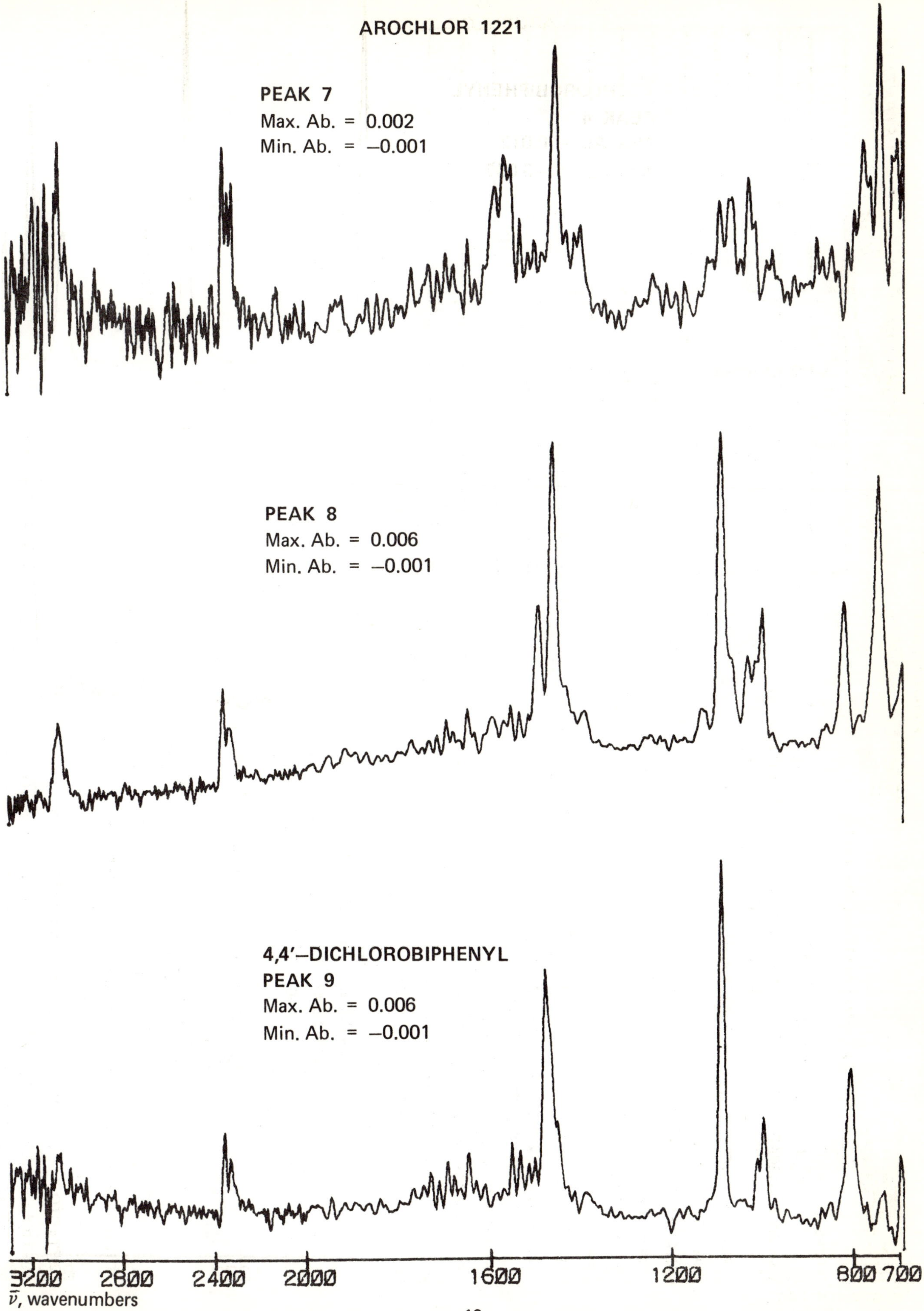

AROCHLOR 1232

GAS CHROMATOGRAPH DATA

TYPE = HP 5880 G.C.
MANUAL SPLITLESS INJECTION

SAMPLE PREPARATION = 50 μl of a saturated solution of the compound in diethyl ether was diluted to 1 ml.

INJECTED AMT. = 2 μl of the diluted solution

DET. TEMP. = 290°C

INJ. TEMP. = 260°C

CARRIER GAS = He

FLOW RATE = ∿ 1.5 ml./min.

PRESSURE = 60 PSI

DETECTOR = FID

COLUMN = J & W, SE–30, FUSED SILICA
30 m. LONG, 0.32 mm. DIA.

FILM THICKNESS = 0.25 microns

OVEN PROGRAM RATES 1 AND 2:

- **INITAL TEMP.** = 35°C
- **INITAL TIME** = 0.40 min.
- **LEVEL 1**
  - **PRGM RATE** = 25.00°C/min.
  - **FINAL TEMP.** = 120°C
  - **FINAL TIME** = 0.00 min.
- **LEVEL 2**
  - **PROGRAM RATE** = 4.00°C/min.
  - **FINAL TEMP.** = 210°C
  - **FINAL TIME** = 20.00 min.

IR DATA

DIGILAB FTS 20 E WITH DIGILAB GC/CT.M. **ACCESSORY**

PURGE = N_2

DETECTOR = LIQUID N_2 COOLED MCT (NARROW BAND)

OPTICAL CELL = GOLD COATED LIGHTPIPE WITH KBr WINDOWS

LIGHTPIPE TEMP. = 270°C

PEAK	RETENTION TIME	PEAK	RETENTION TIME
1	10.56	14	21.17
2	12.90	15	21.36
3	14.52	16	21.93
4	15.22	17	22.10
5	16.29	18	22.28
6	16.68	19	22.68
7	16.98	20	22.83
8	18.71	21	23.16
9	18.80	22	24.11
10	19.39	23	24.28
11	20.17	24	24.41
12	20.62	25	25.01
13	20.96		

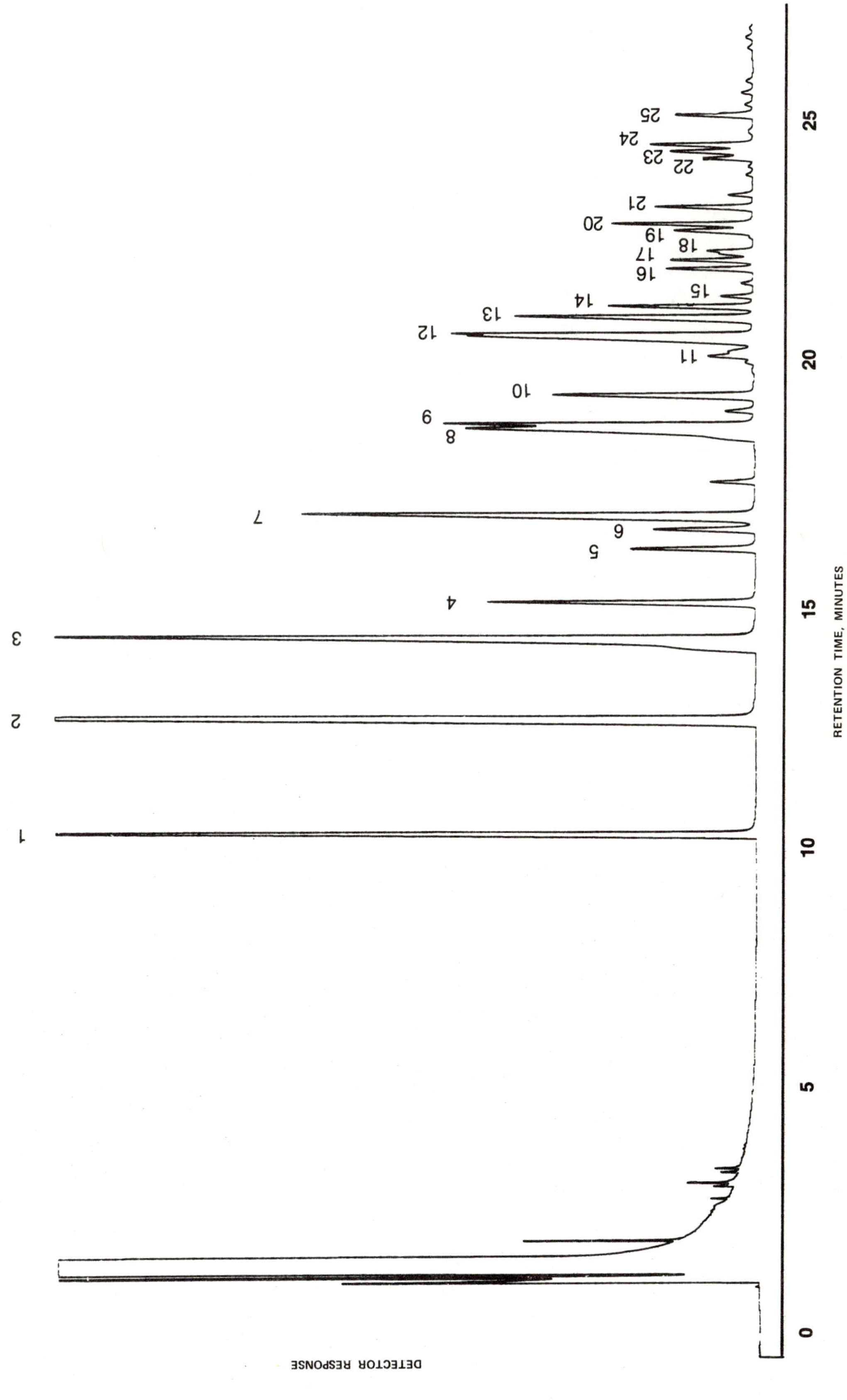

AROCHLOR 1232

AROCHLOR 1232

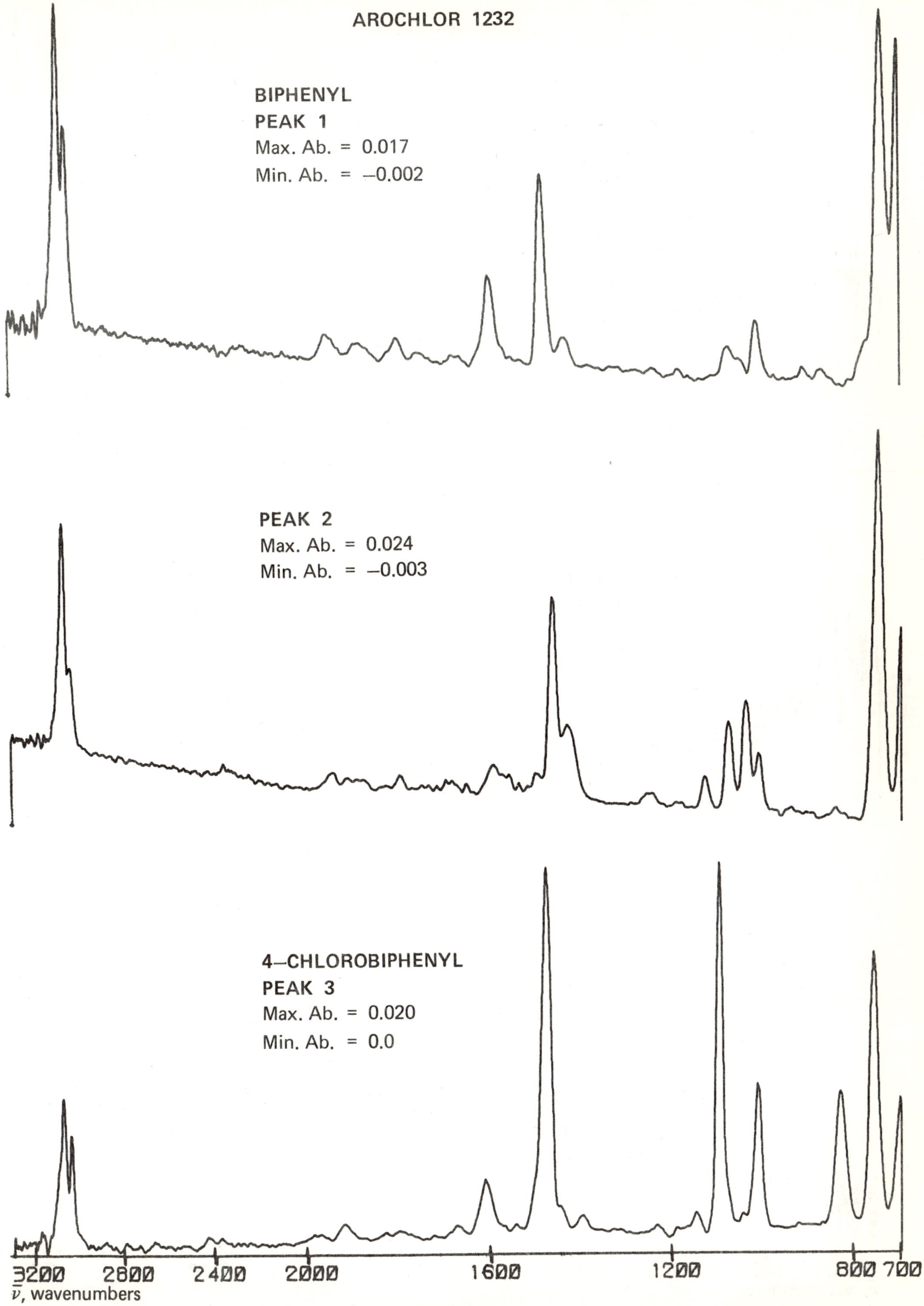

AROCHLOR 1232

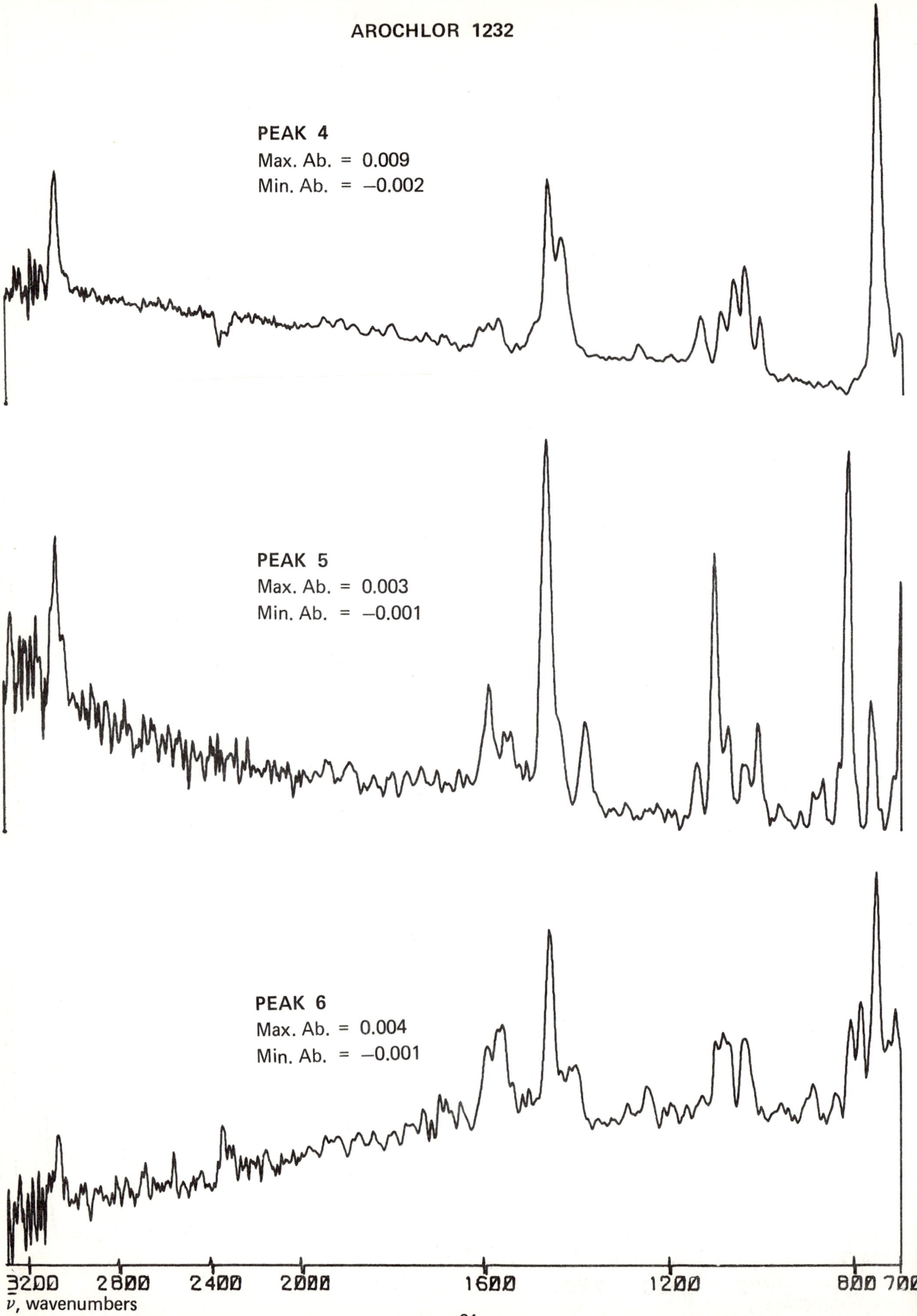

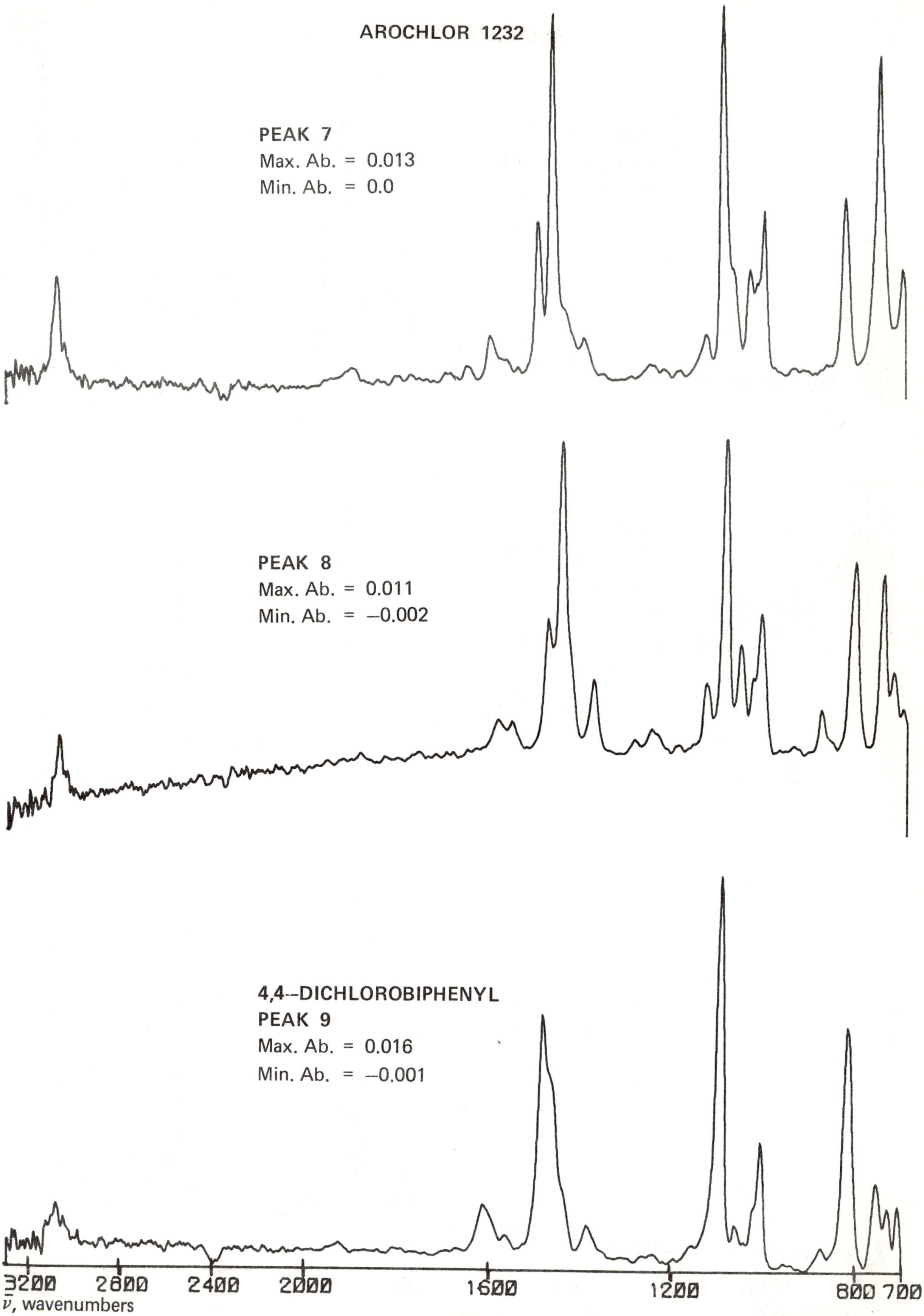
AROCHLOR 1232
PEAK 7
Max. Ab. = 0.013
Min. Ab. = 0.0
PEAK 8
Max. Ab. = 0.011
Min. Ab. = –0.002
4,4–DICHLOROBIPHENYL
PEAK 9
Max. Ab. = 0.016
Min. Ab. = –0.001
3200 2800 2400 2000 1600 1200 800 700
$\bar{\nu}$, wavenumbers

AROCHLOR 1232

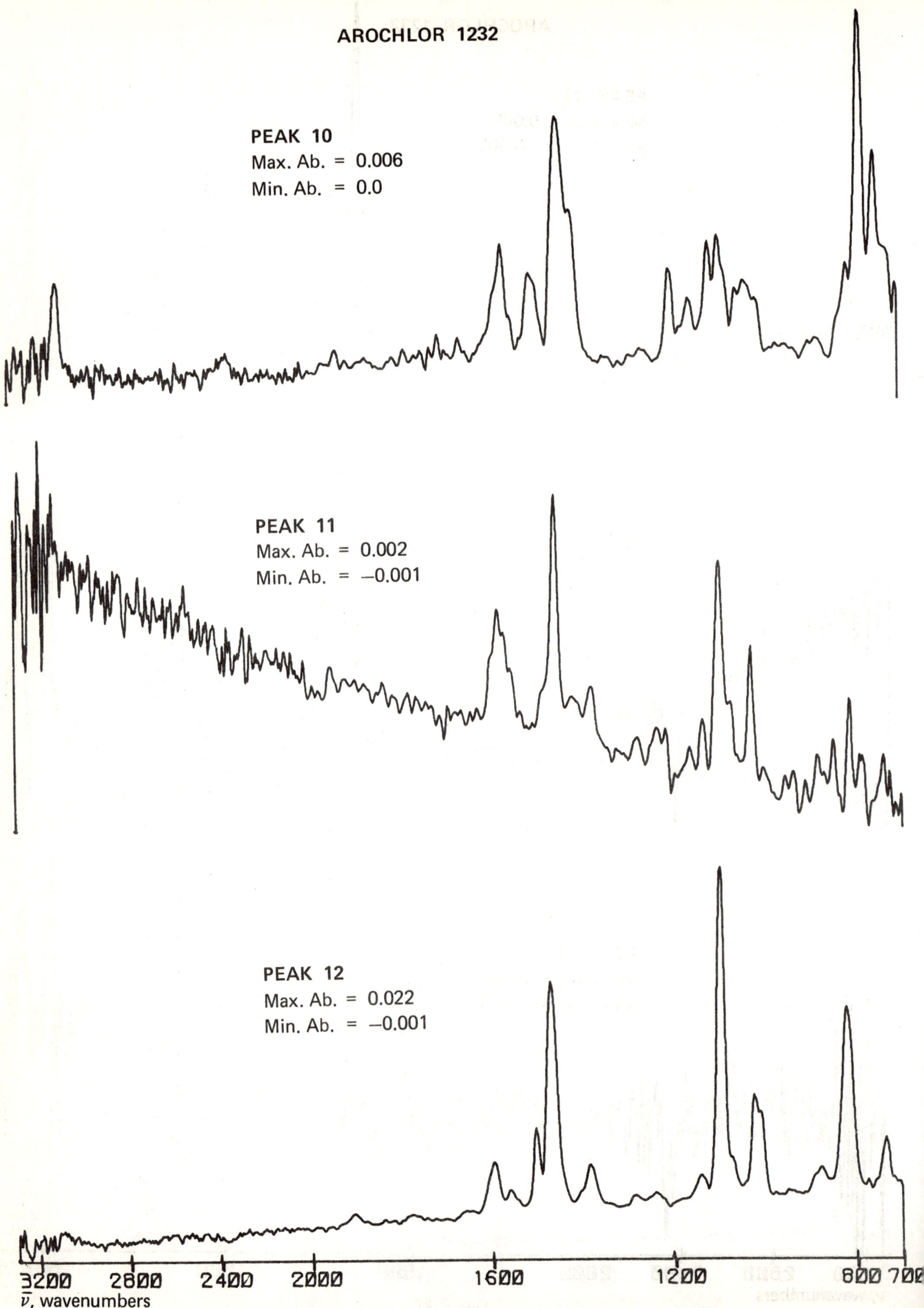

AROCHLOR 1232

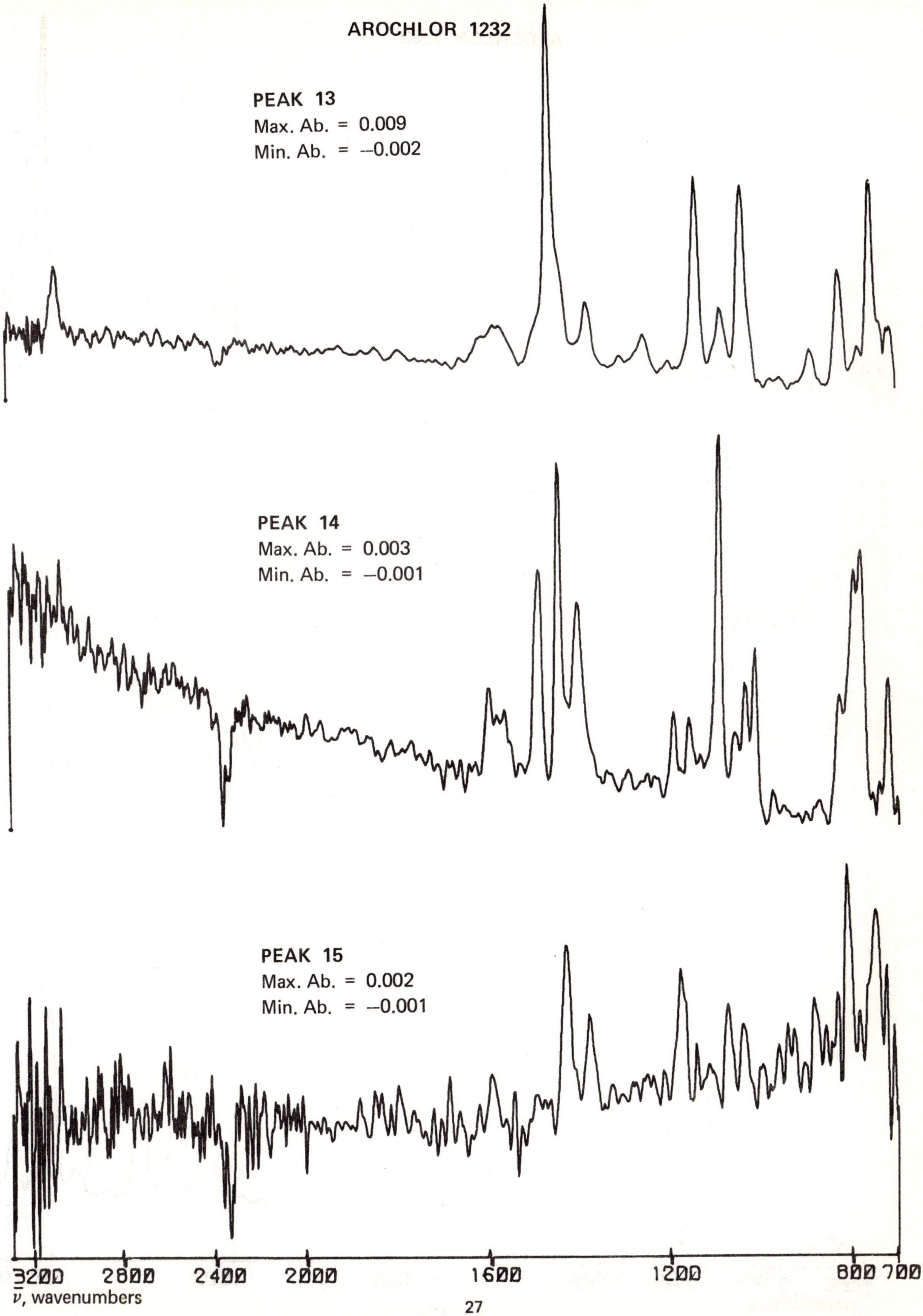

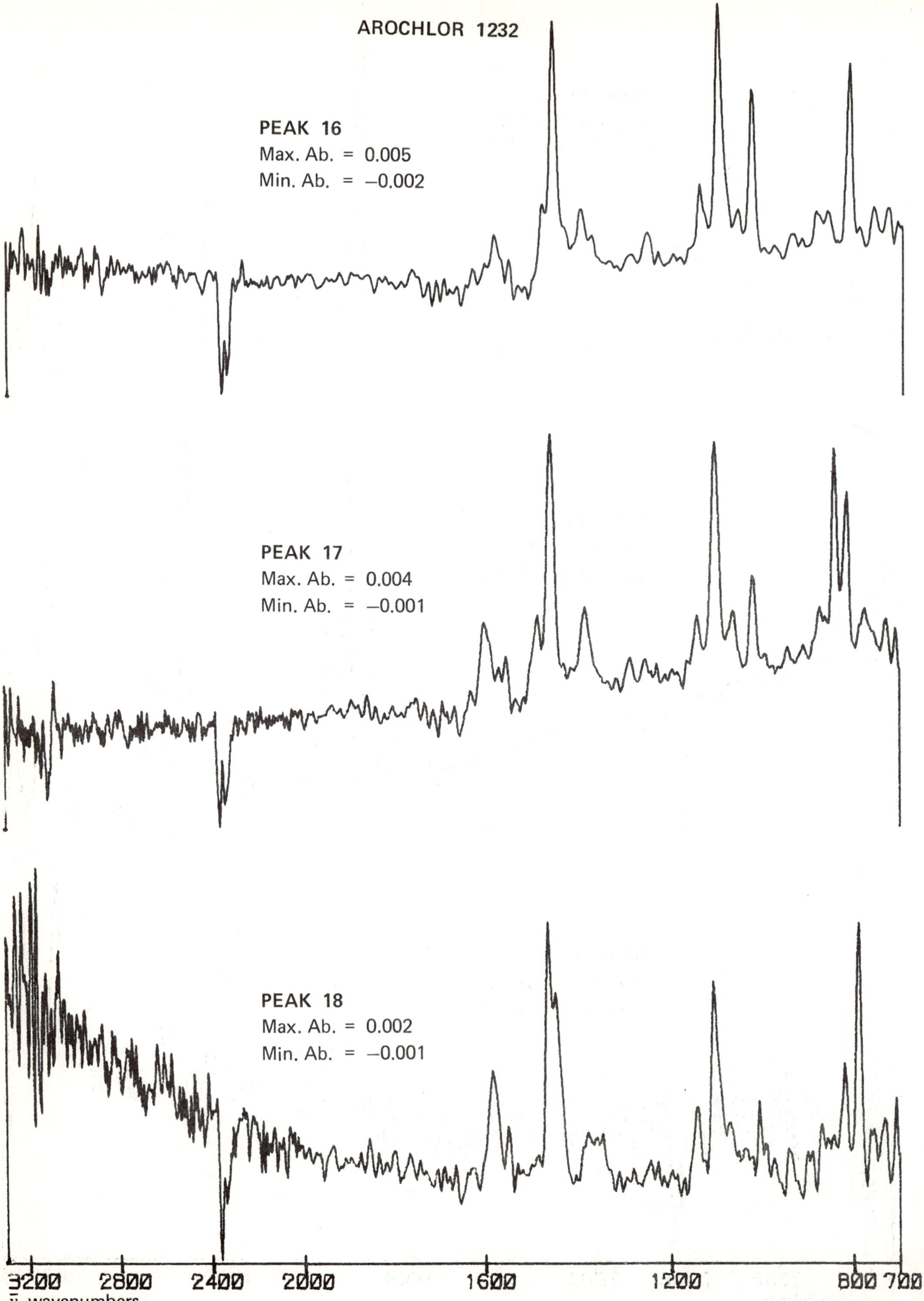
AROCHLOR 1232
PEAK 16
Max. Ab. = 0.005
Min. Ab. = –0.002
PEAK 17
Max. Ab. = 0.004
Min. Ab. = –0.001
PEAK 18
Max. Ab. = 0.002
Min. Ab. = –0.001
3200
2800
2400
2000
1600
1200
800
700
$\bar{\nu}$, wavenumbers

AROCHLOR 1232

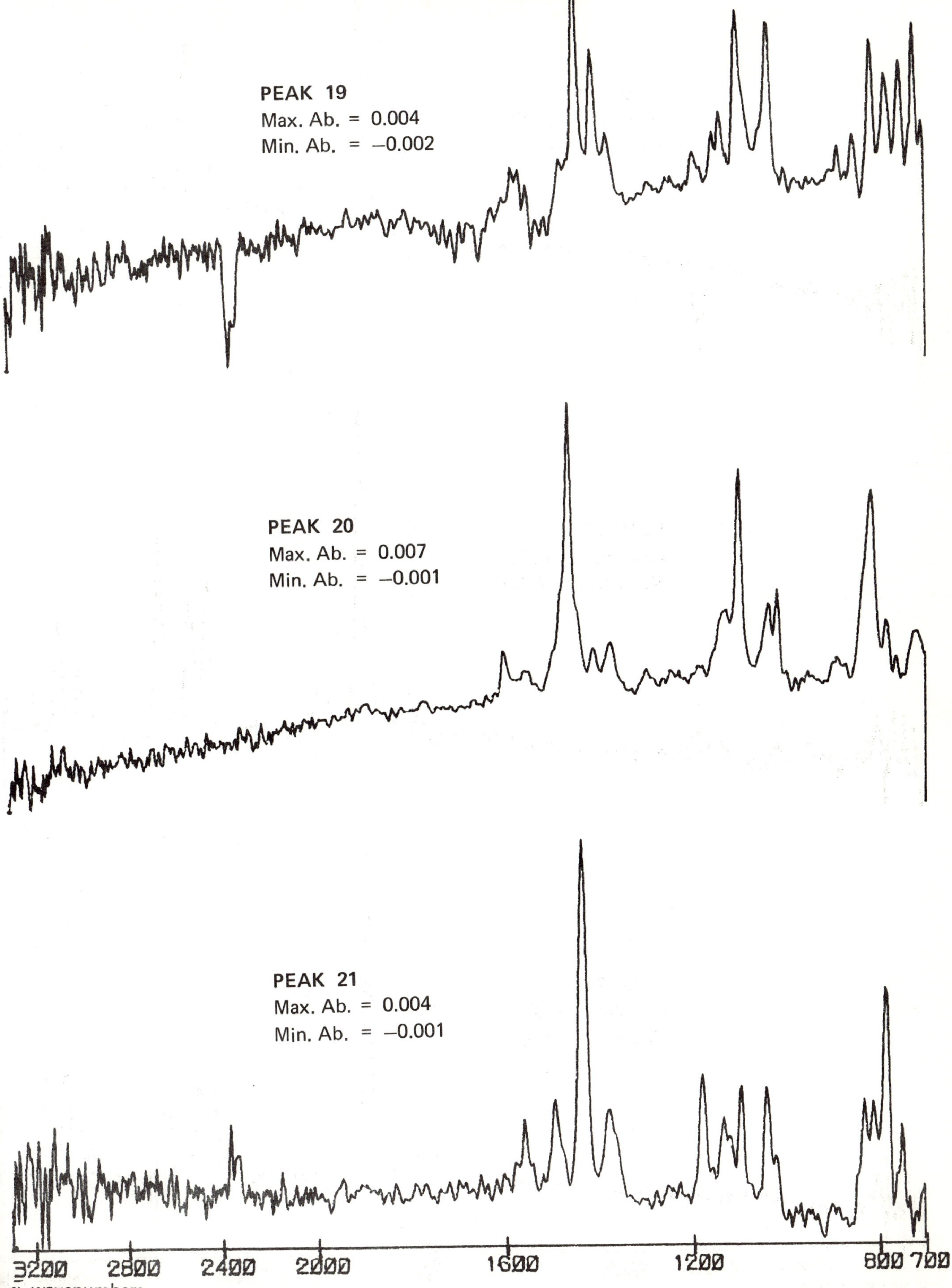

AROCHLOR 1232

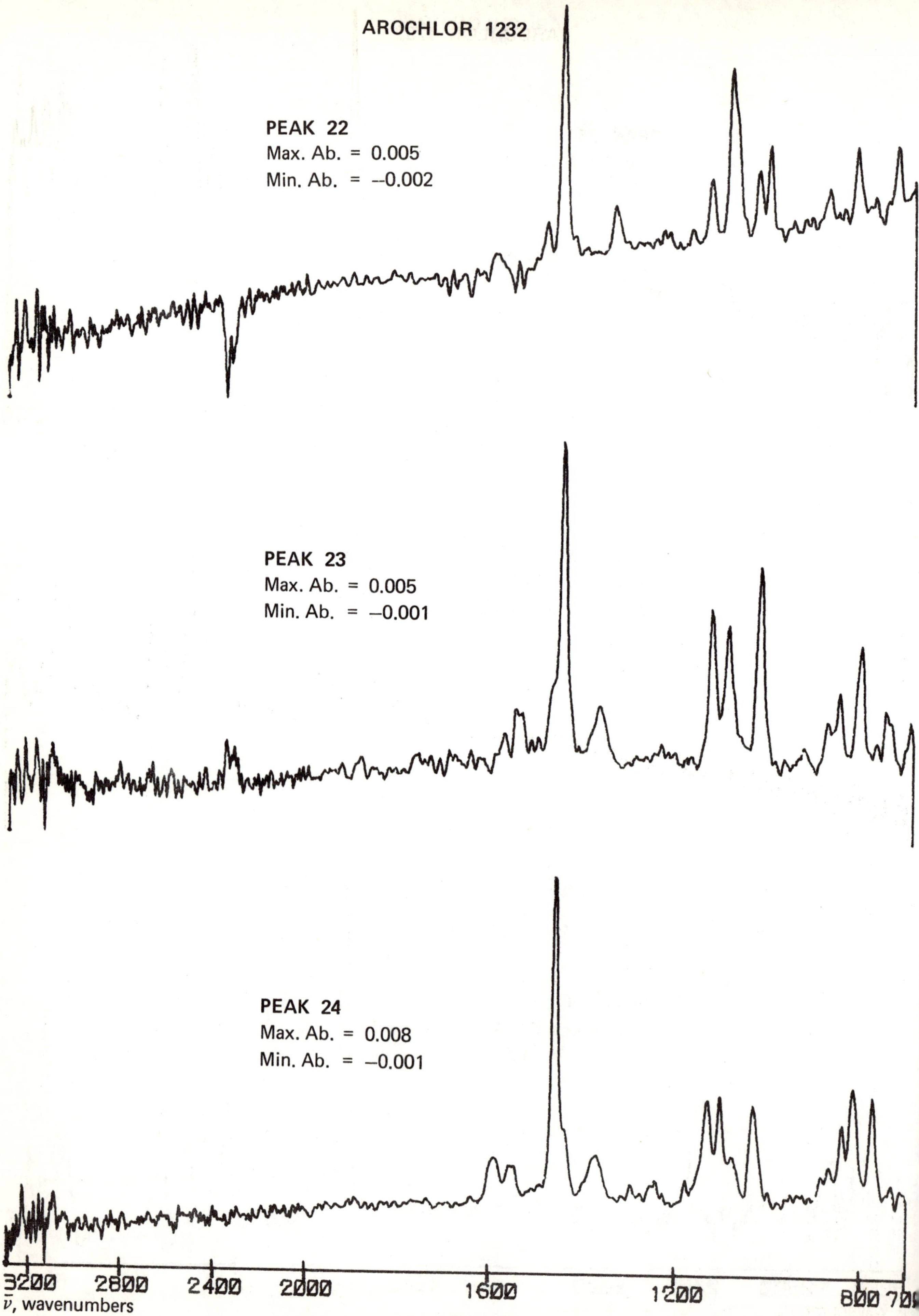

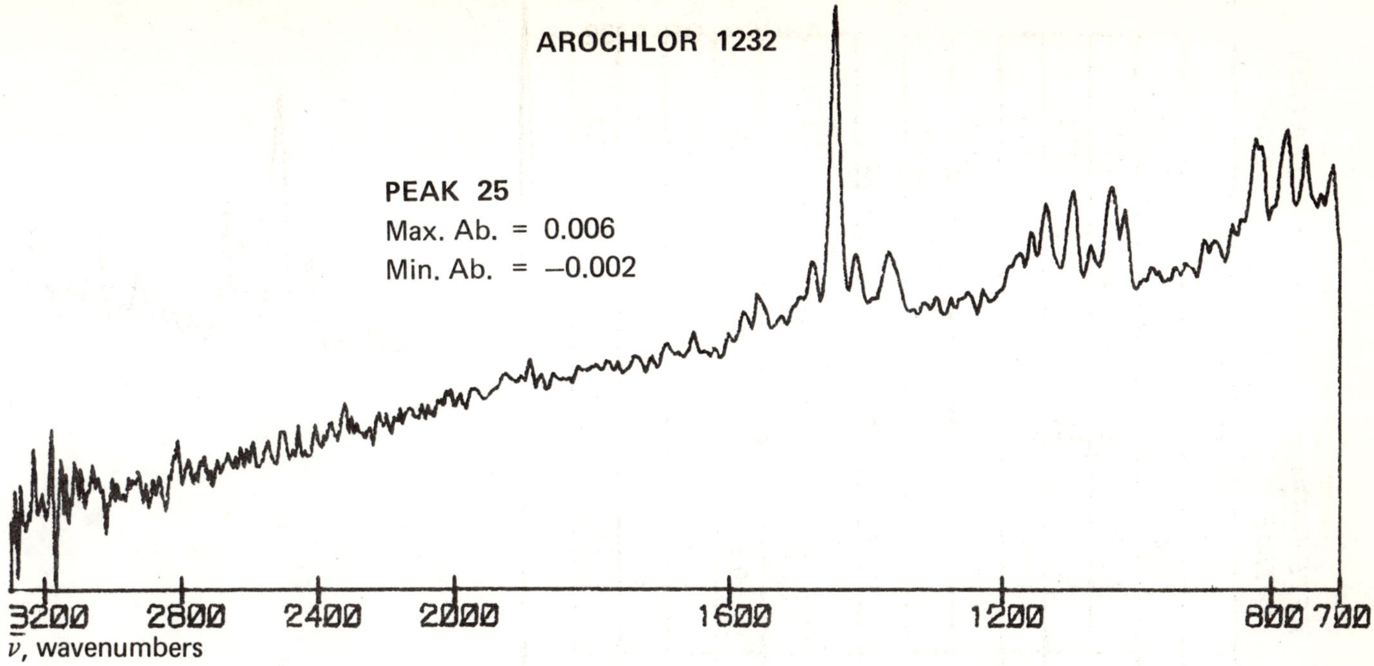
AROCHLOR 1232
PEAK 25
Max. Ab. = 0.006
Min. Ab. = –0.002
3200
2800
2400
2000
1600
1200
800
700
ν̄, wavenumbers

AROCHLOR 1242

GAS CHROMATOGRAPH DATA

TYPE = HP 5880 G.C.
MANUAL SPLITLESS INJECTION

SAMPLE PREPARATION = 50 μl of a saturated solution of the compound in diethyl ether was diluted to 1 ml.

INJECTED AMT. = 2 μl of the diluted solution

DET. TEMP. = 290°C

INJ. TEMP. = 260°C

CARRIER GAS = He

FLOW RATE = ∿ 1.5 ml./min.

PRESSURE = 60 PSI

DETECTOR = FID

COLUMN = J & W, SE–30, FUSED SILICA
30 m. LONG, 0.32 mm. DIA.

FILM THICKNESS = 0.25 microns

OVEN PROGRAM RATES 1 AND 2:

INITAL TEMP. = 35°C

INITAL TIME = 0.40 min.

LEVEL 1

PRGM RATE = 25.00°C/min.

FINAL TEMP. = 120°C

FINAL TIME = 0.00 min.

LEVEL 2

PROGRAM RATE = 4.00°C/min.

FINAL TEMP. = 210°C

FINAL TIME = 20.00 min.

IR DATA

DIGILAB FTS 20 E WITH DIGILAB GC/C$^{T.M.}$ ACCESSORY

PURGE = N_2

DETECTOR = LIQUID N_2 COOLED MCT (NARROW BAND)

OPTICAL CELL = GOLD COATED LIGHTPIPE WITH KBr WINDOWS

PEAK	*RETENTION TIME*	*PEAK*	*RETENTION TIME*
1	15.02	16	21.81
2	16.08	17	21.99
3	16.50	18	22.16
4	16.81	19	22.58
5	17.48	20	22.75
6	18.61	21	23.06
7	18.70	22	23.27
8	18.92	23	24.03
9	19.29	24	24.18
10	20.07	25	24.32
11	20.52	26	24.91
12	20.90	27	25.31
13	21.10	28	26.90
14	21.23	29	28.39
15	21.49		

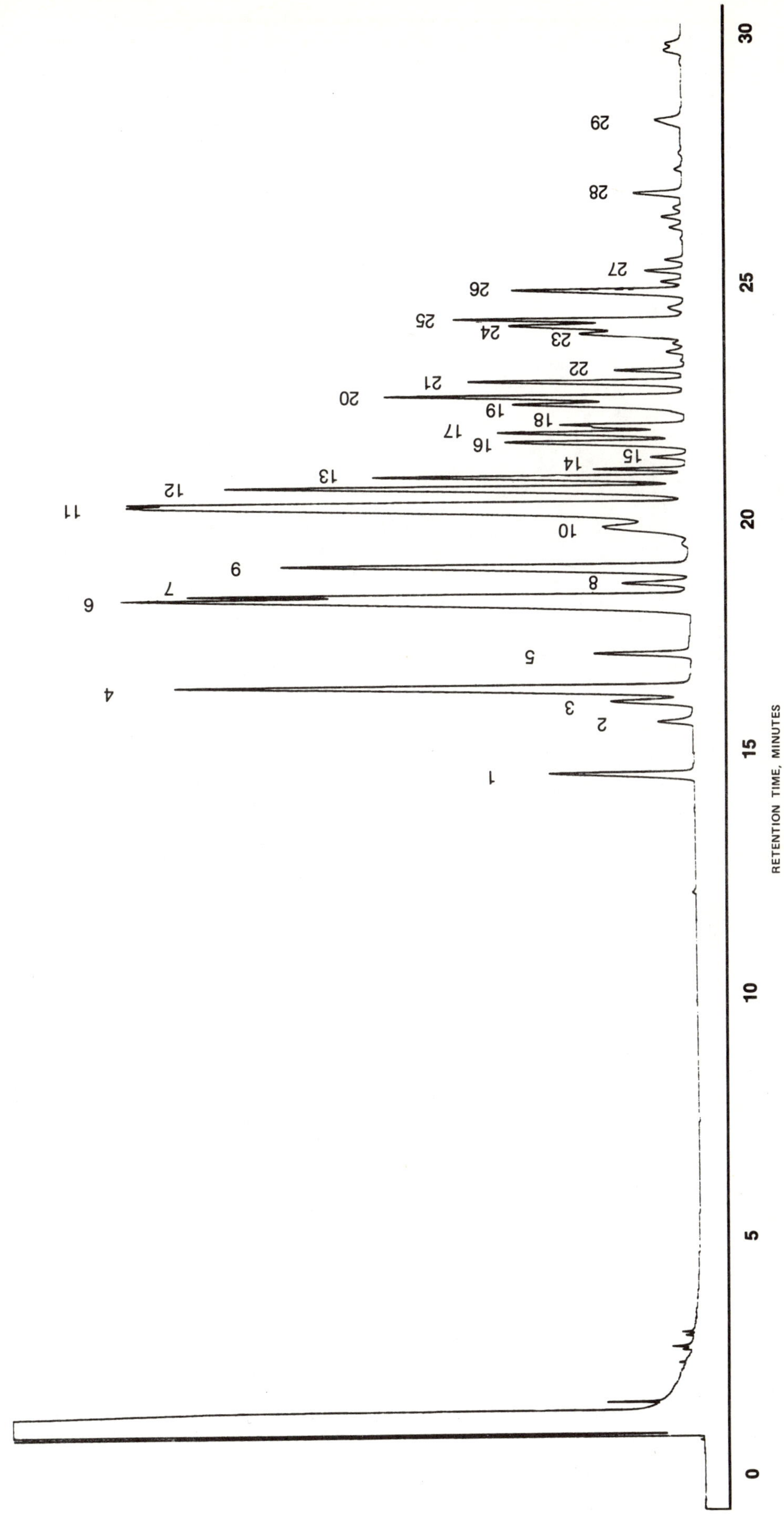

AROCHLOR 1242

AROCHLOR 1242

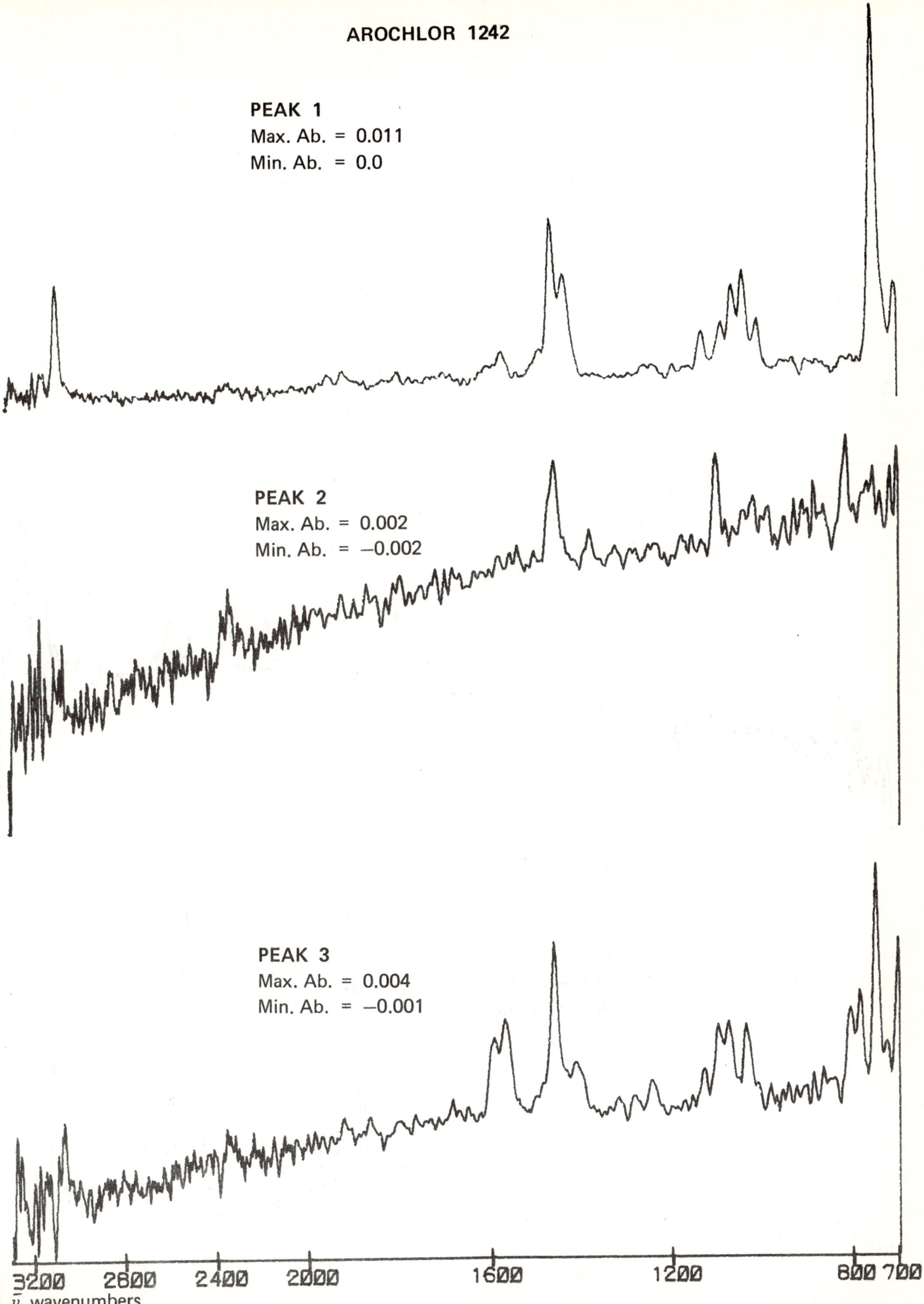

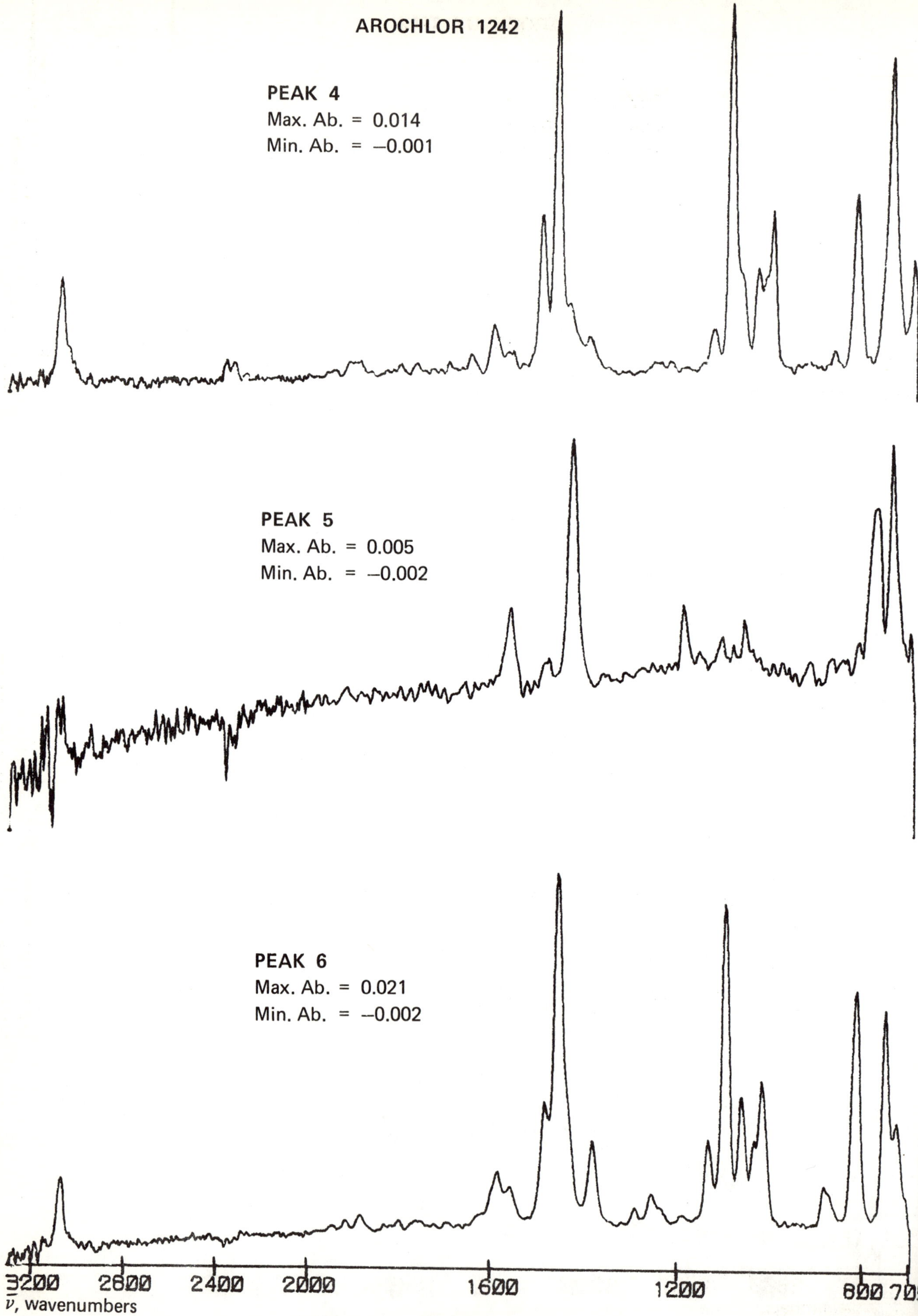
AROCHLOR 1242
PEAK 4
Max. Ab. = 0.014
Min. Ab. = –0.001
PEAK 5
Max. Ab. = 0.005
Min. Ab. = –0.002
PEAK 6
Max. Ab. = 0.021
Min. Ab. = –0.002
3200
2800
2400
2000
1600
1200
800
ν̄, wavenumbers

AROCHLOR 1242

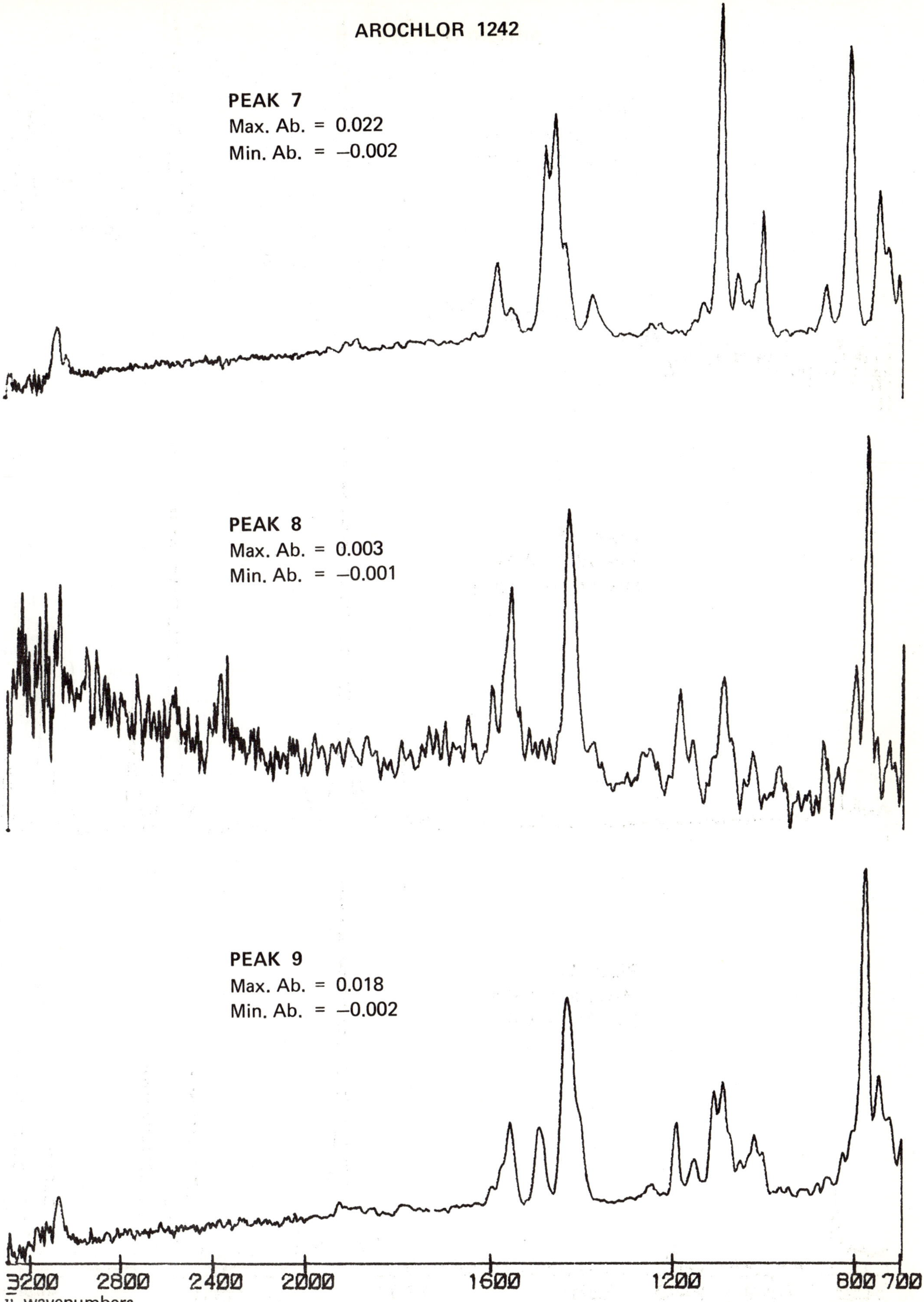

AROCHLOR 1242

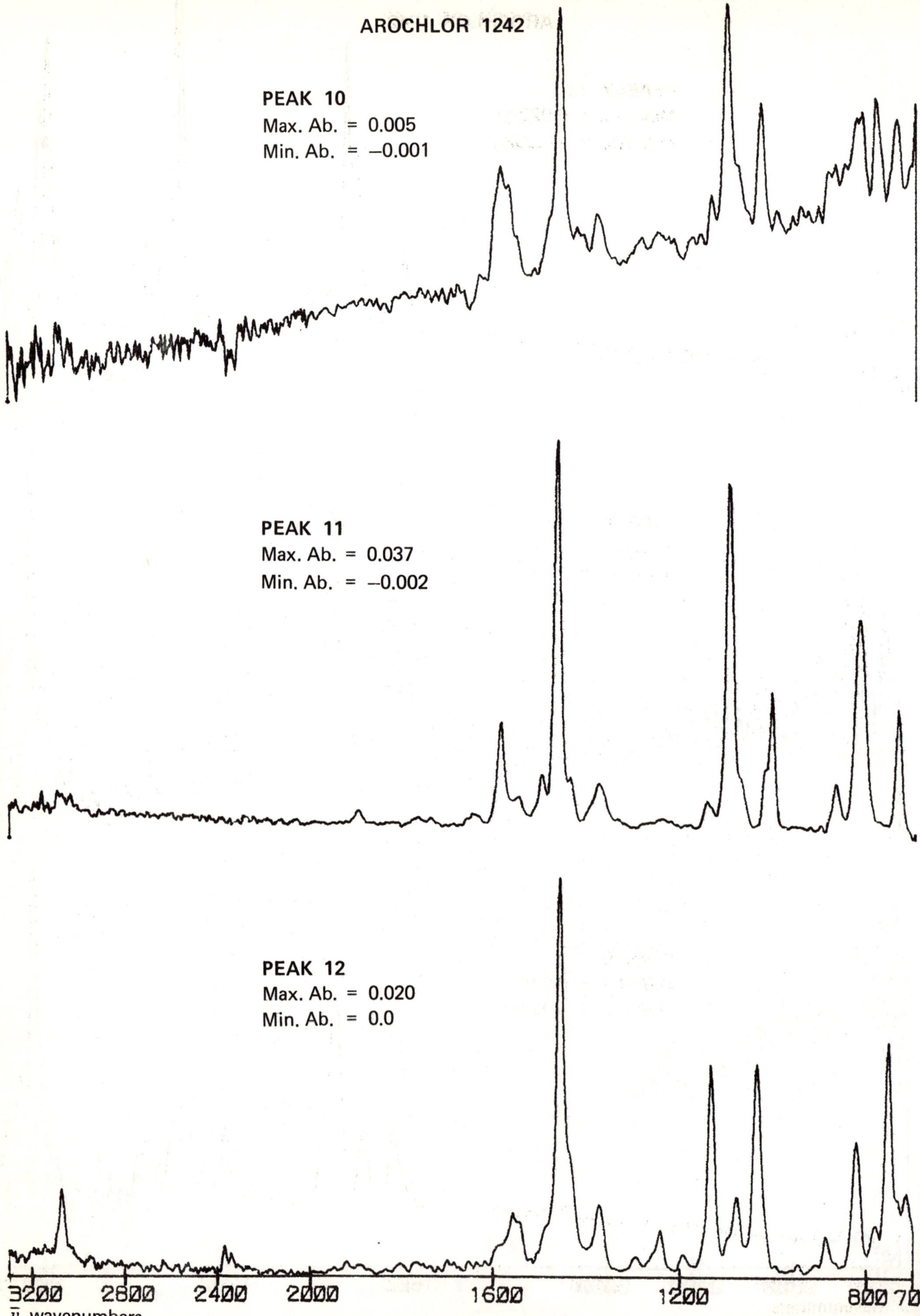

AROCHLOR 1242

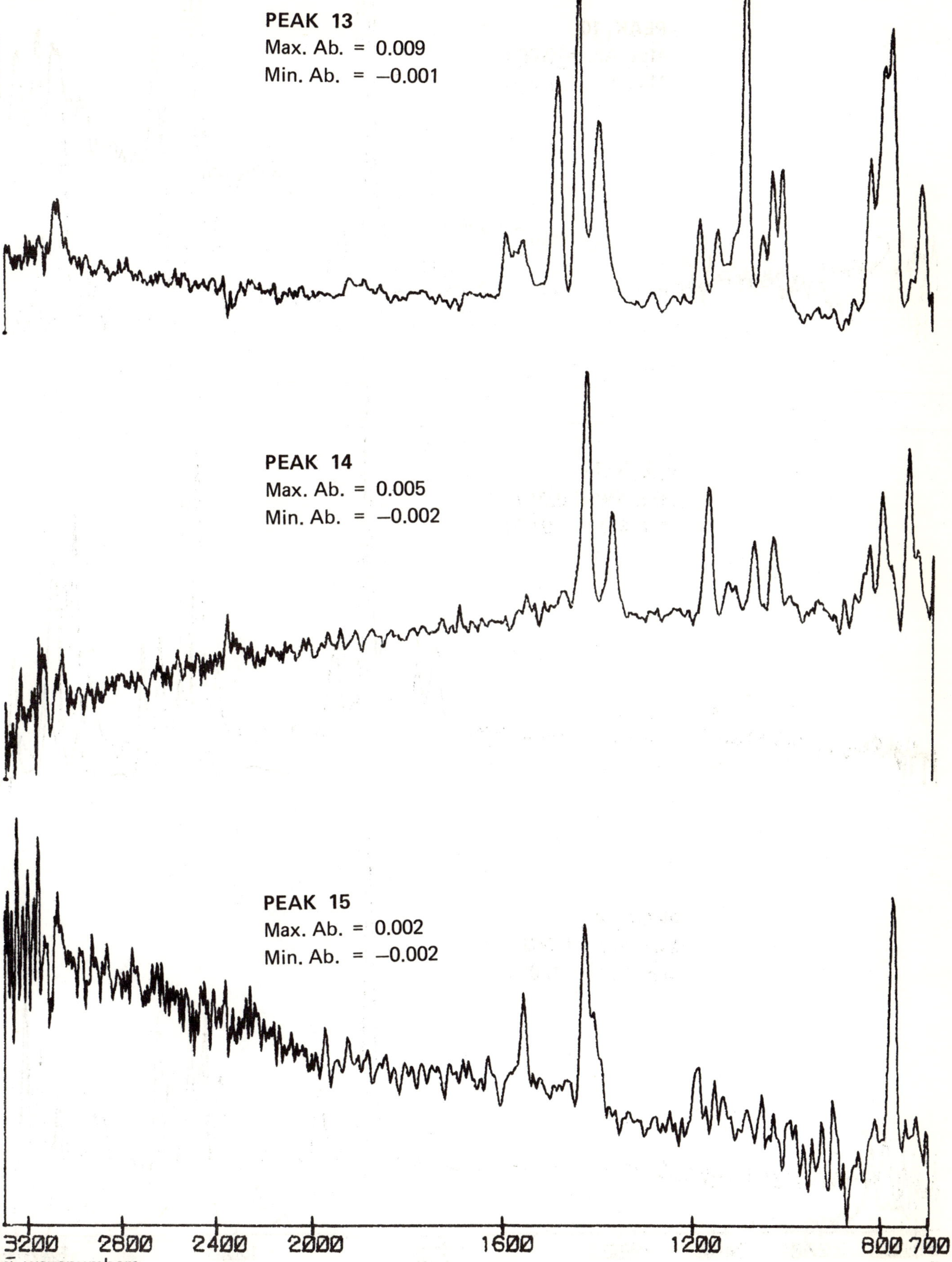

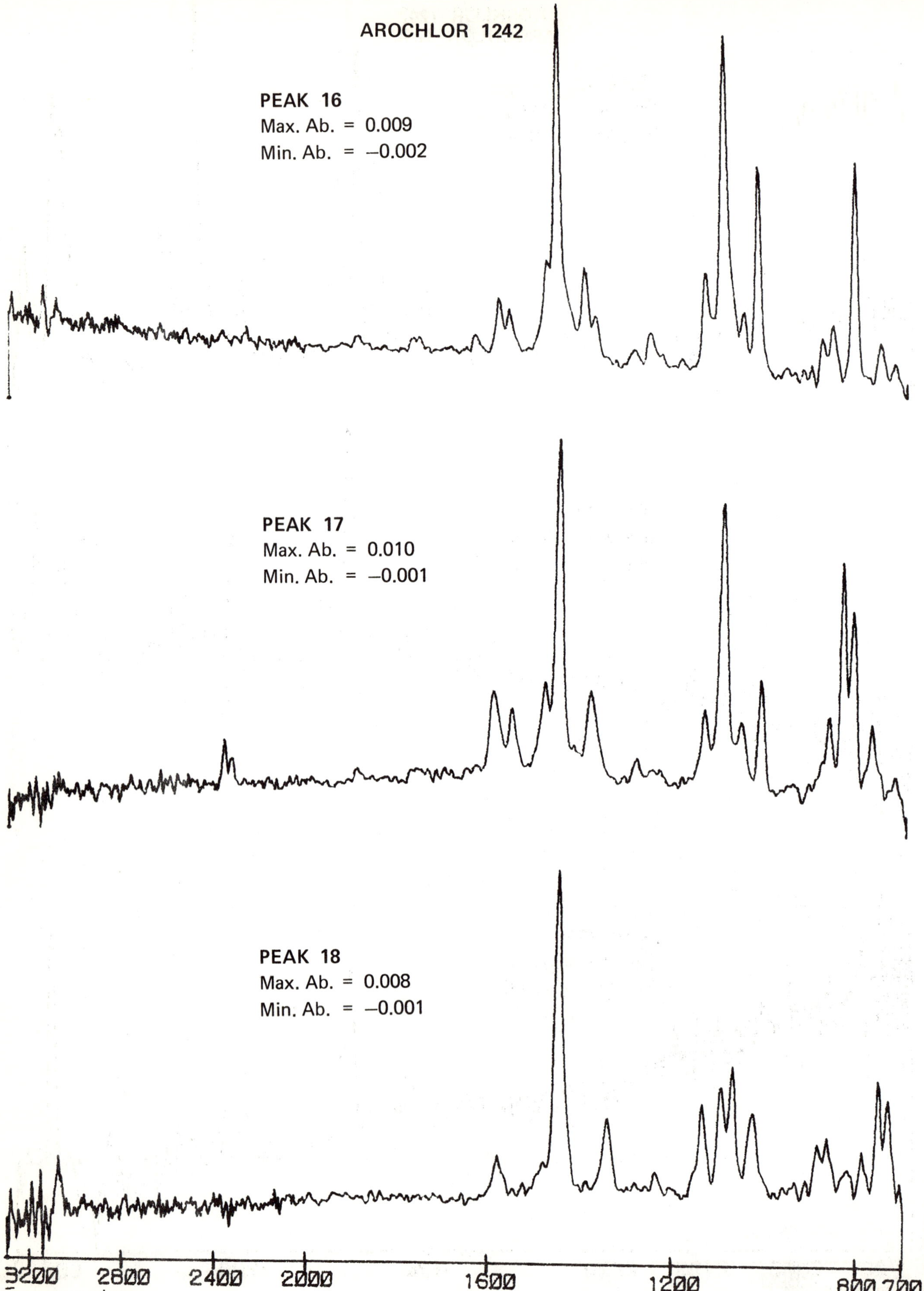
AROCHLOR 1242
PEAK 16
Max. Ab. = 0.009
Min. Ab. = –0.002
PEAK 17
Max. Ab. = 0.010
Min. Ab. = –0.001
PEAK 18
Max. Ab. = 0.008
Min. Ab. = –0.001
3200
2800
2400
2000
1600
1200
800
700
$\bar{\nu}$, wavenumbers

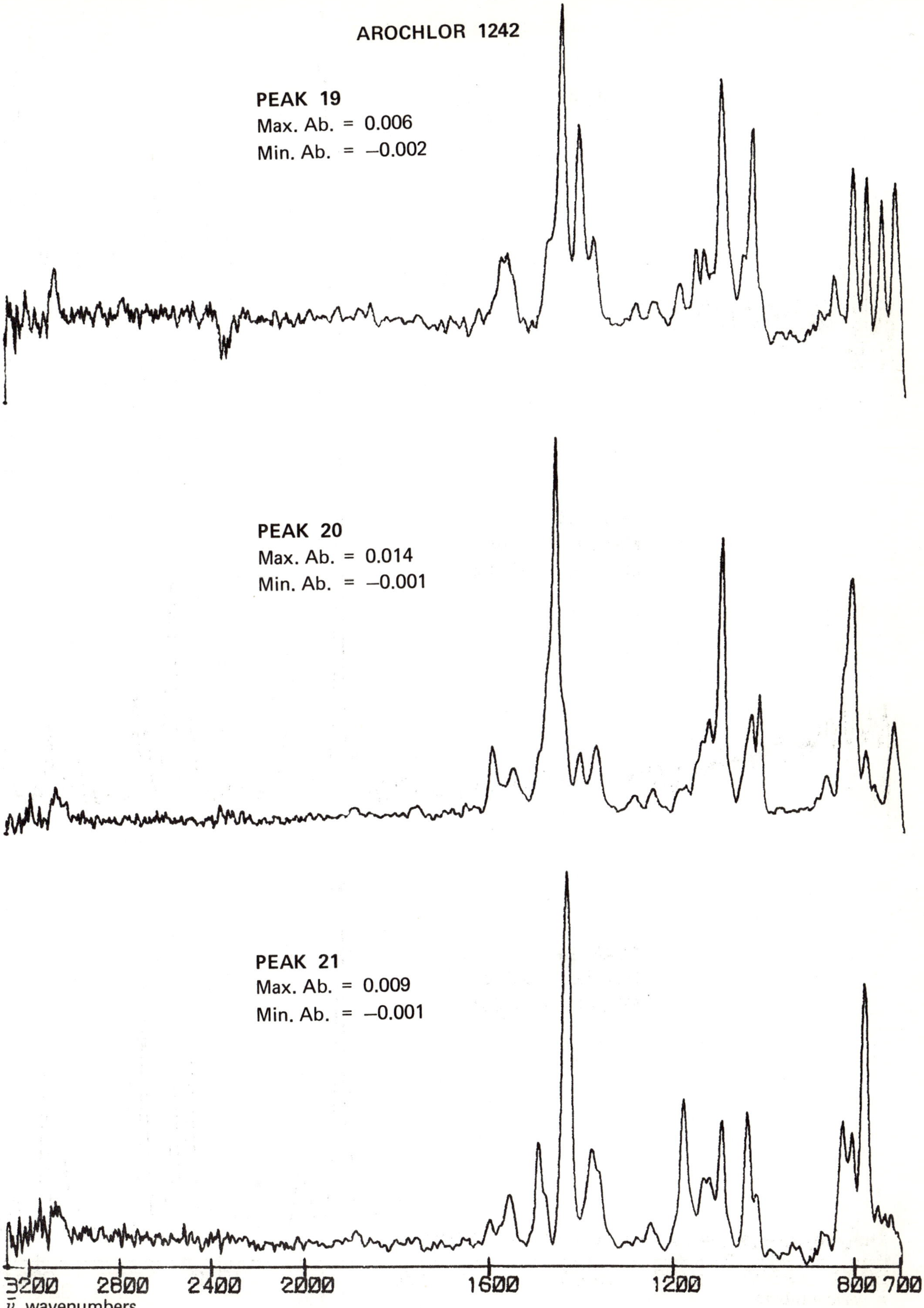
AROCHLOR 1242
PEAK 19
Max. Ab. = 0.006
Min. Ab. = –0.002
PEAK 20
Max. Ab. = 0.014
Min. Ab. = –0.001
PEAK 21
Max. Ab. = 0.009
Min. Ab. = –0.001
3200
2800
2400
2000
1600
1200
800
700
$\bar{\nu}$, wavenumbers

AROCHLOR 1242

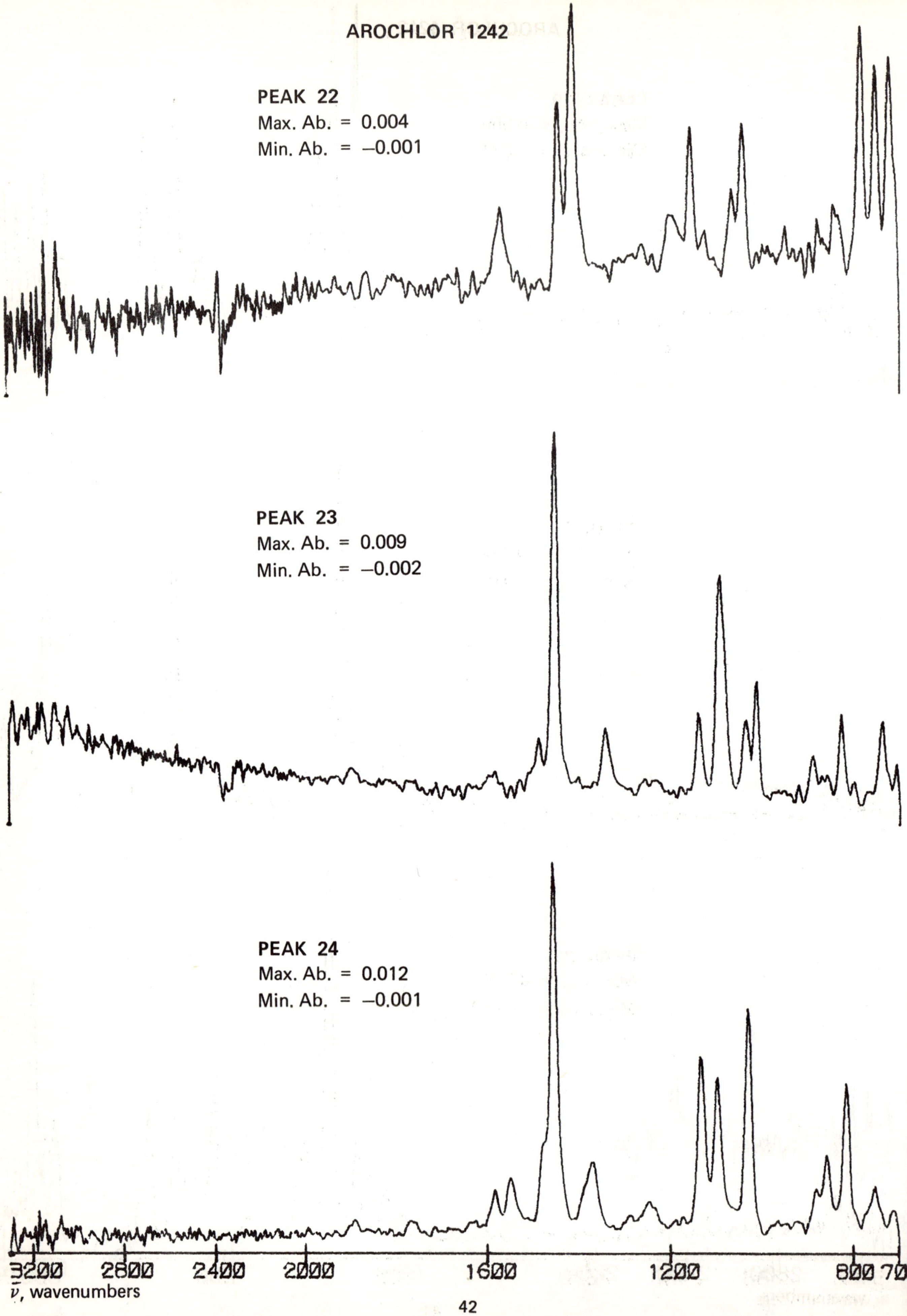

AROCHLOR 1242

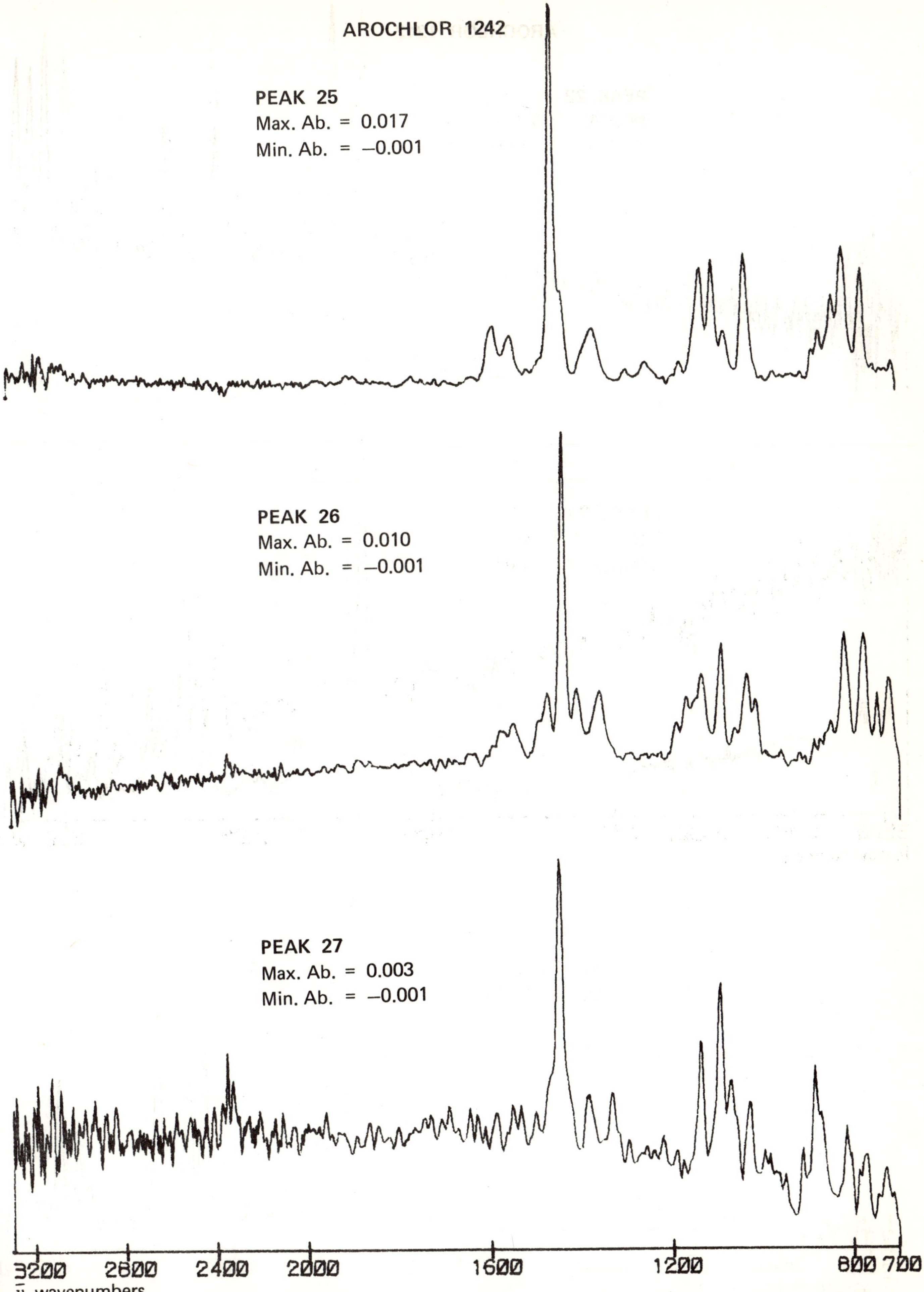

AROCHLOR 1242

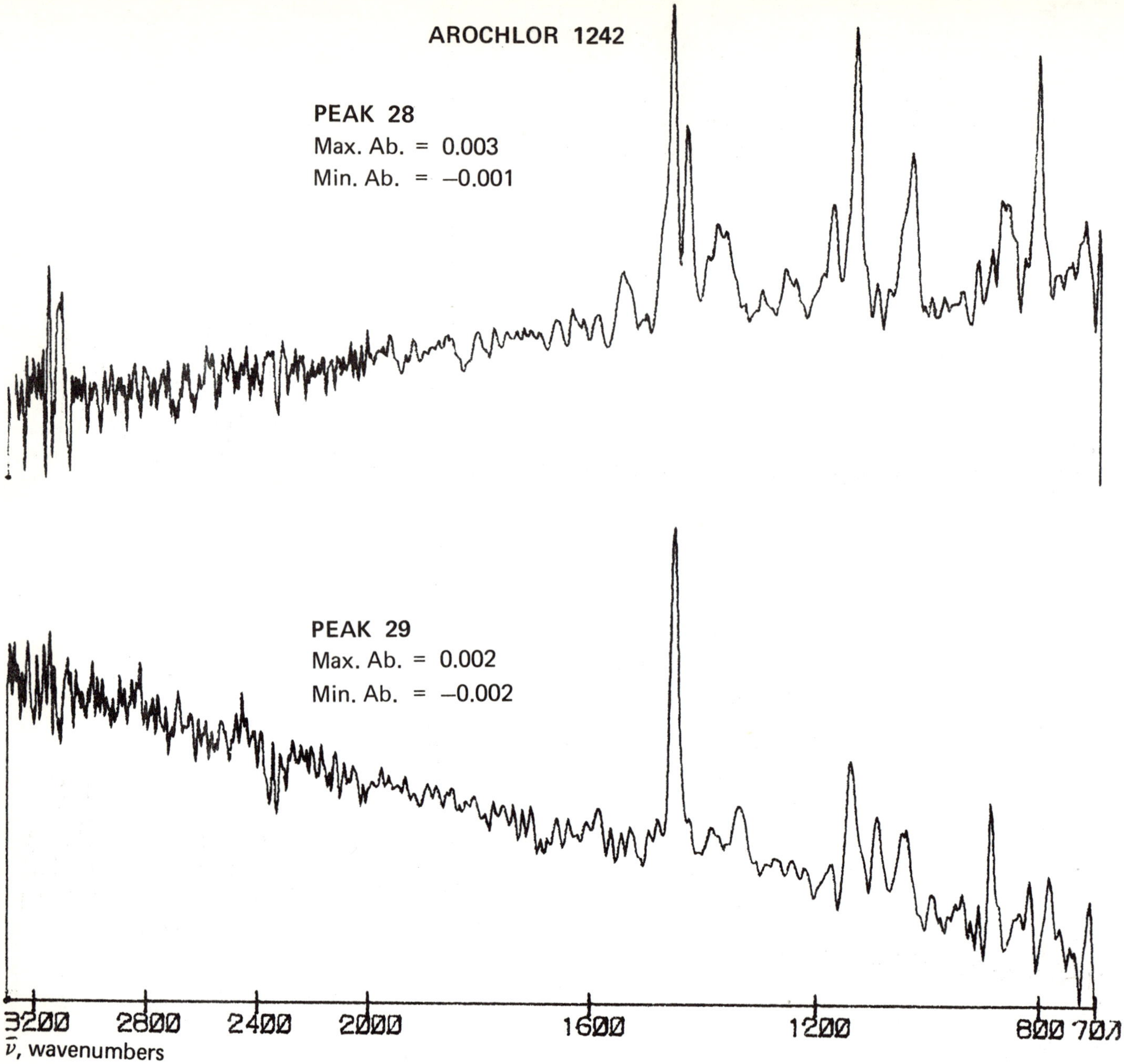

AROCHLOR 1248

GAS CHROMATOGRAPH DATA

TYPE = HP 5880 G.C.
MANUAL SPLITLESS INJECTION

SAMPLE PREPARATION = **50 μl** of a saturated solution of the compound in diethyl ether was diluted to 1 ml.

INJECTED AMT. = 2 μl of the diluted solution

DET. TEMP. = 290°C

INJ. TEMP. = 260°C

CARRIER GAS = He

FLOW RATE = ∿ 1.5 ml./min.

PRESSURE = 60 PSI

DETECTOR = FID

COLUMN = J & W, SE–30, FUSED SILICA
30 m. LONG, 0.32 mm. DIA.

FILM THICKNESS = 0.25 microns

OVEN PROGRAM RATES 1 AND 2:

INITAL TEMP. = 35°C

INITAL TIME = 0.40 min.

LEVEL 1

PRGM RATE = 25.00°C/min.

FINAL TEMP. = 120°C

FINAL TIME = 0.00 min.

LEVEL 2

PROGRAM RATE = 4.00°C/min.

FINAL TEMP. = 210°C

FINAL TIME = 20.00 min.

PEAK	RETENTION TIME	PEAK	RETENTION TIME
1	16.74	16	24.35
2	18.55	17	24.49
3	19.25	18	24.62
4	20.05	19	25.05
5	20.48	20	25.18
6	20.86	21	25.43
7	21.26	22	25.64
8	21.91	23	26.29
9	22.09	24	26.51
10	22.24	25	26.70
11	22.67	26	27.04
12	22.88	27	27.47
13	23.15	28	28.54
14	23.35	29	29.93
15	24.19		

IR DATA

DIGILAB FTS 20 E WITH DIGILAB GC/C[T.M.] ACCESSORY

PURGE = N_2

DETECTOR = LIQUID N_2 COOLED MCT (NARROW BAND)

OPTICAL CELL = GOLD COATED LIGHTPIPE WITH KBr WINDOWS

LIGHTPIPE TEMP. = 270°C

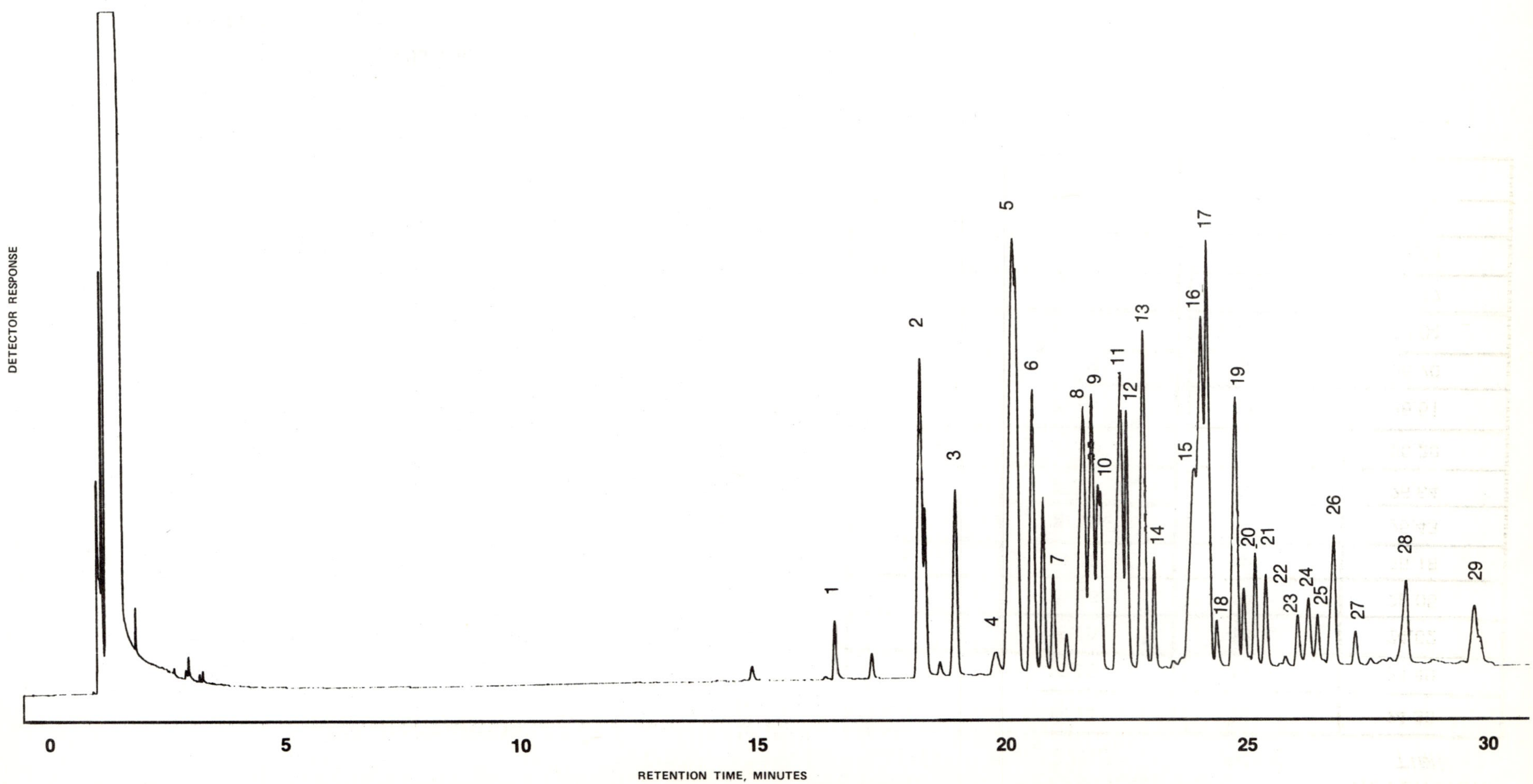

AROCHLOR 1248

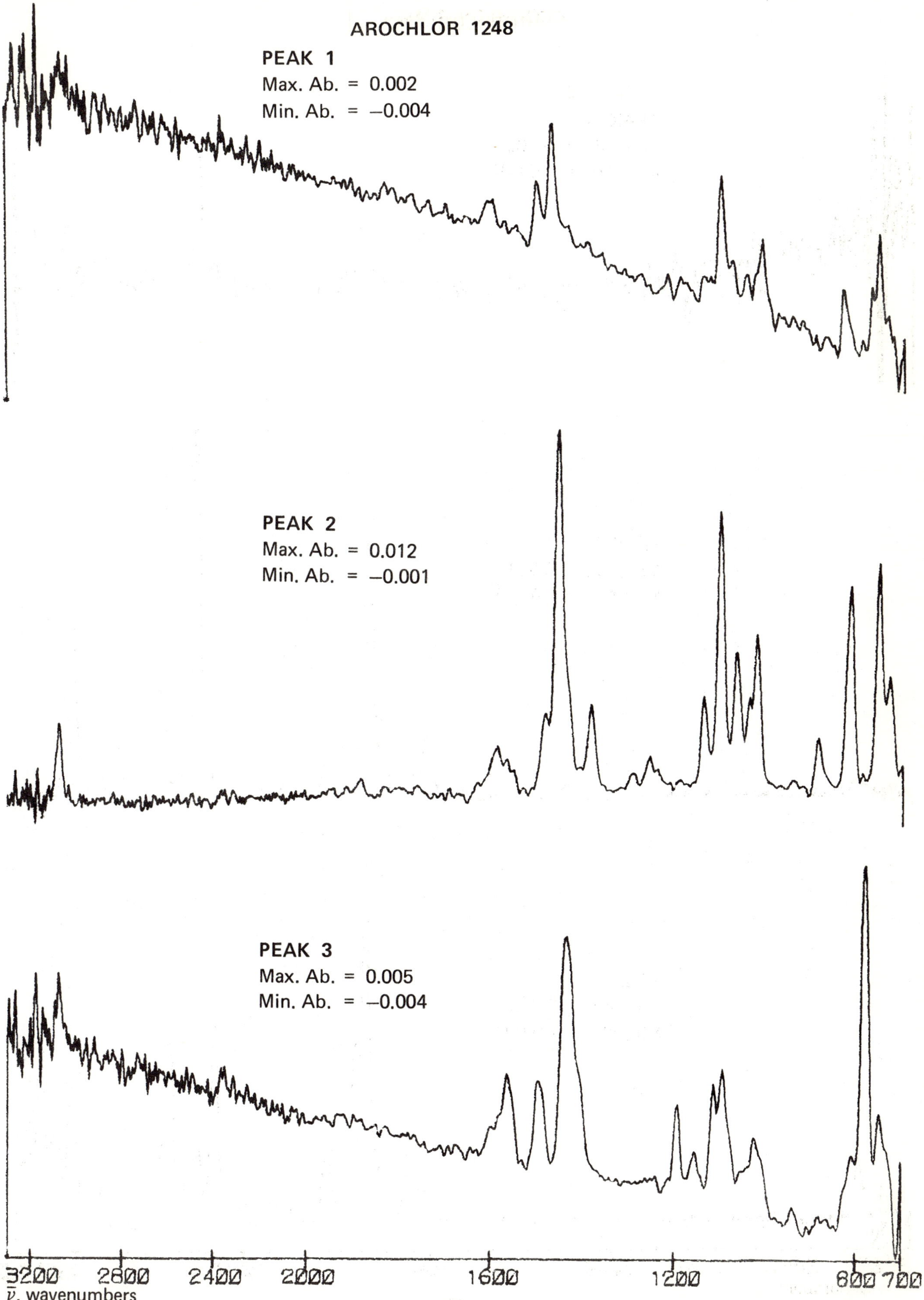
AROCHLOR 1248
PEAK 1
Max. Ab. = 0.002
Min. Ab. = –0.004
PEAK 2
Max. Ab. = 0.012
Min. Ab. = –0.001
PEAK 3
Max. Ab. = 0.005
Min. Ab. = –0.004
3200 2800 2400 2000 1600 1200 800 700
$\bar{\nu}$, wavenumbers

AROCHLOR 1248

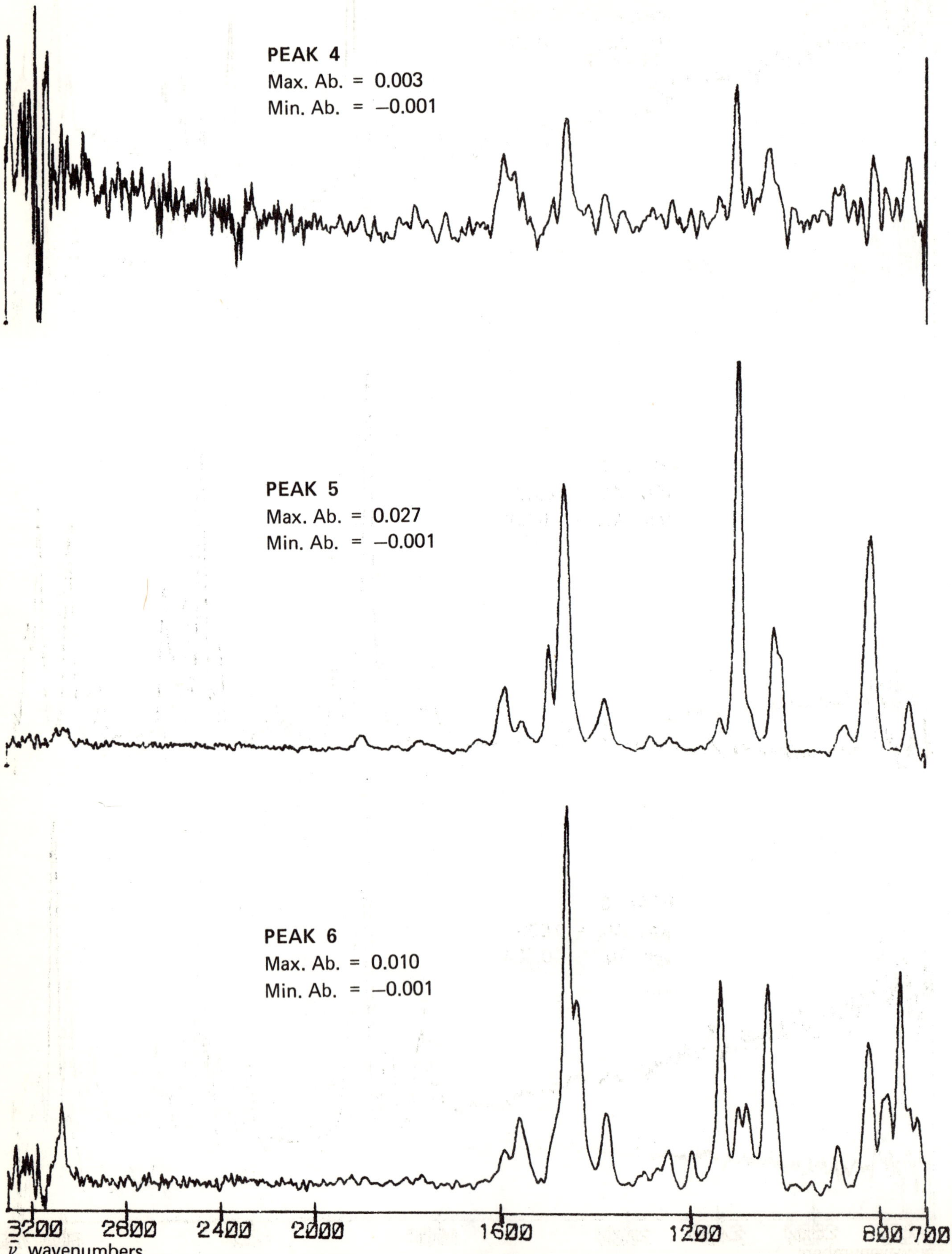

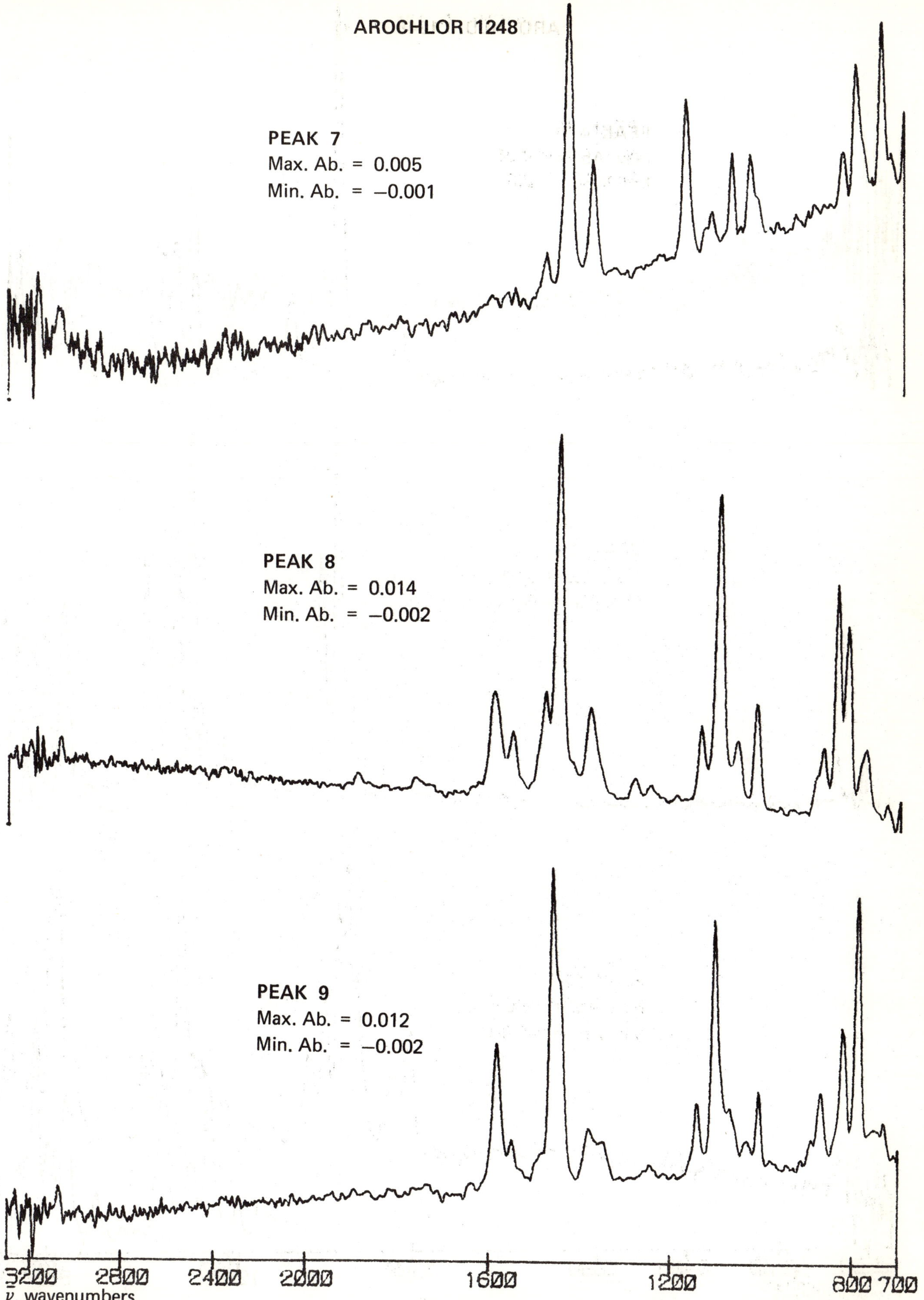
AROCHLOR 1248
PEAK 7
Max. Ab. = 0.005
Min. Ab. = –0.001
PEAK 8
Max. Ab. = 0.014
Min. Ab. = –0.002
PEAK 9
Max. Ab. = 0.012
Min. Ab. = –0.002
3200
2800
2400
2000
1600
1200
800
700
ν, wavenumbers

AROCHLOR 1248

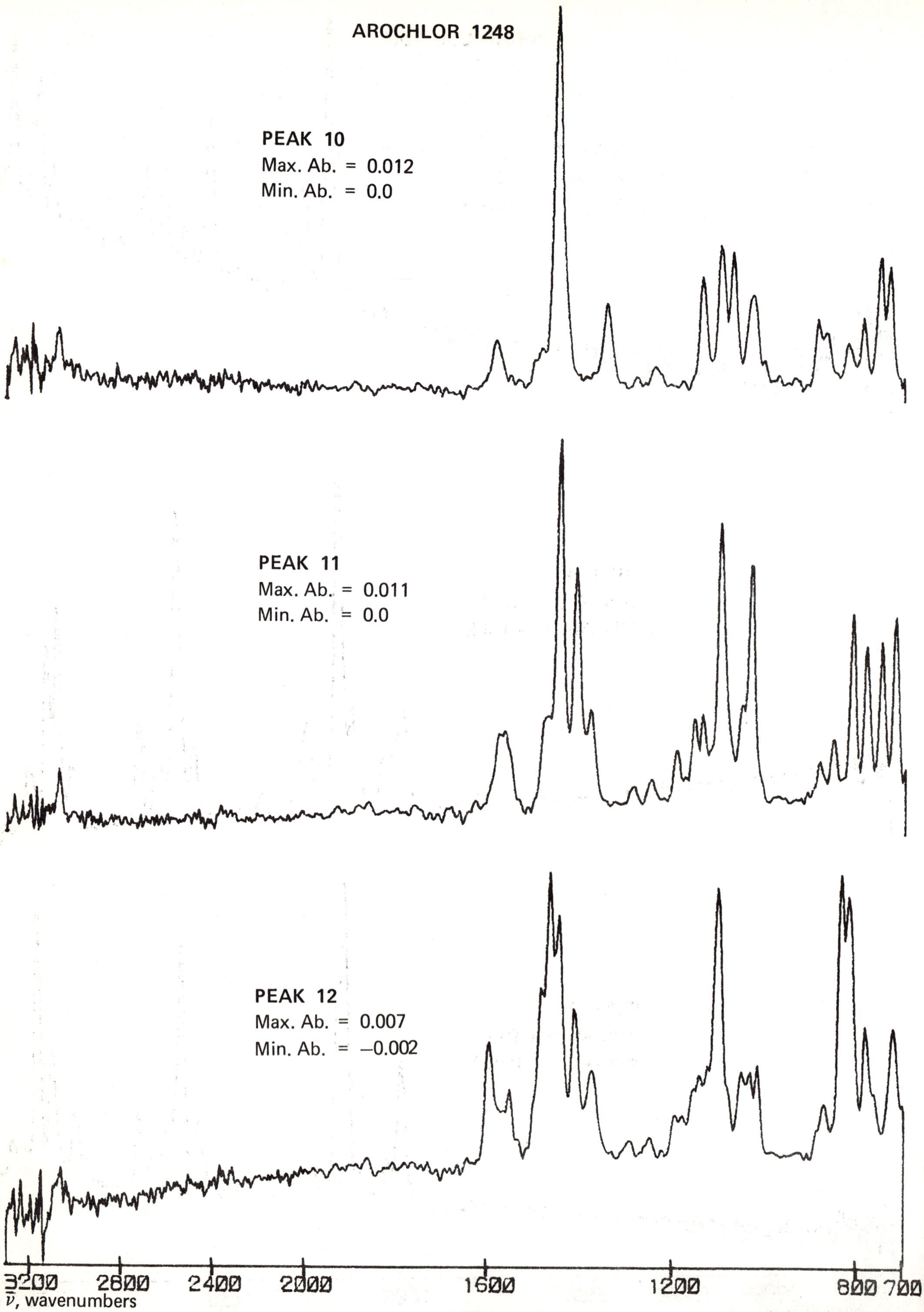

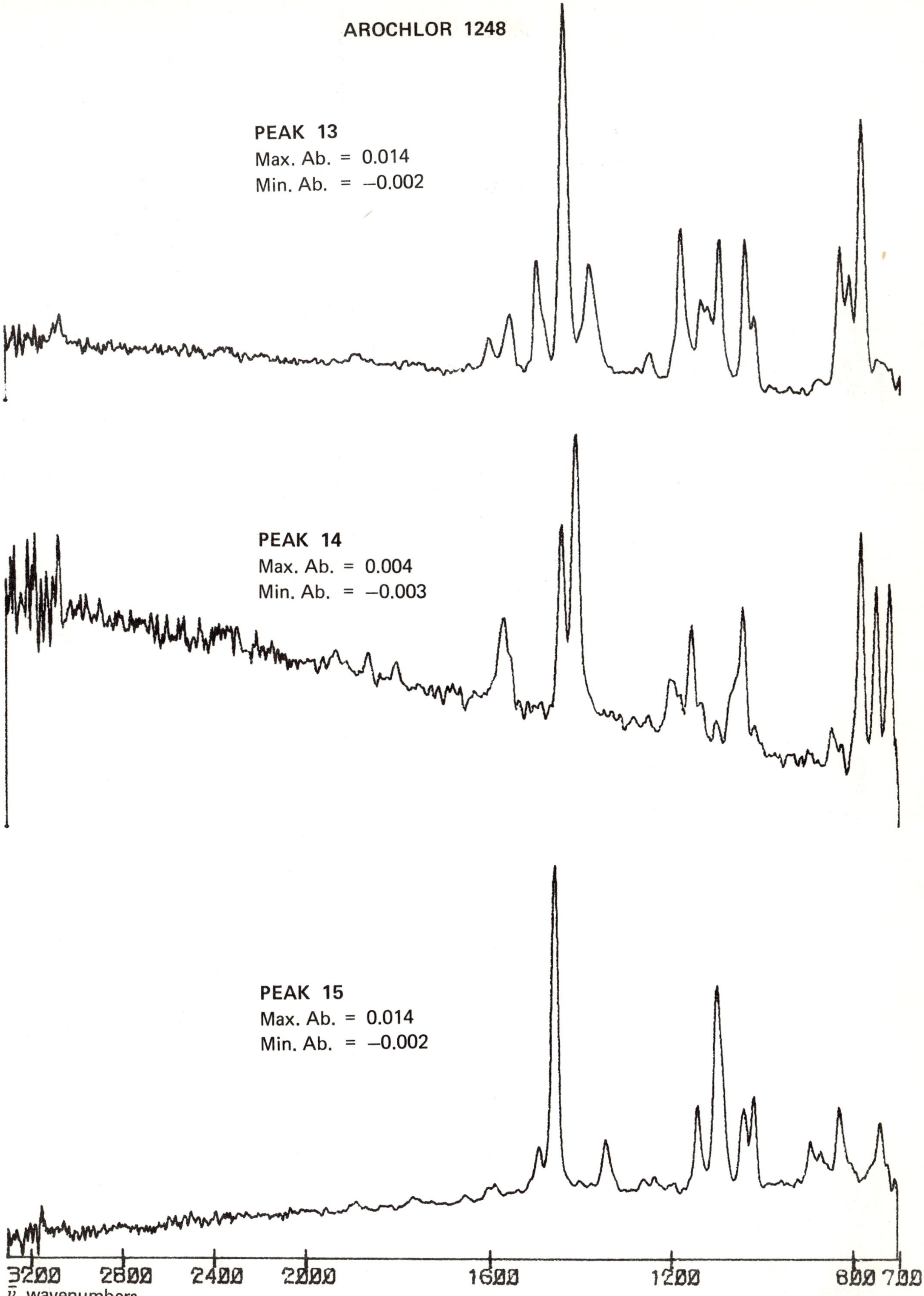
AROCHLOR 1248
PEAK 13
Max. Ab. = 0.014
Min. Ab. = –0.002
PEAK 14
Max. Ab. = 0.004
Min. Ab. = –0.003
PEAK 15
Max. Ab. = 0.014
Min. Ab. = –0.002
3200
2800
2400
2000
1600
1200
800
700
ν̄, wavenumbers

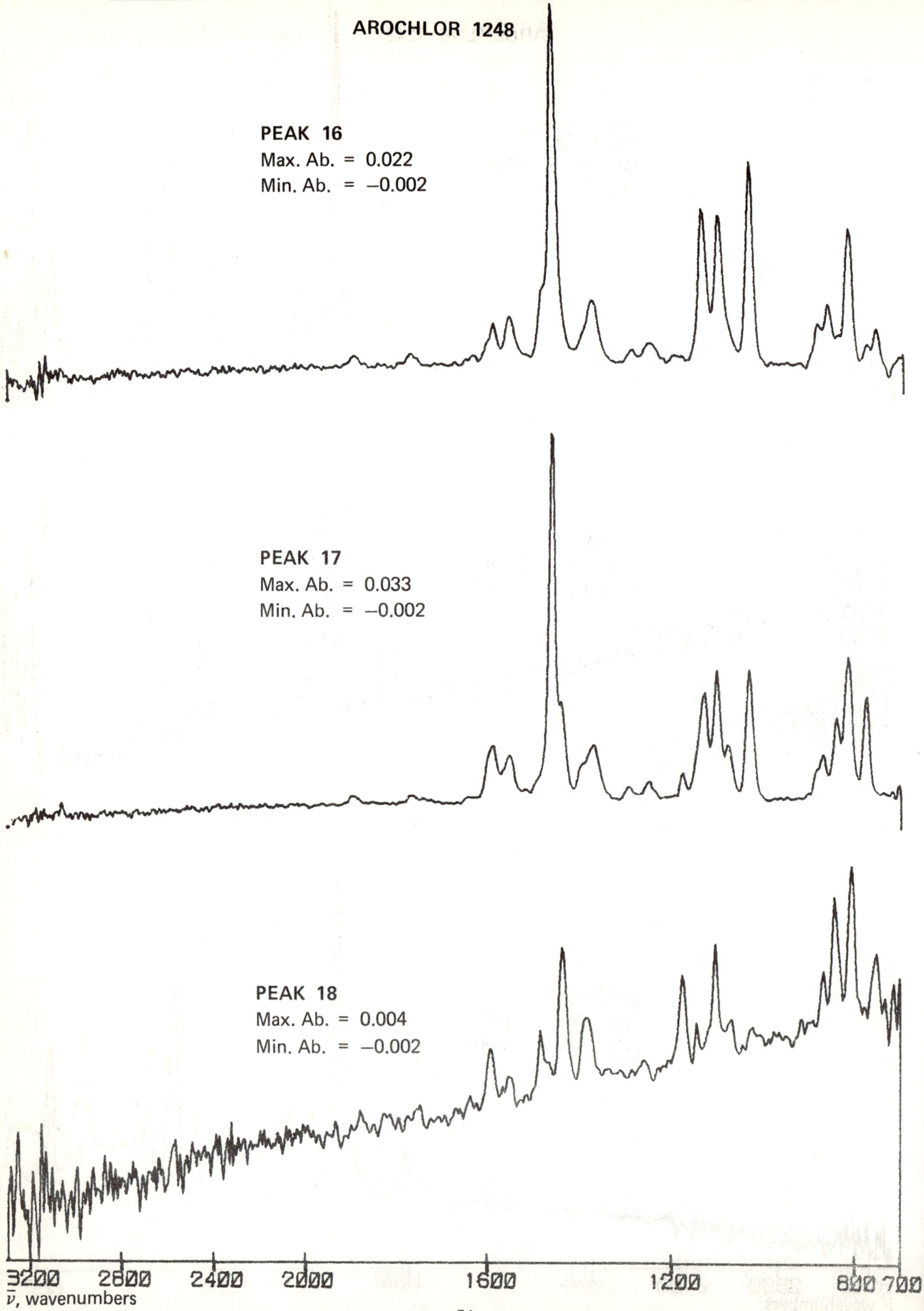
AROCHLOR 1248
PEAK 16
Max. Ab. = 0.022
Min. Ab. = –0.002
PEAK 17
Max. Ab. = 0.033
Min. Ab. = –0.002
PEAK 18
Max. Ab. = 0.004
Min. Ab. = –0.002
3200
2800
2400
2000
1600
1200
800 700
$\bar{\nu}$, wavenumbers

AROCHLOR 1248

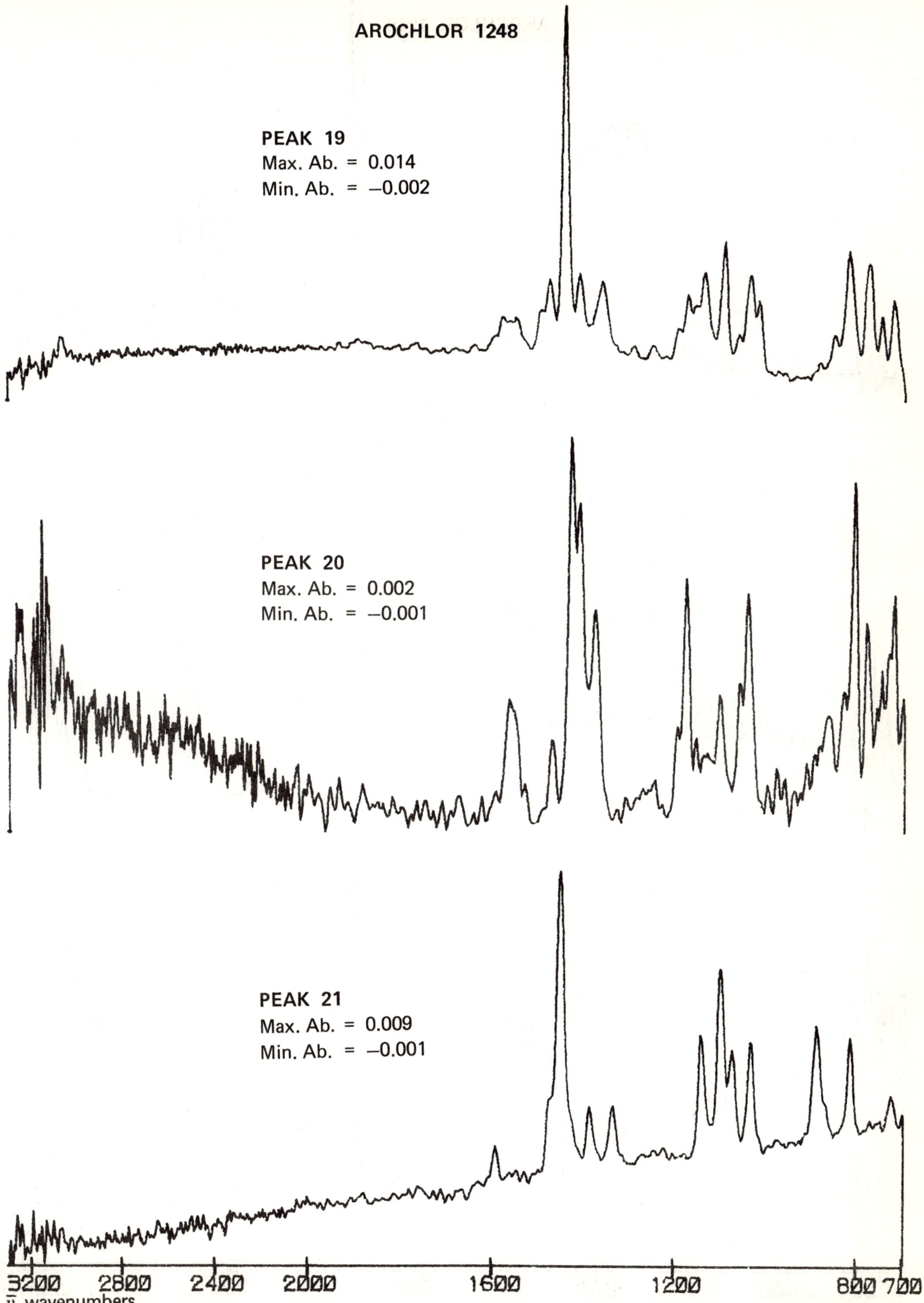

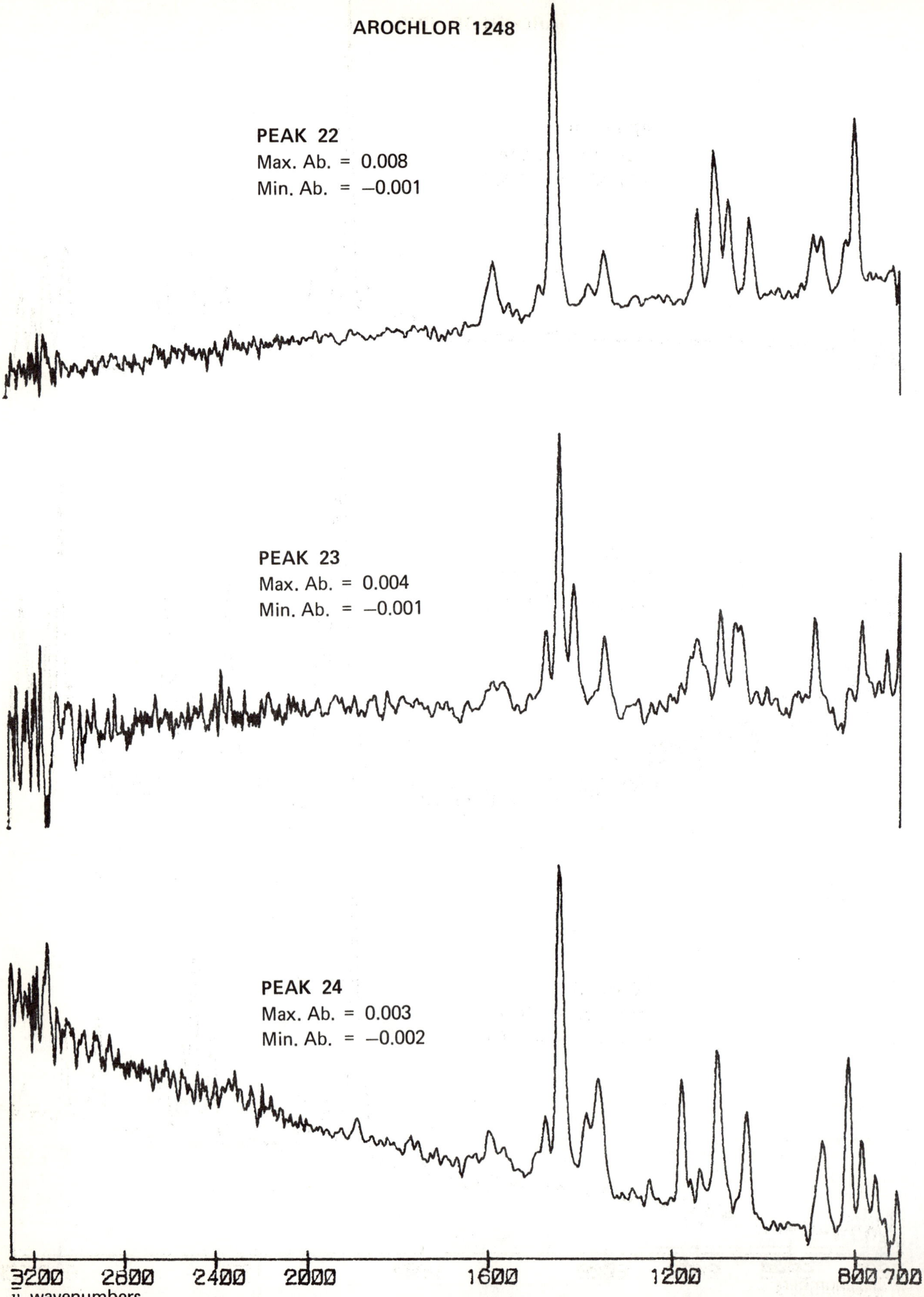
AROCHLOR 1248
PEAK 22
Max. Ab. = 0.008
Min. Ab. = –0.001
PEAK 23
Max. Ab. = 0.004
Min. Ab. = –0.001
PEAK 24
Max. Ab. = 0.003
Min. Ab. = –0.002
3200
2800
2400
2000
1600
1200
800
700
$\bar{\nu}$, wavenumbers

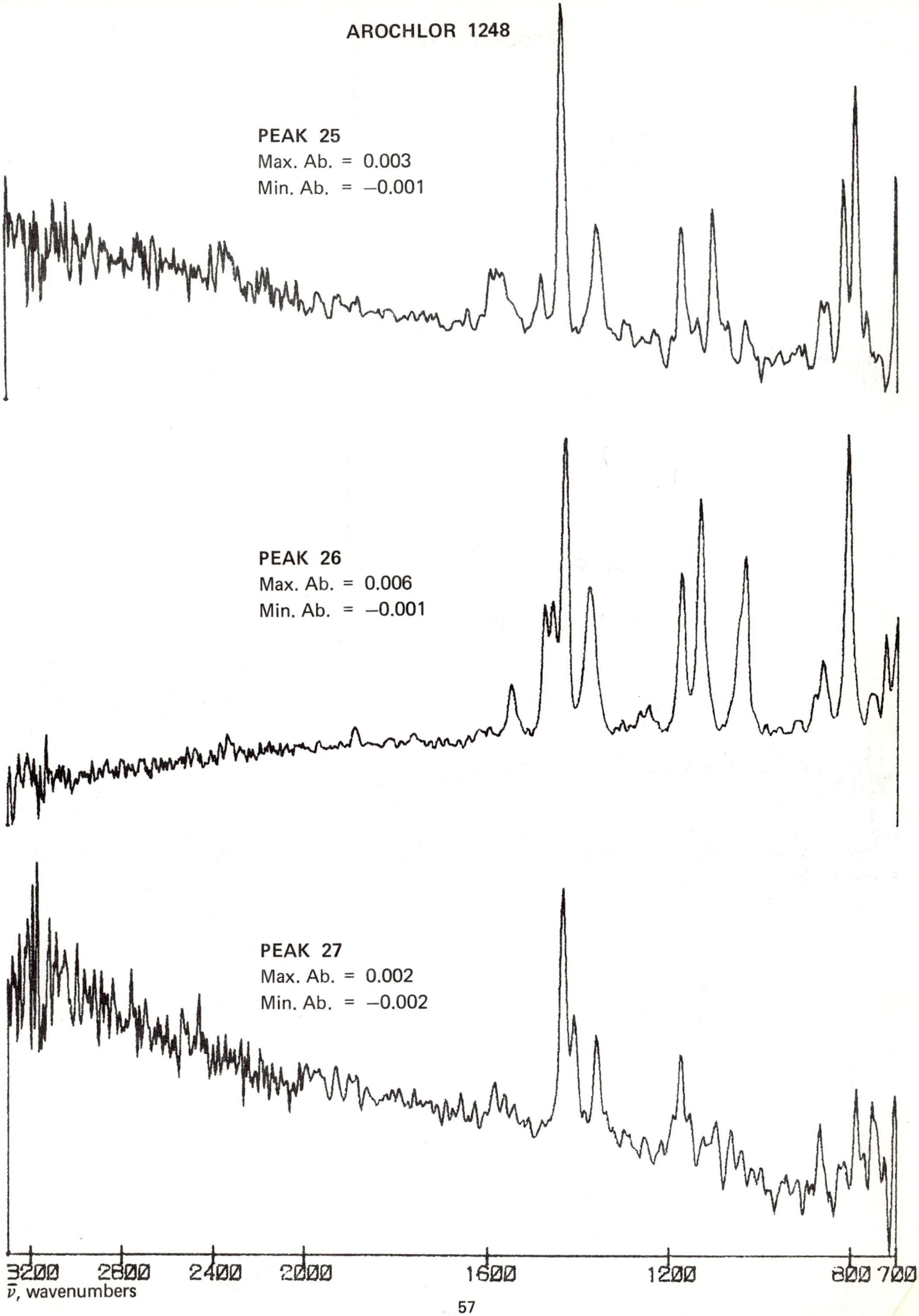
AROCHLOR 1248
PEAK 25
Max. Ab. = 0.003
Min. Ab. = –0.001
PEAK 26
Max. Ab. = 0.006
Min. Ab. = –0.001
PEAK 27
Max. Ab. = 0.002
Min. Ab. = –0.002
3200 2800 2400 2000 1600 1200 800 700
ν̄, wavenumbers

AROCHLOR 1248

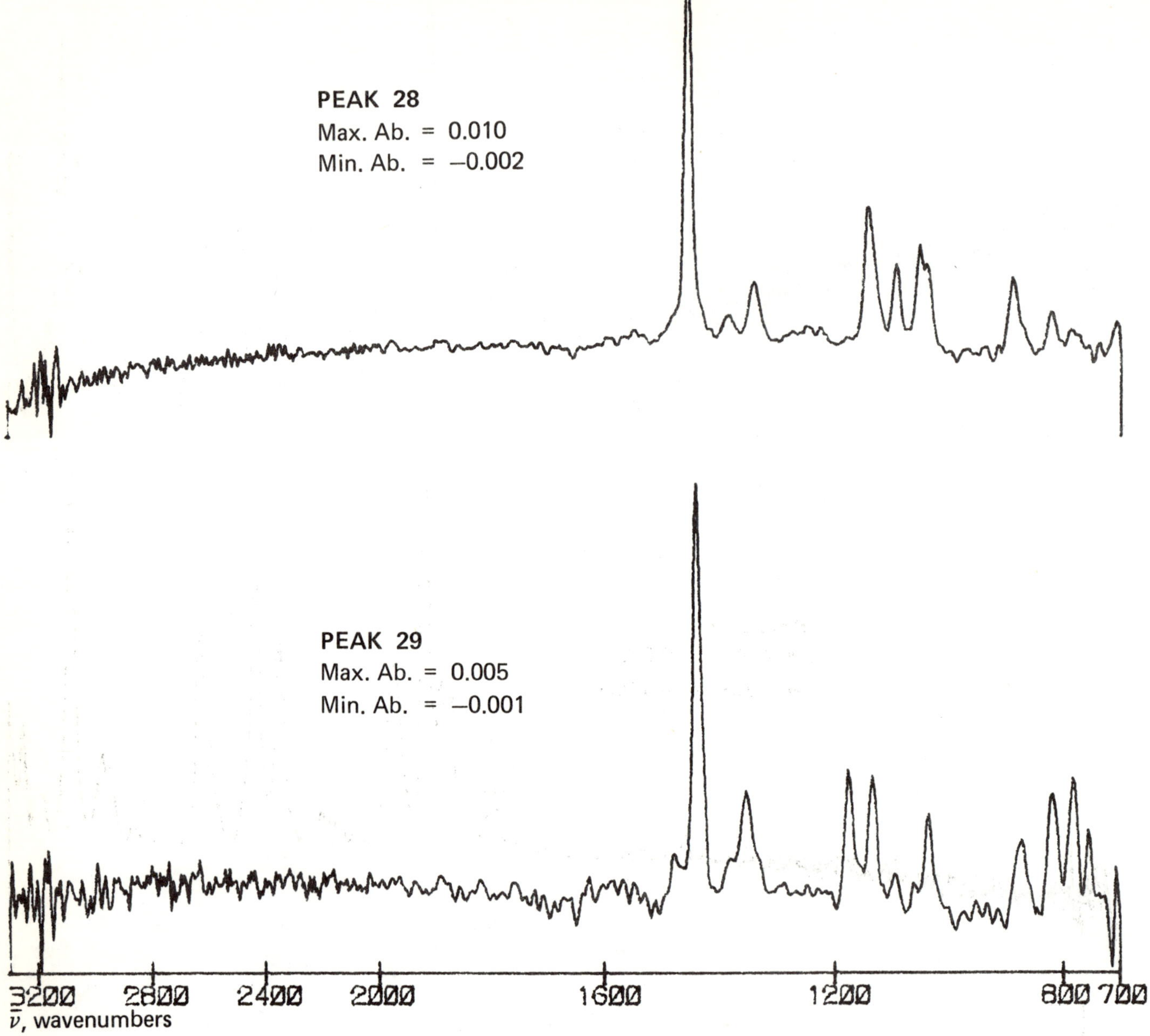

AROCHLOR 1260

GAS CHROMATOGRAPH DATA

TYPE = HP 5880 G.C.
MANUAL SPLITLESS INJECTION

SAMPLE PREPARATION = 50 μl of a saturated solution of the compound in diethyl ether was diluted to 1 ml.

INJECTED AMT. = 2 μl of the diluted solution

DET. TEMP. = 290°C

INJ. TEMP. = 260°C

CARRIER GAS = He

FLOW RATE = ∿ 1.5 ml./min.

PRESSURE = 60 PSI

DETECTOR = FID

COLUMN = J & W, SE–30, FUSED SILICA
30 m. LONG, 0.32 mm. DIA.

FILM THICKNESS = 0.25 microns

OVEN PROGRAM RATES 1 AND 2:

INITAL TEMP. = 35°C

INITAL TIME = 0.40 min.

LEVEL 1

PRGM RATE = 25.00°C/min.

FINAL TEMP. = 120°C

FINAL TIME = 0.00 min.

LEVEL 2

PROGRAM RATE = 4.00°C/min.

FINAL TEMP. = 210°C

FINAL TIME = 20.00 min.

IR DATA

DIGILAB FTS 20 E WITH DIGILAB GC/C[T.M.] ACCESSORY

PURGE = N_2

DETECTOR = LIQUID N_2 COOLED MCT (NARROW BAND)

OPTICAL CELL = GOLD COATED LIGHTPIPE WITH KBr WINDOWS

PEAK	RETENTION TIME	PEAK	RETENTION TIME
1	14.71	28	34.00
2	17.02	29	34.59
3	17.13	30	35.19
4	18.03	31	35.88
5	19.73	32	36.24
6	19.84	33	37.00
7	20.39	34	37.21
8	20.76	35	37.40
9	22.16	36	37.62
10	22.32	37	38.61
11	23.26	38	38.79
12	24.05	39	38.98
13	26.86	40	39.17
14	27.36	41	39.62
15	27.79	42	40.52
16	28.74	43	40.89
17	29.55	44	41.26
18	30.06	45	42.07
19	30.57	46	42.51
20	30.86	47	43.28
21	32.18	48	44.04
22	32.35	49	44.33
23	32.98	50	44.69
24	33.20	51	46.82
25	33.40	52	47.09
26	33.62	53	48.61
27	33.85	54	49.53

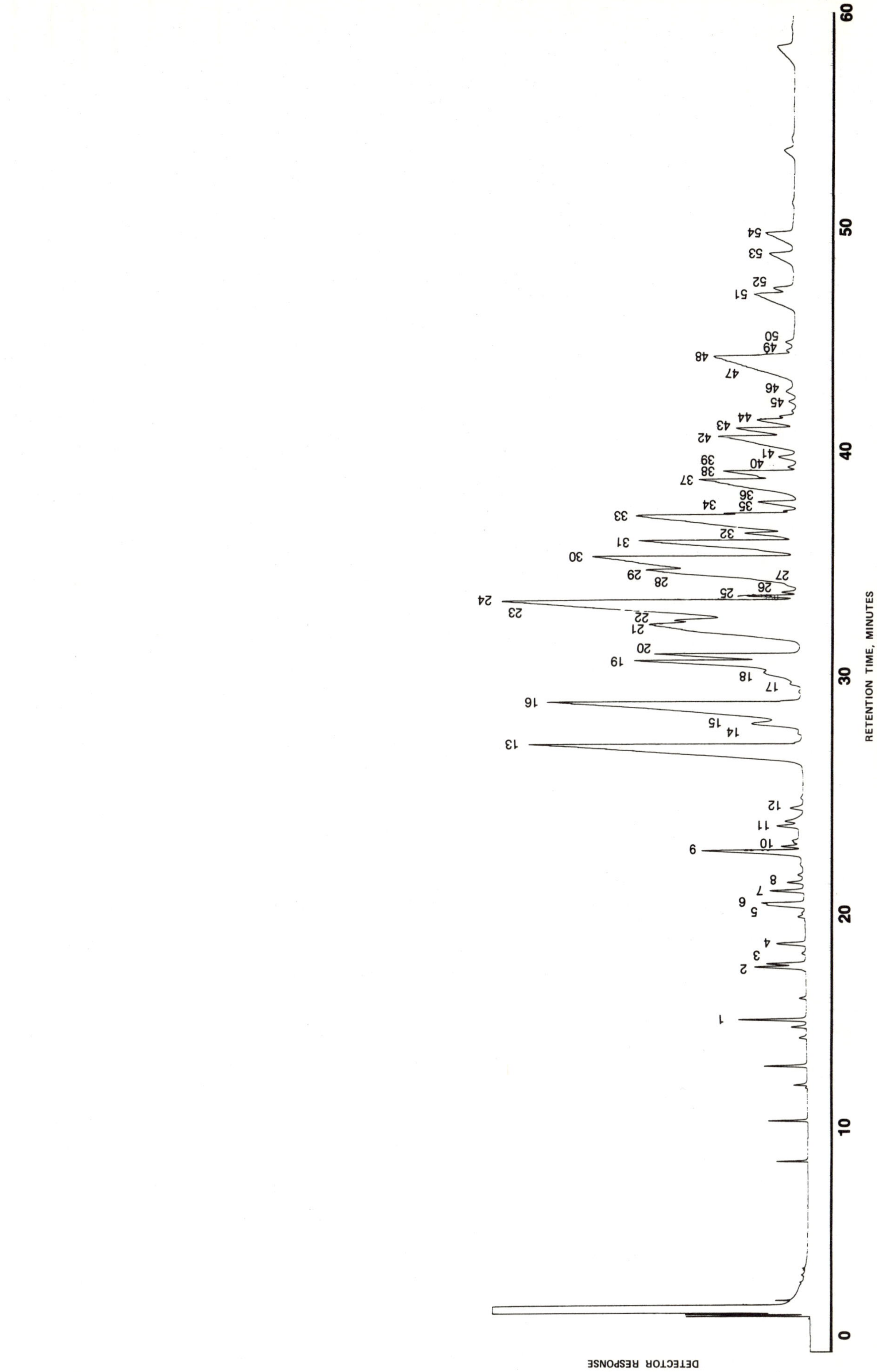

AROCHLOR 1260

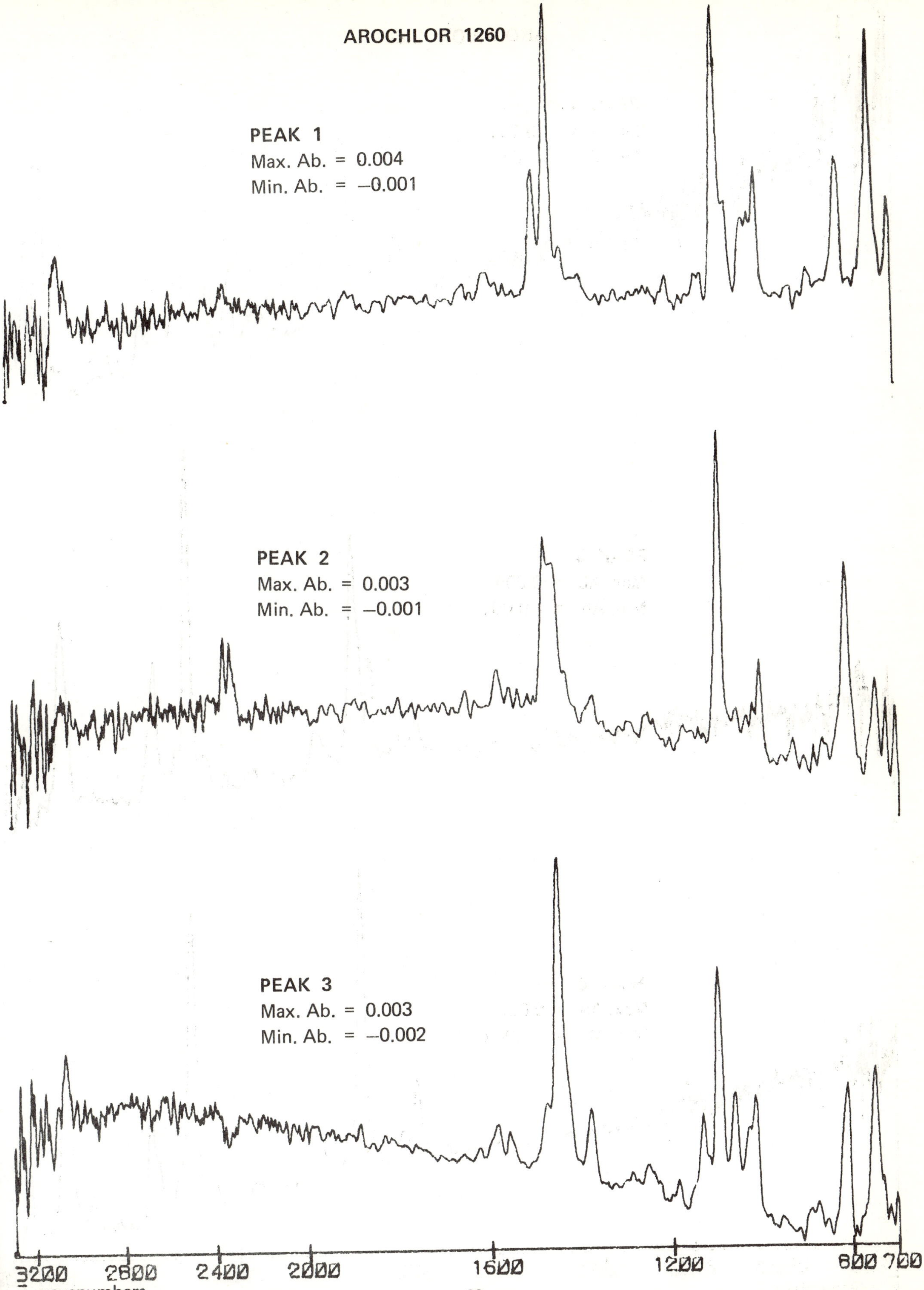

AROCHLOR 1260
PEAK 1
Max. Ab. = 0.004
Min. Ab. = –0.001
PEAK 2
Max. Ab. = 0.003
Min. Ab. = –0.001
PEAK 3
Max. Ab. = 0.003
Min. Ab. = –0.002
3200
2800
2400
2000
1600
1200
800
700
$\bar{\nu}$, wavenumbers

ARROCHLOR 1260

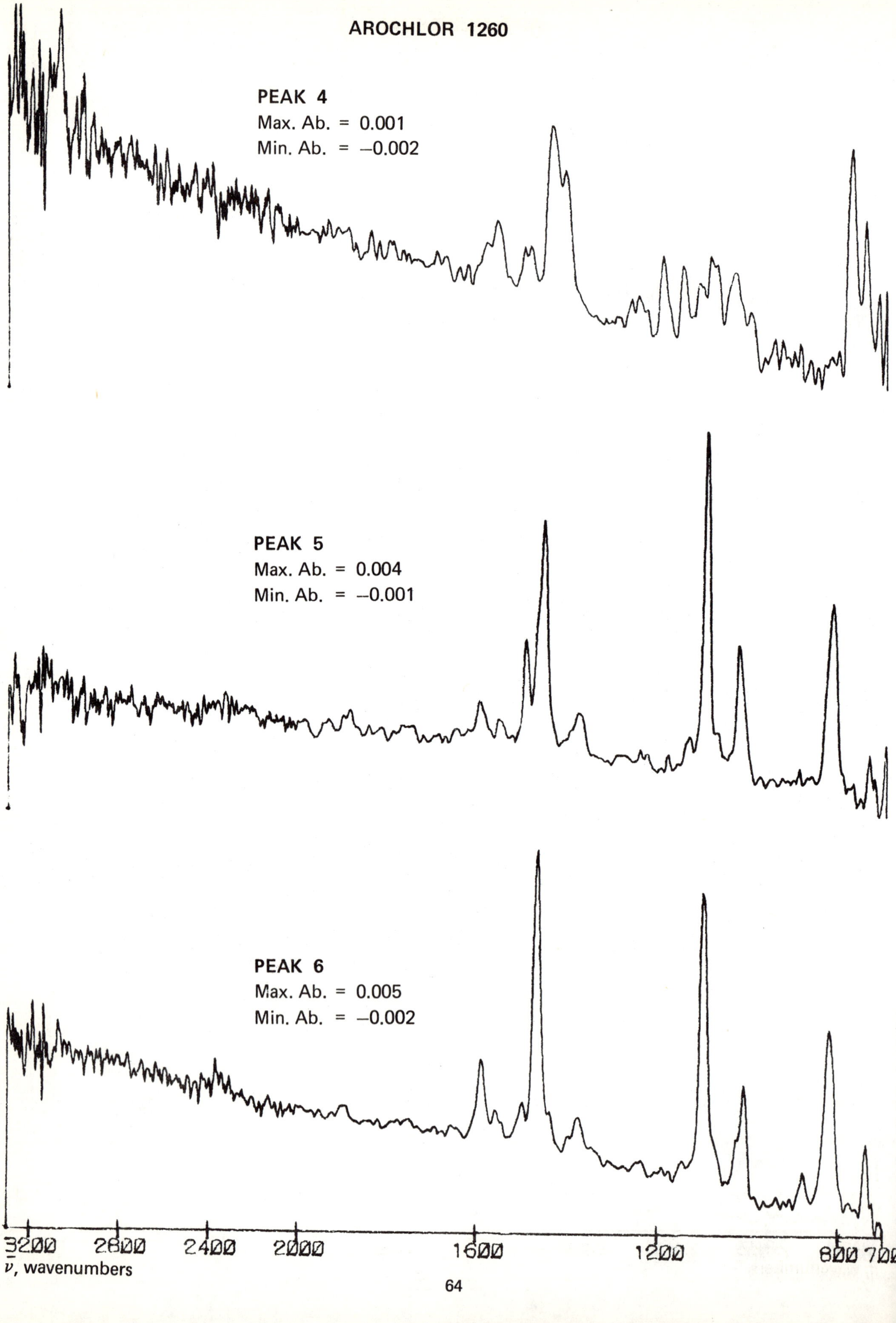

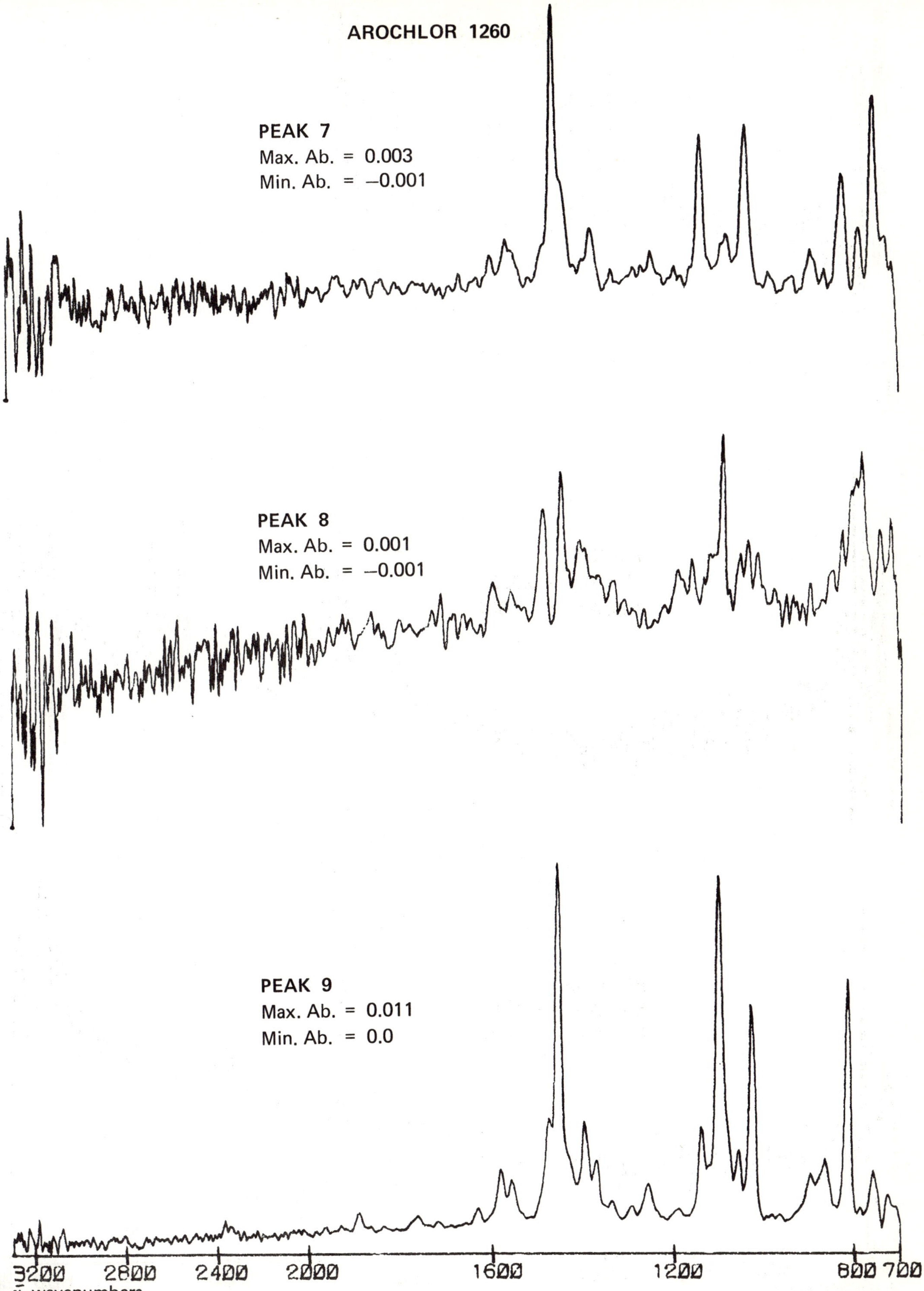
AROCHLOR 1260
PEAK 7
Max. Ab. = 0.003
Min. Ab. = –0.001
PEAK 8
Max. Ab. = 0.001
Min. Ab. = –0.001
PEAK 9
Max. Ab. = 0.011
Min. Ab. = 0.0
3200
2800
2400
2000
1600
1200
800
700
$\bar{\nu}$, wavenumbers

AROCHLOR 1260

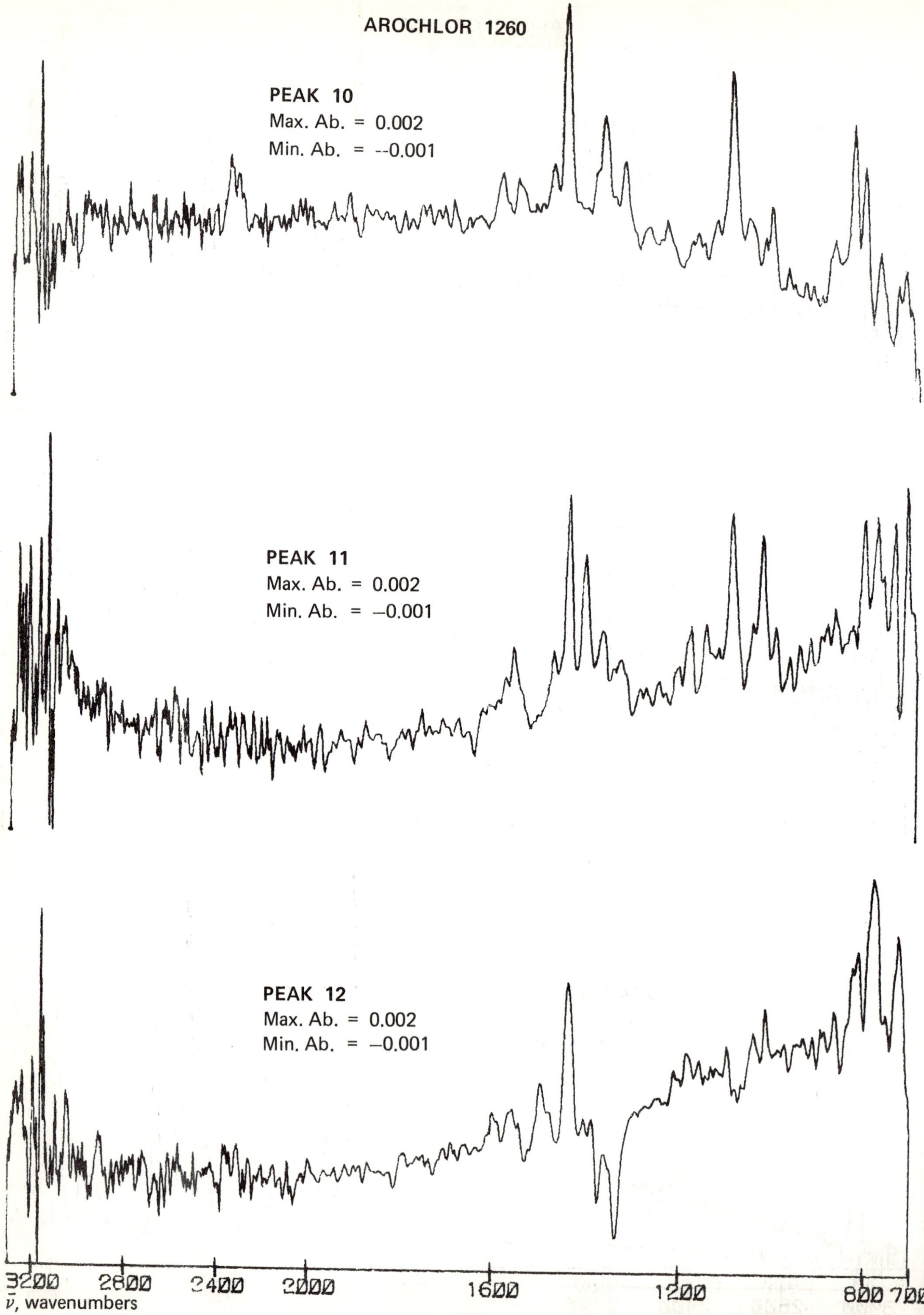

AROCHLOR 1260

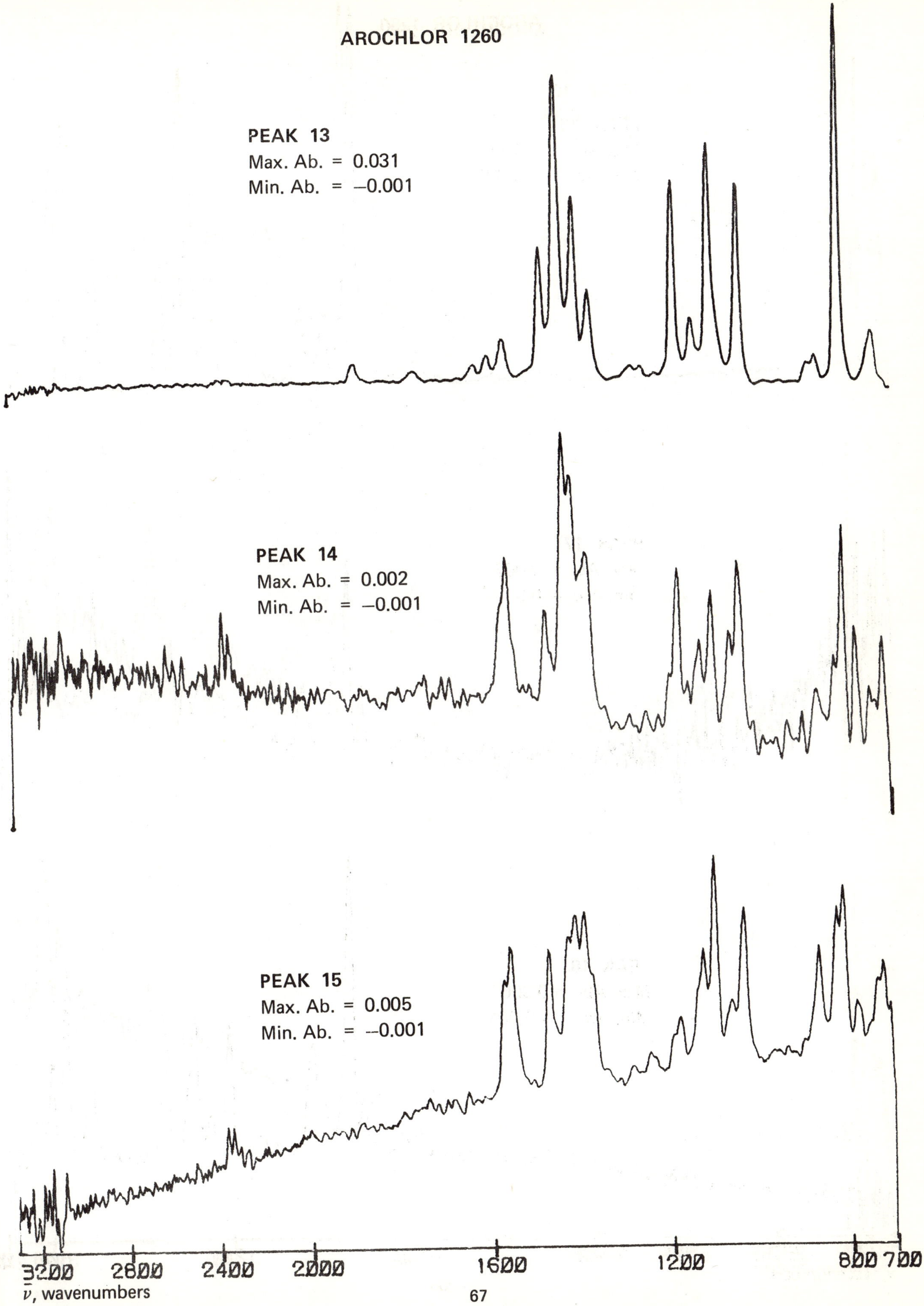

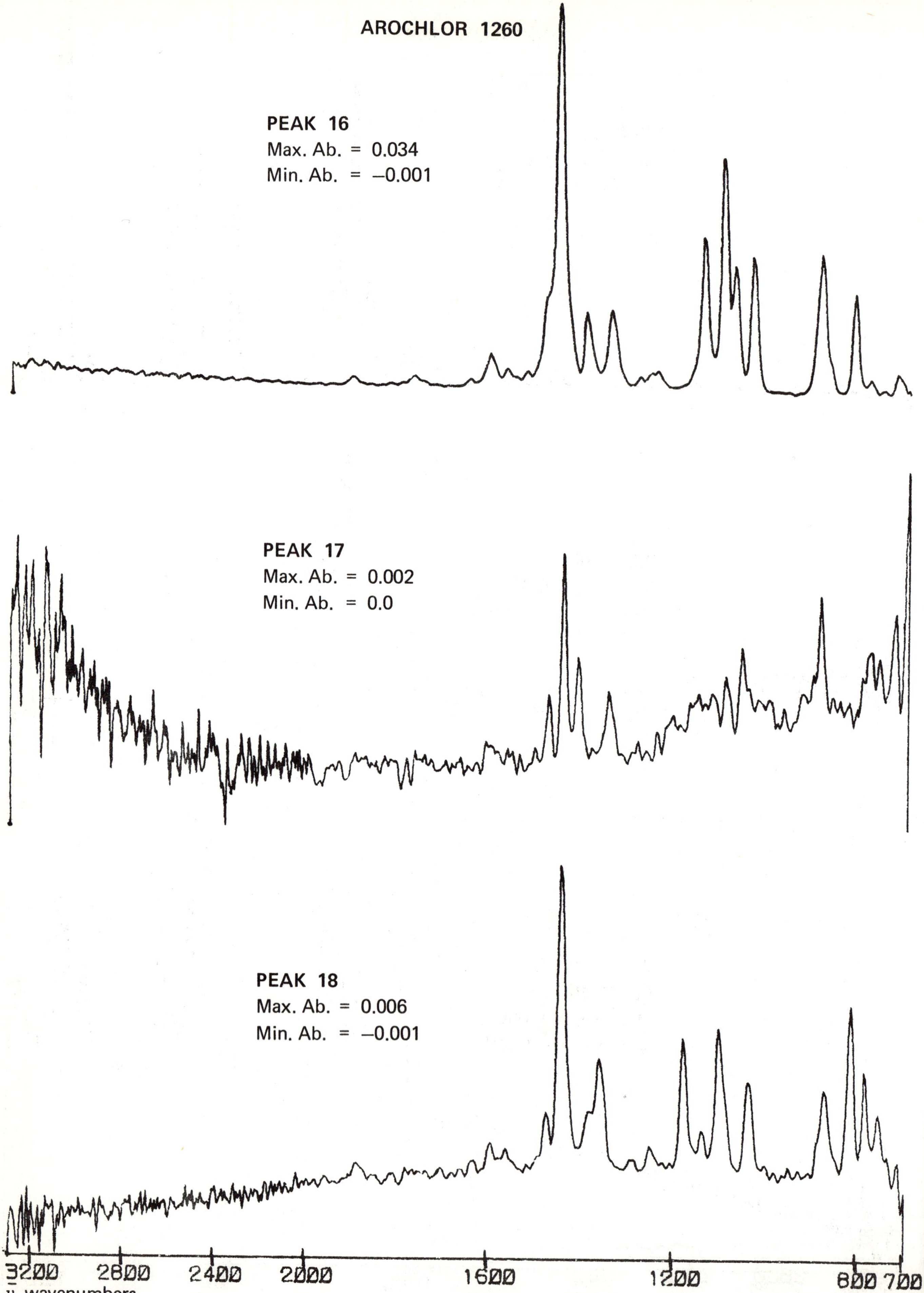
AROCHLOR 1260
PEAK 16
Max. Ab. = 0.034
Min. Ab. = –0.001
PEAK 17
Max. Ab. = 0.002
Min. Ab. = 0.0
PEAK 18
Max. Ab. = 0.006
Min. Ab. = –0.001
3200
2800
2400
2000
1600
1200
800
700
ν, wavenumbers

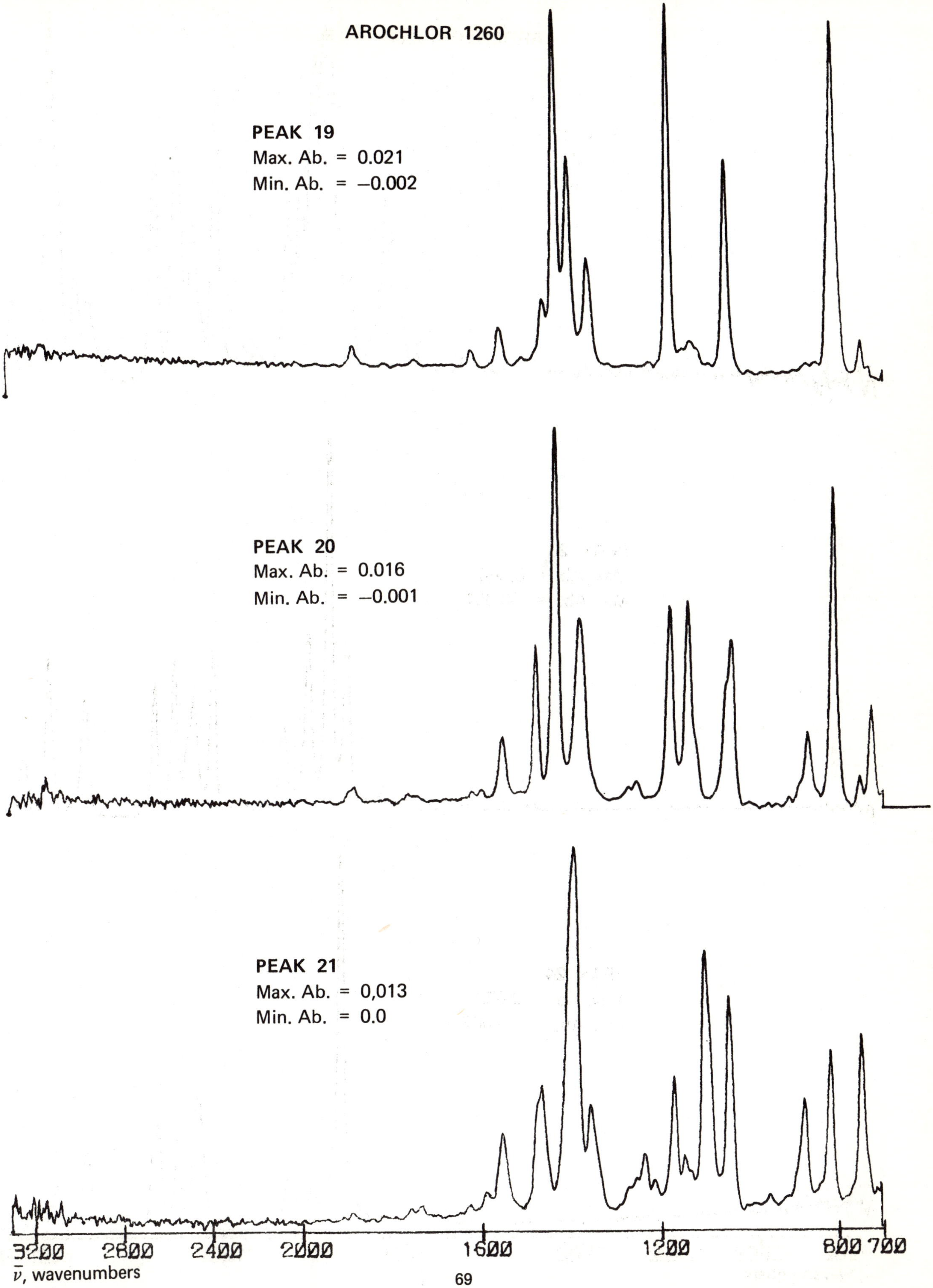

ARCHLOR 1260
PEAK 19
Max. Ab. = 0.021
Min. Ab. = –0.002
PEAK 20
Max. Ab. = 0.016
Min. Ab. = –0.001
PEAK 21
Max. Ab. = 0,013
Min. Ab. = 0.0
3200
2800
2400
2000
1600
1200
800
700
$\bar{\nu}$, wavenumbers

AROCHLOR 1260

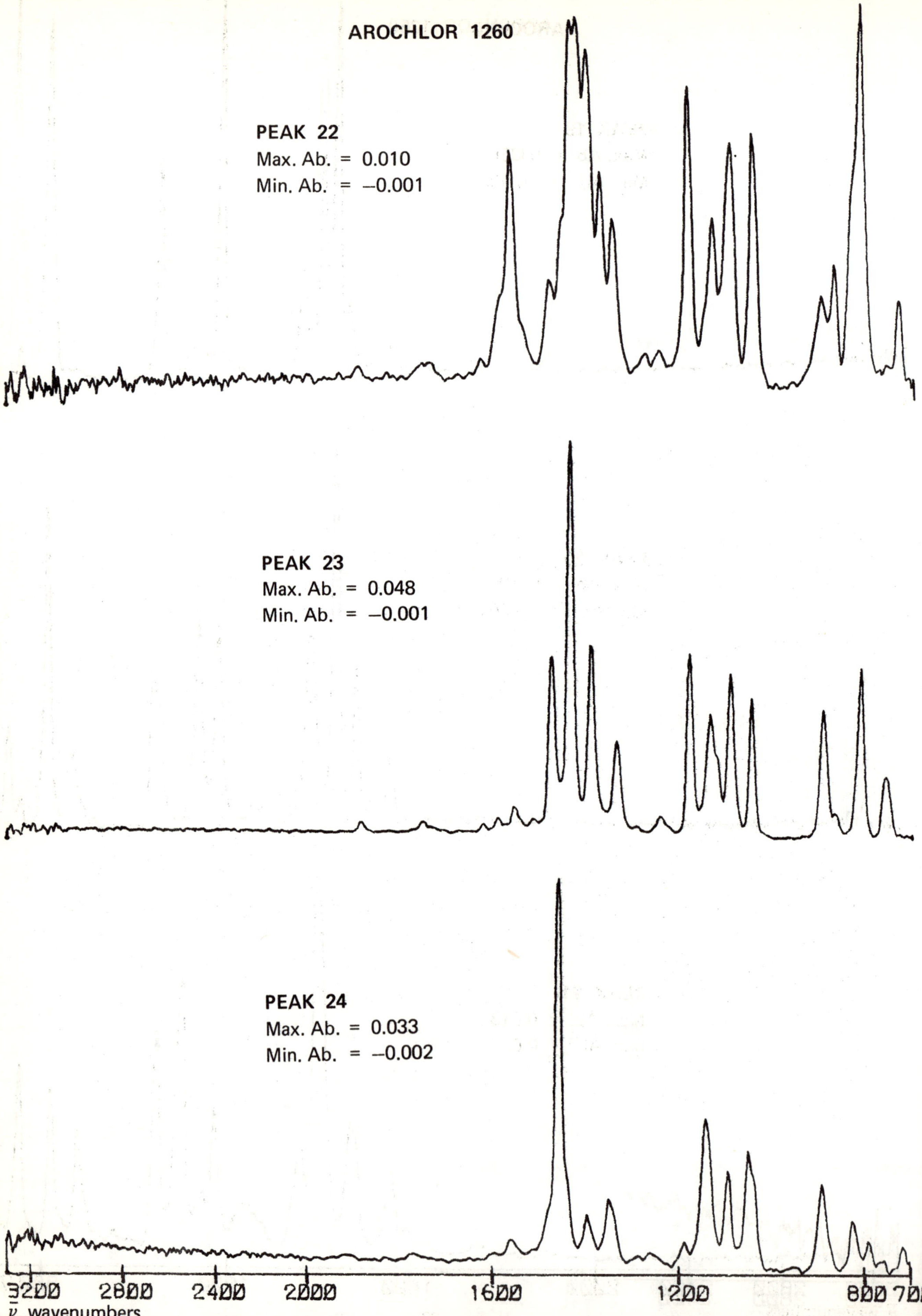

AROCHLOR 1260

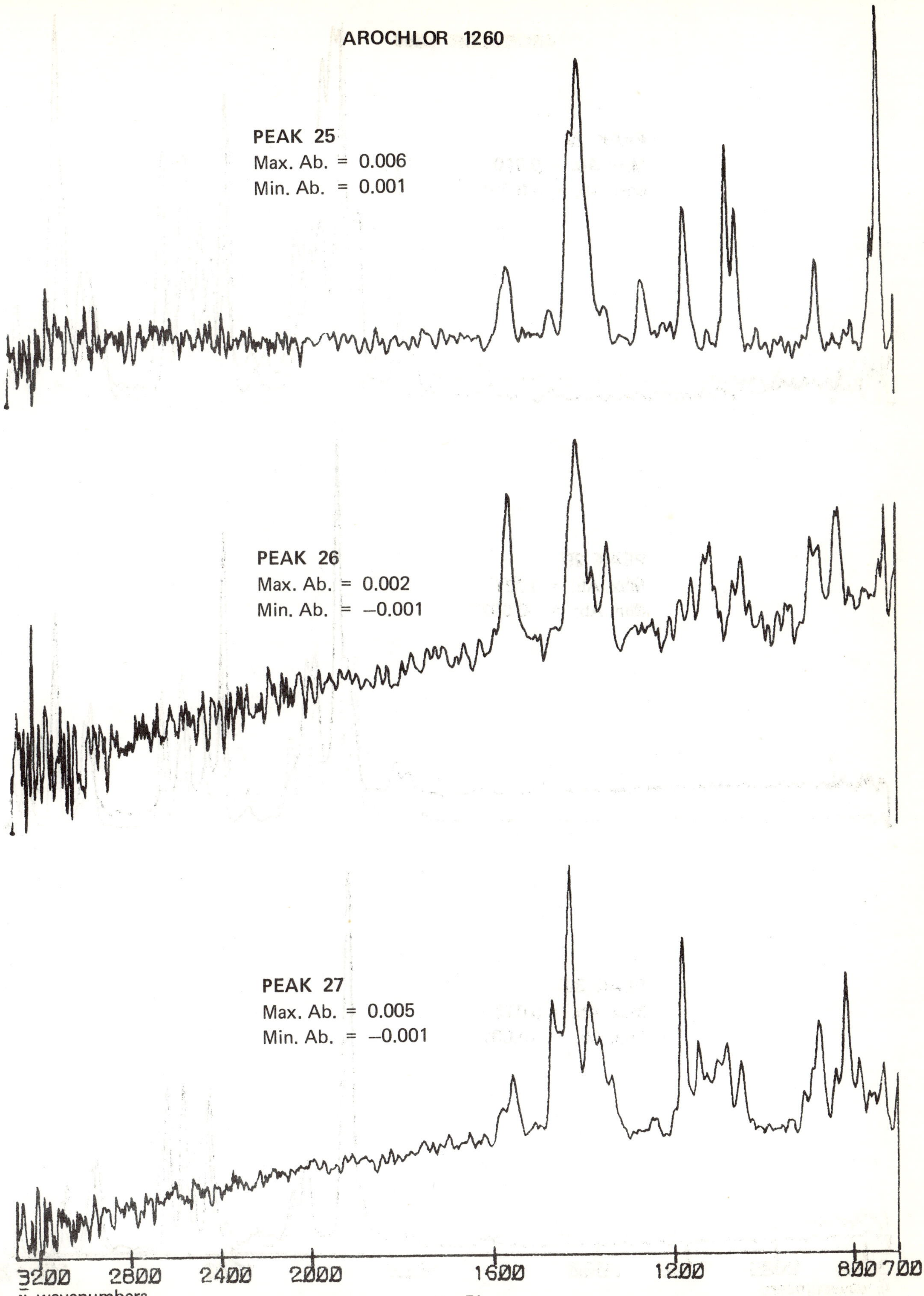

AROCHLOR 1260

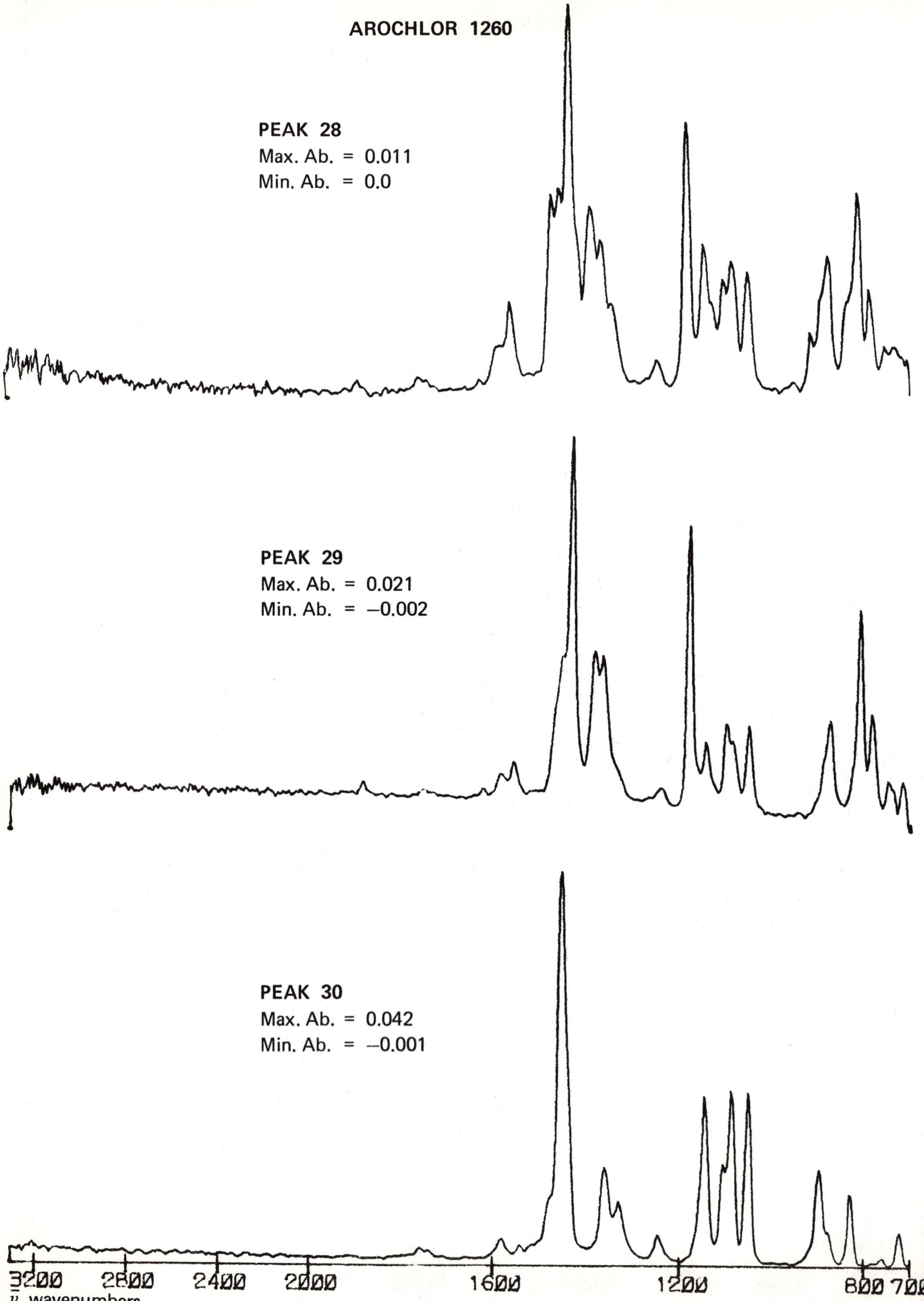

AROCHLOR 1260

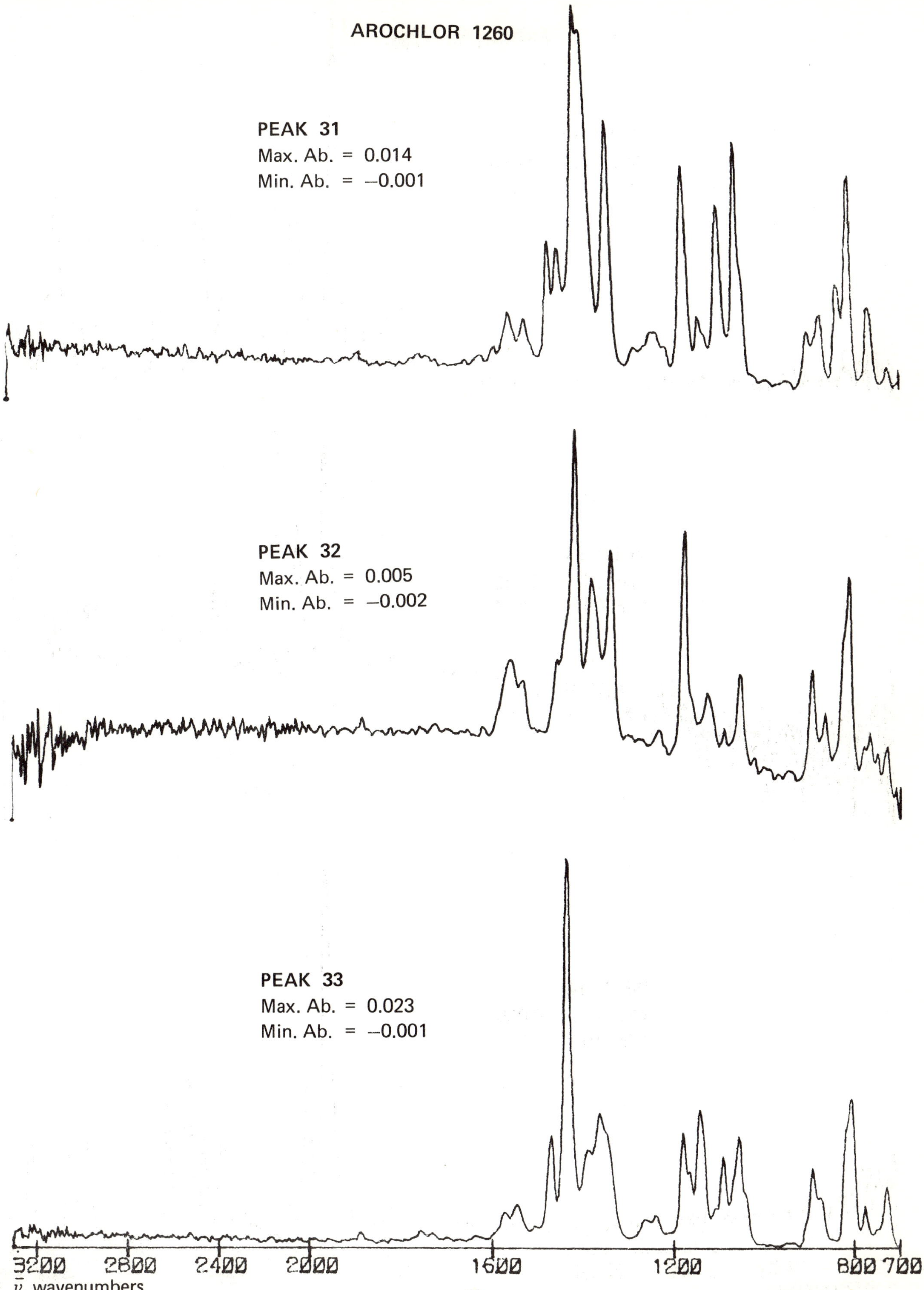

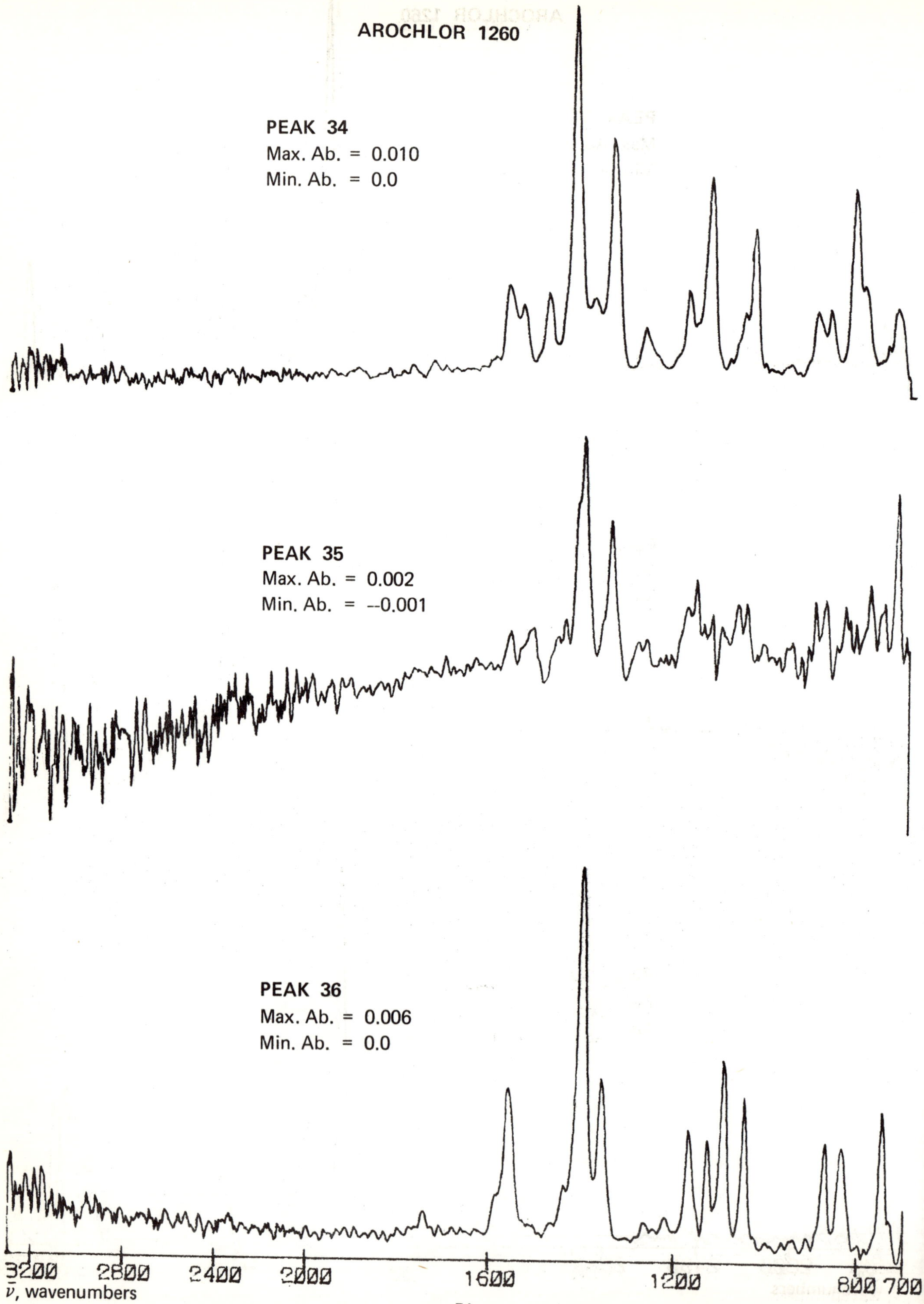
AROCHLOR 1260
PEAK 34
Max. Ab. = 0.010
Min. Ab. = 0.0
PEAK 35
Max. Ab. = 0.002
Min. Ab. = --0.001
PEAK 36
Max. Ab. = 0.006
Min. Ab. = 0.0
3200
2800
2400
2000
1600
1200
800
700
$\bar{\nu}$, wavenumbers

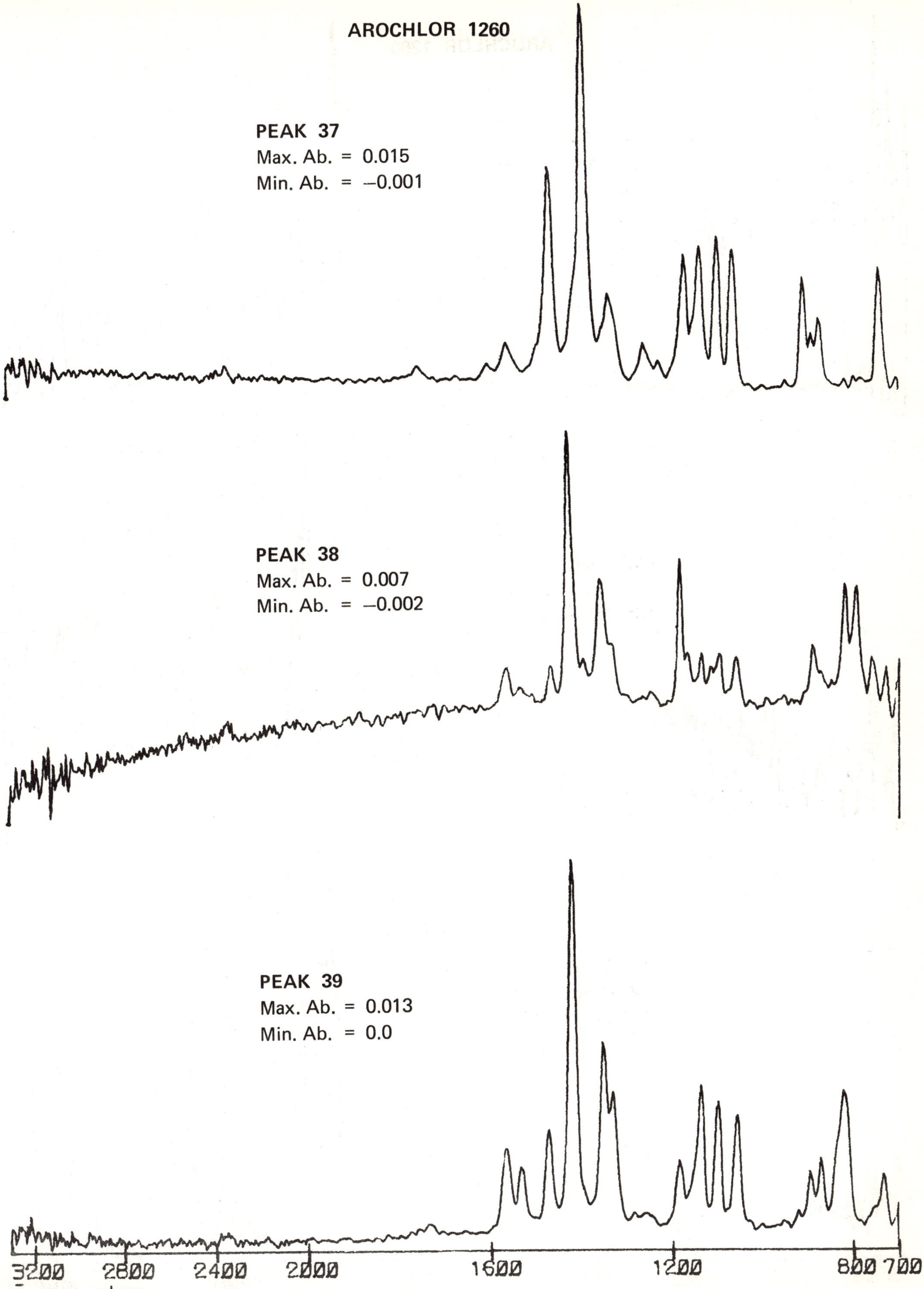
AROCHLOR 1260
PEAK 37
Max. Ab. = 0.015
Min. Ab. = –0.001
PEAK 38
Max. Ab. = 0.007
Min. Ab. = –0.002
PEAK 39
Max. Ab. = 0.013
Min. Ab. = 0.0
3200
2800
2400
2000
1600
1200
800
700
$\bar{\nu}$, wavenumbers

AROCHLOR 1260

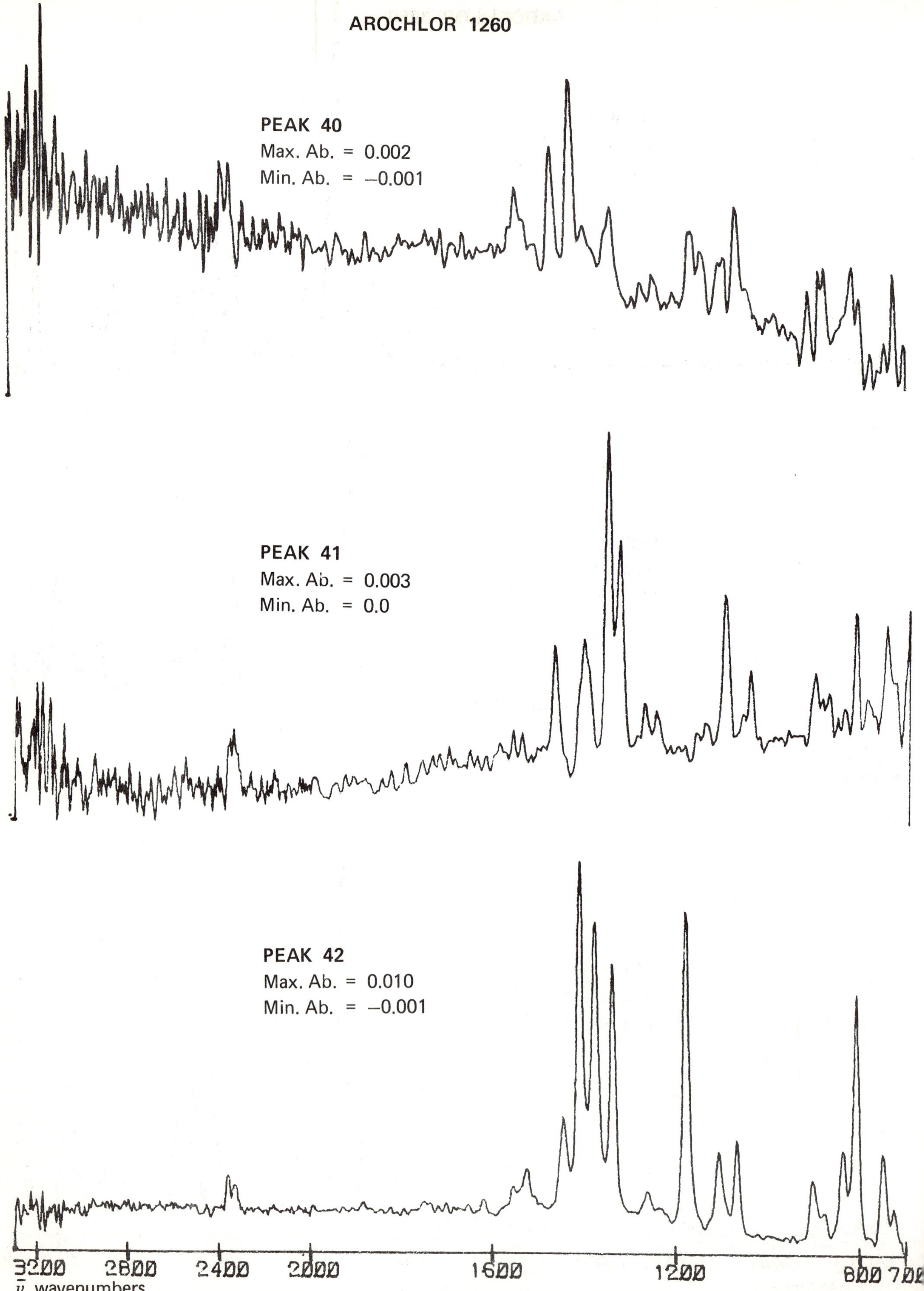

AROCHLOR 1260

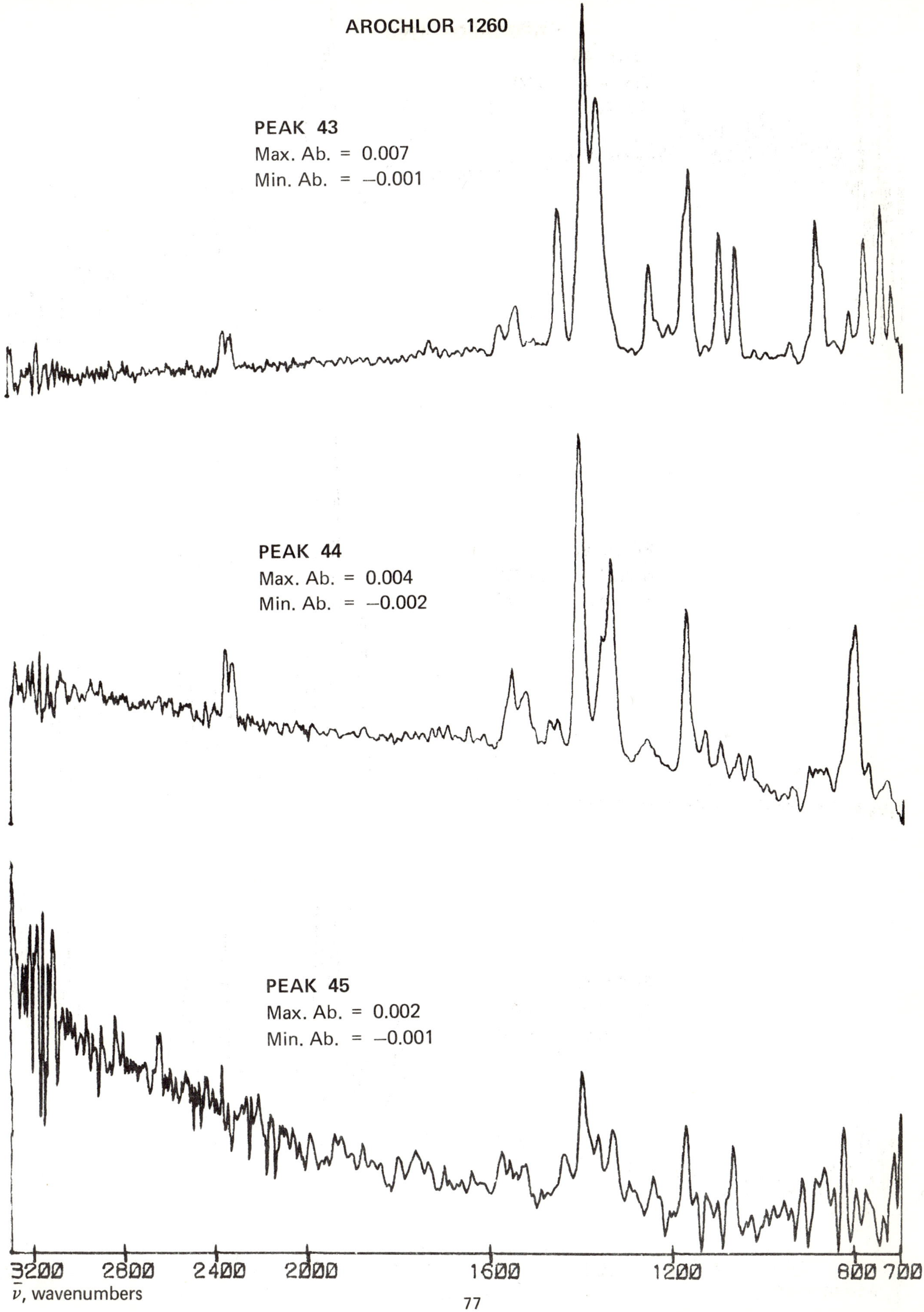

AROCHLOR 1260

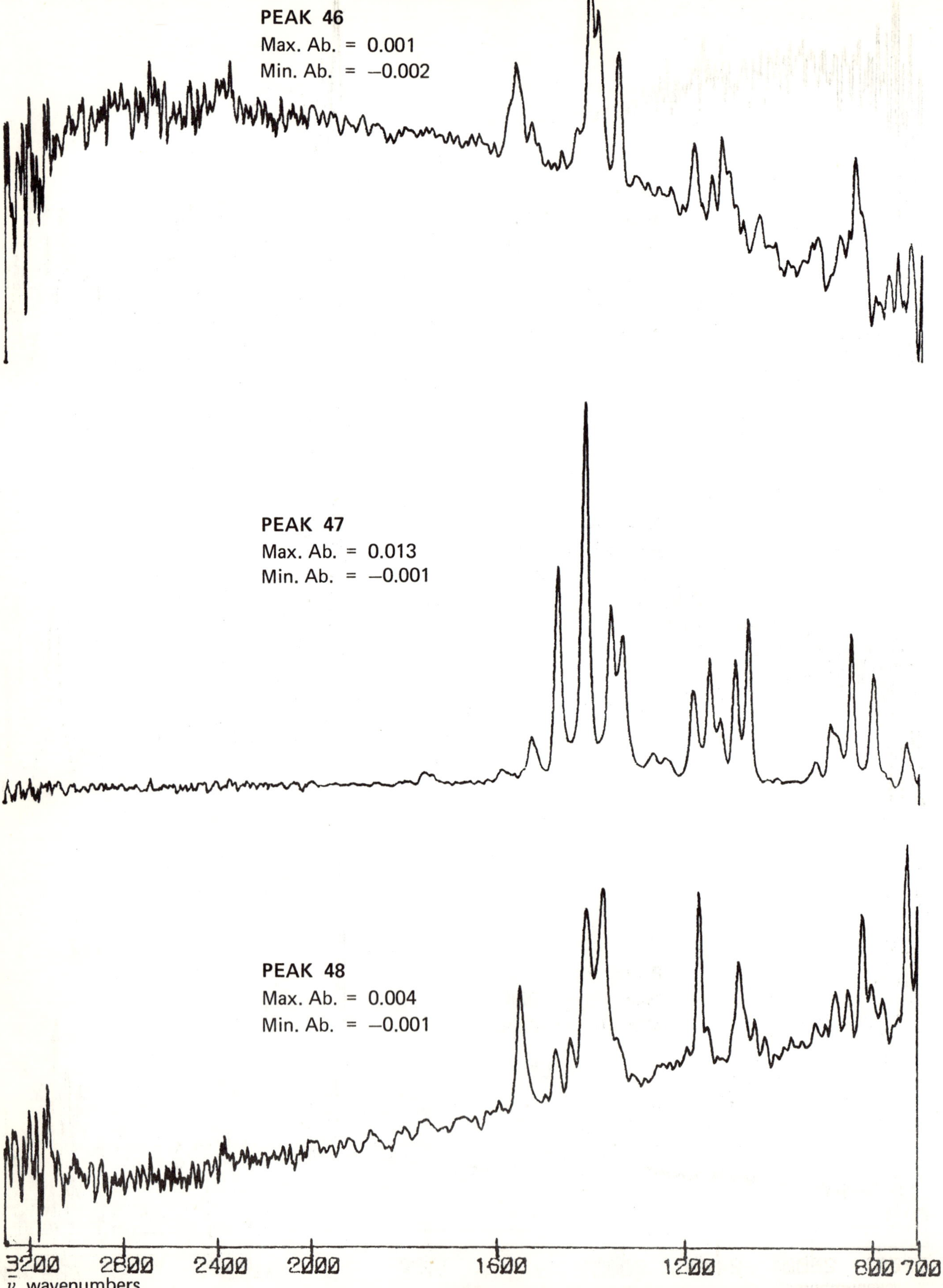

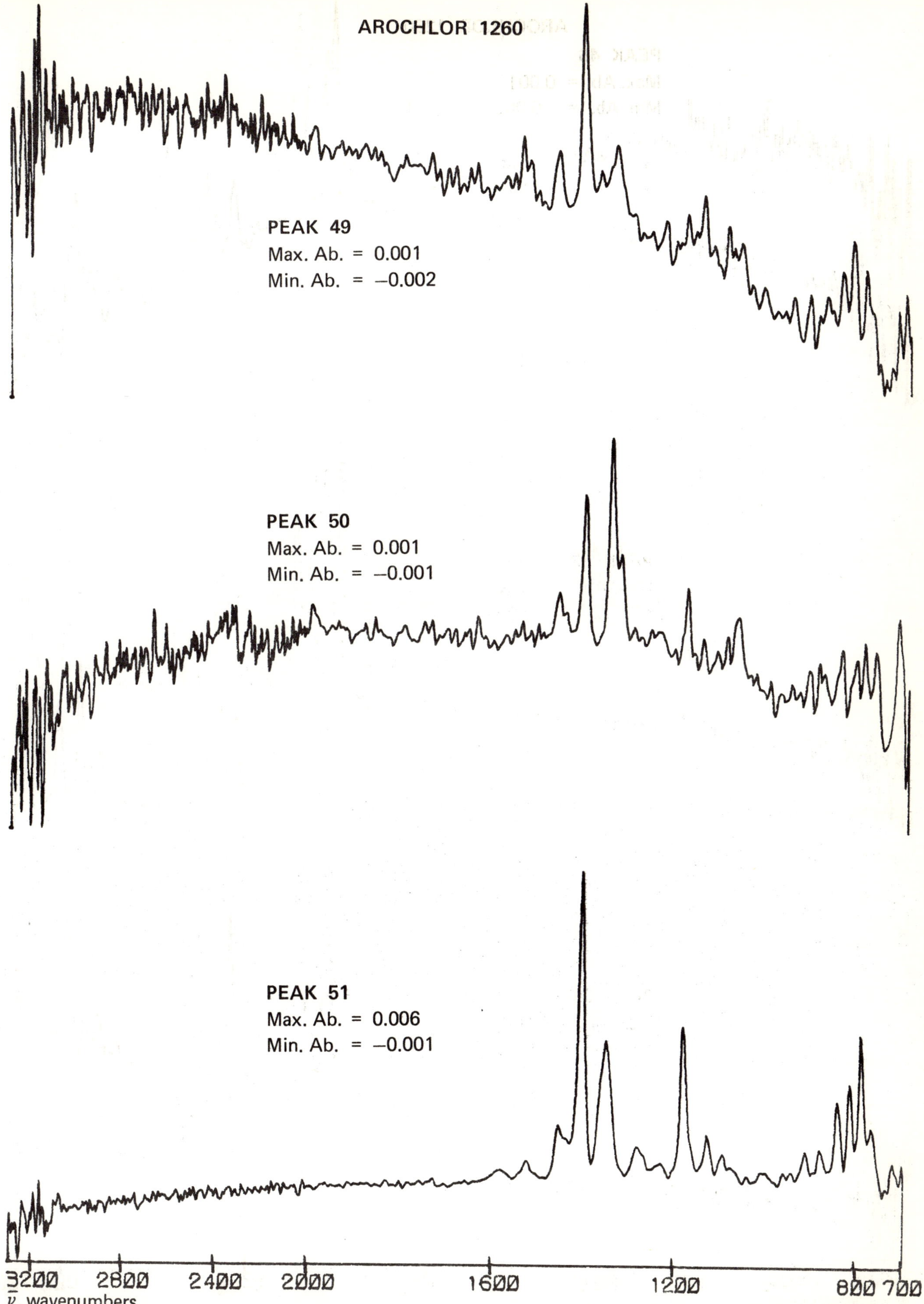
AROCHLOR 1260
PEAK 49
Max. Ab. = 0.001
Min. Ab. = –0.002
PEAK 50
Max. Ab. = 0.001
Min. Ab. = –0.001
PEAK 51
Max. Ab. = 0.006
Min. Ab. = –0.001
3200
2800
2400
2000
1600
1200
800
700
$\bar{\nu}$, wavenumbers

AROCHLOR 1260

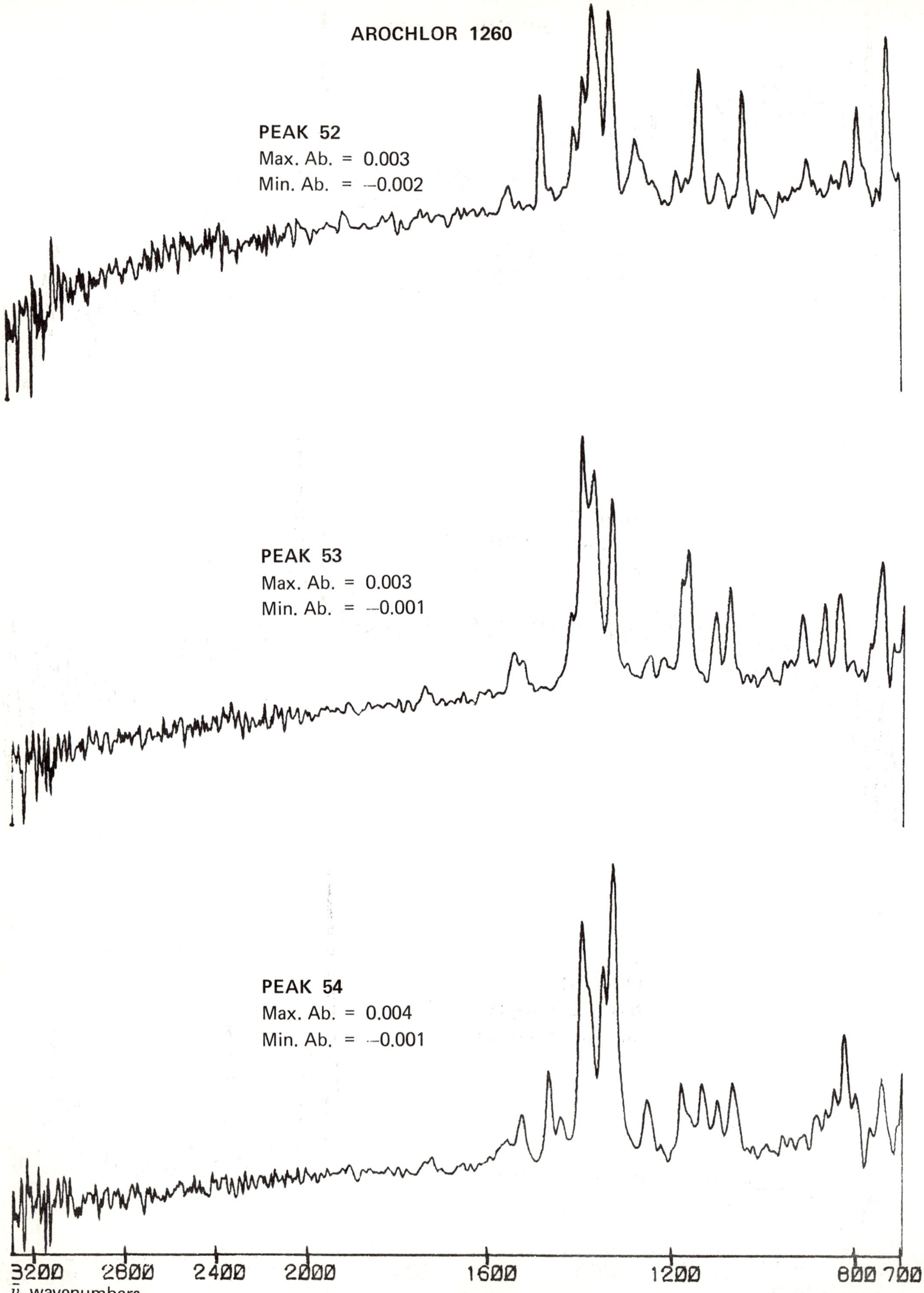

CHLORDAN

GAS CHROMATOGRAPH DATA

TYPE = HP 5880 G.C.
MANUAL SPLITLESS INJECTION

SAMPLE PREPARATION = 50 μl of a saturated solution of the compound in diethyl ether was diluted to 1 ml.

INJECTED AMT. = 2 μl of the diluted solution

DET. TEMP. = 290°C

INJ. TEMP. = 260°C

CARRIER GAS = He

FLOW RATE = ∿ 1.5 ml./min.

PRESSURE = 60 PSI

DETECTOR = FID

COLUMN = J & W, SE–30, FUSED SILICA
30 m. LONG, 0.32 mm. DIA.

FILM THICKNESS = 0.25 microns

OVEN PROGRAM RATES 1 AND 2:

INITAL TEMP. = 35°C

INITAL TIME = 0.40 min.

LEVEL 1

PRGM RATE = 25.00°C/min.

FINAL TEMP. = 120°C

FINAL TIME = 0.00 min.

LEVEL 2

PROGRAM RATE = 4.00°C/min.

FINAL TEMP. = 210°C

FINAL TIME = 20.00 min.

IR DATA

DIGILAB FTS 20 E WITH DIGILAB GC/C[T.M.] ACCESSORY

PURGE = N_2

DETECTOR = LIQUID N_2 COOLED MCT (NARROW BAND)

OPTICAL CELL = GOLD COATED LIGHTPIPE WITH KBr WINDOWS

PEAK	*RETENTION TIME*	*PEAK*	*RETENTION TIME*
1	10.90	15	27.82
2	14.45	16	27.94
3	18.22	17	28.14
4	20.29	18	28.81
5	20.43	19	29.30
6	21.12	20	29.61
7	22.85	21	30.28
8	23.25	22	30.50
9	23.73	23	31.46
10	23.97	24	31.64
11	24.35	25	32.14
12	24.92	26	32.62
13	26.33	27	36.84
14	26.50		

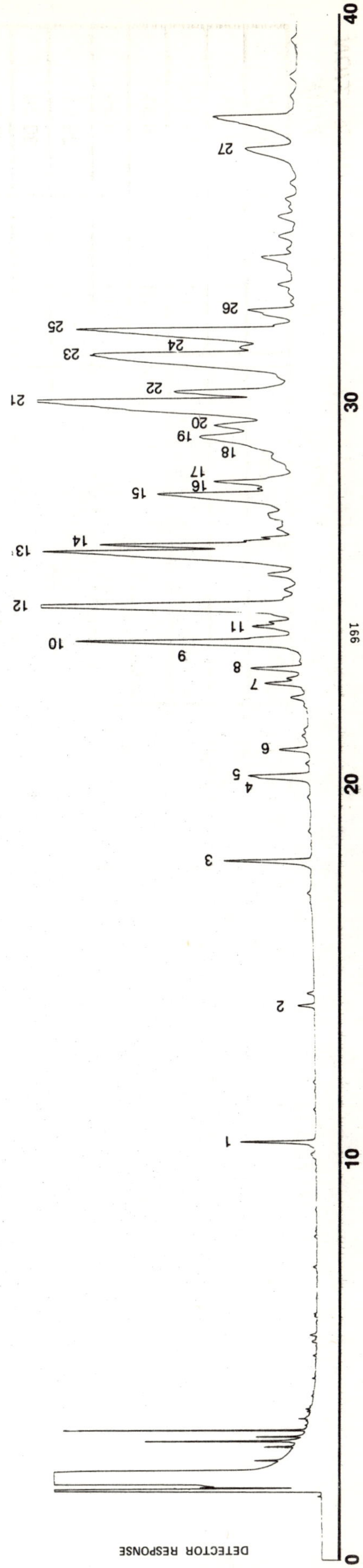

RETENTION TIME, MINUTES

CHLORDAN

CHLORDAN

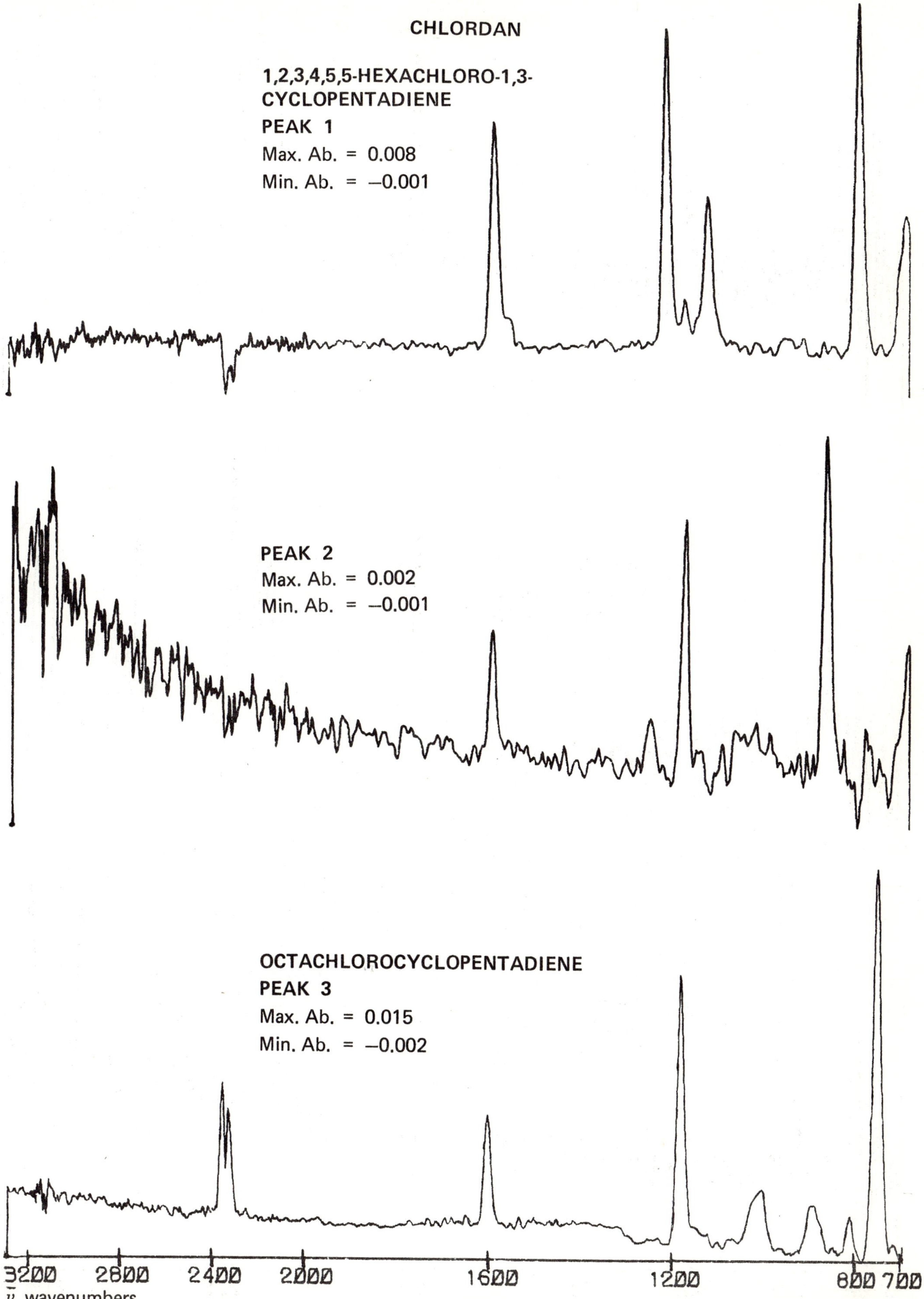
1,2,3,4,5,5-HEXACHLORO-1,3-CYCLOPENTADIENE
PEAK 1
Max. Ab. = 0.008
Min. Ab. = –0.001
PEAK 2
Max. Ab. = 0.002
Min. Ab. = –0.001
OCTACHLOROCYCLOPENTADIENE
PEAK 3
Max. Ab. = 0.015
Min. Ab. = –0.002
3200
2800
2400
2000
1600
1200
800
700
$\bar{\nu}$, wavenumbers

CHLORDAN

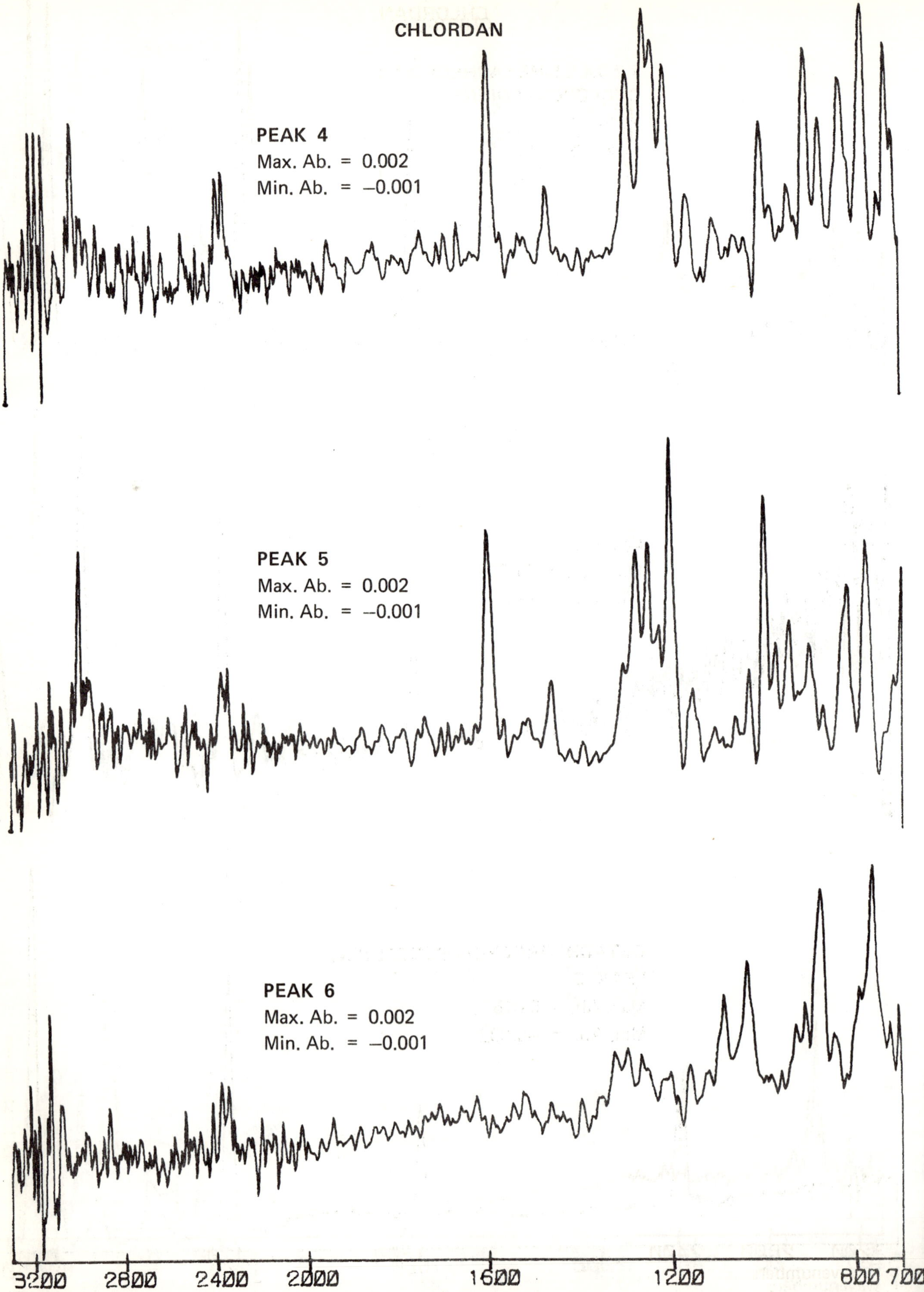

PEAK 4
Max. Ab. = 0.002
Min. Ab. = –0.001
PEAK 5
Max. Ab. = 0.002
Min. Ab. = –0.001
PEAK 6
Max. Ab. = 0.002
Min. Ab. = –0.001
3200
2800
2400
2000
1600
1200
800
700
$\bar{\nu}$, wavenumbers

CHLORDAN

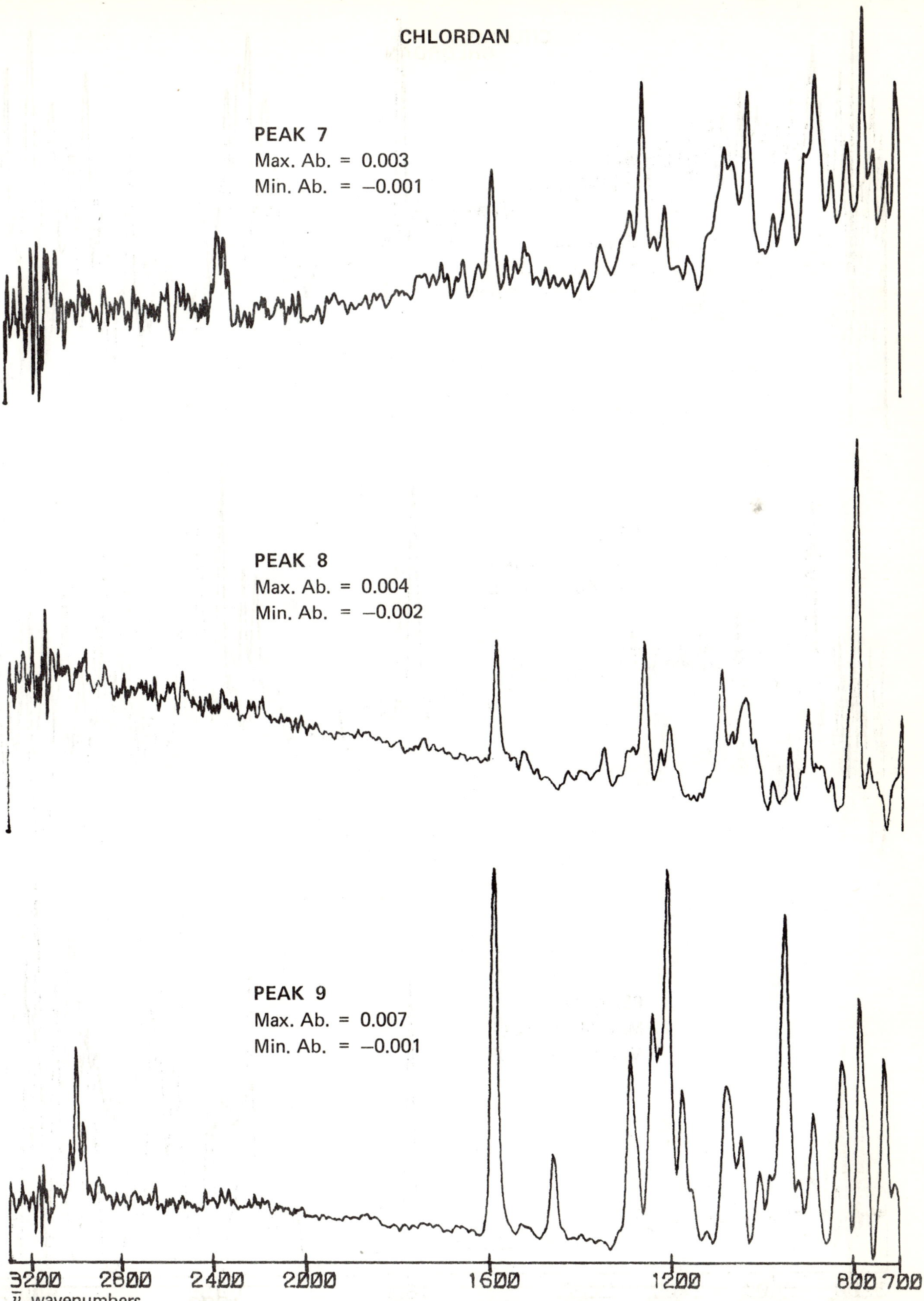
PEAK 7
Max. Ab. = 0.003
Min. Ab. = –0.001
PEAK 8
Max. Ab. = 0.004
Min. Ab. = –0.002
PEAK 9
Max. Ab. = 0.007
Min. Ab. = –0.001
3200
2800
2400
2000
1600
1200
800
700
ν̄, wavenumbers

CHLORDAN

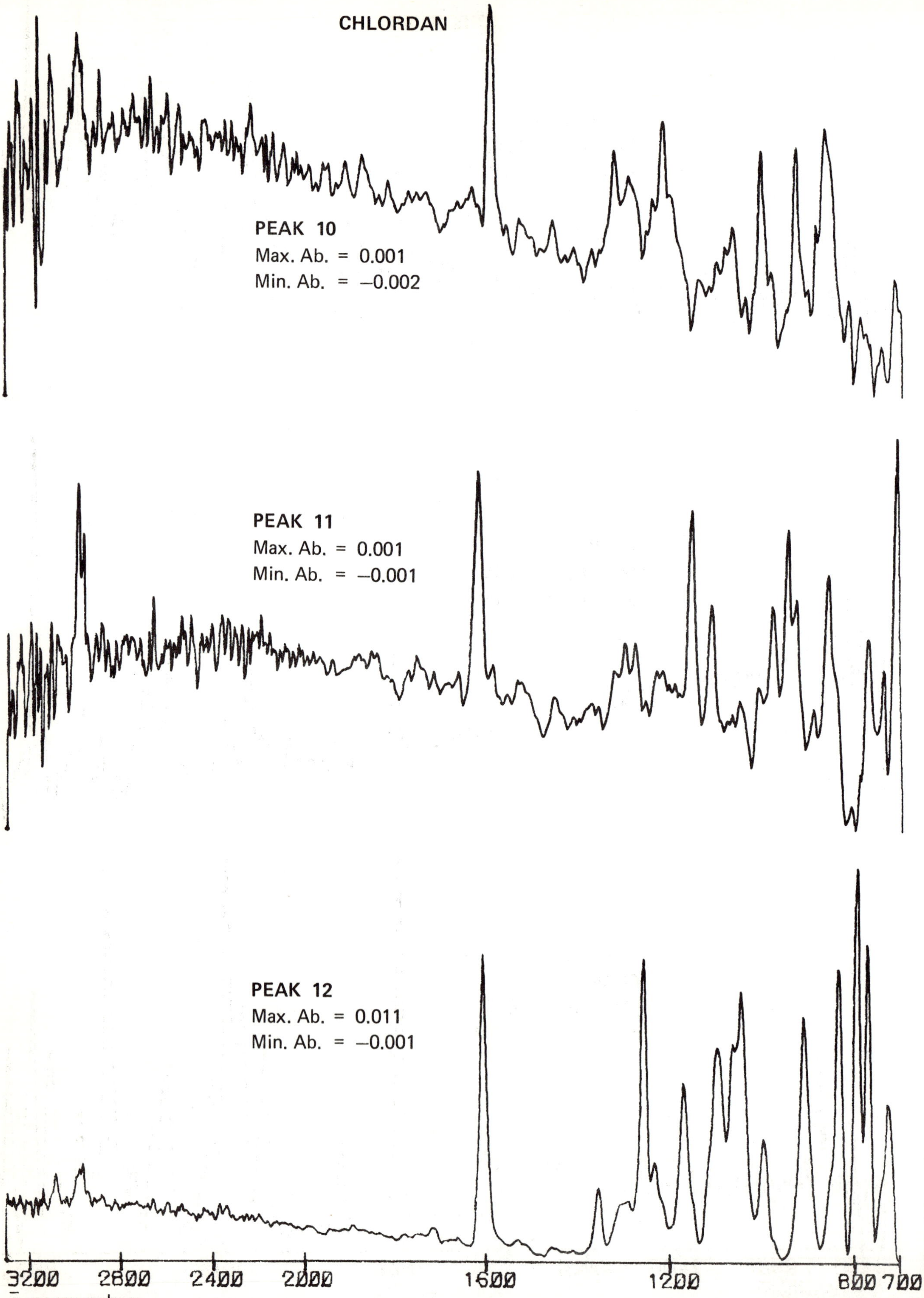
PEAK 10
Max. Ab. = 0.001
Min. Ab. = –0.002
PEAK 11
Max. Ab. = 0.001
Min. Ab. = –0.001
PEAK 12
Max. Ab. = 0.011
Min. Ab. = –0.001
3200
2800
2400
2000
1600
1200
800
700
$\bar{\nu}$, wavenumbers

CHLORDAN

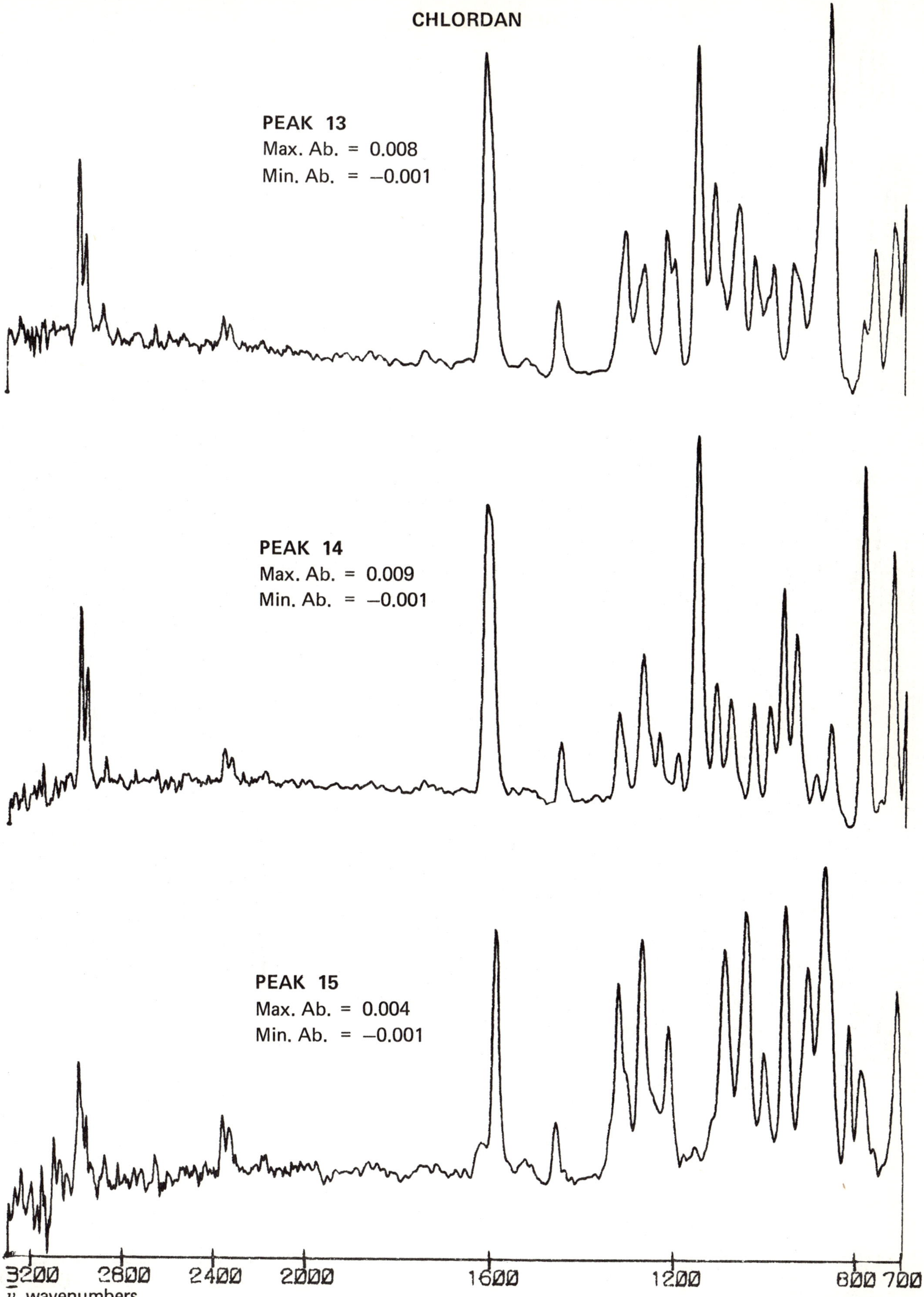

PEAK 13
Max. Ab. = 0.008
Min. Ab. = –0.001
PEAK 14
Max. Ab. = 0.009
Min. Ab. = –0.001
PEAK 15
Max. Ab. = 0.004
Min. Ab. = –0.001
3200
2800
2400
2000
1600
1200
800
700
$\bar{\nu}$, wavenumbers

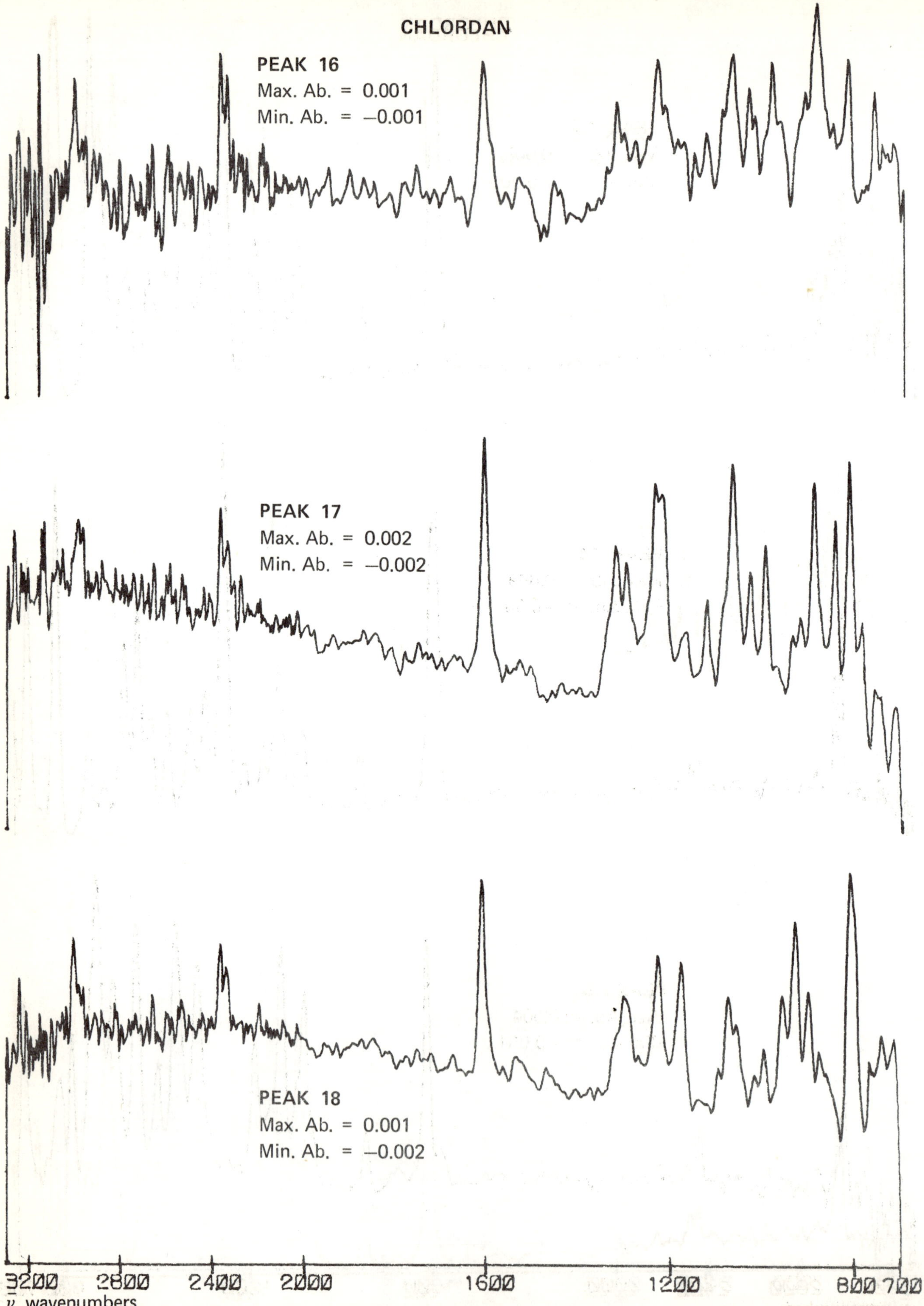

CHLORDAN
PEAK 16
Max. Ab. = 0.001
Min. Ab. = –0.001
PEAK 17
Max. Ab. = 0.002
Min. Ab. = –0.002
PEAK 18
Max. Ab. = 0.001
Min. Ab. = –0.002
3200
2800
2400
2000
1600
1200
800
700
$\bar{\nu}$, wavenumbers

CHLORDAN

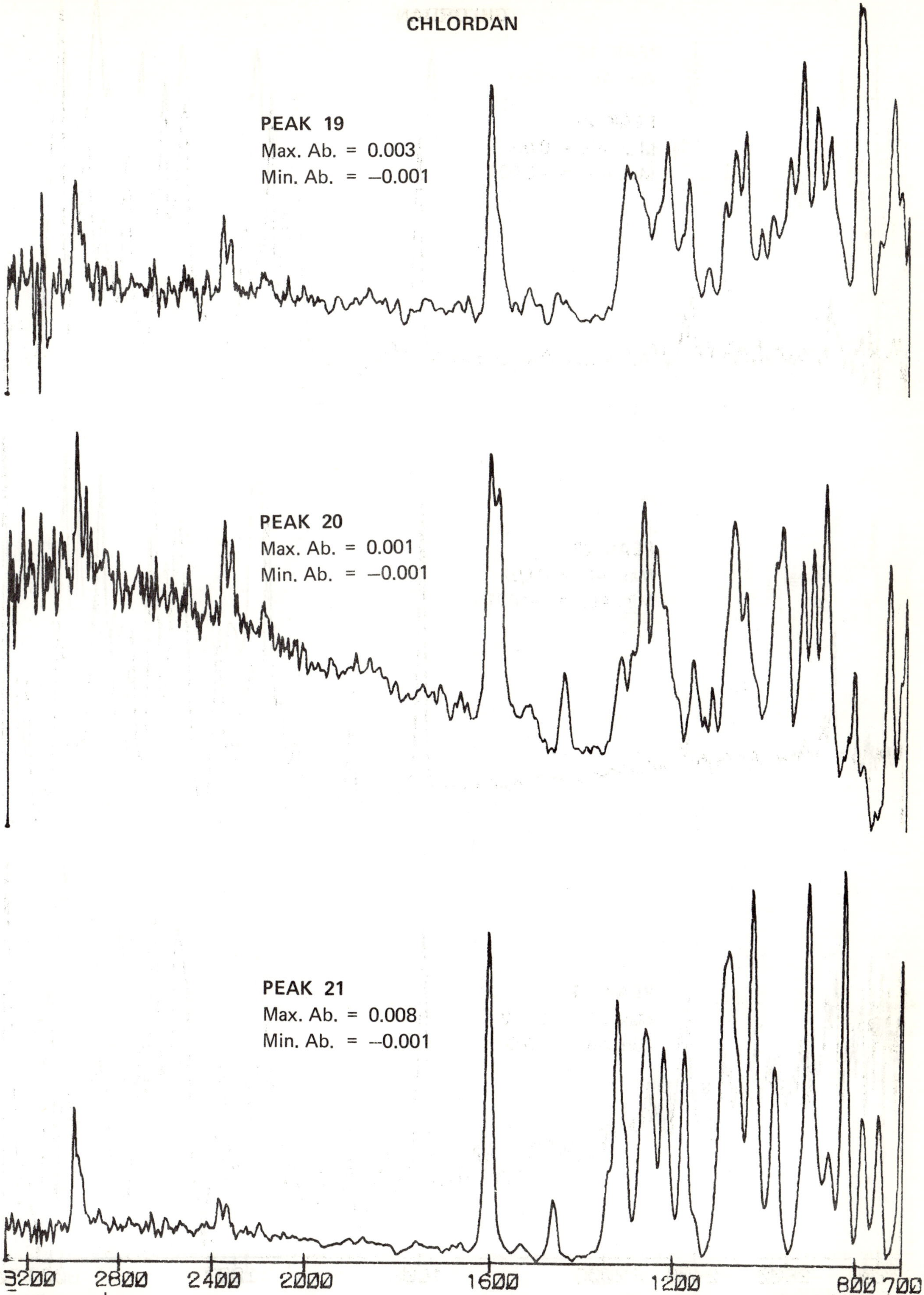
PEAK 19
Max. Ab. = 0.003
Min. Ab. = –0.001
PEAK 20
Max. Ab. = 0.001
Min. Ab. = –0.001
PEAK 21
Max. Ab. = 0.008
Min. Ab. = –0.001
3200
2800
2400
2000
1600
1200
800
700
$\tilde{\nu}$, wavenumbers

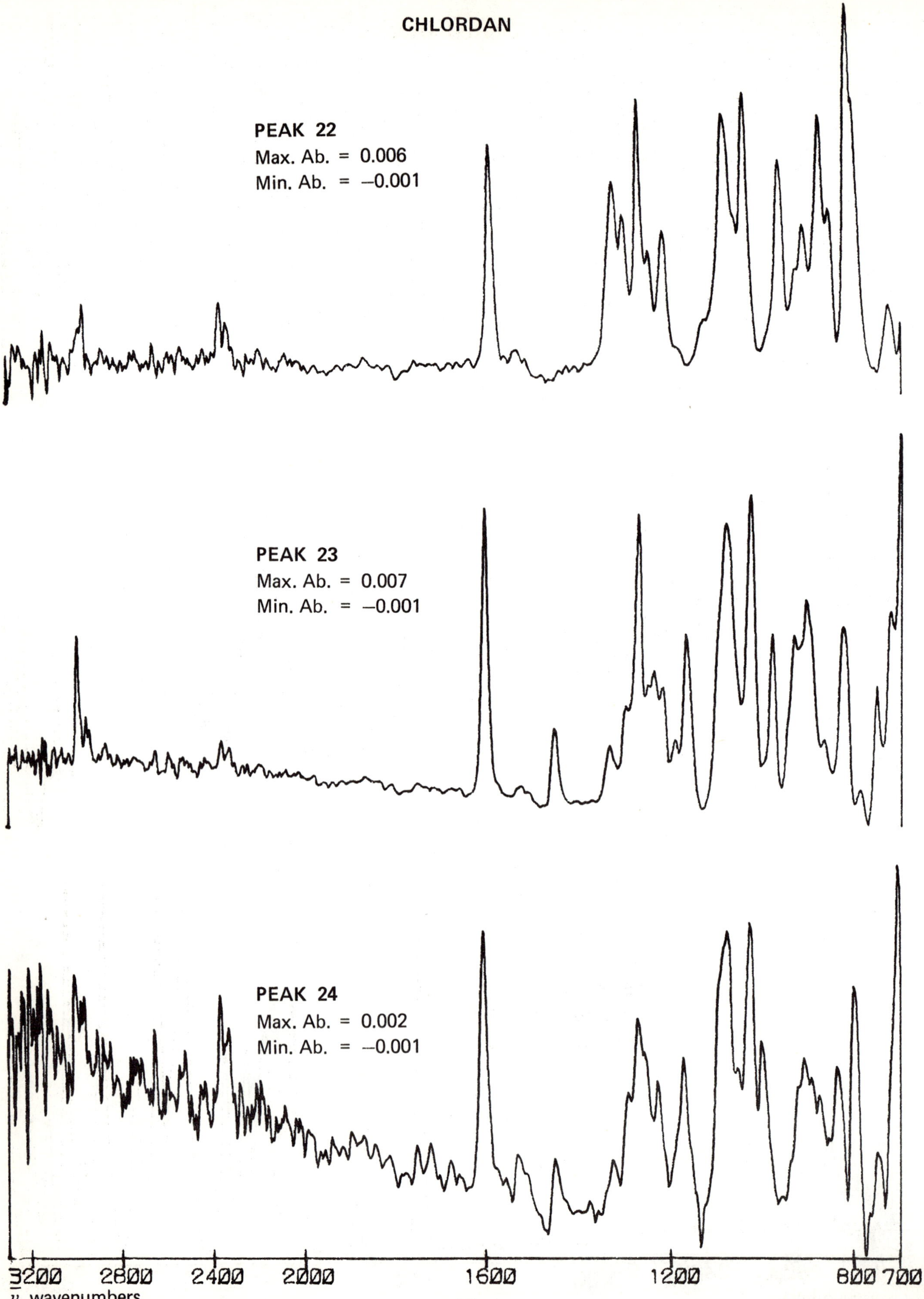
PEAK 22
Max. Ab. = 0.006
Min. Ab. = –0.001
PEAK 23
Max. Ab. = 0.007
Min. Ab. = –0.001
PEAK 24
Max. Ab. = 0.002
Min. Ab. = –0.001
3200
2800
2400
2000
1600
1200
800
700
ν̄, wavenumbers

CHLORDAN

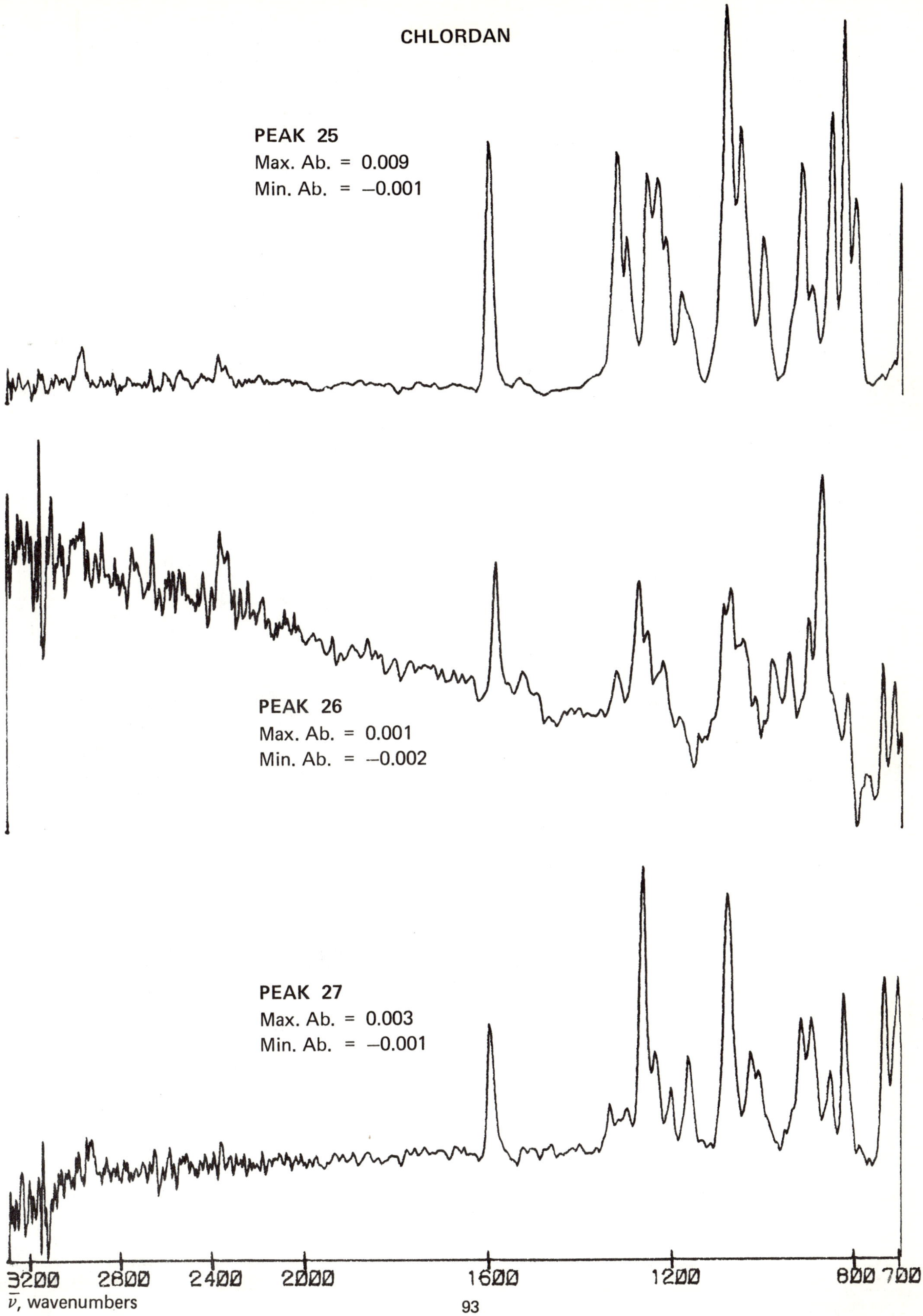
PEAK 25
Max. Ab. = 0.009
Min. Ab. = –0.001
PEAK 26
Max. Ab. = 0.001
Min. Ab. = –0.002
PEAK 27
Max. Ab. = 0.003
Min. Ab. = –0.001
3200
2800
2400
2000
1600
1200
800
700
$\bar{\nu}$, wavenumbers

THE

INFRARED SPECTRA

HANDBOOK OF

PRIORITY POLLUTANTS AND TOXIC CHEMICALS

ALPHABETICAL INDEX

PREFACE

This Alphabetical Index for the *Infrared Spectra Handbook of Priority Pollutants and Toxic Chemicals* contains the book numbers for the 363 compounds appearing in this handbook. It provides easy access to all spectra.

The method of naming the chemical compounds in the Sadtler reference spectra collections closely follows the *Chemical Abstracts* (CA) nomenclature system in operation prior to the publication of the CA Ninth Collective Index. Briefly, this system assigns the parent name on the basis of a predefined priority of functionality (the Sadtler Chemical Class index sequence is comparable to this priority table), and then substituents on the parent compound or derivatives are listed after the parent name in alphabetical sequence. In addition to the systematized names, cross-references to frequently used trivial or generic names are also included in this index.

Since the Sadtler indexing system is totally computerized, certain limitations are imposed in nomenclature symbolism. Deviations which appear in the printed index are as follows:

/	(slash)	--	Read as parenthesis, bracket, or brace
¤	(lozenge)	--	Read as indication of superscripting
<		--	Read as indication of superscripting
PR		--	Read as "prime". Therefore 2PR is read as 2'.
*	(asterisk)	--	Also read as "prime"

Lack of provision for Greek letters, super or subscripts (except as noted above) or lower case letters causes the following changes from normal notation:

A	--	Read as a or α
B	--	Read as b or β
G	--	Read as g or γ
D	--	Read as d or δ
O	--	Read as o, ortho or oxygen
M	--	Read as m or meta
N	--	Read as normal or nitrogen
P	--	Reas as p, para or phosphorus
S	--	Read as s (symmetrical) or sulfur
D & L	--	Still retain their original significance when used in conjunction with carbohydrates.

1/82

1/82

1/82

1/82

1/82

1/82

1/82

1/82

1/82

1/82

NAME BOOK NO.

THE

INFRARED SPECTRA

HANDBOOK OF

PRIORITY POLLUTANTS AND TOXIC CHEMICALS

MOLECULAR FORMULA INDEX

PREFACE

This Molecular Formula Index for the *Infrared Spectra Handbook of Priority Pollutants and Toxic Chemicals* contains the book numbers for the 363 compounds appearing in this handbook. It provides easy access to all spectra.

The arrangement of this Molecular Formula Index is based on the Hill system. Compounds containing carbon are listed first by the number of carbon atoms present, then by the number of hydrogen atoms, and then according to the presence and quality of non-metallic elements following the sequence of Br, Cl, F, I, N, O, P, S and Si. Compounds containing carbon and no hydrogen are indexed at the beginning of each category according to carbon content and arranged within the category by the occurence of the non-metallic elements listed above. Compounds which do not contain carbon are listed on the final pages of the index.

The final column in the formula table, titled M, is used to indicate the presence of other elements not otherwise specified. The presence of one of these elements in a compound is represented by its symbol (NA, K, FE, etc.). If more than one of these elements is present, they are not specified but signified by X2 (for two difference elements), X3 (for three), etc.

For compounds having the same molecular formula, the further arrangement is alphabetical according to chemical name.

There are some special cases which modify the presentation of the molecular formula, they are as follows.

(1) Inorganic acid salts of basic compounds such as amines are found under the formula of the amine alone. However, the anion of "onium" compounds is included in the formula.

(2) Compounds which contain an organic acid salt of a basic component are cross-referenced to the basic component and to the total formula of the compound.

(3) Molecules of hydration are not included in the formula but can be ascertained from the chemical name.

(4) In any case where the number of atoms of an element exceed the space available to list them the highest possible integer is used. For example, for compounds containing 9 or more atoms of a halogen the representation is the maximum available of 9.

MOLECULAR FORMULA INDEX

NAME	C	H	Br	Cl	F	I	N	O	P	S	Si	M	BOOK NO.
ACTINOLITE													217
ALLTOX													351
AMIANTHUS													217
AMOSITE													217
AROCHLOR 1016													28
AROCHLOR 1221													29
AROCHLOR 1232													30
AROCHLOR 1242													31
AROCHLOR 1248													32
AROCHLOR 1260													212
ASBESTOS													217
CHLORINATED CAMPHENE													351
GALLOTANNIC ACID													347
GALLOTANNIN													347
GLYCERITE													347
LEAD NAPHTHENATE													318
NAPHTHENIC ACID													133
NAPHTHENIC ACID, LEAD SALT													318
NATIVE CALCIUM-MAGNESIUM SILICATE													217
PCB-1016													28
PCB-1221													29
PCB-1232													30
PCB-1242													31
PCB-1248													32
PCB-1260													212
PENPHENE													351
POLYCHLORCAMPHENE													351
POLYCHLORINATED BIPHENYL													28
POLYCHLORINATED BIPHENYL													29
POLYCHLORINATED BIPHENYL													30
POLYCHLORINATED BIPHENYL													31
POLYCHLORINATED BIPHENYL													32
POLYCHLORINATED BIPHENYL													212
TANNIC ACID													347
TANNIN													347
TOXAPHENE													351
ALGOFRENE TYPE 2	1			2	2								123
ARCTON 6	1			2	2								123
DICHLORODIFLUOROMETHANE	1			2	2								123
FREON 12	1			2	2								123
GENETRON 21	1			2	2								123
METHANE, DICHLORODIFLUORO-,	1			2	2								123
ALGOFRENE TYPE 1	1			3	1								124
FLUOROTRICHLOROMETHANE	1			3	1								124
FREON 11	1			3	1								124
GENETRON 11	1			3	1								124
METHANE, FLUOROTRICHLORO-,	1			3	1								124
METHANE, TRICHLOROFLUORO-,	1			3	1								124
CARBON TETRACHLORIDE	1			4									63
METHANE, TETRACHLORO-,	1			4									63
CARBON BISULFIDE	1									2			62
CARBON DISULFIDE	1									2			62
CARBON SULFIDE	1									2			62
WEEVILTOX	1									2			62
BROMODICHLOROMETHANE	1	1	1	2									120
METHANE, BROMODICHLORO-,	1	1	1	2									120
METHANE, DICHLOROBROMO-,	1	1	1	2									120
CHLORODIBROMOMETHANE	1	1	2	1									121
METHANE, CHLORODIBROMO-,	1	1	2	1									121
METHANE, DIBROMOCHLORO-,	1	1	2	1									121
BROMOFORM	1	1	3										125
METHANE, TRIBROMO-,	1	1	3										125
TRIBROMOMETHANE	1	1	3										125
CHLOROFORM	1	1		3									64
METHANE, TRICHLORO-,	1	1		3									64
DICHLOROMETHANE	1	2		2									122
METHANE, DICHLORO-,	1	2		2									122
METHANE DICHLORIDE	1	2		2									122
METHYLENE CHLORIDE	1	2		2									122
METHYLENE DICHLORIDE	1	2		2									122
FORMALDEHYDE	1	2						1					111
FORMALIN	1	2						1					111
FORMIC ALDEHYDE	1	2						1					111
METHANAL	1	2						1					111
FORMIC ACID	1	2						2					112
METHANOIC ACID	1	2						2					112
BROM-O-GAS	1	3	1										119

NAME | C | H | Br | Cl | F | I | N | O | P | S | Si | M | BOOK NO.

MOLECULAR FORMULA INDEX

NAME	C	H	Br	Cl	F	I	N	O	P	S	Si	M	BOOK NO.
BROMOMETHANE	1	3	1										119
METHANE, BROMO-,	1	3	1										119
METHOGAS	1	3	1										119
METHYL BROMIDE	1	3	1										119
ROTOX	1	3	1										119
ARTIC	1	3		1									312
CHLOROMETHANE	1	3		1									312
METHANE, CHLORO-,	1	3		1									312
METHYL CHLORIDE	1	3		1									312
THIOCARBAMIDE	1	4					2			1			359
THIOUREA	1	4					2			1			359
THU	1	4					2			1			359
UREA, THIO-,	1	4					2			1			359
METHANETHIOL	1	4								1			128
METHYL MERCAPTAN	1	4								1			128
CARBINAMINE	1	5					1						130
METHANE, AMINO-,	1	5					1						130
METHYLAMINE	1	5					1						130
MONOMETHYLAMINE	1	5					1						130
HYDRAZINE, METHYL-, SULFATE /1 TO 1/	1	6					2						304
ETHYLENE, TETRACHLORO-,	2			4									103
ETHYLENE TETRACHLORIDE	2			4									103
PERCHLOROETHYLENE	2			4									103
PERCLENE	2			4									103
TETRACAP	2			4									103
TETRACHLOROETHYLENE	2			4									103
ETHANE, HEXACHLORO-,	2			6									88
ETHANE HEXACHLORIDE	2			6									88
HEXACHLOROETHANE	2			6									88
PERCHLOROETHANE	2			6									88
PHENOHEP	2			6									88
ACETYLENE TRICHLORIDE	2	1		3									104
ETHYLENE, TRICHLORO-,	2	1		3									104
ETHYLENE TRICHLORIDE	2	1		3									104
TCE	2	1		3									104
TRICHLOROETHYLENE	2	1		3									104
TRICLENE	2	1		3									104
ACETYLENE DICHLORIDE, TRANS-,	2	2		2									102
1,1-DCE	2	2		2									101
1,1-DICHLOROETHYLENE	2	2		2									101
1,2-DICHLOROETHYLENE, TRANS-,	2	2		2									102
ETHENE, 1,1-DICHLORO-,	2	2		2									101
ETHYLENE, 1,1-DICHLORO-,	2	2		2									101
ETHYLENE, 1,2-DICHLORO-, TRANS-,	2	2		2									102
VINYLIDENE CHLORIDE	2	2		2									101
ACETYLENE TETRACHLORIDE	2	2		4									89
BONOFORM	2	2		4									89
CELLON	2	2		4									89
ETHANE, 1,1,2,2-TETRACHLORO-,	2	2		4									89
TETRACHLOROETHANE, 1,1,2,2-,	2	2		4									89
CHLOROETHYLENE	2	3		1									100
ETHENE, CHLORO-,	2	3		1									100
ETHYLENE, CHLORO-,	2	3		1									100
VINYL CHLORIDE	2	3		1									100
ACETIC ACID, CHLORIDE	2	3		1				1					15
ACETYL CHLORIDE	2	3		1				1					15
ETHANOYL CHLORIDE	2	3		1				1					15
AEROTHENE TT	2	3		3									90
CHLORTEN	2	3		3									90
ETHANE, 1,1,1-TRICHLORO-,	2	3		3									90
ETHANE, 1,1,2-TRICHLORO-,	2	3		3									91
ETHANE TRICHLORIDE	2	3		3									91
METHYLCHLOROFORM	2	3		3									90
TRICHLOROETHANE, 1,1,1-,	2	3		3									90
TRICHLOROETHANE, 1,1,2-,	2	3		3									91
VINYL TRICHLORIDE	2	3		3									91
1,2-DIBROMOETHANE	2	4	2										85
DOWFUME W-8	2	4	2										85
DOWFUME W-85	2	4	2										85
DOWFUME 40	2	4	2										85
DOWFUME EDB	2	4	2										85
DOWFUME MC-2	2	4	2										85
ETHANE, 1,2-DIBROMO-,	2	4	2										85
ETHYLENE DIBROMIDE	2	4	2										85
GLYCOL DIBROMIDE	2	4	2										85
1,1-DICHLOROETHANE	2	4		2									86

NAME C H Br Cl F I N O P S Si M BOOK NO.

NAME	C	H	Br	Cl	F	I	N	O	P	S	Si	M	BOOK NO.
1,2-DICHLOROETHANE	2	4		2									87
ETHANE, 1,1-DICHLORO-,	2	4		2									86
ETHANE, 1,2-DICHLORO-,	2	4		2									87
ETHYLENE DICHLORIDE	2	4		2									87
ETHYLIDENE DICHLORIDE	2	4		2									86
GLYCOL DICHLORIDE	2	4		2									87
BCME	2	4		2				1					93
BIS-CME	2	4		2				1					93
CHLOROMETHYL ETHER	2	4		2				1					93
ETHER, BIS/CHLOROMETHYL/,	2	4		2				1					93
AMITROLE	2	4					4						189
AMIZOL	2	4					4						189
TRIAZOLE, 1H-1,2,4-, 3-AMINO-,	2	4					4						189
S-TRIAZOLE, 3-AMINO-,	2	4					4						189
WEEDOCLOR	2	4					4						189
ACETALDEHYDE	2	4						1					3
ACETIC ALDEHYDE	2	4						1					3
ETHANAL	2	4						1					3
ETHYL ALDEHYDE	2	4						1					3
ACETIC ACID	2	4						2					7
ETHANOIC ACID	2	4						2					7
ETHYLIC ACID	2	4						2					7
METHANECARBOXYLIC ACID	2	4						2					7
VINEGAR ACID	2	4						2					7
CHLOROETHANE	2	5		1									84
ETHANE, CHLORO-,	2	5		1									84
ETHYL CHLORIDE	2	5		1									84
KELENE	2	5		1									84
NARCOTILE	2	5		1									84
AZACYCLOPROPANE	2	5					1						107
AZIRIDINE	2	5					1						107
DIMETHYLENIMINE	2	5					1						107
ETHYLENIMINE	2	5					1						107
ACETAMIDE	2	5					1	1					4
ACETIC ACID AMIDE	2	5					1	1					4
ETHANAMIDE	2	5					1	1					4
METHANECARBOXAMIDE	2	5					1	1					4
ACETAMIDE, THIO-,	2	5					1			1			6
ACETOTHIOAMIDE	2	5					1			1			6
ETHANETHIOAMIDE	2	5					1			1			6
TAA	2	5					1			1			6
THIOACETAMIDE	2	5					1			1			6
GUANIDINE, 1-METHYL-3-NITRO-1-NITROSO-,	2	5					5	3					296
GUANIDINE, N-METHYL-N*-NITRO-N-NITROSO-,	2	5					5	3					296
MNG	2	5					5	3					296
MNNG	2	5					5	3					296
DIMETHYLAMINE, N-NITROSO-,	2	6					2	1					80
DMNA	2	6					2	1					80
NITROSAMINE, N,N-DIMETHYL-,	2	6					2	1					80
ETHANE, 1,2-DIHYDROXY-,	2	6						2					106
1,2-ETHANEDIOL	2	6						2					106
ETHYLENE GLYCOL	2	6						2					106
TESCOL	2	6						2					106
METHANESULFONIC ACID, METHYL ESTER	2	6						3		1			127
METHYL METHANESULFONATE	2	6						3		1			127
MMS	2	6						3		1			127
DIMETHYL SULFATE	2	6						4		1			182
METHYL SULFATE	2	6						4		1			182
SULFURIC ACID, DIMETHYL ESTER	2	6						4		1			182
DIMETHYLAMINE	2	7					1						79
DMA	2	7					1						79
ETHANAMINE	2	7					1						98
ETHANE, AMINO-,	2	7					1						98
ETHYLAMINE	2	7					1						98
MONOETHYLAMINE	2	7					1						98
SYM-DIMETHYLHYDRAZINE, DIHYDROCHLORIDE	2	8					2						303
1,2-ETHANEDIAMINE	2	8					2						105
ETHYLENEDIAMINE	2	8					2						105
HYDRAZINE, N,N*-DIMETHYL-, DIHYDRO-CHLORIDE	2	8					2						303
HYDRAZINE, 1,2-DIMETHYL-, DIHYDRO-CHLORIDE	2	8					2						303
HYDRAZINE, 1,1-DIMETHYL-, MONO-HYDROCHLORIDE	2	8					2						302
ACRYLON	3	3					1						19

NAME | C | H | Br | Cl | F | I | N | O | P | S | Si | M | BOOK NO.

NAME	C	H	Br	Cl	F	I	N	O	P	S	Si	M	BOOK NO.
ACRYLONITRILE	3	3					1						19
CARBACRYL	3	3					1						19
PROPENENITRILE	3	3					1						19
1,3-DICHLOROPROPENE	3	4		2									170
PROPENE, 1,3-DICHLORO-,	3	4		2									170
PROPYLENE, 1,3-DICHLORO-,	3	4		2									170
TELONE	3	4		2									170
DALAPON	3	4		2				2					173
2,2-DICHLOROPROPIONIC ACID	3	4		2				2					173
DOWPON	3	4		2				2					173
PROPIONIC ACID, 2,2-DICHLORO-,	3	4		2				2					173
ACROLEIN	3	4						1					16
ACRYLALDEHYDE	3	4						1					16
ALLYL ALDEHYDE	3	4						1					16
ETHYLENE ALDEHYDE	3	4						1					16
PROPENAL	3	4						1					16
ACROLEIC ACID	3	4						2					17
ACRYLIC ACID	3	4						2					17
BETAPRONE	3	4						2					330
EPIHYDRINALDEHYDE	3	4						2					363
ETHYLENECARBOXYLIC ACID	3	4						2					17
GLYCIDALDEHYDE	3	4						2					363
2-OXETANONE	3	4						2					330
1-PROPANAL, 2,3-EPOXY-,	3	4						2					363
PROPENOIC ACID	3	4						2					17
B-PROPIOLACTONE	3	4						2					330
PROPIONALDEHYDE, 2,3-EPOXY-,	3	4						2					363
PROPIONIC ACID, B-HYDROXY-, LACTONE	3	4						2					330
DBCP	3	5	2	1									168
FUMAZONE	3	5	2	1									168
NEMAGON	3	5	2	1									168
PROPANE, 1-CHLORO-2,3-DIBROMO-,	3	5	2	1									168
PROPANE, 1,2-DIBROMO-3-CHLORO-,	3	5	2	1									168
1,2-DICHLOROPROPANE	3	6		2									169
PROPANE, 1,2-DICHLORO-,	3	6		2									169
PROPYLENE DICHLORIDE	3	6		2									169
ACETALDEHYDE, METHYL-,	3	6						1					171
ALLYL ALCOHOL	3	6						1					20
CARBINOL, VINYL-,	3	6						1					20
PROPALDEHYDE	3	6						1					171
PROPANAL	3	6						1					171
2-PROPEN-1-OL	3	6						1					20
PROPIONALDEHYDE	3	6						1					171
PROPIONIC ALDEHYDE	3	6						1					171
ETHANECARBOXYLIC ACID	3	6						2					172
FORMIC ACID, ETHYL-,	3	6						2					172
PROPIONIC ACID	3	6						2					172
PROZOIN	3	6						2					172
PSEUDOACETIC ACID	3	6						2					172
ALDEFORM	3	6						3					191
FORMAGENE	3	6						3					191
MARVOSAN	3	6						3					191
PARAFORMALDEHYDE	3	6						3					191
S-TRIOXANE	3	6						3					191
1,2-OXATHIOLANE, 2,2-DIOXIDE	3	6						3		1			137
1-PROPANESULFONIC ACID 3-HYDROXY-, G-SULTONE	3	6						3		1			137
PROPANESULTONE	3	6						3		1			137
CARBAMIC ACID, ETHYL ESTER	3	7					1	2					60
ETHYL CARBAMATE	3	7					1	2					60
PRACARBAMIN	3	7					1	2					60
URETHANE	3	7					1	2					60
ENU	3	7					3	2					358
UREA, 1-ETHYL-1-NITROSO-,	3	7					3	2					358
ETHYL METHANESULFONATE	3	8						3		1			126
METHANESULFONIC ACID, ETHYL ESTER	3	8						3		1			126
1,3-BUTADIENE, HEXACHLORO-,	4			6									56
HCBD	4			6									56
HEXACHLOROBUTADIENE	4			6									56
2,5-FURANDIONE	4	2						3					116
LYTRON 810	4	2						3					116
LYTRON 820	4	2						3					116
MALEIC ANHYDRIDE	4	2						3					116
TOXILIC ANHYDRIDE	4	2						3					116
ANTAGOTHYROID	4	4					2	1		1			354
DERACIL	4	4					2	1		1			354
PYRIMIDINEDIONE, 2,4/1H,3H/-, 2-THIO-,	4	4					2	1		1			354

NAME C H Br Cl F I N O P S Si M BOOK NO.

1/82

NAME	C	H	Br	Cl	F	I	N	O	P	S	Si	M	BOOK NO.
PIPERAZINE, 1,4-DINITROSO-,	4	8					4	2					164
ACETIC ACID, ETHYL-,	4	8						2					59
BUTANOIC ACID	4	8						2					59
BUTYRIC ACID	4	8						2					59
1,4-DIETHYLENE DIOXIDE	4	8						2					81
DIETHYLENE ETHER	4	8						2					81
P-DIOXANE	4	8						2					81
P-DIOXIN, TETRAHYDRO-,	4	8						2					81
DANA	4	10					2	1					77
DEN	4	10					2	1					77
DENA	4	10					2	1					77
DIETHYLAMINE, N-NITROSO-,	4	10					2	1					77
BASIC LEAD ACETATE	4	10						8				PB	310
LEAD, BIS/ACETATO/TETRAHYDROXYTRI-,	4	10						8				PB	310
LEAD SUBACETATE	4	10						8				PB	310
MONOBASIC LEAD ACETATE	4	10						8				PB	310
BUTANE, 1-AMINO-,	4	11					1						58
BUTYLAMINE	4	11					1						58
DEN	4	11					1						76
DIETHYLAMINE	4	11					1						76
NORVALAMINE	4	11					1						58
C-56	5			6									74
1,3-CYCLOPENTADIENE, 1,2,3,4,5,5-HEXA- CHLORO-,	5			6									74
HEXACHLOROCYCLOPENTADIENE	5			6									74
ARTIFICIAL ANT OIL	5	4						2					113
FURAL	5	4						2					113
2-FURALDEHYDE	5	4						2					113
FURFURAL	5	4						2					113
FURFUROLE	5	4						2					113
ALKIRON	5	6					2	1		1			353
METACIL	5	6					2	1		1			353
MURACIL	5	6					2	1		1			353
PYRIMIDINEDIONE, 2,4/1H,3H/-, 6-METHYL-2-THIO-,	5	6					2	1		1			353
URACIL, 6-METHYL-2-THIO-,	5	6					2	1		1			353
1,3-BUTADIENE, 2-METHYL-,	5	8											114
ISOPRENE	5	8											114
METHACRYLIC ACID, METHYL ESTER	5	8						2					117
METHYL METHACRYLATE	5	8						2					117
MME	5	8						2					117
MONOCITE METHACRYLATE MONOMER	5	8						2					117
PROPENIC ACID, 2-METHYL-, METHYL ESTER	5	8						2					117
FORMALDEHYDE, BIS/2-CHLOROETHYL/ ACETAL	5	10		2				2					118
METHANE, BIS/2-CHLOROETHOXY/-,	5	10		2				2					118
1-NITROSOPIPERIDINE	5	10					2	1					165
PIPERIDINE, N-NITROSO-,	5	10					2	1					165
PIPERIDINE, 1-NITROSO-,	5	10					2	1					165
UREA, 1-BUTYL-1-NITROSO-,	5	11					3	2					356
BENZENE, HEXACHLORO-,	6			6									43
BUNT-CURE	6			6									43
ANTI-CARIE 80	6			6									43
HCB	6			6									43
HEXACHLOROBENZENE	6			6									43
SANOCIDE	6			6									43
CHLOROPHEN	6	1		5				1					152
DOWICIDE G	6	1		5				1					152
DOWICIDE T	6	1		5				1					152
PCP	6	1		5				1					152
PENTACHLOROPHENOL	6	1		5				1					152
PHENOL, PENTACHLORO-,	6	1		5				1					152
WEEDONE	6	1		5				1					152
BENZENE, 1,2,4-TRICHLORO-,	6	3		3									46
TRICHLOROBENZENE, 1,2,4-,	6	3		3									46
DOWICIDE 2S	6	3		3				1					153
OMAL	6	3		3				1					153
PHENACHLOR	6	3		3				1					153
PHENOL, 2,4,6-TRICHLORO-,	6	3		3				1					153
TRICHLOROPHENOL, 2,4,6-,	6	3		3				1					153

MOLECULAR FORMULA INDEX

NAME	C	H	Br	Cl	F	I	N	O	P	S	Si	M	BOOK NO.
BENZENE, M-DICHLORO-,	6	4		2									36
BENZENE, O-DICHLORO-,	6	4		2									37
BENZENE, P-DICHLORO-,	6	4		2									38
BENZENE, 1,3-DICHLORO-,	6	4		2									36
BENZENE, 1,2-DICHLORO-,	6	4		2									37
BENZENE, 1,4-DICHLORO-,	6	4		2									38
CHLOROBEN	6	4		2									37
M-DICHLOROBENZENE	6	4		2									36
O-DICHLOROBENZENE	6	4		2									37
P-DICHLOROBENZENE	6	4		2									38
DIZENE	6	4		2									37
ODCB	6	4		2									37
PARACIDE	6	4		2									38
PDCB	6	4		2									38
SANTOCHLOR 34	6	4		2									38
DCP	6	4		2				1					142
2,4-DICHLOROPHENOL	6	4		2				1					142
PHENOL, 2,4-DICHLORO-,	6	4		2				1					142
BENZENE, M-DINITRO-,	6	4					2	4					39
BENZENE, O-DINITRO-,	6	4					2	4					40
BENZENE, P-DINITRO-,	6	4					2	4					41
BENZENE, 1,3-DINITRO-,	6	4					2	4					39
BENZENE, 1,2-DINITRO-,	6	4					2	4					40
BENZENE, 1,4-DINITRO-,	6	4					2	4					41
M-DINITROBENZENE	6	4					2	4					39
O-DINITROBENZENE	6	4					2	4					40
P-DINITROBENZENE	6	4					2	4					41
DITHANE A-4	6	4					2	4					41
ALDIFEN	6	4					2	5					145
CHEMOX PE	6	4					2	5					145
DINITROPHENOL, BETA-,	6	4					2	5					147
DINITROPHENOL, GAMMA-,	6	4					2	5					146
2,3-DINITROPHENOL	6	4					2	5					144
2,4-DINITROPHENOL	6	4					2	5					145
2,5-DINITROPHENOL	6	4					2	5					146
2,6-DINITROPHENOL	6	4					2	5					147
3,4-DINITROPHENOL	6	4					2	5					148
2,4-DNP	6	4					2	5					145
2,5-DNP	6	4					2	5					146
PHENOL, 2,3-DINITRO-,	6	4					2	5					144
PHENOL, 2,4-DINITRO-,	6	4					2	5					145
PHENOL, 2,5-DINITRO-,	6	4					2	5					146
PHENOL, 2,6-DINITRO-,	6	4					2	5					147
PHENOL, 3,4-DINITRO-,	6	4					2	5					148
SOLFO BLACK B	6	4					2	5					145
SOLFO BLACK 2B	6	4					2	5					145
SOLFO BLACK BB	6	4					2	5					145
BENZENE, CHLORO-,	6	5		1									35
BENZENE CHLORIDE	6	5		1									35
CHLOROBENZENE	6	5		1									35
MCB	6	5		1									35
MONOCHLOROBENZENE	6	5		1									35
PHENYL CHLORIDE	6	5		1									35
O-CHLOROPHENOL	6	5		1				1					141
PHENOL, O-CHLORO-,	6	5		1				1					141
PHENOL, 2-CHLORO-,	6	5		1				1					141
BENZENE, NITRO-,	6	5					1	2					45
CI SOLVENT BLACK 5	6	5					1	2					45
ESSENCE OF MIRBANE	6	5					1	2					45
NITROBENZENE	6	5					1	2					45
OIL OF MIRBANE	6	5					1	2					45
BENZENE, 2-HYDROXYNITRO-,	6	5					1	3					150
BENZENE, 4-HYDROXYNITRO-,	6	5					1	3					151
M-NITROPHENOL	6	5					1	3					149
O-NITROPHENOL	6	5					1	3					150
P-NITROPHENOL	6	5					1	3					151
PHENOL, M-NITRO-,	6	5					1	3					149
PHENOL, O-NITRO-,	6	5					1	3					150
PHENOL, P-NITRO-,	6	5					1	3					151
PHENOL, 3-NITRO-,	6	5					1	3					149
PHENOL, 2-NITRO-,	6	5					1	3					150
PHENOL, 4-NITRO-,	6	5					1	3					151
BENZENE	6	6											33
BENZOL	6	6											33
BENZOLE	6	6											33
CARBON OIL	6	6											33
PHENE	6	6											33
AFICIDE	6	6		6									267
BENZENE HEXACHLORIDE, A-ISOMER	6	6		6									264
BENZENE HEXACHLORIDE, B-ISOMER	6	6		6									265
BENZENE HEXACHLORIDE, D-ISOMER	6	6		6									266
BENZENE HEXACHLORIDE, G-ISOMER	6	6		6									267
BENZENE HEXACHLORIDE /MIXED ISOMERS/	6	6		6									268

NAME | C | H | Br | Cl | F | I | N | O | P | S | Si | M | BOOK NO.

1/82

MOLECULAR FORMULA INDEX

NAME	C	H	Br	Cl	F	I	N	O	P	S	Si	M	BOOK NO.
A-BHC	6	6		6									264
B-BHC	6	6		6									265
D-BHC	6	6		6									266
CYCLOHEXANE, 1,2,3,4,5,6-HEXA- CHLORO-, A-ISOMER	6	6		6									264
CYCLOHEXANE, 1,2,3,4,5,6-HEXA- CHLORO-, B-ISOMER	6	6		6									265
CYCLOHEXANE, 1,2,3,4,5,6-HEXA- CHLORO-, D-ISOMER	6	6		6									266
CYCLOHEXANE, 1,2,3,4,5,6-HEXA- CHLORO-, G-ISOMER	6	6		6									267
CYCLOHEXANE, 1,2,3,4,5,6-HEXA- CHLORO-, /MIXED ISOMERS/	6	6		6									268
KWELL	6	6		6									267
LINDANE	6	6		6									267
A-LINDANE	6	6		6									264
B-LINDANE	6	6		6									265
D-LINDANE	6	6		6									266
BENZENE, HYDROXY-,	6	6						1					140
CARBOLIC ACID	6	6						1					140
PHENOL	6	6						1					140
BENZENE, M-DIHYDROXY-,	6	6						2					176
M-BENZENEDIOL	6	6						2					176
PELAGOL RS	6	6						2					176
RESORCINE	6	6						2					176
RESORCINOL	6	6						2					176
AMINOPHEN	6	7					1						22
ANILINE	6	7					1						22
BENZENAMINE	6	7					1						22
BENZENE, AMINO-,	6	7					1						22
BENZIDAM	6	7					1						22
AMIDON	6	7					3	1					307
ISONIAZID	6	7					3	1					307
ISONICOTINIC ACID, HYDRAZIDE	6	7					3	1					307
4-PYRIDINECARBOXYLIC ACID, HYDRAZIDE	6	7					3	1					307
ZINADON	6	7					3	1					307
HYDRAZINE, PHENYL-, MONOHYDRO- CHLORIDE	6	8					2						305
BENZENE, HEXAHYDRO-,	6	12											72
CYCLOHEXANE	6	12											72
ETHER, BIS/2-CHLOROISOPROPYL/,	6	12		2				1					94
ETHER, BIS/2-CHLORO-1-METHYLETHYL/,	6	12		2				1					94
ACETIC ACID, BUTYL ESTER	6	12						2					8
BUTYL ACETATE	6	12						2					8
BUTYL ALCOHOL, ACETATE	6	12						2					8
BUTYRIC ACID, 2-AMINO-4-/ETHYL- THIO/-, DL-,	6	13					1	2		1			235
BUTYRIC ACID, 2-AMINO-4-/ETHYL- THIO/-, L-,	6	13					1	2		1			236
ETH	6	13					1	2		1			235
ETHIONINE, DL-,	6	13					1	2		1			235
ETHIONINE, L-,	6	13					1	2		1			236
HOMOCYSTEINE, S-ETHYL-, DL-,	6	13					1	2		1			235
HOMOCYSTEINE, S-ETHYL-, L-,	6	13					1	2		1			236
DIPROPYLAMINE, N-NITROSO-,	6	14					2	1					82
DPN	6	14					2	1					82
NITROSAMINE, DIPROPYL-,	6	14					2	1					82
ETHANAMINE, N,N-DIETHYL-,	6	15					1						190
TEN	6	15					1						190
TRIETHYLAMINE	6	15					1						190
BENZONITRILE, 2,6-DICHLORO-,	7	3		2			1						52
CASORON	7	3		2			1						52
2,6-DBN	7	3		2			1						52
DICHLOROBENIL	7	3		2			1						52
2,6-DICHLOROBENZONITRILE	7	3		2			1						52
DU-SPREX	7	3		2			1						52
1,2-BENZISOTHIAZOLIN-3-ONE, 1,1-DI- OXIDE, SODIUM SALT	7	4					1	3		1		NA	228
1,2-BENZISOTHIAZOL-3-OL, 1,1-DIOXIDE, SODIUM SALT	7	4					1	3		1		NA	228
1,2-BENZISOTHIAZOL-3/2H/-ONE, 1,1-DI- OXIDE, SODIUM SALT	7	4					1	3		1		NA	228
SACCHARIN, SODIUM SALT	7	4					1	3		1		NA	228
SODIUM SACCHARIN	7	4					1	3		1		NA	228
BENZENECARBONYL CHLORIDE	7	5		1				1					53
BENZOYL CHLORIDE	7	5		1				1					53
BENZENE, CYANO-,	7	5					1						51
NAME	C	H	Br	Cl	F	I	N	O	P	S	Si	M	BOOK NO.

NAME | C | H | Br | Cl | F | I | N | O | P | S | Si | M | BOOK NO.

NAME	C	H	Br	Cl	F	I	N	O	P	S	Si	M	BOOK NO.
ESTERON	8	5		3				3					13
REDDOX	8	5		3				3					13
T, 2,4,5-,	8	5		3				3					13
WEEDONE	8	5		3				3					13
FANFT	8	5					3	4		1			292
FORMAMIDE, N-/4-/5-NITRO-2-FURYL/- 2-THIAZOLYL/-,	8	5					3	4		1			292
ACETIC ACID, /2,4-DICHLOROPHEN- OXY/-,	8	6		2				3					12
AGROTECT	8	6		2				3					12
O-ANISIC ACID, 3,6-DICHLORO-,	8	6		2				3					26
BANLEN	8	6		2				3					26
BANVEL D	8	6		2				3					26
BENZOIC ACID, 3,6-DICHLORO- 2-METHOXY-,	8	6		2				3					26
2,4-D	8	6		2				3					12
DED WEED	8	6		2				3					12
DICAMBA	8	6		2				3					26
ESTERON	8	6		2				3					12
BENZENE, VINYL-,	8	8											180
CINNAMENE	8	8											180
ETHYLENE, PHENYL-,	8	8											180
STYRENE	8	8											180
STYRON	8	8											180
HYDRAZINE, 1-ACETYL-2-PICOLINOYL-,	8	9					3	2					301
BENZENE, DIMETHYL-,	8	10											360
BENZENE, 1,3-DIMETHYL-,	8	10											193
BENZENE, 1,2-DIMETHYL-,	8	10											194
BENZENE, 1,4-DIMETHYL-,	8	10											195
BENZENE, ETHYL-,	8	10											42
ETHANE, PHENYL-,	8	10											42
ETHYLBENZENE	8	10											42
XYLENE	8	10											360
M-XYLENE	8	10											193
O-XYLENE	8	10											194
P-XYLENE	8	10											195
XYLOL	8	10											360
M-XYLOL	8	10											193
O-XYLOL	8	10											194
P-XYLOL	8	10											195
AZOPHOS	8	10					1	5	1	1			338
METAPHOS	8	10					1	5	1	1			338
METHYL PARATHION	8	10					1	5	1	1			338
NITROX 80	8	10					1	5	1	1			338
PHOSPHOROTHIOIC ACID, O,O-DIMETHYL O-/P-NITROPHENYL/ ESTER	8	10					1	5	1	1			338
2,5-DMP	8	10						1					198
2,6-DMP	8	10						1					199
3,4-DMP	8	10						1					200
3,5-DMP	8	10						1					201
PHENOL, 2,3-DIMETHYL-,	8	10						1					196
PHENOL, 2,4-DIMETHYL-,	8	10						1					197
PHENOL, 2,5-DIMETHYL-,	8	10						1					198
PHENOL, 2,6-DIMETHYL-,	8	10						1					199
PHENOL, 3,4-DIMETHYL-,	8	10						1					200
PHENOL, 3,5-DIMETHYL-,	8	10						1					201
2,3-XYLENOL	8	10						1					196
2,4-XYLENOL	8	10						1					197
2,5-XYLENOL	8	10						1					198
2,6-XYLENOL	8	10						1					199
3,4-XYLENOL	8	10						1					200
3,5-XYLENOL	8	10						1					201
AMINOURACIL MUSTARD	8	11		2			3	2					352
DESMETHYLDOPAN	8	11		2			3	2					352
PYRIMIDINEDIONE, 2,4/1H,3H/-, 5-/BIS/2-CHLOROETHYL/AMINO/-,	8	11		2			3	2					352
URACIL, 5-/BIS/2-CHLOROETHYL/- AMINO/-,	8	11		2			3	2					352
PDMT	8	11					3						188
TRIAZENE, 3,3-DIMETHYL-1-PHENYL-,	8	11					3						188
GLUCOPYRANOSE, 2-DEOXY-2-/3-METHYL- 3-NITROSOUREIDO/-, D-,	8	15					3	7					295
STR	8	15					3	7					295
STREPTOZOTOCIN	8	15					3	7					295
1-BUTANAMINE, N-BUTYL-N-NITROSO-,	8	18					2	1					75
DBN	8	18					2	1					75
DIBUTYLAMINE, N-NITROSO-,	8	18					2	1					75
DISULFOTON	8	19						2	1	3			155
DI-SYSTON	8	19						2	1	3			155
PHOSPHORODITHIOIC ACID, O,O-DIETHYL S-/2-/ETHYLTHIO/ETHYL/ ESTER	8	19						2	1	3			155
SOLVIREX	8	19						2	1	3			155
THIODEMETON	8	19						2	1	3			155

| NAME | C | H | Br | Cl | F | I | N | O | P | S | Si | M | BOOK NO. |

NAME	C	H	Br	Cl	F	I	N	O	P	S	Si	M	BOOK NO.
BLADEX	8	20						7	2				340
ETHYL PYROPHOSPHATE	8	20						7	2				340
PYROPHOSPHORIC ACID, TETRAETHYL ESTER	8	20						7	2				340
TETRAETHYLPYROPHOSPHATE	8	20						7	2				340
TETRON	8	20						7	2				340
ENDOSULFAN	9	6		6				3		1			322
MALIX	9	6		6				3		1			322
6,9-METHANO-2,4,3-BENZODIOXATHIEPIN, 6,7,8,9,10,10-HEXACHLORO-1,5,5A,6,9,9A-HEXAHYDRO-, 3-OXIDE	9	6		6				3		1			322
5-NORBORNENE-2,3-DIMETHANOL, 1,4,5,-6,7,7-HEXACHLORO-, CYCLIC SULFITE	9	6		6				3		1			322
THIODAN	9	6		6				3		1			322
ENDOSULFAN SULFATE	9	6		6				4		1			321
5-NORBORNENE-2,3-DIMETHANOL, 1,4,5,-6,7,7-HEXACHLORO-, CYCLIC SULFATE	9	6		6				4		1			321
4-NQO	9	6					2	3					341
QUINOLINE, 4-NITRO-, 1-OXIDE	9	6					2	3					341
BENZOPYRAN-2-ONE, 2H-1-,	9	6						2					65
CINNAMIC ACID, O-HYDROXY-, D-LACTONE	9	6						2					65
COUMARIN	9	6						2					65
TONKA BEAN CAMPHOR	9	6						2					65
1-AZANAPHTHALENE	9	7					1						175
1-BENZAZINE	9	7					1						175
QUINOLINE	9	7					1						175
AMERCIDE	9	8		3			1	2		1			269
CAPTAN	9	8		3			1	2		1			269
4-CYCLOHEXENE-1,2-DICARBOXIMIDE, N-//TRICHLOROMETHYL/THIO/-,	9	8		3			1	2		1			269
ORTHOCIDE	9	8		3			1	2		1			269
VANGUARD K	9	8		3			1	2		1			269
DICHLORFENIDIM	9	10		2			2	1					357
DIUREX	9	10		2			2	1					357
DIURON	9	10		2			2	1					357
KARMEX	9	10		2			2	1					357
UREA, 3-/3,4-DICHLOROPHENYL/-1,1-DIMETHYL-,	9	10		2			2	1					357
CHLORPYRIFOS	9	11		3			1	3	1	1			157
O,O-DIETHYL O-3,5,6-TRICHLORO-2-PYRIDYL PHOSPHOROTHIOATE	9	11		3			1	3	1	1			157
DOWCO 179	9	11		3			1	3	1	1			157
DURSBAN	9	11		3			1	3	1	1			157
PHOSPHOROTHIOIC ACID, O,O-DIETHYL O-/3,5,6-TRICHLORO-2-PYRIDYL/ ESTER	9	11		3			1	3	1	1			157
2-CYCLOHEXEN-1-ONE, 3,5,5-TRIMETHYL-,	9	14						1					73
ISOPHORONE	9	14						1					73
BLADAN	9	22						4	2	4			336
ETHION	9	22						4	2	4			336
PHOSPHORODITHIOIC ACID, S,S*-METHYLENE O,O,O*,O*-TETRAETHYL ESTER	9	22						4	2	4			336
PHOSPHOTOX E	9	22						4	2	4			336
RODOCID	9	22						4	2	4			336
BICHLORENDO	10			9									315
CYCLOPENTADIENE, HEXACHLORO-, DIMER	10			9									315
DECHLORANE	10			9									315
METHENO-1H-CYCLOBUTA/CD/PENTALENE, 1,3,4-, 1,1A,2,2,3,3A,4,5,5,5A,5B,6-DODECACHLOROOCTAHYDRO-,	10			9									315
MIREX	10			9									315
CHLORDECONE	10			9				1					316
KEPONE	10			9				1					316
MEREX	10			9				1					316
METHENO-2H-CYCLOBUTA/CD/PENTALEN-2-ONE, 1,3,4-, 1,1A,3,3A,4,5,5,5A,5B,6-DECACHLOROOCTAHYDRO-,	10			9				1					316
DICHLONE	10	4		2				2					134
1,4-NAPHTHOQUINONE, 2,3-DICHLORO-,	10	4		2				2					134
PHYGON	10	4		2				2					134

NAME C H Br Cl F I N O P S Si M BOOK NO.

NAME	C	H	Br	Cl	F	I	N	O	P	S	Si	M	BOOK NO.
SANQUINON	10	4		2				2					134
US RUBBER 604	10	4		2				2					134
DRINOX	10	5		7									129
HEPTACHLOR	10	5		7									129
HEPTAGRAN	10	5		7									129
4,7-METHANOINDENE, 1,4,5,6,7,8,8-HEPTA- CHLORO-3A,4,7,7A-TETRAHYDRO-,	10	5		7									129
VELSICOL 104	10	5		7									129
ENT 25,584	10	5		7				1					313
HEPTACHLOR EPOXIDE	10	5		7				1					313
4,7-METHANOINDAN, 2,3-EPOXY-1,4,5,6,7,- 8,8-HEPTACHLORO-3A,4,7,7A-TETRAHYDRO-,	10	5		7				1					313
VELSICOL 53-CS-17	10	5		7				1					313
NAPHTHALENE, B-CHLORO-,	10	7		1									132
NAPHTHALENE, 2-CHLORO-,	10	7		1									132
MOTH BALLS	10	8											131
MOTH FLAKES	10	8											131
NAPHTHALENE	10	8											131
TAR CAMPHOR	10	8											131
CHLORDAN /MIXTURE OF ISOMERS/	10	8		8									314
4,7-METHANOINDAN, 1A,2,4B,5,6,7B,8,8- OCTACHLORO-3AA,4,7,7AA-TETRAHYDRO-,	10	8		8									314
ANTIOXIDANT MB	10	9					1						135
FAST GARNET B	10	9					1						135
FAST SCARLET BASE B	10	9					1						136
A-NAPHTHYLAMINE	10	9					1						135
B-NAPHTHYLAMINE	10	9					1						136
1-NAPHTHYLAMINE	10	9					1						135
2-NAPHTHYLAMINE	10	9					1						136
USAF CB-22	10	9					1						136
BENZENE, 4-ALLYL-1,2-/METHYLENE- DIOXY/-,	10	10						2					34
1,3-BENZODIOXOLE, 5-ALLYL-,	10	10						2					34
SAFROLE	10	10						2					34
SHIKIMOLE	10	10						2					34
AVOLIN	10	10						4					162
DIMETHYL PHTHALATE	10	10						4					162
MIPAX	10	10						4					162
PHTHALIC ACID, DIMETHYL ESTER	10	10						4					162
SOLVARONE	10	10						4					162
AZINOPHOS METHYL	10	12					3	3	1	2			335
GUTHION	10	12					3	3	1	2			335
PHOSPHORODITHIOIC ACID, O,O-DI- METHYL ESTER, S-ESTER WITH 3-/MERCAPTO- METHYL/-1,2,3-BENZOTRIAZIN-4/3H/-ONE	10	12					3	3	1	2			335
BENZENE, 1,2-/METHYLENEDIOXY/- 4-PROPYL-,	10	12						2					44
1,3-BENZODIOXOLE, 5-PROPYL-,	10	12						2					44
SAFROLE, DIHYDRO-,	10	12						2					44
ALKRON	10	14					1	5	1	1			156
ETHYL PARATHION	10	14					1	5	1	1			156
PARATHION	10	14					1	5	1	1			156
PHOSPHOROTHIOIC ACID, O,O-DIETHYL O-/P-NITROPHENYL/ ESTER	10	14					1	5	1	1			156
THIOPHOS	10	14					1	5	1	1			156
ACETIC ACID, /ETHYLENEDINITRILO/- TETRA-,	10	16					2	8					207
EDTA	10	16					2	8					207
ETHYLENEDIAMINETETRAACETIC ACID	10	16					2	8					207
NULLAPON B ACID	10	16					2	8					207
VERSENE	10	16					2	8					207
CYTHION	10	19						6	1	2			181
KARBOFOS	10	19						6	1	2			181
MALATHION	10	19						6	1	2			181
SUCCINIC ACID, MERCAPTO-, DIETHYL ESTER, S-ESTER WITH O,O-DIMETHYLPHOS- PHORODITHIOATE	10	19						6	1	2			181
ZITHIOL	10	19						6	1	2			181
CADMIUM, BIS/DIETHYLDITHIOCARBAM- ATO/-,	10	20					2			4		CD	238
ETHYL CADMATE	10	20					2			4		CD	238
AZODYNE	11	11					5						339
MALLOPHENE	11	11					5						339
PHENAZOPYRIDINE HYDROCHLORIDE	11	11					5						339
PYRIDINE, 2,6-DIAMINO-3-/PHENYL- AZO/-, MONOHYDROCHLORIDE	11	11					5						339
PYRIDIUM	11	11					5						339

NAME C H Br Cl F I N O P S Si M BOOK NO.

NAME	C	H	Br	Cl	F	I	N	O	P	S	Si	M	BOOK NO.
ACENAPHTHYLENE	12	8											2
CYCLOPENTA/DE/NAPHTHALENE	12	8											2
ALDRIN	12	8		6									78
DIMETHANONAPHTHALENE, 1,4:5,8-, 1,2,3,4,10,10-HEXACHLORO-1,4,4A,5,8,8A-HEXAHYDRO-, ENDO-,EXO-,	12	8		6									78
HHDN	12	8		6									78
OCTALENE	12	8		6									78
SEEDRIN	12	8		6									78
DIELDRIN	12	8		6				1					276
DIMETHANONAPHTHALENE, 1,4:5,8-, 6,7-EPOXY-1,2,3,4,10,10-HEXACHLORO-1,4,4A,5,6,7,8,8A-OCTAHYDRO-, ENDO-,-ENDO-,	12	8		6				1					275
DIMETHANONAPHTHALENE, 1,4:5,8-, 6,7-EPOXY-1,2,3,4,10,10-HEXACHLORO-1,4,4A,5,6,7,8,8A-OCTAHYDRO-, ENDO-,-EXO-,	12	8		6				1					276
ENDRIN	12	8		6				1					275
ENDRIN ALDEHYDE	12	8		6				1					279
HEOD	12	8		6				1					276
HEXADRIN	12	8		6				1					275
ILLOXOL	12	8		6				1					276
MENDRIN	12	8		6				1					275
ETHER, P-BROMOPHENYL PHENYL,	12	9	1					1					95
ETHER, P-CHLOROPHENYL PHENYL,	12	9		1				1					97
BIPHENYL, 4-NITRO-,	12	9					1	2					54
4-NITROBIPHENYL	12	9					1	2					54
PNB	12	9					1	2					54
ACENAPHTHENE	12	10											1
ACENAPHTHYLENE, 1,2-DIHYDRO-,	12	10											1
BENZIDINE, 3,3*-DICHLORO-,	12	10		2			2						48
BIPHENYL, 4,4*-DIAMINO-3,3*-DI-CHLORO-,	12	10		2			2						48
BIPHENYLDIAMINE, 4,4*-, 3,3*-DI-CHLORO-,	12	10		2			2						48
DCB	12	10		2			2						48
DIPHENYLAMINE, N-NITROSO-,	12	10					2	1					277
REDAX	12	10					2	1					277
RETARDER 2N	12	10					2	1					277
VULCATARD	12	10					2	1					277
WILTROL N	12	10					2	1					277
BIPHENYL, 4-AMINO-,	12	11					1						55
4-BIPHENYLAMINE	12	11					1						55
ATOXAN	12	11					1	2					242
CARBAMIC ACID, METHYL-, 1-NAPHTHYL ESTER	12	11					1	2					242
CARBARYL	12	11					1	2					242
SEPTENE	12	11					1	2					242
SEVIN	12	11					1	2					242
BARBITURIC ACID, 5-ETHYL-5-PHENYL-, MONOSODIUM SALT	12	11					2	3				NA	218
LUMINAL SODIUM	12	11					2	3				NA	218
PBS	12	11					2	3				NA	218
PHENOBARBITAL, MONOSODIUM SALT	12	11					2	3				NA	218
PHENOBARBITAL SODIUM	12	11					2	3				NA	218
AAB	12	11					3						25
ANILINE, P-/PHENYLAZO/-,	12	11					3						25
ANILINE YELLOW	12	11					3						25
AZOBENZENE, 4-AMINO-,	12	11					3						25
CI SOLVENT YELLOW 1	12	11					3						25
AQUACIDE	12	12	2				2						278
DIAZONIAPHENANTHRENE DIBROMIDE, 8A,10A-, 9,10-DIHYDRO-,	12	12	2				2						278
DIPYRIDO/1,2-A:2*,1*-C/PYRAZINE-DIIUM DIBROMIDE, 6,7-DIHYDRO-,	12	12	2				2						278
DIQUAT DIBROMIDE	12	12	2				2						278
REGLONE	12	12	2				2						278
BENZIDINE	12	12					2						47
BENZIDINE, SULFATE /1 TO 1/	12	12					2						225
BIPHENYL, 4,4*-DIAMINO-,	12	12					2						47
BIPHENYL, 4,4*-DIAMINO-, SULFATE /1 TO 1/	12	12					2						225
BIPHENYLDIAMINE, 4,4*-,	12	12					2						47
BIPHENYLDIAMINE, 4,4*-, SULFATE /1 TO 1/	12	12					2						225
CI AZOIC DIAZO	12	12					2						47

| NAME | C | H | Br | Cl | F | I | N | O | P | S | Si | M | BOOK NO. |

NAME	C	H	Br	Cl	F	I	N	O	P	S	Si	M	BOOK NO.
FAST CORINTH BASE B	12	12					2						47
HYDRAZINE, 1,2-DIPHENYL-,	12	12					2						306
HYDRAZOBENZENE	12	12					2						306
HYDRAZODIBENZENE	12	12					2						306
ENS	12	13					1	4		2			317
1-NAPHTHALENESULFONAMIDE, 4-/ETHYL- SULFONYL/-,	12	13					1	4		2			317
ANOZOL	12	14						4					161
DIETHYL PHTHALATE	12	14						4					161
PALATINOL A	12	14						4					161
PHTHALIC ACID, DIETHYL ESTER	12	14						4					161
SOLVANOL	12	14						4					161
CARBAMIC ACID, METHYL-, 4-/DI- METHYLAMINO/-3,5-XYLYL ESTER	12	18					2	2					241
DOWCO 139	12	18					2	2					241
MEXACARBATE	12	18					2	2					241
3,5-XYLENOL, 4-/DIMETHYLAMINO/-, METHYLCARBAMATE	12	18					2	2					241
ZECTRANE	12	18					2	2					241
IBENZMETHYZINE HYDROCHLORIDE	12	19					3	1					350
NATULAN HYDROCHLORIDE	12	19					3	1					350
PROCARBAZINE, HYDROCHLORIDE	12	19					3	1					350
P-TOLUAMIDE, N-ISOPROPYL-A-/2-METHYL- HYDRAZINO/-, MONOHYDROCHLORIDE	12	19					3	1					350
ALFA-TOX	12	21					2	3	1	1			337
DIAZINON	12	21					2	3	1	1			337
GARDENTOX	12	21					2	3	1	1			337
NUCIDOL	12	21					2	3	1	1			337
PHOSPHOROTHIOIC ACID, O,O-DIETHYL O-/2-ISOPROPYL-6-METHYL-4-PYRIMIDINYL/ ESTER	12	21					2	3	1	1			337
DIPHENYLENEMETHANE	13	10											110
FLUORENE	13	10											110
FLUOREN-2-AMINE	13	11					1						109
FLUORENE, 2-AMINO-,	13	11					1						109
ANILINE, 4,4*-METHYLENEBIS- /2-CHLORO-,	13	12		2			2						23
BENZENAMINE, 4,4*-METHYLENEBIS- /2-CHLORO-,	13	12		2			2						23
CYANASET	13	12		2			2						23
MOCA	13	12		2			2						23
FLUORENE, 2,7-DIAMINO-,	13	12					2						291
FLUORENE-2,7-DIAMINE	13	12					2						291
ANILINE, N-METHYL-P-/PHENYLAZO/-,	13	13					3						211
AZOBENZENE, N-METHYL-4-AMINO-,	13	13					3						211
ALANINE, 3-/P-/BIS/2-CHLOROETHYL/- AMINO/PHENYL/-, L-,	13	18		2			2	2					209
ALKERAN	13	18		2			2	2					209
MELPHALAN	13	18		2			2	2					209
NSC-8806	13	18		2			2	2					209
L-PAM	13	18		2			2	2					209
DDE, P,P*-,	14	8		4									99
ETHYLENE, 1,1-BIS/P-CHLOROPHENYL/- 2,2-DICHLORO-,	14	8		4									99
ANOFEX	14	9		5									290
DDT, P,P*-,	14	9		5									290
DICOPHANE	14	9		5									290
ETHANE, 1,1-BIS/P-CHLOROPHENYL/- 2,2,2-TRICHLORO-,	14	9		5									290
ZERDANE	14	9		5									290
ACARIN	14	9		5				1					224
BENZHYDROL, 4,4*-DICHLORO-A-/TRI- CHLOROMETHYL/-,	14	9		5				1					224
DICOFOL	14	9		5				1					224
ETHANOL, 1,1-BIS/P-CHLOROPHENYL/- 2,2,2-TRICHLORO-,	14	9		5				1					224
KELTHANE	14	9		5				1					224
ANTHRACENE	14	10											27
GREEN OIL	14	10											27
PARANAPHTHALENE	14	10											27
PHENANTHRENE	14	10											139
NAME	C	H	Br	Cl	F	I	N	O	P	S	Si	M	BOOK NO.

NAME	C	H	Br	Cl	F	I	N	O	P	S	Si	M	BOOK NO.
TETRA OLIVE N2G	14	10											27
DDD, P,P*-,	14	10		4									289
DILENE	14	10		4									289
ETHANE, 1,1-BIS/P-CHLOROPHENYL/- 2,2-DICHLORO-,	14	10		4									289
RHOTHANE D-3	14	10		4									289
TDE	14	10		4									289
4-NITROSTILBENE	14	11					1	2					342
STILBENE, 4-NITRO-,	14	11					1	2					342
ACETAMIDE, N-4-BIPHENYLYL-,	14	13					1	1					204
ACETANILIDE, 4*-PHENYL-,	14	13					1	1					204
BIPHENYL, 4-ACETAMIDO-,	14	13					1	1					204
ANILINE, N,N-DIMETHYL-P-/PHENYL- AZO/-,	14	15					3						259
AZOBENZENE, P-/DIMETHYLAMINO/-,	14	15					3						259
BUTTER YELLOW	14	15					3						259
CI SOLVENT YELLOW 2	14	15					3						259
METHYL YELLOW	14	15					3						259
CO-RAL	14	16		1				5	1	1			262
COUMAPHOS	14	16		1				5	1	1			262
COUMARIN, 3-CHLORO-7-HYDROXY- 4-METHYL-, O-ESTER WITH O,O-DIETHYL- PHOSPHOROTHIOATE	14	16		1				5	1	1			262
RESITOX	14	16		1				5	1	1			262
SUNTOL	14	16		1				5	1	1			262
BENZIDINE, 3,3*-DIMETHOXY-,	14	16					2	2					49
BENZIDINE, 3,3*-DIMETHOXY-, DI- HYDROCHLORIDE	14	16					2	2					226
BI-O-ANISIDINE, 4,4*-,	14	16					2	2					49
BI-O-ANISIDINE, 4,4*-, DIHYDRO- CHLORIDE	14	16					2	2					226
CELLITAZOL B	14	16					2	2					49
CI DISPERSE BLACK 6	14	16					2	2					49
CI DISPERSE BLACK 6, DIHYDRO- CHLORIDE	14	16					2	2					226
DIANISIDINE	14	16					2	2					49
BUTYRIC ACID, 4-/P-/BIS/2-CHLORO- ETHYL/AMINO/PHENYL/-,	14	19		2			1	2					237
CHLORAMBUCIL	14	19		2			1	2					237
ELCORIL	14	19		2			1	2					237
LEUKERAN	14	19		2			1	2					237
NSC-3088	14	19		2			1	2					237
AAF	15	13					1	1					5
ACETAMIDE, N-FLUOREN-2-YL-,	15	13					1	1					5
FAA	15	13					1	1					5
FLUORENE, 2-ACETAMIDO-,	15	13					1	1					5
ACETAMIDE, N-FLUOREN-3-YL- N-HYDROXY-,	15	13					1	2					208
ACETOHYDROXAMIC ACID, N-FLUOREN- 3-YL-,	15	13					1	2					208
ANILINE, N,N-DIMETHYL-P-/M-TOLYL- AZO/-,	15	17					3						210
AZOBENZENE, 4-/DIMETHYLAMINO/- 3*-METHYL-,	15	17					3						210
BENZENAMINE, N,N-DIMETHYL- 4-//3-METHYLPHENYL/AZO/-,	15	17					3						210
MDAB	15	17					3						210
ARAMITE	15	23		1				4		1			346
ARATRON	15	23		1				4		1			346
CES	15	23		1				4		1			346
ORTHO-MITE	15	23		1				4		1			346
SULFUROUS ACID, 2-/P-TERT-BUTYL- PHENOXY/-1-METHYLETHYL 2-CHLOROETHYL ESTER	15	23		1				4		1			346
BENZO/JK/FLUORENE	16	10											108
BENZO/DEF/PHENANTHRENE	16	10											174
FLUORANTHENE	16	10											108
IDRYL	16	10											108
PYRENE	16	10											174
ACETAMIDE, N-2-PHENANTHRYL-,	16	13					1	1					203
PHENANTHRENE, 2-ACETAMIDO-,	16	13					1	1					203

NAME | C | H | Br | Cl | F | I | N | O | P | S | Si | M | BOOK NO.

NAME	C	H	Br	Cl	F	I	N	O	P	S	Si	M	BOOK NO.
ETHANE, 1,1-BIS/P-METHOXYPHENYL/- 2,2,2-TRICHLORO-,	16	15		3				2					83
MARLATE	16	15		3				2					83
METHOXCIDE	16	15		3				2					83
METHOXYCHLOR	16	15		3				2					83
METHOXY-DDT	16	15		3				2					83
DIBUTYL PHTHALATE	16	22						4					160
PALATINOL C	16	22						4					160
PHTHALIC ACID, DIBUTYL ESTER	16	22						4					160
POLYCIZER DBP	16	22						4					160
WITCIZER 300	16	22						4					160
EPOXIDE-201	16	24						4					328
7-OXABICYCLO/4.1.0/HEPTANE-3-CARBOX- YLIC ACID, 4-METHYL-, /4-METHYL-7-OXA- BICYCLO/4.1.0/HEPT-3-YL/METHYL ESTER	16	24						4					328
UNOX EPOXIDE 201	16	24						4					328
CI SOLVENT ORANGE 2	17	14					2	1					257
FAT ORANGE II	17	14					2	1					257
2-NAPHTHOL, 1-/O-TOLYLAZO/-,	17	14					2	1					257
OIL ORANGE OPEL	17	14					2	1					257
ACETAMIDE, N,N*-FLUOREN-2,7-YLENE- BIS-,	17	16					2	2					202
2,7-FAA	17	16					2	2					202
FLUORENE, 2,7-DIACETAMIDO-,	17	16					2	2					202
AMUDANE	17	17		1				6					177
FULVICIN	17	17		1				6					177
GRISEOFULVIN	17	17		1				6					177
PONCYL	17	17		1				6					177
SPIRO/BENZOFURAN-2/3H/,1*-/2/CYCLO- HEXENE/-3,4*-DIONE, 7-CHLORO- G*B-METHYL-2*,4,6-TRIMETHOXY-,	17	17		1				6					177
ANILINE, 4,4*-IMIDOCARBONYLBIS- /N,N-DIMETHYL-,	17	21					3						260
AURAMINE	17	21					3						260
CI SOLVENT YELLOW 34	17	21					3						260
WAXOLINE YELLOW O	17	21					3						260
YELLOW PYOCTANINE	17	21					3						260
7H-FURO/3*,2*:4,5/FURO/2,3-C/XANTHEN- 7-ONE, 3A,12C-DIHYDRO-8-HYDROXY- 6-METHOXY-,	18	10						6					294
STERIGMATOCYSTIN	18	10						6					294
BENZ/A/ANTHRACENE	18	12											220
1,2-BENZANTHRACENE	18	12											220
BENZO/A/PHENANTHRENE	18	12											251
1,2-BENZOPHENANTHRENE	18	12											251
CHRYSENE	18	12											251
DIBENZONAPHTHALENE, 1,2:5,6-,	18	12											251
NAPHTHANTHRACENE	18	12											220
TETRAPHENE	18	12											220
6-AMC	18	13					1						250
6-CHRYSENAMINE	18	13					1						250
CHRYSENE, 6-AMINO-,	18	13					1						250
CI ACID RED 26, DISODIUM SALT	18	14					2	7		2		NA	252
2,7-NAPHTHALENEDISULFONIC ACID, 3-HYDROXY-4-/2,4-XYLYLAZO/-, DISODIUM SALT	18	14					2	7		2		NA	252
PONCEAU G	18	14					2	7		2		NA	252
PONCEAU J	18	14					2	7		2		NA	252
PONCEAU R	18	14					2	7		2		NA	252
PONCEAU 2R	18	14					2	7		2		NA	252
PONCEAU BNA	18	14					2	7		2		NA	252
PONCEAU FR	18	14					2	7		2		NA	252
PONCEAU GR	18	14					2	7		2		NA	252
PONCEAU MX	18	14					2	7		2		NA	252
PONCEAU NR	18	14					2	7		2		NA	252
PONCEAU PXM	18	14					2	7		2		NA	252
CI SOLVENT RED 80	18	16					2	3					258
CITRUS RED 2	18	16					2	3					258
2-NAPHTHOL, 1-//2,5-DIMETHOXYPHENYL/- AZO/-,	18	16					2	3					258
EQUILENIN	18	18						2					284

NAME C H Br Cl F I N O P S Si M BOOK NO.

NAME	C	H	Br	Cl	F	I	N	O	P	S	Si	M	BOOK NO.
ESTRA-1,3,5,7,9-PENTAEN-17-ONE, 3-HYDROXY-,	18	18						2					284
ESTRA-1,3,5/10/,6,8-PENTAEN-17-ONE, 3-HYDROXY-,	18	18						2					284
BIO-DES	18	20						2					178
DES	18	20						2					178
EQUILIN	18	20						2					285
ESTRA-1,3,5/10/,7-TETRAEN-17-ONE, 3-HYDROXY-,	18	20						2					285
STILBENEDIOL, 4,4*-, A,A*-DIETHYL-,	18	20						2					178
STILBESTEROL, DIETHYL-,	18	20						2					178
SYNESTRIN	18	20						2					178
CYCLOESTROL	18	22						2					143
ESTRA-1,3,5/10/-TRIEN-17-ONE, 3-HYDROXY-,	18	22						2					287
ESTRIN, KETOHYDROXY-,	18	22						2					287
ESTRONE	18	22						2					287
EXTRA-PLEX	18	22						2					143
FOLLICULIN	18	22						2					287
HEXESTROL	18	22						2					143
KETOHYDROXYESTRIN	18	22						2					287
MENFORMON	18	22						2					287
PHENOL, 4,4*-/1,2-DIETHYLETHYLENE/-DI-, MESO-,	18	22						2					143
STILBESTROL, DIHYDRO-,	18	22						2					143
THEELIN	18	22						2					287
DIHYDROFOLLICULIN	18	24						2					280
DIHYDROTHEELIN	18	24						2					280
ESTRADIOL	18	24						2					280
17A-ESTRADIOL	18	24						2					283
17B-ESTRADIOL	18	24						2					280
ESTRA-1,3,5/10/-TRIENE-3,17B-DIOL	18	24						2					280
ESTRA-1,3,5/10/-TRIENE-3,17A-DIOL	18	24						2					283
ESTRIN, DIHYDROXY-,	18	24						2					280
FOLLICULIN, DIHYDRO-,	18	24						2					280
THEELIN, DIHYDRO-,	18	24						2					280
ESTRA-1,3,5/10/-TRIENE-3,16A,17B-TRIOL	18	24						3					286
ESTRIN, TRIHYDROXY-,	18	24						3					286
ESTRIOL	18	24						3					286
FOLLICULAR HORMONE HYDRATE	18	24						3					286
THEELOL	18	24						3					286
BENZENESULFONIC ACID, DODECYL-,	18	30						3		1			222
DODECYLBENZENESULFONIC ACID	18	30						3		1			222
LAKEWAY SA	18	30						3		1			222
CARBAMIC ACID, DIBUTYLDITHIO-, NICKEL SALT	18	36					2			4		NI	319
NBC	18	36					2			4		NI	319
NICKEL, BIS/DIBUTYLDITHIOCARBAM-ATO/-,	18	36					2			4		NI	319
VANGUARD N	18	36					2			4		NI	319
CI FOOD RED 6, DISODIUM SALT	19	16					2	7		2		NA	256
2,7-NAPHTHALENEDISULFONIC ACID, 3-HYDROXY-4-//2,4,5-TRIMETHYLPHENYL/-AZO/-, DISODIUM SALT	19	16					2	7		2		NA	256
PONCEAU 3R	19	16					2	7		2		NA	256
PONCEAU RN	19	16					2	7		2		NA	256
USACERT RED NO. 1	19	16					2	7		2		NA	256
CI BASIC RED 9, MONOHYDROCHLORIDE	19	17					3						253
PARAFUCHSINE	19	17					3						253
PARAMAGENTA	19	17					3						253
PARAROSANILINE	19	17					3						253
P-TOLUIDINE, A-/P-AMINOPHENYL/-A-/4-IMINO-2,5-CYCLOHEXADIEN-1-YL-IDENE/-, MONOHYDROCHLORIDE	19	17					3						253
BENZYL BUTYL PHTHALATE	19	20						4					158
PALATINOL BB	19	20						4					158
PHTHALIC ACID, BENZYL BUTYL ESTER	19	20						4					158
SANTICIZER 160	19	20						4					158
ANDROLIN	19	28						2					348
ANDROST-4-EN-3-ONE, 17B-HYDROXY-,	19	28						2					348
MERTESTATE	19	28						2					348
TESLEN	19	28						2					348
TESTOSTERONE	19	28						2					348
BENZ/E/ACEPHENANTHRYLENE	20	12											219

NAME | C | H | Br | Cl | F | I | N | O | P | S | Si | M | BOOK NO.

NAME	C	H	Br	Cl	F	I	N	O	P	S	Si	M	BOOK NO.
BENZO/K/FLUORANTHENE	20	12											229
BENZOFLUORANTHENE, 11,12-,	20	12											229
3,4-BENZOFLUORANTHENE	20	12											219
BENZO/A/PYRENE	20	12											232
3,4-BENZOPYRENE	20	12											232
3,4-BP	20	12											232
DIBENZOCARBAZOLE, 3,4:5,6-,	20	13					1						273
7H-DIBENZO/C,G/CARBAZOLE	20	13					1						273
BENZ/A/ANTHRACENE, 7,12-DIMETHYL-,	20	16											221
1,2-BENZANTHRACENE, 9,10-DIMETHYL-,	20	16											221
DBA	20	16											221
DMBA	20	16											221
CI BASIC VIOLET 14, MONOHYDRO- CHLORIDE	20	19					3						254
FUCHSINE	20	19					3						254
ROSANILINE	20	19					3						254
2,4-XYLIDINE, A<4-/P-AMINOPHENYL/- A<4-/4-IMINO-2,5-CYCLOHEXADIEN-1-YL- IDENE/-, MONOHYDROCHLORIDE	20	19					3						254
ESTINYL	20	24						2					323
ESTRADIOL, 17-ETHYNYL-,	20	24						2					323
ETHINYL ESTRADIOL	20	24						2					323
NOVESTROL	20	24						2					323
19-NOR-17A-PREGNA-1,3,5/10/-TRIEN-20- YNE-3,17-DIOL	20	24						2					323
NORETHINDRONE	20	26						2					325
NORETHINYNODREL	20	26						2					327
NORETHISTERONE	20	26						2					325
NORETHYNODREL	20	26						2					327
NORLUTIN	20	26						2					325
19-NOR-17A-PREGN-4-EN-20-YN-3-ONE, 17-HYDROXY-,	20	26						2					325
19-NOR-17A-PREGN-5/10/-EN-20-YN-3-ONE, 17-HYDROXY-,	20	26						2					327
DIBENZ/A,H/ACRIDINE	21	13					1						271
DIBENZACRIDINE, 1,2:5,6-,	21	13					1						271
BENZ/J/ACEANTHRYLENE, 1,2-DIHYDRO- 3-METHYL-,	21	16											243
CHOLANTHRENE, 3-METHYL-,	21	16											243
MC	21	16											243
KWIK-KIL	21	22					2	2					343
4,6-METHANO-6H,14H-INDOLO/3,2,1-IJ/- OXEPINO/2,3,4-DE/PYRROLO/2,3-H/QUINOL- IN-14-ONE, 2,4A,5,5A,7,8,15,15A,15B,- 15C-DECAHYDRO-,	21	22					2	2					343
MOUSE-TOX	21	22					2	2					343
RO-DEX	21	22					2	2					343
STRYCHNINE	21	22					2	2					343
DELTA-MVE	21	26						2					324
MESTRANOL	21	26						2					324
19-NOR-17A-PREGNA-1,3,5/10/-TRIEN-20- YN-17-OL, 3-METHOXY-,	21	26						2					324
CORLUTIN	21	30						2					167
LUTEOSTERONE	21	30						2					167
PREGN-4-ENE-3,20-DIONE	21	30						2					167
PROGESTERONE	21	30						2					167
PROLIDON	21	30						2					167
2-BUTENOIC ACID, 2-METHYL-, 7-//2,3- DIHYDROXY-2-/1-METHOXYETHYL/-3-METHYL- 1-OXOBUTOXY/METHYL/-2,3,5,7A-TETRA- HYDRO-1H-PYRROLIZIN-1-YL ESTER	21	33					1	7					309
LASIOCARPINE	21	33					1	7					309
BENZO/GHI/PERYLENE	22	12											231
BENZOPERYLENE, 1,12-,	22	12											231
BENZO/B/TRIPHENYLENE	22	14											233
DIBENZ/A,C/ANTHRACENE	22	14											233
DIBENZ/A,4/ANTHRACENE	22	14											272
DIBENZANTHRACENE, 1,2:3,4-,	22	14											233
DIBENZANTHRACENE, 1,2:5,6-,	22	14											272
NORETHINDRONE, ACETATE	22	28						3					326
NORETHISTERONE, ACETATE	22	28						3					326

NAME C H Br Cl F I N O P S Si M BOOK NO.

1/82

NAME	C	H	Br	Cl	F	I	N	O	P	S	Si	M	BOOK NO.
NORLUTATE	22	28						3					326
NORLUTIN, ACETATE	22	28						3					326
19-NOR-17A-PREGN-4-EN-20-YN-3-ONE, 17-HYDROXY-, ACETATE	22	28						3					326
AGOVIRIN	22	32						3					349
ANDROST-4-EN-3-ONE, 17B-HYDROXY-, PROPIONATE	22	32						3					349
HORMOTESTON	22	32						3					349
ORETON	22	32						3					349
TESTOSTERONE, PROPIONATE	22	32						3					349
BENZO/RST/PENTAPHENE	24	14											230
DIBENZO/A,I/PYRENE	24	14											230
DIBENZOPYRENE, 1,2:7,8-,	24	14											230
CLINESTROL	24	28						4					179
DIBESTIL	24	28						4					179
STILBENEDIOL, 4,4*-, A,A*-DIETHYL-, DIPROPIONATE, TRANS-,	24	28						4					179
STILBESTEROL, DIETHYL-, DIPROPION-ATE	24	28						4					179
SYNTESTRINE	24	28						4					179
ESTRADIOL, DIPROPIONATE	24	32						4					282
17B-ESTRADIOL, DIPROPIONATE	24	32						4					282
ESTRA-1,3,5/10/-TRIENE-3,17B-DIOL, DIPROPIONATE	24	32						4					282
OVOCYCLIN DIPROPIONATE	24	32						4					282
PROGYNON-DP	24	32						4					282
BIS/2-ETHYLHEXYL/ PHTHALATE	24	38						4					159
DIOCTYL PHTHALATE	24	38						4					163
HATCOL DOP	24	38						4					163
HOWFLEX DOP	24	38						4					159
PHTHALIC ACID, BIS/2-ETHYLHEXYL/ ESTER	24	38						4					159
PHTHALIC ACID, DIOCTYL ESTER	24	38						4					163
POLYCIZER 162	24	38						4					163
PX-138	24	38						4					163
STAFLEX DOP	24	38						4					159
WITICIZER 312	24	38						4					159
ESTRA-1,3,5/10/-TRIEN-17-ONE, 3-HYDROXY-, BENZOATE	25	26						3					288
ESTRONE, BENZOATE	25	26						3					288
DIHYDROFOLLICULIN BENZOATE	25	28						3					281
ESTRADIOL, 3-BENZOATE	25	28						3					281
17B-ESTRADIOL, 3-BENZOATE	25	28						3					281
ESTRA-1,3,5/10/-TRIENE-3,17B-DIOL, 3-BENZOATE	25	28						3					281
FOLLICULIN, DIHYDRO-, BENZOATE	25	28						3					281
PROGYNON BENZOATE	25	28						3					281
CERUBIDIN	27	29					1	10					270
DAUNOMYCIN	27	29					1	10					270
HEXOPYRANOSIDE, LYXO-, 3-ACETYL-6,11-DIOXO-1,2,3,4,6,11-HEXAHYDRO-10-METHOXY-3,5,12-TRIHYDROXY-1-NAPHTH-ACENYL 3-AMINO-2,3,6-TRIDEOXY-, A-L-, /1S,3S/-,	27	29					1	10					270
RUBIDOMYCIN	27	29					1	10					270
RUBOMYCIN C	27	29					1	10					270
AMANIL SKY BLUE	34	22					4	12		4		NA	255
BENZAMINE BLUE	34	22					4	12		4		NA	255
CI DIRECT BLUE 14, TETRASODIUM SALT	34	22					4	12		4		NA	255
NAME	C	H	Br	Cl	F	I	N	O	P	S	Si	M	BOOK NO.

1/82

NAME	C	H	Br	Cl	F	I	N	O	P	S	Si	M	BOOK NO.
DIRECT BLUE 3B	34	22					4	12		4		NA	255
2,7-NAPHTHALENEDISULFONIC ACID, 3,3*-//3,3*-DIMETHYL-4,4*-BIPHENYLYL-ENE/BIS/AZO//BIS/5-AMINO-4-HYDROXY-, TETRASODIUM SALT	34	22					4	12		4		NA	255
BENZENESULFONIC ACID, DODECYL-, CALCIUM SALT /2 TO 1/	36	58						6		2		CA	223
DODECYLBENZENESULFONIC ACID, CALCIUM SALT /2 TO 1/	36	58						6		2		CA	223
MYRJ 45	99	99						99					166
POLYETHYLENE GLYCOL, MONOSTEARATE	99	99						99					166
POLYOXYETHYLENE-8-MONOSTEARATE	99	99						99					166
CADDY				2								CD	239
CADMIUM CHLORIDE				2								CD	239
VI-CAD				2								CD	239
BERYLLIA								1				BE	234
BERYLLIUM OXIDE								1				BE	234
CADMIUM OXIDE								1				CD	240
BUNSENITE								1				NI	320
NICKEL/II/ OXIDE /1 TO 1/								1				NI	320
NICKEL MONOXIDE								1				NI	320
NICKELOUS OXIDE								1				NI	320
NICKEL OXIDE								1				NI	320
ARSENIC SESQUIOXIDE								3				AS	216
ARSENIC TRIOXIDE								3				AS	216
ARSENOUS ACID								3				AS	216
ARSENOUS OXIDE								3				AS	216
WHITE ARSENIC								3				AS	216
CHROME GREEN								3				CR	248
CHROMIC ACID								3				CR	249
CHROMIC ANHYDRIDE								3				CR	249
CHROMIUM/III/ OXIDE /2 TO 3/								3				CR	248
CHROMIUM OXIDE								3				CR	248
CHROMIUM OXIDE								3				CR	249
CHROMIUM TRIOXIDE								3				CR	249
CHROMIUM/VI/ OXIDE /1 TO 3/								3				CR	249
GREEN CINNABAR								3				CR	248
ULTRAMARINE GREEN								3				CR	248
BLOOD STONE								3				FE	297
HEMATITE								3				FE	297
IRON ORE								3				FE	297
RED IRON ORE								3				FE	297
CALCIUM CHROMATE/VI/								4				X2	244
CALCIUM CHROME YELLOW								4				X2	244
CHROMATE OF SODA								4				X2	246
CHROME YELLOW								4				X2	247
CHROMIC ACID, CALCIUM SALT /1 TO 1/								4				X2	244
CHROMIC ACID, DIPOTASSIUM SALT								4				X2	245
CHROMIC ACID, DISODIUM SALT								4				X2	246
CHROMIC ACID, LEAD/2 PLUS/ SALT /1 TO 1/								4				X2	247
COLOGNE YELLOW								4				X2	247
GELBIN								4				X2	244
KINGS YELLOW								4				X2	247
LEAD CHROMATE/VI/								4				X2	247
NEUTRAL POTASSIUM CHROMATE								4				X2	245
NEUTRAL SODIUM CHROMATE								4				X2	246
POTASSIUM CHROMATE/VI/								4				X2	245
SODIUM CHROMATE/VI/								4				X2	246
TARAPACAITE								4				X2	245
YELLOW ULTRAMARINE								4				X2	244
PHOSPHORIC ACID, TRISODIUM SALT, DODECAHYDRATE								4	1			NA	334
SODIUM PHOSPHATE, DODECAHYDRATE								4	1			NA	334
SODIUM PHOSPHATE TRIBASIC								4	1			NA	334
NAME	C	H	Br	Cl	F	I	N	O	P	S	Si	M	BOOK NO.

1/82

PREFACE

This Spec-Finder Index for the ***Infrared Spectra Handbook of Priority Pollutants and Toxic Chemicals*** contains the strongest band of each spectrum for the 363 compounds appearing in this handbook. This is the band which has the greatest intensity by maximum absorption. It provides a method of rapidly locating reference spectra which have an absorption peak similar to an unknown spectrum.

This index is arranged in ascending sequence with the compound number. Since some compounds have the condensed phase and vapor phase represented, both spectra are coded. However, only the compound number is listed and there is no designation that identifies the condensed phase or vapor phase.

CHEMICAL NAME	NUMBER	STRONGEST BAND
NICKEL OXIDE	320	420
BLOOD STONE	297	550
HEMATITE	297	550
IRON ORE	297	550
RED IRON ORE	297	550
DBCP	168	570
FUMAZONE	168	570
NEMAGON	168	570
PROPANE, 1-CHLORO-2,3-DIBROMO-,	168	570
PROPANE, 1,2-DIBROMO-3-CHLORO-,	168	570
DBCP	168	580
FUMAZONE	168	580
NEMAGON	168	580
PROPANE, 1-CHLORO-2,3-DIBROMO-,	168	580
PROPANE, 1,2-DIBROMO-3-CHLORO-,	168	580
1,2-DIBROMOETHANE	85	600
DOWFUME W-8	85	600
DOWFUME W-85	85	600
DOWFUME 40	85	600
DOWFUME EDB	85	600
DOWFUME MC-2	85	600
ETHANE, 1,2-DIBROMO-,	85	600
ETHYLENE DIBROMIDE	85	600
GLYCOL DIBROMIDE	85	600
CHROME GREEN	248	630
CHROMIUM/III/ OXIDE /2 TO 3/	248	630
CHROMIUM OXIDE	248	630
COPPER, /ACETATO/TRIMETAARSENITO-DI-,	261	630
COPPER, BIS/ACETATO/HEXAMETA-ARSENITOTETRA-,	261	630
COPPER ACETATE ARSENITE	261	630
CUPRIC ACETOARSENITE	261	630
GREEN CINNABAR	248	630
PARIS GREEN	261	630
ULTRAMARINE GREEN	248	630
BROMOFORM	125	650
CHLORODIBROMOMETHANE	121	650
ENT 25,584	313	650
HEPTACHLOR EPOXIDE	313	650
METHANE, CHLORODIBROMO-,	121	650
METHANE, DIBROMOCHLORO-,	121	650
METHANE, TRIBROMO-,	125	650
4,7-METHANOINDAN, 2,3-EPOXY-1,4,5,6,7,-8,8-HEPTACHLORO-3A,4,7,7A-TETRAHYDRO-,	313	650
TRIBROMOMETHANE	125	650
VELSICOL 53-CS-17	313	650
BROMOFORM	125	660
CHLORODIBROMOMETHANE	121	660
1,2-DICHLOROPROPANE	169	660
METHANE, CHLORODIBROMO-,	121	660
METHANE, DIBROMOCHLORO-,	121	660
METHANE, TRIBROMO-,	125	660
PROPANE, 1,2-DICHLORO-,	169	660
PROPYLENE DICHLORIDE	169	660
TRIBROMOMETHANE	125	660
BENZENE	33	670
BENZENE HEXACHLORIDE /MIXED ISOMERS/	268	670
BENZOL	33	670
BENZOLE	33	670
CADMIUM OXIDE	240	670
CARBON OIL	33	670
CYCLOHEXANE, 1,2,3,4,5,6-HEXA-CHLORO-, /MIXED ISOMERS/	268	670
1,2-DICHLOROPROPANE	169	670
PHENE	33	670
PROPANE, 1,2-DICHLORO-,	169	670
PROPYLENE DICHLORIDE	169	670
AFICIDE	267	680
BENZENE HEXACHLORIDE, G-ISOMER	267	680
CYCLOHEXANE, 1,2,3,4,5,6-HEXA-CHLORO-, G-ISOMER	267	680
KWELL	267	680
LINDANE	267	680
ALDRIN	78	690
BENZYL CHLORIDE	184	690
A-CHLOROTOLUENE	184	690
1,1-DICHLOROETHANE	86	690
DIMETHANONAPHTHALENE, 1,4:5,8-,1,2,3,4,10,10-HEXACHLORO-1,4,4A,5,8,8A-HEXAHYDRO-, ENDO-,EXO-,	78	690
ETHANE, 1,1-DICHLORO-,	86	690

CHEMICAL NAME	NUMBER	STRONGEST BAND
ETHYLIDENE DICHLORIDE	86	690
HHDN	78	690
OCTALENE	78	690
SEEDRIN	78	690
TOLUENE, A-CHLORO-,	184	690
AROCHLOR 1221	29	700
BENZENE, ETHYL-,	42	700
BENZENE, VINYL-,	180	700
BENZYL CHLORIDE	184	700
A-CHLOROTOLUENE	184	700
CINNAMENE	180	700
1,1-DICHLOROETHANE	86	700
ETHANE, 1,1-DICHLORO-,	86	700
ETHANE, PHENYL-,	42	700
ETHYLBENZENE	42	700
ETHYLENE, PHENYL-,	180	700
ETHYLIDENE DICHLORIDE	86	700
PCB-1221	29	700
POLYCHLORINATED BIPHENYL	29	700
STYRENE	180	700
STYRON	180	700
TOLUENE, A-CHLORO-,	184	700
1,2-DICHLOROETHANE	87	710
ETHANE, 1,2-DICHLORO-,	87	710
ETHYLENE DICHLORIDE	87	710
GLYCOL DICHLORIDE	87	710
AEROTHENE TT	90	720
ANTHRACENE	27	720
BENZENE, METHYL-,	183	720
BROMODICHLOROMETHANE	120	720
CHLORTEN	90	720
1,2-DICHLOROETHANE	87	720
ETHANE, 1,2-DICHLORO-,	87	720
ETHANE, 1,1,1-TRICHLORO-,	90	720
ETHYLENE DICHLORIDE	87	720
GLYCOL DICHLORIDE	87	720
GREEN OIL	27	720
METHACIDE	183	720
METHANE, BROMODICHLORO-,	120	720
METHANE, DICHLOROBROMO-,	120	720
METHANE, PHENYL-,	183	720
METHYLCHLOROFORM	90	720
PARANAPHTHALENE	27	720
TETRA OLIVE N2G	27	720
TOLUENE	183	720
TOLUOL	183	720
TRICHLOROETHANE, 1,1,1-,	90	720
ARTIC	312	730
BENZENE, METHYL-,	183	730
CHLOROMETHANE	312	730
DIPHENYLENEMETHANE	110	730
ETHANE, 1,1,2-TRICHLORO-,	91	730
ETHANE TRICHLORIDE	91	730
FLUORENE	110	730
METHACIDE	183	730
METHANE, CHLORO-,	312	730
METHANE, PHENYL-,	183	730
METHYL CHLORIDE	312	730
PHENANTHRENE	139	730
TOLUENE	183	730
TOLUOL	183	730
TRICHLOROETHANE, 1,1,2-,	91	730
VINYL TRICHLORIDE	91	730
BENZENE, 1,2-DIMETHYL-,	194	740
BENZENE HEXACHLORIDE, B-ISOMER	265	740
B-BHC	265	740
CYCLOHEXANE, 1,2,3,4,5,6-HEXA-CHLORO-, B-ISOMER	265	740
DIBENZ/A,H/ACRIDINE	271	740
DIBENZACRIDINE, 1,2:5,6-,	271	740
DICHLOROMETHANE	122	740
B-LINDANE	265	740
METHANE, DICHLORO-,	122	740
METHANE DICHLORIDE	122	740
METHYLENE CHLORIDE	122	740
METHYLENE DICHLORIDE	122	740
O-XYLENE	194	740
O-XYLOL	194	740
ACETYLENE TETRACHLORIDE	89	750
AROCHLOR 1221	29	750
AROCHLOR 1232	30	750
BENZ/A/ANTHRACENE, 7,12-DIMETHYL-,	221	750
1,2-BENZANTHRACENE, 9,10-DIMETHYL-,	221	750

CHEMICAL NAME	NUMBER	STRONGEST BAND
BENZENE, CYANO-,	51	750
BENZENE, O-DICHLORO-,	37	750
BENZENE, 1,2-DICHLORO-,	37	750
BENZO/K/FLUORANTHENE	229	750
BENZOFLUORANTHENE, 11,12-,	229	750
BENZONITRILE	51	750
BENZO/A/PYRENE	232	750
3,4-BENZOPYRENE	232	750
BENZO/RST/PENTAPHENE	230	750
BONOFORM	89	750
3,4-BP	232	750
CELLON	89	750
CHLOROBEN	37	750
CHLOROFORM	64	750
O-CRESOL	68	750
O-CRESYLIC ACID	68	750
DBA	221	750
DIBENZO/A,I/PYRENE	230	750
DIBENZOPYRENE, 1,2:7,8-,	230	750
O-DICHLOROBENZENE	37	750
DIZENE	37	750
DMBA	221	750
ETHANE, 1,1,2,2-TETRACHLORO-,	89	750
METHANE, TRICHLORO-,	64	750
ODCB	37	750
PCB-1221	29	750
PCB-1232	30	750
PHENOL, O-METHYL-,	68	750
PHENYL CYANIDE	51	750
POLYCHLORINATED BIPHENYL	29	750
POLYCHLORINATED BIPHENYL	30	750
TETRACHLOROETHANE, 1,1,2,2-,	89	750
ALLTOX	351	760
6-AMC	250	760
ARSENIC ACID, MONOPOTASSIUM SALT	215	760
BENZENE, CYANO-,	51	760
BENZENE, DIMETHYL-,	360	760
BENZONITRILE	51	760
BENZO/A/PHENANTHRENE	251	760
1,2-BENZOPHENANTHRENE	251	760
BENZO/B/TRIPHENYLENE	233	760
CARBINAMINE	130	760
CHLORINATED CAMPHENE	351	760
6-CHRYSENAMINE	250	760
CHRYSENE	251	760
CHRYSENE, 6-AMINO-,	250	760
DIBENZ/A,C/ANTHRACENE	233	760
DIBENZANTHRACENE, 1,2:3,4-,	233	760
DIBENZONAPHTHALENE, 1,2:5,6-,	251	760
DRINOX	129	760
HEPTACHLOR	129	760
HEPTAGRAN	129	760
MACQUERS SALT	215	760
METHANE, AMINO-,	130	760
4,7-METHANOINDENE, 1,4,5,6,7,8,8-HEPTA-CHLORO-3A,4,7,7A-TETRAHYDRO-,	129	760
METHYLAMINE	130	760
MONOMETHYLAMINE	130	760
PENPHENE	351	760
PHENOL, 2,3-DIMETHYL-,	196	760
PHENYL CYANIDE	51	760
POLYCHLORCAMPHENE	351	760
POTASSIUM ACID ARSENATE	215	760
POTASSIUM ARSENATE	215	760
POTASSIUM DIHYDROGEN ARSENATE	215	760
TOXAPHENE	351	760
VELSICOL 104	129	760
XYLENE	360	760
2,3-XYLENOL	196	760
XYLOL	360	760
ACARIN	224	770
ACENAPHTHYLENE	2	770
ANTIOXIDANT MB	135	770
BENZENE, 1,3-DIMETHYL-,	193	770
BENZENE HEXACHLORIDE, D-ISOMER	266	770
BENZHYDROL, 4,4*-DICHLORO-A-/TRI-CHLOROMETHYL/-,	224	770
BENZO/JK/FLUORENE	108	770
D-BHC	266	770
CHLOROFORM	64	770
CYCLOHEXANE, 1,2,3,4,5,6-HEXA-CHLORO-, D-ISOMER	266	770
CYCLOPENTA/DE/NAPHTHALENE	2	770

CHEMICAL NAME	NUMBER	STRONGEST BAND
1,3-DICHLOROPROPENE	170	770
DICOFOL	224	770
ETHANOL, 1,1-BIS/P-CHLOROPHENYL/- 2,2,2-TRICHLORO-,	224	770
FAST GARNET B	135	770
FLUORANTHENE	108	770
IDRYL	108	770
KELTHANE	224	770
D-LINDANE	266	770
METHANE, TRICHLORO-,	64	770
A-NAPHTHYLAMINE	135	770
1-NAPHTHYLAMINE	135	770
PROPENE, 1,3-DICHLORO-,	170	770
PROPYLENE, 1,3-DICHLORO-,	170	770
TELONE	170	770
M-XYLENE	193	770
M-XYLOL	193	770
ACENAPHTHENE	1	780
ACENAPHTHYLENE, 1,2-DIHYDRO-,	1	780
BENZ/E/ACEPHENANTHRYLENE	219	780
BENZENE HEXACHLORIDE, A-ISOMER	264	780
3,4-BENZOFLUORANTHENE	219	780
BENZO/JK/FLUORENE	108	780
BENZONITRILE, 2,6-DICHLORO-,	52	780
A-BHC	264	780
CARBINAMINE	130	780
CASORON	52	780
CYCLOHEXANE, 1,2,3,4,5,6-HEXA- CHLORO-, A-ISOMER	264	780
2,6-DBN	52	780
1,1-DCE	101	780
DICHLOROBENIL	52	780
2,6-DICHLOROBENZONITRILE	52	780
1,1-DICHLOROETHYLENE	101	780
DU-SPREX	52	780
ETHANE, HEXACHLORO-,	88	780
ETHANE HEXACHLORIDE	88	780
ETHENE, 1,1-DICHLORO-,	101	780
ETHYLENE, 1,1-DICHLORO-	101	780
FLUORANTHENE	108	780
HEXACHLOROETHANE	88	780
IDRYL	108	780
A-LINDANE	264	780
METHANE, AMINO-,	130	780
METHYLAMINE	130	780
MONOMETHYLAMINE	130	780
MOTH BALLS	131	780
MOTH FLAKES	131	780
NAPHTHALENE	131	780
PERCHLOROETHANE	88	780
PHENOHEP	88	780
TAR CAMPHOR	131	780
VINYLIDENE CHLORIDE	101	780
ACETYLENE TETRACHLORIDE	89	790
ARSENIC SESQUIOXIDE	216	790
ARSENIC TRIOXIDE	216	790
ARSENOUS ACID	216	790
ARSENOUS OXIDE	216	790
BENZENE, CHLORO-,	35	790
BENZENE, 1,4-DIMETHYL-,	195	790
BENZENE CHLORIDE	35	790
BONOFORM	89	790
CARBON TETRACHLORIDE	63	790
CELLON	89	790
CHLOROBENZENE	35	790
DDE, P,P*-,	99	790
DRINOX	129	790
ETHANE, 1,1,2,2-TETRACHLORO-,	89	790
ETHYLENE, 1,1-BIS/P-CHLOROPHENYL/- 2,2-DICHLORO-,	99	790
HEPTACHLOR	129	790
HEPTAGRAN	129	790
MCB	35	790
METHANE, TETRACHLORO-,	63	790
4,7-METHANOINDENE, 1,4,5,6,7,8,8-HEPTA- CHLORO-3A,4,7,7A-TETRAHYDRO-,	129	790
MONOCHLOROBENZENE	35	790
PHENYL CHLORIDE	35	790
TETRACHLOROETHANE, 1,1,2,2-,	89	790
VELSICOL 104	129	790
WHITE ARSENIC	216	790
P-XYLENE	195	790
P-XYLOL	195	790

CHEMICAL NAME	NUMBER	STRONGEST BAND
ARSENIC ACID, LEAD/2 PLUS/ SALT /2 TO 3/	362	800
1-AZANAPHTHALENE	175	800
1-BENZAZINE	175	800
BENZENE, 1,4-DIMETHYL-,	195	800
BENZIDINE, 3,3*-DICHLORO-,	48	800
BIPHENYL, 4,4*-DIAMINO-3,3*-DI-CHLORO-,	48	800
BIPHENYLDIAMINE, 4,4*-, 3,3*-DI-CHLORO-,	48	800
CARBON TETRACHLORIDE	63	800
DCB	48	800
DIBENZOCARBAZOLE, 3,4:5,6-,	273	800
7H-DIBENZO/C,G/CARBAZOLE	273	800
LEAD ARSENATE	362	800
METHANE, TETRACHLORO-,	63	800
NAPHTHALENE, B-CHLORO-,	132	800
NAPHTHALENE, 2-CHLORO-,	132	800
QUINOLINE	175	800
P-XYLENE	195	800
P-XYLOL	195	800
ACETYLENE DICHLORIDE, TRANS-,	102	810
ARSENIC ACID, LEAD/2 PLUS/ SALT /1 TO 1/	214	810
BENZENE, P-DICHLORO-,	38	810
BENZENE, 1,4-DICHLORO-,	38	810
C-56	74	810
1,3-CYCLOPENTADIENE, 1,2,3,4,5,5-HEXA-CHLORO-,	74	810
P-DICHLOROBENZENE	38	810
1,2-DICHLOROETHYLENE, TRANS-,	102	810
ETHYLENE, 1,2-DICHLORO-, TRANS-,	102	810
GYPSINE	214	810
HEXACHLOROCYCLOPENTADIENE	74	810
LEAD ARSENATE	214	810
PARACIDE	38	810
PDCB	38	810
SANTOCHLOR 34	38	810
SCHULTENITE	214	810
SOPRABEL	214	810
ACENAPHTHYLENE	2	820
ACETYLENE DICHLORIDE, TRANS-,	102	820
CYCLOPENTA/DE/NAPHTHALENE	2	820
1,2-DICHLOROETHYLENE, TRANS-,	102	820
ETHYLENE, 1,2-DICHLORO-, TRANS-,	102	820
ACETYLENE TRICHLORIDE	104	840
ALGOFRENE TYPE 1	124	840
ARSENIC ACID, CALCIUM SALT /2 TO 3/	213	840
BENZO/GHI/PERYLENE	231	840
BENZOPERYLENE, 1,12-,	231	840
BENZO/DEF/PHENANTHRENE	174	840
CALCIUM ARSENATE	213	840
CHIP-CAL	213	840
ETHYLENE, TRICHLORO-,	104	840
ETHYLENE TRICHLORIDE	104	840
FLUOROTRICHLOROMETHANE	124	840
FREON 11	124	840
GENETRON 11	124	840
METHANE, FLUOROTRICHLORO-,	124	840
METHANE, TRICHLOROFLUORO-,	124	840
PENCAL	213	840
PYRENE	174	840
TCE	104	840
TRICALCIUM ARSENATE	213	840
TRICHLOROETHYLENE	104	840
TRICLENE	104	840
ALGOFRENE TYPE 1	124	850
AZACYCLOPROPANE	107	850
AZIRIDINE	107	850
1,3-BUTADIENE, HEXACHLORO-,	56	850
DIELDRIN	276	850
DIMETHANONAPHTHALENE, 1,4:5,8-, 6,7-EPOXY-1,2,3,4,10,10-HEXACHLORO-1,4,4A,5,6,7,8,8A-OCTAHYDRO-, ENDO-,-ENDO-,	275	850
DIMETHANONAPHTHALENE, 1,4:5,8-, 6,7-EPOXY-1,2,3,4,10,10-HEXACHLORO-1,4,4A,5,6,7,8,8A-OCTAHYDRO-, ENDO-,-EXO-,	276	850
DIMETHYLENIMINE	107	850
ENDRIN	275	850
ETHYLENIMINE	107	850
FLUOROTRICHLOROMETHANE	124	850

CHEMICAL NAME	NUMBER	STRONGEST BAND
FREON 11	124	850
GENETRON 11	124	850
HCBD	56	850
HEOD	276	850
HEXACHLOROBUTADIENE	56	850
HEXADRIN	275	850
ILLOXOL	276	850
MENDRIN	275	850
METHANE, FLUOROTRICHLORO-,	124	850
METHANE, TRICHLOROFLUORO-,	124	850
NAPHTHALENE, B-CHLORO-,	132	850
NAPHTHALENE, 2-CHLORO-,	132	850
AZACYCLOPROPANE	107	860
AZIRIDINE	107	860
CHROME YELLOW	247	860
CHROMIC ACID, LEAD/2 PLUS/ SALT /1 TO 1/	247	860
COLOGNE YELLOW	247	860
1,1-DCE	101	860
1,1-DICHLOROETHYLENE	101	860
DIMETHYLENIMINE	107	860
2,6-DMP	199	860
ETHENE, 1,1-DICHLORO-,	101	860
ETHYLENE, 1,1-DICHLORO-,	101	860
ETHYLENIMINE	107	860
KINGS YELLOW	247	860
LEAD CHROMATE/VI/	247	860
PHENOL, 2,6-DIMETHYL-,	199	860
VINYLIDENE CHLORIDE	101	860
2,6-XYLENOL	199	860
BENZ/J/ACEANTHRYLENE, 1,2-DIHYDRO-3-METHYL-,	243	870
CHOLANTHRENE, 3-METHYL-,	243	870
MC	243	870
ANTHRACENE	27	880
BENZ/A/ANTHRACENE	220	880
1,2-BENZANTHRACENE	220	880
CHROMIC ACID, DIPOTASSIUM SALT	245	880
GREEN OIL	27	880
NAPHTHANTHRACENE	220	880
NEUTRAL POTASSIUM CHROMATE	245	880
PARANAPHTHALENE	27	880
POTASSIUM CHROMATE/VI/	245	880
TARAPACAITE	245	880
TETRA OLIVE N2G	27	880
TETRAPHENE	220	880
1,3-BUTADIENE, 2-METHYL-,	114	890
CHROMATE OF SODA	246	890
CHROMIC ACID, DISODIUM SALT	246	890
DIBENZ/A,4/ANTHRACENE	272	890
DIBENZANTHRACENE, 1,2:5,6-,	272	890
ISOPRENE	114	890
NEUTRAL SODIUM CHROMATE	246	890
SODIUM CHROMATE/VI/	246	890
CALCIUM CHROMATE/VI/	244	900
CALCIUM CHROME YELLOW	244	900
CHROMIC ACID, CALCIUM SALT /1 TO 1/	244	900
GELBIN	244	900
YELLOW ULTRAMARINE	244	900
BETAPRONE	330	910
BIOXIRANE	57	910
BUTADIENE DIEPOXIDE	57	910
BUTADIENE DIOXIDE	57	910
BUTANE, 1,2:3,4-DIEPOXY-,	57	910
ETHYLENE, TETRACHLORO-,	103	910
ETHYLENE TETRACHLORIDE	103	910
2-OXETANONE	330	910
PERCHLOROETHYLENE	103	910
PERCLENE	103	910
B-PROPIOLACTONE	330	910
PROPIONIC ACID, B-HYDROXY-, LACTONE	330	910
TETRACAP	103	910
TETRACHLOROETHYLENE	103	910
THREITOL, 1,2:3,4-DIANHYDRO-,	57	910
ALGOFRENE TYPE 2	123	920
ARCTON 6	123	920
DICHLORODIFLUOROMETHANE	123	920
FREON 12	123	920
GENETRON 21	123	920
METHANE, DICHLORODIFLUORO-,	123	920
ACETYLENE TRICHLORIDE	104	930
ETHYLENE, TRICHLORO-,	104	930
ETHYLENE TRICHLORIDE	104	930

CHEMICAL NAME	NUMBER	STRONGEST BAND
FORMALDEHYDE	111	930
FORMALIN	111	930
FORMIC ALDEHYDE	111	930
METHANAL	111	930
TCE	104	930
TRICHLOROETHYLENE	104	930
TRICLENE	104	930
BICHROMATE OF SODA	274	940
CHROMIC ACID	249	940
CHROMIC ANHYDRIDE	249	940
CHROMIUM OXIDE	249	940
CHROMIUM TRIOXIDE	249	940
CHROMIUM/VI/ OXIDE /1 TO 3/	249	940
DICHROMIC ACID, DISODIUM SALT	274	940
SODIUM BICHROMATE	274	940
SODIUM DICHROMATE/VI/	274	940
ACTINOLITE	217	950
AMIANTHUS	217	950
AMOSITE	217	950
ASBESTOS	217	950
NATIVE CALCIUM-MAGNESIUM SILICATE	217	950
ACRYLON	19	960
ACRYLONITRILE	19	960
AM-FOL	21	960
AMMONIA	21	960
CARBACRYL	19	960
PROPENENITRILE	19	960
ACRYLON	19	970
ACRYLONITRILE	19	970
AM-FOL	21	970
AMMONIA	21	970
CARBACRYL	19	970
PROPENENITRILE	19	970
LEAD/II/ PHOSPHATE /3 TO 2/	333	980
LEAD ORTHOPHOSPHATE	333	980
PHOSPHORIC ACID, LEAD/2 PLUS/ SALT /2 TO 3/	333	980
TRILEAD PHOSPHATE	333	980
ARAMITE	346	990
ARATRON	346	990
AZINOPHOS METHYL	335	990
CES	346	990
GUTHION	335	990
ORTHO-MITE	346	990
PHOSPHORODITHIOIC ACID, O,O-DIMETHYL ESTER, S-ESTER WITH 3-/MERCAPTOMETHYL/-1,2,3-BENZOTRIAZIN-4/3H/-ONE	335	990
SULFUROUS ACID, 2-/P-TERT-BUTYLPHENOXY/-1-METHYLETHYL 2-CHLOROETHYL ESTER	346	990
ALFA-TOX	337	1000
BLADAN	336	1000
DIAZINON	337	1000
ETHION	336	1000
GARDENTOX	337	1000
NUCIDOL	337	1000
PHOSPHORODITHIOIC ACID, S,S*-METHYLENE O,O,O*,O*-TETRAETHYL ESTER	336	1000
PHOSPHOROTHIOIC ACID, O,O-DIETHYL O-/2-ISOPROPYL-6-METHYL-4-PYRIMIDINYL/ ESTER	337	1000
PHOSPHOTOX E	336	1000
RODOCID	336	1000
DIMETHYL SULFATE	182	1010
DISULFOTON	155	1010
DI-SYSTON	155	1010
HYDRAZINE, METHYL-, SULFATE /1 TO 1/	304	1010
METHYL SULFATE	182	1010
PHOSPHORIC ACID, TRISODIUM SALT, DODECAHYDRATE	334	1010
PHOSPHORODITHIOIC ACID, O,O-DIETHYL S-/2-/ETHYLTHIO/ETHYL/ ESTER	155	1010
SODIUM PHOSPHATE, DODECAHYDRATE	334	1010
SODIUM PHOSPHATE TRIBASIC	334	1010
SOLVIREX	155	1010
SULFURIC ACID, DIMETHYL ESTER	182	1010
THIODEMETON	155	1010
ALLYL ALCOHOL	20	1020
CARBINOL, VINYL-,	20	1020
CHLOROPHOS	331	1020
CROTONIC ACID, 3-HYDROXY-, METHYL ESTER, DIMETHYL PHOSPHATE, TRANS-	71	1020

CHEMICAL NAME	NUMBER	STRONGEST BAND
DANEX	331	1020
MENITE	71	1020
MEVINPHOS	71	1020
PHENOL, 2,3-DIMETHYL-,	196	1020
PHOSDRIN	71	1020
PHOSFENE	71	1020
PHOSPHONIC ACID, /1-HYDROXY- 2,2,2-TRICHLOROETHYL/-, DIMETHYL ESTER	331	1020
2-PROPEN-1-OL	20	1020
TRICHLOROFON	331	1020
TRINEX	331	1020
2,3-XYLENOL	196	1020
AMANIL SKY BLUE	255	1030
ATGARD V	154	1030
BENZAMINE BLUE	255	1030
BLADEX	340	1030
BROMCHLOPHOS	332	1030
BROMEX	332	1030
CI DIRECT BLUE 14, TETRASODIUM SALT	255	1030
DDVP	154	1030
DIBROM	332	1030
DICHLORVOS	154	1030
DIRECT BLUE 3B	255	1030
DISULFOTON	155	1030
DI-SYSTON	155	1030
ETHYL PYROPHOSPHATE	340	1030
NALED	332	1030
2,7-NAPHTHALENEDISULFONIC ACID, 3,3*-//3,3*-DIMETHYL-4,4*-BIPHENYLYL- ENE/BIS/AZO//BIS/5-AMINO-4-HYDROXY-, TETRASODIUM SALT	255	1030
PHOSPHORIC ACID, 1,2-DIBROMO- 2,2-DICHLOROETHYL DIMETHYL ESTER	332	1030
PHOSPHORIC ACID, 2,2-DICHLOROVINYL DIMETHYL ESTER	154	1030
PHOSPHORODITHIOIC ACID, O,O-DIETHYL S-/2-/ETHYLTHIO/ETHYL/ ESTER	155	1030
PYROPHOSPHORIC ACID, TETRAETHYL ESTER	340	1030
SOLVIREX	155	1030
TETRAETHYLPYROPHOSPHATE	340	1030
TETRON	340	1030
THIODEMETON	155	1030
VAPONA	154	1030
ALKRON	156	1040
ETHYL PARATHION	156	1040
PARATHION	156	1040
PHOSPHOROTHIOIC ACID, O,O-DIETHYL O-/P-NITROPHENYL/ ESTER	156	1040
THIOPHOS	156	1040
CROTONIC ACID, 3-HYDROXY-, METHYL ESTER, DIMETHYL PHOSPHATE, TRANS-,	71	1050
HS	300	1050
HYDRAZINE, SULFATE /1 TO 1/	300	1050
HYDRAZINIUM SULFATE	300	1050
HYDRAZONIUM SULFATE	300	1050
MENITE	71	1050
MEVINPHOS	71	1050
PHOSDRIN	71	1050
PHOSFENE	71	1050
ATGARD V	154	1060
DANA	77	1060
DDVP	154	1060
DEN	77	1060
DENA	77	1060
DICHLORVOS	154	1060
DIETHYLAMINE, N-NITROSO-,	77	1060
ETHER, BIS/2-CHLOROISOPROPYL/,	94	1060
ETHER, BIS/2-CHLORO-1-METHYLETHYL/,	94	1060
PHOSPHORIC ACID, 2,2-DICHLOROVINYL DIMETHYL ESTER	154	1060
VAPONA	154	1060
AEROTHENE TT	90	1080
BERYLLIUM SULFATE, TETRAHYDRATE /1 TO 1 TO 4/	344	1080
CHLORTEN	90	1080
ETHANE, 1,1,1-TRICHLORO-,	90	1080
FORMALDEHYDE, BIS/2-CHLOROETHYL/ ACETAL	118	1080
METHANE, BIS/2-CHLOROETHOXY/-,	118	1080
METHYLCHLOROFORM	90	1080
SULFURIC ACID, BERYLLIUM SALT /1 TO 1/, TETRAHYDRATE	344	1080

CHEMICAL NAME	NUMBER	STRONGEST BAND
TRICHLOROETHANE, 1,1,1-,	90	1080
ACETIC ACID, CHLORIDE	15	1090
ACETYL CHLORIDE	15	1090
BERYLLIUM PHOSPHATE	361	1090
DDE, P,P*-,	99	1090
ETHANOYL CHLORIDE	15	1090
ETHER, BIS/2-CHLOROISOPROPYL/,	94	1090
ETHER, BIS/2-CHLORO-1-METHYLETHYL/,	94	1090
ETHYLENE, 1,1-BIS/P-CHLOROPHENYL/- 2,2-DICHLORO-,	99	1090
PHOSPHORIC ACID, BERYLLIUM SALT /2 TO 3/	361	1090
AROCHLOR 1016	28	1100
AROCHLOR 1242	31	1100
BCME	93	1100
BIS-CME	93	1100
CHLOROMETHYL ETHER	93	1100
ETHER, BIS/CHLOROMETHYL/.	93	1100
PCB-1016	28	1100
PCB-1242	31	1100
POLYCHLORINATED BIPHENYL	28	1100
POLYCHLORINATED BIPHENYL	31	1100
BCME	93	1110
BENZIDINE, SULFATE /1 TO 1/	225	1110
BIPHENYL, 4,4*-DIAMINO-, SULFATE /1 TO 1/	225	1110
BIPHENYLDIAMINE, 4,4*-, SULFATE /1 TO 1/	225	1110
BIS-CME	93	1110
CADMIUM SULFATE /1 TO 1/	345	1110
CHLOREX	92	1110
CHLOROMETHYL ETHER	93	1110
DCEE	92	1110
ETHER, BIS/2-CHLOROETHYL/,	92	1110
ETHER, BIS/CHLOROMETHYL/,	93	1110
SULFURIC ACID, CADMIUM SALT /1 TO 1/	345	1110
1,4-DIETHYLENE DIOXIDE	81	1120
DIETHYLENE ETHER	81	1120
P-DIOXANE	81	1120
P-DIOXIN, TETRAHYDRO-,	81	1120
ACETIC ANHYDRIDE	14	1130
ACETIC OXIDE	14	1130
CHLORDECONE	316	1130
CHLOREX	92	1130
DCEE	92	1130
1,4-DIETHYLENE DIOXIDE	81	1130
DIETHYLENE ETHER	81	1130
P-DIOXANE	81	1130
P-DIOXIN, TETRAHYDRO-,	81	1130
ETHANOIC ANHYDRATE	14	1130
ETHER, BIS/2-CHLOROETHYL/,	92	1130
KEPONE	316	1130
MEREX	316	1130
METHENO-2H-CYCLOBUTA/CD/PENTALEN- 2-ONE, 1,3,4-, 1,1A,3,3A,4,5,5,5A,5B,6- DECACHLOROOCTAHYDRO-,	316	1130
BENZENE, M-DIHYDROXY-,	176	1140
M-BENZENEDIOL	176	1140
BICHLORENDO	315	1140
CYCLOPENTADIENE, HEXACHLORO-, DIMER	315	1140
DECHLORANE	315	1140
METHENO-1H-CYCLOBUTA/CD/PENTALENE, 1,3,4-, 1,1A,2,2,3,3A,4,5,5,5A,5B,6- DODECACHLOROOCTAHYDRO-,	315	1140
MIREX	315	1140
PELAGOL RS	176	1140
RESORCINE	176	1140
RESORCINOL	176	1140
BENZENE, M-DIHYDROXY-,	176	1150
M-BENZENEDIOL	176	1150
M-CRESOL	66	1150
M-CRESYLIC ACID	66	1150
PELAGOL RS	176	1150
PHENOL, M-METHYL-,	66	1150
RESORCINE	176	1150
RESORCINOL	176	1150
ALDEFORM	191	1160
BROMODICHLOROMETHANE	120	1160
ETHYL METHANESULFONATE	126	1160
FORMAGENE	191	1160
MARVOSAN	191	1160
METHACRYLIC ACID, METHYL ESTER	117	1160

CHEMICAL NAME	NUMBER	STRONGEST BAND
METHANE, BROMODICHLORO-,	120	1160
METHANE, DICHLOROBROMO-,	120	1160
METHANESULFONIC ACID, ETHYL ESTER	126	1160
METHYL METHACRYLATE	117	1160
MME	117	1160
MONOCITE METHACRYLATE MONOMER	117	1160
PARAFORMALDEHYDE	191	1160
PROPENIC ACID, 2-METHYL-, METHYL ESTER	117	1160
S-TRIOXANE	191	1160
ALDEFORM	191	1170
1,2-BENZISOTHIAZOLIN-3-ONE, 1,1-DIOXIDE, SODIUM SALT	228	1170
1,2-BENZISOTHIAZOL-3-OL, 1,1-DIOXIDE, SODIUM SALT	228	1170
1,2-BENZISOTHIAZOL-3/2H/-ONE, 1,1-DIOXIDE, SODIUM SALT	228	1170
P-CRESOL	70	1170
P-CRESYLIC ACID	70	1170
CYCLOESTROL	143	1170
EXTRA-PLEX	143	1170
FORMAGENE	191	1170
HEXESTROL	143	1170
MARVOSAN	191	1170
PARAFORMALDEHYDE	191	1170
PHENOL, 4,4*-/1,2-DIETHYLETHYLENE/-DI-, MESO-,	143	1170
PHENOL, P-METHYL-,	70	1170
SACCHARIN, SODIUM SALT	228	1170
SODIUM SACCHARIN	228	1170
STILBESTROL, DIHYDRO-,	143	1170
S-TRIOXANE	191	1170
BENZENE, HYDROXY-,	140	1180
BENZENESULFONIC ACID, DODECYL-, CALCIUM SALT /2 TO 1/	223	1180
BIO-DES	178	1180
CARBOLIC ACID	140	1180
CI ACID RED 26, DISODIUM SALT	252	1180
CI FOOD RED 6, DISODIUM SALT	256	1180
CLINESTROL	179	1180
M-CRESOL, 4-CHLORO-,	67	1180
DES	178	1180
DIBESTIL	179	1180
1,2-DIBROMOETHANE	85	1180
3,4-DMP	200	1180
DODECYLBENZENESULFONIC ACID, CALCIUM SALT /2 TO 1/	223	1180
DOWFUME W-8	85	1180
DOWFUME W-85	85	1180
DOWFUME 40	85	1180
DOWFUME EDB	85	1180
DOWFUME MC-2	85	1180
ETHANE, 1,2-DIBROMO-,	85	1180
ETHYLENE DIBROMIDE	85	1180
GLYCOL DIBROMIDE	85	1180
2,7-NAPHTHALENEDISULFONIC ACID, 3-HYDROXY-4-//2,4,5-TRIMETHYLPHENYL/-AZO/-, DISODIUM SALT	256	1180
2,7-NAPHTHALENEDISULFONIC ACID, 3-HYDROXY-4-/2,4-XYLYLAZO/-, DISODIUM SALT	252	1180
PARACHLOROMETA CRESOL	67	1180
PAROL	67	1180
PHENOL	140	1180
PHENOL, 4-CHLORO-3-METHYL-,	67	1180
PHENOL, 3,4-DIMETHYL-,	200	1180
PONCEAU G	252	1180
PONCEAU J	252	1180
PONCEAU R	252	1180
PONCEAU 2R	252	1180
PONCEAU 3R	256	1180
PONCEAU BNA	252	1180
PONCEAU FR	252	1180
PONCEAU GR	252	1180
PONCEAU MX	252	1180
PONCEAU NR	252	1180
PONCEAU PXM	252	1180
PONCEAU RN	256	1180
STILBENEDIOL, 4,4*-, A,A*-DIETHYL-,	178	1180
STILBENEDIOL, 4,4*-, A,A*-DIETHYL-, DIPROPIONATE, TRANS-,	179	1180
STILBESTEROL, DIETHYL-,	178	1180
STILBESTEROL, DIETHYL-, DIPROPIONATE	179	1180

CHEMICAL NAME	NUMBER	STRONGEST BAND
SYNESTRIN	178	1180
SYNTESTRINE	179	1180
USACERT RED NO. 1	256	1180
3,4-XYLENOL	200	1180
ACETONE CYANOHYDRIN	308	1190
ACRYLIC ACID, BUTYL ESTER	18	1190
BUTYL ACRYLATE	18	1190
BUTYL ALCOHOL, ACRYLATE	18	1190
CARBAMIC ACID, METHYLNITROSO-, ETHYL ESTER	61	1190
CARBAMIC ACID, N-METHYL-N-NITROSO-, ETHYL ESTER	61	1190
LACTONITRILE, 2-METHYL-,	308	1190
MNU	61	1190
1,2-OXATHIOLANE, 2,2-DIOXIDE	137	1190
1-PROPANESULFONIC ACID 3-HYDROXY-, G-SULTONE	137	1190
PROPANESULTONE	137	1190
PROPIONITRILE, 2-HYDROXY-2-METHYL-,	308	1190
URETHANE, NITROSOMETHYL-,	61	1190
DIMETHYL SULFATE	182	1200
2,6-DMP	199	1200
ENDOSULFAN	322	1200
ENDOSULFAN SULFATE	321	1200
ETHER, 2-CHLOROETHYL VINYL,	96	1200
GALLOTANNIC ACID	347	1200
GALLOTANNIN	347	1200
GLYCERITE	347	1200
GUANIDINE, 1-METHYL-3-NITRO-1-NITROSO-,	296	1200
GUANIDINE, N-METHYL-N*-NITRO-N-NITROSO-,	296	1200
MALIX	322	1200
6,9-METHANO-2,4,3-BENZODIOXATHIEPIN, 6,7,8,9,10,10-HEXACHLORO-1,5,5A,6,9,9A-HEXAHYDRO-, 3-OXIDE	322	1200
METHYL SULFATE	182	1200
MNG	296	1200
MNNG	296	1200
5-NORBORNENE-2,3-DIMETHANOL, 1,4,5,-6,7,7-HEXACHLORO-, CYCLIC SULFATE	321	1200
5-NORBORNENE-2,3-DIMETHANOL, 1,4,5,-6,7,7-HEXACHLORO-, CYCLIC SULFITE	322	1200
PHENOL, 2,6-DIMETHYL-,	199	1200
SULFURIC ACID, DIMETHYL ESTER	182	1200
TANNIC ACID	347	1200
TANNIN	347	1200
THIODAN	322	1200
2,6-XYLENOL	199	1200
ETHER, 2-CHLOROETHYL VINYL,	96	1210
ACETIC ACID, VINYL ESTER	10	1220
ASTA	329	1220
CYTOXAN	329	1220
ENDOXAN	329	1220
GENOXAL	329	1220
OXAZAPHOSPHORINE, 2H-1,3,2-, 2-/BIS/2-CHLOROETHYL/AMINO/TETRAHYDRO-, 2-OXIDE	329	1220
VINYL ACETATE	10	1220
VINYL A MONOMER	10	1220
ACETIC ACID, AMYL ESTER	9	1230
ACETIC ACID, BUTYL ESTER	8	1230
ACETIC ACID, PENTYL ESTER	9	1230
ACETIC ACID, /2,4-DICHLOROPHEN-OXY/-,	12	1230
AGROTECT	12	1230
AMYL ACETATE	9	1230
BANANA OIL	9	1230
BUTYL ACETATE	8	1230
BUTYL ALCOHOL, ACETATE	8	1230
2,4-D	12	1230
DED WEED	12	1230
2,5-DMP	198	1230
ESTERON	12	1230
PENTYL ACETATE	9	1230
PHENOL, 2,5-DIMETHYL-,	198	1230
2,5-XYLENOL	198	1230
ETHER, P-CHLOROPHENYL PHENYL,	97	1240
ETHANE, 1,1-BIS/P-METHOXYPHENYL/-2,2,2-TRICHLORO-,	83	1250
MARLATE	83	1250
METHOXCIDE	83	1250
METHOXYCHLOR	83	1250

CHEMICAL NAME	NUMBER	STRONGEST BAND
METHOXY-DDT	83	1250
BENZIDINE, 3,3*-DIMETHOXY-, DI-HYDROCHLORIDE	226	1260
BI-O-ANISIDINE, 4,4*-, DIHYDRO-CHLORIDE	226	1260
2-BUTENOIC ACID, 2-METHYL-, 7-//2,3-DIHYDROXY-2-/1-METHOXYETHYL/-3-METHYL-1-OXOBUTOXY/METHYL/-2,3,5,7A-TETRA-HYDRO-1H-PYRROLIZIN-1-YL ESTER	309	1260
CI DISPERSE BLACK 6, DIHYDRO-CHLORIDE	226	1260
LASIOCARPINE	309	1260
ANOZOL	161	1270
BENZYL BUTYL PHTHALATE	158	1270
BIS/2-ETHYLHEXYL/ PHTHALATE	159	1270
DIBUTYL PHTHALATE	160	1270
DICHLONE	134	1270
DIETHYL PHTHALATE	161	1270
HOMFLEX DOP	159	1270
1,4-NAPHTHOQUINONE, 2,3-DICHLORO-,	134	1270
PALATINOL A	161	1270
PALATINOL C	160	1270
PALATINOL BB	158	1270
PHTHALIC ACID, BENZYL BUTYL ESTER	158	1270
PHTHALIC ACID, BIS/2-ETHYLHEXYL/ ESTER	159	1270
PHTHALIC ACID, DIBUTYL ESTER	160	1270
PHTHALIC ACID, DIETHYL ESTER	161	1270
PHYGON	134	1270
POLYCIZER DBP	160	1270
SANQUINON	134	1270
SANTICIZER 160	158	1270
SOLVANOL	161	1270
STAFLEX DOP	159	1270
US RUBBER 604	134	1270
WITCIZER 300	160	1270
WITICIZER 312	159	1270
AVOLIN	162	1280
CERUBIDIN	270	1280
DAUNOMYCIN	270	1280
DIMETHYL PHTHALATE	162	1280
HEXOPYRANOSIDE, LYXO-, 3-ACETYL-6,11-DIOXO-1,2,3,4,6,11-HEXAHYDRO-10-METHOXY-3,5,12-TRIHYDROXY-1-NAPHTH-ACENYL 3-AMINO-2,3,6-TRIDEOXY-, A-L-, /1S,3S/-,	270	1280
MIPAX	162	1280
PHTHALIC ACID, DIMETHYL ESTER	162	1280
RUBIDOMYCIN	270	1280
RUBOMYCIN C	270	1280
SOLVARONE	162	1280
4-NQO	341	1300
QUINOLINE, 4-NITRO-, 1-OXIDE	341	1300
BENZENE, 4-HYDROXYNITRO-,	151	1320
BROM-O-GAS	119	1320
BROMOMETHANE	119	1320
METHANE, BROMO-,	119	1320
METHOGAS	119	1320
METHYL BROMIDE	119	1320
P-NITROPHENOL	151	1320
PHENOL, P-NITRO-,	151	1320
PHENOL, 4-NITRO-,	151	1320
ROTOX	119	1320
ALKRON	156	1330
BENZENE, 2-HYDROXYNITRO-,	150	1330
CARBAMIC ACID, ETHYL ESTER	60	1330
3,4-DINITROPHENOL	148	1330
ETHYL CARBAMATE	60	1330
ETHYL PARATHION	156	1330
O-NITROPHENOL	150	1330
PARATHION	156	1330
PHENOL, 3,4-DINITRO-,	148	1330
PHENOL, O-NITRO-,	150	1330
PHENOL, 2-NITRO-,	150	1330
PHOSPHOROTHIOIC ACID, O,O-DIETHYL O-/P-NITROPHENYL/ ESTER	156	1330
PRACARBAMIN	60	1330
THIOPHOS	156	1330
URETHANE	60	1330
ANTINONIN	69	1340
AZOPHOS	338	1340
BENZENE, P-DINITRO-,	41	1340
BENZENE, 1,4-DINITRO-	41	1340

CHEMICAL NAME	NUMBER	STRONGEST BAND
BENZENE, HEXACHLORO-,	43	1340
BUNT-CURE	43	1340
CAPSINE	69	1340
ANTI-CARIE 80	43	1340
O-CRESOL, 4,6-DINITRO-,	69	1340
P-DINITROBENZENE	41	1340
DITHANE A-4	41	1340
DITROSOL	69	1340
ENS	317	1340
HCB	43	1340
HEXACHLOROBENZENE	43	1340
METAPHOS	338	1340
METHANESULFONIC ACID, METHYL ESTER	127	1340
METHYL METHANESULFONATE	127	1340
METHYL PARATHION	338	1340
MMS	127	1340
1-NAPHTHALENESULFONAMIDE, 4-/ETHYL-SULFONYL/-,	317	1340
4-NITROSTILBENE	342	1340
NITROX 80	338	1340
1,2-OXATHIOLANE, 2,2-DIOXIDE	137	1340
PHOSPHOROTHIOIC ACID, O,O-DIMETHYL O-/P-NITROPHENYL/ ESTER	338	1340
1-PROPANESULFONIC ACID 3-HYDROXY-, G-SULTONE	137	1340
PROPANESULTONE	137	1340
SANDOLIN A	69	1340
SANOCIDE	43	1340
STILBENE, 4-NITRO-,	342	1340
ACETAMIDE, THIO-,	6	1350
ACETOTHIOAMIDE	6	1350
ALDIFEN	145	1350
BENZENE, M-DINITRO-,	39	1350
BENZENE, 1,3-DINITRO-,	39	1350
BENZENE, 4-HYDROXYNITRO-,	151	1350
BIPHENYL, 4-NITRO-,	54	1350
CHEMOX PE	145	1350
M-DINITROBENZENE	39	1350
2,4-DINITROPHENOL	145	1350
2,4-DNP	145	1350
ETHANETHIOAMIDE	6	1350
4-NITROBIPHENYL	54	1350
P-NITROPHENOL	151	1350
PHENOL, 2,4-DINITRO-,	145	1350
PHENOL, P-NITRO-,	151	1350
PHENOL, 4-NITRO-,	151	1350
PNB	54	1350
SOLFO BLACK B	145	1350
SOLFO BLACK 2B	145	1350
SOLFO BLACK BB	145	1350
TAA	6	1350
THIOACETAMIDE	6	1350
1,4-DINITROSOPIPERAZINE	164	1360
FANFT	292	1360
FORMAMIDE, N-/4-/5-NITRO-2-FURYL/-2-THIAZOLYL/-,	292	1360
1-NITROSOPIPERIDINE	165	1360
PIPERAZINE, 1,4-DINITROSO-,	164	1360
PIPERIDINE, N-NITROSO-,	165	1360
PIPERIDINE, 1-NITROSO-,	165	1360
CHLOROPHEN	152	1380
CHLORPYRIFOS	157	1380
O,O-DIETHYL O-3,5,6-TRICHLORO-2-PYRIDYL PHOSPHOROTHIOATE	157	1380
DOWCO 179	157	1380
DOWICIDE G	152	1380
DOWICIDE T	152	1380
DURSBAN	157	1380
ETHYL METHANESULFONATE	126	1380
METHANESULFONIC ACID, ETHYL ESTER	126	1380
METHANESULFONIC ACID, METHYL ESTER	127	1380
METHYL METHANESULFONATE	127	1380
MMS	127	1380
PCP	152	1380
PENTACHLOROPHENOL	152	1380
PHENOL, PENTACHLORO-,	152	1380
PHOSPHOROTHIOIC ACID, O,O-DIETHYL O-/3,5,6-TRICHLORO-2-PYRIDYL/ ESTER	157	1380
WEEDONE	152	1380
AROCHLOR 1260	212	1390
BASIC LEAD ACETATE	310	1390
LEAD, BIS/ACETATO/TETRAHYDROXYTRI-,	310	1390
LEAD SUBACETATE	310	1390

CHEMICAL NAME	NUMBER	STRONGEST BAND
MONOBASIC LEAD ACETATE	310	1390
PCB-1260	212	1390
POLYCHLORINATED BIPHENYL	212	1390
PDMT	188	1400
TRIAZENE, 3,3-DIMETHYL-1-PHENYL-,	188	1400
CADMIUM, BIS/DIETHYLDITHIOCARBAM-ATO/-,	238	1410
CHLORPYRIFOS	157	1410
O,O-DIETHYL O-3,5,6-TRICHLORO-2-PYRIDYL PHOSPHOROTHIOATE	157	1410
DOWCO 179	157	1410
DURSBAN	157	1410
ETHYL CADMATE	238	1410
PHOSPHOROTHIOIC ACID, O,O-DIETHYL O-/3,5,6-TRICHLORO-2-PYRIDYL/ ESTER	157	1410
BENZENE, O-DICHLORO-,	37	1430
BENZENE, 1,2-DICHLORO-,	37	1430
CHLOROBEN	37	1430
O-DICHLOROBENZENE	37	1430
DIZENE	37	1430
ODCB	37	1430
ANILINE, N-METHYL-N-NITROSO-,	24	1440
BENZENAMINE, N-METHYL-N-NITROSO-,	24	1440
BENZONITRILE, 2,6-DICHLORO-,	52	1440
CASORON	52	1440
2,6-DBN	52	1440
DICHLOROBENIL	52	1440
2,6-DICHLOROBENZONITRILE	52	1440
DIMETHYLAMINE, N-NITROSO-,	80	1440
DMNA	80	1440
DU-SPREX	52	1440
NITROSAMINE, N,N-DIMETHYL-,	80	1440
NMA	24	1440
ACETAMIDE, N-FLUOREN-3-YL-N-HYDROXY-,	208	1450
ACETOHYDROXAMIC ACID, N-FLUOREN-3-YL-,	208	1450
AROCHLOR 1242	31	1450
PCB-1242	31	1450
POLYCHLORINATED BIPHENYL	31	1450
ACETIC ACID, /2,4,5-TRICHLOROPHEN-OXY/-,	13	1460
AROCHLOR 1016	28	1460
AROCHLOR 1248	32	1460
BENZENE, 1,2,4-TRICHLORO-,	46	1460
ESTERON	13	1460
PCB-1016	28	1460
PCB-1248	32	1460
POLYCHLORINATED BIPHENYL	28	1460
POLYCHLORINATED BIPHENYL	32	1460
REDDOX	13	1460
T, 2,4,5-,	13	1460
TRICHLOROBENZENE, 1,2,4-,	46	1460
WEEDONE	13	1460
BENZENE, CHLORO-,	35	1470
BENZENE, P-DICHLORO-,	38	1470
BENZENE, 1,4-DICHLORO-,	38	1470
BENZENE CHLORIDE	35	1470
CARBON BISULFIDE	62	1470
CARBON DISULFIDE	62	1470
CARBON SULFIDE	62	1470
CHLOROBENZENE	35	1470
P-DICHLOROBENZENE	38	1470
DOWICIDE 2S	153	1470
ETHER, P-BROMOPHENYL PHENYL,	95	1470
FLUORENE, 2,7-DIAMINO-,	291	1470
FLUORENE-2,7-DIAMINE	291	1470
MCB	35	1470
MONOCHLOROBENZENE	35	1470
OMAL	153	1470
PARACIDE	38	1470
PDCB	38	1470
PDMT	188	1470
PHENACHLOR	153	1470
PHENOL, 2,4,6-TRICHLORO-,	153	1470
PHENYL CHLORIDE	35	1470
SANTOCHLOR 34	38	1470
TRIAZENE, 3,3-DIMETHYL-1-PHENYL-,	188	1470
TRICHLOROPHENOL, 2,4,6-,	153	1470
WEEVILTOX	62	1470
ACETIC ACID, /2,4-DICHLOROPHEN-OXY/-,	12	1480
AGROTECT	12	1480

CHEMICAL NAME	NUMBER	STRONGEST BAND
BENZENE, 1,2-/METHYLENEDIOXY/- 4-PROPYL-,	44	1480
1,3-BENZODIOXOLE, 5-PROPYL-,	44	1480
O-CHLOROPHENOL	141	1480
M-CRESOL, 4-CHLORO-,	67	1480
2,4-D	12	1480
DANA	77	1480
DCP	142	1480
DED WEED	12	1480
DEN	77	1480
DENA	77	1480
2,4-DICHLOROPHENOL	142	1480
DIETHYLAMINE, N-NITROSO-,	77	1480
DIMETHYLAMINE, N-NITROSO-,	80	1480
DIPROPYLAMINE, N-NITROSO-,	82	1480
DMNA	80	1480
DPN	82	1480
ESTERON	12	1480
ETHER, P-BROMOPHENYL PHENYL,	95	1480
NITROSAMINE, N,N-DIMETHYL-,	80	1480
NITROSAMINE, DIPROPYL-,	82	1480
PARACHLOROMETA CRESOL	67	1480
PAROL	67	1480
PHENOL, O-CHLORO-,	141	1480
PHENOL, 2-CHLORO-,	141	1480
PHENOL, 4-CHLORO-3-METHYL-,	67	1480
PHENOL, 2,4-DICHLORO-,	142	1480
SAFROLE, DIHYDRO-,	44	1480
ANILINE, 4,4*-METHYLENEBIS- /2-CHLORO-,	23	1490
ANILINE, N-METHYL-N-NITROSO-,	24	1490
ANOFEX	290	1490
BENZENAMINE, 4,4*-METHYLENEBIS- /2-CHLORO-,	23	1490
BENZENAMINE, N-METHYL-N-NITROSO-,	24	1490
BENZENE, 4-ALLYL-1,2-/METHYLENE- DIOXY/-,	34	1490
BENZENE, 1,2-/METHYLENEDIOXY/- 4-PROPYL-,	44	1490
BENZIDINE, 3,3*-DICHLORO-,	48	1490
1,3-BENZODIOXOLE, 5-ALLYL-,	34	1490
1,3-BENZODIOXOLE, 5-PROPYL-,	44	1490
BIPHENYL, 4,4*-DIAMINO-3,3*-DI- CHLORO-,	48	1490
BIPHENYLDIAMINE, 4,4*-, 3,3*-DI- CHLORO-,	48	1490
CI SOLVENT ORANGE 2	257	1490
CYANASET	23	1490
DCB	48	1490
DDD, P,P*-,	289	1490
DDT, P,P*-,	290	1490
DICOPHANE	290	1490
DILENE	289	1490
DIPHENYLAMINE, N-NITROSO-,	277	1490
ETHANE, 1,1-BIS/P-CHLOROPHENYL/- 2,2-DICHLORO-,	289	1490
ETHANE, 1,1-BIS/P-CHLOROPHENYL/- 2,2,2-TRICHLORO-,	290	1490
FAT ORANGE II	257	1490
MOCA	23	1490
2-NAPHTHOL, 1-/O-TOLYLAZO/-,	257	1490
NMA	24	1490
OIL ORANGE OPEL	257	1490
REDAX	277	1490
RETARDER 2N	277	1490
RHOTHANE D-3	289	1490
SAFROLE	34	1490
SAFROLE, DIHYDRO-,	44	1490
SHIKIMOLE	34	1490
TDE	289	1490
VULCATARD	277	1490
WILTROL N	277	1490
ZERDANE	290	1490
ANILINE, 4,4*-METHYLENEBIS- /2-CHLORO-,	23	1500
AQUACIDE	278	1500
1-AZANAPHTHALENE	175	1500
1-BENZAZINE	175	1500
BENZENAMINE, 4,4*-METHYLENEBIS- /2-CHLORO-,	23	1500
BENZIDINE	47	1500
BIO-DES	178	1500
BIPHENYL, 4,4*-DIAMINO-,	47	1500

CHEMICAL NAME	NUMBER	STRONGEST BAND
BIPHENYLDIAMINE, 4,4*-,	47	1500
BUTYRIC ACID, 4-/P-/BIS/2-CHLORO-ETHYL/AMINO/PHENYL/-,	237	1500
CHLORAMBUCIL	237	1500
CI AZOIC DIAZO	47	1500
CI SOLVENT RED 80	258	1500
CITRUS RED 2	258	1500
CYANASET	23	1500
DES	178	1500
DIAZONIAPHENANTHRENE DIBROMIDE, 8A,10A-, 9,10-DIHYDRO-,	278	1500
1,4-DINITROSOPIPERAZINE	164	1500
DIPYRIDO/1,2-A:2*,1*-C/PYRAZINE-DIIUM DIBROMIDE, 6,7-DIHYDRO-,	278	1500
DIQUAT DIBROMIDE	278	1500
3,4-DMP	200	1500
ELCORIL	237	1500
FAST CORINTH BASE B	47	1500
LEUKERAN	237	1500
MOCA	23	1500
2-NAPHTHOL, 1-//2,5-DIMETHOXYPHENYL/-AZO/-,	258	1500
NSC-3088	237	1500
PHENOL, 3,4-DIMETHYL-,	200	1500
PIPERAZINE, 1,4-DINITROSO-,	164	1500
QUINOLINE	175	1500
REGLONE	278	1500
STILBENEDIOL, 4,4*-, A,A*-DIETHYL-,	178	1500
STILBESTEROL, DIETHYL-,	178	1500
SYNESTRIN	178	1500
3,4-XYLENOL	200	1500
ACETIC ACID, LEAD/2 PLUS/ SALT	205	1510
ALANINE, 3-/P-/BIS/2-CHLOROETHYL/-AMINO/PHENYL/-, L-,	209	1510
ALKERAN	209	1510
BENZIDINE, 3,3*-DIMETHOXY-,	49	1510
BI-O-ANISIDINE, 4,4*-,	49	1510
CARBAMIC ACID, DIBUTYLDITHIO-, NICKEL SALT	319	1510
CELLITAZOL B	49	1510
CI DISPERSE BLACK 6	49	1510
CRESOL	263	1510
CRESYLIC ACID	263	1510
DIANISIDINE	49	1510
ETHANE, 1,1-BIS/P-METHOXYPHENYL/-2,2,2-TRICHLORO-,	83	1510
LEAD ACETATE	205	1510
LEAD DIACETATE	205	1510
LEAD/II/ ACETATE	205	1510
MARLATE	83	1510
MELPHALAN	209	1510
METHOXCIDE	83	1510
METHOXYCHLOR	83	1510
METHOXY-DDT	83	1510
NBC	319	1510
NICKEL, BIS/DIBUTYLDITHIOCARBAM-ATO/-,	319	1510
NSC-8806	209	1510
L-PAM	209	1510
PHENOL, 2,4-DIMETHYL-,	197	1510
SALT OF SATURN	205	1510
VANGUARD N	319	1510
2,4-XYLENOL	197	1510
BENZENE, NITRO-,	45	1520
CI SOLVENT BLACK 5	45	1520
P-CRESOL	70	1520
P-CRESYLIC ACID	70	1520
CYCLOESTROL	143	1520
DINITROPHENOL, BETA-,	147	1520
2,3-DINITROPHENOL	144	1520
2,6-DINITROPHENOL	147	1520
ESSENCE OF MIRBANE	45	1520
EXTRA-PLEX	143	1520
HEXESTROL	143	1520
NITROBENZENE	45	1520
M-NITROPHENOL	149	1520
OIL OF MIRBANE	45	1520
PHENOL, 4,4*-/1,2-DIETHYLETHYLENE/-DI-, MESO-,	143	1520
PHENOL, 2,3-DINITRO-,	144	1520
PHENOL, 2,6-DINITRO-,	147	1520
PHENOL, P-METHYL-,	70	1520
PHENOL, M-NITRO-,	149	1520

CHEMICAL NAME	NUMBER	STRONGEST BAND
PHENOL, 3-NITRO-,	149	1520
STILBESTROL, DIHYDRO-,	143	1520
BENZENE, O-DINITRO-,	40	1530
BENZENE, 1,2-DINITRO-,	40	1530
BENZENE, 2-HYDROXYNITRO-,	150	1530
O-DINITROBENZENE	40	1530
2,4-DINITROTOLUENE	185	1530
2,6-DINITROTOLUENE	186	1530
2,4-DNT	185	1530
O-NITROPHENOL	150	1530
PHENOL, O-NITRO-,	150	1530
PHENOL, 2-NITRO-,	150	1530
PROCASIL	192	1530
PTU	192	1530
PYRIMIDINEDIONE, 2,4/1H,3H/-, 6-PROPYL-2-THIO-,	192	1530
TOLUENE, 2,4-DINITRO-,	185	1530
TOLUENE, 2,6-DINITRO-,	186	1530
URACIL, 6-PROPYL-2-THIO-,	192	1530
ACRYLIC ACID, 2-METHYL-, SODIUM SALT	311	1540
BENZENE, NITRO-,	45	1540
CARBON BISULFIDE	62	1540
CARBON DISULFIDE	62	1540
CARBON SULFIDE	62	1540
CI SOLVENT BLACK 5	45	1540
3,4-DINITROPHENOL	148	1540
ESSENCE OF MIRBANE	45	1540
METHACRYLIC ACID, SODIUM SALT	311	1540
NITROBENZENE	45	1540
M-NITROPHENOL	149	1540
OIL OF MIRBANE	45	1540
PHENOL, 3,4-DINITRO-,	148	1540
PHENOL, M-NITRO-,	149	1540
PHENOL, 3-NITRO-,	149	1540
SODIUM METHACRYLATE	311	1540
WEEVILTOX	62	1540
BENZENE, M-DINITRO-,	39	1550
BENZENE, 1,3-DINITRO-,	39	1550
M-DINITROBENZENE	39	1550
DINITROPHENOL, BETA-,	147	1550
DINITROPHENOL, GAMMA-,	146	1550
2,5-DINITROPHENOL	146	1550
2,6-DINITROPHENOL	147	1550
2,4-DINITROTOLUENE	185	1550
2,6-DINITROTOLUENE	186	1550
2,5-DNP	146	1550
2,4-DNT	185	1550
PHENOL, 2,5-DINITRO-,	146	1550
PHENOL, 2,6-DINITRO-,	147	1550
TOLUENE, 2,4-DINITRO-,	185	1550
TOLUENE, 2,6-DINITRO-,	186	1550
URANIUM, BIS/ACETATO-DIOXO-,	355	1550
URANIUM OXYACETATE	355	1550
URANYL ACETATE	355	1550
ACETIC ACID, NICKEL/2 PLUS/ SALT	206	1560
BARBITURIC ACID, 5-ETHYL-5-PHENYL-, MONOSODIUM SALT	218	1560
BENZENE, O-DINITRO-,	40	1560
BENZENE, 1,2-DINITRO-,	40	1560
O-DINITROBENZENE	40	1560
DINITROPHENOL, GAMMA-,	146	1560
2,5-DINITROPHENOL	146	1560
2,5-DNP	146	1560
LUMINAL SODIUM	218	1560
NICKEL/II/ ACETATE /1 TO 2/	206	1560
NICKELOUS ACETATE	206	1560
PBS	218	1560
PHENOBARBITAL, MONOSODIUM SALT	218	1560
PHENOBARBITAL SODIUM	218	1560
PHENOL, 2,5-DINITRO-,	146	1560
BENZENE, P-DINITRO-,	41	1570
BENZENE, 1,4-DINITRO-,	41	1570
P-DINITROBENZENE	41	1570
2,3-DINITROPHENOL	144	1570
DITHANE A-4	41	1570
PHENOL, 2,3-DINITRO-,	144	1570
BENZENE, M-DICHLORO-,	36	1580
BENZENE, 1,3-DICHLORO-,	36	1580
BUTENEDIOIC ACID, CIS-,	115	1580
BUTYRIC ACID, 2-AMINO-4-/ETHYL-THIO/-, DL-,	235	1580
BUTYRIC ACID, 2-AMINO-4-/ETHYL-THIO/-, L-,	236	1580

CHEMICAL NAME	NUMBER	STRONGEST BAND
CHLOROETHYLENE	100	1580
CI BASIC RED 9, MONOHYDROCHLORIDE	253	1580
CI BASIC VIOLET 14, MONOHYDRO- CHLORIDE	254	1580
M-DICHLOROBENZENE	36	1580
ETH	235	1580
ETHENE, CHLORO-,	100	1580
ETHIONINE, DL-,	235	1580
ETHIONINE, L-,	236	1580
ETHYLENE, CHLORO-,	100	1580
1,2-ETHYLENEDICARBOXYLIC ACID, CIS-,	115	1580
FUCHSINE	254	1580
HOMOCYSTEINE, S-ETHYL-, DL-,	235	1580
HOMOCYSTEINE, S-ETHYL-, L-,	236	1580
MALEIC ACID	115	1580
MALENIC ACID	115	1580
PARAFUCHSINE	253	1580
PARAMAGENTA	253	1580
PARAROSANILINE	253	1580
ROSANILINE	254	1580
P-TOLUIDINE, A-/P-AMINOPHENYL/- A-/4-IMINO-2,5-CYCLOHEXADIEN-1-YL- IDENE/-, MONOHYDROCHLORIDE	253	1580
TOXILIC ACID	115	1580
VINYL CHLORIDE	100	1580
2,4-XYLIDINE, A<4-/P-AMINOPHENYL/- A<4-/4-IMINO-2,5-CYCLOHEXADIEN-1-YL- IDENE/-, MONOHYDROCHLORIDE	254	1580
ANILINE, N-METHYL-P-/PHENYLAZO/-,	211	1590
AZOBENZENE, N-METHYL-4-AMINO-,	211	1590
BENZENE, HYDROXY-,	140	1590
CARBOLIC ACID	140	1590
CHLORDAN /MIXTURE OF ISOMERS/	314	1590
M-CRESOL	66	1590
M-CRESYLIC ACID	66	1590
7H-FURO/3*,2*:4,5/FURO/2,3-C/XANTHEN- 7-ONE, 3A,12C-DIHYDRO-8-HYDROXY- 6-METHOXY-,	294	1590
4,7-METHANOINDAN, 1A,2,4B,5,6,7B,8,8- OCTACHLORO-3AA,4,7,7AA-TETRAHYDRO-,	314	1590
PHENOL	140	1590
PHENOL, M-METHYL-,	66	1590
STERIGMATOCYSTIN	294	1590
THIOCARBAMIDE	359	1590
THIOUREA	359	1590
THU	359	1590
UREA, THIO-,	359	1590
AAB	25	1600
AMINOPHEN	22	1600
AMUDANE	177	1600
ANILINE	22	1600
ANILINE, N,N-DIMETHYL-P-/PHENYL- AZO/-,	259	1600
ANILINE, N,N-DIMETHYL-P-/M-TOLYL- AZO/-,	210	1600
ANILINE, 4,4*-IMIDOCARBONYLBIS- /N,N-DIMETHYL-,	260	1600
ANILINE, P-/PHENYLAZO/-,	25	1600
ANILINE YELLOW	25	1600
ANTINONIN	69	1600
AURAMINE	260	1600
AZOBENZENE, 4-AMINO-,	25	1600
AZOBENZENE, P-/DIMETHYLAMINO/-,	259	1600
AZOBENZENE, 4-/DIMETHYLAMINO/- 3*-METHYL-,	210	1600
BENZENAMINE	22	1600
BENZENAMINE, N,N-DIMETHYL- 4-//3-METHYLPHENYL/AZO/-,	210	1600
BENZENE, AMINO-,	22	1600
BENZIDAM	22	1600
BIPHENYL, 4-AMINO-,	55	1600
4-BIPHENYLAMINE	55	1600
BUTTER YELLOW	259	1600
C-56	74	1600
CAPSINE	69	1600
CI SOLVENT YELLOW 1	25	1600
CI SOLVENT YELLOW 2	259	1600
CI SOLVENT YELLOW 34	260	1600
O-CRESOL, 4,6-DINITRO-,	69	1600
1,3-CYCLOPENTADIENE, 1,2,3,4,5,5-HEXA- CHLORO-,	74	1600
DITROSOL	69	1600
3,5-DMP	201	1600

CHEMICAL NAME	NUMBER	STRONGEST BAND
FULVICIN	177	1600
GRISEOFULVIN	177	1600
HEXACHLOROCYCLOPENTADIENE	74	1600
HYDRAZINE, 1,2-DIPHENYL-,	306	1600
HYDRAZOBENZENE	306	1600
HYDRAZODIBENZENE	306	1600
MDAB	210	1600
METHYL YELLOW	259	1600
PHENOL, 3,5-DIMETHYL-,	201	1600
PONCYL	177	1600
SANDOLIN A	69	1600
SPIRO/BENZOFURAN-2/3H/,1*-/2/CYCLO-HEXENE/-3,4*-DIONE, 7-CHLORO-G*B-METHYL-2*,4,6-TRIMETHOXY-,	177	1600
WAXOLINE YELLOW O	260	1600
3,5-XYLENOL	201	1600
YELLOW PYOCTANINE	260	1600
AMUDANE	177	1610
CORLUTIN	167	1610
3,5-DMP	201	1610
FULVICIN	177	1610
GRISEOFULVIN	177	1610
LUTEOSTERONE	167	1610
PHENOL, 3,5-DIMETHYL-,	201	1610
PONCYL	177	1610
PREGN-4-ENE-3,20-DIONE	167	1610
PROGESTERONE	167	1610
PROLIDON	167	1610
SPIRO/BENZOFURAN-2/3H/,1*-/2/CYCLO-HEXENE/-3,4*-DIONE, 7-CHLORO-G*B-METHYL-2*,4.6-TRIMETHOXY-,	177	1610
3,5-XYLENOL	201	1610
AAB	25	1620
AMINOPHEN	22	1620
AMITROLE	189	1620
AMIZOL	189	1620
ANILINE	22	1620
ANILINE, O-METHYL-,	187	1620
ANILINE, P-/PHENYLAZO/-,	25	1620
ANILINE YELLOW	25	1620
AZOBENZENE, 4-AMINO-,	25	1620
BENZENAMINE	22	1620
BENZENE, AMINO-,	22	1620
BENZIDAM	22	1620
BIPHENYL, 4-AMINO-,	55	1620
4-BIPHENYLAMINE	55	1620
CHLOROETHYLENE	100	1620
CI SOLVENT YELLOW 1	25	1620
ETHENE, CHLORO-,	100	1620
ETHYLENE, CHLORO-,	100	1620
FAST SCARLET BASE B	136	1620
FLUOREN-2-AMINE	109	1620
FLUORENE, 2-AMINO-,	109	1620
B-NAPHTHYLAMINE	136	1620
2-NAPHTHYLAMINE	136	1620
TOLUENE, O-AMINO-,	187	1620
O-TOLUIDINE	187	1620
TRIAZOLE, 1H-1,2,4-, 3-AMINO-,	189	1620
S-TRIAZOLE, 3-AMINO-,	189	1620
UREA, 1-BUTYL-1-NITROSO-,	356	1620
USAF CB-22	136	1620
VINYL CHLORIDE	100	1620
WEEDOCLOR	189	1620
ALKIRON	353	1630
ANILINE, O-METHYL-,	187	1630
AZODYNE	339	1630
1,3-DICHLOROPROPENE	170	1630
FAST SCARLET BASE B	136	1630
IBENZMETHYZINE HYDROCHLORIDE	350	1630
MALLOPHENE	339	1630
METACIL	353	1630
MURACIL	353	1630
B-NAPHTHYLAMINE	136	1630
2-NAPHTHYLAMINE	136	1630
NATULAN HYDROCHLORIDE	350	1630
PHENAZOPYRIDINE HYDROCHLORIDE	339	1630
PROCARBAZINE, HYDROCHLORIDE	350	1630
PROPENE, 1,3-DICHLORO-,	170	1630
PROPYLENE, 1,3-DICHLORO-,	170	1630
PYRIDINE, 2,6-DIAMINO-3-/PHENYL-AZO/-, MONOHYDROCHLORIDE	339	1630
PYRIDIUM	339	1630
PYRIMIDINEDIONE, 2,4/1H,3H/-, 6-METHYL-2-THIO-,	353	1630

CHEMICAL NAME	NUMBER	STRONGEST BAND
TELONE	170	1630
P-TOLUAMIDE, N-ISOPROPYL-A-/2-METHYL- HYDRAZINO/-, MONOHYDROCHLORIDE	350	1630
TOLUENE, O-AMINO-,	187	1630
O-TOLUIDINE	187	1630
URACIL, 6-METHYL-2-THIO-,	353	1630
USAF CB-22	136	1630
AMITROLE	189	1640
AMIZOL	189	1640
TRIAZOLE, 1H-1,2,4-, 3-AMINO-,	189	1640
S-TRIAZOLE, 3-AMINO-,	189	1640
WEEDOCLOR	189	1640
ACETAMIDE, N,N*-FLUOREN-2,7-YLENE- BIS-,	202	1650
DICHLORFENIDIM	357	1650
DIUREX	357	1650
DIURON	357	1650
2,7-FAA	202	1650
FLUORENE, 2,7-DIACETAMIDO-,	202	1650
KARMEX	357	1650
NORETHINDRONE	325	1650
NORETHISTERONE	325	1650
NORLUTIN	325	1650
19-NOR-17A-PREGN-4-EN-20-YN-3-ONE, 17-HYDROXY-,	325	1650
PROCASIL	192	1650
PTU	192	1650
PYRIMIDINEDIONE, 2,4/1H,3H/-, 6-PROPYL-2-THIO-,	192	1650
URACIL, 6-PROPYL-2-THIO-,	192	1650
UREA, 3-/3,4-DICHLOROPHENYL/- 1,1-DIMETHYL-,	357	1650
ACETAMIDE, N-2-PHENANTHRYL-,	203	1660
AGOVIRIN	349	1660
AMIDON	307	1660
ANDROLIN	348	1660
ANDROST-4-EN-3-ONE, 17B-HYDROXY-,	348	1660
ANDROST-4-EN-3-ONE, 17B-HYDROXY-, PROPIONATE	349	1660
2-CYCLOHEXEN-1-ONE, 3,5,5-TRIMETHYL-,	73	1660
HORMOTESTON	349	1660
ISONIAZID	307	1660
ISONICOTINIC ACID, HYDRAZIDE	307	1660
ISOPHORONE	73	1660
KWIK-KIL	343	1660
MERTESTATE	348	1660
4,6-METHANO-6H,14H-INDOLO/3,2,1-IJ/- OXEPINO/2,3,4-DE/PYRROLO/2,3-H/QUINOL- IN-14-ONE, 2,4A,5,5A,7,8,15,15A,15B,- 15C-DECAHYDRO-,	343	1660
MOUSE-TOX	343	1660
ORETON	349	1660
PHENANTHRENE, 2-ACETAMIDO-,	203	1660
4-PYRIDINECARBOXYLIC ACID, HYDRAZIDE	307	1660
RO-DEX	343	1660
STRYCHNINE	343	1660
TESLEN	348	1660
TESTOSTERONE	348	1660
TESTOSTERONE, PROPIONATE	349	1660
ZINADON	307	1660
ACETAMIDE, N-4-BIPHENYLYL-,	204	1670
ACETANILIDE, 4*-PHENYL-,	204	1670
BIPHENYL, 4-ACETAMIDO-,	204	1670
HYDRAZINE, 1-ACETYL-2-PICOLINOYL-,	301	1670
NORETHINDRONE, ACETATE	326	1670
NORETHISTERONE, ACETATE	326	1670
NORLUTATE	326	1670
NORLUTIN, ACETATE	326	1670
19-NOR-17A-PREGN-4-EN-20-YN-3-ONE, 17-HYDROXY-, ACETATE	326	1670
ACETAMIDE	4	1680
ACETIC ACID AMIDE	4	1680
ACROLEIN	16	1680
ACRYLALDEHYDE	16	1680
ALLYL ALDEHYDE	16	1680
AMINOURACIL MUSTARD	352	1680
ARTIFICIAL ANT OIL	113	1680
BENZENECARBOXYLIC ACID	50	1680
BENZOIC ACID	50	1680
BUTENEDIOIC ACID, TRANS-,	293	1680
DESMETHYLDOPAN	352	1680
DICHLONE	134	1680
ETHANAMIDE	4	1680

CHEMICAL NAME	NUMBER	STRONGEST BAND
1,2-ETHENEDICARBOXYLIC ACID, TRANS-,	293	1680
ETHYLENE ALDEHYDE	16	1680
1,2-ETHYLENEDICARBOXYLIC ACID, TRANS-,	293	1680
FUMARIC ACID	293	1680
FURAL	113	1680
2-FURALDEHYDE	113	1680
FURFURAL	113	1680
FURFUROLE	113	1680
METHANECARBOXAMIDE	4	1680
1,4-NAPHTHOQUINONE, 2,3-DICHLORO-,	134	1680
PHYGON	134	1680
PROPENAL	16	1680
PYRIMIDINEDIONE, 2,4/1H,3H/-, 5-/BIS/2-CHLOROETHYL/AMINO/-,	352	1680
RETARDER BA	50	1680
RETARDEX	50	1680
SANQUINON	134	1680
TENN-PLAS	50	1680
URACIL, 5-/BIS/2-CHLOROETHYL/-AMINO/-,	352	1680
US RUBBER 604	134	1680
ACETIC ACID, /ETHYLENEDINITRILO/-TETRA-,	207	1690
ACROLEIC ACID	17	1690
ACRYLIC ACID	17	1690
CORLUTIN	167	1690
2-CYCLOHEXEN-1-ONE, 3,5,5-TRIMETHYL-,	73	1690
EDTA	207	1690
ETHYLENECARBOXYLIC ACID	17	1690
ETHYLENEDIAMINETETRAACETIC ACID	207	1690
ISOPHORONE	73	1690
LUTEOSTERONE	167	1690
NULLAPON B ACID	207	1690
PREGN-4-ENE-3,20-DIONE	167	1690
PROGESTERONE	167	1690
PROLIDON	167	1690
PROPENOIC ACID	17	1690
VERSENE	207	1690
ACETIC ACID, ETHYL-,	59	1700
O-ANISIC ACID, 3,6-DICHLORO-,	26	1700
ANTAGOTHYROID	354	1700
BANLEN	26	1700
BANVEL D	26	1700
BENZOIC ACID, 3,6-DICHLORO 2-METHOXY-,	26	1700
BENZOPYRAN-2-ONE, 2H-1-,	65	1700
BUTANOIC ACID	59	1700
BUTYRIC ACID	59	1700
CINNAMIC ACID, O-HYDROXY-, D-LACTONE	65	1700
COUMARIN	65	1700
DERACIL	354	1700
DICAMBA	26	1700
PYRIMIDINEDIONE, 2,4/1H,3H/-, 2-THIO-,	354	1700
2-THIOURACIL	354	1700
TONKA BEAN CAMPHOR	65	1700
URACIL, 2-THIO-,	354	1700
ATOXAN	242	1710
CARBAMIC ACID, ETHYL ESTER	60	1710
CARBAMIC ACID, METHYL-, 4-/DI-METHYLAMINO/-3,5-XYLYL ESTER	241	1710
CARBAMIC ACID, METHYL-, 1-NAPHTHYL ESTER	242	1710
CARBARYL	242	1710
DOWCO 139	241	1710
ETHANECARBOXYLIC ACID	172	1710
ETHYL CARBAMATE	60	1710
FORMIC ACID, ETHYL-,	172	1710
MEXACARBATE	241	1710
NAPHTHENIC ACID	133	1710
PRACARBAMIN	60	1710
PROPIONIC ACID	172	1710
PROZOIN	172	1710
PSEUDOACETIC ACID	172	1710
SEPTENE	242	1710
SEVIN	242	1710
URETHANE	60	1710
3,5-XYLENOL, 4-/DIMETHYLAMINO/-, METHYLCARBAMATE	241	1710
ZECTRANE	241	1710
AAF	5	1720
ACETALDEHYDE	3	1720

CHEMICAL NAME	NUMBER	STRONGEST BAND
ACETAMIDE, N-FLUOREN-2-YL-,	5	1720
ACETIC ACID	7	1720
ACETIC ALDEHYDE	3	1720
ANOZOL	161	1720
ARTIFICIAL ANT OIL	113	1720
1,2-BENZISOTHIAZOLIN-3-ONE, 1,1-DIOXIDE	227	1720
1,2-BENZISOTHIAZOL-3-OL, 1,1-DIOXIDE	227	1720
1,2-BENZISOTHIAZOL-3/2H/-ONE, 1,1-DI- OXIDE	227	1720
O-BENZOSULFIMIDE	227	1720
DIETHYL PHTHALATE	161	1720
DIOCTYL PHTHALATE	163	1720
ENDRIN ALDEHYDE	279	1720
EQUILENIN	284	1720
EQUILIN	285	1720
ESTRA-1,3,5,7,9-PENTAEN-17-ONE, 3-HYDROXY-,	284	1720
ESTRA-1,3,5/10/,6,8-PENTAEN-17-ONE, 3-HYDROXY-,	284	1720
ESTRA-1,3,5/10/,7-TETRAEN-17-ONE, 3-HYDROXY-,	285	1720
ESTRA-1,3,5/10/-TRIEN-17-ONE, 3-HYDROXY-,	287	1720
ESTRIN, KETOHYDROXY-,	287	1720
ESTRONE	287	1720
ETHANAL	3	1720
ETHANOIC ACID	7	1720
ETHYL ALDEHYDE	3	1720
ETHYLIC ACID	7	1720
FAA	5	1720
FLUORENE, 2-ACETAMIDO-,	5	1720
FOLLICULIN	287	1720
FURAL	113	1720
2-FURALDEHYDE	113	1720
FURFURAL	113	1720
FURFUROLE	113	1720
HATCOL DOP	163	1720
KETOHYDROXYESTRIN	287	1720
MENFORMON	287	1720
METHANECARBOXYLIC ACID	7	1720
NORETHINYNODREL	327	1720
NORETHYNODREL	327	1720
PALATINOL A	161	1720
PHTHALIC ACID, DIETHYL ESTER	161	1720
PHTHALIC ACID, DIOCTYL ESTER	163	1720
POLYCIZER 162	163	1720
19-NOR-17A-PREGN-5/10/-EN-20-YN-3-ONE, 17-HYDROXY-,	327	1720
PX-138	163	1720
SACCHARIN	227	1720
SOLVANOL	161	1720
THEELIN	287	1720
VINEGAR ACID	7	1720
ACETALDEHYDE, METHYL-,	171	1730
ACETIC ACID, AMYL ESTER	9	1730
ACETIC ACID, BUTYL ESTER	8	1730
ACETIC ACID, PENTYL ESTER	9	1730
ACROLEIN	16	1730
ACRYLALDEHYDE	16	1730
ALLYL ALDEHYDE	16	1730
AMERCIDE	269	1730
AMYL ACETATE	9	1730
AVOLIN	162	1730
BANANA OIL	9	1730
BENZYL BUTYL PHTHALATE	158	1730
BIS/2-ETHYLHEXYL/ PHTHALATE	159	1730
BUTYL ACETATE	8	1730
BUTYL ALCOHOL, ACETATE	8	1730
CAPTAN	269	1730
CO-RAL	262	1730
COUMAPHOS	262	1730
COUMARIN, 3-CHLORO-7-HYDROXY- 4-METHYL-, O-ESTER WITH O,O-DIETHYL- PHOSPHOROTHIOATE	262	1730
4-CYCLOHEXENE-1,2-DICARBOXIMIDE, N-//TRICHLOROMETHYL/THIO/-,	269	1730
CYTHION	181	1730
DALAPON	173	1730
DIBUTYL PHTHALATE	160	1730
2,2-DICHLOROPROPIONIC ACID	173	1730
DIHYDROFOLLICULIN BENZOATE	281	1730
DIMETHYL PHTHALATE	162	1730
DOWPON	173	1730

CHEMICAL NAME	NUMBER	STRONGEST BAND
EPOXIDE-201	328	1730
ESTRADIOL, 3-BENZOATE	281	1730
ESTRADIOL, DIPROPIONATE	282	1730
17B-ESTRADIOL, 3-BENZOATE	281	1730
17B-ESTRADIOL, DIPROPIONATE	282	1730
ESTRA-1,3,5/10/-TRIENE-3,17B-DIOL, 3-BENZOATE	281	1730
ESTRA-1,3,5/10/-TRIENE-3,17B-DIOL, DIPROPIONATE	282	1730
ESTRA-1,3,5/10/-TRIEN-17-ONE, 3-HYDROXY-, BENZOATE	288	1730
ESTRONE, BENZOATE	288	1730
ETHYLENE ALDEHYDE	16	1730
FOLLICULIN, DIHYDRO-, BENZOATE	281	1730
FORMIC ACID	112	1730
HOWFLEX DOP	159	1730
KARBOFOS	181	1730
MALATHION	181	1730
METHANOIC ACID	112	1730
MIPAX	162	1730
ORTHOCIDE	269	1730
OVOCYCLIN DIPROPIONATE	282	1730
7-OXABICYCLO/4.1.0/HEPTANE-3-CARBOXYLIC ACID, 4-METHYL-, /4-METHYL-7-OXA-BICYCLO/4.1.0/HEPT-3-YL/METHYL ESTER	328	1730
PALATINOL C	160	1730
PALATINOL BB	158	1730
PENTYL ACETATE	9	1730
PHTHALIC ACID, BENZYL BUTYL ESTER	158	1730
PHTHALIC ACID, BIS/2-ETHYLHEXYL/ ESTER	159	1730
PHTHALIC ACID, DIBUTYL ESTER	160	1730
PHTHALIC ACID, DIMETHYL ESTER	162	1730
POLYCIZER DBP	160	1730
PROGYNON BENZOATE	281	1730
PROGYNON-DP	282	1730
PROPALDEHYDE	171	1730
PROPANAL	171	1730
PROPENAL	16	1730
PROPIONALDEHYDE	171	1730
PROPIONIC ACID, 2,2-DICHLORO-,	173	1730
PROPIONIC ALDEHYDE	171	1730
RESITOX	262	1730
SANTICIZER 160	158	1730
SOLVARONE	162	1730
STAFLEX DOP	159	1730
SUCCINIC ACID, MERCAPTO-, DIETHYL ESTER, S-ESTER WITH O,O-DIMETHYLPHOSPHORODITHIOATE	181	1730
SUNTOL	262	1730
UNOX EPOXIDE 201	328	1730
VANGUARD K	269	1730
WITCIZER 300	160	1730
WITICIZER 312	159	1730
ZITHIOL	181	1730
ACETALDEHYDE, METHYL-,	171	1740
DIOCTYL PHTHALATE	163	1740
HATCOL DOP	163	1740
PHTHALIC ACID, DIOCTYL ESTER	163	1740
POLYCIZER 162	163	1740
PROPALDEHYDE	171	1740
PROPANAL	171	1740
PROPIONALDEHYDE	171	1740
PROPIONIC ALDEHYDE	171	1740
PX-138	163	1740
ACETAMIDE	4	1750
ACETIC ACID, /2,4,5-TRICHLOROPHENOXY/-,	13	1750
ACETIC ACID AMIDE	4	1750
CYTHION	181	1750
ESTERON	13	1750
ETHANAMIDE	4	1750
KARBOFOS	181	1750
MALATHION	181	1750
METHANECARBOXAMIDE	4	1750
REDDOX	13	1750
SUCCINIC ACID, MERCAPTO-, DIETHYL ESTER, S-ESTER WITH O,O-DIMETHYLPHOSPHORODITHIOATE	181	1750
T, 2,4,5-,	13	1750
WEEDONE	13	1750
ZITHIOL	181	1750
ACETALDEHYDE	3	1760

CHEMICAL NAME	NUMBER	STRONGEST BAND
ACETIC ALDEHYDE	3	1760
BENZENECARBOXYLIC ACID	50	1760
BENZOIC ACID	50	1760
CLINESTROL	179	1760
DIBESTIL	179	1760
EPIHYDRINALDEHYDE	363	1760
ETHANAL	3	1760
ETHYL ALDEHYDE	3	1760
GLYCIDALDEHYDE	363	1760
1-PROPANAL, 2,3-EPOXY-,	363	1760
PROPIONALDEHYDE, 2,3-EPOXY-,	363	1760
RETARDER BA	50	1760
RETARDEX	50	1760
STILBENEDIOL, 4,4*-, A,A*-DIETHYL-, DIPROPIONATE, TRANS-,	179	1760
STILBESTEROL, DIETHYL-, DIPROPION- ATE	179	1760
SYNTESTRINE	179	1760
TENN-PLAS	50	1760
AAF	5	1770
ACETAMIDE, N-FLUOREN-2-YL-,	5	1770
O-ANISIC ACID, 3,6-DICHLORO-,	26	1770
BANLEN	26	1770
BANVEL D	26	1770
BENZOIC ACID, 3,6-DICHLORO- 2-METHOXY-,	26	1770
DICAMBA	26	1770
FAA	5	1770
FLUORENE, 2-ACETAMIDO-,	5	1770
2,5-FURANDIONE	116	1770
LYTRON 810	116	1770
LYTRON 820	116	1770
MALEIC ANHYDRIDE	116	1770
TOXILIC ANHYDRIDE	116	1770
ACETIC ACID, ETHYL-,	59	1780
ACROLEIC ACID	17	1780
ACRYLIC ACID	17	1780
BENZENECARBONYL CHLORIDE	53	1780
BENZOPYRAN-2-ONE, 2H-1-,	65	1780
BENZOYL CHLORIDE	53	1780
BUTANOIC ACID	59	1780
BUTYRIC ACID	59	1780
CINNAMIC ACID, O-HYDROXY-, D-LACTONE	65	1780
COUMARIN	65	1780
DALAPON	173	1780
2,2-DICHLOROPROPIONIC ACID	173	1780
DOWPON	173	1780
ETHANECARBOXYLIC ACID	172	1780
ETHYLENECARBOXYLIC ACID	17	1780
FORMIC ACID, ETHYL-,	172	1780
PROPENOIC ACID	17	1780
PROPIONIC ACID	172	1780
PROPIONIC ACID, 2,2-DICHLORO-,	173	1780
PROZOIN	172	1780
PSEUDOACETIC ACID	172	1780
TONKA BEAN CAMPHOR	65	1780
BENZENECARBONYL CHLORIDE	53	1790
BENZOYL CHLORIDE	53	1790
FORMIC ACID	112	1790
METHANOIC ACID	112	1790
ACETIC ACID	7	1800
BUTENEDIOIC ACID, CIS-,	115	1800
ETHANOIC ACID	7	1800
1,2-ETHYLENEDICARBOXYLIC ACID, CIS-,	115	1800
ETHYLIC ACID	7	1800
2,5-FURANDIONE	116	1800
LYTRON 810	116	1800
LYTRON 820	116	1800
MALEIC ACID	115	1800
MALEIC ANHYDRIDE	116	1800
MALENIC ACID	115	1800
METHANECARBOXYLIC ACID	7	1800
TOXILIC ACID	115	1800
TOXILIC ANHYDRIDE	116	1800
VINEGAR ACID	7	1800
ACETIC ACID, CHLORIDE	15	1830
ACETIC ANHYDRIDE	14	1830
ACETIC OXIDE	14	1830
ACETYL CHLORIDE	15	1830
BUTYRIC ACID, 3-HYDROXY-, B-LACTONE	138	1830
B-BUTYROLACTONE	138	1830
ETHANOIC ANHYDRATE	14	1830

CHEMICAL NAME	NUMBER	STRONGEST BAND
ETHANOYL CHLORIDE	15	1830
2-OXETANONE, 4-METHYL-,	138	1830
BUTYRIC ACID, 3-HYDROXY-, B-LACTONE	138	1860
B-BUTYROLACTONE	138	1860
2-OXETANONE, 4-METHYL-,	138	1860
ACETIC ACID, DIAZO-, ETHYL ESTER	11	2120
DAEE	11	2120
DIAZOACETIC ACID, ETHYL ESTER	11	2120
ETHYL DIAZOACETATE	11	2120
SYM-DIMETHYLHYDRAZINE, DIHYDROCHLORIDE	303	2300
HYDRAZINE, N,N*-DIMETHYL-, DIHYDRO- CHLORIDE	303	2300
HYDRAZINE, 1,2-DIMETHYL-, DIHYDRO- CHLORIDE	303	2300
DIMETHYLAMINE	79	2790
DMA	79	2790
DIMETHYLAMINE	79	2800
DMA	79	2800
FORMALDEHYDE	111	2800
FORMALIN	111	2800
FORMIC ALDEHYDE	111	2800
METHANAL	111	2800
BROM-O-GAS	119	2900
BROMOMETHANE	119	2900
1,2-ETHANEDIAMINE	105	2900
ETHYLENEDIAMINE	105	2900
LEAD NAPHTHENATE	318	2900
METHANE, BROMO-,	119	2900
METHOGAS	119	2900
METHYL BROMIDE	119	2900
NAPHTHENIC ACID, LEAD SALT	318	2900
ROTOX	119	2900
BENZENE, HEXAHYDRO-,	72	2920
BERYLLIA	234	2920
BERYLLIUM OXIDE	234	2920
BUTANE, 1-AMINO-,	58	2920
BUTYLAMINE	58	2920
CYCLOHEXANE	72	2920
ENU	358	2920
17A-ESTRADIOL	283	2920
ESTRA-1,3,5/10/-TRIENE-3,17A-DIOL	283	2920
1,2-ETHANEDIAMINE	105	2920
ETHYLENEDIAMINE	105	2920
MYRJ 45	166	2920
NORVALAMINE	58	2920
POLYETHYLENE GLYCOL, MONOSTEARATE	166	2920
POLYOXYETHYLENE-8-MONOSTEARATE	166	2920
UREA, 1-ETHYL-1-NITROSO-,	358	2920
ACRYLIC ACID, BUTYL ESTER	18	2940
BENZENE, HEXAHYDRO-,	72	2940
BUTANE, 1-AMINO-,	58	2940
BUTYL ACRYLATE	18	2940
BUTYL ALCOHOL, ACRYLATE	18	2940
BUTYLAMINE	58	2940
CYCLOHEXANE	72	2940
DELTA-MVE	324	2940
DIHYDROFOLLICULIN	280	2940
DIHYDROTHEELIN	280	2940
ESTRADIOL	280	2940
17B-ESTRADIOL	280	2940
ESTRA-1,3,5/10/-TRIENE-3,17B-DIOL	280	2940
ESTRIN, DIHYDROXY-,	280	2940
ETHANE, 1,2-DIHYDROXY-,	106	2940
1,2-ETHANEDIOL	106	2940
ETHYLENE GLYCOL	106	2940
FOLLICULIN, DIHYDRO-,	280	2940
MESTRANOL	324	2940
MYRJ 45	166	2940
1-NITROSOPIPERIDINE	165	2940
NORVALAMINE	58	2940
PIPERIDINE, N-NITROSO-,	165	2940
PIPERIDINE, 1-NITROSO-,	165	2940
POLYETHYLENE GLYCOL, MONOSTEARATE	166	2940
POLYOXYETHYLENE-8-MONOSTEARATE	166	2940
19-NOR-17A-PREGNA-1,3,5/10/-TRIEN-20- YN-17-OL, 3-METHOXY-,	324	2940
TESCOL	106	2940
THEELIN, DIHYDRO-,	280	2940
BENZENESULFONIC ACID, DODECYL-,	222	2960
1-BUTANAMINE, N-BUTYL-N-NITROSO-,	75	2960
DBN	75	2960
DEN	76	2960
DIBUTYLAMINE, N-NITROSO-,	75	2960

CHEMICAL NAME	NUMBER	STRONGEST BAND
DIETHYLAMINE	76	2960
DIPROPYLAMINE, N-NITROSO-,	82	2960
DODECYLBENZENESULFONIC ACID	222	2960
DPN	82	2960
ETHANAMINE	98	2960
ETHANAMINE, N,N-DIETHYL-,	190	2960
ETHANE, AMINO-,	98	2960
ETHYLAMINE	98	2960
LAKEWAY SA	222	2960
METHANETHIOL	128	2960
METHYL MERCAPTAN	128	2960
MONOETHYLAMINE	98	2960
NAPHTHENIC ACID	133	2960
NITROSAMINE, DIPROPYL-,	82	2960
TEN	190	2960
TRIETHYLAMINE	190	2960
ETHANAMINE	98	2970
ETHANE, AMINO-,	98	2970
ETHYLAMINE	98	2970
MONOETHYLAMINE	98	2970
1-BUTANAMINE, N-BUTYL-N-NITROSO-,	75	2980
DBN	75	2980
DIBUTYLAMINE, N-NITROSO-,	75	2980
ETHANAMINE, N,N-DIETHYL-,	190	2980
HYDRAZINE, PHENYL-, MONOHYDRO-CHLORIDE	305	2980
TEN	190	2980
TRIETHYLAMINE	190	2980
BIOXIRANE	57	3000
BUTADIENE DIEPOXIDE	57	3000
BUTADIENE DIOXIDE	57	3000
BUTANE, 1,2:3,4-DIEPOXY-,	57	3000
CHLOROETHANE	84	3000
ETHANE, CHLORO-,	84	3000
ETHYL CHLORIDE	84	3000
KELENE	84	3000
NARCOTILE	84	3000
THREITOL, 1,2:3,4-DIANHYDRO-,	57	3000
ACETAMIDE, THIO-,	6	3080
ACETOTHIOAMIDE	6	3080
ETHANETHIOAMIDE	6	3080
TAA	6	3080
THIOACETAMIDE	6	3080
HYDRAZINE, 1,1-DIMETHYL-, MONO-HYDROCHLORIDE	302	3240
ESTINYL	323	3280
ESTRADIOL, 17-ETHYNYL-,	323	3280
ETHINYL ESTRADIOL	323	3280
NOVESTROL	323	3280
19-NOR-17A-PREGNA-1,3,5/10/-TRIEN-20-YNE-3,17-DIOL	323	3280
ALLYL ALCOHOL	20	3320
CARBINOL, VINYL-,	20	3320
DIAMIDE	298	3320
DIAMINE	298	3320
HYDRAZINE	298	3320
2-PROPEN-1-OL	20	3320
2,5-DMP	198	3340
ETHANE, 1,2-DIHYDROXY-,	106	3340
1,2-ETHANEDIOL	106	3340
ETHYLENE GLYCOL	106	3340
GLUCOPYRANOSE, 2-DEOXY-2-/3-METHYL-3-NITROSOUREIDO/-, D-,	295	3340
PHENOL, 2,5-DIMETHYL-,	198	3340
STR	295	3340
STREPTOZOTOCIN	295	3340
TESCOL	106	3340
2,5-XYLENOL	198	3340
FLUOREN-2-AMINE	109	3360
FLUORENE, 2-AMINO-,	109	3360
HYDRAZINE, MONOHYDRATE	299	3400
HYDRAZINE HYDRATE	299	3400
ESTRA-1,3,5/10/-TRIENE-3,16A,17B-TRIOL	286	3420
ESTRIN, TRIHYDROXY-,	286	3420
ESTRIOL	286	3420
FOLLICULAR HORMONE HYDRATE	286	3420
THEELOL	286	3420
CADDY	239	3520
CADMIUM CHLORIDE	239	3520
VI-CAD	239	3520